DATE

Methods in Enzymology

Volume 315
VERTEBRATE PHOTOTRANSDUCTION AND
THE VISUAL CYCLE
Part A

METHODS IN ENZYMOLOGY

EDITORS-IN-CHIEF

John N. Abelson   Melvin I. Simon

DIVISION OF BIOLOGY
CALIFORNIA INSTITUTE OF TECHNOLOGY
PASADENA, CALIFORNIA

FOUNDING EDITORS

Sidney P. Colowick and Nathan O. Kaplan

Methods in Enzymology

Volume 315

Vertebrate Phototransduction and the Visual Cycle Part A

EDITED BY

Krzysztof Palczewski

DEPARTMENT OF OPHTHALMOLOGY
UNIVERSITY OF WASHINGTON SCHOOL OF MEDICINE
SEATTLE, WASHINGTON

ACADEMIC PRESS

San Diego London Boston New York Sydney Tokyo Toronto

This book is printed on acid-free paper.

Copyright © 2000 by ACADEMIC PRESS

All Rights Reserved.
No part of this publication may be reproduced or transmitted in any form or by any means, electronic or mechanical, including photocopy, recording, or any information storage and retrieval system, without permission in writing from the Publisher.

The appearance of the code at the bottom of the first page of a chapter in this book indicates the Publisher's consent that copies of the chapter may be made for personal or internal use of specific clients. This consent is given on the condition, however, that the copier pay the stated per copy fee through the Copyright Clearance Center, Inc. (222 Rosewood Drive, Danvers, Massachusetts 01923) for copying beyond that permitted by Sections 107 or 108 of the U.S. Copyright Law. This consent does not extend to other kinds of copying, such as copying for general distribution, for advertising or promotional purposes, for creating new collective works, or for resale. Copy fees for pre-1999 chapters are as shown on the chapter title pages. If no fee code appears on the chapter title page, the copy fee is the same as for current chapters. 0076-6879/00 $25.00

Academic Press
A Harcourt Science and Technology Company
525 B Street, Suite 1900, San Diego, California 92101-4495, USA
http://www.academicpress.com

Academic Press Limited
24-28 Oval Road, London NW1 7DX, UK
http://www.hbuk.co.uk/ap/

International Standard Book Number: 0-12-182216-8

PRINTED IN THE UNITED STATES OF AMERICA
00 01 02 03 04 05 06 MM 9 8 7 6 5 4 3 2 1

Table of Contents

Contributors to Volume 315 xi
Preface. xvii
Volumes in Series xix

Section I. Expression and Isolation of Opsins

1. Heterologous Expression of Bovine Opsin in *Pichia pastoris* — Najmoutin G. Abdulaev and Kevin D. Ridge — 3

2. Baculovirus Expression System for Expression and Characterization of Functional Recombinant Visual Pigments — Corné H. W. Klaassen and Willem J. DeGrip — 12

3. Development of Stable Cell Lines Expressing High Levels of Point Mutants of Human Opsin for Biochemical and Biophysical Studies — Jack M. Sullivan and Michael F. Satchwell — 30

4. Folding and Assembly of Rhodopsin from Expressed Fragments — Kevin D. Ridge and Najmoutin G. Abdulaev — 59

5. Isolation of Isoelectric Species of Phosphorylated Rhodopsin — J. Hugh McDowell, Joseph P. Nawrocki, and Paul A. Hargrave — 70

6. Rhodopsin Trafficking in Photoreceptors Using Retinal Cell-Free System — Dusanka Deretic — 77

Section II. Characterization of Opsins

7. Preparation and Analysis of Two-Dimensional Crystals of Rhodopsin — Gebhard F. X. Schertler and Paul A. Hargrave — 91

8. Domain Approach to Three-Dimensional Structure of Rhodopsin Using High Resolution Nuclear Magnetic Resonance — Arlene D. Albert and Philip L. Yeagle — 107

9. Analysis of Functional Microdomains in Rhodopsin — Steven W. Lin, May Han, and Thomas P. Sakmar — 116

10. Mapping Tertiary Contacts between Amino Acid Residues within Rhodopsin — Mary Struthers and Daniel D. Oprian — 130

11. Energetics of Rhodopsin Photobleaching: Photocalorimetric Studies of Energy Storage in Early and Late Intermediates — Robert R. Birge and Bryan W. Vought — 143

12. Absorption Spectroscopy in Studies of Visual Pigments: Spectral and Kinetic Characterization of Intermediates — JAMES W. LEWIS AND DAVID S. KLIGER — 164

13. Structural Determinants of Active State Conformation of Rhodopsin: Molecular Biophysics Approaches — KARIM FAHMY, THOMAS P. SAKMAR, AND FRIEDRICH SIEBERT — 178

14. pK_a of Protonated Schiff Base of Visual Pigments — THOMAS G. EBREY — 196

15. Assays for Detection of Constitutively Active Opsins — PHYLLIS R. ROBINSON — 207

16. Synthetic Retinals: Convenient Probes of Rhodopsin and Visual Transduction Process — JIHONG LOU, QIANG TAN, ELENA KARNAUKHOVA, NINA BEROVA, KOJI NAKANISHI, AND ROSALIE K. CROUCH — 219

17. Assays for Activation of Opsin by all-*trans*-Retinal — KRISTINA SACHS, DIETER MARETZKI, AND KLAUS PETER HOFMANN — 238

18. Assays for Activation of Recombinant Expressed Opsin by all-*trans*-Retinal — MAY HAN AND THOMAS P. SAKMAR — 251

19. Electrical Approach to Study Rhodopsin Activation in Single Cells with Early Receptor Current Assay — JACK M. SULLIVAN, LIOUBOV BRUEGGEMANN, AND PRAGATI SHUKLA — 268

20. Analysis of Amino Acid Residues in Rhodopsin and Cone Visual Pigments That Determine Their Molecular Properties — HIROO IMAI, AKIHISA TERAKITA, AND YOSHINORI SHICHIDA — 293

21. Phylogenetic Analysis and Experimental Approaches to Study Color Vision in Vertebrates — SHOZO YOKOYAMA — 312

Section III. Proteins That Interact with Rhodopsin

22. Light Scattering Methods to Monitor Interaction between Rhodopsin-Containing Membranes and Soluble Proteins — MARTIN HECK, ALEXANDER PULVERMÜLLER, AND KLAUS PETER HOFMANN — 329

23. Heterogeneity of Rhodopsin Intermediate State Interacting with Transducin — YOSHINORI SHICHIDA, SHUJI TACHIBANAKI, HIROO IMAI, AND AKIHISA TERAKITA — 347

24. Limited Proteolytic Digestion Studies of G Protein–Receptor Interaction — MARIA R. MAZZONI AND HEIDI E. HAMM — 363

25. Monitoring Proton Uptake from Aqueous Phase during Rhodopsin Activation — CHRISTOPH K. MEYER AND KLAUS PETER HOFMANN — 377

26. Use of Peptides-on-Plasmids Combinatorial Library to Identify High Affinity Peptides That Bind Rhodopsin	ANNETTE GILCHRIST, ANLI LI, AND HEIDI E. HAMM	388
27. Purification of Rhodopsin Kinase by Recoverin Affinity Chromatography	CHING-KANG CHEN AND JAMES B. HURLEY	404
28. Heterologous Expression and Reconstitution of Rhodopsin with Rhodopsin Kinase and Arrestin	SHOJI OSAWA, DAYANIDHI RAMAN, AND ELLEN R. WEISS	411
29. Arrestin: Mutagenesis, Expression, Purification, and Functional Characterization	VSEVOLOD V. GUREVICH AND JEFFREY L. BENOVIC	422
30. Mapping Interaction Sites between Rhodopsin and Arrestin by Phage Display and Synthetic Peptides	W. CLAY SMITH AND PAUL A. HARGRAVE	437
31. Characterization of RanBP2-Associated Molecular Components in Neuroretina	PAULO A. FERREIRA	455

Section IV. Transducin and Regulators of G-Protein Signaling

32. Intrinsic Biophysical Monitors of Transducin Activation: Fluorescence, UV/Visible Spectroscopy, Light Scattering, and Evanescent Field Techniques	OLIVER P. ERNST, CHRISTOPH BIERI, HORST VOGEL, AND KLAUS PETER HOFMANN	471
33. Fluorescent Probes as Indicators of Conformation Changes in Transducin on Activation	CHII-SHEN YANG, NIKOLAI P. SKIBA, MARIA R. MAZZONI, TARITA O. THOMAS, AND HEIDI E. HAMM	490
34. $G\alpha_t/G\alpha_{i1}$ Chimeras Used to Define Structural Basis of Specific Functions of $G\alpha_t$	NIKOLAI P. SKIBA, TARITA O. THOMAS, AND HEIDI E. HAMM	502
35. Enzymology of GTPase Acceleration in Phototransduction	CHRISTOPHER W. COWAN, THEODORE G. WENSEL, AND VADIM Y. ARSHAVSKY	524
36. Mutational Analysis of Effectors and Regulators of G-Protein Signaling Interfaces of Transducin	MICHAEL NATOCHIN AND NIKOLAI O. ARTEMYEV	539

Section V. Photoreceptor Protein Phosphatases

37. Isolation and Properties of Protein Phosphatase 2A in Photoreceptors	MUHAMMAD AKHTAR, ALASTAIR J. KING, AND NINA E. M. MCCARTHY	557
38. Purification and Characterization of Protein Phosphate Type 2C in Photoreceptors	SUSANNE KLUMPP AND DAGMAR SELKE	570

39. Photoreceptor Serine/Threonine Protein Phosphatase Type 7: Cloning, Expression, and Functional Analysis
XIZHONG HUANG, MARK R. SWINGLE, AND RICHARD E. HONKANEN 579

Section VI. Photoreceptor Phosphodiesterase and Guanylyl Cyclase

40. Purification and Assay of Bovine Type 6 Photoreceptor Phosphodiesterase and Its Subunits
TERRY A. COOK AND JOSEPH A. BEAVO 597

41. Transcriptional Regulation of cGMP Phosphodiesterase β-Subunit Gene
LEONID E. LERNER AND DEBORA B. FARBER 617

42. Inhibition of Photoreceptor cGMP Phosphodiesterase by Its γ Subunit
ALEXEY E. GRANOVSKY, KHAKIM G. MURADOV, AND NIKOLAI O. ARTEMYEV 635

43. Kinetics and Regulation of cGMP Binding to Noncatalytic Binding Sites on Photoreceptor Phosphodiesterase
RICK H. COTE 646

44. Purification and Autophosphorylation of Retinal Guanylate Cyclase
JEFFREY P. JOHNSTON, JENNIFER G. APARICIO, AND MEREDITHE L. APPLEBURY 673

45. Use of Nucleotide α-Phosphorothioates in Studies of Photoreceptor Guanylyl Cyclase: Purification of Guanylyl Cyclase Activating Proteins
WOJCIECH A. GORCZYCA 689

46. Heterologous Expression and Assays for Photoreceptor Guanylyl Cyclases and Guanylyl Cyclase Activating Proteins
JAMES B. HURLEY AND ALEXANDER M. DIZHOOR 708

47. Spectrophotometric Determination of Retinal Rod Guanylyl Cyclase
GREGOR WOLBRING AND PAUL P. M. SCHNETKAMP 718

48. Calcium-Dependent Activation of Membrane Guanylate Cyclase by S100 Proteins
ARI SITARAMAYYA, NIKOLAY POZDNYAKOV, ALEXANDER MARGULIS, AND AKIKO YOSHIDA 730

49. Characterization of Guanylyl Cyclase and Phosphodiesterase Activities in Single Rod Outer Segments
YIANNIS KOUTALOS AND KING-WAI YAU 742

Section VII. Cyclic Nucleotide-Gated Channels

50. Covalent Tethering of Ligands to Retinal Rod Cyclic Nucleotide-Gated Channels: Binding Site Structure and Allosteric Mechanism
JEFFREY W. KARPEN, MARIALUISA RUIZ, AND R. LANE BROWN 755

51. Using State-Specific Modifiers to Study Rod cGMP-Activated Ion Channels Expressed in *Xenopus* Oocytes — SHARONA E. GORDON — 772

52. Identification and Characterization of Calmodulin Binding Sites in cGMP-Gated Channel Using Surface Plasmon Resonance Spectroscopy — KARL-WILHELM KOCH — 785

53. Determination of Fractional Calcium Ion Current in Cyclic Nucleotide-Gated Channels — STEPHAN FRINGS, DAVID H. HACKOS, CLAUDIA DZEJA, TSUYOSHI OHYAMA, VOLKER HAGEN, U. BENJAMIN KAUPP, AND JUAN I. KORENBROT — 797

54. Modulation of Rod cGMP-Gated Cation Channel by Calmodulin — MARIA E. GRUNWALD AND KING-WAI YAU — 817

Section VIII. Na^+/Ca^{2+}-K^+ Exchanger and ABCR Transporter

55. Purification and Biochemical Analysis of cGMP-Gated Channel and Na^+/Ca^{2+}-K^+ Exchanger of Rod Photoreceptors — ROBERT S. MOLDAY, RENÉ WARREN, AND TOM S. Y. KIM — 831

56. Spectrofluorometric Detection of Na^+/Ca^{2+}-K^+ Exchange — CONAN B. COOPER, ROBERT T. SZERENCSEI, AND PAUL P. M. SCHNETKAMP — 847

57. Purification and Characterization of ABCR from Bovine Rod Outer Segments — JINHI AHN AND ROBERT S. MOLDAY — 864

58. ABCR: Rod Photoreceptor-Specific ABC Transporter Responsible for Stargardt Disease — HUI SUN AND JEREMY NATHANS — 879

AUTHOR INDEX 899

SUBJECT INDEX 935

Contributors to Volume 315

Article numbers are in parentheses following the names of contributors.
Affiliations listed are current.

NAJMOUTIN G. ABDULAEV (1, 4), *Center for Advanced Research in Biotechnology, National Institute of Standards and Technology and University of Maryland Biotechnology Institute, Rockville, Maryland 20850*

JINHI AHN (57), *Department of Biochemistry and Molecular Biology, University of British Columbia, Vancouver, British Columbia V6T 1Z3, Canada*

MUHAMMAD AKHTAR (37), *Division of Biochemistry, University of Southampton, Southampton SO16 7PX, United Kingdom*

ARLENE D. ALBERT (8), *Department of Molecular and Cell Biology, University of Connecticut, Storrs, Connecticut 06269-3125*

JENNIFER G. APARICIO (44), *The Howe Laboratory, Harvard Medical School, Massachusetts Eye and Ear Infirmary, Boston, Massachusetts 02114*

MEREDITHE L. APPLEBURY (44), *The Howe Laboratory, Harvard Medical School, Massachusetts Eye and Ear Infirmary, Boston, Massachusetts 02114*

VADIM Y. ARSHAVSKY (35), *The Howe Laboratory of Ophthalmology, Harvard Medical School, Massachusetts Eye and Ear Infirmary, Boston, Massachusetts 02114*

NIKOLAI O. ARTEMYEV (36, 42), *Department of Physiology and Biophysics, University of Iowa College of Medicine, Iowa City, Iowa 52242-1109*

JOSEPH A. BEAVO (40), *Department of Pharmacology, University of Washington, Seattle, Washington 98195-7280*

JEFFREY L. BENOVIC (29), *Department of Microbiology and Immunology, Kimmel Cancer Institute, Thomas Jefferson University, Philadelphia, Pennsylvania 19107-5566*

NINA BEROVA (16), *Department of Chemistry, Columbia University, New York, New York 10027*

CHRISTOPH BIERI (32), *Laboratory for Physical Chemistry of Polymers and Membranes, Institute of Physical Chemistry, Swiss Federal Institute of Technology, CH-1015 Lausanne, Switzerland*

ROBERT R. BIRGE (11), *Department of Chemistry, Syracuse University, Syracuse, New York 13244-4100*

R. LANE BROWN (50), *Neurological Sciences Institute, Oregon Health Sciences University, Portland, Oregon 97209*

LIOUBOV BRUEGGEMANN (19), *Department of Ophthalmology, State University of New York Health Science Center, Syracuse, New York 13210*

CHING-KANG CHEN (27), *Division of Biology, California Institute of Technology, Pasadena, California 91125*

TERRY A. COOK (40), *Department of Pharmacology, University of Washington, Seattle, Washington 98195-7280*

CONAN B. COOPER (56), *Department of Physiology and Biophysics and Medical Research Council Group on Ion Channels and Transporters, Faculty of Medicine, University of Calgary, Calgary, Alberta T2N 4N1, Canada*

RICK H. COTE (43), *Department of Biochemistry and Molecular Biology, University of New Hampshire, Durham, New Hampshire 03824-3544*

CHRISTOPHER W. COWAN (35), *Department of Biochemistry and Cell and Molecular Biology Graduate Program, Baylor College of Medicine, Houston, Texas 77030*

ROSALIE K. CROUCH (16), *Department of Ophthalmology, Medical University of South Carolina, Charleston, South Carolina 29425*

WILLEM J. DEGRIP (2), *Department of Biochemistry, Institute of Cellular Signalling, University of Nijmegen, NL-6500 HB Nijmegen, The Netherlands*

DUSANKA DERETIC (6), *Departments of Ophthalmology and Visual Sciences, and Anatomy and Cell Biology, Kellogg Eye Center, University of Michigan, Ann Arbor, Michigan 48105*

ALEXANDER M. DIZHOOR (46), *Department of Ophthalmology, Kresge Eye Institute, and Department of Pharmacology, Wayne State University School of Medicine, Detroit, Michigan 48201*

CLAUDIA DZEJA (53), *Institut für Biologische Informationsverarbeitung, Forschungszentrum Jülich, D-52425 Jülich, Germany*

THOMAS G. EBREY (14), *Departments of Cell and Structural Biology, and Biochemistry Center for Biophysics and Computational Biology, School of Molecular and Cellular Biology, University of Illinois, Urbana, Illinois 61801*

OLIVER P. ERNST (32), *Institut für Medizinische Physik und Biophysik, Universitätsklinikum Charité der Humboldt Universität zu Berlin, CCM, D-10098 Berlin, Germany*

KARIM FAHMY (13), *Institut für Biophysik und Strahlenbiologie, Albert-Ludwigs-Universität, D-79104 Freiburg, Germany*

DEBORA B. FARBER (41), *Jules Stein Eye Institute, University of California, School of Medicine, Los Angeles, California 90095-7000*

PAULO A. FERREIRA (31), *Department of Pharmacology, Medical College of Wisconsin, Milwaukee, Wisconsin 53226*

STEPHAN FRINGS (53), *Institut für Biologische Informationsverarbeitung, Forschungszentrum Jülich, D-52425 Jülich, Germany*

ANNETTE GILCHRIST (26), *Department of Molecular Pharmacology and Biological Chemistry, Institute for Neuroscience, Northwestern University, Chicago, Illinois 60611*

WOJCIECH A. GORCZYCA (45), *Department of Microbiology, Ludwik Hirszfeld Institute of Immunology and Experimental Therapy, Polish Academy of Sciences, 53-114 Wroclaw, Poland*

SHARONA E. GORDON (51), *Department of Ophthalmology, University of Washington School of Medicine, Seattle, Washington 98195-6485*

ALEXEY E. GRANOVSKY (42), *Department of Physiology and Biophysics, University of Iowa College of Medicine, Iowa City, Iowa 52242-1109*

MARIA E. GRUNWALD (54), *Department of Molecular and Cell Biology, University of California, Berkeley, California 94720*

VSEVOLOD V. GUREVICH (29), *Ralph and Muriel Roberts Laboratory for Vision Science, Sun Health Research Institute, Sun City, Arizona 85351*

DAVID H. HACKOS (53), *Molecular Physiology and Biophysics Unit, National Institute of Neurological Disorders and Stroke, National Institutes of Health, Bethesda, Maryland 20892*

VOLKER HAGEN (53), *Forschungsinstitut für Molekulare Pharmakologie, D-10315 Berlin, Germany*

HEIDI E. HAMM (24, 26, 33, 34), *Department of Molecular Pharmacology and Biological Chemistry, Institute for Neuroscience, Northwestern University, Chicago, Illinois 60611*

MAY HAN (9, 18), *Department of Molecular Biophysics and Biochemistry, Yale University, New Haven, Connecticut 06520*

PAUL A. HARGRAVE (5, 7, 30), *Departments of Ophthalmology, and Biochemistry and Molecular Biology, University of Florida, Gainesville, Florida 32610*

MARTIN HECK (22), *Institut für Medizinische Physik und Biophysik, Universitätsklinikum Charité der Humboldt Universität zu Berlin, CCM, D-10098 Berlin, Germany*

KLAUS PETER HOFMANN (17, 22, 25, 32), *Institut für Medizinische Physik und Biophysik, Universitätsklinikum Charité der Humboldt Universität zu Berlin, CCM, D-10098 Berlin, Germany*

RICHARD E. HONKANEN (39), *Department of Biochemistry and Molecular Biology, University of South Alabama, Mobile, Alabama 36688*

XIZHONG HUANG (39), *Department of Biochemistry and Molecular Biology, University of South Alabama, Mobile, Alabama 36688*

JAMES B. HURLEY (27, 46), *Department of Biochemistry, Howard Hughes Medical Institute, University of Washington, Seattle, Washington 98195*

HIROO IMAI (20, 23), *Department of Biophysics, Graduate School of Science, Kyoto University, Kyoto 606-8502, Japan*

JEFFREY P. JOHNSTON (44), *The Howe Laboratory, Harvard Medical School, Massachusetts Eye and Ear Infirmary, Boston, Massachusetts 02114*

ELENA KARNAUKHOVA (16), *Department of Chemistry, Columbia University, New York, New York 10027*

JEFFREY W. KARPEN (50), *Department of Physiology and Biophysics, University of Colorado School of Medicine, Denver, Colorado 80262*

U. BENJAMIN KAUPP (53), *Institut für Biologische Informationsverarbeitung, Forschungszentrum Jülich, D-52425 Jülich, Germany*

TOM S. Y. KIM (55), *Department of Biochemistry and Molecular Biology, University of British Columbia, Vancouver, British Columbia V6T 1Z3, Canada*

ALASTAIR J. KING (37), *Walther Oncology Centre, Indianapolis, Indiana 46202*

CORNÉ H. W. KLAASSEN (2), *Department of Biochemistry, Institute of Cellular Signalling, University of Nijmegen, NL-6500 HB Nijmegen, The Netherlands*

DAVID S. KLIGER (12), *Department of Chemistry and Biochemistry, University of California, Santa Cruz, California 95064*

SUSANNE KLUMPP (38), *Institute of Pharmaceutical Chemistry, Universität Marburg, D-35032 Marburg, Germany*

KARL-WILHELM KOCH (52), *Institut für Biologische Informationsverarbeitung, Forschungszentrum Jülich, D-52425 Jülich, Germany*

JUAN I. KORENBROT (53), *Department of Physiology, University of California, San Francisco, California 94143*

YIANNIS KOUTALOS (49), *Department of Physiology and Biophysics, University of Colorado School of Medicine, Denver, Colorado 80262*

LEONID E. LERNER (41), *Jules Stein Eye Institute, University of California, School of Medicine, Los Angeles, California 90095-7000*

JAMES W. LEWIS (12), *Department of Chemistry and Biochemistry, University of California, Santa Cruz, California 95064*

ANLI LI (26), *Department of Molecular Pharmacology and Biological Chemistry, Institute for Neuroscience, Northwestern University, Chicago, Illinois 60611*

STEVEN W. LIN (9), *Laboratory of Molecular Biology and Biochemistry, Rockefeller University, New York, New York 10021*

JIHONG LOU (16), *Department of Chemistry, Columbia University, New York, New York 10027*

DIETER MARETZKI (17), *Institut für Medizinische Physik und Biophysik, Universitätsklinikum Charité der Humboldt Universität zu Berlin, CCM, D-10098 Berlin, Germany*

ALEXANDER MARGULIS (48), *Eye Research Institute, Oakland University, Rochester, Michigan 48309-4480*

MARIA R. MAZZONI (24, 33), *Istituto Polycatthedra di Scienze Biologiche, 56100 Pisa, Italy*

NINA E. M. MCCARTHY (37), *Division of Biochemistry, University of Southampton, Southampton SO16 7PX, United Kingdom*

J. HUGH MCDOWELL (5), *Department of Ophthalmology, University of Florida, Gainesville, Florida 32610*

CHRISTOPH K. MEYER (25), *Institut für Medizinische Physik und Biophysik, Universitätsklinikum Charité der Humboldt Universität zu Berlin, CCM, D-10098 Berlin, Germany*

ROBERT S. MOLDAY (55, 57), *Department of Biochemistry and Molecular Biology, University of British Columbia, Vancouver, British Columbia V6T 1Z3, Canada*

KHAKIM G. MURADOV (42), *Department of Physiology and Biophysics, University of Iowa College of Medicine, Iowa City, Iowa 52242-1109*

KOJI NAKANISHI (16), *Department of Chemistry, Columbia University, New York, New York 10027*

JEREMY NATHANS (58), *Department of Molecular Biology and Genetics, Johns Hopkins University School of Medicine, Baltimore, Maryland 21205*

MICHAEL NATOCHIN (36), *Department of Physiology and Biophysics, University of Iowa College of Medicine, Iowa City, Iowa 52242-1109*

JOSEPH P. NAWROCKI (5), *Laboratory of Bioorganic Chemistry, National Institute of Diabetes and Digestive and Kidney Diseases, Bethesda, Maryland 20892-0805*

TSUYOSHI OHYAMA (53), *Department of Physiology, University of California, San Francisco, California 94143*

DANIEL D. OPRIAN (10), *Department of Biochemistry, Volen Center for Complex Systems, Brandeis University, Waltham, Massachusetts 02454*

SHOJI OSAWA (28), *Department of Cell Biology and Anatomy, University of North Carolina, Chapel Hill, North Carolina 27599-7090*

NIKOLAY POZDNYAKOV (48), *Eye Research Institute, Oakland University, Rochester, Michigan 48309-4480*

ALEXANDER PULVERMÜLLER (22), *Institut für Medizinische Physik und Biophysik, Universitätsklinikum Charité der Humboldt Universität zu Berlin, CCM, D-10098 Berlin, Germany*

DAYANIDHI RAMAN (28), *Department of Cell Biology and Anatomy, University of North Carolina, Chapel Hill, North Carolina 27599-7090*

KEVIN D. RIDGE (1, 4), *Center for Advanced Research in Biotechnology, National Institute of Standards and Technology and University of Maryland Biotechnology Institute, Rockville, Maryland 20850*

PHYLLIS R. ROBINSON (15), *Department of Biological Sciences, University of Maryland Baltimore County, Baltimore, Maryland 21250*

MARIALUISA RUIZ (50), *Department of Physiology and Biophysics, University of Colorado School of Medicine, Denver, Colorado 80262*

KRISTINA SACHS (17), *Institut für Medizinische Physik und Biophysik, Universitätsklinikum Charité der Humboldt Universität zu Berlin, CCM, D-10098 Berlin, Germany*

THOMAS P. SAKMAR (9, 13, 18), *Laboratory of Molecular Biology and Biochemistry, Howard Hughes Medical Institute, Rockefeller University, New York, New York 10021*

MICHAEL F. SATCHWELL (3), *Department of Ophthalmology, State University of New York Health Science Center, Syracuse, New York 13210*

GEBHARD F. X. SCHERTLER (7), *Laboratory of Molecular Biology, Medical Research Council, Cambridge CB2 2QH, England, United Kingdom*

PAUL P. M. SCHNETKAMP (47, 56), *Department of Physiology and Biophysics and Medical Research Council Group on Ion Channels and Transporters, Faculty of Medicine, University of Calgary, Calgary, Alberta T2N 4N1, Canada*

DAGMAR SELKE (38), *Institute of Pharmaceutical Chemistry, Universität Marburg, D-35032 Marburg, Germany*

YOSHINORI SHICHIDA (20, 23), *Department of Biophysics, Graduate School of Science, Kyoto University, Kyoto 606-8502, Japan*

PRAGATI SHUKLA (19), *Department of Ophthalmology, State University of New York Health Science Center, Syracuse, New York 13210*

FRIEDRICH SIEBERT (13), *Institut für Biophysik und Strahlenbiologie, Albert-Ludwigs-Universität, D-79104 Freiburg, Germany*

ARI SITARAMAYYA (48), *Eye Research Institute, Oakland University, Rochester, Michigan 48309-4480*

NIKOLAI P. SKIBA (33, 34), *Department of Molecular Pharmacology and Biological Chemistry, Institute for Neuroscience, Northwestern University, Chicago, Illinois 60611*

W. CLAY SMITH (30), *Departments of Ophthalmology and Neuroscience, University of Florida, Gainesville, Florida 32610*

MARY STRUTHERS (10), *Department of Biochemistry, Volen Center for Complex Systems, Brandeis University, Waltham, Massachusetts 02454*

JACK M. SULLIVAN (3, 19), *Departments of Ophthalmology and Biochemistry, State University of New York Health Science Center, Syracuse, New York 13210*

HUI SUN (58), *Department of Molecular Biology and Genetics, Johns Hopkins University School of Medicine, Baltimore, Maryland 21205*

MARK R. SWINGLE (39), *Department of Biochemistry and Molecular Biology, University of South Alabama, Mobile, Alabama 36688*

ROBERT T. SZERENCSEI (56), *Department of Physiology and Biophysics and Medical Research Council Group on Ion Channels and Transporters, Faculty of Medicine, University of Calgary, Calgary, Alberta T2N 4N1, Canada*

SHUJI TACHIBANAKI (23), *Department of Biology, Graduate School of Science, Osaka University, Toyonaka, Osaka 560-0043, Japan*

QIANG TAN (16), *Department of Chemistry, Columbia University, New York, New York 10027*

AKIHISA TERAKITA (20, 23), *Department of Biophysics, Graduate School of Science, Kyoto University, Kyoto 606-8502, Japan*

TARITA O. THOMAS (33, 34), *Department of Molecular Pharmacology and Biological Chemistry, Institute for Neuroscience, Northwestern University, Chicago, Illinois 60611*

HORST VOGEL (32), *Laboratory for Physical Chemistry of Polymers and Membranes, Institute of Physical Chemistry, Swiss Federal Institute of Technology, CH-1015 Lausanne, Switzerland*

BRYAN W. VOUGHT (11), *Department of Biological Chemistry and Molecular Pharmacology, Harvard Medical School, Boston, Massachusetts 02115*

RENÉ WARREN (55), *Department of Biochemistry and Molecular Biology, University of British Columbia, Vancouver, British Columbia V6T 1Z3, Canada*

ELLEN R. WEISS (28), *Department of Cell Biology and Anatomy, University of North Carolina, Chapel Hill, North Carolina 27599-7090*

THEODORE G. WENSEL (35), *Department of Biochemistry and Cell and Molecular Biology Graduate Program, Baylor College of Medicine, Houston, Texas 77030*

GREGOR WOLBRING (47), *Department of Physiology and Biophysics and Medical Research Council Group on Ion Channels and Transporters, Faculty of Medicine, University of Calgary, Calgary, Alberta T2N 4N1, Canada*

CHII-SHEN YANG (33), *Department of Molecular Pharmacology and Biological Chemistry, Institute for Neuroscience, Northwestern University, Chicago, Illinois 60611*

KING-WAI YAU (49, 54), *Departments of Neuroscience and Ophthalmology, Howard Hughes Medical Institute, Johns Hopkins University School of Medicine, Baltimore, Maryland 21205*

PHILIP L. YEAGLE (8), *Department of Molecular and Cell Biology, University of Connecticut, Storrs, Connecticut 06269-3125*

SHOZO YOKOYAMA (21), *Department of Biology, Syracuse University, Syracuse, New York 13244*

AKIKO YOSHIDA (48), *Eye Research Institute, Oakland University, Rochester, Michigan 48309-4480*

Preface

Molecular methods related to visual transduction and the visual cycle have grown in number and evolved considerably since they were covered in Volume 81 of *Methods in Enzymology* in 1982. During this time the fundamental principles elicited through the study of visual signal transduction have been extended far into the biology of other transduction systems initiated by numerous types of small chemical molecules.

Unprecedented progress has been made in vision research since 1982, and newly developed methods to study vertebrate phototransduction and the visual cycle are summarized in this Volume 315 (Part A) of *Methods in Enzymology* and its companion Volume 316 (Part B). The methods included are very broadly defined. In this volume they appear in sections on characterization of opsins; methods to study properties of proteins that interact with rhodopsin; characterization of photoreceptor protein phosphatases, phosphodiesterase, and guanylyl cyclase; cyclic nucleotide-gated channels; and Na^+/Ca^{2+}-K^+ exchanger and ABCR transporter. In Volume 316 the methods appear in sections on characterization of calcium-binding proteins and calcium measurements in photoreceptor cells; methods to identify posttranslational and chemical modifications; assays to study phototransduction *in vitro* and *in vivo;* characterization of enzymes of the visual cycle; analysis of animal models; and methods to study inherited retinal diseases: from the defective gene to its function and repair.

It is my pleasure to thank the authors, the reviewers, Drs. M. I. Simon and J. N. Abelson (the editors-in-chief), and the publisher who made this project possible. In particular, I would like to thank members of my laboratory for their patience and support during the hectic period of editing these volumes.

KRZYSZTOF PALCZEWSKI

METHODS IN ENZYMOLOGY

VOLUME I. Preparation and Assay of Enzymes
Edited by SIDNEY P. COLOWICK AND NATHAN O. KAPLAN

VOLUME II. Preparation and Assay of Enzymes
Edited by SIDNEY P. COLOWICK AND NATHAN O. KAPLAN

VOLUME III. Preparation and Assay of Substrates
Edited by SIDNEY P. COLOWICK AND NATHAN O. KAPLAN

VOLUME IV. Special Techniques for the Enzymologist
Edited by SIDNEY P. COLOWICK AND NATHAN O. KAPLAN

VOLUME V. Preparation and Assay of Enzymes
Edited by SIDNEY P. COLOWICK AND NATHAN O. KAPLAN

VOLUME VI. Preparation and Assay of Enzymes (*Continued*)
Preparation and Assay of Substrates
Special Techniques
Edited by SIDNEY P. COLOWICK AND NATHAN O. KAPLAN

VOLUME VII. Cumulative Subject Index
Edited by SIDNEY P. COLOWICK AND NATHAN O. KAPLAN

VOLUME VIII. Complex Carbohydrates
Edited by ELIZABETH F. NEUFELD AND VICTOR GINSBURG

VOLUME IX. Carbohydrate Metabolism
Edited by WILLIS A. WOOD

VOLUME X. Oxidation and Phosphorylation
Edited by RONALD W. ESTABROOK AND MAYNARD E. PULLMAN

VOLUME XI. Enzyme Structure
Edited by C. H. W. HIRS

VOLUME XII. Nucleic Acids (Parts A and B)
Edited by LAWRENCE GROSSMAN AND KIVIE MOLDAVE

VOLUME XIII. Citric Acid Cycle
Edited by J. M. LOWENSTEIN

VOLUME XIV. Lipids
Edited by J. M. LOWENSTEIN

VOLUME XV. Steroids and Terpenoids
Edited by RAYMOND B. CLAYTON

VOLUME XVI. Fast Reactions
Edited by KENNETH KUSTIN

VOLUME XVII. Metabolism of Amino Acids and Amines (Parts A and B)
Edited by HERBERT TABOR AND CELIA WHITE TABOR

VOLUME XVIII. Vitamins and Coenzymes (Parts A, B, and C)
Edited by DONALD B. MCCORMICK AND LEMUEL D. WRIGHT

VOLUME XIX. Proteolytic Enzymes
Edited by GERTRUDE E. PERLMANN AND LASZLO LORAND

VOLUME XX. Nucleic Acids and Protein Synthesis (Part C)
Edited by KIVIE MOLDAVE AND LAWRENCE GROSSMAN

VOLUME XXI. Nucleic Acids (Part D)
Edited by LAWRENCE GROSSMAN AND KIVIE MOLDAVE

VOLUME XXII. Enzyme Purification and Related Techniques
Edited by WILLIAM B. JAKOBY

VOLUME XXIII. Photosynthesis (Part A)
Edited by ANTHONY SAN PIETRO

VOLUME XXIV. Photosynthesis and Nitrogen Fixation (Part B)
Edited by ANTHONY SAN PIETRO

VOLUME XXV. Enzyme Structure (Part B)
Edited by C. H. W. HIRS AND SERGE N. TIMASHEFF

VOLUME XXVI. Enzyme Structure (Part C)
Edited by C. H. W. HIRS AND SERGE N. TIMASHEFF

VOLUME XXVII. Enzyme Structure (Part D)
Edited by C. H. W. HIRS AND SERGE N. TIMASHEFF

VOLUME XXVIII. Complex Carbohydrates (Part B)
Edited by VICTOR GINSBURG

VOLUME XXIX. Nucleic Acids and Protein Synthesis (Part E)
Edited by LAWRENCE GROSSMAN AND KIVIE MOLDAVE

VOLUME XXX. Nucleic Acids and Protein Synthesis (Part F)
Edited by KIVIE MOLDAVE AND LAWRENCE GROSSMAN

VOLUME XXXI. Biomembranes (Part A)
Edited by SIDNEY FLEISCHER AND LESTER PACKER

VOLUME XXXII. Biomembranes (Part B)
Edited by SIDNEY FLEISCHER AND LESTER PACKER

VOLUME XXXIII. Cumulative Subject Index Volumes I–XXX
Edited by MARTHA G. DENNIS AND EDWARD A. DENNIS

VOLUME XXXIV. Affinity Techniques (Enzyme Purification: Part B)
Edited by WILLIAM B. JAKOBY AND MEIR WILCHEK

VOLUME XXXV. Lipids (Part B)
Edited by JOHN M. LOWENSTEIN

VOLUME XXXVI. Hormone Action (Part A: Steroid Hormones)
Edited by BERT W. O'MALLEY AND JOEL G. HARDMAN

VOLUME XXXVII. Hormone Action (Part B: Peptide Hormones)
Edited by BERT W. O'MALLEY AND JOEL G. HARDMAN

VOLUME XXXVIII. Hormone Action (Part C: Cyclic Nucleotides)
Edited by JOEL G. HARDMAN AND BERT W. O'MALLEY

VOLUME XXXIX. Hormone Action (Part D: Isolated Cells, Tissues, and Organ Systems)
Edited by JOEL G. HARDMAN AND BERT W. O'MALLEY

VOLUME XL. Hormone Action (Part E: Nuclear Structure and Function)
Edited by BERT W. O'MALLEY AND JOEL G. HARDMAN

VOLUME XLI. Carbohydrate Metabolism (Part B)
Edited by W. A. WOOD

VOLUME XLII. Carbohydrate Metabolism (Part C)
Edited by W. A. WOOD

VOLUME XLIII. Antibiotics
Edited by JOHN H. HASH

VOLUME XLIV. Immobilized Enzymes
Edited by KLAUS MOSBACH

VOLUME XLV. Proteolytic Enzymes (Part B)
Edited by LASZLO LORAND

VOLUME XLVI. Affinity Labeling
Edited by WILLIAM B. JAKOBY AND MEIR WILCHEK

VOLUME XLVII. Enzyme Structure (Part E)
Edited by C. H. W. HIRS AND SERGE N. TIMASHEFF

VOLUME XLVIII. Enzyme Structure (Part F)
Edited by C. H. W. HIRS AND SERGE N. TIMASHEFF

VOLUME XLIX. Enzyme Structure (Part G)
Edited by C. H. W. HIRS AND SERGE N. TIMASHEFF

VOLUME L. Complex Carbohydrates (Part C)
Edited by VICTOR GINSBURG

VOLUME LI. Purine and Pyrimidine Nucleotide Metabolism
Edited by PATRICIA A. HOFFEE AND MARY ELLEN JONES

VOLUME LII. Biomembranes (Part C: Biological Oxidations)
Edited by SIDNEY FLEISCHER AND LESTER PACKER

VOLUME LIII. Biomembranes (Part D: Biological Oxidations)
Edited by SIDNEY FLEISCHER AND LESTER PACKER

VOLUME LIV. Biomembranes (Part E: Biological Oxidations)
Edited by SIDNEY FLEISCHER AND LESTER PACKER

VOLUME LV. Biomembranes (Part F: Bioenergetics)
Edited by SIDNEY FLEISCHER AND LESTER PACKER

VOLUME LVI. Biomembranes (Part G: Bioenergetics)
Edited by SIDNEY FLEISCHER AND LESTER PACKER

VOLUME LVII. Bioluminescence and Chemiluminescence
Edited by MARLENE A. DELUCA

VOLUME LVIII. Cell Culture
Edited by WILLIAM B. JAKOBY AND IRA PASTAN

VOLUME LIX. Nucleic Acids and Protein Synthesis (Part G)
Edited by KIVIE MOLDAVE AND LAWRENCE GROSSMAN

VOLUME LX. Nucleic Acids and Protein Synthesis (Part H)
Edited by KIVIE MOLDAVE AND LAWRENCE GROSSMAN

VOLUME 61. Enzyme Structure (Part H)
Edited by C. H. W. HIRS AND SERGE N. TIMASHEFF

VOLUME 62. Vitamins and Coenzymes (Part D)
Edited by DONALD B. MCCORMICK AND LEMUEL D. WRIGHT

VOLUME 63. Enzyme Kinetics and Mechanism (Part A: Initial Rate and Inhibitor Methods)
Edited by DANIEL L. PURICH

VOLUME 64. Enzyme Kinetics and Mechanism (Part B: Isotopic Probes and Complex Enzyme Systems)
Edited by DANIEL L. PURICH

VOLUME 65. Nucleic Acids (Part I)
Edited by LAWRENCE GROSSMAN AND KIVIE MOLDAVE

VOLUME 66. Vitamins and Coenzymes (Part E)
Edited by DONALD B. MCCORMICK AND LEMUEL D. WRIGHT

VOLUME 67. Vitamins and Coenzymes (Part F)
Edited by DONALD B. MCCORMICK AND LEMUEL D. WRIGHT

VOLUME 68. Recombinant DNA
Edited by RAY WU

VOLUME 69. Photosynthesis and Nitrogen Fixation (Part C)
Edited by ANTHONY SAN PIETRO

VOLUME 70. Immunochemical Techniques (Part A)
Edited by HELEN VAN VUNAKIS AND JOHN J. LANGONE

VOLUME 71. Lipids (Part C)
Edited by JOHN M. LOWENSTEIN

VOLUME 72. Lipids (Part D)
Edited by JOHN M. LOWENSTEIN

VOLUME 73. Immunochemical Techniques (Part B)
Edited by JOHN J. LANGONE AND HELEN VAN VUNAKIS

VOLUME 74. Immunochemical Techniques (Part C)
Edited by JOHN J. LANGONE AND HELEN VAN VUNAKIS

VOLUME 75. Cumulative Subject Index Volumes XXXI, XXXII, XXXIV–LX
Edited by EDWARD A. DENNIS AND MARTHA G. DENNIS

VOLUME 76. Hemoglobins
Edited by ERALDO ANTONINI, LUIGI ROSSI-BERNARDI, AND EMILIA CHIANCONE

VOLUME 77. Detoxication and Drug Metabolism
Edited by WILLIAM B. JAKOBY

VOLUME 78. Interferons (Part A)
Edited by SIDNEY PESTKA

VOLUME 79. Interferons (Part B)
Edited by SIDNEY PESTKA

VOLUME 80. Proteolytic Enzymes (Part C)
Edited by LASZLO LORAND

VOLUME 81. Biomembranes (Part H: Visual Pigments and Purple Membranes, I)
Edited by LESTER PACKER

VOLUME 82. Structural and Contractile Proteins (Part A: Extracellular Matrix)
Edited by LEON W. CUNNINGHAM AND DIXIE W. FREDERIKSEN

VOLUME 83. Complex Carbohydrates (Part D)
Edited by VICTOR GINSBURG

VOLUME 84. Immunochemical Techniques (Part D: Selected Immunoassays)
Edited by JOHN J. LANGONE AND HELEN VAN VUNAKIS

VOLUME 85. Structural and Contractile Proteins (Part B: The Contractile Apparatus and the Cytoskeleton)
Edited by DIXIE W. FREDERIKSEN AND LEON W. CUNNINGHAM

VOLUME 86. Prostaglandins and Arachidonate Metabolites
Edited by WILLIAM E. M. LANDS AND WILLIAM L. SMITH

VOLUME 87. Enzyme Kinetics and Mechanism (Part C: Intermediates, Stereochemistry, and Rate Studies)
Edited by DANIEL L. PURICH

VOLUME 88. Biomembranes (Part I: Visual Pigments and Purple Membranes, II)
Edited by LESTER PACKER

VOLUME 89. Carbohydrate Metabolism (Part D)
Edited by WILLIS A. WOOD

VOLUME 90. Carbohydrate Metabolism (Part E)
Edited by WILLIS A. WOOD

VOLUME 91. Enzyme Structure (Part I)
Edited by C. H. W. HIRS AND SERGE N. TIMASHEFF

VOLUME 92. Immunochemical Techniques (Part E: Monoclonal Antibodies and General Immunoassay Methods)
Edited by JOHN J. LANGONE AND HELEN VAN VUNAKIS

VOLUME 93. Immunochemical Techniques (Part F: Conventional Antibodies, Fc Receptors, and Cytotoxicity)
Edited by JOHN J. LANGONE AND HELEN VAN VUNAKIS

VOLUME 94. Polyamines
Edited by HERBERT TABOR AND CELIA WHITE TABOR

VOLUME 95. Cumulative Subject Index Volumes 61–74, 76–80
Edited by EDWARD A. DENNIS AND MARTHA G. DENNIS

VOLUME 96. Biomembranes [Part J: Membrane Biogenesis: Assembly and Targeting (General Methods; Eukaryotes)]
Edited by SIDNEY FLEISCHER AND BECCA FLEISCHER

VOLUME 97. Biomembranes [Part K: Membrane Biogenesis: Assembly and Targeting (Prokaryotes, Mitochondria, and Chloroplasts)]
Edited by SIDNEY FLEISCHER AND BECCA FLEISCHER

VOLUME 98. Biomembranes (Part L: Membrane Biogenesis: Processing and Recycling)
Edited by SIDNEY FLEISCHER AND BECCA FLEISCHER

VOLUME 99. Hormone Action (Part F: Protein Kinases)
Edited by JACKIE D. CORBIN AND JOEL G. HARDMAN

VOLUME 100. Recombinant DNA (Part B)
Edited by RAY WU, LAWRENCE GROSSMAN, AND KIVIE MOLDAVE

VOLUME 101. Recombinant DNA (Part C)
Edited by RAY WU, LAWRENCE GROSSMAN, AND KIVIE MOLDAVE

VOLUME 102. Hormone Action (Part G: Calmodulin and Calcium-Binding Proteins)
Edited by ANTHONY R. MEANS AND BERT W. O'MALLEY

VOLUME 103. Hormone Action (Part H: Neuroendocrine Peptides)
Edited by P. MICHAEL CONN

VOLUME 104. Enzyme Purification and Related Techniques (Part C)
Edited by WILLIAM B. JAKOBY

VOLUME 105. Oxygen Radicals in Biological Systems
Edited by LESTER PACKER

VOLUME 106. Posttranslational Modifications (Part A)
Edited by FINN WOLD AND KIVIE MOLDAVE

VOLUME 107. Posttranslational Modifications (Part B)
Edited by FINN WOLD AND KIVIE MOLDAVE

VOLUME 108. Immunochemical Techniques (Part G: Separation and Characterization of Lymphoid Cells)
Edited by GIOVANNI DI SABATO, JOHN J. LANGONE, AND HELEN VAN VUNAKIS

VOLUME 109. Hormone Action (Part I: Peptide Hormones)
Edited by LUTZ BIRNBAUMER AND BERT W. O'MALLEY

VOLUME 110. Steroids and Isoprenoids (Part A)
Edited by JOHN H. LAW AND HANS C. RILLING

VOLUME 111. Steroids and Isoprenoids (Part B)
Edited by JOHN H. LAW AND HANS C. RILLING

VOLUME 112. Drug and Enzyme Targeting (Part A)
Edited by KENNETH J. WIDDER AND RALPH GREEN

VOLUME 113. Glutamate, Glutamine, Glutathione, and Related Compounds
Edited by ALTON MEISTER

VOLUME 114. Diffraction Methods for Biological Macromolecules (Part A)
Edited by HAROLD W. WYCKOFF, C. H. W. HIRS, AND SERGE N. TIMASHEFF

VOLUME 115. Diffraction Methods for Biological Macromolecules (Part B)
Edited by HAROLD W. WYCKOFF, C. H. W. HIRS, AND SERGE N. TIMASHEFF

VOLUME 116. Immunochemical Techniques (Part H: Effectors and Mediators of Lymphoid Cell Functions)
Edited by GIOVANNI DI SABATO, JOHN J. LANGONE, AND HELEN VAN VUNAKIS

VOLUME 117. Enzyme Structure (Part J)
Edited by C. H. W. HIRS AND SERGE N. TIMASHEFF

VOLUME 118. Plant Molecular Biology
Edited by ARTHUR WEISSBACH AND HERBERT WEISSBACH

VOLUME 119. Interferons (Part C)
Edited by SIDNEY PESTKA

VOLUME 120. Cumulative Subject Index Volumes 81–94, 96–101

VOLUME 121. Immunochemical Techniques (Part I: Hybridoma Technology and Monoclonal Antibodies)
Edited by JOHN J. LANGONE AND HELEN VAN VUNAKIS

VOLUME 122. Vitamins and Coenzymes (Part G)
Edited by FRANK CHYTIL AND DONALD B. MCCORMICK

VOLUME 123. Vitamins and Coenzymes (Part H)
Edited by FRANK CHYTIL AND DONALD B. MCCORMICK

VOLUME 124. Hormone Action (Part J: Neuroendocrine Peptides)
Edited by P. MICHAEL CONN

VOLUME 125. Biomembranes (Part M: Transport in Bacteria, Mitochondria, and Chloroplasts: General Approaches and Transport Systems)
Edited by SIDNEY FLEISCHER AND BECCA FLEISCHER

VOLUME 126. Biomembranes (Part N: Transport in Bacteria, Mitochondria, and Chloroplasts: Protonmotive Force)
Edited by SIDNEY FLEISCHER AND BECCA FLEISCHER

VOLUME 127. Biomembranes (Part O: Protons and Water: Structure and Translocation)
Edited by LESTER PACKER

VOLUME 128. Plasma Lipoproteins (Part A: Preparation, Structure, and Molecular Biology)
Edited by JERE P. SEGREST AND JOHN J. ALBERS

VOLUME 129. Plasma Lipoproteins (Part B: Characterization, Cell Biology, and Metabolism)
Edited by JOHN J. ALBERS AND JERE P. SEGREST

VOLUME 130. Enzyme Structure (Part K)
Edited by C. H. W. HIRS AND SERGE N. TIMASHEFF

VOLUME 131. Enzyme Structure (Part L)
Edited by C. H. W. HIRS AND SERGE N. TIMASHEFF

VOLUME 132. Immunochemical Techniques (Part J: Phagocytosis and Cell-Mediated Cytotoxicity)
Edited by GIOVANNI DI SABATO AND JOHANNES EVERSE

VOLUME 133. Bioluminescence and Chemiluminescence (Part B)
Edited by MARLENE DELUCA AND WILLIAM D. MCELROY

VOLUME 134. Structural and Contractile Proteins (Part C: The Contractile Apparatus and the Cytoskeleton)
Edited by RICHARD B. VALLEE

VOLUME 135. Immobilized Enzymes and Cells (Part B)
Edited by KLAUS MOSBACH

VOLUME 136. Immobilized Enzymes and Cells (Part C)
Edited by KLAUS MOSBACH

VOLUME 137. Immobilized Enzymes and Cells (Part D)
Edited by KLAUS MOSBACH

VOLUME 138. Complex Carbohydrates (Part E)
Edited by VICTOR GINSBURG

VOLUME 139. Cellular Regulators (Part A: Calcium- and Calmodulin-Binding Proteins)
Edited by ANTHONY R. MEANS AND P. MICHAEL CONN

VOLUME 140. Cumulative Subject Index Volumes 102–119, 121–134

VOLUME 141. Cellular Regulators (Part B: Calcium and Lipids)
Edited by P. MICHAEL CONN AND ANTHONY R. MEANS

VOLUME 142. Metabolism of Aromatic Amino Acids and Amines
Edited by SEYMOUR KAUFMAN

VOLUME 143. Sulfur and Sulfur Amino Acids
Edited by WILLIAM B. JAKOBY AND OWEN GRIFFITH

VOLUME 144. Structural and Contractile Proteins (Part D: Extracellular Matrix)
Edited by LEON W. CUNNINGHAM

VOLUME 145. Structural and Contractile Proteins (Part E: Extracellular Matrix)
Edited by LEON W. CUNNINGHAM

VOLUME 146. Peptide Growth Factors (Part A)
Edited by DAVID BARNES AND DAVID A. SIRBASKU

VOLUME 147. Peptide Growth Factors (Part B)
Edited by DAVID BARNES AND DAVID A. SIRBASKU

VOLUME 148. Plant Cell Membranes
Edited by LESTER PACKER AND ROLAND DOUCE

VOLUME 149. Drug and Enzyme Targeting (Part B)
Edited by RALPH GREEN AND KENNETH J. WIDDER

VOLUME 150. Immunochemical Techniques (Part K: *In Vitro* Models of B and T Cell Functions and Lymphoid Cell Receptors)
Edited by GIOVANNI DI SABATO

VOLUME 151. Molecular Genetics of Mammalian Cells
Edited by MICHAEL M. GOTTESMAN

VOLUME 152. Guide to Molecular Cloning Techniques
Edited by SHELBY L. BERGER AND ALAN R. KIMMEL

VOLUME 153. Recombinant DNA (Part D)
Edited by RAY WU AND LAWRENCE GROSSMAN

VOLUME 154. Recombinant DNA (Part E)
Edited by RAY WU AND LAWRENCE GROSSMAN

VOLUME 155. Recombinant DNA (Part F)
Edited by RAY WU

VOLUME 156. Biomembranes (Part P: ATP-Driven Pumps and Related Transport: The Na,K-Pump)
Edited by SIDNEY FLEISCHER AND BECCA FLEISCHER

VOLUME 157. Biomembranes (Part Q: ATP-Driven Pumps and Related Transport: Calcium, Proton, and Potassium Pumps)
Edited by SIDNEY FLEISCHER AND BECCA FLEISCHER

VOLUME 158. Metalloproteins (Part A)
Edited by JAMES F. RIORDAN AND BERT L. VALLEE

VOLUME 159. Initiation and Termination of Cyclic Nucleotide Action
Edited by JACKIE D. CORBIN AND ROGER A. JOHNSON

VOLUME 160. Biomass (Part A: Cellulose and Hemicellulose)
Edited by WILLIS A. WOOD AND SCOTT T. KELLOGG

VOLUME 161. Biomass (Part B: Lignin, Pectin, and Chitin)
Edited by WILLIS A. WOOD AND SCOTT T. KELLOGG

VOLUME 162. Immunochemical Techniques (Part L: Chemotaxis and Inflammation)
Edited by GIOVANNI DI SABATO

VOLUME 163. Immunochemical Techniques (Part M: Chemotaxis and Inflammation)
Edited by GIOVANNI DI SABATO

VOLUME 164. Ribosomes
Edited by HARRY F. NOLLER, JR., AND KIVIE MOLDAVE

VOLUME 165. Microbial Toxins: Tools for Enzymology
Edited by SIDNEY HARSHMAN

VOLUME 166. Branched-Chain Amino Acids
Edited by ROBERT HARRIS AND JOHN R. SOKATCH

VOLUME 167. Cyanobacteria
Edited by LESTER PACKER AND ALEXANDER N. GLAZER

VOLUME 168. Hormone Action (Part K: Neuroendocrine Peptides)
Edited by P. MICHAEL CONN

VOLUME 169. Platelets: Receptors, Adhesion, Secretion (Part A)
Edited by JACEK HAWIGER

VOLUME 170. Nucleosomes
Edited by PAUL M. WASSARMAN AND ROGER D. KORNBERG

VOLUME 171. Biomembranes (Part R: Transport Theory: Cells and Model Membranes)
Edited by SIDNEY FLEISCHER AND BECCA FLEISCHER

VOLUME 172. Biomembranes (Part S: Transport: Membrane Isolation and Characterization)
Edited by SIDNEY FLEISCHER AND BECCA FLEISCHER

VOLUME 173. Biomembranes [Part T: Cellular and Subcellular Transport: Eukaryotic (Nonepithelial) Cells]
Edited by SIDNEY FLEISCHER AND BECCA FLEISCHER

VOLUME 174. Biomembranes [Part U: Cellular and Subcellular Transport: Eukaryotic (Nonepithelial) Cells]
Edited by SIDNEY FLEISCHER AND BECCA FLEISCHER

VOLUME 175. Cumulative Subject Index Volumes 135–139, 141–167

VOLUME 176. Nuclear Magnetic Resonance (Part A: Spectral Techniques and Dynamics)
Edited by NORMAN J. OPPENHEIMER AND THOMAS L. JAMES

VOLUME 177. Nuclear Magnetic Resonance (Part B: Structure and Mechanism)
Edited by NORMAN J. OPPENHEIMER AND THOMAS L. JAMES

VOLUME 178. Antibodies, Antigens, and Molecular Mimicry
Edited by JOHN J. LANGONE

VOLUME 179. Complex Carbohydrates (Part F)
Edited by VICTOR GINSBURG

VOLUME 180. RNA Processing (Part A: General Methods)
Edited by JAMES E. DAHLBERG AND JOHN N. ABELSON

VOLUME 181. RNA Processing (Part B: Specific Methods)
Edited by JAMES E. DAHLBERG AND JOHN N. ABELSON

VOLUME 182. Guide to Protein Purification
Edited by MURRAY P. DEUTSCHER

VOLUME 183. Molecular Evolution: Computer Analysis of Protein and Nucleic Acid Sequences
Edited by RUSSELL F. DOOLITTLE

VOLUME 184. Avidin–Biotin Technology
Edited by MEIR WILCHEK AND EDWARD A. BAYER

VOLUME 185. Gene Expression Technology
Edited by DAVID V. GOEDDEL

VOLUME 186. Oxygen Radicals in Biological Systems (Part B: Oxygen Radicals and Antioxidants)
Edited by LESTER PACKER AND ALEXANDER N. GLAZER

VOLUME 187. Arachidonate Related Lipid Mediators
Edited by ROBERT C. MURPHY AND FRANK A. FITZPATRICK

VOLUME 188. Hydrocarbons and Methylotrophy
Edited by MARY E. LIDSTROM

VOLUME 189. Retinoids (Part A: Molecular and Metabolic Aspects)
Edited by LESTER PACKER

VOLUME 190. Retinoids (Part B: Cell Differentiation and Clinical Applications)
Edited by LESTER PACKER

VOLUME 191. Biomembranes (Part V: Cellular and Subcellular Transport: Epithelial Cells)
Edited by SIDNEY FLEISCHER AND BECCA FLEISCHER

VOLUME 192. Biomembranes (Part W: Cellular and Subcellular Transport: Epithelial Cells)
Edited by SIDNEY FLEISCHER AND BECCA FLEISCHER

VOLUME 193. Mass Spectrometry
Edited by JAMES A. MCCLOSKEY

VOLUME 194. Guide to Yeast Genetics and Molecular Biology
Edited by CHRISTINE GUTHRIE AND GERALD R. FINK

VOLUME 195. Adenylyl Cyclase, G Proteins, and Guanylyl Cyclase
Edited by ROGER A. JOHNSON AND JACKIE D. CORBIN

VOLUME 196. Molecular Motors and the Cytoskeleton
Edited by RICHARD B. VALLEE

VOLUME 197. Phospholipases
Edited by EDWARD A. DENNIS

VOLUME 198. Peptide Growth Factors (Part C)
Edited by DAVID BARNES, J. P. MATHER, AND GORDON H. SATO

VOLUME 199. Cumulative Subject Index Volumes 168–174, 176–194

VOLUME 200. Protein Phosphorylation (Part A: Protein Kinases: Assays, Purification, Antibodies, Functional Analysis, Cloning, and Expression)
Edited by TONY HUNTER AND BARTHOLOMEW M. SEFTON

VOLUME 201. Protein Phosphorylation (Part B: Analysis of Protein Phosphorylation, Protein Kinase Inhibitors, and Protein Phosphatases)
Edited by TONY HUNTER AND BARTHOLOMEW M. SEFTON

VOLUME 202. Molecular Design and Modeling: Concepts and Applications (Part A: Proteins, Peptides, and Enzymes)
Edited by JOHN J. LANGONE

VOLUME 203. Molecular Design and Modeling: Concepts and Applications (Part B: Antibodies and Antigens, Nucleic Acids, Polysaccharides, and Drugs)
Edited by JOHN J. LANGONE

VOLUME 204. Bacterial Genetic Systems
Edited by JEFFREY H. MILLER

VOLUME 205. Metallobiochemistry (Part B: Metallothionein and Related Molecules)
Edited by JAMES F. RIORDAN AND BERT L. VALLEE

VOLUME 206. Cytochrome P450
Edited by MICHAEL R. WATERMAN AND ERIC F. JOHNSON

VOLUME 207. Ion Channels
Edited by BERNARDO RUDY AND LINDA E. IVERSON

VOLUME 208. Protein–DNA Interactions
Edited by ROBERT T. SAUER

VOLUME 209. Phospholipid Biosynthesis
Edited by EDWARD A. DENNIS AND DENNIS E. VANCE

VOLUME 210. Numerical Computer Methods
Edited by LUDWIG BRAND AND MICHAEL L. JOHNSON

VOLUME 211. DNA Structures (Part A: Synthesis and Physical Analysis of DNA)
Edited by DAVID M. J. LILLEY AND JAMES E. DAHLBERG

VOLUME 212. DNA Structures (Part B: Chemical and Electrophoretic Analysis of DNA)
Edited by DAVID M. J. LILLEY AND JAMES E. DAHLBERG

VOLUME 213. Carotenoids (Part A: Chemistry, Separation, Quantitation, and Antioxidation)
Edited by LESTER PACKER

VOLUME 214. Carotenoids (Part B: Metabolism, Genetics, and Biosynthesis)
Edited by LESTER PACKER

VOLUME 215. Platelets: Receptors, Adhesion, Secretion (Part B)
Edited by JACEK J. HAWIGER

VOLUME 216. Recombinant DNA (Part G)
Edited by RAY WU

VOLUME 217. Recombinant DNA (Part H)
Edited by RAY WU

VOLUME 218. Recombinant DNA (Part I)
Edited by RAY WU

VOLUME 219. Reconstitution of Intracellular Transport
Edited by JAMES E. ROTHMAN

VOLUME 220. Membrane Fusion Techniques (Part A)
Edited by NEJAT DÜZGÜNEŞ

VOLUME 221. Membrane Fusion Techniques (Part B)
Edited by NEJAT DÜZGÜNEŞ

VOLUME 222. Proteolytic Enzymes in Coagulation, Fibrinolysis, and Complement Activation (Part A: Mammalian Blood Coagulation Factors and Inhibitors)
Edited by LASZLO LORAND AND KENNETH G. MANN

VOLUME 223. Proteolytic Enzymes in Coagulation, Fibrinolysis, and Complement Activation (Part B: Complement Activation, Fibrinolysis, and Nonmammalian Blood Coagulation Factors)
Edited by LASZLO LORAND AND KENNETH G. MANN

VOLUME 224. Molecular Evolution: Producing the Biochemical Data
Edited by ELIZABETH ANNE ZIMMER, THOMAS J. WHITE, REBECCA L. CANN, AND ALLAN C. WILSON

VOLUME 225. Guide to Techniques in Mouse Development
Edited by PAUL M. WASSARMAN AND MELVIN L. DEPAMPHILIS

VOLUME 226. Metallobiochemistry (Part C: Spectroscopic and Physical Methods for Probing Metal Ion Environments in Metalloenzymes and Metalloproteins)
Edited by JAMES F. RIORDAN AND BERT L. VALLEE

VOLUME 227. Metallobiochemistry (Part D: Physical and Spectroscopic Methods for Probing Metal Ion Environments in Metalloproteins)
Edited by JAMES F. RIORDAN AND BERT L. VALLEE

VOLUME 228. Aqueous Two-Phase Systems
Edited by HARRY WALTER AND GÖTE JOHANSSON

VOLUME 229. Cumulative Subject Index Volumes 195–198, 200–227

VOLUME 230. Guide to Techniques in Glycobiology
Edited by WILLIAM J. LENNARZ AND GERALD W. HART

VOLUME 231. Hemoglobins (Part B: Biochemical and Analytical Methods)
Edited by JOHANNES EVERSE, KIM D. VANDEGRIFF, AND ROBERT M. WINSLOW

VOLUME 232. Hemoglobins (Part C: Biophysical Methods)
Edited by JOHANNES EVERSE, KIM D. VANDEGRIFF, AND ROBERT M. WINSLOW

VOLUME 233. Oxygen Radicals in Biological Systems (Part C)
Edited by LESTER PACKER

VOLUME 234. Oxygen Radicals in Biological Systems (Part D)
Edited by LESTER PACKER

VOLUME 235. Bacterial Pathogenesis (Part A: Identification and Regulation of Virulence Factors)
Edited by VIRGINIA L. CLARK AND PATRIK M. BAVOIL

VOLUME 236. Bacterial Pathogenesis (Part B: Integration of Pathogenic Bacteria with Host Cells)
Edited by VIRGINIA L. CLARK AND PATRIK M. BAVOIL

VOLUME 237. Heterotrimeric G Proteins
Edited by RAVI IYENGAR

VOLUME 238. Heterotrimeric G-Protein Effectors
Edited by RAVI IYENGAR

VOLUME 239. Nuclear Magnetic Resonance (Part C)
Edited by THOMAS L. JAMES AND NORMAN J. OPPENHEIMER

VOLUME 240. Numerical Computer Methods (Part B)
Edited by MICHAEL L. JOHNSON AND LUDWIG BRAND

VOLUME 241. Retroviral Proteases
Edited by LAWRENCE C. KUO AND JULES A. SHAFER

VOLUME 242. Neoglycoconjugates (Part A)
Edited by Y. C. LEE AND REIKO T. LEE

VOLUME 243. Inorganic Microbial Sulfur Metabolism
Edited by HARRY D. PECK, JR., AND JEAN LEGALL

VOLUME 244. Proteolytic Enzymes: Serine and Cysteine Peptidases
Edited by ALAN J. BARRETT

VOLUME 245. Extracellular Matrix Components
Edited by E. RUOSLAHTI AND E. ENGVALL

VOLUME 246. Biochemical Spectroscopy
Edited by KENNETH SAUER

VOLUME 247. Neoglycoconjugates (Part B: Biomedical Applications)
Edited by Y. C. LEE AND REIKO T. LEE

VOLUME 248. Proteolytic Enzymes: Aspartic and Metallo Peptidases
Edited by ALAN J. BARRETT

VOLUME 249. Enzyme Kinetics and Mechanism (Part D: Developments in Enzyme Dynamics)
Edited by DANIEL L. PURICH

VOLUME 250. Lipid Modifications of Proteins
Edited by PATRICK J. CASEY AND JANICE E. BUSS

VOLUME 251. Biothiols (Part A: Monothiols and Dithiols, Protein Thiols, and Thiyl Radicals)
Edited by LESTER PACKER

VOLUME 252. Biothiols (Part B: Glutathione and Thioredoxin; Thiols in Signal Transduction and Gene Regulation)
Edited by LESTER PACKER

VOLUME 253. Adhesion of Microbial Pathogens
Edited by RON J. DOYLE AND ITZHAK OFEK

VOLUME 254. Oncogene Techniques
Edited by PETER K. VOGT AND INDER M. VERMA

VOLUME 255. Small GTPases and Their Regulators (Part A: Ras Family)
Edited by W. E. BALCH, CHANNING J. DER, AND ALAN HALL

VOLUME 256. Small GTPases and Their Regulators (Part B: Rho Family)
Edited by W. E. BALCH, CHANNING J. DER, AND ALAN HALL

VOLUME 257. Small GTPases and Their Regulators (Part C: Proteins Involved in Transport)
Edited by W. E. BALCH, CHANNING J. DER, AND ALAN HALL

VOLUME 258. Redox-Active Amino Acids in Biology
Edited by JUDITH P. KLINMAN

VOLUME 259. Energetics of Biological Macromolecules
Edited by MICHAEL L. JOHNSON AND GARY K. ACKERS

VOLUME 260. Mitochondrial Biogenesis and Genetics (Part A)
Edited by GIUSEPPE M. ATTARDI AND ANNE CHOMYN

VOLUME 261. Nuclear Magnetic Resonance and Nucleic Acids
Edited by THOMAS L. JAMES

VOLUME 262. DNA Replication
Edited by JUDITH L. CAMPBELL

VOLUME 263. Plasma Lipoproteins (Part C: Quantitation)
Edited by WILLIAM A. BRADLEY, SANDRA H. GIANTURCO, AND JERE P. SEGREST

VOLUME 264. Mitochondrial Biogenesis and Genetics (Part B)
Edited by GIUSEPPE M. ATTARDI AND ANNE CHOMYN

VOLUME 265. Cumulative Subject Index Volumes 228, 230–262

VOLUME 266. Computer Methods for Macromolecular Sequence Analysis
Edited by RUSSELL F. DOOLITTLE

VOLUME 267. Combinatorial Chemistry
Edited by JOHN N. ABELSON

VOLUME 268. Nitric Oxide (Part A: Sources and Detection of NO; NO Synthase)
Edited by LESTER PACKER

VOLUME 269. Nitric Oxide (Part B: Physiological and Pathological Processes)
Edited by LESTER PACKER

VOLUME 270. High Resolution Separation and Analysis of Biological Macromolecules (Part A: Fundamentals)
Edited by BARRY L. KARGER AND WILLIAM S. HANCOCK

VOLUME 271. High Resolution Separation and Analysis of Biological Macromolecules (Part B: Applications)
Edited by BARRY L. KARGER AND WILLIAM S. HANCOCK

VOLUME 272. Cytochrome P450 (Part B)
Edited by ERIC F. JOHNSON AND MICHAEL R. WATERMAN

VOLUME 273. RNA Polymerase and Associated Factors (Part A)
Edited by SANKAR ADHYA

VOLUME 274. RNA Polymerase and Associated Factors (Part B)
Edited by SANKAR ADHYA

VOLUME 275. Viral Polymerases and Related Proteins
Edited by LAWRENCE C. KUO, DAVID B. OLSEN, AND STEVEN S. CARROLL

VOLUME 276. Macromolecular Crystallography (Part A)
Edited by CHARLES W. CARTER, JR., AND ROBERT M. SWEET

VOLUME 277. Macromolecular Crystallography (Part B)
Edited by CHARLES W. CARTER, JR., AND ROBERT M. SWEET

VOLUME 278. Fluorescence Spectroscopy
Edited by LUDWIG BRAND AND MICHAEL L. JOHNSON

VOLUME 279. Vitamins and Coenzymes (Part I)
Edited by DONALD B. MCCORMICK, JOHN W. SUTTIE, AND CONRAD WAGNER

VOLUME 280. Vitamins and Coenzymes (Part J)
Edited by DONALD B. MCCORMICK, JOHN W. SUTTIE, AND CONRAD WAGNER

VOLUME 281. Vitamins and Coenzymes (Part K)
Edited by DONALD B. MCCORMICK, JOHN W. SUTTIE, AND CONRAD WAGNER

VOLUME 282. Vitamins and Coenzymes (Part L)
Edited by DONALD B. MCCORMICK, JOHN W. SUTTIE, AND CONRAD WAGNER

VOLUME 283. Cell Cycle Control
Edited by WILLIAM G. DUNPHY

VOLUME 284. Lipases (Part A: Biotechnology)
Edited by BYRON RUBIN AND EDWARD A. DENNIS

VOLUME 285. Cumulative Subject Index Volumes 263, 264, 266–284, 286–289

VOLUME 286. Lipases (Part B: Enzyme Characterization and Utilization)
Edited by BYRON RUBIN AND EDWARD A. DENNIS

VOLUME 287. Chemokines
Edited by RICHARD HORUK

VOLUME 288. Chemokine Receptors
Edited by RICHARD HORUK

VOLUME 289. Solid Phase Peptide Synthesis
Edited by GREGG B. FIELDS

VOLUME 290. Molecular Chaperones
Edited by GEORGE H. LORIMER AND THOMAS BALDWIN

VOLUME 291. Caged Compounds
Edited by GERARD MARRIOTT

VOLUME 292. ABC Transporters: Biochemical, Cellular, and Molecular Aspects
Edited by SURESH V. AMBUDKAR AND MICHAEL M. GOTTESMAN

VOLUME 293. Ion Channels (Part B)
Edited by P. MICHAEL CONN

VOLUME 294. Ion Channels (Part C)
Edited by P. MICHAEL CONN

VOLUME 295. Energetics of Biological Macromolecules (Part B)
Edited by GARY K. ACKERS AND MICHAEL L. JOHNSON

VOLUME 296. Neurotransmitter Transporters
Edited by SUSAN G. AMARA

VOLUME 297. Photosynthesis: Molecular Biology of Energy Capture
Edited by LEE MCINTOSH

VOLUME 298. Molecular Motors and the Cytoskeleton (Part B)
Edited by RICHARD B. VALLEE

VOLUME 299. Oxidants and Antioxidants (Part A)
Edited by LESTER PACKER

VOLUME 300. Oxidants and Antioxidants (Part B)
Edited by LESTER PACKER

VOLUME 301. Nitric Oxide: Biological and Antioxidant Activities (Part C)
Edited by LESTER PACKER

VOLUME 302. Green Fluorescent Protein
Edited by P. MICHAEL CONN

VOLUME 303. cDNA Preparation and Display
Edited by SHERMAN M. WEISSMAN

VOLUME 304. Chromatin
Edited by PAUL M. WASSARMAN AND ALAN P. WOLFFE

VOLUME 305. Bioluminescence and Chemiluminescence (Part C) (in preparation)
Edited by MIRIAM M. ZIEGLER AND THOMAS O. BALDWIN

VOLUME 306. Expression of Recombinant Genes in Eukaryotic Systems
Edited by JOSEPH C. GLORIOSO AND MARTIN C. SCHMIDT

VOLUME 307. Confocal Microscopy
Edited by P. MICHAEL CONN

VOLUME 308. Enzyme Kinetics and Mechanism (Part E: Energetics of Enzyme Catalysis)
Edited by VERN L. SCHRAMM AND DANIEL L. PURICH

VOLUME 309. Amyloids, Prions, and Other Protein Aggregates
Edited by RONALD WETZEL

VOLUME 310. Biofilms
Edited by RON J. DOYLE

VOLUME 311. Sphingolipid Metabolism and Cell Signaling (Part A)
Edited by ALFRED H. MERRILL, JR., AND Y. A. HANNUN

VOLUME 312. Sphingolipid Metabolism and Cell Signaling (Part B) (in preparation)
Edited by ALFRED H. MERRILL, JR., AND Y. A. HANNUN

VOLUME 313. Antisense Technology (Part A: General Methods, Methods of Delivery, and RNA Studies)
Edited by M. IAN PHILLIPS

VOLUME 314. Antisense Technology (Part B: Applications)
Edited by M. IAN PHILLIPS

VOLUME 315. Vertebrate Phototransduction and the Visual Cycle (Part A)
Edited by KRZYSZTOF PALCZEWSKI

VOLUME 316. Vertebrate Phototransduction and the Visual Cycle (Part B) (in preparation)
Edited by KRZYSZTOF PALCZEWSKI

VOLUME 317. RNA-Ligand Interactions (Part A: Structural Biology Methods) (in preparation)
Edited by DANIEL W. CELANDER AND JOHN N. ABELSON

VOLUME 318. RNA-Ligand Interactions (Part B: Molecular Biology Methods) (in preparation)
Edited by DANIEL W. CELANDER AND JOHN N. ABELSON

VOLUME 319. Singlet Oxygen, UV-A, and Ozone (in preparation)
Edited by LESTER PACKER AND HELMUT SIES

VOLUME 320. Cumulative Subject Index Volumes 290–319 (in preparation)

VOLUME 321. Numerical Computer Methods (Part C) (in preparation)
Edited by MICHAEL L. JOHNSON AND LUDWIG BRAND

VOLUME 322. Apoptosis (in preparation)
Edited by JOHN C. REED

VOLUME 323. Energetics of Biological Macromolecules (Part C) (in preparation)
Edited by MICHAEL L. JOHNSON AND GARY K. ACKERS

VOLUME 324. Branched Chain Amino Acids (Part B) (in preparation)
Edited by ROBERT A. HARRIS AND JOHN R. SOKATCH

Section I

Expression and Isolation of Opsins

[1] Heterologous Expression of Bovine Opsin in *Pichia pastoris*

By NAJMOUTIN G. ABDULAEV and KEVIN D. RIDGE

Introduction

The overproduction of bovine opsin and structurally related receptors in a variety of heterologous systems has so far proven quite difficult and consequently limited both biophysical and structural investigations. Therefore, a highly efficient expression system capable of producing modified opsin or its defined domains in the quantities required for these analyses would be beneficial. Recombinant opsin has been produced by transient and/or stable transfection of COS-1 and human embryonic kidney (HEK) 293 cells,[1-3] RNA injection of *Xenopus laevis* oocytes,[4] in baculovirus-infected Sf9 insect cells,[5] by RNA translation in wheat germ extracts,[6] and recently, in *Saccharomyces cerevisiae*.[7] Although expression in COS-1 and HEK293 cells remain the methods of choice for most structure–function studies on rhodopsin, their use for large-scale production is rather laborious and expensive. The purpose of this chapter is to describe methods for the functional expression of bovine opsin in *Pichia pastoris*,[8] a methylotrophic yeast capable of producing considerable quantities of foreign protein.[9] The focus is on the construction of a suitable expression vector, conditions for

---

[1] D. D. Oprian, R. S. Molday, R. J. Kaufman, and H. G. Khorana, *Proc. Natl. Acad. Sci. U.S.A.* **84,** 8874 (1987).
[2] J. Nathans, C. J. Weitz, N. Agarwal, I. Nir, and D. S. Papermaster, *Vision Res.* **29,** 907 (1989).
[3] P. J. Reeves, R. L. Thurmond, and H. G. Khorana, *Proc. Natl. Acad. Sci. U.S.A.* **84,** 11487 (1996).
[4] H. G. Khorana, B. E. Knox, E. Nasi, R. Swanson, and D. A. Thompson, *Proc. Natl. Acad. Sci. U.S.A.* **85,** 7917 (1988).
[5] J. J. M. Jansen, W. J. van de Ven, W. A. H. M. van Groningen-Luyben, J. Roosien, J. M. Vlak, and W. J. DeGrip, *Mol. Biol. Rep.* **13,** 65 (1988).
[6] S. A. Zozulya, V. V. Gurevich, T. A. Zvyaga, E. P. Shirokova, I. L. Dumler, M. N. Garnovskaya, M. Yu. Natochin, B. E. Shmukler, and P. R. Badalov, *Prot. Eng.* **3,** 453 (1990).
[7] R. Mollaghababa, F. F. Davidson, C. Kaiser, and H. G. Khorana, *Proc. Natl. Acad. Sci. U.S.A.* **93,** 11482 (1996).
[8] N. G. Abdulaev, M. P. Popp, W. C. Smith, and K. D. Ridge, *Prot. Express. Purif.* **10,** 61 (1997).
[9] J. M. Cregg, J. F. Tschopp, C. Stillman, R. Siegel, M. Akong, W. S. Craig, R. G. Buckholtz, K. R. Madden, P. A. Kellaris, G. R. Davis, B. L. Smiley, J. Cruze, R. Torregrossa, G. Velicelebi, and G. P. Thill, *Bio/Technology* **5,** 479 (1987).

stable transformation and expression in *P. pastoris* cells, and the purification and characterization of yeast expressed rhodopsin.

Cloning and Expression of Bovine Opsin in *Pichia pastoris*

Media

Medium A contains yeast extract (10 g/liter), proteose peptone (20 g/liter), and dextrose (200 g/liter). Medium B contains yeast nitrogen base without amino acids (13.4 g/liter), biotin (0.4 mg/liter), and glycerol (10%, v/v). Medium C contains yeast nitrogen base without amino acids (13.4 g/liter), biotin (0.4 mg/liter), and methanol (0.5%, v/v).

Choice and Construction of Expression Vector

The pHIL-S1 plasmid (Invitrogen, Carlsbad, CA) is the preferred vector for expression of the bovine opsin gene in *P. pastoris* because it contains sequences that code for the acid phosphatase (*PHO1*) signal sequence (S), a native *P. pastoris* secretion signal. The rationale behind this choice of vector is that the presence of the secretion signal would likely target the opsin to the plasma membrane. The vector contains a unique *Eco*RI restriction site for cloning downstream from the *P. pastoris* alcohol oxidase 1 (*AOX1*) gene promoter and the secretion signal. Cloning the opsin gene into the *Eco*RI site results in an in-frame insertion with the secretion signal open reading frame. This insertion point is immediately downstream from the secretion signal cleavage site, such that the resulting opsin contains at most three additional amino acids at its amino terminus after proteolytic processing. Therefore, *Eco*RI sites are created at the 5′ and 3′ ends of the bovine opsin gene using the polymerase chain reaction (PCR) and primers 1 (5′-GCGAATTCATGAACGGGACCGAGGGCCCAAA) and 2 (5′-GCGAATTCTTAGGCAGGCGCCACTTGGCTGGTCT). The reaction mixture contains 1 ng of bovine opsin cDNA (provided by Dr. J. Nathans, Johns Hopkins University, Baltimore, MD), 15 pmol of each primer, 200 μM dNTPs, and 1 unit of Thermus brokianus DNA polymerase (Whatman Biometra, Göttingen, Germany) in 10 mM Tris-HCl, pH 8.8, 50 mM KCl, 1.5 mM MgCl$_2$, and 0.1% (v/v) Triton X-100. PCR amplifications are typically 30 cycles at 94° for 30 sec, 55° for 30 sec, and 72° for 2 min. The amplified product is purified by electrophoresis through agarose, digested with *Eco*RI, and subcloned into the *Eco*RI linearized pHIL-S1 plasmid.

Proper insertion and orientation of the opsin gene is confirmed by restriction enzyme analysis and sequencing.

Transformation of P. pastoris

The *P. pastoris* strain GS115 (*his4*) (Invitrogen) is maintained and transformed by electroporation essentially as described in the Manual Version D of the *Pichia* expression kit. Briefly, GS115 cells are grown overnight at 30° in medium A to an A_{600} of 1.4–2.0 and then pelleted by centrifugation at 1500g for 5 min at 4°. The cell pellet is resuspended in 500 ml of ice-cold sterile water and centrifuged again. This same procedure is repeated using 200 ml of ice-cold sterile water and then 1 ml of ice-cold sterile 1 M sorbitol. For transformation, the opsin/pHIL-S1 plasmid (1 μg) is linearized at the unique *Sal*I site and mixed with 40 μl of cells. Immediately after the pulse, 1 ml of ice-cold 1 M sorbitol is added and aliquots (200 μl) of the transformation mixture are spread on dextrose/agar plates lacking histidine. After 16–18 hr at 30°, colonies are isolated and cultured in medium A.

Expression in P. pastoris

Typically, several independent clones are screened and characterized by sodium dodecyl sulfate–polyacrylamide gel electrophoresis (SDS–PAGE) and immunoblot analysis to select for cells with the highest expression levels. The transformed GS115 cells are grown in 5 ml of medium B at 30° to an A_{600} of 10.0. Aliqouts (500 μl) are removed and added to 5 ml of induction medium C and incubated for 48 hr at 30°. A comparison of the YopspHIL-S1–8 clone, which contains a single copy of the opsin gene,[8] with rod outer segment (ROS) opsin is shown in Fig. 1A. Only those cells induced with methanol yield an immunoreactive band which migrates as a single band of ~37 kDa (lane 2), an apparent molecular mass that is slightly lower than that of ROS opsin (lane 3). Based on comparative immunoblot analysis, the yield of protein from this particular clone is ~0.3 mg of opsin per liter of culture. Since 1 liter of yeast grown to an A_{600} of 1.0 yields approximately 40 mg of membrane protein, the expression level of opsin in the YopspHIL-S1–8 clone accounts for ~0.75% of the total membrane protein. For large-scale growth, transformed GS115 cells are grown in 25 ml of medium B at 30° to an A_{600} of 10.0. The cells are harvested by centrifugation at 1500g for 5 min at 4° and resuspended in 250 ml of medium C in a 1-liter capacity shake flask. After further incubation for 2–6 days at

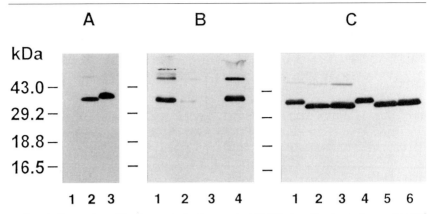

FIG. 1. Expression of bovine opsin in *P. pastoris*. (A) Whole-cell extracts from ~5×10^5 cells before (lane 1) and after (lane 2) induction with methanol (medium C). Lane 3 contains 10 ng of ROS rhodopsin. (B) Fractions obtained from whole-cell extracts after sucrose density sedimentation analysis. Lane 1, membrane containing interface fraction; lane 2, 50% sucrose phase; lane 3, pellet of debris; lane 4, 20 ng of ROS rhodopsin. (C) Effect of glycosidase treatment on yeast expressed and ROS rhodopsins. Lanes 1 and 4, yeast expressed and ROS rhodopsins, respectively; lanes 2 and 5, endoglycosidase H-treated yeast expressed and ROS rhodopsins, respectively; lanes 3 and 6, PNGase F-treated yeast expressed and ROS rhodopsins, respectively. The samples are analyzed by reducing SDS–PAGE, and the proteins visualized by immunoblotting with the rho 1D4 antibody. Positions of molecular size standards are shown at the left in kilodaltons. The immunoreactive bands that migrate above the 43-kDa standard (B) and (C) represent opsin dimers.

30°, the cells are harvested and processed immediately or stored at $-80°$ until use.

Subcellular Localization, Chromophore Formation, and Purification of *P. pastoris* Expressed Rhodopsin

Buffers and Solutions

Buffer A is 7 mM NaH$_2$PO$_4$, pH 6.5, containing 7 mM ethylenediaminetetraacetic acid (EDTA) and 7 mM dithioerythritol. Buffer B is 20 mM NaH$_2$PO$_4$, pH 6.5, containing 140 mM NaCl, 3 mM MgCl$_2$, and 2 mM CaCl$_2$. Buffer C is the same as buffer B except that it also contains 10% (w/v) dodecyl maltoside (DM). Buffer D is 20 mM NaH$_2$PO$_4$, pH 6.5, containing 140 mM NaCl, 2 mM MgCl$_2$, 2 mM adenosine triphosphate (ATP), and 1% (w/v) DM. Buffers A–D are also supplemented with Complete protease inhibitor tablets (Roche Molecular Biochemicals, Mannheim, Germany) and 1 mM phenylmethylsulfonyl fluoride (PMSF). Buffer E is 2 mM NaH$_2$PO$_4$, pH 6.2, containing 0.1% (w/v) DM. Buffer F is the same as buffer E except that it also contains 35 μM c'-TETSQVAPA peptide, which corresponds to the carboxyl-terminal nine amino acids of bovine opsin.

Subcellular Localization and Posttranslational Modifications

To determine whether the opsin is membrane integrated, whole-cell extracts are prepared and layered on a sucrose cushion and centrifuged. Briefly, induced yeast cultures grown to an A_{600} of 1.5–1.6 are centrifuged at 1500g for 5 min at 4°, and the pellet washed once with ice-cold water. The pellet is suspended in buffer A, immediately centrifuged, and resuspended in two pellet volumes of the same buffer. Four pellet volumes of ice-cold acid-washed glass beads (Biospec Products, Bartlesville, OK) are added and the cells disrupted by vigorous mixing for 10 min at 4°. It is important to maintain this cells : buffer : glass beads ratio since optimal cell breakage can be achieved under these conditions. The cell debris is removed by centrifugation at 1500g for 5 min at 4° and the cell lysate layered on a sucrose cushion (50%, w/v, in buffer A) and centrifuged at 120,000g for 1 hr at 4° in a swinging bucket rotor (SW28, Beckman Coulter, Fullerton, CA). Three different fractions are apparent and each is collected by puncturing the tube with an 18-gauge needle and analyzed for opsin content by immunoblotting. Virtually all of the opsin is found in the membrane containing interface fraction located on the border of the buffer/sucrose cushion (Fig. 1B, lane 1). Only a negligible amount of opsin is observed in the slightly turbid sucrose solution, while none is detected in the heavy pellet of debris (Fig. 1B, lanes 2 and 3, respectively). Although this analysis does not show the specific location of the opsin in the cells, the above results indicate that it is integrated into the membrane.

ROS opsin is N-glycosylated at Asn-2 and Asn-15 in the amino-terminal region of the protein. The presence and extent of N-glycosylation on *P. pastoris* expressed opsin can be compared with that of ROS opsin by glycosidase treatment. Typically, crude membrane extracts are incubated with 500 units of endoglycosidase H or PNGase F (Roche Molecular Biochemicals) in 20 mM Tris-HCl, pH 8.0, containing 0.5% (w/v) SDS, 1% (v/v) Nonidet P-40 (NP-40), 5 mM EDTA, and 0.1 mM PMSF for 30 min at 20°. The glycosidase-treated proteins are analyzed by SDS–PAGE and immunoblotting. The yeast expressed apoprotein shows a faster migrating polypeptide after treatment with both endoglycosidase H and PNGase F (Fig. 1C, lanes 1–3). Thus, like ROS opsin (Fig. 1C, lanes 4–6), yeast expressed opsin is high mannose N-glycosylated.

Chromophore Formation and Purification

To test whether the yeast expressed opsin is capable of forming a chromophore with 11-*cis*-retinal, the interface fraction of crude membranes is collected and immediately resuspended in 10 volumes of buffer B. All subsequent manipulations are carried out under dim red light. The crude membrane fraction is incubated with 11-*cis*-retinal (2.5 μM in ethanol) for 2 hr at 4° with gentle agitation. The membranes are collected by centrifugation at 100,000g for 30 min at 4°, the pellets combined, and solubilized in buffer C for 1 hr at 4°. The insoluble material is removed by centrifugation at 100,000g for 1 hr at 4°. The supernatant, containing the solubilized rhodopsin, is purified by immunoaffinity chromatography on rho 1D4-Sepharose.[1] Briefly, to 15 ml of soluble membrane extract is added 0.2 ml of immobilized rho 1D4 (binding capacity = 0.3–0.5 μg rhodopsin/μl resin). The mixture is incubated for 3 hr at 4° with gentle agitation, washed five times with 20 bed volumes of buffer D, and then an additional five times with 20 bed volumes of buffer E. The bound proteins are eluted from the immobilized antibody with buffer F.

Functional Properties of *P. pastoris* Expressed Rhodopsin

Spectral Characterization

UV/visible spectroscopy of the purified yeast expressed pigment shows the characteristic rhodopsin spectrum with absorption maxima at 280 and 500 nm (Fig. 2A). The spectrum shown is obtained from a sample that has been eluted from the immobilized rho 1D4 antibody under conditions of low pH and low ionic strength (buffer F). Here, the ratio of the absorbance at 280 to that at 500 nm is 1.6–1.7, which is characteristic of purified rhodopsin.[10] However, preparations of *P. pastoris* rhodopsin that are eluted from the immunoaffinity matrix with high ionic strength buffers show considerably higher A_{280}/A_{500} ratios (~3–5). This increase is presumably due to the coelution of both correctly folded-regenerated rhodopsin and misfolded opsin that does not bind retinal.[11] The yeast expressed rhodopsin shows a shift in λ_{max} to 380 nm on illumination (>495 nm) and forms the protonated retinyl-Schiff base (λ_{max} 440 nm) on acidification.[12]

---

[10] P. A. Hargrave, *Prog. Ret. Res.* **1,** 1 (1982).
[11] K. D. Ridge, Z. Lu, X. Liu, and H. G. Khorana, *Biochemistry* **34,** 3261 (1995).
[12] G. Wald, and P. K. Brown, *J. Gen. Physiol.* **37,** 189 (1953).

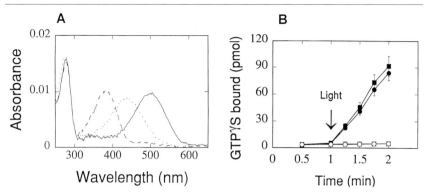

FIG. 2. Spectral and functional characterization of *P. pastoris* expressed rhodopsin. (A) UV/visible spectra of yeast expressed rhodopsin. Spectra are recorded in the dark (continuous line), after illumination (>495 nm) for 10 sec (dashed line), and after acidification to pH ~1.9 (dotted line). (B) Light-dependent activation of G_t by yeast expressed and ROS rhodopsins. Time course for the reaction catalyzed by ROS (circles) and yeast expressed (squares) rhodopsins. Closed symbols are data for the light reaction and open symbols are data for the dark reaction. The results shown are averages ± standard error from two independent determinations.

G-Protein Activation

To test whether the yeast expressed rhodopsin activates G-protein *in vitro*, the activity of the pigment is determined by following the light-dependent binding of [^{35}S]GTPγS by transducin (G_t). G_t is prepared from bovine retina by the method of Fung *et al.*[13] The assay mixture consists of 2 nM rhodopsin, 2 μM G_t, and 4 μM [^{35}S]GTPγS in 10 mM Tris-HCl, pH 7.5, containing 100 mM NaCl, 5 mM MgCl$_2$, 2 mM dithiothreitol (DTT), and 0.012% (w/v) DM. The time course is begun in darkness with the addition of [^{35}S]GTPγS and after a 1-min incubation at 20°, the assay mixture is illuminated (>495 nm) or allowed to remain in darkness. At 15-sec intervals, 35 μl of the 250-μl reaction mixture is removed and filtered through nitrocellulose using a vacuum manifold (Millipore, Bedford, MA). The filters, which retain G_t and its bound nucleotide, are washed four times with 5 ml of 10 mM Tris-HCl, pH 7.5, containing 100 mM NaCl, 5 mM MgCl$_2$, and 2 mM DTT, dried, and analyzed for ^{35}S radioactivity by scintillation counting. As shown in Fig. 2B, the yeast expressed rhodopsin catalyzes light-dependent [^{35}S]GTPγS binding by G_t to ~90% of the level of ROS rhodopsin.

[13] B. K. K. Fung, J. B. Hurley, and L. Stryer, *Proc. Natl. Acad. Sci. U.S.A.* **78**, 152 (1981).

Comments

Several key features of the *P. pastoris* expression system make it an attractive candidate for heterologous expression of opsin. First, the gene of interest is placed under the transcriptional control of the tightly regulated and highly inducible *AOX1* promoter. In some cases, expression of foreign genes from this promoter has yielded gram quantities of the desired protein. Second, the gene of interest can also be fused at its amino terminus to a native *P. pastoris* secretion signal (*PHO1*), a feature that may influence the membrane localization of the expressed protein. Importantly, using a similar vector that lacks the secretion signal fails to produce detectable amounts of membrane-bound opsin. Finally, *P. pastoris* cells possess the necessary machinery for many posttranslational modifications found in higher eukaryotes.

Several lines of evidence suggest that *P. pastoris* expressed opsin is similar to ROS opsin. First, the yeast expressed apoprotein shows a single band of ~37 kDa on SDS–PAGE, an apparent molecular mass that is slightly lower than that of ROS opsin. Second, based on glycosidase sensitivity, the yeast expressed opsin, like ROS opsin, appears to be high mannose N-glycosylated. Although the precise nature of the oligosaccharide moieties on the *P. pastoris* opsin is not evident from this work, it is possible that differences in carbohydrate composition between this and ROS opsin may contribute to the slight difference in apparent molecular mass. Opsin expressed in COS-1 and HEK293 cells is heterogeneously N-glycosylated as evidenced by its tendency to smear on SDS–PAGE.[1-3] On the other hand, opsin preparations from baculovirus-infected Sf9 cells show additional immunoreactive bands that have been attributed to the nonglycosylated form of the protein.[5] Importantly, no heterogeneously glycosylated material or additional immunoreactive bands are apparent in preparations of the yeast expressed opsin. Third, and perhaps most important, a portion of the *P. pastoris* expressed opsin binds exogenously added 11-*cis*-retinal to form the characteristic rhodopsin chromophore. The resulting pigment also undergoes the same light-dependent and acid-induced conformational changes as ROS rhodopsin. Finally, the yeast expressed rhodopsin activates G-protein in a light-dependent manner. Although the reason for the slightly reduced signaling capacity is not clear at the present time, it may also be related to differences in carbohydrate composition since N-glycosylation of bovine opsin appears to be important for the G-protein activating potential of metarhodopsin II.[14]

---

[14] S. Kaushal, K. D. Ridge, and H. G. Khorana, *Proc. Natl. Acad. Sci. U.S.A.* **91**, 4029 (1994).

The yield of correctly folded *P. pastoris* expressed rhodopsin after purification varies from ~10 to 40 μg per liter of shake flask culture (A_{600} = 1.0; ~5 × 10^{10} cells), or roughly 4–15% of the total opsin apoprotein produced. For comparison, an equivalent number of transiently transfected COS-1 cells would produce ~2 mg of rhodopsin,[11] whereas the same number of baculovirus-infected Sf9 cells would yield 3–4 mg of rhodopsin.[6] Although *P. pastoris* cells can be grown and induced for longer periods of time to produce higher milligram amounts of opsin, this does not dramatically change the regenerability of the opsin. These findings suggest that a single copy of the opsin gene is sufficient for high-level expression and that other factors influence the ability of the opsin to adopt a retinal-binding conformation.[8,15]

Identifying additional parameters that lead to increased levels of chromophore formation for opsins expressed in several species of yeast are likely to be mutually beneficial. Such success would allow for the high-level production of mutants and metabolically labeled analogs of opsin, providing an extremely powerful tool for more extensive and detailed structure–function studies.

Acknowledgments

We would like to recognize the contributions of W. Clay Smith and Michael Popp in some aspects of these studies. This work was supported by National Institutes of Health Grant EY11112. Certain commercial materials, instruments, and equipment are identified in this manuscript in order to specify the experimental procedure as completely as possible. In no case does such identification imply a recommendation or endorsement by the National Institute of Standards and Technology nor does it imply that the materials, instruments, or equipment identified are necessarily the best available for the purpose.

[15] C.-H. Sung, B. G. Schneider, N. Agarwal, D. S. Papermaster, and J. Nathans, *Proc. Natl. Acad. Sci. U.S.A.* **88**, 8840 (1991).

[2] Baculovirus Expression System for Expression and Characterization of Functional Recombinant Visual Pigments

By CORNÉ H. W. KLAASSEN and WILLEM J. DEGRIP

Introduction

Visual pigments are a family of closely related proteins mediating the initial event in the visual response in the eye. All members of this family consist of an apoprotein of 350–400 amino acids in a topologic pattern of seven transmembrane segments (TMs), connected by relatively short loops, with the N terminus facing the luminal side of the photoreceptor disk membrane and the C terminus facing the cytoplasm. Responsible for the absorption of visible light is a chromophore (11-*cis*-retinal) covalently attached via a protonated Schiff base linkage to a lysine residue in the seventh TM of the apoprotein.

In the last decade it has become clear that the visual pigments represent a subclass of the superfamily of G-protein-coupled receptors (GPCRs), which all share the 7-TM motif and employ a heterotrimeric G protein for further signal transduction and amplification.[1] This superfamily harbors a large number of other, physiologically prominent receptors (e.g., involved in adrenergic, serotonergic, dopaminergic, and a variety of peptidergic and glycoprotein hormone-induced signaling pathways) and presents important targets for pharmacologic intervention in human disease. About half of the new drugs marketed during the last several years address members of this receptor family. The rod visual pigment rhodopsin is the only representative that can be isolated from native tissue in sufficient amount to allow detailed biochemical and biophysical studies (see elsewhere this volume). In fact, rhodopsin has become a paradigm for the mechanism of receptor activation and the subsequent signal transduction pathways of G-protein-coupled receptors.

Within the general context of the GPCR family, the visual pigments show an intriguing specialization.[2] Their ligand (11-*cis*-retinal) is covalently bound to the receptor and acts as an inverse agonist in the dark state (inactive receptor), thereby reducing the basal activity of the receptor protein to a very low level. Light absorption triggers an isomerization

---

[1] A. G. Beck-Sickinger, *Drug Discov. Today* **1**, 502 (1996).
[2] P. A. Hargrave and J. H. McDowell, *Int. Rev. Cytol.* **137B**, 49 (1992).

reaction, converting the ligand to the all-*trans* state, which acts as a full agonist. Hence, this full agonist is generated very efficiently *in situ*. This photoisomerization reaction has quite unique properties (femtosecond kinetics, high stereospecificity, 0.67 quantum efficiency). Much research effort has been aimed at investigating the mechanism behind this trigger reaction and the subsequent activation of rhodopsin. In addition to being of prime importance for understanding visual pigment function at a molecular level, such studies should also produce general principles for the molecular mechanism of G-protein-coupled receptors.

For detailed studies on the complex structure–function relationships in visual pigments, there obviously is a need for functional expression of recombinant protein, preferably in milligram amounts. Although rhodopsin itself can be purified in large amounts from bovine eyes, the very low abundance of the related cone pigments and their poor thermal stability make it virtually impossible to purify these from a natural source. Furthermore, recombinant proteins from a nonnative cell type often lack the kind of impurities associated with native proteins, which makes them easier to purify. An essential advantage of the ability to produce recombinant proteins of course is that it allows generation of tagged proteins for highly selective purification protocols, as well as site-directed mutants or stable-isotope labeled protein for functional and structural studies. Since the late 1980s, functional expression of rhodopsin and several cone pigments has been achieved in a variety of heterologous cell types.[3–6]

If there is a need for large amounts of recombinant eukaryotic membrane protein (crystallization, biophysical studies), in our view the best option for functional expression presently is the recombinant baculovirus system. Membrane proteins usually contain complex posttranslational modifications (*N*-glycosylation, disulfide bridge formation, thiopalmitoylation, *N*-acetylation) essential for correct folding. Insect cells recognize the corresponding motifs and correctly perform these reactions, except that *N*-glycosylation is restricted to the more simple high-mannose type.[7] Unless a precise composition of the sugar moiety is essential for functioning (e.g., adhesion compounds), this is usually advantageous, because it results in a

---

[3] J. J. M. Janssen, W. J. M. VandeVen, W. A. H. M. VanGroningen-Luyben, J. Roosien, J. M. Vlak, and W. J. DeGrip, *Mol. Biol. Rep.* **13,** 65 (1988).

[4] D. D. Oprian, R. S. Molday, R. J. Kaufman, and H. G. Khorana, *Proc. Natl. Acad. Sci. U.S.A.* **84,** 8874 (1987).

[5] J. Nathans, C. J. Weitz, N. Agarwal, I. Mir, and D. S. Papermaster, *Vision Res.* **29,** 907 (1989).

[6] N. G. Abdulaev, M. P. Popp, W. C. Smith, and K. D. Ridge, *Prot. Express. Purif.* **10,** 61 (1997).

[7] D. L. Jarvis, Z. S. Kawar, and J. R. Hollister, *Curr. Opin. Biotechnol.* **9,** 528 (1998).

much less heterogeneous glycosylation pattern than in mammalian expression systems. The native sugar moieties of visual pigments are trimmed down and relatively small and, in fact, quite similar to the ones generated in insect cells. Furthermore, the expression of recombinant proteins in insect cells has several clear advantages over other expression systems. Insect cells are easy to handle, do not require CO_2 incubators, can easily be grown in monolayer as well as in suspension culture, and the latter is easily scaled up to 10 liters or more. Moreover, the expression levels of functional recombinant proteins in insect cells often exceed those obtained with other eukaryotic expression systems by an order of magnitude.

Consequently, since its introduction in the early 1980s, the recombinant baculovirus system has been developed into a widely used versatile system for the expression of foreign proteins in insect cells. To date, a huge variety of recombinant proteins from prokaryotic as well as eukaryotic origin have successfully been expressed functionally in insect cells including nuclear, cytoplasmic, and secreted proteins. Following our successful expression of rhodopsin in this system,[3] it has also become quite popular for production of integral membrane proteins.

Production of Recombinant Baculovirus

Background

The baculovirus expression system offers a selection of several insect cell lines for heterologous expression and a variety of promoters to drive expression of a foreign cDNA. Three cell lines are used most frequently, two of which were derived from the ovary of the fall armyworm *Spodoptera frugiperda*. These cell lines (Sf9 and Sf21 cells; ATCC, Rockville, MD, CRL-1711 and IPLB Sf21 AE, respectively), which are clonally related, are most popular among baculovirus expressionists because they behave very well in suspension culture. Apart from a minor difference in cell size and in glycosylation potential, these two cell lines behave very similarly. The third cell line derives from *Trichoplusia ni,* and was named High-Five cells (Invitrogen, Carlsbad, CA). This line reportedly yields significantly higher expression levels of, in particular, secreted proteins. We have tested the Sf9 and a *Mamestra brassicae* cell line with regard to expression of bovine rhodopsin, and do not find significant differences in expression level per milligram of protein.[8] Since Sf9 cells are most easily adapted to

---

[8] G. L. J. DeCaluwé, J. VanOostrum, J. J. M. Janssen, and W. J. DeGrip, *Meth. Neurosci.* **15,** 307 (1993).

suspension as well as serum-free culture, all procedures described here pertain to the use of the Sf9 cell line.

Baculovirus offers quite a selection of promoters, from the very strong and very late phase p10 and polyhedrin promoters to the much weaker, very early phase IE-1 promoter (Table I). Most often used because of its power is the polyhedrin promoter. It is activated in the very late phase of infection (starting at 18–24 hr postinfection). Of comparable strength but activated slightly earlier is the p10 promoter. The other promoters are less strong and are activated in much earlier phases of infection, but several are also reported to give satisfactory expression levels of soluble recombinant proteins. Since neither the polyhedrin nor the p10 gene product is required for viral replication, either locus can be used to insert heterologous genes under the promoter of choice. In some instances, the use of an earlier, less strong promoter remarkably resulted in higher expression levels of functional protein compared to using the polyhedrin promoter. The explanation for this phenomenon is usually sought in posttranslational modifications essential for functional expression, which can still be processed efficiently by the cells early after infection. However, despite the fact that rhodopsin also has to undergo several posttranslational modifications (disulfide bond formation, N-glycosylation, thiopalmitoylation, N-acetylation), we have obtained best results using the very late promoters (polyhedrin and p10) to drive the expression of rhodopsin in insect cells. Our results, described later, are all based on the use of the polyhedrin promoter.

Classically, recombinant baculovirus was generated through homologous recombination of a plasmid (transfer vector) containing the cDNA of interest under control of the appropriate promoter, flanked by 1–2 kb of sequence from the polyhedrin locus. The early laborious and prolonged efforts involved with isolating recombinant virus have been markedly accel-

TABLE I
BACULOVIRUS PROMOTERS FOR HETEROLOGOUS PROTEIN EXPRESSION

Promoter	Phase	Strength
Polyhedrin	Very late	+++
P10	Very late	+++
Basic protein	Late	++
PCOR	Late	+
PE 38	Early	+
ME 53	Early	?
IE-1	Immediate early	+

erated by the inclusion of reporter genes[9] and the use of linearized baculovirus genomes,[10] which can only replicate after homologous recombination. These procedures still require the recombination event to occur in insect cells. Lately, a different approach based on transposon-mediated exchange of the cDNA of interest into the polyhedrin locus was developed that is designed for generation of recombinant baculovirus genomes in *Escherichia coli*.[11] This presently is the most rapid procedure, able to generate purified recombinant virus in about 2 weeks. A flowchart showing the procedure for high-level expression of recombinant proteins using the baculovirus expression system is given in Fig. 1.

Culture of Sf9 Cells in Serum-Supplemented Medium

Routinely, insect cell cultures are maintained in serum-supplemented medium. We have always had excellent results with TNM-FH medium. [TNM-FH consists of Grace's insect cell medium supplemented with 3.3 g/liter yeastolate, 3.3 g/liter lactalbumin hydrolyzate, and 5.5 g/liter bovine serum albumin. To complete, add 10% fetal bovine serum (essential) and antibiotics (optional). For suspension cultures add an additional 1 g/liter Pluronic F-68 (Sigma, St. Louis, MO).] Insect cell culturing methods (maintaining cell lines in monolayer as well as suspension cultures, freezing and thawing of insect cells, preparation of culture medium) have been well documented.[12,13] Because we follow these procedures exactly, we refer the reader to these protocols.

Production of Recombinant Baculovirus:
Homologous Recombination, Isolation of Individual Viral Clones and Generation of a High-Titer Viral Stock Solution

Basically, two different procedures are commonly being used to generate a recombinant baculovirus. Both methods rely on a targeted translocation of the gene of interest from a transfer vector to the AcNPV genome. This can be established either in insect cells or in *E. coli*. Both procedures are commercially available and have been optimized to yield recombinant baculoviruses with high frequency [for instance, the Baculogold system from Pharmingen (San Diego, CA) employing linearized baculovirus DNA

---

[9] D. Zuidema, A. Schouten, M. Usmany, A. J. Maule, G. J. Belsham, J. Roosein, E. C. Klinge-Roode, J. W. M. VanLent, and J. M. Vlak, *J. Gen. Virol.* **71**, 2201 (1990).

[10] P. A. Kitts and R. D. Possee, *BioTechniques* **14**, 810 (1993).

[11] V. A. Luckow, S. C. Lee, G. F. Barry, and P. O. Olins, *J. Virol.* **67**, 4566 (1993).

[12] M. D. Summers and G. E. Smith, *Texas Exp. Station Bull. No. 1555* (1987).

[13] D. R. O'Reilly, L. K. Miller, and V. A. Luckow, "Baculovirus Expression Vectors, A Laboratory Manual." W. H. Freeman and Company, New York, 1992.

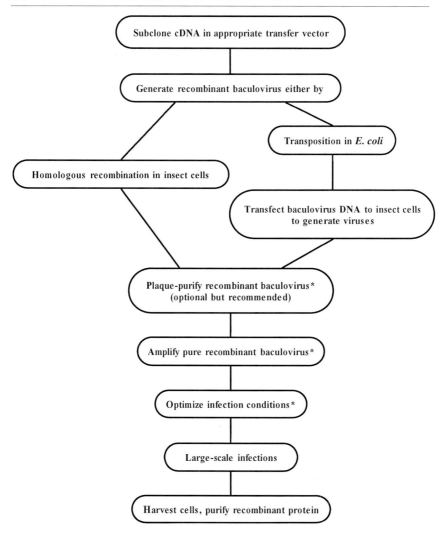

FIG. 1. Flowchart showing the entire procedure from subcloning the cDNA to the production of large amounts of recombinant protein using the baculovirus expression system. Recommended procedures are described in the text for all steps marked with asterisks.

or the Bac-to-bac system from Life Technologies (Rockville, MD) employing transposon-mediated translocation in *E. coli*]. Since both methods are provided with detailed protocols, we do not discuss these procedures here. The choice for either of these methods is based on personal preference rather than on specific advantages or disadvantages of one method over the other. At some point during the procedure, a transfection of viral DNA to insect cells must be performed. From this moment on, viruses will be produced. Although the manufacturer of the commercially available kits may suggest that viral supernatants from a transfection experiment can be used directly for recombinant protein production, we highly recommend that a plaque purification procedure be performed prior to the production of a working stock of virus. It is much more comfortable knowing one is working with a pure recombinant virus rather than having to face the possible presence of viral anomalies, occurring during transfection and improper viral amplification,[14] which may interfere with optimal recombinant protein production. In our hands the following protocol always yields excellent results:

1. In a 6-well culture dish, seed approximately 8×10^5 Sf9 cells in 2 ml of TNM-FH per well. Only take healthy cells, preferentially from a log-phase suspension culture. Gently swirl the plate to distribute the cells evenly and place in incubator at 27°. Allow 15 min or more to attach.

2. Remove the medium and any remaining floating cells from the wells and replace with 600 μl of fresh medium.

3. Prepare 10-fold dilutions of the transfection supernatant (usually within the range of 10^4–10^7). Add 100 μl of each dilution to a well and allow the infection to proceed for approximately 1 hr at 27°.

4. Meanwhile, for each 6-well dish, autoclave 0.225 g of low-temperature gelling agarose (Sea-Plaque agarose, FMC Bioproducts, Rockland, ME) in 7.5 ml of ultrapure water in a disposable polypropylene 50-ml centrifuge tube. After autoclaving, let cool to 37° in a water bath. Prewarm fetal bovine serum and 2× concentrated TNM-FH to 37°.

5. To the agarose, add 6 ml of 2× concentrated TNM-FH, 1.5 ml of fetal bovine serum, and 7.5 μl of antibiotic stock solution (2000× concentrated). Keep at 37°.

6. Remove the medium from the wells and immediately add 2 ml of agarose overlay. Place at room temperature for 30 min to allow the agarose to solidify. Then, place in incubator at 27°. Add an open container filled with distilled water in the incubator to ensure a high humidity.

7. After 5 days, check for the presence of plaques using a bright-field microscope. Plaques can be recognized with the microscope lamp switched off as groups of white cells in a dark background of uninfected cells. With the lights switched on, plaques appear as darker cells in a light background of uninfected cells. Do not use a phase contrast microscope because this

---

[14] T. J. Wickham, T. R. Davis, R. R. Granados, D. A. Hammer, M. L. Shuler, and H. A. Wood, *Biotechnol. Lett.* **13,** 483 (1991).

will obscure the plaques. If no plaques can be identified, continue the incubation for another 1–2 days. If still no plaques can be identified, repeat the procedure using lower dilutions (10^1–10^3).

8. Using a cotton-plugged sterile Pasteur pipette, pick a plug of agarose containing a plaque and expel in 0.5 ml of TNM-FH medium. Pick 3–6 individual plaques. Select only the most isolated plaques from the highest positive dilution. Store overnight at 4° to elute the viruses from the agarose.

9. In a 6-well culture dish, seed 5×10^5 cells in 1.8 ml of TNM-FH per well. Add 200 μl of eluted viruses and incubate at 27° for 6 days.

10. Dislodge the cells from the wells by pipetting. Transfer to a centrifuge tube and pellet the cells for 5 min at 3000g at 27°. Store the supernatant at 4° in a fresh tube and mark the tube as first passage viral supernatant. Check the cells for the presence of the recombinant gene/protein using appropriate detection methods (Western blot, PCR, enzymatic assay, etc.). Any positive clone can be used to prepare a working stock of virus to be used in recombinant protein production. Preferentially, the one with the strongest recombinant protein signal on immunoblots will be used.

11. Take three 175-cm^2 culture flasks, seed 15×10^6 cells in 20 ml TNM-FH in each flask. Dilute 1, 10, and 100 μl, respectively, of first passage viral supernatant from a positive clone in 5 ml TNM-FH and add to the flasks. After 6 days incubation at 27°, transfer the supernatant to a centrifuge tube and remove the cells by centrifugation for 5 min at 3000g at 27°. Filter the supernatant through a 0.2-μm filter and store at 4°. Mark as second passage viral stock.

12. Determine the titer of each stock by agarose overlay assay (steps 1–7 from above). This will usually be around 1×10^8 plaque-forming units (pfu)/ml. If the titers of all supernatants are in the same order of magnitude, the one obtained with the lowest amount of first passage viral supernatant will be used for recombinant protein production. If not, the one with the highest titer will be used.

13. If larger volumes of viral stock have to be prepared, repeat the optimal condition from step 11 with multiple culture flasks or proceed to the use of spinner flasks as follows.

14. Seed 1×10^5 cells/ml in 100 ml TNM-FH in a 250-ml spinner flask (Bellco, Vineland, NJ).

15. Monitor the cell density on a daily basis using a hemocytometer. If the cells have reached approximately 1×10^6 cells/ml (3–4 days), add enough viral supernatant from a second passage viral stock to infect the cells with a multiplicity of infection of 0.1. Continue the incubation for another 3–4 days.

16. Remove the cells by centrifugation for 5 min at 3000g. Filter the supernatant through a 0.2 μm filter and store at 4°. Mark as third passage

viral stock. A viral stock can be stored at 4° for at least 1 year without loss of quality. Prolonged storage methods have been reported in detail before in the literature.[15]

17. Determine the titer by plaque assay (steps 1–7 from above). This will usually be $\geq 1 \times 10^8$ pfu/ml.

Production of Recombinant Visual Pigment

Background

Baculovirus-mediated expression of recombinant protein in insect cells is fairly easily achieved. However, if one aims at large amounts of functional protein, several factors have to be optimized in order to achieve the optimal expression efficiency (volumetric yield of functional protein). The type of promoter and the cell line should be carefully considered, as discussed earlier. Previously we also reported that minimizing the length of the 5' untranslated sequence of the insert can increase the expression level of opsin up to twofold.[8] Additional important parameters are the infection conditions [cellular growth status and multiplicity of infection (MOI)] and the level of functionality of the recombinant protein.

In agreement with reports elsewhere, we have observed that infection of cells in the logarithmic growth phase results in significantly higher expression levels (Table II). Marked effects were also obtained by varying the MOI. While a theoretical study suggested that low MOIs (around 0.1) could improve the yield of recombinant protein,[16,17] the large majority of baculovirus literature uses MOIs in the range of 3–10. We evaluated MOIs in the range of 0.01–1.0 and indeed observed significantly higher volumetric yields at the lower MOIs (Table II). This is largely due to the increase in cell density, since at a low MOI cellular division still can proceed for some time. The net cellular production capacity appears to be quite independent of the MOI.

We emphasize that the absolute expression level is not the best parameter to evaluate expression efficiency. In particular with complex integral membrane proteins, it is commonly observed that only part of the total equivalent of heterologous protein is correctly folded and fully functional. The other, nonfunctional part usually also lacks one or more posttranslational modifications and generally is not correctly targeted, or cannot leave

---

[15] D. L. Jarvis and A. Garcia, Jr., *BioTechniques* **16,** 508 (1994).
[16] P. J. Licari and J. E. Bailey, *Biotechnol. Bioeng.* **37,** 238 (1991).
[17] P. J. Licari and J. E. Bailey, *Biotechnol. Bioeng.* **39,** 432 (1992).

TABLE II
RECOMBINANT RHODOPSIN EXPRESSION LEVELS UNDER VARIOUS CULTURE CONDITIONS[a]

Growth stage at time of infection	MOI	Maximum yield at dpi	Volumetric yield (mg/liter)
Early to mid-log phase	0.01	5–6	3.2
	0.1	4–5	2.4
	1.0	3–4	2.0
Late log phase	0.01	5–6	0.5
	0.1	5–6	1.5
	1.0	5–6	2.0

[a] Data are representative for several similar experiments.

the endoplasmic reticulum (ER). This is still one of the deficits of an heterologous expression system. The percentage functional protein can be as low as 5–10%, hence only monitoring total expression [immunoblot, enzyme-linked immunosorbent assay (ELISA)] can be quite deceptive, and should be complemented by a functional assay (ligand binding and affinity, enzymatic properties, etc.). In the case of the visual pigments, the most straightforward functional assay is the absorbance shift on binding of the ligand, 11-*cis*-retinal, and the photosensitive nature of the resulting absorbance spectrum. Total opsin levels are determined by inhibition ELISA and/or immunoblot, as described elsewhere.[8] In this way we could establish that the majority of the bovine opsin produced in insect cells is functional (60–80%). On the other hand, this situation is much less favorable for cone pigments, where often only 10–20% is functional. Coexpression with a specific cytoplasmic chaperone (RanBP2) has been reported to increase the level of functional red cone pigment in mammalian cells.[18] We are currently evaluating coexpression with ER chaperones to boost functional expression in the baculovirus system.

Finally, the culture medium has to be considered. The classical serum-supplemented media support cellular growth very well and allow storage of cell stocks in liquid nitrogen. However, they are expensive and not suitable if stable-isotope labeling is pursued. For the latter purpose, a minimal medium is preferable, but a satisfactory one has not yet been developed for the baculovirus system. The next best option would be a serum- and protein-free medium. In fact, in the past few years, several culture media have been developed that satisfactorily sustain rapid insect cell growth under serum- and protein-free conditions. In addition, they are

[18] P. A. Ferreira, T. A. Nakayama, W. L. Pak, and G. H. Travis, *Nature* **383,** 637 (1996).

less expensive, which strongly reduces the costs of large-scale insect cell cultures. Moreover, these media have been optimized to yield maximal expression levels of recombinant protein. We have tested several commercially available media with respect to volumetric production of recombinant rhodopsin and best results were obtained with Insect-Xpress (BioWhittaker, Walkersville, MD). However, cells adapted to the serum-free media do not survive very well when stored frozen. For this reason we routinely maintain two cultures of Sf9 insect cells: one culture adapted to serum-supplemented medium (TNM-FH, see earlier discussion) and one adapted to Insect-Xpress. The culture adapted to serum-supplemented medium is used for passage and storage of cellular stocks, for transfection procedures, for plaque purification/quantification assays and for viral stock production. The culture adapted to serum- and protein-free medium is used for recombinant protein production in suspension culture.

Adaptation of Insect Cells to Serum-Free and Protein-Free Culture Conditions

To adapt cells growing in a serum-supplemented medium to serum- and protein-free conditions we have obtained good results with the following protocol:

1. Dislodge the cells from a near confluent monolayer culture in TNM-FH in a 25-cm^2 culture flask (ca. 4 ml) by repeatedly pipetting the medium over the cells. Avoid introducing air bubbles as much as possible. Transfer 0.1 volume of the culture to a fresh 25-cm^2 culture flask containing a 50/50 mixture of serum-supplemented and serum-free medium. Incubate at 27° until confluency (usually 3–4 days).

2. Again transfer 0.1 volume of the culture to a 50/50 mixture of both media.

3. During subsequent similar transfers, the percentage of serum-free medium is gradually increased to 100% (for instance, from 50% to 75%, 90%, 95%, and finally 100% serum-free medium). The cells are passed twice in each new mixture. As the percentage of serum-free medium increases, it may become more difficult to dislodge the cells from the culture flask since they tend to adhere more strongly. It also will take somewhat more time for the culture to reach confluency (4–5 days) since the doubling time in serum-free medium (ca. 24 hr) is longer than in serum-supplemented medium (ca. 18 hr).

4. Once the cells have been passed in 100% serum-free medium, increase the size of the culture flasks to 175 cm^2 (25 ml). Make sure to transfer at least 2.5 ml of old culture to a new culture flask.

5. Gather enough cells from 175-cm^2 culture flasks to seed a 100-ml culture in a 250-ml spinner flask with at least 5×10^5 cells/ml. From this moment on the cell density is monitored on a daily basis and the maximal cell density should reach at least 5×10^6 cells/ml.

6. On reaching the stationary phase of growth, the culture is transferred to a new spinner flask with an initial density of at least 5×10^5 cells/ml. If lower amounts of cells are being used, the culture may display a lag phase of up to 3 days before they start to divide on a regular base. However, this is only observed in serum-free medium. In a serum-supplemented medium, initial seeds may go as low as 1×10^5 cells/ml.

7. Cells fully adapted to growth in serum-free medium can be maintained for at least 6 months. Discard the culture when the maximal cell density no longer reaches satisfactory levels and/or when recombinant protein production levels are compromised. We did not succeed in reviving adapted cells from frozen cultures in serum- and protein-free medium, hence a subsequent serum-free culture starts again with adaptation of a TNM-FH culture.

Finding Optimal Infection Conditions

Many different parameters affecting culture conditions can be changed in attempts to maximize recombinant protein expression levels. We highly recommend that at least the following two parameters be evaluated: MOI and duration of infection. The following protocol has been proven to be very effective in our hands:

1. In three 250-ml spinner flasks, seed 5×10^5 cells/ml in 100 ml Insect-Xpress medium. Incubate at 27°, stir at approximately 100 rpm. Use only a stirrer with a nonheating surface or take all necessary precautions to prevent the stirrer from heating up the culture (standard laboratory stirrers tend to heat up during continuous operation).

2. Monitor the cell density on a daily basis by counting with a hemocytometer. If the cells have reached no more than 30–40% of their maximal density, infect with a MOI of 0.01, 0.1, and 1.0 by slowly adding the necessary amount of viral stock solution (usually around 0.02, 0.2, and 2.0 ml of viral stock, respectively).

3. Every consecutive day, for up to 6 days postinfection (dpi), monitor the cell density and take a 10-ml sample. Pellet the cells by centrifugation for 5 min at 3000g, remove the supernatant, and store the cell pellet at −80° until further analyzed.

4. Measure the functional expression level in the various samples using appropriate procedures, to determine the optimal conditions for recombi-

nant protein production. These conditions can also be used for large-scale experiments. For visual pigments we routinely monitor functional levels by ligand binding (11-*cis*-retinal) to generate their specific absorbance spectrum (regeneration).

Functional Analysis of Recombinant Visual Pigment by Ligand Binding (Regeneration)

Monitoring functional expression levels is done on a small scale. The same procedure is used for batch regeneration of recombinant apopigment (opsin) into photoactive visual pigment. Because ligand binding substantially stabilizes these proteins, large-scale regenerations are performed immediately after harvesting of the cells, and the regeneration mix can be stored at $-80°$ until further purification. The following protocol is used for large-scale regeneration, but can simply be downscaled for smaller samples:

1. Resuspend the cell pellet at a density equal to 1×10^8 cells/ml in ice-cold 6 mM 1,4-piperazinediethanesulfonic acid (PIPES)/Na$^+$, 10 mM ethylenediaminetetraacetic acid (EDTA), 5 mM dithioerythritol (DTE), and 5 μM leupeptin, pH 6.5. Take into account that the cell pellet itself will accommodate approximately 40% of the final volume.

2. Homogenize the cells using 6 strokes in a motor-driven Potter–Elvehjem tube at 150 rpm on ice.

3. Centrifuge for 20 min at 40,000g at 4° to pellet the membranes. Remove the supernatant as completely as possible.

4. Resuspend the membranes at a density equal to 2×10^8 cells/ml in 20 mM PIPES/Na$^+$, 140 mM NaCl, 10 mM KCl, 3 mM MgCl$_2$, 2 mM CaCl$_2$, 1 mM EDTA, and 5 μM leupeptin, pH 6.5. Again take into account that the membranes themselves will accommodate approximately 40% of the final volume.

5. While stirring, slowly add dodecylmaltoside (Anatrace, Maumee, OH) to a final concentration of 0.5 mM. Continue stirring under a gentle stream of argon for another 20–30 min.

6. All of the following manipulations will be performed in the dark or under dim red light (>630 nm).

7. From a 20 mM stock solution of 11-*cis*-retinal in hexane, put 1 μl for each 1×10^8 cells in a 1.5-ml polypropylene reaction tube. Slowly dry the hexane with a gentle stream of argon. Then dissolve the 11-*cis*-retinal in two to three times the original volume of dimethylformamide. The 11-*cis*-retinal was a gift from Dr. Rosalie Crouch (Dept. of Ophthalmology, Medical University of South Carolina, Charleston, SC).

8. While stirring under a gentle stream of argon, slowly add the 11-*cis*-retinal to the membrane suspension, obtained in step 5.

9. Pour 20- to 25-ml aliquots of the suspension in disposable 50-ml centrifuge tubes (e.g., Falcon tubes), replace the gas phase with argon and wrap in aluminum foil.

10. Attach to a rotary wheel and gently rotate for another 60 min at ambient temperature, then either continue with extraction and purification of the rhodopsin or store at $-80°$ in a light-tight container.

Optimized Procedure for Baculovirus-Based Expression of Recombinant Visual Pigments

A typical example of an optimization procedure for bovine rhodopsin using the protocol described earlier is already shown in Table II. Under optimal conditions, typically functional rhodopsin expression levels of 15–20 pmol/10^6 cells ($\sim 10^7$ copies/cell) can be achieved, which is equivalent to 3–4 mg of rhodopsin per liter culture. These levels were reached following 5–6 days incubation after infection of the cells with a multiplicity of infection of only 0.01.

Special Effects

Large-Scale Production (≥ 20 mg Recombinant Protein per Batch)

As already mentioned, one of the advantages of the recombinant baculovirus/insect cell expression system is that culture volumes can be easily scaled up. The optimal conditions derived from small-scale experiments (Sf9 cells in Insect-Xpress medium, infection in logarithmic growth phase, MOI of 0.01, cells harvested at 6 dpi) could be directly adapted to large-scale suspension culture (10-liter bioreactor, Applikon, Schiedam, The Netherlands). Now, the low MOI is a big advantage, as no excessively large volumes of viral inoculate have to be used. Bioreactor expression yields very similar functional expression levels of bovine rhodopsin as 100-ml spinner cultures, and allows us to generate recombinant rhodopsin in amounts of 30–40 mg in a single batch. For cone pigments, actually a significant improvement in functional expression level was obtained in the bioreactor (two- to threefold). Our current explanation is that the more controlled conditions in the bioreactor, in particular the constant oxygen supply, improve the ability of the cells to process these much less stable membrane proteins.

Stable-Isotope Labeling

Biophysical techniques like solid-state (ss) nuclear magnetic resonance (NMR) spectroscopy and Fourier transform infrared (FTIR) difference spectroscopy are potentially very powerful in resolving aspects of structure and mechanism of membrane proteins at atomic detail. However, NMR analysis requires stable-isotope labeling of the protein (^{13}C and/or ^{15}N), and peak assignment in the complex FTIR spectra of biomembranes also appears to be feasible only with the help of stable-isotope labeling. We approached this issue on the basis of our experience with expression of rhodopsin under protein- and serum-free conditions (see earlier discussion). A customized version of the Insect-Xpress medium was prepared with strongly reduced concentrations of specified amino acids. Following supplementation of these amino acids, where required in a stable-isotope labeled form, this medium was used for expression of rhodopsin under the same conditions applied for the standard medium. Essentially the same functional levels of recombinant rhodopsin were obtained as with the standard medium. Incorporation of labeled amino acid in cellular protein was analyzed by gas chromatography–mass spectrometry (GC-MS) following pronase treatment and derivatization of the liberated amino acids.[19] Label incorporation levels in rhodopsin paralleled those in total cellular protein. Because the customized medium still requires a low molecular weight protein extract, which presents a background of unlabeled amino acids, full labeling of specific amino acid residues could not yet be achieved. Nevertheless, both for an essential amino acid (either L-[^{15}N$_2$] or L-[4,4,5,5-^{2}H$_4$]lysine) and for a nonessential amino acid ([ring-^{2}H$_4$] or [1-^{13}C$_1$]tyrosine) label incorporation in the range 60–75% could be attained, which readily permits NMR as well as FTIR identification. In fact, sufficient amounts already were produced in the bioreactor to allow the first ss-NMR analysis of a eukaryotic membrane protein. It can already be concluded from these first results that unique structural information is available through these approaches.[19–21]

Purification and Reconstitution

To purify recombinant rhodopsin, a number of affinity procedures have been reported in the literature. Immunoaffinity purification is based on the availability of monoclonal antibody 1D4 with a well-defined C-terminal epitope.[4] The corresponding peptide is used to elute bound protein. Lectin affinity chromatography is based on the affinity of concanavalin A for the N-linked glycosyl moieties on the protein. This approach is less efficient

---

[19] C. H. W. Klaassen, P. H. M. Bovee-Geurts, J. Raap, K. J. Rothschild, and W. J. DeGrip, in preparation (2000).

[20] F. DeLange, C. H. W. Klaassen, S. E. Wallace-Williams, P. H. M. Bovee-Geurts, X. Liu, W. J. DeGrip, and K. J. Rothschild, *J. Biol. Chem.* **273**, 23735 (1998).

[21] A. F. L. Creemers, C. H. W. Klaassen, P. H. M. Bovee-Geurts, R. Kelle, U. Kragl, J. Raap, W. J. DeGrip, J. Lugtenburg, and H. J. M. DeGroot, *Biochemistry* **38** (1999).

since several other glycoproteins of viral origin are produced in insect cells on baculovirus infection, and complete purification of the recombinant rhodopsin from those contaminants cannot be achieved in a single step. We finally opted for immobilized metal affinity chromatography (IMAC).

To allow IMAC, the protein has to be extended with a polyhistidine tag of at least 6 residues. Despite this modification of the protein, we favor the use of IMAC for several reasons. Compared to the other two methods, the IMAC matrix (we use Superflow from Qiagen, Hilden, Germany) allows relatively high flow rates and has a very high binding capacity (up to 3 mg rhodopsin/ml bed volume, unpublished observations), which are important advantages in large-scale procedures. Furthermore, IMAC has good selectivity, yielding purification factors of ≥500-fold, and can be used both batchwise and in columns.

Overall, rapid purification protocols could be established using this approach,[22] which is a clear benefit when working with unstable pigments. The histidine tag was placed at the C terminal of rhodopsin. N-terminal constructs do not allow satisfactory purification. The additional His tag induces a slight downshift of the pK_a of the metarhodopsin I $\rightleftharpoons$ II equilibrium, but this does not really affect the essential features of the signaling function of rhodopsin.[22,23] IMAC-based procedures for purfication of recombinant visual pigments have been reported in detail elsewhere and are not discussed here.[22-24] Note that the procedure allows considerable adaptation to satisfy specific protein requirements. For instance, to achieve purification of the quite unstable cone pigments we switched to CHAPS as a detergent instead of dodecylmaltoside, and included retina lipids in wash and elution buffers to increase the thermal stability of these pigments.[24] Detergents, and detergent–lipid mixtures as well, strongly destabilize visual pigments and perturb their functional properties. For that reason we always reconstitute purified visual pigments in a suitable lipid environment (proteoliposomes) for functional analysis and long-term storage. For the lipid environment we have opted for a bovine total retina lipid extract. This can easily be obtained in large quantities from a native source and closely approaches the native lipid environment of visual pigments, which is characterized by an unusually high level of polyunsaturated fatty acids (ca. 45 mol% of 22:6n3).

A variety of methods have been developed during the last several de-

---

[22] J. J. M. Janssen, P. H. M. Bovee-Geurts, M. Merkx, and W. J. DeGrip, *J. Biol. Chem.* **270**, 11222 (1995).

[23] C. H. W. Klaassen, P. H. M. Bovee-Geurts, G. L. J. DeCaluwé, and W. J. DeGrip, *Biochem. J.* **342**, 293 (1999).

[24] P. M. A. M. Vissers, P. H. M. Bovee-Geurts, M. D. Portier, C. H. W. Klaassen, and W. J. DeGrip, *Biochem. J.* **330**, 1201 (1998).

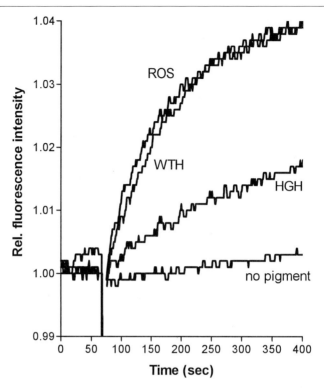

FIG. 2. G-protein activation assay demonstrating the ability of the recombinant His-tagged visual pigments to bind and activate the G-protein transducin (ROS, native bovine rhodopsin from rod outer segments; WTH, recombinant wild-type histidine-tagged bovine rhodopsin; HGH, wild-type histidine-tagged human green cone pigment). The assay was performed according to Vissers et al.[24] using equimolar amounts of each visual pigment.

cades for reconstitution of detergent-solubilized membrane proteins into proteoliposomes. The most popular ones include detergent removal by dialysis or dilution, by adsorption to hydrophobic beads, or by gel filtration. None of these techniques can be successfully applied to every detergent. To overcome this problem, we have developed a generic reconstitution procedure that is based on selective removal of detergents by cyclodextrin inclusion.[25] We have meanwhile successfully applied this technique for reconstitution of rod and cone pigments into retina lipids using a variety of detergents (dodecylmaltoside, nonylglucoside, CHAPS, Triton X-100, zwittergent TM-314, LDAO). For technical details of this procedure refer to ref. 25.

[25] W. J. DeGrip, J. VanOostrum, and P. H. M. Bovee-Geurts, *Biochem. J.* **330,** 667 (1998).

Functional Characterization of Recombinant Baculovirus Expressed Rhodopsin

A variety of assays are commonly in use to study structural as well as functional aspects of the visual pigments. These include the obvious UV/visible spectrophotometric assays to characterize the properties and kinetics of the photointermediate products, FTIR, and Raman spectroscopy for the overall and localized structural protein and chromophore changes, ultrafast transient laser spectroscopy for time-resolved kinetics, ss-NMR analysis for high-resolution structural constraints, and also G-protein binding/activation assays to analyze signal transduction capacity (see elsewhere in this volume). These assays have all been reported in detail before and are not discussed in this chapter.

We have analyzed recombinant visual pigments using several of the assays mentioned above and as far as we can tell there are no significant differences between wild-type and purified recombinant protein with respect to their functional properties.[22,24] In addition, baculovirus mediated expression was successfully used to produce functional visual pigments that could not be characterized in the native state before.[26,27] In light of this issue, we would like to emphasize that adding a His tag to the C terminus does not compromise the capacity of rhodopsin to bind and activate the G-protein transducin, as has already been demonstrated for His-tagged human green cone pigment[24] (Fig. 2).

Conclusions

We have demonstrated the potential of the baculovirus expression system for the functional expression of recombinant visual pigments. Given the additional features possible with this system, like large-scale production and isotope labeling, at present we think this is the most promising expression system allowing researchers to investigate both structural and functional aspects of membrane proteins. However, there is no reason to assume that the methods we have developed and the results we have obtained are only applicable to the visual pigment family of membrane proteins. Hence, we believe that similar results could be obtained with a variety of other, in particular secreted and membrane, proteins. We have provided ready-to-go protocols that have always worked well in our hands as guidelines for other researches considering the use of this expression system.

---

[26] S. E. Wilkie, P. M. A. M. Vissers, D. Das, W. J. DeGrip, J. K. Bowmaker, and D. M. Hunt, *Biochem. J.* **330,** 541 (1998).

[27] Z. K. David-Gray, J. W. H. Janssen, W. J. DeGrip, E. Nevo, and R. G. Foster, *Nature Neurosci.* **1,** 655 (1998).

[3] Development of Stable Cell Lines Expressing High Levels of Point Mutants of Human Opsin for Biochemical and Biophysical Studies

By Jack M. Sullivan and Michael F. Satchwell

Introduction

Structure–function studies of vertebrate visual pigments have employed heterologous expression of transfected wild-type and mutated rhodopsin genes in tissue cultured cells.[1-4] Absorption, infrared, and electron spin resonance spectroscopies, as well as rapid proton uptake measurements, have been used to study the biophysical properties of expressed visual pigments. Biochemical studies of transducin and phosphodiesterase activation have also been conducted. Numerous studies have contributed to an expanse of knowledge during the last decade on spectral tuning, biochemical activation, and structural organization.[5,6]

Most structure–function studies on expressed rhodopsin pigments have utilized transient transfection techniques with plasmids delivered by calcium phosphate, DEAE-dextran, lipid-complexing agents, or electroporation into cultured COS or HEK293 kidney cells. Such approaches are subject to considerable expense in materials and labor in both the repeated preparation of high-quality supercoiled plasmid and in lengthy transfection protocols, which are often subject to low transfection efficiencies and significant variability in yield. Due to these challenges, a major component of time allocation in expressed rhodopsin studies unfortunately ends up being devoted to preparation of the protein rather than experimental effort on the structure–function problem. Studies of expressed visual pigments often require moderate to high levels (hundreds of micrograms to milligrams) of purified protein for spectroscopic, biochemical, and biophysical analysis. To partially alleviate these challenges we have invested considerable upfront effort to develop stable cell lines that constitutively express high levels ($\approx 10^6$ molecules/cell) of mutant opsins in the suspension-adapted

---

[1] D. D. Oprian, R. S. Molday, R. J. Kaufman, and H. G. Khorana, *Proc. Natl. Acad. Sci. U.S.A.* **84,** 8874 (1987).
[2] J. Nathans, C. J. Weitz, N. Agarwal, I. Nir, and D. S. Papermaster, *Vision Res.* **29,** 907 (1989).
[3] J. Nathans, *Biochemistry* **29,** 937 (1990).
[4] D. D. Oprian, *Methods Neurosci.* **15,** 301 (1993).
[5] H. G. Khorana, *J. Biol. Chem.* **267,** 1 (1992).
[6] T. P. Sakmar, *Prog. Nucl. Acid Res. Mol. Biol.* **59,** 1 (1998).

Copyright © 2000 by Academic Press
All rights of reproduction in any form reserved.
0076-6879/00 $30.00

HEK293S cell line.[7] Without additional transfection, visual pigments can be rapidly regenerated with 11-*cis*-retinal and purified for structure–function analysis. To date we have generated constitutively expressing stable cell lines of wild-type human opsin, and the following human rod opsin mutants: D83N, G90D, E113Q, E113Q/E134Q (double mutant), E122Q, E134Q, H211F, and C187Y. High-level expression cell lines were generated for D83N, G90D, E122Q, E134Q, and H211F mutant pigments, constituting the bulk of potential intramembrane residues capable of proton-transfer reactions. Stable expression clones of E113Q and E113Q/E134Q (not shown) were also generated, but at lower expression levels.

Stable cell lines expressing mutant human opsin pigments are a useful resource for structure–function studies requiring high levels of visual pigment because each cell of a stable transfection clone expresses a uniform and quantifiable level of opsin protein. They are of considerable utility in our time-resolved experiments on rhodopsin charge motion analysis.[8–11] Since HEK293S cells can be grown in suspension, anticipated experimental needs for opsin can be managed by simple scaleup of cell growth to produce the required amount of opsin protein. Cells can be grown in liter quantities in spinner flasks, roller bottles, or even bioreactors for time-resolved spectroscopic or biochemical techniques (e.g., absorption, Fourier transform infrared spectroscopy) requiring hundreds of micrograms to milligrams of visual pigments. These or similar cell lines expressing high yields of recombinant rhodopsin protein should facilitate the extension of structure–function studies to the time-resolved domain of biophysical analysis. A preliminary report of these results has been presented.[12]

Materials, Methods, and Results

Materials

Restriction and other enzymes are obtained from New England Biolabs (Beverly, MA) or Life Technologies (Gaithersburg, MD) and used according to the manufacturers' recommendations. pBlueScriptII-KD(−) is obtained from Stratagene (La Jolla, CA). pCIS is obtained from Genentech

---

[7] B. W. Stillman and Y. Gluzman, *Mol. Cell. Biol.* **5**, 2051 (1985).
[8] J. M. Sullivan, *Rev. Sci. Inst.* **69**, 527 (1998).
[9] P. Shukla and J. M. Sullivan, *Invest. Ophthal. Vis. Sci.* **39**, S974 (1998).
[10] J. M. Sullivan and P. Shukla, *Biophys. J.* **77**, 1333 (1999).
[11] J. M. Sullivan, L. Brueggemann, and P. Shukla, *Methods Enzymol.* **315** [19] (1999) (this volume).
[12] M. F. Satchwell and J. M. Sullivan, *Invest. Ophthal. Vis. Sci.* **39**, S958 (1998).

(South San Francisco, CA). pCDNA3 (neomycin resistance, neo^R) and pCDNA3.1 (hygromycin resistance, hygro^R) are obtained from Invitrogen (Carlsbad, CA). pSV2-neo is obtained from the American Type Culture Collection (Manassas, VA). Oligonucleotides are prepared by GenoSys (The Woodlands, TX) or at the DNA Core Facility at the University of Michigan. The 1D4 monoclonal antibody is prepared at the National Cell Culture Center (Cellex Biosciences, Minneapolis, MN). HEK293S cells were a kind gift of Dr. Bruce Stillman (Cold Spring Harbor Laboratories, NY).[7] Culture media and additives are obtained from Life Technologies. G418 (Geneticin) and hygromycin are obtained from Life Technologies. *n*-Dodecyl-β-D-maltopyranoside (DM) is obtained from Anatrace (Maumee, OH). Cyanogen bromide-activated Sepharose 4B is obtained from Pharmacia (Piscataway, NJ). All other chemicals are of reagent grade and obtained from Sigma (St. Louis, MO).

Engineering Mutant Human Rhodopsin Genes

The human rod opsin cDNA[13] containing the coding region (1044 bp), a 21-bp 5'-untranslated region, and a 3'-untranslated region extending through the first polyadenylation site is directionally subcloned between the *Cla*I and *Xba*I sites of a modified pBlueScriptII-KS(−) plasmid. A unique *Hpa*I mutation is engineered into an intergenic region of the plasmid by oligonucleotide-dependent site-specific mutagenesis[14] while selecting with a silent suppression of a unique *Xmn*I site in the ampicillin-resistance gene of the plasmid. Site-specific mutagenesis is used to create opsin point mutations in the plasmid. In this method a mutagenic primer anneals to a particular stretch of the coding region of the opsin cDNA to prime the desired mutation while the selection primer switches a unique restriction site (*Hpa*I) in an intergenic region of pBS to a different unique recognition sequence (*Afl*II) or deletes the site. Both primers are designed to anneal to the same strand of pBlueScriptII-KS(−) with human opsin cDNA (pBS-opsin) and are optimized for tight binding of the 3' end, a minimum of internal secondary structure, and a low probability of hybridization to alternative sequences in the vector. In addition, the mutagenic oligonucleotide is designed to make the missense mutation in the rod opsin cDNA and to silently (with respect to the opsin codon information) add or suppress a neighboring rare-cutting restriction enzyme recognition motif; this involves at most one additional nucleotide change. These silent mutations maintain the amino acid sequence information of the neighboring codons.

[13] J. Nathans and D. S. Hogness, *Proc. Natl. Acad. Sci. U.S.A* **81,** 4851 (1984).
[14] W. P. Deng and J. A. Nickoloff, *Anal. Biochem.* **200,** 81 (1992).

TABLE I
Oligonucleotide Sequences for Mutagenic Primers[a]

Mutant	Restriction enzyme	Add/ suppress	Mutagenic oligonucleotide sequence
D83N	EarI	−	5'-GCCACCTAGGACCAT(GAAtAG)**GTt**AGCCACGGC-3'
G90D	BsmFI	+	5'-GAGGGTGCTGGTGAA(Gt**C**cC)CTAGGACC-3'
E113Q	PstI	+	5'-GGCAAAGAAGCC(**CTg**CAg)ATTGCATCCTGTGGGCCC-3'
E122Q	BsrI	+	5'-CAGGGCAA**TTg**ACCGC(CCAGt)GTGGCAAAGAAGCCC-3'
E134Q	BsrBI	−	5'-CACCACCACaTA(CC**GCTg**)GATGGCCAG-3'
C187Y	PstI	−	5'-GCGTGTAGTAGTCGATTCC**AtA**CGAGCA(CTgAG)GCCCTCGG-3'
H211F	AflIII	−	5'-GGGGATGGTGAA**Gaa**GACCACGA(ACATaT)AGATGAC-3'

[a] The mutagenic codon programs both the missense mutation and the silent addition or deletion of a rare or unique restriction cleavage site. All oligonucleotides are antisense with respect to the rhodopsin coding sequence. The codon affected by the missense mutation is shown in boldface type. The nucleotides altered are shown in lowercase. The restriction site recognition sequence modified silently with respect to the coding sequence is enclosed in parentheses. A single nucleotide change accomplished the addition (+) or suppression (−) of a restriction enzyme recognition motif. In rare cases the missense mutation results directly in the addition or loss of a restriction cleavage motif (e.g., E134Q). The oligonucleotide used to make E134Q incorporates a silent codon variant that was found in sequencing.

Mutagenic and selection primers are designed according to established rules[14] and tested for complementarity to the targeted binding site in pBS-opsin using the Bestfit program of GCG (Genetics Computer Group, University of Wisconsin, Madison). Primer designs are energetically optimized for stable and site-specific annealing. The oligonucleotides used to program the desired mutations for each clone as well as the restriction sites altered are shown in Table I.

Primers are 5'-end-phosphorylated with adenosine triphosphate (ATP) (nonradioactive) and T4 polynucleotide kinase, annealed to heat-denatured pBS-opsin plasmid (0.25 pmol), and T4 DNA polymerase and T4 DNA ligase are used to complete and ligate secondary strands using dNTPs. Initially we used a 100-fold molar excess of both mutagenic and selection primers relative to the plasmid backbone [25 pmol (Mut): 25 pmol (Sel)]. In later experiments we doubled the concentration of the mutagenic primer (50 pmol) in order to increase the probability of mutagenic primer annealing and to increase the number of selected clones (HpaI → AflII) that were positive for the desired mutation. The plasmid pool is then transfected into chemically competent *Escherichia coli* BMH-71-18 (mutS, mismatch repair defective) (Clontech, Palo Alto, CA) and the next day the pooled plasmid preparation is digested to effective completion with HpaI, which targets only wild-type plasmid for linearization. Linearization of the wild-type pool reduces, but does not eliminate, the probability of transfection into competent bacteria (*E. coli*, INVα-F', Invitrogen, Carlsbad, CA), thus enriching for the growth of select clones (*Afl*II$^+$/*Hpa*I$^-$). Plasmid DNA is isolated from INVα-F' colonies and analyzed by restriction analysis and DNA sequencing[15] for the proper genotype. Restriction digests are run on 1% agarose gels in 1× Tris–acetate–EDTA with ethidium bromide staining, and photographed using ethidium fluorescence.[16] The *Hpa*I/*Afl*II selection switch has been a convenient means to reduce, but not eliminate, WT background in our hands. Of note, plasmids that are already *Afl*II$^+$ can be reverted back to *Hpa*I by a reverse switching selection oligonucleotide such that repeated rounds of mutagenesis can be operationally used to make multiple mutations in a single opsin gene. The *Hpa*I → *Afl*II conversion oligonucleotide used is 5'-GGCGGTAATACGCTTAAGCACA-GAATCA-3'. To switch back from *Afl*II → *Hpa*I the following oligonucleotide can be used: 5'-GCGGTAATACGGTTAACCACAGAATCAGGG-3'. With this added engineering we are able to rapidly identify INVα-F' clones that harbor plasmids with the desired opsin sequence by simple restriction enzyme digestion rather than employing sequencing as the screening tool.

Successful annealing of the mutagenic primer is measured by the alteration of the restriction digest due to the silently altered restriction recognition sequence. For example, in the generation of the G90D cDNA, the silent addition of a *Bsm*FI restriction site is programmed with the desired missense mutation in the mutagenic oligonucleotide. Figure 1 shows a double restriction digest of plasmid clones (D1 → D5) and control plasmid (C1 → C2) (pBS-opsin) with *Hpa*I and *Xba*I. Control plasmid does not have a *Hpa*I site and is linearized with *Xba*I (4706 bp). D1, D2, D4, and D5 show two bands with *Hpa*I/*Xba*I double digest (2188 and 2518 bp) indicating wild-type plasmids where the *Hpa*I site did not convert to *Afl*II (selection oligonucleotide did not anneal). D3 shows linearization of the vector backbone (4706 bp) on double digest apparently due to a single *Xba*I cleavage with the *Hpa*I site switched to *Afl*II. All isolates are also double digested with *Afl*II and *Xba*I. Controls and clones D1, D2, D4, and D5 only linearize (4706 bp) due to *Xba*I digestion without *Afl*II cleavage. D3 shows two bands indicating the presence of both *Xba*I and *Afl*II sites (2186 and 2520 bp) and supporting the conclusion of annealing of the selection primer in D3. All clones were also digested with *Bsm*FI. Annealing of the mutagenic primer is expected to make the G90D mutation and add another *Bsm*FI site. C1, C2, D1, D2, D4, D5 isolates demonstrate a wild-

---

[15] F. Sanger, S. Nicklen, and A. R. Coulson, *Proc. Natl. Acad. Sci. U.S.A.* **74**, 5463 (1977).
[16] J. Sambrook, E. F. Fritsch, and T. Maniatis, *in* "Molecular Cloning: A Laboratory Manual," 2nd ed., Chap. 6. Cold Spring Harbor Laboratory Press, Plainview, New York, 1989.

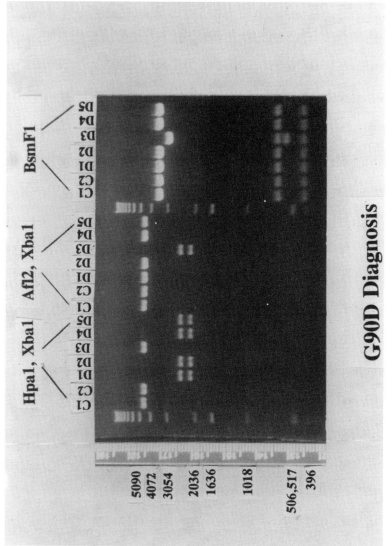

FIG. 1. Screening and confirmation of mutagenic plasmid clones. Isolated INV α-F' clones were used to grow minipreparations of plasmid DNA. Plasmids were restriction digested to completion with *XbaI* and *HpaI*, *XbaI* and *Afl*II, or *Bsm*FI. Fragments were separated on 1% agarose TAE gels and stained with ethidium bromide.

type plasmid digestion pattern with *Bsm*FI. D3, however, shows the appearance of an additional band of 556 bp and a shortening of the slowest migrating *Bsm*FI band (4200 bp → 3644 bp). This demonstrates the addition of a *Bsm*FI site within the 4200-bp fragment and indicates the annealing of the mutagenic primer. Thus, simple restriction analysis gives a diagnosis for clone D3 as likely to be the desired genotype (*Hpa*I$^-$, *Afl*II$^+$, *Bsm*FI^{add}). Clones with the desired restriction sites are then sequenced with a regional primer to confirm the mutation and to rule out anomalous changes. The opsin cDNA is then completely sequenced over the entire coding region using four uniformly spaced primers to rule out additional spurious mutations. Sequencing is conducted by the dideoxynucleotide chain termination method using [α-^{35}S]dATP.[15] After confirmation, the complete cDNA clones were directionally subcloned into pCDNA3 (5'*Kpn*I/3'*Xba*I) (neo^R), pCIS[17] (5'*Cla*I/3'*Xba*I), or pCDNA3.1 (5'*Kpn*I/3'*Xba*I) (hygro^R). In all expression plasmids opsin cDNAs are expressed from the human cytomegalovirus (CMV) early immediate promoter. The opsin mRNA transcribed from pCIS contains an intron for splicing, whereas opsin mRNA expressed from pCDNA3 or pCDNA3.1 contains no intron and expression levels are enhanced by a human immunodeficiency virus long terminal repeat (HIV LTR) enhancer element just upstream and overlapping the transcription start site of the CMV promoter. Expression plasmid stocks are maintained in INVα-F' and production of bulk expression plasmids for transfections is conducted according to standard techniques (Qiagen, Santa Clarita, CA; Promega, Madison, WI). It is useful to mention that we have identified clones with both the desired selection and mutational screening enzyme genotypes but which had unusual rearrangements (deletions), identified by sequencing, which apparently occurred by partial sequence complementarity of the mutagenic primer to both the target region and another region of the plasmid.

Development of Cell Lines Constitutively Expressing Human Rod Opsin Visual Pigments

HEK293S cells are grown to subconfluence (70–80%) in 10-cm dishes in Dulbecco's modified Eagle's medium (DMEM)/F12 medium with 10% heat-inactivated calf serum plus 2 mM L-glutamine and antibiotics (penicillin/streptomycin) at 37° in 5% CO_2. Supercoiled plasmid, 10–20 μg per 10-cm plate, is used to transfect log-phase cultures of HEK293S cells using the slow calcium phosphate technique.[18,19] The appropriate amount of con-

---

[17] C. M. Gorman, D. R. Gies, and G. McCray, *DNA Prot. Eng. Tech.* **2,** 3 (1990).
[18] C. Chen and H. Okayama, *Mol. Cell. Biol.* **7,** 2745 (1987).
[19] C. Chen and H. Okayama, *BioTechniques* **6,** 632 (1988).

centrated plasmids is first mixed in a sterile 0.5-ml microcentrifuge tube. For every 1.0 ml final volume of transfection solution, 0.5 ml of 2× BBS transfection buffer is first diluted in the appropriate amount of sterile water in a sterile 15-ml conical tube, the plasmid mix is added next and swirled, and finally 0.05 ml of 0.25 M $CaCl_2$ is added to initiate the precipitation reaction. The 2× BBS is 50 mM N,N-bis[2-hydroxyethyl]-2-aminoethanesulfonic acid (BES) (Sigma), 280 mM NaCl, and 1.5 mM Na_2HPO_4 at final titrated pH 6.95 (store at $-20°$). After mixing plasmids with buffer and $CaCl_2$, the solution is incubated for 1 hr at room temperature until birefringent crystals are visible in the phase contrast microscope. We interpret these "cross crystals" to represent the early development of calcium phosphate precipitate. At this point the transfection solution is added (1.0 ml/10-cm plate) with precipitation allowed to occur over a 16- to 20-hr period at 3% (v/v) CO_2 in air at 37°. When pCIS expression plasmids are used during preliminary experiments, pSV2-neo[20] is cotransfected (ratio of pCIS-opsin:pSV2-neo is 10–20:1) to introduce neomycin resistance (neo^R).

During the study we switched to sole use of the pCDNA3 vector to express both opsin and neo^R because our results indicated a significant improvement in the number of neo^R clones that were opsin$^+$ with pCDNA3-opsin constructs. This occurs presumably because both characteristics are expressed from the same plasmid with pCDNA3, and although they are not expressed from the same mRNA they are nonetheless structurally linked during the integration process. In both cases the neo^R expression cassette is controlled by a strong simian verus 40 (SV40) promoter. Given the amount of time required to generate and screen clones it is desirable to minimize the number of opsin$^-$/neo$^+$ clones.

After transfection, the precipitate-containing medium is aspirated and fresh medium added for a 48-hr period to allow expression of neomycin phosphotransferase II from the neo^R expression cassette. Media containing 500 μg/ml of the neomycin analog G418 is then added to the cells and replaced every 2–3 days as untransfected cells begin to die off due to G418 toxicity. Over a period of 2–3 weeks in the continuous presence of drug, small G418-resistant colonies become identifiable. As colonies grow to macroscopic visibility they are individually harvested and grown as separate clones in 6-well tissue culture plates. To minimize the time needed to screen individual clones, we add a poly(L-lysine)-coated glass coverslip to each well and then process the coverslips for primary screening leaving the live remainder of the clone for tissue culture passage. Thus, clone isolation and immunocytochemical tests for opsin expression are managed on the first

---

[20] P. J. Southern and P. Berg, *J. Molec. Appl. Genet.* **1,** 327 (1982).

passage. Later in the study pCDNA3.1-opsin plasmids expressing hygro^R are used to supertransfect existing opsin clones that are G418 resistant. When introducing pCDNA3.1-opsin constructs (hygro^R) into established opsin-expressing neo^R lines, slow calcium phosphate transfections are conducted and hygromycin (200 μg/ml) challenge is initiated 2 days after supertransfection.

Immunocytochemistry

Cells are grown on poly(L-lysine)-coated coverslips and fixed and permeabilized in 100% ice-cold methanol for 5 min, after which the methanol is aspirated and the samples air dried. Samples are rehydrated with phosphate-buffered saline (PBS) (in mM) (NaCl 137, KCl 2.7, Na$_2$HPO$_4$ 5.6, KH$_2$PO$_4$ 1.47) plus 0.02% sodium azide and then incubated with the carboxyl-terminal 1D4 mouse anti-bovine opsin monoclonal antibody[21] (1 μg/ml final concentration) in PBS plus 10% goat serum for 1 hr at room temperature, washed with PBS, and then incubated with secondary fluorescein isothiocyanate (FITC)-labeled horse anti-mouse IgG (Vector Laboratorics, Burlingame, CA) at a final concentration of 1 μg/ml in PBS plus 10% (v/v) horse serum for 1 hr at room temperature. The 1D4 recognition epitope of bovine rod opsin is conserved in human rod opsin. Coverslips were washed with distilled water, inverted onto a drop of 2% (w/v) 1,4-diazabicyclo[2.2.2]octane (DABCO, fluorescent antifade agent) in 80% glycerol on a clean slide and the edges of the slide were sealed with fast drying polymer. Cells were photographed on a Nikon Microphot-FX fluorescent microscope through FITC filters. A high-level wild-type human expression cell line[2] (wild-type standard control) and a negative control (untransfected HEK293S cells[7]) are included for comparison. Clones are examined for the quantity of cellular fluorescence on a scale of 0 (background, like HEK293S cells), 1$^+$ (faint fluorescence), 2$^+$ (moderate), 3$^+$ (bright), and 4$^+$ (intense and comparable to the wild-type standard line). Several fields are examined to achieve the qualitative estimate of expression levels, and fields are only examined once to avoid any bleaching of the FITC label. Each clone is also carefully examined for uniformity of expression to ensure that each clone is not a mixed population of cells having markedly varying levels of expression. Mixed clones are not further evaluated, although uniform producers could have been isolated from such cultures by limiting dilution subcloning. Each clone is also evaluated for the predominant location of the expressed opsin molecules (plasma membrane versus intracellular membranes).

[21] R. S. Molday and D. MacKenzie, *Biochemistry* **22**, 653 (1983).

High-level clones (3–4⁺) are identified by immunocytochemistry for all mutants generated (Fig. 2), and these are further evaluated for opsin levels and ability to regenerate visual pigment. One opsin expressing clone of the E113Q/E134Q double mutant is also identified but not evaluated (not shown). The wild-type standard line and untransfected HEK293S cells are shown as positive and negative controls (Figs. 2A and B). The faint background fluorescence in untransfected cells is due to flavinoid fluorescence.[22] We maintain G418 selection for several passages after primary clonal isolation and then release it with later immunocytochemical follow-up to ensure that the clones are stably producing opsin on the basis of integrated transgenes. Episomally replicating plasmids dilute with further cell divisions in the absence of selection under these conditions. In all mutant clones except C187Y most of the opsin partitioned to the plasma membrane (Fig. 2J). C187Y is unlikely to fold and traffic correctly because C187 is involved in the critical disulfide bond found in normal rhodopsin.[23]

Quantification of Opsin Production in Selected High-Level Producing Clones

Mutant opsin expression cell lines are grown to near confluence in 10-cm dishes and washed from the plate after a 3-min incubation in PBS containing 5 mM EDTA. Cell number is quantified spectrophotometrically by measurement of absorbance of the cell suspension at 800 nm.[24] Previous calibration of spectroscopic versus hemocytometer measurements of HEK293S cell numbers allow generation of a reference linear regressed data set so that OD_{800} measurements can be simply used to assay cell number and concentration. Cells are gently centrifuged into a pellet in a table-top clinical centrifuge at room temperature ($500g$) for 5 min, washed once with PBS, and solubilized in 0.1% DM in PBS for 4 hr at 4°. Nuclei are pelleted at $16000g$ at 4° for 1 hr and the supernatants centrifuged at $43,000g$ for 2 hr at 4° to remove residual cellular debris and polysomes. Cell extracts are stored at −80° until use. Protein concentration in cellular detergent extracts is quantitated using Coomassie Brilliant Blue G-250 binding[25] (Bio-Rad, Hercules, CA). Slot immunoblots are used for quantification of opsin levels because the total amount of opsin can be ascertained without concerns over losses during immunoaffinity purification. Slot blots

---

[22] M. J. Zylka and B. J. Schnapp, *BioTechniques* **21,** 220 (1996).
[23] S. S. Karnik, T. P. Sakmar, H.-B. Chen, and H. G. Khorana, *Proc. Natl. Acad. Sci. U.S.A.* **85,** 8459 (1988).
[24] W. A. Mohler, C. A. Charlton, and H. M. Blau, *BioTechniques* **21,** 260 (1996).
[25] M. M. Bradford, *Anal. Biochem.* **72,** 248 (1976).

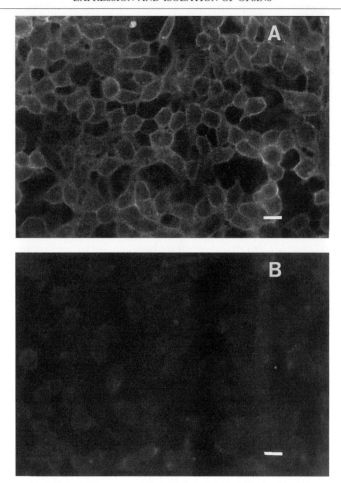

FIG. 2. Immunocytochemical screening of opsin clones. Primary wild-type and mutant clones were screened by indirect immunofluorescence relative to the wild-type standard control and untransfected HEK293S cells. Representative 3–4+ clones from several stable transfections are shown: (A) wild-type standard, (B) HEK293S cells, (C) wild-type-2, (D) D83N-20, (E) G90D-22, (F) E113Q-8, (G) E122Q-8, (H) E134Q-7, (I) E134Q-4 (hygro^R), (J) C187Y-2, and (K) H211F-4. The E134Q-4 (hygro^R) clone resulted from a secondary transfection of E134Q-7 with pCDNA3.1-(hygro)-E134Q plasmid and selection in 200 μg/ml of hygromycin. The primary localization of immunoreactive opsin appeared to be at the level of the plasma membrane in all but the C187Y mutant. In C187Y, immunoreactive opsin was essentially entirely localized in intracellular membranes (endoplasmic reticulum, Golgi) with little to no plasma membrane staining. Bar: 10 μm (A–K).

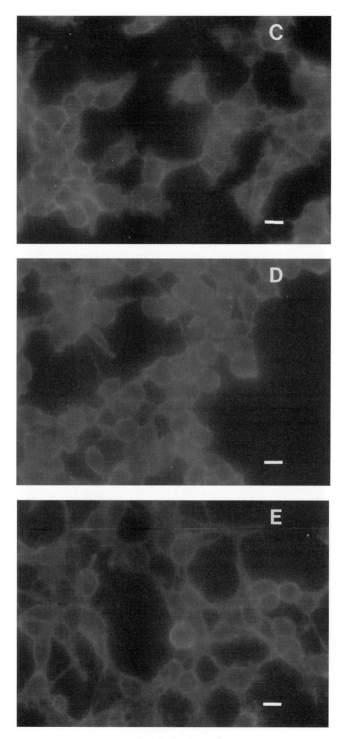

FIG. 2. (*continued*)

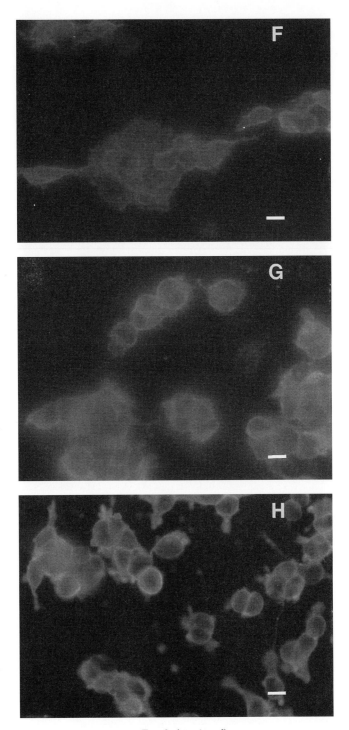

FIG. 2. (*continued*)

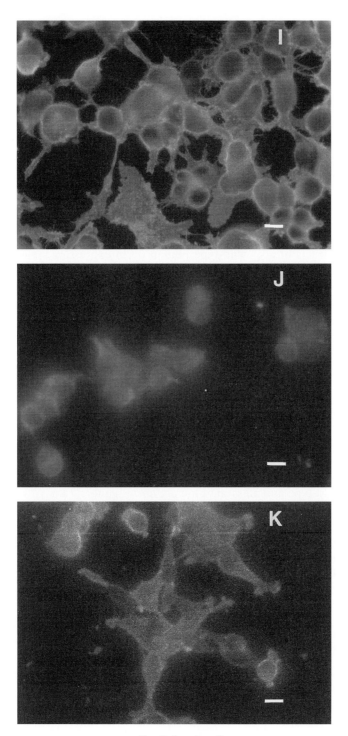

FIG. 2. (*continued*)

are used rather than Western blots so that all forms of the opsin protein can be simultaneously measured in a single band without losses due to electroblotting transfer inefficiencies. Dimers and trimers of aggregated opsin form during Western analysis,[26,27] which complicates densitometry since the opsin in each of three bands must be measured and summed (not shown). Samples are diluted in 0.1% DM in PBS to the same total protein concentration such that each sample is loaded onto the slot apparatus in the same volume to achieve 10 μg of total protein presented to the membrane. Preliminary experiments with the wild-type standard line indicate that intense immunoreactive banding can be observed with 10 μg total protein loading, while the blots fail to percolate more than 10–20 μg of total protein. We chose to load 10 μg of total protein to ensure detection sensitivity over a sufficient range to quantitate opsin from clones with low or high expression levels. Extracts of a 4$^+$ wild-type human opsin expressing standard clone and extracts of untransfected HEK293S cells are added as positive and negative controls. At room temperature cell extracts are drained by gravity onto a nitrocellulose membrane (NitroBind, 0.45 μm pore size, Micron Separations Inc., Westboro, MA) on a slot-blot apparatus (Bio-Rad).[28,29] The membrane is blocked in 5% (w/v) nonfat milk in 50 mM Tris-HCl (pH 8), 100 mM NaCl, 0.05% (v/v) Tween 20 (TBST) for 1 hr at room temperature, and then incubated with the 1D4 opsin monoclonal antibody[21] (600 ng/ml) in TBST overnight at 4°. The blot is washed 4 × 5 min in TBST, incubated with the secondary horseradish peroxidase-conjugated horse anti-mouse IgG (125 ng/ml) (Vector Laboratories) in TBST for 1 hr at room temperature, and washed 4 × 5 min in TBST. The blot is then incubated in a 97.5 mM Tris-HCl solution (pH 7.5) containing 800 μg/ml diaminobenzidine, 400 μg/ml NiCl$_2$, and 0.09% H$_2$O$_2$ for 5–10 min to allow for color development. Developed membranes are air dried on Whatman (Clifton, NJ) paper. Wild-type and all mutant human opsins generated have an identical 1D4 epitope at the carboxy terminus of the polypeptide. Qualitatively, clones were evaluated relative to staining of the wild-type standard, which produces about 6 × 10^6 wild-type opsins/cell as evaluated by immunoaffinity purification and spectrophotometry following 11-*cis*-retinal regeneration. In quantitative experiments, known quantities (as determined by spectrophotometry, extinction at 493 nm, 40,000 M^{-1} cm^{-1}) of immunoaffinity-purified WT human rhodopsin in 0.1% DM in PBS are diluted in PBS and

---

[26] C.-H. Sung, B. G. Schneider, N. Agarwal, D. S. Papermaster and J. Nathans, *Proc. Natl. Acad. Sci. U.S.A.* **88,** 8840 (1991).
[27] C.-H. Sung, C. M. Davenport, and J. Nathans, *J. Biol. Chem.* **268,** 26645 (1993).
[28] W. N. Burnette, *Anal. Biochem.* **112,** 195 (1981).
[29] J.-Y. Xu, M. K. Gorny, and S. Zolla-Pazner, *J. Immunol. Meth.* **120,** 179 (1989).

applied to the membrane with the panel of high-producing mutant clonal extracts. In this manner the known quantities of purified WT opsin standards can be quantified in parallel to cell extracts from unknown mutant clones. Figure 3A shows a representative blot that brings together all of our high-producing wild-type and mutant clones, relative to wild-type standard cells and untransfected cells, with a standard series of wild-type rhodopsin, immunoaffinity purified from the wild-type standard cell line. The level of expression varies in different clones. The lowest levels of expression hare found in E113Q clones and the highest level occurred in one G90D clone. No opsin was detected in the untransfected HEK293S cells. C187Y and E113Q/E134Q (not shown) clones have not yet been quantified for opsin.

To investigate whether secondary transfection with additional opsin expression plasmid programming an alternative drug resistance marker would enhance the levels of opsin protein, we supertransfect pCDNA3-E134Q (hygro^R) into the neo^R E134Q-7 clone and select with hygromycin (200 μg/ml) while maintaining G418 selection (500 μg/ml). We reasoned that stable transfection of additional opsin expressing transgenes would increase mutant opsin protein levels, under the assumption that cellular transcription and translation machinery are not saturated.[30] Again, the level of expression varies, and while many of the clones appear to express about the same or less than the parent clone, two clones appear to express greater amounts of E134Q opsin (raw data not shown, see Table III later). Thus, supertransfection and secondary drug selection is a means of enhancing expression levels in particular clones.

The amount of opsin deposited in each slot on the developed membrane is semiquantitatively measured by reflectance-mode measurements on a scanning densitometer (Hoefer Inc., San Francisco, CA; model GS 300) and signals are acquired using GS365W software (Hoefer). Bands are scanned orthogonal to their long axis (slot dimensions: 0.75 $\times$ 7 mm) on the face of the membrane to which the samples were applied. Unknown bands are measured relative to known amounts of purified wild-type human rhodopsin added to the membrane.[31] Files are converted to ASCII output and then analyzed in Origin 4.1 (MicroCal Software, Northampton, MA). Scans of developed blot membranes result in single symmetric Gaussian-like peaks that are individually integrated relative to neighboring baseline to "weigh" the amount of immunoreactive opsin in each slot band (data not shown). Quantitative assessments of mutant cell line opsin content relative to purified opsin standards, or relative comparisons of opsin content in a mutant

---

[30] R. F. Santerre, N. E. Allen, J. N. Hobbs, Jr., R. Nagaraja, and R. J. Schmidt, *Gene* **30,** 147 (1984).

[31] C. A. Dennis-Sykes, W. J. Miller, and W. J. McAleer, *J. Biol. Standard.* **13,** 309 (1985).

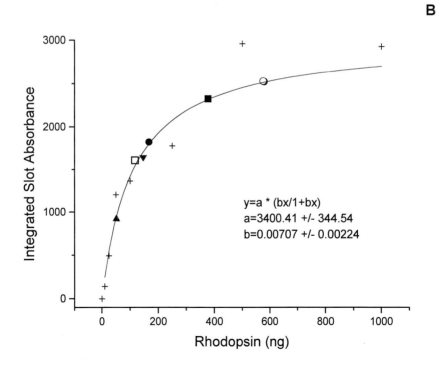

line compared to wild-type or other control line, are made with respect to individual developed membranes. Quantities of opsin with respect to the standard curves are normalized with respect to protein content of the cell extracts and the cell numbers from which the extracts were derived. A graph of the integrated density versus known amounts of purified WT rhodopsin protein is generated from the neo^R slot blot, and the location of the mutant clones relative to this curve is shown in Fig. 3B. To quantitate results a rectangular hyperbolic function is fit to the standard data to allow comparison of the mutant clones relative to the standard wild-type series. At and above 250 ng of purified rhodopsin (integrated slot absorbance >1500) the intensity of opsin immunoreactivity begins to saturate. The saturation found in the standards above 250 ng may result from several mechanisms—the saturation of the protein binding capacity of the membrane under each well (≈ 5.25 μg) due to the combination of rhodopsin and the 1D4 peptide, a competition of the matrix-bound opsin and peptide for the primary antibody (1D4), or saturation of steps in the blot development process (e.g., limiting primary or secondary antibody). To obtain measures for the opsin quantity of mutant clones, the scan weight from each blot is compared to the standard curve to estimate the total amount of immunoreactive opsin protein present in the applied sample. For clones with absorbance values in areas of saturation of the standard curve the quantification of opsin levels represents the minimum amount present. The

FIG. 3. Opsin quantification in high-level-producing clones. (A) Slot blot showing a series of purified wild-type opsin standards of known quantity (left-hand side). Mutant and wild-type clones with 3–4$^+$ opsin immunofluorescence were analyzed (10 μg total protein) in parallel on the same blot (right-hand side). Wild-type is the standard positive control line. The slot-blot photograph was obtained from a digital imaging station using a flatbed scanner and dye sublimation printer (Kodak, Rochester, NY). (B) The relationship between integrated scan absorbance versus the quantity of purified wild-type opsin loaded onto the blot shown in (A). A rectangular hyperbolic curve was fit to the wild-type standard data (+) ("a" is saturation level of hyperbola and "b" is slope characteristic). Given the integrated slot absorbance of an unknown band, the amount of immunoreactive opsin in the particular unknown clone can be determined from the wild-type standard curve. The wide range of standards was needed to cover the expression levels of the diverse clones identified and analyzed on the same membrane. With the blot development method and the practical limitations of sample volumes, the relationship of known opsin quantity to measured density saturates. This likely reflects the saturation of binding sites for opsin protein on the nitrocellulose matrix or competition of matrix-bound 1D4 elution peptide with the 1D4 monoclonal antibody. Symbols for the highest producing clones are as follows: WT-2 (■), D83N-11 (●), E113Q-8 (▲), E122Q-8 (▼), E134Q-12 (□), H211F-3 (○). G90D-22, the highest producing clone of G90D opsin, had absorbance (3896) which was in excess of the largest amount of wild-type standard added (1000 ng) and, by the fitted curve, would correspond to a wild-type standard quantity of greater than 3200 ng.

individual mutant clones are also internally compared to the slot density of the wild-type standard cell extract on the blot by taking a ratio of band absorbances. In all cases, results are normalized to the amount of total protein added to the slot and to the number of cells which contributed to the sample added ($\approx 2.1 \times 10^7$ cells/10-cm dish at 95% confluence). Finally, the average number of wild-type or mutant opsins present per cell is calculated. We then calculate the total amount of opsin obtainable in a linear scaleup assuming growth of 10^9 cells in a 1-liter suspension flask. These results for the mutant opsins are presented in Table II and for the super-transfected hygro^R E134Q clones in Table III.

D83N clones express comparable levels of opsin compared to the wild-type standard and would generate a minimum of 400–500 μg of opsin/10^9 cells. Clone G90D-22 is the highest expression clone we have isolated to date and would generate a minimum of about 3 mg of opsin/10^9 cells. E113Q clones are the lowest expressing clones and would express on the order of 100 μg of opsin/10^9 cells. E122Q clones are similar to D83N and would produce at a minimum about 400–500 μg opsin/10^9 cells. The highest producing E134Q primary neo^R clone would produce around 400 μg opsin/10^9 cells. Two hygro^R/neo^R clones would be expected to increase the E134Q

TABLE II
QUANTIFICATION OF OPSIN EXPRESSION LEVELS IN CELL LINES (NEO^R)[a]

Cell line	Integrated slot absorbance	Absorbance relative to wild-type standard	ng Opsin/μg total protein	μg Protein from 2.1×10^7 cells	μg Opsin/ 2.1×10^7 cells	μg Opsin/10^9 cells	Opsin #/cell
D83N-11	1791.5	0.812	15.8	65.3	1.032	491.4	7.39×10^5
D83N-20	1771.1	0.803	15.4	54.2	0.835	397.6	5.99×10^5
G90D-18	729.8	0.331	3.9	49.8	0.194	92.4	1.39×10^5
G90D-22	3896.4	1.766	100.0[b]	60.3	6.030	2871.4	4.32×10^6
E113Q-8	896.8	0.406	5.1	57.9	0.295	140.5	2.11×10^5
E113Q-14	358.4	0.162	1.7	49.4	0.084	40.0	6.02×10^4
E122Q-7	1580.7	0.716	12.3	57.1	0.702	334.3	5.03×10^5
E122Q-8	1639.6	0.743	13.2	63.8	0.842	401.0	6.04×10^5
E134Q-7	615.5	0.279	3.1	59.4	0.184	87.6	1.32×10^5
E134Q-12	1627.7	0.738	13.0	55.0	0.715	340.5	5.12×10^5
H211F-3	2513.9	1.139	40.1	51.2	2.053	977.6	1.47×10^6
H211F-4	1594.1	0.722	12.5	62.3	0.779	371.0	5.59×10^5
Wild-type-2	2313.0	1.048	30.1	72.5	2.182	1039.0	1.56×10^6
Wild-type-6	1171.0	0.531	7.4	56.0	0.414	197.1	2.97×10^5
Wild-type standard	2206.4	1.000	26.1	73.0	1.905	907.1	1.37×10^6

[a] Integrated opsin absorbance bands were compared with the absorbance curve of known wild-type opsin standards, and then normalized to the amount of total cellular protein and the cell number that contributed to the amount of cellular protein loaded onto the blots. For standard rhodopsin quantities above 250 ng the corresponding blot absorbance measurements (above 1500) are approaching saturation so the amount of opsin computed for cell lines with absorbances above 1500 is a minimum estimate. Estimates of the quantity of opsin expected in 10^9 cells and the number of opsin molecules per cell are shown. A comparison to the wild-type standard line is included.
[b] Estimate.

TABLE III
QUANTIFICATION OF OPSIN EXPRESSION LEVELS IN SUPERTRANSFECTED HYGRO^R/NEO^R
E134Q CELL LINES (NEO^R)[a]

Sample	Integrated slot absorbance	Absorbance relative to E134Q #7	Absorbance relative to wild-type standard
Wild-type standard	3139.0	1.707	1.000
E134Q #7	1838.4	1.000	0.586
Hygro #1	969.9	0.528	0.309
Hygro #2	993.1	0.540	0.316
Hygro #3	1796.4	0.977	0.572
Hygro #4	2683.0	1.459	0.855
Hygro #5	1402.7	0.763	0.447
Hygro #6	1006.0	0.547	0.320
Hygro #7	1684.9	0.917	0.537
Hygro #8	1712.2	0.931	0.545
Hygro #9	727.6	0.396	0.232
Hygro #10	1435.2	0.781	0.457
Hygro #11	1093.6	0.595	0.349
Hygro #12	1073.6	0.584	0.342
Hygro #13	879.3	0.478	0.280
Hygro #14	973.8	0.530	0.310
Hygro #15	613.4	0.334	0.195
Hygro #16	2105.8	1.145	0.671
Hygro #17	1586.8	0.863	0.506
Hygro #18	681.2	0.371	0.217

[a] Opsin absorbance bands of the supertransfected clones (labeled hygro) were compared with the absorbance of known wild-type opsin standards from the curve. Again, for standard rhodopsin quantities above 250 ng the blot intensities are saturated and the amount of opsin computed for cell lines with these quantities is thus a minimum estimate. Relative quantities of absorbance for hygro clones relative to the parent E134Q-7 cell line and the wild-type standard are shown.

harvest level between 20 and 50% (500–600 μg/10^9 cells). The H211F clones would express at least 400 and 1000 μg/10^9 cells. Finally, one wild-type clone (WT-2) would generate about 1 mg/10^9 cells, comparable to the wild-type "standard" line.

1D4 Immunoaffinity Purification and Spectroscopic Properties of Expressed Rhodopsins

Selected clones of mutants and wild-type are grown to about 1×10^8 cells in two 90–95% confluent, 15-cm polystyrene plates, harvested in PBS/ 5 mM EDTA, counted, washed, and concentrated in 2 ml of PBS. Cells are regenerated with 50 μM 11-*cis*-retinal for 3 hr at 4° in the dark, pelleted, and solubilized in 1.0% DM for 1.5 hr at 4°. All procedures are carried out in darkness or under dim far-red light (Kodak safelight, filter 2, No. 1521525, Kodak Inc.). DM leaves nuclei intact, which are sedimented at 43,000g for 2 hr at 4° and the supernatants transferred for binding to preequilibrated 1D4-Sepharose-4B resin prepared as described.[1,4] Rhodopsin is allowed to bind to the resin overnight at 4° with gentle agitation. The supernatant is removed and the resin washed in PBS plus 0.1% DM until protein bound nonspecifically to the resin is minimized in spectrophotometric analyses (A_{280}). Rhodopsin is then eluted from the resin using 50–100 μM of COOH-terminal competing epitope peptide (14-mer) in PBS plus 0.1% DM, and quantified by absorption spectrophotometry against a blank of peptide elution buffer.

Wild-type, D83N, G90D, E122Q, E134Q, and H211F rhodopsins are spectroscopically evaluated by bleaching with light to form Meta-II/retinal and by formation of an acid-denatured ground state. Difference spectra are generated to demonstrate the transitions between spectral states.[32] Dark–light difference spectroscopy is used to characterize the formation of a visual pigment by way of protonated Schiff base linkage of 11-*cis*-retinal with opsin and the bleaching characteristics in comparison to wild-type human rhodopsin. Absorption spectroscopy is performed on both dark-adapted and light-exposed pigment preparations to determine if the dark-adapted peak (≈ 493 nm) disappears concurrent with the appearance of a Metarhodopsin-II state (A_{380}) or free retinal (bleaching). Pigments are bleached for 10 sec using light from a 300-W slide projector filtered through a 450-nm-long pass filter. Ground state pigments are also acid-denatured by the addition of 1 μl of 3 N HCl per 75 μl of sample to a final concentration of 40 mM in order to trap a protonated Schiff base linkage to the protein with absorbance at 440 nm.[33] Absorption spectroscopy is performed on both pre- and post-acid-denatured samples to determine if the ground state pigment would denature to a 440-nm absorbing species, indicating a protonated Schiff base attachment of retinal to the opsin protein.[33] Spectra are obtained on a DU-640 spectrophotometer (Beckman, Columbia, MD)[3,32] and analyzed off-line using Origin 4.1 software.

In all pigments tested, addition of acid to dark-adapted rhodopsin promotes quantitative conversion to a ≈ 440-nm absorbing band of a protonated Schiff base in the absolute spectra which blue shifts slightly in the difference spectra (Fig. 4, left-hand side). Presentation of intense light (>450 nm) for a few seconds at room temperature in wild-type, D83N, E122Q, E134Q, and H211F rhodopsins promotes quantitative appearance of a ≈ 380-nm

---

[32] J. Nathans, *Biochemistry* **29**, 9746 (1990b).
[33] Y. Kito, T. Suzuki, M. Azuma, and Y. Sekoguti, *Nature* **218**, 955 (1968).

absorbing band (Fig. 4, right-hand side). The 380-nm band is consistent with Metarhodopsin-II formation and the slow hydrolysis of all-*trans*-retinal from the deprotonated Schiff base. G90D bleaches through a blue-shifted intermediate at 15° as in bovine rhodopsin[34] and appears to have a thermally unstable ground state even at 4° (shown as an absolute spectrum in Fig. 4, right-hand side). We found G90D to be temperature labile in the dark even after prolonged incubation at 4°. The dark spectrum shows two peaks, one around 500 nm (ground state) and another around 440 nm. After clamping the temperature at 15°, light is presented and the 500-nm peak disappears with quantitative appearance of a blue-shifted intermediate ($\approx$440 nm), which appears related to the same population found in dark-adapted rhodopsin. At higher temperatures the 440-nm peak bleaches with the gradual appearance of a peak around 380 nm. Of note, we were unable to grow the largest producing E113Q clone to sufficient cell numbers in 15-cm plates to generate a measurable pigment yield after detergent extraction. For example, it typically took 3–4 weeks to grow near-confluent 15-cm plates of E113Q after even heavy seeding. Thus, high yields require not only high levels of opsin expressed per cell but also reasonable cellular mitotic rates to increase the overall biomass.

Discussion

The powerful CMV enhancer/promoter was used in this study to drive expression of human rod opsins in cultured transformed human embryonic kidney cells. HEK293S cells express the leftmost region of the adenovirus 5 genome (e.g., E1A, E1B genes), which may affect opsin expression levels since the E1A protein is thought to transactivate the CMV promoter.[17,35] While single-cell expression levels for stably transformed cells are generally lower than after transient transfection, the presence of a predictable amount of opsin in every cell avoids the problem of variable transfection efficiency and ensures previously quantified amounts of opsin starting material for structure–function experiments. The time investment needed to produce high-level stable expression lines is beneficial when the anticipated needs of a given mutant pigment are great. When mutant pigments with interesting properties are identified in transient transfection experiments, and the experimental studies are expected to be extensive, then we would recommend development of constitutively expressing lines.

All clones are primarily selected in G418, conditionally dependent on expression of neomycin phosphotransferase. Selection is dominant but not

[34] V. R. Rao, G. B. Cohen, and D. D. Oprian, *Nature* **367,** 639 (1994).
[35] C. M. Gorman, D. Gies, G. McCray, and M. Huang, *Virology* **171,** 377 (1989).

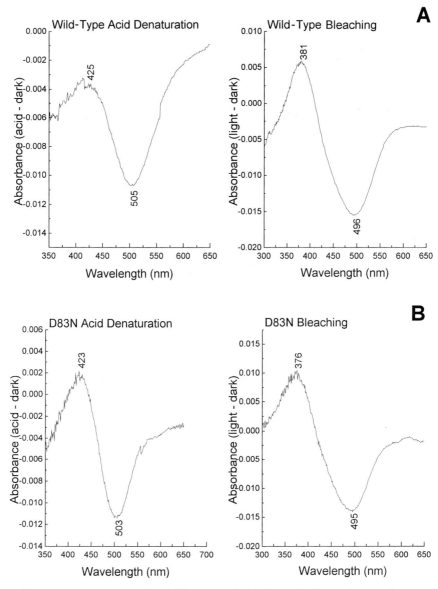

FIG. 4. Ground state properties and formation of Metarhodopsin-II in wild-type and mutant rhodopsins. (A) Wild-type (B) D83N, (C) G90D, (D) E122Q, (E) E134Q, (F) H211F. (Left-hand side) Difference spectra of the dark-adapted purified rhodopsins (after acid—before acid) during acid trapping. The ground state absorbance (around 500 nm) appears as a negative band in both wild-type and all mutants examined. The quantitative appearance of a band around 440 nm in the absolute absorbance spectra following acid treatment was indicative

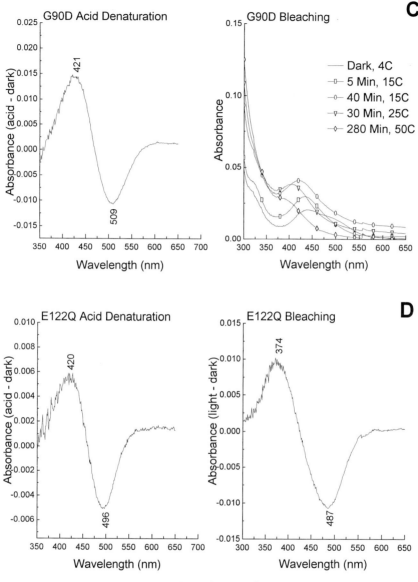

FIG. 4. (*continued*)
of the presence of a protonated Schiff base in wild-type and all mutants tested. The positive acid-trapped band is slightly blue shifted in the difference spectrum relative to 440 nm due to absorbance in the corresponding region of the spectrum prior to acid treatment (not shown). (Right-hand side) The bleaching characteristics were examined by difference spectra on the same samples (light–dark) following exposure to bleaching light (longpass cutoff 450 nm) for 10 sec. The negative peak around 500 nm represents the dark-adapted absorption band,

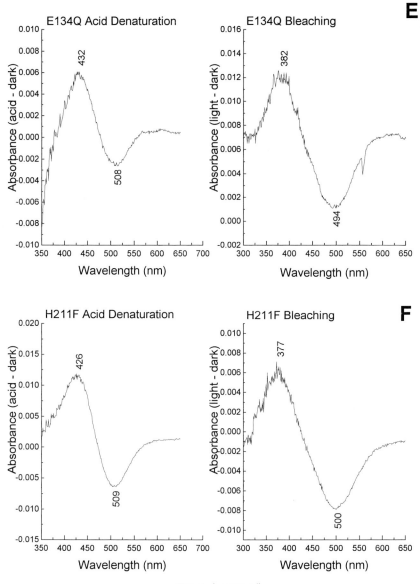

Fig. 4. (*continued*)

whereas the peak around 380 nm represents the Metarhodopsin-II absorbing intermediate (380 nm) or the appearance of all-*trans*-retinal (374 nm) due to Schiff base hydrolysis. The G90D bleaching spectrum has absolute scales, showing both the anomalous bleaching characteristics and the temperature dependence of the process. The symbols for the different spectra shown are labeled in the inset for G90D. The small step in the spectra at 560 nm is due to a spectrophotometer filter change.

explicitly tied to the production of the opsin protein because the neo^R or $hygro^R$ mRNAs and the opsin mRNAs are expressed from independent but physically linked transgenes in pCDNA3 or pCDNA3.1, respectively. Therefore, quantities of opsin vary greatly from clone to clone. Clonal variance of opsin expression likely depends on many factors including the number of stably integrated transgenes, genomic positional effects on expression, and the effects of opsin protein on cell survival. Some opsins might express to low levels in stable clones because proteins are toxic and only low levels might permit cell survival during G418 selection. For example, the E113Q clones with the best rhodopsin levels were slow growing in cell culture and the one clone of E113Q/E134Q, which was also opsin positive on an immunocytochemical level, was also slow growing. The levels of expression for given mutants probably reach maximum levels that depend on the nature of the mutation and the limits of HEK293S cell expression machinery.

Clones that are drug resistant do not necessarily express detectable levels of opsin because the opsin transgene could be deleted or positionally suppressed after integration. In these experiments there was no coupled relationship between neo^R and opsin expression as would occur with a dicistronic mRNA expressing both opsin and neo^R. The yield of opsin expressing neo^R clones was higher when neo^R and opsin were expressed from the same plasmid where the genes are physically linked. Therefore, it may be possible to increase further the number of opsin$^+$ neo^R clones by linearizing expression plasmid in a region remote from either expression cassette prior to transfection. This may minimize deletion or rearrangement events in the opsin transgene during genomic integration. We found that an immunocytochemical assay is an effective and rapid means to screen significant numbers of clones to identify those that are likely to be high-level producers. We identified clones that express to the level of about 10^6 opsins/cell. Stable supertransfection of opsin expression plasmids containing a different drug resistance marker (hygromycin) yielded some clones with greater levels than the parent clone. Surprisingly, some supertransfected clones produced less opsin for unclear reasons. To date we have routinely used standard levels of selection drug dosages to promote growth of G418 (500 μg/ml) or hygromycin (200 μg/ml) resistant clones. Due to the expense of G418 we did not try to select clones in higher levels of G418 (e.g., 3.0 mg/ml) as was reported in a study[36] using HEK293S cells and, as in this study, an opsin transgene expressed from a CMV promoter. However, the neo^R was expressed from a relatively weak promoter. Neo^R selec-

---

[36] P. J. Reeves, R. L. Thurmond, and H. G. Khorana, *Proc. Natl. Acad. Sci. U.S.A.* **93**, 11487 (1996).

tion at high drug concentrations was thought to select for high-level expression wild-type bovine opsin clones (e.g., 2 mg/10^9 cells) that resulted from integration events in chromosomal regions of high transcriptional activity. Four mutant opsin expression constructs altered in the cytoplasmic loops were also used to generate stable lines using high G418 concentrations and 0.3–1 mg/liter expression clones were isolated. We also found that high levels of opsin expression could also be obtained in wild-type (e.g., WT-2) and some point mutant clones (e.g., G90D-22) which should yield opsin levels comparable to those found in the other study[36] during scaleup (10^9 cells) even if the neo^R cassette in pCDNA3 and pSV2-neo is expressed from the strong SV40 promoter and special enriched growth media are not used. High levels of wild-type bovine opsin can also be obtained from baculovirus transgene delivery under control of the polyhedrin promoter.[37] However, these transient methods require production of virus, are conducted in insect cell lines which may not mimic mammalian cellular machinery, and stable cell lines cannot apparently be generated by this approach.

In the development of stable mammalian cell lines by plasmid transfection, sampling frequency plays a major role in detecting the highest level producing clones. We routinely screened between 30 and 50 clones (colonies) for each mutant type in a single transfection. While this may appear large, it is highly unlikely that we have identified the highest producers in the population. For this reason we have saved transfected cell populations for future selection and screening. Moreover, even though neo^R and hygro^R are "dominant" selectable markers, the level of resistance enzyme expression in a given cell will depend in part on the plasmid transgene load introduced. Thus, use of higher concentrations of selection drug should promote death of all cells except for those expressing the highest levels of drug resistance enzyme. This would likely increase the probability of identifying the highest producing opsin clones from physically linked transgenes. A more efficient method to screen for the highest producing opsin clones would be with fluorescence-activated cell sorting. This would require application of an extracellular (e.g., N-terminal) expression tag which could be efficiently recognized in live cells, or the use of an anti-opsin antibody that recognized the N terminus or extracellular loops. N-Terminal expression tags of rhodopsin might affect folding or expression.[38] We did try the B6-30N opsin monoclonal antibody,[39] which labels live

---

[37] J. J. M. Janssen, W. J. M. van de Ven, W. A. H. M. van Groningen-Luyben, J. Roosien, J. M. Vlak, and W. J. DeGrip, *Mol. Biol. Rep.* **13,** 65 (1988).

[38] J. Borgijin and J. Nathans, *J. Biol. Chem.* **269,** 14715 (1994).

[39] G. Adamus, Z. S. Zam, A. Arendt, K. Palczewski, J. H. McDowell, and P. A. Hargrave, *Vision Res.* **31,** 17 (1991).

Xenopus laevis photoreceptors.[40,41] However, we were unable to detect opsin in the WT standard line using indirect immunofluorescence. Since wild-type opsin expresses outside-out in the HEK293S cell line,[2] the lack of B6-30N immunoreactivity is likely to have some relationship to the nature or composition of the HEK293S cell surface since the immunologic epitopes in frog and human differ by only one amino acid. If a fluorescence probe against extracellular components of opsin could be identified, then cell sorting techniques would be a means of screening several orders to magnitude more clones than is practical by the plate selection techniques. However, this would not work well for those opsin mutants that fail to express well to the HEK293S plasma membrane (e.g., C187Y) as is commonly found in human rod opsin mutants that cause autosomal dominant retinitis pigmentosa.[26,27,42]

Spectroscopic properties of expressed human rhodopsins are consistent with other investigations on human and bovine wild-type rhodopsin and D83N, G90D, E122Q, E134Q, and H211F bovine mutants.[3,32,34,43–47] In small-scale experiments we did not purify sufficient E113Q pigment for spectroscopic analysis. Unique findings are the apparent sensitivity of the G90D pigment to undergo formation of a blue-shifted absorbing state in the dark during storage at 4°. This observation was not previously reported.[34] Otherwise, pigments had the expected spectroscopic properties indicating again that HEK293S cells provide an expression environment similar to both COS cells, where most of the bovine pigments have been expressed, and to photoreceptors.[36,46,47]

Summary and Conclusions

Stable HEK293S cell lines expressing high levels of normal and mutant human rod opsins were generated. Cellular expression is uniform across a population. Secondary overexpression of the same opsin transgene linked to a different drug selection marker (hygro^R) yielded expression clones with increased opsin levels compared to the neo^R parent strain. Wild-type and mutant human opsins regenerate with native chromophore and demonstrate spectroscopic properties consistent with previous reports of

---

[40] G. Adamus, *Methods Neurosci.* **15**, 151 (1993).
[41] J. C. Besharse and M. G. Wetzel, *J. Neurocytol.* **24**, 371 (1995).
[42] J. E. Richards, K. M. Scott, and P. A. Sieving, *Ophthalmology* **102**, 669 (1995).
[43] T. P. Sakmar, R. R. Franke, and H. G. Khorana, *Proc. Natl. Acad. Sci. U.S.A.* **86**, 8309 (1989).
[44] G. B. Cohen, D. D. Oprian, and P. R. Robinson, *Biochemistry* **31**, 12592 (1992).
[45] G. B. Cohen, T. Yang, P. R. Robinson, and D. D. Oprian, *Biochemistry* **32**, 6111 (1993).
[46] C. J. Weitz and J. Nathans, *Neuron* **8**, 465 (1992).
[47] C. J. Weitz and J. Nathans, *Biochemistry* **32**, 14176 (1993).

bovine opsin mutants. HEK293S cells can be grown in larger scale suspension culture (10^9 cells/liter) or in roller bottles (10^8 cells/bottle) to facilitate milligram-order preparations of purified pigments. These cell lines should be useful in any time-resolved spectroscopic or biophysical experiments that require either uniform cellular levels of opsin protein or regenerable pigment, or large amounts of purified visual pigment. They should also be useful in experiments where uniform constitutive levels of a given mutant human visual pigment are needed in each cell. These and similar types of constitutive or inducible cell lines may also be useful for studying mechanisms of human cell death that occur by mutations in the human rod opsin gene.

Acknowledgments

Supported by the National Eye Institute (EY11384 to J.M.S.), a Career Development Award from the Foundation Fighting Blindness (Harry J. Hocks Memorial Fund award), the Department of Ophthalmology (SUNY Health Science Center, NY), and Research to Prevent Blindness. We thank Genentech (South San Francisco, CA) for providing the pCIS plasmid, Dr. Jeremy Nathans (Johns Hopkins Medical Center, Baltimore, MD) for the wild-type human opsin cDNA and the wild-type "standard" human opsin cell line, Dr. Robert Molday (University of British Columbia, Victoria) for access to the 1D4 monoclonal antibody, Dr. Paul Hargrave (University of Florida, Gainesville) for providing the B6-30N antibody, Dr. Margaret Maimone (SUNY Health Science Center) for introducing us to the slow calcium phosphate transfection method, Dr. Barry Knox (SUNY Health Science Center) for providing the diaminobenzidine protocol for slot/Western blot development, a protocol for 1D4 immunoaffinity purification of rhodopsin pigments, and for supplies of 1D4-Sepharose and competing 1D4 epitope peptide, and Dr. Edward Shillitoe (SUNY Health Science Center) for making the Hoefer densitometer available for our use. We acknowledge the contributions of Kathleen M. Scott, Stephen Wrzesinski, Rachel C. Janssen, and Steven T. Jansen in the development of mutant human rod cDNAs. This work was assisted by the Wisconsin Package Version 8, Genetics Computer Group (GCG), Madison, WI. We especially thank Drs. Paul A. Sieving and Julia E. Richards (Department of Ophthalmology, University of Michigan) for encouraging the early phases of this work which occurred in their laboratories.

[4] Folding and Assembly of Rhodopsin from Expressed Fragments

By Kevin D. Ridge and Najmoutin G. Abdulaev

Introduction and Principles

The dim-light photoreceptor rhodopsin is a prototypical member of the family of seven transmembrane helix receptors that are coupled to guanine nucleotide-binding proteins (G-proteins). Although chemical aspects of the membrane-embedded chromophore environment, the cytoplasmic sites of interaction for proteins involved in visual transduction and desensitization, and the intradiscal posttranslational modifications of rhodopsin have been the subject of active investigation, the relative contributions of these three regions to the overall folding and assembly of the photoreceptor have received considerably less attention. An understanding of how rhodopsin adopts its tertiary structure is important not only to clarify details of the folding and assembly process, but also to gain insight into the severe visual impairments occurring as an immediate consequence of natural mutations affecting opsin structure and function.

One approach that has been used successfully to study the mechanism of protein folding and assembly is to use fragments of a polypeptide. It is well known that certain polypeptide fragments behave as independent folding domains,[1–3] and membrane protein fragments have been functionally recombined *in vitro* and *in vivo*.[4–11] The ability to form native-like rhodopsin complexes from expressed polypeptide fragments would provide opportunities for the study of a number of important aspects of bovine

---

[1] D. B. Wetlaufer, *Adv. Prot. Chem.* **34,** 61 (1981).
[2] H. Taniuchi, G. R. Parr, and M. A. Juillerat, *Methods Enzymol.* **131,** 185 (1986).
[3] P. Schimmel and J. A. Landro, *Curr. Opin. Struct. Biol.* **3,** 549 (1993).
[4] K.-S. Huang, H. Bayley, M.-J., Liao, E. London, and H. G. Khorana, *J. Biol. Chem.* **256,** 3802 (1981).
[5] J.-L. Popot, S.-E. Gerchman, and D. M. Engelman, *J. Mol. Biol.* **198,** 655 (1987).
[6] W. Stühmer, F. Conti, H. Suzuki, X. Wang, M. Noda, N. Yahagi, H. Kubo, and S. Numa, *Nature (London)* **339,** 597 (1989).
[7] B. K. Kobilka, T. S. Kobilka, K. Daniel, J. W. Regan, M. G. Caron, and R. J. Lefkowitz, *Science* **240,** 1310 (1989).
[8] E. Bibi and H. R. Kaback, *Proc. Natl. Acad. Sci. U.S.A.* **87,** 4325 (1990).
[9] C. Berkower and S. Michaelis, *EMBO J.* **10,** 3777 (1991).
[10] R. Maggio, Z. Vogel, and J. Wess, *FEBS Lett.* **319,** 195 (1993).
[11] J. D. Groves and M. J. A. Tanner, *J. Biol. Chem.* **270,** 9097 (1995).

opsin folding and assembly. They could offer insights into which regions of the polypeptide chain contain sufficient information to fold independently, insert into a membrane, and assemble to form a retinal-binding opsin. They could also give information about the locations of topologic determinants, interhelical interactions within the opsin apoprotein, and the minimum structure required for the binding of retinal. Of considerable interest is the prospect of examining the structural consequences of both site-directed and natural mutations on the folding and assembly of various segments of the opsin polypeptide chain. The purpose of this chapter is to describe methods for investigating whether expressed complementary bovine opsin fragments fold and assemble *in vivo*.[12-14] The focus is on the construction of opsin gene fragments, their transient expression in COS-1 cells, the biochemical characterization of the expressed fragments, and their implications to the overall mechanism of opsin biosynthesis.

Construction of Opsin Gene Fragments

The points of separation in most of the opsin fragments this laboratory has examined correspond to proteolytic cleavage sites or intron/exon boundaries in rhodopsin.[15-16] These sites are of interest because they may reflect the boundaries of independent structural or folding domains in the encoded protein. Additionally, gene fragments have also been generated where naturally occurring stop codon mutations found in retinitis pigmentosa result in premature termination of the polypeptide chain.[17,18] The actual points of separation for some of these fragments are shown in Fig. 1 and occur in all three regions of the protein. All of the opsin gene fragments are constructed by restriction fragment replacement of the synthetic bovine opsin gene[19] in the pMT3 expression vector (Fig. 2). The pMT3 vector contains the simian virus 40 (SV40) origin of replication and

---

[12] K. D. Ridge, S. S. J. Lee, and L. L. Yao, *Proc. Natl. Acad. Sci. U.S.A.* **92,** 3204 (1995).

[13] K. D. Ridge, S. S. J. Lee, and N. G. Abdulaev, *J. Biol. Chem.* **271,** 7860 (1996).

[14] K. D. Ridge, T. Ngo, S. S. J. Lee, and N. G. Abdulaev, *J. Biol. Chem.* **274,** 21437 (1999).

[15] P. A. Hargrave, J. H. McDowell, E. Smyk-Randall, E. D. Siemiatkowski-Juszcak, T. Cao, A. Arendt, and H. Kuhn, in "Membrane Proteins: Proceedings of the Membrane Symposium (S. C. Goheen, ed.), p. 81, 1987.

[16] J. Nathans and D. S. Hogness, *Cell* **34,** 807 (1983).

[17] J. P. Macke, C. M. Davenport, S. G. Jacobson, J. C. Hennessey, F. Gonzalez-Fernandez, B. P. Conway, J. Heckenlively, R. Palmer, I. H. Maumenee, P. Sieving, W. Good, and J. Nathans, *Am. J. Hum. Genet.* **53,** 80 (1993).

[18] P. J. Rosenfeld, G. S. Cowley, T. L. McGee, M. A. Sandberg, E. L. Berson, and T. P. Dryja, *Nature Genet.* **1,** 209 (1992).

[19] L. Ferretti, S. S. Karnik, H. G. Khorana, M. Nassal, and D. D. Oprian, *Proc. Natl. Acad. Sci. U.S.A.* **83,** 599 (1986).

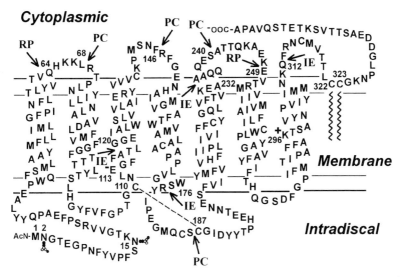

FIG. 1. Secondary structure model of bovine rhodopsin showing the points of polypeptide chain discontinuity studied in this laboratory. The seven transmembrane α-helical segments are lettered A–G, and the membrane–solvent boundaries are shown approximately by the interrupted horizontal lines. The points of polypeptide chain discontinuity examined so far are indicated by arrows. PC, Proteolytic cleavage sites; I/E, intron/exon boundaries; RP, positions of naturally occurring stop codons found in retinitis pigmentosa.

is therefore capable of replicating in COS-1 cells (see later discussion). Transfected cells, therefore, will contain multiple copies of the plasmid that can produce ample quantities of the fragment(s) encoded by the inserted gene(s). The fragments are placed between the adenovirus major late pro-

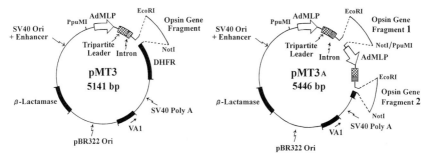

FIG. 2. Vectors used for expression of the opsin fragments in COS-1 cells.

moter (AdMLP) and the SV40 poly(A) addition site on an *Eco*RI/*Not*I fragment. The amino-terminal fragments are prepared by replacing the appropriate restriction fragment with a synthetic oligonucleotide duplex containing a termination codon after the last encoded amino acid. To facilitate expression of the complementary carboxyl-terminal fragments, they are constructed by replacing the appropriate restriction fragment with a synthetic oligonucleotide duplex containing a CCACC consensus sequence and a Met codon to provide a translation initiation site.

For those sets of complementary fragments which show the capacity to form noncovalent complexes (see later discussion), a pMT3-based expression vector containing each of the complementary gene fragments under the control of a separate AdMLP is used (Fig. 2). Initially, the gene encoding the amino-terminal fragment is isolated as a *Ppu*MI/*Not*I fragment so that it contains a copy of the AdMLP. In practice, the procedure involves first linearizing the vector with *Not*I, filling in the *Not*I overhang with the Klenow fragment of DNA polymerase I, and then digesting the vector with *Ppu*MI. The gene encoding the carboxyl-terminal fragment is first linearized with *Ppu*MI, filled in with the Klenow enzyme, and then digested with *Not*I. Both fragments encoding the amino- and carboxyl-terminal portions of opsin downstream from an AdMLP are then ligated to a pMT3 vector digested with *Ppu*MI and *Not*I.

Expression of Opsin Gene Fragments in COS-1 Cells

The COS-1 cell line was isolated from African green monkey kidney cells (CV-1), which were stably transformed with an origin-defective mutant of SV40 viral DNA.[20] They contain a copy of the complete early region of SV40 DNA integrated into the CV-1 genome and produce the SV40 T antigen that is required for viral replication. Thus, any transiently transfected plasmid DNA that contains the SV40 origin of replication, such as pMT3, is able to replicate in COS-1 cells. The optimal method for transfection of mammalian cells with plasmid DNA is dependent on the cell type. Oprian *et al.*[21] have established procedures for the culturing of COS-1 cells in Dulbecco's modified Eagle's medium containing 10% (v/v) fetal calf serum and transient transfection of COS-1 cells with the vector containing the synthetic opsin gene using the DEAE-dextran procedure. These procedures have been successfully adapted for the expression of the opsin gene fragments in COS-1 cells with only slight modifications. The COS-1 cells are plated at a density of $1.25-1.5 \times 10^7$ cells per 150×25-

---

[20] Y. Gluzman, *Cell* **23**, 175 (1981).
[21] D. D. Oprian, R. S. Molday, R. J. Kaufmann, and H. G. Khorana, *Proc. Natl. Acad. Sci. U.S.A.* **84**, 8874 (1987).

mm culture dish and transfected 10–14 hr later with 12.5–15 μg of each plasmid DNA or the vector containing both opsin fragments. In our experience, maintaining a ratio of 1 μg of plasmid DNA per 1×10^6 cells gives optimal transfection results. The transfected cells are harvested with a rubber policeman 55–72 hr after transfection and washed with phosphate-buffered saline (PBS: 10 mM NaH$_2$PO$_4$, pH 7.0, containing 150 mM NaCl). The cells are pelleted at 2000g for 5 min at 4° in a microcentrifuge and processed immediately or stored at $-80°$.

Cellular expression of the opsin polypeptide fragments is examined by protein immunoblotting of whole-cell detergent extracts with anti-rhodopsin monoclonal antibodies. Cells from one plate ($\sim 2 \times 10^7$) are solubilized in 1 ml of PBS, containing 1% (w/v) dodecylmaltoside (DM), and 0.1 mM phenylmethylsulfonyl fluoride (PMSF) for 1 hr at 4°. Insoluble material is pelleted at 16,000g for 5 min at 4° in a microcentrifuge and the supernatant is removed. Aliquots of protein (~ 25 μg) from the supernatant are analyzed by nonreducing sodium dodecyl sulfate (SDS)/ Tris–glycine polyacrylamide gel electrophoresis (PAGE) and electroblotted onto poly(vinyl difluoride) membranes (Millipore, Bedford, MA). For some of the smaller opsin fragments (<10 kDa), the proteins are analyzed by nonreducing SDS/Tris–tricine PAGE. Immunoreactive protein is detected using the anti-rhodopsin rho 4D2, rho 1D4, B6-30N, T13-34L, or K42-41L primary antibodies,[22–24] and horseradish peroxidase-conjugated goat anti-mouse immunoglobulin G (IgG) as the second antibody (Promega, Madison, WI). The protein bands are visualized using the enhanced chemiluminescence detection system (Amersham Pharmacia Biotech, Piscataway, NJ). In most cases, stable expression of both the amino- and carboxyl-terminal fragment is detected. This suggests that the fragments are tightly folded and not degraded by the quality control system that eliminates misfolded proteins from the cell. However, in some cases, expression of the amino-terminal fragment is only detected on coexpression with a carboxyl-terminal fragment.[13,14]

Characterization of Expressed Opsin Fragments

Membrane Integration Assay

To assay for insertion of the fragments into the membrane, transfected cells are subjected to hypotonic lysis and the crude membranes extracted

---

[22] R. S. Molday and D. McKenzie, *Biochemistry* **22,** 653 (1983).
[23] D. Hicks and R. S. Molday, *Exp. Eye Res.* **42,** 55 (1986).
[24] G. Adamus, Z. S. Zam, A. Arendt, K. Palczewski, J. H. McDowell, and P. A. Hargrave, *Vision Res.* **31,** 17 (1991).

with alkali. Crude membranes are prepared by resuspending 4×10^7 cells in 2 ml of 15 mM Tris-HCl, pH 7.5, containing 2 mM MgCl$_2$, 2 mM dithiothreitol (DTT), 0.1 mM PMSF, and a Complete protease inhibitor tablet (Roche Molecular Biochemicals, Mannheim, Germany). After 45 min at 4°, the suspension is passed through a 20-gauge needle and centrifuged at 100,000g for 30 min at 4°. The pellet is homogenized and layered on a 20%/50% (w/v) sucrose gradient and centrifuged at 120,000g in a swinging bucket rotor (SW 50.1, Beckman Coulter, Fullerton, CA) for 45 min at 4°. The crude membranes, which band at the border of the 20%/50% sucrose cushion, are collected by puncturing the tube with an 18-gauge needle and washed by centrifugation at 100,000g for 30 min at 4°. The crude membranes are resuspended in 100 mM Na$_2$CO$_3$, pH 11.0, and incubated with gentle agitation for 1 hr at 20°. The membranes are collected by centrifugation at 100,000g for 30 min at 4° and equivalent amounts of protein (~25 μg) from the supernatant and pellet are examined by SDS/Tris–glycine or SDS/Tris–tricine PAGE followed by immunoblotting. Many of the fragments examined so far are resistant to extraction with alkali, suggesting that they are membrane integrated. Two notable exceptions include fragments composed of the amino-terminal region and the first transmembrane helix.[13]

Cleavage of Oligosaccharides with Glycosidases

Rod outer segment (ROS) rhodopsin is N-glycosylated at Asn-2 and Asn-15 in the amino-terminal region of the protein. In many cases, the singly expressed amino-terminal opsin fragments show two distinct polypeptides after SDS–PAGE and immunoblotting. To test whether this is the result of heterogenous N-glycosylation, the opsin fragments are treated with glycosidases. Typically, the whole-cell detergent extracts are incubated with 500 units of endoglycosidase H or PNGase F (Roche Molecular Biochemicals) in 20 mM Tris-HCl, pH 8.0, containing 0.5% (w/v) SDS, 1% (v/v) Nonidet P-40 (NP-40), 5 mM ethylenediaminetetraacetic acid (EDTA), and 0.1 mM PMSF for 30 min at 20°. The glycosidase treated proteins are analyzed by SDS–PAGE and immunoblotting. In most cases, a faster migrating single polypeptide is observed after treatment of the amino-terminal polypeptides with endoglycosidase H, indicating differences in their extent of high mannose N-glycosylation. Notably, coexpression of the amino-terminal fragment with its complementary carboxyl-terminal fragment often yields a noncovalent complex where the amino-terminal polypeptide migrates as an elongated band with a sharp leading edge and a trailing smear (Fig. 3). Here, the amino-terminal fragment shows sensitivity only to PNGase F. This difference results from further processing

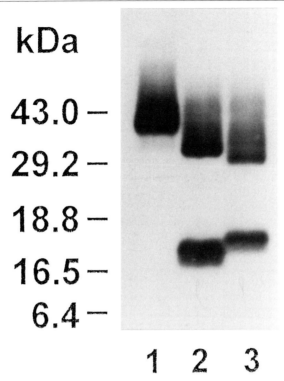

FIG. 3. Immunoblot analysis of rhodopsin-like complexes formed from two complementary fragments. Purified wild-type (lane 1) and fragment rhodopsins (lanes 2 and 3) were analyzed by SDS–PAGE, detected with the rho 4D2 and rho 1D4 antibodies, and visualized by chemiluminescence.

to the complex carbohydrate form and is characteristic of COS-1 cell-expressed wild-type opsin (Fig. 3, lane 1).

Quantitation of Expression Levels

The expression levels of the amino- and carboxyl-terminal fragments can be determined from enzyme-linked immunosorbent assays (ELISA). For this purpose, whole-cell detergent extracts containing the opsin fragments are serially diluted in detergent extracts obtained from cells transfected with the pMT3 vector lacking the opsin fragment gene(s). After drying the samples for 12–16 hr at 37° in microtiter plates (Nalge Nunc, Rochester, NY), the unbound proteins are washed away with PBS, pH 7.2, containing 0.1% (w/v) bovine serum albumin (BSA) and 0.05% (v/v) Tween

20. The wells are then blocked with 10 mM KH$_2$PO$_4$, pH 7.4, containing 2.5% (w/v) BSA, 0.5% (w/v) gelatin, and 0.02% (w/v) NaN$_3$ for 6 hr at 20°. Immunoreactive protein is detected by incubating the samples with the above-mentioned primary antibodies in blocking buffer for 16 hr at 4°. After washing away excess primary antibody, horseradish peroxidase-conjugated goat anti-mouse IgG (Promega) is added as the second antibody for 1 hr at 20°. Excess peroxidase-conjugated IgG is washed away before adding the 3,3′,5,5′-tetramethylbenzidine peroxidase substrate (Kirkegaard and Perry Laboratories, Gaithersburg, MD). The reactions are monitored using an automated ELISA reader equipped with a 650-nm filter (Dynex Technologies, Middlesex, UK). A known amount of photobleached ROS rhodopsin (3.2–400 ng) solubilized in 1% (w/v) DM is used as the standard for quantitation. The absorbance at 650 nm is linear over this range of protein concentration. Typically, the expression levels range from 0.6 to 6 μg opsin fragment per 1×10^7 cells.

Chromophore Formation and Purification

To test whether the coexpressed complementary fragments are capable of forming a chromophore with 11-*cis*-retinal, freshly harvested transfected cells are resuspended in PBS, pH 7.0 and incubated with 5 μM 11-*cis*-retinal for 3 hr at 4° in the dark with gentle agitation. The retinal reconstituted proteins are solubilized with 1% (w/v) DM in PBS, pH 7.2, containing 0.1 mM PMSF, and purified on immobilized rho 1D4 antibody.[21] The binding capacity of the immobilized antibody is typically 0.3–0.5 μg rhodopsin/μl of resin. The matrices are washed five times with 20 bed volumes of 20 mM Tris-HCl, pH 8.0, containing 1 M NaCl, 2 mM adenosine triphosphate (ATP), 2 mM MgCl$_2$, and 0.1% (w/v) DM and then an additional five times with 20 bed volumes of 2 mM NaH$_2$PO$_4$, pH 6.0, containing 0.1% (w/v) DM or PBS, pH 7.2, containing 0.1% (w/v) DM. The bound proteins are eluted from the immobilized antibody with 35 μM c′-tetsquapa peptide (a synthetic peptide corresponding to the carboxyl-terminal nine amino acids of opsin) in 2 mM NaH$_2$PO$_4$, pH 6.0, containing 0.1% (w/v) DM or PBS, pH 7.2, containing 0.1% (w/v) DM.

Functional Properties of Rhodopsin–Fragment Complexes

Spectral Properties

Spectra are recorded on a UV/visible spectrophotometer from 650 to 250 nm at a scan speed of 200 nm per minute. Some representative spectra of rhodopsin-like pigments formed on coexpression of two or three complementary fragments are shown in Fig. 4. Here, the ratio of the absorbance at 280 nm to that of 500 nm is 1.6–1.7, which is characteristic of purified rhodopsin. These preparations are obtained by eluting the protein from the immunoaffinity matrix under conditions of low ionic strength. However, preparations of coexpressed fragments, which are eluted from the immunoaffinity matrix in high ionic strength buffers, show higher A_{280}/A_{500} ratios (~2–3). This difference appears to result from the coelution of correctly folded–regenerated rhodopsin fragment complexes with the carboxyl-terminal fragment that has not formed a complex. The stability of the chromophore in the dark can be measured by adding neutral hydroxylamine to a final concentration of 100 mM. Like ROS or wild-type COS-1 cell rhodopsin, the rhodopsin fragment complexes are stable to hydroxylamine in the dark, indicating that the retinyl-Schiff base is not accessible to hydrolytic attack.

To examine whether the rhodopsin fragment complexes undergo the same light-dependent and acid-induced conformational changes as wild-type rhodopsin, the pigments are illuminated for 10 sec with a 150-W light

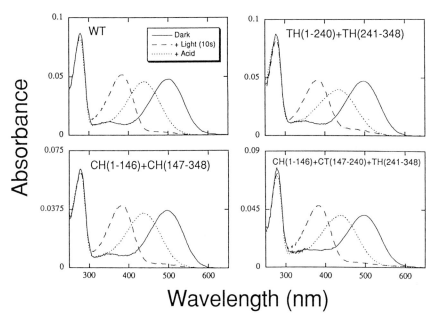

FIG. 4. UV/visible spectra of rhodopsin-like complexes formed from two or three complementary fragments.

source through a >495-nm-long pass filter and acid denatured by adjusting the pH to 1.9 with 2 M H_2SO_4. As shown in Fig. 4, rhodopsin fragment complexes formed from two or three fragments show the characteristic shift in λ_{max} to 380 nm on illumination and form the protonated retinyl-Schiff base ($\lambda_{max} \sim 440$ nm) on acidification. Thus, formation and decay of the 380-nm Metarhodopsin-II intermediate appears to be similar to that of wild-type rhodopsin.

Although the ability of complementary opsin fragments to form noncovalent complexes *in vivo* provides important information about the mechanism of opsin folding and assembly, it is also desirable to show that these fragments retain the capacity to regenerate a chromophore with 11-*cis*-retinal *in vitro*. These experiments are done by illuminating the purified pigments for 10 sec, recording a spectrum, adding a 3–4 molar excess of 11-*cis*-retinal, and recording a series of spectra at 3- to 5-min intervals. Because free retinal ($\lambda_{max} \sim 380$ nm) contributes a small amount of absorbance at 500 nm, the absorbance values at 540 nm (where retinal does not absorb) are compared before photolysis and after addition of 11-*cis*-retinal. Although wild-type opsin can be regenerated to greater than 90%, the fragment complexes typically show less than 60% regeneration. This is presumably due, in part, to the lack of a covalent connection between the two or three fragment complexes.

G-Protein Activation Assays

To test whether the rhodopsin fragment complexes activate G-protein *in vitro*, the activity of the rhodopsin fragment complexes is determined by following the light-dependent binding of [^{35}S]GTPγS by transducin (G_t). G_t is prepared from bovine retina by the method of Fung *et al.*[25] The assay mixture consists of 2 nM rhodopsin fragment complex, 2 μM G_t, and 4 μM [^{35}S]GTPγS in 10 mM Tris-HCl, pH 7.5, containing 100 mM NaCl, 5 mM $MgCl_2$, 2 mM DTT, and 0.012% (w/v) DM. The time course is begun in darkness with the addition of [^{35}S]GTPγS, and after a 1-min incubation at 20°, the assay mixture is illuminated (>495 nm) or allowed to remain in darkness. At 15-sec intervals, 35 μl of the 250-μl reaction mixture is removed and filtered through nitrocellulose with the aid of a vacuum manifold (Millipore). The filters, which retain G_t and its bound nucleotide, are washed four times with 5 ml of 10 mM Tris-HCl, pH 7.5, containing 100 mM NaCl, 5 mM $MgCl_2$, and 2 mM DTT, dried, and analyzed for ^{35}S radioactivity by scintillation counting. Not surprisingly, many of the rhodopsin fragment complexes with polypeptide chain discontinuities in the cyto-

[25] B. K. K. Fung, J. B. Hurley, and L. Stryer, *Proc. Natl. Acad. Sci. U.S.A.* **78**, 152 (1981).

plasmic region exhibit a significantly reduced capacity to activate G_t. This is presumably due to disruption of the cytoplasmic binding site(s) for G_t.

Remarks

Examining the contributions of various segments of the opsin polypeptide chain to the *in vivo* folding and assembly process has offered considerable insights into the steps involved in opsin biosynthesis and stability. For example, virtually all of the singly expressed fragments examined to date fold to a conformation that allows for membrane integration. The fact that many of these opsin polypeptide fragments are competent for endoplasmic reticulum (ER) translocation is consistent with the conclusion that bovine opsin contains multiple internal signal and stop–transfer sequences.[26] Following their entry into the ER lumen, the amino-terminal polypeptide fragments are high mannose *N*-glycosylated while the carboxyl-terminal polypeptide fragments presumably undergo no modification(s). As proposed by the two-stage model for integral membrane protein folding and assembly,[5] it is likely that the folding of these polypeptide fragments begins with the stable formation of the α helices prior to insertion. Further, based on a recent model for the structural organization of the α helices in rhodopsin,[27] it is reasonable to suggest that the complementary fragments associate through side-to-side interactions once appropriately inserted in the membrane. Whether this can occur spontaneously, or requires the participation of various chaperones involved in cellular protein folding and assembly, deserves further study. Once these noncovalent complexes have formed, they presumably exit the ER and are transported to the Golgi for further cellular processing. This is suggested by the presence of complex carbohydrates on the amino-terminal fragments in these rhodopsins. It is possible that structural determinants located within the COOH-terminal region of the opsin polypeptide, or the presence of a complete opsin structure, may be required for normal cellular trafficking.

Several factors may influence the stability of the rhodopsin fragment complexes once they are formed and isolated in detergent solution. In the dark, the chromophore may link the polypeptide fragments with a covalent attachment to one via Lys-296, and noncovalent interactions with the other(s). Added stability is probably afforded by the electrostatic interaction between the protonated retinyl-Schiff base and its Glu-113 counterion. Here, the retinal moiety serves to stabilize interactions between the complexed fragments. On photoexcitation, however, the stability of these rho-

---

[26] M. Friedlander and G. Blobel, *Nature* (*London*) **318**, 338 (1985).
[27] J. M. Baldwin, G. F. X. Schertler, and V. M. Unger, *J. Mol. Biol.* **272**, 144 (1997).

dopsins appears to be largely dependent on the extent of protein–protein interactions.

Acknowledgments

We would like to acknowledge the contributions of Tony Ngo and Stephen S. J. Lee to many aspects of these studies. This work was supported by National Institutes of Health Grant EY11112. Certain commercial materials, instruments, and equipment are identified in this manuscript in order to specify the experimental procedure as completely as possible. In no case does such identification imply a recommendation or endorsement by the National Institute of Standards and Technology nor does it imply that the materials, instruments, or equipment identified are necessarily the best available for the purpose.

[5] Isolation of Isoelectric Species of Phosphorylated Rhodopsin

By J. Hugh McDowell, Joseph P. Nawrocki, and Paul A. Hargrave

Introduction

Rhodopsin is the photoreceptor protein of rod cells. On activation by light, rhodopsin initiates transduction through activation of the G protein, transducin, which ultimately leads to the perception of light.[1] Once activated, rhodopsin continues to activate visual transduction until the rhodopsin is deactivated by a series of reactions. The first of these reactions is the phosphorylation of rhodopsin mediated by rhodopsin kinase.[2–4] The phosphorylated, photoactivated rhodopsin can continue to activate transducin, although at a somewhat reduced level,[5] until arrestin binds to the phosphorylated, photoactivated rhodopsin.[6] Arrestin remains bound until the all-*trans*-retinal is removed, relaxing the activation state of rhodopsin. A phosphatase then removes the phosphate groups[7] and finally 11-*cis*-retinal is added to the opsin, regenerating rhodopsin.

A number of studies have addressed the question of how many phosphates are required for arrestin to bind *in vivo* as well as how effective are the lower phosphorylation extents in reducing activation of transducin.

---

[1] K. Palczewski, *Invest. Ophthalmol. Visual Sci.* **35,** 3577 (1994).
[2] H. Kühn and W. J. Dreyer, *FEBS Lett.* **20,** 1 (1972).
[3] D. Bownds, J. Dawes, J. Miller, and M. Stahlman, *Nature* **237,** 125 (1972).
[4] R. N. Frank, H. D. Cavanagh, and K. R. Kenyon, *J. Biol. Chem.* **248,** 596 (1973).
[5] A. Sitaramayya, *Biochemistry* **25,** 5460 (1986).
[6] K. P. Hofmann, A. Pulvermüller, J. Buczylko, P. Van Hooser, and K. Palczewski, *J. Biol. Chem.* **267,** 15701 (1992).
[7] K. Palczewski, J. H. McDowell, S. Jakes, T. S. Ingebritsen, and P. A. Hargrave, *J. Biol. Chem.* **264,** 15770 (1989).

While *in vivo* experiments suggest that phosphorylation of only two sites is sufficient to induce action by arrestin,[8] *in vitro* studies have been less conclusive and have been hampered by an inability to produce well-defined species of phosphorylated rhodopsin. We describe here the isolation of phosphorylated rhodopsins that are phosphorylated at 1, 2, and 3 residues.

Preparation of Rod Outer Segments

All procedures are performed under dim red light unless indicated otherwise. Rod outer segments (ROS) are prepared following the method of Wilden and Kühn as described earlier.[9] The concentration of the ROS is adjusted to 1 mg/ml rhodopsin in 100 mM potassium phosphate, pH 7.0, 1 mM MgCl$_2$, 1 mM dithiothreitol (DTT), and 0.1 mM ethylenediaminetetraacetic acid (EDTA). Aliquots of the ROS are frozen in liquid nitrogen and stored at $-75°$ until use.

Phosphorylation of Rod Outer Segments

In preparing the phosphorylated rhodopsin, often one or two of the phosphorylated species of rhodopsin is most desired in a particular preparation. To maximize the production of the desired phosphorylated species, a 3-mg aliquot of the ROS preparation is used to determine the time course of the production of the variously phosphorylated species of rhodopsin. For these experiments, the ROS in the buffer given is diluted to 0.5 mg/ml rhodopsin and brought to 0.6 mM in adenosine triphosphate (ATP). The ROS are then sonicated in a small sonic cleaning bath for a couple of minutes and exposed to white light (a 150-W flood lamp at about 15 cm) with stirring while contained in a 25° water-jacketed beaker. Timed aliquots (500 μl) are removed and made 10 mM in EDTA to stop the phosphorylation reaction. When all samples have been taken, a threefold molar excess of 11-*cis*-retinal is added to each sample in the dark. The retinal is dissolved in ethanol and added to yield a final concentration of ethanol less than 1%.

The phosphorylated, regenerated ROS membranes are collected by centrifugation (48,400g, 15 min) and dissolved in 100 μl of 10 mM dodecylmaltoside. The phosphorylated rhodopsin is analyzed by isoelectric focusing (IEF) essentially as described by Adamus *et al.*[10] The gel is prepared as

---

[8] H. Ohguro, J. P. VanHooser, A. H. Milam, and K. Palczewski, *J. Biol. Chem.* **270**, 14259 (1995).

[9] J. H. McDowell, in "Photoreceptor Cells" (P. A. Hargrave, ed.), p. 123. Academic Press, Orlando, Florida, 1993.

[10] G. Adamus, Z. S. Zam, J. H. McDowell, G. P. Shaw, and P. A. Hargrave, *Hybridoma* **7**, 237 (1988).

recommended by the manufacturer. A mixed bed deionizer such as Amberlite MB-3 (2.5 g) is added to a solution containing 24.25 g acrylamide, 0.75 g N,N'-methylenebisacrylamide in 250 ml water. The solution is stirred for 1 hr, then filtered to remove the resin. To pour the gel (230 × 115 × 1 mm), 153 mg dodecylmaltoside is dissolved in 5 ml water in a vacuum flask taking care to prevent foaming. In a 25-ml graduated cylinder, 4 ml of glycerol is mixed with 15 ml of the previously given acrylamide solution, 1.14 ml of Pharmalyte 2.5–5 and 0.76 ml of Pharmalyte 5–8 and finally adjusted to 25 ml with water. This is carefully mixed with the dodecylmaltoside in the vacuum flask and degased under house vacuum for 10 min, again avoiding foaming. Freshly made ammonium persulfate (200 μl of 22.8 mg/ml) and 25 μl of N,N,N',N'-tetramethylethylenediamine are added and the gel is cast between silanized glass plates, one plate treated with Bind-Silane (Pharmacia, Piscataway, NJ) and one treated with Repel-Silane ES (Pharmacia). The gel is usually allowed to cure overnight before use. The samples are loaded using a sample applicator strip. Glutamic acid (0.04 M) is used as the anode solution and 1 M NaOH is the cathode solution. IEF is performed at 25-W constant power for 2 hr on a flatbed electrophoresis apparatus cooled by circulating coolant at 4°.

IEF analysis of the time course of phosphorylation is shown in Fig. 1. This analysis allows selecting a time of phosphorylation for a larger size ROS preparation that maximizes the production of the desired form of phosphorylated rhodopsin. The 1, 2, and 3 phosphorylated forms of rhodopsin are maximally produced in a reaction time of 5–20 min for most ROS preparations. The larger scale phosphorylation is carried out for the time selected using the same conditions as given earlier. The phosphorylated opsin is then regenerated by scaling up the regeneration procedure given earlier.

The yield of regeneration is checked and a second aliquot of 11-*cis*-retinal is added if the regeneration is less than 90%, again keeping the total ethanol concentration at less than 1%.

Affinity Purification on Concanavalin A-Agarose

The phosphorylated rhodopsin (75 mg) is then submitted to affinity chromatography using concanavalin A-agarose (ConA-agarose) essentially as described earlier.[11] The phosphorylated, regenerated ROS are collected by centrifugation and solubilized in either 50 mM octyl glucoside or 6 mM dodecylmaltoside in 50 mM Tris–acetate, 1 mM CaCl$_2$, 1 mM MgCl$_2$, 1 mM MnCl$_2$, pH 6.9, at a concentration of rhodopsin of about 1 mg/ml. The phosphorylated rhodopsin solution is centrifuged to remove any insoluble

---

[11] B. J. Litman, *Methods Enzymol.* **81**, 150 (1982).

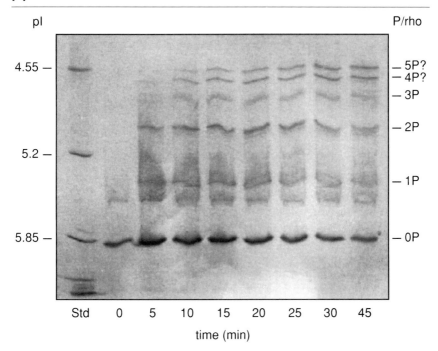

FIG. 1. Isoelectric focusing of the time course of phosphorylation of rhodopsin. IEF and phosphorylation were carried out as described in the text. The time of phosphorylation for each sample is given at the bottom of each lane. The pI values of the standards are given at the left of the gel going from higher pI values at the bottom to lower pI at the top.

material before loading on the ConA-agarose column (pre-equilibrated in buffer with detergent). For 100 mg of rhodopsin, a column with a 100-ml bed volume of ConA-agarose is used. The column is washed with 2 column volumes of buffer with detergent to remove the excess retinal and then eluted with 250 mM methyl α-D-mannopyranoside in buffer with detergent collecting 5-ml fractions. The rhodopsin is pooled (approximately 90 ml) based on OD_{498} and dialyzed versus two changes of 4 liters of 10 mM imidazole, pH 7.4. If octylglucoside is used in the chromatography, the pooled rhodopsin is adjusted to 6 mM dodecylmaltoside to prevent precipitation during dialysis.

Separation of Rhodopsin from Phosphorhodopsin

Rhodopsin is separated from phosphorhodopsin following the method of Andersson and Porath.[12] A chelating column is prepared using a 25-ml

[12] L. Andersson and J. Porath, *Anal. Biochem.* **154,** 250 (1986).

bed volume of iminodiacetic acid epoxy-activated Sepharose 6B (Sigma no. I 4510). The column is washed with 10 mM FeCl$_3$ until iron is visible in the effluent. The column is washed with 10 mM imidazole, pH 5.0, until the effluent is colorless. The column is then equilibrated with 3–5 column volumes of 10 mM imidazole, 6 mM dodecylmaltoside, pH 5.0. The dialyzed rhodopsin is carefully adjusted to pH 5.0 with HCl and applied to the chelating column collecting fractions and monitoring OD$_{500}$ of the effluent. Unphosphorylated rhodopsin does not bind to the column, so the column is washed with 10 mM imidazole, 6 mM dodecylmaltoside, pH 5.0, until a baseline at OD$_{498}$ is reached. The column is then eluted with 10 mM imidazole, 6 mM dodecylmaltoside, pH 7.4, collecting fractions and monitoring at OD$_{498}$. Figure 2 shows IEF analysis of these fractions. An unphosphorylated, unbleached ROS sample was applied to lane 1. In lanes 2 and 3 (Fig. 2), the solubilized phosphorylated and regenerated ROS were applied before and after ConA-agarose purification respectively. The phosphorhodopsins that bound to and were eluted from the chelating column were applied to lane 4 and contain no unphosphorylated rhodopsin. In some preparations the amount of the higher phosphorylated forms, presumably

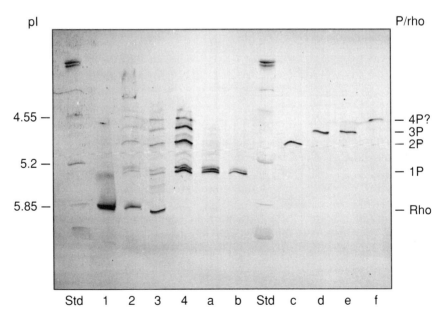

FIG. 2. Isoelectric focusing of forms of phosphorylated rhodopsin. Lane 1, Unphosphorylated ROS; lane 2, phosphorylated ROS; lane 3, ConA-agarose purified phosphorylated ROS; lane 4, phosphorylated rhodopsins isolated by chelating chromatography; lanes labeled a through f are samples from the labeled pools of the chromatofocusing in Fig. 3.

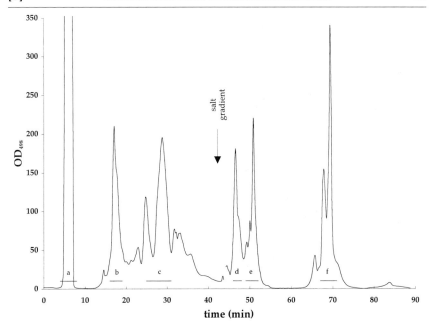

FIG. 3. Chromatofocusing of phosphorhodopsins. Phosphorhodopsins were isolated and chromatofocused as described in the text. After chromatofocusing a linear salt gradient was applied to the column as noted by the arrow. The OD_{498} of the effluent was monitored during both procedures. Peaks were pooled as noted and analyzed by IEF (Fig. 2).

5 phosphates per rhodopsin or more, were reduced after purification on ConA-agarose and/or chelating chromatography (data not shown).

Isolation of Various Isoelectric Forms of Phosphorylated Rhodopsin

Phosphorhodopsins are separated by chromatofocusing on a Mono P column (HR 5/20, Pharmacia) using modifications of previous procedures.[13,14] The Mono P column is first equilibrated with 10 mM imidazole, 0.5 mM DTT, pH 7.0, and then washed with 2–3 column volumes of 10 mM imidazole, 0.5 mM DTT, 6 mM dodecylmaltoside, pH 7.0. Polybuffer 74 (Pharmacia, 1:12.5 dilution) is made with 0.5 mM DTT, 6 mM dodecylmaltoside and adjusted to pH 4.5 with HCl. This buffer, 2.5 ml, is applied to the column. The phosphorhodopsins pooled from the chelating column

[13] P. P. Philippov, V. Y. Arshavsky, and A. M. Dizhoor, *Sov. Sci. Rev. D. Physicochem. Biol.* **9,** 243 (1990).

[14] G. Adamus, A. Arendt, P. A. Hargrave, T. Heyduk, and K. Palczewski, *Arch. Biochem. Biophys.* **304,** 443 (1993).

are applied to the column at a flow rate of 1 ml/min. The column is developed with 40 ml of Polybuffer 74 adjusted to pH 4.5 as done earlier. A linear salt gradient is then applied from 0 to 0.4 M NaCl in 10 mM imidazole, 0.5 mM DTT, 6 mM dodecylmaltoside, pH 7.0. The column effluent is monitored for OD_{500} and 1-ml fractions are collected. Fractions containing OD_{500} peaks are pooled as shown in Fig. 3. The pools are analyzed by IEF as shown in Fig. 2. The phosphorylation extent of each peak is based on further analysis of the rhodopsin's carboxyl terminal peptide from these pools for mono-, di-, and triphosphorylated rhodopsins.[15] The tetra- and pentaphosphorylated species (noted in Figs. 1 and 2) are tentatively assigned to the species with more acidic pI values. Monophosphorylated rhodopsin clearly shows as a doublet in samples prepared under these conditions. Monophosphorylation under these conditions occurs at either Ser-338 or Ser-343.[15] Ohguro et al.[16] found that changing the conditions of phosphorylation could alter the ratio of phosphorylation at these sites. Diphosphorylated rhodopsin is phosphorylated at both sites and the triphosphorylated form is probably also phosphorylated at Thr-336.[17] The localization of sites for the higher phosphorylated forms has not been determined.

Conclusion

In conclusion, mono-, di-, tri-, and tetraphosphorylated rhodopsin have been isolated and can be used to explore further the effect of phosphorylation on the interactions of rhodopsin with other proteins involved in visual transduction.

Acknowledgments

This work was supported by NIH grants EY06625 and EY06626 and by an unrestricted departmental award and Senior Scientific Investigator Award (to P.A.H.) from Research to Prevent Blindness.

---

[15] J. H. McDowell, J. P. Nawrocki, and P. A. Hargrave, *Biochemistry* **32**, 4968 (1993).
[16] H. Ohguro, R. S. Johnson, L. H. Ericsson, K. A. Walsh, and K. Palczewski, *Biochemistry* **33**, 1023 (1994).
[17] H. Ohguro, K. Palczewski, L. H. Ericsson, K. A. Walsh, and R. S. Johnson, *Biochemistry* **32**, 5718 (1993).

[6] Rhodopsin Trafficking in Photoreceptors Using Retinal Cell-Free System

By DUSANKA DERETIC

Introduction

In retinal photoreceptor cells rhodopsin is delivered from the site of its synthesis and processing to the site of its function, the rod outer segments (ROS), by post-Golgi carrier membranes that have unique protein and lipid composition.[1–3] Sorting into these membranes of the appropriate proteins and lipids destined for delivery to the ROS is critical for the maintenance of photoreceptor polarity and functional integrity of retinal rods. Valuable information on the properties of post-Golgi membrane carriers was gained by morphologic analysis and biochemical fractionation of the photoreceptor cells.[1,4,5] However, these methods have limitations to the extent that only trafficking in intact cells can be monitored. This precludes any interference with rhodopsin trafficking by membrane-impermeable agents that could potentially interfere with these processes and help reveal underlying molecular mechanisms. To circumvent this problem, we have developed a cell-free system derived from frog retinas that reconstitutes formation of rhodopsin-bearing post-Golgi carrier membranes from the trans-Golgi network (TGN) *in vitro*.

Processes taking place in the sorting compartment of the cell, at the TGN, can be readily studied in cell-free membrane traffic systems. These are either based on perforated cells or on cell homogenates and fractions derived from these homogenates. An invaluable contribution to our understanding of the molecular mechanisms that regulate membrane trafficking came from the cell-free system that reconstitutes intra-Golgi transport.[6] This system made possible the identification of membrane receptors termed SNAREs (SNAP receptors).[7] These are receptors for the general soluble

---

[1] D. Deretic and D. S. Papermaster, *J. Cell Biol.* **113,** 1281 (1991).
[2] D. Deretic, B. Puleo Scheppke, and C. Trippe, *J. Biol. Chem.* **271,** 2279 (1996).
[3] E. B. Rodriguez de Turco, D. Deretic, N. G. Bazan, and D. S. Papermaster, *J. Biol. Chem.* **272,** 10491 (1997).
[4] D. Deretic and D. S. Papermaster, *J. Cell Sci.* **106,** 803 (1993).
[5] D. Deretic, L. A. Huber, N. Ransom, M. Mancini, K. Simons, and D. S. Papermaster, *J. Cell Sci.* **108,** 215 (1995).
[6] W. E. Balch, W. G. Dunphy, W. A. Braell, and J. E. Rothman, *Cell* **39,** 405 (1984).
[7] T. Sollner, S. W. Whiteheart, M. Brunner, H. Erdjument-Bromage, S. Geromanos, P. Tempst, and J. E. Rothman, *Nature* **362,** 318 (1993).

fusion proteins NSF and SNAPs. SNAREs are highly conserved among species and regulate membrane budding, docking, and fusion events in organisms as diverse as yeast and humans.[7,8] SNAREs were shown to represent the minimal fusion complex,[9] and further structural analysis of this complex[10–12] has revealed common motifs in SNARE protein complexes and the viral fusion proteins, providing insight into a more general mechanism for the regulation of membrane fusion (reviewed by Skehel and Wiley[13]).

Trafficking from the TGN has been studied in several cell-free systems.[14–18] These studies revealed that budding of post-Golgi carrier membranes from the TGN is highly regulated by an interplay of protein–lipid interactions. Protein and lipid phosphorylation were found to play essential roles in cell-free post-Golgi membrane trafficking.[19,20] Membrane budding from the TGN is regulated by proteins that belong to the large family of GTPases,[15,18,21] while membrane scission is regulated by specific isoforms of GTPase dynamin[22] and by protein kinase C.[23] In a cell-free system derived from PC12 cells post-Golgi membrane budding critically depends on the high phosphatidylinositol/phosphatidylcholine (PI/PC) ratio maintained by PI/PC transfer protein (PI-TP).[17] PI-TP may also be essential for budding of rhodopsin-bearing post-Golgi membranes, because we have recently reported that frog photoreceptor TGN membranes display the highest DHA-PI/DHA-PC ratio.[3] In *Drosophila* mutation in the gene encoding membrane-associated PI-TP causes retinal degeneration B

[8] J. E. Rothman and G. Warren, *Curr. Biol.* **4,** 220 (1994).
[9] T. Weber, B. V. Zemelman, J. A. McNew, B. Westermann, M. Gmachl, F. Parlati, T. H. Sollner, and J. E. Rothman, *Cell* **92,** 759 (1998).
[10] P. I. Hanson, R. Roth, H. Morisaki, R. Jahn, and J. E. Heuser, *Cell* **90,** 523 (1997).
[11] M. A. Poirier, W. Xiao, J. C. Macosko, C. Chan, Y. K. Shin, and M. K. Bennett, *Nature Struct. Biol.* **5,** 765 (1998).
[12] R. B. Sutton, D. Fasshauer, R. Jahn, and A. T. Brunger, *Nature* **395,** 347 (1998).
[13] J. J. Skehel and D. C. Wiley, *Cell* **95,** 871 (1998).
[14] S. A. Tooze and W. B. Huttner, *Cell* **60,** 837 (1990).
[15] S. M. Jones, J. R. Crosby, J. Salamero, and K. E. Howell, *J. Cell Biol.* **122,** 775 (1993).
[16] J. Salamero, E. S. Sztul, and K. E. Howell, *Proc. Natl. Acad. Sci. U.S.A.* **87,** 7717 (1990).
[17] M. Ohashi, K. Jan de Vries, R. Frank, G. Snoek, V. Bankaitis, K. Wirtz, and W. B. Huttner, *Nature* **377,** 544 (1995).
[18] J. P. Simon, I. E. Ivanov, B. Shopsin, D. Hersh, M. Adesnik, and D. D. Sabatini, *J. Biol. Chem.* **271,** 16952 (1996).
[19] M. Ohashi and W. B. Huttner, *J. Biol. Chem.* **269,** 24897 (1994).
[20] S. M. Jones and K. E. Howell, *J. Cell Biol.* **139,** 339 (1997).
[21] S. A. Tooze, U. Weiss, and W. B. Huttner, *Nature* **347,** 207 (1990).
[22] S. M. Jones, K. E. Howell, J. R. Henley, H. Cao, and M. A. McNiven, *Science* **279,** 573 (1998).
[23] J. P. Simon, I. E. Ivanov, M. Adesnik, and D. D. Sabatini, *J. Cell Biol.* **135,** 355 (1996).

($rdgB$).[24] A dominant mutant of $rdgB$ significantly reduces the steady-state level of rhodopsin, indicating its important role in rhodopsin biogenesis.[25] Therefore, identification of important regulatory components in cell-free assays derived from various cells is likely to provide increasingly useful information about common mechanisms that govern protein trafficking and contribute to our understanding of how these processes are affected in diseased cells, including retinal photoreceptors.

This chapter focuses on the experimental details of a frog retinal cell-free assay that reconstitutes post-Golgi trafficking of rhodopsin. This cell-free system is a modification of an assay described by Tooze and Huttner.[14,26] The frog retinal cell-free system reconstitutes the ATP-, GTP-, and cytosol-dependent physiologic process of post-Golgi membrane budding as judged by the appropriate morphology, topology, and protein composition of budded membranes and by the retention of resident proteins in the TGN.[2] Using this system we have identified the C-terminal five amino acids as the domain that directs intracellular trafficking of rhodopsin.[27] Because mutations that cluster within the five C-terminal amino acids of rhodopsin cause particularly severe forms of autosomal dominant retinitis pigmentosa (ADRP),[28,29] our findings using this cell-free assay clearly suggest a possible underlying molecular mechanism: these mutations most likely result in abnormal post-Golgi membrane formation and mistargeting of mutant rhodopsin.

Pulse Labeling of Newly Synthesized Proteins in Cultured Isolated Frog Retinas

The kinetics of rhodopsin trafficking through the post-Golgi compartment of photoreceptor cells has been well established in southern leopard frogs, *Rana berlandieri*.[1] Because these frogs are well suited for biochemical and morphologic studies we use the same species for the cell-free membrane traffic assay. Frogs are maintained at 22° under dim white light in a 12-hr

---

[24] T. S. Vihtelic, M. Goebl, S. Milligan, J. E. O'Tousa, and D. R. Hyde, *J. Cell Biol.* **122,** 1013 (1993).

[25] S. C. Milligan, J. G. Alb, Jr., R. B. Elagina, V. A. Bankaitis, and D. R. Hyde, *J. Cell Biol.* **139,** 351 (1997).

[26] S. A. Tooze and W. B. Huttner, *Methods Enzymol.* **219,** 81 (1992).

[27] D. Deretic, S. Schmerl, P. A. Hargrave, A. Arendt, and J. H. McDowell, *Proc. Natl. Acad. Sci. U.S.A.* **95,** 10620 (1998).

[28] M. A. Sandberg, C. Weigel DiFranco, T. P. Dryja, and E. L. Berson, *Invest. Ophthalmol. Vis. Sci.* **36,** 1934 (1995).

[29] A. V. Cideciyan, D. C. Hood, Y. Huang, E. Banin, Z. Y. Li, E. M. Stone, A. H. Milam, and S. G. Jacobson, *Proc. Natl. Acad. Sci. U.S.A.* **95,** 7103 (1998).

light–dark cycle and fed mealworms. Because frog rhodopsin synthesis is maximal in the afternoon the light cycle is shifted as previously described.[30] Frogs are dark adapted for 2 hr before the experiment. Dark adaptation causes the retraction of pigment epithelium, which facilitates retinal isolation. All experiments are conducted under dim red light because subcellular organelle distribution has been well established for the dark-adapted rhodopsin-containing membranes.[1]

Procedure

Frogs *R. berlandieri* (100–250 g, Rana Co., Brownsville, TX) are sacrificed under dim red light (Kodak, Rochester, NY, Adjustable Safelight Lamp, model B, with Kodak Safelight Filters 1A) by decapitation using a small animal decapitator (PGC Scientific, Gaithersburg, MD) and retinas are isolated as described.[31] Seven retinas are incubated per 15 ml of oxygenated media in 50 ml polycarbonate Erlenmeyer flasks (Nalgene, Rochester, NY). The composition of the medium for retinal cultures has been previously described.[30] [^{35}S]-Express protein labeling mixture (1000 Ci/mmol, DuPont/NEN, Boston, MA) is added to a final concentration of 25 μCi per retina and retinal cultures are incubated at 22° for 1 hr on the benchtop orbital shaker at 25 rpm. After isotope incorporation, retinas are rinsed in a petri dish with 34% (w/w) ice-cold sucrose homogenizing medium containing protease inhibitors and transferred to a high-speed polycarbonate tube with fresh medium on ice (0.3 ml per retina). Preparation of stock solutions and sucrose homogenizing media has been previously described.[30]

Removal of ROS from Frog Retina and Preparation of Photoreceptor-Enriched Postnuclear Supernatant from Radiolabeled Cells

Due to the complexity of the retinal tissue, biosynthetic organelles of the rod inner segments are accessible only on removal of the ROS from the neural retina. To avoid extensive damage of retinal tissue ROS are removed from the retina by low shear forces and separated from the remainder of the retina by flotation on high-density sucrose. Biosynthetic compartments from photoreceptor cell inner segments are then isolated from the remainder of the retina by rehomogenization of the retinal pellet in 0.25 M sucrose under conditions of slightly greater shear. The 0.25 M sucrose homogenizing medium used in these experiments is modified from that

---

[30] D. Deretic and D. S. Papermaster, *in* "Methods for the Study of Photoreceptor Cells," Vol. 15 (P. A. Hargrave, ed.), p. 108. Rockefeller University Press, New York, 1993.

[31] D. S. Papermaster, *Methods Enzymol.* **96,** 609 (1983).

previously described[30] so that NaCl is omitted to avoid the aggregation of membranes during *in vitro* incubation.[26] This homogenization step preferentially releases photoreceptor inner segment membranes, while the rest of the retina is relatively unbroken in large fragments. Retinal fragments and nuclei are then sedimented at low speed. This pellet contains ~85% of retinal tissue, but less than 10% of the newly synthesized rhodopsin.[32] The postnuclear supernatant (PNS), therefore, is highly enriched in photoreceptor organelles involved in the biosynthesis of the ROS proteins. Preparation of postnuclear supernatant and the retinal subcellular fractionation procedure is schematically outlined in Fig. 1.

Procedure

ROS are sheared by five passes through a 14-gauge pipetting needle by gentle aspiration with a 10-ml syringe. The homogenate in 34% sucrose homogenizing medium is overlaid with 1 ml of 1.10 g/ml sucrose solution[30] and centrifuged at 16,000 rpm ($31,000g_{av}$), for 20 min at 4° in a JA 25.50 rotor in a Beckman Avanti J-25 centrifuge (Beckman Instruments Inc., Palo Alto, CA). ROS are collected from the 1.10/34% sucrose interface and removed. After removal of crude ROS and excess 34% sucrose homogenizing medium, the retinal pellet is rehomogenized in 0.25 M sucrose in 10 mM Tris–acetate, pH 7.4, containing 1 mM magnesium acetate and protease inhibitors (modified 0.25 M sucrose homogenizing medium[30]) (0.15 ml/retina). Retinal pellets are homogenized in 1-ml aliquots, with five passes of a loose-fitting Teflon–glass homogenizer (Thomas AA Tissue Grinder, Thomas Scientific, Swedesboro, NJ; with a clearance increased to 150 μm) driven by a variable speed stirrer, model GT21 (G. K. Heller, Thomas Scientific) with motor controller at setting 30, in the cold room, under dim red light. Homogenized aliquots are pooled and collected in a 13-ml polycarbonate tube, and the homogenate is centrifuged at 3300 rpm ($1300g_{av}$), for 4 min at 4° in a JA 25.50 rotor to pellet the nuclei and unbroken retinal fragments. The postnuclear supernatant obtained after this centrifugation is the starting material for the cell-free reaction.

In vitro Incubation of Photoreceptor-Enriched Postnuclear Supernatant and Cell-Free Formation of Post-Golgi Carrier Membranes

Radiolabeled PNS are incubated in the presence of an ATP-regenerating system for 2 hr of chase *in vitro*, a chase period found to be optimal for radiolabeling of post-Golgi membranes *in vivo*.[1] After 1 hr of pulse

---

[32] D. S. Papermaster, C. A. Converse, and J. Siu, *Biochemistry* **14,** 1343 (1975).

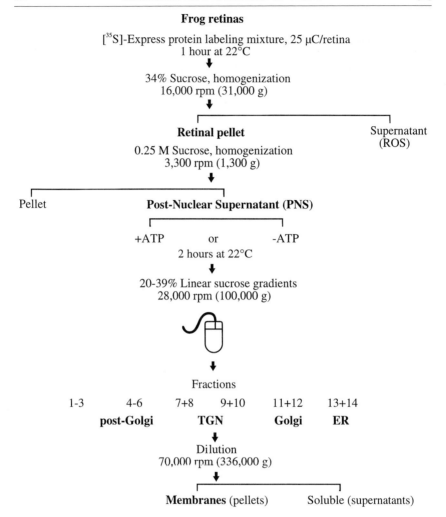

FIG. 1. Schematic representation of the cell-free assay for rhodopsin trafficking. Following radiolabeling in cultured retinas, ROS are sheared from the retina and floated on 34% sucrose. The retinal pellet is rehomogenized in 0.25 M sucrose and retinal fragments and nuclei are removed by low-speed centrifugation. PNS is incubated in the presence of an ATP-regenerating or an ATP-depleting system for 2 hr of cell-free chase. Subcellular compartments are then fractionated on a linear 20–39% sucrose gradient and membrane fractions are analyzed for the presence of radiolabeled rhodopsin.

labeling *in vivo* rhodopsin is predominantly found in the Golgi, with only a small fraction remaining in the endoplasmic reticulum (ER).[1] The kinetics of the transfer of rhodopsin from the ER to the Golgi complex and its export from the Golgi indicates that the rate of transport through the Golgi is significantly slower than the ER-to-Golgi transport.[1] Therefore, the majority of radiolabeled protein is localized in the Golgi at the onset of a typical *in vitro* chase experiment.[2] Under such conditions, rhodopsin trafficking during the cell-free chase predominantly reflects intra-Golgi transport and post-Golgi membrane formation.

Procedure

The standard cell-free assay is as follows: To 1 ml of PNS in 0.25 M sucrose (obtained from seven radiolabeled retinas) 115 μl of 10× concentrated buffer stock solution is added to give a final concentration of 25 mM HEPES–KOH, pH 7.0, 25 mM KCl, and 2.5 mM magnesium acetate. The 10× concentrated buffer stock solution (250 mM HEPES–KOH, pH 7.0, 250 mM KCl, and 15 mM magnesium acetate) is designed to provide optimal conditions to achieve transport *in vitro,* as described by Balch *et al.*[6] The use of low-ionic-strength medium is important because incubation of the PNS in the presence of salt causes membrane aggregation, which interferes with subsequent subcellular fractionation. This was also observed by Tooze and Huttner.[26] The cell-fee transport assay is initiated by the addition of 50 μl of an ATP-regenerating system[33] and by transfer to 22°. The ATP-regenerating system contains equal volumes of 100 mM ATP (Na$^+$ salt, neutralized with NaOH), 800 mM creatine phosphate, and 4 mg/ml creatine phosphokinase in 50% w/v glycerol. To investigate the energy requirements of the post-Golgi membrane formation, an ATP-depleting system is added to the PNS. The ATP-depleting system contains 10 mg/ml hexokinase in 250 mM D-glucose.[33] ATP, creatine phosphate, creatine phosphokinase (800 U/mg) and hexokinase (450 U/mg) are purchased from Boehringer-Mannheim (Roche Molecular Biochemicals, Indianapolis, IN). In the standard assay for cell-free post-Golgi membrane formation all the indicated amounts are for one cell-free reaction. Depending on the number of cell-free reactions to be performed, these amounts need to be multiplied accordingly. Samples are routinely incubated for 2 hr without agitation. The assay is terminated by the addition of 0.2 M ethylenediaminetetraacetic acid (EDTA) to a final concentration of 3 mM and the assay mixture is loaded on linear sucrose gradients. In addition to termination of the trafficking assay, EDTA also disrupts divalent cation cross-linking of unrelated membranes and permits their separation on sucrose gradients.

[33] J. Davey, S. M. Hurtley, and G. Warren, *Cell* **43**, 643 (1985).

To examine the cytosol requirements of the cell-free trafficking PNS is subfractionated into membrane-enriched and cytosolic fractions. One milliliter of PNS in 0.25 M sucrose (obtained from 14 retinas) is overlaid on a step sucrose gradient containing 1 ml of 0.5 M sucrose and 3 ml of 49% sucrose. Gradients are spun for 1 hr at 28,000 rpm (75,000g_{av}) at 4° in a SW 50.1 rotor in an Optima L-90K ultracentrifuge (Beckman). Soluble proteins that remain in 0.25 M sucrose are further centrifuged at 70,000 rpm (336,000g_{av}) for 30 min in a 70.1 Ti rotor (Beckman) to remove residual membranes, and supernatants containing cytosolic proteins are collected. Membrane proteins that enter the 0.5 M sucrose layer are collected and divided in half. To one-half of the membranes 0.5 ml of 5 mg/ml bovine serum albumin (BSA) in 0.25 M sucrose is added (−cytosol) and to the other half 0.5 ml of cytosolic proteins is added back (+cytosol), restoring photoreceptor-enriched PNS to the original ratio of cytosol:membranes. Reconstituted PNS (±cytosol) is assayed for post-Golgi membrane formation after addition of 10× buffer stock and an ATP-regenerating system as described earlier. To study the role of specific cytosolic factors in post-Golgi membrane biogenesis, complete cytosol in these experiments can be further subfractionated, either biochemically or by immunodepletion.

The standard cell-free assay is modified in the samples where preincubation with antibodies, purified proteins, or synthetic peptides is used to perturb the function of the regulatory components. For example, to test the role of small GTPases of the Rab family in post-Golgi membrane formation radiolabeled retinal PNS is preincubated with 200 μM GDP and 2 μM recombinant Rab GDI (GDP dissociation inhibitor).[2] PNS is preincubated for 1 hr at 22° with 200 μM GDP to displace the bound GTP on membrane-bound Rab proteins and this is followed by 30 min at 30° with 2 μM recombinant bovine Rab GDI to remove membrane-associated GDP-bound Rab proteins. Control samples are preincubated with the buffer under identical conditions. The role of the cytoplasmic domain of rhodopsin is tested by preincubation with 100 μg of anti-rhodopsin C-terminal antibody (or its Fab fragments) for 30 min on ice before 10× stock solution and an ATP-regenerating system are added and trafficking in the *in vitro* assay is initiated.[2] Synthetic peptides are also useful reagents to test the role of participating proteins in post-Golgi trafficking. We have used synthetic peptides that correspond to the C-terminal sequence of rhodopsin to establish the role of this domain in rhodopsin trafficking.[27] PNS was preincubated for 30 min at 22° with 50 μM peptide before initiation of *in vitro* trafficking. Although peptides are powerful tools for the study of membrane trafficking, certain peptides have a propensity to assume a conformation of an amphipathic α helix and act as detergents in cell-free assays, thus completely solubilizing membrane compartments and inhibiting trafficking steps in a nonspecific fashion.[34] It is important to examine the secondary structure and the solubility of the peptides before they are introduced into a cell-free assay.

Subcellular Fractionation of Postnuclear Supernatant on Sucrose Density Gradients Following Cell-Free Formation of Post-Golgi Membranes

To follow the kinetics of movement of radiolabeled rhodopsin from the Golgi to the TGN and to monitor the cell-free formation of post-Golgi membranes, we use equilibrium sucrose gradient centrifugation. Because of their unique low buoyant density, rhodopsin-bearing post-Golgi membranes formed *in vivo* can be enriched >85% and separated away from the Golgi and the TGN on such gradients.[1,4] The rhodopsin-bearing post-Golgi membrane carriers formed *in vitro* are indistinguishable from those formed *in vivo* by their protein composition, topology, and morphology and by a distinct profile of membrane-associated small GTPases of the Rab family.[2] Based on their identical buoyant density they can be separated away from the donor compartment, the TGN, at the end of the cell-free reaction on the same gradients as post-Golgi membranes formed *in vivo*.

Procedure

On completion of cell-free post-Golgi membrane formation, assay mixtures are overlaid on 10-ml linear 20–39% (w/w) sucrose gradients in 10 mM Tris–acetate, pH 7.4, and 0.1 mM MgCl$_2$, above a 0.5-ml cushion of 49% (w/w) sucrose in the same buffer. Preparation of sucrose solutions has been previously described.[30] Gradients are prepared as follows: a 0.5-ml cushion of 49% sucrose is prepared in the 13.5-ml ultraclear tube for the SW40 rotor; 5 ml of 39% sucrose is pipetted into the near chamber and 5 ml of 20% sucrose into the far chamber of the gradient maker (Pharmacia, Piscataway, NJ) that is connected to an Auto Densi-Flow fractionator (Labconco, Kansas City, MO), which, in the "deposit" mode, automatically deposits a linear gradient on the underlying cushion. Gradients are centrifuged at 28,000 rpm (100,000g_{av}) at 4° for 15 hr in an SW40 rotor in an Optima L-90K ultracentrifuge (Beckman). Fractions (0.9 ml) are collected from the top of the gradient using the Auto Densi-Flow fractionator (in the "remove" mode) and a fraction collector (Gilson FC 205, Middleton,

---

[34] P. J. Weidman and W. M. Winter, *J. Cell Biol.* **127**, 1815 (1994).

WI). A 50-μl aliquot of each fraction is used to measure the refractive index. The remainder is diluted with 10 mM Tris–acetate, pH 7.4, and centrifuged at 70,000 rpm (336,000g_{av}) for 30 min in a 70.1 Ti rotor (Beckman). Six different cell-free assays are routinely performed (a control and five experimental conditions), and subcellular fractions are pooled according to the kinetics of their acquisition of radiolabeled rhodopsin as described[2] as follows: pool 1, fractions 1–3; pool 2, 4–6; pool 3, 7–8; pool 4, 9–10; pool 5, 11–12; and pool 6, 13–14. Pooled fractions are then diluted with 10 mM Tris–acetate, pH 7.4, and centrifuged at 70,000 rpm (336,000g_{av}) for 30 min in a 70.1 Ti rotor (Beckman). Pellets are resuspended in 10 μl/ retina of 10 mM Tris acetate, pH 7.4, and aliquoted for further analysis.

Biochemical Analysis

To determine the efficiency of the cell-free post-Golgi membrane, formation aliquots from each fraction are subjected to scintillation counting (10 μl, equivalent to 1 retina) and sodium dodecyl sulfate–polyacrylamide gel electrophoresis (SDS–PAGE) (20 μl, equivalent to two retinas). Dried SDS gels are subjected to quantitative analysis of ^{35}S-labeled rhodopsin in retinal subcellular fractions in a PhosphorImager (Molecular Dynamics, Sunnyvale, CA), while the images of the gels are generated by autoradiography at $-85°$ using Kodak BioMax MR film. An example of such analysis that illustrates the ATP-dependent transfer of radiolabeled rhodopsin to the post-Golgi fraction during the cell-free chase is shown in Fig. 2. To control for the possible fragmentation of the TGN the retention of the TGN markers in the TGN at the end of the reaction is determined by assaying the distribution of sialyltransferase activity across the sucrose density gradient as described[4] and as shown in Fig. 2. To ascertain the presence of a distinct set of membrane-associated small GTPases of the Rab family [^{32}P]GTP overlays and immunoblotting with specific antibodies are performed as described[5] using the enhanced chemiluminescence (ECL) Western blotting detection system (Amersham, Piscataway, NJ).

Morphologic Analysis

For electron microscopic analysis, membranes from the sucrose gradient fractions are pelleted for 1 hr at 50,000 rpm (230,000g_{av}) in a Beckman SW50.1 rotor with adapters to obtain a small pellet. Pellets are fixed with 2% glutaraldehyde in 120 mM cacodylate, pH 7.4, containing 3% sucrose, for 30 min on ice, postfixed with OsO$_4$ stained with uranyl acetate, and embedded in 2% agarose as described.[1] Blocks of membranes in agarose are dehydrated in ethanol and embedded in Epon. Thin sections along the axis of sedimentation are examined in a Philips CM-100 electron micro-

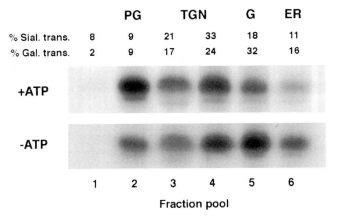

FIG. 2. Rhodopsin trafficking to the post-Golgi membranes in the retinal cell-free system. Following pulse labeling in the isolated retinas and cell-free chase in the presence or absence of ATP, aliquots corresponding to radiolabeled membrane proteins from two retinas are separated by SDS–PAGE and autoradiographed. In the absence of ATP (−ATP), rhodopsin remains in the Golgi (G) during *in vitro* chase, indicating that the assay is ATP dependent.[2] On addition of ATP (+ATP), radiolabeled rhodopsin exits the Golgi and appears in the TGN and in low-density post-Golgi fractions (PG). The distribution of Golgi and TGN membranes has been determined by their galactosyltransferase or sialyltransferase activities, respectively.[1,2,4] Activities of pooled membrane fractions are indicated. ER, Endoplasmic reticulum. [Reproduced from D. Deretic *et al., Proc. Natl. Acad. Sci. U.S.A.* **95,** 10620 (1998), with permission.]

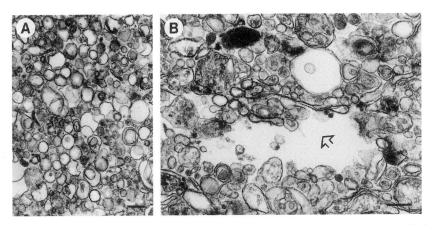

FIG. 3. Post-Golgi membranes formed in the cell-free system are morphologically identical to those formed *in vivo* and differ from Golgi-enriched fraction. Electron micrographs of the thin sections through the pellet obtained after cell-free trafficking assay. (A) A population of vesicular structures ~300 nm in diameter is a major component of post-Golgi membrane fraction. (B) Golgi-enriched fraction is very heterogeneous. Occasional profiles that may represent cross sections of the Golgi cisternae can be seen (open arrow). In addition this fraction contains mitochondria, smooth membranes, and multivesicular bodies. Bar: 0.3 μm. [Reproduced from D. Deretic *et al., J. Biol. Chem.* **271,** 2279 (1996), with permission.]

scope. Electron microscopic analysis of the post-Golgi membranes formed in the cell-free assay is shown in Fig. 3A. These membranes have an appearance of vesicular structures with ~300-nm diameter and without a visible coat, indistinguishable from the membranes formed *in vivo* (shown in ref. 1). They are significantly larger than the vesicles involved in intra-Golgi trafficking and less uniform in size. Analysis of post-Golgi transport intermediates in living COS cells has indicated that these are large pleiomorphic tubular structures, rather than small vesicles.[35] Likewise, the majority of rhodopsin-bearing post-Golgi membrane carriers may have similar appearance *in vivo*, but fragment into smaller vesicular structures upon homogenization and preparation for EM analysis. The Golgi-enriched fraction examined after *in vitro* post-Golgi membrane budding is very heterogeneous (Fig. 3B) but profiles that may represent cross sections of Golgi cisternae can be seen.

Conclusion

The retinal cell-free system provides access to photoreceptor biosynthetic membranes at the critical step when newly synthesized proteins and lipids are incorporated into rhodopsin-bearing post-Golgi membranes. The development of this cell-free system to study rhodopsin trafficking has allowed new insights into these processes at the molecular level. Using this system we found that post-Golgi trafficking to the ROS depends on the presence of membrane-bound Rab proteins and on the access of cytoplasmic proteins to the carboxyl-terminal domain of rhodopsin in the TGN membranes.[2] We further mapped the domain that directs intracellular trafficking using synthetic peptides as competitive inhibitors and found that the five C-terminal residues QVS(A)PA play a critical role in regulating rhodopsin sorting into specific post-Golgi membranes from the TGN.[27] This cell-free system should be useful to identify other factors involved in the regulation of post-Golgi trafficking of rhodopsin and its associated proteins and lipids to the ROS, and will serve as a basis for further modifications for the studies of intracellular trafficking in retinal photoreceptors.

Acknowledgments

I thank Dr. David Papermaster, Belen Puleo-Scheppke, Claudia Trippe, Nancy Ransom, and Sonia Schmerl for invaluable help throughout the course of this work. Supported by NIH grants EY-12421 and EY-6891 and the Core Grant for Vision Research EY-07003. D. Deretic is a recipient of the Career Development Award from Research to Prevent Blindness.

---

[35] K. Hirschberg, C. M. Miller, J. Ellenberg, J. F. Presley, E. D. Siggia, R. D. Phair, and J. Lippincott-Schwartz, *J. Cell. Biol.* **143**, 1485 (1998).

Section II

Characterization of Opsins

[7] Preparation and Analysis of Two-Dimensional Crystals of Rhodopsin

By Gebhard F. X. Schertler and Paul A. Hargrave

Rhodopsin: A Model G-Protein-Coupled Receptor

It is through cell-surface proteins of the G-protein-coupled receptor (GPCR) family that cells sense alterations in their environment and effect biochemical changes in the cell interior. At the present time more than 800 GPCRs have been cloned from a variety of species, from fungi to humans. As much as 3.4% of the genome of *Caenorhabditis elegans* consists of genes for such receptors.[1] GPCRs all share some degree of sequence similarity and are likely to have a similar arrangement of their seven-transmembrane helices.[2,3] They are of such pharmacologic importance that in the past 20 years more than 100 new drugs have been registered that "activate" or "inhibit" GPCRs.[4]

Rhodopsin, the receptor for dim light in the vertebrate retina, is one of the most intensively studied of the GPCRs.[5] Such receptors are normally present in only a small number of copies per cell, but rhodopsin can be obtained in hundreds of milligrams from a single vertebrate retina. Rhodopsin is held tightly in a nonsignaling conformation by its covalently bound chromophore 11-*cis*-retinal, which functions as a receptor antagonist (inverse agonist). Rhodopsin's extracellular domain is relatively rigid, which may help to reduce spontaneous activation of the receptor in the absence of light.[6–8] Vertebrate rhodopsin is one of the most stable and detergent tolerant of the GPCRs, another factor making it an ideal candidate for structural investigations.

---

[1] The *C. elegans* Sequencing Consortium, *Science* **282,** 2012 (1998).
[2] J. M. Baldwin, *EMBO J* **12,** 1693 (1993).
[3] J. M. Baldwin, G. F. X. Schertler, and V. M. Unger, *J. Mol. Biol* **272,** 144 (1997).
[4] J. M. Stadel, S. Wilson, and D. J. Bergsma, *Trends Pharmacol. Sci.* **18,** 430 (1997).
[5] T. P. Sakmar, *in* "Progress in Nucleic Acid Research and Molecular Biology" (K. Moldave, ed.), Vol. 59, p. 1. Academic Press, San Diego, 1998.
[6] G. F. X. Schertler, *Eye* **12,** 504 (1998).
[7] H. G. Khorana, *J. Biol. Chem.* **267,** 1 (1992).
[8] V. M. Unger, P. A. Hargrave, J. M. Baldwin, and G. F. X. Schertler, *Nature* **389,** 203 (1997).

Comparison of Electron Crystallography with Other Structural Methods

Protein structures may be determined by X-ray crystallography, nuclear magnetic resonance (NMR), or electron microscopy. Application of NMR is currently limited to proteins of less than 35 kDa, and is therefore not practical for application to the entire molecule of rhodopsin, which is 40 kDa. X-ray crystallography has been used to solve the structures of thousands of proteins, but requires well-ordered three-dimensional (3-D) crystals. To date only a handful of structures of membrane proteins have been solved by this approach: photosynthetic reaction center,[9] porin,[10] bc1 complex,[11] and cytochrome oxidase.[12,13] This is due to the fact that it has proven difficult to obtain sufficiently large and well-ordered crystals from the detergent complexes of membrane proteins. Efforts have been made to obtain 3-D crystals of rhodopsin.[14,15] The prospects for application of these methods to rhodopsin have been reviewed.[16]

For the technique of electron crystallography, two-dimensional (2-D) crystals are required, and it is usually easier to prepare these than it is to obtain 3-D crystals. An additional advantage of 2-D crystals is that both amplitudes and phases can be obtained from images of the same crystal directly, whereas X-ray diffraction requires that phases must be determined separately.

Because electrons interact more strongly with matter than do X rays, smaller and thinner samples such as a monomolecular layer of protein molecules can be studied. Unfortunately, this strong interaction of electrons is also damaging to biologic specimens, requiring strategies to reduce beam damage, for example, use of low temperatures and low beam intensities.[17] The low doses of electrons required to avoid damage limits the amount of information obtained from a single molecule. This requires a larger assembly such as a 2-D crystal for proteins the size of rhodopsin. Even nonideal 2-D crystals can yield valuable information since computer processing of

---

[9] J. Deisenhofer and H. Michel, *Science* **245**, 1463 (1989).
[10] M. S. Weiss, A. Kreusch, E. Schiltz, U. Nestel, W. Welter, J. Wechesser, and G. E. Schulz, *FEBS Lett.* **280**, 379 (1991).
[11] D. Xia, C. A. Yu, H. Kim, J. Z. Xian, A. M. Kachurin, L. Zhand, L. Yu, and J. Deisenhofer, *Science* **277**, 60 (1997).
[12] C. Ostermeier, S. Iwata, B. Ludwig, and H. Michel, *Nature Struct. Biol.* **2**, 842 (1995).
[13] T. Tsukihara, H. Aoyama, E. Yamashita, T. Tomizaki, H. Yamaguchi, K. Shinzawaitoh, R. Nakashima, R. Yaono, and S. Yoshikawa, *Science* **272**, 1136 (1996).
[14] W. J. DeGrip, J. Van Oostrum, and G. L. J. De Caluwé, *J. Crystal Growth* **122**, 375 (1992).
[15] E. V. Yurkova, V. V. Demin, and N. G. Abdulaev, *Biomed. Sci.* **1**, 585 (1990).
[16] P. A. Hargrave, *Behav. Brain Sci.* **18**, 403 (1995).
[17] R. Henderson, *Quart. Rev. Biophys.* **28**, 171 (1995).

images enables removal of some of the disorder present in the crystal lattice.[18–20]

Electron crystallography has been successful in determination of high-resolution structures of bacteriorhodopsin[21] and light-harvesting complex.[22] Many other proteins have had their structures determined by electron crystallography to lower resolution: rhodopsin,[23] halorhodopsin,[24] aquaporin,[25] gap junction,[26,27] sarcoplasmic reticulum Ca^{2+}-ATPase,[28] and plasma membrane H^{+}-ATPase.[29]

Preparation of Two-Dimensional Crystals

Methods for preparation of 2-D crystals of membrane proteins have been described in detail in several reviews.[18,30–33] We describe the methods that have been successfully applied to rhodopsin. This includes purification of rhodopsin, the formation of 2-D crystals, the search for such crystals by electron microscopy, and a 3-D electron density map produced by the study of such crystals. The experimental steps necessary to obtain a 3-D map starting with 2-D crystals are outlined in Fig. 1. Electron microscopy and computational steps are described in detail in several reviews.[18,34–36]

[18] G. F. X. Schertler, in "Structure–Function Analysis of G-Protein-Coupled Receptors" (J. Wess, ed.), pp. 233–287. Wiley-Liss, Inc., New York, 1999.
[19] W. Chiu, *Annu. Rev. Biophys. Biomol. Struct.* **22**, 233 (1993).
[20] R. Henderson, J. M. Baldwin, K. H. Downing, J. Lepault, and F. Zemlin, *Ultramicroscopy* **19**, 147 (1986).
[21] R. Henderson, J. M. Baldwin, T. A. Ceska, F. Zemlin, E. Beckmann, and K. H. Downing, *J. Mol. Biol.* **213**, 899 (1990).
[22] W. Kühlbrandt, D. N. Wang, and Y. Fujiyoshi, *Nature* **367**, 614 (1994).
[23] A. Krebs, C. Villa, P. C. Edwards, and G. F. X. Schertler, *J. Mol. Biol.* **282**, 991 (1998).
[24] W. A. Havelka, R. Henderson, and D. Osterhelt, *J. Mol. Biol.* **247**, 726 (1995).
[25] T. Walz, T. Hirai, K. Murata, J. B. Heymann, K. Mitsuoka, Y. Fujiyoshi, B. L. Smith, P. Agre, and A. Engel, *Nature* **387**, 624 (1997).
[26] V. M. Unger, N. M. Kuman, N. B. Gilula, and M. Yeager, *Nature Struct. Biol.* **4**, 39 (1997).
[27] V. M. Unger, N. M. Kumar, N. B. Gilula, and M. Yeager, *Science* **283**, 1176 (1999).
[28] P. J. Zhang, C. Toyoshima, K. Yonekura, N. M. Green, and D. L. Stokes, *Nature* **392**, 835 (1998).
[29] M. Auer, G. A. Scarborough, and W. Kuhlbrandt, *Nature* **392**, 840 (1998).
[30] A. Engel, A. Hoenger, A. Hefti, C. Henn, R. C. Ford, J. Kistler, and M. Zulauf, *J. Struct. Biol.* **109**, 219 (1992).
[31] B. K. Jap, M. Zulauf, T. Scheybani, A. Hefti, W. Baumeister, U. Aebi, and A. Engel, *Ultramicroscopy* **46**, 45 (1992).
[32] W. Kühlbrandt, *Quart. Rev. Biophys.* **25**, 1 (1992).
[33] J. L. Rigaud, G. Mosser, J. J. Lacapere, A. Olofsson, D. Levy, and J. L. Ranck, *J. Struct. Biol.* **118**, 226 (1997).
[34] M. Yeager, V. M. Unger, and A. K. Mitra, *Methods Enzymol.* **294**, 135 (1999).
[35] Y. Fujiyoshi, *Adv. Biophys.* **35**, 25 (1998).
[36] L. A. Amos, R. Henderson, and P. N. T. Unwin, *Prog. Biophys. Mol. Biol.* **39**, 183 (1982).

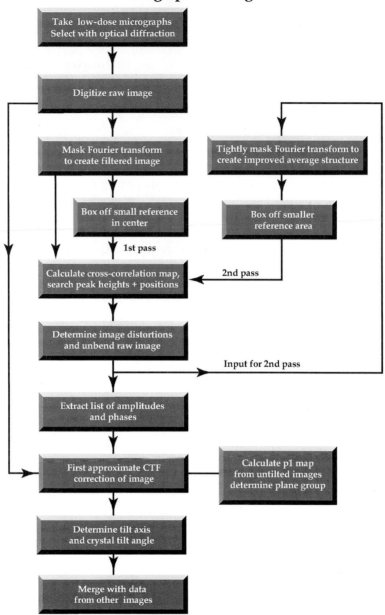

FIG. 1. Flowchart for the electron crystallography of 2-D crystals. Steps are presented for obtaining a structure by a combination of electron microscopy, image processing, and electron crystallography. (Reproduced with permission from G. F. X. Schertler, *in* "Structure–Function Analysis of G-Protein-Coupled Receptors," pp. 283–287. Wiley-Liss, New York, 1999.)

Crystallization of Membrane Proteins in the Cell Membrane

Membrane proteins only rarely exist in cell membranes in crystalline form. Bacteriorhodopsin occurs in crystalline form in purple membrane patches in halobacteria; halorhodopsin[37,38] and gap junctions[39] can form crystalline arrays in the cell when they are overexpressed. Density gradient centrifugation can be utilized to purify such membranes based on the fact that their high density of protein gives them a unique density compared to that of other cellular membranes. However, most membrane proteins exist in lower concentration along with other membrane proteins and they must be purified by solubilization in detergent and purified by chromatography in the presence of detergent.

Purification of Frog Rhodopsin

Retinas are obtained from dark-adapted adult *Rana catesbiana* following guidelines of the Association for Research in Vision and Ophthalmology. Dissected retinas are quick frozen in liquid nitrogen, and stored in light-tight tubes at −80°. Pooled retinas from several dozen frogs are thawed, homogenized, and submitted to discontinuous sucrose density gradient centrifugation.[40] Rod outer segments (ROS) band at the interface of density 1.115/1.135 and are harvested, diluted with phosphate buffer, and pelleted by centrifugation (40,000g, 20 min at 4°). Yields are generally ~25–60 μg rhodopsin per retina. Amounts of rhodopsin are conveniently determined by solubilization of an aliquot of ROS in a detergent such as lauryldimethylamine oxide (LDAO) and measuring absorption at 498 nm.[40]

ROS membranes prepared as described have been used directly for some crystallization experiments. In other experiments the ROS were first suspended in 5% Ficoll, let stand in the cold to swell in order to lyse the plasma membrane and free the small amount of cytoplasmic proteins. Disk membranes containing rhodopsin were then isolated by centrifugation. This disk membrane preparation method was originally developed using bovine ROS[41] from which satisfactory yields were obtained, but it gives rather low yields when applied to frog ROS.

---

[37] W. A. Havelka, R. Henderson, J. A. W. Heymann, and D. Oesterhelt, *J. Mol. Biol.* **234,** 837 (1993).
[38] J. S. W. Heymann, W. A. Havelka, and D. Oesterhelt, *Mol. Microbiol.* **7,** 623 (1993).
[39] M. Yeager, *Acta Cryst.* **D50,** 1 (1994).
[40] J. H. McDowell, *in* "Photoreceptor Cells" (P. A. Hargrave, ed.), p. 123. Academic Press, Orlando, Florida, 1993.
[41] H. G. Smith, G. W. Stubbs, and B. J. Litman, *Exp. Eye Res.* **20,** 211 (1975).

Purification of Bovine Rhodopsin

Eyes are obtained from freshly slaughtered cattle at a slaughterhouse, placed in a light-tight container, and taken to a dark room where the retinas are dissected under dim red light. Alternatively, retinas prepared as above are obtained frozen from a commercial source (Lawson Co., Lincoln, NE). Rod outer segments are prepared by vigorous shaking or homogenization of 200 retinas in 180 ml sucrose-containing buffer [45% sucrose in buffer A: 100 mM potassium phosphate, pH 7.0, containing 1 mM MgCl$_2$, 0.5 mM dithiothreitol, and 0.1 mM ethylenediaminetetraacetic acid (EDTA)]. This is followed by centrifugation to obtain a crude ROS preparation (centrifuge in a fixed-angle rotor for 5 min at 3000g at 4°). The supernatant is filtered through gauze into a 1-liter measuring cylinder and diluted 1:1 with buffer A. Crude ROS are obtained by centrifugation 7 min at 4400g. The supernatants are discarded and each pellet is suspended in about 1 ml of sucrose–buffer A (d 1.105). The ROS are purified by discontinuous density gradient centrifugation.[40,42]

Gradients are prepared in six 50-ml polycarbonate centrifuge tubes by overlayering 18 ml of sucrose–buffer (d 1.135) with 17 ml sucrose–buffer (d 1.115), and distributing the suspended crude ROS equally among the tubes. The tubes are centrifuged in a swinging bucket rotor for 1 hr at 27,000g and 4° (e.g. Beckman JS13 rotor at 13,000 rpm) and the rotor allowed to decelerate with no brake. The gradients are viewed with dim red backlighting and the ROS band at the 1.115–1.135 interface is collected using a syringe fitted with a long cannula. The ROS are diluted 1:1 with buffer A and harvested by centrifugation at 39,000g and 4°. Pellets are stored at −70°.

Such ROS contain about 85% rhodopsin along with small amounts of other membrane proteins and a variety of soluble ROS proteins important in the visual transduction process. To further purify rhodopsin, the ROS membranes are dissolved in a buffer containing LDAO, centrifuged, and the soluble rhodopsin-containing solution is submitted to concanavalin A column chromatography.[43]

Induction of Two-Dimensional Crystals by Selective Extraction of Membranes

Disk membranes contain a high concentration of rhodopsin in a matrix of lipid rich in unsaturated fatty acid side chains, at a ratio of 70 lipids per

---

[42] D. S. Papermaster and W. J. Dreyer, *Biochemistry* **13**, 2438 (1974).
[43] W. J. DeGrip, *Methods Enzymol.* **81**, 197 (1982).

rhodopsin molecule.[44] Corless and co-workers discovered that extraction of these membranes with polyoxyethylene sorbitan (Tween) detergents induced the formation of 2-D crystals of frog rhodopsin in frog disk membranes.[45] We further optimized the conditions (Tween/protein ratio, pH, buffer composition) and were able to improve the reproducibility, yield, and quality of the crystals. Combining Tween 20 and Tween 80 yielded additional improvement.[46]

The process of 2-D protein crystal formation is not completely understood. Bovine disk membranes incubated under the same conditions with Tween detergents fail to produce stable rhodopsin crystals.[47] It is likely that the detergents are extracting lipid from the membranes, thus enhancing protein–protein interaction resulting in crystal formation. In most experiments we obtain crystals with randomized rather than the native orientation of the protein, suggesting that membrane fusion has occurred.[46] Similar methods using Tween[26] or deoxycholate[48,49] extraction, or lipase treatment[50] have been employed to reduce membrane lipid content and induce crystallization or improve the 2-D order.

Experimental

Frog disk membranes or ROS membranes are mixed with various Tween detergents (Tween 20, 40, 60, and 80; Sigma Chemical Co., St. Louis, MO) at a variety of molar ratios to induce the formation of 2-D arrays of rhodopsin.[45] Membranes are gently mixed with detergent at room temperature in foil-wrapped 1-ml Eppendorf centrifuge tubes (Fischer Scientific Co., Pittsburgh, PA) on a rotating mixer (e.g., Fischer Scientific Co., specimen tube rotator). Rhodopsin is present at a final concentration of 1 mg/ml, Tween detergents at a molar ratio of 1:500 to 1:1000. Membrane samples (25 μg) are removed at various times (16–30 hr) and diluted with 1 ml buffer without detergent. Membranes are harvested by centrifugation and gently resuspended to a concentration of ~0.1 mg/ml by vortexing in ~200 μl buffer. Two examples of membrane crystals obtained by these methods are presented in Fig. 2.

---

[44] B. J. Litman and D. C. Mitchell, in "Rhodopsin and G-Protein Linked Receptors" (A. G. Lee, ed.), pp. 1–32. JAI Press, Stamford, Connecticut, 1996.
[45] J. M. Corless, D. R. McCaslin, and B. L. Scott, *Proc. Natl. Acad. Sci. U.S.A.* **79,** 1116 (1982).
[46] G. F. X. Schertler and P. A. Hargrave, *Proc. Natl. Acad. Sci. U.S.A.* **92,** 11578 (1995).
[47] E. A. Dratz, J. F. Van Breemen, K. M. P. Kamps, W. Keegstra, and E. F. J. Van Bruggen, *Biochim. Biophys. Acta* **832,** 337 (1985).
[48] R. M. Glaeser, J. S. Jubb, and R. Henderson, *Biophys. J.* **48,** 775 (1985).
[49] I. N. Tsygannik and J. M. Baldwin, *Eur. Biophys. J.* **14,** 263 (1987).
[50] B. K. Jap, *J. Mol Biol.* **199,** 229 (1988).

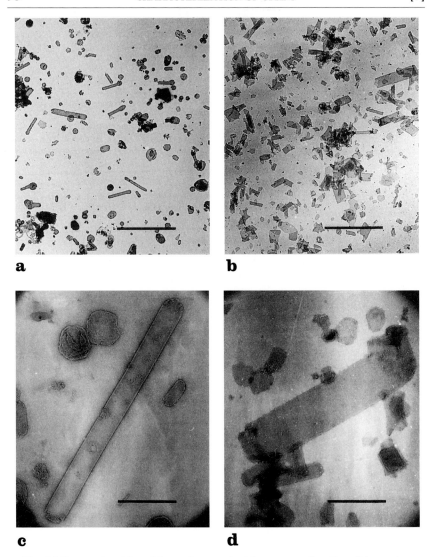

FIG. 2. Micrography of frog disk membranes after Tween extraction. Frog disk membranes were incubated for 16 hr with Tween 80 or a mixture of Tween 80 and Tween 20 at a molar ratio of 1000:1 in TES buffer (25 mM TES, 100 mM NaCl, 0.1 mM dithiothreitol, pH 7.5), washed three times, resuspended in TES buffer at a rhodopsin concentration of 1 mg/ml, and stained with 1% uranyl acetate. Crystals obtained by extraction with Tween 80 that contain rhodopsin in the $p2$ crystal form as shown in (a) and (c). Crystals in the $p22_12_1$ crystal form are shown in (b) and (d). Bars: 5 μm (a and b) or 0.5 μm (c and d). [Reproduced with permission from G. F. X. Schertler and P. A. Hargrave, *Proc. Natl. Acad. Sci. U.S.A.* **92,** 11578 (1995).]

Induction of Two-Dimensional Crystals by Reconstitution

The aim of a reconstitution experiment is to achieve as high a protein concentration as feasible in order to induce crystallization to form a single large membrane.[18,30–33] Electron diffraction experiments need coherent crystalline areas larger than 1 μm in diameter. Smaller membranes can only be studied by imaging.

Experience has shown that conditions that use the smallest ratio of lipid to protein that still give extended membrane sheets are the most successful. The halobacteria purple membrane contains as little as 10 lipids per bacteriorhodopsin molecule.[51] When membrane proteins are chromatographically purified in detergent they become partially delipidated and may contain less than 5 lipids per protein molecule.[18,30] Therefore it is often necessary to add additional lipid before the detergent is removed by dialysis. Differences in amount and type of lipid present probably account for much of the variability seen in reconstitution experiments. Other factors are the nature, stability, and purity of the protein as well as the presence of stabilizing agents such as cholesterol, glycerol, and sugars. The nature and concentration of the detergent have a strong influence on the kinetics of detergent removal that leads to reconstitution of the protein with lipid. Other parameters include salt concentration and pH, which affect the aggregation state and stability of the protein, and divalent metal ion concentration (Mg^{2+}, Ca^{2+}), which can interact with lipid headgroups and induce membrane fusion. Aggregation of 2-D crystals can be influenced by buffer conditions and by additives such as 2-propanol. These parameters must be optimized empirically in order to find a suitable protocol for each protein.

Experimental

Method 1: Reconstitution with Added Lipids

Rhodopsin is purified by concanavalin A chromatography in LDAO, exchanged into buffer containing 0.2% (w/v) N-octyltetraoxyethylene (C_8E_4; Bachem, Bubendorf, Switzerland) by chromatography on Sephadex G-50, followed by ion-exchange chromatography (S-Sepharose fast flow, Pharmacia Piscataway, NJ).[52,53] Freshly prepared rhodopsin is reconstituted into soybean phosphatidylcholine (Sigma) under dim red light. Rhodopsin concentration is 1 mg/ml in 20 mM HEPES, pH 7.0, 100 mM NaCl, 10 mM

---

[51] N. Grigorieff, T. A. Ceska, K. H. Downing, J. M. Baldwin, and R. Henderson, *J. Mol. Biol.* **259**, 393 (1996).
[52] G. F. X. Schertler, C. Villa, and R. Henderson, *Nature* **362**, 770 (1993).
[53] V. M. Unger and G. F. X. Schertler, *Biophys. J.* **68**, 1776 (1995).

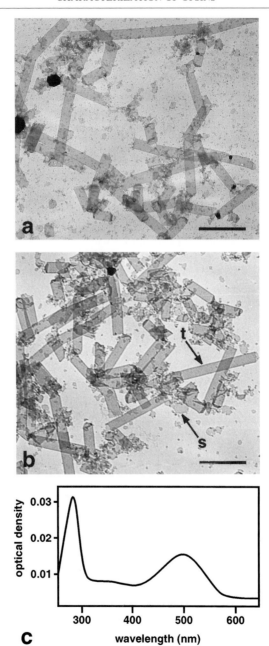

$MgCl_2$, 1 mM dithiothreitol, 3 mM NaN_3 at 20°, and lipid concentration varied from 3–30 lipids per rhodopsin molecule. The detergent (0.2% C_8E_4) was removed by dialysis.

Method 2: Reconstitution with Natural Lipids

Rhodopsin, partially delipidated by concanavalin A chromatography in LDAO, is concentrated to ~3 mg/ml and adjusted to 1 mg/ml with a buffer containing 20 mM HEPES, 10 mM $MgCl_2$, 100 mM NaCl, 3 mM NaN_3, and 0.5% LDAO, pH 7.0.[23] This solution is then dialyzed in ~1-ml aliquots using Slide-A-Lyzers (Pierce, Rockford, IL). Dialysis is carried out at 18° for 11 days against the same buffer (containing 2 mM dithiothreitol but no detergent), with daily buffer changes. The membrane suspension obtained is purified by centrifugation on a continuous sucrose gradient, from 15 to 45% sucrose. Crystals of bovine rhodopsin are found only in the fraction of density 1.166 g/ml. The concentration and optical properties of the reconstituted rhodopsin are determined spectroscopically following resolubilization of an aliquot in dialysis buffer containing 1% LDAO (Fig. 3).

Screening for Two-Dimensional Crystals

Electron microscopy is the most informative method for determining the formation of 2-D crystals.

Experimental

Copper–rhodium electron microscope grids (400 mesh, Graticules Ltd., Tonbridge, UK) are carbon coated according to standard procedures,[21] and activated by glow discharging in the presence of amylamine immediately before use. Such treatment enhances adsorption of protein to the grid. The membrane suspension (~1 μl) is applied under dim red illumination and allowed to adsorb to the grid prior to gentle blotting to remove excess liquid. The specimen is then directly stained for 10 sec with 1% uranyl

---

FIG. 3. Negatively stained 2-D crystals of bovine rhodopsin. (a) Before purification and (b) following purification by sucrose density gradient centrifugation. Crystals were obtained only in the fraction with density 1.166. Tubular crystals had a diameter of ~0.2 μm and a length of up to 5 μm. Single layers (0.2–0.5 μm in diameter) were rare. Bar: 1 μm. A single layer crystal (**s**) and tubular crystal (**t**) are indicated. (c) Spectroscopic characterization of the fraction of the crystalline suspension with density 1.166 g/ml, after dissolving in dialysis buffer containing 1% LDAO. [Reproduced with permission from A. Krebs *et al.*, *J. Mol. Biol.* **282**, 991 (1998).]

acetate (~10 μl) without an additional wash step, to avoid loss of sample. The grid is gently blotted on filter paper and dried.

Taking Overview Pictures

Overview pictures are electron micrographs taken at low magnification (1000–3000×) and strong defocus (300 μm) to enhance image contrast. The contrast achieved by such a defocused diffraction mode is essential because a single-layer membrane is a low-contrast object and easily missed. Approximately one-quarter of a grid square is imaged (Figs. 2a and 2b). Such overview pictures provide a good idea of the bulk properties of the sample, such as the presence of precipitate, vesicles, tubes, and other objects and their relative amounts. Such pictures document the results of reconstitution experiments and are an effective and rapid method to find the most promising conditions. Negatively stained sample grids are then selected for imaging at higher magnification followed by optical diffraction. All photographs are taken at the same magnification (~30,000×) and defocus level.[18] Two examples of such negatively stained crystals are shown in Figs 2c and 2d.[46]

Optical Diffraction

An early indication of the success of a crystallization experiment can be obtained by examination of negatives of pictures recorded from the crystals by optical diffraction. If there is a crystal imaged, the image acts as a diffraction grating for light. Examination by laser diffractometer provides indications of crystal symmetry elements and the plane group of the crystal lattice (Fig. 4).

Vitrification of Specimens for Cryomicroscopy

Membrane preparations that are promising following preliminary examination of stained samples are prepared for cryomicroscopy. The sample is rapidly frozen in a thin film of water, adsorbed to the carbon film of the microscope grid. The membrane suspension (1 μl) is applied to the microscopy grid under dim red illumination, blotted gently on filter paper, and the grid rapidly plunged into liquid ethane. If the freezing rate is rapid enough, ice crystals do not form and the specimen is embedded in an amorphous glass-like water layer. Grids containing the frozen membrane specimens are then rapidly transferred to liquid nitrogen for storage and later examination on the cold stage of an electron microscope.

The vitreous state of water is stable at temperatures below $-160°$ under the high vacuum of the electron microscope.

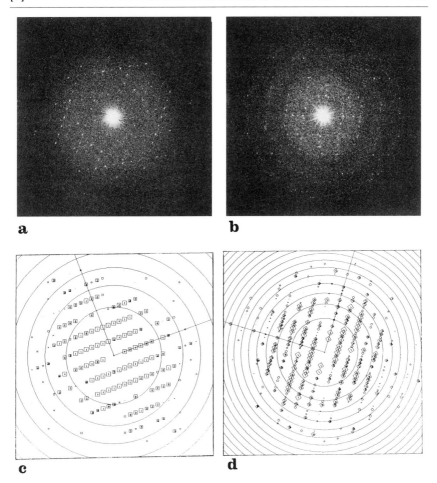

FIG. 4. Optical diffraction of frog rhodopsin crystals. Crystals were rapidly frozen in liquid ethane. Electron micrographs were taken at liquid nitrogen temperature with a Gatan 626 cold stage and inspected by optical diffraction. In (a) the diffraction spots can be indexed on two independent p2 lattices with $a = 32$ Å, $b = 83$ Å, and $\gamma = 91°$ for both layers. (b) Diffraction pattern of a frog $p22_12_1$ crystal. Two lattices can be indexed with $a = 40$ Å, $b = 146$ Å, and $\gamma = 90°$. Digitized areas were corrected for distortions in the crystal lattice. C shows the output for a p2 crystal, and D for a $p22_12_1$ crystal. [Reproduced with permission from G. F. X. Schertler and P. A. Hargrave, *Proc. Natl. Acad. Sci. U.S.A.* **92,** 11578 (1995).]

Three-Dimensional Structure of Rhodopsin

It is beyond the scope of this article to describe in detail how one uses electron microscopy to take images from frozen membrane samples and to process these images to obtain a 3-D structure for the membrane protein. This is the subject of several reviews.[18,20,36] Using these methods, projection maps and 3-D structures have been determined for rhodopsin from frog, squid, and bovine retinas (Table I). This has given the first direct evidence for the seven-helix construction of rhodopsin and for the orientation of the helices with respect to one another.

Rhodopsin molecules are visualized by this method to have planar dimensions of 28×39 Å and to be approximately 64 Å in height. A solid model of rhodopsin has been constructed from contour cross sections (Fig. 5, see color insert). Slices through the electron density map parallel to the plane of the membrane show the arrangement of helices at various positions within the membrane. Seven main density peaks representing the seven transmembrane helices of rhodopsin can be clearly resolved.[8] Using information from rhodopsin and from other G-protein-linked receptors, the map has been interpreted and the seven density peaks have been assigned to individual helices within the protein sequence.[2,52]

When individual slices of the map taken at different positions within the membrane are compared (Fig. 5B), it is possible to note differences between the helices and in their relationships. In the section taken closer to the cytoplasmic surface, at 12 Å above the center of the membrane, the helices are packed tightly together. In contrast, in the section closer to the intradiskal surface, at -8 Å, the helices are more spread apart, and the molecule has a larger cross-sectional area. There is space present encompassed by helices 3, 4, 5, 6, and 7, that must be the binding site for rhodopsin's light-sensitive ligand, 11-*cis*-retinal.

Four of the helices (4, 6, and 7) appear in nearly the same positions at the different levels within the membrane, and are perpendicular to the plane of the membrane through most of their distance. The other helices are more tilted. Helix 3 shows the greatest tilt, with an inclination of 30°. It is buried within the molecule and therefore has extensive contact with other helices. On the cytoplasmic surface helix 3 closes the retinal binding pocket, and separates the vertical helices 4, 6, and 7. Because of the limited vertical resolution of the map it is not possible to resolve where cysteine-110 in helix 3 makes its important stabilizing disulfide linkage to cysteine-187 in the loop connecting helices 3 and 4.

Less electron density is observed on the cytoplasmic side in comparison with the extracellular side, suggesting that the cytoplasmic loops are more loosely packed than are the extracellular loops. Electron density extends furthest on the cytoplasmic side corresponding to helix 6, in agreement with independent results from EPR measurements.[54] Portions of the cytoplasmic

---

[54] C. Altenbach, K. Yang, D. L. Farrens, Z. T. Farahbakhsh, H. G. Khorana, and W. L. Hubbell, *Biochemistry* **35**, 12470 (1996).

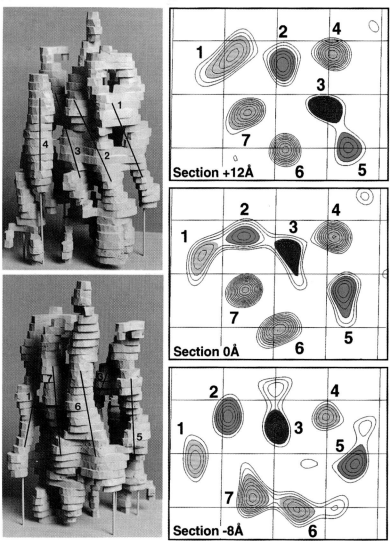

FIG. 5. The seven helices of rhodopsin. (A) The structure of frog rhodopsin obtained by electron cryomicroscopy. Two views of a solid model of the rhodopsin electron density map. Upper left: a view from helix 2 toward helix 6. Lower left: a view from helix 6 toward helix 3. The model was constructed from 33 contour sections 2 Å apart. The cytoplasmic side is at the top and the intradiskal or extracellular side is at the bottom. The central sections of the seven transmembrane helices are marked with lines starting at section +12 Å at the top and ending at section − 8 Å. The corresponding sections are shown in (B). The peaks representing the seven helices are interpreted according to the sequence assignment[2] to the projection map of rhodopsin.[52] (B) Three slices through the best part of the electron density map of frog rhodopsin.[8] In each of these sections peaks can be seen for each of the seven transmembrane helices. The section closer to the cytoplasmic side is at $z = +12$ Å from the center, and the last section at $z = -8$ Å from the center of the map. The least tilted helices (4, 6, and 7) are colored gray, and the most tilted helices (1, 2, 3, and 5) are each a different color. The grid spacing is 10 Å with lines parallel to the **a** and **b** axes (+**b** is horizontal to the right and +**a** points toward the bottom of the figure). [Modified from V. M. Unger et al., Nature **389**, 203 (1997)].

TABLE I
Rhodopsin Crystals Investigated by Electron Cryomicroscopy

Rhodopsin crystals	Bovine projection	Bovine 3-D structure	Frog projection	Squid projection	Frog projection	Frog 3-D structure	Bovine projection
Important steps	Soybean phosphatidyl-choline[a]	Soybean phosphatidyl-choline[a]	Tween 80/Tween 20[b]	Squid photoreceptor lipids[a]	Tween 80[b]	Tween 80[b]	Lauryldimethylamine oxide[c]
Ratio lipid/rhodopsin	10:1 to 50:1	3:1 to 30:1		50:1			
Number of images	8	16	4	8	7	49	14
Number of crystalline images	13	23	8	11	14	60	20
Image tilt	0–8°	Up to 30°				Up to 45°	Up to 6.9°
Crystal form	$p222_1$	$p222_1$	$p22_12_1$	$p222_1$	$p2$	$p2$	$p22_12_1$
a	43 ± 3%	43 ± 3%	40 ± 3%	44 ± 1.7%	32 ± 2%	32 ± 2%	60.6 ± 1.2%
b	140 ± 3%	140 ± 3%	146 ± 3%	131 ± 2.5%	83 ± 2%	83 ± 2%	86.3 ± 1.8%
Angle	90 ± 3%	90 ± 3%	90 ± 3%	90	91 ± 2%	91 ± 2%	89.5 ± 0.6%
Area/molecule	1505 Å^2	1505 Å^2	1460 Å^2	1441 Å^2	1328 Å^2	1328 Å^2	1307 Å^2
Number of reflections	1755 (to 7 Å)	1314 (to 7.5 Å)	1763 (to 5 Å)	1431 (to 7 Å)	1546 (to 5 Å)	3378 (to 7.5 Å)	3548 (to 5 Å)
Number of unique reflections	109 (to 7 Å)	86 (to 7.5 Å)	190 (to 5 Å)	105 (to 7 Å)	154 (to 5 Å)		179 (to 5 Å)
Phase residual	29.7° (to 7 Å)	26.6° (to 7.5 Å)	33° (to 5 Å)	32° (to 7 Å)	27.5° (to 5 Å)	22.6° (to 7.5 Å)	13.4° (to 5 Å)
Achieved resolution	9.0 Å	9.5 Å in plane 47 Å normal	7.0 Å	8.0 Å	6.0 Å	7.5 Å in plane 16.5 Å normal	5 Å
Reference	52	53	46	60	46	8	23

[a] Lipid reconstitution of delipidated protein.
[b] Selective extraction.
[c] Lipid reconstitution of partially delipidated protein.
"a" is the length of the a axis in that crystal form, and "b" is the corresponding length of the b axis.

surface loops have been demonstrated to interact with the cytoplasmic proteins rhodopsin kinase, transducin, and arrestin.

As mentioned earlier, an important result of these studies is the demonstration that the helix arrangement close to the cytoplasmic surface of rhodopsin is significantly more compact than that near the extracellular surface (in Fig. 5B, compare section +12 Å with section −8 Å). The formation of photoactivated rhodopsin (metarhodopsin-II) has been shown to be associated with the movement of helices.[55–57] This is consistent with the finding that the formation of metarhodopsin-II occurs with an increase of the volume of the protein,[58] and with the formation of a binding site for transducin and other cytoplasmic proteins.

Conclusions and Future Prospects

A new crystal form of rhodopsin has been generated by using natural lipids in reconstitution experiments.[23] The crystals are better ordered than those previously obtained,[52] and there is always the possibility of further optimizing conditions of crystallization to yield larger and better ordered arrays. Some images from currently produced membranes contain information to a resolution beyond 5 Å. The projection structure obtained to date is a reasonable starting point for further 3-D analysis. Use of a field emission microscope as compared to a conventional electron microscope has provided superior images.[18] Promising diffraction patterns have been obtained from rhodopsin crystals using a cooled charge-coupled device (CCD) camera as the detector,[59] giving highest resolution spots at 3.5 Å.[23] In addition, superior methods of data analysis have enabled improved map construction. The best diffraction patterns obtained to data suggest that it may be possible to obtain high-resolution data from two-dimensional crystals of rhodopsin.

Acknowledgments

We would like to thank the MRC Laboratory of Molecular Biology for its generous support of this work over a period of years. This work was also supported in part by grant EY06226

---

[55] D. L. Farrens, C. Altenbach, K. Yang, W. L. Hubbell, and H. G. Khorana, *Science* **274**, 768 (1996).

[56] S. P. Sheikh, T. A. Zvyaga, O. Lichtarge, T. P. Sakmar, and H. R. Bourne, *Nature* **383**, 347 (1996).

[57] Z. T. Farahbakhsh, K. D. Ridge, H. G. Khorana, and W. L. Hubbell, *Biochemistry* **34**, 8812 (1995).

[58] A. A. Lamola, T. Yamane, and A. Zipp, *Biochemistry* **13**, 738 (1974).

[59] A. R. Faruqi, H. N. Andrews, and R. Henderson, *Nucl. Instrum. Methods* **367**, 408 (1995).

[60] A. Davies, G. F. X. Schertler, B. E. Gowan, and H. R. Saibil, *J. Struct. Biol.* **117**, 36 (1996).

from the National Institutes of Health, an unrestricted departmental award and a Senior Scientist Award (to P.A.H.) from Research to Prevent Blindness, Inc. P.A.H. is Francis N. Bullard Professor of Ophthalmology. The authors thank V. Unger, A. Krebs, P. Edwards, C. Villa, R. Henderson, and J. H. McDowell who contributed greatly to the studies described.

[8] Domain Approach to Three-Dimensional Structure of Rhodopsin Using High-Resolution Nuclear Magnetic Resonance

By ARLENE D. ALBERT and PHILIP L. YEAGLE

Introduction

The determination of the structure of integral membrane proteins such as rhodopsin has presented substantial challenges. Direct structural determination by nuclear magnetic resonance (NMR) has been successful for soluble proteins and for peptides. However, integral membrane proteins are not amenable to this technique due to their size and solubility in aqueous solution. Recent work on the largely helical protein, myohemerythrin, has suggested it is feasible to elucidate protein structure by piecing together structural domains. The structures of peptides that represent these domains exhibited the same secondary structure (as determined by NMR solution methods) as found in the structure of the intact protein.[1] Secondary structures (helices and loops) are stabilized by short-range interactions. When the protein consists primarily of a bundle of helices, peptides coding for the helices or the loops connecting the helices exhibit the same structures as are observed in the native protein.

These data suggest that the structure of the whole protein can be reduced to a summation of the individual structural domains. In this approach, structures are experimentally determined for individual peptides which encode the domains of secondary structure like α helices or β turns. The sequences of these domains are chosen such that they have significant overlap with the sequence of domains adjacent in the primary sequence. These regions of overlap allow the individual domains of the protein to be assembled into a protein structure that contains all of the secondary structural elements. The structure that emerges at this point is stabilized exclusively by short-range interactions. Subsequent addition of long-range experimentally determined interactions constitutes the constraints which then

---

[1] H. J. Dyson, G. Merutka, J. P. Waltho, R. A. Lerner, and P. E. Wright, *J. Mol. Biol.* **226**, 795 (1992).

define the proper packing of the elements of secondary structure into the structure of the whole protein.

This approach is suitable for an integral membrane protein like rhodopsin. The basic structure of a bundle of seven transmembrane helices has now been established by the elegant low-resolution electron diffraction work of Shertler et al.[2] The helices include approximately three-quarters of the rhodopsin sequence. Because helices are stabilized by short-range interactions, they can be expected to form independently of the remainder of the protein. These helices are connected by six loops, three facing the cytoplasm and three on the opposite face. The literature suggests that certain types of turns are stable as peptides in solution.[3] Thus it is reasonable to predict that the loops of rhodopsin are also stabilized by short-range interactions. Each of these portions of rhodopsin could form structures independently of the remainder of the protein. The observation that peptides corresponding to cytoplasmic loops of rhodopsin exhibit biological activity in solution[4–6] further supports the proposition that the structures of these loop peptides in solution are closely related to the structures expressed in the whole protein. Therefore, in the case of rhodopsin, built around a bundle of seven transmembrane helices, the protein structure can be considered to consist of smaller domains each of whose structure is stabilized by short-range interactions and fully coded by the local primary sequence.

Methodology

Peptide Length

To obtain relevant structural information about the native protein by examining structures of portions of the protein as individual peptides, the size and positioning of these peptides in the sequence of the protein must be chosen carefully. Our studies have led us to the following guidelines in the choice of peptides:

1. In the case of a membrane protein like rhodopsin, the peptides can be chosen as loops (which may be soluble in water), transmembrane helices

---

[2] G. R. X. Schertler, C. Villa, and R. Henderson, *Nature* **362,** 770 (1993).
[3] A.-S. Yang, B. Hitz, and B. Honig, *J. Mol. Biol.* **259,** 873 (1996).
[4] B. Konig, A. Arendt, J. H. McDowell, M. Kahlert, P. A. Hargrave, and K. P. Hofmann, *Proc. Natl. Acad. Sci. U.S.A.* **86,** 6878 (1989).
[5] D. J. Takemoto, L. J. Takemoto, J. Hansen, and D. Morrison, *Biochem. J.* **232,** 669 (1985).
[6] D. J. Takemoto, D. Morrison, L. C. Davis, and L. J. Takemoto, *Biochem. J.* **235,** 309 (1986).

[which can be studied in dimethyl sulfoxide (DMSO), in detergent or phospholipid micelles, or in $CDCl_3/CD_3OH$], amino terminus and carboxyl terminus (either one of which, or both, may be soluble in water). With respect to the loops, size is an important variable. Experience has shown that the choice of small peptides (15–18 amino acids) will reveal only the structure of the turn in these loops. To see some evidence of the helices that connect to these turns, it is necessary to synthesize larger peptides. Inclusion of enough sequence to code for two turns of helix (as predicted from an analysis of the primary sequence) on each side of the loop has been shown to be sufficient to reveal the helix–turn–helix motif of these turns in the solution structures. Although it is important to have peptides of sufficient length to visualize the structural elements, peptides that extend too far into the hydrophobic core create solubility problems that even the use of DMSO does not eliminate. Hydrophobic peptides from transmembrane helices of 35 residues and longer appear to be unstable in DMSO and must be studied in micelles. Hydrophobic peptides up to 20 residues appear to be stable in DMSO.

2. To be able to "fit" the structure of the chosen peptides together to construct the whole protein, the peptides chosen need to overlap each other in sequence. Experience has now shown that this overlap should be substantial. The ends of the peptides tend to be disordered in the solution structures. Therefore the peptides need to be long enough to produce sufficient ordered structure in the overlap region to be able to accurately superimpose (overlap) structures from adjacent sequences with confidence. In practical terms, these needs appear to require an overlap region of at least seven amino acids in each of the peptide ends.

Initial Structural Analysis

The peptides are synthesized by solid-phase peptide synthesis. The most readily available structural information on such peptides is from the methods of solution structure determination by NMR. Therefore the conditions under which the solubility of each peptide can be maintained must be determined. In the case of rhodopsin, all the peptides corresponding to the cytoplasmic loops and the carboxyl terminus of this receptor are soluble in aqueous solution. Interestingly the loops on the opposite face of rhodopsin are not soluble in water. As an alternative, these loops were dissolved in DMSO and structures were obtained. Likewise the transmembrane helices of rhodopsin were synthesized and because of their hydrophobic nature were dissolved in both DMSO and in $CDCl_3/CD_3OH$. The peptides were more stable in DMSO. Subsequent structure determinations by NMR can also be undertaken in perdeuterated dodecylphosphocholine micelles to verify that the structures of the transmembrane helices are the same in a more membrane-like environment.

A useful start to the structural analysis is the use of circular dichroism (CD) to obtain a measure of the secondary structure content of the peptide. This analysis can be performed for the peptides soluble in water, but not for the peptides dissolved in DMSO, which has too high a UV cutoff to permit the analysis of secondary structure. As an example, CD analysis of the structure of the 43-amino-acid peptide representing the carboxyl terminal of rhodopsin (residues 303–348) compared well with the actual secondary structure content after the solution structure was determined. Therefore even though this peptide is modest in length, CD analysis can provide a reasonable measure of secondary structure in the peptide. Note also that this structure also agreed with a Fourier transform infrared (FTIR) analysis of the secondary structure content of the carboxyl-terminal region of rhodopsin.[7]

Solution Structure Determination

The peptides used in these studies are all synthesized. Therefore all nuclei are in natural abundance in these peptides. Consequently only two-dimensional homonuclear ^{1}H NMR techniques can be used to determine the solution structure of the peptides from the sequence of rhodopsin. These techniques are well known so will not be described in detail here, but a summary follows.

All NMR spectra in water are recorded on a Bruker AMX-600 spectrometer at 5° or 10° in 1 mM phosphate buffer, pH 5, and including 0.1 mM α-mercaptoethanol for peptides containing cysteine. All DMSO spectra are recorded at 25°. Standard pulse sequences and phase cycling are employed to record in H$_2$O (10% D$_2$O), double quantum filtered (DQF) COSY, HOHAHA,[8,9] and NOESY (150- to 400-ms mixing times).[10] All spectra are accumulated in a phase-sensitive manner using time-proportional phase incrementation for quadrature detection in F1. Chemical shifts are referenced to internal methanol or the water peak in DMSO. Spectra are analyzed using Felix (MSI).

The sequence-specific assignment of the ^{1}H NMR spectrum is carried

---

[7] A. M. Pistorius and W. J. DeGrip, *Biochem. Biophys. Res. Commun.* **198**, 1040 (1994).
[8] L. Braunschweiler and R. R. Ernst, *J. Magn. Reson.* **53**, 521 (1983).
[9] A. Bax and D. G. Davis, *J. Magn. Reson.* **65**, 355 (1985).
[10] A. Kumar, R. R. Ernst, and K. Wuthrich, *Biochem. Biophys. Res. Commun.* **95**, 1 (1980).

out using standard methods.[11] The CαH chemical shifts of the peptide can be analyzed according to the scale of Wishart et al.[12] to test for the presence of elements of secondary structure. In our experience, this analysis is only modestly successful in small peptides, compared to the considerable success for larger proteins. Connectivities from the NOESY map are consistent with the elements of secondary structure found in the structure determination.

Assigned NOE cross peaks are segmented using a statistical segmentation function and characterized as strong, medium, and weak corresponding to upper bound distance range constraints of 2.7, 3.7, and 5.0 Å, respectively. Lower bounds between nonbonded atoms are set to the sum of their van der Waals radii (1.8 Å). Pseudoatom corrections are added to interproton distance restraints where necessary.[13] Distance geometry calculations are carried out using the program DIANA[14] within the SYBYL 6.4 package (Tripos Software Inc., St. Louis, MO). First-generation DIANA structures, 150 in total, are optimized with the inclusion of three REDAC cycles. Energy refinement calculations (restrained minimizations/dynamics) are carried out on the best distance geometry structures using the SYBYL program implementing the Kollman all-atom force field. Statistics on structures are obtained from Xplor. These calculations are performed on a Silicon Graphics R10000 computer. Imaging and overlay of the resulting structures and construction of the transmembrane helices of rhodopsin are performed on a PowerMac with MacImdad (Molecular Applications Group, Palo Alto, CA).

Structural Results

The structures of rhodopsin peptides are shown in Fig. 1 (see color insert). Only the general features of these structures are described because details have been published elsewhere.[15–21] All loops connecting helices

[11] G. C. K. Roberts, in "The Practical Approach Series" (D. Rickwood and B. D. Hames, eds.), p. IRL Press, Oxford, 1993.
[12] D. S. Wishart, B. D. Sykes, and F. M. Richards, *Biochemistry* **31,** 1647 (1992).
[13] K. Wüthrich, M. Billeter, and W. J. Braun, *J. Mol. Biol.* **169,** 949 (1983).
[14] P. Guntert, W. Braunk, and K. Wuthrich, *Mol. Biol.* **217,** 517 (1991).
[15] A. D. Albert, P. L. Yeagle, J. L. Alderfer, M. Dorey, T. Vogt, N. Bhawasar, P. A. Hargrave, J. H. McDowell, and A. Arendt, *Biochim. Biophys. Acta* **1416,** 217 (1999).
[16] P. L. Yeagle, J. L. Alderfer, and A. D. Albert, *Biochemistry* **36,** 9649 (1997).
[17] A. D. Albert, P. L. Yeagle, J. L. Alderfer, M. Dorey, T. Vogt, N. Bhawasar, P. A. Hargrave, J. H. McDowell, and A. Arendt, *Peptides: Chemistry, Structure and Biology,* in press (1997).
[18] P. L. Yeagle, J. L. Alderfer, and A. D. Albert, *Biochemistry* **36,** 3864 (1997).
[19] P. L. Yeagle, J. L. Alderfer, and A. D. Albert, *Mol. Vis.* **2,** http://www.emory.edu/molvis/v2/yeagle (1996).
[20] P. L. Yeagle, J. L. Alderfer, and A. D. Albert, *Biochemistry* **34,** 14621 (1995).
[21] P. L. Yeagle, J. L. Alderfer, and A. D. Albert, *Nature Struct. Biol.* **2,** 832 (1995).

form stable turns as peptides in solution. These include the three cytoplasmic loops and the three loops on the opposite face, as well as the amino terminal and the carboxyl terminal. All domains on the cytoplasmic face are soluble in aqueous solution. Some of these peptides formed clear macroscopic gels with time in water, but these gels did not inevitably hinder the collection of NMR data. The NMR data for these peptides was obtained at 5° or 10° because the structures proved to be unstable at higher temperatures. Surprisingly the domains on the opposite face were not stable as a solution in water for the length of time required to collect the NMR data, but proved to be soluble and stable in DMSO. Likewise the transmembrane helices formed stable structures in solution when DMSO was used. The NMR data for these peptides were collected at 30°.

As expected, the structures of these peptides, particularly the smaller ones, tended to be frayed at the ends. The most ordered regions were in the middle of the peptides. In the case of the third cytoplasmic loop of rhodopsin, one side of the loop remains disordered in the final structure. The same observation is made in the structure of the third cytoplasmic loop of the parathyroid hormone (PTH) receptor, which is remarkably structurally homologous to the third cytoplasmic loop of rhodopsin.[22] Because of the disordered ends of these peptides, it was necessary to work with longer regions of rhodopsin peptides to ensure that a sufficient piece of ordered structure could be obtained to assemble the protein as described next.

Cytoplasmic Surface of Rhodopsin

A structure for the entire cytoplasmic surface of rhodopsin can be assembled from the individual cytoplasmic structural domains of rhodopsin. Two methods can be used to achieve this. The first method combines the low-resolution information on the structure of the transmembrane domain of rhodopsin[2] with the structures of the cytoplasmic loops. A model of the transmembrane domain was constructed, using the helix assignment and rotational orientation proposed by Baldwin,[23] which has found considerable experimental support.[24–31] A structure of the cytoplasmic face of rhodopsin

---

[22] D. F. Mierke, M. Royo, M. Pellegrini, H. Sun, and M. Chorev, *J. Am. Chem. Soc.* **118**, 8998 (1996).
[23] J. M. Baldwin, *EMBO J.* **12**, 1693 (1993).
[24] S. P. Sheikh, T. A. Zvyaga, O. Lichtarge, T. P. Sakmar, and H. R. Bourne, *Nature* **383**, 347 (1996).
[25] W. Zhou, C. Flanagan, J. A. Ballesteros, K. Konvicka, J. S. Davidson, H. Weinstein, R. P. Millar, and S. C. Sealfon, *Mol. Pharmacol.* **45**, 165 (1994).
[26] H. Yu, M. Kono, T. D. McKee, and D. D. Oprian, *Biochemistry* **34**, 14963 (1995).
[27] C. E. Elling, S. M. Nielsen, and T. W. Schwartz, *Nature* **374**, 74 (1995).
[28] C. E. Elling and T. W. Schwartz, *EMBO J.* **15**, 6213 (1996).

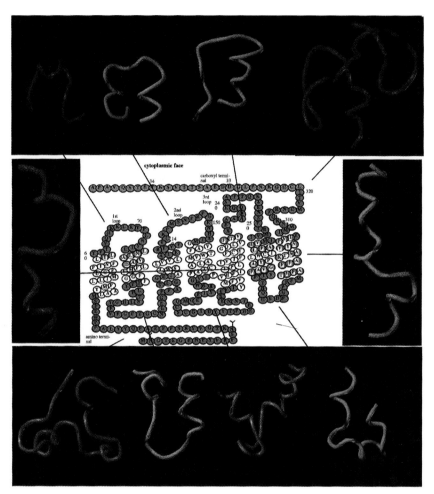

FIG. 1. The structures of peptides determined by solution NMR that correspond to the regions of rhodopsin indicated by the lines.

is obtained by docking the experimentally determined structures of the four cytoplasmic domains to the transmembrane domain. Docking is achieved by overlap of the helix in the structure of the cytoplasmic domain with the appropriate helix of the transmembrane domain. (A useful animation showing this docking of the carboxy terminus can be seen at http://www.emory.edu/molvis/v2/yeagle.) This docking organizes the cytoplasmic domains relative to each other. The energy of the structure, without the transmembrane helices, was then minimized within Sybyl.

The second method for determining the structure cytoplasmic surface did not require use of the transmembrane helices to organize the structures relative to each other. Rather, the structure of the complex consisting of all four domains of rhodopsin from the cytoplasmic face associated in solution was determined by two-dimensional ^{1}H NMR in water. Details of the structure determination were previously reported.[16] Several long-range interactions were discovered in the NOESY map of the complex and used to fold the complex together. The structure was obtained by simulated annealing and is similar to the structure obtained by docking described earlier, except there is a greater exposure of the β sheet of the carboxyl-terminal domain in the latter structure. This is particularly interesting in that it is a region that contains the rhodopsin kinase phosphorylation sites.

As has been described elsewhere,[16] this structure obtained for the cytoplasmic face of rhodopsin agrees well with the known site-to-site distance measurements made on the intact molecule. This remarkable agreement with available data supports the validity of the domain approach to the determination of the structure of rhodopsin. In the remainder of this article the expansion of this domain approach from the cytoplasmic face to the structure of the whole molecule is described.

Comments

Experiments have now shown that it is possible to divide all of rhodopsin into small domains consisting of loops, transmembrane helices, or the amino or carboxyl termini of the protein, and to obtain useful structural information from each domain. All of these small domains are dominated by elements of secondary structure, that are stabilized by short-range interac-

---

[29] T. Mizobe, M. Maze, V. Lam, S. Suryanarayana, and B. K. Kobilka, *J. Biol. Chem.* **271**, 2387 (1996).

[30] D. L. Farrens, C. Altenbach, K. Yang, W. L. Hubbell, and H. G. Khorana, *Science* **274**, 768 (1996).

[31] K. Yang, D. L. Farrens, C. Altenbach, Z. T. Farahbakhsh, W. L. Hubbell, and H. G. Khorana, *Biochemistry* **35**, 14040 (1996).

tions. For the most part the secondary structures observed are α helix or β turn. The former is the dominant structure in the transmembrane helices and the amino terminal, whereas the latter is found in several of the loops that exhibit a helix–turn–helix motif. Both α helices and β turns are stabilized by short-range interactions ($i \rightarrow i + 1$; $i \rightarrow i + 2$; $i \rightarrow i + 3$; $i \rightarrow i + 4$). There are many other examples of relatively small peptides exhibiting these secondary structures in solution.[32–34]

If the peptides of the parent protein are appropriately designed, the ends of peptides representing adjacent regions of the primary sequence of rhodopsin overlap. This would allow the combination of all of the structural elements into a structural whole. When the structures of the loops derived from the NMR studies are combined with the transmembrane helices derived from the electron diffraction studies, a relatively compact structure is produced. This result is obtained in the complete absence of long-range interactions and using the assumption that the transmembrane helices are straight. This model for rhodopsin cannot adequately describe the native protein. However, it does suggest that the turns themselves are an important factor in determining the spatial location of the transmembrane helices. It is reasonable then to expect that the addition of long-range interactions as well as refined transmembrane helical structures would result in a more accurate structure.

Five of the transmembrane helices of rhodopsin contain at least one proline. Although modeling of G-protein receptors assumes normal, straight helical structure, the presence of the proline suggests that these transmembrane helices contain kinks. In fact, the structures actually determined for the peptides with the sequences of these transmembrane helices of rhodopsin (as well as other transmembrane proteins) are kinked by the proline.[35,36] These kinks have significant ramifications for the structure of the transmembrane region of rhodopsin.

A complete structure of rhodopsin also requires the inclusion of long-range interactions to describe the appropriate three-dimensional packing of the smaller structural domains. In principle, this is achieved by combining all of the short-range experimental constraints that stabilize the individual

---

[32] F. J. Blanco and L. Serrano, *Eur. J. Biochem.* **230**, 634 (1994).
[33] N. Goudreau, F. Cornille, M. Duchesne, F. Parker, B. Tocqué, C. Garbay, and B. P. Roques, *Nature Struct. Biol.* **1**, 898 (1994).
[34] M. Adler, M. H. Sato, D. E. Nitecki, J.-H. Lin, D. R. Light, and J. Morser, *J. Biol. Chem.* **270**, 23366 (1995).
[35] J.-P. Berlose, O. Convert, A. Brunissen, G. Chassaing, and S. Lavielle, *Eur. J. Biochem.* **225**, 827 (1994).
[36] C. F. Snook, G. A. Wolley, G. Oliva, V. Pattabhi, S. P. Wood, T. L. Blundell, and B. A. Wallace, *Structure* **6**, 783 (1998).

structures of the small domains with experimental long-range constraints obtained from other methods on the intact protein in a simulated annealing process.

Other experimental methods provide long-range constraints that define the relationship among the elements of secondary structure represented in the small domains. Many of these constraints are obtained as single site-to-site distances. The groups of Hubbell and Khorana have placed spin labels in pairs in specific locations of rhodopsin and used the dipolar interactions between the unpaired electrons to determine distances between the two sites on the intact protein.[30,31,37-40] This approach has led to several long-range constraints for rhodopsin. Using paramagnetic enhancement of relaxation between a phosphate on the C terminus and a spin label on Cys-140, another long-range distance constraint was obtained.[15] The counterion to the shift base and the disulfide bond between Cys-110 and Cys-185 provide more long-range constraints. These long-range constraints can be combined with all the short-range constraints described earlier through a simulated annealing process to fold the structure of the entire protein.

This approach uses experimentally determined high-resolution structures for each domain of the protein in combination with long-range distance measurements to provide an alternative approach to the determination of the three-dimensional structure of rhodopsin and other G-protein-coupled receptors. Unlike a modeling approach, all structural elements are experimentally determined. Although a crystal structure is anticipated for rhodopsin, it is less likely for other G-protein receptors. Therefore, this technique will complement the crystallography as well as provide an alternate means of structural determination for other receptors. This structural information is important to determine the molecular mechanism of signal transduction initiated by light absorption by rhodopsin. Furthermore, the development of retinitis pigmentosa by mutations in rhodopsin primary sequence can be investigated by structural studies on peptides synthesized to reflect the specific mutations.

---

[37] Z. T. Farahbakhsh, K. D. Ridge, H. G. Khorana, and W. L. Hubbell, *Biochemistry* **34,** 8812 (1995).
[38] C. Altenbach, K. Yang, D. L. Farrens, Z. T. Farahbakhsh, H. G. Khorana, and W. L. Hubbell, *Biochemistry* **35,** 12470 (1996).
[39] K. Yang, D. L. Farrens, W. L. Hubbell, and H. G. Khorana, *Biochemistry* **35,** 12464 (1996).
[40] R. Langen, K. Kai, H. G. Khorana, and W. L. Hubbell, *Biophys. J.* **74,** A290 (1998).

[9] Analysis of Functional Microdomains of Rhodopsin

By STEVEN W. LIN, MAY HAN, and THOMAS P. SAKMAR

Introduction

The activity of rhodopsin and visual pigments as light receptors requires communication between residues in the transmembrane (TM) helical domain and surface loop domain. Residues within the TM domain perform two primary functions: (1) They create the binding pocket for 11-*cis*-retinal and regulate its spectral properties and (2) they participate in intramolecular interactions that control the tertiary structure and dynamics of the receptor. Residues on the cytoplasmic surface regulate the interaction of the receptor to its G protein, transducin (G_t), and other cytoplasmic proteins. Previous site-directed mutagenesis studies of the TM domain have focused mainly on elucidating residues located near the chromophore or on investigating the effects of mutations associated with autosomal-dominant retinitis pigmentosa on the expression and assembly of rhodopsin. More recently, studies using multiple-site replacements as well as multiple substitutions at a single site have probed for structure–function relationships and interactions between different sites. The analysis of biochemical and physical data from these mutants has benefited from the application of molecular modeling methods. This integrated biophysical approach has facilitated interpretation and has provided valuable insight into the role of certain TM residues and regions in the mechanism of rhodopsin activity.

Mutagenesis of Functional Residues in Transmembrane Domain

Chromophore Binding Pocket and Spectral Tuning

One of the primary functions of opsin is to tune the spectral response of 11-*cis*-retinal to a particular region of the electromagnetic spectrum. This is accomplished by enclosing the retinal in a low dielectric pocket supplemented by interactions with side chains of key residues (Fig. 1). Protonation of the Schiff base enhances electron delocalization along the polyene and causes the λ_{max} to move from 380 toward 440 nm. The +1 charged protonated Schiff base (PSB) species is stabilized by the glutamate counterion at position 113[1–3] which, unlike in solution, resides near the

---

[1] T. P. Sakmar, R. R. Franke, and H. G. Khorana, *Proc. Natl. Acad. Sci. U.S.A.* **86,** 8309 (1989).

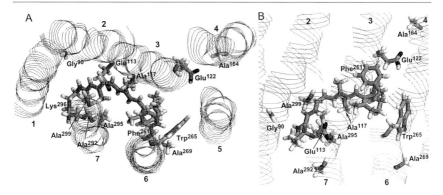

Fig. 1. Computer model of rhodopsin [Shieh *et al., J. Mol. Biol.* **269,** 373 (1997)]. Residues in the retinal-binding pocket and amino acid sites that are involved in spectral tuning of color visual pigments are shown. (A) View from the cytoplasmic surface. The TM helices are numbered clockwise 1 to 7. (B) Side view (cytoplasm is at the top).

C-12 of retinal.[4,5] This unusual arrangement may also be supported by the interaction of the PSB group with water molecules nearby.[6,7] The weaker, long-range electrostatic interaction between the PSB and its counterion may explain the majority of the opsin shift to 500 nm seen in rhodopsin.

The molecular basis of opsin shift between green- and red-sensitive visual pigments[8–12] and between green- and blue-sensitive visual pigments[13] has been addressed by mutagenesis. The opsin shift can be accounted for by the aggregate interactions of several residues with the chromophore. One theme that emerges is the role of long-range electrostatic interaction of dipolar residues. These groups are capable of inducing λ_{max} shifts to

---

[2] J. Nathans, *Biochemistry* **29,** 9746 (1990).
[3] E. A. Zhukovsky and D. D. Oprian, *Science* **246,** 928 (1989).
[4] M. Han, B. S. DeDecker, and S. O. Smith, *Biophys. J.* **65,** 899 (1993).
[5] R. R. Birge, L. P. Murray, B. M. Pierce, H. Akita, V. Balogh-Nair, L. A. Findsen, and K. Nakanishi, *Proc. Natl. Acad. Sci. U.S.A.* **82,** 4117 (1985).
[6] S. Nishimura, H. Kandori, and A. Maeda, *Photochem. Photobiol.* **66,** 796 (1997).
[7] H. Deng, L. Huang, R. Callender, and T. Ebrey, *Biophys. J.* **66,** 1129 (1994).
[8] M. Neitz, J. Neitz, and G. H. Jacobs, *Science* **252,** 971 (1991).
[9] T. Chan, M. Lee, and T. P. Sakmar, *J. Biol. Chem.* **267,** 9478 (1992).
[10] A. B. Asenjo, J. Rim, and D. D. Oprian, *Neuron* **12,** 1131 (1994).
[11] S. W. Lin, Y. Imamoto, Y. Fukada, Y. Shichida, T. Yoshizawa, and R. A. Mathies, *Biochemistry* **33,** 2151 (1994).
[12] G. G. Kochendoerfer, Z. Wang, D. D. Oprian, and R. A. Mathies, *Biochemistry* **36,** 6577 (1997).
[13] S. W. Lin, G. G. Kochendoerfer, K. S. Carroll, D. Wang, R. A. Mathies, and T. P. Sakmar, *J. Biol. Chem.* **273,** 24583 (1998).

either longer or shorter wavelength depending on their location in the vicinity of the ionone ring or the Schiff base, respectively. The λ_{max} of the human green cone pigment or rhodopsin may be shifted ~30 nm to the red if the amino acids at position 164, 261, and 269 on TM helices 4 and 6 are replaced with hydroxyl-bearing residues. Mutagenesis of rhodopsin also indicates that dipolar effects from serines substituted in at positions 90, 292, 295, a cysteine at position 299, and a tyrosine at position 265 could account for close to 50% of the λ_{max} difference between the human blue cone pigment and rhodopsin (~420 versus 500 nm). Further blue shift down to ~438 nm seems to require a stronger interaction of the PSB group with the counterion environment, which is indirectly induced in part by the substitutions of Gly and Leu at the 117 and 122 positions, respectively.

Tryptophan Absorbance Change as Probe of Receptor Activation

Tryptophans are commonly found in the interior of proteins[14] and at membrane interfaces of many integral membrane proteins.[15,16] They are believed to contribute to the stabilization of protein structure. In G-protein-coupled receptors (GPCRs) of the rhodopsin-like receptor family, which include hormone-binding receptors such as serotonin receptors, muscarnic receptors, and chemokine receptors, several tryptophan molecules reside in the putative TM regions. A specific Trp residue is conserved in virtually all of these receptors in the aromatic motif (F or Y XXX W XX F or Y) present in TM helix 6.[17]

A tryptophan is an amenable target for absorption and fluorescence spectroscopy because its indole side chain serves as a good reporter group for the polarity or hydrophobicity of its surroundings. Early absorption studies on rhodopsin had detected changes in near-UV absorbance between ~240 and 320 nm on photoactivation, consistent with absorbance by Trp residues.[18,19] It was presumed that the environment of at least one Trp residue was becoming more polar in the activated metarhodopsin-II (meta-II) state relative to unphotolyzed rhodopsin. Rhodopsin contains a total of five Trp residues at positions 35, 126, 161, 175,

---

[14] G. E. Schulz and R. H. Schirmer, *in* "Principles of Protein Structure." Springer-Verlag, New York, 1979.
[15] G. von Heijne, *Annu. Rev. Biophys. Biomol. Struct.* **23,** 167 (1994).
[16] R. A. F. Reithmeier, *Curr. Opin. Struct. Biol.* **5,** 491 (1995).
[17] J. M. Baldwin, G. F. X. Schertler, and V. Unger, *J. Mol. Biol.* **272,** 144 (1997).
[18] C. N. Rafferty, C. G. Mullenberg, and H. Shichi, *Biochemistry* **19,** 2145 (1980).
[19] M. Chabre and J. Breton, *Vision Res.* **19,** 1005 (1979).

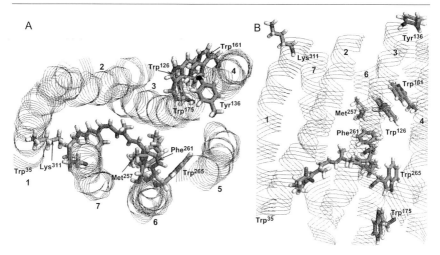

FIG. 2. Model of rhodopsin showing location of tryptophan residues studied using UV absorption spectroscopy. Sites that were studied by insertion Trp mutagenesis (Tyr-136, Met-257, Phe-261, and Lys-311) are also included. (A) View from the cytoplasmic surface. (B) Side view (cytoplasm is at the top). Trp-35, which lies at the extracellular interface of TM helix 1, is not shown.

and 265 (Fig. 2). Secondary structure and hydrophobicity analyses place Trp-126, Trp-161, and Trp-265 in the membrane bilayer region of the predicted TM helices 3, 4, and 6, respectively. To investigate if any of the UV-difference features could be ascribed to tryptophan(s), UV-difference absorption spectroscopy was performed on recombinant rhodopsins in which each of the five tryptophans had been mutated to either phenylalanine or tyrosine.[20]

Absorption Spectroscopy on Rhodopsin Mutants

Synthetic oligonucleotides encoding the desired codon alteration(s) were cloned into the rhodopsin gene located on the pMT vector. The mutant opsin was expressed in COS cells, regenerated with 11-*cis*-retinal, solubilized in dodecylmaltoside (DM), and purified by an immunoaffinity procedure as described in Han and Sakmar.[21] Solubilized mutant rhodopsin was eluted from the antibody-conjugated resin in sodium phosphate buffer (1 m*M*, pH 6, 0.1% DM) supplemented with the antigenic peptide. Absorption spectroscopy was performed on a Perkin-Elmer (Norwalk, CT) λ-19

[20] S. W. Lin and T. P. Sakmar, *Biochemistry* **35,** 11149 (1996).
[21] M. Han and T. P. Sakmar, *Methods Enzymol.* **315** [18] (1999) (this volume).

spectrophotometer. The $A_{280}/A_{\lambda max}$ ratio of purified samples typically ranged between ~1.8 and 2.5. An approximately 100-μl aliquot of sample (50 mM Tris-HCl, pH 6.8, 0.15 M NaCl, 0.1% DM) was transferred into a quartz cuvette maintained at ~4° to prolong the lifetime of the meta-II species. Absorbance spectra between 240 and 340 nm were recorded from samples in the dark and after irradiation with a tungsten–halogen lamp (150 W, ~15 sec). Each absorption spectrum was normalized to unity at the visible λ_{max} peak since the measured extinction coefficients of the mutant pigments are similar ($< \pm 15\%$ variation). A photobleaching difference spectrum was obtained by subtracting the "dark" sample spectrum from the "irradiated" sample spectrum (L–D).

UV-Difference Spectra

The substitution of Trp-126 and Trp-265 specifically resulted in the decrease of the difference bands at 294 and 302 nm (Fig. 3). Thus, the alterations of Trp-126 and Trp-265 environments on meta-II formation are responsible for the near-UV spectral features. Control experiments on rhodopsin solubilized in digitonin and on the E113A/A117E pigment establish these UV absorbance signals as specific to the receptor activation step.[20] A much smaller difference signal is observed in the W161F mutant, but the spectra of W35F and W175F appear essentially identical to the spectrum of native rhodopsin. Therefore, the environment of Trp-161 experiences weak perturbation, but those of Trp-35 and Trp-175 are indistinguishable in the two receptor states as probed by this method.

The differential absorbance spectrum of an individual Trp residue in rhodopsin and meta-II may be recovered by subtracting the difference spectrum of the Trp mutant from that of native rhodopsin (i.e., compute the double-difference spectrum) (Fig. 3). The presence of a negative difference band at ~294 nm indicates that Trp residues at 126 and 265 transfer from a nonpolar or hydrophobic environment to a more hydrophilic one in meta-II. Solvent-dependence studies on Trp absorption further suggest that difference extinction at ~302 nm can be interpreted in terms of a weakening hydrogen bonding of the indole group. Based on the band magnitudes, Trp-265 experiences a larger perturbation of its local environment.

Once the origin of the UV-difference bands in native rhodopsin has been determined, it is possible to probe for environmental change at another location by monitoring for absorbance signal from a Trp residue that has been inserted into the sequence by mutagenesis. The environments of Tyr-136 in TM helix 3, Met-257 and Phe-261 in TM helix 6, and Lys-311 in TM helix 7 were studied in this manner. The difference spectra of Y136W, M257W, F261W, and K311W mutants and the associated double-difference

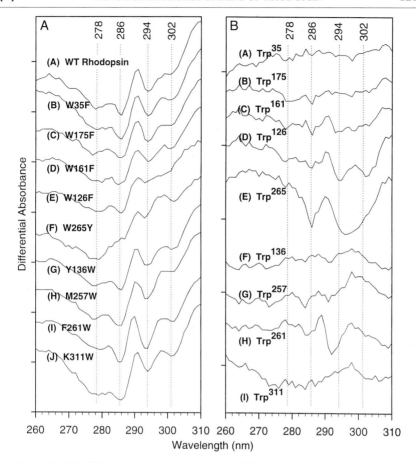

FIG. 3. **(A)** UV-difference spectra of rhodopsin and Trp mutant pigments. (A) Rhodopsin. (B) W35F. (C) W175F. (D) W161F. (E) W126F. (F) W265Y. (G) Y136W. (H) M257W. (I) F261W. (J) K311W. Each spectrum was generated by subtracting the spectrum of the mutant pigment in the dark from the spectrum recorded after irradiation with a tungsten–halogen lamp (L–D). **(B)** UV-difference spectra of native Trp residues in rhodopsin. (A) Trp-35. (B) Trp-175. (C) Trp-161. (D) Trp-126. (E) Trp-265. The double-difference spectrum was generated by subtracting the difference spectrum of the Trp mutant from that of rhodopsin. UV-difference spectra of Trp residue substituted into rhodopsin. (F) Trp-137. (G) Trp-257. (H) Trp-261. (I) Trp-311. Each of these double-difference spectra was generated by subtracting the difference spectrum of native rhodopsin from that of the Trp mutant. The spectrum can be attributed to an absorbance change from the inserted Trp residue if it is assumed that the extra Trp residue does not alter the native Trp bands.

spectra are presented in Fig. 3. The local environments at the 136, 257, and 311 positions are not altered significantly in meta-II, at least to the degree that is measurable by this method, since their difference spectra appear relatively flat.[22] This may be due to the fact that Tyr-136, Met-257, and Lys-311 are located closer to the cytoplasmic surface and solvent (Fig. 2) and, therefore, experience relatively smaller dielectric change locally. By contrast, the Phe-261 environment becomes less hydrophobic because only a negative difference band at 292 nm is present in the Trp-261 spectrum. Perturbation at this site is not unexpected since Phe-261 lies one helix turn from Trp-265 and also comprises the chromophore-binding pocket. Overall, the Trp absorption data indicate that regions of TM helices 3 and 6 nearer the chromophore pocket become more polar in the active conformation, and that the magnitude of the environmental change becomes smaller for sites closer to the cytoplasmic surface. Other techniques such as fluorescence,[23] resonance Raman,[12,24] and/or Fourier transform infrared (FTIR) spectroscopies may be utilized to characterize the Trp environments in more detail.

In summary, UV absorbance studies on Trp residues have been able to detect concerted activating structural changes in the TM regions of helices 3 and 6 that are coupled to changes shown by other biophysical methods[25-27] to occur near the cytoplasmic surface. UV-difference spectroscopy of Trp mutants is a simple method that can yield information that complements studies of surface residues using a combination of site-directed mutagenesis and exogenously added environmental probe molecules. Moreover, it can extend structural analysis to sites in the interior that may not be readily accessible and/or reactive to these probes.

Coupling of Chromophore and Protein Dynamics

The *cis–trans* photoisomerization reaction of the retinal chromophore serves as a binary switch to translate the event of photon absorption to a

---

[22] However, there are indications of small positive deflections in the spectra of Trp-257 and Trp-311, which represent an increase in hydrophobic character at these positions in meta-II. The environment of Val-250 has also been shown to be more hydrophobic in meta-II [T. D. Dunham and D. L. Farrens, *J. Biol. Chem.* **274,** 1683, (1999)].

[23] M. Nakagawa, S. Kikkawa, T. Iwasa, and M. Tsuda. *Biophys. J.* **72,** 2320 (1997).

[24] S. Hashimoto, H. Takeuchi, M. Nakagawa, and M. Tsuda, *FEBS Lett.* **398,** 239 (1996).

[25] D. L. Farrens, C. Altenbach, K. Yang, W. L. Hubbell, and H. G. Khorana, *Science* **274,** 768 (1996).

[26] S. P. Sheikh, T. A. Zvyaga, O. Lichtarge, T. P. Sakmar, and H. R. Bourne, *Nature* **383,** 347 (1996).

[27] T. D. Dunham and D. L. Farrens, *J. Biol. Chem.* **274,** 1683 (1999).

distinct molecular state that can be recognized by the opsin moiety. This recognition induces structural alterations such as changes in hydrogen bonding of Asp-83, Glu-122,[28,29] Trp-126, and Trp-265, Schiff base deprotonation, and neutralization of Glu-113.[30] It also triggers movements, at the minimum, of TM helices 3 and 6, which renders the cytoplasmic region active toward G_t. One prerequisite for the occurrence of TM helix 3 motion appears to be the breaking of the salt bridge between the PSB group of retinal and Glu-113.[31] A detailed map of surface residues that undergo alterations in their environments has been determined and evidence of gross motion of TM helices 3 and 6 has been obtained, but the specific details of how the isomerization process initiates these gross structural changes have yet to be understood. Specifically, "what elements or microdomain in the transmembrane region respond or sense the *trans* state of retinal in the pocket?" This question was specifically addressed by analyzing the biochemical properties of recombinant rhodopsins with substitutions on TM helices 3 and 6.

Methods

Mutagenesis and opsin purification, regeneration with retinal, and reconstitution are performed as described.[21] Mutant pigments are characterized using standard assays. UV/visible spectrophotometry is performed on detergent-purified pigment to measure the λ_{max} value and the $A_{280}/A_{\lambda_{max}}$ ratio. Reactivity of the mutant pigment to hydroxylamine (HA) is monitored by UV/visible spectroscopy. Briefly, a spectrum of purified pigment is recorded in darkness, and after addition of a freshly prepared aliquot of 1 M HA (pH 7, neutralized by NaOH addition) to a final concentration of 25 mM in the assay buffer [50 mM Tris-HCl, pH 6.9, 0.1 M NaCl, 0.1% (w/v) dodecylmaltoside], successive spectra are recorded over time. The half-time of decay of pigment in the presence of HA ($T_{1/2HA}$) is obtained from a single-exponential fit of the plot of λ_{max} absorbance over time. GTPγS filter-binding assay for G_t activation by pigment is performed as described.[21] Dark activity is assayed by measuring the level of G_t activation in the dark by mutant opsin in COS membranes incubated with 11-*cis*-retinal.

---

[28] K. Fahmy, F. Jäger, M. Beck, T. A. Zvyaga, T. P. Sakmar, and F. Siebert, *Proc. Natl. Acad. Sci. U.S.A.* **90,** 10206 (1993).
[29] P. Rath, L. L. J. DeCaluwé, P. H. M. Bovee-Geurts, W. J. DeGrip, and K. J. Rothschild, *Biochemistry* **32,** 10277 (1993).
[30] F. Jäger, K. Fahmy, T. P. Sakmar, and F. Siebert, *Biochemistry* **33,** 10878 (1994).
[31] E. A. Zhukovsky, P. R. Robinson, and D. D. Oprian, *Science* **251,** 558 (1991).

Interaction of Retinal with TM Helix 3 at the Glycine-121 Position

Glycine-121 is located on TM helix 3 and is strictly conserved in the alignments of virtually all visual opsins sequenced to date.[32] Inspection of models for the retinal–opsin complex suggested this location as a possible contact site between retinal and the opsin.[33] A set of amino acids (A, S, V, T, I, L, W) with increasing side-chain volume was substituted at this position to test this hypothesis.

The mutant pigments displayed size-dependent disruption of normal chromophore–protein interactions and binding-pocket structure (Table I). First, the amount of mutant pigment that could be purified is inversely correlated to the size of the amino acid at the 121 position. The yield of the G121A pigment was comparable to that of wild-type rhodopsin processed in parallel. On the other hand, no stable pigment of G121W could be regenerated and purified in detergent. Second, the λ_{max} values of the Gly-121 mutants displayed progressive blue shifts as the residue was changed from Ala to Trp, indicating a perturbation of the chromophore–protein interactions. For example, the λ_{max} of the G121L pigment is at 475 nm. A linear positive correlation exists between the λ_{max} value of the pigment expressed in wavenumbers and the side-chain volume.[33] Third, the stability of the mutant pigment in the presence of HA is lowered in the presence of a bulkier group at the 121 site. The Schiff base in the wild-type rhodopsin is unreactive to hydrolysis by HA in the dark ($T_{1/2HA} > 600$ min). However, Gly-121 mutants display $T_{1/2HA}$ values that generally decrease with larger 121 substitution. They range from 180 min for G121A to less than 1 min for G121I. Residues of similar size, but with tertiary β carbons such as G121V and G121I caused accelerated Schiff base hydrolysis, for instance, compared to G121L. These data indicate that the added volume at the 121 site interferes with proper binding of retinal in the opsin pocket and also suggest direct steric interaction between 11-*cis*-retinal and TM 3 at position 121 (Fig. 4). In addition to structural perturbations, the Gly-121 pigments display elevated levels of dark activity of G_t activation (Fig. 5). The dark activities of the mutants range from 2.3% in G121A up to 30% in G121W.[33] The level of dark activity also shows size dependence that is similar to that observed for other phenotypes and is linearly correlated to the size at the 121 position.[34] This correspondence implies that the functional activity of the receptor may be related to direct steric interaction between the retinal

---

[32] Alignments of visual opsin sequences can be found at www.gpcr.org/7tm/seq [Horn *et al.*, *Nucl. Acids Res.* **26**, 277 (1998)].

[33] M. Han, S. W. Lin, S. O. Smith, and T. P. Sakmar, *J. Biol. Chem.* **271**, 32330 (1996).

[34] M. Han, M. Groesbeek, T. P. Sakmar, and S. O. Smith, *Proc. Natl. Acad. Sci. U.S.A.* **94**, 13442 (1997).

TABLE I
BIOCHEMICAL PROPERTIES OF MUTANT PIGMENTS AND ANALOG RHODOPSINS

Mutant	Approximate yield[a]	λ_{max}[b] (nm)	Hydroxylamine reactivity ($T_{1/2HA}$)[c] (min)	Dark activity[d] (%)
Wild-type	100%	500	>600	0.8 ± 0.3
G121A	H	498 ± 0.7	175 ± 53	2.3 ± 1.3
G121S	H	497 ± 0.4	43 ± 2	5.0 ± 2.6
G121T	M	483 ± 0.6	9.2 ± 1.7	5.6 ± 1.8
G121V	H	477 ± 0.5	2.6 ± 0.8	7.8 ± 3.7
G121I	L	475 ± 2.2	<0.5	21.1 ± 4.8
G121L	L	475 ± 0.3	6.4 ± 0.4	14.6 ± 2.5
G121W	N	(461)[e]	n/a[f]	30.0 ± 6.2
F261A	H	500 ± 0.4	>600	1.7 ± 0.1
F261V	H	502 ± 0.7	>600	2.5 ± 0.4
F261W	M	501 ± 0.6	>600	0.5 ± 0.1
G121L/F261A	H	482 ± 0.3	11.1 ± 1.3	4.4 ± 1.1
G121W/F261A	M	467	0.43 ± 0.01	18 ± 2
G121L/F261V	M	477 ± 0.9	1.5 ± 0.4	27.0 ± 8.2
G121L/F261W	vL	478 ± 0.7	0.43 ± 0.01	12.4 ± 4.1
I219A	H	486 ± 0.8	>600	0.6 ± 0.4
M257A	H	501 ± 0.2	>600	3 ± 1
V258A	H	500 ± 0.3	>600	1 ± 1
W265Y	H	485 ± 0.9	>600	2 ± 2
G121L/I219A	N	n/a	n/a	34 ± 8
G121L/M257A	L	473 ± 1.5	n/a[g]	79 ± 5
G121L/V258A	L	473	12.7 ± 8.1	14 ± 3
G121L/W265Y	vL	477 ± 1.5	n/a[g]	32 ± 3
Wild-type + 11-cis-9-ethylretinal	N	n/a	n/a	5.4 ± 1.2[h]
Wild-type + 11-cis-9-propylretinal	N	n/a	n/a	11.2 ± 5.3[h]

[a] Approximate yield of dodecylmaltoside reconstituted pigment relative to the yield of wild-type rhodopsin purified in parallel. H, 75–100%; M, 50–75%; L, 25–50%; vL, <25%; N, no detectable amount.
[b] Visible λ_{max} of dodecylmaltoside reconstituted pigment.
[c] Decay of λ_{max} in the presence of 25 mM hydroxylamine (HA) at 20°. A half-time of decay was calculated from the best fit to a single-exponential function.
[d] G_t activation in the dark by mutant opsin in COS cell membranes incubated with 11-cis-retinal. Dark activity is reported as a percentage of G_t activation by the same opsin + 11-cis-retinal in COS cell membranes under continuous illumination (light activity) assayed in parallel.
[e] No stable G121W pigment could be purified. The λ_{max} is an estimate based on an extrapolation from the λ_{max} of G121W/F261A pigment [Han et al., J. Biol. Chem. **271**, 32337 (1996)].
[f] n/a, not applicable.
[g] Unstable in the absence of HA.
[h] G_t activation in the dark by wild-type opsin in COS cell membranes incubated with retinal analog. Dark activity is reported as a percentage of G_t activation by the same opsin + retinal analog in COS cell membranes under continuous illumination (light activity).

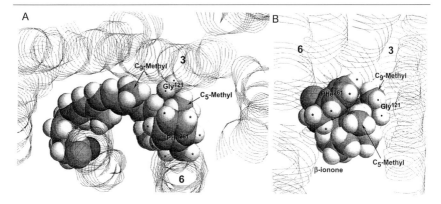

Fig. 4. Space-filling model of rhodopsin depicting the packing of 11-*cis*-retinal between TM helices 3 and 6. The C-9 methyl group and the β-ionone ring contact the opsin at Gly-121 and Phe-261. The methyl groups attached to C-5 in the β-ionone ring and to C-9 in the polyene are labeled. For clarity, hydrogen atoms of Phe-261 and Gly-121 are labeled with the symbols * and •, respectively. (A) View from the cytoplasmic surface. (B) Edge-on view from the ionone ring looking in toward the Schiff base. The β-ionone ring is sandwiched between Gly-121 and Phe-261. The cytoplasm is at the top.

and TM helix 3 at position 121. No phenotypes analogous to those seen in Gly-121 mutants were observed in control studies involving Gly-114 and Ala-117.[33]

Interaction between Glycine-121 and C-9 Methyl Group of Retinal

Mapping the precise sites of receptor–ligand interaction can be achieved by studying the binding and efficacy of ligand analogs onto the receptor or receptor mutants. This basic strategy has been applied to rhodopsin to determine the portion of retinal that is interacting with Gly-121. The 9-methyl group of retinal has been implicated in many studies to be necessary for normal activation of receptor.[35,36] A model of retinal–opsin complex based on NMR constraints[37] and an opsin model by Baldwin[38] suggest the 9-methyl group of retinal as the likely

[35] U. M. Ganter, E. D. Schmid, D. Perez-Sala, R. R. Rando, and F. Siebert, *Biochemistry* **28,** 5954 (1989).
[36] Morrison D. F., T. D. Ting, V. Vallury, Y. K. Ho, R. K. Crouch, D. W. Corson, N. J. Mangini, and D. R. Pepperberg, *J. Biol. Chem.* **270,** 6718 (1995).
[37] M. Han and S. O. Smith *Biochemistry* **34,** 1425 (1995).
[38] J. M. Baldwin, *EMBO J.* **12,** 1693 (1993).

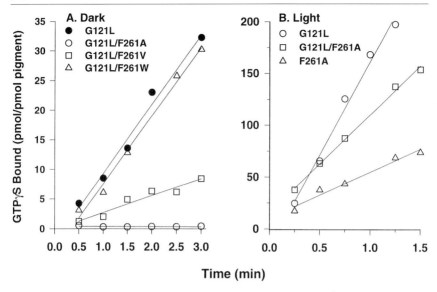

FIG. 5. Mutual functional rescues of G121L and F261A through steric compensation. (A) The dark activity of G121L and double mutants G121L/F261A, V, W measured in 0.01% DM. F261A mutation completely suppresses the dark activity of G121L mutant. A 30-nM pigment is used for measuring dark activity. (B) The light activity of single-site mutants G121L, F261A, and the double-site mutant G121L/F261A. The G121L mutation can partially rectify the lowered light activation of G_t by the F261A pigment. A 3-nM pigment is used in each assay.

site of interaction with Gly-121. This hypothesis predicts that volume increase at the C-9 position of retinal should augment dark activity, mirroring the effect of bulky substitutions at the 121 position. To test this, the dark activities of 24 opsin–retinal complexes obtained by combining eight opsins (wild-type and G121A, S, T, V, L, I, W) with three C-9 retinal analogs (11-*cis*-retinal, 11-*cis*-9-ethylretinal, and 11-*cis*-9-propylretinal) were measured. Wild-type opsin indeed becomes active in the dark in the presence of 11-*cis*-9-ethyl- or -propylretinals, analogous to activation of Gly-121 mutants by native 11-*cis*-retinal (Table I). The combination of mutant opsins and retinal analogs (24 complexes) yields a general trend of synergistic activation of the receptor by increasing the molecular volume at both position 121 and retinal C-9.[34] These data are consistent with the notion that receptor activation in the absence of light is caused by exaggerated steric interaction between the side chain at position 121 and the C-9 moiety of retinal. They also suggest that the 9-methyl group of retinal is involved in functional interaction with TM helix 3 at Gly-121 in the ground state of rhodopsin.

Interaction between TM Helices 3 and 6 at Glycine-121 and Phenylalanine-261 Mediated by Retinal

Short of crystal structure analysis, complementary mutagenesis is one of the most effective ways to establish specific interactions between two or more sites in a protein. The basic strategy is to find a second site whose mutation can reverse or complement the gain or loss of function caused by a substitution at the first site. This strategy was applied to map other opsin sites that potentially interacted with Gly-121. In light of evidence for concerted TM changes in TM helix 3 and 6, [20]Met-257, Val-258, Phe-261, and Trp-265 were mutated since they were predicted in structural models[38,39] to lie on the face of TM helix 6 opposing TM helix 3. Additionally, Ile-219 on TM helix 5, which could be in proximity to a Leu residue at position 121 according to a molecular model of rhodopsin,[11] was considered.

The majority of structural and functional defects observed in the G121L and G121W mutants were specifically rescued by the substitution of Phe-261 by a smaller residue, such as Ala or Val (Table I). First, structural integrity in the G121L and G121W opsins was improved by the F261A mutation. G121L/F261A and G121W/F261A mutants could be successfully regenerated into pigments and purified at a much higher yield than the respective single-site Gly-121 mutants. This effect was not observed with a F261W mutation. Second, the magnitude of the λ_{max} blue shift in G121L pigment (25 nm) is reduced to ~18 nm by F261A second-site mutation. Third, the F261A mutation can suppress the dark activity caused by the G121L mutation (Fig. 5). The reversal of abnormalities resulting from Gly-121 mutations follows a trend that is also size dependent. The partial rescue elicited by the Val substitution is weaker compared to that by Ala residue, and no rescue is observed with a F261W substitution although this single-site mutant is spectrally and functionally similar to the wild-type. These results imply that a reduction of steric volume at the 261 site can compensate for the increase at the 121 position. Conversely, bulkier substitution at the 121 position is able to correct the defect caused by a decrease at the 261 position. For instance, the G121L/F261A substitution rescues the lowered light-dependent G_t activation of the F261A mutant (Fig. 5). Second-site

---

[39] G. F. X. Schertler, C. Villa, and R. Henderson, *Nature* **362,** 770 (1993).

substitutions at positions 219, 257, 258, or 265 were ineffective in reversing the G121L phenotypes (Table I). The steric complementarity between Gly-121 and Phe-261 is likely to be mediated by the retinal chromophore, because the Phe-261 substitutions partially reverse properties of opsins containing the retinal chromophore. These results suggest that 11-*cis*-retinal binds between Gly-121 on TM helix 3 and Phe-261 on TM helix 6 (Fig. 4).

The Gly^{121}-retinal-Phe^{261} moiety forms a microdomain in the interior of rhodopsin whose structure is important for (1) correct ligand binding, (2) stabilizing the inactive receptor state (i.e., suppressing dark activity or noise), and (3) efficient receptor activation in response to photoisomerization. Because an aromatic residue at position 261 is highly conserved among GPCRs (82% conservation),[17] similar TM helix 3–ligand–TM helix 6 interaction may be of general importance in receptor activation.

Hydrogen-Bonding Network of Water Molecules inside Opsin

Changes in the interactions of internal water molecules are correlated with conformational transition of rhodopsin. Much of this evidence has been derived from alteration in the IR absorbance of O–H stretching modes between ~3500 and 3600 cm^{-1} between rhodopsin and its primary photoproduct, bathorhodopsin, detected by FTIR spectroscopy. At least three water molecules in distinct environments have been detected.[6] Based on the analysis of E113Q mutant, it was concluded that one water molecule resides in the binding pocket and is hydrogen bonded to the Glu-113 counterion and the PSB group.[40] The O–H stretching band of the second water molecule is perturbed by mutation of Gly-120 or Asp-83, and to a lesser degree by mutation at Gly-121, but none at all by mutation at Glu-122.[41] This suggests that this water bridges the two sites on TM helices 2 and 3 by hydrogen bonding to carboxylic carbonyl of Asp-83 and the peptide carbonyl of Gly-120. The structure of these water molecules in the active receptor conformation, that is, meta-II, has yet to be determined, and the functional role of internal water interactions in the receptor activation process requires further study.

---

[40] T. Nagata, A. Terakita, H. Kandori, D. Kojima, Y. Shichida, and A. Maeda, *Biochemistry* **36**, 6164 (1997).
[41] T. Nagata, A. Terakita, H. Kandori, Y. Shichida, and A. Maeda, *Biochemistry* **37**, 17216 (1998).

Acknowledgments

We thank Juergen Isele for assistance with absorbance measurements, Ethan Marin for constructing the gene for the K311W mutant, and Steven. O. Smith for helpful discussions. Support for this work was provided by the Howard Hughes Medical Institute, the Allene Reuss Memorial Trust, and the Charles H. Revson Foundation. M.H. is currently a Helen Hay Whitney Foundation fellow.

[10] Mapping Tertiary Contacts between Amino Acid Residues within Rhodopsin

By MARY STRUTHERS and DANIEL D. OPRIAN

Introduction

Cysteine scanning mutagenesis and oxidative cross-linking have been successfully used to map tertiary interactions in transmembrane helices of the bacterial chemotactic receptors, a class of homodimeric membrane-bound receptors.[1-4] We have adapted this method to the study of the monomeric G-protein-coupled receptor (GPCR), rhodopsin.[5] The method is based on the principle that disulfide bond formation occurs when the introduced cysteines are proximal to each other in the three-dimensional structure of the protein. Thus, a tertiary contact in the native structure between the original residues at these positions is suggested by disulfide bond formation in the corresponding double cysteine mutants. This method can be used to verify and refine current models of GPCRs by providing information regarding the extent of helical structure within a proposed transmembrane region, as well as the proximity and relative tilt of the helical regions of GPCRs.

Split Receptors

Key to the practical application of cysteine scanning mutagenesis and oxidative cross-linking to the monomeric protein rhodopsin is the ability

---

[1] J. J. Falke and D. E. Koshland, Jr., *Science* **237**, 1596 (1987).
[2] G. F. Lee, G. G. Burrows, M. R. Lebert, D. P. Dutton, and G. L. Hazelbauer, *J. Biol. Chem.* **269**, 29920 (1994).
[3] A. A. Pakula and M. I. Simon, *Proc. Natl. Acad. Sci. U.S.A.* **89**, 4144 (1992).
[4] S. A. Chervitz, C. M. Lin, and J. J. Falke, *Biochemistry* **34**, 9722 (1995).
[5] H. Yu, M. Kono, T. D. McKee, and D. D. Oprian, *Biochemistry* **34**, 14963 (1995).

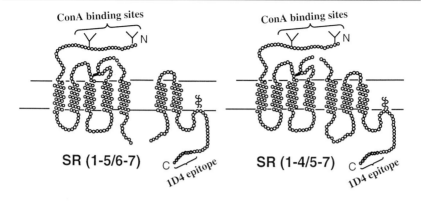
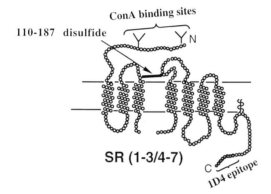

FIG. 1. Schematic representation of the structure of split rhodopsins. The 1D4 epitope is represented by shaded circles and the carbohydrates at positions 2 and 15 are represented by branched lines. The palmitoyl groups at positions 322 and 323 are shown as crooked lines. The 110–187 disulfide is represented by a boldface line between these residues.

to express rhodopsin as two noncovalently associated fragments.[5,6] The split receptors we have developed for this purpose are illustrated in Fig. 1. Although this method may be used on full-length receptor mutants, detection of the disulfide would require labor-intensive techniques such as protease digestion and fragment analysis, which do not lend themselves to the investigation of the numerous double cysteine mutants required to map tertiary interactions. The production of a split receptor consisting of noncovalently associated N- and C-terminal rhodopsin fragments (each of which contains a single cysteine mutation) allows for the utilization of a

[6] K. D. Ridge, S. S. Lee, and L. L. Yao, *Proc. Natl. Acad. Sci. U.S.A.* **92,** 3204 (1995).

facile gel shift assay for cross-linking. The formation of a disulfide between the introduced cysteines creates a covalent bond between the two fragments and can therefore be readily detected by the decrease in mobility of the cross-linked double mutant in nonreducing sodium dodecyl sulfate–polyacrylamide gel electrophoresis (SDS–PAGE) as shown schematically in Fig. 2. This simple assay is amenable to the rapid analysis of numerous samples simultaneously. Concanavalin A (ConA), a lectin that binds to the mannose-containing carbohydrates at positions 2 and 15 of rhodopsin, is used to detect the N-terminal fragments of split rhodopsins in Western blot analyses. Because the split rhodopsin mutants are purified by immunoaffinity chromatography using immobilized antibodies to the C-terminal 1D4 epitope, probing the Western blot with ConA ensures that only split receptors consisting of fully associated N- and C-terminal fragments are detected.

An additional advantage of using split receptors for mapping tertiary interactions in rhodopsin is the ease with which a large number of double mutants can be constructed from combinations of N- and C-terminal fragments, each containing single cysteine mutations. In a cysteine scanning mutagenesis experiment, single cysteine mutations within a proposed transmembrane region within one fragment (a cysteine scan) are paired individually with cysteine substitutions in a separate transmembrane (TM) region in the complementary fragment. Using split receptors, the total number of mutant plasmids that need to be constructed is greatly reduced. For example, a cysteine scan of 10 residues in TM 3 paired with each of 10 cysteine substitutions at target positions in TM 5 would only require the construction of 10 mutants each in the N- and C-terminal gene fragments. Coexpression of various combinations of these 20 single mutant gene fragments would produce the desired 100 double mutant proteins. In contrast, the same experiment performed in a full-length rhodopsin construct would require the construction of all 100 double cysteine mutant genes.

Disulfide Bond Formation

Two methods have been used to induce disulfide bond formation in our studies of rhodopsin. The first is based on the cysteine scanning and oxidative cross-linking studies performed on the bacterial chemotactic receptors[1] and utilizes $Cu(phen)_3^{2+}$ to catalyze the oxidation of cysteines to cystine in proteins.[7] However, $Cu(phen)_3^{2+}$ is a particularly strong oxidant and has the potential to form disulfides in nonnative conformations arising from minor local dynamic movements of the protein. We therefore use a second milder, and more specific, oxidation strategy. This second method involves

---

[7] K. Kobashi, *Biochim. Biophys. Acta* **158**, 239 (1968).

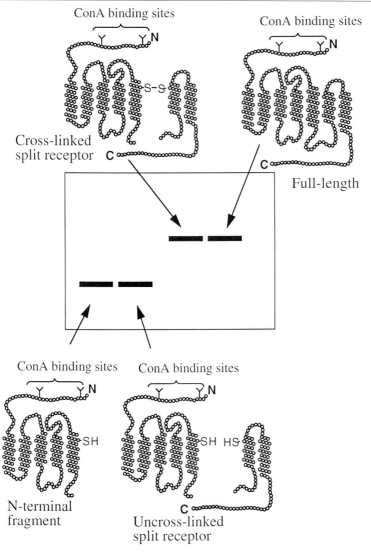

FIG. 2. Schematic illustration of the results of ConA blot analysis of cross-linked and unlinked mutant split receptors using the SR(1-5/6-7) receptor as an example.

"spontaneous" oxidation of double cysteine split rhodopsin mutants by ambient oxygen at slightly elevated pH (pH 8.0). Double cysteine split receptor mutants are purified in the predominantly reduced state (pH 6.0) and the pH is raised to promote disulfide bond formation. This mild oxidation protocol places a more stringent requirement on residue proximity for disulfide bond formation than the use of $Cu(phen)_3^{2+}$.

We have used these two oxidation methods together to scan helical regions of rhodopsin, including TM 3 and 5 on the cytoplasmic side and TM 5 and 6 on the extracellular side.[8,9] An initial cysteine scan using $Cu(phen)_3^{2+}$ is followed by repeating the scan using the more mild spontaneous cross-linking method. Cysteine mutants that are observed to completely cross-link with $Cu(phen)_3^{2+}$ but not with ambient oxygen at high pH are likely to be more distal than those that cross-link under both conditions. The two methods therefore complement each other and allow the identification of tertiary interactions between transmembrane helices.

Mapping tertiary interactions in rhodopsin using cysteine scanning mutagenesis and oxidative cross-linking involves seven general steps, the procedures for which are described here:

1. *Construction of the cysteine mutants in an appropriate mammalian expression vector containing the gene for the rhodopsin fragment.* We use a pMT3-based vector[10] and introduce cysteines into our constructs by standard cassette mutagenesis,[5] taking advantage of the multiple restriction sites within a synthetic opsin gene.[11] However, any standard DNA methodology can be used for the construction of the desired mutants.

2. *Small-scale (one 100-mm plate of COS-1 cells) transfection, expression, and purification of double cysteine mutants for the desired scan.* This scale is used for samples analyzed only by SDS–PAGE and Western blot.

3. *Disulfide cross-linking of the mutant split receptors using both $Cu(phen)_3^{2+}$ and air oxidation in turn.* This step may be repeated with varying oxidation times and incubation temperatures to obtain information concerning the relative rates of disulfide bond formation within a cysteine scanning experiment.

4. *Detection of disulfide bond formation between fragments by SDS–PAGE and ConA Western blots of cross-linking reactions.*

5. *Verification that the cross-links identified in step 4 are due to specific disulfide bond formation between the engineered cysteines.* For this purpose, reversibility of cross-linking in double cysteine mutants is demonstrated by treating the cross-linked split mutant with dithiothreitol (DTT). Additionally, the absence of similar cross-linking reactions in the corresponding single cysteine mutants is verified.

[8] H. Yu and D. D. Oprian, *Biochemistry* **38,** 12033 (1999).
[9] M. Struthers, H. Yu, M. Kono, and D. D. Oprian, *Biochemistry* **38,** 6597 (1999).
[10] R. R. Franke, T. P. Sakmar, D. D. Oprian, and H. G. Khorana, *J. Biol. Chem.* **263,** 2119 (1988).

6. *Large-scale (ten 100-mm COS-1 plates) transfection, reconstitution, and purification of double mutants that exhibit significant cross-linking for functional analysis.*
7. *Assessment of the spectral properties and transducin activation by both the reduced and cross-linked forms of the double cysteine mutants identified in step 4.*

Procedures

Expression, Reconstitution, and Purification of Split Receptors

We have developed three split rhodopsin constructs for use in these studies,[5] which are shown in Fig. 1. Each of the six fragments has been constructed from a synthetic rhodopsin gene[5,11] and cloned in a pMT3-based expression vector. Each construct includes a Kozak consensus sequence immediately 5' of the initiator methionine. The wild-type fragments consist of the following sequences: SR(1-3), Met-1 to Pro-142; SR(4-5), Met-143 to Ala-348; SR(1-4) Met-1 to Pro-194; SR(5-7) His-195 to Ala-358 preceded by an initiator methionine; SR(1-5), Met-1 to Ser-240; SR(6-7) Ala-241 to Ala-358 preceded by an initiator methionine.

Production of the noncovalently associated mutant split rhodopsin is accomplished by coexpression of both the N-terminal and C-terminal gene fragments in COS-1 cells.[5] Transfection of DNA into COS-1 cells is accomplished by the DEAE-dextran method.[12] Equal amounts of plasmid DNA for each fragment are used (2 μg of each plasmid for each 100-mm plate of cells). The procedure for one 100-mm plate of COS cells is as follows: COS cells are grown to 90% confluency using Dulbecco's modified Eagle's medium containing D-glucose (4.5 g/liter), streptomycin (100 mg/ml), penicillin (100 mg/ml), 2 mM L-glutamine, and 10% heat-inactivated calf serum (medium A). After washing the cells twice with 5 ml of medium B (medium A without calf serum), the COS cells are incubated for 6 hr at 37° in 5 ml of transfection cocktail consisting of 2 μg of each plasmid, 3.5 ml of medium B, 0.5 ml of 1 M Tris buffer (pH 7.0), 0.5 ml of DEAE-dextran solution (2.5 mg/ml in medium B) and 0.5 ml of 1 mM chloroquine. The cells are then washed once with 5 ml of medium B and treated for 3 min at room temperature with 2 ml of 10% (v/v) dimethyl sulfoxide (DMSO) in 21 mM HEPES (pH 7.0), 137 mM NaCl, 5 mM KCl, 0.7 mM Na$_2$HPO$_4$, and 5.5

---

[11] L. Ferretti, S. S. Karnik, H. G. Khorana, M. Nassal, and D. D. Oprian, *Proc. Natl. Acad. Sci. U.S.A.* **83,** 599 (1986).

[12] D. D. Oprian, *Meth. Neurosci.* **15,** 301 (1993).

mM dextrose. Aspiration of the DMSO solution is followed immediately by an overlay of 5 ml of 0.1 mM chloroquine in medium A and incubation for 2 hr at 37°. Finally the plate is washed twice in 5 ml of medium B followed by incubation with 10 ml of medium A at 37°. The medium is changed 24 hr posttransfection and the cells are harvested 60–72 hr posttransfection, washed with PBS, and either used immediately or frozen (−80°).

Split receptors are reconstituted by suspension of the harvested cells expressing the desired mutant opsins in PBS and incubation with 11-*cis*-retinal (5 μM, 1.0 ml/100-mm plate) for 2 hr at 4° in the dark. All subsequent purification steps and manipulation of the purified receptors prior to denaturation are performed under dim red light. Solubilization in 1.0% (w/v) *n*-dodecyl-β-D-maltoside (DM, Calbiochem, La Jolla, CA) is followed by immunoaffinity chromotography using the anti-rhodopsin antibody 1D4,[13] which binds to the C-terminal 8 residues of rhodopsin. The antibody is coupled to Sepharose 4B resin using the method of Cuatrecasas.[14] Purifications are done on two scales, and the procedures used for purification differ slightly to accommodate the difference in scale. For cysteine scanning experiments in which numerous split receptor mutants are reconstituted and analyzed using only SDS–PAGE and Western blotting, small-scale reconstitutions of single 100-mm plates of COS-1 cells expressing mutant split receptors are performed. In this case the purification procedure is performed in microcentrifuge tubes. After solubilization with 1% DM, the postnuclear supernate (16,000g, 3 min) containing the mutant split receptors is incubated with 1D4–Sepharose resin for 2 hr at 4°. Following this incubation, the resin is isolated by centrifugation in a microcentrifuge (16,000g, 1 min) and aspiration of the supernate above the resin. The resin is washed eight times with 100–200 μl of 0.1% DM in phosphate-buffered saline (PBS) by centrifugation of the sample and aspiration of the wash solution. The protein is eluted by incubation of the washed resin with 40 μl of peptide I (a synthetic peptide corresponding to the C-terminal 18 residues of rhodopsin) at a concentration of 0.18 mg/ml in 0.1% (w/v) DM in PBS at room temperature for 30 min. Following centrifugation, the purified split receptor in the supernatant is removed by pipetting taking care to avoid the resin beads. One 100-mm plate of COS-1 cells typically yields enough mutant split receptor (1 μg) to perform one cross-linking experiment and an appropriate control (i.e., two lanes on an SDS–PAGE gel as described later).

Purification of mutant split receptors is performed on a larger scale (5–10 plates of transfected COS cells) when functional analysis is desired.

[13] R. S. Molday and D. MacKenzie, *Biochemistry* **22**, 653 (1983).
[14] P. Cuatrecasas, *J. Biol. Chem.* **245**, 3059 (1970).

Cells are harvested and the purification procedure performed in 15-ml disposable clinical centrifuge tubes. After the harvested cells have been incubated with 11-cis-retinal and solubilized with 1% DM the postnuclear supernate (1380g, 3 min) containing the mutant split receptors is incubated with 1D4–Sepharose resin for 2 hr at 4°. After this incubation, the resin containing the bound receptor is transferred to a spin column consisting of a 1-ml syringe barrel that has been plugged with glass wool and placed in a 15-ml disposable centrifuge tube. All remaining solution is removed from the resin by centrifugation of the spin column in a clinical table-top centrifuge (1380g for 30 sec). The resin in the spin column is then washed eight times with 1 ml 0.1% (w/v) DM (in PBS; or 10-fold the total volume of resin). The wash solution is removed by centrifugation (1380g for 30 sec). Elution of the receptor is accomplished by incubation of the resin in the spin column (after blocking the end of the syringe with parafilm) with peptide I (2- to 4-fold the volume of resin used) for 30 min at room temperature. The supernatant containing the split receptor is isolated by centrifugation. Yields of the split receptors vary but are typically 1 μg for SR(1-3/4-7), 3 μg for SR(1-4/5-7), and 1.5 μg for SR(1-5/6-7) per 100-mm plate of COS-1 cells (as quantitated by the absorbance at 500 nm).[5] The pH of the PBS solutions used in reconstitution depends on the method of cross-linking used (see next section).

Disulfide Cross-Linking

The cross-linking of purified mutant split receptors in 0.1% (w/v) DM in PBS is initiated by the addition of Cu(phen)$_3^{2+}$ to a final concentration of 3 mM from a stock solution (15 mM CuSO$_4$, 45 mM phenanthroline in 10 mM phosphate buffer, pH 7.0, and 30% glycerol) as a modification of the procedure described by Lee et al.[2] When Cu(phen)$_3^{2+}$ is used as the oxidant, mutant split receptors are reconstituted and purified at pH 7.0. Reactions are incubated in the dark at either 25° or 37° for 30 min to 1 hr. The cross-linking reactions are quenched by addition of SDS loading buffer containing final concentrations of 12.5 mM N-ethylmaleimide to block free sulfhydryls, 12.5 mM EDTA to chelate Cu^{2+}, 60 mM Tris (pH 6.8), 2% (w/v) SDS, 6% (w/v) sucrose, and 0.005% (w/v) bromphenol blue.

The milder "spontaneous" cross-linking strategy involves a jump in pH from 6.0 to 8.0. Mutant split receptors are incubated with 11-cis-retinal, solubilized, and bound to 1D4–Sepharose resin in PBS (10 mM sodium phosphate, 150 mM NaCl) at pH 6.0. This is followed by washing of the resin and elution of the mutant receptor in a pH 6.0 buffer containing peptide I, 2 mM sodium phosphate, 150 mM NaCl, and 0.1% (w/v) DM. The pH is raised to 8.0 by the addition of 100 mM dibasic sodium phosphate,

150 NaCl, and the split receptors are incubated in the dark at 25°. Disulfide bond formation is quenched by the addition of SDS load buffer containing NEM [final concentrations of 12.5 mM N-ethylmaleimide, 60 mM Tris (pH 6.8), 2% (w/v) SDS, 6% (w/v) sucrose, and 0.005% (w/v) bromphenol blue].

In both procedures, cross-linking reactions are initiated immediately after elution from the 1D4–Sepharose resin. Because the rates of disulfide bond formation depend on the structural context of the target cysteines, the times required for complete cross-linking vary. Typically, reactions are allowed to proceed for 30 min at 25° for $Cu(phen)_3^{2+}$-induced cross-linking and for 2 hr at 25° for cross-linking using ambient oxygen at pH 8.0. Quenched cross-linking reactions are either analyzed immediately by SDS–PAGE and Western blot or stored overnight at 4° and analyzed the following day.

Detection of Cross-Linking by Western Blot Using Concanavalin A

Quenched cross-linking reactions are analyzed by SDS–PAGE using a 12% polyacrylamide gel. Proteins are transferred to a nitrocellulose membrane (Schleicher & Schuell, Keene, NH; 0.2-μM pore size) using a semidry transfer apparatus (Bio-Rad, Richmond, CA). The blot is then incubated at 37° for 1 hr in blocking buffer: 5% (w/v) bovine serum albumin (BSA, Sigma, St. Louis, MO) in ConA buffer (50 mM HEPES buffer, pH 7.0, 100 mM NaCl, 1 mM $CaCl_2$, 1 mM $MnCl_2$) followed by overnight incubation with ConA–biotin [Sigma, 0.001% (w/v) ConA–biotin in blocking buffer]. After washing three times in 0.05% (w/v) Nonidet P-40 (NP-40, Sigma) in ConA buffer, the blot is incubated for 1 hr at room temperature in a 1 : 2500 dilution of strepavadin-horseradish peroxidase conjugate (Amersham, Arlington Heights, IL) in 0.05% (w/v) NP-40 in ConA buffer. After washing three times with 0.05% (w/v) NP-40 in ConA buffer and once with 10 mM Tris, pH 7.4, 138 mM NaCl, proteins in the blot are visualized using the enhanced chemiluminescence (ECL) detection kit (Amersham) following manufacturer protocols. A cross-link between introduced cysteines is indicated by a decrease in the electrophoretic mobility of the N-terminal fragment of the split mutant such that it migrates with the same apparent molecular mass as full-length, wild-type rhodopsin (Fig. 2). Confirmation that a cross-link observed in this manner is due to a disulfide is obtained by observing the collapse of this higher molecular weight species to a band with the mobility expected for the N-terminal fragment alone on addition of 2-mercaptoethanol (BME) or dithiothreitol (DTT) to the load buffer prior to electrophoresis. In this case, a vacant lane of the gel must separate the BME or DTT treated sample from other samples because BME or DTT may leak into the immediately adjacent lanes and result in partial reduction of any samples in those lanes.

Assessment of Spectral Properties of Double Cysteine Mutants

Dark-state absorbance spectra of cross-linked split receptor mutants identified in the cysteine scanning experiments described earlier are recorded from larger scale reconstitutions of these mutants. Spectra are recorded with a 100-μl sample using a microcuvette (Helma, Germany) and a spectrophotometer modified for darkroom use (Hitachi, Japan). Spectra of cross-linked mutants obtained by the pH jump strategy can be directly recorded after the cross-linking reaction and compared (after correcting for dilution) with the spectrum of the original reduced sample (taken immediately after purification). For split receptors treated with $Cu(phen)_3^{2+}$, which has a significant UV/visible absorbance spectrum, the excess reagent is removed by loading the cross-linking reaction on a spin column containing Sephadex G-50 resin (bed volume is 10 times the cross-linking reaction volume) and centrifuging at approximately 1500g in a swinging bucket clinical centrifuge for 3 min. The absorbance spectrum of the eluted sample is then recorded. The spin column is prepared by hydrating the resin in 0.1% DM in PBS at room temperature for 1 hr followed by transfer to a polystyrene column (Pierce, Rockford, IL) in a disposable centrifuge tube. The resin is washed three times with 0.1% (w/v) DM in PBS ($\sim$1500g, 3 min). Removal of $Cu(phen)_3^{2+}$ on spin columns typically results in a slight dilution of the protein sample.

Assessment of Transducin Activation by Double Cysteine Mutants

Transducin is purified from bovine retinae as described elsewhere[15,16] with some modification. The purified protein is dialyzed against 10 mM Tris buffer (pH 7.5) containing 50% (v/v) glycerol, and 1 mM MgCl$_2$. The ability of purified split receptors to catalyze the exchange of GDP for radiolabeled [^{35}S]-GTPγS in transducin is examined using a filter binding assay.[17] A solution of [^{35}S]GTPγS solution is prepared by dilution of 2 μl of radioactive [^{35}S]GTPγS (NEN, 0.25 mCi/20 μl) into 80 μl of 80 μM nonlabeled GTPγS (Boehringer Mannheim, Germany). The assay reaction mixture consists of 10 μl of 10$\times$ assay buffer (100 mM Tris, pH 7.5, 50 mM MgCl$_2$, 1 mM EDTA, 1 M NaCl), 10 μl of 50 nM rhodopsin, 4 μl of 80 μM [^{35}S]GTPγS, 10 μl of 25 μM transducin, and 68 μl of distilled water for a total reaction of 100 μl. The reaction is performed in the dark and initiated by addition of the [^{35}S]GTPγS. A series of 10-μl aliquots is spotted on prewetted nitrocellulose filters (Millipore, Bedford, MA, HAWP, 0.45

---

[15] M. Wessling-Resnick and G. L. Johnson, *J. Biol. Chem.* **262**, 3697 (1987).
[16] W. Baehr, E. A. Morita, R. J. Swanson, and M. L. Applebury, *J. Biol. Chem.* **257**, 6452 (1982).
[17] E. A. Zhukovsky, P. R. Robinson, and D. D. Oprian, *Science* **251**, 558 (1991).

μm) in a vacuum apparatus and the filters are washed three times with 4 ml of 1× assay buffer. The first four aliquots are taken in the dark, the reaction is illuminated for 20 sec with >450-nm light and four additional aliquots are taken. The nitrocellulose filters are transferred to scintillation vials and 10 ml of scintillation fluid added.

$Cu(phen)_3^{2+}$ must also be removed by Sephadex G-50 spin column treatment before the assay because it inhibits transducin activation. The concentration of mutant and wild-type receptors is determined by absorption spectroscopy at 500 nm, based on the assumption that the extinction coefficient for the mutants does not differ from that of wild-type rhodopsin. The appropriate dilution of the receptor into 0.1% DM in PBS is then performed to generate the 50-nM solution required for the assay.

If an engineered cross-link in the mutant rhodopsin is observed to inhibit activation of transducin, verification that this inhibition is due to the disulfide is obtained by assaying the split receptor after reduction of the disulfide with DTT. Typically, treatment of the mutant split receptor in 0.1% DM with high concentrations of DTT (10–20 mM) overnight at 25° results in complete disulfide reduction. Such DTT treatment has been observed to have a slightly activating effect on both wild-type and mutant split receptors.[8,9] Therefore, comparative analyses of transducin activation by the wild-type split receptor and single cysteine mutants should also be performed as controls.

Practical Considerations

The original cysteine scanning and oxidative cross-linking experiments that studied the bacterial chemotactic receptors used $Cu(phen)_3^{2+}$ as an oxidation reagent.[1] Although the mechanism of cysteine oxidation by this reagent is not well understood it presumably involves the coupled reduction of the Cu^{2+} species to Cu^{1+} (the latter being readily oxidized by ambient oxygen back to Cu^{2+}).[7] Oxidation of cysteine appears to proceed primarily to the disulfide stage,[7] although higher oxidation states of sulfur such as sulfinic or sulfonic acid have been proposed in studies of the D-galactose chemosensory receptor in order to account for the lack of reactivity of some cysteines after treatment of mutant proteins with this reagent.[18] We have not observed a significant occurrence of such side reactions in our studies of rhodopsin.

Treatment of a series of consecutive cysteine mutations in a region of proposed helical structure with $Cu(phen)_3^{2+}$ typically produces select rap-

---

[18] C. L. Careaga and J. J. Falke, *J. Mol. Biol.* **226**, 1219 (1992).

idly formed cross-links which follow a pattern expected for secondary structure. However, one problem we have noted is that often in such a series all positions exhibit a slow reactivity for disulfide bond formation. The select rapidly formed cross-links most likely result from the proximity of these positions in the native structure, while the ubiquitous slower reactions are likely due to disulfide bond formation in nonnative conformations of the protein that result from minor local conformational fluctuations. In contrast to treatment with $Cu(phen)_3^{2+}$, oxidation by ambient oxygen at pH 8.0 of a series of consecutive cysteine mutations typically cross-links only a specific subset of residues and does not exhibit the slower reactions observed with the copper reagent. There is a general correspondence between the rapidly formed cross-links obtained with $Cu(phen)_3^{2+}$ and the cross-links observed with the milder method. Additionally, the use of ambient oxygen at high pH is unlikely to produce higher oxidation states of sulfur in the absence of added metal.

An additional problem with using $Cu(phen)_3^{2+}$ as an oxidant is the cross-linking of native cysteines at positions 140 and 222 in rhodopsin by this reagent.[19] This is a particular problem for experiments performed in SR(1-4/5-7) or SR(1-3/4-7) since formation of the Cys^{140}-Cys^{222} disulfide creates a cross-link between the two fragments. Cysteine scanning experiments using $Cu(phen)_3^{2+}$ must therefore be performed in a C140S and C222S background. A disulfide between Cys-140 and Cys-222 is not detected in experiments performed with SR(1-5/6-7) because the disulfide is contained within the N-terminal fragment. These native cysteines have not been observed to cross-link by the milder spontaneous air oxidation method.

Use of the SR(1-3/4-7) split receptor for cysteine scanning and oxidative cross-linking is also complicated by the native 110–187 disulfide bond between the two fragments (Fig. 1). If this split receptor is denatured in the absence of a thiol modification reagent such as NEM, a disulfide exchange reaction occurs with the net result of formation of a nonnative 185–187 disulfide and a free thiol at position 110, producing two noncovalently linked fragments.[20] Denaturation in the presence of NEM results in modification of Cys-185, retention of the native 110–187 disulfide, and, thus, a covalent cross-link between the two fragments in wild-type SR(1-3/4-7). Therefore cysteine scanning mutagenesis and oxidative cross-linking might be possible using this split receptor if cross-linked proteins are denatured in the absence of NEM, although we have not yet attempted such studies.

[19] H. Yu, M. Kono, and D. D. Oprian, *Biochemistry* **38,** 12028 (1999).
[20] M. Kono, H. Yu, and D. D. Oprian, *Biochemistry* **37,** 1302 (1998).

Discussion of Method

There are several issues to be aware of when using this method. Since cross-linking may be influenced by factors other than residue proximity, two engineered cysteines may be close to each other in the tertiary structure of rhodopsin and yet not afford a cross-link using this method. Therefore, negative results are inconclusive and should be viewed only in the context of multiple scanning experiments. It is best to obtain a complete set of internally consistent cross-linking data, that is, if the region under investigation is part of a transmembrane helix then a cysteine scan of this region would be expected to produce a helical pattern of cross-linking (cross-links only at positions i and $i + 4$, for example) with respect to a single position in a neighboring helix. Observation of such a pattern, while not conclusive, implies that a discrete secondary structure is present and therefore that the cross-links observed are not due to random structural fluctuations. Additionally, the effect of cross-linking on the structure of the protein should be monitored by absorption spectroscopy. Since native dark-state rhodopsin has a characteristic absorbance maximum at 500 nm, observation of this absorbance signature in a cross-linked mutant implies that the secondary structure suggested by the cross-links is part of the native dark-state structure. Transducin activation by cross-linked mutant receptors also indicates that the engineered disulfide does not adversely affect the protein structure or the structural rearrangement required for receptor activation. If cross-linking of a split receptor has no effect on the spectral properties but inhibits transducin activation, then the disulfide may either restrict the structural rearrangement required for receptor activation or disturb the local structure of the protein and affect transducin activation without disrupting the retinal binding pocket. The latter possibility must be considered especially if the engineered disulfide is on the cytoplasmic side of rhodopsin, which interacts directly with transducin and is distal from the retinal binding pocket.

The relative rates of disulfide bond formation have been used as an indicator of the distance between the introduced cysteines in some studies[2,21] and can often provide additional structural information. As previously mentioned, a number of factors in addition to distance can contribute to the rate of disulfide bond formation such as disulfide bond geometry and access to solvent and oxygen, although interresidue distance frequently plays the dominant role. An additional problem that has been noted in our investigations is the considerable variability observed in cross-linking rates, particularly for cysteine mutants in the cytoplasmic region of rhodopsin.

---

[21] B. L. Stoddard, J. D. Bui, and D. E. Koshland, Jr., *Biochemistry* **31,** 11978 (1992).

Although absolute rates appear to be variable, the relative cross-linking rates generally vary much less between experiments. Thus, conclusions based on cross-linking rates should be restricted to comparisons within a single scanning experiment.

[11] Energetics of Rhodopsin Photobleaching: Photocalorimetric Studies of Energy Storage in Early and Later Intermediates

By ROBERT R. BIRGE and BRYAN W. VOUGHT

Introduction

The conversion of light energy into stored internal energy represents a critical component of the primary event of light-transducing proteins. In 1979, Alan Cooper used photocalorimetry to show that about 60% of the energy of the absorbed photon is stored in the primary photoproduct of rhodopsin. This observation was unexpected, and prompted important revisions in our modeling of the primary photochemical event. More recent work has provided a useful perspective on the energetics of the entire photobleaching sequence, and it is now possible to map the entire enthalpy surface. This chapter examines the energetics of the photobleaching sequence and compares and contrasts the viability of optoacoustic spectroscopy and photocalorimetry as used to measure energy storage in proteins. We include a discussion of the measurement and interpretation of the energetics of both visual rhodopsin and bacteriorhodopsin, and through comparison of these two proteins, gain insight into the overall mechanism that each protein utilizes to convert stored energy into function. We conclude with an examination of a rhodopsin side reaction that may represent the principal source of intrinsic photoreceptor noise.

Methods and Procedures of Photoacoustic Calorimetry

Laser-induced optoacoustic spectroscopy (LIOAS) measures protein energetics by observing the sound wave that propagates through a sample in response to the absorption of energy from a laser pulse. Under ideal conditions, this technique can determine the enthalpy storage (ΔH), the change in molecular volume (ΔV), and the quantum yield of photochemistry (Φ_{st}) that accompany a photoactivated event (for recent reviews, see Refs.

1–5). First we outline the theory and then discuss the experimental methods. We end with a discussion of the application of photoacoustic spectroscopy to the study of rhodopsin and bacteriorhodopsin.

Theory of Photoacoustic Spectroscopy

The theoretical foundations of this method have been discussed in detail.[6–10] The fundamental equation is based on an analysis of the energy balance[4]:

$$E_{ph} = \Phi_f E_f + \alpha E_{ph} + \Phi_{st} E_{st} \tag{1}$$

where E_{ph} is the energy added to the sample due to laser excitation and the right-hand side partitions the energy among three processes, fluorescence ($\Phi_f E_f$), prompt radiationless decay (αE_{ph}), and photochemistry ($\Phi_{st} E_{st}$). The fluorescence term describes the energy loss due to emission of light from the excited singlet state manifold where Φ_f is the fluorescence quantum yield and E_f is the fluorescence energy. For retinal proteins, this term is usually neglected because the fluorescence quantum yield is very small ($\Phi_f \approx 10^{-5}$).[11] The second term is the amount of heat generated from the laser flash that is faster than the time resolution of the detection. This time, τ_a, is equal to the time it takes a sound wave to travel to the detector and equals D/ν, where D is the laser beam diameter, and ν is the speed of

---

[1] N. S. Foster, S. T. Autrey, J. E. Amonette, J. R. Small, and E. W. Small, *American Laboratory* **31**, 965 (1999).

[2] S. E. Braslavsky, *in* "Progress in Photothermal and Photoacoustic Science and Technology," Vol. 3, "Life and Earth Sciences" (A. Mandelis and P. Hess, eds.), pp. 1–15. SPIE—The International Society for Optical Engineering, Bellingham, Washington, 1997.

[3] P. J. Schulenberg and S. E. Braslavsky, *in* "Progress in Photothermal and Photoacoustic Science and Technology," Vol. 3, "Life and Earth Sciences" (A. Mandelis and P. Hess, eds.), pp. 57–81. SPIE—The International Society for Optical Engineering, Bellingham, Washington, 1997.

[4] S. E. Braslavsky and G. E. Heibel, *Chem. Rev.* **92**, 1381 (1992).

[5] S. E. Braslavsky and K. Heihoff, *in* "CRC Handbook of Organic Photochemistry" (J. C. Scaiano, ed.), Vol. 1, pp. 327–355. CRC Press, Boca Raton, Florida, 1989.

[6] L. J. Rothberg, J. D. Simon, M. Bernstein, and K. S. Peters, *J. Am. Chem. Soc.* **105**, 3464 (1983).

[7] J. E. Rudzki, J. L. Goodman, and K. S. Peters, *J. Am. Chem. Soc.* **107**, 7849 (1985).

[8] K. Heihoff, S. E. Braslavsky, and K. Schaffner, *Biochemistry* **26**, 1422 (1987).

[9] P. J. Schulenberg, M. Rohr, W. Gärtner, and S. E. Braslavsky, *Biophys. J.* **66**, 838 (1994).

[10] D. Zhang and D. Mauzerall, *Biophys. J.* **71**, 381 (1996).

[11] A. G. Doukas, M. R. Junnarkar, R. R. Alfano, R. H. Callender, T. Kakitani, and B. Honig, *Proc. Natl. Acad. Sci. U.S.A.* **81**, 4790 (1984).

sound in the solvent (τ_a ranges from 100 to 800 ns).[4,5,12,13] The fraction of the sample that promptly converts the light to heat is α. The third term contains Φ_{st}, the quantum yield of photochemistry (isomerization) and E_{st}, the energy stored during that photochemical event.

The signal from the piezoelectric detector, H, is equal to Eq. (2)[6,7]:

$$H = K\alpha E_{ph}(1 - 10^{-A}) \tag{2}$$

where A is the absorbance of the sample, and K is a constant characteristic of the sample geometry and response from the electronics. A reference sample can eliminate K from Eq. (2), [as in Eq. (3)].[8,9,14]

$$\varphi = \frac{H^S}{H^R} = \alpha + \left(\frac{\Phi \Delta V_R}{E_{ph}}\right)\left(\frac{c_p \rho}{\beta}\right) \tag{3}$$

The ratio of signal magnitudes, φ_i, is a function of H^S, the optoacoustic signal from the sample, and H^R, the detector signal of the reference. Heat capacity at constant pressure, c_p, density, ρ, and thermal expansivity, β, of the sample are dependent on temperature and solvent.

Experimental Methods in Photoacoustic Spectroscopy

Figure 1 shows a schematic of a typical photoacoustic spectrometer.[5,7] Femtosecond to nanosecond laser pulses photoactivate the sample at a repetition rate of 0.1–3 Hz. Fractions of each pulse are used to trigger the recorder with a photodiode and monitor the laser intensity with an energy meter. A calibrated pinhole is used to set the diameter of the laser, which, in turn, determines the effective acoustic time ($\tau_a = D/\nu$, where D is the laser beam diameter, and ν is the speed of sound in the solvent). The heat created by the absorption of light in the cuvette is detected by a piezoelectric detector clamped to the side of the cuvette. Lead-zinc-titanate (PZT) and β-polyvinylidene difluoride (PVF$_2$) are the most commonly used piezoelectric elements.[5,8] The signal is then amplified and recorded. An average of 100–300 laser pulses are averaged for each photoacoustic waveform.

A reference sample is used to eliminate K in Eq. (2). Reference compounds should have similar absorption spectra (λ_{max} and optical density) to the sample being used and α should equal 1. That is, all of the light absorbed should be converted to heat, none of the energy should contribute

---

[12] M. Rohr, W. Gärtner, G. Schweitzer, A. R. Holzwarth, and S. E. Braslavsky, *J. Phys. Chem.* **96**, 6055 (1992).
[13] S. L. Logunov, M. A. El-Sayed, L. Song, and J. K. Lanyi, *J. Phys. Chem.* **100**, 2391 (1996).
[14] T. Gensch, J. M. Strassburger, W. Gaertner, and S. E. Braslavsky, *Isr. J. Chem.* (in press).

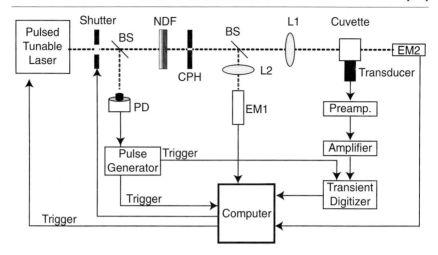

FIG. 1. A schematic diagram of a photoacoustic spectrometer. The computer opens the shutter to select a single pulse from a tunable laser. A fraction of the beam is diverted with a beam splitter (BS) to a photodiode (PD), which triggers the transient digitizer to start recording. A neutral density filter (NDF) and calibrated pinhole aperture (CPH) adjust the beam intensity to a few microjoules per pulse. A second beam splitter diverts a small part of the beam to an energy meter (EM1). A lens (L2) is used to expand the laser to the area of the energy meter probe. Most of the laser intensity passes through a second lens (L1) before illuminating the sample in the cuvette. A second energy meter (EM2) is used to monitor the transmitted laser intensity. The absorbance of the sample is computed from each laser flash to ensure that no unwanted photochemistry is generated. A piezoelectric transducer converts the sound wave to a voltage that is amplified prior to measurement via the transient digitizer.

to fluorescence or energy storage. Dyes such as bromocresol purple or inorganic salts (cobalt chloride, copper chloride) are examples. Care must be taken to avoid multiphoton absorption[15] or concentration effects associated with high concentrations of the reference compound (see later discussion). To determine α and ΔV, waveforms are recorded from 4° to 30° to take advantage of the fact that the term $(c_p\rho/\beta)$ is temperature dependent [see Eqs. (2)–(4)]. At 4°, the "magic temperature," β is zero. However, the signal (amplitude) decreases with temperature, which may prevent measurements at 4°. The amplitude of the sample signal divided by the amplitude of the reference signal, H^S/H^R, is plotted against $(c_p\rho/\beta)$ at different temperatures. The resultant slope is equal to $\Phi\Delta V_R/E_{ph}$ and the y intercept equals α [Eq. (3)].

[15] S. L. Logunov and M. A. El-Sayed, *J. Phys. Chem. B* **101**, 6629 (1997).

Applications to Bacteriorhodopsin and Visual Pigments

A majority of photoacoustic studies have investigated bacteriorhodopsin because this protein is light stable and can be studied conveniently at ambient temperature. Because the K and L intermediates of bacteriorhodopsin are close in energy, energy storage in the K intermediate has been determined by modifying Eq. (1) to yield Eq. (4)[12,13,15–17]:

$$1 - \alpha \approx \Phi E_K / E_{ph} \tag{4}$$

where E_K is the energy stored in the K intermediate. By plotting the amplitude of the sample signal divided by the amplitude of the reference signal, H^S/H^R, against $(c_p\rho/\beta)$ at different temperatures, the slope equals $(\Phi \Delta V_R/E_{ph})$ and the y intercept equals α [Eq. (3)]. Plugging α into Eq. (4), along with the E_{ph} from the laser and $\Phi = 0.65$,[18] E_K, the energy stored in the K intermediate, can be determined. The large number of studies on bacteriorhodopsin have provided a useful perspective on potential sources of error inherent in photoacoustic spectroscopy. If either the sample[19,20] or the reference[15] undergoes multiphoton absorption, it can significantly alter the calculated energy storage.[15] This problem becomes very relevant when picosecond or subpicosecond excitation is used. If a large concentration of a reference dye is required to generate an optical density equal to the sample, this situation will not only increase the probability of two-photon absorption but also generate a difference between the temperature dependence of the term $(c_p\rho/\beta)$ [Eq. (3)] between the sample and reference. The result in both cases is to overestimate the energy storage. In addition, the assignment of the volume change is not trivial (see discussion in Ref. 10). If assigned incorrectly, the volume change will lead to a corresponding error in the assignment of energy storage. These sources of error have been discussed with respect to the measurement of energy storage in bacteriorhodopsin, and have led to reassignments of the enthalpy of storage in the primary photoproduct from previous values near $\Delta H = 160 \pm 30$ kJ/mol[12,13,17] to smaller values of 45 ± 10 kJ/mol.[10,15] The latter values are in good agreement with the earlier photocalorimetric assignment of 49 ± 14 kJ/mol.[21]

---

[16] S. L. Logunov, M. A. El-Sayed, and J. K. Lanyi, *Biophys. J.* **71**, 1545 (1996).

[17] S. L. Logunov, M. A. El-Sayed, and J. K. Lanyi, *Biophys. J.* **70**, 2875 (1996).

[18] R. Govindjee, S. P. Balashov, and T. G. Ebrey, *Biophys. J.* **58**, 597 (1990).

[19] R. R. Birge, L. P. Murray, B. M. Pierce, H. Akita, V. Balogh-Nair, L. A. Findsen, and K. Nakanishi, *Proc. Natl. Acad. Sci. U.S.A.* **82**, 4117 (1985).

[20] R. R. Birge and C. F. Zhang, *J. Chem. Phys.* **92**, 7178 (1990).

[21] R. R. Birge, T. M. Cooper, A. F. Lawrence, M. B. Masthay, C. F. Zhang, and R. Zidovetzki, *J. Am. Chem. Soc.* **113**, 4327 (1991).

Energy storage in the early rhodopsin intermediates has been studied by modifying Eq. (3) into the following two relationships[14,22,23]:

$$\varphi E_{ph} = q + (\Phi \Delta V_R)(c_p \rho / \beta) \quad (5a)$$
$$\Delta H = (E_{ph} - q)/\Phi \quad (5b)$$

where $q = \alpha E_{ph}$. Rhodopsin is more complicated than bacteriorhodopsin because an equilibrium between bathorhodopsin (batho) and the blue-shifted intermediate (BSI) forms at these temperatures.[24] Deconvolution of the spectra is necessary to extract information on the batho and BSI states.[14,23] We will compare the results with the photocalorimetric studies later.

Methods and Procedures of Cryogenic Photocalorimetry

Cryogenic photocalorimetry provides a direct method of measuring enthalpy changes associated with light-induced processes.[25–27] The photocalorimeter measures the heat generated in response to the absorption of a laser pulse, or a gated, wavelength-selected light flux. The enthalpy of any intermediate can be measured provided the intermediate of interest can be trapped at reduced temperature. By using the photostationary state as an internal standard, analysis of the data is relatively straightforward, provided the photochemistry and the quantum yields are assigned accurately. Our current pulsed laser photocalorimeter is shown in Figs. 2 and 3. We describe the operation of the photocalorimeter with specific reference to the measurement of energy storage in the primary events of rhodopsin (R) and isorhodopsin (I). These two pigments form a trinary photochemical system with a common photochemical intermediate, bathorhodopsin (B)[27]:

$$\text{Rhodopsin(R)} \underset{h\nu}{\overset{h\nu}{\rightleftarrows}} \text{bathorhodopsin(B)} \underset{h\nu}{\overset{h\nu}{\rightleftarrows}} \text{isorhodopsin(I)}$$

A pulsed tunable laser (Coherent Infinity Nd:YAG with an internal optical parametric oscillator) sends a single 4-ns, 5- to 20-mJ pulse of monochro-

---

[22] K. Marr and K. S. Peters, *Biochemistry* **30**, 1254 (1991).
[23] J. M. Strassburger, W. Gärtner, and S. E. Braslavsky, *Biophys. J.* **72**, 2294 (1997).
[24] S. J. Hug, J. W. Lewis, C. M. Einterz, T. E. Thorgeirsson, and D. S. Kliger, *Biochemistry* **29**, 1475 (1990).
[25] A. Cooper, *Methods Enzymol.* **88**, 667 (1982).
[26] T. M. Cooper, H. H. Schmidt, L. P. Murray, and R. R. Birge, *Rev. Sci. Instrum.* **55**, 896 (1984).
[27] G. A. Schick, T. M. Cooper, R. A. Holloway, L. P. Murray, and R. R. Birge, *Biochemistry* **26**, 2556 (1987).

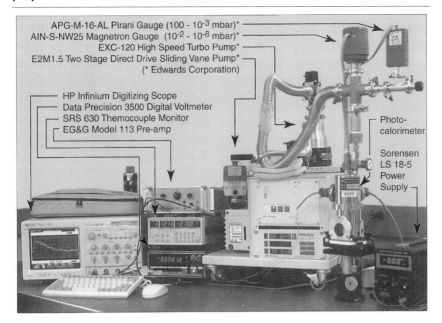

FIG. 2. A view of the primary components of a pulsed laser cryogenic photocalorimeter. The laser and the shutter system are not shown. The calorimeter is housed inside the evacuated chamber within a liquid nitrogen Dewar (DCD-300, Janis Research Co., Wilmington, MA) of the type normally used to cool infrared detectors. A schematic diagram is shown in Fig. 3.

matic radiation into a cooled sample cell surrounded by Peltier receivers (Fig. 3). The concentration of the protein sample is adjusted so that at least 95% of the excitation light is absorbed by the reaction mixture (both sample and photostationary state). The heat produced by the decay of the excited state species back to the ground state species via radiationless decay and photochemistry is measured by a pair of Peltier receivers directly attached to the sample cell (Fig. 3). The heat flow is measured as an output voltage and, after amplification, is integrated to give the total heat produced:

$$I_\nu = \int_{t_1}^{t_1+3\tau} V(t)\,dt \qquad (6)$$

Here $V(t)$ is the observed voltage response of the Peltier receivers and the integration is performed from the time the laser is fired (t_1) to the end of effective heat flow, $t_1 + 3\tau$, where τ is the photocalorimeter time constant. The photocalorimeter shown in Figs. 2 and 3 has a time constant of about 30 sec.

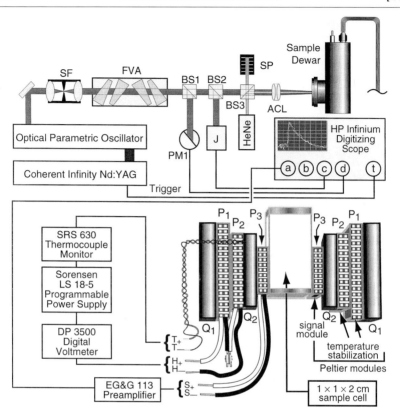

Fig. 3. A schematic diagram of a pulsed laser cryogenic photocalorimeter. The sample solution is degassed and sealed inside the sample cell. The back surface of the sample cell is sputtered with silver so that light making it through the sample is reflected back into the sample for a second pass through the solution. The Dewar is evacuated, liquid nitrogen is then added and the sample cell is cooled down to ~77 K. Two small Peltier receivers on both sides of the sample cell monitor the heat (signals on S_+ and S_-), and four larger Peltier modules maintain a constant "bath" temperature of 80–200 K via current supplied to P_+/P_- terminals using thermocouples to monitor the temperature and a proportional temperature controller. The liquid nitrogen chamber is connected directly to the end plates, Q1. The bath temperature is defined as the temperature of the two beryllium–copper slabs, Q2, which are held at a constant temperature by the four Peltier modules, P_1 and P_2. Two small Peltier receivers on both sides of the sample cell monitor the heat (signals on S_+ and S_-). The wiring connections to the Peltier modules are identical on both sides, but are only shown for the left set for clarity.

The energetics of the reaction are related to the calorimeter signal by[26-28]:

$$I_\nu = KN\Delta H^\lambda_{\text{photon}}\Gamma^\lambda, \qquad (7)$$

where K is the efficiency of the calorimeter, N is the number of moles of sample excited by the laser pulse,

$$N = 8.3593462 \times 10^{-12} \text{ mol } E\{\text{mJ/pulse}\} \lambda\{\text{nm}\} \qquad (8)$$

$\Delta H^\lambda_{\text{photon}}$ is the energy per mole of photons for wavelength λ,

$$\Delta H^\lambda_{\text{photon}} = \frac{119{,}624 \text{ kJ mol}^{-1}}{\lambda\{\text{nm}\}} = \frac{28{,}591 \text{ kcal mol}^{-1}}{\lambda\{\text{nm}\}} \qquad (9)$$

and Γ^λ is the dimensionless molecular response function.[26-28] Braces indicate the unit assignment for the independent variables. The molecular response function is interpreted as the photochemical light-to-heat conversion efficiency of the (nonemitting) photochemical system ($\Gamma^\lambda = 1$ for a nonphotochemical absorber; $\Gamma^\lambda < 1$ for an endothermic reaction; $\Gamma^\lambda > 1$ for an exothermic reaction). These assignments and the relationship in Eq. (7) are applicable only to solutions that absorb all of the laser pulse and are nonfluorescent. For reaction mixtures in which the existing radiation initiates more than one photochemical reaction, Γ^λ will contain contributions from each of the individual reaction paths. In the present example we are examining the case where three solutes may be present simultaneously with photochemical interconversions described by the following quantum yields:

$$R \underset{\phi_2}{\overset{\phi_1}{\rightleftharpoons}} B \underset{\phi_4(\lambda)}{\overset{\phi_3}{\rightleftharpoons}} I$$

The quantum yields and the photochemical processes are depicted schematically in Fig. 4. Note that Φ_4 is wavelength dependent (see later discussion).

The functional form of Γ^λ for the photoequilibrium shown in Eq. (10) may be derived by considering individually the contributions from each of the processes depicted in Fig. 4 [see derivation of Eq. (8) in Ref. 27]:

$$\Gamma^\lambda = 1 + (\alpha^\lambda_B \Phi_2 - \alpha^\lambda_R \Phi_1) \frac{\Delta H_{\text{BR}}}{\Delta H^\lambda_{\text{photon}}} + (\alpha^\lambda_B \Phi_3 - \alpha^\lambda_I \Phi_4) \frac{\Delta H_{\text{BI}}}{\Delta H^\lambda_{\text{photon}}} \qquad (11)$$

where ΔH_{BR} is the enthalpy of bathorhodopsin minus the enthalpy of rhodopsin, ΔH_{BI} is the enthalpy of bathorhodopsin minus the enthalpy of

---

[28] R. R. Birge and T. M. Cooper, *Biophys. J.* **42**, 61 (1983).

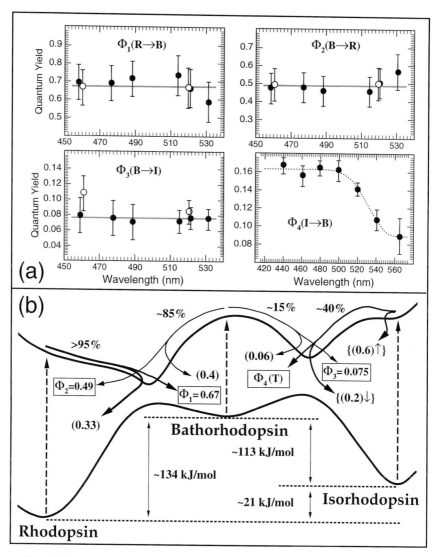

FIG. 4. (a) Temperature effect on the principal quantum yields and (b) a schematic representation of the ground and first excited singlet state surfaces connecting rhodopsin, bathorhodopsin, and isorhodopsin using a simplified (linearized) reaction coordinate. The quantum yield data are at 77 K (solid circles) and 70 K (open circles) and are from Refs. 67–69. The potential surfaces are based on all-valence electron molecular orbital calculations including single and double configuration interaction[41] and were adjusted to reflect the experimental transition energies and ground state enthalpies.[27,31] Absolute quantum yields of photoisomerization are displayed at the tips of the arrows indicating the processes.[68,69] Values given in parentheses (●) are predicted by using semiempirical molecular dynamics theory to calculate the reverse/forward yield ratios and multiplying these values by the experimental forward yields (shown without parentheses). Values listed in braces {●} are ambient temperature quantum yields which display temperature dependence. The arrows indicate the effect that lowering the temperature will have on a given value. (Adapted and updated from Refs. 68 and 69.)

isorhodopsin, and the α_i^λ values are the partition functions for excitation of the ith species:

$$\alpha_i^\lambda = \frac{\varepsilon_i(\lambda)[i]}{\varepsilon_R(\lambda)[R] + \varepsilon_B(\lambda)[B] + \varepsilon_I(\lambda)[I]} \quad (12)$$

where $\varepsilon_i(\lambda)[i]$ is the molar absorptivity for component i at wavelength λ. In Eq. (8) ΔH_{BR} and ΔH_{BI} will both be positive for the case where $E^\circ_B > E^\circ_I > E^\circ_R$ (E°_i is the ground-state energy for component i). When the mixture reaches a photostationary equilibrium, the molecular response function, Γ^λ, equals unity because:

$$(\alpha_B^\lambda \Phi_2 - \alpha_R^\lambda \Phi_1) \frac{\Delta H_{BR}}{\Delta H_{photon}^\lambda} = -(\alpha_B^\lambda \Phi_3 - \alpha_I^\lambda \Phi_4) \frac{\Delta H_{BI}}{\Delta H_{photon}^\lambda} \quad (13)$$

This relationship describes a situation in which all the light energy put into the system is returned as heat because no net photochemistry takes place. This observation provides a convenient means by which the photocalorimeter may be calibrated. An experimental value for Γ^λ may therefore be determined by

$$\Gamma^\lambda_{obsvd} = \frac{(I_\nu/N)_{sample}}{(I_\nu/N)_{ctrl}} \quad (14)$$

where $(I_\nu/N)_{ctrl}$ is the laser intensity–normalized calorimeter signal obtained for the photostationary state mixture. It is important to measure both the sample and control signals over a range of pulse energies to verify a linear relationship between I_ν and N.

Inherent in Eq. (13) is the assumption that K, the photocalorimeter efficiency factor [see Eq. (8)], is invariant to changes in the prepared solute versus photostationary states. This will not be the case if the photostationary state does not absorb >80% of the laser excitation. It is noted that formation of the stationary state invariably reduces the optical density of the solution, but if one starts with a sufficient solute concentration, this is normally not a problem. It is also important to maintain the same temperature for all experiments, because the responsivity of the Peltier receivers is temperature dependent and thus K [Eq. (7)] is temperature dependent.

Photocalorimetric Studies of Visual Pigments

Rhodopsin and its intermediates have been studied extensively by using photocalorimetry.[25,27,29–32] Two approaches have been used. The pioneering studies of Alan Cooper and co-workers utilized Peltier-based photocalorim-

---

[29] A. Cooper and C. A. Converse, *Biochem.* **15**, 2970 (1976).
[30] A. Cooper, *Nature (London)* **282**, 531 (1979).
[31] A. Cooper, *FEBS Lett.* **100**, 382 (1979).
[32] A. Cooper, *FEBS Letters* **123**, 324 (1981).

eters with irradiation carried out using filtered Xe/Hg arc lamps. An excellent discussion of their methods and procedures may be found in Ref. 25. Subsequent pulsed laser studies based on the methods and procedures outlined here were in good agreement with the original studies and the results are shown in Fig. 4.[27] Enthalpies of later intermediates in the rhodopsin bleaching sequence appear to be rather sensitive to experimental conditions, and this sensitivity complicates comparative examinations of technique-based differences. Energy storage in bathorhodopsin is predicted by photocalorimetry to be either ~146 kJ/mol[30] or ~135 kJ/mol.[27] Although a low-temperature photoacoustic study[33] is consistent with these values, more recent LIAOS studies suggest that bathorhodopsin stores ~207 kJ/mol.[14] The differential reverses when comparing latter intermediates. While Cooper predicted that lumirhodopsin stores ~109 kJ/mol based on photocalorimetry,[32] photoacoustic measurements yield much lower values of ~16 kJ/mol[22] or ~85 kJ/mol.[23]

Energetics of Bacteriorhodopsin Photocycle

We briefly overview the energetics of the bacteriorhodopsin photocycle first because this system provides a useful perspective on the less well understood energetics of the rhodopsin photobleaching sequence. The combination of photocalorimetric,[21] photoacoustic,[34] and kinetic studies[35,36] have provided a detailed picture of the energetics in the photocycle of light-adapted bacteriorhodopsin. The results are shown in Fig. 5. The primary event stores 49 ± 14 kJ/mol.[21] This process is completed in a few picoseconds, and involves an all-*trans* to 13-*cis* photoisomerization of the chromophore and little, if any, protein relaxation. It has been argued that roughly one-half of the energy is stored in charge separation and the other half in conformational distortion.[28] As noted later, energy storage in rhodopsin appears to be dominated by conformational distortion. It is reasonable to assume that the change in entropy during the primary event is negligible. The L → M_1 reaction involves the transfer of the chromophore imine proton to a nearby aspartic acid counterion (Asp-85), and generates a significant change in the electrostatic environment within the chromophore binding site and the lower proton channel. As can be seen by reference to Fig. 5, the K → L → M_1 dark reactions are driven by modest enthalpy changes with little contribution from entropy. Deprotonation of the chro-

---

[33] F. Boucher and R. M. Leblanc, *Photochem. Photobiol.* **41,** 459 (1985).
[34] D. R. Ort and W. W. Parson, *Biophys. J.* **25,** 355 (1979).
[35] G. Varo and J. K. Lanyi, *Biochem.* **30,** 5016 (1991).
[36] G. Varo and J. K. Lanyi, *Biochemistry* **30,** 5008 (1991).

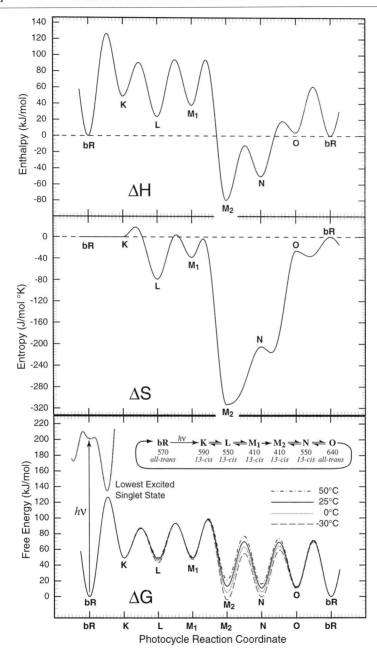

FIG. 5. The enthalpy, entropy, and free-energy surfaces associated with the photocycle of light-adapted bacteriorhodopsin. The plots are based on the data presented in Refs. 35 and 36 updated to reflect the revised K intermediate enthalpy.[21]

mophore, however, induces a large change in the structure of the protein, and the $M_1 \to M_2$ reaction is characterized by a significant decrease in both enthalpy and entropy. Note that M_2 has an enthalpy about 80 kJ/mol lower than bR.[34] The remaining portion of the photocycle involving $M_2 \to N \to O \to bR$ is thus entropy driven. A side reaction allowing N to directly return to bR is an important step not represented in Fig. 5.

Energetics of Rhodopsin Photobleaching

We do not have sufficient kinetic and calorimetric data to analyze the free-energy surface of rhodopsin. At present, we can only provide a perspective on the enthalpy surface, and this requires adoption of the following simplified reaction scheme:

$$\text{Rhodopsin(R)} \xrightarrow{h_\nu} \text{bathorhodopsin(B)} \underset{k_{32}}{\overset{k_{23}}{\rightleftarrows}}$$

$$\text{blue-shifted intermediate(BSI)} \xrightarrow{k_{34}}$$

$$\text{lumirhodopsin(L)} \xrightarrow{k_{45}} \text{meta-I(MI)} \underset{k_{65}}{\overset{k_{56}}{\rightleftarrows}}$$

$$\text{meta-II(MII)} \xrightarrow{k_{67}} \text{opsin + retinal} \quad (15)$$

This scheme ignores some of the potential contributions of meta-III as well as back reactions involving lumirhodopsin and meta-I. Table I presents an overview of the various experimental data, as well as the preferred values based on our analysis of the literature. A plot of the enthalpy surfaces is shown in Fig. 6. It is important to note that the enthalpy surface is known with lower confidence than that for bacteriorhodopsin (Fig. 5). Nevertheless, we can clearly see a significant difference in the energetics between these two light-transducing proteins. The surprising similarities between the spectroscopic properties of the intermediates in the two proteins does not translate into any significant energetic similarities. Note the sharp decrease in the enthalpy of the M_2 state relative to the M_1 state in bacteriorhodopsin as compared to the increase in enthalpy in meta-II versus meta-I. To the extent comparisons are possible, it appears that meta-I is more similar to M_2 and that the meta-I $\to$ meta-II $\to$ opsin + retinal portion of the photobleaching sequence is primarily entropy driven, as was the case of the latter stages of the bacteriorhodopsin photocycle.

The blue-shifted intermediate (BSI) remains somewhat controversial because it is not possible to trap this intermediate at reduced temperatures,

TABLE I
ENTHALPIES OF PRINCIPAL INTERMEDIATES IN RHODOPSIN PHOTOBLEACHING
AT pH 7–8

Intermediate	ΔH (kJ/mol)		
	Photocalorimetry	Photoacoustics	Preferred
Bathorhodopsin	135[a]	207[b]	135
BSI	140–155[c]	135[b]	145
Lumirhodopsin	110[d]	80–85[b]	105
Meta I	70[d]	—	70
Meta II	115[d]	—	115
Opsin + retinal	92[d]	—	90

[a] Error of ±4 kJ/mol. G. A. Schick, T. M. Cooper, R. A. Holloway, L. P. Murray, and R. R. Birge, *Biochemistry* **26,** 2556 (1987).
[b] Errors reported in the range of 50–150 kJ/mol, with errors increasing for later intermediates. T. Gensch, J. M. Strassburger, W. Gaertner, and S. E. Braslavsky, *Isr. J. Chem.* (in press).
[c] Based on estimates relative to photocalorimetric data. S. J. Hug, J. W. Lewis, C. M. Einterz, T. E. Thorgeirsson, and D. S. Kliger, *Biochemistry* **29,** 1475 (1990).
[d] Errors are about 10% except for ΔH of lumirhodopsin, which is ~20%. A. Cooper, *FEBS Lett.* **123,** 324 (1981).

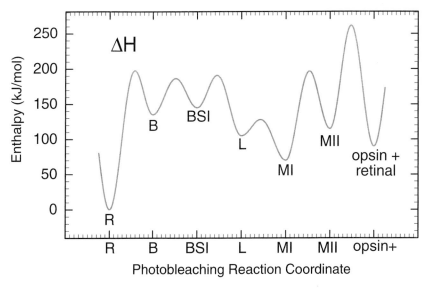

FIG. 6. The enthalpy surface associated with the photobleaching sequence of rhodopsin at pH 7–8 based on the data of Table I (preferred column) and the activation energies compiled in Ref. 32.

in contrast to the other intermediates shown in Fig. 6. Rather, this intermediate can only be deduced by transient absorption techniques and kinetic deconvolution.[24] Examination of Fig. 6 provides a rather straightforward explanation of the reason this intermediate would be difficult to trap thermally. If we assume the free-energy surface mirrors the enthalpy surface at the early stages of the bleaching sequence, BSI has an energy larger than both the preceding and subsequent intermediates. As the temperature is lowered, the bathorhodopsin and BSI equilibrium will shift in favor of the former intermediate. Thus, it is not surprising that this state has eluded detection until recently.

The large energy storage in the primary event remains a subject of active study and discussion. A number of theoretical studies have tackled this issue.[37-41] Early studies emphasized the importance of charge separation, with suggestions that 40% or more of the energy was stored via the motion of the positively charged chromophore away from a negatively charged counterion.[37,39] NMR studies suggest that charge separation may play a much smaller role than was originally predicted,[42] and the most recent theoretical study emphasizes the role of chromophore distortion in the bathochromic shift and the storage of energy.[40] We know from two-photon studies that the binding site of rhodopsin is neutral.[19] The principal counterion near the protonated chromophore has now been identified as Glu-113.[43,44] There is strong evidence to suggest that the primary photochemical event in rhodopsin proceeds along a barrierless excited state surface, and is propagated along that surface via an electrostatic interaction between the chromophore and Glu-113 (see discussion in Ref. 45). These observations suggest that at least 20% of the energy storage is due to charge separation or dipole–dipole interactions within the protein.[41]

---

[37] B. Honig, T. Ebrey, R. H. Callender, U. Dinur, and M. Ottolenghi, *Proc. Natl. Acad. Sci. U.S.A.* **76,** 2503 (1979).
[38] A. Warshel and N. Barboy, *J. Am. Chem. Soc.* **104,** 1469 (1982).
[39] R. R. Birge and L. M. Hubbard, *J. Am. Chem. Soc.* **102,** 2195 (1980).
[40] A. Bifone, H. J. M. de Groot, and F. Buda, *J. Phys. Chem. B* **101,** 2954 (1997).
[41] J. R. Tallent, E. Q. Hyde, L. A. Findsen, G. C. Fox, and R. R. Birge, *J. Am. Chem. Soc.* **114,** 1581 (1992).
[42] S. O. Smith, J. Courtin, H. J. M. de Groot, R. Gebhard, and J. Lugtenburg, *Biochemistry* **30,** 7409 (1991).
[43] T. P. Sakmar, R. R. Franke, and H. G. Khorana, *Proc. Natl. Acad. Sci. U.S.A.* **86,** 8309 (1989).
[44] E. A. Zhukovsky and D. D. Oprian, *Science* **246,** 928 (1989).
[45] J. A. Stuart and R. R. Birge, in "Biomembranes" (A. G. Lee, ed.), Vol. 2A, pp. 33–140. JAI Press, London, 1996.

Origin and Energetics of Photoreceptor Noise

The human visual system can detect faint stars at night and distinguish objects in direct sunlight for an effective operating range of about 10 log units of light intensity. At high light intensities the key limitation of visual function is the bleaching of photoreceptor pigments. At low light levels the key limiting factor is noise in the photoreceptors. We can reliably detect pulses of light that send roughly 100 photons through the pupil and activate approximately 10–20 rhodopsin molecules in as many rod photoreceptors. We see such dim flashes of light against a background of noise caused by each photoreceptor eliciting false signals ("noise") that are indistinguishable from the signals triggered by single photon absorptions.[46] The mechanism responsbile for photoreceptor noise remains a subject of debate.[47–53] Because the noise scales in proportion to the amount of rhodopsin in the photoreceptor,[54] it is now generally accepted that a side reaction within the rhodopsin protein is responsible.

Baylor and co-workers have assigned the thermodynamic properties of the thermally activated dark processes in toad rod outer segments: (E_a = 92 ± 7 kJ/mol, $\Delta G^\ddagger$ = 133 ± 9 kJ/mol, $\Delta H^\ddagger$ = 90 ± 7 kJ/mol, $\Delta S^\ddagger$ = −148 ± 23 kJ/mol-K).[46] Studies on other animals yield comparable, though slightly higher, activation energies (for a review, see Ref. 54). A comparison with denaturation activation energies measured for cattle rhodopsin (E_a = ~400 kJ/mol), frog rhodopsin (E_a = ~190 kJ/mol), and squid rhodopsin (E_a = ~300 kJ/mol) indicates that protein denaturation is not the origin of the dark signal.[55] Measurements of thermal isomerization of 11-*cis*-retinal in 1-propanol solution, however, appear to offer a more compatible set of thermodynamic properties (E_a = 94 kJ/mol, $\Delta G^\ddagger$ = 123 kJ/mol, $\Delta H^\ddagger$ = 91 kJ/mol, $\Delta S^\ddagger$ = −90 kJ/mol-K).[56] Comparison of the latter measurements on 11-*cis*-retinal with those observed by Baylor on rod segments prompted some investigators to propose that thermal isomerization of "the chromophore" is responsible for dark activation of rhodopsin.[46–48] However, this hypothesis is not consistent with the energetics of ground state isomerization

---

[46] D. A. Baylor, G. Matthews, and K. W. Yau, *J. Physiol.* **309**, 591 (1980).
[47] A. C. Aho, K. Donner, C. Hyden, L. O. Larsen, and T. Reuter, *Nature* **334**, 348 (1988).
[48] H. B. Barlow, *Nature* **334**, 296 (1988).
[49] R. B. Barlow, Jr. and E. Kaplan, *Biol. Bull.* **177**, 323 (1989).
[50] R. B. Barlow, Jr. and T. H. Silbaugh, *Invest. Ophthalmol. Vis. Sci. Suppl.* **30**, 61 (1989).
[51] R. R. Birge, *Biochim. Biophys. Acta* **1016**, 293 (1990).
[52] K. Fahmy and T. P. Sakmar, *Biochemistry* **32**, 7229 (1993).
[53] R. R. Birge, *Biophys. J.* **64**, 1371 (1993).
[54] R. R. Birge and R. B. Barlow, *Biophys. Chem.* **55**, 115 (1995).
[55] R. Hubbard, *J. Gen. Physiol.* **42**, 259 (1958).
[56] R. Hubbard, *J. Biol. Chem.* **241**, 1814 (1966).

of the protein bound chromophore (Fig. 6). The protein bound chromophore is not 11-*cis*-retinal, but the protonated Schiff base of 11-*cis*-retinal. The ground state barrier to isomerization of the protein bound chromophore is estimated to be $\Delta H^\ddagger = 188 \pm 13$ kcal mol^{-1} (Fig. 6). We can establish a lower limit of $\Delta H^\ddagger \geq 177$ kJ/mol based on the relative enthalpy of bathorhodopsin ($\Delta H_{BR} = 135 \pm 4$ kJ/mol)[27] plus the activation enthalpy of the bathorhodopsin → lumirhodopsin dark reaction ($\Delta H^\ddagger = 42 \pm 8$ kJ/mol)[57] and assuming additive errors. In contrast, the activation energies for thermal activation are all less than 150 kJ/mol, and a majority are less than 126 kJ/mol. Thus, thermal (ground state) isomerization of the native (protonated) chromophore cannot be responsible for thermal activation of the protein.

Another possibility is that there is an equilibrium within the rhodopsin binding site coupling protonated versus unprotonated chromophores. The experiments of Longstaff *et al.* have demonstrated that deprotonation of the Schiff base of retinal is *obligate* for rhodopsin activation.[58] It is thus possible that deprotonation of the Schiff base is *sufficient* for activation. More precisely, deprotonation generates a form of rhodopsin that is interpreted by transducin as activated (i.e., R*), and the cascade is initiated. (Deprotonation would break the above-mentioned salt bridge.) The observation that the thermal noise signals have amplitudes identical to the light-activated signals requires that the rhodopsin molecule with an unprotonated chromophore have a lifetime identical to activated rhodopsin. This is unlikely due to the instability of the R_d species. Recent site-directed mutagenesis studies by Fahmy and Sakmar provide more explicit evidence that deprotonation is not sufficient for activation.[59] These investigators replaced the primary counterion of rhodopsin, the glutamate (E) residue at position 113, with glutamine (Q) and regenerated the mutant opsin with 11-*cis*-retinal. The E113Q substitution dramatically decreases the pK_a of the Schiff base proton and generates a pigment containing a mixture of protonated and unprotonated retinyl chromophores with absorption maxima at 490 and 380 nm, respectively. The key observation is that the unprotonated (380-nm) species does not activate transducin, which indicates that deprotonation is not sufficient to activate rhodopsin.

We have proposed an alternative mechanism that accommodates the available experimental evidence.[51,53,60] The first step is deprotonation of

---

[57] K. H. Grellmann, R. Livingston, and D. Pratt, *Nature* **193**, 1258 (1962).
[58] C. Longstaff, R. D. Calhoon, and R. R. Rando, *Proc. Natl. Acad. Sci. U.S.A.* **83**, 4209 (1986).
[59] K. Fahmy and T. P. Sakmar, *Biochemistry* **32**, 9165 (1993).
[60] R. B. Barlow, R. R. Birge, E. Kaplan, and J. R. Tallent, *Nature* **366**, 64 (1993).

the 11-*cis* protonated Schiff base chromophore. The second step is thermal 11-*cis* to 11-*trans* isomerization of the (unprotonated Schiff base) chromophore. The reaction scheme and the adiabatic potential energy surfaces generated by using MNDO/AM1 and INDO-PSDCI molecular orbital theory are shown in Fig. 7. The calculated adiabatic activation energies are as follows:

$$R(11-cis\text{-PSB}) \underset{\Delta E_{-1} \approx 42 \text{ kJ/mol}}{\overset{\Delta E_1 \approx 113 \text{ kJ/mol}}{\rightleftharpoons}} R_d(11-cis\text{-SB})$$

$$\xrightarrow{\Delta E_2 \approx 96 \text{ kJ/mol}} B_d(11-trans\text{-SB}) \quad (16)$$

These energetics would produce an apparent (experimental) Arrhenius-like activation energy of approximately 100–105 kcal mol^{-1} (see derivation in Ref. 54), which is consistent with the experimental noise activation energies (see earlier comments). We note that the B_d species generated via the above mechanism would be virtually identical to the R* generated via the light-induced photobleaching sequence, and that both species would decay to form all-*trans*-retinal and opsin. Thus, the two-step mechanism given here is viable with respect to observed energetics as well as observed thermal photoreceptor noise signals.

If the mechanism given is correct, and the binding site is accessible for titration from the aqueous bulk medium, then changes in extracellular pH should affect the ratio of unprotonated versus protonated chromophores within the binding site. Recent experiments by Callender and co-workers[61] on bovine rhodopsin are consistent with the studies of Lisman and Strong[62] on *Limulus* and indicate that the binding site of rhodopsin in both vertebrate and invertebrate photoreceptors is accessible to changes in extracellular pH. If the pH were decreased, then the equilibrium would favor the protonated species, and the rate of thermal activation processes would decrease. Experiments on *Limulus*, in which extracellular pH could be monitored and adjusted while simultaneously monitoring dark noise, are consistent with this model.[60]

Nature has designed the rhodopsin binding site in order to minimize intrinsic photoreceptor noise. We can write the noise rate as proportional to two factors:

$$k_{tot} = \frac{k_1 k_2}{k_{-1}} \propto k_2 \, 10^{-pK_a^{PSB}} \quad (17)$$

---

[61] H. Deng, L. Huang, R. Callender, and T. Ebrey, *Biophys. J.* **66**, 1129 (1994).
[62] J. E. Lisman and L. A. Strong, *J. Gen. Physiol.* **73**, 219 (1979).

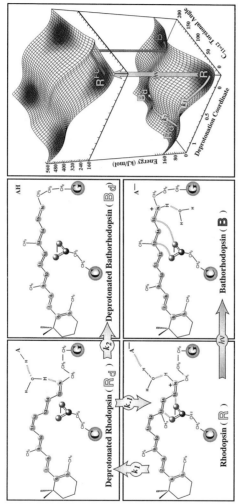

FIG. 7. Molecular schematics of the photochemical and thermal pathways of activation of rhodopsin. The membrane-spanning helix to which the chromophore is attached via Lys-296 is labeled G. The primary counterion, Glu-113, is attached to helix C.[44,70] Only one water molecule is shown for simplicity, and the symbol A represents one or more negatively charged amino acids outside of the binding site which ultimately accepts the proton. [Note that the rhodopsin (R) binding site is neutral,[19] and thus the acceptor group must be separated from the chromophore by a larger distance than inferred in this figure.] Spectroscopic and theoretical studies indicate that the glutamic acid counterion interacts primarily with the chromophore in the region C12—C13 = C14—C15 region of the chromophore and that at least one water molecule is hydrogen bonded to the imine proton.[9,51,61,69,71] The calculated energy surfaces in the ground and excited state associated with both the primary photochemical event and the dark reaction responsible for photoreceptor noise are shown in the insert at right[60] (Adapted and updated from Refs. 54 and 72.)

where

$$R \underset{k_{-1}}{\overset{k_1}{\rightleftharpoons}} R_d \overset{k_2}{\longrightarrow} B_d \qquad (18)$$

It follows from Eq. (17) that a high pK_a of the protonated Schiff base chromophore is important to the biological control of photoreceptor noise. Model protonated retinyl-Schiff base chromophores in solution exhibit pK_a values around 7.[63–66] The retinyl chromophore in bacteriorhodopsin exhibits a pK_a of ~13.[63] The pK_a of the protonated Schiff base chromophore of rhodopsin is 16 or greater.[65] Thus, if the chromophore of rhodopsin had the same pK_a as that observed in bacteriorhodopsin, the visual pigment would have a dark noise rate 3 to 4 orders of magnitude larger than observed. It is likely that the adjustment of the pK_a in rhodopsin is intimately related to the natural selection of a photoreceptor protein exhibiting minimal dark noise.

Acknowledgments

The authors thank Drs. Silvia Braslavsky, Thomas Ebrey, Mostafa El-Sayed, Barry Knox, and David Mauzerall for interesting and helpful discussions. We also thank Dr. Braslavsky for providing us with a preprint of Ref. 14 prior to publication. The research in our group discussed in this chapter was funded in part by NIH grant GM-34548.

---

[63] S. Druckmann, M. Ottolenghi, A. Pande, J. Pande, and R. H. Callender, *Biochemistry* **21**, 4953 (1982).
[64] M. Sheves, A. Albeck, N. Friedman, and M. Ottolenghi, *Proc. Natl. Acad. Sci. U.S.A.* **83**, 3262 (1986).
[65] G. Steinberg, M. Ottolenghi, and M. Sheves, *Biophys. J.* **64**, 1499 (1993).
[66] C. Sandorfy, L. S. Lussier, H. L. Thanh, and D. Vocelle, in "Biophysical Studies of Retinal Proteins" (T. G. Ebrey, H. Frauenfelder, B. Honig, and K. Nakanishi, eds.), pp. 247–251. University of Illinois Press, Urbana, 1987.
[67] T. Suzuki and R. H. Callender, *Biophys. J.* **34**, 261 (1981).
[68] R. R. Birge and R. H. Callender, in "Biophysical Studies of Retinal Proteins" (T. Ebrey, H. Frauenfelder, B. Honig, and K. Nakanishi, eds.), pp. 270–281. University of Illinois Press, Urbana, 1987.
[69] R. R. Birge, C. M. Einterz, H. M. Knapp, and L. P. Murray, *Biophys. J.* **53**, 367 (1988).
[70] T. P. Sakmar, R. R. Franke, and H. G. Khorana, *Proc. Natl. Acad. Sci. U.S.A.* **86**, 8309 (1989).
[71] H. Kakitani, T. Kakitani, H. Rodman, and B. Honig, *Photochem. Photobiol.* **41**, 471 (1985).
[72] R. Barlow, R. Birge, E. Kaplan, and J. Tallent, *Nature* **366**, 64 (1993).

[12] Absorption Spectroscopy in Studies of Visual Pigments: Spectral and Kinetic Characterization of Intermediates

By JAMES W. LEWIS and DAVID S. KLIGER

Introduction

UV–vis spectroscopy is a reliable tool for the characterization of discrete intermediates that occur in protein function. It is particularly suited to visual pigments, where the chromophore is intimately involved with the activation mechanism. Further, UV–vis spectroscopy can be relatively easily performed on the extremely short timescales involved in visual pigment function. As a consequence, much has been learned about visual pigment photointermediates. These results have been reviewed elsewhere.[1–3] Here we describe fundamental problems encountered in applying time-resolved UV–vis spectroscopy to visual pigments and discuss operational approaches to overcoming those problems.

After absorption of a single photon, visual pigments proceed with a high quantum yield through a series of intermediates. Of particular interest are the mechanisms involved in the formation and decay of the intermediate that activates transducin.[4] The understanding of such signaling intermediates is a primary goal of visual pigment research. To achieve this, intermediates have been studied in pigments with modified chromophores and/or site-specific mutations. Historically, insight has also come from investigation of pigments isolated from a variety of different species and photoreceptor types, some of which are available in very limited quantities. Here, to simplify discussion, techniques are described in terms of the best characterized visual pigment, bovine rhodopsin (rho). Application of these techniques to other visual pigments requires appropriate modification for the specific properties (λ_{max}, decay times, etc.) of those pigments.

---

[1] G. G. Kochendoerfer and R. A. Mathies, *Isr. J. Chem,* **35,** 211 (1995).
[2] D. S. Kliger and J. W. Lewis, *Isr. J. Chem.* **35,** 289 (1995).
[3] K. P. Hofmann, *Photobiochem. Photobiophys,* **13,** 309 (1986).
[4] O. G. Kisselev, J. Kao, J. W. Ponder, Y. C. Fann, N. Gautam, and G. R. Marshall, *Proc. Natl. Acad. Sci. U.S.A.* **95,** 4270 (1998).

Primary Photolysis

High signal-to-noise (S/N) ratio measurements are required to understand subtle aspects of visual pigment function. As discussed in the next section, there are limits to noise reduction for optical measurements, making it important to maximize signal size. Because the absorbance signal due to the presence of a photointermediate is proportional to its concentration, an intense actinic pulse is required to initiate visual pigment activation in order to produce as much of the photointermediates as possible. Under such conditions, sequential absorption of two or more photons becomes possible. While the quantum yield of these secondary photolyses is generally lower than for the primary photolysis of rhodopsin, and some secondary processes do not lead to new species (i.e., rho + $h\nu \to$ bathorhodopsin + $h\nu \to$ rho), the wavelength and duration of the photolysis pulse should be chosen to minimize secondary photolysis. For rho the 7-ns, 477-nm pulse from a dye laser containing C-480 pumped by the third harmonic of a Nd:YAG laser is ideal. This wavelength is chosen primarily to reduce photolysis of bathorhodopsin (batho, $\lambda_{max} \approx 535$ nm), but it also helps to photolyze any of the 9-*cis* isomer, isorhodopsin (iso), which forms due to secondary photolysis of batho.

Only laser sources provide short enough pulses for the study of the early photointermediates and, as just described, their narrow bandwidth can be useful in mitigating the effects of secondary photolysis. However, laser light has other properties that need to be considered in experimental design. For example, secondary photolysis can be aggravated if the sample is irradiated with the spatially nonuniform beam of a typical dye laser. Combinations of cylindrical and diverging lenses can be used to produce relatively uniform illumination of the sample, allowing those parts of the beam that are most inhomogeneous to be discarded. Another significant property of laser light is its linear polarization. Because visual pigment molecules rotate little during the 7-ns pulse duration, only about one-third of the transition dipoles are oriented so as to allow absorption of the laser light. This both limits the signal achievable in absorbance measurements and results in anisotropic distribution of the photoproduct absorbance. Care must be taken so that when rotational relaxation occurs the associated absorbance changes are not mistaken for photointermediate decay. Laser polarization must also be taken into account when modeling the effects of secondary photolysis. This is necessary because reorientation during the laser pulse makes the batho formed much more liable to secondary photolysis than is predicted by an isotropic model of batho absorbance.[5,6]

[5] D. S. Kliger, J. S. Horwitz, J. W. Lewis, and C. M. Einterz, *Vis. Res.* **24**, 1465 (1984).
[6] J. W. Lewis, C. M. Einterz, S. J. Hug, and D. S. Kliger, *Biophys. J.* **56**, 1101 (1989).

Monitoring Absorbance Difference Spectra

Probe Light Sources

The noise level in a well-designed absorbance measurement is proportional to the reciprocal square root of the number of probe beam photons detected. Thus a bright probe beam reduces noise levels. The photolabile nature of rho and, to a lesser degree, its photointermediates puts important limits on probe beam flux, making it critically important to achieve the proper balance between precise measurement and significant photolysis by the probe beam. This balance is nearly impossible to strike with a continuous probe beam. After primary photolysis, the absorbance of the sample changes rapidly at early times and then more slowly later, with the consequence that the duration of individual measurements needs to be short at first, but then could be lengthened later as the signal becomes relatively constant for longer periods. This means that with a continuous probe beam it is possible to measure late absorbance changes more precisely than is justified by the precision of early measurements. This is a serious problem since the extra precision is accomplished by exposing the sample to unnecessary probe beam photolysis. Visual pigment photoreactions span more than 13 orders of magnitude in time, with typical studies ranging over 2 to 5 decades, making this precision change very significant. In typical single-wavelength kinetic experiments, a particularly troublesome aspect of the artifactual species produced by probe beam photolysis is that their concentrations will depend on the monitoring wavelength(s). This makes it very difficult to estimate their concentrations and correct for their presence in analysis of the data. In summary then, the photons in a continuous probe beam may not arrive when they are needed and, particularly for monitoring slow processes in visual pigments, may cause artifacts.

A better approach is to measure the absorbance at all wavelengths simultaneously using a pulse of white light from a flash lamp. A system that has proven to be very effective is shown in Fig. 1. Since the flash lamp need only be fired shortly before the time at which the absorbance spectrum is measured, accumulation of artifactual species is minimized. The exact time at which the absorbance spectrum is measured depends on the duration of the gate pulse applied to a gated optical multichannel analyzer (OMA), which is synchronized to occur near the peak output of the flash lamp. A stable flash lamp such as the EG&G FXQ-856 is essential since measurements are carried out in single-beam mode (i.e., the I_0 of Beer's law is measured using a different flash from I). For such a lamp, pulse-to-pulse reproducibility of 0.3% is routinely achieved. The resulting limit in precision is comparable to that produced by the single-scan, 16,000 photoelectron/

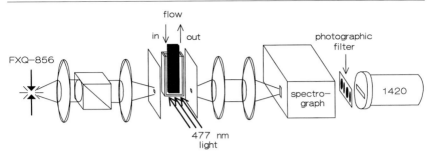

FIG. 1. Schematic of an apparatus for measuring time-resolved difference spectra. Rhodopsin samples pumped into the optical path contained under the opaque cuvette insert are photolyzed with a pulse of vertically polarized, 477-nm light. Absorbance changes are probed with light from the flash lamp that is collimated before it passes through the polarizer and then is refocused onto an opaque mask, which confines the beam to the photolyzed region under the cuvette insert. Transmitted light is then refocused onto the slit of a spectrograph which disperses it across the channels of the OMA detector. Response at all detector channels is balanced using a piece of photographic film previously exposed to the flash lamp and then developed.[7] Since the spectrograph transmits light in second order as well as in first order, where necessary order blocking filters (not shown) must be used with this system.

channel dynamic range of the Princeton Applied Research Model 1420 (R option) microchannel plate-intensified diode array detector (similar to the Princeton Instruments IRY 700 or the Andor DH520). Accurate synchronization of the laser pulse, flash lamp, and OMA gate is accomplished under computer control by use of digital delay generator such as the Stanford Research Systems DG-535. When all else is optimized, the S/N ratio across the spectrum is only limited by the intensity variation due to spectral lines in the flash lamp output. This factor can be substantially eliminated by using a simple photographic filter to balance the intensity seen by all detector channels.[7] In typical applications the probe flash itself bleaches ~0.25% of the rho present. Further, half of this bleaching occurs after the OMA gate and thus does not affect the measured absorbance.

Probe Beam Polarization

Probe beam polarization must be carefully considered within the context of the specific experimental objectives. Depending on whether rhodopsin is in detergent or membrane suspension, absorbance transients due to rotational diffusion are expected to occur on the 100-ns or 5-μs timescale, respectively.[8] Observation of much slower reactions are thus expected to

---

[7] J. W. Lewis, S. J. Hug, and D. S. Kliger, *Rev. Sci. Instrum.* **58**, 2342 (1987).
[8] R. A. Cone, *Nature New Biol.* **236**, 39 (1972).

have almost no dependence on probe beam polarization. Hence, in at least some experiments elaborate precautions to prevent rotational diffusion artifacts are probably unnecessary, especially if the signal of interest is relatively large.

Even on the relevant rotational timescales, choice of probe polarization at the magic angle, 54.7°, dramatically reduces observation of absorbance changes due to rotation of, as opposed to evolution of, photointermediates. Of course, care must be taken with this approach in unusual situations[9] or where very small components of the signal are considered, but it is operationally easier than the more cumbersome method of measuring $A_\parallel$ (the absorbance with parallel actinic and probe polarizations) and $A_\perp$ (the absorbance with perpendicular actinic and probe polarizations) separately at each time delay and computing $\overline{A} \equiv [A_\parallel + 2A_\perp]/3$, which is rotationally invariant if cylindrical symmetry prevails. It is important to note that even the latter method is not infallible since in membrane suspensions of rhodopsin actinic beam deviations usually destroy the cylindrical symmetry that is required for $\overline{A}$ to be completely free of rotational contributions.[10] Thus, simple steps that reduce rotational contributions are usually worthwhile taking, but even when elaborate precautions are taken, a rotational origin for small contributions must always be considered. For example, in cases of two-dimensional rotational diffusion, such as rhodopsin in membrane, a slight degree of anisotropy persists even after the 2-D rotation is complete[11] and can be detected in very careful measurements. In practice, it is a good idea to determine rotational diffusion times for samples using linear dichroism in order to identify time regions where rotational diffusion might be responsible for changes seen in absorption measurements.

Signal Averaging Considerations

Because rhodopsin photolysis is irreversible the sample must be replaced after each measurement so that all measurements are performed on samples with the same composition. Flow cells are therefore required, and since they must be manipulated in the dark, it is prudent to use a relatively simple commercial cell with flow channeled through an insert made from unbreakable material. Designs have been developed with a 1-cm path length[12] (24-μl volume) and 2-mm path length[13] (1-μl volume). For either

[9] J. W. Lewis and D. S. Kliger, *Photochem. Photobiol,* **54,** 963 (1991).
[10] S. Jäger, I. Szundi, J. W. Lewis, T. L. Mah, and D. S. Kliger, *Biochemistry* **37,** 6998 (1998).
[11] R. J. Cherry and R. E. Godfrey, *Biophys. J.* **36,** 257 (1981).
[12] J. W. Lewis, J. Warner, C. M. Einterz, and D. S. Kliger, *Rev. Sci. Instrum.* **58,** 945 (1987).
[13] J. W. Lewis and D. S. Kliger, *Rev. Sci. Instrum.* **64,** 2828 (1993).

design a computer-controlled pump can be constructed from a linear stepper motor, such as the Airpax L92121-P2 (Thomson Industries, Port Washington, NY) driving the appropriate size glass syringe. Temperature control can be achieved either using a thermostatted insert made from copper or via a conventional thermostatted jacket, but in either case care must be taken that the sample flow is slow enough that samples have time to equilibrate to the correct temperature. To speed this process when the measurement temperature is more than a few degrees away from ambient, it is useful to have several sample volumes of preheated (or precooled) tubing preceding the optical path. Samples should be degassed at a temperature no lower than that at which measurements will be conducted in order to prevent bubbles from forming in the optical path during thermal equilibration.

The sequence of data collection should be designed to preclude any systematic difference either between data collected at different time delays or between I_0 and $I(\Delta t)$ measurements at a particular time delay. For example, the power of the actinic laser pulse might steadily drift over the course of an experiment. If data were collected with the time delay steadily increasing as the experiment progressed, the effect of the power drift might be attributed to sample kinetics. Consequently, it is important to repeat measurements at all time delays and intersperse them in a pattern designed to detect drifts and, if possible, cancel their effects on the averaged data.

Similarly, the pattern of pumping and laser photolysis must be carefully considered to preclude systematic differences in conditions prevailing during I_0 and $I(\Delta t)$ measurement. Consider the case shown in Fig. 2 of an I_0 measurement using four probe flashes in the absence of actinic pulses. While the sample must obviously be changed after each probe flash of an $I(\Delta t)$ measurement when laser photolysis is occurring, it is also advisable to pump a sample volume between each of the probe flashes of the I_0 measurement in order to prevent the bleaching due to the four probe flashes from accumulating. However, pumping during I_0 collection can cause another problem since sample exchange is never more than about 95% effective at practical levels of flow. As shown in the second row of Fig. 2, during the measurement of $I(\Delta t)$ this results in laser photolysis of a solution containing only 99% unbleached rhodopsin (assuming a typical laser bleach of 20%) rather than the >99.75% unbleached rho sampled by the probe pulses of an I_0 measurement that includes pumping (row 3 of Fig. 2). Thus, pumping alone during I_0 measurement can cause a larger error in this case than omitting pumping during I_0 collection entirely. A better procedure is to introduce laser photolysis pulses into the I_0 measurement cycle shortly *after* each probe flash pulse and before pumping. As shown by the fourth row of Fig. 2, this causes the composition of the solution in the cuvette to remain constant for both I_0 and $I(\Delta t)$ as it should in order for Beer's law

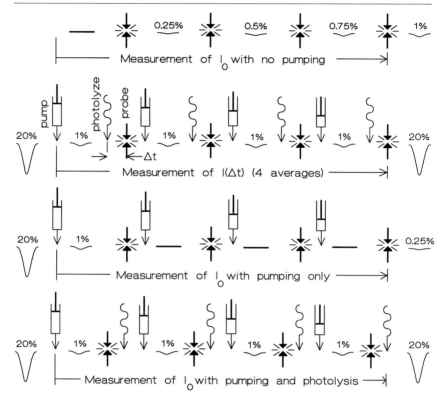

FIG. 2. Measurement of I_0 and $I(\Delta t)$, the light intensities transmitted by a sample immediately prior to and at a time Δt after photolysis by a laser pulse, respectively. The top row shows I_0 measurement beginning with 100% unphotolyzed rhodopsin using four probe flashes with no pumping. The spectrum of rhodopsin bleached by the preceding probe flashes is shown schematically and given in percent. The second row shows the more complicated sequence required to measure $I(\Delta t)$ including pumping and laser photolysis. Explicitly incorporated is the approximately 95% efficiency of sample replacement. This accounting process requires the bleached rhodopsin content of the cell to be specified immediately before data collection begins, and here a 20% bleach caused by a single laser pulse is assumed. Again the bleached rhodopsin content of the cell is shown immediatley prior to photolysis by the individual laser pulses. It is this composition that should be probed by an I_0 measurement in order for the Beer's law absorbance $\equiv -\log[I(\Delta t)/I_0]$ to correctly reflect the difference spectra of intermediates formed by the laser pulse. The third row shows that pumping alone during I_0 collection actually increases the discrepancy. The bottom row shows how laser pulses after pump/probe cycles can recreate the conditions needed for accurate I_0 measurement. An important practical requirement met by the sequences in rows 2 and 4 is that they begin and end with the cell in the same state. This property is obviously needed for continuous data acquisition.

to be correctly interpreted. Imperfections in sample flow could lead to artifactual photolysis signals if the bleaching product that remains after pumping (1%) absorbs laser light and has significant photolysis signals, but this is usually not the case.

Other steps to preclude systematic $I(\Delta t)$ and I_0 differences include varying whether $I(\Delta t)$ is measured before or after I_0 at a particular time delay. This becomes important if there is any slow kinetic process (such as metarhodopsin-III formation) that might proceed appreciably during the time that an absorbance spectrum computed from $I(\Delta t)$ and I_0 is being viewed or stored. Analyses similar to that shown in Fig. 2 can be conducted to show that the effect of any such constant delay prior to collection of I and I_0 will cancel when $I(\Delta t)$ is collected first half the time. As might be expected, if slow kinetic processes occur after photolysis for a particular sample, it is important to avoid pauses in data collection that would allow substantial concentrations of the slow photoproduct to build up in the cell. The presence of pumping during I_0 measurement serves to prevent what would otherwise amount to an extended pause in pumping.

Measurement of Bleach

Time-resolved absorption measurements on visual pigments are useful because a photointermediate's λ_{max} reports on the chromophore's environment at that point in the activation pathway. Further, the shape of the absorbance bands that are observed gives information about the homogeneity of the species present as a function of time. None of this information can be directly determined from difference spectra, and interpretation requires that difference spectra be converted to absolute spectra by adding back the bleach, which is the spectrum of the visual pigment that was present during the I_0 measurement and absent during measurement of I. In contrast to photochemical systems that cycle, for visual pigments the bleach can often be determined directly. A procedure for doing this has been described elsewhere,[14] and it can be modified as needed after consideration of such factors as the particular visual pigment system's detergent stability and/or formation of isopigment.

Global Fitting of Time-Dependent Spectra

The set of time-dependent absorption difference spectra, $\{\Delta A(\lambda, t)\}$ collected as just described can be well fit by a sum of exponential decays:

---

[14] A. Albeck, N. Friedman, M. Ottolenghi, M. Sheves, C. M. Einterz, S. J. Hug, J. W. Lewis, and D. S. Kliger, *Biophys. J.* **55**, 233 (1988).

$$\Delta a(\lambda, t) \equiv b_1(\lambda) + b_2(\lambda) \exp(-t/\tau_1) + b_3(\lambda) \exp(-t/\tau_2) + \ldots \quad (1)$$

Such a functional form applies to any mechanism that contains time-independent rates (i.e., kinetic schemes that are first order or pseudo first order). Numerical aspects of fitting are handled well by commercial analysis software packages such as Matlab (Mathworks, Natick, MA) and only an overview is given here. Two types of parameters appear in Eq. (1): a large number of linear parameters contained in the b spectra, $b_i(\lambda)$, and a much smaller number of nonlinear parameters, the τ_i, which are the reciprocals of the apparent, or "observed," rates. Although the vast majority of the parameters are linear, those can always be uniquely fit (using least squares) if the τ_i are known. However, the τ_i, being nonlinear parameters, require a more elaborate, indeterminate search, which can be defeated by noise in the data. Singular value decomposition (SVD) is a technique that can be used to speed the linear part of the fitting by reducing the number of parameters that must be fit during each iteration of the nonlinear search.[15,16] In the process of doing this, SVD produces a set of basis spectra (not to be confused with the b spectra) equal in number to the difference absorption spectra, $\{\Delta A(\lambda, t)\}$, originally transformed. The basis spectra can be used in linear combinations to express the $\{\Delta A(\lambda, t)\}$ and have the important property that the relatively few basis spectra contributing the most to the $\{\Delta A(\lambda, t)\}$ tend to be related to the signal of interest while the remaining basis spectra that contribute relatively little to $\{\Delta A(\lambda, t)\}$ tend to represent only noise. If desired, dramatic noise reduction in the data is possible by omitting the majority of the insignificant basis spectra from the fitting process. Care must be taken not to overdo this procedure since omission of significant basis spectra causes the resulting b spectra to be distorted.

In practice, a data set is fit with more and more τ_i's until finally enough terms are included in Eq. (1) to give a set of residuals $\{r(\lambda, t) \equiv \Delta A(\lambda, t) - \Delta a(\lambda, t)\}$ that exhibits no spectral features above the noise level in the data set. An example of this process is shown in Fig. 3. Confidence in the τ_i values and b spectra obtained from this process can come only after repeated experiments where similar τ_i values and b spectra are obtained. If reproducibility cannot be achieved, it often signifies the presence of an additional component that is not being fit in the individual data sets because the noise level is too high. Further experiments should then be conducted to allow more (and hopefully more reproducible) parameters to be fit. Similar improvements can sometimes be achieved by restricting the wavelength range that is fit. OMA detectors can provide data over a wide spectral

---

[15] R. I. Shrager and R. W. Hendler, *Anal. Chem.* **54**, 1147 (1982).
[16] R. W. Hendler and R. I. Shrager, *J. Biochem. Biophys. Meth.* **28**, 1 (1994).

range (typically ~400 nm), and this often includes spectral regions with little or no absorbance change. Noise in those regions interferes with fitting and in marginal cases a better fit can be obtained by excluding them before fitting is attempted. However, collection of data in regions where no photochemical changes occur (near 700 nm for rhodopsin) can be used to correct for small baseline offsets caused by probe flash fluctuations. Optimum fits can often be obtained by initially using such data to correct for offsets and then excluding it from the final fitting process.

Determination of Possible Kinetic Schemes

Only for the simplest kinetic schemes do the parameters of Eq. (1) correspond to microscopic properties of the reaction mechanism. For example, if no time-dependent kinetic changes are detected, the data can be fit using only $b_0(\lambda)$. This corresponds to the case where photolysis of the sample produces a stable photoproduct, and $b_0(\lambda)$ gives the difference between that product's spectrum and the bleached material that went to form it. Even such a scheme is arguably not unique since, if the product's spectrum deduced from the data is unexpectedly complex, a branched scheme might be considered:

$$B \longleftarrow A \longrightarrow B'$$

versus

$$A \longrightarrow B$$

Expectations about the standard characteristics of intermediate spectra are an important tool in analysis and need to be taken advantage of if maximum insight into mechanisms is to be achieved.[17,18] As more terms in Eq. (1) can be fit to a data set, ever larger numbers of kinetic schemes become possible. The mechanism fitting problem does not consist of finding a single scheme that fits the data. Several may fit reasonably well, and the actual task is better described as eliminating schemes that do not fit the data. Once these are eliminated further experiments can be conducted to discriminate within those that fit. However, initially all possible schemes need to be tested. Consideration of all possible schemes is vastly more time consuming than collecting data and global fitting combined, particularly since the latter steps can be highly automated.

Originally, progress in fitting kinetic schemes was slow because each candidate scheme was approached using the closed form algebraic solution.

---

[17] J. F. Nagle, L. A. Parodi, and R. H. Lozier, *Biophys. J.* **38,** 161 (1982).
[18] I. Szundi, J. W. Lewis, and D. S. Kliger, *Biophys. J.* **73,** 688 (1997).

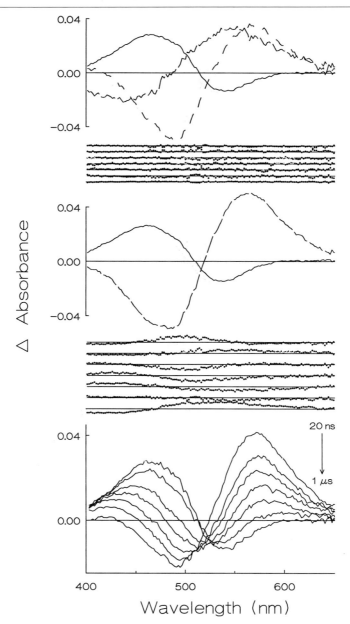

This procedure was able to resolve the correct scheme for batho decay through the blue-shifted intermediate (BSI) to form lumirhodopsin (lumi):

$$\text{rho} \rightsquigarrow \text{batho} \rightleftharpoons \text{BSI} \rightarrow \text{lumi} \ldots$$

from competing schemes:

$$\text{lumi} \leftarrow \text{batho}_f \rightsquigarrow \text{rho} \rightsquigarrow \text{batho}_s \rightarrow \text{lumi}'$$

$$\text{rho} \rightsquigarrow \text{batho} \rightarrow \text{BSI} \rightarrow \text{lumi}$$

and

$$\text{rho} \rightsquigarrow \text{batho} \rightarrow \text{lumi}$$
$$\updownarrow$$
$$\text{BSI}$$

for the case of the two exponential processes occurring from 20 ns to 1 μs after rhodopsin photolysis.[19] With considerably more effort the three exponential processes involved in lumi decay from 1 μs to 100 ms were also solved with an algebraic approach.[20] However, currently much more rapid progress is possible by extending the numerical methods used in the fitting process to the scheme determination step.[18] This allows the testing of arbitrary schemes essentially independent of their algebraic complexity, enabling a much more thorough and rapid exploration of the scheme "space" to be conducted.

[19] S. J. Hug, J. W. Lewis, C. M. Einterz, T. E. Thorgeirsson, and D. S. Kliger, *Biochemistry* **29,** 1475 (1990).
[20] T. E. Thorgeirsson, J. W. Lewis, S. E. Wallace-Williams, and D. S. Kliger, *Biochemistry* **32,** 13861 (1993).

FIG. 3. Difference spectra for early rhodopsin intermediates fit with one and two exponentials. Individual 24-μl samples of rhodopsin in 2% octylglucoside suspension ([rho] = 25 μM) were photolyzed at a temperature of 21° using a fluence of 0.08 mJ/mm². The bottom panel shows absorbance changes that were monitored at magic angle in a 1-cm path length at 7 time delays ranging from 20 ns to 1 μs with an OMA gate of 10 ns. Above the data (middle panels) are shown the residuals and b spectra from a single exponential fit which gave τ = 141 ns for the dashed b spectrum (solid line is b_0, the time-independent spectral change corresponding to the difference between the final product on the timescale of the measurement and the original sample). In the topmost panels are b spectra and residuals for a two exponential fit of the data which gave τ_1 = 36 ns (solid line) and τ_2 = 215 ns (dashed line). The fastest process is the decay of batho toward an equilibrated mixture of batho and BSI. The slow process is the decay of that mixture to lumi.[19]

Applications

Figure 4 shows typical data for spectral changes resulting from slower steps in the activation reactions of rhodopsin in membrane suspension collected using the larger scale apparatus.[12] These data were of sufficiently high S/N to detect not only the three exponential decays associated with the decay of lumi but also an additional small amplitude fourth component whose lifetime was in the 5- to 10-μs range associated with rotational diffusion of rhodopsin in membrane suspension. The b spectrum of this fast component was consistent with this assignment and the assignment was confirmed by analysis of linear dichroism data.[10] The amplitude of this fastest component in $\overline{A}$ is more variable than the components associated with decay of rho photointermediates, presumably because the actinic beam deviations that break down cylindrical symmetry and allow rotational diffusion to produce changes in $\overline{A}$ depend on sample turbidity, which varies considerably.

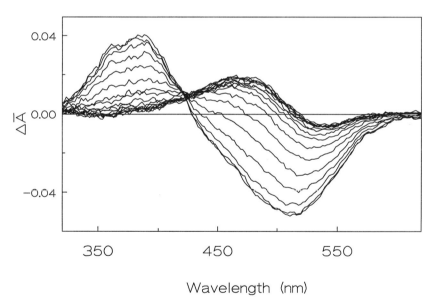

FIG. 4. Difference absorbance spectra measured with 15 mg of rhodopsin. Difference spectra were recorded with probe polarization parallel and perpendicular to the laser polarization and $\overline{A}$ was calculated. Individual 24-μl samples of rhodopsin in membrane suspension ([rho] = 5.4 μM) were photolyzed at a temperature of 20° using a fluence of 0.08 mJ/mm^2. Absorbance changes were monitored using a 1-cm path length at 18 time delays ranging from 1 μs (smallest amplitude) to 75 ms (largest amplitude) after photolysis. The OMA gate duration was 200 ns for parallel and 300 ns for perpendicular data collection. The sample was sonicated before use as described previously.[20]

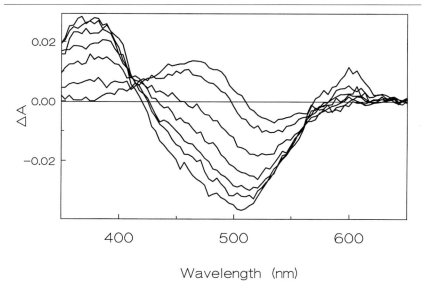

FIG. 5. Difference absorption spectra measured with 225 μg of rhodopsin in the presence of the pH indicator bromcresol purple. Difference spectra were recorded directly using magic angle polarization. Individual 1-μl samples of rhodopsin in lauryl maltoside (0.25%) suspension ([rho] = 25 μM) were photolyzed and the resulting absorbance changes were measured in a 2-mm path length at 7 times ranging from 10 μs to 5 ms. The increased absorbance at late times near 600 nm is due to deprotonation of bromcresol purple ([bromcresol purple] = 250 μM) as one 380-nm absorbing intermediate takes up a proton to form R*, another 380-nm absorbing intermediate.

Analysis of data similar to that in Fig. 4 indicates that there are two intermediates with similar absorption spectra, peaking near 380 nm, formed after photolysis of rhodopsin.[20] Subsequent work[10] indicated that while the earlier of these intermediates, meta-I_{380}, is apparently homogeneous, the one forming later, meta-II, is heterogeneous, occurring in unprotonated, meta-II'_a, and protonated, meta-II_b, forms. Such a system of isospectral intermediates provides a challenge to time-resolved optical methods but, as has been demonstrated for rhodopsin, one that is not beyond the capabilities of modern spectral detectors and matrix analytical methods. However, given the difficulties it is important that efforts be made to increase the information content of time-resolved absorbance techniques.

Figure 5 shows data collected with a much smaller amount of rhodopsin. The amount used is in the range obtainable from COS cell expression of rhodopsin mutants.[21] The data shown in Fig. 5 also demonstrate how time-

[21] S. Jäger, J. W. Lewis, T. A. Zvyaga, I. Szundi, T. P. Sakmar, and D. S. Kliger, *Biochemistry* **36**, 1999 (1997).

resolved spectral measurements can be used to simultaneously monitor rhodopsin photochemical processes and related events disclosed by an indicator dye. Detailed descriptions of such measurements are the subject of another article in this volume.[22] The data in Fig. 5 show clear evidence for proton uptake by one of the late, 380-nm absorbing rho intermediates.[10,23] This is demonstrated by the fact that the 590-nm absorbance of the deprotonated form of the indicator dye, bromcresol purple, does not appreciably increase during the formation of the first two 380-nm absorbing species, meta-I_{380} and meta-II'_a, and is delayed until the later meta-II_b appears. The data demonstrate the practicality of simultaneous study of rhodopsin photochemistry combined with proton uptake and release measurements using small sample sizes. As such they typify the wide range of studies that are currently possible.

Acknowledgments

The research described in this article was supported by a grant (EY00983) from the National Institutes of Health. We also gratefully acknowledge the contributions of our collaborators to this work.

[22] K. P. Hofmann, *Meth. Enzymol.* **315** [17] (1999) (this volume).
[23] I. Szundi, T. L. Mah, J. W. Lewis, S. Jäger, O. P. Ernst, K. P. Hofmann, and D. S. Kliger. *Biochemistry* **37**, 14237 (1998).

[13] Structural Determinants of Active State Conformation of Rhodopsin: Molecular Biophysics Approaches

By KARIM FAHMY, THOMAS P. SAKMAR, and FRIEDRICH SIEBERT

Introduction

G-protein-coupled receptors (GPCRs) are integral membrane proteins that couple to cytosolic heterotrimeric G proteins. Members of the rhodopsin family of GPCRs are unique in that they contain a bound ligand, which is converted to the active form by photoisomerization. The elucidation of the activation mechanism of GPCRs presents a major challenge in receptor research. Despite extensive efforts, details of how the binding of an agonist ligand, or the photoisomerization of the bound chromophore, induces conformational changes at the cytosolic surface necessary for activation of the G protein are still not understood. Because no structural information at molecular resolution is available for any GPCR, it is not possible to outline the precise signaling pathway from ligand binding (or retinal isomerization) to the activated structure. Nevertheless, in the case of rhodopsin, considerable progress has been made in developing satisfactory structural models for the transmembrane domain of the receptor.

The primary basis for this molecular modeling was the determination of a three-dimensional electron density map of rhodopsin at 7.5-Å resolution in the approximate plane of the membrane bilayer. At this resolution, the seven transmembrane helices can be appreciated and their tilts with respect to an arbitrary membrane normal can be measured.[1] The comparison of the primary structures of a large number of related GPCRs allowed the identification of residues in the seven helices that were most likely to be facing the lipid phase as opposed to those involved in helix–helix contacts or in the ligand binding site.[2,3] Solid-state nuclear magnetic resonance (NMR) investigations provided information on the geometry of the retinal binding site.[4] These data and constraints obtained by optimizing hydrogen-bonding interactions[5] or from the results of site-directed mutagenesis[6] allowed the development of refined molecular models of rhodopsin structure. The models are not precise enough to deduce reliably the protein conformational changes involved in the activation of the pigment from the structural differences between 11-*cis*- and all-*trans*-retinylidene chromophores. However, site-directed EPR spin-labeling experiments and the introduction of reversible cross-links between transmembrane helices 3 and 6 demonstrated that the activation process involves an increase of the distance between the cytoplasmic ends of these two helices.[7,8] In addition, the spin-labeling experiments demonstrated conformational changes of the cytosolic loops that had been postulated from earlier studies.[9–11] However, it is still not precisely known how these changes are evoked by the isomerization of the

---

[1] V. M. Unger, P. A. Hargrave, J. M. Baldwin, and G. F. X. Schertler, *Nature* **389**, 203 (1997).

[2] J. M. Baldwin, *EMBO J.* **12**, 1693 (1993).

[3] J. M. Baldwin, G. F. Schertler, and V. M. Unger, *J. Mol. Biol.* **12**, 144 (1997).

[4] M. Han and S. O. Smith, *Biochemistry* **34**, 1425 (1995).

[5] I. D. Pogozheva, A. L. Lomize, and H. I. Mosberg, *Biophys. J.* **70**, 1963 (1997).

[6] T. Shieh, M. Han, T. P. Sakmar, and S. O. Smith, *J. Mol. Biol.* **269**, 373 (1997).

[7] D. L. Farrens, C. Altenbach, K. Yang, W. L. Hubbell, and H. G. Khorana, *Science* **274**, 768 (1996).

[8] S. P. Sheikh, T. A. Zvyaga, O. Lichtarge, T. P. Sakmar, and H. R. Bourne, *Nature (London)* **383**, 347 (1996).

[9] C. Altenbach, K. Yang, D. L. Farrens, Z. T. Farahbakhsh, H. G. Khorana, and W. W. Hubbell, *Biochemistry* **35**, 12470 (1996).

[10] Z. T. Farahbakhsh, K. D. Ridge, H. G. Khorana, and W. L. Hubbell, *Biochemistry* **34**, 8812 (1995).

chromophore and subsequent Schiff base deprotonation in metarhodopsin-II (MII).

To elucidate the information pathway leading to the active state conformation of the receptor, molecular changes in the interior ligand-binding domain must be detected and amino acids that are directly or indirectly linked to these changes have to be identified. Fourier transform infrared (FTIR)-difference spectroscopy is an excellent tool to detect such internal molecular changes. In combination with site-directed mutagenesis, FTIR-difference spectroscopy of recombinant mutant pigments can also identify the essential contributing amino acids. In short, FTIR-difference spectroscopy can monitor molecular vibrations of groups within rhodopsin (e.g., amino acid side chains, peptide backbone, and chromophore) as it changes from inactive to active conformation. The molecular vibrations are probes of changes in chemical structure and environment. The study of the activation mechanism of rhodopsin with FTIR spectroscopy always involves the measurement of a difference spectrum between rhodopsin and one of its photoproducts. Therefore, characteristics of both of these species (rhodopsin and a particular photoproduct) are reflected in the data.

Examples are described in this chapter that illustrate how FTIR-difference spectroscopy in combination with site-directed mutagenesis has contributed to our understanding of the activation mechanism of rhodopsin. The main focus is on molecular changes in the interior of the protein. We emphasize studies of the photoproduct of primary biochemical interest, MII, which has been shown to be the active state of the receptor responsible for G_t activation.[12,13] Earlier photoproducts are described only in a few select cases. We also demonstrate that FTIR spectroscopy, in combination with the attenuated total reflection (ATR) sampling technique is equally well suited to monitor pH-dependent molecular changes, and to study protein–protein interactions, in particular, the interaction of activated rhodopsin with G_t. Original literature will be cited for experimental details. The general approach of infrared spectroscopy has been reviewed recently.[14,15]

Preparation of Recombinant Mutant Pigments for FTIR Study

Bovine rhodopsin can be studied in its natural environment, the rod outer segment disk membrane, purified in detergent solution, or reconsti-

---

[11] J. F. Resek, Z. T. Farahbakhsh, W. L. Hubbell, and H. G. Khorana, *Biochemistry* **32**, 12025 (1993).
[12] D. Emeis, H. Kühn, J. Reichert, and K. P. Hofmann, *FEBS Lett.* **143**, 29 (1982).
[13] C. Longstaff, R. D. Calhoon, and R. R. Rando, *Proc. Natl. Acad. Sci. U.S.A.* **83**, 4209 (1986).
[14] F. Siebert, *Isr. J. Chem.* **35**, 309 (1995).
[15] F. Siebert, *Mikrochim. Acta (Suppl.)* **14**, 43 (1997).

tuted into natural or synthetic phospholipids. Recently, methods have been developed to study recombinant expressed bovine pigments in detergent or after reconstitution into lipid vesicles.

Opsin genes, or opsin mutant genes, are expressed in COS-1 cells and the pigments resulting from reconstitution with 11-*cis*-retinal are purified from the cell membranes by immunoaffinity adsorption as previously described.[16,17] The addition of 11-*cis*-retinal and all subsequent steps are carried out under dim red light. Recombinant pigments are characterized by UV–visible spectroscopy and G_t activation assays as previously described.[18,19] To reconstitute the recombinant pigments into phospholipid vesicles, the purified pigments are obtained in 0.5 mM sodium phosphate (pH 6.5) and 1.5% (w/v) octylglucoside detergent solution. Phosphastidylcholine (L-α-lecithin) from fresh egg yolk is lyophilized and resuspended in 1 mM sodium phosphate (pH 6.5). The phospholipid suspension is mixed with a purified rhodopsin sample in a molar ratio of 100:1 (phospholipid:pigment) and kept on ice for at least 1 hr. The suspension is incubated for 10 min in a bath sonicator and then dialyzed against 1 mM sodium phosphate (pH 6.5) for 24 hr at 8° with a 20,000 molecular weight cutoff membrane in a microdialyzer unit (Pierce, Rockford, IL) in flow-through mode. Vesicles are pelleted by centrifugation at 80,000g (4°, 16 hr). The vesicles are then washed and resuspended in a small volume of 1 mM sodium phosphate (pH 6.5). The concentration of the pigment in the vesicles is determined by UV–visible spectroscopy.

FTIR Spectroscopy Investigation of Active State of Rhodopsin

Spectral Features Characterizing the Active State

Figure 1 shows FTIR-difference spectra of bovine rhodopsin and selected recombinant pigments. The upper four spectra have been published previously[20]; the lower spectrum is from M. Beck.[21] Negative bands are due to the dark state; positive bands to the active state MII. As shown before, the mutant pigments D83N/E122Q and H211N on illumination form normal MII-like states that have the ability to activate G_t. The upper

---

[16] R. R. Franke, T. P. Sakmar, R. M. Graham, and H. G. Khorana, *J. Biol. Chem.* **267**, 14767 (1992).
[17] D. D. Oprian, R. S. Molday, R. J. Kaufman, and H. G. Khorana, *Proc. Natl. Acad. Sci. U.S.A.* **84**, 8874 (1987).
[18] K. Fahmy and T. P. Sakmar, *Biochemistry* **32**, 7229 (1993).
[19] K. C. Min, T. A. Zvyaga, A. M. Cypess, and T. P. Sakmar, *J. Biol. Chem.* **268**, 9400 (1993).
[20] M. Beck, T. P. Sakmar, and F. Siebert, *Biochemistry* **37**, 7630 (1998).
[21] M. Beck, Ph.D. Dissertation, Albert-Ludwigs-Universität, Freiburg, 1998.

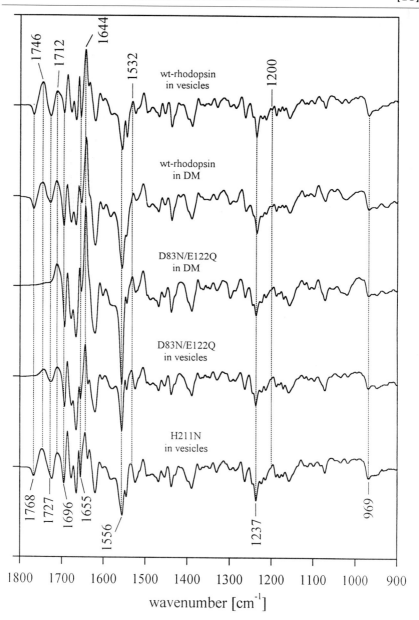

FIG. 1. Rhodopsin → metarhodopsin-II FTIR-difference spectra of (from top to bottom) wild-type (wt) bovine rhodopsin in vesicles,[20] wild-type rhodopsin in dodecylmaltoside detergent,[20] mutant pigment D83N/E122Q in dodecylmaltoside detergent,[20] double mutant D83N/E122Q in vesicles,[20] and mutant pigment H211N in vesicles.[21]

spectrum [wild-type (wt) rhodopsin in vesicles] is obtained from bovine rhodopsin reconstituted into vesicles prepared from egg lecithin. We have demonstrated that this spectrum is indistinguishable from that obtained from rod outer segment disk membranes.[20] The MII spectra are characterized, in contrast to those of earlier photointermediates, by large changes in the amide-I and amide-II spectral ranges. The amide-I (1690–1620 cm^{-1}) and amide-II (1570–1500 cm^{-1}) spectral regions represent mainly the C=O stretching vibration, and the NH bending coupled to the C–N stretching vibration of the peptide groups, respectively. The magnitude of changes in these spectral regions indicates significant changes of the protein backbone in converting from rhodopsin to MII.[22] These changes in the peptide backbone structure are congruent with the conformational changes required to switch the inactive rhodopsin into MII to allow G_t activation.

Bands of the 11-*cis* chromophore in the dark state can be seen at 1237 and 969 cm^{-1}. The 1237-cm^{-1} band represents a delocalized mode involving C–C stretching, CH bending, and NH bending vibrations of the Schiff base region up to C-14.[23] The 969-cm^{-1} band represents the C-11–C-12 hydrogen-out-of-plane (HOOP) mode of the chromophore that is twisted by intrinsic steric hindrance.[24] Compared with a protonated retinylidene Schiff base, the infrared absorption strength of the chromophore linked by an unprotonated Schiff base is considerably reduced.[25] Therefore, the larger positive bands in the spectra of Fig. 1 must be assigned to molecular vibrations of groups of the protein.

The bands (negative and positive) between 1780 and 1700 cm^{-1}, and the negative and positive bands at 1655 and 1644 cm^{-1}, respectively, have been identified as special features characterizing the transition to the active MII state. The absorption in the former spectral range is caused by the C=O stretching vibration of specific protonated internal carboxyl groups (see later discussion), whereas, the latter two bands reflect backbone changes of the protein (amide-I vibrations). The correlation of these spectral features with the active state was possible by comparison of the spectra of the MI and MII states[22] and by studying rhodopsin pigments with artificial chromophores that caused defects in the G_t activation capability.[22,26]

The comparison of the spectra of rhodopsin in lipid vesicles and in dodecylmaltoside (DM) detergent (wt rhodopsin in DM) shows a great

---

[22] U. M. Ganter, E. D. Schmid, D. Perez-Sala, R. R. Rando, and F. Siebert, *Biochemistry* **28**, 5954 (1989).
[23] U. M. Ganter, W. Gärtner, and F. Siebert, *Biochemistry* **27**, 7480 (1988).
[24] U. M. Ganter, W. Gärtner, and F. Siebert, *Eur. Biophys. J.* **18**, 295 (1990).
[25] F. Siebert and W. Mäntele, *Biophys. Struct. Mech.* **6**, 147 (1980).
[26] F. Jäger, S. Jäger, O. Kräutle, N. Friedman, M. Sheves, K. P. Hofmann, and F. Siebert, *Biochemistry* **33**, 7389 (1994).

deal of similarity. Nevertheless, two differences between the spectra of the pigments in vesicles and detergent are remarkable. First, the difference band at $1727(-)/1745(+)$ cm^{-1} is smaller for the sample in detergent solution, and for this reason reveals a shoulder at 1735 cm^{-1}. Second, the amide-I bands have larger intensities in the detergent solubilized sample. This detergent effect on the amide-I bands intensities can be explained if orientational effects, in the case of rhodopsin in vesicles, reduce band intensities for modes with preferential absorption perpendicular to the membrane plane.[27] The spectral deviation in the $1727(-)/1745(+)$ cm^{-1} difference band is discussed later.

Because the bands in the spectral range between 1780 and 1700 cm^{-1}, which are mainly caused by protonated carboxyl groups, are characteristic of the active state, the effects of site-directed mutations of aspartic and glutamic acid residues are of special interest. In the double mutant D83N/E122Q, Asn and Gln replace the membrane-embedded carboxyl groups Asp-83 on helix 2 and Glu-122 on helix 3, respectively. The FTIR-difference spectra of D83N/E122Q in detergent and in vesicles are shown in Fig. 1. The negative band at 1768 cm^{-1} is absent and the band intensity at 1745 cm^{-1} is reduced in the vesicle preparation (D83N/E122Q in vesicles). Furthermore, if the mutant is prepared in detergent (D83N/E122Q in DM), the differential band at $1727(-)/1745(+)$ cm^{-1} disappears. Thus, in combination with measurements of the single mutants D83N and E122Q, it was concluded that protonated Asp-83 causes a differential band at $1768(-)/1750(+)$ cm^{-1} and protonated Glu-122 causes a differential band at $1727(-)/1745(+)$ cm^{-1} with a negative shoulder at 1735 cm^{-1}.[27] Although these conclusions are valid for rhodopsin in detergent, a certain modification of this interpretation has to be made based on the spectrum of the double mutant measured in lipid vesicles.[28] This spectrum still reveals a differential band at $1726(-)/1743.5(+)$ cm^{-1}, even though both Asp-83 and Glu-122 have been replaced. Since this band is not influenced by ^{2}H$_2$O, it has been concluded that it cannot be caused by a protonated carboxyl group since deuteration causes a downshift of the C=O stretch by 2–10 cm^{-1}. Thus, the $1726(-)/1743.5(+)$ cm^{-1} band has been assigned to the ester C=O stretching mode of the phospholipids, and preliminary experiments with ^{13}C isotopic labeling of the phospholipid carbonyl carbon support this identification.

Mutation of His-211 has been reported to influence the equilibrium between the MI and MII photoproducts. Based on biochemical character-

[27] K. Fahmy, F. Jäger, M. Beck, T. A. Zvyaga, T. P. Sakmar, and F. Siebert, *Proc. Natl. Acad. Sci. U.S.A.* **90**, 10206 (1993).
[28] M. Beck, F. Siebert, and T. P. Sakmar, *FEBS Lett.* **436**, 304 (1998).

ization of site-directed mutant pigments, His-211 was proposed to take part in the proton uptake process involved in the formation of the MII active state.[29] However, as noted earlier, the H211N mutation does not appear to affect the MI/MII equilibrium nor to reduce the capability of MII to activate G_t. Therefore, His-211 does not directly participate in the proton uptake process that is concomitant with the formation of MII.[30] Nevertheless, as is evident from the lower FTIR-difference spectrum in Fig. 1, the H211N mutation causes spectral alterations around 1725 cm^{-1}. In studies of MI-like spectra of the H211N mutant pigment, we could show that the mutation causes replacement of the double-peak structure at 1727 and 1734 cm^{-1} assigned to two C=O stretching modes of protonated Glu-122 in the dark state (see earlier discussion) by a single band at 1721 cm^{-1}.[20] The superposition with the lipid band identified earlier causes the spectral feature in the spectrum of the H211N mutant in Fig. 1. We have further demonstrated that the unusual downshift of the C=O stretch of Glu-122 in wild-type MI to 1701 cm^{-1} is reverted into an upshift to 1729 cm^{-1} in the MI-like spectrum of H211N.

The MI-like spectrum of H211N also displays appreciable spectral changes between 1250 and 1150 cm^{-1} that are probably not caused by the disappearance of histidine modes. Therefore, we have concluded that in wild-type rhodopsin there is a special interaction between Glu-122 and an additional group, which is modified by the His-211 mutation.[20] From these findings we have concluded that the infrared spectra of these mutant pigments can only be explained by a close interaction between helices 3 and 5 in the neighborhood of these residues. According to the recent electron density map deduced from electron cryomicroscopy,[1] and according to an α-carbon model of GPCRs,[3] such a juxtaposition of transmembrane helices 3 and 5 is possible. Furthermore, the data strongly support two recent models that have proposed interactions between His-211 and Glu-122.[5,6]

Molecular Changes Leading to Receptor Activation as Deduced from FTIR Spectroscopy

FTIR-difference spectroscopy studies of rhodopsin, recombinant expressed opsins, and mutant opsins have identified several changes in molecular structure that lead from inactive rhodopsin to the active MII:

1. *Backbone changes:* Structural changes in the protein backbone cause the strong amide-I bands at 1656 cm^{-1} (negative) and 1645 cm^{-1} (positive) in the transition to MII.

---

[29] C. J. Weitz and J. Nathans, *Neuron* **8**, 465 (1992).
[30] G. B. Cohen, D. D. Oprian, and P. R. Robinson, *Biochemistry* **31**, 12592 (1992).

2. *Asp-83:* The protonated carboxylic acid side chain of Asp-83 undergoes an environmental change from a very hydrophobic surrounding in the dark state to moderate hydrogen-bonding conditions in MII. This residue is conserved in many GPCRs, although its precise role is not unequivocally settled.
3. *Glu-122:* In the dark state, the protonated carboxylic acid side chain of Glu-122 is in a special environment that causes the splitting of the C=O stretching vibration. The band at 1735 cm^{-1} shifts to 1725 cm^{-1} in the presence of ^{2}H$_2$O, superimposing onto the band at 1727 cm^{-1}, which is largely unaffected (it may also be shifted to 1725 cm^{-1}).[22] We have proposed that Glu-122 exhibits heterogeneous vibrational modes.

In one state, the C=O stretch shows the common 10-cm^{-1} deuteration-induced downshift, indicating strong coupling with the OH bending mode. Therefore, the movement of the hydrogen must be mainly within the OCO plane. In the other state, the C=O stretching band is almost completely decoupled from the OH bending mode, which could be caused by a mainly out-of-plane character of the movement of the hydrogen. In MII, only a single band is observed at 1745 cm^{-1} that shifts down to 1739 cm^{-1}. Therefore, not only is the coupling behavior altered, but also the uncoupled C=O stretching vibration is upshifted from 1727 to 1739 cm^{-1}.

Influence of Schiff Base Environment

The formation of the active state of rhodopsin is coupled to the deprotonation of the Schiff base. Mutagenesis studies identified Glu-113 in bovine rhodopsin to be the negatively charged counterion that stabilizes the positive charge of the protonated Schiff base.[31–33] The Schiff base counterion is the obvious leading candidate to fulfill the role of Schiff base proton acceptor during the conversion of MI to MII. For example, in bacteriorhodopsin, FTIR investigations identified the primary counterion of the protonated Schiff base, Asp-85, to be the proton acceptor in the Schiff base deprotonation reaction.[34–36] Furthermore, mutations that remove the

---

[31] J. Nathans, *Biochemistry* **29**, 9746 (1990).
[32] T. P. Sakmar, R. R. Franke, and H. G. Khorana, *Proc. Natl. Acad. Sci. U.S.A.* **86**, 8309 (1989).
[33] E. A. Zhukovsky and D. D. Oprian, *Science* **246**, 928 (1989).
[34] M. S. Braiman, T. Mogi, T. Marti, L. J. Stern, H. G. Khorana, and K. J. Rothschild, *Biochemistry* **27**, 8516 (1988).
[35] K. J. Rothschild, M. S. Braiman, Y. He, T. Marti, and H. G. Khorana, *J. Biol. Chem.* **265**, 16985 (1990).
[36] K. Fahmy, O. Weidlich, M. Engelhard, J. Tittor, D. Oesterhelt, and F. Siebert, *Photochem. Photobiol.* **56**, 1073 (1992).

negative charge at position 113, or replace Lys-296 at the site of the Schiff base chromophore linkage, constitutively activate the mutant opsins even in the absence of chromophore.[30] The role of the environment around the Schiff base, which can be altered by mutation at one or more sites, on receptor function can be addressed by FTIR-difference spectroscopy methods. Selected FTIR spectroscopy experiments designed to elucidate the Schiff base environment and its role in receptor activation are described later.

The FTIR MII-difference spectra of wild-type rhodopsin, and mutants E113A, E113A/A117E, G90D, and G90D/E113A are shown in Fig. 2. All measurements were performed on detergent-solubilized samples. Therefore, the lipid band described earlier does not contribute. The comparison of the first two spectra allowed the identification of Glu-113 as the net proton acceptor for the Schiff base deprotonation reaction.[37] Despite the neutralization of the counterion by mutation, a pigment with a protonated Schiff base can still be obtained provided the pH is low enough and a solute anion such as Cl^- is present.[32,38] The spectrum of mutant E113A (Fig. 2) was obtained at pH 4.2 in the presence of 800 nmol NaCl. Under these conditions a Cl^- ion substitutes for the carboxylate Schiff base counterion. The spectrum of the acid form of E113A is very similar to that of wild-type rhodopsin. Nevertheless, a few deviations are obvious. In the spectrum of the mutant, the band normally observed at 1712 cm^{-1} has very little intensity and its position is shifted to 1715 cm^{-1}; the band at 1656 cm^{-1} has lower intensity and the band at 1638 cm^{-1} has larger intensity. In addition, the band at 1238 cm^{-1} is missing in the spectrum of E113A. These spectral differences in the E113A mutant can be explained most directly if it is assumed that replacement of Glu-113 also removes the proton acceptor group.[36]

The lack of a band at 1712 cm^{-1} in the spectrum of E113A can be explained by the removal of the protonated carboxylic acid side chain of Glu-113. Thus, the 1712-cm^{-1} band is identified to be the C=O stretching vibration of Glu-113 in rhodopsin. A protonated carboxyl group cannot cause the residual band at 1715 cm^{-1} since it is not downshifted by deuteration. The spectral changes at 1656 and 1638 cm^{-1} could be due to the shift of the C=N stretching mode from 1656 cm^{-1} in wild-type rhodopsin to 1638 cm^{-1} in the mutant. However, it must be noted that these modes, especially the 1656-cm^{-1} band, are always highly variable (e.g., compare the spectra of two wild-type rhodopsin samples in detergent presented in Figs. 1 and 2).

[37] F. Jäger, K. Fahmy, T. P. Sakmar, and F. Siebert, *Biochemistry* **33**, 10878 (1994).
[38] T. P. Sakmar, R. R. Franke, and H. G. Khorana, *Proc. Natl. Acad. Sci. U.S.A.* **88**, 3079 (1991).

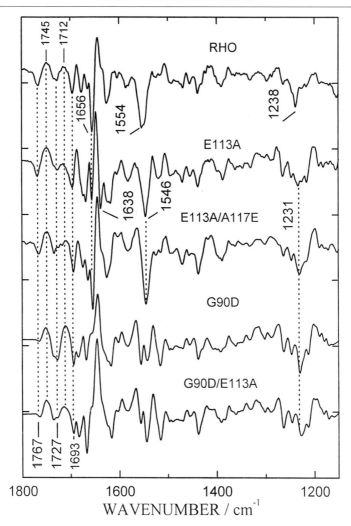

FIG. 2. Rhodopsin → metarhodopsin-II FTIR-difference spectra of (from top to bottom) wild-type rhodopsin[37] and of the mutants E113A,[37] E113A/A117E,[46] G90D,[50] and G90D/E113A.[49] All measurements were performed in dodecylmaltoside detergent.

The negative band at 1238 cm^{-1} of rhodopsin is caused by a complex mode involving C-12–C-13 and C-14–C-15 stretching vibrations and N–H, C_{14}–H, and C_{15}–H bending vibrations. Experiments and theoretical models suggest that the Schiff base is hydrogen bonded to water molecule(s).[23,26,39–41] Such a special interaction will certainly influence the N–H bending vibration, and therefore the 1238-cm^{-1} chromophore mode as well. Indeed, FTIR experiments of the E113Q mutant in the spectral range above 2500 cm^{-1} have indicated that a water molecule sensitive to the chromophore isomerization state is removed by the Glu-113 replacement, thereby causing a N–H stretching mode typical of chloride salts of protonated retinylidene Schiff base model compounds.[42]

Interestingly, it also has been shown that the photoproduct created directly from the unprotonated Schiff base species of the E113Q mutant is capable of activating G_t with high quantum yield.[43] Although this result may bear on the mechanism of activation of UV-sensitive visual pigments in general, it is remarkable that the proton transfer from the Schiff base to Glu-113 per se seems not to be required for attaining the active state. Thus, it could be that either neutralization of the Schiff base or neutralization of the residue at position 113 would be sufficient for receptor activation. To try to distinguish between these two possibilities, a double-replacement mutant was constructed in which the counterion was moved by one helix turn so that the protonated Schiff base could be stabilized without necessarily providing a proton acceptor group at the proper orientation. Biochemical studies of this double-replacement mutant (E113A/A117E) showed that its active photoproduct did in fact contain a protonated Schiff base. Thus, the conformation of the E113A/A117E mutant that could activate G_t contained a charged Schiff base, but a neutral residue at position 113 due to the mutation.[44,45] Therefore, it was of particular interest to monitor the protonation states of carboxyl groups and the molecular changes leading to the active state of the E113A/A117E mutant by FTIR spectroscopy.

The comparison of the spectrum of the E113A/A117E mutant[46] with that of wild-type rhodopsin (Fig. 2) again shows significant similarities. As expected, the absorbance around 1712 cm^{-1} attributed to protonation of Glu-113 is considerably reduced since this residue is replaced by alanine and the Schiff base is still protonated. The small residual band at 1710 cm^{-1} does not shift down on deuteration and cannot, therefore, be identified as a protonated carboxyl group. Since the other bands attributed to protonated

---

[39] U. M. Ganter, E. D. Schmid, and F. Siebert, *J. Photochem. Photobiol. B* **2**, 417 (1988).
[40] C. N. Rafferty and H. Shichi, *Photochem. Photobiol.* **33**, 229 (1981).
[41] R. R. Birge and C.-F. Zhang, *J. Chem. Phys.* **92**, 7178 (1990).
[42] T. Nagata, A. Terakita, H. Kandor, D. Kojima, Y. Shichida, and A. Maeda, *Biochemistry* **36**, 6164 (1997).
[43] K. Fahmy and T. P. Sakmar, *Biochemistry* **32**, 9165 (1993).
[44] T. A. Zvyaga, K. C. Min, M. Beck, and T. P. Sakmar, *J. Biol. Chem.* **268**, 4661 (1993).
[45] T. A. Zvyaga, K. Fahmy, and T. P. Sakmar, *Biochemistry* **33**, 9753 (1994).
[46] K. Fahmy, F. Siebert, and T. P. Sakmar, *Biochemistry* **33**, 13700 (1994).

Asp-83 and Glu-122 are very similar in the two spectra, protonation changes of the new counterion Glu-117 can be excluded. The spectral differences around 1550 cm^{-1} can be best explained by the superposition of the C=C stretching modes of the chromophore in the dark and in photoproduct states. The C=C stretching modes of the chromophore in the photoproduct states are upshifted due to the change in the visible absorption maximum from 490 nm (dark) to 474 nm (photoproduct). In wild-type rhodopsin, the positive band of MII is practically missing since it represents a state with an unprotonated Schiff base. As in the spectrum of the E113A mutant, the negative band at 1238 cm^{-1} is influenced. However, in this region of the spectrum a clear band can be still discerned at 1231 cm^{-1}. Thus, the environment of the Schiff base in the dark state is altered by the shift of the counterion by one helix turn, but the alteration is different than that observed if the counterion is removed. The clear band at 1231 cm^{-1} could indicate that the water molecule is still present, which might be required to stabilize the protonated Schiff base both in the dark and photoproduct states. At higher pH, a species with deprotonated Schiff base is rapidly formed. Biochemical studies have indicated that this state is inactive. The results of FTIR spectroscopic characterization of this inactive photoproduct of the mutant pigment clearly show that many of the spectral characteristics of the active state are less developed or even absent.[46]

As a last example in which the environment of the Schiff base has been altered we discuss a mutation responsible for congenital night blindness in humans, which is thought to be caused by excessive biochemical noise in the photoreceptor rod cell. The mutation has been identified to be a replacement of the neutral residue Gly-90 (helix 2) by a potentially negative residue Asp.[47] *In vitro* studies of this mutant indicated that the active photoproduct contains predominantly a protonated Schiff base. In the absence of the chromophore, the mutant opsin also exhibits constitutive activity. In the double mutant G90D/E113Q, the protonated Schiff base is stabilized by Asp-90.[48] These findings suggest that Asp-90 also assumes the counterion function in the G90D mutant, although this has not been directly demonstrated by these experiments, nor has the counterion of the photoproduct state been identified. According to the examples discussed previously, such questions may be addressed by FTIR spectroscopy. In addition, since such experiments provide some information on protein conformational changes, they may provide insights regarding the origin of the dark activity displayed by the G90D mutant.

---

[47] P. A. Sieving, J. E. Richards, F. Naarendorp, E. L. Binghan, K. Scott, and M. Alpern, *Proc. Natl. Acad. Sci. U.S.A.* **92,** 880 (1995).
[48] V. R. Rao, G. B. Cohen, and D. D. Oprian, *Nature* **367,** 639 (1994).

The last two spectra in Fig. 2 provide information about the transitions of the mutants G90D and G90D/E113A to their active states.[49] Despite the great similarity to the spectrum of wild-type rhodopsin, several deviations are noteworthy. In the spectrum of the mutant G90D, the bands around 1727, 1712, and 1693 cm^{-1} are much more pronounced, an effect that is largely reverted by the second-site mutation E113A. In both spectra, the characteristic negative bands at 1767 (caused by Asp-83) and 1656 (amide-I) cm^{-1} have considerably reduced intensities. Finally, as seen in the case of the E113A/A117E mutant, the fingerprint mode at 1238 cm^{-1}, which is sensitive to the Schiff base environment, is shifted to 1231 cm^{-1}. The reversion of the intensity increase at 1727 cm^{-1} by the second-site E113A mutation suggests an environmental change of Glu-113 in the G90D mutant. In contrast to the case in wild-type rhodopsin, Glu-113 may be protonated in both the dark and photoproduct states of mutants G90D and G90D/E113A. This possibility is confirmed by low-temperature FTIR spectroscopy of the dark, isorhodopsin, and bathorhodopsin states of these mutants. The band of the dark state at 1727 cm^{-1} is abolished by the additional Glu-113 mutation.[49] The larger intensity of the negative band around 1695 cm^{-1} could be explained by additional amide-I spectral changes. However, since it is also reverted by the additional mutation, this indicates that Asp-90 is also partially protonated in the dark state. Therefore, both Glu-113 and Asp-90 exist in equilibrium to provide the negative charge that serves as the Schiff base counterion. Only if the residue at position 113 is neutralized by mutation does Asp-90 adopt the full function of Schiff base counterion. The residual band at 1712 cm^{-1} in the spectrum of the double mutant indicates that the photoproduct also contains, to a small extent, a deprotonated Schiff base, which allows a fractional protonation of Asp-90. This is in agreement with UV–visible spectral investigations. The shifted fingerprint mode at 1231 cm^{-1} clearly indicates that the interaction of the Schiff base with the counterion(s) is changed, as was the case in the E113A/A117E mutant. Again, it appears plausible that a water molecule is still present to stabilize the protonated Schiff base.

Additional information related to the altered Schiff base environment is obtained from low-temperature FTIR studies.[50] A pronounced HOOP band at 921 cm^{-1} assigned to the uncoupled C11-HOOP vibration characterizes the bathorhodopsin state of wild-type rhodopsin.[51] The decoupling from the C12-HOOP mode is explained by a special interaction with coun-

---

[49] T. A. Zvyaga, K. Fahmy, F. Siebert, and T. P. Sakmar, *Biochemistry* **35,** 7536 (1996).

[50] K. Fahmy, T. A. Zvyaga, T. P. Sakmar, and F. Siebert, *Biochemistry* **35,** 15065 (1996).

[51] I. Palings, E. M. M. van den Berg, J. Lugtenburg, and R. A. Mathies, *Biochemistry* **28,** 1498 (1989).

terion Glu-113, located according to solid-state NMR experiments close to C-12.[4] This is in agreement with the observation that the HOOP mode for the bathorhodopsin of the E113A mutant is located at 931 cm^{-1}. In the spectra of the G90D and G90D/E113A mutants, the band is further shifted to 937 cm^{-1}.[49] This band shift further supports the conclusions about the drastically altered electrostatic environment of the chromophore in these mutants. Finally, the reduced intensities of the negative bands at 1767 and 1656 cm^{-1} cannot be explained by incomplete activation, since both mutants can fully activate G_t. However, the reduced band intensities are evidence that these two mutants exhibit some characteristics of the active state already in the dark. The dark states of mutants G90D and G90D/E113A display reduced intensities in bands that are otherwise indicative of the transition from a completely inactive to an active state. From these investigations, speculations on the origin of the dark activity can be made. It appears possible that the "active" properties of the dark state cause the dark activity and dark noise. In addition, the altered electrostatic environment may reduce the thermodynamic barrier for thermal isomerization of the chromophore, thereby increasing dark activity and dark noise.[50]

FTIR Investigations with ATR Sampling Technique

Thus far, the potential of FTIR-difference spectroscopy for the elucidation of intramolecular processes during photoactivation of rhodopsin has been well documented. However, only recently has this technique been applied more systematically to investigate physical interactions of rhodopsin with its aqueous environment. The strong IR absorption of water restricts transmissive infrared spectroscopy to the study of thin hydrated films, thereby preventing IR-absorption measurements of molecular interactions that require diffusion of ligands in a bulk aqueous phase. FTIR-difference spectroscopy in combination with ATR circumvents this problem.[52,53] The ATR method exploits the interaction of an evanescent infrared field with a sample that is intimately attached to the surface of an internal reflection element (IRE) (i.e., an IR-transmissive crystal). Depending on the number of reflections, the absorption by the adsorbed material corresponds typically to an effective sample thickness of only about 1 μm, irrespective of the actual dimensions of the sample.[54] Therefore, the adsorbed sample can be brought into contact with bulk water, which will be sensed by the IR field

---

[52] U. P. Fringeli, in "Internal Reflection Spectroscopy" (F. M. Mirabella, Jr., ed.), p. 255. Marcel Dekker, New York, 1992.
[53] J. Heberle and C. Zscherp, Appl. Spectrosc. **50**, 588 (1996).
[54] N. J. Harrick, Appl. Spectrosc. **37**, 573 (1983).

only within the effective penetration depth, and relevant solutes can be added easily during spectral recordings.

The ATR technique has been used to measure the light-dependent increase of the H/^{2}H exchange rate of peptide NH groups in rhodopsin,[55] demonstrating that photoisomerization exposes previously inaccessible peptide bonds. Likewise, the pH and ionic strength dependence of the MI/MII photoproduct equilibrium has been determined in this way under much more well-defined conditions than in hydrated films.[56] The pH sensitivity of the MI/MII photoproduct equilibrium indicates that, in addition to the internal proton transfer to Glu-113, proton exchange reactions between external groups and the aqueous phase regulate receptor conformation. Since FTIR-difference spectroscopy provides a direct assessment of conformational changes, their detection is independent of chromophore absorption, which classically defines the MI and MII states. Thus, the receptor conformation can be studied also under conditions where no visible absorption changes are evoked by pH, as in dark rhodopsin. The ATR technique is particularly advantageous in this case, because proton-induced, rather than light-induced, absorption changes have to be recorded.

Figure 3 shows the infrared absorption changes of dark rhodopsin adsorbed on a surface of a trapezoidal IRE (3-cm^2) made of ZnSe. Absorption decreases at 1402 and 1561 cm^{-1} and increases at 1712 cm^{-1} during a pH change from 7.5 to 5.5. The decreases at 1402 and 1561 cm^{-1} correspond to the loss of symmetric and antisymmetric COO$^-$ stretching modes and increases at 1712 cm^{-1} correspond to the appearance of C=O stretching vibrations, as expected on titratation of carboxylic acid groups. The shoulder around 1740 cm^{-1} in the broad absorption increase between 1700 and 1800 cm^{-1} suggests that heterogeneous populations of hydrogen-bonded COOH groups exist. Small absorption changes in the amide-I region around 1660 cm^{-1} may indicate that a minor protein conformational change is induced by acidification. In contrast, a prominent absorption increase at 1645 cm^{-1} accompanies the pH change when the sample has been illuminated at pH 7.5 (i.e., when MI had been predominantly formed before titration). The additional band corresponds well with the marker band of the MII state,[57] which is generated from the MI form during acidification.[58] This experiment demonstrates that (1) MI/MII equilibrium is not preformed in the dark and (2) proton uptake is coupled to a specific conformational change in

---

[55] P. Rath, W. J. DeGrip, and K. J. Rothschild, *Biophys. J.* **74,** 192 (1998).
[56] F. Delange, M. Merkx, P. H. M. Bovee-Geurts, A. M. A. Pistorius, and W. J. DeGrip, *Eur. J. Biochem.* **243,** 174 (1997).
[57] K. Fahmy, F. Siebert, and T. P. Sakmar, *Biophys. Chem.* **56,** 171 (1995).
[58] J. H. Parkes and P. A. Liebman, *Biochemistry* **23,** 5054 (1984).

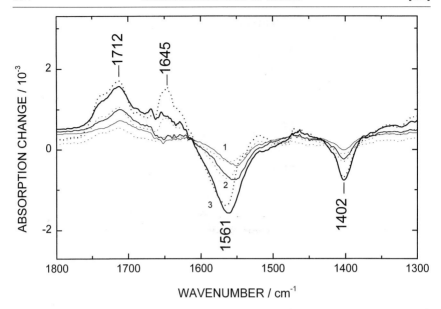

FIG. 3. ATR–FTIR spectra of bovine rhodopsin at various pH values are presented. Absorption changes accompanying acidification from pH 7.5 to 5.5 in the dark (solid lines) and after illumination (1 min, 150-W projector equipped with Schott GG495 filter) at pH 7.5 (broken lines). Absorption changes were determined relative to the initial IR absorption (pH 7.5). The pH change was performed by dialysis[62] (10 mM phosphate buffer, 150 mM NaCl) over 15 min and was complete by the third trace of each series (determined by pH electrode measurement at the end of the experiment). The first and second pair of traces were recorded at 5-min intervals during acidification. Experiments were carried out at 5°. Absorption changes are scaled to equal amounts of rhodopsin determined from the magnitude of the light-induced absorption changes recorded before and after acidification of illuminated and dark membranes, respectively.

the photoactivated receptor. Thus, the photoreaction can be viewed as a light-induced transition from a pH-insensitive to a pH-sensitive receptor state.

The conserved Glu-134 in the second cytoplasmic loop of rhodopsin is a prime candidate for a functional role in coupling proton uptake to a conformational transition.[43,59] However, previous FTIR experiments on recombinant rhodopsin have not identified corresponding vibrational modes.[60,61] In a recent ATR–FTIR study, the spectral characteristics of

[59] S. Arnis, K. Fahmy, K. P. Hofmann, and T. P. Sakmar, *J. Biol. Chem.* **269**, 23879 (1994).
[60] P. Rath, L. L. J. DeCaluwé, P. H. M. Bovee-Geurts, W. J. DeGrip, and K. J. Rothschild, *Biochemistry* **32**, 10277 (1993).
[61] G. L. J. DeCaluwé, P. H. M. Bovee, P. Rath, K. J. Rothschild, and W. J. DeGrip, *Biophys. Chem.* **56**, 79 (1995).

the MII–G_t complex formation were described.[62] Prominent IR spectral features are caused by the interaction of the G_t C terminus with rhodopsin. These interactions include protonation of a carboxylic acid group on the surface of rhodopsin when complexed with G_t. In contrast to transmissive FTIR spectroscopy of the complex,[63] the more native-like conditions of the ATR experiment allow the identification of its C=O stretching vibration between 1730 and 1740 cm^{-1} and also provide a high signal-to-noise ratio for the amide-I bands. Corresponding experiments with rhodopsin mutants in complex with G_t may allow verification of the generally postulated protonation of the conserved cytoplasmic Glu (Asp) residue at the cytoplasmic border of transmembrane helix 3 in GPCRs.

Conclusions

The examples discussed in this article clearly demonstrate how the method of FTIR-difference spectroscopy, in combination with site-directed mutagenesis, provides detailed insight into the activation mechanism of rhodopsin. An inherently sensitive spectroscopic technique that is capable of detecting molecular changes of single amino acid residues attains high specificity in combination with molecular biological methods. In addition, the use of the ATR–FTIR technique allows the identification of pH-induced structural changes and of protonation changes of external carboxyl groups. Furthermore, studies of rhodopsin–G_t interaction are now possible using the ATR–FTIR method.

Mutational effects may provide direct information on the role of particular amino acid residues. However, the interpretation of the spectral effects of mutations may be complicated in many cases, especially if key functionally important residues are replaced, or if mutations cause indirect effects. Specificity of IR band assignments without interfering with structure and function can only be obtained by the technique of site-directed isotopic labeling.[64] However, this technology is not yet available for membrane proteins that require expression in mammalian cell lines.

Experiments on rhodopsin by other groups have emphasized other important aspects of the photoactivation process. The involvement of water

---

[62] K. Fahmy, *Biophys. J.* **75**, 1306 (1998).

[63] S. Nishimura, J. Sasaki, H. Kandori, T. Matsuda, Y. Fukada, and A. Maeda, *Biochemistry* **35**, 13276 (1996).

[64] S. Sonar, C.-P. Lee, M. Coleman, N. Patel, X. Liu, T. Marti, H. G. Khorana, U. L. RajBhandary, and K. J. Rothschild, *Struct. Biol.* **1**, 512 (1994).

molecules in the rhodopsin photoreaction has been reported.[42,65,66] The change of a SH-stretching band of a cysteine has been shown to be characteristic of the activation process.[67] Finally, the expression of rhodopsin in the *Spodoptera frugiperda* insect cells has allowed the incorporation of isotopically labeled amino acids and the identification of tyrosine structural changes that may correlate with the movement of helices that occur during rhodopsin activation.[68]

Acknowledgments

This work was supported by Deutsche Forschungsgemeinschaft, grant Si-278/16-1, the Fonds der Chemischen Industrie (to F.S.) and the Howard Hughes Medical Institute. We thank students and colleagues who have contributed to the work described in this chapter. In particular, we thank Dr. Mareike Beck and Dr. Tatyana A. Zvyaga.

[65] A. Maeda, Y. J. Ohkita, J. Sasaki, Y. Shichida, and T. Yoshizawa, *Biochemistry* **32**, 12033 (1993).
[66] S. Nishimura, H. Kandori, M. Nakagawa, M. Tsuda, and A. Maeda, *Biochemistry* **36**, 864 (1997).
[67] P. Rath, P. H. M. Bovee-Geurts, W. J. DeGrip, and K. J. Rothschild, *Biophys. J.* **66**, 2065 (1994).
[68] F. Delange, C. H. W. Klaassen, S. E. Wallace-Williams, P. H. M. Bovee-Geurts, X.-M. Liu, W. J. DeGrip, and K. J. Rothschild, *J. Biol. Chem.* **273**, 23735 (1998).

[14] pK_a of the Protonated Schiff Base of Visual Pigments

By THOMAS G. EBREY

Introduction

"Retinene," a compound closely related to vitamin A, was identified as the chromophore of the visual pigment by Wald in the late 1930s. Soon thereafter Morton and co-workers showed that retinene was the aldehyde of vitamin A, retinal, and then proposed that the retinal was attached to the apoprotein via a Schiff base linkage (reviewed in Ref. 1). Next, Hubbard and Kropf proposed that the Schiff base must be protonated in order to account for the ability of the retinal chromophore to have its absorption spectrum shifted from that of free retinal in solution (380 nm) or that of a Schiff base of retinal (ca. 370 nm) to that of rhodopsin (ca. 500 nm).

[1] R. Morton, *in* "Handbook of Sensory Physiology," Vol. VII/1, "Photochemistry of Vision" (H. Dartnall, ed.), p. 33. Springer-Verlag, Berlin, 1972.

Indeed, model protonated Schiff bases do absorb much nearer to 500 nm than unprotonated Schiff bases (450 versus 370 nm); however, they still are far to the blue of rhodopsin (500 nm) let alone that of many long-wavelength-sensitive cone pigments, ca. 570 nm. Hubbard and Kropf's proposal was that the π electrons in the polyene chain were further delocalized by external charges and polar groups of the apoprotein to shift the spectrum further to the red. This seems, in principle, correct (reviewed in Ref. 2).

Firm evidence that the retinal was attached as a Schiff base to a lysine side chain of the opsin was provided by Bownds[3] and Akhtar et al.[4] Direct evidence that the bond normally is a *protonated* Schiff base comes from resonance Raman studies of rhodopsin.[5] This technique allows the selective detection of those vibrations of the pigment just due to the chromophore. In the resonance Raman spectrum of visual pigments a vibration at ca. 1655 cm^{-1} was observed, which is where the Schiff base vibration of model protonated Schiff bases is expected. Moreover, a very compelling experiment is that of suspending rhodopsin in D_2O; since the proton on the Schiff base is easily exchangeable, then if rhodopsin's chromophore is a protonated (deuterated) Schiff base, the new molecular species will be C=ND, which will have a different vibrational frequency because of the change of mass when H is replaced by D. In model compounds, this change of solvents from H_2O to D_2O and the subsequent deuteration shifts the Schiff base vibration from about 1656 to 1630 cm^{-1} (Ref. 5). When bovine rhodopsin is resuspended in D_2O, only one vibration shows a major change: The 1655 cm^{-1} shifts to 1622 cm^{-1}, demonstrating that its chromophore is a protonated Schiff base.

There are five distinct visual pigment families in vertebrates: the mid/long-wavelength-sensitive (M/LWS), short-wavelength-sensitive 1 (SWS1), short-wavelength-sensitive 2 (SWS2), rhodopsin 1 (RH1), and rhodopsin 2 (RH2) families (reviewed in Ref. 6). As noted earlier, bovine rhodopsin, a member of the RH1 family, has a protonated Schiff base. The chicken LWS visual pigment, P571 ("iodopsin"), has also been shown[7] to have a protonated Schiff base as has the toad SWS2 visual pigment P430 (the

---

[2] M. Sheves and M. Ottolenghi, *in* "Chemistry and Biology of Synthetic Retinoids" (M. Dawson and W. Okamura, eds.), p. 99. CRC Press, Boca Raton, Florida, 1990.
[3] D. Bownds, *Nature* **216,** 1178 (1965).
[4] M. Akhtar, D. Blosse and P. Dewhurst, *Biochem. J.* **110,** 693 (1968).
[5] A. Oseroff and R. Callender, *Biochemistry* **13,** 4243 (1974).
[6] T. Ebrey and Y. Koutalos, *Recent Prog. Vision Ophthal.*, submitted for publication.
[7] S. Lin, Y. Yamamoto, Y. Fukuda, Y. Shichida, T. Yoshizawa, and R. Mathies, *Biochemistry* **33,** 2151 (1994).

"green" rod pigment[8]). So far no members of the other two visual pigment families have been studied. All known members of the RH2 family absorb from 465 to 510 nm and so are also expected to be protonated Schiff bases, but the SWS1 family members absorb from 355 to 425 nm (the human "blue" cone) and conceivably be unprotonated Schiff bases. However, Pande et al.[9] showed that an ultraviolet absorbing visual pigment from an invertebrate, the owlfly, probably has a protonated Schiff base. Thus, it is possible that all members of the SWS1 family have protonated Schiff bases.

Consequences of a High pK_a for the Protonated Schiff Base of Visual Pigments

Three interesting points arose from these important early results. First, it is known that after light absorption by rhodopsin, the pigment bleaches through a series of photointermediates and one of the most striking changes in the absorption spectrum is during the meta-I to meta-II transition. Here the spectrum of the intermediate changes from ca. 470 to ca. 380 nm. Mathews and co-workers[10] suggested that this probably is due to the deprotonation of the Schiff base, and again resonance Raman experiments showed conclusively that meta-I is a protonated Schiff base and meta-II is a deprotonated Schiff base and so the Schiff base does deprotonate in the meta-I to meta-II transition.[11] Second, the observation that the deuteron would exchange for a proton on the Schiff base meant that the Schiff base was in some way accessible to the external, bathing solution. Further studies by Deng et al.[12] showed that the exchange takes place very quickly, in less than 3 ms. Finally, it was shown that while the pK_a of model Schiff base compounds is usually around 7 (see, e.g., Ref. 13) rhodopsin must have a much higher pK_a since it retains its color up to at least pH 9 (Ref. 14) and even pH 11 (Refs. 15 and 16, and see later discussion).

With this background, we can suggest at least two reasons why the pK_a of the visual pigment Schiff base is of such great interest. First, the pK_a of the protonated Schiff base must be at least 1 and preferably 2 or more units higher than the pH of the pigment's environment so that the pigment's chromophore will stay protonated in the photoreceptor cell and thus be

---

[8] G. Loppnow, B. Barry, and R. Mathies, *Proc. Natl. Acad. Sci. U.S.A.* **86,** 1515 (1989).
[9] A. Pande, H. Deng, P. Rath, R. Callender, and J. Schwemer, *Biochemistry* **26,** 7426 (1987).
[10] R. Mathews, R. Hubbard, P. Brown, and G. Wald, *J. Gen. Physiol.* **47,** 215 (1963).
[11] A. Doukas, B. Aton, R. Callender, and T. Ebrey, *Biochemistry* **17,** 2430 (1978).
[12] H. Deng, L. Huang, R. Callender, and T. Ebrey, *Biophys. J.* **66,** 1129 (1994).
[13] H. Deng and R. Callender, *Biochemistry* **26,** 7418 (1987).
[14] C. Radding and G. Wald, *J. Gen. Physiol.* **39,** 923 (1956).
[15] Y. Koutalos, *Biophys. J.* **61,** 272 (1992).
[16] G. Steinberg, M. Ottolenghi, and M. Sheves, *Biophys. J.* **64,** 1499 (1993).

able to catch the photons in the "visible" region of the spectrum. Taking the intracellular pH as 7.4, the pK_a should be greater than 9. Second, there is a great deal of evidence (reviewed in Refs. 17 and 18) that a vertebrate visual pigment is required to deprotonate its Schiff base after light absorption in order to reach its activating state, essentially a subspecies of meta-II that occurs early in meta-II's lifetime (before phosphorylation and binding of arrestin). Thus not only does a visual pigment have to initially have an unusually high pK_a, but one of the major consequences of light absorption must be to change the pK_a of the Schiff base so that it becomes low enough to allow the Schiff base to deprotonate.

There is quite possibly a third reason why the pK_a of visual pigments is of such great interest and especially why it is so high (see later discussion). Birge and Barlow[19,20] have put forth a quite provocative hypothesis: that the high pK_a of the Schiff base of visual pigments is required to lower the rate of thermal isomerization of the chromophore. Otherwise a high rate of thermal isomerization would lead to a high and disruptive level of dark noise in the photoreceptor cells. The notion that the thermal isomerization of rhodopsin could be a major source of thermal noise in photoreceptors was raised initially by Barlow[21] for vertebrates and Srebro and Behbehani[22] for invertebrates. This idea was probably put forth in its most cogent form by Baylor et al.[23] in their discussion of the thermal noise records they obtained from vertebrate (toad) rods. The assumption that a thermal isomerization has the same effect on the photoreceptor cell as a photoisomerization seems plausible. There are ca. 3×10^9 rhodopsin molecules in a toad rod so even a very small rate constant for thermal isomerization of a visual pigment molecule could lead to many isomerizations per minute for the photoreceptor cell. Having a very low rate of thermal isomerization probably is a great advantage for a visual pigment, at least in the most sensitive photoreceptors, the rods. Rods are capable of detecting a single photon and are part of a visual system whose threshold is at a very small number of photons absorbed.[24] This low rate of thermal isomerization would require that the barrier for thermal isomerization be high. Even in photoreceptors

---

[17] T. Sakmar, *Prog. Nucl. Acid Res. Mol. Biol.* **59,** 1 (1998).
[18] M. Nakagawa, T. Iwasa, M. Kikkawa, M. Tsuda, and T. Ebrey, *Proc. Natl. Acad. Sci. U.S.A.* **96,** 6189 (1999).
[19] R. Birge, *Biophys. J.* **64,** 1371 (1993).
[20] R. Birge and R. Barlow, *Biophys. Chem.* **55,** 115 (1995).
[21] H. Barlow, in "Photophysiology" (A. Giese, ed.), Vol. 2, p. 163. Academic Press, New York (1964).
[22] R. Srebro and M. Behbehani, *J. Physiol.* **224,** 349 (1972).
[23] D. Baylor, G. Mathews, and K. W. Yao, *J. Physiol.* **309,** 591 (1980).
[24] S. Hecht, S. Shlaer, and M. Pirenne, *J. Gen. Physiol.* **25,** 819 (1942).

like cones that function at higher light levels, thermal noise could be an important limitation in setting the signal-to-noise ratio. Birge and Barlow[19,20] presented calculations suggesting that if a proton were to dissociate from the protonated Schiff base, the barrier to thermal isomerization would be lowered. Thus they proposed that it is important that the pK_a be high, to reduce the probability of the Schiff base deprotonating, and so reduce the amount of time the pigment is in the deprotonated state that they had suggested has a higher rate of thermal isomerization of the chromophore. However, there is a problem with this proposal. The deprotonation of the Schiff base would remove some kinds of electrostatic barriers to thermal isomerization. But deprotonation of the Schiff base, as noted earlier, leads to a chromophore that will absorb at shorter wavelengths; this means that the π electrons are less delocalized and so the partial single bonds go closer to being essential single bonds, whereas the double bonds go more toward being essential double bonds. This will raise the barrier to thermal isomerization about the double bonds. Thus deprotonation of the Schiff base will probably make it harder, not easier, to thermally isomerize about the 11–12 double bond.

However, there is a variation of the Birge and Barlow hypothesis that still requires a high pK_a for the protonated Schiff base in order to lower the probability of thermal isomerization about the 11–12 bond. There is very good evidence that the *protonation* of the *counterion* to the Schiff base will delocalize the π electrons, causing both a red shift of the absorption spectrum and a lowering of the barrier to thermal isomerization about the double bonds. Thus the protonation of the counterion will increase the rate constant for thermal isomerization.[25] The lower the pK_a of the counterion, the less likely it will be to become transiently protonated, and thus to be in a state that has a greatly enhanced probability for thermal isomerization of the chromophore. The evidence for these statements comes initially from model compound studies of Sheves and co-workers.[26,27] A closer example is our study of the rate of thermal isomerization of the protonated Schiff base of the retinal chromophore of bacteriorhodopsin, a pigment built in a very similar way to rhodopsin. We showed that protonation of its counterion, Asp-85, increased the rate of thermal isomerization of its chromophore by at least three orders of magnitude.[28,29] Our conclusion that the pro-

---

[25] S. Balashov, R. Govindjee, M. Kono, E. Imasheva, E. Lukashev, T. Ebrey, R. Crouch, D. Menick, and Y. Feng, *Biochemistry* **32,** 10331 (1993).

[26] T. Baasov and M. Sheves, *Biochemistry* **25,** 5249 (1986).

[27] A. Albeck, N. Livnah, H. Gottlieb, and M. Sheves, *J. Am. Chem. Soc.* **114,** 2400 (1992).

[28] S. Balashov, R. Govindjee, E. Imasheva, S. Misra, T. Ebrey, Y. Feng, R. Crouch, and D. Menick, *Biochemistry* **34,** 8820 (1995).

[29] S. Balashov, E. Imasheva, T. Ebrey, N. Chen, D. Menick, and R. Crouch, *Biochemistry,* **36,** 8671 (1997).

tonation of Asp-85 reduces the barrier for thermal isomerization by at least 4.5 kcal/mol (Ref. 25) agrees with the calculations of Logunov and Schulten[30]; they calculated that protonation of the counterion Asp-85 would lower the barrier for thermal isomerization by almost 7 kcal/mol.

Can we relate the high pK_a of the Schiff base to a low pK_a of its counterion? To some extent, such a correspondence does rely on the precise mechanism by which the pK_a of the Schiff base is regulated, but the studies of Sheves and co-workers[31,32] show that such a correlation is not only possible, but likely for retinal proteins. In bacteriorhodopsin and visual pigments it seems plausible that the high pK_a of the Schiff base can be correlated with a low pK_a for the counterion. Only in bacteriorhodopsin do we have any data; here the pK_a of the Schiff base is high, about 13, while that of Asp-85 is low, 2.6 (Ref. 33). The pK_a of another part of the complex counterion of bacteriorhodopsin, Asp-212, is even lower. Its exact value is not known, but it must be below 1, because titrations find that at pH values close to 1, Asp-85 has long been protonated, while Asp-212 still has not.[34]

Experimental Determination of the pK_a of Visual Pigments

General Considerations

The most straightforward method used to measure the pK_a of the Schiff base of a visual pigment is to measure its absorption spectrum as a function of pH. Since the absorption spectrum is expected to show a large shift to the blue as the Schiff base becomes deprotonated, a plot of the absorbance either at the absorption maximum or at the wavelength where the deprotonated Schiff base is expected to absorb, ca. 360 nm, will be the titration curve (see Fig. 1). Its midpoint will be the pK_a. There is one very important caution that one must be aware of: the pigment could be denaturing or unfolding as the pH is increased. Then the absorption change will not be due just to the deprotonation of the Schiff base but would be set by the pigment denaturing or unfolding and subsequently the Schiff base environment now being altered to something close to that of a model compound, allowing its deprotonation at the elevated pH values that caused the denaturation. As noted earlier, the pK_a of a model Schiff base of retinal will be around 7, so the pigment would be sure to deprotonate if it were

---

[30] I. Logunov and K. Schulten, *J. Am. Chem. Soc.* **118,** 9727 (1996).
[31] Y. Gat and M. Sheves, *J. Am. Chem. Soc.* **115,** 3772 (1993).
[32] I. Rousso, N. Friedman, M. Sheves, and M. Ottolenghi, *Biochemistry* **34,** 12059 (1995).
[33] R. Jonas and T. Ebrey, *Proc. Natl. Acad. Sci. U.S.A.* **88,** 149 (1991).
[34] G. Metz, F. Siebert, and M. Engelhard, *FEBS Lett.* **303,** 237 (1992).

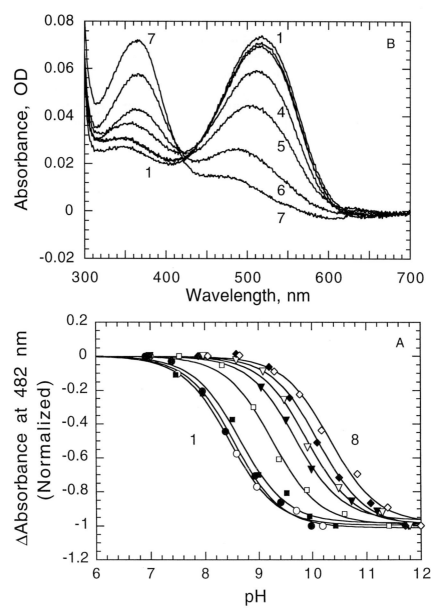

FIG. 1. (A) The absorption spectrum of gecko P521 mid-wavelength-sensitive visual pigment soublized in 0.2% dodecylmaltoside, in 50 mM KCl at pH 7.2, 8.2, 8.6, 9.3, 9.6, 10.1, and 10.5, traces 1 through 7, respectively. (B) pH titration of the gecko P521 pigment in 2% digitonin at constant ionic strength, varying the concentration of KCl. Potassium acetate was used to adjust the ionic strength to 3 M. The concentration of Cl$^-$ is 0 mM, 1 mM, 10 mM, 50 mM, 500 mM, 1 M, 2 M, and 3 M for curves 1 throught 8, respectively. [From C. Yuan *et al.*, *Biochemistry* **38**, 4649 (1999).]

denatured or unfolded by pH values of 8 or above. One way to check on the possible denaturation would be if the titration was reversible, but this would not detect unfolding. Sheves and co-workers[35] came up with an innovative solution to this problem. They synthesized a group of chemically modified retinals in which the substituent groups on the retinal lowered the *intrinsic* pK of the retinal Schiff base.[35] For instance, a Schiff base of all-*trans*-retinal with n-butylamine had its pK at 7.3, but the modified retinal, all-*trans*-13-trifluoromethylretinal, when combined with n-butylamine to form a Schiff base, had a much lower pK_a of 1.8. This difference is due to the increased electronegativity of the trifluoro group attached to position 13 of the chain. Because this alteration is an intrinsic property of the modified retinal, it would be expected to roughly maintain its difference in pK_a with retinal when either are incorporated into a pigment. This seems to be born out in the case of bacteriorhodopsin.[35] In this case standard titrations found that the pK_a of the Schiff base of this pigment is about 13.3. However, when the all-*trans*-13-trifluororetinal is used to regenerate a pigment, this bacteriorhodopsin analog has a much lower pK_a of 8.0. The difference in pK values between the model chromophores, 5.5, is remarkably close to the difference in pK_a values of the Schiff bases of the pigments, 5.3.

pK_a of the Schiff Base of Bovine Rhodopsin

Steinberg *et al.*[16] used the retinal analog method to determine a lower limit for the pK_a of bovine rhodopsin. They tried to titrate the Schiff base of bovine rhodopsin, but could only see evidence that the pigment was denaturing at pH values above about 11. Similar results were seen with the 11-*cis*-13-trifluororetinal-based pigment. Because this pigment is expected to have its Schiff base pK at 5.5 pH units lower than that with the native retinal chromophore, Steinberg *et al.* inferred that the native pigment must have a pK_a greater than 11 + 5.5 = 16.5. For a conservative estimate, the pK_a of the native rhodopsin is taken to be greater than 15. Taking the pK_a of a model Schiff base as about 7, then the pigment has raised the pK_a by at least 8 pH units, a surprisingly large number. There seems no obvious reason for the pK_a to be this high (as opposed to pH 8 or 9) except for the argument given earlier that a high pK_a of the protonated Schiff base is closely related to a low pK_a for the counterion and this is part of the mechanism that keeps the rate of thermal isomerization of rhodopsin's chromophore low, in order to lower the dark noise of the photoreceptor.

[35] M. Sheves, A. Albeck, N. Friedman, and M. Ottolenghi, *Proc. Natl. Acad. Sci. U.S.A.* **86**, 3262 (1986).

pK_a of the Schiff Base of Mid/Long-Wavelength-Sensitive Visual Pigments

Kojima et al.[36] showed that the gecko visual pigment P521 is a member of the M/LWS visual pigment family. Most members of this family are found in cone-shaped photoreceptors (reviewed in Ref. 6), but the gecko example is found in rods. Like most members of this family, the absorption spectrum shifts to the red (from 495 to 521 nm) when chloride is added to the bathing media. Wang et al.[37] have shown that the chloride binding site is formed by two amino acid residues, a lysine and a histidine, in the loop between helices 4 and 5. We[38] found that the pK_a of the Schiff base of P521, unlike rhodopsin, is low enough to be readily titratable (Fig. 1). In 100 mM KCl, the pK_a is 9.9, more than 5 pH units below that of bovine rhodopsin. To ensure that this pK_a was not due to the pigment denaturing in a pH-dependent way, we not only showed that the titration was at least partially reversible, but that substituting 9-*cis*-14-fluororetinal for the natural chromophore lowered the pK_a of the titration by 1.1 pH units compared with the gecko 9-*cis* pigment. The 9-*cis* isomer had to be used because it was difficult to synthesize the 11-*cis* isomer of 14-fluororetinal. For model Schiff bases, the pK_a of 9-cis-14-fluororetinal was 2.4 units lower than that for native retinal. Steric effects due to the bulkier group at the 14 position actually work to increase the pK_a (by ca. 0.7 units), so the artificial chromophore lowers the pK_a in the pigment by almost 2 pH units, close to the value seen in the model compounds. This provides excellent evidence that the pK_a of the native pigment that we measured was the true pK_a of the gecko P521 Schiff base.

In a later study, we showed[39] that the presence of chloride made a large difference in the pK_a of the Schiff base. Without chloride, the pK_a was ca. 8.4; at physiologic chloride concentrations, the pK_a was about 9.7; but at 2 M chloride the pK_a could reach as high as 10.5 (Fig. 1). Thus not only does chloride shift the absorption spectrum, it also raises the pK_a. This study also showed that the surface potential, which in some cases can greatly affect the proton concentration close to the membrane, making the bulk (measured) pH quite different from the surface pH, must be small and it does not have much influence on the apparent pK_a.

---

[36] D. Kojima, Y. Okano, Y. Shichida, T. Yoshizawa, and T. Ebrey, *Proc. Natl. Acad. Sci. U.S.A.* **89,** 9181 (1992).

[37] Z. Wang, A. Asenjo, and D. Oprian, *Biochemistry* **32,** 2125 (1993).

[38] J. Liang, G. Steinberg, N. Livnah, M. Sheves, T. Ebrey, and M. Tsuda, *Biophys. J.* **67,** 848 (1994).

[39] C. Yuan, O. Kuwata, S. Balashov, and T. Ebrey, *Biochemistry* **38,** 4649 (1999).

Some preliminary data[40,41] have been reported for the human middle-wavelength-sensitive ("green," P531) and long-wavelength-sensitive ("red," P568) cone pigments. Unfortunately they must be studied after solubilization in detergent and so there may be an influence of the lipid/detergent environment of the pigment. Nevertheless, the tentative values reported are ca. 9, reasonably close to the gecko P521 pK_a and even further away from the bovine rhodopsin value.

pK_a of the Schiff Base of Octopus Rhodopsin

Koutalos et al.[42] made a preliminary measurement of the pK_a of the Schiff base of octopus rhodopsin. There was a small surface potential effect, and the pK_a at high salt (small surface potential) was ca. 10.5. Using 9-cis-14-fluororetinal to change the intrinsic pK of the Schiff base, Liang et al.[38] showed that the pK_a of the artificial pigment was 6.8, about 3.5 pH units below that of the octopus 9-cis pigment, somewhat larger than the difference seen for the model compound of 2.2 pH units. The conclusion is that the value measured for the native 11-cis pigment, 10.5, is indeed the authentic value for the pK_a of the Schiff base of the octopus rhodopsin.

Factors Controlling the pK_a of Schiff Bases

Some of the factors controlling the pK_a of the Schiff base of retinal-based pigments have been discussed in Refs. 31, 32, 42, and 43. The main variable is the nature of the environment of the Schiff base. This includes not only the amino acid residues that are close by, but also chloride in the case of the gecko pigment, and probably a quite important small molecule, water. In bacteriorhodopsin water has been shown to form hydrogen bonds spanning the Schiff base with the counterion.[44,45] This hydrogen bonding with both the counterion and the protonated Schiff base will alter the pK_a values of these groups. Of the amino acid residues in vertebrate pigments almost certainly the most important influence on the pK_a of the Schiff base is the counterion, Glu-113. This is seen most dramatically in mutants in which the glutamic at 113 is changed to a neutral polar residue like

---

[40] Z. Wang and D. Oprian, cited in G. Kochendoerfer, Z. Wang, D. Oprian, and R. Mathies, *Biochemistry* **36**, 6577 (1997).
[41] C. Yuan, O. Kuwata, and T. Ebrey, unpublished observations (1998).
[42] Y. Koutalos, T. Ebrey, H. Gilson, and B. Honig, *Biophys. J.* **58**, 493 (1990).
[43] R. Sampogna and B. Honig, *Biophys. J.* **66**, 1341 (1994).
[44] A. Maeda, J. Sasaki, Y. Yamazaki, R. Needleman, and J. Lanyi, *Biochemistry* **33**, 1713 (1994).
[45] H. Luecke, H.-T. Richter, and J. Lanyi, *Science* **280**, 1934 (1998).

glutamine. For bovine rhodopsin the pK_a drops by at least 8 pH units to ca. 7 (Refs. 46–48), close to the pK_a of the Schiff base of a model compound like n-butylamine retinal. A similar drop in the pK_a of the Schiff base is seen in bacteriorhodopsin when a key component of the pigment's complex counterion, Asp-85, is changed from an aspartic acid to an asparagine.[49]

These findings have two potentially quite important consequences for visual pigments. First, evidence from model compounds does suggest that in a pigment, the tendency of the Schiff base to lose its proton is most directly related to the tendency of the counterion to pick up a proton. That is, if the charge groups do not change their geometry, then a high pK_a for the protonated Schiff base can be related to a low pK_a for the counterion. In bacteriorhodopsin the effect on each other's pK_a can be seen in theoretical calculations of the charge–charge effects of the protonated Schiff base on the counterion(s) and vice versa (see Ref. 43). Second, the preceding observations allow a very qualitative explanation for the relatively low pK_a for the Schiff base of octopus rhodopsin. A recent study[18] has concluded that although the position occupied by the anionic Glu-113 in vertebrate pigments is usually occupied by a tyrosine in invertebrate pigments, this tyrosine is *not* in its anionic, ionized form but rather is in its neutral form. In addition, no alternative anionic species could be found in octopus rhodopsin, leading to the conclusion that the "counterion" site of octopus rhodopsin is neutral. Since making the counterion neutral in bovine rhodopsin (and bacteriorhodopsin) greatly lowers the pK_a, this is a simple and reasonable explanation for the relatively low pK_a of octopus rhodopsin compared to, say, bovine rhodopsin. Incidentally, this probably is not the explanation for the low pK_a of gecko P521. This pigment has a glutamic acid at position 113 and preliminary FTIR experiments by us[50] suggest that this group acts in the same way as it does in bovine rhodopsin: light leads to the formation of metarhodopsin-II by transferring a proton from the Schiff base to the Glu-113 (Ref. 51): when the meta-II of gecko P521 is formed, a carboxylic acid is also found to become protonated; this presumably is Glu-113, indicating that it is not protonated in the initial unphotolyzed state of the pigment.

[46] E. Zhukovsky and D. Oprian, *Science* **246**, 928 (1989).
[47] T. Sakmar, R. Franke, and H. G. Khorana, *Proc. Natl Acad. Sci. U.S.A.* **86**, 8309 (1989).
[48] J. Nathans, *Biochemistry* **29**, 9746 (1990).
[49] S. Subramaniam, S. Marti, and H. G. Khorana, *Proc. Natl. Acad. Sci. U.S.A.* **87**, 1013 (1990).
[50] A. Maeda, S. Balashov, and T. Ebrey, unpublished data (1998).
[51] K. Fahmy, F. Jaeger, M. Beck, T. Zvyaga, T. Sakmar, and F. Siebert, *Proc. Natl. Acad. Sci. U.S.A.* **90**, 10206 (1995).

Summary

The pK_a of bovine rhodopsin is greater than 15; that of the long-wavelength-sensitive gecko P521 pigment ranges from 8.4 to 10.5 depending on chloride concentration; and that of octopus, an invertebrate, is 10.5. These pK_a values are much higher than are needed just to maintain the Schiff base in its protonated state in the photoreceptor cell. The high pK_a of the Schiff base may be at least partially related to a low pK_a of its counterion, which would lower the frequency of thermal isomerization of the chromophore and thus lower the dark noise in the photoreceptor cell. After light absorption, the high pK_a of the protonated Schiff base of a vertebrate visual pigment must get lowered enough to allow it to deprotonate, a required step in vertebrate visual excitation. This deprotonation step is not required in invertebrate visual excitation.[18]

Acknowledgments

I would like to thank Sergei Balashov, Akio Maeda, and Mudi Sheves for many enlightening discussions. This work was supported by NIH grant EY01323.

[15] Assays for Detection of Constitutively Active Opsins

By PHYLLIS R. ROBINSON

Introduction

The visual pigment rhodopsin is the best studied member of the large family of G-protein-coupled receptors. Initially, bovine rhodopsin served as a model system for this family of receptors with seven transmembrane helices because it was available in large quantities for *in vitro* biochemical studies and could be easily assayed spectrophotometrically. The capability to perform functional *in vitro* assays was aided by the ability to purify significant quantities of the rod-specific G protein, transducin, and the effector enzyme, cGMP phosphodiesterase. Today rhodopsin still serves as a model system for the study of G-protein-coupled receptors because functional rhodopsin can be expressed in a heterologous expression system. The availability of the rhodopsin gene, the ability to construct site-directed rhodopsin mutants, and the ability to assay recombinant rhodopsin function *in vitro* have allowed us to study the role of specific sites in the function of rhodopsin.

In the course of these studies, we discovered mutations in rhodopsin that result in constitutively active molecules[1–3] (see Ref. 4 for a review). These rhodopsin mutants can signal or activate transducin in the absence of light and in the absence of the chromophore, 11-*cis*-retinal. Because the rhodopsin molecule is comprised of both a chromophore and a protein moiety, we refer to the apoprotein in the absence of chromophore as opsin. Since our initial discovery of these agonist-independent opsin mutants, others have described additional mutations that result in constitutive activity.[5–8] The study of these mutants has given us insights into the structure of rhodopsin and the mechanism of receptor activation. These studies clearly demonstrate the power of this experimental approach.

This chapter describes the procedures we use to determine if rhodopsin mutants are constitutively active. This includes an *in vitro* reconstitution assay in which the ability of a mutant rhodopsin to activate transducin is measured. Because the activation of transducin by rhodopsin is pH dependent, the rate of transducin activity is assayed under a variety of pH conditions. This enables us to construct a pH profile of the rate of transducin activation by various mutants and to determine how a particular mutant affects the protonation step that is essential for rhodopsin activation. When possible, constitutively active rhodopsin mutants are also assayed spectrophotometrically to determine the effect of a particular mutation on the spectral properties of rhodopsin.

Methods and Results

Materials

11-*cis*-retinal is the generous gift of the NIH. Dodecyl-β-D-maltoside is from Calbiochem (La Jolla, CA). The buffers MES, HEPES, PAPS, and Tris are obtained from Sigma (St. Louis, MO). Bovine retinas are purchased either from J. A. Lawson Co. (Lincoln, NE) or Schenk Packing Company (Stanwood, WA). The monoclonal antibody rho 1D4, which is specific for

---

[1] P. R. Robinson, G. B. Cohen, E. A. Zhukovsky, and D. D. Oprian, *Neuron* **9,** 719 (1992).
[2] G. B. Cohen, D. D. Oprian, and P. R. Robinson, *Biochemistry* **31,** 12592 (1992).
[3] G. B. Cohen, T. Yang, P. R. Robinson, and D. D. Oprian, *Biochemistry* **32,** 6111 (1993).
[4] V. R. Rao and D. D. Oprian. *Annu. Rev. Biophys. Biomol. Struct.* **25,** 287 (1996).
[5] T. P. Dryja, E. L. Berson, V. R. Rao, and D. D. Oprian, *Nature Genet.* **4,** 280 (1993).
[6] V. R. Rao, G. B. Cohen, and D. D. Oprian, *Nature* **367,** 639 (1994).
[7] M. Han, S. W. Lin, M. Minkova, S. O. Smith, and T. P. Sakmar, *J. Biol. Chem.* **271,** 32337 (1996).
[8] M. Han, S. O. Smith, and T. P. Sakmar, *Biochemistry* **37,** 8253 (1998).

the C terminus of rhodospin,[9] is purified from hybridoma culture medium (National Cell Culture Center, Minneapolis, MN). The hybridoma cells are the generous gift of R. S. Molday (University of British Columbia, Vancouver, Canada). [^{35}S]GTPγS (1156 Ci/mmol) is from NEN (Beverly, MA) and the unlabeled GTPγS (tetralithium salt) is from Boehringer Mannheim (Indianapolis, IN). LipofectAMINE is from Life Technologies, Inc. (Rockville, MD). Peptide I (DEASTTVSKTETSQVAPA) is purchased from the American Peptide Co., Inc. (Sunnyvale, CA).

Mutagenesis of the Bovine Rhodopsin Gene

Mutations are created in a synthetic bovine rhodopsin gene[10,11] using either the method of cassette mutagenesis[1,12] or the transformer mutagenesis system (Clontech, Palo Alto, CA).[13] The synthetic bovine rhodopsin gene used in these experiments has been cloned into a mammalian expression/shuttle vector that is a derivative of pMT2,[14] which contains a prokaryotic origin of replication, an ampicillin resistance gene for selection in *Escherichia coli,* simian virus (SV40) origin of replication, and an adenovirus-2 major late promoter. Cassette mutagenesis utilizes the 30 unique restriction sites that were engineered into the synthetic gene.[10] The transformer mutagenesis system, on the other hand, uses oligonucleotide-directed mutagenesis. Two oligonucleotides are synthesized, one to create the desired mutation in the rhodopsin gene and a second to mutate a unique selection restriction site located in the plasmid outside of the rhodopsin gene. We use the transformer mutagenesis system to make mutants that are difficult to construct with cassette mutagenesis. Mutations in the rhodopsin gene are confirmed by the dideoxy method of sequencing.[15]

Expression of Rhodopsin and Rhodopsin Mutants

Rhodopsin and rhodopsin mutants are transiently expressed in COS-1 cells. COS cells, derived from African green monkey kidney cells, are an attractive heterologous expression system for the study of rhodopsin and

---

[9] R. S. Molday and D. MacKenzie, *Biochemistry* **22,** 653 (1983).

[10] L. Ferretti, S. S. Karnik, H. G. Khorana, M. Nassal, and D. D. Oprian, *Proc. Natl. Acad. Sci. U.S.A.* **83,** 599 (1986).

[11] D. D. Oprian, R. S. Molday, R. J. Kaufman, and G. H. Khorana, *Proc. Natl. Acad. Sci. U.S.A.* **84,** 8874 (1987).

[12] E. A. Zhukovsky, and D. D. Oprian, *Science* **246,** 928 (1989).

[13] J. I. Fasick and P. R. Robinson, *Biochemistry* **37,** 433 (1998).

[14] R. J. Kaufman, L. C. Wasley, M. V. Davies, R. J. Wise, D. I. Israel, and A. J. Dorner, *Mol. Cell Biol.* **9,** 1233 (1989).

[15] F. Sanger, S. Nicklen, and A. R. Coulsen, *Proc. Natl. Acad. Sci. U.S.A.* **74,** 5463 (1977).

its mutants because of the high levels of functional rhodopsin that can be expressed. COS cells are routinely maintained in the laboratory using the methods described by Oprian.[16] We employ two methods for the transfection of the COS-7 cells. One method utilizes DEA–dextran–DNA and a dimethyl sulfoxide (DMSO) exposure. This method has been thoroughly described by Oprian.[16] The second method is a cationic liposome-mediated transfection for which we use LipofectAMINE from Life Technologies. For each plate of COS cells to be transfected we use 50 μl of the LipofectAMINE solution, and 10 μg of DNA. Typically we use 8–10 plates for the expression of each rhodopsin mutant. When COS cells are 80–100% confluent ($\sim$0.8–1 $\times$ 10^7 cells per plate), they are washed 2 times with medium B [Dulbecco's modified Eagle's medium containing D-glucose (4.5 g/liter), and a supplement of 2 mM L-glutamine]. Then 5 ml of medium B is added to the plates and 1.5 ml of medium B containing 50 μl of the LiofectAMINE solution and 10 μg of DNA. After the cells have incubated with the DNA cocktail overnight for a total of 18 hr, this medium is removed and replaced with 10 ml of medium A [medium B with 10% (v/v) heat-inactivated calf serum]. The cells are now allowed to incubate for an additional 48 hr prior to being harvested. We find that the cationic liposome-mediated-transfection is easier, and has a higher yield of expressed rhodopsin than the DEA–dextran method. The one disadvantage of using LipofectAMINE is its expense.

Preparation of COS Cell Membranes

To determine if a rhodopsin mutant is constitutively active, we measure transducin activation by the opsin in the absence of chromophore. However, we find that constitutively active rhodopsin mutants are unstable when solubilized in detergent and purified in the absence of chromophore. To assay for constitutive activity in a transducin activity assay we use COS membranes containing the expressed opsin mutant or purified opsins reconstituted into lipid vesicles (see later discussion). Membranes are prepared from COS cells that are harvested 72 hr after transfection with DNA in the case of a DEA–dextran transfection and 48 hr after transfection when cationic liposomes are used. Cells are harvested from 100-mm tissue culture plates by first removing the tissue culture medium, and washing the cells twice with 5 ml of 10 mM sodium phosphate buffer (pH 7.0) containing 150 mM NaCl. Then 1 ml of this phosphate buffer is added to each plate and the cells are removed by scraping with a rubber policeperson. The

[16] D. D. Oprian, in "Methods in Neurosciences: Photoreceptor Cells," pp. 301–306. Academic Press, Inc., San Diego, 1993.

transfected cells are collected into a tube and pelleted in a clinical centrifuge (1000g for 5 min at room temperature). These cells are washed twice with the isotonic phosphate buffer. To help with the lysis, cells are frozen on dry ice and thawed. All procedures are now carried out on ice or at 4°. Cells from 10 transfected plates are lysed in 15 ml of a hypotonic buffer containing 10 mM Tris buffer (pH 7.4) and 7 μg/ml phenylmethylsulfonyl fluoride. These cells are further lysed by forcing the suspension through a 25-gauge needle four times. The lysed cell homogenate is then layered onto 20 ml of a solution containing 37% (w/v) sucrose, 10 mM Tris buffer (pH 7.4), 150 mM NaCl, 1 mM MgCl$_2$, 1 mM CaCl$_2$, and 0.1 mM EDTA. This two-step discontinuous sucrose gradient is then centrifuged in a Beckman SW 28.1 rotor at 15,000 rpm (33,000g_{av}) for 20 min at 4°. Membranes are then collected from the interface using a syringe and long blunt needle. The membranes are diluted 10-fold into the hypotonic lysis buffer. The membranes are collected by centrifugation for 45 min in a Beckman 50.2 Ti rotor at 33,000 rpm (100,000g_{av}). The resulting membrane pellet is resuspended in 1 ml of 10 mM Tris buffer (pH 7.4), 150 mM NaCl, 1 mM MgCl$_2$, 1 mM CaCl$_2$, and 0.1 mM EDTA (100 μl/plate). This membrane suspension is then divided into 50- to 100-μl aliquots and frozen on dry ice and stored at $-80°$. Once an aliquot is thawed for an experiment, it is discarded.

Purification of Opsins and Reconstitution into Lipid Vesicles

The procedure that we use to purify and reconstitute opsin mutants into lipid vesicles for use in functional assays was modified from one originally developed by Rim and Oprian.[17] Cells that have been transfected with mutant opsin genes are harvested and washed twice with an isotonic phosphate buffer. They are then solubilized in 50 mM Tris buffer, pH 7.0, containing 140 mM NaCl, 1 mM dithiothreitol (DTT), 1% CHAPS, and 10 mg/ml asolectin for 40 min at 4°. The soluble fraction is separated from the nuclear pellet by centrifugation in a clinical centrifuge (1000g for 5 min at room temperature). Expressed proteins are purified by immunoaffinity chromatography using the bovine rhodopsin antibody, 1D4, as described by Oprian et al.[11] Briefly, the supernatant containing solubilized membrane proteins is removed and incubated with the antibody column, 1D4–Sepharose 4B matrix (300 μl of gel suspension) for 30 min. This suspension is then centrifuged in a clinical centrifuge for 1 min and the supernatant is removed. The beads are then transferred to a small 1-ml column made from a 1-ml syringe. The column is then washed 10 times with 1-ml portions of the solubilization buffer. After the final wash, the bottom of the column

[17] J. Rim and D. D. Oprian, *Biochemistry* **34**, 11938 (1995).

is sealed with Parafilm and the opsin is eluted from the immunoaffinity beads by incubation with 300 µl of solubilization buffer containing 50 µM peptide I (DEASTTVSKTETSQVAPA) for 30 min and subsequent centrifugation. This procedure is repeated with a second 300 µl of the elution solution. The purified receptor in a CHAPS asolectin solution is then applied to a 1 × 10-cm Sephadex G-50 column to remove the detergent and form lipid vesicles containing the purified opsin. The Sephadex column had been equilibrated with a 10 mM Tris buffer (pH 7.4) containing 1 mM $MgCl_2$, 1 mM EDTA, and 1 mM DTT. The vesicles are eluted from the column using the equilibration buffer under gravity flow. Then 500-µl fractions are collected and assayed for vesicles by use of light scattering. The turbid fractions containing the vesicles are pooled and concentrated 10-fold by centrifugation in a Centricon-30 (Amicon, Danvers, MA). The vesicles are then stored at 4°.

Preparation of 11-cis-Retinal Schiff Base Complexes

EtSB and nPrSB, the Schiff base complexes of 11-cis-retinal with ethylamine and n-propylamine, respectively, are prepared in ethanol solution as described by Zhukovsky et al.[18] To prepare nPrSB, 13 mM 11-cis-retinal is reacted with a 10-fold excess of n-propylamine in ethanol at 4° overnight in the dark. EtSB is prepared by reacting 4 mM 11-cis-retinal with a 10-fold excess of ethylammonium hydrochloride in the presence of 4 mM triethylamine at 4° overnight in the dark. The reaction is monitored using a spectrophotometer and monitoring the change in absorbance from 380 nm for the aldehyde to 355 nm for the unprotonated Schiff base.

Quantitation of Expressed Rhodopsin

Rhodopsin concentration is estimated from immunoblots using the rhodopsin 1D4 monoclonal antibody according to the methods of Burnette.[19] In most cases the binding of the primary antibody is visualized using an alkaline phosphatase-conjugated secondary antibody. In some experiments when we are interested in a more precise quantitation of the rhodopsin concentration, we use a ^{35}S-labeled goat anti-mouse-IgG antibody (Amersham, Oslo, Norway) as the secondary antibody. Blots are then quantitated using a Molecular Dynamics PhosphorImager with ImageQuant v4.2 software (Sunnyvale, CA).

[18] E. A. Zhukovsky, P. R. Robinson, and D. D. Oprian, *Science* **251**, 558 (1991).
[19] W. N. Burnette, *Annal. Biochem.* **112**, 195 (1981).

Purification of Transducin

Purification of transducin is a key component in the assay of constitutively active rhodopsin mutants. We have found that successful purification of transducin is dependent on finding a source of retinas that are removed and frozen quickly after the animal is sacrificed. Failure to use these rapidly frozen retinas may result in the purification of inactive transducin. The other consideration that is essential for these assays is to ensure that the transducin preparation is pure and does not have any rhodopsin contamination. Even trace amounts of rhodopsin can interfere with the accuracy of the transducin activation assay described here.

Transducin is purified from bovine retinas according to the procedure of Wessling-Resnick and Johnson[20] and then subjected to ion-exchange chromatography on DE52-cellulose as described by Baehr et al.[21] As a final step, the protein is dialyzed overnight against 10 mM Tris buffer (pH 7.5) containing 50% (v/v) glycerol, 1 mM MgCl$_2$, and 1 mM DTT. Dialysis against glycerol results in the concentration of the G protein; Bradford analysis is performed on the dialyzed preparation and concentrations of up to 70 μM are common, with yields of approximately 1 ml. (We typically start the preparation with 100 bovine retinas.) The transducin preparation is then monitored for purity by SDS–PAGE and for rhodopsin contamination with immunoblot analysis with the rhodopsin monoclonal antibody 1D4. Functional assays are also used to detect the presence of rhodopsin contamination and to determine the specific activity of our transducin preparation. Incubation with the chromophore 11-*cis*-retinal should show no light-dependent activity.

Assay for Activation of Transducin

Rhodopsin mutants are assayed for their ability to activate transducin catalytically by following the binding of [^{35}S]GTPγS to transducin using a filter binding assay.[12] The binding of radiolabeled GTPγS to activated G protein is determined by washing aliquots of the reaction through nitrocellulose filters that bind G protein/GTPγS but not free GTPγS. The assays are conducted with rhodopsin or the opsin mutants in COS cell membranes or lipid vesicles instead of in detergent-solubilized solution. The reason for this is that we have found that opsin, the apoprotein form of rhodopsin, is relatively unstable and does not survive purification in detergent. For light-dependent experiments, membranes are incubated in the dark with 20

---

[20] M. Wessling-Resnick and G. L. Johnson, *J. Biol. Chem.* **262,** 3697 (1987).
[21] W. Baehr, E. A. Morita, R. J. Swanson, and M. I. Applebury, *J. Biol. Chem.* **257,** 6452 (1982).

μM chromophore for 1 hr at 4° and then diluted 10-fold into the final reaction mixture.

Typically our 100-μl reaction mixtures contain 5 nM COS cell expressed rhodopsin or a rhodopsin mutant, 20 mM buffer at the desired pH (see below buffers used to measure pH dependence of the reaction; if we are not assaying for the pH dependence of activation then we routinely use 20 mM Tris, pH 7.5), 150 mM NaCl (in some experiments the NaCl concentration will vary from 150 to 185 mM to keep ionic strength constant), 1 mM MgCl$_2$, 1 mM CaCl$_2$, 0.1 mM EDTA, 1 mM DTT, 2 μM transducin. The reaction is then initiated by the addition of 3 μM [^{35}S]GTPγS (5 Ci/mmol). Then 10-μl aliquots are taken at 30-sec intervals and washed with 15 ml of reaction buffer through nitrocellulose filters (Schleicher & Schuell, Keene, NH) using a Millipore collection manifold. Filters are added to 5 ml of scintillation fluid (Amersham, BCS) and allowed to shake vigorously for $\geq$30 min before counting. The data are plotted as the amount of GTPγS bound as a function of time, and from these data the rate of the reaction is determined (see Fig. 1). Under conditions of our assay, the reaction rate is directly proportional to the rhodopsin or opsin concentration.

pH Dependence of Transducin Activation

We routinely characterize the activation of transducin by constitutively active opsin mutants by determining a pH–rate profile for activation. Several different buffers are used to cover the pH range from 4.9 to 9.5. MES (pK_a 6.1) is used for pH 4.9–6.7, HEPES (pK_a 7.5) is used for pH from 6.7 to 8.1, and TAPS (pK_a 8.4) is used for pH 8.1–9.5. Membranes or vesicles containing mutant opsins are diluted 10-fold in a buffer of the desired pH and then assayed in the transducin activity assay described earlier.

The pH–rate profiles for all constitutively active rhodopsin mutants, as well as for light-activated wild-type rhodopsin, are bell shaped (see Fig. 2 and Refs. 2 and 3). We refer to the lower apparent pK_a as pK_{a1} and the higher apparent pK_a as pK_{a2}.

We assume that the uptake of a proton with the ionization constant of K_{a2} is necessary to activate opsin mutants and wild-type rhodopsin while the further uptake of another proton with the ionization constant of K_{a1} inactivates opsin or light-activated wild-type rhodopsin. To determine the values of the apparent pK_a values the data are simulated with Eq. (1):

$$\text{Observed rate} = V_{\max}H/[(H + K_{a2})][K_{a1}/(H + K_{a1})] + T \quad (1)$$

where H is the proton concentration and the expression $H/(H + K_{a2})$ is the fraction of the protein that is protonated at site 2 at proton concentration

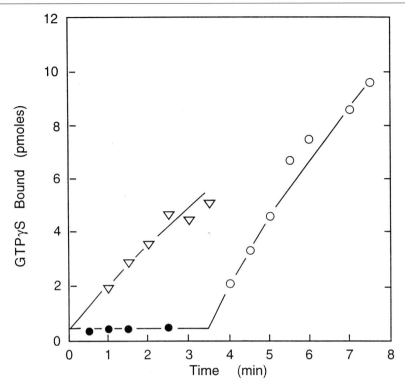

FIG. 1. Transducin activation by Lys296Gly in the absence of chromophore and the inhibition of constitutive activity by chromophore in the dark. Open triangles, reaction of transducin with Lys296Gly in the absence of added chromophore; closed circles, dark reaction with Lys296Gly preincubated with 100 μM nPrSB; open circles, dark samples after exposure to light at 3.5 min.

H. Similarly, the expression $K_{a1}/(H + K_{a1})$ is the fraction of the protein that is deprotonated at site 1. Therefore, the fraction of the protein that is active is the fraction that is protonated at site(s) 2 and deprotonated at site(s) 1 and expressed in Eq. (2).

$$[H/(H + K_{a2})][K_{a1}/(H + K_{a1})] = \text{fraction of active protein} \quad (2)$$

It should be noted that although we can calculate the net uptake of protons for each transition, we do not know how many ionizable groups are actually involved. The term T has been added to account for the pH-dependent activation of transducin that occurs in the absence of opsin. (See Ref. 2 for a complete discussion of the derivation of this expression.)

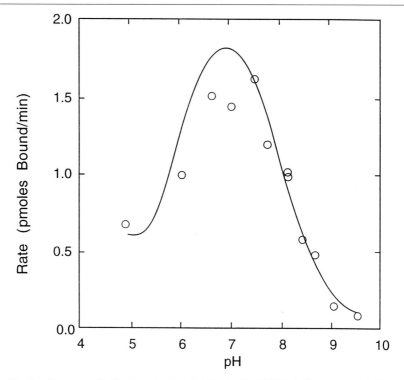

Fig. 2. pH–rate profile for transducin activation by Lys296Gly in the absence of chromophore. Rates were determined from the linear portion of each reaction time course. The solid line is simulated as described in the text. pK_{a1} is 5.9 and pK_{a2} is 8.0.

Purification of Constitutively Active Mutants for Spectral Analysis

Depending on the site mutated to create a constitutive-active mutant, it is possible to characterize the spectral properties of the visual pigment that has been reconstituted with chromophore. In the mutants where Lys-296 has been mutated it is possible to form a reconstituted visual pigment using a Schiff base complex of 11-*cis*-retinal.[18] We purify the reconstituted visual pigments from COS cells with an immunoaffinity column using[9] bovine rhodopsin monoclonal antibody, 1D4, as described in detail elsewhere.[11,16] Harvested transfected cells are resuspended in 1 ml/plate of 10 mM sodium phosphate buffer (pH 7.0) containing 150 mM NaCl and incubated with 20–40 μM of the appropriate chromophore in the dark.[11,16,18] Proteins are solubilized from cell membranes using dodecylmaltoside (DM). After incubation with the immunoaffinity column, the sample is eluted into a 0.1% DM solution (300 μl) and assayed spectrophotometrically.

Purified pigments eluted from the immunoaffinity matrix are maintained at 6° in a water-jacketed cuvette holder. A Hitachi model U-3300 dual-path spectrophotometer is used to record absorption spectra before and after bleaching with a 175-W fiber optic light source. Visual pigment samples are bleached for 10 min in the presence of 50 mM hydroxylamine (pH 7.0). In some experiments, difference curves are then calculated from pre- and postbleach spectra. To assign λ_{max}, dark-adapted or difference spectra were compared with template spectra for rhodopsin using a least-squares procedure.[22,23]

Assay for Rhodopsin Kinase Activity Using Constitutively Active Opsin Mutants

Light-activated wild-type rhodopsin not only binds and activates transducin, but it binds to and is a substrate for rhodopsin kinase. To further characterize some constitutively active rhodopsin mutants, we assay them for their ability to act as a substrate for purified rhodopsin kinase.

Rhodopsin kinase is purified and its activity determined as previously described.[24–26] Wild-type rhodopsin and constitutively active rhodopsin mutants expressed in COS cell membranes are assayed for their ability to act as a substrate for purified rhodopsin kinase. COS cell membranes are assayed in the absence of chromophore or if possible after regeneration with 11-*cis*-retinal or the appropriate 11-*cis*-retinal Schiff base complex. Membranes (50 nM opsin) are incubated with 200 μM chromophore for 1 hr in the dark at room temperature. These membranes are then concentrated 10-fold by centrifugation and resuspension in a buffer containing 50 mM BTP (pH 7.5) and 1 mM MgCl$_2$. Membranes are added in the dark to a reaction mixture containing 5 μl [γ-^{32}P]ATP (1.6 mM 1200 cpm/pmol) and 1 μg of purified rhodopsin kinase. This mixture is then incubated at 20° in the dark or under the illumination of a 150-W lamp. Then 20-μl aliquots are removed at the desired time and the kinase reaction is terminated by the addition of 5 μl of 25 mM EDTA/100 mM dodecyl-β-D-maltoside in 50 mM BTP. Quenched samples are then analyzed by SDS–PAGE. Radiolabeled proteins are visualized either using autoradiography

---

[22] G. D. Bernard, *J. Opt. Soc. Am. A* **4**, 123 (1987).
[23] T. W. Cronin, *J. Comp. Phys. A* **172**, 339 (1989).
[24] J. Buczyiko, C. Gutmann, and K. Palczewski. *Proc. Natl. Acad. Sci. U.S.A.* **88**, 2568 (1991).
[25] Palczewski, K., ed., in "Methods in Neurosciences: Photoreceptor Cells," pp. 217–225. Academic Press, Inc., San Diego, 1993.
[26] H. Ohguro, M. Rudnicka-Nawrot, J. Buczylko, X. Zhao, J. A. Taylor, K. A. Walsh, and K. K. Palczewski, *J. Biol. Chem.* **271**, 5215 (1996).

or using a Molecular Dynamics PhosphorImager with ImageQuant v4.2 software.

Mutations that Disrupt Salt Bridge Between Lys-296 and Glu-113 Resulting in Constitutive Activation of Opsin

The power of combining a heterologous expression system for rhodopsin with *in vitro* biochemical assays is illustrated by our discovery that the apoprotein of the mutant Lys296Gly activates transducin in our *in vitro* assay. The ligand-independent or constitutive activation has the same specific activity as we measure for wild-type rhodopsin. This constitutive activity of the mutant Lys296Gly can be inhibited by incubation in the dark with the Schiff base complex, nPrSB. The reconstituted pigment can then be activated by illumination (see Fig. 1). We have further characterized the constitutive activity of this mutant by assaying its activity as a function of pH (see Fig. 2). The bell-shaped curve suggests that activation depends on the uptake of a proton with the ionization constant of K_{a2}. We have made a series of mutations at position 296 and characterized the pH dependence of each.[3] The determination of the pK_{a2} value for each of the 12 mutants is an indication of the degree to which the active state is favored. Plotting these data as a function of both the size and the hydrophobicity of the substituted amino acid reveals that an inverse correlation exists between the pK_{a2} of the substituted amino acid and the size of the amino acid.

We have also analyzed the mutant Glu113Gln in a similar fashion. Glu-113 is the counterion to the protonated Schiff base nitrogen. We hypothesized that the interaction of this amino acid with the Lys-296 in the apoprotein might be involved in maintaining the protein in an inactive position. This mutant is also constitutively active as an apoprotein. The addition of the chromophore 11-*cis*-retinal suppresses this constitutive activity. The pK_{a2} value for this mutant is 6.8. These types of data have led us to speculate that the transition to the active state of rhodopsin involves the breaking of a salt bridge between Lys-296 and the counterion Glu-113 (Refs. 1 and 2).

Acknowledgments

I thank M. Brannock and L. Newman for reading the manuscript. Part of this research was funded by National Institutes of Health grant EY 10205-01.

[16] Synthetic Retinals: Convenient Probes of Rhodopsin and Visual Transduction Process

By Jihong Lou, Qiang Tan, Elena Karnaukhova, Nina Berova, Koji Nakanishi, *and* Rosalie K. Crouch

Introduction

All the visual pigment proteins are integral membrane proteins, crossing the membrane seven times with the majority of the protein being composed of hydrophobic amino acids within the membrane. This has made these proteins difficult to study by many standard biophysical techniques. However, each of these opsins has a chromophore, generally 11-*cis*-retinal or one of a few closely related structures, that is attached to the protein via a protonated Schiff base linkage. The site of attachment is at a lysine in the center of the seventh helix. *In vitro*, exposure of these proteins to light in the presence of hydroxylamine results in the complete detachment of the retinal, and washing the membranes with bovine serum albumin (1%) removes any residual retinal. This generates the opsin apoprotein, which can reform rhodopsin on the addition of 11-*cis*-retinal in the dark. This provides a most convenient tool for the examination of rhodopsin structure and function as the protein has also been found to accommodate various retinal isomers and analogs, forming pigments or complexes which can then be tested for functionality and physical properties.

A large number of retinals have been synthesized and tested with opsins from various species (see a recent review[1]). Most of these experiments have been carried out *in vitro* although a few studies have shown that retinal analogs can be successfully incorporated into some species including vitamin A-deficient rats.[2] Very little work to date has been conducted on the cone opsins due to problems in obtaining reasonable quantities of the proteins. However, with the development of both improved expression systems and purification techniques, these studies are now being initiated on the cone opsins as well. Interestingly, the physiologic recordings conducted to date on isolated rod and cone photoreceptors utilizing retinal analogs indicate that there are some significant differences in the interactions of these respective opsins with their chromophores.[3] Recently studies have been initiated

---

[1] K. Nakanishi and R. K. Crouch, *Isr. J. Chem.* **35,** 253 (1996).
[2] R. Crouch, B. R. Nodes, J. I. Perlman, D. R. Pepperberg, H. Akita, and K. Nakanishi, *Invest. Ophthalmol. Vis. Sci.* **25,** 419 (1984).
[3] V. J. Kefalov, M. C. Cornwall, and R. K. Crouch, *J. Physiol.* (in press) (1999).

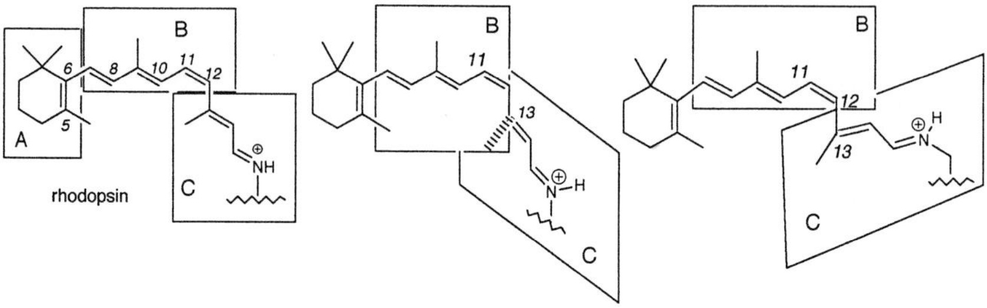

FIG. 1. The chromophore in rhodopsin is not planar, that is, planes A, B, and C are not coplanar.

on combining the use of retinal derivatives with opsins containing specific site mutations, which is indeed a powerful method of exploring the structural boundaries of the chromophore binding site.[4]

The constraints of the binding site of rod rhodopsin have been extensively examined using retinal analogs and from these studies the following generalizations can be made: (1) All isomers (including di- and tri-*cis* isomers), except for 13-*cis* and all-*trans,* can be accommodated; (2) the methyl at C-9 on the polyene chain is critical to the absorption properties and activity of the protein; and (3) the binding cavity has little tolerance for additional bulk in the region of the cyclohexyl ring; however, the ring itself is not essential for pigment formation if one methyl group is present, corresponding to the C-1 or C-6 methyl. The exact location of the chromophore within the protein has been probed by photoaffinity cross-linking studies and these results are discussed elsewhere in this volume. Due to the lack of significant quantities of the cone opsins, little is understood about the cone opsin-binding site.

We describe here some recent findings on rod rhodopsin using analogs of retinal that seek (1) to explain the unique circular dichroism (CD) spectrum of rhodopsin, (2) to define the absolute conformation around the 12-*s-trans* bond of the chromophore in bovine rhodopsin, and (3) to address the control of the activity of the protein by the retinal–protein interactions.

Circular Dichroism Spectrum of Rhodopsin

11-*cis*-Retinal is twisted around the C-6/C-7 and C-12/C-13 single bonds due to steric interactions between 5-CH$_3$/8-H and 13-CH$_3$/10-H, respectively. Namely, planes A/B and B/C (Fig. 1) are not coplanar. Then what is the absolute sense of twist between these planes? In rhodopsin, light

[4] M. Han, M. Groesbeek, S. O. Smith, and T. P. Sakmar, *Biochemistry* **37,** 538 (1998).

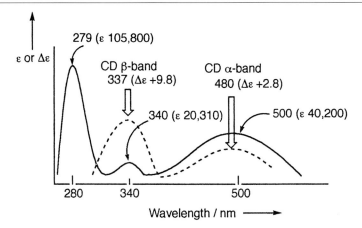

FIG. 2. UV–vis (solid line, ε) and CD (dashed line, Δε) of bovine rhodopsin in 23 mM octylglucoside solution (pH 7.0).

induces isomerization of the 11-*cis*-retinylidene chromophore to its all-*trans* form. To accomplish this transformation, the C-11/C-12 double bond is required to rotate 180°. Taking the steric interaction between 13-CH_3 and 10-H into account, isomerization should involve a rotation with the 13-CH_3 moving away from the adjacent 10-H.[5]

Elucidating the absolute extent of twist around the 6-*s-cis* and 12-*s-trans* bonds of the retinal chromophore in rhodopsin is central for clarifying the changes in the chromophore–receptor interaction along the visual transduction pathway. In addition, the twists together with other factors, for example, the protonated Schiff base, the counteranion distance, and the electrostatic charge distribution within the binding site, dictate the regulations of absorption maxima of various visual pigments.[6,7]

The nonplanar conformation of the retinal chromophore also accounts for the unique CD spectrum of rhodopsin. Native rhodopsin exhibits two positive Cotton effects in its CD spectrum at 480 nm (Δε = +2.8, α-band) and 337 nm (Δε = +9.8, β-band), respectively (Fig. 2). A rhodopsin incorporating the retinal analog ret5 with a five-membered ring bridging C-10 and C-13, in which planes B/C are coplanar, exhibits CD with negligible α-band.[8] In contrast, the pigment formed from a retinal analog in which planes A/B are kept coplanar by a five-membered ring bridging C-8 and

---

[5] G. G. Kochendoerfer and R. A. Mathies, *Isr. J. Chem.* **35**, 211 (1995).
[6] K. Nakanishi, *Am. Zool.* **31**, 479 (1991).
[7] K. Nakanishi, *Pure Appl. Chem.* **63**, 161 (1991).
[8] Y. Fukada, Y. Shichida, T. Yoshizawa, M. Ito, A. Kodama, and K. Tsukida, *Biochemistry* **23**, 5826 (1984).

C-5, showed a weak β-band.[9,10] Based on these observations, Ito and coworkers have assigned the origin of α- and β-band to distortions around the 12-*s-trans* and 6-*s-cis* bonds, respectively.[11] Although it is generally accepted that the two positive Cotton effects (CE) reflect the interaction between the twisted chromophore and its protein environment, the absolute senses of twist around the 6-*s-cis* and 12-*s-trans* bonds, or the absolute conformation of the retinylidene chromophore in rhodopsin, remains to be established.

The absolute twist around the 12-*s*-bond has been determined by two independent approaches: (1) chiroptical spectroscopy based on excitoncoupled CD and (2) bioorganic studies based on incorporation of an 11,12-cyclopropyl retinal analog. The twist around the 6-*s* bond is under investigation.

Chiroptical Spectroscopy

The approach was to modify the retinal structure so that the absolute sense of twist around the 12-*s* bond could be detected spectroscopically. Exciton-coupled CD has proven to be an extremely versatile and sensitive method for stereochemical studies.[12] However, since the method depends on the through-space chiral coupling between two or more isolated chromophores, it cannot be applied to the native retinal system, which consists of a single twisted polyene system. Saturation of the 11-ene breaks the chromophore into two exciton conjugated systems, a triene and a diene, which might interact or couple through space within the protein binding site and thus give rise to an exciton couplet reflecting the C-12/C-13 twist. As shown in Fig. 3, a positive helicity between B/C planes would yield a positive couplet, and vice versa.

11,12-Dihydrorhodopsins

Figure 4 shows the UV of 11,12-dihydroretinal (2H-ret) (**1**) and its protonated Schiff base (PSB) with *n*-butylamine (2H-PSB) (**2**) in methanol, and the UV (A) and CD (B) of 11,12-dihydrorhodopsin (2H-Rh) (**3**) in octylglucoside solution. The pigment incorporating 11,12-dihydroretinal,

---

[9] M. Ito, T. Hiroshima, K. Tsukida, Y. Shichida, and T. Yoshizawa, *J. Chem. Soc. Chem. Commun.* 1443 (1985).

[10] Y. Katsuta, M. Sakai, and M. Ito, *J. Chem. Soc. Perkin Trans. I*, 2185 (1993).

[11] M. Ito, Y. Katsuta, Y. Imamoto, Y. Shichida, and T. Yoshizawa, *Photochem. Photobiol.* **56**, 915 (1992).

[12] K. Nakanishi, N. Berova, and R. W. Woody, "Circular Dichroism: Principles and Applications." VCH, New York, 1994.

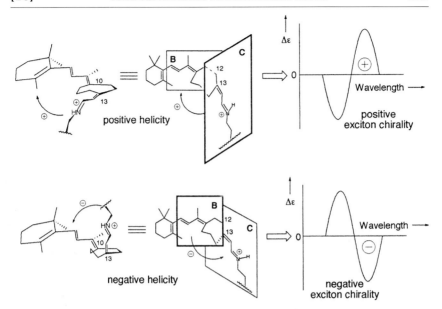

**Positive helicity in B / C planes gives positive couplet.
Negative helicity in B / C planes gives negative couplet.**

Fig. 3. Expected signs of exciton couplet resulting from the two directions of twist between planes B and C.

2H-Rh (**3**) absorbs at 279 nm (ε 27,000) and exhibits a *negative bisignate* CD, 295 nm ($\Delta\varepsilon$ − 1.2) and 275 nm ($\Delta\varepsilon$ + 2.5), A value −3.7 (Fig. 4). This bisignate nature of the CD is in contrast to native rhodopsin, λ_{max} 500 nm, which exhibits two positive CD Cotton effects at 480 nm ($\Delta\varepsilon$ + 2.8, α-band) and 337 nm ($\Delta\varepsilon$ + 9.8, β-band, Fig. 2). This bisignate CD is thus interpreted as being due to the coupling between the triene and diene (protonated Schiff base, PSB) moieties and reflects the absolute sense of twist between planes B and C.

11-*Cis*-Locked Seven-Membered Ring 11,12-Dihydrorhodopsins (2H-Rh7)

The similarity in the CD and UV–vis spectra of the pigment incorporating ret7 (**4**), that is, Rh7, with native rhodopsin indicated that the conformations of the seven-membered chromophore and the native chromophore were alike.[13] Moreover, because saturation of the double bond at the center

---

[13] K. Nakanishi, A. H. Chen, F. Derguini, P. Franklin, S. Hu, and J. Wang, *Pure Appl. Chem.* **66**, 981 (1994).

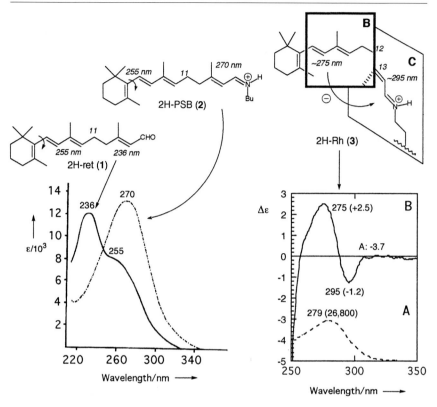

FIG. 4. 11,12-2H-ret (**1**). UV of 11,12-2H-ret and its protonated Schiff base in methanol and UV/CD of 11,12-2H-Rh in octylglucoside solution.

of the polyene chain increases the chromophoric flexibility within the binding site, a chain was bridged across C-10 and C-13 to fix the C-11/C-12 as *cis* in the seven-membered ring analogs **5** and **6** (2H-ret7). Figure 5 shows 13-*trans*-2H-ret7 (**5**), its PSB (**7**) with *n*-butylamine, and their respective UV spectra in methanol. The enal and nonplanar triene moieties in 2H-ret7 (**5**) absorb at 244 and 270 nm (shoulder) in methanol, respectively (Fig. 5). In the protonated Schiff base with *n*-butylamine, the absorption maximum of the enal is shifted to 272 nm, whereas the triene absorption stays at 270 nm.

Figure 6 shows the UV, CD, and projected structures of 13-*trans*-2H-Rh7 (**8**) and 13-*cis*-2H-Rh7 (**9**), that is, the rhodopsin analogs incorporating **5** or **6**, respectively. They absorb maximally at 283 nm (ε 27,000) and 284 nm (ε 29,000), respectively.

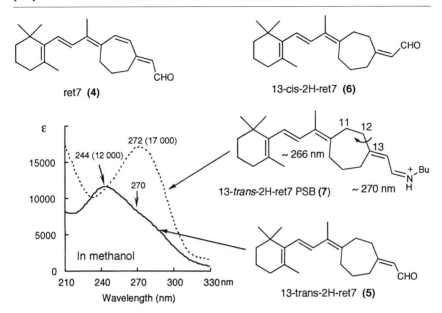

Fig. 5. Structures of ret7 (**4**), 13-*trans*-2H-ret7 (**5**), and 13-*cis*-2H-ret7 (**6**). UV of 13-*trans*-2H-ret7 (**5**) and 13-*trans*-2H-ret7 PSB (**7**).

13-*trans*-2H-Rh7 (**8**) exhibits a *negative split Cotton effect* CD at 297 nm ($\Delta\varepsilon$ − 6.5) and 273 nm ($\Delta\varepsilon$ + 5.2), that is, A value (amplitude) −11.7. Namely, the triene absorption originally at 270 nm is red-shifted to ca. 275 nm arising from its more planar shape in the protein, whereas the 272 nm maximum of the PSB group is also red-shifted to 295 nm due to its interaction with the global electrostatic charge within the binding site. This is further supported by the larger A value of −14.2 for the pigment 13-*cis*-2H-Rh7 (**9**) as compared to the A value of −11.7 for the pigment 13-*trans*-2H-Rh7 (**8**) (Fig. 5). The projection angle of the two interacting chromophores (through the C-13/C-10 axis), is ca. 60° and ca. 20°, respectively, for 13-*cis* and 13-*trans* isomers. In agreement with the known fact that the A value of bisignate CD curves is maximal around 70°,[12] it follows that the A value of the pigment derived from the 13-*cis* isomer is larger.

The amplitude of 2H-Rh is smaller than those of the 2H-ret7-derived pigments, probably due to the more flexible nature of the side chain. Nevertheless, the fact that the CD spectra of all these dihydro pigments are negative bisignate curves shows that the triene and PSB moieties constitute a negative couplet. Thus it is concluded that the C-12/C-13 bond of the

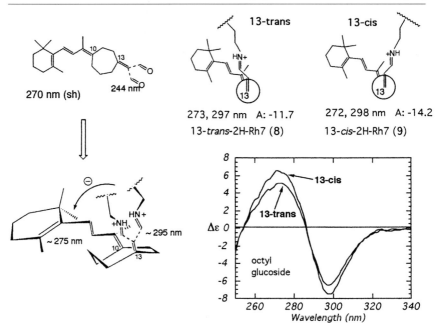

FIG. 6. Projected structures of 13-*trans*-2H-Rh7 (**8**) and 13-*cis*-2H-Rh7 (**9**) and their CD.

chromophore is twisted in a negative sense as depicted in Fig. 4 (for 2H-Rh) and Fig. 11 (given later).[14]

Absolute Conformation of 12-*s-trans* Bond of Retinal in Rhodopsin; 11,12-Cyclopropylretinal and Derived Pigment

The exciton-coupled circular dichroic studies of 11,12-dihydrorhodopsin pigments just described led to the absolute conformation around the 12-*s-trans* bond of the retinal chromophore in bovine rhodopsin.[14] Namely, the negative CD couplets of 11,12-dihydrorhodopsin pigments indicate that planes B and C are oriented as shown in Fig. 7.

This negative helicity agrees with the theoretical calculation by Kakitani and co-workers[15] and the results from the solid-state nuclear magnetic resonance (NMR) studies by Smith and co-workers.[16,17] However, Buss *et*

[14] Q. Tan, J. Lou, B. Borhan, E. Karnauknova, N. Berova, and K. Nakanishi, *Angew. Chem. Int. Ed. Engl.* **36,** 2089 (1997).
[15] H. Kakitani, T. Kakitani, and S. Yomosa, *J. Phys. Soc. Japan* **42,** 996 (1977).
[16] M. Han and S. O. Smith, *Biochemistry* **34,** 1425 (1995).
[17] M. Han and S. O. Smith, *Biochemistry* **36,** 7280 (1997).

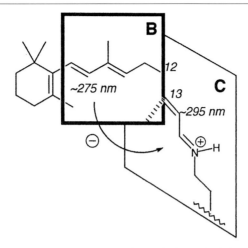

FIG. 7. Absolute sense of twist around C-12/C-13 single bond of the retinal chromophore in rhodopsin based on negative CD couplets of 11,12-dihydrorhodopsin pigments.

al. proposed a positive twist around the 12-*s-trans* bond in retinal on the basis of semiempirical and nonempirical calculations of the 11-*cis*-retinylidene chromophore.[18] To resolve the discrepancy between these theoretical calculations and to verify the assignments obtained from exciton-coupled CD and solid-state NMR data,[14,16,17] further investigation of the retinal conformation in rhodopsin was carried out. The following work represents the first case of enantioselective binding of a chiral retinal analog to bovine opsin, that is, one of the enantiomers forms a pigment with opsin while the other does not bind to opsin.[19] This bioorganic study unambiguously establishes the chirality of the retinal binding site in rhodopsin.

11-*cis*-Seven-membered ring-locked retinal analogs with a cyclopropyl ring incorporated to the C-11/C-12 bond (**10** and **11**) have been designed for opsin binding studies. The rationale in choosing a seven-membered-ring locking the retinal into 11-*cis* conformation lies in the earlier observation that the UV–vis and CD spectra of rhodopsin pigment containing 11-*cis*-cycloheptatrienylidene retinal or ret7 (**4**) closely resemble those of the native rhodopsin.[20] Incorporation of a cyclopropyl ring into the C-11/C-12 bond introduces a chiral element to the retinal chromophore and induces

---

[18] V. Buss, K. Kolster, F. Terstegen, and R. Vahrenhorst, *Angew. Chem. Int. Ed. Engl.* **37**, 1893 (1998).
[19] J. Lou, M. Hashimoto, N. Barrow, and K. Nakanishi, *Org. Lett.* **1**, 51 (1999).
[20] H. Akita, S. P. Tanis, M. Adams, V. Balogh-Nair, and K. Nakanishi, *J. Am. Chem. Soc.* **102**, 6370 (1980).

a predetermined rigid twist around the C-12/C-13 bond. The two enantiomeric cyclopropyl retinal analogs adopt opposite twists around the C-12/C-13 bond, and hence it is expected that the enantiomer with the right geometry will bind preferably to opsin, while binding of the other enantiomer will be less favored or will not proceed at all.

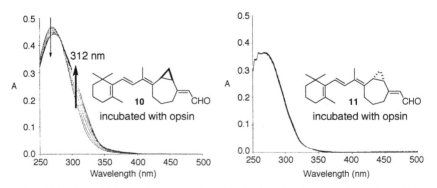

Binding Studies with Opsin

Both cyclopropyl retinal analogs **10** and **11** absorb maximally at 266 nm in methanol, their *n*-butylamine Schiff bases absorb at 260 nm, while the protonated Schiff bases absorb at 284 nm in methanol. Retinals **10** and **11** were incubated with bovine opsin solubilized in CHAPSO/HEPES buffer in a UV sample cuvette and the progress of binding was monitored by UV–vis spectra. The concentrations of opsin and retinal analog in the two binding experiments were about the same. Generally, the binding of a retinal analog to opsin results in a red-shifted absorption band in UV that originates from formation of the protonated Schiff base within the retinal binding pocket of rhodopsin, and this is what was observed in the binding of **10** to opsin. A new band absorbing at 312 nm was formed during a 1-hr incubation at room temperature, indicating formation of cyclopropylrhodopsin pigment (Fig. 8), while for **11**, no change in UV was observed, clearly indicating that **11** does not bind to opsin (Fig. 8). The optical purity

FIG. 8. Binding of **10** to opsin was monitored by UV; no binding was observed for **11**.

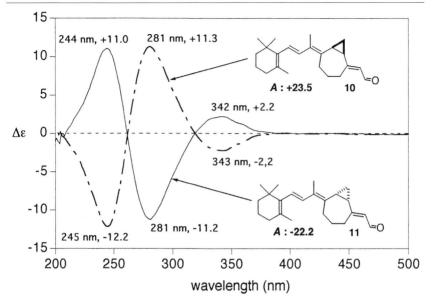

FIG. 9. CD of cyclopropylretinal analogs **10** and **11** in methanol.

of both enantiomers is about the same as seen by similar intensities of the Cotton effects in the CD (Fig. 9).

Conformational Analysis of the Cyclopropylretinal Analogs

The binding experiments of **10** and **11** with opsin clearly demonstrated the chiral preference of the retinal binding site. Analog **10** binds to opsin and forms a pigment absorbing at 312 nm, whereas **11** does not bind to opsin at all. Thus the conformation of **10** should represent the shape of the retinal binding pocket in rhodopsin. Molecular modeling by MacroModel using MM3 force field yielded the lowest energy conformations of **10** and **11** as shown in Fig. 10. The twists around C-12/C-13 single bond in **10** and **11** are opposite to each other, and for **10**, the analog that binds to opsin, there is a negative twist around the C-12/C-13 bond. This negative twist is in agreement with the above-mentioned conclusion based on exciton-coupled CD of 11,12-dihydrorhodopsin.

The pigment formed from **10** absorbs maximally at 312 nm, which is red-shifted by more than 25 nm as compared to the λ_{max} of the pigment formed from the seven-membered ring-locked 11,12-dihydroretinal analog **5** (which absorbs maximally at 285 nm).[14] Such a red shift in absorption maximum is due to the partial conjugation effect resulting from the cyclo-

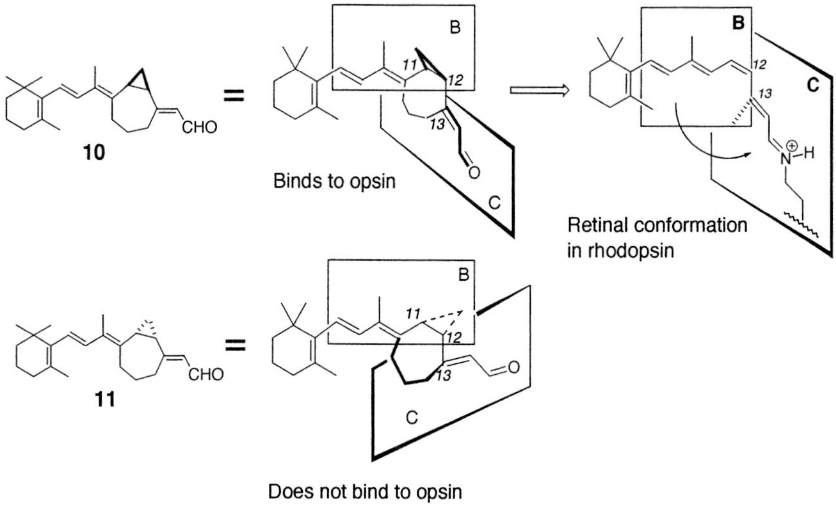

FIG. 10. Conformation of cyclopropylretinal analogs, MacroModel using MM3 force field.

propyl ring. Namely, the CD of the cyclopropyl rhodopsin pigment (not shown) does not originate from exciton coupling, which requires zero or negligible molecular orbital overlaps. Theoretical calculation of the chiroptical data of the cyclopropylretinal analogs is currently under way.

Two independent studies, chiroptical and bioorganic, show that the absolute sense of twist between planes B and C is negative as depicted in Fig. 11. Figure 11 also incorporates the conclusion of earlier photoaffinity labeling studies with 3-diazo-4-keto-11-*cis*-retinal, which showed that C-3 of the ionone ring is close to Trp-265 located around the middle of helix F.[21]

Role of 13-Methyl Group in Control of Rhodopsin Activity

The removal of the 13-methyl group from the polyene side chain has a remarkable effect on the properties of rhodopsin. As noted earlier, in the 11-*cis* isomer of retinal, steric hindrance occurs between the C-10 hydrogen and the C-13 methyl (Fig. 11) and polyene side chain twists to accommodate this interaction. Removal of the 13-methyl relieves this strain. Early reports on the pigment formed with 11-*cis*-13-demethylretinal showed that phosphodiesterase was activated in the absence of light.[22] In collaboration

[21] H. Zhang, K. Lerro, T. Yamamoto, T. Lien, L. Sastry, M. Gawinowicz, and K. Nakanishi, *J. Am. Chem. Soc.* **116**, 10165 (1994).
[22] T. G. Ebrey, M. Tsuda, G. Sassenrath, J. L. West, and W. H. Waddell, *FEBS Lett.* **116**, 217 (1980).

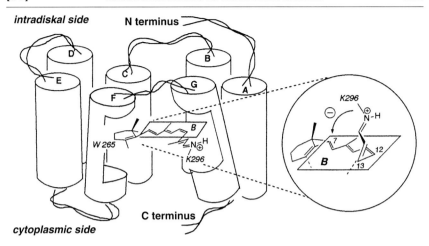

FIG. 11. Location of chromophore and the negative twist around C-12/C-13 bond in rhodopsin ground state.

with Palczewski, Saari, and colleagues, biochemical studies yielded the surprising result that on the addition of 11-*cis*-13-demethylretinal to the apoprotein opsin, rhodopsin kinase is activated (Fig. 12). Over time, this activity decreases. The native chromophore, 11-*cis*-retinal, shows no activity in the same time range. The all-*trans* isomer of the 13-demethylretinal, like the all-*trans* isomer of retinal itself, combines with opsin to form a fully active state, with phosphorylation occurring at the same sites as in rhodopsin itself. The pigment formation between 11-*cis*-13-demethylretinal and opsin is quite slow (one-ninth the rate of rhodopsin). The complex formed between the 11-*cis*-13-desmethyl retinal and the opsin places the protein into a conformation that evidently is recognized by the rhodopsin kinase as a metarhodopsin-II conformation. This results in activation of rhodopsin kinase and phosphorylation of the pigment. If the pigment is allowed to fully regenerate (24 hr) before exposure to rhodopsin kinase, no activation is obtained until the pigment is photolyzed and has similar activity to photolyzed rhodopsin. Therefore, either the 13-methyl group has a critical interaction with the protein, which facilitates the Schiff base formation, or the change in the conformation of the polyene chain as a result of the relief of the strain due to the interaction between the 10-hydrogen and the 13-methyl places the retinal in a conformation that is no longer optimum for interaction with the critical lysine.

The use of this retinal derivative has been critical to the understanding that retinal has a most precise role in the control of the activity of this G-protein receptor. A photon of light of a specific wavelength isomerizes

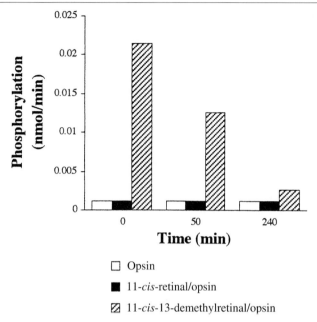

FIG. 12. Levels of phosphorylation of opsin–retinal complexes.

the critical 11-*cis* double bond to the all-*trans* conformation, bringing the receptor from its inactive state to the active state. 11-*cis*-Retinal therefore acts as an inverse agonist, locking the receptor in an inactive conformation, which of course is critical in keeping the receptor "quiet" when not receiving light. The light-induced isomerization of the chromophore would appear to be designed to break the glutamate-113/lysine-296 Schiff base salt bridge and change the conformation of the receptor to allow maximal interaction with the G protein. In the all-*trans* conformation, the retinal is acting as an agonist, activating the receptor. At least two forms of the rod rhodopsin containing the all-*trans* conformation are proposed to have activity: (1) the metarhodopsin-II (or R*) form, which is very effective in activating the transduction process (see recent review) and (2) a second form of metarhodopsin-II, proposed to be phosphorylated and bound to arrestin, having an activity of 10^{-5} that of R*.[23] The apoprotein opsin with no ligand in the binding site also has some basal activity, estimated at 10^{-7} that of R*.[24] While this is a decidedly low level,[25] it is quite meaningful on the physiologic scale, as has been demonstrated using both isolated rods and cones. A

---

[23] C. S. Leibrock, T. Reuter, and T. D. Lamb, *Eye* **12**, 511 (1998).
[24] M. C. Cornwall, and G. L. Fain, *J. Physiol.* **480**, 261 (1994).
[25] T. J. Melia, Jr., C. W. Cowan, J. K. Angleson, and T. G. Wensel, *Biophys. J.* **73**, 3182 (1997).

number of laboratories have also determined that combination of the apoprotein opsin with all-*trans*-retinal forms a complex that has activity resembling that of the photolyzed rhodopsin as measured by phosphodiesterase,[26] phosphorylation by rhodopsin kinase, and arrestin binding.[27] The activity of this complex does not depend on the formation of a Schiff base.[28] The role of such a complex in the physiologic process is as yet unclear.

Retinal is central to the entire visual process because it is the ligand of the visual pigments and is critical for the control of the activity of these pigments. The molecule itself is quite small (molecular weight <300) but it very precisely controls the conformation of these proteins, which in turn determines the interaction of these pigments with the various proteins involved in the transduction cascade. 11-*cis*-Retinal can be thought of in the classical pharmacologic sense as being a reverse agonist, locking rhodopsin in the inactive conformation.

Opsin	→	rhodopsin	→	metarhodopsin-II
no chromophore		11-*cis*-retinal		all-*trans*-retinal
(*partially active*)		(*inactive*)		(*active*)

The apoprotein opsin, which is lacking the chromophore and therefore this "lock" in the inactive conformation, can assume conformations that are partially active. The role of light in the process is to isomerize the 11-*cis* bond of retinal to the all-*trans* isomer, placing rhodopsin into the active meta-II conformation. In this form, the chromophore is acting as an agonist. The results reported here, as well as work in other laboratories, demonstrate that complexes formed with all-*trans* isomers of retinal and this desmethyl analog can also place the opsin protein into a meta-II conformation and thus mimic the action of light on native rhodopsin. In addition, there are retinals analogs which act as antagonists as they block the binding site but do not activate the pigment. The large body of literature on studies of pigments formed with retinal analogs has been fundamental to establishing this critical role of retinal in orchestrating the activity of this G-protein receptor.

General Procedures

General Information

Chemicals including solvents are purchased from Aldrich (Milwaukee, WI), Fisher (Suwanee, GA), and Sigma (St. Louis, MO). Water for bio-

---

[26] Y. Fukada and T. Yoshizawa, *Biochim. Biophys. Acta* **675,** 195 (1981).
[27] K. P. Hofmann, A. Pulvermuller, J. Buczylko, P. Van Hooser, and K. Palczewski, *J. Biol. Chem.* **267,** 15701 (1992).
[28] S. Jager, K. Palczewski, and K. P. Hofmann, *Biochemistry* **35,** 2901 (1996).

chemical procedures is double distilled and deionized. Dark-adapted retinas are obtained from W. Lawson Co. (Lincoln, NE). Centrifugation is performed on a Du Pont Sorvall RC-58 refrigerated superspeed centrifuge or a Beckman L8-M ultracentrifuge using appropriate rotors. Anhydrous solvents are either purchased or prepared by distillation over appropriate dehydrating reagents. Glassware for anhydrous purposes is flame-dried. The progress of reactions is checked by thin-layer chromatography using 250-μm precoated silica gel TLC plates from EM Separation Technology, which are visualized by phosphomolybdic staining or UV light. Flash column chromatography is carried out using 32-63 mesh silica gel from ICN. All isomers of retinal analogs are isolated by HPLC on a YMC-Pack SIL column with a size of 250 × 10 mm ID. All UV–vis spectra are recorded on a Perkin-Elmer (Norwalk, CT) Lambda 6 UV–vis spectrophotometer, and CD spectra on a JASCO J-720 spectropolarimeter, 1-cm light path cell.

Synthesis of Retinal Analogs

1. Retinal derivatives are synthesized by several laboratories. A general overall scheme employing the Horner reaction is shown in Fig. 13. Several different agents are employed both for the reduction steps

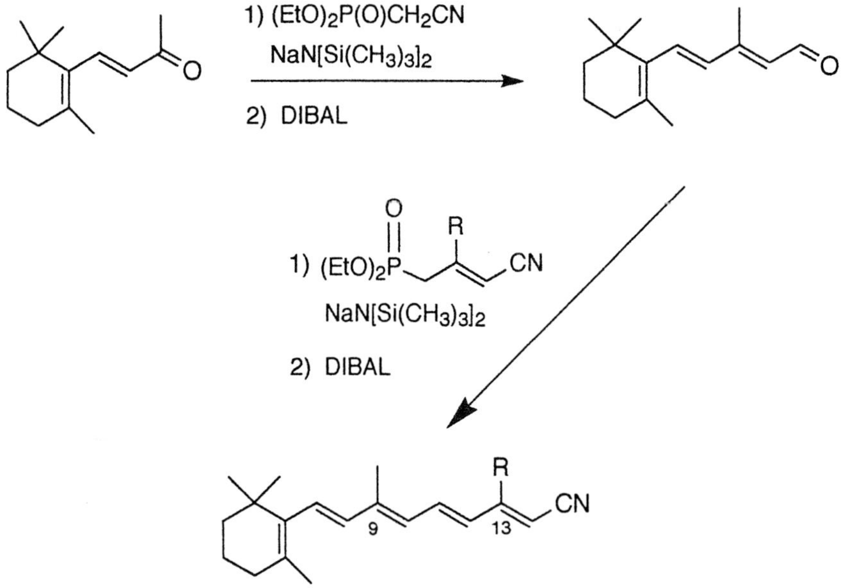

FIG. 13. General scheme for synthesis of retinal analogs.

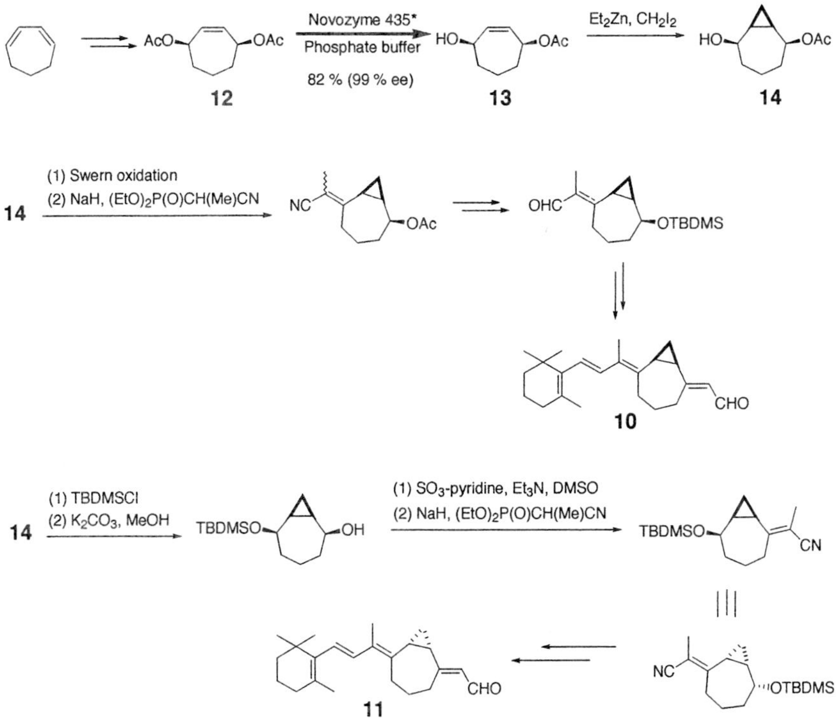

FIG. 14. Synthesis of cyclopropylretinal analogs.

as well as oxidizing agents. Other types of Wittig-based schemes are also employed (see Ref. 1 for citations to the individual preparations).

2. The cyclopropyl retinal analogs are synthesized as enantiomerically pure isomers following the route summarized in Fig. 14. The key step involves enantioselective hydrolysis of diacetate **12** by lipase Novozyme 435 to give monoacetate **13** in 82% yield (99% ee).[29] The Simmons-Smith reaction of **13** gives cyclopropyl alcohol **14**, which can be converted to both **10** and **11** through separate reaction sequences as shown in Fig. 14.

Preparation of Schiff Bases

n-Butylamine (300 μl) is added to a mixture of 0.3 μmol of retinal analog and 5 mg anhydrous K_2CO_3. After stirring for 24 hr at room temperature, *n*-

[29] C. R. Johnson and S. J. Bis, *Tetrahedron Lett.* **33**, 7287 (1992).

butylamine is removed *in vacuo*, the residue is stirred with 1 ml hexane and the K_2CO_3 filtered off. The hexane is removed *in vacuo*. The remaining Schiff base is dissolved in 1 ml absolute methanol, treated with 0.1 ml of methanolic HCl, and the UV measured.

ROS Preparation

Bovine rod outer segments (ROS) are isolated according to a standard procedure with modifications. All operations are performed at 4° under dim red light (>680 nm). Sucrose solutions are prepared with isotonic buffer (10 mM Tris, pH 8.0, 60 mM KCl, 30 mM NaCl, 2 mM MgCl$_2$, 1 mM dithiothreitol). The frozen retinas (~100) are defrosted overnight and gently shaken in 80 ml 35% sucrose solution for 1 min and centrifuged at 5000 rpm in a Sorvall SS-34 rotor for 10 min. The pellet is resuspended in 80 ml 35% (w/v) sucrose solution and centrifuged similarly. The two supernatants are combined, diluted with isotonic buffer to 26% (w/v) sucrose concentration, and centrifuged at 15,000 rpm in a Sorvall SS-34 rotor for 30 min. The pellet is resuspended in 26% (w/v) sucrose solution, and added gently to the top of a 26–35% (w/v) discontinuous sucrose gradient by a Pasteur pipette. After centrifugation at 23,000 rpm for 45 min in a Beckman SW28 swinging bucket rotor, the ROS are found at the interface of the gradient. The ROS are collected with a syringe fitted with a flat-tipped needle and washed three times with isotonic buffer. For the experiments with rhodopsin kinase, the membranes are also washed with 500 mM NaCl (35 min; 15,000 rpm) and then with 10 mM 1,3-bis[tris(hydroxymethyl)methylamino]propane (BTP), pH 7.5, containing 50 mM NaCl (35 min; 15,000 rpm). This procedure removes the soluble and membrane-associated proteins. The membranes are stored frozen in 10 mM BTP, pH 7.5, with 50 mM NaCl at −20°.

ROS Bleaching and Opsin Preparation

The ROS are suspended in 67 mM phosphate buffer (pH 7.0) containing 0.1 M hydroxylamine to a final concentration of 0.5–2 OD/ml. At 0° the suspension is illuminated by a projector equipped with a 470-nm cutoff filter for 30 min, or under room light for 2 hr. The bleaching is complete when the red color changes into pale yellow, and the absorption at 500 nm vanishes. Excess hydroxylamine and oxime are removed by washing with 5% (w/v) bovine serum albumin (BSA) in 0.02 M Tris buffer (pH 7.0) three times followed by three washes with 67 mM phosphate buffer (pH 7.0).

Preparation and Characterization of Artificial Pigments

All procedures are carried out at 25° in the dark. Opsin (1 OD) suspended in 426 μl 67 mM phosphate buffer (pH 7.0) is added to two screw-capped vials with magnetic stir bars. To one vial, 1.5 molar equivalence of retinal analog in 5 μl ethanol is added. To the other vial, 5 μl ethanol is added as a control reference. Both mixtures are stirred for 5 hr and then centrifuged at 25,000 rpm for 15 min. In some cases the pellets are washed with 1% BSA to remove excess chromophore. The pellets are either suspended in phosphate buffer or dissolved in 1 ml 23 mM octylglucoside solution (67 mM phosphate buffer, pH 7.0) and centrifuged at 25,000 rpm for 10 min. The supernants are analyzed by UV and CD measurements. The amount of opsin in each supernatant is determined from the A_{500} value after reconstitution with 11-*cis*-retinal.

Activation of Rhodopsin

Several assays for the measurement of the activation of rhodopsin or rhodopsin analog pigments have been reported in the literature. Activation of rhodopsin kinase provides one convenient monitor for level of activation of rhodopsin. Further, this method provides the potential of determining if the activated conformation of the analog pigment is identical to that of the native rhodopsin as the sites of phosphorylation can be determined.[30] The assay has been described in detail elsewhere.[31] Rhodopsin kinase is expressed in insect *Spodoptera frugiperda* cells and purified by chromatography on DEAE-cellulose. The activity is measured by the P_i transferred per minute per milligram rhodopsin or analog pigment using [γ-^{32}P]ATP. The reaction is terminated by the addition of potassium phosphate buffer (250 mM, pH 7.2) containing 200 mM EDTA, 100 mM KF, 5 mM adenosine, and 200 mM KCl.

Acknowledgments

The studies were supported by NIH grants GM36564 to K.N. and EY04939 to R.K.C. and by the Foundation to Prevent Blindness. The rhodopsin kinase activation experiments were conducted by Drs. Palczewski, Buczylko, and Saari at the University of Washington.[32]

---

[30] D. I. Papac, J. E. Oatis, Jr., R. K. Crouch, and D. R. Knapp, *Biochemistry* **32**, 5930 (1993).

[31] K. Palczewski, *Methods Neurosci.* **15**, 217 (1993).

[32] J. Buczylko, J. C. Saari, R. K. Crouch, and K. Palczewski, *J. Biol. Chem.* **271**, 20621 (1996).

[17] Assays for Activation of Opsin by all-*Trans*-Retinal

By KRISTINA SACHS, DIETER MARETZKI, and KLAUS PETER HOFMANN

Introduction

Rhodopsin, once triggered by cis–trans photoisomerization of 11-*cis*-retinal, retains all-*trans*-retinal (*atr*) covalently bound to Lys-296 in Schiff base linkage. The deprotonated Schiff base form of this photoproduct, the metarhodopsin-II conformation (R*) catalytically activates the G protein transducin (G_t) at very high rate (>1000 G_t s^{-1}) for signal transduction in the visual cascade. Deactivation of the photoactivated receptor in rod cells occurs in the so-called rhodopsin cycle via phosphorylation by rhodopsin kinase (RK), arrestin binding, and reduction of *atr* by the retinol dehydrogenase. The resulting phosphorylated opsin (P-opsin) can be dephosphorylated by cellular phosphatases[1] (see Hofmann *et al.*[1]; reviewed by Helmreich and Hofmann[2]). Opsin by itself has a low catalytic rate for G_t activation *in vitro*. A salt bridge between Lys-296 and Glu-113, forces the protein into the inactive conformation.[3] Melia *et al.*[4] estimated the opsin to be 10^{-6} as active as metarhodopsin-II (R*) at neutral pH. Higher activities of opsin found in G_t activation assays eventually depend on residual retinal derivatives in the preparations.[5]

Addition of *atr* to opsin enhances its activity by the formation of noncovalent complexes[3,6,7] and reversible "pseudo photoproducts,"[1] which interact with G_t, arrestin, and RK by different mechanisms than metarhodopsin-II.[6,8] The ligand-like noncovalent receptor–agonist complex between opsin and *atr* can activate G_t on the order of 10^{-2} as well as R* at physiologically relevant 1:1 molar ratio. These complexes may thus adopt a light-independent signaling state.[6] Such active complexes could be functional during

---

[1] K. P. Hofmann, A. Pulvermüller, J. Buczylko, P. Van Hooser, and K. Palczewski, *J. Biol. Chem.* **267**, 15701 (1992).
[2] E. J. Helmreich and K. P. Hofmann, *Biochim. Biophys. Acta* **1286**, 285 (1996).
[3] G. B. Cohen, D. D. Oprian, and P. R. Robinson, *Biochemistry* **31**, 12592 (1992).
[4] T. J. Melia, Jr., C. W. Cowan, J. K. Angelson, and T. G. Wensel, *Biophys. J.* **73**, 3182 (1997).
[5] A. Surya, K. W. Foster, and B. E. Knox, *J. Biol. Chem.* **270**, 5024 (1995).
[6] S. Jäger, K. Palczewski, and K. P. Hofmann, *Biochemistry* **35**, 2901 (1996).
[7] J. Buczylko, J. C. Saari, R. K. Crouch, and K. Palczewski, *J. Biol. Chem.* **271**, 20621 (1996).
[8] K. Palczewski, S. Jäger, J. Buczylko, R. K. Crouch, D. L. Bredberg, K. P. Hofmann, M. A. Asson-Batres, and J. C. Saari, *Biochemistry* **33**, 13741 (1994).

continuous illumination of the retina and play a role in the physiologic phenomenon of "bleaching desensitization."[9]

These interactions of opsin may arise *in vivo* after the spontaneous decay of the metarhodopsin-II intermediate of bleached rhodopsin by hydrolysis of the deprotonated Schiff base linkage between Lys-296 and *atr*.[10] As long as the shut-off reactions in the bleached condition are not completed by regeneration with 11-*cis*-retinal, opsin or P-opsin and *atr* remain in a state capable of light-independent catalytic activity in the visual cascade. Different mechanisms of desensitization are possible due to the presence of free opsin in rod outer segments[9] or the formation of noncovalent opsin/*atr* complexes[6,8] in favor of a considerably high gain in G_t turnover. Thus, a secondary light-independent signaling does exist, originating from opsin and its complexes with *atr* (see also Ref. 11).

In this chapter, we discuss methods and data related to *in vitro* analysis of the apoprotein opsin and its interaction with the photolyzed chromophore *atr*.

Membrane and Protein Preparations

Membrane-Bound Opsin

Large amounts of bovine opsin are easily prepared by modifications of a procedure described previously by Surya *et al.*[5] Aliquots of bovine rod outer segments (ROS) or washed ROS membranes (ca. 20 mg of rhodopsin) are thawed and vigorously mixed in 100 ml ice-cold buffer, 10 mM sodium phosphate, pH 7.0, containing 10 mM hydroxylamine and fatty-acid-free bovine serum albumin (BSA, 2%) from Sigma (St. Louis, MO). The suspension is then bleached for 10 min on ice using orange light. Membranes are pelleted by centrifugation (30,000g, 20 min), and are resuspended in 10 mM sodium phosphate buffer, pH 6.5, containing 5 M urea (50 ml), with a glass/glass Potter homogenizer. After 10 min of incubation on ice, 50 ml phosphate buffer (10 mM, pH 6.5) containing fatty-acid-free BSA (2%) is added to the suspension, which is vigorously mixed before centrifugation. Retinaloxime is extracted from the membrane-bound opsin by four washes with the same buffer containing fatty-acid-free BSA. Four washes are carried out with the buffer alone to remove BSA. Finally, the membranes are

---

[9] J. Jin, R. K. Crouch, D. W. Corson, B. M. Katz, E. F. MacNicol, and M. C. Cornwall, *Neuron* **11**, 513 (1993).
[10] G. Wald, *Science* **162**, 230 (1968).
[11] C. S. Leibrock, T. Reuter, and T. D. Lamb, *Vision Res.* **34**, 2787 (1994).

washed with isotonic BTP buffer {containing [1,3-bis(trishydroxymethyl)-methylamine]propane, 50 mM, pH 7.5, 130 mM NaCl, 5 mM MgCl$_2$} and are resuspended in the same buffer containing 0.3 M sucrose.

Phosphorylated Opsin

Rhodopsin phosphorylation is carried out following conditions similar to those described previously by Kühn and Wilden.[12]

ROS are resuspended in potassium phosphate buffer (0.1 M, pH 7.4), containing adenosine triphosphate (ATP, 3 mM), MgCl$_2$ (1.5 mM), and guanosine triphosphate (GTP, 100 μM), at a concentration of 25 μM rhodopsin. Samples (20 ml) are sonicated and illuminated with orange light for 30 min at 30° in a water bath.

Phosphorylation is then stopped by cooling on ice and sixfold dilution with phosphate buffer, pH 7.0, containing hydroxylamine (10 mM) and sodium fluoride (10 mM). Membranes were pelleted by centrifugation (30,000g, 20 min) and treated with urea and fatty-acid-free BSA as described for the preparation of opsin, except that all solutions contained sodium fluoride (10 mM). A stoichiometry of 2-3 phosphates/opsin is determined under these conditions using radioactive [γ-^{32}P]ATP as a tracer in aliquots of the preparation.

Methylation Procedure

Rhodopsin membranes are reductively methylated in the dark, employing procedures similar to those described by Longstaff and Rando,[13] to prevent random Schiff base formation between *atr* and peripheral amines. Specifically, two rounds of the methylation procedure are performed using 20 mM formaldehyde (added from a stock solution of 0.2 M formaldehyde freshly prepared by hydrolysis of paraformaldehyde) and 0.2 M NaCNBH$_3$ (10 times more than used in the original procedure[13]). Permethylated rhodopsin membranes (PM-rhodopsin) are desalted using a Sephadex G-25 column. PM-opsin is prepared from bleached PM-rhodopsin as described for opsin.[6]

Arrestin, Rhodopsin Kinase, and Transducin

Arrestin is purified from frozen, dark-adapted bovine retinas as described by Palczewski and Hargrave[14] and Palczewski *et al.*[15] Rhodopsin

[12] H. Kühn and U. Wilden, *Methods Enzymol.* **81**, 489 (1982).
[13] C. Longstaff and R. R. Rando, *Biochemistry* **24**, 8137 (1985).
[14] K. Palczewski and P. A. Hargrave, *J. Biol. Chem.* **266**, 4201 (1991).
[15] K. Palczewski, A. Pulvermüller, J. Buczylko, C. Gutman, and K. P. Hofmann, *FEBS Lett.* **295**, 195 (1991).

kinase is prepared and activity is measured using opsin membranes as described by Palczewski et al.[16] and Pulvermüller et al.[17] G_t was isolated from washed bovine ROS membranes by elution with GTP following the procedure described previously.[18]

Methods and Applications

Spectrophotometry

A two-wavelength spectrophotometer (Shimadzu UV 3000) is used to measure the differences of absorption at 450 and 417 nm. Quartz cuvettes of 2-mm path length containing 400-μl suspensions of phosphorylated opsin or opsin membranes (5 μM) are placed in the measuring and reference beam to record the spectra of photoproducts (see Fig. 1). For the fast regeneration kinetics of opsin, a Hewlett-Packard HP 8452 diode array spectrophotometer was used at a reference wavelength of 600 nm.

Regeneration of Opsin with 11-cis-Retinal

Opsin or P-opsin membranes are incubated with 11-cis-retinal (dissolved in ethanol) at 2:1 or 1:1 molar ratio for 2 hr at room temperature and 16 hr at 0° under completely dark conditions. The membranes are pelleted by centrifugation and washed twice with isotonic BTP buffer, pH 7.5, containing fatty-acid-free BSA (2%), followed by two washes with the buffer only.

Regeneration kinetics of opsin with 11-cis-retinal is measured as described previously.[6] After addition of 5 μM 11-cis-retinal to 2.5 μM opsin in 50 mM BTP buffer, pH 7.5, containing 130 mM NaCl, spectra are recorded at chosen time points at 20° and regeneration is measured as a rise of the 500-nm absorption.

G_t Activation Assay

Intrinsic G_t Fluorescence

G_t activation is measured by the intrinsic tryptophan fluorescence assay for G proteins.[19,20] The assay allows measurement of the GDP-for-GTPγS

---

[16] K. Palczewski, J. Buczylko, P. Van Hooser, S. A. Carr, M. J. Huddleston, and J. W. Crabb, J. Biol. Chem. **267,** 18991 (1992).

[17] A. Pulvermüller, K. Palczewski, and K. P. Hofmann Biochemistry **32,** 8220 (1993).

[18] M. Heck and K. P. Hofmann, Biochemistry **32,** 8220 (1993).

[19] T. Higashijima, K. M. Ferguson, P. C. Sternweis, E. M. Ross, M. D. Smigel, and A. G. Gilman, J. Biol. Chem. **262,** 752 (1987).

[20] W. J. Phillips and R. A. Cerione, J. Biol. Chem. **263,** 15498 (1988).

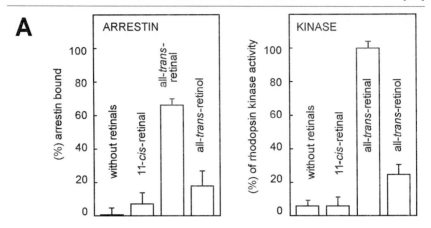

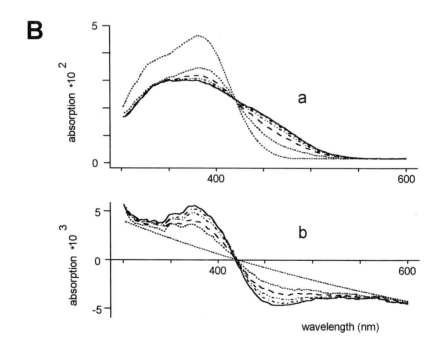

or GDP-for-GTP exchange in the $G_t\alpha$ subunit by a proportional fluorescence change of the dissociated $G\alpha$–GTPγS or $G\alpha$–GTP protein as described previously.[6,21] A version of this assay has also been used to study purified recombinant rhodopsins in detergent solution.[22]

Assays are carried out in 600-μl volumes of a suspension of 0.2 μM receptor protein membranes, which are reconstituted with 0.5 μM G_t in isotonic BTP buffer, pH 7.5. The samples are continously stirred in the cuvette at 20°. The membrane preparations are sonicated before the measurements (Branson Sonifier S 125, 80 W, 10 sec) to obtain uniform size particles without aggregation. The sample is excited at 300 nm and fluorescence emission is recorded at 340 nm using a SPEX Fluorolog fluorimeter.

[21] O. P. Ernst, K. P. Hofmann, and T. P. Sakmar, *J. Biol. Chem.* **270**, 10580 (1995).
[22] K. Fahmy and T. Sakmar, *Biochemistry* **32**, 7229 (1993).

FIG. 1. Reversible binding of arrestin to phosphorylated opsin and effect on phosphorylation of opsin in the presence of retinoids by formation of pseudo photoproducts. (A) Binding of arrestin to phosphorylated opsin and phosphorylation of opsin by rhodopsin kinase: effect of *atr*. Arrestin binding was determined by a centrifugation assay. Phosphorylated opsin (30 μM) was mixed with purified arrestin (5 μM) in 10 mM HEPES buffer, pH 7.5, containing 100 mM NaCl (total volume of 60 μl). Samples were incubated for 5 min at 30° without and with retinoids, and centrifuged using an air-fuge (Beckmann TLX, 180,000g) for 2 min. Supernatant was stored for SDS–PAGE and the pellet was washed with 180 μl of the HEPES buffer (centrifuged for 2 min at 180,000g). Supernatant and pellet were solubilized in 55 μl of 1% SDS, containing 0.1% 2-mercaptoethanol, followed by SDS–PAGE analysis using Laemmli's procedure. Binding of arrestin was assayed by the amount of protein either bound to the membrane pellet or present in the supernatant using the Bradford method. Purified rhodopsin kinase was added to opsin membranes in the presence of [γ-^{32}P]ATP and the incorporation of ^{32}P into opsin was measured in the dark without and with retinoids. [See K. Palczewski, J. Buczylko, P. van Hooser, S. A. Carr, M. J. Huddleston, and J. W. Crabb, *J. Biol. Chem.* **267**, 18991 (1992), and A. Pulvermüller, K. Palczewski, and K. P. Hofmann, *Biochemistry* **32**, 8220 (1993).] (B) Formation of pseudo photoproducts: (a) formation of 380-nm product (M_{380}) and 470-nm product (M_{470}) from phosphorylated opsin and *atr*. (b) Stabilization of M_{380} by arrestin at the expense of M_{470}. The measurements (absorbance differences) employed a two-wavelength spectrophotometer (Shimadzu UV 3000) as described earlier. The sample and reference cuvettes were filled with the phosphorylated opsin membrane suspensions (400 μl, 5μM P-opsin) and the starting agents. Reactions were started by adding *atr* (4 μl, final concentration 10 μM) to the P-opsin sample and alcohol to the reference (a), by adding arrestin (final concentration 5 μM) to the P-opsin sample and arrestin buffer to the reference, both after equilibration with *atr* (b). All spectra were plotted against a baseline taken after equilibration. In each set, (a) and (b), the solid line represents the final spectrum; in B (b), the dotted line represents calculated absorption due to the turbidity change by arrestin in the sample. The turbidity caused a curved baseline. The dotted line fits the equation $E = a\lambda^{-2} + b$, where E is the absorbance (turbidity) of the sample at λ (wavelength in nm), and a and b are fit parameters. All spectra were recorded at 2.5-min intervals, in 100 mM HEPES buffer, pH 7.5, at 8°.

The slit of the excitation double monochromator is kept open at 0.4 mm to reduce the effect of 300-nm light on the sample. After sample equilibration for 10 min GTPγS is injected into the cuvette to give a final concentration of 5 μM, and the fluorescence change over time is recorded. The positive fluorescence change indicates the transition of the G protein into its active conformation.

In the case of dark-adapted or 11-*cis*-retinal-generated receptor membrane preparations, immediately before the measurement, rhodopsin is illuminated with orange light for 20 sec to produce meta-II. For the opsin measurements, the protein is preincubated with retinoid for 30 min before the reaction is initiated with GTPγS or GTP to a final concentration of 5 μM.

To obtain the catalytic G_t activation rates k_{on}, the traces recorded by the fluorescence assay were fit by the function

$$C(t) = C_{max}(1 - e^{-kt})$$

as described previously.[6] C_{max} is the maximal amplitude of the fluorescence increase and k is the activation rate in s^{-1} derived from the initial slope of the linear approximation of the early phase of the recorded curve. (For limitations of the fluorescence assay, see later discussion and chapter by Ernst *et al.* in this volume.[22a])

all-*trans*-Retinal Promoting Opsin Binding to Arrestin, RK, and G_t

We have demonstrated that free *atr*, added exogenously to opsin membranes, can react with the opsin apoprotein to form pseudo photoproducts that are spectrally similar to the photoinduced metarhodopsin intermediates (meta-I, -II, -III). Figure 1A shows that *atr* enhances significantly the binding of arrestin to P-opsin and phosphorylation of opsin by purified rhodopsin kinase. Evidently, arrestin binds to P-opsin/*atr* and kinase binds to opsin/*atr* complexes, in which *atr* is retained by noncovalent interactions.

Spectral analysis of the reaction between *atr* and P-opsin (2:1 molar ratio) shows the formation of 380-nm product (M_{380}) and 470-nm product (M_{470}) as can be seen in Fig. 1B (a). A reversible binding of *atr* forms pseudo photoproducts with P-opsin. Stabilization of M_{380} at the expense of M_{470} is observed by adding arrestin to the P-opsin sample (Fig. 1B, part b).

Figure 2A shows that G_t fails to stabilize the pseudo photoproducts after equilibration with *atr*. However, *atr* recombines with opsin as well as P-opsin, forming activating species of the receptor. P-opsin and opsin form similar active complexes with *atr*, as seen in G_t activation assays (Fig.

[22a] O. P. Ernst, C. Bieri, H. Vogel, and K. P. Hofmann, *Methods Enzymol.* **315** [32], 1999 (this volume).

2B). *Atr* does not compete with 11-*cis*-retinal incorporation into opsin membranes in the regeneration procedure (Fig. 2C). Regeneration of P-opsin with 11-*cis*-retinal in the presence of *atr* is also undisturbed (data not shown). On the other hand, β-ionone at increasing amounts competes with 11-*cis*-retinal at the regeneration of PM-opsin as previously reported.[6]

We examined the mechanism by which *atr* activates opsin.[6] To exclude other amines except active site Lys-296 from formation of Schiff bases, we reductively methylated rhodopsin (PM-rhodopsin), which we then bleached to generate PM-opsin. Using the fluorescence assay for G_t activation we found that *atr* interacted with PM-opsin, producing a noncovalent complex that catalyzes nucleotide exchange (GTPγS uptake) by G_t (Fig. 2D). This is intriguing evidence for noncovalent complex formation between *atr* and opsin. Other experiments also show that the opsin/*atr* complex does not require a stable Schiff base linkage for activity.[23] These findings are in contrast to those obtained by others with the opsin mutant M257A. It formed an active complex in COS cell membranes that was stable enough to be purifed in dodecylmaltoside,[24,25] indicating the formation of a covalent bond (see also Han and Sakmar in this volume[25a]).

Figure 3 (*inset*) shows that opsin regenerated with submolar 11-*cis*-retinal catalyzes GTPγS uptake by G_t already at the maximum rate if only about 20% is in the photoactivated form. This result shows that the assay is rate limited (see later discussion). When the opsin/*atr* complex with increasing *atr* molar ratio is used instead of photolyzed rhodopsin (Fig. 3), a higher concentration is necessary to reach an apparent "maximum" G_t activation rate. Metarhodopsin-II activity saturates at 0.2 molar 11-*cis*-retinal : opsin ratio (Fig. 3, *inset*) independent of the opsin : G_t ratio at lower rates. Therefore, the fluorescence assay is fast enough to assess *in vitro* the G_t activation by opsin/*atr* complexes, which reaches considerably high activity at the physiologically important 1 : 1 stoichiometry for light-independent reactions of G_t (Fig. 3). The conditions and limitations of this assay are described next.

Limitations of Fluorescence Assay

We have shown that the fluorescence assay for G_t activation using permethylated disk membranes does not distinguish between 20 and 100%

---

[23] A. Surya and B. E. Knox, *Exp. Eye Res.* **66**, 599 (1998).
[24] M. Han, S. W. Lin, M. Minkova, S. O. Smith, and T. P. Sakmar, *J. Biol. Chem.* **2712**, 32337 (1996).
[25] M. Han, S. O. Smith, and T. P. Sakmar, *Biochemistry* **37**, 8253 (1998).
[25a] M. Han and T. P. Sakmar, *Methods Enzymol.* **315** [18], 1999 (this volume).

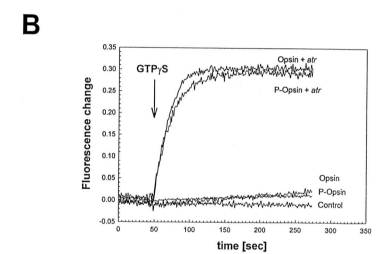
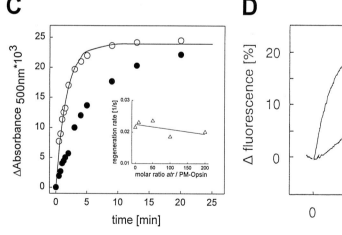
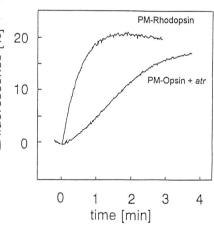

photoactivated rhodopsin.[6] When opsin is regenerated with increasing 11-cis-retinal at the molar ratio from 0 to 1, the G_t activation curves (Fig. 3, inset) could be fit using a single exponential for first-order kinetics. The catalytic activation rates k were plotted as a function of the molar ratio (Fig. 3) and again this plot fit by a single exponential. The G_t activation saturates at ~0.2 stoichiometry of 11-cis-retinal/opsin regenerated and photoactivated rhodopsin (Fig. 3, inset). This is explained by the fact that, to make R* limiting with the densely packed, fully bleached disk membranes, the membrane association of G_t is the rate-limiting step. This occurs at higher than 0.2 stoichiometric ratio of photoisomerized retinal bound to opsin at Lys-296, as shown previously using permethylated opsin membranes.[6] The activation of G_t is faster than its association with the membrane. This idea was also supported by light scattering measurements.[26]

The more general problem behind this result is that of intrinsically slow in vitro assays, which do not allow the assessment of the true G_t activation capacity by light-activated rhodopsin. It is worthwhile to mention that in functioning photoreceptor cells of isolated bovine retinas, the catalyic rate per photoactivated rhodopsin does not depend on the amount of bleached rhodopsin in the membranes.[27]

[26] K. P. Hofmann, in "Handbook of Experimental Pharmacology," Vol. 108/II, "GTPases in Biology" (B. F. Dickey and L. Birnbaumer, eds.), p. 267. Springer-Verlag, Berlin, 1993.
[27] M. Kahlert, D. R. Pepperberg, and K. P. Hofmann, Nature **345**, 537 (1990).

FIG. 2. Light-independent activation of transducin catalyzed by opsin and P-opsin, and by the complexes with atr. (A) Absence of pseudo photoproduct stabilization in an equilibrated mixture of opsin, atr, and transducin. Opsin membrane suspensions (400 µl, 5 µM) were mixed with atr (10 µM) and equimolar transducin (G_t). Reaction was started by adding GTPγS (final concentration 25 µM) to the opsin sample and buffer to the reference, both after equilibration with atr and G_t. Conditions and records were made as described in Fig. 1B. (B) The fluorescence assay was employed to measure the rate of G_t activation by receptor-catalyzed $G_t\alpha$/GTPγS uptake. Time course for the reaction by opsin or P-opsin samples (0.2 µM) of membrane preparations reconstituted with G_t (0.5 µM) and their activation promoted by preincubation with atr at 1:2 molar ratio. G_t activity is initiated by injection of GTPγS to a final concentration of 5 µM. Traces of G_t activation at pH 7.5 are relative changes of the fluorescence emission at 340 nm. Control represents G_t preparation without receptor. The data fit the function $C(t) = C_{max}(1 - e^{-kt})$. (C) Regeneration of native opsin (2.5 µM, closed circles) and PM-opsin (2.5 µM, open circles) in the presence of 11-cis-retinal (5 µM). Inset: Rate constants of PM-opsin regeneration in the presence of increasing amounts of atr expressed as the atr/PM-opsin ratio. (D) Activation of G_t by the PM-opsin/atr complex and by regenerated photolyzed PM-rhodopsin. The reaction was initiated by the addition of 0.08 µM atr to 0.04 µM PM-opsin or by illumination of PM-rhodopsin (0.04 µM) in the presence of 0.4 µM G_t and 20 µM GTPγS.

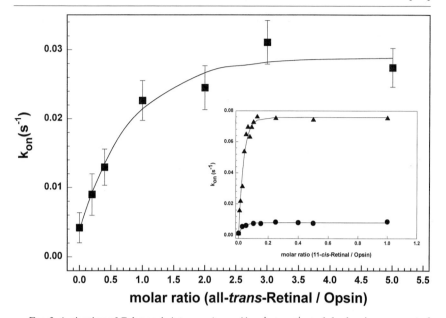

FIG. 3. Activation of G_t by opsin/atr complex and by photoactivated rhodopsin, regenerated substoichiometrically with 11-*cis*-retinal. Rhodopsin was regenerated from opsin membranes (*inset*), and catalytic activation of G_t was in both cases triggered by injection of GTPγS. The activation rates k_{on} are given for opsin/atr complex at different molar retinoid ratio produced by 0.2 μM opsin and 0.5 μM G_t. The measurements are carried out as described in Fig. 2B, but opsin is incubated with increasing concentrations of *atr*. *Inset:* Activation rates k_{on} for submolar regenerated and photoactivated rhodopsin, produced by 0.5 (triangles) or 0.05 (circles) μM opsin with 11-*cis*-retinal and 1 or 0.5 μM G_t, respectively, are plotted as a function of the molar ratio retinal/opsin = 0–1. The reaction was initiated after illumination (30 sec) by addition of 5 μM GTPγS in isotonic BTP buffer at pH 7.5 and 20°. All data could be fit by the given single exponential for first-order kinetics, yielding the relative activation rates k of each point given.

In situ the reaction is usually limited by R* in the form of a few photoactivated receptor molecules in the disk membrane. The kinetic analysis is based on

$$G^*(t) = k[R^*][G_t]$$

where G_t denotes the total concentration of activatable G protein in the GDP-binding form.

Only in the absence of slow reaction steps is the initial rate of $G^*(t)$ a measure of the catalytic efficiency of the R*–G_t interaction.

In another chapter by Heck *et al.* in this volume,[27a] we present data obtained with the light scattering dissociation signal. The kinetic fidelity of this monitor, which is only applicable to flash-induced activation, is on the order of 1 ms.

We note that fluorescence assays in detergent[22] or filter-binding assays in membranes with much lower density of expressed rhodopsin[3,28] yield reliable results.

pH Dependency of all-*trans*-Retinal Activation of Opsin

The pH/rate profiles of G_t or RK activation (Fig. 4) reflect the degree of coupling between the retinal binding site and the interaction domain for a partner protein. "Forced protonation" appears to be a key requirement for forming an active species that can allow formation of a catalytically active conformation. Energy is required to protonate a residue at a pH higher than its endogenous pK_a, and this is expressed in the relative position of the right part of the bell-shaped pH/activity profiles (Fig. 4). There is a relative shift to the right for meta-II, as compared with the all-*trans*-retinal–opsin complex by ca. 2 pH units, both for G_t and RK, under saturating conditions. This may reflect that in the light-activated meta-II state, the coupling between the retinal binding site and the interaction domain is stronger. All pK_a values are higher than that of isolated Glu (pK_a 4.3). This was explained by Glu-Glu pairing.[7]

The left part of the pH/rate profiles, which is seen for G_t and RK, reflects the necessary deprotonation with a pK_a of the order of 5.5–6.0. It may indicate a pH dependence of the ground state or proton release during formation of the active state, with a possible role for proton release in the early MI_{380} product.[29]

Summary

The data collected with the techniques discussed in this chapter suggest significant differences between the active conformation(s) of the opsin/*atr* complex, which are reversibly formed in the dark, and the active conformation (R*) of the meta-II photoproduct. First, there is good evidence for noncovalent opsin/*atr* complexes with considerable activity (although covalent binding of *atr* is found in mutant opsins.

[27a] M. Heck, A. Pulvermüller, and K. P. Hofmann, *Methods Enzymol.* **315** [22], 1999 (this volume).
[28] M. Han, S. W. Lin, S. O. Smith, and T. P. Sakmar, *J. Biol. Chem.* **271**, 32330 (1996).
[29] I. Szundi, T. L. Mah, J. W. Lewis, S. Jäger. O. P. Ernst, K. P. Hofmann, and D. S. Kliger, *Biochemistry* **37**, 14237 (1998).

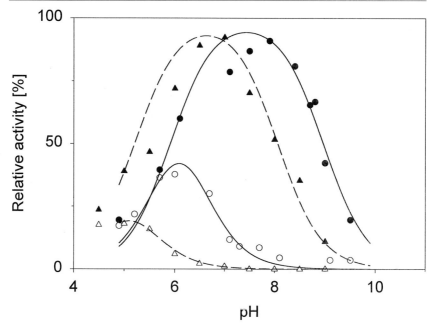

FIG. 4. pH/rate profiles of rhodopsin-induced activities in transducin and rhodopsin kinase The data points are measured relative (in %) to nucleotide exchange by G_t (circles) and of phosphorylation by RK (triangles), both stimulated by interaction with light-activated rhodopsin or the opsin/*atr* complex. Closed symbols, light-induced formation of active rhodopsin; open symbols, light-independent opsin/*atr* complex. The solid and broken lines are fits with the function: % rate = $100 [H/(H + K_2)] [K_1/(H + K_1)]$, where H is the free H^+ concentration, and K_1 and K_2 are proton dissociation constants. Solid lines belong to G_t, data from G. B. Cohen, D. D. Oprian, and P. R. Robinson, *Biochemistry* **31**, 12592 (1992); broken lines to RK, data from J. Buczylko, J. C. Saari, R. K. Crouch, and K. Palczewski, *J. Biol. Chem.* **271**, 20621 (1996).

Even more intriguing, all-*trans*-retinal in an amount that saturates the activity of the opsin/*atr* complex toward G_t does not measurably inhibit the access of 11-*cis*-retinal to the light-sensitive binding site during regeneration (Fig. 2C). On the other hand, forced protonation at or near Glu-134 appears to be an integral mechanism for both the meta-II and the opsin-like activities (Fig. 4).

Thus, it is not inconceivable that these two activities of the receptor arise from two fundamentally different conformations, one meta-II-like and one opsin-like. They would be similar with respect to the G_t (or RK) protein–protein interaction but different in their mode of retinal–protein interaction.

Acknowledgments

We wish to thank Prof. T. P. Sakmar for stimulating discussions and for reviewing the manuscript. The expert technical assistance of Jana Engelmann is gratefully acknowledged. Work was supported by grants from the Deutsche Forschungsgemeinschaft (to K.P.H. and to D.M., MA 1784/1-2).

[18] Assays for Activation of Recombinant Expressed Opsins by all-*trans*-Retinals

By May Han and Thomas P. Sakmar

Introduction

Rhodopsin, a member of a large family of G-protein-coupled receptors (GPCRs) with seven transmembrane helices, is responsible for dim-light vision in vertebrates. It has been a subject of intensive physiologic, biochemical, and biophysical investigations for a century.[1] Rhodopsin contains an 11-*cis*-retinal chromophore covalently bound to K296 on transmembrane (TM) helix 7 via a protonated Schiff base linkage. E113 on TM helix 3 serves as the counterion to the positively charged 11-*cis*-retinylidene protonated Schiff base. Absorption of a photon isomerizes the 11-*cis*-retinal to its all-*trans* isomer. This event triggers conformational changes in rhodopsin, which in turn activate the visual G protein, transducin (G_t). Members of the GPCR family share many structural and functional similarities and receive a variety of extracellular stimuli including photons, odorants, neurotransmitters, and peptide hormones. They regulate many critical cellular processes. Malfunctions of GPCRs cause a variety of human diseases and disorders, accounting for the fact that more than 60% of the identified targets for current nonantibiotic drugs are members of the GPCR family.

Rhodopsin as a photoreceptor molecule is one of nature's most impressive high-gain sensors: the energy threshold for human photon detection was measured to be 6×10^{-22} kcal, corresponding to five to seven photons reaching the retina.[2] Rhodopsin, as a highly specialized photon receptor, has distinctive features that differ from other GPCRs. These differences have brought it special attention, but are also responsible for rhodopsin sometimes being excluded from general discussions of GPCRs. First, it is the only receptor of the family with which the ligand forms a covalent bond

---

[1] W. Kuhne, "On the Photochemistry of the Retina and on the Visual Purple." Macmillan, London, 1878.
[2] S. Hecht, S. Shlaer, and M. H. Pirenne, *J. Gen. Physiol.* **25,** 819 (1942).

in its binding pocket. However, it has been shown that the covalent bond is not required for ligand incorporation nor receptor activation for K296 replacement mutants.[3] Second, the activation of rhodopsin requires two events, ligand binding and photon absorption. The natural agonist of rhodopsin, all-*trans*-retinal (ATR), is converted from the inverse agonist 11-*cis*-retinal through photolysis within the receptor. More evidence has accumulated to show that opsin, when combined with exogenously added ATR, can also partially activate, or interact with G_t, rhodopsin kinase and arrestin.[4–6] Moreover, some opsin mutants can be activated fully in the presence of ATR in the absence of light.[7–9] These results imply that the highly specialized receptor is, in fact, very similar to other GPCRs. The special features of rhodopsin as a photoreceptor molecule are only necessary to carry out its physiologic function as a photon receptor, but not indispensable for receptor activation.

The observation that ATR can partially activate the receptor (termed ATR activity), analogous to G-protein activation of other GPCRs in the presence of agonist, has been documented by several groups, in both the native rod outer segment (ROS)[5,6,10] and recombinant COS cell membrane systems.[8,9,11] The advantages of using a recombinant heterologous expression system to study receptor activation are as follows: (1) since the vision is highly specialized in the photoreceptor cells of the retina, fewer cellular components in COS cells are capable of cross reacting with rhodopsin and G_t, making results easier to interpret; and (2) in a recombinant system, one can obtain information by slightly perturbing the receptor through mutagenesis. The disadvantage is that ROS contains highly specialized membranes, which COS cell membranes may not fully mimic. But the difference can be minimized, if not eliminated, by always normalizing the data to activities measured for wild-type rhodopsin under the same conditions. In the following sections, methods employed to measure the receptor activation by ATR in COS cell membranes are

---

[3] E. A. Zhukovsky, P. R. Robinson, and D. D. Oprian, *Science* **251**, 558 (1991).
[4] K. P. Hofmann, A. Pulvermüller, J. Buczylko, P. Van Hooser, and K. Palczewski, *J. Biol. Chem.* **267**, 15701 (1992).
[5] K. Palczewski, S. Jäger, J. Buczylko, R. K. Crouch, D. L. Bredberg, K. P. Hofmann, M. A. Asson-Batres, and J. C. Saari, *Biochemistry* **33**, 13741 (1994).
[6] S. Jäger, K. Palczewski, and K. P. Hofmann, *Biochemistry* **35**, 2901 (1996).
[7] T. P. Sakmar, R. R. Franke, and H. G. Khorana, *Proc. Natl. Acad. Sci. U.S.A.* **86**, 8309 (1989).
[8] M. Han, S. W. Lin, M. Minkova, S. O. Smith, and T. P. Sakmar, *J. Biol. Chem.* **271**, 32337 (1996).
[9] M. Han, S. O. Smith, and T. P. Sakmar, *Biochemistry* **37**, 8253 (1998).
[10] A. Surya and B. E. Knox, *Exp. Eye Res.* **66**, 599 (1998).
[11] G. B. Cohen, D. D. Oprian, and P. R. Robinson, *Biochemistry* **31**, 12592 (1992).

described in detail. Some results obtained through such measurements are presented. The interpretation of these results may provide insights into understanding the structure–function relationships of visual receptor activation.

Methods

Heterologous Expression of Recombinant Opsin

Opsin genes are expressed in COS-1 cells as previously described[12] except that a lipofection procedure is employed using LipofectAMINE (Gibco-BRL, Gaithersburg, MD) in place of the DEAE–dextran transfection procedure. Briefly, 60 μl of LipofectAMINE is incubated with 6 μg of the expression vector for 30 min at room temperature in 800 μl of Dulbecco's modified Eagle's medium (DMEM). The mixture is then added to a 100-mm tissue culture plate of ~70% confluent COS cells. After a 5-hr incubation at 37°, the mixture is replaced with DMEM supplemented with 10% serum and the cells are fed again with fresh media 18 hr later. The transient expression is terminated 40 hr after the start of the transfection, timed from the moment LipofectAMINE was added to the cells. The cells are washed with 5 ml of phosphate-buffered saline (PBS) and then scraped from the plate in 1 ml of PBS using a rubber policeman. The cells are immediately subjected to membrane preparation as described next without the addition of chromophore. Alternatively, the cells are incubated with 5 μl of 11-*cis*-retinal (7.5 mM) or ATR (4.7 mM) at room temperature for 1 hr and then subject to immunoaffinity purification in dodecylmaltoside (DM) detergent buffer.

Preparation of Membranes from Transfected COS-1 Cells

Cells from one 100-mm culture plate are washed in 1 ml PBS, collected in a 1.5-ml microfuge tube and spun for 1 min in a tabletop centrifuge. The supernatant fraction is thoroughly removed and the pellet is resuspended in 750 μl of hypotonic buffer [10 mM Tris-HCl, pH 6.8, 1 mM EDTA, 0.1 mM phenylmethylsulfonyl fluoride (PMSF)], and forced through a 26-gauge needle three times. The lysate is layered onto 750 μl of a 37.7% (w/v) sucrose solution in buffer A (20 mM Tris-HCl, pH 6.8, 150 mM NaCl, 1 mM MgCl$_2$, 1 mM CaCl$_2$, 10 mM EDTA, 0.1 mM PMSF) in an 11- × 34-

---

[12] R. R. Franke, T. P. Sakmar, R. M. Graham, and H. G. Khorana, *J. Biol. Chem.* **267**, 14767 (1992).

mm TLS-55 centrifuge tube and spun at 22,000 rpm for 20 min at 4° in an Optima-TL Tabletop Ultracentrifuge (Beckman, Palo Alto, CA). The interface band containing the membrane fraction is collected with a 1-ml syringe with a 25-gauge needle, and the volume is brought to 3.5 ml with buffer A in a 13- × 51-mm TLA-100.3 centrifuge tube. The membranes are resuspended and then spun at 60,000 rpm for 30 min at 4°. The pellet is again resuspended and washed with 3.5 ml of buffer A. The pellet is homogenized in 0.5 ml buffer A by passing the suspension through a 30-gauge needle. Aliquots (80 μl) are flash-frozen in liquid nitrogen and stored at −80°. Each membrane aliquot is used once immediately after thawing. For each series of transfections, wild-type opsin is transfected and manipulated in parallel with mutant opsin samples. This serves as an internal standard in order to account for experimental variation among different transfections. All data presented are the statistical average of at least three measurements from at least two different transfections. The total protein content of each sample is determined by the modified Bradford method (Bio-Rad, Hercules, CA).

Immunoaffinity Purification of Covalent Opsin–Retinal Complex

All manipulations are carried out at 4° under dim red light unless otherwise specified. Rhodopsin monoclonal antibody 1D4[13] (National Cell Culture Center) is immobilized on CNBr-activated Sepharose 4B (Pharmacia, Piscataway, NJ) COS cells expressing opsin genes are spun down after incubation with 11-*cis*-retinal or ATR. The cell pellet is solubilized in 1 ml solubilization buffer (50 mM Tris-HCl, pH 6.8 at room temperature, 100 mM NaCl, 1 mM CaCl$_2$, 0.1 mM PMSF, 1% w/v DM) for 1 hr with occasional vortexing. Insoluble material is removed by centrifugation at 60,000 rpm for 30 min in polyallomer microfuge tubes in a TLA-100.3 rotor. The supernatant fraction is incubated with 150 μl 1D4 antibody resin slurry for 2 hr with rotation. The resin is then washed four times with 1 ml wash buffer (50 mM Tris-Cl, pH 6.8 at room temperature, 100 mM NaCl, 0.1% DM). Rhodopsin is then eluted with 400 μl elution buffer (wash buffer supplemented with 0.18 mg/ml 1D4 peptide TETSQVAPA) by incubation for 2 hr at 4°. Typical yields for wild-type rhodopsin from one 10-cm plate are 10–20 μg, quantified by UV–vis spectroscopy (λ-19, Perkin-Elmer, Norwalk, CT), assuming $\varepsilon = 42.7 \times 10^3 \, M^{-1} \, cm^{-1}$ at 500 nm. Some mutants may have lowered yields.

---

[13] D. MacKenzie, A. Arendt, P. Hargrave, J. H. McDowell, and R. S. Molday, *Biochemistry* **23**, 6544 (1984).

Filter-Binding Assay of Receptor-Catalyzed GTPγS Uptake by G_t

A radionucleotide filter-binding assay, which monitors the guanine-nucleotide exchange by G_t, using a nonhydrolyzable GTP analog, GTPγS, is carried out as described earlier.[14] The filter-binding assay is based on the property that G_t and its bound nucleotide are retained on nitrocellulose filters, whereas free nucleotide passes through. The assay mixture consists of (in final concentration) 2 μM purified G_t, and 20 μM [^{35}S]GTPγS in assay buffer [50 mM Tris-HCl, pH 7.2, 100 mM NaCl, 4 mM MgCl$_2$, 1 mM dithiothreitol (DTT)]. For samples purified in DM, pigment concentration is determined before the assay by UV–vis spectrophotometry. Purified pigment (100 nM) in 0.1% DM is used to provide a final concentration of 10 nM, and [DM] is kept at 0.01%. This is very important because the ability of G_t to pick up GTPγS greatly decreases when [DM] is higher, and activity becomes essentially undetectable at 0.1% DM.[15] Either purified pigment in DM or a COS cell membrane suspension (5 μl) is added to 45 μl of reaction mixture followed by thorough gentle mixing. The time course is started by turning on a 150-W projection lamp (Dolan-Jenner) equipped with a 495-nm long-pass filter. After a given time interval, a 7-μl aliquot is removed from the reaction mixture and quickly applied onto a nitrocellulose filter (25-mm diameter, Schleicher & Schuell, Keene, NH) fixed on a vacuum manifold (Millipore, Bedford, NY). The filter is washed twice with 3 ml of ice-cold reaction buffer. After six aliquots are taken, the filters are removed from the manifold and allowed to dry on blotting paper. They are then dissolved in 4.5 ml of scintillation counting fluid (Ready Safe, Beckman, Fullerton, CA). The GTPγS bound to the filter is quantified by scintillation counting. The G_t activation rate is determined by linear-regression analysis of the increasing amount of radioactivity retained on successive filters. Experiments are conducted under the condition that G_t is not significantly consumed (<2%) in order to ensure that the initial rate is being measured and the amount of G_t is not limiting. For reactions where G_t is activated at a measurable level, a data set is used only when the time points display a good linear correlation with $R^2 \geq 0.95$ and a Y intercept close to 0.

The activity assay of mutant opsins in COS cell membranes is performed as follows. Membrane aliquots are thawed at room temperature, and 1 μl of 11-cis-retinal (7.5 mM), ATR (4.7 mM), or other retinoids (5–10 mM) in an ethanolic solution is incubated with 30 μl of membrane suspension in the dark at room temperature for 1 hr before assaying at room tempera-

---

[14] K. C. Min, T. A. Zvyaga, A. M. Cypess, and T. P. Sakmar, *J. Biol. Chem.* **268,** 9400 (1993).
[15] M. Han, M. Groesbeek, S. O. Smith, and T. P. Sakmar, *Biochemistry* **37,** 538 (1998).

ture. A 5-μl aliquot of the sample is mixed with 45 μl of the reaction mixture as described earlier. Light activity refers to light-dependent G_t activation caused by samples incubated with 11-*cis*-retinal. ATR activity refers to the G_t activation caused by opsin samples incubated with ATR assayed under dim red light without illumination. Opsin activity refers to any basal activity of a mutant in the absence of chromophore assayed under ambient room light. For ATR and opsin activities, the reaction starts the moment the membranes are added to the reaction mixture.

All assays are performed at room temperature. For reactions with very fast kinetics, such as most of the photoactivated rhodopsin experiments, data points are taken at 15, 25, 35, 45, 55, and 65 sec after the reaction. For reactions with slower kinetics, such as most opsin activities, samples are taken every 30 sec or 1 min.

Fluorescence Assay of G_t Activation

Spectrofluorimetric assay of G_t activation can be carried out to determine precise kinetic rate constants.[16] Active transducin α-subunit concentration in a preparation of holotransducin can be precisely determined by measuring rhodopsin-catalyzed nucleotide-induced fluorescence increase on addition of different amounts of GTPγS. The assay allows measurement of the rate of GTPγS or GTP uptake in the $G_{t\alpha}$ subunit by monitoring the increase in its intrinsic fluorescence. The fluorescence assay employed was similar to one described previously.[17,18] A version of this assay has been used to study purified recombinant rhodopsins in detergent solution.[16] Fluorescence was measured with a specially modified SPEX-Fluorolog II spectrofluorometer in signal/reference mode with excitation at 300 nm (2-nm bandwidth) and emission at 345 nm (12-nm bandwidth). Signal integration time was 2 sec. The reaction mixture (1.6 ml) containing 10 mM Tris-Cl (pH 7.4), 100 mM NaCl, 2 mM MgCl$_2$, and 0.01% DM, was cooled to 10° and stirred continuously using a magnetic cuvette stirrer set at maximum speed. The addition of rhodopsin and nucleotide was done by injecting 50 μl of the appropriate solution into the cuvette with a gastight syringe kept at 10°. The sample was continuously illuminated in the cuvette with 543-nm light from a HeNe laser (Melles-Griot, Rochester, NY) connected to the sample compartment by a fiber optic light guide. Stray light was efficiently blocked from reaching the detector by a double monochromator. Under assay conditions, the rate of photobleaching to form a 380-nm species characteristic of metarhodopsin-II (MII) was complete in less than 15 seconds.

[16] K. Fahmy and T. P. Sakmar, *Biochemistry* **32**, 7229 (1993).
[17] P. M. Guy, J. G. Koland, and R. A. Cerione, *Biochemistry* **29**, 6954 (1990).
[18] W. J. Phillips and R. A. Cerione, *J. Biol. Chem.* **263**, 15498 (1988).

Exogenously Supplied all-*trans*-Retinal to Activate Wild-Type Opsin

The ATR chromophore, when converted from 11-*cis*-retinal by photon absorption in the ligand-binding site, is the natural agonist of opsin. The activation of G_t by rhodopsin under these conditions can be defined as full activation (100%) for a given set of assay conditions. However, when added exogenously to membranes containing opsin, ATR only partially (14% of the light-induced activity of rhodopsin) activates the aporeceptor in our recombinant COS cell system. No covalent bond is established between the ATR and opsin, in contrast to the case for native R* (activated receptor), MII.[6] In this context, the exogenous ATR also functions as an agonist to opsin, but only with partial efficiency. This is not surprising given that the physiologic function of rhodopsin is to detect photons rather than ATR. Opsin binds 11-*cis*-retinal preferentially over ATR, and ATR does not compete for 11-*cis*-retinal incorporation.[6] However, the ligand preference for different retinal isomers can be altered by mutagenesis of residues in the retinal-binding site.[19]

Several groups have established conditions for receptor activation by exogenously added ATR. Although all of the data reported were normalized to rhodopsin light activity assayed under similar conditions, the reported results are drastically different. Palczewski and co-workers used a carefully controlled ROS system, where the opsin–ATR complex was measured to exhibit 3% G_t activation compared to that of photoactivated rhodopsin.[6] Sakmar and co-workers, on the other hand, by using a recombinant COS cell expression system, recorded that opsin–ATR exhibited 14% relative activation (Table I).[8] These data agree with a similar study conducted by Oprian and co-workers, where 16% activation was measured.[11] Surya and Knox, however, reported 200% ATR activity relative to photoactivated rhodopsin.[10] The discrepancy may be due to the differences in the conditions under which the experiments were conducted. The first three experiments compared initial rates measured under unsaturated conditions where a negligible portion of G_t was consumed by the end of the measurement. A pH difference is the most likely factor that contributes to the difference between these reports, 3% (measured at pH 8.0) in Jäger *et al.*[6] versus 14–16% (pH 6.8) in Han *et al.*[8] and Cohen *et al.*,[11] respectively. The pH has been shown to greatly influence the absolute opsin–ATR activity as well as its relative activity normalized to photoactivated rhodopsin, since the two activating species have different pH profiles.[11] The latter experiments by Surya and Knox were done under saturation conditions

---

[19] M. Han, S. W. Lin, S. O. Smith, and T. P. Sakmar, *J. Biol. Chem.* **271**, 32330 (1996).

where as much as 50% of the G_t was consumed during the measurements.[10] The results thus obtained may reflect different properties from those measured under unsaturated conditions.

all-*trans*-Retinal Acting as Full Agonist to Constitutively Active Opsin Mutants

The COS cell recombinant system allows investigation of the ATR activation of rhodopsin mutants. Due to the noncovalent nature of the ATR–opsin complex, the G_t activation is usually assayed in COS cell membranes. During the last few years, we examined more than 50 rhodopsin mutants for various and sometimes unrelated purposes.[8,9,19] When all of the data are compiled, an interesting correlation between basal opsin activity and ATR activity emerges as shown in Fig. 1. ATR activity increases rapidly as the relative basal opsin activity increases. A plateau in ATR activity of nearly 100% is reached when basal activity is $\geq$10%.

Constitutive activity (i.e., elevated basal activity) of GPCRs, defined as receptor activation in the absence of ligand, has been recognized as an important feature of these receptors since it was first reported to occur in mutant α_{1B}-adrenergic receptors[20] and later in opsin[21] as well as many other GPCRs. Certain wild-type receptors also have intrinsically high levels of basal activity, which may be important for their physiologic function.[22,23] Constitutive activation can result from overexpression of either the receptor[24] or the G protein.[25] Rhodopsin has evolved a unique mechanism to minimize receptor basal activity. The chromophore 11-*cis*-retinal, which acts as a potent pharmacologic inverse agonist to opsin, is covalently bound to the active site of rhodopsin. In dark-adapted photoreceptor cells, all of the receptors are bound to 11-*cis*-retinal to warrant extremely low noise. Mutant opsins that are constitutively active *in vitro* have been found in certain retinal diseases.[26]

The relationship between constitutive activity and agonist ligand binding and activation can be described in terms of the two-state model of GPCR

---

[20] S. Cotecchia, S. Exum, M. G. Caron, and R. J. Lefkowitz, *Proc. Natl. Acad. Sci. U.S.A.* **87**, 2896 (1990).
[21] P. R. Robinson, G. B. Cohen, E. A. Zhukovsky, and D. D. Oprian, *Neuron* **9**, 719 (1992).
[22] M. Tiberi and M. G. Caron, *J. Biol. Chem.* **269**, 27925 (1994).
[23] D. P. Cohen, C. N. Thaw, A. Varma, M. C. Gershengorn, and D. R. Nussenzveig, *Endocrinology* **138**, 1400 (1997).
[24] R. A. Bond, P. Leff, T. D. Johnson, C. A. Milano, H. A. Rockman, T. R. McMinn, S. Apparsundaram, M. F. Hyek, T. P. Kenakin, L. F. Allen, and R. J. Lefkowitz, *Nature* **374**, 272 (1995).
[25] E. S. Burstein, T. A. Spalding, and M. R. Brann, *Mol. Pharmacol.* **51**, 312 (1997).
[26] V. R. Rao and D. D. Oprian, *Annu. Rev. Biophys. Biomol. Struct.* **25**, 287 (1996).

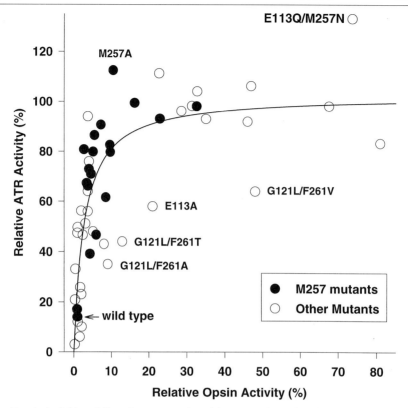

FIG. 1. Activities of 50 opsin mutants plotted for comparison with the two-state model of GPCR activation. Each of the opsin mutants was assayed in COS cell membranes for the ability to activate G_t. The level of activity in the presence of ATR is plotted as a function of the basal opsin activity. Numerical values are presented in Table I. The so-called two-state model can describe the relationship between opsin activity and ATR activity. The model predicts the following relationship between ATR activity and opsin activity: ATR activity = opsin activity/[$(1 - k)$(opsin activity) + k], where k is a constant representing the relative affinity of a given ligand to R and R*, the inactive and active forms of the receptor, respectively. The curve represents the best fit of this equation to the experimental data points where k is assumed to remain constant for each mutant. Thus, the solid line represents the idealized relationship between ATR activity and opsin activity predicted by the two-state model. A majority of the opsin mutants follow the curve predicted by the model, with a k of 0.029 and a standard error of 19%.

function first proposed by Lefkowitz and co-workers.[20,27] The results presented in Fig. 1 are a quantitative experimental test of the two-state model of GPCR function as described by Leff.[28] Namely, the equilibrium constant L ($L = [R]/[R^*]$, where [R] and [R*] are the concentration of aporeceptor

[27] R. J. Lefkowitz, S. Cotecchia, P. Samama, and T. Costa, *Trends Pharmacol. Sci.* **14,** 303 (1993).
[28] P. Leff, *Trends Pharmacol. Sci.* **16,** 89 (1995).

in the inactive and the active states, respectively) is related to the level of basal opsin activity listed in Table I by the formula 1/(opsin activity) − 1. We showed previously that at the concentration of ATR (157 μM) used to assay ATR activity, opsin is fully saturated with the ligand.[19] The ATR activity observed then reflects the maximum fraction of receptor in the active form (f_{R*}). The two-state model predicts that the maximum partition of the active receptor is related to the equilibrium constant L by the formula $f_{R*} = 1/(1 + LK_{A*}/K_A)$, where K_{A*} and K_A are the equilibrium constants of receptor bound to ligand A in its active and inactive states (K_{A*} = [R*][A]/[AR*], K_A = [R][A]/[AR]).[28] The two-state model predicts that the increased basal activity (i.e., a decrease in L) will result in increased ATR activity (maximum f_{R*}), which would eventually reach a saturation level of 1 when L is small, or basal activity high enough. If one and the same constant $k = K_{A*}/K_A$ (the relative affinity of a given ligand to R and R*) is assumed for all of the mutants, then the correlation between ATR activity and opsin basal activity can be described by the following function: ATR activity = opsin activity/[(1 − k) (opsin activity) + k)]. This equation is plotted as a curve in Fig. 1. Most of the mutants tested follow the ideal curve very well (k of 0.029 and a standard error of 19%). A k value of 0.029 means that the ATR ligand has ~34-fold higher affinity for the active form of the receptor than the inactive form.

Much higher ATR activities than predicted are likely due to decreased k values (i.e., increased affinity to R* relative to R) for some mutants compared to that of wild-type opsin. Two mutants with much higher ATR activities than predicted are M257A and G113Q/M257N. Likewise, much lower than predicted ATR activities are likely due to increased k values relative to that of opsin. Two mutants with much lower ATR activities than predicted are E113A and double mutants G121L/F261A, T or V. These deviations (Fig. 1) may be related to each other, and may reflect some mechanistic difference in these mutants from the others that behave according to model predictions.

The pharmacologic characteristics of most of the large number (~50) of mutant opsins that we have characterized to date generally support the so-called two-state model of GPCR function.[27,28] It is also clear from these data that the model is an oversimplification of the dynamics involved in receptor conformational changes.[29]

all-*trans*-Retinal Forming Covalent Schiff Base Bond with Constitutively Active Opsin

It has been shown convincingly that ATR does not form a covalent bond with wild-type opsin.[6] However, Sakmar *et al.*[7] have shown that E113A

---

[29] K. Fahmy, F. Siebert, and T. P. Sakmar, *Biophys. Chem.* **56,** 171 (1995).

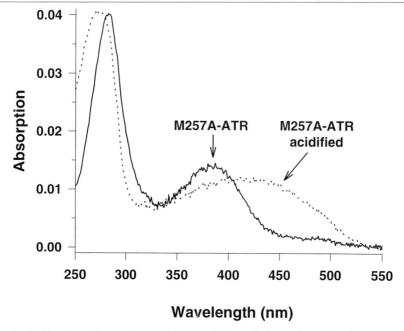

FIG. 2. The absorption spectrum of M257A–ATR covalent complex. COS cells expressing M257A opsin are incubated with ATR and purified in DM. A pigment with a λ_{max} at 380 nm is observed (solid line). A shift toward 440 nm is observed when 10 μl of 1 M HCl is added to 90 μl of M257A–ATR pigment solution (dotted line), indicating a covalent bond between M257A and ATR. Wild-type opsin treated in parallel does not give rise to absorption at 380 nm.

opsin can form a covalent bond with ATR that is stable enough to sustain the DM detergent solubilization and purification. Recently, we have shown that mutants M257A, M257Y, and double mutant G121L/M257A can also form a stable covalent bond with ATR. This is illustrated in Fig. 2 using M257A as an example. Wild-type opsin–ATR complex purified in parallel, however, does not show absorbance at 380 nm. Furthermore, the detergent-purified ATR–opsin complex can efficiently activate G_t (data not shown). E113 is the counterion of the protonated Schiff base; therefore, it is conceivable that a neutralizing mutation E113A can drastically alter the Schiff base environment. M257 is not involved in the retinal-binding site at all, since none of the 18 M257 mutants shows any significant change in λ_{max}.[9] However, mutation of M257 facilitates the formation of a covalent bond with ATR. Despite the very different nature of the mutations in E113A and M257A or Y with respect to the retinal-binding site, each of these mutant opsins displays high constitutive activity and full ATR activity when assayed in the membranes (Table I).

TABLE I
G_t ACTIVATION BY OPSIN MUTANTS IN COS CELL MEMBRANES

Mutant[a]	Opsin[b] (%)	all-*trans*-Retinal[b] (%)	11-*cis*-Retinal/light[b,c] (%)
F261W	0.2 ± 0.0	3.1 ± 1	120 ± 11
G121T	0.3 ± 0.1	21 ± 3	82 ± 18
G121V	0.4 ± 0.1	33 ± 5	94 ± 21
V258A	0.6 ± 0.1	13 ± 2	189 ± 27
M257L	0.8 ± 0.1	17 ± 5	112 ± 14
G121S	0.9 ± 0.2	47 ± 6	82 ± 1
wild-type	0.9 ± 0.2 (5)	14 ± 5 (4)	100
F261A	1.0 ± 0.2	12 ± 4	79 ± 7
G121I	1.0 ± 0.4	50 ± 8	81 ± 14
G121A	1.6 ± 0.3	26 ± 2	101 ± 2
G121L	1.9 ± 0.3	56 ± 9 (5)	99 ± 12 (5)
F261T	1.9 ± 0.5	23 ± 11	81 ± 16
G121L/V258A	2.0 ± 0.3	54 ± 3	218 ± 57
W265Y	2.0 ± 0.8	10 ± 3	70 ± 3
G121W	2.3 ± 0.4	47 ± 6 (6)	76 ± 12 (6)
M257Q	2.6 ± 0.3	81 ± 3	94 ± 2
E134Q	3.0 ± 1.2	51 ± 26	98 ± 14
M257F	3.3 ± 0.7	67 ± 3	113 ± 6
M257G	3.6 ± 0.2	66 ± 8	72 ± 7
M257C	3.9 ± 0.6	73 ± 10	110 ± 26
G121L/W265Y	3.9 ± 0.8	76 ± 11	89 ± 14
M257W	4.2 ± 0.9	39 ± 11	69 ± 6
M257I	4.4 ± 1.1	71 ± 20	109 ± 5
G121L/F261W	4.8 ± 1.2	48 ± 5	86 ± 19 (6)
M257H	5.0 ± 0.5	80 ± 9	85 ± 4
M257P	5.8 ± 1.3	47 ± 14	14 ± 1
M257E	6.4 ± 0.9	86 ± 2	90 ± 6
M257K	7.1 ± 3.0	91 ± 12	96 ± 8
F261V	7.6 ± 1.6	43 ± 11	99 ± 14

Although not all of the mutants that are constitutively active and fully activatable by ATR have been tested for their ability to form a covalent bond with ATR, we highly suspect that this correlation may be true in general. This observation implies that in constitutively active mutants, exogenously added ATR can covalently bind to opsins in the same way as in MII, where ATR was converted from 11-*cis*-retinal by photolysis. It is possible that due to certain conformational restraints, exogenously added ATR cannot be accommodated stably into the retinal-binding site in wild-type opsin. These constraints can only be overcome by the high-energy input of the photon absorption (~57 kcal/mol for 500-nm photon). In

TABLE I (continued)

Mutant[a]	Opsin[b] (%)	all-*trans*-Retinal[b] (%)	11-*cis*-Retinal/light[b,c] (%)
M257T	8.5 ± 0.8	62 ± 5	98 ± 13
G121L/F261A	9.4 ± 3.6	35 ± 11	105 ± 14
M257V	9.6 ± 1.6	83 ± 13	81 ± 6
M257D	9.7 ± 0.7	80 ± 3	60 ± 9
M257A	10.5 ± 1.0	112 ± 16	102 ± 3
G121L/F261T	13 ± 2	44 ± 13	119 ± 26 (6)
M257S	16.2 ± 2.7	99 ± 16	87 ± 9
E113A	21 ± 6	58 ± 12	100 ± 16
M257N	22.9 ± 7.4	93 ± 10	72 ± 9
E113Q	23 ± 0.8	111 ± 17	97 ± 18
E113Q/E134Q	31 ± 11	98 ± 5	78 ± 6
E134Q/F261V	28 ± 2	96 ± 10	69 ± 4
M257Y	32.6 ± 2.0	98 ± 6	101 ± 3
E113Q/F261V	33 ± 4	104 ± 8	78 ± 8
E113A/G121L	35 ± 8	93 ± 3	96 ± 20
G121L/M257A	46 ± 4	92 ± 12	91 ± 3
E134Q/M257A	47 ± 5	106 ± 12	60 ± 3
G121L/F261V	48 ± 5	64 ± 10	94 ± 4
E113Q/M257A	68 ± 8	98 ± 11	73 ± 11
E113Q/M257N	74 ± 11	133 ± 7	74 ± 6
E113Q/M257Y	81 ± 7	83 ± 10	78 ± 14

[a] Opsins are arranged in order of increasing opsin activity.
[b] Values indicate the ability of opsin alone, opsin with all-*trans*-retinal in the dark, and opsin with 11-*cis*-retinal in the light to activate G_t. Values are given as mean ± S.E., $n = 3$, unless otherwise specified. Activities are normalized to the 11-*cis*-retinal/light activity of the corresponding purified mutant measured in parallel. Data are assembled from previously published work.[8,9,19]
[c] Light activities in the presence of 11-*cis*-retinal are normalized to the 11-*cis*-retinal/light activity of wild-type opsin.

constitutively active mutants, the receptor is predisposed to its activated conformation, where at least some of the restraints are already released. ATR can then bind in the correct orientation to form the covalent bond. In the case of E113A, the restraint relieved is likely to be the salt bridge between E113 and K296.[11,30] In the case of M257 mutants, the restraint relieved is likely to be the packing interaction between M257 and the NPXXY motif on TM helix 7.[9] Relieving both constraints is predicted to result in a receptor with high opsin activity and high ATR activity as observed for E113Q/M257A, N, Y (Table I).

[30] G. B. Cohen, T. Yang, P. R. Robinson, and D. D. Oprian, *Biochemistry* **32**, 6111 (1993).

all-*trans*-Retinal as One Extreme of Continuum of Retinoid Compounds

Like other GPCRs, opsin can interact with a spectrum of chemically related ligands. As many as 100 or more retinoids have been used to probe the retinal-binding site of rhodopsin to yield many important insights into our current understanding of the receptor (reviewed by Nakanishi and Crouch[31] and references therein). Many retinoids when combined with wild-type opsin have been shown to activate rhodopsin kinase.[5,32] Some retinoids can also relieve bleach desensitization.[33] all-*trans*-Retinal can be viewed as one end of a continuum of retinoids, being the natural (when generated from photolysis) and the most efficient agonist of opsin.

Rhodopsin is specially evolved to be a low-noise, high-gain sensory system. When G_t activation by opsin is probed with a series of related retinoids, only 13-*cis*- and all-*trans*-retinals exhibit significant activity (Fig. 3, open bars). Therefore, G_t activation by opsin–retinoid complexes has not been as instructive as for rhodopsin kinase and bleaching adaptations. However, the system can be probed to reveal much more information by choosing certain mutations. G121 on TM helix 3 has been shown to interact with the 9-methyl group of retinal, and the interaction is important for receptor activation.[34] The G121L mutant has exaggerated steric interactions with retinal due to the bulky leucine substitution. Two series of retinoids, representing different elements of the native retinal ligand, are tested for their ability to activate G121L opsin (Fig. 3, solid bars). A progression of effects is observed, correlating well with the chemical components of the retinoids.

The first series consists of β-cyclocitral, 9-demethyl (dm)-β-ionone and β-ionone, 11-*cis*-9-dm-retinal, and 11-*cis*-retinal (Fig. 3A). Neither of these compounds significantly activates wild-type opsin although they are likely to bind. This finding is consistent with the finding that β-ionone, 11-*cis*-9-dm-retinal, and 11-*cis*-retinal are capable of relieving the desensitization on bleaching of cone cells.[33,35] In contrast, the G121L mutant opsin can be significantly activated by all, except for β-cyclocitral, which defines these

---

[31] K. Nakanishi and R. K. Crouch, *Isr. J. Chem.* **35**, 253 (1995).

[32] J. Buczylko, J. C. Saari, R. K. Crouch, and K. Palczewski, *J. Biol. Chem.* **271**, 20621 (1996).

[33] J. Jin, R. K. Crouch, D. W. Corson, B. M. Katz, E. F. MacNichol, and M. C. Cornwall, *Neuron* **11**, 513 (1993).

[34] M. Han, M. Groesbeek, T. P. Sakmar, and S. O. Smith, *Proc. Natl. Acad. Sci. U.S.A.* **94**, 13442 (1997).

[35] D. W. Corson, M. C. Cornwall, E. F. MacNichol, S. Tsang, F. Derguini, R. K. Crouch, and K. Nakanishi, *Proc. Natl. Acad. Sci. U.S.A.* **91**, 6958 (1994).

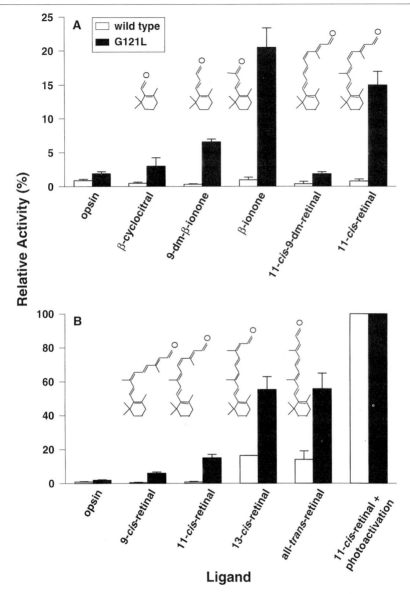

FIG. 3. A variety of retinoids can activate G121L opsin as determined by GTPγS uptake by G_t in COS cell membranes. (A) β-ionone and 9-dimethyl analogs. (B) Retinal isomers. All measurements are carried out under dim red light, except for those of opsin activities (ambient light) and photoactivated rhodopsin after 11-*cis*-retinal regeneration.

compounds as partial agonists to G121L opsin. Pairwise comparison can help dissect the important elements for receptor activation. The β-ionone and 11-*cis*-retinal exhibit much higher activities when compared to their 9-dm counterparts. This implies that the 9-methyl group plays a very important role in receptor activation. Because β-ionone causes a slightly higher agonist activity than 11-*cis*-retinal, the polyene segments from C-10 to the Schiff base do not contribute to the partial agonism of the 11-*cis*-retinylidene chromophore. On the contrary, it may serve to reduce the activity of the receptor when in the *cis* configuration. This notion is supported by the comparison between 9-dm-β-ionone and 9-dm-11-*cis*-retinal. The agonist efficiency of β-cyclocitral, which can be viewed as β-ionone further shortened by removing C-8 and C-9, is lower than that of 9-dm-β-ionone.

The second series of retinoids consists of four retinal isomers: 9-*cis*-, 11-*cis*-, 13-*cis*-, and all-*trans*-retinal (Fig. 3B). This series reveals the strong correlation between the ability to activate G121L opsin and shape of the isomer, which can be characterized by the distance L between the center of the cyclohexenyl ring and last carbon atom C-15.[36] The order is all-*trans* (photoactivated from 11-*cis*) > all-*trans* = 13-*cis* > 11-*cis* > 9-*cis* (i.e., the greater the L, the higher the activation). It is noteworthy that for both wild-type opsin and G121L, 13-*cis*-retinal and all-*trans*-retinal behaved nearly identically. Although the all-*trans* conformation is the only chromophore normally formed after illumination of the 11-*cis*-retinal in the binding site of rhodopsin, both opsin and G121L opsin do not seem to distinguish between the two isomers when added exogenously.

In summary, G_t activation by G121L opsin complexed with a spectrum of retinoids suggests that the 9-methyl group and the β-ionone-like portion of the retinal are very important for R* formation. The polyene chain from C-10 to C-15 activates the receptor very efficiently when in *trans* configuration with the β-ionone portion, and plays little, or even an inhibitory, role in R* formation when in *cis*. The longer the ligand isomer, measured by L, the more efficient is its agonism.

Implications of all-*trans*-Retinal Opsin Interaction in Visual Desensitization

Visual desensitization refers to the decreased sensitivity of visual perception that cannot be solely accounted for by the decrease in the amount of visual pigment available after light exposure. The primary biochemical entity that causes this fascinating phenomenon has been of great contro-

---

[36] H. Matsumoto and T. Yoshizawa, *Vis. Res.* **18,** 607 (1978).

versy, since its discovery more than 50 years ago.[37] The three possible etiologies, which may not be mutually exclusive, are (1) empty opsin, (2) opsin–ATR noncovalent complex, and (3) opsin–ATR covalent complex. More and more compelling evidence suggests that the ATR–opsin complex generated by bleaching and subsequent decay of MII could be involved in desensitization. Saari et al.[38] have shown that ATR is indeed accumulated in photoreceptor cells during constant illumination. As discussed earlier, ATR is capable of, and probably is, forming a complex with opsin. Leibrock and Lamb[39] also showed that in toad rod cells, hydroxylamine exposure could relieve desensitization. Hydroxylamine can react with retinylidene Schiff base as well as retinal, thereby disrupting both the covalent and noncovalent complex between ATR and opsin. It has no effect on free opsin and unbleached receptor. These two studies, among others, seem to suggest that a covalent or noncovalent opsin–ATR complex is the cause of desensitization.

However, Crouch and co-workers have demonstrated that it is likely to be the presence of free opsin that is responsible.[33] They have shown that desensitization can be relieved by β-ionone delivered by liposomes or ethanol. The effect could then be washed away and the cell restored to its original desensitized state, which highly suggests free opsin being the cause. ATR has been shown not to compete with β-ionone for interacting with opsin.[6] It is possible that ATR can be readily displaced by β-ionone, but is not washed away by the superfusion system due to its higher hydrophobicity. Therefore, it can again form the ATR–opsin complex to resume desensitization after β-ionone is washed away. If this is indeed the case, then the results strongly support that the noncovalent opsin–ATR complex is responsible for desensitization. Some more conclusive experiments may soon resolve the mystery, but it is undoubtedly true that the ATR–opsin complex does exist during constant illumination and may play important roles in the visual signal transduction pathway.

Acknowledgments

Support for the work carried out in the laboratory of T.P.S was provided by the Howard Hughes Medical Institute, the Allene Reuss Memorial Trust, and the Charles H. Revson Foundation. M.H. is currently a Helen Hay Whitney Foundation fellow. We are grateful to K. P. Hofmann for many illuminating discussions and Carsten Schubert for commenting on the manuscript. We also thank Tanya Zvyaga, Steven O. Smith, and Steven W. Lin.

---

[37] S. Hecht, C. Haig, and A. M. Chase, *J. Gen. Physiol.* **20,** 831 (1937).
[38] J. C. Saari, G. G. Garwin, J. P. Van Hooser, and K. Palczewski, *Vis. Res.* **38,** 1325 (1998).
[39] C. S. Leibrock and T. D. Lamb, *J. Physiol.* **501,** 97 (1997).

[19] Electrical Approach to Study Rhodopsin Activation in Single Cells with Early Receptor Current Assay

By JACK M. SULLIVAN, LIOUBOV BRUEGGEMANN, and PRAGATI SHUKLA

Introduction

Rhodopsin is the most well-understood G-protein-coupled receptor. As an integral membrane protein with seven transmembrane domains it resides in the disks and plasma membranes of rod photoreceptor cells in the vertebrate retina. Rhodopsin is constituted by the covalent interaction of its ligand, 11-*cis*-retinal (11cRet), with the apoprotein, opsin, by way of a protonated Schiff base (PSB) linkage with lysine-296 (K296; single-letter chemical abbreviations are used for amino acids: C = cysteine; D = aspartate; E = glutamate, F = phenylalanine; G = glycine; H = histidine; K = lysine; N = asparagine; Q = glutamine; Y = tyrosine). The K296 side chain and the ligand binding pocket are buried near the middle of the membrane in a presumably lower dielectric environment.[1] To tune the spectral absorbance characteristics of the pigment and stabilize the cationic PSB charge, the pigment has a counterion[2] to the PSB which in vertebrate visual pigments is a charged glutamate carboxylate ion side chain at $E113^-$. The ground state of rhodopsin is incredibly stable in the dark-adapted rod photoreceptor with a half-life on the order of 1000 years for a single molecule at normal body temperature. That is, only about one molecule of rhodopsin activates without light every 20 sec in the entire rod photoreceptor despite the large copy number of the pigment ($\approx 10^8$) in the outer segment.[3] This stability permits the low dark noise that allows rods to signal detection of single photons to the vertebrate visual apparatus through a high-gain transduction machinery. Despite this incredible dark stability, rhodopsin is an efficient molecular machine converting photon energy into conformational transitions that activate the biochemical transduction cascade in the photoreceptor. The activation probability of rhodopsin is about 0.7 after a single photon encounter.[4] The combination of a low noise and highly efficient activation mechanism is an evolutionary accomplishment that remains a challenge for vision scientists to understand.

[1] J. M. Baldwin, *EMBO J.* **12**, 1693 (1993).
[2] H. G. Khorana, *J. Biol. Chem.* **267**, 1 (1992).
[3] K.-W. Yau, G. Matthews, and D. A. Baylor, *Nature* **279**, 806 (1979).
[4] R. R. Birge, *Biochim. Biophys. Acta* **1016**, 293 (1990).

On a femtosecond timescale photon absorption leads to *cis*-to-*trans* isomerization of the chromophore and charge separation in the ligand binding pocket associated with efficient photon energy uptake by the pigment.[4,5] The early spectral intermediates bathorhodopsin$_{538\text{ nm}}$ (nanosecond lifetime) and lumirhodopsin$_{497\text{ nm}}$ (microsecond lifetime) are associated with red and blue shifts of the absorption dipole indicating successive changes in local conformation of the chromophore environment.[6] These early transitions are thought to be largely steric in nature.[7] On a millisecond timescale the critical biochemical events occur in the rhodopsin pigment. The environment around the Schiff base becomes supportive to deprotonation of the Schiff base, whereupon the E113$^-$ counterion becomes protonated.[8] Schiff base deprotonation is associated with the metarhodopsin-I$_{468\text{ nm}}$ (meta-I) to metarhodopsin-II$_{380\text{ nm}}$ (meta-II) spectral transition that reports the beginning of the cascade of conformational activation which is now thought to be largely electrostatic in nature.[7,9] Specifically, Schiff base deprotonation in the buried ligand binding pocket precedes the uptake of two protons on the remote cytoplasmic surface of the wild-type protein.[10–12] Meta-II formation is catalyzed in its equilibrium with meta-I (pK_a 6.8) by acidification on the cytoplasmic surface despite the fact that deprotonation of the Schiff base drives the meta-I to meta-II transition. In contrast to wild-type rhodopsin the mutant E134Q has no proton uptake and E134D takes up protons at a slower rate.[12] E134 is a highly conserved residue in rhodopsins and other G-protein-coupled receptors and is an important link in the millisecond-order appearance of the transducin binding domain in the cytoplasmic loops of rhodopsin.[9] During meta-II formation helix 3 moves inward (toward the cytoplasm) relative to the interacting helix 6, and inward motion of helix 7 also occurs.[13–15] Other intramembrane amino acids capable of proton transfer reactions influence meta-II formation (e.g., D83, E122, H211).[16,17] The importance of proton exchange reactions and α-helical

[5] R. W. Schoenlein, L. A. Peteanu, R. A. Mathies, and C. V. Shank, *Science* **254**, 412 (1991).
[6] J. W. Lewis and D. S. Kliger, *J. Bioenerg. Biomembr.* **24**, 201 (1992).
[7] T. Shieh, M. Han, T. P. Sakmar, and S. O. Smith, *J. Mol. Biol.* **269**, 373 (1997).
[8] F. Jager, K. Fahmy, T. P. Sakmar, and F. Siebert, *Biochemistry* **33**, 10878 (1994).
[9] K. Fahmy, F. Siebert, and T. P. Sakmar, *Biophys. Chem.* **56**, 171 (1995).
[10] J. H. Parkes and P. A. Liebman, *Biochemistry* **23**, 5054 (1984).
[11] S. Arnis and K. P. Hofmann, *Proc. Natl. Acad. Sci. U.S.A.* **90**, 7849 (1993).
[12] S. Arnis, K. Fahmy, K. P. Hofmann, and T. P. Sakmar, *J. Biol. Chem.* **269**, 23879 (1994).
[13] Z. T. Farahbakhsh, K. Hideg, and W. L. Hubbell, *Nature* **262**, 1416 (1993).
[14] D. L. Farrens, C. Altenbach, K. Yang, W. L. Hubbell, and H. G. Khorana, *Science* **274**, 768 (1996).
[15] N. Abdulaev and K. D. Ridge, *Proc. Natl. Acad. Sci. U.S.A.* **95**, 12854 (1998).
[16] C. J. Weitz and J. Nathans, *Neuron* **8**, 465 (1992).
[17] C. J. Weitz and J. Nathans, *Biochemistry* **32**, 14176 (1993).

movements (macrodipoles) to the millisecond-order process of rhodopsin activation led us to apply electrical techniques to the study of this process.

The early receptor potential (ERP) is due to a charge redistribution in rhodopsin during its activation.[18] The small fast depolarizing R_1 phase is thought to be associated with early photochemical processes and energy uptake,[19] while the slower and much larger hyperpolarizing R_2 phase occurs on a millisecond timescale overlapping meta-II formation.[20] Unique ERP waveforms originate on flash photolysis of late spectral intermediates (meta-I, meta-II, meta-III) suggesting that distinct electrical states occur on the timescale of the spectral metarhodopsin intermediates.[18] The molecular mechanism of the ERP signal remains to be established. Prior investigations indicated that both intramembrane charge transfer and proton uptake contributed to the R_2 signal, and that a local positive electrostatic potential accumulates near the cytoplasmic surface during meta-II formation.[21–23] These studies suggest that the biochemically relevant conformational changes in rhodopsin have a significant electrostatic component.

Early receptor current (ERC) measurements of rhodopsin activation were made in isolated vertebrate rod and cone photoreceptors.[24,25] The ERC is the direct measure of charge transfer and does not explicitly depend on the electrical charging time constant of the membrane. The ERC has opposite polarity to the ERP. In the ERC the R_1 current is fast and inward directed, whereas the R_2 current is slow and outward directed. In the standard electrophysiologic convention, R_1 is consistent with an inward (toward the cytoplasm of the cell) cationic charge motion, whereas R_2 is consistent with an outward cationic charge motion. ERP and ERC measurements have been widely used in studies of proton exchange and pumping in bacteriorhodopsin where protein mutagenesis and analog chromophores have advanced mechanistic understanding of structure–function relationships during the photocycle.[26] Therefore it was reasonable to propose that similar structure–function problems could be addressed in an expression system of vertebrate rhodopsin provided several experimental challenges

---

[18] R. A. Cone and W. L. Pak, *in* "Handbook of Sensory Physiology: Principles of Receptor Physiology" (W. R. Lowenstein, ed.), p. 345. Springer-Verlag, Berlin and New York, 1971.
[19] H. W. Trissl, *Biophys. Struct. Mech.* **8**, 213 (1982).
[20] J. D. Spalink and H. Stieve, *Biophys. Struct. Mech.* **6**, 171 (1980).
[21] N. Bennett, M. Michel-Villaz, and Y. Dupont, *Eur. J. Biochem.* **111**, 105 (1980).
[22] M. Lindau and H. Ruppel, *Photobiochem. Photobiophys.* **9**, 43 (1985).
[23] D. S. Cafiso and W. L. Hubbell, *Biophys. J.* **30**, 243 (1980).
[24] S. Hestrin and J. I. Korenbrot, *J. Neurosci.* **10**, 1967 (1990).
[25] C. L. Makino, W. R. Taylor, and D. A. Baylor, *J. Physiol.* **442**, 761 (1991).
[26] S. Moltke, M. P. Heyn, M. P. Krebs, R. Mollaaghababa, and H. G. Khorana, *in* "Structures and Functions of Retinal Proteins" (J. L. Rigaud, ed.), Vol. 221, Colloque INSERM, p. 201. John Libbey Eurotext Ltd., 1992.

were met: a high-level expression system, a wide-bandwidth, low-noise recording technique, efficient regeneration with chromophore, and the delivery of flashes of sufficient strength to activate significant numbers of regenerated rhodopsin molecules. We have solved these problems for the human rod visual pigment and are now conducting a structure–function study of millisecond-order electrical events in both wild-type and mutant pigments regenerated with 11cRet and analog chromophores.[27] Our success depends on adapting the highly sensitive and low-noise patch clamp electrophysiologic techniques to the challenge of rhodopsin activation. To simply state the utility of the method, we can resolve rhodopsin charge motions from a single cell expressing on the order of a picogram of regenerated visual pigment with submillisecond temporal resolution! This is at least 10^7-fold more sensitive than other time-resolved spectroscopic techniques currently being applied to study rhodopsin activation.[27] Moreover, the "vector" of protein activation space sampled with electrophysiology is orthogonal to the membrane plane and oriented to measure electrical events that could correspond to the proton exchange reactions or bulk movements of α helices that are now known to play important roles in the biochemical phase of rhodopsin activation.

In this chapter we describe the evolution of our methods to record ERCs from heterologously expressed opsin cDNAs in cells that were regenerated with 11cRet. We discuss the essential variables and potential pitfalls and demonstrate our success to record large ERC signals from both wild-type and mutant rhodopsins that are in fact larger than signals reported in photoreceptors.

Materials and Methods

Materials

pBlueScript-II-KS(−) (pBS) is obtained from Stratagene (La Jolla, CA). pCIS is obtained from Genentech (South San Francisco, CA). pCDNA3 (neomycin resistance, neo^R) is obtained from Invitrogen (Carlsbad, CA). pSV2neo is obtained from the American Type Culture Collection (Manassas, VA). Oligonucleotides are synthesized by GenoSys (The Woodlands, TX). The 1D4 monoclonal antibody is prepared at the National Cell Culture Center (Cellex Biosciences, Minneapolis, MN). HEK293S cells were a kind gift of Dr. Bruce Stillman (Cold Spring Harbor Laboratories).[28]

---

[27] J. M. Sullivan and P. Shukla, *Biophys. J.* **77**, 1333 (1999).
[28] B. W. Stillman and Y. Gluzman, *Mol. Cell. Biol.* **5**, 2051 (1985).

Culture media and additives are obtained from Life Technologies (Gaithersburg, MD). G418 (Geneticin) is obtained from Life Technologies. Cell culture tested polyethylene glycol (PEG) is from Boehringer Mannheim (Indianapolis, IN). 11cRet was a kind gift of Dr. Rosalie Crouch (Medical University of South Carolina) and the National Eye Institute. Fatty-acid-free bovine serum albumin (FAF-BSA) is obtained from Sigma (St. Louis, MO). All other chemicals are of reagent grade and obtained from Sigma, Acros (Pittsburgh, PA), or Aldrich (Milwaukee, WI).

Site-Specific Mutagenesis

The human wild-type rod opsin cDNA is directionally cloned into the pBlueScriptII-KS(−) vector (pBS-opsin). This cDNA clone contains a 21-nucleotide stretch of the 5' untranslated region, the complete coding region (1044 bp), and a long 683-bp 3' untranslated tail that extends 311 bp beyond the first putative polyadenylation signal.[29] Oligonucleotide-directed site-specific mutagenesis is conducted in pBS-opsin according to established techniques[30] as described extensively in an accompanying article in this volume.[31] cDNAs are completely sequenced in the pBS-opsin vector to confirm the mutation and rule out spurious ones before directional subcloning into CMV expression vectors, either pCIS[32] or pCDNA3 for cellular transfections.

Cellular Transfections and Development of Stable Opsin Expressing Cell Lines

HEK293S cells are grown to subconfluence (70–80%) in 10-cm dishes in Dulbecco's modified Eagle's medium (DMEM)/F12 (Life Technologies, Gaithersburg, MD) with 10% heat-inactivated calf serum plus 2 mM L-glutamine and antibiotics at 37° in 5% (v/v) CO_2. Then 10–20 μg of supercoiled plasmids (pCIS-opsin plus pSV2-neo or pCDNA3-opsin) are transfected per 10-cm plate according to the slow calcium phosphate method[33] with precipitation allowed to occur over a 16- to 20-hr period at 3% CO_2 at 37°. To make stable cell clones, drug selection is in G418 (500 μg/ml) as detailed in the accompanying paper in this volume.[31] Transient transfections are conducted with the slow calcium phosphate method with-

---

[29] J. Nathans and D. S. Hogness, *Proc. Natl. Acad. Sci. U.S.A.* **81**, 4851 (1984).
[30] W. P. Deng and J. A. Nickoloff, *Anal. Biochem.* **200**, 81 (1992).
[31] J. M. Sullivan and M. F. Satchwell, *Methods Enzymol.* **315** [3], 1999 (this volume).
[32] J. Nathans, C. J. Weitz, N. Agarwal, I. Nir, and D. S. Papermaster, *Vis. Res.* **29**, 907 (1989).
[33] C. Chen and H. Okayama, *Mol. Cell. Biol.* **7**, 2745 (1987).

out G418 selection or with electroporation in 0.4-cm cuvettes (capacitance 71 μF, load resistance ∞, capacitor charge voltage 350 V, Electroporator II, Invitrogen).[34]

Immunocytochemistry

Cells grown on poly(L-lysine)-coated coverslips are fixed and permeabilized in 100% ice-cold methanol. The primary antibody is a mouse anti-bovine monoclonal directed toward the carboxyl terminus of rhodopsin (1D4)[35] (human and bovine epitopes are identical) and the secondary antibody is a fluorescein isothiocyanate (FITC)-coupled horse anti-mouse IgG monoclonal antibody. Details of our procedure are presented in the accompanying article in this volume.[31] For immunocytochemistry of giant cells and in tests of transfection efficiency, single cells are first grown on polystyrene dishes (see later discussion). Giant cells are photographed with Hoffman interference optics (Olympus, Lake Success, NY) under transmitted light while on the tissue culture dish. A stage micrometer is photographed to allow measurements of cell diameters. Giant cells are then transferred to poly(L-lysine)-coated coverslips for attachment and processed for immunocytochemistry the following day. For transient transfection experiments cells are transfected in tissue culture plates (calcium phosphate) or in cuvettes (electroporation) prior to transfer to poly(L-lysine)-coated coverslips. Immunofluorescent photomicroscopy is conducted on the Olympus scope fitted with a standard epifluorescent attachment and an FITC dichroic cube. Hoffman contrast and FITC fluorescence images are photographed on Kodak (Rochester, NY) Ektachrome slide file (ASA 400).

Cellular Fusion and Pigment Regeneration

Initially we grow stable expressing cell lines on poly(L-lysine)-coated coverslips in six-well polystyrene dishes. Nearest neighbor cells are then fused by transient exposure ($\approx$2 min) to 50% PEG in 75 mM HEPES (pH 8).[36] We have more recently turned to growth of cells on 10-cm polystyrene dishes and then fusing with PEG on dishes. This method generates giant cells of more uniform size with fewer single cells. Once fused we avoid the use of trypsin to detach giant cells from plates which will be used for electrophysiology because there are potential cleavage sites in the opsin protein.[15] Instead, giant cells on dishes are harvested with phosphate-buf-

---

[34] M. E. Jurman, L. M. Boland, Y. Liu, and G. Yellen, *BioTechniques* **17**, 876 (1994).
[35] R. S. Molday and D. MacKenzie, *Biochemistry* **22**, 653 (1983).
[36] M. F. Sheets, J. W. Kyle, S. Krueger, and D. A. Hanck, *Am. J. Physiol.* **271**, C1001 (1996).

fered saline (PBS) (in mM: 137 NaCl, 2.7 KCl, 5.6 Na$_2$HPO$_4$, 1.47 KH$_2$PO$_4$) containing 5 mM disodium EDTA and transferred to poly(L-lysine)-coated coverslips. Prior to chromophore regeneration and recording, fused giant cells are transferred to poly(L-lysine)-coated coverslips for attachment. Single fused spherical giant cells on coverslip fragments are then selected for recordings. Fused giant kidney cells do not divide but maintain active membrane potentials.[37]

In initial experiments we regenerated cells by the addition of concentrated ethanolic stocks of 11cRet directly to a physiologic "regeneration buffer" (140 NaCl, 5.4 KCl, 1.8 CaCl$_2$, 1.0 MgCl$_2$, 10 glucose, 10 HEPES, pH 7.2). However, we learned that the solubility of retinoids in aqueous solutions could limit partition into cellular membranes. We became aware of a means to solubilize 11cRet and regenerate visual pigment from bleached opsin in rod outer segments by the addition of 2% (w/v) FAF-BSA to our regeneration buffer.[38] We now routinely use this regeneration technique. Giant cells are dark adapted in regeneration buffer containing 2% FAF-BSA, 25 μM 11cRet (added as a concentrated ethanolic stock), and 0.025% (v/v) α-D-tocopherol as an antioxidant. Cells are regenerated in the darkroom for at least 30 min at room temperature in an opaque black box. Coverslips are dipped in recording buffer (without 11cRet) and placed in the recording chamber.

Whole-Cell ERC Recording

Whole-cell electrodes are fashioned by two-stage pulls on a vertical apparatus (Model 730, Kopf Ind. Inc, Tujunga, CA) using borosilicate glass (Kimax 51, Kimble Glass, Toledo, OH).[39] Electrode resistances of 2–8 MΩ are made leading to series resistances of between 5 and 20 MΩ. Electrodes are coated with black Sylgard (Model 173, Dow Chemical, Midland, MI) to prevent light from the flash from striking the Ag|AgCl electrode in the pipette. Electrodes are not fire polished. In these experiments we fill pipettes (intracellular) with a solution type commonly used in ion channel gating current experiments that included (in mM): 70 tetramethylammonium hydroxide (TMA-OH), 70 tetramethylammonium fluoride, 70 2-[N-morpholino]ethanesulfonic acid (MES-H), 10 EGTA–CsOH, 10 HEPES–CsOH, pH 6.5).[36] The extracellular solution is similar (in mM): 140 TMA-OH, 140

---

[37] U. Kersting, H. Joha, W. Steigner, B. Gassner, G. Gstaunthaler, W. Pfaller, and H. Oberleithner, *J. Membr. Biol.* **111,** 37 (1989).

[38] J. H. McDowell, *Methods Neurosci.* **15,** 123 (1993).

[39] O. P. Hamill, A. Marty, E. Neher, B. Sakmann, and F. J. Sigworth, *Pflugers Arch.* **391,** 85 (1981).

MES-H, 1.0 CaCl$_2$, 2.0 MgCl$_2$, 5.0 HEPES-NaOH, pH 7.0. Once in the bath, electrode capacitance is compensated electronically. Gigohm seals (>10^9 Ω) are regularly and easily obtained from giant HEK293S cells expressing rhodopsins, and whole-cell recording (WCR) is readily obtained by applying pulses of suction. Cells are quite large, typically 60–120 μm, and the cellular membrane capacitance is typically greater than 100 pF, but whole-cell capacitance and series resistance are not compensated because the ERC is a capacitative current. Although this would not affect the amplitude of the ERC signals it does affect the fast kinetics of R$_1$ or the R$_1$ to R$_2$ transition.[27] However, we are predominantly interested in the R$_2$ signal and its relaxation kinetics and thus select large giant cells for recordings in order to obtain larger and more stable signals over time. Data are acquired with an Axopatch 1C amplifier with a resistive CV-4 headstage (Axon Instruments Inc., Foster City, CA), filtered through an eight-pole Bessel filter tuned to 5-kHz cutoff (Model LPF 902B, Frequency Devices, Haverhill, MA), and digitized in the pCLAMP 5.51 environment using CLAMPEX (Axon Industries) at 10 kHz on an AT clone computer. This digitization rate undersamples the very fast components of the signal but allows us to acquire 100 ms of ERC data with each flash in order to fully record the stretched tail of the R$_2$ relaxation in the wild-type pigment. Flashes are generated with a novel microsecond microbeam apparatus designed around a short arc xenon flash tube.[40] In these experiments flash stimuli are about 4×10^8 photons/μm^2 and spectrally tuned with a 70-nm bandpass three-cavity interference filter centered on 500 nm (Omega Inc., Brattleboro, VT). Data are analyzed using Origin 4.1 (MicroCal, Northampton, MA).

Results

Adaptation of a High-Level Opsin Expression System for ERC Studies

Charge motion resulting from rhodopsin activation is linearly proportional to the amount of pigment activated in a single flash over a broad range of flash intensities.[18] Calculations done on the basis of the amount of regenerable wild-type opsin expressed in human cell lines,[32] and the amount of charge that moves with each rhodopsin molecule,[18] led to initial estimates that ERC signals might be recordable in single cells. However, the fraction of opsin expressed to the plasma membrane would need to be high to record ERC signals. To record signals from a significant portion of

---

[40] J. M. Sullivan, *Rev. Sci. Instrum.* **69,** 527 (1998).

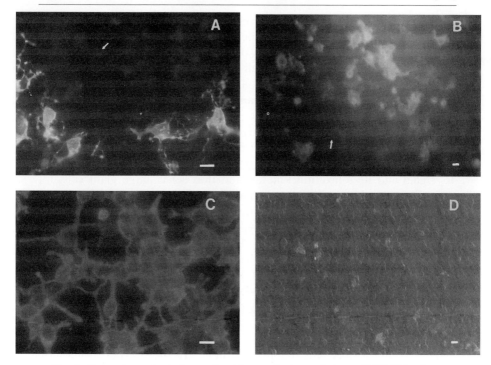

Fig. 1. Comparison of transient and stable transfection techniques. HEK293S cells were transiently transfected with wild-type human opsin expression plasmid by (A) electroporation or (B) the slow calcium phosphate method and processed for immunocytochemistry in comparison to a stably transfected cell line for wild-type human opsin (WT-6) (C). In the immunocytochemistry of transiently transfected cells, representative cells not expressing opsin are indicated by single arrows in (A) and (B). The field of cells (Hoffman image) from the calcium phosphate transfection (B) is shown in (D). Bars: 20 μm.

cells, transfection efficiency should be high. Therefore, we examined the uniformity of opsin expression level following various means of plasmid transfection. Wild-type and mutant human rod opsins were expressed from a cassette containing the human cytomegalovirus (CMV) early intermediate promoter/enhancer and a cDNA containing the complete coding region, a 21-bp 5' untranslated region containing a Kozak consensus sequence, and a 3' untranslated region extending through the first polyadenylation site.[29] Figure 1 shows fields of HEK293S cells transiently transfected with wild-type expression plasmid by the electroporation or slow calcium phosphate methods in comparison to a stably transfected wild-type opsin expressing cell line that we developed (WT-6).[31] Both electroporation and calcium phosphate lead to low transfection efficiencies such that at most 15–20%

of cells express wild-type opsin as detected by opsin immunocytochemistry. In transient transfections opsin expression levels varied considerably. Some cells expressed very high levels of opsin-dependent immunofluorescence, and some expressed little fluorescence over the background which results from autofluorescence of flavinoids in the untransfected HEK293S cells.[41] In contrast, every cell in the stably transfected wild-type cell line expresses opsin at very uniform and moderately intense levels across the population and almost all of the opsin immunofluorescence emanates from the plasma membrane. Thus, while some transiently transfected cells express higher levels of opsin compared to cells from a high-producing stable line, electrophysiology has practical sampling limits of at most a few tens of cells on a given day. We initially anticipated that low transient transfection efficiencies would make the cellular ERC success rate low and the variability of transient expression levels would predictably lead to considerable variance in the levels of ERC signals. In contrast, every cell of a stable constitutively expressing line should support measureable ERC signals with little variance, provided that the regenerable opsin levels were sufficiently high to yield ERC charge motions above background cellular noise on flash photolysis. We decided to initially pursue ERC recordings in cells that were stably transfected. Single-cell ERC signals were recordable but the amplitudes were small, making quantitative analysis challenging.[27,40,42] We therefore sought means of increasing ERC signal strength while continuing work with a high producing stable wild-type human opsin expression line.

The ERC is a light-activated capacitative current across the plasma membrane. On flash photolysis charge motion results in an oriented population of rhodopsin molecules and is linearly dependent on the number of rhodopsin molecules activated. Assuming independent activation of each rhodopsin molecule, to increase the peak ERC current and total charge (integral of the current) required increased numbers of rhodopsin molecules resident in the plasma membrane. In fact, cone photoreceptors have larger ERC signals than rods because plasma membrane rhodopsin levels in cones are about 10^8 molecules per cell or about an order of magnitude larger than in rods. Electrophysiology puts significant constraints on the levels of expression needed to record ERCs because signals are measured in a single cell and the currents resulting from rhodopsin activation must be larger than the band of whole-cell noise in order to be minimally detectable. Quantitative ERC analysis (e.g., action spectra, relaxation kinetics, pH dependence, temperature) requires high signal-to-noise ERC recordings with rhodopsin charge motions well distinguished from the whole-cell noise

---

[41] M. J. Zylka and B. J. Schnapp, *BioTechniques* **21**, 220 (1996).
[42] J. M. Sullivan, *Invest. Ophthal. Vis. Sci.* **37**, S811 (1996).

band. One way to increase signals is to enhance opsin expression levels in each cell. We invested considerable time to generate stable expression clones of wild-type rhodopsin which demonstrate wild-type human opsin levels even greater than an existing high level producing standard line ($\approx 6 \times 10^6$ rhodopsin molecules/cell).[31,32] However, for single drug selections wild-type opsin expression levels in new lines were comparable to, but not greater than, that in the wild-type standard line that we used for single-cell ERC measurements. Regardless of the line, immunocytochemistry demonstrated that wild-type human opsin partitioned almost exclusively to the plasma membrane in HEK293S cells (Fig. 1).

Two mutant E134Q human opsin cell lines were also generated for ERC recordings (see later discussion) and expressed about 0.3×10^6 opsins/cell and at least 0.51×10^6 opsins/cell.[31] Secondary transfections with additional E134Q human opsin expression plasmid containing an independent drug selection marker cassette (hygromycin) allowed identification of clones with higher expression levels for the E134Q mutant ($\approx 0.8 \times 10^6$ opsins/cell), but we have not examined these secondary lines in ERC recordings. The sequential transfection/selection approach is a means of progressively building on opsin expression levels in an established expression line. It is unlikely that stable expression clones we have developed have saturated the HEK293S cellular expression machinery because one does not see an opsin band in Coomassie blue stained denaturing polyacrylamide gels in comparison to untransfected cells (not shown). However, the time investment to generate ever increasingly higher opsin expression lines is impractical for most laboratories. With a rapid method to select expression clones (e.g., fluorescence activated cell sorting) this process would become more feasible but we have not yet identified monoclonal antibodies that will avidly detect opsin in live HEK293S cells.

Another means to increase plasma membrane opsin levels is to increase cell size and thus the plasma membrane surface area available for expression. In *Xenopus* oocytes, which are more than 1 mm in diameter, approximately 10^{11} molecules/cell can be expressed from injected cRNA.[43] Moreover, oocytes are routinely used to record ionic channel gating currents which, like ERCs, are also protein conformation-dependent signals.[44] Sheets *et al.*[36] recently demonstrated a means to record Na$^+$ channel gating currents in HEK293 cells. Untransfected cells were first fused with PEG and then transfected with plasmids with a Na$^+$ channel cDNA under control of a CMV promoter. They were able to record Na$^+$ channel gating currents.

[43] H. G. Khorana, B. E. Knox, E. Nasi, R. Swanson, and D. A. Thompson, *Proc. Natl. Acad. Sci. U.S.A.* **85,** 7917 (1988).
[44] F. Bezanilla and E. Stefani, *Methods Enzymol.* **293,** 331 (1998).

Using a similar rationale we fused stable high-level expression human wild-type cells to generate giant cells with greater surface area for expressing opsin. Figure 2 shows results of PEG fusion of wild-type expressing HEK293S cells on polystyrene dishes. Fusion results in giant cells with spherical contours that are clearly several-fold larger in volume than that of single cells (arrow). PEG-induced chemical fusion is an isovolumic process[37] and thus knowledge of the spherical cellular diameters of giant cells and single unfused cells can allow estimates of the numbers of cells that fused. When giant cells spread out on the dish or poly(L-lysine)-coated coverglass surface, nuclei are more visible and the largest of cells contain hundreds of discernible nuclei. Fused giant cells expressing wild-type rod opsin were examined by immunocytochemistry. Like unfused cells (Fig. 1), fused giant cells made from the wild-type standard line have opsin localized predominantly in the plasma membrane. The increase in cell size leads to increased levels of plasma membrane opsin in the giant cells in proportion to the number of cells that participated in the fusion event. Spherical giant cells that have not formed extensive substrate attachments demonstrate surface plasma membrane opsin labeling. Well-attached giant cells demonstrate attachment feet with immunoreactive opsin clearly in the plasma membrane. In Fig. 2, note the large number of nuclei in each of the two fused giant cells that have formed a giant cell pair. The PEG fusion technique, applied to single cells stably expressing uniform amounts of plasma membrane opsin, is a simple and effective technique to isolate giant cells (e.g., 60–120 μm in diameter) with amplified levels of plasma membrane opsin. Thus, even cell lines that do not express as intensively as wild-type can be fused to generate single giant cells with comparable levels of opsin. We anticipated that fused giant cells would support larger ERC signals for quantitative analysis and allow biophysical study of remaining structure–function problems in rhodopsin activation using electrophysiologic tools.

Instrumentation Needed for Rhodopsin ERC Studies

In addition to a high-level expression system the design and assembly of an advanced electrophysiologic recording apparatus was necessary. Because ERC signals are similar to ionic channel gating currents, an apparatus to record small and fast signals with high fidelity and low noise is a requirement. An intense light pulse is needed to activate a large portion of the rhodopsin molecules with each flash to obtain high-level signals. In addition the light-sensitive nature of the protein of interest requires that experiments be conducted in darkness. Several important criteria of the instrumentation are presented here. We use a patch-clamp apparatus (Axopatch 1C) with a resistive headstage to record ERC signals, although a capacitative headstage

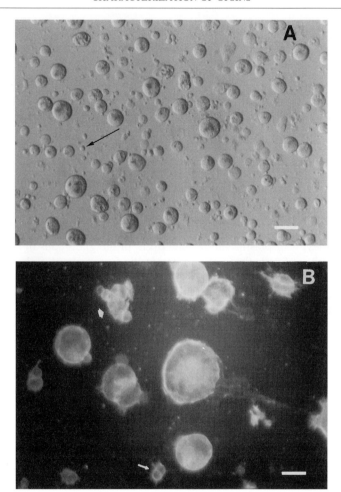

FIG. 2. Cell fusion generates giant cells that express plasma membrane opsin. (A) Fused wild-type human giant cells imaged by Hoffman contrast optics. Bar: 100 μm. Arrowhead (black) points to a single cell. (B) Immunocytochemistry of a field of spherical fused wild-type giant cells that contain surface plasma membrane immunoreactive opsin. A pair of single cells is indicated (single arrow) as well as a cluster of unfused or partially fused single cells (broad arrowhead). Bar: 50 μm. (C) Touching pair of fused giant cells that are well attached to the poly(L-lysine)-coated coverslip. Note the immunoreactive opsin-containing footpads that splay out from each cell indicating the presence of opsin in the plasma membrane (single white arrows) and the large number of nuclei present in each cell (broad black arrow). Bar: 50 μm.

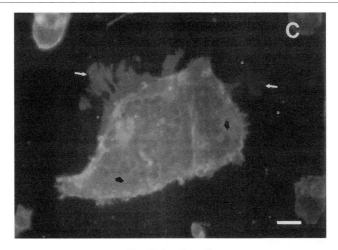

FIG. 2. (continued)

could also be used.[25] Routinely, we measure the series resistance but do not compensate the series resistance/cell capacitance because compensation changes the shape or even polarity of the ERC waveform. This would be expected because the ERC is a capacitative current. We use very uniform electrodes with low access resistance (2 MΩ), which results in uniform series resistances between 5 and 15 MΩ. For accurate measurements of R_1 peak or the R_1 to R_2 transition, cell size can affect the kinetics and must be considered, whereas R_2 relaxation kinetics is independent of cell size.[27] The patch-clamp technique[39] requires gigaohm seals to be formed, and then the WCR technique to be achieved, whereupon the charge motions of the ensemble of plasma membrane rhodopsin molecules activated by each flash can be measured as a capacitative current with respect to the membrane bilayer. We record data using the patch-clamp amplifier in coordination with an AT clone computer running pCLAMP (version 5.51) with a Scientific Solutions LabMaster interface card. While this is currently sufficient for our needs more advanced systems are commercially available. A schematic of the apparatus is shown in Fig. 3.

The inverted microscope (Nikon Diaphot) is housed in a grounded continuous aluminum Faraday cage coated on the inside with black felt to efficiently suppress environmental electrical noise and prevent stray light reflections (Fig. 3). The entire setup is housed in a darkroom with flat black wall painting and equipped with a darkroom far red light. Care was taken to block light emission from diodes on instrumentation. The microscope/table pair was mounted on an air isolation table to allow vibrationally stable recordings over hours. Final positioning of patch pipettes is controlled with

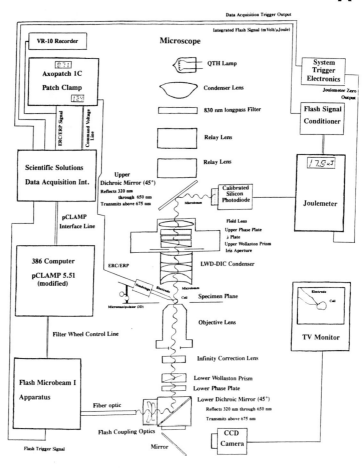
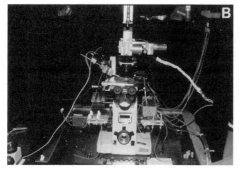

a fine hydraulic micromanipulator. The recording chamber was machined from Plexiglas and allows bulk bath perfusion from a peristalsis pump. We can locally perfuse cells with chromophores or other solutions using a pipette and a gravity-dependent delivery system. The tungsten light source for the microscope is remotely housed outside the Faraday cage, filtered through an infared long-pass filter (>830 nm), and relayed to the microscope condenser using lenses. A long working distance condenser is used with differential interference contrast optics. High-throughput objective lenses are used and a custom-designed dichroic mirror sits below the objective turret and allows reflective coupling of the flash output (near-UV–visible, tunable) from an orthogonally mounted fiber optic onto the optical axis while transmitting the infared light for visualizing the cells. A charge-coupled device camera is used to image the cells from the phototube output of the microscope and is transmitted to a TV monitor that sits above

FIG. 3. ERC recording apparatus and electro-optical delivery system. (A) Schematic of the electro-optical components of the ERC recording, optical delivery, and measurement system. Boxes indicate electrical units and lines indicate analog or digital connections. Optical components indicated are not drawn to scale. (B) Photograph of the microscope and attachments in the Faraday cage. The metallic vertical rail is used to mount the microscope light source on the top of the cage (not shown) and the two optical mounts attached to the rail contain the relay lenses (one shown). Between the lower relay lens and the long working distance differential interference contrast condenser is a metallic mount that contains the upper dichroic 45° mirror, which is positioned on or off the optical axis by an integrated motor and worm drive. To the right of the upper dichroic mirror is a mount (cylindrical) for the silicon photodiode, which measures light flashes with the joulemeter. The small box beneath the condenser is the CV-4 headstage, which converts picoampere signals to millivolts through a 500-MΩ resistor. The lower 45° dichroic mirror mounts on a manual positioning rod slide beneath the objective turret and contains additional optics to couple the flash to the objective (not seen). The fiber optic mount of the flash delivery system attaches directly to the lower aspect of the objective turret (not seen). The CCD camera is attached to the left lower camera port of the microscope. The set of tubes on a vertical rail at the upper right drapes an array of tubes to a box that contains microvalves that are digitally regulated to control microperfusion flow through a multibarrel pipette which is attached to a manipulator next to the recording chamber. (C) Dual-beam optical table in which a single halogen lamp (150 W), controlled by a tightly regulated constant current supply, illuminates two optical paths each of which has (in order) a neutral density wedge filter, coupling optics, a shutter, a monochromator, and coupling optics that image the output slits of the monochromator into the separate inputs of a dual leg/common arm glass fiber optic for transmission to the microscope. Pieces of paper sit just in front of the fiber optic chucks on the two legs of the fiber to show the colored beams emanating from the two monochromators. The common arm mounts in an adapter in the cylindrical pod in (B) and replaces the photodiode to project light through the condenser into the specimen plane. This table allows independent wavelength, intensity, and temporal control of two different stimuli. [(A) reprinted with permission from J. M. Sullivan, *Rev. Sci. Instrum.* **69,** 527 (1998). Copyright 1998 American Institute of Physics.]

the Faraday cage and is covered with a long-pass red filter with spectral transmission properties outside rhodopsin absorbance. Above the condenser of the microscope and below the last relay lens is positioned another dichroic mirror with matched properties to the one in the cube below the objective. The second dichroic allows the condenser to couple the flash microbeam spot in the specimen plane onto the detector surface of a photodiode for measurement by a joulemeter. The position of the photodiode can be replaced with a fiber optic originating from an optical table where two monochromators, neutral density wheels, and shutters allow dual-channel delivery of low-intensity step light stimuli that could be used to bias the pigment population into particular states (at lower temperatures) prior to giving the flash.

The flash apparatus is a custom device designed to deliver high-intensity (10^8 photon/μm^2), brief (≈ 14-μs) monochromatic flashes to the specimen plane of the microscope where the cells are parafocally imaged.[40] Spot sizes in the specimen plane are small and inversely related to the objective power. Bands of stimulation are controlled by an interference filter wheel. Flash light originates from a short-arc xenon flash tube and is collimated, filtered, and reimaged into a small core fiber optic that delivers light to the epifluorescent attachment of the microscope. Fiber optic delivery allows high-intensity flash delivery without polluting the recording apparatus with the electromagnetic discharge of the xenon tube and triggering elements. Flash stimulation and ERC recording are coordinated through additional homemade electronics that zero the joulemeter and trigger pCLAMP. pCLAMP then triggers the flash tube and activates data recording of the ERC through the interface board. A short-pulse, wavelength-tunable (dye) laser should offer equivalent performance in comparison to our apparatus.

Rhodopsin Early Receptor Currents Recorded in Rod Photoreceptors and Giant HEK293 Cells

A representative ERC recorded from a rod photoreceptor of tiger salamander under WCR conditions and bright flash photolysis[25] is shown in Fig. 4. Also shown are ERCs elicited by bright flash photolysis and measured in three fused giant HEK293S-WT cells of different size under WCR conditions. The giant cells were generated by fusion of a constitutively transfected wild-type line (WT-12). In both rod photoreceptors and giant HEK293-WT cells a definite inward R_1 current is noted immediately coincident with the flash. R_1 peaks within 1 ms. The outward R_2 current follows, peaks in several milliseconds, and is followed by a relaxation tail over 50–70 ms. The amount of charge (integral of the respective current waves) is much larger for the R_2 signals in both photoreceptors and the fused giant

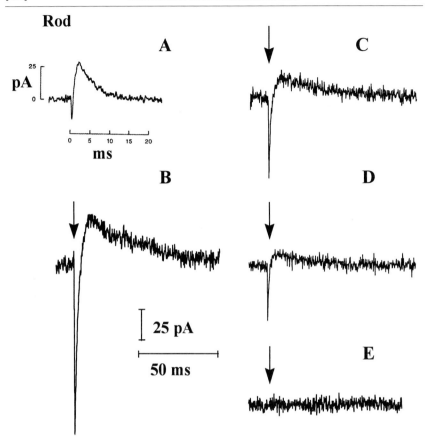

FIG. 4. Comparison of ERCs from amphibian rod photoreceptors and expressed wild-type human rod rhodopsin in HEK293 Cells. (A) Representative ERC measured under WCR from a red rod photoreceptor of *Ambystoma tigrinum* by flash photolysis at 500 nm at 6.5 × 10^7 photon μm^2 centered around 500 nm (30-nm bandpass) and with a 300-μs duration. Just before the flash there was 73% of the dark-adapted rhodopsin quantity. (B–D) ERCs recorded on the first flash from several fused giant cells expressing wild-type human opsin regenerated with 11cRet, The giant cell diameters and membrane capacitance were (B) 86 μm and 168 pF, (C) 52 μm and 70 pF, and (D) 41 μm and 47.5 pF. (E) Trace from a giant fused HEK293S cell (30 μm and 40 pF) that was generated by fusion of untransfected cells and regenerated with 11cRet. No ERC is elicited by flash photolysis. In (B–E) flashes were band limited (full-width half-maximum = 70 nm) centered around 500 nm with an intensity in the specimen plane of $\approx 4 \times 10^8$ photon/μm^2 over a flash duration of approximately 14 μs. [(A) reprinted with permission from C. L. Makino, W. R. Taylor, and D. A. Baylor, *J. Physiol.* **442**, 761 (1991). Copyright 1991 Physiological Society, London.]

cell expression system. The ERCs recorded from the three giant cells of different size occurred on the first flash following dark adaptation with 11cRet complexed to FAF-BSA. Both the R_1 and R_2 signals scale with the size of the cell. ERC signal amplitudes scale with the size of the giant cell because more plasma membrane opsin is available for regeneration in proportion to the number of cells participating in the fusion event. Signal levels are also dependent on the intensity of the flash at 500 nm, near the peak absorption wavelength of human rod rhodopsin ($\approx$493 nm), and the amount of unbleached rhodopsin remaining in the cell.[27] The similarities of the ERC waveforms and kinetics between native photoreceptors and the HEK293S-WT expression system are clear. This indicates that absorption of light by rhodopsin leads to charge motions that are similar in human and amphibian visual pigments that are A_1 and largely A_2 pigments, respectively.[25] Charge motions occur on different timescales during rhodopsin activation and are associated with conformational or chemical transitions that result in net redistribution of charge.

Also shown in Fig. 4 is a representative trace from a fused giant cell made from untransfected HEK293S cells and regenerated in the same way with 11cRet. No ERC results from flash photolysis. Thus, appropriately oriented rhodopsin (outside-out) must be present in the plasma membrane of the giant cell in order to record capacitative ERCs with the same polarity and waveform shape as are found in photoreceptors. ERCs from unfused single wild-type cells were smaller but qualitatively similiar to those recorded from photoreceptors (not shown).[27,40,42] Because giant cell size is controlled experimentally, ERC signals can be obtained that are much larger than those found in single rod photoreceptors that contain a finite quantity of plasma membrane rhodopsin ($\approx 10^7$ molecules).[25]

Electrical Structure–Function Approach to the ERC R_2 Signal

Large ERCs are generated on flash photolysis of giant cells expressing wild-type opsin regenerated with 11cRet. Consistency at recording ERCs from giant cells depends on the rhodopsin regeneration protocol with 11cRet complexed to albumin and efficient chromophore solubilization that aids prompt loading into intracellular membranes.[45] A surprising yet experimentally useful outcome of the regeneration protocol is that the ERCs in cells can be extinguished by successive flashes and, on dark adapatation without added chromophore, sufficient pigment regenerates to support ERC signals during secondary series of flash photolytic challenges using flashes in the visible band. Thus, primary chromophore loading into

[45] P. Shukla and J. M. Sullivan, *J. Gen. Physiol.* **114**, 609 (1999).

cells is sufficient to regenerate plasma membrane rhodopsin pigment after opsin is bleached. Cells can be cycled several times by simple dark adaptation. One difference with respect to the first flash series is that the R_1 signal is not seen on secondary flash series under the recording conditions stated (Figs. 5A and 5B). This presents an experimental convenience in that it allows the R_2 signal to be studied in isolation for straightforward comparison of wild-type with mutant R_2 signals, which occur on a millisecond timescale commensurate with meta-II formation.[45] The loss of the R_1 signal is under investigation. The recovery of ERC signals on dark adaptation depends on an intracellular source of chromophore that is established on primary regeneration.[45]

The R_2 phase of the ERC correlates with the timing of the meta-I $\rightleftharpoons$ meta-II equilibrium measured spectroscopically.[20] The later phases of rhodopsin activation have many features that indicate that electrostatic energy contributes to the conformation changes including proton exchange reactions and α-helical movements that constitute macrodipole transitions. Schiff base deprotonation initiates the sequence of molecular events that result in the transducin binding state of rhodopsin on a millisecond timescale. The later phases of these events include proton uptake into the cytoplasmic surface of rhodopsin. The highly conserved E134 side chain is critically involved in regulating the important configuration of the transducin docking domain. When this side chain is replaced in E134Q the apoprotein acquires greater background biochemical activity, the regenerated pigment demonstrates partial agonist activity in darkness consistent with incomplete movements of α helices, and during light activation spectral meta-II is formed and transducin is activated even more effectively than in wild-type protein although no proton uptake is observed.[12,46] Therefore, the E134Q pigment serves as a useful example to test whether the ERC approach can be used to study structure–function properties of the millisecond scale of activation consistent with the R_2 signal.

Figures 5C and 5D show ERCs recorded from a fused giant cell expressing E134Q. The first flash on the primary and secondary flash series is shown. We have been able to record large ERCs with giant cells made from the E134Q-12 cell line.[31] Curiously, an R_1 signal was always missing in E134Q on the first flash series even in cells that had peak R_2 amplitudes as large or larger than R_2 in wild-type giant cells that manifested large R_1 signals. The R_2 signal of E134Q is brief in comparison to wild-type R_2. Previous work indicates that at least two kinetic states contribute to the wild-type R_2 relaxation, whereas only a single state is necessary to explain

[46] J.-M. Kim, C. Altenbach, R. L. Thurmond, H. G. Khorana, and W. L. Hubbell, *Proc. Natl. Acad. Sci. U.S.A.* **94**, 14273 (1997).

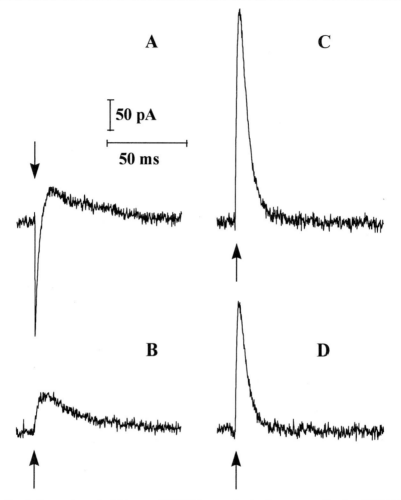

FIG. 5. Comparison of wild-type and E134Q human rhodopsins expressed in giant HEK293S cells. (A) ERC from a fused giant HEK293S cell expressing wild-type human rhodopsin regenerated with 11cRet. This ERC was elicited on the first flash of the first bleaching series and demonstrates both R_1 and R_2 signals. (B) ERC elicited from the same cell as in (A) on the first flash of the fifth secondary bleaching series lacks R_1 but a similar R_2 signal. Note the slow stretched relaxation of the R_2 current in the wild-type pigment. (C) ERC from a fused giant HEK293S cell expressing E134Q human rhodopsin regenerated with 11cRet (line E134Q-12). This ERC was elicited on the first flash of the first bleaching series and demonstrates only the R_2 signal. (D) ERC elicited on the first flash of the second secondary bleaching series in the same cell as in (C) also demonstrates only the R_2 signal. In E134Q the R_2 signals are large but brief in comparison to the wild-type signals. In (A) and (B) the cell diameter was 86 μm and the capacitance was 168 pF. In (C) and (D) the cell diameter was 81 μm and the capacitance was 113.7 pF. Flash strength was 4×10^8 photon/μm^2 (500 nm).

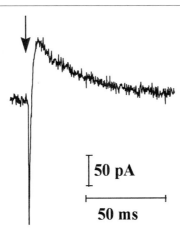

FIG. 6. Transiently transfected cells can have large ERC signals. This ERC trace was recorded from a giant cell that was fused by PEG on the fourth day following slow calcium phosphate transfection of single HEK293S cells with pCDNA3-WT plasmid. This ERC (first flash, initial series) was recorded on the fifth day following transfection. Both R_1 and R_2 signals are measurable and signal amplitudes are very high, even larger in comparison to fused giant cells made from stably transfected cells (e.g., see Fig 4B). Cell diameter was 80 μm and capacitance was >100 pF. Flash strength was 4×10^8 photon/μm^2 (500 nm).

the E134Q signal.[27,45] These findings provide evidence that the R_2 and perhaps R_1 signals can be investigated in expressed mutant visual pigments in relationship to the normal process of the charge motion in the wild-type protein.

Structure–function investigations of ERC charge motions in mutant pigments would be greatly assisted if ERC signals could be recorded from transiently transfected cells in addition to constitutively expressing stable cell lines. One could rapidly screen mutant pigment constructs in transiently transfected cells, whereas the development and characterization of stable cell lines requires significant time investment.[31] Transient transfection techniques are now being applied in ERC experiments. The slow calcium phosphate transient transfection technique was used to transfect pCDNA3-WT opsin into HEK293S cells. Cells were fused on the fourth day following transfection and ERC recordings were conducted on the fifth and sixth days. Figure 6 shows first flash ERC from a fused giant cell formed from transiently transfected cells. Both R_1 and R_2 signals were found with the same waveform as in fused giant cells made from the stable wild-type cell line. We found that about 10–20% of giant cells made from calcium phosphate transfected single cells had quantifiable ERC currents. That the frequency of ERCs appears to follow the transient transfection efficiency (see Fig. 1) suggests that individual transfected or untransfected individual cells grow to form local populations that are then fused by PEG. Also, cell fusion would act to suppress variance in expression levels in individual cells. ERC recording in transiently transfected cells is exciting and, if the success frequency can be increased, may become a useful approach.

Discussion

We developed methods to record high-amplitude ERC signals from cells expressing wild-type or mutant opsins regenerated with 11cRet, applying modern cellular electrophysiologic tools to the problem of biochemical activation of rhodopsin. Whole-cell recording allows explicit control of several sources of energy across the plasma membrane in which oriented rhodopsin molecules reside, including intracellular and extracellular pH, transmembrane voltage, surface potential, and temperature. The sensitivity of gigaohm seal WCR allows electrical measurements of rhodopsin activation in single fused giant cells containing on the order of one picogram of regenerated rhodopsin. Morever, ERC data are currently acquired with submillisecond temporal resolution, which allows investigation of charge motions in normal or mutant rhodopsins on a timescale faster than that of meta-II formation. The most sensitive time-resolved approach previously applied to study rhodopsin conformational activation to meta-II has been time-resolved electron spin resonance spectroscopy, which requires on the order of 100 μg of expressed and purified visual pigments to which spin probes are covalently attached at engineered cysteine residues.[13,14] The ERC approach improves on the sensitivity of spin-label time-resolved techniques by about 10^7-fold. ERC data can be acquired from an ensemble of rhodopsin molecules in a single giant cell in comparison to heterologously expressed visual pigment from more than twenty 15-cm plates of transfected COS cells. By applying extremely sensitive "patch-clamp" electrophysiologic methods developed by Hamill et al.[39] to the problem of rhodopsin activation, we have developed a tool that is likely to prove useful in furthering knowledge of conformational transitions during meta-II formation.

We have presented characteristics of a cellular system needed to study ERCs of expressed visual pigments. We chose the HEK293S cells over others (e.g., COS) for several reasons. First, HEK293S is a suspension-adapted line that can be used for both single-cell electrophysiology and large-scale growth in experiments requiring large quantities of rhodopsin. Second, unlike HEK293S cells constitutively expressing opsin, COS cells overexpress proteins to such high (toxic) levels that generation of stable cell lines is not feasible. Third, COS cells are rarely used to express ionic channels for single-cell electrophysiologic studies. While COS is an excellent biosynthetic protein factory, it is a poor physiologic environment because overexpression makes cell membranes leaky with increased noise and makes high-fidelity ERC measurements difficult. Transiently transfected COS cells or similar HEK293T cells (constitutive expression of simian virus 40 large T antigen) could be useful for ERC studies, but high-efficiency transfection methods and appropriate timing of experiments for healthy nonleaky cells would likely be necessary. We showed that ERC signals can be recorded from giant cells formed after calcium phosphate transfection of single cells. While it will be difficult to estimate how much opsin is contributing to the signal for thermodynamic analysis in transiently transfected versus stable cell lines, transient transfection could be very useful to rapidly screen new rhodopsin mutants for changes in ERC waveform or state kinetics and to identify which mutants warrant the investment needed to generate and quantify stable expression lines.

The instrumentation and expertise needed to conduct ERC experiments is common in cellular electrophysiologic laboratories, but is uncommon in biochemistry laboratories. Standard patch-clamp setups require extensive modifications to accomplish ERC recordings given the challenges of adding light control, delivery, and measurement to the apparatus as well as conducting experiments in darkness. Our previous work describes the accomplishment of a novel microsecond intense flash microbeam apparatus designed and fabricated for this work.[40] We presented the basic instrumental issues in this work but suitable reviews on low-noise patch-clamping techniques and similar gating current measurements are available.[44,47,48]

Although the expression-ERC technique is extremely sensitive, it has several limitations. First, the need for high-level expression to the HEK293 plasma membrane indicates that many mutant rod opsins made in the laboratory or that cause retinitis pigmentosa may not be investigated with this technique because they do not traffic well to the cell surface.[49,50] Cell fusion may allow recording from some of those mutants when at least a significant fraction of opsin partitions to the plasma membrane. Second, the plasma membrane opsin must be able to rapidly regenerate with 11cRet or another appropriate chromophore. Some mutant opsins might express to the plasma membrane but would be unable to form visual pigments with 11cRet (e.g., K296G),[51] and some mutant opsins might not regenerate with 11cRet rapidly (e.g., G121L).[52] Third, if stable cell lines are used, then the

---

[47] R. A. Levis and J. L. Rae, *Methods Enzymol.* **207,** 14 (1992).
[48] R. A. Levis and J. L. Rae, *Methods Enzymol.* **293,** 218 (1998).
[49] C.-H. Sung, B. G. Schneider, N. Agarwal, D. S. Papermaster, and J. Nathans, *Proc. Natl. Acad. Sci. U.S.A.* **88,** 8840 (1991).
[50] C.-H. Sung, C. M. Davenport, and J. Nathans, *J. Biol. Chem.* **268,** 26645 (1993).
[51] E. A. Zhukovsky, P. R. Robinson, and D. D. Oprian, *Science* **251,** 558 (1991).
[52] M. Han, S. W. Lin, S. O. Smith, and T. P. Sakmar, *J. Biol. Chem.* **271,** 32330 (1996).

high-level plasma membrane opsin expression should be associated with rapid cell growth. Mutant opsin clones that grow poorly, perhaps because of the nature of the mutation, would make these studies more difficult.

Flash photolysis of wild-type and E134Q pigments yields ERCs with very different waveforms and a distinct simplification in E134Q. These observations may be associated with known properties of the E134Q pigment studied by other methods.[12,46] We made quantitative measurements of ERC R_2 kinetics in E134Q and D83N mutant pigments with respect to wild-type.[45] The E134Q ERC is currently under extensive investigation.[53] The differences we observe to date in ERCs of rhodopsin mutants indicate that electrical studies of mutant pigments will likely contribute important structure–function knowledge on rhodopsin activation. An investigation of many rhodopsin mutants is anticipated. Electrical measurements have been essential in structure–function investigation of proton pumping in bacteriorhodopsin.[26,54]

Summary and Conclusions

The ERC is a conformation-dependent charge motion similar to the gating currents of ionic channels. Both the waveforms and bandwidth of ERCs and ionic channel gating currents are similar, providing support to the initial suggestion that the ERP was a kind of gating current.[55] In ionic channels the electrostatic field promotes motion of α-helical elements that stimulate large-scale molecular events that promote opening of the ionic pore. In ionic channels gating currents of expressed channel mutants has contributed significantly to understanding the mechanism of activation.[44] Given the known role of electrical processes to rhodopsin activation, the ERC approach applied to mutant and wild-type visual pigments is likely to lead to a fuller understanding of the mechanism of conformational activation. This method is currently well suited to investigate the later phases of rhodopsin activation that are thought to be electrostatic in nature.[7,9] We anticipate that ERC studies will make significant contributions to understanding how the breakdown of the electrostatic interaction between the PSB and its counterion is initiated and propagated to induce the proton uptake on the cytoplasmic surface of the pigment and the shaping of the transducin docking domain. We encourage collaboration to apply the ERC methodology to interesting mutant pigments and retinal analogs. We expect that the ERC methodology can soon be applied to understand rapid charge

[53] L. Brueggemann and J. Sullivan, work in progress.
[54] H. W. Trissl, *Photochem. Photobiol.* **51,** 793 (1990).
[55] H.-W. Trissl, A. Darszon, and M. Montal, *Proc. Natl. Acad. Sci. U.S.A.* **74,** 207 (1977).

displacements associated with photochemistry (i.e., R_1), the effects of transduction proteins on R_2, and the measurement of electrical processes during cone visual pigment activation.

Acknowledgments

Supported by the National Eye Institute (EY11384 to J.M.S.), a Career Development Award from the Foundation Fighting Blindness, the Department of Ophthalmology (SUNY Health Science Center), and Research to Prevent Blindness. We especially thank Dr. Paul Sieving (Department of Ophthalmology, University of Michigan) who provided laboratory space and resources that catalyzed the early stages of this project. We thank Dr. Hugh McDowell (University of Florida at Gainesville) who shared experience and protocols on fatty-acid-free bovine serum albumin regeneration of opsin and Dr. Michael Sheets (Northwestern University) who shared experience and protocols regarding ion channel gating current measurements in HEK293 cells. We thank Dr. Gary Yellen (Massachusetts General Hospital, Boston) for providing a protocol used to transfect HEK293 cells by electroporation. We thank Dr. Clint Makino (Harvard Medical School) for kindly providing the rod ERC trace shown in Fig. 4A. We thank Drs. Makino, Denis Baylor, and W. R. Taylor and the Physiological Society (London, UK) for providing permission to republish these data. We thank the American Institute of Physics for permission to republish the recording schematic shown in Fig. 3A. We thank Dr. Gary L. Trick for making the monochromators available to us. We thank Genentech (South San Francisco, CA) for providing the pCIS plasmid for our use, Dr. Bruce Stillman (Cold Spring Harbor Laboratory, NY) for providing the suspension-adapted HEK293S cell line, Dr. Jeremy Nathans (Johns Hopkins Medical Center, Baltimore, MD) for the wild-type human opsin cDNA and a wild-type human opsin HEK293S cell line, and Dr. Robert Molday (University of British Columbia, Victoria), for access to the 1D4 monoclonal antibody. We acknowledge the participation of Michael F. Satchwell in some of the immunocytochemistry experiments.

[20] Analysis of Amino Acid Residues in Rhodopsin and Cone Visual Pigments That Determine Their Molecular Properties

By HIROO IMAI, AKIHISA TERAKITA, and YOSHINORI SHICHIDA

Introduction

Most vertebrate retinas contain two types of photoreceptor cells, rods and cones, which are responsible for twilight (scotopic) and daylight (photopic) vision, respectively. Rods are more sensitive to light than cones, while cones display rapid photoresponse and rapid adaptation compared to rods.[1,2] Because both cells have signal transduction proteins whose functions are similar but whose amino acid sequences are different from each other, the difference in photoresponse patterns between rods and cones should originate from the different properties of these proteins. Thus investigations of the differences in molecular properties of these proteins between rods and cones are important for our understanding of the molecular mechanisms that underlie physiologic differences between rods and cones.

Among the signal transduction proteins, visual pigment receives a photon signal from the environment using a light-absorbing chromophore, 11-cis-retinal, and transfers the light signal to the retinal G protein, transducin, by binding to it and catalyzing the GDP-GTP exchange reaction on it.[3] Several lines of evidence have revealed that the pigment cycle initiated by light can be related to the physiologic response of photoreceptor cells. Thus an investigation of the differences in molecular properties between rod and cone visual pigments is one of the initial approaches for elucidating the differences in photoresponse patterns between rods and cones. In line with this, we have investigated the molecular properties of cone visual pigments from chicken retinas and compared them with those of the rod visual pigment rhodopsin. The results showed that cone visual pigments exhibit faster regeneration from 11-cis-retinal and opsin[4,5] and faster decay of the physiologically active intermediate (meta-II) than rhodopsin.[5-7] These results correlated well with the functional difference between rods and cones. Furthermore, we have identified the amino acid residue at position 122 (Fig. 1) that regulates not only the regeneration rate but also the thermal decay rate of meta-II.[8] Here we show the methods for the large-scale purification of cone visual pigments from chicken retinas and for the spectroscopic and biochemical analyses of the molecular properties of cone visual pigments, followed by identification of amino acid residues using modified proteins expressed in artificial systems.

Preparation of Chicken Rhodopsin, Cone Pigments, and Their Mutants

Buffers

Buffer A: 50 mM N-2-hydroxyethylpiperazine-N'-2-ethanesulfonic acid (HEPES), 0.6% (w/v) 3-[(3-cholamidopropyl)dimethylam-

---

[1] K.-W. Yau, *Invest. Opthalmol. Vis. Sci.* **35,** 9 (1994).
[2] D. Baylor, *Proc. Natl. Acad. Sci. U.S.A.* **93,** 560 (1996).
[3] Y. Shichida and H. Imai, *Cell Mol. Life Sci.* **54,** 1299 (1998).
[4] G. Wald, P. K. Brown, and P. H. Smith, *J. Gen. Physiol.* **38,** 623 (1955).
[5] Y. Shichida, H. Imai, Y. Imamoto, Y. Fukada, and T. Yoshizawa, *Biochemistry* **33,** 9040 (1994).

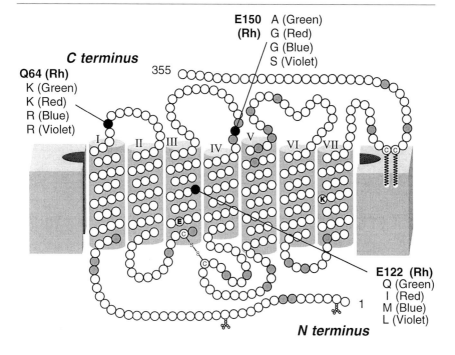

- ○ Each Cone Pigment ≠ Rhodopsin
- ● All Cone Pigments ≠ Rhodopsin

FIG. 1. Amino acid positions at which the residues are different in electrical properties between chicken rhodopsin and cone visual pigments. The transmembrane topography is based on the model of P. A. Hargrave, J. H. McDowell, D. R. Curtis, J. K. Wang, E. Juszczak, S.-L. Fong, J. K. M. Rao, and P. Argos, *Biophys. Struct. Mech.* **9,** 235 (1983). Based on the amino acid sequences of all the visual pigments investigated so far, the common property of cone visual pigments is that they have many basic amino acid residues, whereas rhodopsin has acidic residues. Amino acid positions expressed with white circles show that rhodopsin and four types of cone visual pigments have residues similar in electrical properties. The gray or black circles indicate the positions at which some of the cone pigments or almost all of the cone pigments have residues electrically different from those of rhodopsin, respectively. We, therefore, designed and expressed site-directed mutants of rod and cone pigments at three positions (64, 122, 150) shown by the black circles. The residues of rhodopsin replaced are denoted by single letters and numbered using the bovine rhodopsin numbering system. Corresponding residues of chicken cone pigments are also denoted. [Reproduced with permission from H. Imai, D. Kojima, T. Oura, S. Tachibanaki, A. Terakita, and Y. Shichida, *Proc. Natl. Acad. Sci. U.S.A.* **94,** 2322 (1997).]

monio]-1-propane sulfonate (CHAPS), 0.8 mg/ml L-α-phosphatidylcholine from egg yolk (PC), 140 mM NaCl, 1 mM MnCl$_2$, 1 mM CaCl$_2$

Buffer B: 50 mM HEPES, 0.6% CHAPS, 0.8 mg/ml PC, 10 mM NaCl, 1 mM MnCl$_2$, 1 mM CaCl$_2$

Buffer C: 50 mM HEPES, 0.6% CHAPS, 0.8 mg/ml PC, 20% (w/v) glycerol, 10 mM NaCl

Buffer D: 50 mM HEPES, 0.6% CHAPS, 0.8 mg/ml PC, 20% (w/v) glycerol, 140 mM NaCl

Buffer E: 50 mM HEPES, 0.75% CHAPS, 1 mg/ml PC, 140 mM NaCl

Buffer P: 50 mM HEPES, 140 mM NaCl

Buffer Pm: 50 mM HEPES, 140 mM NaCl, 3 mM MgCl$_2$

Each buffer is supplemented with 1 mM dithiothreitol (DTT), 0.1 mM phenylmethylsulfonyl fluoride (PMSF), 1 µg/ml aprotinin, 1 µg/ml leupeptin, adjusted to pH 6.5 at 4° with a small amount of NaOH solution (5 N) before use.

Preparation of Visual Pigments from Chicken Retinas

Rod and cone pigments are extracted from fresh chicken retinas by a mixture of CHAPS and PC and purified using affinity and ion-exchange chromatographies[7,9] (Fig. 2). The photoreceptor outer segments are isolated from about 2000 chicken retinas using a conventional sucrose flotation method, from which visual pigments are extracted with buffer E by homogenizing with a Teflon homogenizer (30 strokes). The concentrations of CHAPS and PC (0.75% and 1 mg/ml, respectively) are suitable for extraction of the visual pigments, but cone visual pigments are unstable in these conditions. Thus lowering the CHAPS and PC concentration of the extract to 0.6% and 0.8 mg/ml, respectively (buffer A), is necessary to stabilize the cone pigments after the extraction procedure.[9] It is possible to extract visual pigments with buffer P containing dodecylmaltoside (0.5 or 1%) as a detergent. However, about 70% of the red-sensitive cone visual pigments will be denatured during the extraction procedure, and most of the rhodopsin will be intact.[9a]

---

[6] Y. Shichida, T. Okada, H. Kandori, Y. Fukada, and T. Yoshizawa, *Biochemistry* **32**, 10832 (1993).

[7] H. Imai, Y. Imamoto, T. Yoshizawa, and Y. Shichida, *Biochemistry* **34**, 10525 (1995).

[8] H. Imai, D. Kojima, T. Oura, S. Tachibanaki, A. Terakita, and Y. Shichida, *Proc. Natl. Acad. Sci. U.S.A.* **94**, 2322 (1997).

[9] T. Okano, Y. Fukada, I. D. Artamonov, and T. Yoshizawa, *Biochemistry* **28**, 8848 (1989).

[9a] H. Imai and Y. Shichida, unpublished data (1998).

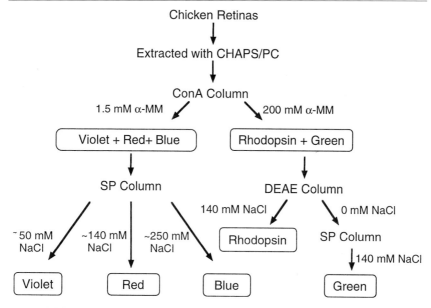

FIG. 2. Flowchart of the large-scale preparation procedure of native rod and cone pigments from chicken retina.

The extract is applied to a concanavalin A (ConA)-Sepharose (Pharmacia, Piscataway, NJ) column (16 × 300 mm) that has been equilibrated with buffer A. After washing with more than 8 bed volumes of buffer A to remove unbound materials including oil droplet components, a mixture of chicken violet, red, and blue is eluted with buffer B supplemented with 1.5 mM methyl-α-D-mannoside. The column is then washed with buffer B supplemented with 5 mM methyl-α-D-mannoside to remove a small amount of these pigments still bound to the column, and a mixture of chicken green and rhodopsin is eluted with buffer B supplemented with 200 mM methyl-α-D-mannoside. Glycerol is added to the eluates to make a final concentration of 20% (w/v) to stabilize the pigments.[9]

For further purification, the fractions containing chicken violet, red, and blue are applied to a sulfophenyl (SP)-Sepharose (Pharmacia) column (16 × 50 mm) that has been equilibrated with buffer C. Chicken violet, red, and blue are separated by using a linear gradient of NaCl (100–500 mM) in buffer A.[10] To separate chicken green from rhodopsin, the fraction containing chicken green and rhodopsin is applied to a diethylaminoethyl (DEAE)-Sepharose (Pharmacia) column (16 × 300 mm) that has been

[10] T. Okano, Thesis, Kyoto University, Kyoto, Japan (1993).

equilibrated with buffer C. Under our experimental conditions, chicken green passed through the column, but rhodopsin adsorbed on the column. The passed through fraction containing chicken green is further applied to an SP-Sepharose column (16 × 50 mm), from which chicken green is eluted with buffer D. Finally, each fraction is dialyzed against buffer D to adjust the buffer condition. Dialysis is inevitable, especially when the ion-exchange column is used as the final step of purification. In our hands, about 50, 30, 3, 3, and 1 mg of purified rhodopsin, chicken red, green, blue, and violet are obtained from 2000 chicken retinas.

Opsin (protein moiety of visual pigment) is prepared as follows[11]: The pigment is first adsorbed on the Con A-Sepharose column, followed by washing with buffer A supplemented with 50 mM hydroxylamine. Then the pigment is irradiated with a yellow light (>480 nm) and the column immediately washed with 10 bed volumes of buffer A to remove retinal oxime produced from photoisomerized retinal chromophore and hydroxylamine and unreacted hydroxylamine. Then the opsin is eluted with buffer D supplemented with 200 mM methyl-α-D-mannoside. Because cone opsins are much more unstable than the opsin of rhodopsin, the irradiation should be performed after adsorbing the pigments on the column.

Expression of Rod and Cone Pigments and Their Mutants

The coding regions of chicken rhodopsin and cone pigments are derived from a cDNA clone[12] and attached with the *Hin*dIII and *Eco*RI sites to the 5' and 3' ends respectively, then subcloned into pBlueScript-II-KS$^+$ (Stratagene, La Jolla, CA). Point mutations are introduced using the kits Sculptor TM (Amersham Pharmacia) or Quickchange (Stratagene). All coding regions are sequenced by the dideoxy termination method. Each of the fragments is recloned into the expression vector pUSRα.[13] Plasmids are purified by means of a large-scale CsCl gradient or ion-exchange column (Concert, GIBCO BRL) and transfected into 293S cells[14] using the calcium phosphate method.[15] Wild-type and mutant rhodopsins are harvested 72 hr after the transfection, whereas cone pigments are harvested 48 hr after the transfection because the amounts of expressed pigments are reduced during incubation from 48 to 72 hr. After harvesting, the cell mem-

[11] Y. Fukada, T. Okano, Y. Shichida, T. Yoshizawa, A. Trehan, D. Mead, M. Denny, A. E. Asato, and R. S. H. Liu, *Biochemistry* **29,** 3133 (1990).
[12] T. Okano, D. Kojima, Y. Fukada, Y. Shichida, and T. Yoshizawa, *Proc. Natl. Acad. Sci. U.S.A.* **89,** 5932 (1992).
[13] S. Kayada, O. Hisatomi, and F. Tokunaga, *Comp. Biochem. Physiol.* **110,** 599 (1995).
[14] J. Nathans, *Biochemistry* **29,** 937 (1990).
[15] C. M. Gorman, D. R. Gies, and G. McCray, *DNA Prot. Eng. Techniques* **2,** 3 (1990).

brane is isolated by means of sucrose flotation methods. If necessary, 11-*cis*-retinal is added to the opsin solution and incubated >3 hr at 4°. Then the regenerated pigments are extracted with buffer E. Because the CHAPS contained in buffer E is one of the detergents whose extraction efficiency is low, repeated extraction is recommended. If the extraction of cone pigments is performed without addition of 11-*cis*-retinal, the amount of acquired pigments is significantly reduced (~30%). As in the case of native pigments, extraction with dodecylmaltoside results in denaturing of the red-sensitive cone visual pigments by more than 80%.

Spectroscopy

Absorption spectra of visual pigments are recorded on a recording spectrophotometer (MPS-2000, Shimadzu, Kyoto, Japan) interfaced to a personal computer (PC9801RA NEC, Tokyo, Japan). The spectrophotometer used in the experiments is one of the typical spectrophotometers having probe lights that are very weak, so that little bleaching of the visual pigment in the sample induced by the probe light occurs. In fact, our experiments indicate that less than 1.5% of bleaching is observed after 100 recordings of the spectra at wavelengths from 750 to 330 nm. When necessary, the sample is irradiated with light from a 1-kW tungsten–halogen lamp (Philips, Tokyo, Japan) through a specially designed optical setup including shutters and a movable mirror operated by a rotary solenoid,[16] by which one can irradiate the sample without transferring the optical cell containing the sample from the sample compartment of the spectrophotometer. This kind of setup is necessary for recording spectral changes of less than 0.01 absorbance unit without disturbing the baseline. Other special methods for recording spectra are described in the following sections.

Assay for Determination of Regeneration Rates of Pigments

Regeneration processes of visual pigments can be monitored by the change in the absorption spectra after the mixing of opsin and 11-*cis*-retinal. 11-*cis*-retinal is prepared by irradiation of all-*trans*-retinal (Sigma, St. Louis, MO) dissolved in acetonitrile with light from a fluorescent lamp (20 W) for 16 hr, purified by means of high-performance liquid chromatography (HPLC)[17] and dissolved in ethanol. A sample cell holder connected to a thermostatic circulator (RTE-220, Neslab, Tokyo, Japan) is installed in the

[16] Y. Shichida, S. Tachibanaki, H. Imai, and A. Terakita, *Methods Enzymol.* **315** [23], 1999 (this volume).
[17] A. Maeda, Y. Shichida, and T. Yoshizawa, *J. Biochem.* **83**, 661 (1978).

sample compartment of the spectrophotometer to keep the sample at a constant temperature. Nitrogen gas is sprayed on the surface of the sample cell to inhibit the surface from clouding. The retinal solution is added to the opsin solution in the sample cell holder followed by incubation at this temperature until the changes in absorbance due to the regeneration of the pigment are saturated. Compared to the amount of opsin, a >10 times molar excess of retinal is necessary to make the reaction as a quasi-first-order reaction. The wavelength for monitoring should be >530 nm because of the formation of artificial retinylidene Schiff bases, which are formed randomly with contaminating protein or phospholipids.[18] In wild-type rhodopsin, regeneration occurs in a few minutes under the conditions and can be monitored with the wavelength scanning mode of spectrophotometers (e.g., MPS-2000, Shimadzu). On the other hand, regeneration of cone pigments and rhodopsin mutants mimicking cone pigments occurs much (tens to hundreds of times) faster than wild-type rhodopsin; therefore, some adaptation of the experimental conditions are required.

First, the time scan mode of the spectrophotometer is useful for monitoring these fast reactions. If one records the absorbance change before and after the addition of retinal, one can monitor the regeneration processes with a time resolution of seconds. Second, the temperature can be varied; the lower the temperature of the sample, the slower the regeneration reaction. The regeneration reaction of rhodopsin mutants E122Q and E122I could be monitored by time scan mode at 2°. Figure 3A shows the regeneration processes of wild-type, E122Q, and E122I mutants of chicken rhodopsin monitored by the change in absorbance at 530 nm. It is clearly shown that rhodopsin mutants E122Q and E122I regenerate much faster than wild-type rhodopsin, whereas the other mutants derived from the amino acid replacements at positions 64 and 150 regenerate with rates similar to that of wild-type rhodopsin.[8] In the case of the quasi-first-order reaction, the experimental data should be simulated by a single exponential curve. The change in absorbance at 530 nm of the wild-type, E122Q, and E122I mutants of chicken rhodopsin fit exponential curves with time constants of 26 (wild-type), 1.3 (E122Q), and 0.94 (E122I) min.

Although the regeneration processes of rhodopsin mutants can be monitored under the condition of the quasi-first-order reaction, the processes of native and wild-type cone pigments are one order faster and thus cannot be monitored under this condition. Therefore, concentrations of the retinal and opsin should be reduced and equimolar portions of retinal and opsin should be mixed. Under this condition, the course of regeneration, that is, the two-molecule reaction, should be simulated by a hyperbolic curve.

[18] H. Matsumoto, K. Horiuchi, and T. Yoshizawa, *Biochim. Biophys. Acta* **501,** 257 (1978).

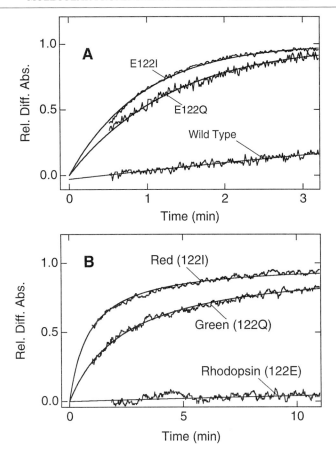

FIG. 3. Regeneration processes of rhodopsin, cone pigments, and rhodopsin mutants. (A) Regeneration of wild-type, E122Q, and E122I rhodopsins monitored by the change in absorbance at 530 nm. 11-*cis*-Retinal solution (500 μM) in the ethanol (5 μl) was added to the respective opsin solution (220 μl) at 2°, and the increase in absorbance at 530 nm due to the regeneration of pigment was recorded. In the panel, the maximal absorbance due to the full regeneration is normalized. Solid curves are the fitted single exponential curves with time constants of 26 (wild-type), 1.3 (E122Q), and 0.94 (E122I) min. [Reproduced with permission from H. Imai, D. Kojima, T. Oura, S. Tachibanaki, A. Terakita, and Y. Shichida, *Proc. Natl. Acad. Sci. U.S.A.* **94,** 2322 (1997).] (B) Time courses of regeneration of chicken green, red, and rhodopsin. 11-*cis*-Retinal dissolved in 5 ml ethanol (7.5 mM) was added to 250 ml of each opsin solution (150 nM) at 2°, followed by incubation at this temperature until the changes in absorbance due to the regeneration of the pigments were saturated. Wavelengths for monitoring are 530 nm in chicken green and rhodopsin, and 570 nm in chicken red. Solid curves are the fitted hyperbolic curves with time constants of 2.5 min, 1 min, and 4 hr for chicken green, red, and rhodopsin, respectively. [Reproduced with permission from Y. Shichida, H. Imai, Y. Imamoto, Y. Fukada, and T. Yoshizawa, *Biochemistry* **33,** 9040 (1994).]

Figure 3B shows the regeneration processes of chicken green, red, and rhodopsin after addition of 11-*cis*-retinal to the respective opsin solutions at 2°. The changes in absorbance fit the hyperbolic curves (solid lines) with time constants of 2.5 min, 1 min, and 4 hr for chicken green, red, and rhodopsin, respectively. The combined results clearly show that regeneration of the cone pigments occurs hundreds of times faster than that of rhodopsin, and the replacement of an amino acid residue at position 122 changes the regeneration rate of rhodopsin closer to those of the cone pigments. These results suggest that the amino acid residue is one of the major determinants of the regeneration rate.

Analysis of Thermal Reaction of Meta Intermediates

To monitor the thermal behavior of spectrally distinctive intermediates of rod and cone pigments, we have developed time-resolved low-temperature spectroscopy in which the spectral change was recorded for a long time from a single pigment sample at a continuous low temperature.[19] In addition, we have developed a biochemical assay in which the thermal reaction of the active state for transducin could be monitored.[20] In this section, we introduce the process in which the differences in thermal reaction of the intermediates between rod and cone pigments have been found by a combination of these methods.

Thermal Behavior of Meta Intermediates Monitored by Spectroscopic Analysis

As already described, some spectrophotometers (e.g., MPS-2000, Shimadzu) use a very weak probe light that causes little bleaching of the visual pigment during scanning. Using this type of spectrophotometer, one can record the spectral change in a visual pigment sample again and again after irradiation in order to monitor the thermal behavior of the intermediates. One disadvantage experienced with this is that type of spectrophotometer recording the spectra is slow (measured in minutes). However, most chemical reactions slow down as the temperature decreases. Thus, one can obtain spectral data even when the reaction is too fast to record at room temperature, if the temperature of the sample is lowered.[21] An advantage is that a large amount of spectral data can be obtained with a good signal-to-noise ratio compared to a flash photolysis experiment. For example,

---

[19] H. Imai, T. Mizukami, Y. Imamoto, and Y. Shichida, *Biochemistry* **33**, 14351 (1994).
[20] H. Imai, A. Terakita, S. Tachibanaki, Y. Imamoto, T. Yoshizawa, and Y. Shichida, *Biochemistry* **36**, 12773 (1997).
[21] T. Yoshizawa and Y. Shichida, *Methods Enzymol.* **81**, 333 (1982).

>100 spectral data can be obtained from a 4- to 50-μg single visual pigment sample without concentration.

To cool the sample in the sample compartment of the spectrophotometer, some special equipment is required. For experimental conditions above 0°, the measurement system is identical to that used in monitoring the regeneration rates. For experimental conditions below 0°, an optical cryostat (e.g., CF-1204, Oxford Instruments, Oxford, England) should be used. The sample temperature is regulated to within 0.1° by a temperature controller (ITC-4, Oxford) attached to the cryostat. Regarding the sample conditions, more than an equal volume of glycerol or an equivalent material should be added to the sample to inhibit freezing under temperature conditions above $-40°$. More than twice the volume of glycerol should be added to the sample to inhibit freezing under the temperature conditions below $-80°$. The wavelength of the irradiated light was selected with a glass cutoff filter or an interference filter.

Under a typical condition, the amount of rhodopsin in the sample bleached by the 30-sec irradiation (>570 nm) was about 50%, which can be estimated by complete bleaching after each experiment.[19] This fact suggests that multiple photoreactions occur during the irradiation. To confirm that low-temperature irradiation did not form retinal isomers other than all-*trans*, 11-*cis*, and 9-*cis* forms, the chromophores should be extracted from the irradiated sample after the experiments, and their isomeric composition should be analyzed by an HPLC (YMC-A0123, Yamamura, Tokyo, Japan) method.[22]

As an example of differences in the thermal reaction of meta intermediate of rod and cone pigments, we present spectral data on the thermal behavior of the intermediates of chicken green and rhodopsin after irradiation of these pigments at temperatures of $-20°$ and $-10°$ (Fig. 4). After the irradiation of a rhodopsin sample with orange light (>570 nm) for 30 sec at $-20°$ (Fig. 4A, top), an increase in absorbance at about 380 nm and a decrease at about 500 nm were observed. Because the absorption maxima of lumi meta-I, meta-II, and meta-III intermediates are located ~500, ~480, ~380, ~480 nm, respectively, this reaction represents the conversion from a thermal equilibrium state between lumirhodopsin and meta-I to meta-II.[19] The conversion from meta-II to meta-III (Fig. 4A, bottom, and 4B, top) and from meta-III to all-*trans*-retinal plus opsin (Fig. 4B, bottom) was then observed.

The experimental results for chicken green (Figs. 4C and 4D) are somewhat different from those of rhodopsin. First, at $-20°$, the early spectral change (Fig. 4C, top) was similar to that in rhodopsin (Fig. 4A). It was followed by a broad absorbance decrease (Fig. 4C, bottom). The absorbance

---

[22] Y. Imamoto, T. Yoshizawa, and Y. Shichida, *Biochemistry* **35**, 14599 (1996).

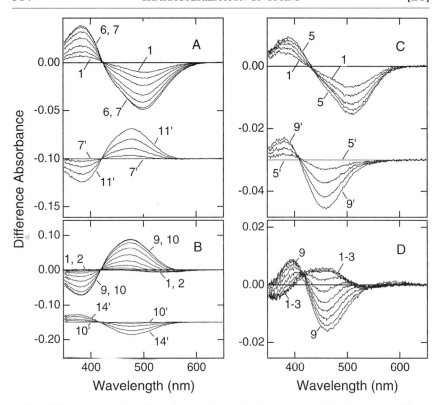

FIG. 4. Thermal reactions of meta intermediates of chicken green and rhodopsin at −20° and −10° monitored by low-temperature time-resolved spectroscopy. The pigment/56% glycerol mixture was cooled to each temperature and irradiated with orange light (>570 nm) for 30 sec. Absorption spectra were recorded immediately (0 min), 2, 4, 8, 16, 32, 64, 128, 256, 512, 1024, 2048, 4096, 8192, and 15,424 min after the irradiation. (A) *Top:* Difference spectra calculated by subtracting the spectrum recorded immediately after irradiation from those recorded at 2–128 min after irradiation of the rhodopsin sample at −20° (curves 1–7). *Bottom:* Difference spectra calculated by subtracting the spectrum recorded at 128 min after irradiation from those recorded at 128–2048 min after irradiation (curves 7'–11'). (B) *Top:* Difference spectra calculated by subtracting the spectrum recorded immediately after irradiation from those recorded at 2–1024 min after irradiation of the rhodopsin sample at −10° (curves 1–10). *Bottom:* Difference spectra calculated by subtracting the spectrum recorded at 1024 min after irradiation from those recorded at 1024–15,424 min after irradiation of the rhodopsin sample at −10° (curves 10'–14'). (C) *Top:* Difference spectra calculated by subtracting the spectrum recorded immediately after irradiation from those recorded at 2–32 min after irradiation of the chicken green sample at −20° (curves 1–5). *Bottom:* Difference spectra calculated by subtracting the spectrum recorded at 32 min after irradiation from those recorded at 32–512 min after irradiation (curves 5'–9'). (D) Difference spectra calculated by subtracting the spectrum recorded immediately after irradiation from those recorded at 2–512 min after irradiation of the chicken green sample at −10° (curves 1–9). [Reproduced with permission from H. Imai, Y. Imamoto, T. Yoshizawa, and Y. Shichida, *Biochemistry* **34**, 10525 (1995).]

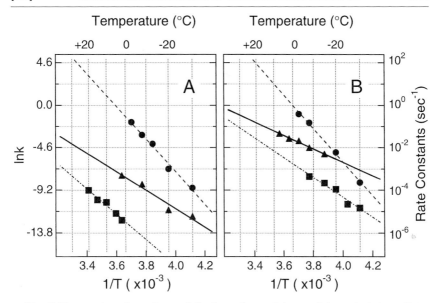

FIG. 5. Temperature dependence of the formation and decay of the meta intermediates of (A) chicken rhodopsin and (B) green. Apparent rate constants of meta-II formation, meta-II decay, and meta-III decay (circles, triangles, and squares, respectively) are plotted in an Arrhenius manner, and the experimental points are fitted with straight lines (dashed, solid, and dashed-dotted lines, respectively) by the least squares method. [Reproduced with permission from H. Imai, Y. Imamoto, T. Yoshizawa, and Y. Shichida, *Biochemistry* **34**, 10525 (1995).]

at about 380 nm increased in this phase, and no spectral changes were observed when the sample was further warmed to 20°; therefore, this phase would be due to the dissociation process of the intermediate into all-*trans*-retinal and opsin. In addition, these spectral changes were similar in shape to that in the decay process of meta-III of rhodopsin (Fig. 4B, bottom). Therefore, the phase can be assigned to the decay of meta-III of chicken green. Although the formation process of meta-III was not observed at −20°, it was clearly observed at −10° (Fig. 4D, curves 1–3) and −15° (data not shown). Thus it is suggested that the formation time constant of meta-III becomes similar to that of meta-II when the temperature is lowered to hide the apparent spectral change representing the meta-II to meta-III transition. In fact, the temperature-dependence slopes of the rates of these reactions are different in the Arrhenius plots, whereas those of the corresponding processes are similar between rhodopsin and green (Fig. 5). In addition, a singular value decomposition (SVD) analysis followed by global fitting analysis[23] clearly separates the spectral changes in the chicken green

[23] T. Mizukami, Thesis, Kyoto University, Kyoto, Japan (1994).

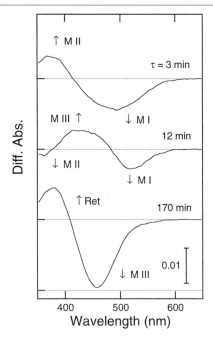

FIG. 6. The *b*-spectra representing components of the spectral change of the chicken green sample at −20° (Fig. 4C).

sample at −20°. Namely, the spectral changes are separated as three *b*-spectra with time constants of 3, 12, and 170 min, respectively. Strictly speaking, the *b*-spectrum exhibits a spectral component with certain time constants calculated from the initial spectrum as a baseline and it does not exhibit the spectral change between two specific intermediates. However, the similar spectra between meta-I and -III and those between meta-II and all-*trans*-retinal, in addition to the considerable differences among three time constants, enable us to infer that the spectral changes observed in chicken green are separated as the conversion from meta-I to meta-II (Fig. 6, top), meta-II to meta-III (Fig. 6, middle), and the decay of meta-III (Fig. 6, bottom). For a detailed analysis of the spectral changes by means of SVD and global fitting, the reader is referred to other papers.[16,24]

When the absorbance changes of 380 nm at −10° were plotted, it was clearly shown that the decay rate of meta-II of green was faster than that

[24] E. T. Thorgeirsson, J. W. Lewis, S. T. Wallace-Williams, and D. S. Kliger, *Biochemistry* **32**, 13861 (1993).

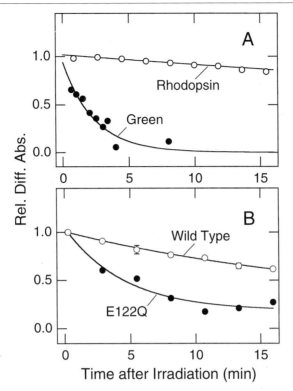

FIG. 7. Meta-II decay rates of chicken green and rhodopsin and mutant rhodopsin. (A) Absorbances at 380 nm of the native sample at $-8°$ are plotted against the incubation time after irradiation. Solid curves are the fitted single exponential curves with the time constants of 88 (rhodopsin) and 1.7 (chicken green) min. (B) Absorbances at 380 nm of the wild-type and E122Q mutant rhodopsin samples at $2°$ are plotted against the incubation time after irradiation. Solid curves are the fitted single exponential curves with time constants of 160 (wild-type) and 3.3 (E122Q) min.

of the corresponding process of rhodopsin (Fig. 7A). Because meta-II of rhodopsin is the intermediate activating transducin,[25–27] the difference in the lifetime of the intermediate is important for estimating the efficiency in photosignal transduction in rod and cone photoreceptor cells. Thus which amino acid residue(s) are responsible for the difference in the lifetime of meta-II intermediates is of interest. As described earlier, the amino acid residue at position 122 is one of the regulators of the regeneration rate of

[25] Y. Fukada and T. Yoshizawa, *Biochim. Biophys. Acta* **675**, 195 (1981).
[26] N. Bennet, M. Michel-Villaz, and H. Kuhn, *Eur. J. Biochem.* **127**, 97 (1982).
[27] J. Kibelbek, D. C. Mitchell, J. M. Beach, and B. J. Litman, *Biochemistry* **30**, 6761 (1991).

the pigments. As in the case of the regeneration reaction, interestingly, replacement of the residue caused acceleration of meta-II decay (Fig. 7B), whereas the replacement of the other residues showed little effect.[8] Thus, it is demonstrated that this residue is one of the major regulators of the meta-II decay rate.

Thermal Behavior of Meta Intermediates Monitored by Biochemical Assay

Because it is well known that meta-II of rhodopsin and chicken red activate the retinal G protein, transducin,[25-28] the decay process of the meta intermediates should be monitored by biochemical assays such as GTPase activity and GTPγS binding of transducin. For this purpose, the purified pigments were incorporated into PC liposome by dialysis at 2° for 18 hr with five buffer changes with 300-fold volumes of buffer Pm. Transducin was prepared from the fresh bovine retina by the method previously reported.[29]

We measured the amount of transducin activated by the irradiated pigments as the GTPase activities of transducin (Fig. 8). The pigment suspension was irradiated with orange light (blue, >520 nm; rhodopsin and green, >570 nm; red, >640 nm) for 30 sec at 2°. After the indicated preincubation time in the dark at this temperature, 20-μl aliquots were removed and transferred to the reaction solution (50 μl; final concentration is 0.1 μM pigments, 1.5 μM transducin, 2.4 μM [γ-^{32}P]GTP precooled at 2°). The mixture was then incubated at 25° for 2 min for the GTPase reaction, and the reaction was stopped with EDTA solution. The released P_i was collected by addition of activated charcoal and assayed by a liquid scintillation counter (LS6000IC, Beckman, Palo Alto, CA). The activity of the pigment without irradiation was measured and subtracted as background. Aliquots of irradiated pigments (100 μl) were completely bleached with yellow light (>480 nm) in the presence of hydroxylamine (10 mM) to estimate the amounts of bleached pigments during the assay. The average and standard deviations were calculated for each point from at least three independent experiments. The activity immediately after the irradiation can be measured from the sample mixed before the irradiation and was confirmed to be almost identical to that of pigment added to the reaction mixture after the irradiation. The linear relationship between the active pigment concentration in the sample and GTPase activity should be confirmed.

Figure 9 shows the experimental result of the assay of rod and cone pigments (Fig. 9A) and their mutants (Fig. 9B). When the pigments were

[28] T. Okada, T. Matsuda, H. Kandori, Y. Fukada, T. Yoshizawa, and Y. Shichida, *Biochemistry* **33**, 4940 (1994).
[29] Y. Fukada, H. Ohguro, T. Saito, T. Yoshizawa, and T. Akimo, *J. Biol. Chem.* **264**, 5937 (1989).

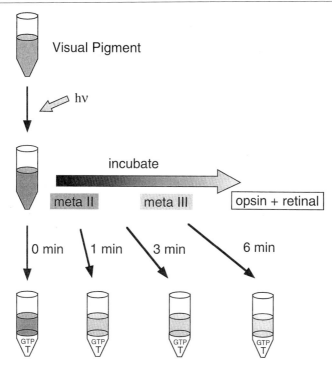

Fig. 8. Schematic drawing of the assay for monitoring the thermal reaction of the active state for transducin.

irradiated in the presence of transducin and GTP, the differences in the activities induced by the pigments were within 10% of each other. Therefore, one can conclude that the catalytic turnover rates in transducin activation were similar in these pigments, which is consistent with the previous study using chicken red.[30] On the other hand, the activities induced by the irradiated cone pigments were greatly reduced as the preincubation time increased, while that by irradiated rhodopsin was not. These results demonstrated that decay of the active state of cone pigment is much faster than that of rhodopsin. In addition, a close relationship is strongly suggested between the decay rate of the meta intermediates monitored by spectroscopic measurement (Fig. 7) and the decay rate of the active state for transducin (Fig. 9). This fact gives a clue to the identity of the intermediate(s) that activate transducin.

The decay kinetics of the active state of cone pigments monitored by the GTPase activity could be separated into two components. The fast compo-

[30] Y. Fukada, T. Okano, I. D. Artamonov, and T. Yoshizawa, *FEBS Lett.* **246,** 69 (1989).

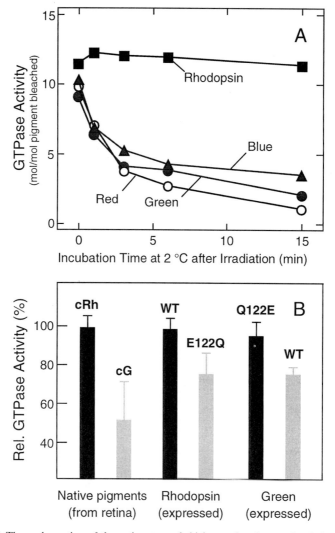

FIG. 9. Thermal reaction of the active state of chicken rod and cone visual pigments and their mutants monitored by GTPase activity assay. (A) Thermal decay of the active state of chicken rod and cone visual pigments monitored by time-resolved GTPase assay. The solutions of chicken rhodopsin (square), red (open circle), green (closed circle), and blue (triangle) were irradiated at 2° with orange light for 30 sec. [Reproduced with permission from H. Imai, A. Terakita, S. Tachibanaki, Y. Imamoto, T. Yoshizawa, and Y. Shichida, *Biochemistry* **36**, 12773 (1997).] (B) Change in the activation of transducin by native, wild-type, and mutant rhodopsin and chicken green. The pigments 6 min after irradiation were added to the reaction mixture containing transducin, and the extent of GTPase activity was measured. The activities relative to those immediately after irradiation are plotted. The standard deviations are estimated from four independent experiments using different preparations. [Reproduced with permission from H. Imai, D. Kojima, T. Oura, S. Tachibanaki, A. Terakita, and Y. Shichida, *Proc. Natl. Acad. Sci. U.S.A.* **94**, 2322 (1977).]

nent is largely complete after a 3-min incubation, whereas the slow component still remains after a 15-min incubation. The spectroscopic measurements revealed that the meta-II to meta-III transition occurs within several minutes, followed by the decay of the equilibrium state between meta-II and -III (Fig. 4). Note that the decay kinetics of meta-II and -III in the detergent solution present in this study are almost identical with those observed in PC liposome (data not shown). Therefore, the fast component observed by the biochemical method reflects the conversion process of meta-II to the equilibrium state between meta-II and -III. The slow component reflects the decay process of the equilibrium state between meta-II and -III to all-*trans*-retinal and opsin. Thus one can conclude that the intermediate that activates transducin should be meta-II even in cone pigments.

It has been shown that the decay rate of meta-II is regulated by the residue at position 122; therefore, we examined the thermal decay rate of the active state of mutants of rhodopsin and cone pigments by biochemical assay. As in the case of native pigments, the absolute GTPase activities were similar when both pigments were irradiated in the presence of transducin and GTP, suggesting that the catalytic turnover rates to transducin activation are not changed by the mutation. However, the thermal reactions of the active state for transducin were changed by the mutation. In Fig. 9B, the activities of transducin that were induced by addition of pigments at a preincubation time of 6 min after the irradiation relative to those just after irradiation are compared. Although the activity of wild-type rhodopsin hardly changed, that of rhodopsin mutant E122Q, which mimics chicken green, was significantly reduced. Furthermore, the reverse mutant Q122E of chicken green reverses the decrease in transducin activity shown by the wild-type chicken green. Therefore, it has been shown that the amino acid residue at position 122 really changes the thermal reaction of active state to transducin as predicted by the thermal reaction of the meta-II intermediate monitored by spectroscopic analysis.

Conclusions

We have described methods for the analysis of molecular properties of cone visual pigments that are different from those of rhodopsin and the process of analysis of the amino acid residue responsible for the difference. It is of interest that the single amino acid residue at position 122 regulates two independent properties of the pigments such as regeneration rate and decay rate of meta-II. However, we could not accomplish full conversion of the rod and cone pigments, possibly because of the presence of other amino acid residue(s) in addition to this residue. In addition, other molecular properties like sensitivity for hydroxylamine and thermal stability of the parent pigment and opsin were not changed by the mutation of the residue. It would be important to identify all amino acid residues responsible for these molecular properties.

Acknowledgments

We thank Prof. J. Nathams for his gift of 293S cell line, and Prof. A. Tokunaga for providing a pUSRα expression vector and experimental guidance. This work was supported in part by Grants-in-Aid for Scientific Research from the Japanese Ministry of Education, Science, Sports and Culture.

[21] Phylogenetic Analysis and Experimental Approaches to Study Color Vision in Vertebrates

By SHOZO YOKOYAMA

Introduction

Visual pigments in the retinas are classified into five evolutionarily distinct groups: (1) rhodopsin (RH1); (2) RH1-like (RH2); (3) short-wavelength-sensitive (SWS1); (4) SWS1-like (SWS2); and (5) long-wavelength-sensitive or middle-wavelength-sensitive (LWS/MWS) pigment clusters.[1-5] The wavelengths of maximal absorption (λ_{max}) of RH1, RH2, SWS1, SWS2, and LWS/MWS pigments are given by 490–500, 470–510, 360–420, 430–460, and 510–570 nm, respectively.[4,6] Visual pigments also exist in the pineal glands and other tissues and are evolutionarily related to the retinal pigments.[4] The six groups of pigments have been generated by at least five gene duplication events, followed by nucleotide substitutions in the coding regions.[4]

The molecular mechanisms of color vision in vertebrates may be elucidated in three steps: (1) cloning and characterization of the retinal opsin genes, (2) identification of amino acid changes that are potentially important in changing the λ_{max} of the pigments, and (3) determination of the actual effects of these mutations identified in the second step on the shifts in the

---

[1] S. Yokoyama, *Mol. Biol. Evol.* **11,** 32 (1994).
[2] S. Yokoyama, *Mol. Biol. Evol.* **12,** 53 (1995).
[3] S. Yokoyama, *Genes Cells* **1,** 784 (1996).
[4] S. Yokoyama, *Annu. Rev. Genet.* **31,** 3150 (1997).
[5] S. Yokoyama and R. Yokoyama, *Annu. Rev. Ecol. Syst.* **27,** 543 (1996).
[6] S. Yokoyama and F. B. Radlwimmer, *Mol. Biol. Evol.* **15,** 560 (1998).

λ_{max} of visual pigments. The cloning and molecular characterization of the retinal and nonretinal opsin genes can be accomplished using standard recombinant DNA methods.[7-12] The purpose of this chapter is to describe the principles and methods of the second and third steps of the analyses.

Phylogenetic Analyses of Pigment Sequences

Identification of critical amino acid changes that may shift the λ_{max} of a visual pigment is based on the fact that vertebrates have adapted to various photic environments by modifying, often extensively, their visual systems.[5] Thus, studying the evolutionary changes of visual pigments, we should be able to identify such potentially important amino acid changes. To establish associations between specific amino acid substitutions and the directions of the λ_{max} shift of pigments, the knowledge of the phylogenetic relationships of various pigments becomes essential.

Construction of Phylogenetic Trees

To construct a phylogenetic tree of visual pigments, their amino acid sequences must be aligned. The currently available amino acid sequences of various pigments have been aligned using a multiple alignment program of clustal W[13] and adjusted further visually to increase their similarity.[1] The aligned sequences are available by requesting them from syokoyam@mailbox.syr.edu. Based on such aligned amino acid sequences, the evolutionary tree can be constructed using different statistical methods: (1) distance matrix method, (2) maximum parsimony method, and (3) maximum likelihood method. Readers may consult Nei[14] for more information on this and other methods. Computer programs also exist to help conduct phylogenetic analyses. Web sites provide information on them. Sites include http://onyx.si.edu/PAUP/ (for PAUP), http://www.gcg.com/ (for GCG package), http://evolution.genetics.washington.edu/phylip.html (for PHYLIP), http://www.bio.psu.edu/faculty/nei/imeg (for MEGA/METREE),

---

[7] J. Sambrook, E. F. Fritsch, and T. Maniatis, "Molecular Cloning: A Laboratory Manual," 2nd ed. Cold Spring Harbor Laboratory Press, New York, 1989.
[8] J. Nathans and D. S. Hogness, *Cell* **34,** 807 (1983).
[9] J. Nathans, D. Thomas, and D. S. Hogness, *Science* **232,** 193 (1986).
[10] S. Kawamura and S. Yokoyama, *Gene* **182,** 213 (1996).
[11] S. Kawamura and S. Yokoyama, *Vis. Res.* **37,** 1867 (1997).
[12] S. Yokoyama and H. Zhang, *Gene* **202,** 89 (1997).
[13] J. D. Thompson, D. G. Higgins, and T. J. Gibson, *Nucleic Acids Res.* **22,** 4673 (1994).
[14] M. Nei, "Molecular Evolutionary Genetics." Columbia University Press, New York, 1987.

and http://phylogeny.arizona.edu/macclade/macclade.html (for Mac-Clade).[15] In the following, I describe only distance matrix methods.

UPGMA Method

The simplest distance matrix method is the unweighted pair-group method with arithmetic mean (UPGMA). In this method, the number of amino acid substitutions (K) is estimated for each pair of amino acid sequences. If the number of amino acid substitutions follows a Poisson process, K is estimated from $K = -\ln(1 - p)$, where p is the proportion of different amino acids between a pair of pigment sequences. Or, it may be estimated using an empirical formula $K = -\ln(1 - p - p^2/5)$.[16]

To illustrate the algorithm, let us consider four pigment sequences 1, 2, 3, and 4 with the distance matrix shown in Fig. 1A, where K_{ij} denotes the number of amino acid substitutions between pigments i and j with i, $j = 1, 2, 3,$ and 4. Suppose that K_{34} is smallest among all K_{ij} values. Then, pigments 3 and 4 are considered to be most closely related and are clustered with a branching point at distance $K_{34}/2$ (Fig. 1A). In this method, we assume that the number of amino acid substitutions per site per year is constant, producing two branches of equal length. Then, new distances between these grouped pigments and the other pigments, $K_{1(34)}$ and $K_{2(34)}$, are calculated by taking averages (Fig. 1B). We again search for the smallest value within the new distance matrix. Assume that $K_{2(34)}$ is smallest among three. Then, pigment 2 is clustered with the pigment group (34). The branch lengths of pigment 2 and the pigment group (34) are given by $K_{2(34)}/2$ (Fig. 1B). The remaining pigment 1 is the last to be clustered with the group (234). The branch length of pigment 1 and that of the pigment group (234) are given by $K_{1(234)}/2$ (Fig. 1C). Because of the rate constancy of amino acid substitution, the UPGMA tree has a root and the location of the common ancestor can be located (branch point a in Fig. 1C).

As an example, let us consider the aligned amino acid sequences of six RH1 pigments in Fig. 2. The Poisson-corrected K values vary from 0.084 to 0.281 (Table I). From this distance matrix, we can obtain the UPGMA tree in Fig. 3. This tree suggests that the cavefish and European eel pigments are most distantly related to the other pigments. This part of Fig. 3 does not agree with the accepted organismal tree, where the jawless fish lamprey is most distantly related. This discrepancy occurred because the assumption of the rate constancy among different RH1 pigment lineages does not hold.

---

[15] J. A. Lake and J. E. Moore, *Trends Guide Bioinformatics,* **22** (1998).
[16] M. Kimura, "The Neutral Theory of Molecular Evolution." Cambridge University Press, Cambridge, England, 1983.

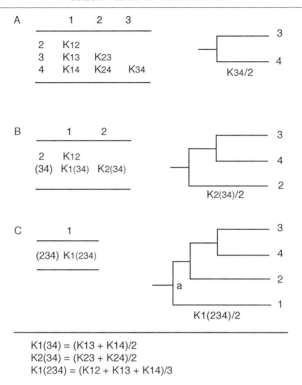

$K1(34) = (K13 + K14)/2$
$K2(34) = (K23 + K24)/2$
$K1(234) = (K12 + K13 + K14)/3$

FIG. 1. Distance matrices and steps for constructing the UPGMA tree. Point a denotes the root of the tree.

In this particular case, the wrong tree topology has been obtained because of a slow evolutionary rate of lamprey pigment.[17]

NJ Method

To alleviate this problem, we can relax the assumption of the rate constancy. One such method for doing so is the neighbor-joining (NJ) method.[18] The principle of this method is to sequentially find a pair of sequences (neighbors) that minimizes the total branch length of the tree. The topology of the NJ tree based on the K values in Table I is given in Fig. 4. This tree does not have a root. However, if we assume that point a in Fig. 4 is the root, then the topology of this tree becomes identical to that of the UPGMA tree (Fig. 3). When we add more distantly related

[17] H. Zhang and S. Yokoyama, *Gene* **191**, 1 (1997).
[18] N. Saitou and M. Nei, *Mol. Biol. Evol.* **4**, 268 (1987).

```
Lamprey    MNGTEGDNFYVPFSNKTGLARSPYEYPQYYLAEPWKYSAL   40
Cavefish   ......PY....M..A..VV............P..A.AC.   40
Eur. Eel   ......P...I.M..I..VV...F...........A.TI.   40
Chicken    ......QD....M.....VV...F............F...   40
Human      ......P.........A..VV...F...........QF.M.  40
Dolphin    ......L............VV...F...........QF.V.  40
Anc. 1     MNGTEGPNFYVPMSNKTGVVRSPFEYPQYYLAEPWKYSAL
Anc. 2     MNGTEGPNFYVPMSNITGVVRSPFEYPQYYLAEPWAYSIL
Anc. 3     MNGTEGPNFYVPMSNKTGVVRSPFEYPQYYLAEPWKFSAL
Anc. 4     MNGTEGPNFYVPFSNKTGVVRSPFEYPQYYLAEPWQFSVL

Lamprey    AAYMFFLILVGFPVNFLTLFVTVQHKKLRTPLNYILLNLA   80
Cavefish   .................Y..IE................   80
Eur. Eel   .....T...L.......Y..IE................   80
Chicken    .....M...L.......Y..I.................V  80
Human      .....L..VL...I...Y.....................  80
Dolphin    .....L..VL...I...Y.....................  80
Anc. 1     AAYMFFLILLGFPVNFLTLYVTIQHKKLRTPLNYILLNLA
Anc. 2     AAYMFFLILLGFPVNFLTLYVTIEHKKLRTPLNYILLNLA
Anc. 3     AAYMFLLILLGFPVNFLTLYVTIQHKKLRTPLNYILLNLA
Anc. 4     AAYMFLLIVLGFPINFLTLYVTVQHKKLRTPLNYILLNLA

Lamprey    MANLFMVLFGFTVTMYTSMNGYFVFGPTMCSIEGFFATLG  120
Cavefish   V.D....FG...T....L.......RLG.NL......F.  120
Eur. Eel   V......FG...T.V....H......E.G.NL...Y....  120
Chicken    V.D....FG...T...........V.G.Y..........  120
Human      V.D.....G...S.L...LH........G.NL.......  120
Dolphin    V......FG...T.L...LHA.......G.NL.......  120
Anc. 1     VADLFMVFGGFTTTMYTSMNGYFVFGPTGCNLEGFFATLG
Anc. 2     VADLFMVFGGFTTTMYTSMNGYFVFGPTGCNLEGFFATLG
Anc. 3     VADLFMVFGGFTTTMYTSMNGYFVFGPTGCNLEGFFATLG
Anc. 4     VADLFMVFGGFTTTLYTSLHGYFVFGPTGCNLEGFFATLG

Lamprey    GEVALWSLVVLAIERYIVICKPMGNFRFGNTHAIMGVAFT  160
Cavefish   .INS..C....S...WV.V....S.....EN.........  160
Eur. Eel   ..IS...........WV.V....S.....EN.....L...  160
Chicken    ..I.........V..V.V....S.....EN.........S 160
Human      ..I............V.V....S.....EN.........  160
Dolphin    ..I............V.V....S.....EN.....L.L.  160
Anc. 1     GEIALWSLVVLAIERYVVVCKPMSNFRFGENHAIMGVAFT
Anc. 2     GEISLWSLVVLAIERWVVVCKPMSNFRFGENHAIMGVAFT
Anc. 3     GEIALWSLVVLAIERYVVVCKPMSNFRFGENHAIMGVAFT
Anc. 4     GEIALWSLVVLAIERYVVVCKPMSNFRFGENHAIMGVAFT
```

FIG. 2. Aligned amino acid sequences of 6 RH1 pigments. Dots indicate identical amino acids with those of the Lamprey pigment. Gaps, indicated by dashes, are introduced to maximize sequence similarity. Anc. 1–Anc. 4 show the ancestral amino acid sequences at nodes 1–4 in Fig. 4. Seven transmembrane domains[40] are boxed.

```
Lamprey    WIMALACAAPPLVGWSRYIPEGMQCSCGPDYYTLNPNFNN  200
Cavefish   .F.....TV....................I....RAEG...  200
Eur. Eel   ....NS..M...F................V.....K.EV..  200
Chicken    ....M.......F................I.....K.EI..  200
Human      .V.........A.........L.....I.....K.EV..   200
Dolphin    ....M....A..................I....SRQEV..  200
Anc. 1     WIMALACAAPPLVGWSRYIPEGMQCSCGIDYYTLKPEINN
Anc. 2     WIMALACAVPPLVGWSRYIPEGMQCSCGIDYYTLKPEINN
Anc. 3     WIMALACAAPPLVGWSRYIPEGMQCSCGIDYYTLKPEINN
Anc. 4     WIMALACAAPPLVGWSRYIPEGMQCSCGIDYYTLKPEVNN

Lamprey    ESYVVYMFVVHFLVPFVIIFFCYGRLLCTVKEAAAAQQES  240
Cavefish   ..F.I........T.LFV.T......V.......Q....  240
Eur. Eel   ..F.I...I...S..LT..S......V.......Q....  240
Chicken    ..F.I.......MI.LAV......N.V.......Q....  240
Human      ..F.I.......TI.MI.......Q.VF......Q....  240
Dolphin    ..F.I.......TI.L........Q.VF......Q....  240
Anc. 1     ESFVIYMFVVHFLIPLVIIFFCYGRLVCTVKEAAAQQQES
Anc. 2     ESFVIYMFVVHFLVPLVIISFCYGRLVCTVKEAAAQQQES
Anc. 3     ESFVIYMFVVHFMIPLVIIFFCYGRLVCTVKEAAAQQQES
Anc. 4     ESFVIYMFVVHFTIPLVIIFFCYGQLVFTVKEAAAQQQES

Lamprey    ASTQKAEKEVTRMVVLMVIGFLVCWVPYASVAFYIFTHQG  280
Cavefish   ET..R..R......I..F.AY....L.....SWW...N..  280
Eur. Eel   ET..R..R.......I...A............W......  280
Chicken    .T............II...A..I.............N..  280
Human      .T............II...A..I................  280
Dolphin    .T............II..VA..I................  280
Anc. 1     ATTQKAEKEVTRMVIIMVIAFLVCWVPYASVAFYIFTHQG
Anc. 2     ETTQRAEREVTRMVIIMVIAFLVCWVPYASVAWYIFTHQG
Anc. 3     ATTQKAEKEVTRMVIIMVIAFLICWVPYASVAFYIFTHQG
Anc. 4     ATTQKAEKEVTRMVIIMVIAFLICWVPYASVAFYIFTHQG

Lamprey    SDFGATFMTLPAFFAKSSALYNPVIYILMNKQFRNCMITT  320
Cavefish   .E..PI...V........SI.......CL.....H.....  320
Eur. Eel   .T..PV...V.S.......I...L...CL.S.........  320
Chicken    ....PI...I.........I.......V............  320
Human      .N..PI...I........A.I.......M..........L..  320
Dolphin    ....PI...I.S......SI.......M..........L..  320
Anc. 1     SDFGPIFMTIPAFFAKSSAIYNPVIYIVMNKQFRNCMITT
Anc. 2     SDFGPIFMTVPAFFAKSSAIYNPVIYICLNKQFRNCMITT
Anc. 3     SDFGPIFMTIPAFFAKSSAIYNPVIYIVMNKQFRNCMITT
Anc. 4     SDFGPIFMTIPAFFAKSSAIYNPVIYIMMNKQFRNCMLTT
```

FIG. 2. (*continued*)

```
Lamprey    LCCGKNPLGDDE-SGASTSKTEVSSVSTSPVSPA 353
Cavefish   .......FEEE.GASTTA....A....S--.... 352
Eur. Eel   .F.....FQEE.GASTTA....A....S--.... 352
Chicken    ..........ED-TS.G--...T......Q.... 351
Human      I............-AS.TV....T.-----Q.A.. 348
Dolphin    ....R........-ASTTA....T.-----Q.A.. 348
Anc. 1     LCCGKNPLGDDE ASAT    KTETS     VSPA
Anc. 2     LCCGKNPFEEEE ASTT    KTEAS     VSPA
Anc. 3     LCCGKNPLGDDE ASAT    KTETS     VSPA
Anc. 4     LCCGKNPLGDDE ASAT    KTETS     VAPA
```

FIG. 2. (*continued*)

TABLE I
K VALUES OF RH1 PIGMENTS

Source	Lamprey	Cavefish	Eur. eel	Chicken	Human
Cavefish	0.281				
European eel	0.262	0.177			
Chicken	0.194	0.217	0.189		
Human	0.214	0.240	0.200	0.133	
Dolphin	0.228	0.243	0.193	0.143	0.084

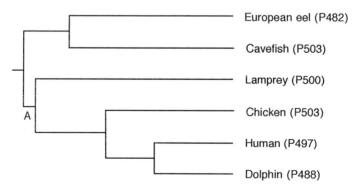

FIG. 3. The UPGMA tree of 6 RH1 pigments. Point A denotes the branch between the common ancestor of vertebrate pigments and that of lamprey and land animals' pigments. The numbers after P refer to λ_{max} values.

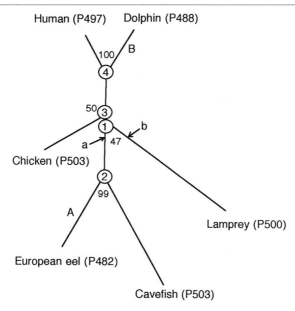

FIG. 4. The NJ tree of six RH1 pigments. Values beside branches are bootstrap values. Points a and b denote potential roots; points A and B are branches leading to European eel (P482) and dolphin (P488) pigments, respectively. The circled numbers denote ancestral pigments.

pigments (known as an outgroup) in the analysis, the NJ tree has the root at point b in Fig. 4, showing that the lamprey RH1 pigment is in fact most distantly related to the rest of the RH1 pigments.[4]

To test the reliability of the tree topology, we can resample the data under study so that the confidence level of the tree topology can be evaluated. This method is known as a bootstrap analysis.[19] Suppose that n amino acids for each pigment are analyzed. We randomly resample n sites from the data with replacement and then reconstruct a tree. This process is repeated a large number of times, for example, 1000 times. We then evaluate how often certain clusters occur among 1000 trials. The dolphin and human pigment cluster and the European eel and cavefish pigment cluster have bootstrap supports of 100 and 99%, respectively, and are highly reliable. But the phylogenetic positions of chicken and lamprey pigments have about 50% of bootstrap supports and are not reliable (Fig 4). Because of these low bootstrap supports, we cannot determine the phylogenetic positions of

[19] J. Felsenstein, *Evolution* **39**, 783 (1985).

lamprey, fish, chicken, and mammalian pigments unambiguously. A short length of branch A in Fig. 3 leads to the same conclusion.

Critical Amino Acid Substitutions

To determine the direction of amino acid changes, we have to infer the amino acid sequences of all ancestral pigments. Ancestral amino acid sequences of the pigments can be inferred using a distance-based Baysian method,[20] parsimony method,[21] and likelihood-based Baysian method.[22] These methods utilize various models of amino acid substitution.[23,24] For example, the amino acid substitution models may be based on empirical amino acid substitution data for many different proteins, whereas the equal-input model assumes that the probability of change of one amino acid to another is proportional to the frequency of the amino acids in the extant sequences.[25]

Example

Applying the likelihood-based Baysian method to the aligned amino acid sequences of the RH1 pigments in Fig. 2, the amino acid sequences of the ancestral pigments 1, 2, 3, and 4 in Fig. 4 were inferred (Fig. 2). Using these amino acids, we can determine amino acid substitutions at all branches in Fig. 4. Because the RH1 pigments generally have λ_{max} values at about 500 nm,[4] it is most likely that the λ_{max} values of European eel (P482) and dolphin (P488) pigments are blue-shifted. Being closely related to a "typical" RH1 cavefish (P503) pigment, the blue shift in the λ_{max} values of the European eel (P482) must have occurred at branch A in Fig. 4. Similarly, the blue shift in the λ_{max} of dolphin (P488) pigment must have occurred at branch B. We can identify 25 and 13 amino acid substitutions at branches A and B, respectively (Table II). Among these, D83N, V157L, and A292S are common to the two branches and, therefore, are associated with the blue shift of the λ_{max} value. Interestingly, these changes are located in the transmembrane domains (Fig. 2), where the opsin and 11-*cis*-retinal interact. Thus, these amino acid substitutions become excellent candidates

---

[20] J. Zhang and M. Nei, *J. Mol. Evol.* **44,** S139 (1997).
[21] W. P. Maddison and D. R. Maddison, "McClade: Analysis of Phylogeny and Character Evolution," Version 3. Sinauer, Sunderland, Massachusetts, 1992.
[22] Z. Yang, S. Kumar, and M. Nei, *Genetics* **141,** 1641 (1995).
[23] D. T. Jones, W. R. Taylor, and J. M. Thornton, *Comput. Appl. Biosci.* **8,** 275 (1992).
[24] M. O. Dayhoff, R. M. Schwartz, and B. C. Orcutt, in "Atlas of Protein Sequence and Structure" (M. O. Dayhoff, ed.), p. 345. National Biomedical Research Foundation, Washington, DC, 1978.
[25] M. Nei, J. Zhang, and S. Yokoyama, *Mol. Biol. Evol.* **14,** 611 (1997).

TABLE II
AMINO ACID SUBSTITUTIONS IN EUROPEAN EEL AND DOLPHIN

Branch[a]	Amino acid substitutions
European eel (A)	V11I, S38T, F46T, D83N, M95V, N100H, P107E, V157L, L165H, A166S, V168M, V173F, I189V, I198V, V209I, L213S, V217T, I255V, D282T, I286V, A292S, V304L, K310S, C322F, E329Q
Dolphin (B)	P7L, D83N, G101A, V157L, F159L, L165M, P170A, L194S, K195R, P196Q, I259V, A292S, A299S

[a] In Fig. 4.

that might have caused the blue shifts in λ_{max} values of the European eel (P482) and dolphin (P488) pigments. Note that D83N and A292S in bovine RH1 pigment can shift the λ_{max} value ~5 nm[26–28] and ~10 nm[28,29] toward blue, respectively. Thus, it appears that both D83N and A292S can contribute significantly to the blue shifts in the λ_{max} values of European eel (P482) and dolphin (P488) pigments. The actual effects of D83N, V157L, and A292S on the λ_{max} values of the two pigments have yet to be tested.

Functional Tests of Mutant Pigments

Regeneration of Wild-Type and Mutant Pigments

To obtain RNA, the protocol of Chomczynski and Sacchi[30] can be followed. A retina or whole eye is homogenized in 0.5 ml GuSCN buffer (4 M guanidium thiocyanate, 25 mM sodium citrate, pH 7, and 0.5% N-lauroylsarcosine) and 3 μl 2-mercaptoethanol. Total RNA is extracted with 0.5 ml phenol/chloroform ratio, precipitated in 2 volumes of 100% ethanol, and washed with 80% (v/v) ethanol. From the total RNA, the opsin cDNA can be cloned by reverse transcriptase–polymerase chain reaction (RT-PCR) amplification and subcloned into an expression vector pMT5.[31]

The pMT5 is composed of three segments: (1) a segment containing the simian virus 40 (SV40) origin of replication and enhancer, the adenovirus major late promoter, a majority of the adenovirus tripartite leader

[26] J. Nathans, *Biochemistry* **29,** 937 (1990).
[27] J. Nathans, *Biochemistry* **29,** 9746 (1990).
[28] J. I. Fasick and P. R. Robinson, *Biochemistry* **37,** 432 (1998).
[29] C. Sun, J. P. Macke, and J. Nathans, *Proc. Natl. Acad. Sci. U.S.A.* **94,** 8860 (1997).
[30] P. Chomczynski and N. Sacchi, *Anal. Biochem.* **162,** 156 (1987).
[31] L. Ferretti, S. S. Karnik, H. B. Khorana, M. Nassal, and D. D. Oprian, *Proc. Natl. Acad. Sci. U.S.A.* **83,** 599 (1986).

sequences, and an intervening sequence; (2) a segment containing the dihydrofolate reductase coding region, the SV40 polyadenylation sequence, and sequences encoding the adenovirus VA1 RNA; and (3) a segment containing the bacterial origin of replication and the bacterial β-lactamase gene conferring ampicillin resistance derived from the pUC8 plasmid. The pMT5 contains the cloning sites *Eco*RI and *Sal*I. The sequences in the large fragment of the *Eco*RI/*Sal*I-digested pMT5 are necessary for expression in cultured COS-1 cells. The COS-1 cell is a simian cell that is permissive for SV40 replication and supports the replication of a recombinant DNA molecule containing a SV40 origin of replication. The last 15 codons of synthetic bovine RH1 cDNA, between *Sal*I and *Not*I sites, in the pMT5 encode the rho 1D4 epitope sequence necessary for the antibody-binding for the purification of the opsin protein.[32] These extra 15 amino acids do not affect their spectral properties.[33]

The forward primer for the RT-PCR amplification should contain about 6 nucleotides 5' to the *Eco*RI site (GAATTC) followed by CACC and about 20 nucleotides of the coding region, starting with the initiation codon. The first 6 nucleotides and the overlapping Kozak sequence (CCACC)[34] facilitate *Eco*RI digestion and transcription, respectively. Six nucleotides 5' to the *Sal*I site (GTCGAC) in the reverse primers are followed by about 20 nucleotides of the reverse complement sequence from the nucleotide at the second position of the last codon.[35] Using the forward and reverse primers, cDNA is reverse-transcribed for 1 hr at 42° and for 5 min at 95°, and then PCR is carried out for 30 cycles at 94° for 45 sec, 55° for 1.5 min, and 72° for 2 min. The PCR products may be subcloned into pBlueScript-(SK$^-$) (Stratagene, La Jolla, CA), which can be readily sequenced.

The *Eco*RI/*Sal*I-digested pMT5 fragments can be ligated with the desired *Eco*RI/*Sal*I-digested wild-type opsin cDNA fragments. The resulting plasmid DNAs are purified through ultracentrifugation in CsCl density gradient[7] and are transiently expressed in COS-1 cells using 12.5–19 μg DNA per 13-cm plate containing 1.9–2.2 × 10^7 cells.[36,37]

A mutant pigment may be created by the Kunkel *et al.* method.[38] As an example, we may assume that our goal is to change tyrosine (TAC) to

---

[32] R. S. Molday and D. MacKenzie, *Biochemistry* **22,** 653 (1983).
[33] D. D. Oprian, A. B. Asenjo, N. Lee, and S. L. Pelletier, *Biochemistry* **30,** 11367 (1991).
[34] M. Kozak, *Nucleic Acids Res.* **12,** 857 (1984).
[35] S. Yokoyama, F. B. Radlwimmer, and S. Kawamura, *FEBS Lett.* **423,** 155 (1998).
[36] D. D. Oprian, R. S. Molday, R. J. Kaufman, and H. G. Khorana, *Proc. Natl. Acad. Sci. U.S.A.* **84,** 8874 (1987).
[37] S. S. Karnik, T. P. Sakmar, H.-B. Chen, and H. G. Khorana, *Proc. Natl. Acad. Sci. U.S.A.* **85,** 8459 (1988).
[38] T. A. Kunkel, J. D. Roberts, and R. A. Zakuor, *Methods Enzymol.* **154,** 367 (1987).

phenylalanine (TTC).[39] We begin with the sequence we want to mutagenize by cloning into M13 mp18. We grow the recombinant clone using *Escherichia coli* CJ236 (*dut⁻ ung⁻* F) cells with 0.25 μg/ml uridine and harvest single-stranded (sense-strand) phage DNA. To this strand, we can anneal an oligonucleotide (19-mer, say) that is complementary to the region of interest, except for a single central base change, such as an A for a T. This primer will be extended to form a full-length antisense strand *in vitro* and the double-stranded heteroduplex DNA will be used to transform *E. coli* TG1 cells. The uracil-substituted template strand is degraded in the *ung*⁺ TG1 cells, leaving the newly replicated strand primed with the mutagenic oligonucleotide to serve as a template. The same strand will now bear the mutagenized TTC codon in place of the original TAC codon.

Site-directed mutagenesis can also be easily performed following the protocols provided by commercially available mutagenesis kits such as the QuickChange kit from Stratagene (La Jolla, CA). This particular method uses two complementary primers (21-mers), each of which contains the desired central base change, to anneal to 10 ng of the plasmid DNA of interest. *PfuTurbo* DNA polymerase is used to extend the product in a thermal cycler by heating for 30 sec at 95° followed by 12 cycles of amplification at 95° for 30 sec, 55° for 1 min, and 68° for 8 min. This product is treated with 10 units of *Dpn*I, which is specific for methylated and hemimethylated DNA, to digest the parental DNA template, and to select for mutation-containing synthesized DNA. The newly synthesized DNA incorporating the desired mutation is then transfected into *E. coli* XL1-Blue supercompetent cells. The majority of colonies produced contain the desired mutant opsins. Using this protocol, we can also construct mutant opsins with multiple nucleotide changes and insertion or deletion of single and multiple nucleotides.

Regeneration, Purification, and Determination of λ_{max} Values of Pigments

Transfected cells are harvested 63 hr after addition of 12.5 μg of DNA per plate. COS-1 cells from 20 plates are washed once in 50 mM HEPES (pH 6.6), 140 mM NaCl, 3 mM MgCl$_2$ (buffer Y1) at 37°. Cells are scraped off and suspended in ice-cold buffer Y2 (buffer Y1 with 10 μg/ml each aprotinin and leupeptin). Visual pigments are regenerated by resuspending cells with 5 μM of 11-*cis*-retinal for 3 hr at 0–4° in the dark or in dim red light (Kodak No. 2 safelight filter), solubilizing in buffer Y2 with 1% dodecylmaltoside and 20% (w/v) glycerol for 1.5 hr, and binding to 1D4

---

[39] R. Yokoyama, B. E. Knox, and S. Yokoyama, *Invest. Ophthalmol. Vis. Sci.* **36,** 939 (1995).

Sepharose[32] overnight. The 11-*cis*-retinal can be obtained from the Storm Eye Institute (Medical University of South Carolina, SC). The regenerated pigments are washed in 10 ml of buffer W2 [buffer Y2 with 0.1% *n*-dodecyl-maltoside and 20% (w/v) glycerol] twice and 10 ml and 1 ml of buffer W1 (buffer W2 without aprotinin and leupeptin) once and five times, respectively. The pigments are further eluted from the 1D4 Sepharose by two to four successive incubations with 40 μM of peptide I (the last 15 amino acids of bovine RH1 pigment, obtained from the Cell Culture Center, Minneapolis, MN) in buffer W1. The eluted pigments are concentrated using Centricon C-30 (Amicon, Danvers, MA).

The λ_{max} values of visual pigments are recorded at 20° using a spectrophotometer. For light bleaching experiments, Kodak Wratten Gelatin Filter No. 3 is used to cut off wavelengths shorter than 440 nm from a 60-W room lamp. Usually, visual pigments are bleached by a 60-W room lamp with 440-nm cutoff filter. However, SWS1 pigments are bleached by a 366-nm ultraviolet light illuminator.[35] The data are analyzed with Sigmaplot software (Jandel Scientific, San Rafael, CA).

Bovine RH1 Mutant

To generate mutant pigments, vision scientists have been using the synthetic bovine RH1 cDNA of 1057 bp in length extensively, which contains 28 unique restriction sites that are, on average, 60 bp apart.[31] Thus, replacement of specific restriction fragments by synthetic counterparts containing the desired nucleotide changes also permits specific mutagenesis in any part of the cDNA. For example, using this method, an amino acid change A292S was incorporated in the bovine RH1 opsin cDNA and the mutant pigment was regenerated using cultured cells. The absorption spectra of the wild-type and mutant pigments are shown in Fig. 5. The dark spectra of the two pigments have two peaks at 270 nm and at around 500 nm. The first peak is detected by proteins other than visual pigments, whereas the second peak shows the λ_{max} value of the visual pigments. The λ_{max} values can also be evaluated from dark–light difference spectra (Fig. 5, inset). The λ_{max} values of the wild-type pigment (499 nm) and the pigment with A292S (490 nm) agree well with the previous results.[28,29,41]

Summary

To elucidate the molecular mechanisms of vertebrate color vision, it is essential to establish associations between amino acid substitutions and the

---

[40] P. A. Hargrave, J. H. McDowell, D. R. Curtis, J. K. Wang, E. Jaszczack, S. L. Fong, J. K. Mohanna Rao, and P. Argos, *Biophys. Structural Mech.* **9**, 235 (1983).
[41] S. Yokoyama and G. Yu, unpublished results.

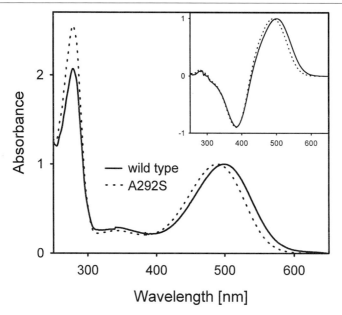

FIG. 5. Absorbance spectra of bovine RH1 pigments in the dark and the dark–light difference spectra (inset). The solid line and dashed line represent measurements of the wild-type pigment and A292S mutant, respectively.

directions of λ_{max} shifts of visual pigments. In this way, we can identify critical amino acid changes that may be responsible for λ_{max} shifts of visual pigments. In this process, we may consider only highly conserved residues, simply because the evolutionary conservation often implies functional importance. Using such an "evolutionary model"[2] as a convenient tool in designing mutagenesis experiments, we can test specific hypotheses on the molecular mechanisms that are responsible for color vision in vertebrates.

Virtually any vertebrate opsin cDNA can be expressed in COS cells, reconstituted with 11-*cis*-retinal, and the λ_{max} values of the regenerated pigments can be measured rather easily. By constructing mutant pigments with desired amino acid changes and conducting the *in vitro* assay and comparing their λ_{max} values with those of corresponding wild-type pigments, we can elucidate the molecular mechanisms of λ_{max} shifts—and color vision—of vertebrates rigorously.

Acknowledgments

I thank Ruth Yokoyama for comments on the manuscript. This work was supported by USPHS grant GM-42379.

Section III

Proteins that Interact with Rhodopsin

[22] Light Scattering Methods to Monitor Interactions between Rhodopsin-Containing Membranes and Soluble Proteins

By MARTIN HECK, ALEXANDER PULVERMÜLLER, and KLAUS PETER HOFMANN

Introduction

Many physicochemical processes are accompanied by changes of scattered light. An everyday example is the condensation of atmospheric water to small liquid droplets, leading to opaque fog. In cell biology and biochemistry, light scattering changes accompany, for example, the growth or osmotic swelling of bacteria,[1] the swelling of liposome bilayers due to osmotic flow of water,[2] or the aggregation of liposomes in surfactant research.[3] The generality of the light scattering phenomenon makes it difficult to interpret a measured change in terms of a specific process. In favorable cases, however, physicochemical and biochemical reactions are readily measured and, with the help of adequate controls, unambiguously identified by this simple technique. Combined with chemical probing, kinetic light scattering (KLS) provides a real-time monitor of specific molecular changes.

Measurements of light-induced changes of the scattered light of rod outer segment and disk membrane preparations date back to the 1970s. These early investigations focused on analyzing physical processes such as disk shrinkage[4] or osmotic swelling.[5] Kühn et al.[6] were the first to recognize that two of the measured changes reflect important steps in the rhodopsin-catalyzed activation of the G protein, namely, its binding to rhodopsin ("binding signal") and its subsequent dissociation ("dissociation signal"). The physicochemical background of both these observations is that G_t, unlike other G proteins, is in equilibrium between a membrane-bound and a soluble form.[7] The gain of mass of a membrane, when the G_t holoprotein

---

[1] A. L. Koch, *Biochim. Biophys. Acta* **51**, 429 (1961).
[2] M. Jansen and A. Blume, *Biophys. J.* **68**, 997 (1995).
[3] A. Meyboom, D. Maretzki, P. A. Stevens, and K. P. Hofmann, *J. Biol. Chem.* **272**, 14600 (1997).
[4] K. P. Hofmann, A. Schleicher, D. Emeis, and J. Reichert, *Biophys. Struct. Mech.* **8**, 67 (1981).
[5] D. G. McConnell, C. N. Rafferty, and R. A. Dilley, *J. Biol. Chem.* **243**, 5820 (1968).
[6] H. Kühn, N. Bennett, M. Michel Villaz, and M. Chabre, *Proc. Natl. Acad. Sci. U.S.A.* **78**, 6873 (1981).
[7] A. Schleicher and K. P. Hofmann, *J. Membr. Biol.* **95**, 271 (1987).

is bound from solution, and the loss of mass with dissociation of (mainly) the α subunit produces large and readily measurable changes in light scattering.

Based on the same principle, other interaction partners of active rhodopsin, including arrestin and Tween-solubilized rhodopsin kinase, can also be followed by KLS. One decisive advantage of KLS is that crucial irregularities of the sample such as aggregation of the membranes immediately show up in the baseline scattering. The fine dispersion of the membrane particles results in short diffusion paths of the interacting proteins, so that after calibration of the mass shifts by centrifugation, KLS offers real-time information and continuous recording with good sensitivity and fidelity.

This chapter focuses on interactions that go along with changes of membrane association. Attempts have been made to extend the applicability of KLS to other processes, including intact photoreceptors.[8-10] We will give one example, namely, the so-called PDE signal, which monitors G_t–PDE interaction on isolated membranes, but does not reflect a change of membrane association.[11] In another chapter of this volume,[12] the KLS technique is compared with other techniques with emphasis on the activation of the G protein, G_t. Finally, we note that the application of KLS is not limited to interactions in the visual system, although it reaches its full sensitivity in light-sensitive systems. Caged substances can provide the trigger in other systems (see, for example, Meyboom et al.[13]).

Theoretical Background

Light Scattering by Membrane Vesicles

A comprehensive treatment of theories and approaches used for the description of light scattering can be found in Kerker.[14] Light scattering (LS) of particles with properties of disk vesicles (size in the range of $\lambda/20$ to $\lambda/2$ and a refractive index only slightly different from the solvent) can be described by the classical Rayleigh–Gans (or Rayleigh–Debye) approxi-

---

[8] T. M. Vuong, M. Chabre, and L. Stryer, *Nature* **311,** 659 (1984).
[9] D. R. Pepperberg, M. Kahlert, A. Krause, and K. P. Hofmann, *Proc. Natl. Acad. Sci. U.S.A.* **85,** 5531 (1988).
[10] M. Kahlert, D. R. Pepperberg, and K. P. Hofmann, *Nature* **345,** 537 (1990).
[11] M. Heck and K. P. Hofmann, *Biochemistry* **32,** 8220 (1993).
[12] O. P. Ernst, C. Bieri, H. Vogel, and K. P. Hofmann, *Methods Enzymol.* **315** [32], 1999 (this volume).
[13] A. Meyboom, D. Maretzki, P. A. Stevens, and K. P. Hofmann, *Biochim. Biophys. Acta,* in press (1999).
[14] M. Kerker, *Phys. Chem.* **16,** 1 (1969).

mation (see, e.g., van de Hulst[15] and Kerker[14]). In graphic terms, this theory neglects the different wavelength within and outside the scattering particle. The scattering intensity at angle θ is written as a product of two terms (including a cos term that accounts for randomly polarized incident light):

$$I(\theta) = (1 + \cos^2 \theta) I_\alpha P(\theta) \tag{1}$$

The term I_α depends on the polarizability α of the particle. It is the sum of the scattering by the individual Rayleigh dipoles contained in the particle, independent of the scattering angle; $P(\theta)$ is the particle-scattering function (or form factor), a term dependent on the particle shape and size as a consequence of destructive interference of scattered light from different parts of the particle. Generally, $P(\theta)$ is equal to unity in the forward direction ($\theta = 0°$), but decreases as θ becomes larger. Although analytical solutions for $P(\theta)$ of particle shapes, including osmotically swollen disks,[16] are available, it is often not possible to give a general interpretation of the scattering curves for complex biological structures (see later discussion).

A more recent development is quasielastic light scattering, a technique that allows us to extract kinetic parameters such as diffusion coefficients from an analysis of the correlation in time between incident light quanta and the scattered quanta.

Kinetic Light Scattering

Kinetic light scattering (KLS) combines the light scattering monitor with a relaxation technique. Generally, a trigger induces processes in the particles that are seen as a change in light scattering. The light trigger offers the advantage that side effects such as convection of the sample suspension or deterioration of the membrane particles are minimized.

The angular dependence of the relative scattering intensity change ($\Delta I/I$; "LS signal") is defined as a "difference scattering curve"[17] in analogy to the difference spectrum derived from the spectral dependence of an absorption change. According to Eq. (1) it can help to interpret a signal physically. More importantly, a conclusive assignment of a LS signal to the underlying biochemical reaction is usually achieved by comparative studies with other techniques, such as simultaneously measured absorption changes[7] and centrifugation assays[18] (see later discussion). The additional

[15] H. C. van de Hulst, "Light Scattering by Small Particles." John Wiley & Sons, New York, 1957.
[16] W. F. Hoffman, T. Norisuye, and H. Yu, *Biochemistry* **16**, 1273 (1977).
[17] K. P. Hofmann and D. Emeis, *Biophys. Struct. Mech.* **8**, 23 (1981).
[18] A. Pulvermüller, K. Palczewski, and K. P. Hofmann, *Biochemistry* **32**, 14082 (1993).

kinetic information contained in the triggered scattering change considerably reduces the interpretation problem and gives first insight into mechanistic details. A routine analysis may include variation of temperature[7] and pH[19] dose–response relationships,[11,18] and probing by biochemical modifications or site-directed mutagenesis.[20]

KLS Signals from Triggered Membrane Association of Proteins

An extensively used application of the KLS method is the real-time monitoring of light-induced binding of soluble proteins to disk membranes. In suspensions of isolated disk membranes reconstituted with the purified protein, the 840-nm incident light beam is almost exclusively scattered by the disks; the comparatively small soluble proteins do not measurably contribute to the overall scattering in the angular range used.

We consider the scattering intensity change when the soluble protein binds to the disk surface. Substituting I_α [Eq. (1)] according to the Rayleigh theory, the scattering intensity at the scattering angle θ is proportional to

$$I(\theta) \sim (1 + \cos^2 \theta)[(n/n_0)^2 - 1]^2 V^2 P(\theta) \qquad (2)$$

where n is the refractive index of the particle, n_0 the refractive index of the solvent, and V the volume of the particle. The $P(\theta)$ term is in the case of disks only dependent on the disk radius and nearly independent of the disk.[4] The relative scattering intensity change ($\Delta I/I$) is therefore independent of the scattering angle when the soluble protein is associated. This holds with good approximation even for slightly irregular disk-like vesicles, as long as the association of protein mass does not lead to a shape change. On the assumption that the soluble protein and the disk vesicle have equal refractive indices, protein association has an effect only on the volume V. When we further assume a homogeneous density of the disks and the soluble protein, V can be replaced by the particle mass M in Eq. (2). This substitution leads to an equation relating the relative mass change with the scattering intensity change:

$$dI/I = 2dM/M \qquad (3)$$

For the expected small changes, the relative intensity changes $\Delta I/I$ is predicted to be twice the relative mass change $\Delta M/M$. The experimental test of Eq. (3), using the transducin binding signal, has confirmed the proportionality but with a factor of 1.2, instead of 2.[7]

---

[19] O. G. Kisselev, C. K. Meyer, M. Heck, O. P. Ernst, and K. P. Hofmann, *Proc. Natl. Acad. Sci. U.S.A.* **96**, 4898 (1999).

[20] O. P. Ernst, K. P. Hofmann, and T. P. Sakmar, *J. Biol. Chem.* **270**, 10580 (1995).

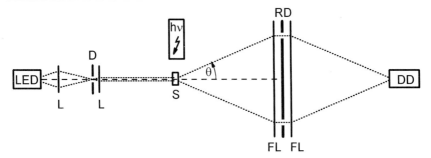

FIG. 1. Schematic diagram of a kinetic light scattering setup. LED, light emitting diode (λ = 840 nm); L, lens; D, diaphragm; S, sample; θ, scattering angle; FL, fresnel lens; RD, ring diaphragm; DD, detector (for details see text).

Measuring Protocols and Devices

Standard Spectrophotometer

A standard spectrophotometer is often adapted to measure light scattering (see, for example, Kühn et al.[6]). One may either observe the scattered light at right angle to the incident beam ($\theta = 90°$) or measure the turbidity ($\theta = 0°$). The latter is the decrease of the transmitted light by scattering, and thus equal to the scattering over all angles. Although elaborate biochemical controls allow interpretation of such data, the resolution (signal-to-noise ratio) is in most cases not satisfying.

Light Scattering Setup

Light-scattering changes are measured in a setup similar to the one previously described by Hofmann and Emeis[17] (Fig. 1). The measuring wavelength has to be chosen in the near infrared to avoid interference by both absorption changes and bleaching of rhodopsin. A light-emitting diode (Hitachi HLP 60R, $\lambda = 840$ nm, 0.4×0.4 mm) with a high luminous density is first focused on a small diaphragm and then, through a second lens via a nearly parallel beam (diameter about 4 mm), into the measuring cuvette. The scattered light is collected by two Fresnel lenses and focused on a solid-state photodetector (Centronics OSD 100 5-T). Using different ring diaphragms positioned between the Fresnel lenses the scattering angle θ is selected, in the standard configuration $\theta = 16° \pm 2°$. For particles of the dimension of disk vesicles, $\Delta I/I$ is measured with good signal-to-noise ratio in the range $\theta = 6°–25°$. LS signals are recorded with 100-μs to 100-

ms dwell time of the A/D converter (Nicolet 400, Madison, WI); the cutoff frequency is routinely set to 0.2 of the dwell time.

All measurements are performed in 10-mm path cuvettes, with 300-μl volumes. Reactions are triggered by flash photolysis of rhodopsin with a green flash (500 ± 20 nm), attenuated by appropriate neutral density filters. The flash intensity is quantified by the amount of rhodopsin bleached and expressed in terms of the mole fraction of photoexcited rhodopsin (R*/R).

Although experimental tests of Eq. (3) yield a reasonable approximation, a rigorous quantitative evaluation of light scattering signals by means of scattering theory is limited by several factors, which depend on properties of the membrane samples. These factors include the following:

1. The polydispersity of disk vesicles, which are, depending on the method of preparation, inhomogeneous in size and shape. Electron micrographs of the disks used here show vesicles in a size range of 100–1000 nm with an average size of about 400 nm. Contamination by large particles (e.g., vesicle aggregates) must be avoided because they produce a large elevation of baseline scattering at small angles, thus artefactually lowering $\Delta I/I$.

2. The concentration is chosen low enough so that most of the quanta are scattered only once when passing through the sample. The onset of multiple scattering is a function not only of particle concentration but also of shape and size of the particles. With disk vesicles in a 10 mm path cuvette, multiple scattering begins at 1 μM rhodopsin.

Membrane and Protein Preparations

Rhodopsin-Containing Membranes

Rod outer segments (ROS) are isolated from fresh, dark-adapted bovine retinas as described.[21] Washed disk membranes (WM) are prepared by removing the soluble and membrane-associated proteins by repetitive washes with a low ionic strength buffer.[22] Phosphorylated opsin is prepared according to Wilden and Kühn,[23] yielding phosphorylated WMs by regeneration with 11-*cis*-retinal. Excess 11-*cis*-retinal is removed by washing the membranes with NH_2OH.[24] The membrane suspensions are stored at $-80°$

---

[21] D. S. Papermaster, *Methods Enzymol.* **81,** 48 (1982).
[22] H. Kühn, *Curr. Top. Membr. Transport* **15,** 171 (1981).
[23] U. Wilden and H. Kühn, *Biochemistry* **21,** 3014 (1982).
[24] K. P. Hofmann, A. Pulvermüller, J. Buczylko, P. Van Hooser, and K. Palczewski, *J. Biol. Chem.*, **267,** 15701 (1992).

in 20 mM 1,3-bis[tris(hydroxymethyl)methylamino]propane (BTP), pH 7.5, 130 mM NaCl, 1 mM MgCl$_2$. After thawing, the suspensions are rigorously shaken with glass beads (2-mm diameter), to dissolve aggregates.

Rhodopsin Kinase

Unphosphorylated rhodopsin kinase is purified as described by Pulvermüller et al.[18] The preparation yield rhodopsin kinase contaminated with an approximately equal amount of arrestin. Note that the concentration of Tween 80 is reduced to 0.004% (2.5-fold critical micellar concentration). This detergent concentration is a compromise between the reduction of the hydrophobic interaction of the rhodopsin kinase with the membrane and the negative effect of the solubilization of the membrane by the detergent.

Arrestin

Arrestin is purified as described by Palczewski et al.[25] with the following modifications. The crude retinal extract was loaded on a DEAE-Sephacel column (Pharmacia, Piscataway, NJ, 1 × 30 cm). After washing the column as described, arrestin is eluted with 60 mM NaCl in 10 mM HEPES, pH 7.5, at a flow rate of 0.5 ml/min and directly applied on a HiTrap heparin-Sepharose column (Pharmacia, 5 ml). Arrestin is eluted by a linear gradient (0–2 mM) of InsP$_6$ in 10 mM HEPES, pH 7.5, and 100 mM NaCl. InsP$_6$ was then removed by rechromatography on a HiTrap heparin-Sepharose column as described.

Transducin and Phosphodiesterase

These proteins are purified as described by Heck and Hofmann[11] with the following modification. Crude PDE is chromatographed on a TSK gel-heparin-5PW column (TosoHaas, GmbH, Stuttgart, Germany; 0.75 × 7.5 cm). Elution is carried out with a linear NaCl gradient (0–600 mM) in 20 mM BTP, pH 7.5, 1 mM MgCl$_2$, and 1 mM dithiothreitol (DTT) at a flow rate of 1 ml/min.

Light-Induced Binding of Soluble Proteins to Disk Membranes

Light Scattering Signals

After absorption of a photon and relaxation via intermediates, rhodopsin adopts a conformation that is recognized by several key proteins of the

---

[25] K. Palczewski, A. Pulvermüller, J. Buczylko, C. Gutmann, and K. P. Hofmann, *FEBS Lett.* **295**, 195 (1991).

visual cascade, including the G protein transducin (G_t); rhodopsin kinase (RK), which phosphorylates photoactivated rhodopsin; and arrestin (also known as 48 k or S-antigen). The light-activated form of rhodopsin (R*) is related to the metarhodopsin (meta) intermediates of the rhodopsin reaction sequence.

In contrast to arrestin, both RK and G_t are posttranslationally modified by N-terminal myristoylation and C-terminal farnesylation. The hydrophobic modifications result in a light-independent interaction of both these proteins with the disk membrane. Although purified RK is fully solubilized by the necessary Tween treatment (see Methods section), a significant (however variable) fraction of G_t is membrane bound under the conditions of the experiment (see later section), complicating the analysis of light-induced membrane binding.

As described earlier, light-induced transition of a soluble protein to the disk membrane can be monitored as an increase of the scattered light. Binding signals of RK, arrestin, and G_t are shown in Fig. 2. The fast negative component of the LS transient (Fig. 2A, inset) represents the N-signal, which relates directly to the conformational changes of rhodopsin on photoexcitation.[26] It is subtracted point by point from the original record, to yield the resulting binding signal. Binding of the proteins was studied using both unphosphorylated and (pre)phosphorylated rhodopsin, respectively. The results presented in Fig. 2 show that RK, but not arrestin, interacts with unphosphorylated R*. Thus, the presence of arrestin in the rhodopsin kinase preparation does not affect the rhodopsin kinase binding signal. [Note that, despite the presence of rhodopsin kinase, R* remains unphosphorylated in these experiments, due to the lack of adenosine triphosphate (ATP).] Phosphorylation of R* dramatically reduces the affinity for RK and switches to tight binding of arrestin (binding of arrestin contained in the RK preparation is much slower compared to RK binding and is therefore not seen on the timescale used in Fig. 2A). Binding of G_t to R* is much less influenced by the phosphorylation confirming the notion that additional binding of arrestin (or RK) is required for shutoff of the receptor.

Centrifugation Assay

Light-induced binding of the proteins to disk membranes was also investigated using the centrifugation assay.[18,27] Disk membranes were incubated with the purified proteins and pelleted by centrifugation. The amount of

[26] K. P. Hofmann, R. Uhl, W. Hoffmann, and W. Kreutz, *Biophys. Struct. Mech.* **2,** 61 (1976).
[27] H. Kühn, *Nature* **283,** 587 (1980).

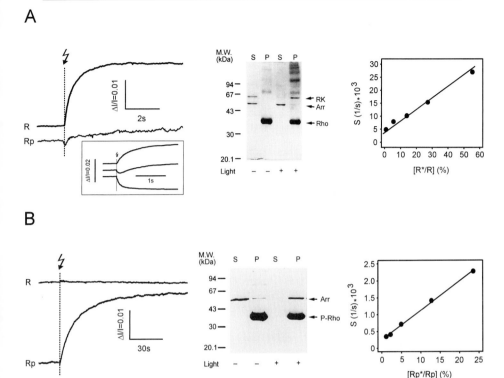

FIG. 2. KLS monitoring of the light-induced binding of soluble proteins to disk membranes. *Left-hand side:* Flash-induced light-scattering binding signals measured on suspensions of washed disk membranes reconstituted with (A) purified rhodopsin kinase, (B) arrestin, or (C) transducin. (Note the different timescales.) Binding of the proteins was studied using disk membranes containing either unphosphorylated (R, upper traces) or (pre)phosphorylated rhodopsin (R_p, lower traces), respectively. All signals are corrected for control signals measured without added protein as indicated in the inset. *Inset:* Point-by-point subtraction of the control without RK ("N-signal," lower trace) from the original record (middle trace), yields the RK binding signal (upper trace). Measuring conditions: 3 μM rhodopsin and 1 μM RK, arrestin or G_t. The flash illuminated 35% of rhodopsin in the RK experiments and 23.5% in the experiments with arrestin and G_t. *Middle:* Centrifugation assay of the light-dependent binding of (A) purified rhodopsin kinase, (B) arrestin, or (C) transducin. Washed disk membranes were reconstituted with the purified protein and kept either in the dark (labeled −) or illuminated under white light (labeled +). The membranes were then pelleted at 15,000g for 10 min at room temperature and washed twice with buffer. The first supernatant (labeled S) and the pellet (labeled P) were analyzed using SDS–PAGE. Measuring conditions: 5 μM rhodopsin and 1 μM RK, arrestin, or G_t. *Right-hand side:* Dependence of the initial slopes of the binding signals of (A) rhodopsin kinase, (B) arrestin, or (C) transducin on the mole fraction of light-activated rhodopsin (R*/R).

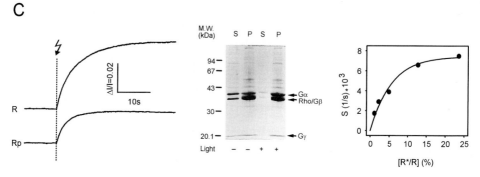

Fig. 2. (*continued*)

protein either bound to the membrane pellet or present in the supernatant was analyzed by sodium dodecyl sulfate–polyacrylimide gel electrophoresis (SDS–PAGE). Figure 2 shows that the photoactivation of rhodopsin leads to the translocation of the soluble proteins from the supernatant to the membrane fraction, confirming the interpretation of the LS signals. In the case of G_t, both the light-independent and the light-induced binding are seen in the gels.

Bimolecular Analysis

A comparison of the binding signals shown in Fig. 2, including dose–rate curves (Fig. 2, right), allows a first characterization of the underlying reactions.

The simplest model for the binding signal would be that the scattering change is a direct real-time monitor of complex formation of R* and the soluble protein (X). This assumption implies that after the fast formation of R* the binding signal follows a bimolecular reaction scheme:

$$R^* + X \underset{k_{\text{off}}}{\overset{k_{\text{on}}}{\rightleftharpoons}} R^*X \qquad (4)$$

where the initial rate depends on both initial concentrations $[R^*]_0$ and $[X]_0$ (the back reaction is negligible at $t = 0$):

$$d[R^*X]/dt|_{t=0} = k_{\text{on}}[R^*]_0[X]_0 \qquad (5)$$

A necessary condition for Eq. (5) to describe the reaction correctly is that the initial slope of the signal $S = d(\Delta I/I)/dt$ is proportional to $[R^*]$ at fixed $[X]_0$.

The experimentally determined initial slopes of RK binding signals indeed show the predicted linear dependence on R* (Fig. 2A, right). This is evidence that the fast binding signals seen with rhodopsin kinase (at least in a Tween solubilized form) reflect direct interaction with R*, without an intervening interaction with the membrane.[18] Although an analogous analysis of arrestin binding signals (Fig. 2B, right) yields the same dose–rate curve, enzyme digestion experiments[28] have shown that the interaction with R* triggers a conformational transition in arrestin, which, as the rate-limiting step, explains the slowness of the binding signal and the substantial activation energy[29] (see later discussion).

In the case of transducin, visual inspection of the data (Fig. 2C) suggests that the interaction between transducin and the photoexcited receptor does not follow the simple bimolecular reaction scheme [Eq. (4)]. We will see later that it proceeds via an intermediary membrane binding step.[7]

In summary, this first analysis shows that different kinetics can in some cases give immediate evidence for significant differences in the mechanism (as is evident from the difference between G_t versus RK or arrestin). However, in less favorable cases (as in the case of arrestin versus RK), a closer analysis including other techniques is required.

Rhodopsin Kinase

KLS allows one to determine basic parameters such as the dissociation constant (K_D) or the bimolecular rate constant (k_{on}) of the binding reaction. The following analysis of the interaction of RK with R* is just one example. Analogous calculations were applied to R*–arrestin[29] and G_t–PDE[11] interactions.

Dissociation Constant

At a constant concentration of R* and varying protein concentration, the LS signal amplitude increases and exhibits saturation (Fig. 3). As stated earlier, interaction between R* and RK can be described by a simple bimolecular reaction [Eq. (4)]. To fit the data, the mass action law is used, together with the conservation of the total amount of receptor ($[R^*]_{tot}$) and kinase ($[RK]_{tot}$):

$$[R^*RK] = \frac{1}{K_D}([R^*]_{tot} - [R^*RK])([RK]_{tot} - [R^*RK]) \qquad (6)$$

---

[28] K. Palczewski, A. Pulvermüller, J. Buczylko, and K. P. Hofmann, *J. Biol. Chem.* **266**, 18649 (1991).
[29] A. Schleicher, H. Kühn, and K. P. Hofmann, *Biochemistry* **28**, 1770 (1989).

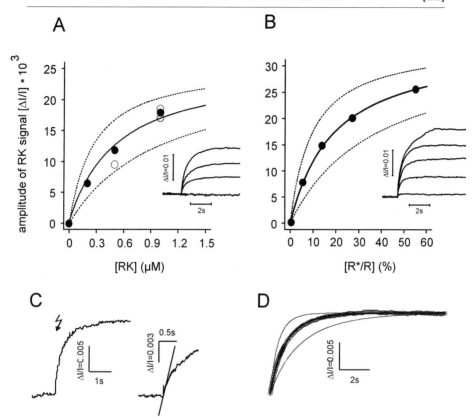

FIG. 3. Analysis from KLS binding signals of rhodopsin kinase binding to light-activated rhodopsin. (A) Saturation of the RK binding signal as a function of the RK concentration. The plot shows final amplitudes of RK binding signals from a first flash on a fresh sample; closed circles and open circles identify different RK preparations. The best fit curve (solid line) of the experimental data calculated according to Eq. (7) yields a dissociation constant for the R*–RK interaction of $K_D = 0.5\ \mu M$. Upper and lower dashed lines represent curves for K_D values of 0.25 and 1 μM, respectively. *Inset:* Binding signals for 0, 0.25, 0.5, and 1 μM RK, respectively. Measuring conditions: 5 μM R, flash excitation R*/R = 35%, 100 mM BTP buffer, pH 7, 20°; Tween 80 was maximally 0.0043%. (B) Saturation of the RK binding signal as a function of the mole fraction of R*. The best fit curve (solid line) of the experimental data calculated according to Eq. (7) yields a dissociation constant for the R*–RK interaction of $K_D = 0.3\ \mu M$. Upper and lower dashed lines represent curves for K_D values of 0.25 and 1 μM, respectively. *Inset:* Signals with flash excitation of R*/R = 0.8, 5.5, 13.8, 27.1 and 55.2%. Measuring conditions as in (A) (Tween 80 was maximally 0.0077%). (C) Analysis of the initial slope of the RK binding signal (1 μM RK, 1.8 μM R*). The initial slope obtained from the binding signal is enlarged on the right. This simple analysis yields a detergent depended on-rate of $k_{on} = 1\ \mu M^{-1}\ s^{-1}$. The Tween 80 concentration was maximally 0.01%. (D) Bimolecular reaction fit of the binding signal (1 μM RK, 2.75 μM R*). The solid line represents best fit according to Eq. (8), yielding an on-rate for the R*–RK interaction of $k_{on} = 0.8\ \mu M^{-1}\ s^{-1}$. Dotted lines are the kinetics with twice or half the on-rate assumed.

The solution is:

$$[R^*RK] = \frac{-b - \sqrt{b^2 - 4ac}}{2a} \quad (7)$$

where $a = 1$, $b = -(K_D + [R^*]_{tot} + [RK]_{tot})$, and $c = [R^*]_{tot}[RK]_{tot}$. Equation (7) was used to model the relative amplitudes ($\Delta I/I$) of the binding signal arising from the titration with both $[RK]_{tot}$ and $[R^*]_{tot}$. A scaling factor (α) relates the measured values to corresponding concentration units ($\Delta I/I = \alpha[R^*RK]$). In the curves shown in Fig. 3, K_D and α were allowed to vary.

The K_D obtained for the R*–RK complex was estimated to be 0.5 μM (Fig. 3A). When a fixed concentration of RK was titrated with varying [R*], the K_D was estimated to be 0.3 μM (Fig. 3B). Thus the K_D values for the two approaches are close (0.3 and 0.5 μM) but not identical. This indicates that the LS monitor is not strictly stoichiometric for one or both of the reaction components. Possible artifacts include a slight variation of R* binding affinity with increasing concentrations in the membrane.

Bimolecular Rate Constant

The bimolecular rate constant for the formation of the complex between R* and RK can either be obtained from the initial slope of the binding signal [Eq. (5), see Fig. 3C] or by applying a bimolecular reaction fit [Eq. (8), see Fig. 3D] to the signal.

The experiment in Fig. 3C was performed with $[RK]_{tot} = 1\ \mu M$, $[R^*]_{tot} = 1.8\ \mu M$. The initial concentrations of the proteins immediately after the first flash are $[R^*] = [R^*]_{tot}$, $[RK] = [RK]_{tot}$, and $[R^*RK] = 0$. From a calibration that assumes multiple flashes lead to virtually complete kinase binding, the concentration of kinase bound after 0.15 sec was estimated to be 0.25 μM. Using Eq. (5) it follows that, 0.25 μM/0.15 sec = $k_{on} \times 1.8\ \mu M \times 1\ \mu M$, yielding $k_{on} = 1\ \mu M^{-1}\ s^{-1}$. The variation of k_{on} (0.5 to 1 $\mu M^{-1}\ s^{-1}$ in four determinations) correlates with the different detergent concentrations in the sample (0.0043% Tween 80 for the fastest and 0.016% Tween for the slowest recorded rates, respectively).

Based on the reversible bimolecular reaction scheme [Eq. (4)] an analytical solution for the time course of R*–RK complex formation can be obtained (for details, see Pulvermüller et al.[30]):

$$[R^*RK] = \frac{[R^*RK]_\infty [R^*]_{tot}[RK]_{tot}[\exp(Zk_{on}t) - 1]}{[R^*RK]_\infty^2[\exp(Zk_{on}t)] - [R^*]_{tot}[RK]_{tot}} \quad (8)$$

[30] A. Pulvermüller, D. Maretzki, M. Rudnicka Nawrot, W. C. Smith, K. Palczewski, and K. P. Hofmann, *Biochemistry* **36**, 9253 (1997).

where $Z = ([R^*RK]_\infty^2 - [R^*]_{tot}[RK]_{tot})/[R^*RK]_\infty$, and $[R^*RK]_\infty$ represents the concentration of the complex after an infinitely long time and was determined from K_D. Equation (8) was used to model the RK binding signals, yielding $k_{on} = 0.6–1\ \mu M^{-1}\ s^{-1}$ (Fig. 3D), which is in good agreement with the results obtained earlier.

Arrestin

The reaction between arrestin and light-activated phosphorylated rhodopsin (R_p^*) is about 30 times slower than the reaction between rhodopsin kinase and rhodopsin (see Fig. 2). The different kinetics (and the high energy of activation[29]) is explained by a reaction model with at least two forms of arrestin[28]:

$$R_p^* + A_i \overset{1}{\rightleftharpoons} R_p^* + A_b \overset{2}{\rightleftharpoons} R_p^*A_b \qquad (9)$$

According to this model, arrestin exists in an equilibrium between the forms A_i and A_b. While A_i is incompatible with all signaling states of rhodopsin, A_b can bind to phosphorylated rhodopsin. In its free form, the A_i/A_b equilibrium is strongly shifted to A_i. Stabilization of A_b by binding to photolyzed and phosphorylated rhodopsin shifts the equilibrium to this interactive form. In contrast to the transducin–rhodopsin interaction (see later discussion), both forms of arrestin (A_i and A_b) are soluble. Consequently, in the coupled equilibrium [Eq. (9)], only step 2 gives rise to a binding signal, while the preequilibrium [reaction 1 in Eq. (9)] results in slowing of the monitored binding reaction. Since the rate constants of the preequilibrium can be integrated in the on-rate of the actual binding step, the kinetics of the overall reaction apparently follows a simple bimolecular reaction scheme (see earlier comments). Accordingly, the initial slopes of the binding signals depend linearly on the R* concentration (Fig. 2B, right). Extrapolation to zero [R*] shows a small intercept, which may reflect the small amount of immediately interactive A_b present at the time of flash activation. Interestingly a similar behavior is seen for RK binding to R* (Fig. 2A, right) suggesting that a comparable equilibrium of two RK conformations may exist.

Transducin

The catalytic activation of transducin involves a sequence of interactions between the light-activated receptor and the G_t–holoprotein (see also the chapter by Ernst et al.[12] in this volume). Upon binding, the affinity of G_t

for GDP becomes very low and the nucleotide dissociates.[31] An "empty site" complex of G_t with R* (R*$G_{t[empty]}$) is formed, which remains stable in the absence of nucleotide. Binding of GTP to the $G_{t\alpha}$ subunit within the R*$G_{t[empty]}$ complex enables a conformational change that eventually leads to dissociation of the active $G_{t\alpha}$GTP subunit from R* and $G_{t\beta\gamma}$.

Analysis of the R*–G_t interaction is complicated by the interaction of G_t with the membranes in the dark. Although it is assumed that essentially all G_t is bound to the disk membranes *in situ*, on dilution a significant fraction of G_t dissolves into solution. Under the conditions of the experiment (i.e., 1000-fold diluted) about 50% of G_t is bound to the membrane (the exact percentage varies with different types of disks and G_t preparations; furthermore it depends on pH, ionic strength, and divalent cations). This light-independent membrane binding is directly seen in SDS–PAGE (Fig. 2C).

In the following we analyze the binding of G_t to R* in the absence of added nucleotide. The scattering changes related to the fast activation of G_t in the presence of GTP are dealt with also.

Figure 2C shows typical G_t-binding signals as observed on suspensions of disk membranes containing rhodopsin or prephosphorylated rhodopsin, respectively. Centrifugation experiments performed under comparable conditions confirm the light-induced transition of G_t from solution to the membranes (Fig. 2C). The amplitude of the binding signal is proportional to the R* concentration and hence to the amount of G_t that binds to R*.[22] Its time course, however, is markedly slower than that of R*G_t complex formation, which can be simultaneously monitored by spectroscopic measurement of the extra metarhodopsin II formed ("extra MII"[32]; see Ernst *et al.*[12]). This demonstrates that the binding signal is not a real-time monitor of the R*G complex formation itself. The slowed but stoichiometric expression of the R*G_t complex formation in the scattering signal is explained by a two-step model [Eq. [10]], which takes the membrane binding of G_t into account.[7] In this model, membrane associated G_t (G_{tmb}) rapidly binds to R* as monitored by the extra MII formation. The depletion of the G_{tmb} pool induces binding of previously soluble G_t (G_{tsol}) to the membrane and eventually to R*. The significant activation energy (44 kJ/mol) suggests that the transition to the membrane is accompanied by a significant (rate-limiting) conformational change of G_t:

$$R^* + G_{tsol} \underset{}{\overset{1}{\rightleftharpoons}} R^* + G_{tmb} \underset{}{\overset{2}{\rightleftharpoons}} R^*G_t \qquad (10)$$

[31] N. Bennett and Y. Dupont, *J. Biol. Chem.* **260**, 4156 (1985).
[32] D. Emeis and K. P. Hofmann, *FEBS Lett.* **136**, 201 (1981).

Note that the reaction scheme is analogous to the reaction model of the interaction between arrestin and R_p^* [Eq. (9)]. In both schemes an equilibrium between an interactive and a noninteractive species precedes the actual complex formation. In contrast to arrestin, G_t binds to the membrane in the first step of the reaction sequence. Accordingly, reaction 1 gives rise to the binding signal, whereas the fast R^*G_t complex formation on the membrane (reaction 2) is not monitored by light scattering, consistent with the nonlinear dependence of the initial slopes on R^* concentration (see Fig. 2C). See Schleicher and Hofmann[7] for details.

The kinetic analysis of the binding signals yields both the rate of the membrane association of G_{tsol} and the dissociation constant (K_D) of the interaction.[7] The rate constants of forward (k_{on}) and backward (k_{off}) reaction were determined as $k_{on} = 2 \times 10^6\ M^{-1}\ s^{-1}$ and $k_{off} = 0.67\ s^{-1}$. When the concentration of membrane binding sites is tentatively equated with the rhodopsin concentration, a value of $K_D = 10^{-5}\ M$ is obtained. Due to the high overall concentrations, membrane binding is most likely not rate limiting for G_t activation *in vivo*. The G_t activation rate *in vitro*, however, is affected by the relatively slow membrane binding of the soluble fraction of G_t, which (particularly at low membrane concentrations) can lead to an underestimation of the activation rate in assays that intrinsically contain the membrane binding step (such as intrinsic $G_{t\alpha}$ fluorescence; see Ernst et al.[12]).

Other Light-Induced LS Signals

The amplitudes (i.e., final levels) of the LS signals discussed so far are stoichiometric to R^*. Other LS signals found in the literature saturate at very low R^* and were shown to be related to R^*-catalyzed G_t activation.

Dissociation Signal

In the presence of GTP, the formation of the $R^*G_{t[empty]}$ complex is rapidly followed by binding of GTP to the empty nucleotide site on the $G_{t\alpha}$ subunit and the subsequent dissociation of $G_{t\alpha}$GTP from both R^* and $G_{t\beta\gamma}$. Released R^* is free to activate further molecules of G_t, resulting in amplification of the light signal.

Activation of G_t is accompanied *in vitro* by a complete release of the $G_{t\alpha}$ GTP subunit and a partial release of the $G_{t\beta\gamma}$ subunit from the disk surface. Accordingly, activation of G_t can be KLS monitored in real time (i.e., with a fidelity of <5 ms) by probing the resulting loss of mass of the disk vesicles. The scattering change (dissociation signal; Fig. 4) is of opposite polarity to the binding signal.

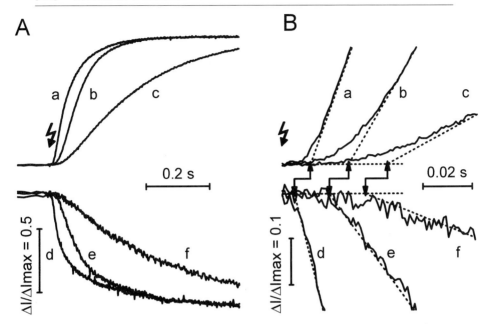

FIG. 4. Kinetic analysis of dissociation signal and PDE signal. (A) PDE signals (traces a–c) and dissociation signals (traces d–f) induced by flashes of intensity $R^*/R = 8.4 \times 10^{-2}$ (a, d), 2×10^{-3} (b, e), and 3×10^{-4} (c, f). The signals were normalized to their final amplitude for better kinetic comparison. For intense flashes the N-signal was substracted (see Fig. 2A, inset). Measuring conditions: 3 μM rhodopsin (washed membranes), 0.4 μM purified transducin, 1 mM GTP; (a, b, c) 0.3 μM purified PDE, (d, e, f) no PDE. (B) Replot of the signals shown in (A) on expanded timescales. Two parameters were used to characterize the kinetics of the signals, namely, (1) the slope of tangent at the inflection point (dashed lines) normalized to the final amplitude, and (2) the delay time that is, the time after the flash where this tangent intersects the baseline (arrows). Delays belonging to the same intensity are indicated by horizontal lines.

It is important to note that only the fraction of G_t that is present on the membranes at the time of the flash is monitored in the dissociation signal. The slower (transient) binding and activation of the soluble fraction are not reflected in light scattering on the short timescale. Consequently, the amplitude of the dissociation signal, by comparison with the binding signal allows estimation of the sizes of the [G_{tsol}] and [G_{tmb}] pools.

PDE Signal

The effector of the rod photoreceptor, the cGMP phosphodiesterase (PDE), consists of a catalytically active PDE$_{\alpha\beta}$ dimer that is inhibited in the dark by two PDE_γ subunits.[33] By contrast with the catalytic R*–G interaction, activation of PDE results from the successive binding of two $G_{t\alpha}GTP$ molecules to one molecule of PDE. Like G_t, the PDE is peripherally bound to the disk membranes by weak interactions, leading to a partial release of the enzyme into the solution at low overall concentrations.[34,35] Analysis of PDE activation is also complicated by the finding that the activation mechanism of soluble PDE differs from that of membrane bound PDE (for a review, see Hofmann and Heck[36]). For such a complex system, KLS can by useful as a complementary approach, since it probes an endogenous physically predominant step in a complicated sequence of events and measures exclusively the membrane-bound interaction mode.

Figure 4 shows typical dissociation signals evoked in suspensions of washed disk membranes by a weak flash, and in the presence of GTP and purified transducin. When purified PDE is added to an otherwise identical sample, the same flash evokes a different LS change (PDE signal; Fig. 4). We discuss this signal because it shows a real-time KLS monitor of the interaction between light-activated G_t and the membrane-bound PDE. The assignment to the reaction was made by the following observations: (1) preactivation of PDE by purified $G_{t\alpha}GTP_\gamma S$ gradually suppresses the signal, (2) the peak amplitude of the signal is stoichiometric with respect to both G* and PDE, and (3) the rise and delay times fit into a kinetic model for a reaction of activated G_t (Fig. 4). The PDE signal shows that KLS can provide a real-time monitor for a biochemical reaction even in cases in which the physical origin is not yet understood.

Titration of the PDE signal yields an apparent dissociation constant for the interaction of active G_t with the first binding site on membrane bound PDE of $K_D \leq 2.5$ nM.

Kinetics of Dissociation Signal and PDE Signal

The time course of G_t activation by R* is monitored with high fidelity by the dissociation signal. The formation of the partially active $G_{t\alpha}GTP$–PDE complex on the membrane surface, as monitored by the PDE signal, is delayed but otherwise identical in its time course. Two parameters characterize the kinetics of the signals: (1) the slope of the tangent at the inflection

---

[33] P. Deterre, J. Bigay, F. Forquet, M. Robert, and M. Chabre, *Proc. Natl. Acad. Sci. U.S.A.* **85,** 2424 (1988).
[34] P. Catty, C. Pfister, F. Bruckert, and P. Deterre, *J. Biol. Chem.* **267,** 19489 (1992).
[35] J. A. Malinski and T. G. Wensel, *Biochemistry* **31,** 9502 (1992).
[36] K. P. Hofmann, and M. Heck, in "Biomembranes II" (A. G. Lee, ed.), p. 141. JAI Press, Greenwich, Connecticut, 1996.

point, normalized to the final amplitude; and (2) the delay time, that is, the time after the flash where this tangent intersects the baseline (Fig. 4B).

The slope of the dissociation signal yields the lower limit for the rate at which membrane-bound G_t is activated. At low R*, the dissociation signal and the PDE signal display the same slope, because the rate at which active G_t is formed limits its interaction with PDE. With increasing R*, both signals are accelerated in parallel. For large fractional bleaches (R*/R > 1%) the slope of the PDE signal saturates, indicating that now the $G_{t\alpha}$GTP–PDE interaction becomes rate limiting.[11]

The delay of both the dissociation signal (t_{diss}) and the PDE signal (t_{PDE}) shortens with increasing R* (Fig. 4B). The PDE signal delay approaches 10 ms for sufficiently high R*, which gives an upper limit for the overall reaction time between rhodopsin photolysis and formation of the $G_{t\alpha}$GTP–PDE complex. Analogous evaluation of the dissociation signal yields an upper limit of 5 ms for the $G_{t\alpha}$GTP formation. The relative delay $t_{diss} - t_{PDE} = 5–8$ ms can be directly seen in Fig. 4B. It sets an upper limit to the time an average $G_{t\alpha}$GTP, once formed, spends in diffusion, coupling to PDE, and formation of the complex. The actual process could well be faster because we cannot ignore that part of the observed delay is due to the monitored structural change. In any case it fits nicely into the irreducible total delay of 7 ms of the electrical response evoked by very intense flashes.[37] It is remarkable that this reaction, measured on a 1000-fold diluted, reconstituted system, has the same speed as in the functioning photoreceptor cell. Although the membranes are diluted, the interacting proteins seem to retain their density and collisional efficiency on the membrane.

[37] W. H. Cobbs and E. N. Pugh, Jr., *J. Physiol. Lond.* **394**, 529 (1987).

[23] Heterogeneity of Rhodopsin Intermediate State Interacting with Transducin

By Yoshinori Shichida, Shuji Tachibanaki, Taku Mizukami, Hiroo Imai, and Akihisa Terakita

Introduction

Rhodopsin, the visual pigment present in rod photoreceptor cells, is a prototypical G-protein-coupled receptor (GPCR) that receives a light signal from the outer environment. In contrast to a large variety of receptors that are activated by diffusible agonist ligands, rhodopsin has the 11-*cis*-retinal chromophore as a light-activated ligand in its protein moiety. The advantages of the studies on rhodopsin in comparison with those on other GPCRs are that rhodopsin can be synchronously activated by light even at a freezing temperature. The chromophore acts as an intrinsic spectroscopic probe to monitor the protein structural changes so that each intermediate state after absorption of the photon is characterized by its absorption spectrum. Thus, rhodopsin is one of the model receptors whose activation mechanism can be elucidated at submolecular or atomic resolution.

On absorption of light, rhodopsin fades in color from red to pale yellow. This phenomenon has been referred to as *photobleaching*. So far, at least six intermediate states have been identified in the photobleaching process of rhodopsin.[1] Among the intermediate states, the state named metarhodopsin-II (meta-II) was thought to be a crucial state that binds to and activates transducin. The first implication that meta-II could be an active state of rhodopsin came from the biochemical evidence that phosphodiesterase was activated even at low temperature where meta-II did not convert to the subsequent intermediate, metarhodopsin-III (meta-III).[2] Then, spectroscopic measurements indicated that a large amount of meta-II accumulates in the presence of transducin and the accumulation was abolished by guanosine 5'-O-(3-thiotriphosphate) (GTPγS).[3,4] These findings also demonstrated that the state interacting with transducin is unequivocally identified by detecting the change of its formation and decay kinetics in the presence of transducin.

We have developed a technique called time-resolved low-temperature spectroscopy that provides both high time-resolution and wide wavelength-range recording.[5,6] We applied this technique to investigate how many intermediate states interact with transducin and showed that chicken[7] and bovine[8] rhodopsins have a new transducin-interacting state that appears between formally identified metarhodpsin-I (meta-I, now referred to as meta-Ia) and meta-II. The intermediate state named meta-Ib can form a complex with transducin but induce no GDP–GTP exchange reaction in transducin, while the exchange reaction occurs in the meta II–transducin complex. Here we show the method for the spectroscopic analysis to identify

[1] Y. Shichida and H. Imai, *Cell. Mol. Life Sci.* **54**, 1299 (1998).
[2] Y. Fukada and T. Yoshizawa, *Biochim. Biophys. Acta* **675**, 195 (1981).
[3] D. Emeis, H. Kühn, J. Reichert, and K. P. Hofmann, *FEBS Lett.* **143**, 29 (1982).
[4] K. P. Hofmann, *Biochim. Biophys. Acta* **810**, 278 (1985).
[5] H. Imai, T. Mizukami, Y. Imamoto, and Y. Shichida, *Biochemistry* **33**, 14351 (1994).
[6] H. Imai, Y. Imamoto, T. Yoshizawa, and Y. Shichida, *Biochemistry* **34**, 10525 (1995).
[7] S. Tachibanaki, H. Imai, T, Mizukami, T. Okada, Y. Imamoto, T. Matsuda, F. Fukada, A. Terakita, and Y. Shichida, *Biochemistry* **36**, 14173 (1997).
[8] S. Tachibanaki, H. Imai, A. Terakita, and Y. Shichida, *FEBS Lett.* **425**, 126 (1998).

the intermediate states interacting with transducin by using chicken rhodopsin as an example.

Preparation of Rhodopsin and Transducin

Purification of Rhodopsin

Rhodopsin is extracted from chicken retinas by a mixture of 3-[(3-cholamidopropyl)dimethylammonio]-1-propane sulfonate (CHAPS) and L-α-phosphatidylcholine from egg yolk (PC) and purified by means of column chromatography.[5,7,9] The purified rhodopsin in buffer A [0.6% (w/v) CHAPS, 0.8 mg/ml PC, 20% (w/v) glycerol, 50 mM N-2-hydroxyethylpiperazine-N'-2-ethanesulfonic acid (HEPES), 140 mM NaCl, 2 mM $MgCl_2$, 1 mM dithiothreitol (DTT), 0.1 mM phenylmethylsulfonyl fluoride (PMSF), 4 μg/ml leupeptin, 50 KIU/ml aprotinin, pH 7.5, at 4°] is stored at −80° until use. The concentration of rhodopsin in the sample is estimated by its absorbance at the maximum (503 nm, molar extinction coefficient 40,500; Ref. 10).

Purification of Transducin

Transducin is purified from fresh bovine retinas according to the method reported by Fukada et al.[11] Briefly, transducin is extracted from bovine rod outer segment (ROS) membranes with a hypotonic buffer B [5 mM tris(hydroxymethyl)aminomethane (Tris), 0.5 mM $MgCl_2$, 0.1 mM PMSF, 4 μg/ml leupeptin, 50 KIU/ml aprotinin, 1 mM DTT, pH 7.2, at 4°] supplemented with 100 μM guanosine triphosphate (GTP). Then it is applied to a Blue Sepharose (Pharmacia, Piscataway, NJ) column that has been equilibrated with buffer B. Transducin α subunit (Tα) binds to the column, whereas its βγ subunit (Tβγ) flows through the column. Thus Tα is eluted from the column with buffer C [10 mM 3-(N-morpholino)propanesulfonic acid (MOPS), 600 mM NaCl, 2 mM $MgCl_2$, 0.1 mM PMSF, 4 μg/ml leupeptin, 50 KIU/ml aprotinin, 1 mM DTT, pH 7.5, at 4°]. The eluted fractions containing Tβγ are applied to a DEAE-Toyopearl 650S (Tosoh, Tokyo, Japan) column from which Tβγ is eluted with buffer C. Tα and Tβγ fractions are then applied to a gel filtration column, Superdex 75-pg HiLoad 26/60

---

[9] T. Okano, Y. Fukada, I. D. Artamonov, and T. Yoshizawa, *Biochemistry* **28**, 8848 (1989).
[10] T. Okano, Y. Fukada, Y. Shichida, and T. Yoshizawa, *Photochem. Photobiol.* **56**, 995 (1992).
[11] Y. Fukada, T. Matsuda, K. Kokame, T. Takao, Y. Shimonishi, T. Akino, and T. Yoshizawa, *J. Biol. Chem.* **269**, 5163 (1994).

(Pharmacia), from which they are eluted with buffer C. Tα is eluted as a single peak, whereas T$\beta\gamma$ is eluted as two peaks, which contain farnesylated/ nonmethylated and farnesylated/methylated forms of γ subunits, respectively. It is known that the latter form of Tγ is dominant *in vivo* and we therefore use the latter fractions for the experiments. The fractions contain >95% of methylated Tγ, which is estimated by a previously described method.[12] Then equal amounts of Tα and T$\beta\gamma$ are mixed and the buffer exchanged for 50 mM HEPES, 140 mM NaCl, 2 mM MgCl$_2$, 1 mM DTT, 4 μg/ml leupeptin, 50 KIU/ml aprotinin (pH 7.5 at 4°) using a gel filtration column (PD10, Pharmacia Biotech), followed by concentration to about 150 μM by ultrafiltration membrane (Centricon-30, Amicon, Danvers, MA). Note that Tα prepared under our experimental conditions is a GDP-bound form[13–16] and can form heterotrimer with T$\beta\gamma$.[11] The sample is stored at −80° until use.

Time-Resolved Low-Temperature Spectroscopy

Methods

Low-temperature absorption spectroscopy can be used to determine the number of intermediate states present in the photobleaching process of rhodopsin.[17] In this method, the rhodopsin sample is cooled to a temperature where an intermediate state produced by irradiation of rhodopsin is spectroscopically trapped. This method is useful in surveying the bleaching process of rhodopsin and several distinctive intermediates, such as bathorhodopsin (batho), lumirhodopsin (lumi), metarhodopsins-I,-II, and -III, have been identified.[1] Furthermore, in combination with other spectroscopic techniques such as circular dichroism (CD),[18] resonance Raman, and Fourier transform infrared[19] spectroscopies, this low-temperature trapping method elucidates detailed structural changes of the chromophore and the changes in interaction between the chromophore and nearby proteins.

---

[12] H. Ohguro, Y. Fukada, T. Yoshizawa, T. Saito, and T. Akino, *EMBO J.* **10,** 3669 (1991).
[13] C. Kleuss, M. Pallast, S. Brendel, W. Rosenthal, and G. Schults, *J. Chromatog.* **407,** 281 (1987).
[14] A. Yamazaki, M. Tatsumi, D. C. Torney, and M. W. Bitensky, *J. Biol. Chem.* **262,** 9316 (1987).
[15] J. P. Noel, H. E. Hamm, and P. B. Sigler, *Nature* **366,** 654 (1993).
[16] D. G. Lambright, J. P. Noel, H. E. Hamm, and P. B. Sigler, *Nature* **369,** 621 (1994).
[17] T. Yoshizawa and Y. Shichida, *Methods Enzymol* **81,** 333 (1982).
[18] T. Yoshizawa and Y. Shichida, *Methods Enzymol* **81,** 634 (1982).
[19] K. Fahmgi, T. Sakmar, and F. Siebert, *Methods Enzymol.* **315** [13], 1999 (this volume).

However, because of experimental limitations, this method has not been applied to the kinetic measurements of the intermediate states and these have been studied by flash (or laser) photolysis techniques.[20] The flash photolysis techniques are suitable for monitoring the kinetics of the intermediate states but the signal-to-noise ratio of the obtained data is low compared to the low-temperature trapping method. Thus one must prepare a large amount of sample to integrate and average the obtained data especially for a photobleachable pigment such as rhodopsin. We have applied a low-temperature technique to investigate the kinetics of rhodopsin intermediate using a conventional spectrophotometer by which spectral changes observed on the millisecond timescale at room temperature were monitored on the minute timescale by cooling the sample at a relatively low temperature. As a result, more than 100 spectra with good signal-to-noise ratio were recorded from a 4- to 50-μg single rhodopsin sample during the incubation of the sample up to 3 days after the irradiation. The most important point is that one can monitor spectral changes using a single sample to be free from a dispersion of the experimental data due to the exchange of the sample.

The experimental setup is schematically illustrated in Fig. 1. Absorption spectra are recorded on a recording spectrophotometer (MPS-2000, Shimadzu, Kyoto, Japan) interfaced to a personal computer (PC9801RA NEC, Tokyo, Japan). The spectrophotometer used in the experiments is one of the typical spectrophotometers having a probe light that is very weak, so that little bleaching of the rhodopsin in the sample induced by the probe light occurs. In fact, our experiments indicate that less than 1.5% of bleaching is observed after 100 recordings of the spectra at wavelengths from 750 to 330 nm. A flow-type optical cryostat (CF1204, Oxford, UK) with a 1-cm light path optical cell is used to record the spectral changes for more than 3 days at low temperatures. The sample temperature is regulated to within $0.1°$ by a temperature controller (ITC4, Oxford, UK) attached to the cryostat. The sample is irradiated with light from a 1-kW tungsten–halogen lamp (Philips, Tokyo, Japan) through the specially designed optical setup including shutters and a movable mirror operated by a rotary solenoid (Fig. 1). This setup enables us to irradiate the sample in the cryostat without removing the cryostat from the spectrophotometer. The wavelength of the irradiation light is selected with a glass cutoff filter (Toshiba, Tokyo, Japan). A 5-cm water layer is placed in front of the light source to remove heat from the irradiation light. Thermal reactions of intermediates initiated by irradiation of rhodopsin sample at $-35°$ or $-25°$ are monitored by recording absorption spectra with intervals of 2.5–30 min until the reactions were almost satu-

[20] J. W. Lewis and D. S. Kliger, *Methods Enzymol.* **315** [12], 1999 (this volume).

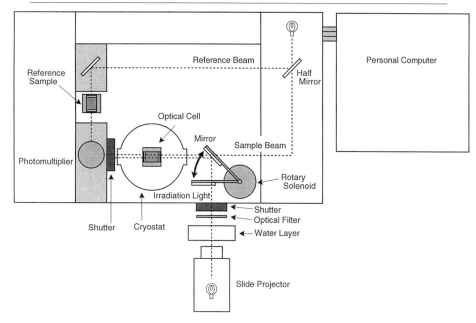

FIG. 1. A schematic diagram of the time-resolved low-temperature spectroscopy apparatus.

rated. All the procedures including irradiation of the sample and recording the spectra are computerized. The recording of each absorption spectrum in the wavelength region from 750 to 330 nm required 66 sec.

To investigate which intermediates interact with transducin, the effects of transducin and guanosine GTPγS on the thermal reactions of rhodopsin intermediates were examined. GTPγS is a good reagent for studying rhodopsin–transducin interaction because it is known that GTPγS is hardly hydrolyzable and can abolish the coupling between rhodopsin intermediate meta-II and transducin.[3,4] The following three samples in buffer A supplemented with 50% glycerol were prepared: Sample A contained 3.01 μM rhodopsin, sample B contained 3.01 μM rhodopsin and 9.60 μM transducin, sample C contained 3.01 μM rhodopsin, 9.60 μM transducin, and 421 μM GTPγS. Fifty percent glycerol was used to prevent the sample from freezing during low-temperature spectroscopy. Each glycerol-containing sample shows high viscosity and should be carefully mixed.

Estimation of Amount of Photoconverted Pigment

The amount of rhodopsin photoconverted to the intermediates by irradiation at low temperature ($-35°$ or $-25°$) is estimated as follows: The irradiated sample at $-35°/-25°$ is warmed to 20° and 1 M hydroxylamine is added to the sample at a final concentration of 10 mM, followed by incubation at this temperature until intermediates produced by the irradiation are completely converted to retinal oxime and opsin. Then the sample is cooled to $-35°/-25°$ to record the spectrum. To bleach the residual rhodopsin and the small amount of isorhodopsin present in the irradiated sample, the sample is irradiated with yellow light (>500-nm light) at 0° and the spectrum recorded at $-35°/-25°$. The differences spectrum between these spectra at wavelengths longer than 500 nm is then simulated with the separately obtained spectra of rhodopsin and isorhodopsin.[5] The total amount of intermediates produced is calculated by subtracting the amount of residual rhodopsin and isorhodopsin from that of the original rhodopsin. Under our experimental conditions, about 40% of rhodopsin in the sample is always photoconverted to the intermediates, indicating the irradiation system is reproducible.

Collected spectra

Figure 2 shows the absorption spectra of the rhodopsin intermediates in three samples measured by time-resolved low-temperature spectroscopy, namely, samples which contained (A) only rhodopsin, (B) rhodopsin and transducin, and (C) rhodopsin, transducin, and GTPγS. Each sample was cooled to $-35°$ or $-25°$, irradiated with >570-nm light for 30 sec, and the spectral changes due to the thermal reactions of intermediates recorded. The spectral changes obtained were then expanded as shown in the lower panels of Figs. 2a–f, in which the changes are represented by the difference spectra calculated by subtracting the spectrum recorded immediately after the irradiation from those recorded at later times after the irradiation.

Irradiation of each sample at $-35°$ formed a mixture containing mainly lumi[5] and caused a blue shift of absorption spectrum with a slight increase in absorbance. Subsequent incubation at this temperature resulted in decrease of absorbance at about 520 nm and increase in absorbance at about 440 nm (Figs. 2a–c). The irradiation at $-25°$ also caused a blue shift of spectrum with increase of absorbance and the shift of maximum was slightly larger than that observed at $-35°$. Subsequent incubation resulted in an increase in absorbance at about 380 nm with a concurrent decrease in absorbance at about 500 nm (Figs. 2d–f). The final product formed by the incubation was meta-II because it displayed a large absorbance at about 380 nm. At these two temperatures, the thermal reaction of rhodopsin can be recorded from lumi to meta-II.

The addition of transducin in the rhodopsin sample causes formation of a larger amount of meta-II as reported previously.[3,4] The difference

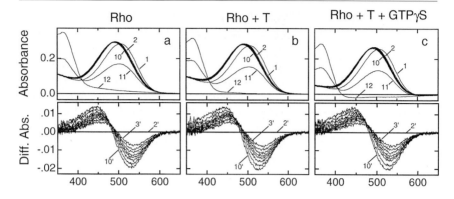
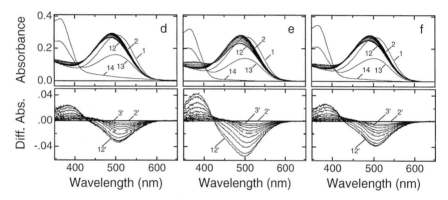

FIG. 2. Effects of transducin and/or GTPγS on the thermal reactions of rhodopsin intermediates at various temperatures. (a–c) Three samples, each of which contained (a) rhodopsin [sample A], (b) rhodopsin + transducin [sample B], or (c) rhodopsin + transducin + GTPγS [sample C] were cooled to −35° (curves 1) and irradiated with >570-nm light for 30 sec (curves 2), followed by incubation at this temperature for 2.5, 5, 10, 20, 40, 80, 160, and 320 min (curves 3–10). Then the samples were warmed to 20° and 1 M hydroxylamine was added to the sample at a final concentration of 10 mM. After recording the spectra at −35° (curves 11), they were irradiated with >500-nm light at 0° and the spectra were measured at −35°. (d–f) Three samples, each of which contained (d) rhodopsin [sample A], (e) rhodopsin + transducin [sample B], or (f) rhodopsin + transducin + GTPγS [sample C], were cooled to −25° (curves 1) and irradiated with >570-nm light for 30 sec (curves 2), followed by incubation at this temperature for 2.5, 5, 10, 20, 40, 80, 160, 320, 640, and 1280 min (curves 3–12). Then the samples were warmed to 20° and 1 M hydroxylamine was added to the sample at a final concentration of 10 mM. After recording the spectra at −25° (curves 13), they were irradiated with >500-nm light at 0° and the spectra were measured at −25° (curves 14). In the lower panel of each figures, difference spectra obtained by subtracting the spectrum recorded immediately after the irradiation from the spectra recorded at later times after the irradiation are presented. Each spectrum is the sum of two spectra recorded by independent experiments.

spectrum obtained by irradiation of rhodopsin in the presence of transducin at $-25°$ (curve 12' in Fig. 2e) displayed absorbance at 380 nm significantly larger than that obtained in the absence of transducin (curve 12' in Fig. 2d), showing the formation of extra meta-II. On the other hand, the presence of GTPγS caused little enhancement of the formation of meta-II even in the presence of transducin (curve 12' in Fig. 2f), suggesting that GTPγS can abolish the interaction between meta-II and transducin. The detection of extra meta-II indicates that the coupling of rhodopsin intermediate(s) with transducin is monitored under these experimental conditions.

These fine time-resolved recordings show several important points that suggest the existence of a new intermediate state between meta-I and meta-II; at each temperature ($-35°$ or $-25°$), the spectral changes due to the thermal reactions of intermediates did not form an isosbestic point, and the positive and negative maxima of the difference spectrum shifted to the blue as the time of incubation increased, indicating that at least two conversion processes took place during the incubation at each temperature. Thus, we have to carefully analyze these spectral changes in order to identify rhodopsin intermediate state(s) interacting with transducin.

Spectral Analyses

The spectral changes due to the thermal reactions of intermediates were analyzed by one of the mathematical methods using the matrix transformation procedures, singular value decomposition (SVD), and global exponential fitting with computer programs originally written on a NEC PC9821 V10 computer, according to the theory reported previously[21–24] in order to obtain b-spectra and apparent time constants (Fig. 3).

First, a set of difference spectra reflecting the spectral change was calculated by subtracting the spectrum recorded 2.5 min after the irradiation from those recorded at later times after the irradiation. Then the difference spectra whose recording times after the irradiation were near the $2.5 \times 2^{0.3i}$ min ($i = 1, 2, \ldots, n$) were selected from the set of difference spectra. After the number of wavelength points (421 points) of the spectra was reduced to 84 points by averaging over each 5 points, these spectra were subjected to the spectral analysis.

[21] G. H. Golub and C. Reinsch, *Numer. Math.* **14**, 403 (1970).
[22] S. J. Hug, J. W. Lewis, C. M. Einterz, T. E. Thorgeirsson, and D. S. Kliger, *Biochemistry* **29**, 1475 (1990).
[23] E. R. Henry and J. Hofrichter, *Methods Enzymol.* **210**, 129 (1992).
[24] T. E. Thorgeirsson, J. W. Lewis, S. E. Wallace-Williams, and D. S. Kliger, *Biochemistry* **32**, 13861 (1993).

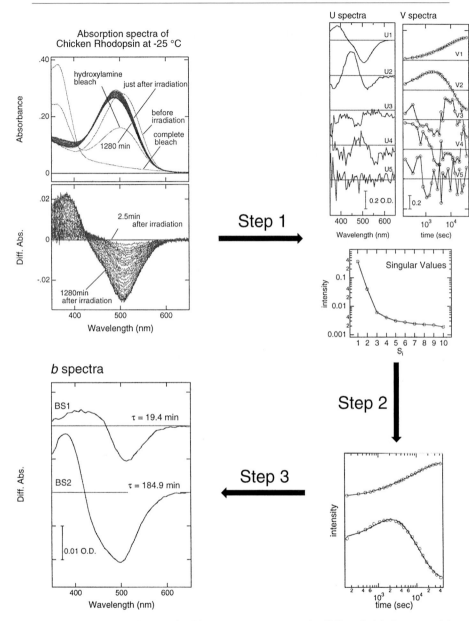

FIG. 3. Analyses of time-resolved low-temperature spectra by SVD and global exponential fitting. The time-resolved low-temperature spectra were recorded after irradiation of rhodopsin with >570-nm light for 30 sec at $-25°$ (upper left panel). From the difference spectra calculated by subtracting the spectrum recorded 2.5 min after the irradiation from those recorded at later times after the irradiation, an $84 \times n$ matrix was made and it was transferred to a product

The difference spectra expressed by an $84 \times n$ matrix were transferred to a product of three matrices **U**, **S**, and $\mathbf{V^T}$ by means of SVD (step 1 in Fig. 3). **U** is an $84 \times n$ matrix whose columns are called basis spectra. **S** is an $n \times n$ diagonal matrix whose diagonal elements are singular values. **V** is an $n \times n$ matrix describing the composition of the original matrix in terms of the basis spectra. This mathematical analyses can provide an estimation of how many spectrally distinct intermediates are needed to reproduce the spectral change within the noise of the experiments. Then the *b*-spectra representing spectral changes and their apparent time constants were calculated after global exponential fitting of the rows of $\mathbf{V^T}$ with exponential time functions (step 2 in Fig. 3). The number of exponential time functions was estimated by the number of *meaningful* singular values and basis spectra. In the case of Fig. 3, meaningful singular values are S1 and S2. Note that opposite signed *b*-spectra are presented in Figs. 4, 5, and 6 in order to be easily compared with the spectral changes observed by the low-temperature experiments.

The spectral changes observed at $-35°$ were expressed by two *b*-spectra. None of the *b*-spectra reflect the formation of meta-II because they show no absorbance peaks below 400 nm (Figs. 4a–c, upper and lower curves). This observation pointed out the presence of at least two intermediates in the conversion process from lumi to meta-II. The spectral properties also show that the intermediates have absorption maxima located between those of lumi and meta-II. Two intermediates between lumi and meta-II are referred to as meta-Ia and meta Ib.[7] The apparent time constant for the first *b*-spectrum was significantly smaller (18 times) than that for the second *b*-spectrum, indicating that the two conversion processes were well separated. Thus, it can be simply summarized that the first *b*-spectrum reflects the decay of lumi to meta-Ia and the second *b*-spectrum the decay of meta-Ia to meta-Ib.[25]

---

[25] Strictly, the first and the second *b*-spectra reflect the decay of lumi to the quasi-equilibrium state of lumi and its following intermediate (meta-Ia) and that of the quasi-equilibrium state to an equilibrium state of lumi, meta-Ia, and meta-Ib, respectively. Note that these explanations are valid only when two conversion processes are well separated. For the detailed analysis of the conversion processes in terms of reaction kinetics, the reader is referred to other papers[7,24] and reviews.[20,23]

---

of three matrices **U**, **S**, and $\mathbf{V^T}$ by means of SVD (step 1). The rows of $\mathbf{V^T}$ were selected on the basis of the meaningful singular values (S1 and S2) and were fitted with exponential time functions by means of global exponential fitting (step 2). The *b*-spectra were then calculated by using the selected columns of **U**, singular values and the matrix produced by global fitting of the rows of $\mathbf{V^T}$ (step 3).

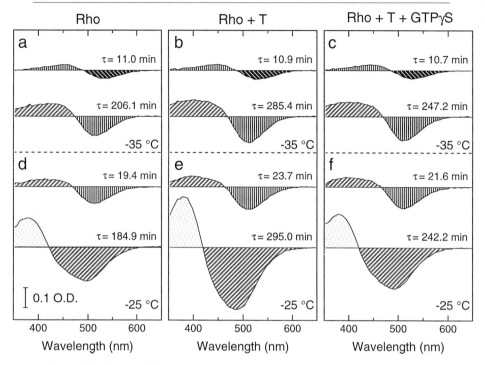

Fig. 4. Calculation of b-spectra from the spectral changes observed at −35° and −25°. Samples contained (a, d) only rhodopsin [sample A], (b, e) rhodopsin + transducin [sample B], and (c, f) rhodopsin + transducin + GTPγS [sample C]. In each panel, upper and lower curves are the first and second b-spectra, respectively. Time constants for the respective b-spectra and the measured temperatures are shown in the panels. The b-spectra are normalized so that they represent the changes induced by the photoreaction of 1.0 absorbance of rhodopsin.

The spectral changes observed at −25° were also expressed by two b-spectra (Figs. 4d–f) with significant difference in the apparent time constant. The first b-spectra at −25° (Figs. 4d–f, upper curves) were similar in shape to the second b-spectra at −35° (Figs. 4a–c, lower curves), suggesting that the state formed at the later stage at −35° is similar to that formed at the early stage at −25°, namely, the reaction meta-Ia → meta-Ib. The second b-spectrum also roughly showed the formation of meta-II from meta-Ib (Figs. 4d–f, lower curves).

Identification of Intermediate States Interacting with Transducin

The b-spectra of the sample containing only rhodopsin were compared with the spectra of the sample containing rhodopsin and transducin and

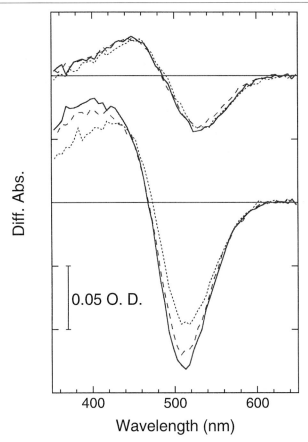

Fig. 5. The b-spectra obtained from the spectral changes observed at $-35°$. Upper and lower curves represent the first and second b-spectra. The b-spectra of the samples containing only rhodopsin [sample A], rhodopsin + transducin [sample B], and rhodopsin + transducin + GTPγS [sample C] are represented by dotted, solid, and dashed curves, respectively. The b-spectra are normalized so that they represent the changes induced by the photoreaction of 1.0 absorbance of rhodopsin.

the sample containing rhodopsin, transducin, and GTPγS to identify the intermediate states whose thermal reaction is affected by transducin and/ or GTPγS (Figs. 4 and 5). The b-spectra and their apparent time constants reflecting the conversion from lumi to meta-Ia were similar among each other (Fig. 5, upper curves), indicating that transducin as well as GTPγS does not affect the conversion process from lumi to meta-Ia. So it is concluded that either lumi or meta-Ia does not interact with transducin.

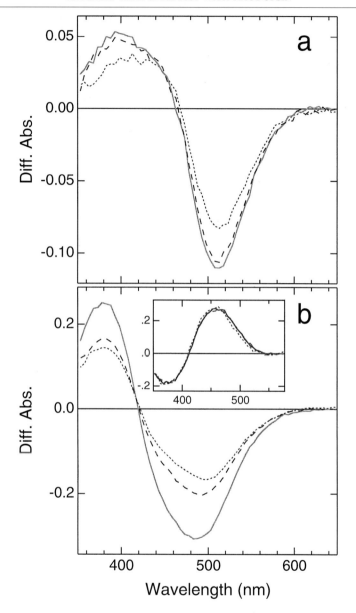

The shape of the b-spectrum reflecting the conversion from meta-Ia to meta-Ib (Fig. 5, lower curves, and Fig. 6a) changed when transducin was present. Because transducin does not interact with lumi or meta-Ia, meta-Ib is the intermediate that interacts with transducin. The interesting observation was that the b-spectrum obtained in the presence of both transducin and GTPγS (lower dashed curve in Fig. 5 and dashed curve in Fig. 6a) was similar in shape to that obtained in the presence of only transducin (lower solid curve in Fig. 5 and solid curve in Fig. 6a), but different from that obtained in the absence of transducin (lower dotted curve in Fig. 5 and dotted curve in Fig. 6a). These results indicated that the addition of GTPγS had no effect on binding of transducin to meta-Ib. Therefore, it can be concluded that meta-Ib interacts with transducin but does not induce the GDP–GTP exchange reaction in transducin.

The presence of transducin caused the decrease of the negative ~510-nm peak and the increase of the positive absorption between 400 and 440 nm. This spectral effect roughly suggested that the addition of transducin enlarged the formation of meta-Ib, that is, the interaction of transducin stabilized meta-Ib.[26]

On the other hand, as already inferred from the raw data of spectral changes (Figs. 2d–f), the b-spectra reflecting the formation of meta-II were greatly affected by transducin (Fig. 6b). Transducin enhanced the accumulation of meta-II and GTPγS abolished the enhancement, demonstrating that transducin can form a complex with meta-II and dissociates from meta-II through the GDP–GTP exchange reaction.

The comparison of b-spectrum in view of the GTPγS effect highlighted two stepwise reactions of G protein with rhodopsin, that is, transducin binds to meta-Ib state and GTP–GDP exchange occurs in the meta-II state.

Estimation of Absorption Maximum of Intermediates

The absorption maximum of the intermediate state gives information about partial molecular characteristics of the intermediate. To estimate the

---

[26] This speculation should be confirmed by analyses of rate constants.[7]

FIG. 6. The b-spectra obtained from the spectral changes observed at $-25°$. The b-spectra of the samples containing only rhodopsin [sample A], rhodopsin + transducin [sample B], and rhodopsin + transducin + GTPγS [sample C] are represented by dotted, solid, and dashed curves, respectively. (a) The first b-spectra. (b) The second b-spectra. *Inset:* Normalized difference spectra between meta-Ib and meta-II calculated from the b-spectra obtained from sample A (dotted curve), sample B (solid curve), and sample C (dashed curve). The b-spectra are normalized so that they represent the changes induced by the photoreaction of 1.0 absorbance of rhodopsin.

absorption spectrum of meta-Ib, the following calculations can be performed using the first (BS1) and second b-spectra (BS2) at $-25°$. The conversion process at $-25°$ was expressed by two b-spectra, while four intermediates (lumi, meta-Ia, meta-Ib, and meta-II) appeared in the process. Strictly, these facts indicate the presence of a quasi-equilibrium state between lumi and meta-Ia although it was mentioned that the first b-spectra at $-25°$ mainly reflected the decay of meta-Ia to meta-Ib. Thus the b-spectra are expressed by a linear combination of the spectra of these intermediates as follows:

$$BS1 = a_1(\alpha\varepsilon_{lumi} + \beta\varepsilon_{meta-Ia}) + b_1\varepsilon_{meta-Ib} + c_1\varepsilon_{meta-II}$$
$$BS2 = a_2(\alpha\varepsilon_{lumi} + \beta\varepsilon_{meta-Ia}) + b_2\varepsilon_{meta-Ib} + c_2\varepsilon_{meta-II}$$

where ε_{lumi}, $\varepsilon_{meta-Ia}$, $\varepsilon_{meta-Ib}$, $\varepsilon_{meta-II}$ are the absorption spectra of lumi, meta-Ia, meta-Ib, and meta-II intermediates, and a_i, b_i, and c_i ($a_i + b_i + c_i = 0$; $i = 1, 2$) are the mole fractions of the respective intermediates. Constants α and β are derived from the equilibrium constant between lumi and meta-Ia. The BS1 exhibited a positive maximum at the shorter wavelength (Fig. 6a). This fact indicates that the absorption spectrum of meta-Ib is blue shifted from those of lumi and meta-Ia, because BS1 reflects mainly the conversion from a mixture of lumi and meta-Ia to meta-Ib. Therefore, both the BS1 and BS2 at longer wavelengths originate from only lumi and meta-Ia. Then we calculated the ratio of a_1 to a_2 from absorbances of the b-spectra at a wavelength region from 555 to 580 nm, where the ratios of absorbance were constant, while at shorter wavelengths they increased. The difference spectrum between meta-Ib and meta-II was calculated by subtracting the BS2 from the BS1 after normalizing the BS2 with the ratio. The calculated spectrum is shown in the inset of Fig. 6b. Because meta-II has no absorbance at wavelength regions longer than 420 nm, the absorption maximum of meta-Ib can be estimated to be 460 nm. This fact suggested that a retinal Schiff base was still protonated in meta-I but deprotonated in meta-II state. The meta-Ib and meta-II are well distinguished by the characteristics of their absorption spectrum.

Conclusion

Our method, which combined time-resolved low-temperature spectroscopy with a mathematical method (SVD) demonstrated heterogeneous intermediate states interacting with transducin, meta-Ib, and meta-II. Our method shows not only the heterogeneity but also the two-step interaction of rhodopsin with transducin, which means the activation of transducin by photoactivated rhodopsin. A similar two-step interaction has been also found in bovine rhodopsin by using the same method. Unlike chicken meta-

Ib, bovine meta-Ib was detected only by detailed kinetics analysis of the bleaching process, but it was stabilized by transducin and visualized in the observed spectral changes as in Fig. 2.[8] From the effect of GTPγS, it was also revealed that bovine meta-Ib induced no GDP–GTP exchange reaction in transducin. Thus meta-Ib is a common intermediate of vertebrate rhodopsin and transducin is activated, in two steps, by meta-Ib and meta-II.

Although our experiments were performed at relatively low temperature ($-25°$ or $-35°$), this method is powerful enough to examine the interaction between rhodopsin and transducin.

Acknowledgments

This work was supported in part by Grants-in-Aid for Scientific Research from the Japanese Ministry of Education, Science, Sports and Culture.

[24] Limited Proteolytic Digestion Studies of G Protein–Receptor Interactions

By MARIA R. MAZZONI and HEIDI H. HAMM

Introduction

The recent resolution of crystal structures of Gα subunits in both active[1–3] and inactive[4] conformations as well as of isolated Gβγ subunit[5] and Gαβγ heterotrimers[6,7] has provided a fundamental context for understanding their activation mechanism and ability to interact with heptahelical receptors. Elucidation of Gα-subunit structure has allowed us to define the mechanistic basis for data obtained by biochemical,[8–12] mutagenic, or

---

[1] J. P. Noel, H. E. Hamm, and P. B. Sigler, *Nature* **366,** 654 (1993).
[2] D. E. Coleman, A. M. Berghuis, E. Lee, M. E. Linder, A. G. Gilman, and S. R. Sprang, *Science* **265,** 1405 (1994).
[3] R. K. Sunahara, J. J. G. Tesmer, A. G. Gilman, and S. R. Sprang, *Science* **278,** 1943 (1997).
[4] D. G. Lambright, J. P. Noel, H. E. Hamm, and P. B. Sigler, *Nature* **369,** 621 (1994).
[5] J. Sondek, A. Bohm, D. G. Lambright, H. E. Hamm, and P. B. Sigler, *Nature* **379,** 369 (1996).
[6] D. G. Lambright, J. Sondek, A. Bohm, N. P. Skiba, H. E. Hamm, and P. B. Sigler, *Nature* **379,** 311 (1996).
[7] M. A. Wall, D. E. Coleman, E. Lee, J. A. Iniguez-Lluhi, B. A. Posner, A. G. Gilman, and S. R. Sprang, *Cell* **83,** 1047 (1995).
[8] H. E. Hamm, Deretic D., A. Arendt, P. A. Hargrave, B. Koenig, and K. P. Hoffmann, *Science* **241,** 832 (1988).
[9] M. M. Rasenick, M. Watanabe, M. B. Lazarevic, G. Hatta, and H. E. Hamm, *J. Biol. Chem.* **269,** 21519 (1994).
[10] E. L. Martin, S. Rens-Damiano, P. J. Schatz, and H. E. Hamm, *J. Biol. Chem.* **271,** 361 (1996).
[11] M. R. Mazzoni and H. E. Hamm, *J. Biol. Chem.* **271,** 30034 (1996).
[12] A. Gilchrist, M. R. Mazzoni, B. Dineen, A. Dice, J. Linden, W. R. Proctor, C. R. Lupica, T. V. Dunwiddie, and H. E. Hamm, *J. Biol. Chem.* **273,** 14912 (1998).

chimeric[13-18] studies of receptor-interacting regions on G proteins. The main model system characterized by these investigations[8,10,11,15] is the light receptor, rhodopsin (Rh), and the rod G protein, transducin or G_t. Light activation of rhodopsin leads to the formation of metarhodopsin-II (Rh*), which, like an agonist-activated receptor, is able to interact with and in turn activate heterotrimeric G_t. Metarhodopsin-II catalyzes a conformational change in the $G\alpha_t$ subunit with opening of the nucleotide binding pocket, a decrease of GDP affinity, and the consequential nucleotide release. In the "empty-pocket" state, G_t binds with high affinity to metarhodopsin-II. This high-affinity interaction is disrupted when GTP replaces GDP, leading to a further conformational change of $G\alpha_t$ that causes a decrease of the subunit affinity for both rhodopsin and $G\beta\gamma$. In the absence of guanine nucleotides, the molecular basis of the high-affinity interaction between receptors and G proteins can be examined in great detail using this model system. In fact, metarhodopsin-II tightly bound to G_t shows a functional behavior similar to hormone or neurotransmitter receptors in the high-affinity state for agonist ligands.

A variety of studies have implicated the carboxyl terminus of the $G\alpha$ subunits in mediating receptor G-protein selectivity.[8,10,12-16] However numerous evidence now supports a model in which other $G\alpha$ regions[8,9,11,17,18] and perhaps some parts of the $G\beta\gamma$ subunits[19-21] cooperate in forming more complex interaction sites for receptors. In the attempt to delineate regions on $G\alpha_t$ and $G\beta_t$ involved in interactions with rhodopsin, we have studied the limited tryptic digestion of G_t in the presence of the inactive or light-activated receptor.[11] This approach is possible because the tryptic digestion pattern of G_t is well known[22,23] and the cleavage sites have been character-

[13] B. R. Conklin, Z. Farfel, K. D. Lusting, D. Julius, and H. R. Bourne, *Nature* **363**, 274 (1993).
[14] J. Liu, B. R. Conklin, N. Blin, J. Yun, and J. Wess, *Proc. Natl. Acad. Sci. U.S.A.* **92**, 11642 (1995).
[15] R. Onrust, P. Herzmark, P. Chi, P. D. Garcia, O. Lichtarge, C. Kingsley, and H. R. Bourne, *Science* **275**, 381 (1997).
[16] E. Kostenis, B. R. Conklin, and J. Wess, *Biochemistry* **36**, 1487 (1997).
[17] C. Lee, A. Katz, and M. I. Simon, *Mol. Pharmacol.* **47**, 218 (1995).
[18] H. Bae, K. Anderson, L. A. Flood, N. P. Skiba, H. E. Hamm, and S. G. Graber, *J. Biol. Chem.* **272**, 32071 (1997).
[19] J. Taylor, G. Jacob-Mosier, R. Lawton, M. VanDort, and R. Neubig, *J. Biol. Chem.* **271**, 3336 (1996).
[20] O. Kisselev, A. Promin, M. Ermolaeva, and N. Gautam, *Proc. Natl. Acad. Sci. U.S.A.* **92**, 9102 (1995).
[21] H. Yasuda, M. Lindorfer, K. Woodfork, J. Fletcher, and J. Garrison, *J. Biol. Chem.* **271**, 18588 (1996).
[22] B. K.-K. Fung and C. R. Nash, *J. Biol. Chem.* **258**, 10503 (1983).
[23] M. R. Mazzoni, J. A. Malinski, and H. E. Hamm, *J. Biol. Chem.* **266**, 14072 (1991).

ized.[24] The limited tryptic digestion patterns of both G_i and G_o are also defined.[24,25] Here, we describe the use of limited tryptic digestion of G_t in the presence of rod outer segment (ROS) membranes or phospholipid vesicles to analyze the molecular mechanisms of G_t's interaction with rhodopsin and membranes, and the conformational state Rh* · Gα–empty$\beta\gamma$. This methodological approach can be adapted to other receptor–G-protein systems because they share similar molecular mechanisms of interaction and activation.

General Considerations

Limited Proteolytic Patterns of G-Protein Subunits

A protein can be cleaved chemically or enzymatically to generate various internal peptides. The number of peptides produced depends on whether a protein is cleaved completely at many sites or at a limited number of sites. Analysis of limited proteolytic patterns is a powerful way to probe the conformation of the tissue purified or the *in vitro* synthesized Gα or G$\beta\gamma$ subunits. Various proteolytic enzymes have been utilized under native conditions to digest G proteins.[22–29] However, the tryptic cleavage products of Gα and G$\beta\gamma$ are the best characterized.[22–26]

Both Gα and G$\beta\gamma$ subunits have more than 30 potential tryptic cleavage sites each. Nevertheless, in the native molecules, the number of sites available is extremely limited. Whereas the Gγ subunit is apparently uncleaved by trypsin, the Gβ subunit has only one tryptic cleavage site (Arg-129) accessible in the native molecule that splits it into two fragments of ~23 and 14 kDa[22,23,25] (Fig. 1). The tryptic cleavage pattern of Gα subunits depends on whether a guanine nucleoside diphosphate or triphosphate is bound to the active site. When a nonhydrolyzable guanine nucleoside triphosphate analog, such as guanosine 5'-O-(3-thiotriphosphate) (GTPγS), is bound to the Gα subunits, a tryptic cleavage site at an Arg residue (Arg-204 in Gα_t) in the switch II region is protected.[22,23,25] In this condition, a 2-kDa amino-terminal peptide is cleaved from Gα_i, Gα_o, and Gα_t,[22,23,25]

---

[24] J. B. Hurley, M. I. Simon, D. B. Teplow, J. D. Robishaw, and A. G. Gilman, *Science* **226**, 860 (1984).
[25] J. W. Winslow, J. R. Van Amsterdam, and E. J. Neer, *J. Biol. Chem.* **261**, 7571 (1986).
[26] E. J. Neer, L. Pulsifer, and L. G. Wolf, *J. Biol. Chem.* **263**, 8996 (1988).
[27] S. E. Navon and B. K.-K. Fung, *J. Biol. Chem.* **262**, 15746 (1987).
[28] M. Pines, P. Gierschik, G. Milligan, W. Klee, and A. Spiegel, *Proc. Natl. Acad. Sci. U.S.A.* **82**, 4095 (1985).
[29] M. R. Mazzoni and H. E. Hamm, *J. Prot. Chem.* **12**, 215 (1993).

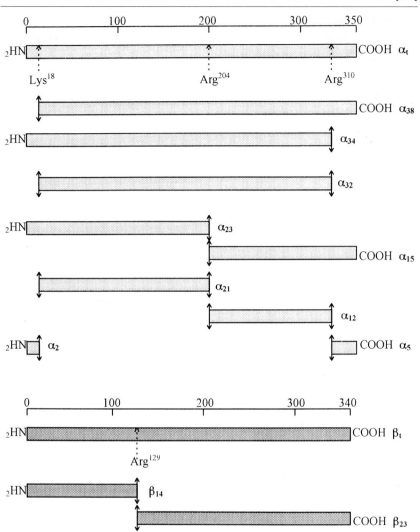

FIG. 1. Origin of fragments generated by limited tryptic digestion of G_t. The dashed lines represent the cleavage sites available during the limited tryptic digestion of $G\alpha_t$–GDP and $G\beta_t$. The number that follows α or β indicates the size of the fragment in kilodaltons.

while tryptic cleavage of $G\alpha_s$–GTPγS subunits generates stable fragments that are 1 kDa smaller than the native molecules.[30] Removal of the amino terminus from $G\alpha_o$–GTPγS (39 kDa) and $G\alpha_i$–GTPγS (41 kDa) produces stable 37- and 39-kDa fragments, respectively.[25] The $G\alpha_t$–GTPγS subunit

[30] T. H. Hudson, J. F. Roeber, and G. L. Johnson, *J. Biol. Chem.* **256**, 1459 (1981).

(39 kDa) is cleaved at Lys-18 generating a 38- to 37-kDa fragment that is digested further at the carboxyl terminus (Arg-310) producing stable 32- and 5-kDa fragments[22,23] (Fig. 1). In the GDP-bound form, the 37- and 38-kDa fragments of $G\alpha_o$ and $G\alpha_i$ are cleaved further to relatively stable 25- to 26-kDa and 17- to 18-kDa polypeptides. The latter contain the carboxyl terminus.[25] The 32-kDa fragment of $G\alpha_t$–GDP is also digested to two smaller and stable polypeptides, 21 and 12 kDa (Fig. 1).[22,23] During the time course of limited tryptic digestion of heterotrimeric G_t, other transient fragments are produced from $G\alpha_t$–GDP (Fig. 1).[11,23] A 34-kDa fragment appears as consequence of initial cleavage at Arg-310 rather than Lys-18 and two smaller polypeptides of relative molecular mass, 23 and 15 kDa, are also generated (Fig. 1).[11,23]

As described here and clearly shown in Fig. 1, the origin of $G\alpha_t$ proteolytic fragments is well characterized; therefore, an analysis of their time course can be useful to study functional properties of this subunit as well as interactions with other proteins. For example, the accessibility to a cleavage site (Glu-21) for *Staphylococcus aureus* V8 protease on $G\alpha_t$ amino terminus is reduced by the presence of the $G\beta\gamma_t$ subunit, which binds to this region.[27] If the interacting protein is a poor substrate for limited tryptic digestion, this technique may become an important tool for this type of investigation. In the case of native rhodopsin, 13 cleavage sites are available for tryptic digestion on intradiskal and cytoplasmic surfaces of the protein.[31] However, treatment of urea-stripped ROS membranes with trypsin for 30 min at 30° mainly cleaves rhodopsin at Lys-339, releasing the carboxyl-terminal 9 residues.[32,33]

Identification of Proteolytic Fragments

After limited digestion, the following experimental steps are separation and identification of proteolytic fragments. Separation can be achieved either by sodium dodecyl sulfate–polyacrylamide gel electrophoresis (SDS–PAGE) or reversed-phase high performance liquid chromatography (RP-HPLC). Polypeptides separated by gel electrophoresis can be detected by Coomassie blue staining or electroblotted to an immobilizing membrane and probed with specific antibodies. Figure 2 shows examples of Coomassie blue stained gels. When the proteolytic pattern of a protein is examined in the presence of membranes and/or another protein, fragment identification may be difficult by gel staining (Fig. 2B–D). Thus, a specific recognition of polypeptides is needed and can be obtained by means of antibodies (see Fig. 3).

[31] H. G. Khorana, *J. Biol. Chem.* **267,** 1 (1992).
[32] J. L. Miller and E. A. Dratz, *Vis. Res.* **24,** 1509 (1984).
[33] N. M. Greene, D. S. Williams, and A. C. Newton, *J. Biol. Chem.* **270,** 6710 (1995).

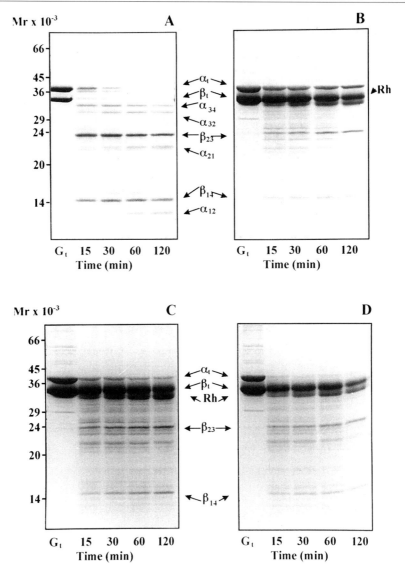

Fig. 2. Time course of limited tryptic digestion of G_t bound to urea-washed ROS membranes containing Rh, Rh*, or Rh* plus GDPβS. Bovine G_t (0.54 mg) is incubated in 1 ml of buffer in the presence urea-washed ROS membranes (33 μM Rh) with or without 100 μM GDPβS in the dark or light for 10 min. The samples are centrifuged and the pellets resuspended in buffer containing 25% glycerol to a final volume of 0.34 ml. Digestion with TPCK-trypsin is carried out at a trypsin to G_t (w/w) ratio of 1:25 in ice. As control, soluble G_t (0.54 mg) is

Numerous antisera against short peptides that represent sections of the primary amino acid sequences of specific Gα and Gβ subunits are now available.[34,35] An antiserum that recognizes a major fragment produced during limited tryptic digestion of the Gα subunit is suitable for this purpose. The use of various antisera that are specific for different sections of the subunit is recommended. Table I shows amino acid sequences of peptides used for generation of antisera that have been utilized to detect Gα_t polypeptides in immunoblots. Examples of typical immunoblots are shown in Fig. 3. To obtain quantitative measures from Coomassie blue stained gels or immunoblots the density of polypeptide bands can be measured using a densitometer.

Alternatively, proteolytic fragments can be separated by reversed-phase HPLC and their identify analyzed by tandem mass spectroscopy as recently described for limited tryptic digestion of Gα_o isoforms.[36] Another possible means of detection is by the use of matrix-assisted laser desorption mass spectroscopy (MALDI).[36]

Experimental Procedures

Materials

L-1-Tosylamido-2-phenylethyl chloromethyl ketone-treated trypsin (TPCK-treated trypsin) is purchased from Worthington Biochem. Corp. (Freehold, NJ). GDP, guanosine 5'-O-(2-thiodiphosphate) (GDPβS), and 1-chloro-3-tosylamido-7-amino-2-heptanone hydrochloride (TLCK) are products of Boehringer Mannheim Biochem. Corp. (Indianapolis, IN). The

---

[34] G. Milligan, *Methods Enzymol.* **237,** 268 (1994).
[35] A. N. Pronin and N. Gautam, *Methods Enzymol.* **237,** 482 (1994).
[36] W. E. McIntire, J. Dingus, K. L. Schey, and J. D. Hildebrandt, *J. Biol. Chem.* **273,** 33135 (1998).

---

similarly treated with TPCK-trypsin. At the indicated time points, duplicate aliquots are removed from the incubation mixtures, and the reaction is stopped by adding TLCK to a final concentration of 40 μg/ml. The proteolytic fragments are separated by electrophoresis on SDS–polyacrylamide gels (12.5%) that are stained with Coomassie blue. (A) Time course of limited tryptic digestion of soluble G_t; (B) time course of limited tryptic digestion of G_t in the presence of Rh*; (C) time course of limited tryptic digestion of G_t in the presence of Rh* plus GDPβS; (D) time course of limited tryptic digestion of G_t in the presence of Rh. Molecular weight standards are indicated, as are the size and the origin of the fragments. G_t, control at time 0. [Data from M. R. Mazzoni and H. E. Hamm, *J. Biol. Chem.* **271,** 30034 (1996), with permission.]

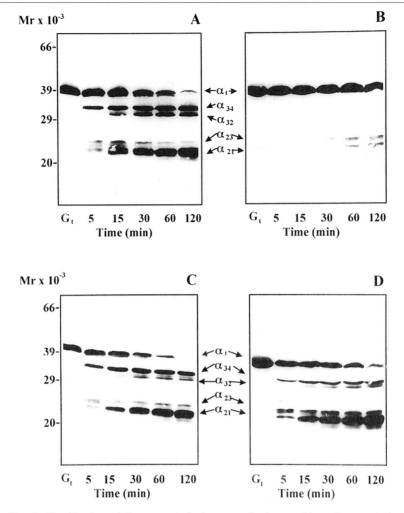

FIG. 3. Identification of $G\alpha_t$ proteolytic fragments by immunoblot using a polyclonal antibody. Bovine G_t (0.55 mg) is incubated in 1 ml of buffer with SLUV or urea-washed ROS membranes (36 μM Rh) in the dark or light for 10 min. The samples are centrifuged and the pellets resuspended in buffer containing 25% glycerol to a final volume of 0.34 ml. Tryptic digestion conditions are the same as described for Fig. 2. As control, soluble G_t (0.54 mg) is also treated with TPCK-trypsin. At the indicated time points, duplicate aliquots are removed from the incubation mixtures, and the reactions are stopped by adding TLCK to a final concentration of 40 μg/ml. The proteolytic fragments are separated by electrophoresis on SDS–polyacrylamide gels (12.5%) and blotted onto nitrocellulose. Blots are incubated in a phosphate-buffered solutions/milk buffer containing a polyclonal antibody (antiserum dilution 1 : 8,000) which recognizes $G\alpha_t$(195–206) sequence. (A) Time course of limited tryptic digestion of soluble $G\alpha_t$; (B) time course of limited tryptic digestion of $G\alpha_t$ in the presence of Rh*;

LumiGLO substrate kit and peroxidase-labeled antibodies to rabbit or mouse immunoglobulin G (IgG) are obtained from Kirkegaard & Perry Laboratories, Inc. (Gaithersburg, MD). Nitrocellulose (0.1 μM) is from Schleicher & Schuell (Keene, NH). All other reagents are of the best grade available.

ROS membranes are prepared from freshly collected bovine retinas as described previously.[23] ROS stripped of peripheral proteins are obtained by washing ROS membranes after limited digestion with 4 M urea.[37] Aliquots of both ROS membrane preparations are stored in the dark at $-80°$ until needed. Concentration of rhodopsin in octylglucoside-solubilized (1%) samples is determined spectrophotometrically at 500 nm (ε = 42,700 M^{-1} cm^{-1}).[38]

Phospholipid vesicles are prepared with highly purified neutral phosphatidylcholine (PC). A relatively homogeneous population of large unilamellar vesicles (LUV) is obtained performing several cycles of freezing and thawing followed by extrusion through sized polycarbonate filters as described.[39] Sucrose-loaded LUV (SLUV) are prepared following the method described by Kim et al.[40] and resuspended in 10 mM 3-(N-morpholino)propanesulfonic acid (MOPS), pH 7.5, 200 mM NaCl, 2 mM MgCl$_2$ and 0.1 mM EDTA. The phospholipid concentration is determined as reported by Malinski and Wensel.[39] In the final SLUV preparation, the total phospholipid concentration is approximately 4.9 mM which corresponds to that present in ROS membranes containing 82 μM rhodopsin.

Heterotrimeric G_t is prepared by GTP elution of hypotonically washed ROS membranes as described[41] and stored in 40% (v/v) glycerol as aliquots at $-20°$. G_t concentration is determined by the Coomassie blue binding method,[42] using bovine serum albumin (BSA) as a standard (Pierce, Rockford, IL).

---

[37] G. Yamanaka, F. Eckstein, and L. Stryer, *Biochemistry* **24,** 8094 (1985).
[38] K. Hong and W. L. Hubbell, *Proc. Natl. Acad. Sci. U.S.A.* **69,** 2617 (1972).
[39] J. A. Malinski and T. G. Wensel, *Biochemistry* **31,** 9502 (1992).
[40] J. Kim, P. J. Blackshear, J. D. Johnson, and S. McLaughlin, *Biophys. J.* **67,** 227 (1994).
[41] L. Stryer, J. B. Hurley, and B. K.-K. Fung, *Methods Enzymol.* **96,** 617 (1983).
[42] M. M. Bradford, *Anal. Biochem.* **72,** 248 (1976).

(C) time course of limited tryptic digestion of Gα_t in the presence of PL; (D) time course of limited tryptic digestion of Gα_t in the presence of Rh. Molecular weight standards are indicated, as are the size and origin of the fragments. G_t, control at time 0. [Data from M. R. Mazzoni and H. E. Hamm, *J. Biol. Chem.* **271,** 30034 (1996), with permission.]

TABLE I
Anti-Gα_i Peptide Antibodies Used in Immunoblots for Detection of Gα_t Proteolytic Fragments

Designation	Code number[a]	Peptide sequence	Gα_i amino acids
Gα_{common}[b]	1398	GAGESGKSTIVK	40–51
Gα_i/Gα_o/Gα_t[c]	8645	FDVGGQRSERKK	199–211
G$\alpha_{i1/2}$[d]	116	IKNNLKDCGLF	344–354

[a] Antisera raised against synthetic peptides corresponding to segments of Gα_i amino acid sequence are provided by Dr. D. Manning (Department of Pharmacology, University of Pennsylvania, PA). All three antisera give analog results in immunoblots. However, antiserum 8645 shows the best resolution of major Gα_t fragments in combination with a low background.

[b] This polyclonal antibody is directed against an amino acid sequence that is common to Gα_s/Gα_i/Gα_o/Gα_t. Some residues of this region form coordinate interactions with the triphosphate portion of GTPγS and Mg^{2+} in activated Gα subunits.[1–3]

[c] The polyclonal antibody is directed against an amino acid sequence that is common to Gα_i/Gα_o/Gα_t. The sequence is part of switch II region[1,2] and contains some residues that take contact with Gβ residues in heterotrimeric G proteins.[6,7]

[d] This polyclonal antibody is directed against the carobxyl-terminal sequence of G$\alpha_{i1/2}$, which diverges for only one amino acid from that of Gα_t (IKENNLKDCGLF). The antibody also recognizes the Gα_t subunit.[11]

Table I shows a list of rabbit anti-Gα antisera used to identify specific Gα_t fragments in immunoblots. These antisera raised against synthetic peptides corresponding to segment of Gα_i sequence are generously provided by Dr. D. Manning (Department of Pharmacology, University of Pennsylvania, PA). Other polyclonal antisera with different specificity are available from various sources. Antisera 1398 and 8645 recognize two distinct internal sequences that are common to both Gα_i and Gα_t, whereas antiserum 116 has been raised to an 11 amino acid synthetic peptide representing G$\alpha_{i1/2}$ carboxyl terminus. This anti-G$\alpha_{i1/2}$ antiserum cross-reacts with the Gα_t subunit, which shows only one amino acid difference in its carboxyl terminus (G$\alpha_{i1/2}$ sequence, IKNNLKDCGLF; Gα_t sequence, IKENLKDCGLF). A monoclonal antibody (MAb 4A) that binds to Gα_t and two transient fragments (34 and 23 kDa) in immunoblots[23] is also utilized in some experiments. The main portion of the MAb 4A epitope appears to be localized in the amino-terminal region of Gα_t.[23,43]

Rhodopsin–G$_t$ Interaction

The interaction reaction between urea-washed ROS membranes and G$_t$ is performed in either 1.5-ml Eppendorf or 2-ml-thick walled polycarbonate

[43] S. E. Navon and B. K. Fung, *J. Biol. Chem.* **263**, 489 (1988).

tubes. When G_t is incubated with SLUVs, polycarbonate tubes are always used since an ultracentrifugation step is required. The composition of the incubation buffer is as follows: 10 mM MOPS, pH 7.5, 200 mM NaCl, 2 mM MgCl$_2$, and 1 mM dithiothreitol (DTT). DTT is added fresh to the buffer the same day of the experiment. Freshly prepared stock solutions of either GDP or GDPβS are made in the same buffer.

1. G_t (~0.54 mg) is diluted in the incubation buffer (~0.95 ml) and mixed by vortexing. In some experiments, the guanine nucleoside diphosphate (GDP or GDPβS) is also included to obtain a final concentration of 100 μM in 1 ml of the reaction mixture. This nucleotide concentration is maintained throughout the assay.
2. The reaction is started by the addition of dark urea-washed ROS membrane or SLUV suspensions and vortexing. The final volume of the reaction mixture is 1 ml. The concentrations of rhodopsin and G_t are 34 and 6.8 μM with a G_t:rhodopsin molar ratio of 1:5. In experiments with SLUVs a final concentration of 2.2 mM PC is required to obtain a similar phospholipid concentration to that in samples with ROS membranes.
3. Samples are incubated for 10 min at room temperature under ambient light (light-activated rhodopsin, Rh*) or in the dark (dark rhodopsin, Rh). All manipulations of dark samples are performed under dim red light (Kodak, Rochester, NY, filter 1).
4. After incubation, samples in Eppendorf tubes are centrifuged at 12,000 rpm for 5 min at room temperature in a microfuge. When the incubation reaction is carried out in 2 ml polycarbonate tubes, samples are centrifuged at 50,000 rpm for 30 min at 4° in a Beckman (Palo Alto, CA) fixed angle TL-100.2 ultracentrifuge rotor. Sample ultracentrifugation is required to achieve SLUV precipitation. ROS membrane pellets are also more compact after ultracentrifugation.
5. Supernatants are aspirated and placed in Eppendorf tubes that are kept in ice. Aliquots of these samples are treated with equal volumes of 2× electrophoresis sample buffer[44] and loaded on SDS–PAGE to evaluate the amount of soluble G_t.
6. Pellets are resuspended in 10 mM MOPS, pH 7.5, 200 mM NaCl, 2 mM MgCl$_2$, 1 mM DTT containing 25% glycerol (buffer A) at a G_t concentration of approximately 1.6 mg/ml and kept in ice. This protein concentration is estimated on the basis of the initial amount of G_t.

Experiments performed in ambient light intend to investigate the interactions between Rh* and G_t in the state of "empty pocket" (no guanine nucleotide added) or in the Gα_t–GDP bound form (GDP or GDPβS added). Thus, rhodopsin is light activated when urea-washed ROS membranes are bleached during the incubation with G_t under ambient light. On the other hand, to study the interactions between Rh and G_t, all manipulations of the samples containing urea-washed ROS membranes are carried out under dim red light or in the dark.

Limited Proteolysis of G_t

A stock solution of TPCK-treated trypsin (2 mg/ml) is prepared in 0.1 mM HCl and stored at 4°. A diluted solution (0.06 mg/ml) in buffer A is made fresh each time from the stock solution. To block the proteolytic reaction, TLCK at a final concentration of 40 μg/ml is used. A stock solution of TLCK (18 mg/ml) is prepared fresh in distilled water.

1. Duplicate aliquots are removed from ROS membrane or SLUV suspensions containing bound G_t (~1.6 mg/ml) and placed in Eppendorf tubes with equal volumes of buffer A in the presence of TLCK (40 μg/ml). After incubation at room temperature for 5 min, 2× electrophoresis sample buffer[44] is added and the aliquots that represent controls at time 0 are immediately frozen.
2. An equal volume of the diluted TPCK-treated trypsin solution (0.06 mg/ml) is added to the G_t samples (~1.6 mg/ml). In the digestion mixture the final concentration of G_t is 0.8 mg/ml and the trypsin : G_t (w/w) ratio is 1 : 25.
3. Limited proteolysis is carried out by incubating samples in ice.
4. At various incubation times (i.e., 5, 15, 30, 60, and 120 min) duplicate aliquots are removed and incubated with TLCK in Eppendorf tubes for 5 min in ice. This short incubation is required to obtain a complete inactivation of the enzyme before denaturation of G_t. After addition of 2× electrophoresis sample buffer,[44] the aliquots are immediately frozen using dry ice with 2-propanol and stored at $-80°$ until used. To prevent formation of rhodopsin aggregates these sample are not boiled.

As control experiment (Figs. 2 and 3A), soluble G_t (1.6 mg/ml) in buffer A is proteolyzed with TPCK-treated trypsin under identical conditions as described earlier but in the absence of membranes or phospholipid vesicles. When the limited proteolysis of G_t bound to bleached ROS membranes is investigated in the presence of the guanine nucleotide, fresh GDP or GDPβS is added before stating digestion to keep the nucleotide concentration at 100 μM.

[44] U. K. Laemmli, *Nature* **227**, 680 (1970).

Sodium Dodecyl Sulfate–Polyacrylamide Gel Electrophoresis

Separation of proteolytic fragments by SDS–PAGE is carried out in 12.5% (w/v) acrylamide slabs according to Laemmli.[44] Frozen samples are thawed at room temperature and gently resuspended. Aliquots (~9 μg of G_t/lane) of each sample are loaded in the wells of duplicate gels. A mixture of molecular weight standards (M_r ranging from 97,400 to 14,000) dissolved in Laemmli's sample buffer[44] is always loaded in the first well of the slab gel. When electrophoresis is followed by electroblotting, fluorescent molecular weight markers (Sigma, St. Louis, MO) are used. Following gel electrophoresis, proteins and polypeptide fragments are either stained with Coomassie blue (Fig. 2) or electroblotted to nitrocellulose (0.1 μm) essentially as described by Towbin et al.[45]

Immunoblotting

Polyacrylamide gels are removed from the electrophoresis cell and proteins are electroblotted.[45]

After transfer, the nitrocellulose is examined under UV light to detect the fluorescent molecular weight markers. Their sites of migration are marked. The nitrocellulose is then incubated in 10 mM NaH_2PO_4, pH 7.4, 0.9% NaCl (PBS) containing 3% (w/v) low-fat dried milk and 0.2% (v/v) Tween 20 (PBS/milk) at room temperature and under continuous gentle shaking. After 30 min, the nitrocellulose is incubated in PBS/milk containing an anti-$G\alpha_t$ antiserum (1:8000 dilution) or MAb 4A (50 μg/ml) for 1 hr at room temperature. The immunoblots are washed four times (10 min each) with PBS/milk and then incubated in PBS/milk containing peroxidase-labeled second antibody (1:10,000 dilution) for 1 hr at room temperature. The washing step is repeated as described earlier, followed by two washes with PBS and one with distilled water. The immunoblots are incubated in LumiGLO substrate for ~1 min at room temperature and then exposed to Kodak (Rochester, NY) XAR-2 film for a few seconds. Figure 3 shows immunoblots obtained using the anti-$G\alpha_t$(195–206) antiserum (antiserum code, 8645) as primary antibody. On following the limited tryptic digestion of $G\alpha_t$ this polyclonal antibody gives better results than the other two antisera or monoclonal antibody 4A since it recognizes major transient (α_{38}, α_{34}, α_{32} and α_{23}) and final (α_{21}) fragments. To obtain a quantitative estimation, the density of the polypeptide bands is measured using a densitometer. The integrated area of each band is divided by the combined integrated areas of all bands present in the sample lane. Thus,

---

[45] H. Towbin, T. Staehelin, and J. Gordon, *Proc. Natl. Acad. Sci. U.S.A.* **76**, 4350 (1979).

a quantitative analysis of $G\alpha_t$ disappearance and fragment production is carried out allowng the evaluation of tryptic digestion rates.

Conclusions

Analyzing the time course of limited tryptic digestion of G_t in the presence and absence of dark or light bleached urea-washed ROS membranes, we have defined the regions of $G\alpha_t$ involved in the interaction with the receptor, rhodopsin. The accessibility of $G\alpha_t$ to proteolysis is changed dramatically when G_t is in the high affinity interaction with Rh*. In this condition, all cleavage sites (Lys-18, Arg-204, and Arg-310) are substantially protected (Figs. 2B and 3B). The presence of the guanine nucleoside diphosphate partially reverts the protection at Arg residues (Fig. 2C). This result suggests that in the "empty pocket" state the $G\alpha_t$ subunit forms a tight complex with $G\beta\gamma_t$ and Rh*. In agreement with other evidences,[8] the protection at Arg-310, which is located in a surface-exposed loop of $G\alpha_t$ ($\alpha 4/\beta 6$ region)[1,6] is likely the consequence of direct Rh* binding. On the other hand, the cleavage site at Arg-204 located in the switch II appears be protected for an indirect effect. The $G\beta\gamma_t$ subunit that interacts with the switch II region in the heterotrimeric G_t,[6] may cooperate in producing this protection.

This investigation has also allowed us to elucidate an aspect of the interaction between $G\alpha_t$ amino terminus and membrane phospholipids. Our data show that the cleavage rate at Lys-18 is substantially decreased in the presence of either urea-washed ROS membranes or phospholipid vesicles as compared to digestion of soluble G_t (Figs. 2 and 3). The effect may be the direct or indirect consequence of the presence of phospholipids since the $G\beta\gamma_t$ subunit binds to the amino-terminal α helix of $G\alpha_t$ in the heterotrimeric G_t protein.[6] In fact, the presence of $G\beta\gamma_t$ partially protects cleavage sites for chymotrypsin[29] and *S. aureus* V8 protease[27] in the $G\alpha_t$ amino terminus. The interaction of both subunits with phospholipid may reinforce this protective effect. The presence of ROS membranes also seems to reduce the cleavage rate of $G\beta_t$ at Arg-129 (Fig. 2).

Even considering some limitations, analysis of limited proteolytic digestion patterns of G proteins may represent a useful approach to study their molecular mechanisms of interaction with heptahelical receptors. In particular, this method can be adopted to investigate the molecular basis for G_i-protein activation by a wide variety of receptors.

[25] Monitoring Proton Uptake from Aqueous Phase during Rhodopsin Activation

By CHRISTOPH K. MEYER and KLAUS PETER HOFMANN

Introduction

Proton transfer reactions are involved in key steps of the light activation of rhodopsin. Deprotonation of the Schiff base linking Lys-296 and the retinal chromophore and protonation of the counterion Glu-113 mark one of the key steps in transducing the energy of the photon absorbed into the conformational changes necessary for a transition into the catalytically active state of the receptor. Spectroscopically, this is accompanied by a shift of the visible absorption maximum to 380 nm, hence defining the metarhodopsin-II (MII) intermediate[1,2] (see also the chapters by Ernst et al.,[2a] other chapters of this volume and a recent review[2b]).

MII is in a temperature- and pH-dependent equilibrium with the species preceding it in the activation reaction, metarhodopsin-I (MI).[3,4] The deprotonation of the Schiff base is accompanied by the net uptake of a proton from the aqueous phase.[5,6] This explains why the *de*protonation of the Schiff base is favored at acidic pH, and shows that at least two proton accepting groups are involved in this process. Indeed, as demonstrated for rhodopsin solubilized with the detergent *n*-dodecylmaltoside (DM), the deprotonation of the Schiff base and proton uptake are sequential steps in the activation reaction.[7] This led to a distinction between two states of MII, called MIIa and MIIb, both of which have deprotonated Schiff bases (with the accompanying 380-nm absorbance maximum), and MIIb arising from MIIa through uptake of a proton from the aqueous phase. Recent work has further supported this model.[8,9]

---

[1] D. Emeis, H. Kühn, J. Reichert, and K. P. Hofmann, *FEBS Lett.* **143**, 29 (1982).
[2] J. Kibelbeck, D. C. Mitchell, J. M. Beach, and B. J. Litman, *Biochemistry* **30**, 6761 (1991).
[2a] O. P. Ernst, C. Bieri, H. Vogel, and K. P. Hofmann, *Methods Enzymol.* **315** [32], 1999 (this volume).
[2b] K. P. Hofmann, *in* "Rhodopsins and Phototransduction," Novartis Foundation, Symp. 224. Wiley, New York, 1999.
[3] R. G. Matthews, R. Hubbard, P. K. Brown, and G. Wald, *J. Gen. Physiol.* **47**, 215 (1963).
[4] J. H. Parkes and P. A. Liebman, *Biochemistry* **23**, 5054 (1984).
[5] J. K. Wong and S. E. Ostroy, *Arch. Biochem. Biophys.* **154**, 1 (1973).
[6] N. Bennett, *Eur. J. Biochem.* **111**, 99 (1980).
[7] S. Arnis and K. P. Hofmann, *Proc. Natl. Acad. Sci. U.S.A.* **90**, 7849 (1993).
[8] S. Jäger, I. Szundi, J. W. Lewis, T. L. Mah, and D. S. Kliger, *Biochemistry* **37**, 6998 (1998).
[9] I. Szundi, T. L. Mah, J. W. Lewis, S. Jäger, O. P. Ernst, K. P. Hofmann, and D. S. Kliger, *Biochemistry* **37**, 14237 (1998).

TABLE I
COMMONLY USED INDICATOR DYES

Dye	pH range	Absorption maximum (nm)	
		Acidic	Basic
Bromocresol purple	5.2–6.8	430	595
Cresol red	7.0–8.8	430	575

In the protonation assay described in this article, the uptake or loss of a proton to the aqueous phase is monitored as the signature event of conformational or other changes occurring in the proteins. More direct evidence for the processes or conformations involved, e.g., from infrared spectroscopy is currently accumulating.[10] However, proton uptake measurements are in some cases the only technique presently available for the investigation of the molecular mechanisms.

Optical Spectroscopy Using pH Indicator Dyes

In samples with a small buffer capacity, proton uptake or release will change the pH of the aqueous solution of the sample. By adding a pH indicator dye (see Table I), one can monitor these changes of pH using time-resolved absorption spectroscopy. The advantages of this technique include good time resolution (milliseconds) and the fact that important parameters of the assay such as rhodopsin concentration and the fraction of rhodopsin bleached per flash are easily determined.

General Considerations

Depending on the pH of the sample, a certain percentage of the indicator dye molecules in the sample will be protonated. If, following activation, protons are released into or removed from the sample solution, this will lead to a change in the pH of the sample. As a consequence, the percentage of protonated indicator dye molecules will change, leading to a change in the absorption spectrum of the sample.

Some of the experimental consequences follow:

1. For a good signal-to-noise ratio with this technique, the pH change of the solution following activation has to be maximized, requiring the

---

[10] F. Siebert, *Isr. J. Chem.* **35,** 309 (1995).

buffering capacity of the sample to be low. As a result, all pH buffers such as MES, HEPES, BTP, etc., should not be present in the sample (the resulting pH instability of the sample will be dealt with later). As a further step, degassing the solutions to be used will reduce the buffering by CO_2.

2. The contribution of rhodopsin bleaching to the absorption changes recorded in the sample has to be eliminated to determine the true pH-induced dye response following light activation. This is done by measuring a strongly pH buffered but otherwise identical sample, which will eliminate the pH-induced contribution to the absorption changes recorded. (As a side effect, this measurement is valuable to detect dye artifacts that can arise with some dyes or other components added to the system.)

3. Calibration of the indicator dye response is necessary because (a) the dye response depends on the sample pH and (b) the many protonatable groups present in the sample influence the change in pH after light activation in a way that is difficult to predict quantitatively. As a result, the light-induced dye response is compared with the dye response caused by the addition of a defined (small) amount of protons (HCl).

4. The technique monitors net proton uptake or release effects. This characteristic is important to keep in mind when performing experiments on systems containing multiple active protonatable groups, such as rhodopsin, or complex systems containing rhodopsin and some ligand, e.g., peptides (see later discussion): Observed protonation behavior will be a superposition of deprotonations and/or protonations of these groups, which can reside either on rhodopsin or on the ligand.

Instrumentation

We use a modified version of the custom-made two-wavelength spectrophotometer described by Emeis et al.[1] In brief, the light from a 150-W halogen light bulb is sent through two monochromators (Jobin-Yvon), one of which is always tuned to 380 nm. The other one is tuned to the red absorption maximum of the pH indicator dye (or close to it). Custom-made silicon photodiode light detectors and amplifiers record changes in the transmitted light intensity and send their signals to a Nicolet Model 2090-IIIA two-channel transient recorder. The photoreaction is triggered using a flash lamp (EG&G Electro Optics). For the conversion of changes in the transmitted light intensity to changes in indicator dye absorption, see the Appendix. Another important piece of equipment is a glass pH electrode small enough to allow measurement of the pH directly in the measuring cuvette.

Preparation of Detergent-Solubilized Rhodopsin

Because the use of pH indicator dyes with rhodopsin incorporated into the native disk membranes has proved difficult,[10a] we use rhodopsin that has been purified from hypotonically washed disk membranes in detergent (e.g., DM) using concanavalin A affinity chromatography.[11] The buffer used during solubilization of disk membranes and washing of the column is 20 mM BTP, pH 7.5, 130 mM NaCl, 1 mM MgCl$_2$ plus detergent. For the elution from the column, the concentration of BTP is decreased to 4 mM and the dialysis to remove the methylmannoside used for elution is performed against 130 mM NaCl, 0.03% DM, with no buffer added.

As a result of the solubilization in detergent, the Schiff base deprotonation and proton uptake steps of rhodopsin activation are kinetically decoupled, allowing the proton uptake reaction to be studied separately, either as a function of pH and temperature or using ligands in 'extra MIIb' experiments.

Sample Preparation

Because no pH buffer is present in the sample and the solution is no longer saturated with CO_2, the pH of the samples is very sensitive to handling and exposure to air. The following procedure has proved to yield reproducible and reliable results.

Solutions

    130 mM NaCl (filtered through 0.2 μM membrane and degassed)
    10% (w/v) DM
    Indicator dye (e.g., 1 mM cresol red, filtered and degassed)
    10 mM HCl and 10 mM NaOH (degassed)
    35–40 μM DM-solubilized rhodopsin in 0.03% DM

Procedure

All operations are performed under dim red light. The HCl or NaOH is used to set the desired pH of the measurement, and necessary amounts have to be determined through experience. The solutions are combined into the measurement cuvette, with rhodopsin added last, such that the final concentrations are as follows:

    130 mM NaCl
    0.03% DM

[10a] S. Arnis, unpublished data.
[11] W. DeGrip, *Methods Enzymol.* **81**, 197 (1982).

24 μM cresol red
4 μM rhodopsin

A small magnetic stirrer is placed into the cuvette and the exact pH is measured using a microelectrode (Microelectrodes, Inc.). It has proven important to prepare the sample in the measuring cuvette, because transferring the sample, e.g., from a sample tube, will shift the pH uncontrollably by up to 0.5 pH units. The opening of the cuvette is only slightly larger than the microelectrode shaft to reduce uptake of atmospheric CO_2.

It is important to cap the cuvette, because influx of CO_2 will cause a large drift in the signal, especially during the calibration procedure (see below).

Determining Pure pH Signal

The absorption spectra of many commonly used pH indicator dyes, e.g., cresol red (CR) and bromocresol purple (BCP), overlap with the absorption spectrum of rhodopsin. This means that the time course of absorption changes recorded at, e.g., 584 nm for CR will contain contributions from the bleaching of rhodopsin in addition to the pH-dependent signal of interest.

To extract the pure pH signal, we perform measurements of identical samples adjusted to the same pH, except that 65 mM MES or HEPES is added to the 130 mM NaCl solution. This abolishes the pH-dependent contributions to the signal, leaving only the component due to the bleaching of rhodopsin. Then, the pure pH signal is the difference of the unbuffered measurement minus the buffered measurement.

Calibrating Indicator Dye Response

The magnitude of the pure pH indicator signal, besides being a function of the number of protons released or taken up by the sample following activation, will vary according to several parameters, among them the titration curve of the dye, the exact buffering capacity of the sample at the pH of the measurement, the amount of rhodopsin bleached per flash, etc.

It is convenient to determine the amount of rhodopsin bleached per flash by repeatedly flashing the same sample and performing a linear regression of the logarithms of the signal amplitudes. It has been noted that this method is only approximately correct if the flash induces MI and MII present in the sample to photoregenerate to the dark state of rhodopsin.[12] In the experimental setup described here, the occurrence of photoregeneration is readily visible in the recording of absorbance at 380 nm and can be avoided by a judicious choice of colored filters for the activating flash. In this respect

[12] J. H. Parkes and P. A. Liebman, *Biophys. J.* **66,** 80 (1994).

it is helpful that for DM-solubilized rhodopsin, the MI–MII equilibrium is shifted completely to MII with an absorbance spectrum that is well separated from that of dark rhodopsin.

The other factors influencing the signal amplitude have to be determined by performing a calibration with a known amount of protons, i.e., HCl (all other conditions, such as intensity of the spectrometer lamp being the same). We have developed the following protocol to accomplish this task, which can be performed on every sample prior to flash activation.

Solutions

130 mM NaCl (filtered and degassed)
50 μM HCl (diluted from the stock solution using 130 mM NaCl, filtered and degassed)
100 μM HCl (prepared likewise)

Procedure

The measurement is a long timescale recording of adding equal volumes of the solutions given to the sample. To prevent unnecessary bleaching of the sample by the measuring light, a neutral density filter with OD = 1 is placed into the pH light path and the 380-nm light path is blocked.

For every solution, a syringe (Hamilton) is loaded with the same volume (e.g., 10 μl) of the solution and inserted through a small hole in the cap of the measuring cuvette. The measurement is started, and at some point the solution is injected. The sample is mixed using the small magnetic stirrer in the cuvette.

The reliability of the calibration can be judged by looking at a superposition of the three traces that are recorded in this way. After the addition of the solutions, the traces should be equidistant (see Results section).

Calculation

The number of protons taken up or released by every illuminated molecule of rhodopsin can now be calculated as follows (for a validation of this procedure please see the Appendix).

If ΔI_f is the change in transmitted light intensity following the activating flash, and $b\%$ is the percentage of rhodopsin bleached per flash, then

$$\Delta I_{f\text{Total}} = \Delta I_f (100/b\%)$$

would be the signal amplitude, if all rhodopsin were bleached in a single flash. Let ΔI_{HCl} be the change in transmitted intensity due to the addition of the acid, as determined in the calibration, corrected for the neutral

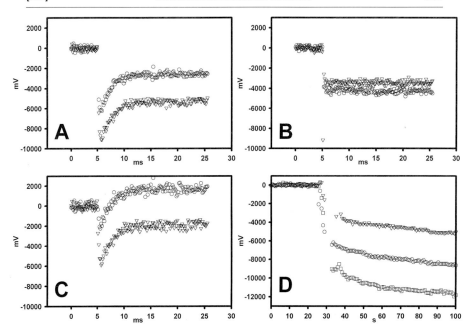

FIG. 1. Flash-induced changes of transmitted light intensity, sample prepared as described in text using cresol red as indicator dye. (A) Unbuffered measurement, (B) buffered measurement, and (C) difference of (A) and (B). Measurements at pH 7.4 (circles) and pH 7.8 (triangles). (D) Calibration, at pH 7.8, with 10 μl of 130 mM NaCl with 0, 50, and 100 μM HCl added (top to bottom trace). All measurements at 22°.

density filter that may be present in the light path during calibration, and let c_{HCl} be the final concentration of the added HCl when it has been added to the cuvette. Then full bleaching of rhodopsin would correspond to the addition (or removal) of

$$c_{HCl}(\Delta I_{fTotal}/\Delta I_{HCl})$$

of HCl. Then the value of interest, protons taken up or released per illuminated rhodopsin molecule, is calculated to be

$$H^+/R^* = (c_{HCl}/c_{Rho})(\Delta I_{fTotal}/\Delta I_{HCl}) \tag{1}$$

if c_{Rho} is the concentration of rhodopsin in the cuvette.

Results

In Fig. 1 we show measurements of proton transfer by rhodopsin recorded at pH 7.4 (circles) and pH 7.8 (triangles) using the pH indicator

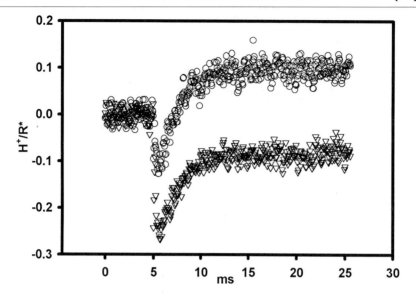

FIG. 2. H$^+$/R* at pH 7.4 (circles) and pH 7.8 (triangles), calculated using Eq. (1) and the data from Figs. 1C and 1D, plus a similar calibration at pH 7.4. One sees a fast deprotonation [I. Szundi, T. L. Mah, J. W. Lewis, S. Jäger, O. P. Ernst, K. P. Hofmann, and D. S. Kliger, *Biochemistry* **37,** 14237 (1998)] followed by incomplete protonation of illuminated rhodopsin. The shift from *net* uptake (pH 7.4) to *net* release (pH 7.8) is visible.

dye cresol red (CR). The traces shown are the unbuffered measurements (Fig. 1A), the buffered measurements (Fig. 1B), and the difference of the two measurements, adjusted for rhodopsin bleaching (Fig. 1C). These are the pure pH-induced indicator signals. Their relative amplitude will be influenced by the calibration measurements. All traces shown are the average of four flashes on one sample, each flash bleaching 8–10% of rhodopsin. Figure 1D shows a typical calibration, recorded at pH 7.8, performed with 10 μl of 130 mM NaCl with 0, 50, and 100 μM HCl added (top to bottom trace). The calibration was repeated at pH 7.4 (data not shown). Traces shown are corrected for the neutral density filter used during calibration.

In Fig. 2 we show the measurements after the pure pH signal has been converted to protons per illuminated rhodopsin using the calibration measurement and Eq. (1).

Dynamics of Proton Transfer Processes

The interpretation of the converted protonation signal is straightforward once the system has reached equilibrium, because then equilibrium thermodynamics can be used.

The dynamic aspects of proton translocations are much more complex and have been treated in detail, e.g., by Gutman and Nachliel.[13] Apart from the biologically significant reactions in or between the proteins (with the reaction rates of interest to us), three main processes are involved in detecting pH changes using pH indicator dyes: (1) deprotonation and/or protonation of the reactive groups presented by the proteins, (2) proton transfer from the proteins to the indicator, and (3) deprotonation or protonation of the indicator molecules. The reaction rate respective timescales for these processes given by Gutmann and Nachliel are $k_{diss} = 10^5-10^{10}/$sec for the dissociation of an acid and complexation of its proton by water, and $k_{ass} \approx 10^{10}/(M\ s)$ for the diffusion-controlled association of a proton with a base. (For the concentrations used in our experimental setup one arrives at 10^4-10^5/sec, 2 to 3 orders of magnitude faster than the dynamics of the protein involved.) The time needed for diffusion of the protons can be estimated, using the diffusion constant of protons in water ($D_{H^+} = 9\ 10^{-5}\ cm^2/s$) and the concentrations of the reactants involved, to be on the order of tenths of microseconds. In this context it may be interesting to note that *intra*molecular proton transfer reactions, such as that of the Schiff base proton, usually occur on the timescale of picoseconds or less once the participating groups have reached a favorable orientation, distance, and arrangement of the electric fields. In graphic terms, the rate of these processes is limited by waiting for the conformation suited for the actual transfer process to occur.[13]

For rhodopsin and in the current article, protonation studies have been confined to pH-indicator dyes in solution with rhodopsin. Studies on bacteriorhodopsin purple membrane have compared protonation dynamics of solution indicator dyes with those of dyes covalently bound to the bacteriorhodopsin molecule.[14] The results add a new level of complexity because they show that appearance of the proton at the protein surface and its release into the aqueous phase are separated by ca. 700 μs (at 22°) during which the proton diffuses along the membrane, confined to the hydration layer on its surface. Whether this is also the case for detergent-solubilized rhodopsin will be the subject of future investigations. Protonation sites sequestered in pockets of the protein would be another source for major delays between the protein events and the reaction of the indicator dye.[15–17]

---

[13] M. Gutman and E. Nachliel, *Biochim. Biophys. Acta* **1015**, 391 (1990).
[14] J. Heberle and N. A. Dencher, *Proc. Natl. Acad. Sci. U.S.A.* **89**, 5996 (1992).
[15] K. Fahmy and T. P. Sakmar, *Biochemistry* **32**, 7229 (1993).
[16] S. Arnis and K. P. Hofmann, *Biochemistry* **34**, 9333 (1995).
[17] J.-M. Kim, C. Altenbach, R. L. Thurmond, H. G. Khorana, and W. L. Hubbell, *Proc. Natl. Acad. Sci. U.S.A.* **94**, 14273 (1997).

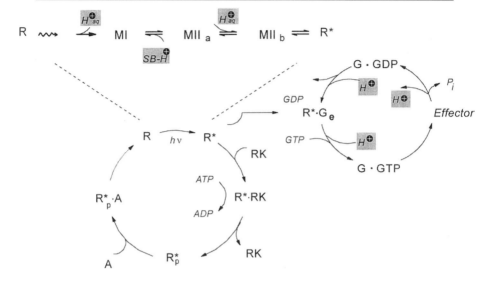

FIG. 3. The first steps of the visual cascade showing their signature protons. For details, see text.

Other Proton-Dependent Processes in Visual Cascade

Protons are involved in many of the processes of the visual cascade, in varying roles (Fig. 3). In 1986, Schleicher and Hofmann showed that proton uptake accompanies the formation of the complex between active rhodopsin and G_t.[18] A proton is liberated when $G_{t\alpha}$ hydrolyzes GTP to GDP,[19] and similarly, the catalytic hydrolyzation of cGMP by the visual effector protein phosphodiesterase (PDE) releases a proton into the medium. This last effect has been used to investigate PDE catalytic activity in the well-known cGMP hydrolysis proton assay.[20] In this case, measuring proton release with a pH electrode is advantageous to avoid residual rhodopsin activation by the measuring light necessary for the indicator dye methods.[20] Since protonation experiments are measurements of *net* protonation effects, in these complex systems with several reaction partners one sometimes encounters the difficulty of ascribing the proton bound or released to one of the reaction partners involved. Consequently, Schleicher and Hofmann could only state that the *complex* of rhodopsin and G_t takes up the proton and could not specify the binding site.

[18] A. Schleicher and K. P. Hofmann, *Z. Naturforsch.* **40c**, 400 (1985).
[19] M. Heck, unpublished data.
[20] P. A. Liebman and A. T. Evanczuk, *Methods Enzymol.* **81**, 532 (1982).

Appendix

As stated in the text, to achieve a higher resolution with the analog-to-digital converter, our instrument records changes ΔI in the transmitted light intensity, rather than the absolute values I of the intensity. For the convenience of the reader, we give a short derivation of the equivalence of the quantity ΔI and the absorbance change ΔA more commonly employed.

Lambert–Beer's law states that the absorbance A is related to the concentration c of the absorbing substance and its absorption coefficient ε by the following formula:

$$A = c\varepsilon d$$

so that the change ΔA in absorbance due to change Δc in the concentration of the absorbing substance is

$$\Delta A = \Delta c \varepsilon d$$

If I is the transmitted intensity before the flash, then

$$I = I_0 e^{-A}$$

The change in absorbance will result in a change of the transmitted intensity according to

$$I + \Delta I = I_0 e^{-(A+\Delta A)}$$

Dividing by the formula above and rearranging terms we find that

$$1 + \Delta I/I = e^{-\Delta A}$$

We now take the natural logarithm of both sides. Since ΔI is very small compared to I, one can expand the left side in a power series according to $\ln(1 + x) = x + \ldots$ *(if $x \ll 1$; under our experimental conditions $x < 0.1$)* and we find that

$$\Delta I/I = -\Delta A$$

Inserting this into the expression given earlier, we have

$$\Delta I = -\Delta c \varepsilon d I$$

This result also validates our calibration procedure because I is the same for the calibration and the flash-induced measurement.

[26] Use of Peptides-on-Plasmids Combinatorial Library to Identify High-Affinity Peptides That Bind Rhodopsin

By ANNETTE GILCHRIST, ANLI LI, and HEIDI E. HAMM

Introduction

The transduction of many signals from extracellular to intracellular environments requires the actions of heterotrimeric G proteins. The carboxyl-terminal region of Gα subunits represents an important site of interaction between heterotrimeric G proteins and their cognate receptors.[1-3] Within this region mutations,[4-6] covalent modification by pertussis toxin-catalyzed ADP-ribosylation,[7] or antibody binding[8] uncouple the receptor from heterotrimeric G protein. Synthetic peptides corresponding to the last 11 residues from Gα_t can bind the photoactivated receptor rhodopsin, resulting in stabilization of the activated metarhodopsin-II state.[9] Similarly, the carboxyl-terminal peptide from Gα_s can mimic the ability of Gα_s to evoke high-affinity binding to β-adrenergic receptors.[10] Furthermore, the carboxyl-terminal peptide from Gα_i can negatively modulate agonist binding, compete with heterotrimeric G protein for binding to the adenosine A$_1$ receptor, and block signal transduction events *in vitro*[11] and *in vivo*.[12] The carboxyl-terminal residues of Gα proteins not only provide the molecular basis for receptor-mediated activation of G proteins, they also play an important role in determining the fidelity of receptor activation.[13-15]

---

[1] J. Wess, *Pharmacol. Ther.* **80**, 231 (1998).
[2] H. Hamm, *J. Biol. Chem.* **273**, 669 (1998).
[3] H. E. Hamm and A. Gilchrist, *Curr. Opin. Cell Biol.* **8**, 189 (1996).
[4] S. Osawa and E. R. Weiss, *J. Biol. Chem.* **270**, 31052 (1995).
[5] P. Garcia, R. Onrust, S. Bell, T. Sakmar, and H. Bourne, *EMBO J.* **14**, 4460 (1995).
[6] R. Sullivan, D. Kunze, and M. Kroll, *Blood* **87**, 648 (1996).
[7] R. E. West, J. Moss, M. Vaughan, T. Lui, and T. Y. Lin, *J. Biol. Chem.* **260**, 14428 (1985).
[8] W. Simonds, P. Goldsmith, C. Woodard, C. Unson, and A. Spiegel, *FEBS Lett.* **249**, 189 (1989).
[9] H. E. Hamm, D. Deretic, A. Arendt, P. A. Hargrave, B. Koenig, and K. P. Hofmann, *Science* **241**, 832 (1988).
[10] M. M. Rasenick, M. Watanabe, M. B. Lazarevic, G. Hatta, and H. E. Hamm, *J. Biol. Chem.* **269**, 21519 (1994).
[11] A. Gilchrist, M. Mazzoni, B. Dineen, A. Dice, J. Linden, T. Dunwiddie, and H. E. Hamm, *J. Biol. Chem.* **273**, 14912 (1998).
[12] A. Gilchrist, M. Bünemann, A. Li, M. M. Hosey, and H. E. Hamm, *J. Biol. Chem.* **274**, 6610 (1999).
[13] C. Fong, D. Bahia, S. Rees, and G. Milligan, *Mol. Pharmacol.* **54**, 249 (1998).

The association between G-protein-complex receptors (GPCR) and G protein is a transient event. Following ligand stimulation the GPCR becomes activated, conformational changes in the receptor lead to activation of the G protein, with subsequent decreased affinity of Gα for GDP, dissociation of the GDP and replacement with GTP. Once GTP is bound, Gα assumes its active conformation and dissociates from the receptor.[16–18] The transitory nature of receptor–G protein interactions indicates that nature has evolved for the association to be of a sufficiently high affinity, but not necessarily the highest possible affinity. To identify peptides that can bind to the GPCR rhodopsin, with higher affinity than the native sequence, we have screened a biased combinatorial peptide library based on the carboxyl-terminus of Gα_t.[19]

The use of combinatorial libraries has greatly increased in recent years. A number of methods have been developed to display a large collection of peptides such that identification of the displayed peptides can be made through determination of attached DNA sequences. The strategies for the construction of soluble and solid phase-bound chemically or biologically generated peptide libraries, as well as the techniques used to screen them, have been described in great detail elsewhere.[20–23] Combinatorial peptide libraries have been used for the detection of epitopes, as well as the identification of peptide antagonists.[24–26]

---

[14] B. Conklin, P. Herzmark, S. Ishida, T. Voyno-Yasenetskaya, Y. Sun, Z. Farfel, and H. Bourne, *Mol. Pharmacol.* **50,** 885 (1996).

[15] R. B. Conklin, Z. Farfel, K. D. Lustig, D. Julius, and H. R. Bourne, *Nature* **363,** 274 (1993).

[16] H. Bourne, *Curr. Opin. Cell Biol.* **9,** 134 (1997).

[17] S. Rens-Domiano and H. Hamm, *FASEB J.* **9,** 1059 (1995).

[18] E. Weiss, D. Kelleher, C. Woon, S. Soparkar, S. Osawa, L. Heasley, and G. Johnson, *FASEB J.* **2,** 2841 (1988).

[19] E. L. Martin, S. Rens-Domiano, P. J. Schatz, and H. E. Hamm, *J. Biol. Chem.* **271,** 361 (1996).

[20] M. Zwick, J. Shen, and J. Scott, *Curr. Opin. Biotechnol.* **9,** 427 (1998).

[21] F. al-Obeidi, V. Hruby, and T. Sawyer, *Mol. Biotechnol.* **9,** 205 (1998).

[22] J. Scott and L. Craig, *Curr. Opin. Biotechnol.* **5,** 40 (1994).

[23] M. Needels, D. Jones, E. Tate, G. Heinkel, L. Kochersperger, W. Dower, R. Barrett, and M. Gallop, *Proc. Natl. Acad. Sci. U.S.A.* **90,** 10700 (1993).

[24] H. Hiemstra, P. van Veelen, N. Schloot, A. Geluk, K. van Meijgaarden, S. Willemen, J. Leunissen, W. Benckhuijsen, R. Amons, R. de Vries, B. Roep, T. Ottenhoff, and J. Drijfhout, *J. Immunol.* **161,** 4078 (1998).

[25] S. Cwirla, P. Balasubramanian, D. Duffin, C. Wagstrom, C. Gates, S. Singer, A. Davis, R. Tansik, L. Mattheakis, C. Boytos, P. Schatz, D. Baccanari, N. Wrighton, R. Barrett, and W. Dower, *Science* **276,** 1696 (1997).

[26] S. Yanofsky, D. Baldwin, J. Butler, F. Holden, J. Jacobs, P. Balasubramanian, J. Chinn, S. Cwirla, E. Peters-Bhatt, E. Whitehorn, E. Tate, A. Akeson, T. Bowlin, W. Dower, and R. Barrett, *Proc. Natl. Acad. Sci. U.S.A.* **93,** 7381 (1996).

In this paper we describe the use of a "peptide-on-plasmid" combinatorial library to identify high-affinity peptides from the carboxyl terminus of $G\alpha_t$ that bind light-activated or dark-adapted rhodopsin. The peptide-on-plasmids technique represents a unique system by which the high-affinity bond between LacI and lacO is exploited. In this technique, a library of peptides produced via degenerate polymerase chain reaction (PCR) are fused at the carboxyl terminus of LacI, and expressed via a plasmid vector carrying the fusion gene. Following transcription and translation, the LacI–peptide fusion protein binds back to the encoding plasmid, which contains a lacO DNA binding sequence. Thus, a stable LacI–peptide–plasmid complex is formed and can be screened by affinity purification to an immobilized receptor.[27]

We have selected this approach for several reasons including (1) the fusion of the peptide at the carboxyl terminus of the presenting protein, which mimics its normal presentation; (2) the ease of setting up the technique in the laboratory; and (3) the possibility that the high-affinity peptides may be useful tools to block receptor–G protein interaction.

Library Construction

The vector used for library construction is pJS142 (Fig. 1A). It has a linker sequence between the LacI and the biased peptide, as well as restriction sites for cloning the library oligonucleotide (Fig. 1B). The oligonucleotide synthesized to encode the mutagenesis library is synthesized with 70% of the correct base and 10% of each of the other bases. This mutagenesis rate leads to a biased library such that there is approximately a 50% chance that each of the 11 codons will be the appropriate amino acid, and a 50% chance that it will be another amino acid. In addition, four random NNK (where N denotes A, C, G, and T; and K denotes G and T) codons were synthesized at the 5' end of the sequence to make a total of 15 randomized codons. Construction of the biased peptide library of peptides has been described in detail previously.[19,27] Using this protocol, a library with greater than 10^9 independent clones per microgram of vector used in the ligation has been constructed based on the K341R derivative of the native 340–350 carboxyl-terminal sequence of $G\alpha_t$.

The *Escherichia coli* strain used for panning is ARI814, and has the following genotype: Δ(*srl-recA*) *endA1 nupG lon-11 sulA1 hsdR17* Δ(*ompT-fepC*)266 Δ*clpA319::kan* Δ*lacI lacZU118*. The strain contains the *hsdR17* allele that prevents restriction of unmodified DNA introduced by transformation or transduction. The *ompT-fepC* deletion removes the gene

[27] P. J. Schatz, M. G. Cull, E. L. Martin, and C. M. Gates, *Methods Enzymol.* **267**, 171 (1996).

A) map of pJS142 vector

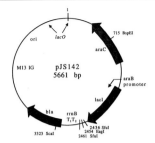

B) cloning sites at 3' end of lacI gene

```
  ❖ lacI ❖ ❖ ⋇ ❖ ❖ ❖ ❖ ❖ ❖ linker ❖ ❖ ❖ ⋇
  XhoI                           StuI  SfiI     HpaI  EagI   SfiI        MscI        SalI
  L   E   S   G   Q   V   V   H   G   E   Q   V   G   G   E   A   S   G   A   V   N   G   R   G   L   A   G   Q   *
5'- CTCGAGAGCGGGCAGgtggtgcatggggagcaggtgggtggtgagGCCTCCGGGGCCGTTAACGGCCGTGGCCTAGCTGGCCAATAAgtcgac
    GAGCTCTCGCCCGTCcaccacgtaccc ctcgtccacccaccactcCGGAGGCCCCGGCAATTGCCGGCACCGGSTCGACCGGTTATTcagctg
```

C) vector sequence after library addition

```
  ❖ lacI ❖ ❖ ⋇ ❖ ❖ ❖ ❖ ❖ ❖ ❖ linker ❖ ❖ ❖ ❖ ❖ ❖ ❖ ⋇ library
  XhoI                             StuI  BspEI                      MscI        SalI
  L   E   S   G   Q   V   V   H   G   E   Q   V   G   G   E   A   S   G   G   G   Xn   *
CTCGAGAGCGGGCAGgtggtgcatggggagcaggtgggtggtgagGCCTCCGgaggtggt(NNK)ₙtaactaagtaaagcTGGCCAATAAGTCGAC
GAGCTCTCGCCCGTCcaccacgtaccc ctcgtccacccaccactcCGGAggcctccacca(NNK)ₙattgattcattTCGACCGGTTATTcagctg
```

D) enzymes used to cut pJS142

Enzyme	Recognition site
BspEI	t/ccgaa
EagI	c/ggccg
ScaI	agt/act
SfiI	ggccnnnn/nggcc

FIG. 1. (A) Map of pJS142 vector. The 5661-bp pJS142 vector includes an *araC* gene to permit positive and negative regulation of the *araB* promoter driving expression of the *lacI* fusion gene, a *bla* gene to permit selection on ampicillin, two *lacO_s* sequences, the M13 phage intragenic region (*M13IG*) to permit rescue of single-stranded DNA, a plasmid replication origin (*ori*), and a *rrnB* transcriptional terminator. In addition, the restriction enzyme sites for *Sfu*I, *Eag*I, and *Sca*I are noted. (B) Cloning sites at 3' end of the *lacI* gene. Sequence of the cloning region at the 3' end of the *lacI* gene, including the *Sfi*I and *Eag*I restriction enzyme sites that are used for library construction. (C) Vector sequence after library addition. Sequence of the cloning region at the 3' end of the *lacI* gene following ligation of the annealed library oligonucleotides into the *Sfi*I restriction enzyme sites. Cloning of the annealed library oligonucleotides into pJS142 creates a *Bsp*EI restriction site near the beginning of the coding region of the library. The oligonucleotide synthesized to encode the mutagenesis library are represented as $(NNK)_n$, where n is the number of codons, in our case, 15. Digestion with *Bsp*EI and *Sca*I allows the purification of a 900-bp DNA fragment that is subcloned into pELM3. (D) Enzymes used to cut pJS142. The restriction enzymes used to introduce the annealed library oligonucleotides into pJS142 (*Sfi*I and *Eag*I), as well as those needed to cut the pJS142 vector (*Bsp*EI and *Sca*I) listed. Digestion of the pELM3 vector with *Age*I and *Sca*I allows efficient cloning of the *Bsp*EI–*Sca*I fragment from the pJS142 library. Solidus (/) indicates where the restriction enzyme cuts in the recognition site.

encoding the OmpT protease, which digests peptides between paired basic residues. The *lon-11* and *clpA* mutations also limit proteolysis as they prevent expression of ATP-dependent, cytoplasmic proteases. The deletion of the *lacI* gene prevents expression of wild-type lac repressor, which would compete with the fusion constructs for binding to the *lacO* sites on the plasmid. The *lacZ* mutation prevents waste of the cell's metabolic resources to make β-galactosidase in the absence of the repressor. The *endA1* mutation eliminates a nuclease that has deleterious effects on affinity purification, and the *recA* deletion prevents multimerization of plasmids through RecA-catalyzed homologous recombination. This strain was selected for its robust growth properties and yields of electrocompetent cells.[27] To test the efficiency we use 1 picogram of a pBlueScript plasmid (Stratagene, La Jolla, CA), and typically the cells yield transformation efficiencies of 2×10^{10} colonies per milligram of DNA.

Procedure for Production of Electrocompetent Cells

1. Start a single colony in 10 ml SOP [per liter: 20 g Bacto-tryptone; 10 g Bacto-yeast extract; 5 g NaCl; 2.5 g K_2HPO_4 (anhydrous); 1 g $MgSO_4 \cdot 7H_2O$, autoclave], grow overnight.
2. Add 1 ml of the overnight to 500 ml SOP, grow until OD_{600} reads 0.6–0.8.
3. Place cells immediately in ice-water bath for >15 min. (All subsequent washes should be done on ice with buffers and centrifuge rotors at or below 4°.)
4. Centrifuge at 4000g for 15 min at 4°.
5. Resuspend cell pellet in 500 ml of 10% glycerol, rest on ice for 30 min.
6. Centrifuge at 4000g for 15 min at 4°.
7. Resuspend cell pellet in 500 ml of 10% glycerol.
8. Centrifuge at 4000g for 15 min at 4°.
9. Resuspend cell pellet in 20 ml 10% glycerol.
10. Centrifuge at 5000g for 10 min at 4°.
11. Resuspend cell pellet in 1 ml 10% glycerol.
12. Aliquot the cells into 100- to 200-μl portions and quick freeze using dry ice and 2-propanol.

Procedure for Library Amplification

1. Chill sterile 0.1-cm electrode gap cuvette (Bid-Rad, Hercules, CA) on ice.
2. Thaw aliquot of electrocompetent ARI814 cells.

3. Transfer 40 μl of ARI814 cells into a chilled microcentrifuge tube, add 2 μl of library DNA, and mix.
4. Transfer cell–plasmid mixture into cuvettes. We set a Bio-Rad *E. coli* Pulsar to 1.8 kV, 25-μF capacity, and the pulser controller unit to 200 mΩ.
5. Apply one pulse (time constant should be 4–5 ms).
6. Immediately add 1 ml of SOC (per liter: 20 g Bacto-tryptone; 5 g Bacto-yeast extract; 0.5 g NaCl; 10 ml 0.25 M KCl; adjust to pH 7 with NaOH, after autoclaving add 1 ml 1 M glucose; 0.5 ml 1 M MgCl$_2$), transfer to a labeled 17- × 100-mm polystyrene tube and shake for 1 hr at 37°.
7. Remove and plate cells 100 μl undiluted to 10^{-6} dilution on LB-Amp (ampicillin) plates. Counts of these plates will yield the size of the library.
8. The remaining ~900 μl should be added to 500 ml of LB/Amp and grown at 37° with vigorous shaking until A_{600} reaches 0.5–0.8.
9. Place the cells in an ice–water bath for >10 min. (All subsequent washes should be done on ice with buffers and centrifuge rotors at or below 4°.)
10. Centrifuge at 5000g for 10 min at 4°.
11. Resuspend the cell pellet in 300 ml WTEK buffer (50 mM Tris, pH7.5, 10 mM EDTA, 100 mM KCl).
12. Centrifuge at 5000g for 10 min at 4°.
13. Resuspend the cell pellet in 150 ml TEK buffer (10 mM Tris, pH 7.5, 0.1 mM EDTA, 100 mM KCl).
14. Centrifuge at 5000g for 10 min at 4°.
15. Resuspend the cell pellet in 10 ml HEK buffer (35 mM HEPES, 0.1 mM EDTA, 50 mM KCl; adjust to pH 7.5 with KOH). Aliquot into cryovials, 2 ml per vial, and store at −70°.

Panning Protocols

The peptide-on-plasmid library is released from the ARI814 cells by gentle enzymatic digestion of the cell wall using lysozyme. After pelleting the cell debris, the lysate is then added directly to the immobilized receptor for affinity purification (panning). The panning is carried out in low-salt buffers because high-salt concentrations destabilize the LacI-*lacO* complex, and could lead to peptides becoming associated with the incorrect plasmid. For this same reason, the panning buffers also contain lactose, which causes the LacI to bind more tightly to *lacO*.

Procedure for Panning

1. Coat wells with 100 μl purified receptor, in our case, urea-washed rod outer segments (ROS), allow to shake gently for 1 hr at 4°. The receptor should be at concentration of approximately 1 μg/well in HEK buffer with 1 mM dithiothreitol (DTT). In the first round we typically coat 24 wells of a 96-well plate with receptor. From the second round on, we use 8 wells coated with receptor and 8 wells without receptor. Comparison of the number of plasmids from these samples gives an indication of whether receptor-specific clones are being enriched during the panning.
2. Block wells for nonspecific binding by adding an equal volume HEKL (35 mM HEPES, 0.1 mM EDTA, 50 mM KCl, 0.2M α-lactose; adjust to pH 7.5 with KOH, warm buffer to dissolve lactose if necessary) with 1 mM DTT and 2% bovine serum albumin (BSA); allow to shake gently for 1 hr at 4°. (We usually alternate blocking agents after two rounds such that rounds 1 and 2 are blocked with 1% BSA, rounds 3 and 4 are blocked with 1% nonfat dry milk.)
3a. While blocking, prepare fresh 10 mg/ml lysozyme in cold HE buffer (35 mM HEPES, 0.1 mM EDTA; adjust to pH 7.5 with KOH), store on ice until needed.
3b. Thaw the library (2-ml aliquot, step 15, Procedure for Library Amplification) and add to 6 ml lysis buffer [4.25 ml HE, 1 ml 50% glycerol, 750 μl 10 mg/ml protease-free BSA in HE, 10 μl 0.5 M DTT, 6.25 μl 0.2 M phenylmethylsulfonyl fluoride (PMSF)], keep on ice.
3c. Add 150 μl lysozyme solution to the library/lysis buffer, invert gently several times, and incubate on ice for no more than 2 min. The extent of lysis is evidenced by an increase in viscosity that can be observed by noting the slow migration of bubbles to the top of the tube after mixing.
3d. End the lysis by adding 2 ml 20% lactose and 250 μl 2 M KCl and mix by inverting several times.
3e. Immediately centrifuge library/lysis buffer/lactose/KCl mixture at 13,000g for 15 min at 4°.
3f. Transfer the supernatant to a new tube. Save 0.1% in a labeled tube (PRE).
4. Wash the plate with immobilized receptor 4× with HEKL/1% BSA. For rounds 1 and 2 we add 200 μl of the crude lysate to each well and allow to shake gently for 1 hr at 4°. In round 3 we reduce the amount of library to 100 μl. In all rounds after 3 we dilute the lysate 1:10 in HEKL/BSA. In all rounds after 2 we also include 200 μM native peptide to chase off peptides that bind with lower affinity.

5. Wash plate 4× with cold HEKL/1% BSA.
6. Add 200 μl 0.1 mg/ml bulk DNA (sonicated salmon sperm DNA) in HEKL/1% BSA; allow to shake gently for 30 min at 4°.
7. Wash plate 4× with cold HEKL.
8. Wash plate 2× with cold HEK.
9. Elute by adding 50 μl/well of 1 mM IPTG/0.2 M KCl in HE; shake vigorously at room temperature for 30 min.
10. Combine all eluents, (−) or (+) receptor, into a microcentrifuge tube. In the first round we actually use three microcentrifuge tubes for the (−) or (+) receptor wells. Bring the volume of the PRE sample (step 3f) up to match the volume of the (−) or (+) receptor samples and precipitate it in parallel with the other samples. Add 1/10 volume of 5 M NaCl, mix, then add 1 μl 20 mg/ml glycogen per microcentrifuge tube as carrier and mix thoroughly.
11. Precipitate plasmids with an equal volume of 2-propanol at room temperature, mix thoroughly.
12. Centrifuge 15 min at 13,000g.
13. Carefully aspirate supernatant and wash with 500 μl of 80% (v/v) cold ethanol.
14. Centrifuge 10 min at 13,000g.
15. Resuspend plasmid DNA in sterile doubly distilled H_2O; 200 μl for the PRE sample, and 4 μl for the (−) or (+) receptor samples. The plasmid DNA is stored at −20°.

Procedure for Amplification of Affinity Purified Plasmids

1. Chill three sterile 0.1-cm electrode gap cuvettes on ice.
2. Thaw an aliquot of electrocompetent ARI814 cells.
3. Transfer 40 μl of ARI814 cells into each of three chilled microcentrifuge tube, add 2 μl of plasmid (−), (+), or (PRE) to each and mix.
4. Transfer the cell–plasmid mixtures into cold cuvettes.
5. Apply one pulse. We use a BioRad *E. coli* pulsar set to 1.8 kV, 25-μF capacity, and the pulser controller unit to 200 mΩ, the time constant should be 4–5 ms.
7. Immediately add 1 ml of SOC, mix by pipetting and transfer the cell–plasmid–SOC mixture to a labeled 17- × 100-mm polystyene tube and incubate at 37°, shaking, for 1 hr.
8. Prewarm a 1-liter flask with 200 ml of LB-Amp to 37°.
9. Remove an aliquot of cells from the (−), (+), or (PRE) tubes, plate 100 μl undiluted to 10^{-6} dilution on LB-Amp plates. Counts of the PRE plates will indicate library diversity, while comparison of the

(−) and (+) plates will indicate whether receptor specific clones are being enriched by the panning procedure.

10. The remaining 900 μl for the (+) receptor tube should be added to the prewarmed LB-Amp media and grown at 37°, shaking until A_{600} is 0.5.
11. Place the cells in an ice water bath for >10 min. (All subsequent washes should be done on ice with buffers and centrifuge rotors at or below 4°.)
12. Centrifuge at 5000g for 6 min at 4°.
13. Resuspend the cell pellet in 100 ml WTEK buffer.
14. Centrifuge at 5000g for 6 min at 4°.
15. Resuspend the cell pellet in 50 ml TEK buffer.
16. Centrifuge at 5000g for 6 min at 4°.
17. Resuspend the cell pellet in 4 ml HEK buffer. Aliquot into 2 cryovials, 2 ml per vial, and store at −70°. Use one tube for the next round, and keep the other as a backup.
18. In the last round, several clones are selected from the (+) receptor plates and grown up overnight in LB-Amp media. DNA is purified using any standard minipreparation protocol (we use Qiagen Spin-prep kits, Valencia, CA) and sequenced using a 19-bp reverse primer, which is homologous to the vector at a site 56 bp downstream from the TAA stop codon that terminates the random region of a library (GAAAATCTTCTCTCATCCG).

Preparation of LacI Lysates

1. Several clones are selected from the plates of the final round (step 9, Procedure for Amplification of the Affinity Purified Plasmids) and grown overnight in 5 ml LB-Amp media; 37°; shaking.
2. 300 μl of the overnight culture is diluted into 3 ml LB-Amp to make ELISA lysate; 300 μl is added to an equal volume of 50% glycerol and stored in labeled microcentrifuge tubes at −70°, the remaining 4.5 ml is used to make DNA that is stored at −20°.
3. The 3.3 ml culture is allowed to shake for 1 hr at 37°.
4. Expression is induced by adding 33 μl 20% arabinose (0.2% final concentration), shake at 37° for 2–3 hr.
5. Centrifuge at 4000g for 5 min.
6. Resuspend pellet in 3 ml cold WTEK buffer.
7. Centrifuge at 4000g for 5 min.
8. Resuspend cell pellet in 1 ml cold TEK, transfer to 1.5-ml microcentrifuge tubes.
9. Centrifuge at 13,000g, 2 min, and aspirate supernatant.

TABLE I
Light Rhodopsin High-Affinity Sequences

G_t 340-350 (K341R)	Library sequence									
	X X X X	I	R	E	N	L	K	D	C	G	L	F
Clone												
8	F V P D	L	L	E	N	L	R	D	C	G	M	F
9	T R F A	L	Q	Q	V	L	K	D	C	G	L	L
10	A L D Y	I	C	E	N	L	K	D	C	G	L	F
18	K Q R N	M	L	E	N	L	K	D	C	G	L	F
23	T G G R	V	L	E	D	L	K	S	C	G	L	F
24	K G Q A	M	L	K	N	L	K	D	C	G	M	F
%Identity		17	0	67	67	100	83	83	100	100	83	83

10. Resuspend cell pellet in 1 ml lysis buffer, incubate on ice, 1 hr. (To prepare 50 ml lysis buffer mix 42 ml HE, 5 ml 50% glycerol, 3 ml 10 mg/ml BSA made in HE, 750 μl 10 mg/ml lysozyme made in HE, and 62.5 μl 0.2 M PMSF.)
11. Add 110 μl 2 M KCl (final concentration 0.2 M), invert to mix.
12. Centrifuge at 13,000g, 15 min, 4°.
13. Transfer clear crude lysate (~0.9 ml supernatant) to a new tube. Store at −70°.

We have screened both light-activated rhodopsin[19] and dark-adapted rhodopsin in this manner. Six of the sequences obtained using light-activated rhodopsin were 100 to 1000 fold more potent than native sequence at binding rhodopsin and are noted in Table I (for a more complete listing please refer to the original paper[19]). When the G_t library is used to pan light-activated rhodopsin we observed that the residues L344, C347, and G348 were always invariant. We also observed that in each of the highest affinity sequences the positive residue at position 341 (R341) was changed to a neutral residue. When the G_t library is used to pan dark-adapted rhodopsin (Table II) we noted that (1) the L344, C347, and G348 residues were no longer invariant (L344 present in 62.5% of sequences, C347 present in 25% of sequences, G348 present in 75% of sequences), and (2) that the residue at position 341 was nearly always invariant (75%). Thus, it appears that the conformation of the receptor in its dark-adapted state allows it to bind to a different set of peptide analogs than the light-activated receptor. In addition, it appears that in the light-activated receptor, the last 7 amino acids are the most important (344–350), whereas the first 6 amino acids (340–345) are more important for dark-adapted rhodopsin binding.

TABLE II
Dark Rhodopsin Sequences

G_t 340-350 (K341R)	Library sequence										
	X X X X	I	R	E	N	L	K	D	C	G	L	F
Clone												
2	I M E C	I	R	E	K	W	K	D	L	A	L	F
3	D F W H	V	R	D	N	L	K	N	C	F	L	F
7	P G M H	I	G	E	Q	I	E	D	C	G	P	F
17	I R H T	I	R	N	N	L	K	R	Y	G	M	F
21	L K A W	I	R	E	N	L	K	D	L	G	L	V
26	G L F K	I	R	E	N	F	K	Y	L	G	L	W
33/37	R K L T	S	L	E	I	L	K	D	W	G	L	F
41	R P K L	I	R	G	T	L	K	G	W	G	L	F
%Identity		75	75	63	50	63	88	50	25	75	75	75

Selection of Higher Affinity Peptides

In the normal peptide-on-plasmids protocol, native peptide is added at the same time as the library lysates (step 4, Procedure for Panning). To identify peptides with even higher affinity than those originally identified we screened light-activated ROS for sequences that bind the receptor in the presence of a high-affinity peptide identified in the first screening (peptide 8; LLENLRDCGMF). By including 100 μM of this peptide sequence that is 1000-fold better at binding rhodopsin than native sequence we have now identified sequences with presumably even higher affinity for light-activated rhodopsin (Table III).

Use of the high-affinity peptide (peptide 8) rather than the native peptide in rounds 3 and 4 has given us several clones that both bind rhodopsin with high affinity and stabilize it in its active form, metarhodopsin-II. Comparison of the sequences from Table I and Table III indicate that when the G_t library is used to pan light-activated rhodopsin we observed that the residues L344, C347, and G348 were invariant whether or not native peptide or peptide 8 is included. As was observed previously, when peptide 8 was included the positive residue at position 341 was changed to a neutral residue in each of the sequences. We also noted differences when peptide 8 was used to compete against the locI-peptide fusion proteins in the later rounds. With inclusion of peptide 8 there was a higher incidence for isoleucine at position 340 (17% with native peptide versus 71% with peptide 8) such that a residue previously being selected against now appears to be selected for. Also seen was that the glutamine at position 342 was selected against (67% with native peptide versus 29% with peptide 8). Thus, it appears this refinement of the peptide-on-plasmid method will not only be successful at obtaining potent analog peptide sequences but will also allow

TABLE III
Light Rhodopsin High-Affinity Sequences

G_t 340-350 (K341R)	Library sequence										
	X X X X	I	R	E	N	L	K	D	C	G	L	F
	Competing sequence										
Peptide 8		L	L	E	N	L	R	D	C	G	M	F
Clone												
3	V G R S	I	L	E	N	L	K	D	C	G	L	L
7	T S K P	M	L	D	N	L	K	D	C	G	L	F
8	G Y L Q	I	V	K	N	L	E	D	C	G	L	F
10	X F R X	I	R	D	N	L	K	D	C	G	L	F
13	S F Q S	I	S	K	N	L	R	D	C	G	L	L
17	L N S D	I	L	Q	N	L	K	D	C	G	L	F
19	V T A L	M	L	D	N	L	K	A	C	G	L	F
%Identity		71	0	29	100	100	71	86	100	100	86	71

us to further our understanding of the structural framework that underlies the sites of contact between Gα and receptor.

Transfer into MBP Vector

The binding properties of the peptides encoded by the individual clones are examined following several rounds of panning using an enzyme-linked immunosorbant assay (ELISA) that detects receptor-specific binding by the LacI–peptide fusion protein. LacI is normally a tetramer and the minimum functional DNA binding species is a dimer. Thus, the peptides are displayed multivalently on the fusion protein leading to binding to the immobilized receptor in a cooperative fashion. This cooperative binding permits the detection of binding events of quite low intrinsic affinity. The sensitivity of the assay is an advantage in that initial hits of low affinity (such as what is observed with the dark adapted rhodopsin) can be identified, but the disadvantage is that the signal in the ELISA does not correlate with the intrinsic affinity of the peptides. To permit testing in an ELISA where signal strength is better correlated with affinity the sequences of interest from a population of clones (i.e., the last round) are fused in frame with the gene encoding maltose binding protein (MBP).[28]

The cloning of the library into pJS142 creates a *Bsp*EI restriction site near the beginning of the random coding region of the library (Fig. 1C).

---

[28] M. Zwick, L. Bonnycastle, K. Noren, S. Venturini, E. Leong, C. R. Barbas, C. Noren, and J. Scott, *Anal. Biochem.* **264,** 87 (1998).

Digestion with *Bsp*EI and *Sca*I (Fig. 1D) allows the purification of a 900-bp DNA fragment that is subcloned into pELM3, a vector that directs the MBP fusion protein to the cytoplasm, a reducing environment. Alternatively, the fragment can be cloned into pELM15, a vector that directs the MBP fusion protein to the periplasm—an oxidizing environment. pELM3 and pELM15 are simple modifications of the pMALc2 and pMALp2 vectors, respectively, available commercially from New England Biolabs (Beverly, MA). Digestion of pELM3 with AgeI and ScaI allows efficient cloning of the *Bsp*EI–*Sca*I fragment from the pJS142 library.

Procedure for Subcloning into MBP Vectors

1. Digest pELM3 with *Age*I (New England Biolabs, digest is performed at room temperature). Separate cut from uncut vector on a 0.7% agarose gel, and purify DNA using the Qiagen Extract-a-gel kit.
2. Digest the *Age*I cut vector DNA with *Sca*I (New England Biolabs). Separate the 5.6-kb MBP vector fragment away from the 1.0-kb fragment on a 1% agarose gel, and purify the 5.6-kb MBP vector DNA.
3. During the final affinity purification round (step 10, Procedure for Amplification of the Affinity Purified Plasmids), set aside a 20-ml portion from the 200-ml amplification culture before harvesting the cells. Allow the 20-ml portion to grow to saturation (usually overnight), then prepare DNA from the cells (Qiagen midi-prep kit).
4. Digest the pJS142 plasmid DNA with *Bsp*EI and *Sca*I. Separate the 0.9-kb peptide-encoding fragment from the 3.1 and 1.7 vector fragments on a 1% agarose gel, and purify the 0.9-kb peptide-encoding fragment DNA.
5. Ligate different ratios of the 5.6-kb MBP vector fragment and the peptide-encoding 0.9-kb fragment (typically we try 1:2; 1:1; 2.5:1; 5:1; 10:1). We make our own ligase buffer such that it contains 0.4 mM ATP, and all ligations are done at 14°, overnight.
6. Inactivate the T4 DNA ligase by increasing the temperature to 65° for 10 min.
7. To lower the background, digest the ligation mix with *Xba*I. Isopropanol precipitate the ligation mix using 1 μl of glycogen as carrier. Wash once with 80% ethanol and resuspend the pellet in 20 μl sterile H$_2$O.
8. Transform ARI814 cells (described in detail in Procedure for Library Amplification, steps 1–7) with 1 μl of the precipitated *Xba*I digested ligation mix. After allowing the cells in 1 ml SOC to shake for 1 hr at 37°, spread 100 μl on an LB-Amp plate.

Preparation of MBP Crude Lysates

1. Several clones are selected from the plates (step 8, Procedure for Subcloning into MBP Vectors) and grown overnight in 5 ml LB-Amp media; 37°, shaking.
2. 300 μl of the overnight culture is diluted into 3 ml LB-Amp to make ELISA lysate; 300 μl is added to an equal volume of 50% glycerol and stored in labeled microcentrifuge tubes at −70°; the remaining 4.5 ml is used to make DNA that is stored at −20°.
3. The 3.3 ml culture is allowed to shake for 1 hr at 37°.
4. Expression is induced by adding 9.9 μl 100 mM IPTG (0.3 mM final concentration), shake 37°, 2–3 hr.
5. Centrifuge at 5000g, 5 min.
6. Resuspend pellet in 3 ml cold WTEK buffer.
7. Centrifuge at 5000g, 5 min.
8. Resuspend cell pellet in 1 ml cold TEK, transfer to 1.5-ml microcentrifuge tube.
9. Centrifuge at 13,000g, 2 min, and aspirate supernatant.
10. Resuspend cell pellet in 1 ml lysis buffer, incubate on ice, 1 hr. (To prepare 50 ml lysis buffer mix 42 ml HE, 5 ml 50% glycerol, 3 ml 10 mg/ml BSA made in HE, 750 μl 10 mg/ml lysozyme made in HE, and 62.5 μl 0.2 M PMSF.)
11. Add 110 μl 2 M KCl (final concentration 0.2 M), invert to mix.
12. Centrifuge at 13,000g, 15 min, 4°.
13. Transfer clear crude lysate (~0.9 ml supernatant) to a new tube. Store at −70°.

MBP–Peptide Fusion Protein Purification

1. Inoculate 200 ml LB/Amp with 1 ml of an overnight culture, shake at 37°.
2. Grow until the OD_{600} is 0.5
3. Induce protein expression with 150 μl 1 M isopropylthiogalactoside (IPTG) (final concentration 0.3 mM), continue shaking at 37° for 2 hr.
4. Centrifuge culture at 5000g for 20 min.
5. Resuspend the cell pellet in 5 ml of column buffer (10 mM Tris, pH 7.4, 200 mM NaCl, 1 mM EDTA, 1 mM DTT), add 16.25 μl 0.2M PMSF. If necessary, stop here and store resuspended cell pellet at −70°. The cells should then be slowly thawed in cold water.
6. Place cells in ice bath and sonicate in short pulses of <15 sec. (We set our Fisher Scientific 55 Sonic Dismembrator to 40% constant time, output 5, repeating 5 times with a 1-min total duration.) If

necessary, the release of protein can be monitored using a standard protein assay.
7. Centrifuge at 9000g for 30 min.
8. Save supernatant and dilute to 100 ml using column buffer. (Should be approximately 2.5 mg/ml protein.)
9. Pour 7.5 ml amylose resin in a Bio-Rad disposable column.
10. Wash with 8 column volumes of column buffer.
11. Load diluted crude extract by gravity flow (~1 ml/min for a 2.5-cm column).
12. Wash with 8 column volumes of column buffer.
13. Elute the fusion protein with 10 mM maltose in column buffer. These can be collected as fractions, but we generally find this unnecessary. The protein generally elutes in the first 10 ml.
14. Concentrate the protein using Amicon (Danvers, MA) Centriplus 30 columns.
15. Aliquot and store concentrated proteins at $-70°$.

ELISA Methods

The individual clones are initially examined by producing large amounts of the LacI–peptide fusion proteins for use in an ELISA to determine the binding properties of the analog peptides encoded by the individual clones. However, as the LacI–peptides are displayed multivalently, the cooperative binding leads to signals in the ELISA that do not necessarily correlate with the intrinsic affinity of the peptides. To permit testing in an ELISA where signal strength is better correlated with affinity the DNA from the last round is fused in frame with the gene encoding MBP.[28] Once the sequences have been transferred into the monomeric fusion protein MBP, they can be overexpressed in *Escherichia coli* and utilized as either crude lysates (see Preparation of MBP Crude Lysates) or further purified using an amylose resin column (see MBP-peptide Fusion Protein Purification) for use in an ELISA. The crude lysates (step 13, Preparation of MBP Crude Lysates) or purified fusion proteins (step 15, MBP–peptide Fusion Protein Purification are placed in wells containing immobilized receptor. After blocking, the wells are washed and probed with an anti-MBP antibody, followed by a goat anti-rabbit HRP conjugated antibody. The A_{450} is measured on a microtiter plate reader and those samples with an absorbance of at least two standard deviations above background are considered) "positive."

Procedure for ELISA Using Crude Lysate

1. Coat microtiter wells with the receptor (we use 1 μg/well) diluted in a final volume of 100 μl in HEK with 1 mM DTT, allow to shake at 4° for 1 hr.

2. Block receptor with 1% BSA in HEK with 1 mM DTT by adding 100 μl 2% BSA in HEK with 1 mM DTT to wells with receptor, allow to shake at 4° for >30 min.
3. Dilute the crude lysates (step 13, Procedure for the Preparation of MBP ELISA Lysates) 1/50 in HEK with 1 mM DTT. Wash plate 4× with HEK with 1 mM DTT and add 100 μl diluted lysate per well, allow to shake at 4° for 1 hr.
4. Wash plate 4× with PBS/0.05% Tween and add 100 μl of diluted primary antibody made up in PBS (1/1000 dilution of rabbit anti-MBP; New England Biolabs), allow to shake at 4° for 30 min.
5. Wash plate 4× with PBS/0.05% Tween and add 100 μl diluted secondary antibody made up in PBS (1/7500 dilution goat anti-rabbit antibody conjugated to HRP; KPL) allow to shake at 4° for 30 min.
6. Wash plate 4× with PBS/0.05% Tween. Add 100 μl HRP substrate (KPL, Gaithersburg, MD) and allow color to form, approximately 20–30 min. Stop reaction by addition of 100 μl 2 N sulfuric acid. Read plate at OD$_{450}$. If the reaction occurs too quickly (<10 min, or if the backgrounds are too high) we repeat the ELISA using 1/100 or 1/200 dilutions of the crude lysates.

Procedure for ELISA Using Purified Fusion Protein

1. Coat microtiter wells with the receptor (we use 1 μg/well) diluted to a final volume of 100 μl in HEK/DTT, allow to shake at 4° for 1 hr.
2. Block receptor with 1% BSA in HEK/DTT by adding 100 μl 2% BSA in HEK/DTT, allow to shake at 4° for >30 min.
3. Dilute the purified MBP–Peptide Fusion Protein (step 15, MBP–peptide Fusion Protein Purification) in HEK/DTT to final concentrations from 0.2–120 nM. Wash plate 4× with HEK/DTT and add 100 μl diluted MBP–peptide fusion protein/well, allow to shake at 4° for 1 hr.
4. Wash plate 4× with PBS/0.05% Tween and add 100 μl of diluted primary antibody made up in PBS (1/3000 dilution of rabbit anti-MBP; New England Biolabs), allow to shake at 4° for 30 min.
5. Wash plate 4× with PBS/0.05% Tween and add 100 μl diluted secondary antibody made up in PBS (1/10,000 dilution goat anti-rabbit antibody conjugated to HRP; KPL) allow to shake at 4° for 30 min.
6. Wash plate 4× with PBS/0.05% Tween. Add 100 μl HRP substrate (KPL) and allow color to form, approximately 20–30 min. Stop reaction by addition of 100 μl 2 N sulfuric acid. Read plate at OD$_{450}$.

The positive purified proteins can be further tested by measuring their ability to be competed off of the immobilized receptor using peptide ($G_t\alpha$340-350K341R), LacI-$G_t\alpha$340-350K341R, or heterotrimer G_t. Use of the competitive ELISA allows one to calculate IC_{50} values for the binding of individual MBP fusion protein to the immobilized receptor.[19] Additionally, the potentially important peptide sequences can then by synthesized and used for ELISA or in receptor binding studies.[11,19]

Conclusion

The peptide-on-plasmid method provides an efficient means of identifying specific and potent peptides capable of regulating receptor–G protein signal transduction. We have shown that the method can define important determinants in the interaction between the carboxyl terminus of $G\alpha_t$ and rhodopsin.[19] Because of the similarity between $G\alpha_t$ and $G\alpha_i$ we have tested the $G\alpha_t$ peptide analogs for their ability to bind other G-coupled receptors.[11] Once the high-affinity analog peptides have been determined, they can be used as dominant negative inhibitors of receptor–G protein interactions.[12] Thus, this strategy lays a structural framework by which therapeutic agents can be developed that specifically interfere with G protein signaling by a given GPCR.

Acknowledgment

This work was supported by grants EY06062 (H.E.H.) and EY10291 (H.E.H.), by a Distinguished Investigator Award from the National Alliance for Research on Schizophrenia and Depression (H.E.H.), and by postdoctoral training grant HL07829 (A.G.) from the National Institutes of Health.

[27] Purification of Rhodopsin Kinase by Recoverin Affinity Chromatography

By CHING-KANG CHEN and JAMES B. HURLEY

Introduction

Recoverin is a retina and pineal-specific Ca^{2+}-binding protein.[1–3] It represents a family of neuronal specific calcium sensors (the NCS fam-

---

[1] A. H. Milam, D. M. Dacey, and A. M. Dizhoor, *Vis. Neurosci.* **10,** 1 (1993).
[2] H. W. Korf, B. H. White, N. C. Schaad, and D. C. Klein, *Brain Res.* **595,** 57 (1992).
[3] A. M. Dizhoor, S. Ray, S. Kumar, G. Niemi, M. Spencer, D. Brolley, K. A. Walsh, P. P. Philipov, J. B. Hurley, and L. Stryer, *Science* **251,** 915 (1991).

ily),[4] which includes neurocalcin, hippocalcin, visinin, frequenin, VILIP, NCS-1, Ce-NCS and guanylyl cyclase activating proteins.[5,6] The N terminus of recoverin is heterogeneously acylated with one of four types of fatty acyl residues, including myristoyl (C14:0), C14:1, C14:2, and C12:0 moieties.[7] Recoverin inhibits rhodopsin kinase (RK)-catalyzed rhodopsin phosphorylation *in vitro* in a Ca^{2+}-dependent manner.[8] A direct and Ca^{2+}-dependent interaction of recoverin and RK underlies such inhibitory effect.[8] Recoverin also interacts with rod outer segment (ROS) membranes *in vitro* in a Ca^{2+}-dependent manner through a mechanism called *calcium myristoyl switch*.[9,10] The acylated N terminus of recoverin is sequestered within a hydrophobic pocket when it is in a Ca^{2+}-free state. When recoverin binds Ca^{2+}, the acylated N terminus is released from the hydrophobic pocket and serves as an anchor for membrane association.[11–13]

The function of this N-terminal acylation (N-acylation) has been shown in several experiments in which the activities of recombinant myristoylated and nonacylated recoverin were compared. N-acylation reduces the affinity of recoverin for Ca^{2+} and introduces cooperativity.[14] It is also required for Ca^{2+}-dependent binding of recoverin to ROS membranes.[9,10] N-acylation is not required for the interaction of recoverin with RK, but it introduces cooperativity and enhances the ability of recoverin to inhibit rhodopsin phosphorylation *in vitro*.[8]

Immobilized recoverin can be used as an affinity matrix for rapid isolation of native RK from bovine ROS extracts and recombinant RK from *Spodoptera frugiperda* (Sf9) cell extracts.[8] Unlike most published methods,[15–17] our procedures exclude the use of detergents. As a result, RK

---

[4] P. Nef, in "Guidebook to the Calcium Binding Proteins" (M. R. Celio, T. Pauls, and B. Schwaller, eds.), pp. 94–98. Oxford University Press, New York, 1996.

[5] K. Palczewski, I. Subbaraya, W. A. Gorczyca, B. S. Helekar, C. C. Ruiz, H. Ohguro, J. Huang, X. Zhao, J. W. Crabb, R. S. Johnson, K. A. Walsh, M. P. Gray-Keller, P. B. Detwiler, and W. Baehr, *Neuron* **13,** 395 (1994).

[6] A. M. Dizhoor, E. V. Olshevskaya, W. J. Henzel, S. C. Wong, J. T. Stults, I. Ankoudinova, and J. B. Hurley, *J. Biol. Chem.* **270,** 25200 (1995).

[7] A. M. Dizhoor, L. H. Ericsson, R. S. Johnson, S. Kuman, E. V. Olshevskaya, S. Zozulya, T. A. Neubert, L. Stryer, J. B. Hurley, and K. A. Walsh, *J. Biol. Chem.* **267,** 16033 (1992).

[8] C.-K. Chen, J. Inglese, R. J. Lefkowitz, and J. B. Hurley, *J. Biol. Chem.* **270,** 18060 (1995).

[9] A. M. Dizhoor, C.-K. Chen, E. Olshevskaya, V. V. Sinelnikova, P. Phillipov, and J. B. Hurley, *Science* **259,** 829 (1993).

[10] S. Zozulya and L. Stryer, *Proc. Natl. Acad. Sci. U.S.A.* **89,** 11569 (1992).

[11] J. B. Ames, R. Ishima, T. Tanaka, J. I. Gordon, L. Stryer, and M. Ikura, *Nature* **389,** 198 (1997).

[12] J. B. Ames, T. Tanaka, M. Ikura, and L. Stryer, *J. Biol. Chem.* **270,** 30909 (1995).

[13] R. E. Hughes, P. S. Brzovic, R. E. Klevit, and J. B. Hurley, *Biochemistry* **34,** 11410 (1995).

[14] J. B. Ames, T. Porumb, T. Tanaka, M. Ikura, and L. Stryer, *J. Biol. Chem.* **270,** 4526 (1995).

[15] K. Cha, C. Bruel, J. Inglese, and H. G. Khorana, *Proc. Natl. Acad. Sci. U.S.A.* **94,** 10577 (1997).

purified by this method can be concentrated to >1 mg/ml and stored frozen at $-70°$ for several months without loss of activity. We routinely obtained 300–500 μg purified RK from a liter culture of Sf9 cells. The specific activity of the RK preparation is typically 1200 units/mg when assayed at room temperature with 1 mM adenosine triphosphate (ATP) and fully bleached rhodopsin in urea-stripped ROS membranes (one unit is defined by 1 nmol of phosphate transfer onto rhodopsin per minute). When reconstituted with urea-stripped ROS membranes, recombinant RK isolated by this method catalyzes high-gain phosphorylation in which photoactivation of one rhodopsin molecule causes incorporation of up to several hundred phosphates into the total rhodopsin pool.[8]

Methods

Production of Nonacylated and Myristoylated Recoverin in *Escherichia coli*

Bacterial strain BL21(DE3), which harbors the plasmid pET11a-mr21,[18] is cultured at $37°$ in Luria–Bertani (LB) broth with ampicillin (75 μg/ml). Production of nonacylated recoverin is induced by 1 mM isopropylthiogalactoside (IPTG) when the culture has reached an OD_{600} of 0.4 and the cells are harvested at OD_{600} of 1.2. For production of acylated recoverin, plasmid pBB131, which carries yeast protein N-myristoyltransferase[19] must also be transformed into BL21(DE3)/pET11a-mr21. The resulting BL21(DE3)/pET11a-mr21/pBB131 strain is cultured at $37°$ in LB containing ampicillin (75 μg/ml) and kanamycin (50 μg/ml). Myristate (Fluka, Ronkonkoma, NY) is added to the culture at OD_{600} of 0.3. Production of myristoylated recoverin is induced at 1 mM IPTG when the culture reaches OD_{600} of 0.4 and the cells are harvested at OD_{600} of 1.2. Harvested cells can be stored at $-20°$ for up to 2 years.

To isolate recombinant recoverin, the frozen cells are resuspended in NLB buffer (100 mM Tris-HCl, pH 8.0, 50 mM NaCl, 2 mM $MgCl_2$, 200 μM $CaCl_2$, 5% glycerol) and lysed by French press (16,000 psi) or by

---

[16] K. Palczewski, *Methods Neurosci.* **15**, 217 (1993).
[17] J. Rim and D. D. Oprian, *Biochemistry* **34**, 11938 (1995).
[18] S. Ray, S. Zozulya, G. A. Niemi, K. M. Flaherty, D. Brolley, A. M. Dizhoor, D. B. McKay, J. Hurley, and L. Stryer, *Proc. Natl. Acad. Sci. U.S.A.* **89**, 5705 (1992).
[19] R. J. Duronio, E. Jackson-Machelski, R. O. Heuckeroth, P. O. Olins, C. S. Devine, W. Yonemoto, L. W. Slice, S. S. Taylor, and J. I. Gordon, *Proc. Natl. Acad. Sci. U.S.A.* **87**, 1506 (1990).

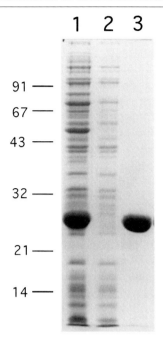

FIG. 1. Purification of recombinant recoverin by phenyl-Sepharose chromatography. Coomassie-stained gel showing BL21(DE3)/pET11a-mr21/pBB131 extracts before chromatography (lane 1), flow-through (lane 2), and EGTA eluate (lane 3) from a phenyl-Sepharose column. Recoverin interacts with phenyl-Sepharose in a Ca^{2+}-dependent manner and can be readily purified to homogeneity in one step.

repeated sonication. The lysate is spun at 50,000g for 30 min at room temperature and the supernatant incubated at 65° for 30 min. The supernatant is recentrifuged at 50,000g for 30 min to remove denatured proteins and applied to a 1.5- × 15-cm phenyl-Sepharose 4B column (Pharmacia, Piscataway, NJ) equilibrated with NLB buffer. The column is then washed extensively with NLB buffer until no further protein elutes. Bound recoverin is eluted with 5 mM EGTA in NLB buffer. As shown in Fig. 1, recombinant recoverin eluted from this column is nearly homogeneous. The typical yield of recoverin from 1 liter of culture is about 20–50 mg. Purified recoverin should be dialyzed extensively against autoclaved Milli-Q water (Millipore, Bedford, MA) and kept frozen at −20° at a concentration of >5 mg/ml.

Immobilization of Recoverin

CNBr-activated Sepharose CL-4B (Sigma, St. Louis, MO) is first washed with 50 mM HCl for 30 min then incubated with 100 mM sodium borate (pH 8.5) at room temperature for 1 hr. The beads are then washed with SBB buffer (100 mM NaHCO$_3$, pH 8.3, 200 mM NaCl). CaCl$_2$ is included at a final concentration of 1 mM. Recoverin is then added to the slurry at a final concentration of 1.5 mg/ml beads. The slurry is gently shaken at 4° for 6 hr. Excess activated beads are blocked by adding Tris-HCl buffer (pH 8.0) to a final concentration of 100 mM and incubating for another 2 hr. Coupling efficiency is typically >99%. Immobilized recoverin should be stored at 4° in ROS-Ca$_1$ buffer (20 mM MOPS, pH 7.0, 30 mM NaCl, 60 mM KCl, 2 mM MgCl$_2$, 1 mM DTT, 1 mM CaCl$_2$, 200 μM PMSF) in the presence of 0.02% sodium azide.

Isolation of Ca^{2+}-Dependent Recoverin Binding Proteins from Bovine Retinal Extracts

Fifty frozen bovine retinas (Excel, Kansas City, MO) are resuspended and shaken in 100 ml ROS buffer (ROS-Ca$_1$ buffer without 1 mM CaCl$_2$) containing 47% sucrose and then centrifuged at 30,000g at 4° for 30 min. The supernatant is diluted with an equal volume of ROS buffer and centrifuged at 30,000g at 4° for 30 min. To remove residual ROS membranes, the supernatant is spun again at 100,000g at 4° for an additional 90 min. All preparative work for retinal extracts should be done in a dark room under infrared illumination. The supernatant is applied to a 1- × 10-cm immobilized recoverin column at a flow rate of 1 ml/min and the column washed extensively with ROS buffer until no protein can be detected by Bradford assay. Bound proteins are then eluted by 5 mM EGTA in ROS buffer. As shown in Fig. 2, interphotoreceptor retinoid binding protein (IRBP), RK, and tubulin typically associate with myristoylated recoverin. Only RK and tubulin bind to nonacylated recoverin.

Immobilized Nonacylated Recoverin as Affinity Matrix for RK Purification

Because RK is nearly quantitatively depleted from retinal extracts by immobilized nonacylated recoverin and because bound RK is eluted when Ca^{2+} is removed, we develop a simple procedure that uses immobilized nonacylated recoverin to purify RK from retinal extracts. As shown in Fig. 3, this procedure includes two additional high-salt washes to remove bound tubulin prior to elution of RK by EGTA. The immobilized nonacylated recoverin column can be reused after regeneration by washing with 100 mM glycine (pH 2.5) and preequilibration by ROS buffer. A recoverin column can be reused at least 10 times. However, repetitive use results in loss of immobilized recoverin and eventually reduces the yield. Because

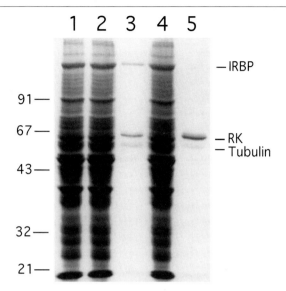

FIG. 2. RK is a Ca^{2+}-dependent recoverin binding protein; 12% SDS–PAGE followed by Coomassie staining. Lane 1, 20 μg retinal extract; lane 2, 20 μg flow-through from a myristoylated recoverin column; lane 3, 0.1 μg proteins eluted by EGTA from a myristoylated recoverin column; lane 4, 20 μg flow-through from a nonacylated recoverin column; lane 5, 0.1 μg proteins eluted by EGTA from a nonacylated recoverin column. Calmodulin and bovine serum albumin (BSA) columns were used as controls. IRBP binds only to myristoylated recoverin, RK binds only to recoverin columns, and tubulin binds both to recoverin and calmodulin columns (data not shown).

recombinant recoverin is easy to produce, we recommend using freshly prepared immobilized recoverin for each RK purification to achieve maximal purity and reproducibility. The yield of RK by this scheme has ranged from 10 to 20 μg/50 retinas.

Expression and Purification of RK in Sf9 Cells

Recombinant autographa California nuclear polyhedrosis virus (ACMNPV) for expressing bovine RK in Sf9 cells has been described previously.[8] Sf9 cells are maintained on 75-cm^2 plates in Grace's insect medium supplemented with 10% (v/v) heat inactivated fetal calf serum (GIBCO, Grand Island, NY) and 1% Pluronic F68. Two days prior to virus infection, Sf9 cells are inoculated into a 500-ml suspension culture in a 1-liter spinner flask with a concentration of 0.75 × 10^6 cells/ml. Cells are infected with recombinant RK virus (MOI = 5) at a concentration of 2 × 10^6 cells/ml. Media containing cells infected with

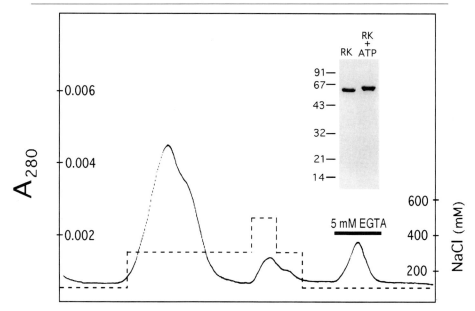

FIG. 3. Immobilized nonacylated recoverin column as an affinity matrix for rapid RK isolation. Protein extracts from 50 frozen bovine retinas were prepared and applied to a 1- × 10-cm nonacylated recoverin column at a flow rate of 1 ml/min as described in the text. One-milliliter fractions were collected. Only the chromatogram after the extensive ROS-Ca$_1$ wash is shown. *Inset:* Pooled EGTA eluate analyzed by Comassie-stained 12% SDS–PAGE. Incubation of the eluted RK with 10 μM ATP at 37° for 10 min causes the apparent molecular weight of RK to shift from 64,000 to 66,000, indicating that autophosphorylation has occurred.

RK virus are supplemented with 4 mM mevalonic acid lactone. Infected Sf9 cells are harvested by centrifugation between 60 and 72 hr after infection.

To purify recombinant RK, 10^9 cells are sonicated in ROS-Ca$_1$ buffer. The lysate is spun at 30,000g for 30 min at 4° to remove cell debris, and the supernatant is applied to a 1- × 10-cm immobilized nonacylated recoverin column. Conditions used for RK purification are identical to the ones used for native RK shown in Fig. 3. The yield of recombinant RK ranges from 300 to 500 μg per liter culture of Sf9 cells. Purified RK should be concentrated by ultrafiltration (YM10 membrane, Amicon, Danvers, MA) to >1 mg/ml and kept in −70° in TM buffer (50 mM Tris-HCl, pH 7.5, 2 mM MgCl$_2$).

[28] Heterologous Expression and Reconstitution of Rhodopsin with Rhodopsin Kinase and Arrestin

By Shoji Osawa, Dayanidhi Raman, and Ellen R. Weiss

Introduction

Rhodopsin, the light receptor of the vertebrate rod cell, mediates a rapid but controlled response to light through its interactions with transducin (the rod cell G protein), rhodopsin kinase (also known as GRK1), and arrestin. The phosphorylation of rhodopsin by rhodopsin kinase and the subsequent binding of arrestin mediate desensitization, a rapid turnoff mechanism that limits the lifetime of photoactivated rhodopsin. Seven serine and threonine residues at the rhodopsin COOH terminus can serve as substrates *in vitro* for phosphorylation by rhodopsin kinase. Rhodopsin kinase belongs to a unique family of G-protein-coupled receptor kinases (GRKs) that phosphorylate only activated receptors. This may be explained by the observation that the cytoplasmic loops of light-activated rhodopsin stimulate rhodopsin kinase activity, promoting phosphorylation of the rhodopsin COOH terminus.[1] Although phosphorylation at multiple sites can support arrestin binding,[2] only one or two sites are thought to be necessary for arrestin binding *in vivo*.[1] The interaction of arrestin with the phosphorylated COOH terminus of rhodopsin has been shown to induce a conformational change in arrestin that promotes its stable binding to a site presumed to include the cytoplasmic loops.[3] Therefore, multiple domains of rhodopsin participate in its interactions with both rhodopsin kinase and arrestin. Our laboratory is interested in identifying these domains as a step toward understanding the complex process of desensitization. This chapter describes procedures developed to express wild-type and site-directed mutants of rhodopsin in HEK293 cells and to measure the ability of these recombinant proteins to be phosphorylated by rhodopsin kinase and to bind arrestin.

Expression of Bovine Rhodopsin in HEK293 Cells

The cDNA for bovine rhodopsin,[4] obtained from Dr. Jeremy Nathans (Johns Hopkins University, Baltimore, MD), was truncated to remove part

---

[1] K. Palczewski, *Eur. J. Biochem.* **248**, 261 (1997).
[2] L. Zhang, C. D. Sports, S. Osawa, and E. R. Weiss, *J. Biol. Chem.* **272**, 14762 (1997).
[3] R. Sterne-Marr and J. L. Benovic, *Vit. Hormones* **51**, 193 (1995).
[4] J. Nathans and D. S. Hogness, *Cell* **34**, 807 (1983).

of the 5' and 3' noncoding domains using the restriction enzyme SmaI, creating a 1.3-kb fragment. Truncation of the cDNA significantly enhances the level of rhodopsin expression in cultured mammalian cells. After the addition of HindIII linkers, the cDNA is ligated into the vector pcDNAI/Amp (Invitrogen, Carlsbad, CA). The construct is transfected into HEK293 cells using DEAE-dextran following procedures similar to those described previously.[5]

HEK293 cells are plated at approximately 75% confluence in 10-cm dishes in Dulbecco's modified Eagle's medium (DMEM)/F12 medium containing 10% fetal calf serum (complete medium) 1 day prior to transfection. On the day of transfection, the cells are rinsed once with phosphate-buffered saline (PBS; 4.3 mM $Na_2HPO_4 \cdot 7H_2O$, 1.4 mM KH_2PO_4, 2.7 mM KCl, 137 mM NaCl) and incubated with 4 ml of DMEM/F12 containing either 10% (v/v) NuSerum (Collaborative Biomedical Products) or 2.5% newborn calf serum. This prevents heavy protein precipitation during the transfection. Each 10-cm dish of cells is cotransfected with 4 μg of pcDNAI/Amp-rhodopsin and 2 μg of pRSV-TAg (T antigen; a gift from Dr. Jeremy Nathans). Transfection with pRSV-TAg increases the level of rhodopsin expression in the HEK293 cell line. The solutions are prepared as described next.

The transfection buffer is composed of 25 mM Tris-HCl, pH 7.4, 137 mM NaCl, 5 mM KCl, 1.4 mM Na_2HPO_4, 10 mM $CaCl_2$, and 5 mM $MgCl_2$. To prepare 100 ml of the transfection buffer, mix 10 ml of solution A (250 mM Tris-HCl, pH 7.4, 1.37 M NaCl, 50 mM KCl, 14 mM Na_2HPO_4) and 1 ml of solution B (1 M $CaCl_2$, 0.5 M $MgCl_2$) with 89 ml of distilled, deionized water and filter sterilize. The DNA needed for transfection is resuspended at a concentration of 0.2 mg/ml in the transfection buffer. A 10 mg/ml solution of DEAE-dextran (500,000 MW; Pharmacia, Piscataway, NJ) and a 10 mM solution of chloroquine are also prepared in the transfection buffer and filter sterilized.

For each dish, 20 μl of pcDNAI/Amp-rhodopsin and 10 μl of pRSV-TAg are mixed with 200 μl of the DEAE-dextran solution at room temperature. Chloroquine (40 μl) is also added to the DNA/DEAE-dextran solution and the entire mixture is added dropwise to the plate, swirling slowly to distribute evenly. The cells are incubated for 3 hr in a 5% (v/v) CO_2 incubator at 37°, then rinsed with 2 ml PBS and incubated in 4 ml 10% dimethyl sulfoxide (DMSO) in PBS for 2 min. Because prolonged exposure to DMSO is highly toxic to the cells, the DMSO solution is removed immediately and each dish is gently rinsed with 2–4 ml complete medium. Care must be taken during this step not to dislodge the cells, which are very loosely attached. The cells are refed with 10 ml complete medium

[5] E. R. Weiss, S. Osawa, W. Shi, and C. D. Dickerson, *Biochemistry* **33**, 7587 (1994).

and incubated in a 5% CO_2 incubator at 37° for approximately 3 days before harvesting.

Preparation of Plasma Membranes

Plates of transfected cells are rinsed with 2 ml ice-cold PBS, scraped in 2 ml PBS, then transferred to a chilled 50-ml conical centrifuge tube. Typically, the cells from 5–10 plates are pooled in one centrifuge tube. The cells are pelleted by centrifugation at 400g (Sorvall, Newtown, CA; H-1000B, 1,500 rpm) for 5 min at 4°. The cell pellet is frozen at −80° to improve cell disruption, thawed on ice, and resuspended in 18 ml of 0.25 M sucrose in buffer containing 0.1 M sodium phosphate, pH 6.5, 1 mM EDTA, 1 mM dithiothreitol (DTT), 2 μg/ml aprotinin, and 1 μg/ml leupeptin. After Dounce homogenization (approximately 20 strokes), the mixture is layered over a 20-ml cushion consisting of 1.1 M sucrose in the same buffer. The homogenates are centrifuged at 103,900g (Beckman, Fullerton, CA; SW 28, 24,000 rpm) for 30 min. The membranes are collected at the interface between the 0.25 M and 1.1 M sucrose layers, diluted approximately ninefold with 50 mM HEPES, pH 6.5, 2 mM EDTA, 1 mM DTT and centrifuged at 41,200g (Beckman 70 Ti, 20,000 rpm) for 20 min to remove the sucrose. The pellet containing the membranes is resuspended by Dounce homogenization in 50 mM HEPES, pH 6.5, 140 mM NaCl, 3 mM $MgCl_2$, 2mM EDTA, 1 mM DTT and stored in aliquots at −80°. Approximately 100 μg of membrane protein per plate of cells is obtained using these methods.

The level of rhodopsin expression is measured by Western blot analysis. Typically, 10 μg of HEK293 cell membrane protein is subjected to 10% sodium dodecyl sulfate–polyacrylamide gel electrophoresis (SDS–PAGE) and compared with a standard curve of urea-stripped rod outer segment (ROS) membranes ranging from 50 to 300 ng of rhodopsin (Fig. 1). The urea-stripped ROS membranes are prepared from dark-adapted bovine retinas as described,[6] except that the ROS are isolated using a 25%/30% sucrose step gradient rather than a 25%/35% gradient. The level of rhodopsin in the urea-stripped ROS membranes is determined after solubilization in 1.5% octylglucoside by absorbance at 498 nm using a molar extinction coefficient of 42,700 M^{-1} cm^{-1}.[7] The gel also contains a lane of membranes isolated from nontransfected cells that is used to measure nonspecific binding of the anti-rhodopsin antibody. After electrophoretic transfer of the protein, the nitrocellulose membrane is incubated with an anti-rhodopsin monoclonal antibody, R2-15N (a gift from Drs. Grazyna Adamus and Paul

---

[6] M. Wessling-Resnick and G. L. Johnson, *J. Biol. Chem.* **262**, 3697 (1987).
[7] K. Hong and W. L. Hubbell, *Biochemistry* **12**, 4517 (1973).

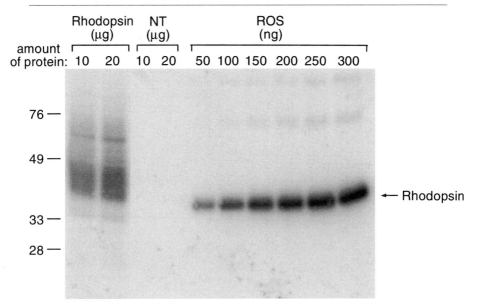

FIG. 1. Western blot analysis of bovine rhodopsin transiently expressed in HEK293 cells. Membranes prepared from nontransfected (NT) or rhodopsin-transfected HEK293 cells (rhodopsin) and urea-stripped ROS membranes were subjected to 10% SDS–PAGE, transferred to nitrocellulose membrane and blotted with the rhodopsin monoclonal antibody, R2-15N. The level of antibody binding was detected using I^{125}-labeled protein A and visualized by scanning with a Molecular Dynamics PhosphorImager. Numbers on the left side represent molecular size markers in kilodaltons (Bio-Rad prestained markers).

Hargrave, University of Florida, Gainesville, FL) at a concentration of 2 ng/μl of blocking buffer composed of 10 mM Tris-HCl, pH 7.4, 150 mM NaCl, 5% nonfat dry milk, and 0.1% Tween 20. This antibody recognizes the NH$_2$-terminal 15 amino acids of the rhodopsin polypeptide.[8] The level of antibody binding is detected with ^{125}I-labeled protein A (0.1 μCi/ml in blocking buffer). A Molecular Dynamics PhoshorImager (Sunnyvale, CA) is used to quantify the level of expression. Because rhodopsin is heterogeneously glycosylated when expressed in HEK293 cells, it appears as a series of multiple bands on Western blots (Fig. 1). Therefore, the entire lane above 35 kDa is used for quantification. Although expression varies for different transfections and different mutants, amounts of rhodopsin equaling 1.5–3% of the total membrane protein are typical. The level of rhodopsin expressed in these membranes can also be measured spectrophotometrically

[8] G. Adamus, Z. S. Zam, A. Arendt, K. Palczewski, J. H. McDowell, and P. A. Hargrave, *Vis. Res.* **31,** 17 (1991).

after reconstitution with 11-*cis*-retinal using methods described by Nathans.[9] The amount of rhodopsin estimated by Western blot analysis correlates well with the amount of functional rhodopsin determined by absorbance.[5]

Assay for Phosphorylation of Rhodopsin by Rhodopsin Kinase

An assay method was developed to measure the ability of rhodopsin kinase to phosphorylate rhodopsin mutants expressed in HEK293 cell membranes. Wild-type and mutant rhodopsin may be expressed at different levels, which may affect the rate of phosphorylation (see later discussion). Therefore, it is critical that each reaction mixture contain equal amounts of rhodopsin and total protein. To achieve this, the membranes are adjusted so that 1 μg of rhodopsin is used in each reaction. Nontransfected cell membranes are added to samples where necessary to equalize the amount of total protein. Before phosphorylation, the membranes must be reconstituted with 11-*cis*-retinal. Under Eastman Kodak (Rochester, NY) No. 2 safelights (dark red), membranes sufficient for duplicate samples are diluted into 0.5–1.0 ml phosphorylation buffer containing 20 mM Tris-HCl, pH 7.5, 2 mM EDTA, 6 mM MgCl$_2$, 1 mM DTT, 50 mM NaF, aprotinin (2 μg/ml), leupeptin (1 μg/ml), and 14 μM 11-*cis*-retinal. The samples are rocked in the dark at room temperature for 60 min. Because 11-*cis*-retinal is absorbed nonspecifically to the membranes, it is important to use at least 7 nmol/200 μg membrane protein to achieve maximum binding to rhodopsin. After reconstitution with 11-*cis*-retinal, the membranes are centrifuged at 12,000g (Beckman TLA, 14,000 rpm) for 10 min at 4°. The pellet is resuspended in phosphorylation buffer and membranes corresponding to 1 μg of rhodopsin are diluted into a final reaction volume of 250 μl containing phosphorylation buffer, 150 μM [γ-^{32}P]ATP (50 μCi/ml), and approximately 60 μl of rhodopsin kinase extracted from dark-adapted bovine retinas using 200 mM Na–HEPES, pH 8.0, 20 mM EDTA, and 2 mM DTT as described.[10] This amount of rhodopsin kinase is approximately the amount extracted from one bovine retina. The reaction is initiated by exposure to fluorescent room light at 30°. To compare the ability of wild-type and mutant rhodopsin to be phosphorylated by rhodopsin kinase, the reaction time is typically 8 min. The level of phosphorylation in the dark and phosphorylation of nontransfected cell membranes are included as controls. The reaction is terminated by incubating the assay mixture on ice and dilution with 500 μl of ice-cold 10 mM ATP neutralized with 0.1 M Tris-HCl, pH 7.5. The mixture is centrifuged for 15 min at 38,500g (Beckman TLA-45, 25,000 rpm) at 4° to remove unincorporated radioactive ATP.

[9] J. Nathans, *Biochemistry* **29**, 937 (1990).
[10] D. J. Kelleher and G. L. Johnson, *J. Biol. Chem.* **265**, 2632 (1990).

The pellets can be stored frozen at −80° for later use or resuspended immediately for immunoprecipation of rhodopsin.

The monoclonal antibody R2-15N is used to immunoprecipitate phosphorylated rhodopsin from HEK293 cell membranes. Microfuge tubes (1.5 ml) are pretreated with 10% γ-globulin-free horse serum in Tris-buffered saline (10 mM Tris-HCl, pH 7.5, 150 mM NaCl; TBS) for at least 1 hr, rocking at room temperature to reduce the nonspecific binding of protein to the walls. If 0.02% sodium azide is added, the tubes can be stored at 4° for future use. The blocking solution is removed just prior to the assay. Protein A-Sepharose CL-4B (Pharmacia, Piscataway, NJ) beads are also pretreated with 10% γ-globulin-free horse serum in TBS for 1 hr on a rocker, centrifuged at 600g (Sorvall H-1000B, 1,700 rpm) for 2 min and washed twice by centrifugation with TBS. The beads are stored as a 50% slurry in TBS containing 0.02% sodium azide. For immunoprecipitation of phosphorylated rhodopsin, HEK293 cell membranes are solubilized in 250 μl TBS containing 2 mM $MgCl_2$, 2 μg/ml aprotinin, 1 μg/ml leupeptin, 1 mM DTT, 0.1 mM EDTA, 1.5% octylglucoside, 50 mM NaF and incubated for 1 hr at room temperature on a rocker. The mixture is centrifuged at 4° for 15 min at 125,000g (Beckman TLA-45, 45,000 rpm). The supernatant is transferred to a pretreated microcentrifuge tube and incubated with the R2-15N monoclonal antibody at a ratio of 6 μg per microgram of rhodopsin at room temperature. After a 1-hr incubation, 50 μl of the pretreated protein A-Sepharose CL-4B bead slurry is added and the mixture is incubated for an additional 30 min at room temperature on a rocker. After centrifugation at 600g (Sorvall H-1000B, 1700 rpm) for 2 min at 4°, the beads are washed by centrifugation three times in TBS containing 0.1% sodium deoxycholate and once in 10 mM Tris-HCl, pH 6.8. The immune complexes are solubilized from the beads by incubation in 100 μl Laemmli buffer[11] for 15–30 min at room temperature on a rocker. The beads are pelleted by centrifugation as described earlier and the supernatant subjected to 10% SDS–PAGE. Phosphorimage analysis is used to quantify the level of phosphorylation of rhodopsin in the dark and in the light. The phosphorylation of nontransfected cell membranes is used as a measure of background and subtracted from values obtained for membranes expressing rhodopsin.

The stoichiometry of phosphorylation, which is determined by excising the bands and measuring incorporated radioactivity by liquid scintillation spectroscopy, is approximately 0.5–1.0 mole of phosphate per mole of rhodopsin.[12] We have observed that the rate of rhodopsin phosphorylation in HEK293 cell membranes is much slower than that observed in urea-stripped ROS membranes. The HEK293 cell membranes appear to suppress

---

[11] U. K. Laemmli, *Nature* **227**, 680 (1970).
[12] W. Shi, S. Osawa, C. D. Dickerson, and E. R. Weiss, *J. Biol. Chem.* **270**, 2112 (1995).

the activity of rhodopsin kinase, perhaps accounting for the low stoichiometry observed in our assay. As shown in Fig. 2, phosphorylation increases progressively but begins to plateau after 60 min. This is similar to the time course of phosphorylation observed for rhodopsin expressed in COS-1 cells.[13] The level at which the reaction reaches a plateau increases proportionately with the amount of rhodopsin kinase in the assay, suggesting that this does not represent saturation of the substrate, but rather a time-dependent inactivation of an essential component of the assay. The most likely candidate is metarhodopsin-II, the active form of rhodopsin, which decays progressively in a temperature sensitive manner (see later section).

Assay for Arrestin Binding to Rhodopsin

To measure arrestin binding, the membranes containing rhodopsin must first be phosphorylated by rhodopsin kinase. As described earlier, the amount of rhodopsin and the amount of total protein are adjusted so that they are the same in each reaction mixture. HEK293 cell membranes containing 1 μg rhodopsin are incubated with 11-*cis*-retinal as described earlier, centrifuged at 12,000g (Beckman TLA-45, 14,000 rpm) at 4°, and resuspended in 100 μl of a buffer containing 20 mM Tris-HCl, pH 7.5, 2 mM EDTA, 6 mM MgCl$_2$, 1 mM DTT, 10 mM NaF, aprotinin (2 μg/ml), and leupeptin (1 μg/ml). Approximately 30 μl of rhodopsin kinase extracted from bovine ROS as described earlier (equivalent to the amount isolated from 0.5 retina) and ATP at a final concentration of 2 mM are added to the membranes and incubated for 1 hr at 30° under fluorescent room light. After three 1-ml washes in the same buffer by centrifugation at 38,500g (Beckman TLA-45, 25,000 rpm) for 15 min at 4°, the membranes are resuspended in 1 ml of 30 mM HEPES, pH 7.5, 2 mM MgCl$_2$, 150 mM potassium acetate, 1 mM DTT, 10 mM NaF, aprotinin (2 μg/ml), leupeptin (1 μg/ml) and incubated with 14 μM 11-*cis*-retinal for 1 hr at room temperature in the dark to regenerate any rhodopsin that has decayed during the phosphorylation reaction. The membranes are again pelleted by centrifugation at 38,500g (Beckman TLA-45, 25,000 rpm) for 15 min at 4° and resuspended at a rhodopsin concentration of 40 ng/μl in the buffer described earlier. The membranes can be stored at $-80°$ at this stage for later use.

The bovine arrestin cDNA[14] in pG2S6-I, a pGEM-based vector that contains an idealized 5' untranslated region to promote high-level *in vitro* expression,[15] was a generous gift from Dr. Vsevolod Gurevich (Sun Health

---

[13] S. Bhattacharya, K. D. Ridge, B. E. Knox, and H. G. Khorana, *J. Biol. Chem.* **267**, 6763 (1992).
[14] G. J. Wistow, A. Katial, C. Craft, and T. Shinohara, *FEBS Lett.* **196**, 23 (1986).
[15] V. V. Gurevich and J. L. Benovic, *J. Biol. Chem.* **267**, 21919 (1992).

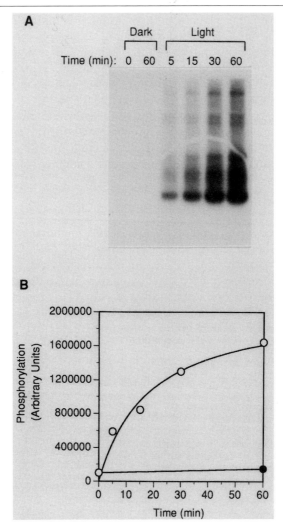

FIG. 2. Phosphorylation of bovine rhodopsin expressed in HEK293 cells by rhodopsin kinase. (A) Time course of rhodopsin phosphorylated in the dark and in light. HEK293 cell membranes containing rhodopsin were phosphorylated by rhodopsin kinase, immunoprecipitated with the rhodopsin monoclonal antibody, R2-215N, and subjected to 10% SDS–PAGE. The radioactivity was visualized by autoradiography. (B) Quantitation of the gel from (A) using a Molecular Dynamics PhosphorImager. Open circles, samples phosphorylated in the light; closed circles, samples phosphorylated in the dark. [Reprinted from W. Shi, S. Osawa, C. D. Dickerson, and E. R. Weiss, *J. Biol. Chem.* **270**, 2112 (1995).]

Research Institute, Sun City, AZ). The plasmid is linearized with HindIII and transcribed *in vitro* using SP6 RNA polymerase (New England Biolabs) in the presence of the cap, P^1-5′-(7-methyl)guanosine-P^3-5′-guanosine triphosphate [m^{7}G(5′)ppp(5′)G; Boehringer Mannheim] to increase mRNA stability and translation efficiency. Approximately 2–4 μg of the synthesized RNA is translated *in vitro* using 35 μl of rabbit reticulocyte lysate (micrococcal nuclease-treated; Promega, Madison, WI), 2 μl of 1 mM amino acid mix (methionine-free), 1 μl RNAsin, and 4 μl [^{35}S]methionine (final concentration = 1200 μCi/ml) in a 50-μl volume at 30° for 1 hr according to procedures supplied by the manufacturer. The reaction mixture is centrifuged at 1100g (Sorvall H-1000B, 2250 rpm) through a Bio-Spin 6 column (Bio-Rad Laboratories, Hercules, CA) into a microfuge tube containing 5 μl of 300 mM HEPES, pH 7.5, 20 mM MgCl$_2$ and 1.5 M potassium acetate (a 10× concentration of the arrestin binding buffer described in the next paragraph) to remove unincorporated [^{35}S]methionine and exchange the buffer. The amount of *in vitro*-translated protein synthesized using these procedures is measured as incorporation of [^{35}S]methionine into a hot trichloroacetic acid (TCA)-insoluble fraction.[16] The reaction mixture (2.5 μl) is spotted onto 1- × 1-cm 3MMChr paper (Millipore, Bedford, MA). The paper is boiled in a beaker containing 5% TCA for 10 min, then rinsed three times with 5% TCA and four times with 95% ethanol at room temperature. The filters are dried and the level of radioactivity is quantified by liquid scintillation spectroscopy. To calculate the specific activity of the preparation, the concentration of nonradioactive methionine must be obtained from Promega for each batch of reticulocyte lysate. The concentration of methionine is approximately 5 μM for most batches of reticulocyte lysate. Taking into account that bovine arrestin has 8 methionines, this value can be used to calculate the yield of *in vitro*-translated arrestin. Typically, 800 fmol are obtained from a 50-μl reaction. Approximately 10 fmol of *in vitro*-translated arrestin is used in the arrestin binding assay for each sample.

The arrestin binding assay is a modification of procedures originally developed by Gurevich and Benovic.[15,17] It is performed in a reaction volume of 35 μl. The arrestin (10 fmol) is diluted into 25 μl of ice-cold arrestin binding assay buffer containing 30 mM HEPES, 2 mM MgCl$_2$ and 150 mM potassium acetate, pH 7.5. Ten microliters of HEK293 cell membranes containing 0.4 μg of phosphorylated, 11-*cis*-retinal-regenerated rhodopsin are added to the arrestin in the dark at 4°. The mixture is incubated in the light at 4° for 5 min, then transferred to a 37° water bath

[16] F. J. Bollum, *Methods Enzymol.* **12**, 169 (1968).
[17] V. V. Gurevich and J. L. Benovic, *J. Biol. Chem.* **268**, 11628 (1993).

A

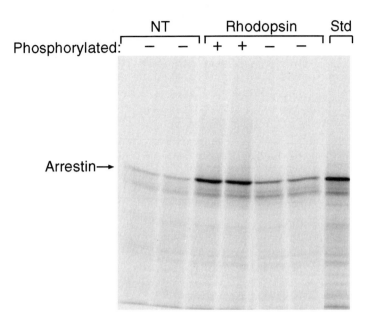

B

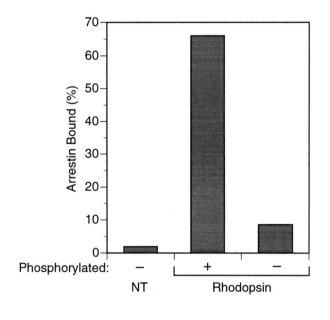

for 5 min to allow arrestin to bind rhodopsin. To terminate the reaction, the assay mixture is placed on ice and 200 μl of ice-cold arrestin binding buffer is added. The diluted sample is loaded onto a 200 μl cushion of ice-cold 0.2 M sucrose in arrestin binding buffer and immediately centrifuged at 2° at 125,000g (Beckman TLA-45, 45,000 rpm) to separate free arrestin from arrestin bound to rhodopsin. The length of centrifugation time has been varied from 3 to 30 min without significant changes in the level of arrestin bound to rhodopsin. The pellets are washed once with 200 μl of ice-cold arrestin binding buffer, solubilized in Laemmli buffer, and analyzed by 10% SDS-PAGE and phosphorimage analysis. A sample consisting of 10 fmol of *in vitro*-translated arrestin is applied to the gel as a standard so that the amount of arrestin bound to rhodopsin can be estimated. Arrestin binding to nontransfected cell membranes is used as a measure of nonspecific binding. Approximately 40–65% of the arrestin in the assay binds to HEK293 cell membranes containing wild-type rhodopsin under these conditions (Fig. 3B).

Our laboratory has used this method to characterize the phosphorylation sites on rhodopsin that play a role in arrestin binding *in vitro*.[2] We have measured an apparent K_D of approximately 0.74 nM for arrestin binding to HEK293 cell membranes expressing rhodopsin. However, the active form of rhodopsin, metarhodopsin-II, is known to decay in a temperature-sensitive manner. For example, detergent-extracted rhodopsin isolated from COS-1 cells decays with a half-time of 18 min at 20°.[18] Therefore, equilibrium binding studies cannot accurately be performed. Nevertheless, the value that we reported is significantly lower than the previously published value of 50 nM, obtained from light scattering studies[19] and not very different from the value of 2 nM reported for the binding of visual arrestin to the β_2-adrenergic receptor.[20] This may be due in part to the ability of arrestin to stabilize metarhodopsin-II in a manner similar to transducin,[19]

---

[18] T. Sakamoto and H. G. Khorana, *Proc. Natl. Acad. Sci. U.S.A.* **92**, 249 (1995).
[19] A. Schleicher, H. Kühn, and K. P. Hofmann, *Biochemistry* **28**, 1770 (1989).
[20] V. V. Gurevich, S. B. Dion, J. J. Onorato, J. Ptasienski, C. M. Kim, R. Sterne-Marr, M. M. Hosey, and J. L. Benovic, *J. Biol. Chem.* **270**, 720 (1995).

FIG. 3. Arrestin binding to bovine rhodopsin expressed in HEK293 cells. (A) Nontransfected (NT) and rhodopsin-transfected HEK293 cell membranes (Rhodopsin) were incubated with [^{35}S]methionine-labeled, *in vitro*-translated arrestin, centrifuged to isolate the arrestin/rhodopsin complexes, and subjected to 10% SDS-PAGE. A lane of arrestin representing 10 fmol (Std) was also applied to the gel. The samples were analyzed using a Molecular Dynamics PhosphorImager. –, Nonphosphorylated; +, phosphorylated. (B) Quantitation of the polyacrylamide gel shown in (A). Labels are as described in (A). The label Arrestin Bound (%) refers to the fraction of arrestin in the assay that binds to rhodopsin.

allowing the isolation of arrestin/rhodopsin complexes without appreciable dissociation during centrifugation. These observations suggest that this arrestin binding assay is a reasonable method for determining relative affinities for rhodopsin mutants expressed in cultured cells.

[29] Arrestin: Mutagenesis, Expression, Purification, and Functional Characterization

By VSEVOLOD V. GUREVICH and JEFFREY L. BENOVIC

Introduction

In the photoreceptor cell the activation of rhodopsin by a photon of light initiates two cascades of events: signal transduction and signal termination. The visual amplification cascade (rhodopsin → transducin → cGMP phosphodiesterase) has long served as an archetypal G-protein-coupled receptor signaling system.[1] A single light-activated rhodopsin (Rh*) can sequentially activate hundreds of transducin molecules, each of which in turn activates a cGMP phosphodiesterase resulting in the hydrolysis of thousands of cGMP molecules. Thus, the potential for signal amplification is enormous, providing for very high sensitivity. However, because relatively modest local changes in the cGMP concentration are sufficient for a full cellular response, the signaling machinery needs to be turned off as soon as the signal goes through. A parallel chain of events brings this about.[2] The initial step in this turn off is the recognition and phosphorylation of Rh* by the enzyme rhodopsin kinase. Rhodopsin phosphorylation attenuates its ability to activate transducin and increases its affinity for the retinal protein, arrestin. Arrestin binds to phosphorylated Rh* (P-Rh*) and effectively shuts down the phototransduction cascade by blocking further transducin activation. Arrestin appears to stay bound to P-Rh* until it decays into phosphoopsin after which arrestin dissociates and phosphoopsin is dephosphorylated by a type IIA protein phosphatase.

Retinal arrestin, also termed 48-kDa protein or S-antigen, was initially identified as a protein that binds to disk membranes following light activation of rhodopsin.[3] Arrestin binding and its ability to terminate signaling require rhodopsin phosphorylation.[3] In fact, although arrestin can indepen-

---

[1] P. A. Hargrave and J. H. McDowell, *FASEB J.* **6,** 2323 (1992).
[2] U. Wilden, *Biochemistry* **34,** 1446 (1995).
[3] U. Wilden, S. W. Hall, and H. Kuhn, *Proc. Natl. Acad. Sci. U.S.A.* **83,** 1174 (1986).

dently recognize both activation and phosphorylation states of rhodopsin, it demonstrates an amazing selectivity toward only one functional form, P-Rh*.[4,5] This selectivity is the key to the function of arrestin because it ensures that both arrestin binding to rhodopsin and its subsequent dissociation from phosphoopsin occur with the appropriate timing to effectively regulate phototransduction. The methods employed to elucidate the molecular mechanism enabling arrestin to recognize its proper target and explore arrestin–rhodopsin interaction are the focus of this chapter.

Expression of Wild-Type and Mutant Arrestins

A systematic structure–function analysis of any functional region of a protein requires the construction and analysis of a considerable number of mutants. A simple, rapid, and cost-effective expression system is a prerequisite for such studies. Because the primary function of arrestin is to bind to rhodopsin, the ability to produce radiolabeled arrestin for binding analysis would be especially advantageous. Functional expression in an *in vitro* translation system satisfies these requirements. Indeed, this system has been successfully used to express 10–20 different mutants in a few days.[4] Commercially available [^{3}H]leucine with a specific activity of 140–150 Ci/mmol enables the production of arrestin (which has 34 leucines) with specific activities up to 5000 Ci/mmol. Different ratios of cold and hot leucine in the translation mix yield proteins with specific activities within the 0–5000 Ci/mmol range. We found the use of proteins with specific activities of 100–400 Ci/mmol (i.e., 200–800 dpm/fmol) most practical for functional analysis. Typical expression yields (50–200 pmol/ml) are sufficient for binding studies. Moreover, recombinant arrestin is the only radiolabeled protein in the translation mix, so that it can be used without further purification.[4,5] A simple and sensitive direct binding assay[4,5] suitable for quantitative analysis of arrestin binding to any function form of rhodopsin has been developed on the basis of this expression system.

Arrestin Expression in Cell-Free Translation System

In vitro Transcription

We have recently described all of the methods we employ for *in vitro* transcription in considerable detail.[6] Briefly, purified plasmids with arrestin

[4] V. V. Gurevich and J. L. Benovic, *J. Biol. Chem.* **268**, 11628 (1993).
[5] V. V. Gurevich and J. L. Benovic, *J. Biol. Chem.* **267**, 21919 (1992).
[6] V. V. Gurevich, *Methods Enzymol.* **75**, 382 (1996).

cDNAs under control of an SP6 promoter are linearized downstream of the stop codon. Then 30 μg/ml of linearized DNA is incubated in 120 mM HEPES–K, pH 7.5, 2 mM spermidine, 16 mM $MgCl_2$, 40 mM dithiothreitol (DTT), 3 mM each of adenosine triphosphate (ATP), guanosine triphosphate (GTP), cytidine triphosphate CTP, and uridine triphosphate (UTP), 2.5 U/ml inorganic pyrophosphatase, 200 U/ml RNasin, and 1500 U/ml SP6 RNA polymerase at 38° for 2 hr. mRNA is then precipitated by addition of 0.4 volume of 9 M LiCl, 10 min incubation on ice, and centrifugation at 8000g for 10 min at 4°. The pellet is then washed with 1 ml of 2.5 M LiCl (4°) followed by 1 ml of 70% (v/v) ethanol (room temperature). The final pellet is dissolved in 1 volume of diethyl pyrocarbonate (DEPC)-treated water and reprecipitated by addition of 0.1 volume of 3 M sodium acetate and 3.3 volumes of ethanol. mRNAs in this suspension are stable for several years at −80°.

In vitro Translation

Arrestin translation using rabbit reticulocyte lysate (RRL) is carried out with a few modifications of the original method,[7] as described in detail.[8] Briefly, translation reactions contain 70% (v/v) RRL, 120 mM potassium acetate, 30 mM creatine phosphate, 200 μg/ml creatine kinase, 200 U/ml RNasin or Prime RNase inhibitor, 0.1 μg/ml pepstatin, 0.1 μg/ml leupeptin, 0.1 mg/ml soybean trypsin inhibitor, 5 mM cAMP, 50 μM of 19 unlabeled amino acids and 40–50 μM of [^{14}C]leucine (14,000–35,000 dpm/ml). When [^{3}H]leucine is used, 800,000–1,000,000 dpm/ml is added along with [^{14}C]leucine to a final concentration of 40–50 μM (final specific activity of 5–15 Ci/mmol leucine). Translation is carried out using 50–150 μg/ml of uncapped mRNA at 22.5° for 2 hr. Translation is optimal when the construct contains an "idealized" 5′ untranslated region.[4,5] Following translation, 1 mM ATP and 1 mM GTP are added followed by incubation for 7 min at 37° ("runoff"). The samples are then cooled on ice and all aggregated proteins are pelleted by centrifugation in a TLA 100.1 rotor (Beckman, Fullerton, CA) at 100,000 rpm for 60 min at 4°. Supernatants can be used for the binding assay either directly or after gel filtration on a Sephadex G-25 column equilibrated with 50 mM Tris-Cl, pH 7.5, 50 mM potassium acetate, 2 mM EDTA. Protein synthesis is determined using the amount of [^{3}H]- or [^{14}C]leucine incorporated into a hot trichloroacetic acid (TCA)-insoluble fraction and/or into the respective band after resolution by sodium dodecyl sulfate–polyacrylamide gel electrophoresis (SDS–PAGE).[4,5] These values

[7] R. J. Jackson and T. Hunt, *Methods Enzymol.* **96**, 50 (1983).
[8] V. V. Gurevich, M. J. Orsini, and J. L. Benovic, in "Receptor Biochemistry and Methodology," Vol. IV, p. 157, 1999.

usually agree within 10% of each other. The radioactivity in the arrestin band along with the specific activity is then used to determine the yields.

Stock Solutions

The solutions listed here are kept in aliquots at −80° (stable for several years), except creatine phosphokinase and RNasin, which are kept at −20°. Except for RRL, all solutions can be refrozen several times without any adverse effects. DEPC-treated water is used throughout the procedure.

RRL (commercial or prepared as described[7])

Mix of 19 amino acids (-Leu) (20×), 1 mM each (prepared out of 20 mM solutions of individual amino acids as described[7])

1 M creatine phosphate in water

Mix of protease inhibitors (100×, contains 10 μg/ml each pepstatin and leupeptin and 10 mg/ml soybean trypsin inhibitor) in water

0.2 M cAMP in water

2 M Potassium acetate, pH 7.4, in water

[^{14}C]Leucine (from NEN, Boston, MA; concentrated on Speed-Vac to 10 mCi/ml)

[^{3}H]Leucine (from NEN, 5 mCi/ml in water)

Creatine phosphokinase (CPK), 20 mg/ml in 50% glycerol. [We use CPK Type I from Sigma (St. Louis, MO). Dissolve at 40 mg/ml in 100 mM Tris-HCl, pH 7.4, 1 mM dithiothreitol (DTT), then mix with equal volume of glycerol. This solution is stable for 3–6 months at −20°.]

RNasin, 40 U/μl (commercially available from Promega, Madison, WI, Boehringer Mannheim, etc.), or Prime RNase inhibitor, 30 U/μl (5 Prime, 3 Prime, Inc., Boulder, CO)

mRNA Preparation

We store all mRNAs at −80° as a suspension in 75% ethanol. An appropriate aliquot of mRNA suspension is transferred to a microcentrifuge tube, mRNA is pelleted by centrifugation for 5–10 min, the pellet is washed with 1 ml of 70% (v/v) ethanol by vortexing and recentrifugation, and allowed to dry (all steps at room temperature).

Translation

For a standard 200-μl translation add (on ice, in the following order) 140 μl RRL, 10 μl amino acid mix (final concentrations 50 μM each), 6 μl 1 M creatine phosphate (30 mM final), 2 μl protease inhibitor mix, 5 μl 0.2 M cAMP (5 mM final), 11 μl 2 M potassium acetate (110 mM final), 0.8 μl concentrated [^{14}C]leucine (supplies 30–35 μM leucine), 2.2 μl [^{3}H]leucine (supplies 0.3–0.35 μM leucine), vortex briefly (avoid frothing), then add 1 μl RNase inhibitor and 2 μl creatine kinase, and vortex again. Dissolve dried mRNA in 20 μl DEPC-treated water, immediately add to the translation mix, and vortex thoroughly. Incubate 2 hr at 22–23°, add 4 μl of a solution containing 50 mM ATP and GTP, incubate 7 min at 37°, cool on ice, and spin in a TLA 100.1 rotor (Beckman) at 100,000 rpm for 60 min at 4°. Remove supernatant (carefully avoiding dark red pellet), aliquot, and freeze at $-80°$. Translated arrestins are stable at $-80°$ for several months, and 3–5 freeze–thaw cycles do not appreciably reduce their activity. It is also a good idea to translate a sample (50–100 μl) with water added instead of mRNA for use as a control.

Yield Calculation

To calculate the yield of product, the specific activity of leucine in the translation mix and the number of leucine residues per molecule of protein needs to be determined. While most batches of RRL contain ~5–7 μM cold leucine, the easiest way to precisely determine the concentration of endogenous cold leucine is to run three identical analytical-scale (20–50 μl) translations in the presence of 30, 60, and 90 μM [^{14}C]leucine. It can be safely assumed that these translations give identical yields, and by comparing the amounts of TCA-insoluble radioactivity in the three samples (see later discussion), the actual concentration of endogenous leucine can be calculated. We prefer to use [^{14}C]leucine instead of cold leucine to increase precision since certain errors in weighing out and serial dilutions of cold leucine are virtually unavoidable. In addition, the higher energy of ^{14}C also helps to visualize the arrestin band after electrophoresis and increases the efficiency of counting. Arrestins labeled with ^{14}C alone are "hot" enough (12–15 Ci/mmol, i.e., 27–33 dpm/fmol for various arrestins) for some applications, although the direct binding assay (see later discussion) usually requires higher specific activities (100–400 dpm/fmol).

To determine the concentration of [^{14}C] and [^{3}H]leucine in the translation mix, we dilute a 2-μl aliquot 10-fold with water, and count 5-μl aliquots in triplicate using a dual-label disintegrations per minute (dpm) mode. To determine the amount of protein-incorporated radioactivity, we dilute a 2-μl aliquot of translation mix 10-fold after the incubation. (It is a good idea to do so both before and after high-speed centrifugation to enable determination of the percentage of soluble product.) We then apply 5 μl of diluted mix to Whatman (Clifton, NJ) 3MM paper in duplicate (1- × 1-cm square per application). The paper is dried for 3–5 min, immersed in ice-cold 10% TCA for 10–15 min, boiling 5% TCA for 10 min, rinsed briefly with ethanol (to remove water) and diethyl ether (to remove TCA), and then dried and cut into individual squares. Each square is put into a scintillation vial containing 0.6 ml of 1% SDS with 50 mM NaOH. After the protein dissolves (30–50 min at room temperature), 5 ml of water-miscible scintillation fluid is added (we use ScintiSafe Econo 2 from Fisher, Pittsburgh, PA) and the contents are mixed and counted.

Calculate the total concentration of leucine in the translation mix (endogenous + [^{14}C] + [^{3}H], it should be in the 30- to 50-μM range) and total radioactivity (in dpm) per microliter of translation mix, and from these two values calculate the specific activity of leucine (in dpm/fmol). Specific activities of arrestins can then be calculated by multiplying the specific activity of leucine by the number of leucine residues in the protein (e.g., 34 in bovine visual arrestin). Dividing the total protein-incorporated (TCA-insoluble) radioactivity (dpm per microliter of translation mix with the value from the control sample subtracted) by the specific activity (dpm/fmol) of the arrestin gives the yield in fmol/μl. We usually get 70–150 fmol/μl for wild-type bovine visual arrestin.

Electrophoresis and Fluorography

It is always a good idea to run all translated samples on a gel prior to any functional experimentation. Arrestins usually give a single band and there should be no bands in the control sample, but it is necessary to ascertain for every translation reaction that all the protein-incorporated radioactivity is indeed in full-length arrestin. We add 2 μl of each translation mix to 18 μl of SDS sample buffer, mix, and run on a standard 10% SDS–polyacrylamide gel as described.[4] The gels are stained with Coomassie Blue G-250, destained for 30 min, and then soaked in 20% 2,5-diphenyloxazole in glacial acetic acid for 10 min. The fluorochrome is then precipitated by washing the gel with two or three changes of water and the gels are dried. Fluorographs are exposed at $-80°$ overnight using Fuji X-ray film. Labeled protein bands also can be excised and counted in a liquid scintillation counter in order to double-check the yields.

Troubleshooting

It is advisable to keep a "standard" mRNA that has produced high yields in the past for troubleshooting. If the yields decrease approximately twofold, a fresh CPK solution should help (without CPK the yields decrease two- to threefold). If the yield drops sharply (below 10 fmol/μl), most likely either the RRL or mRNA is bad. The RRL can be checked using the "standard" mRNA since each freeze–thaw cycle of RRL reduces its activity by 20–30%. Sometimes the [^{3}H]leucine is the source of the problem. If [^{3}H]leucine is more than 1 year old, try translating with [^{14}C]leucine alone

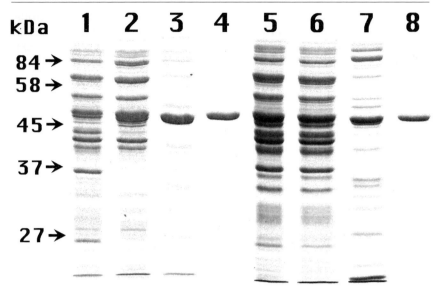

FIG. 1. Expression and purification of arrestins in *E. coli*. Visual arrestin (lanes 1–4) and arrestin (R175E) (lanes 5–8) were expressed in BL21 cells and purified, as described in the text. Aliquots for electrophoresis were taken before ammonium sulfate precipitation (lanes 1 and 5; 15 μg protein/lane), before heparin-Sepharose chromatography (lanes 2 and 6; 15 μg protein/lane), from pooled heparin-Sepharose fractions (lanes 3 and 7; 5 μg protein/lane), and from pooled Q-Sepharose fractions (lanes 4 and 8; 1 μg protein/lane), precipitated with methanol, dissolved in 18 μl of SDS sample buffer, subjected to SDS–PAGE, and stained with Coomassie G-250. The positions of molecular weight markers are indicated by arrows.

($[^{14}C]$leucine is stable at $-80°$ for at least 5 years). Several additional problems, which are discussed in detail elsewhere,[7] may arise if $[^{35}S]$methionine is used. If a smear or more than one protein band is obtained, the mRNA is most likely degraded. In general, since so many components go into the translation reaction, it is always easier to throw away all the "old" aliquots and use new ones.

Expression and Purification of Arrestins From *Escherishia coli*

Detailed *in vitro* characterization of arrestin's ability to quench the activation of transducin and cGMP phosphodiesterase and experiments involving injection of arrestin mutants into photoreceptor cells and crystallization require milligram quantities of purified arrestin. To obtain such quantities, these proteins were expressed in *E. coli* (strain BL21) using the pTrcHisB vector (Invitrogen, Carlsbad, CA). This system yields fully functional arrestins that can be purified by two sequential chromatographic steps (Fig. 1).

All arrestin open reading frames (ORFs) were excised with NcoI and HindIII and subcloned into NcoI/HindIII digested pTrcHisB (these constructs do not retain the His tag). In some cases (e.g., bovine visual arrestin) it was necessary to introduce an extra codon (Ala) between the normal first and second codons in order to create an NcoI site and an optimal Kozak[9] consensus for the initiator codon. The insertion of Ala between Met-1 and Lys-2 does not affect the functional characteristics of arrestin.[5] BL21 cells are transfected with an appropriate arrestin construct and plated. An individual colony is then used to inoculate 10 ml of LB containing 100 μg/ml of ampicillin (LB/Amp) and the cells are grown at 250 rpm at 30° to an OD_{600} of ~0.2–0.3. Glycerol is then added to 10% (v/v) and cells are aliquoted, frozen, and stored at −80° until needed. Cells are stable for years.

Cell Growth

Inoculate 160 ml LB/Amp in a 500-ml flask with an arrestin/BL21 colony or frozen cells. Grow at 30°, 250 rpm, for 6–8 hr (to OD_{600} of ~0.1–0.4). Divide the culture between 4 × 1.1 liter of LB/Amp containing 20–30 μM isopropylthiogalactoside (IPTG) in 2.8-liter flasks, and grow at 30°, 250 rpm, for 12–16 hr. Alternatively, for arrestin proteins with relatively low expression levels and/or with *in vivo* proteolysis problems, grow four 1.1-liter cultures for 10–11 hr without IPTG, then add 20–30 μM IPTG and grow an additional 3–5 hr.

Lysis and Precipitation

The following protocol applies to 4.5 liters of culture. Pellet cells in four 1-liter bottles for 20 min, 5000 rpm at 4°. Carefully discard supernatant, let the bottles stand for 1 min upside down, and wipe away all traces of supernatant using Kimwipes. Immediately add 37.5 ml ice-cold lysis buffer [50 mM Tris/HCl, pH 8.0, 5 mM EGTA, 1 mM DTT (from 1 M stock in water), 2 mM benzamidine (from 1 M stock in water), 1 mM PMSF (from fresh 1 M stock in ethanol), 10 μM leupeptin (from 40 mM stock in water), 0.7 μg/ml pepstatin A (from 1.4 mg/ml stock in methanol), and 5 μg/ml chymostatin (from 10 mg/ml stock in DMSO)] to each bottle (150 ml total). All stock solutions are kept at −20°. Resuspend the cells thoroughly by vortexing and pipetting, carefully avoiding frothing. Then add 1.5 ml of fresh 3 mg/ml lysozyme (dissolved in lysis buffer) to each bottle and incubate 30–40 min on ice. Freeze the cells at −80°. Cells can be kept frozen at this point for weeks. If you intend to proceed immediately, freeze the

---

[9] M. Kozak, *Microbiol. Rev.* **47,** 1 (1983).

cells for 20 min, then thaw at 4° for 20–30 min. After the suspension is almost completely thawed, add an additional 110 ml of lysis buffer per bottle, carefully resuspend (making sure that all frozen cell paste is fully thawed), sonicate the suspension for 2 min (it should become quite viscous because of released DNA), then add 1.2 ml 1 M $MgCl_2$ (to final concentration of >7 mM) and 1 ml of 3 mg/ml DNase II (Sigma) to each bottle, mix thoroughly and incubate on ice for 40 min. Transfer the suspension to 250-ml bottles and pellet cell debris in a GSA rotor at 12,000 rpm for 60–90 min. Transfer supernatant (~600 ml) (carefully avoiding pellet) to a 1-liter beaker. Under continuous gentle stirring (to avoid frothing) at 4°, add 210 g (to 0.35 g/ml) of ammonium sulfate in four to five portions over a 20- to 40-min period. Let stirring continue until all ammonium sulfate is dissolved and then pellet the precipitated protein in a GSA rotor at 12,000 rpm for 60–90 min. Carefully remove supernatant and floating material, and thoroughly wipe the walls clean with Kimwipes. Cover the tube with Parafilm and aluminum foil and store at −80° until needed (stable for at least a month or two).

Heparin-Sepharose Chromatography

Dissolve the pelleted protein from 4.5 liters of cells in 200 ml of ice-cold column buffer (10 mM Tris/HCl, pH 7.5, 2 mM EDTA, 2 mM EGTA, 2 mM benzamidine, 1 mM PMSF, 10 μM leupeptin, 0.7 μg/ml pepstatin A, and 5 μg/ml chymostatin) (CB). It takes 20–40 min of gentle shaking (avoid frothing). Pellet the insoluble material (GSA rotor, 12,000 rpm, 60–90 min) and carefully determine the volume of the supernatant to estimate the volume of the initial pellet. (It is usually 12–18 ml and contains 1200–1600 mg of protein.) Filter the supernatant through a 0.8-μm Millipore (Bedford, MA) filter. (Do not use negative pressure!) Dilute the filtrate with CB to 65 times the volume of the initial pellet for wild-type visual arrestin. Because CB has a very low ionic strength, the addition of CB to the filtrate sometimes results in precipitation of some protein that may clog the column. Thus, it is advisable to either dilute the sample slowly with continuous stirring or dilute the sample during loading onto the column using a gradient mixer and an appropriate ratio of filtrate and CB.

Load the diluted filtrate onto a 25-ml column of heparin-Sepharose, equilibrated with CB/0.1 M NaCl, at 1.5–4.5 ml/min (this step takes 7–14 hr and is often convenient to perform overnight). Wash the column with ~200 ml of CB/0.1 M NaCl (takes 1–2 hr) until the baseline stabilizes. Elute with a 400-ml linear gradient (CB/0.1 M NaCl → CB/0.4 M NaCl) and collect 10-ml fractions. The major arrestin peak elutes in 8–12 fractions between 200 and 240 mM NaCl. Since some wild-type arrestins and mutants elute at higher salt, it is usually safer to initially use a gradient from 0.1 to 1 M NaCl when purifying an unknown wt or mutant arrestin.

Q-Sepharose Chromatography

Pool arrestin-containing fractions (total protein 15–60 mg, ~20–70% arrestin), concentrate to ~10 ml (Amicon, Danvers, MA, YM30 membrane), and filter through a 0.8-μm Millipore filter. Estimate the concentration of NaCl in this pool and then load the sample at 1 ml/min onto a 10-ml Q-Sepharose column (equilibrated with CB) while diluting the sample (using a gradient mixer) with CB to a final NaCl concentration of ~10 mM (takes ~3 hr). It is important that the protein be diluted just before it is pumped onto the column because arrestins aggregate at low-salt concentrations in solution but seem fine while bound to the resin. Wash the column with 40–80 ml of CB and then elute with a 400-ml linear gradient (CB → CB/0.1 M NaCl). Because some arrestins elute at higher salt, use of a gradient to 0.5 M NaCl will elute any arrestin from Q-Sepharose. Arrestin (>95% purity) elutes in a peak at about 60 mM. Pool the arrestin-containing fractions and concentrate to 0.5–3 mg/ml (expect 2–5 mg per liter of bacterial culture), filter through a 0.8-μm Millipore filter, aliquot, and freeze at $-80°$. Frozen arrestins are stable for at least 2 years and two to three freeze–thaw cycles do not appreciably reduce their activity.

Electrophoresis and Western Blot Analysis

We use methanol precipitation to prepare samples for electrophoresis. Add 0.27 ml of methanol to a 30-μl aliquot of the appropriate fraction at room temperature, vortex, spin in a tabletop microcentrifuge for 5 min, wash the pellet with 1 ml of 90% methanol by vortexing and centrifugation, dry the pellet, and dissolve in 20 μl of SDS sample buffer. The sample is electrophoresed in a 15-lane 0.75-mm 10% Mini-PROTEAN gel (Bio-Rad, Hercules, CA), using 2 μl for Western analysis and the remainder for Coomassie blue staining. Arrestin is clearly visible in the crude soluble protein on a Coomassie-stained gel (Fig. 1), while some mutants become apparent only in the heparin-Sepharose fractions. For the detection of arrestins from most species we use the mouse monoclonal antibody F4C1, which was produced and characterized by Donoso.[10] Proteins are transferred to a polyvinylidene fluoride (PVDF) membrane for 45–60 min using a Mini Trans-Blot unit (Bio-Rad), and the blot is then blocked with 5% dry nonfat milk in TBS/0.05% Tween (TBST) for 30 min, washed 3–5 times

---

[10] L. A. Donoso, D. S. Gregerson, L. Smith, S. Robertson, V. Knospe, T. Vrabec, and C. M. Kalsow, *Curr. Eye Res.* **9**, 343 (1990).

with TBST, incubated with primary (F4C1) and secondary (proxidase-conjugated goat anti-mouse antibody, Boehringer Mannheim) (both at 1:5000 dilution in TBST supplemented with 2% BSA) for 1 hr each, followed by 5–7 brief and 2–3 longer 5-min washes each, and then detected by ECL (enhanced chemiluminescence) using SuperSignal (Pierce, Rockford, IL). Working solutions of antibodies can be kept frozen at $-20°$, and reused up to 10–15 times without a substantial change in sensitivity. When necessary, the solutions can be spiked with fresh antibody (1–2 μl per 10 ml) and reused. If the antibodies are used in a solution containing dry milk instead of BSA, the autoradiograms will generally not be as clean and the solutions cannot be frozen and reused. Note that the arrestin band (especially in the case of mutants with relatively low expression levels) can be identified best by comparison of the Coomassie-stained gel with the corresponding Western blot. On Western, 0.5–2 ng of any arrestin gives a sharp visible band at 5- to 20-sec exposures, while 20–30 ng or more gives virtually the same kind of "fat" overloaded band at any exposure. In our hands the signal is more or less linear in a narrow range between 0.2 and 5 ng. Therefore, after the identification of the arrestin band, the fractions should be pooled *solely* on the basis of the pattern revealed by Coomassie staining.

Potential Problems and Variations

1. The expression plasmids should be grown in *E. coli* at 30°, beginning with the construction, in order to avoid selection against effective constructs. Arrestin expression at 37° and/or at high IPTG (e.g., 1 mM) results in expression levels up to 50–60% of total protein. However, under these conditions virtually all of the arrestin appears in the insoluble fraction (inclusion bodies). While this protein could presumably be used for immunization, all attempts to renature the insoluble arrestin are usually futile. Even a small percentage of denatured arrestin in a sample yields a substantially disturbed and often misleading elution pattern, and should be avoided. Therefore, it is crucial to filter all samples prior to chromatography and storage.

2. Even a brief exposure of any arrestin to low-salt conditions (<50 mM) results in its partial denaturation and aggregation. Denatured/aggregated protein should always be eliminated from the sample by ultracentrifugation and/or filtration through a 0.8-μm Millipore filter.

3. We have successfully used similar procedures (with appropriate variations of salt gradients) for the expression and purification of wild-type and mutant visual and nonvisual arrestins from a number of species. This method usually yields 95–98% pure arrestin (as a rule, the higher the expression

level, the higher the purity of the final product) that can be further purified, if necessary, using S-Sepharose chromatography.[11]

Direct Binding Assay

To characterize arrestin interaction with rhodopsin, it is useful to compare its binding to four different functional forms of rhodopsin[4,5]: dark phosphorylated (P-Rh), light-activated phosphorylated (P-Rh*), dark unphosphorylated (Rh), and light-activated unphosphorylated (Rh*). We termed arrestin binding to these four functional forms "selectivity profile," reserving the term "specificity" for the relative affinity of various arrestins for activated phosphorylated forms of different receptors.[12] For high-affinity binding, arrestin requires that rhodopsin be both activated and phosphorylated. This selectivity is ensured by a sequential multisite mechanism of arrestin binding.[4,5,11] Arrestin independently recognizes the activation and phosphorylation state of rhodopsin, but its binding to P-Rh* is at least 10 times higher than to dark P-Rh or Rh*. To rationalize these facts we hypothesized that arrestin has two *primary binding sites:* an *activation-recognition site,* which interacts with parts of rhodopsin that change conformation upon light-activation, and a *phosphorylation-recognition site,* binding to phosphates on phosphorhodopsin. Arrestin works as a coincidence detector, i.e., when both primary sites are simultaneously engaged (which can happen only with P-Rh*), arrestin binds and undergoes a transition into a high-affinity receptor-binding state[11,13] involving the mobilization of a potent *secondary binding site.*[4]

On the basis of the expression of radiolabeled *in vitro* translated arrestins, we developed a simple and sensitive direct binding assay. This assay takes advantage of the expression of arrestins with high specific activity, which provides sufficient sensitivity to study high-affinity arrestin binding to P-Rh*, as well as low-affinity interactions with other forms of rhodopsin[4,5] (Fig. 2).

Visual Arrestin Binding to Rhodopsin

In vitro translated tritiated visual arrestin can be used without further purification, because it is the only labeled protein in the translation mix.

---

[11] J. A. Hirsh, C. Schubert, V. V. Gurevich, and P. B. Sigler, *Cell* **97,** 257 (1999).
[12] V. V. Gurevich, S. B. Dion, J. J. Onorato, J. Ptasienski, C. M. Kim, R. Sterne-Marr, M. M. Hosey, and J. L. Benovic, *J. Biol. Chem.* **270,** 720 (1995).
[13] A. Schleicher, H. Kuhn, and K. P. Hofmann, *Biochemistry* **28,** 1170 (1989).

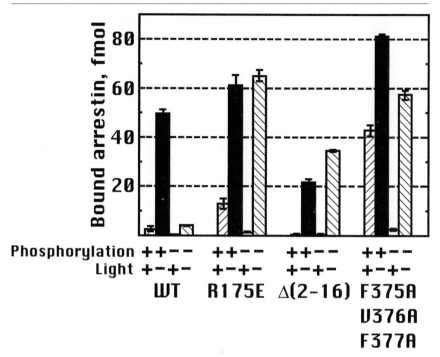

FIG. 2. Selectivity profiles of wild-type arrestin and mutants with different types of "constitutive activity."[4,14,15] Arrestin proteins (100 fmol/assay) were incubated with indicated forms of rhodopsin (0.3 μg/assay) for 5 min at 37°; specific binding was determined as described in the text. Means ± S.D. of two experiments performed in duplicates are presented.

We dilute arrestin for the experiment (50–100 fmol per assay, for 1–2 nM final concentration) with 50 mM Tris-HCl, pH 7.5, 50 mM potassium acetate, 0.5 mM MgCl$_2$ (RB buffer) containing 0.5 mM DTT to 2–4 fmol/μl, and add 25 μl of this solution per Eppendorf tube. Under dim red light, Rh and P-Rh are diluted with the same buffer to 12 μg/ml, and distributed between tubes (25 μl/assay, i.e., 0.3 μg or ~7.5 pmol per tube). Arrestin is then incubated with the appropriate form of rhodopsin in the dark or with illumination at 37° for 5 min. Exact timing is crucial because of the rapid decay of light-activated rhodopsin. The samples are then immediately cooled on ice and bound and free arrestins are separated by gel filtration at 4° (under dim red light for dark rhodopsin) on a 2-ml Sepharose CL-2B column equilibrated with 10 mM Tris-HCl, pH 7.5, 100 mM NaCl (100/10). The membrane-containing fraction elutes in the void volume between

[14] V. V. Gurevich, *J. Biol. Chem.* **273,** 15501 (1998).
[15] V. V. Gurevich and J. L. Benovic, *Mol. Pharmacol.* **51,** 161 (1997).

0.5 and 1.1 ml and is counted in a liquid scintillation counter. This is accomplished by having columns in a 10 × 10 rack that can be directly placed over a standard 10 × 10 box of scintillation vials. Samples are loaded on the columns and allowed to soak in. The column is then washed with 100 μl of 100/10, 400 μl of 100/10, and then eluted into scintillation vials using 600 μl 100/10. It is crucial that the columns for dark samples be run under dim red light. While this method readily demonstrates the exquisite selectivity of arrestin for binding to P-Rh*, arrestin binding to other functional forms of rhodopsin can also be studied (Fig. 2). Nonspecific arrestin binding (using 0.3 μg of liposomes) is determined and subtracted. While the columns for this assay can be reused many times, after each experiment the columns should be washed with 4 × 3 ml of 100/10, and stored capped at 4° in 100/10 buffer.

Receptor Phosphorylation Levels Required for High-Affinity Arrestin Binding

When β-adrenergic receptor kinase (GRK2) is used for receptor phosphorylation, a stoichiometry of phosphorylation of ~2 mol/mol appears to be necessary and sufficient for high-affinity arrestin binding to several different receptors that have been tested.[12] Increasing phosphorylation levels to 5–7 mol/mol does not change the net arrestin binding nor apparent arrestin affinity. When rhodopsin kinase (GRK1) is used to phosphorylate rhodopsin, only 1 phosphate per rhodopsin appears to be sufficient for arrestin binding.[16]

Arrestin Affinity

Keep in mind that the direct binding assay never reaches equilibrium since light-activated rhodopsin rapidly decays into opsin (half-life <10 min even at 0°), whereas due to its size (and hence slow diffusion) arrestin needs 20–40 min to reach equilibrium (as established in similar direct receptor-binding assays for nonvisual arrestins). Thus, the true arrestin affinity for rhodopsin cannot be measured, although differences in arrestin binding to various functional forms of rhodopsin and in the binding of various mutants to the same functional form of rhodopsin most likely directly reflect changes in affinity. One of the ways to determine the relative affinity of two arrestins for P-Rh* is to express these proteins in *E. coli*, purify them, and measure their relative efficacy in competing out radiolabeled wild-type arrestin binding.[5] Such analysis yields K_d values of 18–28

---

[16] J. G. Krupnick, V. V. Gurevich, and J. L. Benovic, *J. Biol. Chem.* **272**, 18125 (1997).

nM for arrestin binding to P-Rh*.[5] An alternative "extra Meta-II assay"[13,17] has yielded similar values that range from 50 nM[13] to 20 nM[17] for wild-type arrestin binding to P-Rh*. Note, however, that in our competition assay the apparent IC_{50} is always conspicuously close to the concentration of rhodopsin used (which was 25 nM in the experiments mentioned earlier), suggesting that the actual K_d is lower (i.e., the actual affinity is much higher). The fact that the affinity of visual arrestin for phosphorylated activated β_2-adrenergic receptor ($K_d \sim 2$ nM) and m2 muscarinic cholinergic receptor ($K_d \sim 7$ nM)[12] is higher than previously reported for rhodopsin also supports this notion. Recently[18] a subnanomolar affinity of arrestin for P-Rh* was reported (see also Osawa et al., this volume[19]). Generally speaking, to determine the true arrestin affinity for P-Rh* one needs to use a constitutively active rhodopsin mutant (such as K296E), to make the binding equilibrium possible and all the conventional techniques (Scatchard analysis, etc.) applicable. However, the competition assay can be effectively used to estimate relative affinities of various arrestins for rhodopsin.

Assay Variations and Troubleshooting

The direct binding assay is very flexible and most of the parameters can be changed depending on the purpose of a particular experiment. The optimum temperature for mammalian proteins is 37°, while for amphibian proteins it is 25° (the optimum time is still 5 min, which appears to be the best compromise between the time necessary for arrestin binding and the rate of Rh* decay). The amount of rhodopsin in the assay can be lowered to 0.05 μg/assay (25 nM) to increase the sensitivity of the competition assay, or increased to 0.6–2.4 μg/assay (300–600 nM) to study arrestin binding to nonpreferred forms of rhodopsin (dark P-Rh, Rh, or Rh*). At 1.2–2.4 μg rhodopsin virtually all wild-type arrestin present binds to P-Rh*, so that these high concentrations can be used to assess the percentage of functional arrestin in the total expressed protein (it is usually >85%). The concentration of radiolabeled arrestin can also be varied (at least between 0.1 and 5 nM). (This can be done without substantially compromising assay sensitivity if the specific activity of the arrestin is adjusted accordingly.) The two major problems that can be encountered are as follows: (1) With some arrestins all or most of the protein may be nonfunctional (because of misfolding, denaturation, and/or proteolysis). (2) Some mutants yield high nonspecific binding. The first problem is easy to detect if the

[17] A. Pulvermuller, D. Maretzki, M. Rudnicka-Nawrot, W. C. Smith, K. Palczewski, and K. P. Hofmann, *Biochemistry* **36,** 9253 (1997).
[18] L. Zhang, C. D. Sports, S. Osawa, and E. R. Weiss, *J. Biol. Chem.* **272,** 14762 (1997).
[19] S. Osawa, V. Gurevich, and E. R. Weiss, *Methods Enzymol.* **315** [28], 1999 (this volume).

proteins are run on a gel, where even modest (<20%) amounts of proteolytic products clearly indicate that the protein does not fold properly and/or denatures. In some cases doing the "run-off" and subsequent binding assay at 22° helps, but we were never able to express certain bovine visual arrestin mutants and one of the salamander arrestins in a functional form by *in vitro* translation (although some of these "troublesome" proteins express very well in *E. coli*). Nonspecific "binding" actually consists of at least two major components: aggregated arrestin that elutes on Sepharose 2B together with rhodopsin-containing membranes and arrestin bound to membranes themselves. With the majority of wild-type and mutant arrestins, most of the nonspecific binding is due to aggregation (i.e., roughly the same levels of nonspecific "binding" is observed in the presence or absence of liposomes), and overall nonspecific binding is less than 1–2% of total arrestin present in the assay. However, some mutants and splice variants, especially those with COOH-terminal truncations, demonstrate substantial binding to liposomes. Because such behavior usually correlates with a high propensity to aggregate, we believe that the phospholipid membranes simply facilitate the denaturation of these proteins. Decreasing the incubation temperature to 20–25° and increasing the ionic strength of the column buffer (e.g., using 100 mM NaCl, 10 mM Tris-HCl, pH 7.5, instead of 20 mM Tris-HCl) usually helps, but in some cases the proteins themselves are so unstable and prone to aggregation that it cannot be helped.

[30] Mapping Interaction Sites between Rhodopsin and Arrestin by Phage Display and Synthetic Peptides

By W. CLAY SMITH and PAUL A. HARGRAVE

Introduction

Protein–protein interactions form the underlying basis for activation of the components of the phototransduction cascade. Photoactivated rhodopsin interacts with transducin, leading to transducin activation, which then binds and activates cGMP phosphodiesterase. The inactivation process also hinges on protein–protein interactions, starting with the binding of rhodopsin kinase to photoactivated rhodopsin (R*), which then phosphorylates R*. Phosphorylated, activated rhodopsin (R*P) is bound by arrestin, which sterically blocks access of transducin to R*P, thereby quenching the rhodopsin component of phototransduction.

Much has been learned about phototransduction using classical and innovative approaches to studying these protein interactions. For example, synthetic peptides matching the sequence of the cytoplasmic loops of rhodopsin were used to demonstrate that transducin interacts with multiple sites on R*, including loop 3–4, loop 5–6, and a small loop in the carboxy-terminal tail.[1] In a different approach, antibodies against specific regions of rhodopsin kinase were used to show that the amino-terminal region of rhodopsin kinase binds to R*, but that blocking the amino-terminal terminal part of the protein does not block the kinase domain.[2] Molecular biology has provided a different set of tools for dissecting these interactions. For example, site-directed mutagenesis of rhodopsin has been used to demonstrate that loop 3–4 of rhodopsin contributes to the activation of transducin,[3] and that arrestin likely binds to loop 5–6 with some contribution of loops 1–2 and 3–4.[4,5]

In this chapter, we show how two different techniques—phage display and synthetic peptides—can be used to study protein–protein interactions. We attempt to give broad principles regarding the use of these techniques followed by specific application of these tools to the mapping of regions of arrestin that bind to rhodopsin.

Use of Phage Display

Phage display has rapidly emerged during the 1990s as a powerful method to study protein–protein interactions. George Smith first proposed the idea that the filamentous bacteriophage could be used as an expression host to produce fusions of foreign peptides with any one of several phage coat proteins, and that these fusion proteins should be accessible to other potential ligands.[6] Consequently, phage displaying a potential ligand for a substrate could be enriched over ordinary phage by affinity purification, a process known as *panning*.

The filamentous phage are a class of *Escherichia coli* phage that contain a single strand of DNA encapsulated by 8–10 coat proteins (Fig. 1). The two proteins most commonly used to display proteins are the major coat

---

[1] B. König, A. Arendt, J. H. McDowell, M. Kahlert, P. A. Hargrave, and K. P. Hofmann, *Proc. Natl. Acad. Sci. U.S.A.* **86**, 6878 (1989).
[2] K. Palczewski, J. Buczylko, L. Lebioda, J. W. Crabb, and A. S. Polans, *J. Biol. Chem.* **268**, 6004 (1993).
[3] R. R. Franke, B. König, T. P. Sakmar, H. G. Khorana, and K. P. Hofmann, *Science* **250**, 123 (1990).
[4] J. G. Krupnick, V. Gurevich, T. Schepers, H. E. Hamm, and J. L. Benovic, *J. Biol. Chem.* **269**, 3226 (1994).
[5] D. Raman, S. Osawa, and E. R. Weiss, *Invest. Ophthalmol. Vis. Sci.* **39**, S954 (1998).
[6] G. P. Smith, *Science* **228**, 1315 (1985).

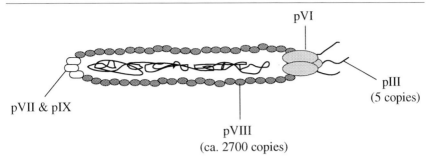

FIG. 1. Stereotypical structure of the M13 filamentous phage. The single-stranded DNA is surrounded by approximately 10 different types of coat proteins. Proteins pVIII and pIII are most commonly modified for use in phage display.

protein (pVIII), which is present at approximately 2700 copies/particle, and the minor coat protein (pIII), present at 3–5 copies/particle. Protein III is responsible for binding to the mating pilus of *E. coli* and initiating the infection process. The selection of which coat protein to use for a phage display experiment is usually dictated by the goal of the experiment and by the size of the polypeptide being displayed. Generally, the smaller the protein being displayed, the more tolerant the phage will be of multiple copies. In addition, proteins displayed as fusions with the major coat protein are expressed in many more copies than those fused with pIII, thus giving the opportunity to select for protein interactions that might have low affinity but high avidity. If selection of high affinity binders is desired, fusions with pIII are preferred, since the maximum number of fusions would be five on each phage. Display of multiple copies of the fusion protein is known as polyvalent display. Selection of high-affinity binders can be further enhanced by the use of the phagemid/helper phage system. In this system, the fusion is made with gene III on a phagemid vector and coexpressed with helper phage that provide an excess of coat proteins (including pIII). Under these conditions, phage are produced that on average will display 0–1 copies of the fusion protein (monovalent display). The methods described in the following sections utilize this system since we attempted to bias our results toward isolating regions of arrestin that bind to rhodopsin with high affinity.

The strength of phage display lies in its ability to rapidly screen very large (10^8–10^9) pools of potential protein ligands against a selected substrate. This substrate can be virtually any protein or nonprotein substance that can be used as an affinity matrix. The ability to use any substance as a target is a decided advantage over two-hybrid systems which require that both the library and the target (also known as the "bait") be not only proteins, but also proteins that can be localized to the nucleus. Readers are directed to several recent reviews for further background information on phage display.[7–11]

Vector Selection

General Considerations

The choice of vectors is dictated by the goal of the experiment, such as whether a researcher wants to screen a random peptide library, or target sequences from a specific protein. If random peptides are to be screened, several commercial peptide libraries are available for those interested in identifying short-peptide ligands for their substrate (e.g., 7-mer and 12-mer libraries from New England BioLabs, Beverly, MA). For those interested in constructing their own peptide libraries or screening sequences from a specific protein, several companies market phagemid vectors with the geneIII DNA included in the vector [e.g., pCANTAB 5E (Pharmacia-Amersham, Piscataway, NJ), M13KE (New England BioLabs), and pSKAN-8 (Display Systems Biotech, Vista, CA)]. A number of other companies sell generalized phagemid vectors that can be modified for use in phage display by the addition of the geneIII or geneVIII cDNA's (e.g., Life Technologies, Rockville, MD, and Stratagene, La Jolla, CA).

Specific Application

For the display of arrestin fragments, we chose to use the pCANTAB 5E vector provided by Pharmacia as part of its recombinant antibody phage kit. This vector is provided linearized at *Sfi*I and *Not*I sites in the geneIII DNA. To accept random blunt fragments of arrestin cDNA, the pCANTAB 5E vector was modified by the introduction of a multiple cloning site including an *Eco*RV blunt site. Complementary oligonucleotides were synthesized that created *Sfi*I–*Xho*I–*Sac*I–*Eco*RV–*Sac*I–*Not*I sites, annealed, and ligated into pCANTAB 5E to create the pCAN-ECO253 vector used in these studies.

Insert Preparation

General Considerations

The key to insertion of fragments is to maximize the diversity of the inserts. The first consideration is to maintain proper reading frames. Be-

[7] E. M. Phizicky and S. Fields, *Microbiol. Rev.* **59,** 94 (1995).
[8] A. Bradbury and A. Cattaneo, *Trends Neurosci.* **18,** 243 (1995).
[9] B. A. Katz, *Annu. Rev. Biophys. Biomol. Struct.* **26,** 27 (1997).
[10] D. R. Wilson and B. B. Finlay, *Can. J. Microbiol.* **44,** 313 (1998).
[11] B. K. Kay, A. V. Kurakin, and R. Hyde-DeRuyscher, *Drug Disc. Today* **3,** 370 (1998).

cause the insert is being introduced into the middle of either geneIII or geneVIII, reading frames must be maintained on both ends. Consequently, use of directional insertion is preferred, because random insertion will give only 1/18 in the proper direction and proper reading frame (1/2 in correct direction, 1/3 in the proper 5' reading frame, and 1/3 in the proper 3' reading frame). Directional cloning works well for preparing libraries of short peptides since synthesis of random oligonucleotides is used to prepare the inserts, and these can be synthesized with the appropriate 5' and 3' adapter sequences.

However, for the insertion of randomly generated fragments, such as of a specific cDNA, it is difficult to obtain any type of directional or reading frame orientation. Consequently, it is essential that the primary library be made as diverse as possible so that there is adequate representation of all portions of the gene sequence. Two approaches are generally taken for the preparation of random blunt fragments of a specific cDNA. The first uses DNase I digestion, which in the presence of Mn^{2+} cleaves double-stranded DNA nonspecifically to give blunt fragments.[12] The concentration of DNase I needs to be adjusted to give the desired amount of cleavage, but a concentration of 0.1 U/μg DNA in 50 mM Tris and 10 mM MnCl$_2$ (pH 8.0) for 1–5 min (25°) is a good starting point.

The second approach uses mechanical cleavage of the cDNA through sonication. This method produces more staggered breaks in the DNA than does DNase I digestion, but in our hands appears to yield a more random representation of the cDNA. Figure 2 shows a comparison of the arrestin cDNA fragmented by sonication (lanes 2–6) and by DNase I (lanes 7–12). Both appear qualitatively similar; however, the phage library we produced using DNase-treated cDNA showed an inherent bias toward the amino-terminal portion of arrestin, whereas sonication produced a more uniform library (see Library Evaluation section).

Specific Application

For the production of the arrestin phage library, 1.3 μg of pCAN-ECO253 was linearized with 20 U *Eco*RV (12 hr at 37°) and treated with calf-intestinal alkaline phosphatase (5 U, 2 hr, 37°) to reduce vector self-ligation. The background of uncut and unphosphorylated phagemid was confirmed to be acceptably low (i.e., 50 ng of self-ligated vector yielded less than 10 colonies from standard chemical transformation). Arrestin cDNA was prepared for insertion by sonication for 2 min at 45% of maximum output (Microzoon Systems Cell Disruptor with 2-mm probe; Heat Systems Ultrasonics Inc., Farmingdale, NY). This length of sonication was

---

[12] S. Anderson, *Nucl. Acid Res.* **9,** 3015 (1981).

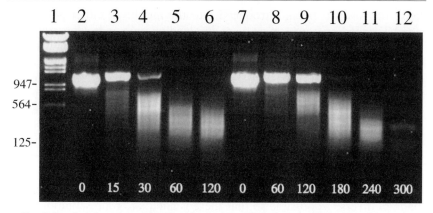

FIG. 2. Random fragmentation of bovine arrestin cDNA. Full-length cDNA was fragmented by sonication (lanes 2–6) or by treatment with 0.1 U/μg DNase I (lanes 7–12) for the indicated length of time (in seconds). Sonication was performed with the sample in an ice–water slurry, and DNase I digestion performed at room temperature. DNA samples subjected to 2 min of sonication were used for constructing the arrestin phage library. Lane 1, lambda DNA standard cut with EcoRI and HindIII.

empirically determined to give the maximum amount of DNA in the 200- to 300-bp range, which targets polypeptides of 67–100 amino acids. The sonicated DNA was ligated into pCAN-ECO253, using 12 reactions containing 50 ng of phagemid, and approximately 80 ng of sonicated DNA. Competent TG1 E. coli cells were prepared,[13] and each ligation was used to transform 1 ml of competent cells, plating transformed cells on SOB plates[14] with 2% (w/v) glucose and 50 μg/ml carbenicillin. This method yielded 3.8 × 10^3 independent clones with a background of 5% nonrecombinant.

Library Evaluation

General Considerations

Because the ultimate success of panning is dependent on having the potential binder in the library, it is a good idea to have some assessment of the quality of the library. In the case of random peptide libraries, 10–20 randomly picked colonies should be sequenced to verify the diversity of the peptides (in this small sample size there should be no identical peptides).

[13] S. R. Kushner, in "Genetic Engineering" (H. W. Boyer and S. Nicosia, eds.), p. 17. Elsevier/North Holland, Amsterdam, 1978.
[14] J. Sambrook, E. F. Fritsch, and T. Maniatis, "Molecular Cloning: A Laboratory Manual." Cold Spring Harbor Laboratory, Cold Spring Harbor, New York, 1989.

For gene fragment libraries, the best evaluation is one that assesses the expression properties of the library. This analysis requires a panel of antibodies directed against various epitopes across the polypeptide in order to demonstrate that each part of the protein is adequately represented in the primary library and that there is not an inherent bias to the library. Unfortunately, few laboratories have available such a diverse panel of antibodies. In the absence of these antibodies, an alternative approach is to ascertain that all portions of the cDNA are well represented.

For this analysis, colony lifts of the primary library are prepared and then probed with specific portions of the cDNA in order to determine that each part of the cDNA is present in the library. This type of analysis can also be used to determine if there is an inherent bias to the library.

Specific Application

Because antibodies against arrestin are available for only a limited range of epitopes (most targeted toward the carboxy terminus), we evaluated our phage library on the basis of the cDNA fragments that had been inserted into geneIII. Defined portions of arrestin cDNA 100–200 bp in length were prepared by polymerase chain reaction (PCR) that covered the entire coding portion of arrestin. These PCR products were then labeled with [α-^{32}P]dCTP (Oligolabelling Kit, Pharmacia, Piscataway, NJ), and hybridized to duplicate colony lifts containing 1500 colonies/lift in aqueous Denhardt's solution following standard DNA/DNA hybridization procedures.[15] The lifts were then exposed to X-ray film, and positive colonies counted for each probe (Table I). Each portion of the arrestin cDNA hybridized to some colonies, indicating that all portions of the arrestin cDNA are present in the phage library. Furthermore, when the number of positive colonies for each probe is normalized to the length of the probe, it is clear that each portion of the arrestin cDNA is represented approximately equally within the library (within one standard deviation from the mean) with the exception of 938–1107, which is slightly underrepresented in the library, and 1090–1212, which is slightly overrepresented in the library.

Panning

General Considerations

Panning provides a method to selectively enrich for the phage that recognize a potential target substrate. Normally, the enrichment provided

---

[15] F. M. Ausubel, R. Brent, R. E. Kingston, D. M. Moore, J. G. Seidman, J. A. Smith, and K. Struhl, "Current Protocols in Molecular Biology." Wiley, New York, 1998.

TABLE I
ASSESSMENT OF QUALITY OF ARRESTIN PHAGE LIBRARY BY COLONY HYBRIDIZATION[a]

Portion of arrestin coding region	Fragment length (bp)	No. of colonies	No. of colonies/bp
1–150	150	132	0.88
68–249	182	120	0.66
246–410	165	131	0.79
409–612	204	181	0.89
547–762	216	156	0.72
653–837	185	147	0.79
810–935	126	108	0.86
938–1107	170	106	0.62
1090–1212	123	153	1.24
$x \pm$ S.D.			0.83 ± 0.18

[a] Colony lifts were probed with nine portions of the arrestin cDNA, spanning the entire length of the cDNA, and the number of colonies to which each probe hybridized counted. The final column indicates the number of colonies averaged by the length of each probe.

by a single round of panning is on the order of 10^2–10^4, with 10^3 being quite common. Consequently, multiple cycles must usually be performed in order to be able to identify the specific binders from nonspecific background. In a library that contains 10^{8-9} original clones, this typically means three to five rounds of panning.

The basis for panning of any phage library is to have available some mechanism for separating the phage bound to the substrate from those that do not bind. The most common method employed in the literature is to use substrate coated on plastic dishes such as tissue culture flasks or tubes (e.g., Nunc Immunotubes, Fisher Scientific, Pittsburgh, PA). Under these circumstances, unbound phage can be simply aspirated from the surface, and the substrate rinsed. However, if it is important that a conformation be preserved in the substrate, then an alternative approach must be used since coating on plastic will denature a protein substrate. One approach taken by several workers is to biotinylate the substrate, thereby allowing for capture of the substrate by streptavidin-coated beads.[16,17] Commercial biotinylation kits are available (e.g., Pharmacia, Piscataway, NJ) as well as several suppliers of streptavidin coated beads (e.g., Dynal, Lake Success, NY). Importantly, it should be determined empirically if biotinylation influences the binding properties in which the researcher is interested.

A final consideration when planning the panning process is the method used to elute the bound phage. Because the phagemid/helper phage system

[16] S. P. Parmley and G. P. Smith, *Gene* **73,** 305 (1988).
[17] R. E. Hawkins, S. J. Russell, and G. Winter, *J. Molec. Biol.* **226,** 889 (1992).

produces monovalently displayed pIII fusion polypeptides on average, the remaining pIII proteins are wild type and can directly infect *Escherichia coli*. Consequently, bacteria can be added to the substrate containing the bound phage and the phage will infect the bacteria. Alternatively, the phage can be eluted from the substrate by mild acid treatment and then added to *E. coli* for infection. The former method has the advantage of simplicity, and does not run the risk of damaging phage particles by acid treatment (although the filamentous phage are quite resistant to pH extremes of pH 2–12). The direct infection method, however, has two inherent disadvantages. First, direct infection by the addition of bacteria to the bound phage may potentially lose phage that are bound to the substrate with very high affinity, since the affinity of the wild-type pIII is in the nanomolar K_d range. Second, direct infection requires that the phage obtained in each round of panning must be amplified by the bacterial host before being used in the subsequent panning cycle. This reamplification means that there is the potential for the introduction of a bias due to differential growth rates because of the different pIII fusions.

Acid elution circumvents the first problem, and can also avoid the second problem if the acid-eluted phage are then used directly in subsequent panning cycles without reamplification. The disadvantage of using the phage directly in subsequent pannings is that if the yield of phage from the previous cycle is low then the total phage recovered after three to four cycles of panning may be vanishingly small. Depending on the yield of phage from the initial panning cycles, it may be necessary to infect and amplify after the first or second panning before proceeding to the next round of panning. Consequently, the optimal approach may use a combination of acid elution and direct infection.

Controls

Probably the single most important consideration in successful application of phage display is the use of every conceivable mechanism to reduce background due to binding of the phage to some reaction component other than the target substrate. First, all surfaces on the reaction vessel should be blocked. If the target substrate is applied to plates or tubes, then the unbound surfaces should be blocked after the substrate is applied. We routinely block the microfuge tubes used for panning overnight with 1% (w/v) bovine serum albumin (BSA; Sigma, Fraction V) or 1% (w/v) nonfat powdered milk in phosphate-buffered saline (PBS: 137 mM NaCl, 2.7 mM KCl, 10 mM Na$_2$HPO$_4$, 1.8 mM KH$_2$PO$_4$, pH 7.4). BSA or milk protein should also be included in the panning reaction to help minimize nonspecific binding. However, the phage could bind to these proteins as well; consequently, the phage library should be preincubated with whichever blocking protein is being used to remove the portion of phage that could bind to these blocking agents.

Second, if any material other than the target substrate is present in the panning solution, adequate controls must be used to prevent the phage from binding to this material. For example, in solution-phase panning, the target is often labeled with biotin to allow for capture of the target–phage complex by streptavidin-coated beads. The phage library should be preincubated with both biotin and the streptavidin beads prior to incubation with the target (usually 1 hr at 4° is sufficient). Finally, phage tend to aggregate. Therefore, all phage-containing solutions should be filtered (0.45-μm filter) to remove these aggregates; single phage will easily pass through this pore size.

Specific Application

We wanted to pan our arrestin phage library against R*P with the goal being to isolate the predominant region of arrestin that bound to R*P. Because the conformation of rhodopsin may be an important constraint for identifying the particular binding region we needed rhodopsin in its undenatured state so that we could prepare photoactivated and phosphorylated rhodopsin against which to pan our library. Consequently, we could not use rhodopsin-coated tubes. However, because rhodopsin comprises 95% of the protein in rod disk membranes,[18] we were able to use disk membrane preparations and could sediment these membranes in order to separate the arrestin-phage that bound to R*P from those that did not.

Because the success of phage display hinges on the panning process, we have provided a detailed protocol that we used so other users may be better able to adapt the process to their study.

Protocol

1. The primary phage library is cultured in 40 ml 2× YT media (10 g/liter yeast extract, 17 g/liter tryptone) containing 50 μg/ml carbenicillin (Sigma, St. Louis, MO) with 2% (w/v) glucose.
2. When the culture reaches OD_{600nm} of 0.6, M13K07 helper phage (Promega, Madison, WI) are added at a multiplicity of infection of 5 (using the approximation of 5×10^8 cells/unit OD_{600nm}).
3. After allowing the helper phage to infect the cells (1 hr, 37° with gentle shaking), the media is exchanged to 2× YT containing both

---

[18] W. Krebs and H. Kühn, *Exp. Eye Res.* **25**, 511 (1977).

50 μg/ml carbenicillin and 50 μg/ml kanamycin to select for bacteria that contain both the phagemid and helper phage. Phage are allowed to be secreted into the media 12–16 hr (37°). The yield of phage is usually on the order to 10^{10}–10^{13} particles.
4. The culture supernatant is collected, and phage precipitated by the addition of 1/10 volume 20% (w/v) polyethylene glycol (PEG) 8000/ 2.5 M NaCl (60 min, 4°).
5. The phage are collected by centrifugation (10,000 rpm, 10 min, 4°) resuspended in 1 ml PBS containing 1% (w/v) BSA (the BSA serves to help block nonspecific binding), and filtered (0.45 μm) to remove any residual bacteria and aggregates of phage.
6. Disk membranes containing 200 μg of phosphorylated rhodopsin[19] are added to the phage, exposed to white light (2 min) to photoactivate the RP, and allowed to incubate with gentle mixing (60 min, 4°).
7. The membrane/phage mixture is centrifuged (16,000g, 10 min, 4°), and the pellet resuspended in 1 ml PBS containing 0.1% (v/v) Tween 20. This washing with detergent is repeated once more, and then twice more with PBS without detergent.
8. Following the final wash, the phage are acid eluted with 400 μl PBS, pH 4.0 containing 0.5 M NaCl (15 min at room temperature), then neutralized with 400 μl PBS, pH 11.0.
9. The eluted phage are mixed with 800 μl 2× YT with 2% BSA, to which 200 μg RP is added and the panning process repeated.
10. Following each panning cycle, an aliquot of the phage is used to infect TG1 *E. coli* cells to monitor the number of phage eluted with each cycle. For titration, TG1 cells are grown to OD_{600nm} of 0.3, a serial titration of phage added (usually 10^{-1}–10^{-5}), and the infected cells plated on 2× YT agar plates with 2% glucose and 50 μg/ml carbenicillin. Each panning cycle should give approximately 10^2- to 10^3-fold enrichment of phage that bind to the substrate. If the number of infectious phage particles drops below 10^4, the phage should be reamplified as in step 1 before proceeding to the next cycle of panning.
11. After completing four panning cycles, isolated colonies from the plates are cultured and phagemid DNA prepared following standard minipreparation protocols for the isolation of plasmid DNA (e.g., Wizard Miniprep kits, Promega, Madison, WI). Because the TG1 strain of *E. coli* is not an endonuclease-deficient strain (EndA$^-$),

---

[19] J. Puig, A. Arendt, F. L. Tomson, G. Abdulaeva, R. Miller, P. A. Hargrave, and J. H. McDowell, *FEBS Lett.* **362**, 185 (1995).

proteases should be included in the sample preparation buffers in order to improve the quality of DNA for sequencing.

We sequenced the phagemid DNA manually using the dideoxy termination method[20] in the presence of [^{35}S]dATP, although automatic sequencing works as well. It is sufficient to sequence across the insertion site from both directions to determine which portion of the arrestin cDNA was cloned into the geneIII DNA and to ensure that proper frame orientation is maintained. Any inserts that are not in frame with the geneIII open reading frame should be discarded from the data. If only a few colonies are obtained after panning, it is an easy matter to sequence all clones. However, if several thousand colonies are obtained, then only a sample of the colonies can be reasonably sequenced.

We obtained 2.4×10^3 colonies from phage panned against R*P in disk membranes and 300 colonies from phage panned against red blood cell membranes. We sequence 200 colonies from the phage panned against R*P and 50 colonies from phage panned against our control membranes.

Control Considerations

In our experiment rhodopsin was provided in disk membranes prepared from rod outer segments. This provided the added component of the lipid bilayer in which the rhodopsin was embedded. The best control would have been the same lipid membrane with rhodopsin removed. However, this material is quite difficult to prepare in the quantities required for the panning experiments. Instead, we chose an alternative approach in which membranes were collected from lysed red blood cells.[21] These membranes were used in a parallel panning experiment. From the small number of colonies obtained in this control experiment we were able to conclude that the contribution due to nonspecific binding to membranes was likely to be insignificant.

Data Analysis

General Considerations

Data analysis is relatively straightforward since it involves simply a compilation and alignment of the sequences obtained that bound to the substrate. For random peptide sequences, the goal is to identify one or a few residues that emerge as a consensus residue in the same location of

---

[20] F. Sanger, S. Nicklen, and A. R. Coulson, *Proc. Natl. Acad. Sci. U.S.A.* **74,** 5463 (1977).
[21] D. Hanahan and J. Ekholm, *Methods Enzymol.* **31,** 168 (1974).

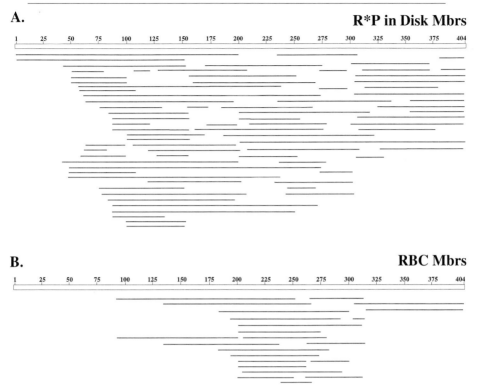

FIG. 3. Fragments of arrestin that were displayed on phage that bound (A) to disk membranes containing R*P or (B) to membranes prepared from red blood cells (RBCs). The arrestin polypeptide is indicated as an open bar with amino acid positions numbered above. Each displayed peptide that bound to the substrate is indicated as a single line.

the peptide (usually present in 50–70% of the isolated peptides). For screens of a specific protein, the goal is to isolate the minimal epitope from the overlapping peptides.

Specific Application

Figure 3 shows a compilation of the arrestin sequences that bound to R*P in disk membranes (Fig. 3A) and those that bound to red blood cell membranes (Fig. 3B). Under ideal circumstances, a minimal epitope will be revealed that can be used for further studies. In our experiment, the entire molecule of the protein was represented. Perhaps this reflects the fact that multiple portions of arrestin contribute to the binding to rhodopsin.

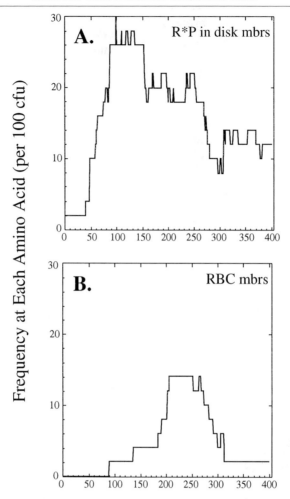

FIG. 4. Frequency histogram representing the number of times [per 100 colony-forming units (cfu)] any given amino acid was displayed on a phage that bound to R*P in (A) disk membranes or (B) to membranes from RBCs.

However, if these data are converted to a frequency histogram (Fig. 4), simply calculating the number of times any particular amino acid of arrestin was present in a phage that bound to R*P, it is clear the most prevalent epitope was 90–140, with additional peaks at 160–210 and 240–270. Note that for the primary peak of binding (90–140) there was virtually no binding to red blood cell membranes, indicating that it is unlikely that the phage displaying this region of arrestin were binding nonspecifically to the lipid membrane.

Data Confirmation

General Considerations

One critical feature of phage display is that like other library screening techniques, such as yeast two-hybrid or cDNA expression library screening, all potential positives must be tested to eliminate false-positives and establish biological relevance. This can be approached using any number of methods that are appropriate for the particular system, but usually include heterologous expression to produce protein for measuring binding parameters, receptor activation/inactivation assays, or binding competition assays.

Specific Application

For our arrestin phage display, we chose to express a portion of the principal binding region as a fusion with glutathione S-transferase and demonstrate that this fusion could function as a competitor with arrestin for binding to R*P. We first expressed the region of arrestin spanning amino acids 95–140 as a carboxyl-terminal fusion protein with glutathione S-transferase. The arrestin cDNA corresponding to this region was amplified by PCR, adding 5' EcoRI and 3' SalI restriction sites with synthetic oligonucleotide primers, and cloned into the pGEX-4T-1 vector (Pharmacia) at the EcoRI and SalI sites. This recombinant plasmid was then introduced into BL21 E. coli cells and used to express soluble protein, inducing with 1 mM IPTG (4 hr, 37°). The expressed protein was purified over glutathione-agarose (Pharmacia) following the manufacturer's recommended procedures. We were able to obtain approximately 2 mg/liter of pure protein.

We then assayed the ability of this protein to compete for the binding of arrestin to R*P. Arrestin (10 μg) was mixed with R*P (40 μg) contained in disk membranes in PBS in the presence of increasing amounts (0–1 mM) of 95–140/GST fusion protein, and 1 μM glyceraldehyde 3-phosphate dehydrogenase in a 150-μl reaction volume. (The dehydrogenase serves as an internal control for quantitation.) Following incubation (2 min, 4°), the reaction mixture was centrifuged (24,000g, 30 min, 4°) to sediment the disk membranes along with any bound arrestin. The amount of arrestin and R*P used was empirically determined to give complete binding of the arrestin in the absence of fusion protein. The reaction supernatant was then subjected to sodium dodecyl sulfate–polyacrylamide gel

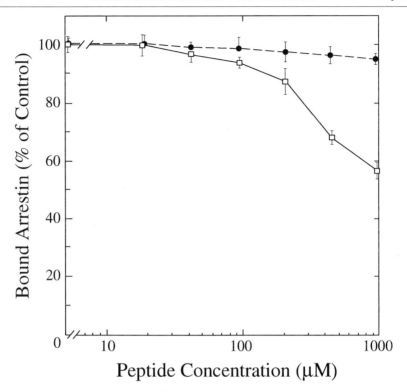

FIG. 5. Binding inhibition of arrestin to R*P using 95–140/GST fusion protein. Incubations containing 1.4 μM arrestin, 8 μM R*P in disk membranes, and the indicated amount of fusion protein (open squares) or glutathione S-transferase alone (closed circles) in the presence of 1 μM glyceraldehyde 3-phosphate dehydrogenase were prepared as described. The amount of arrestin competitively released into the supernatant was measured by SDS–PAGE, stained with Coomassie blue, and quantified by scanning densitometry. The curves represent the means ($\pm$ S.D.) from three experiments.

(SDS–PAGE) electrophoresis to determine if any arrestin was displaced by competition into this fraction. The amount of arrestin present in the reaction supernatant was quantified by scanning densitometry (versions of NIH Image are available for free downloading from NIH's website, http://rsb.info.nih.gov/nih-image/), normalizing the samples to the internal glyceraldehyde 3-phosphate control. Figure 5 shows that as the concentration of 95–140/GST is increased, there is an increasing amount of unbound arrestin. Note that GST alone has no influence on the binding of arrestin to R*P. If this competition is projected, the IC_{50} for the inhibition is projected to be in the low millimolar range. This experiment provides corroboration of the data obtained using phage display.

Use of Synthetic Peptides

Protein–protein interactions form the basis of many key events in biological processes. It is often possible to test synthetic peptides from the sequence of one of the interacting proteins to determine whether they compete in a protein–protein binding reaction, and thus determine the sites of interaction between the proteins. This technique has been successfully applied in several studies of protein interaction of interest in the study of the biochemistry of vision. Some peptides representing sequences displayed on the cytoplasmic surface of rhodopsin compete for the binding of transducin to R*, thus helping delineate binding sites on rhodopsin for transducin.[22] Synthetic peptides corresponding to two regions near the carboxyl terminus of the transducin α subunit compete with transducin for the binding to R*, thus suggesting part of the regions of transducin that interact with R*.[23] The third cytoplasmic loop peptide of rhodopsin competes for the binding of arrestin to R*P, suggesting this loop as a major site of interaction between the two proteins.[24] And the interaction site by which guanylate cyclase becomes activated by its activating protein has been shown to be within the catalytic domain of guanylate cyclase, by peptide competition.[25] Synthetic peptides have also been used to map epitope specificities of libraries of anti-rhodopsin monoclonal antibodies.[26,27]

Specific Application

For this research overlapping peptides 20 amino acids in length were synthesized using RAMPS Multiple Peptide Synthesis System (DuPont, Wilmington, DE). Most peptides were synthesized using "Wang" Carboxylate Resin cartridges with 0.1 mmol of first 9-fluoroenylmethoxycarbonyl amino acids on p-alkoxybenzyl alcohol resin (DuPont). Sequential amino acids were coupled according to the RAMPS protocol using HOBt esters prepared *in situ* from Fmoc amino acids. Cleavage and deprotection from resin was carried out by incubating the resin for 3 hr at room temperature

---

[22] B. König, A. Arendt, J. H. McDowell, M. Kahlert, P. A. Hargrave, and K. P. Hofmann, *Proc. Natl. Acad. Sci. U.S.A.* **86,** 6878 (1989).

[23] H. E. Hamm, D. Deretic, A. Arendt, P. A. Hargrave, B. Koenig, and K. P. Hofmann, *Science* **241,** 832 (1988).

[24] J. G. Krupnick, V. V. Gurevich, T. Schepers, H. E. Hamm, and J. L. Benovic, *J. Biol. Chem.* **269,** 3226 (1994).

[25] I. Sokal, F. Haeseleer, A. Arendt, E. T. Adman, P. A. Hargrave, and K. Palczewski, *Biochemistry* **38,** 1387 (1999).

[26] R. S. Hodges, R. J. Heaton, J. M. R. Parker, L. Molday, and R. S. Molday, *J. Biol. Chem.* **263,** 11768 (1988).

[27] G. Adamus, Z. S. Zam, A. Arendt, K. Palczewski, J. H. McDowell, and P. A. Hargrave, *Vis. Res.* **31,** 17 (1991).

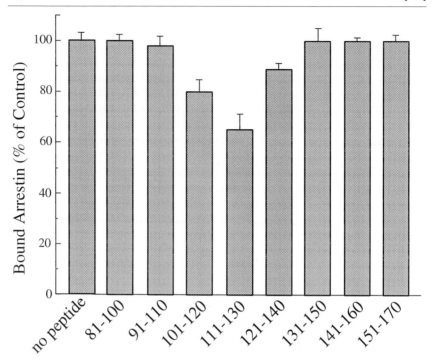

FIG. 6. Binding competition of arrestin to phosphorylated photoactivated rhodopsin using synthetic arrestin peptides. Photoactivated disk membranes containing 8 μM R*P were incubated with 1.4 μM arrestin in the presence of 700 μM synthetic peptides. Following centrifugation, inhibition of arrestin binding was measured by analyzing the amount of arrestin remaining in the supernatant using SDS–PAGE. Arrestin was quantified by scanning densitometry, normalizing arrestin to an internal standard (glyceraldehyde 3-phosphate dehydrogenase). The means ($\pm$ S.D.) from three experiments are shown.

in 90:5:5 (v/v) trifluoroacetic acid (TFA):H$_2$O:ethanedithiol. For purification, peptides were precipitated from solution using cold diethyl ether, lyophilized, and further purified over Bio-Gel P-2 (Bio-Rad, Hercules, CA) in 5% (v/v) acetic acid. Quality was assessed by amino acid analyses and analytical HPLC. The peptides synthesized corresponded to arrestin sequence 81–100, 91–110, 101–120, 111–130, 120–140, 131–150, 141–160, and 151–170.

The peptides were tested in a binding competition assay, using the same methodology as delineated earlier for testing the effect of 95–140/GST fusion on arrestin binding to R*P except that increasing concentrations of peptides (0–700 μM) were used instead of the fusion protein. Figure 6 shows the amount of arrestin binding that was inhibited by each peptide. These results show some effect on binding by peptides 101–120, 111–130, and 121–140, with the largest effect by 111–130. These data clearly implicate the region 101–130 as containing a binding domain, which corresponds nicely with the phage display data. Because 111–130 has limited solubility in aqueous buffers, this peptide was resynthesized to include Lys-109 and Lys-110, as well as to substitute Cys-128 with serine (i.e., 109–130/Cys128 Ser). Using this peptide, we were able to completely block the binding of arrestin to R*P with an IC_{50} of 1.1 mM. The relatively high IC_{50} likely indicates that arrestin has multiple points of contact with R*P, which is consistent with the observations from other laboratories.[28–30] Note that a peptide containing the same amino acid composition, but ordered randomly (i.e., KDYLVPLSSNAPLYLGTFPFTK) was not an effective competitor.

In conclusion, the use of synthetic peptides identifies a binding region in arrestin for R*P that is composed of, at least in part, amino acids 111–130. These data are consistent with those obtained using phage display. Consequently, the synthetic peptides are useful as additional corroboration of the phage display study, but also provide a more narrow definition of the binding region.

Acknowledgment

This research was supported by a Career Development Award from the Research to Prevent Blindness (RPB) Foundation to W.C.S., a Senior Scientific Investigator award from RPB to PAH, grants from the National Eye Institute (EY06225, EY06226, EY08571), and an unrestricted grant to the Department of Ophthalmology from RPB.

[28] V. V. Gurevich and J. L. Benovic, *J. Biol. Chem.* **268**, 11628 (1993).
[29] V. V. Gurevich and J. L. Benovic, *J. Biol. Chem.*, **270**, 6010 (1995).
[30] K. Palczewski, J. Buczylko, N. R. Imami, J. H. McDowell, and P. A. Hargrave, *J. Biol. Chem.* **266**, 15334 (1991).

[31] Characterization of RanBP2-Associated Molecular Components in Neuroretina

By PAULO A. FERREIRA

Introduction

In search of proteins that mediate the biogenesis of vertebrate G-protein-coupled light receptors, opsins, we have cloned several products[1]

[1] P. Ferreira, J. Hom, and W. Pak, *J. Biol. Chem.* **270**, 23179 (1995).

from bovine retinas with homology to the transcript of *ninaA* gene of *Drosophila*.[2-4] NinaA is required for the biogenesis of a subclass of light receptors.[5,6] A subset of bovine cyclophilin-related products identified in our screen[1] is made up orthologs of the human[7,8] and mouse[9] transcripts of Ran-binding protein 2 (RanBP2). They are predominantly expressed in the retina.[1] Recent results from our laboratory indicate that RanBP2 is a scaffold protein containing multiple and well-defined structural modules. These mediate, singly or in combination, the recruitment of protein machines and substrates involved in protein biogenesis.[10-12]

To dissect the role of RanBP2 in retinal function, we began structure–function analysis of each structural module of RanBP2[1,7,8,9] because they likely mirrored distinct functional domains. Identification of molecular partners interacting specifically with each of these domains will provide strong and direct clues of the overall role of RanBP2 and their associated components in retinal function. Among several methods currently available to detect protein–protein interactions, we initially chose protein affinity chromatography to carry out screening of retinal extract proteins that interact specifically with RanBP2.

Several reasons led us to choose this screening method. First, it was possible to generate large quantities of protein constructs containing domains of RanBP2 or subdomains thereof. We chose to produce protein constructs, whereas a given RanBP2 domain is fused downstream of glutathione S-transferase (GST). The 26-kDa GST moiety is highly soluble and stable, properties highly desirable in protein purification. To this end, we used an engineered GST-expression vector containing thrombin cleavage and glycine kinker sequences between GST and the fused protein of interest

---

[2] R. Stephenson, J. O'Tousa, N. Scavarda, L. Randall, and W. L. Pak, in "The Biology of Photoreception" (D. J. Cosens and D. Vince-Price, eds.), pp. 447–501. Cambridge University Press, Cambridge, UK, 1983.

[3] B.-H. Shieh, M. Stamnes, S. Seavello, G. Harris, and C. S. Zuker, *Nature* **338**, 67 (1989).

[4] S. Schneuwly, R. Shortridge, D. Larrivee, T. Ono, M. Ozaki, and W. L. Pak, *Proc. Natl. Acad. Sci. U.S.A.* **86**, 5390 (1989).

[5] D. Larrivee, S. Conrad, R. Stephenson, and W. L. Pak, *J. Gen. Physiol.* **78**, 521 (1981).

[6] M. Stamnes, B.-H. Shieh, L. Chuman, G. Harris, and C. S. Zuker, *Cell* **65**, 219 (1991).

[7] N. Yokoyama, N. Hayashi, T. Seki., N. Pante, T. Ohba, K. Nishii, K. Kuma, T. Hayashida, T. Miyata, U. Aebi, M. Fukui, and T. Nishimoto, *Nature* **376**, 184 (1995).

[8] J. Wu, M. Manutis, D. Kraemer, G. Blobel, and E. Coutavas, *J. Biol. Chem.* **270**, 14209 (1995).

[9] N. Wilken, J.-L. Senecal, U. Sheer, and M.-C. Dabauvalle *Eur. J. Cell Biol.* **68**, 211 (1995).

[10] P. Ferreira, T. Nakayama, W. Pak, and G. Travis, *Nature* **383**, 637 (1996).

[11] P. Ferreira, T. Nakayama, and G. Travis, *Proc. Natl. Acad. Sci. U.S.A.* **94**, 1556 (1997).

[12] P. Ferreira, C. Yunfei, D. Schick, and R. Roepman, *J. Biol. Chem.* **273**, 24676 (1998).

(gift from J. E. Dixon).[13] The thrombin cleavage site allowed the rapid purification of the domain of interest from GST. The glycine kinker increased the efficiency of thrombin cleavage and degree of conformational freedom between GST and the fused protein. This latter feature may also prevent potential steric hindrance of a protein substrate to the GST-fused moiety. Second, GST-fused proteins are bound to a glutathione S-agarose matrix at concentrations well above the K_d of most protein–protein interactions.[13,14] Weak protein interactions within relevant physiologic range (e.g., 10^{-4}–$10^{-5} M$) may thus, be detected. Third, incubation of a given GST-fused protein with tissue extracts permits one to screen all extracts' proteins equally and purify single binding protein species or tethered protein complexes from these extracts. Solubilized extracts may be prepared under different array of conditions since binding of GST to the cognate matrix is extremely strong. Fourth, this method may allow detection of protein interactions that are transient in nature by co-incubation of GST-fused proteins with cofactors required to lock-in the fused or extract protein(s) in an active (binding) state conformation. Finally, specific association of extract proteins with a GST-fused protein may be further tested by coincubation of the GST-fusion construct with excess of counterpart cleaved (competing) construct and/or comparative analysis with other GST-fused unrelated proteins. The former method developed and extensively explored by us, has proven to be very powerful in screening and identifying retinal extract proteins specifically binding to GST constructs. It also prevents the use of highly stringent washes leading to nonspecific disruption of biologically significant protein–protein interactions. Additional variations to this method may be introduced for specific purposes by using metabolic labeled extracts, and/or cross-linking agents to lock-in transient protein interactions to the GST-fused moiety. Also, a candidate protein approach may be taken to search for proteins binding to GST-fused constructs by Western blot analysis of coprecipitates of binding reactions with antibodies against cognate proteins.[10,11] These techniques further enhance the sensitivity of the pull-down affinity assays.

Here, we describe two examples of methods we used to identify Ran-GTPase and 19S cap subunits of the proteasome, as components of the RanBP2 macroassembly complex in the neuroretina. These techniques should be readily applied in the screening and identification of components comprising other macroassembly complexes in the neuroretina and other tissues.

---

[13] K. Guan and J. Dixon, *Anal. Biochem.* **192,** 262 (1990).
[14] D. Smith and K. Johnson, *Gene,* **67,** 31 (1988).

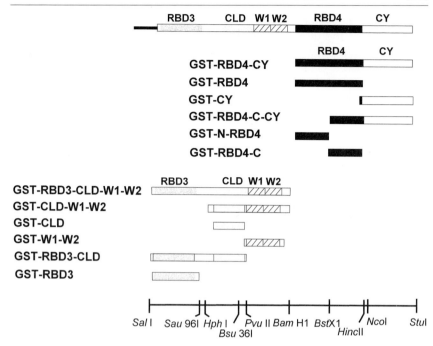

FIG. 1. Schematic representation of GST-fusion RanBP2 constructs. Constructs were derived from clone, CY15,[1] containing the C-terminal half of RanBP2 (top) and were generated as GST-fused and cleaved proteins. These constructs contained single or combination of RanBP2 structural modules, Ran-binding domain 3 (RBD3), cyclophilin-like domain (CLD), tandem modules (W1W2), Ran-binding domain 4 (RBD4), and cyclophilin (CY). This report focuses on data collected from the characterization of RBD4 and CLD modules. Restriction map of CY15 clone used for construction of GST-fused constructs.[1] The HincII site was only used for the generation of GST-CY.

Procedures

Purification of GST-Fused and Cleaved RanBP2 Protein Constructs

As shown in Fig. 1, several GST-fused protein constructs have been made containing multiple, single, or subdomains of RanBP2.[1,10,12] These constructs have been subcloned by standard cloning procedures into the bacterial expression vector, pGEX-KG,[13] as described elsewhere.[1,10,12] Plasmid constructs are electroporated into *Escherichia coli* host cells, XL1-Blue (Stratagene, La Jolla, CA). Transformants are screened for expression of GST-fused proteins by growing them overnight in 2 ml of Luria–Bertani (LB) (or 2× YT) medium followed by inoculation, in duplicate, of 2 ml of

2× YT medium with 100 µl of overnight cultures.[13,15] Cells are grown for 3–4 hr at 37° ($A_{600} \sim 1.0$). Then single cultures are induced with 0.2–1 mM isopropyl-1-thio-β-D-galactopyranoside (IPTG) (Roche Molecular Biochemicals, Indianapolis, IN). Noninduced and induced cultures are grown for an additional 3–4 hr. Aliquots of 80 µl of noninduced and induced cultures are loaded and analyzed on sodium dodecyl sulfate–polyacrylamide gel electrophoresis (SDS–PAGE)[16] after staining with Coomassie Brilliant Blue R-250 [0.05% (w/v) Brilliant Blue R-250, 25% (v/v) 2-propanol, 10% (v/v) acetic acid] and destaining [25% (v/v) 2-propanol, 10% (v/v) acetic acid]. *E. coli* transformants overexpressing GST-fusion proteins are selected for large-scale expression and purification. In general, optimal expression conditions have to be determined empirically. These may vary by adjusting parameters such as temperature of growth, strain of *E. coli* host cells (e.g., host cells that are protease deficient, overexpressing the *lac* repressor, reduced plasmid copy number, etc.) and time of growth and/or induction.

Large-scale preparation of GST-fusion proteins was carried out by growing overnight 50-ml cultures in 2× YT medium and inoculation of these into 500–1000 ml 2× YT.[13,14] Induction and growth of expressing cultures are carried out as previously described.[13] Bacterial cells are harvested, frozen at −80°, resuspended in 10 ml of PBST lysis buffer [1× PBS, pH 7.4, 5 mM EDTA, pH 7.5, 0.1% 2-mercaptoethanol, 1% Triton X-100, and cocktail of protease inhibitors (Boehringer Mannheim)], loaded and passed twice in a French press cell at 1200 lb/in.^{2}.[13] Lysates are centrifuged at 10,000g and supernatants (∼9 ml) are loaded into Bio-Rad (Hercules, CA) columns with 2 ml of preswollen and equilibrated glutathione *S*-agarose beads (Sigma, St. Louis, MO, or Amersham Pharmacia Biotech, Piscataway, NJ). These beads have an adsorption capacity of at least 8 mg/ml[14] and the amounts of expressed protein can be as high as 15 mg/liter of culture. Lysates are incubated and shaken gently for 45 min to 1 hr at 4°.[13] Eluates are discarded and beads washed five times with 15 ml (7–8 bead volumes) of PBST buffer containing 5 mM benzamidine (Sigma) and 1 mM phenylmethylsulfonyl fluoride (PMSF) (Sigma).[13]

To cleave fused protein moieties from GST, glutathione *S*-agarose beads are equilibrated once with 15 ml of thrombin cleavage buffer (50 mM Tris, pH 8.0, 150 mM NaCl, 2.5 mM CaCl$_2$, and 0.1% 2-mercaptoethanol) and incubated twice with 3–4 ml of cleavage buffer containing thrombin (Sigma) at ∼3 µg/ml for ∼30–40 min at 25° with gentle shaking.[13] Thrombin could

---

[15] J. Sambrook, E. Fritsh and T. Maniatis, "Molecular Cloning: A Laboratory Manual." Cold Spring Harbor Laboratory Press, Cold Spring Harbor, New York, 1989.
[16] U. K. Laemmli, *Nature,* **221,** 680 (1970).

be effectively inactivated by adding Pefabloc SC (Boehringer Mannheim) to eluent containing cleaved protein at a final concentration of 1–20 mM. Alternatively, thrombin can be removed by desorption onto benzamidine Sepharose 6B (Amersham Pharmacia Biotech) according to the manufacturer instructions. Finally, eluates are concentrated and the buffer exchanged twice with a storage buffer (50 mM Tris, pH 8.0, 0.1% 2-mercaptoethanol, 10% glycerol) in Centricons (Millipore, Bedford, MA) with appropriate molecular mass cutoffs. Proteins are stored at $-80°$ in aliquots to avoid freeze–thaw cycles.

Elution of GST-fused proteins is carried out by equilibration of glutathione S-agarose beads with 15 ml of elution buffer (50 mM Tris, pH 7.5, 0.1% 2-mercaptoethanol) and elution of GST-fused protein with elution buffer containing 10 mM reduced glutathione (Sigma) at $4°$.[13] Concentration of GST-fused proteins, removal of glutathione, and buffer exchange are carried out in Centricons (Millipore) as previously described. Finally, protein yields are determined according to the Bradford assay against a bovine serum albumin (BSA) standard (Sigma).[17] Purified proteins are then resolved on SDS–PAGE,[16] Coomassie Blue, and/or silver stained for potential degradation and purity analysis. Figure 2 shows some recombinant constructs (Fig. 1) at different steps of purification.

We generally regenerate glutathione S-agarose beads by stripping them once with 3–4 volumes of 6 M GnHCl (or three times with 0.5% SDS, $1 \times$ PBS, pH 7.4) followed by equilibration three times in PBST buffer. Finally, DnaK, a bacterial host protein of about 70 kDa, sometimes copurifies in small amounts with GST-fused constructs.[18] If necessary, this protein can be separated from GST-fusion proteins after purification by ion-exchange chromatography.[18]

Preparation of Retinal Extracts

Bovine retinal extracts are prepared by grinding 20 ($\sim$10 g) fresh frozen bovine retinas (Pel-Freez Biological, Rogers, AR, or local slaughter house) to fine powder on dry ice followed by homogenization (30–40 strokes) in a glass homogenizer with 30 ml of cold retinal homogenization buffer {0.75% 3-[(3-cholamidopropyl)dimethylammonio]-2-hydroxy-1-propane sulfonate] (CHAPS), 2 mM Tris-HCl, pH 6.8, 250 mM NaCl, 2 mM 2-mercaptoethanol, 0.02% NaN$_3$, 5% (v/v) glycerol and cocktail of protease inhibitors (Boehringer Mannheim)}.[10–12] Where applicable, proteasome inhibitors, lactacystine, and MG-132 (BIOMOL, Plymouth

[17] M. M. Bradford, *Anal. Biochem.* **72**, 248 (1976).
[18] Amersham Pharmacia Biotech, *Science Tools,* **2**(2), 26 (1997).

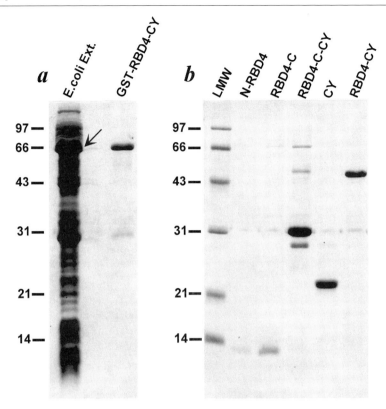

Fig. 2. Expression and purification of recombinant GST-fused and cleaved proteins. (a) Coomassie-stained SDS–PAGE analysis of crude and purified GST–RBD4–CY. First lane is a 10-μl aliquot of bacterial lysates (10 ml) obtained from a 1-liter culture expressing GST-RBD4-CY. Open arrow shows expressed protein. Second lane is an aliquot (~4 μg) of GST-RBD4-CY purified protein. (b) Coomassie-stained SDS–PAGE analysis of purified and thrombin-cleaved (free) RBD4-CY and fragments therof. Protein amounts loaded ranged from 0.5 to 6 μg. LMW, low molecular mass markers (in kDa). Nomenclature of constructs are shown in Fig. 1.

Meeting, MA), are added to the homogenization buffer at final concentrations of 20 nM and 20 μM, respectively.[12] Retinal homogenates are centrifuged at 10,000g and 4° for 20 min. Retinal supernatants are collected and precleared with 0.5 ml of swollen glutathione S-agarose beads and 500 μg of recombinant glutathione S-transferase for about 30 min at 4°.[10,11] Protein extract yields are determined according to the Bradford method (Bio-Rad, Richmond, CA).[17]

Ran-Binding Domain 4 (RBD4) Associated Specifically with 27-kDa GTP-Binding Protein

To screen for retinal proteins that specifically associate with RanBP2, we express single or combinations of well-defined domains of RanBP2 fused to GST. We begin with RBD4-CY and RBD4 domains of RanBP2. To this end, we conduct SDS–PAGE analysis of coprecipitates of pull-down analytical reactions of retinal extracts incubated with these GST-fused domains. When applicable, some of these reactions are also coincubated with 10-fold molar excess of the counterpart cleaved protein or nonhydrolyzable nucleotide analogs, respectively, to test binding specificity or nucleotide-dependent binding of retinal proteins to GST-fused constructs.[10,11] Typical analytical binding assays are done by incubating about 2 μM of GST-fused protein with 2–3 mg of CHAPS-solubilized retinal extracts containing 1 mM Pefabloc SC (Boehringer Mannheim) and, when applicable, 5 to 10-fold molar excess of unfused (cleaved) recombinant protein and/or 300 μM of nucleotide analogs, GTPγS, GDPβS, ATPγS, or ADPβS (Sigma).[10–12] Incubation reactions are carried out for about 1 hr in a nutator. Then, 60 μl of a slurry of 50% glutathione S-agarose beads in incubation buffer (50 mM Tris-HCl, pH 7.5, 100 mM NaCl, 2 mM CaCl$_2$, 2 mM MgCl$_2$, 0.5% CHAPS) is added to binding reactions and incubations are continued for another hour. Glutathione S-agarose beads are precipitated by quick centrifugation (2–3 sec at 10,000g) and washed four times with 1 ml of Cold (4°) washing buffer (50 mM Tris-HCl, pH 7.5, 100 mM NaCl, 2 mM CaCl$_2$, 2 mM MgCl$_2$, 0.2% Triton X-100). Retinal coprecipitates of binding reactions are resuspended in SDS sample buffer,[15,16] boiled for 3–5 min, and fractionated by electrophoresis in 13.5% SDS–polyacrylamide gels. Gels were run at constant current (~50 mA) on a Hoefer SE600 electrophoresis apparatus (Amersham Pharmacia Biotech).

Retinal coprecipitates are visualized by a modified silver-staining procedure of SDS–PAGE gels.[19] Gels are rinsed in deionized water, fixed overnight (>2 hr) in about 400 ml of fixing solution [50% (v/v) methanol, 10% (v/v) glacial acetic acid, 10% (v/v) glycerol] and rinsed two to three times with 500 ml of deionized water with gentle agitation for 1 hr each time. Then, gels are incubated in developing solution under agitation for about 15–20 min or until bands became visible. Developing solution is prepared by mixing (under constant stirring) equal volumes of freshly prepared solutions of 0.2% silver nitrate, 1% (w/v) tungstosilicic acid, and 0.3% (w/v) formaldehyde with 5% (v/v) disodium carbonate anhydrous. Staining reactions are stopped by transferring and emerging gels in a 5% (v/v) glacial acetic acid solution under agitation for 5–10 min followed by a rinse in

---

[19] M. Gottlieb and M. Chavko, *Anal. Biochem.* **165**, 33 (1987).

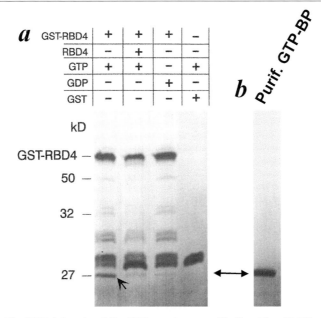

FIG. 3. The RBD4 domain of RanBP2 associates specifically with a 27-kDa protein in a GTP-dependent manner. (a) Silver-stain SDS–PAGE analysis of GST–RBD4 coprecipitates from retinal reactions. A protein with an apparent mass of 27 kDa bound to GST–RBD4 in the presence of GTPγS but not to GDPβS or free RBD4. Bands just below GST–RBD4 protein were degradation products of GST–RBD4. (b) Aliquot (5% v/v) of purified RBD4-binding protein from bovine retinas. Purif. GTP-BP, purified GTP- and RBD4-binding protein.

deionized water. If overdeveloping occurs, gels are destained with a 1× solution of Kodak Rapid Fix (Kodak, Rochester, NY).[20] Destaining is stopped with a solution of 5× hypoclearing agent (Kodak) followed by incubation in 50% (v/v) methanol for 5 min.[20] If necessary, complete destaining of the gels may be carried out by this procedure and these may be then restained with Coomassie blue.

As shown in Fig. 3a, SDS–PAGE (13.5%) analysis of analytical binding reactions provided evidence that RBD4 bound a retinal protein with an apparent molecular mass of around 27 kDa. This binding was GTP dependent because association of the 27-kDa protein to GST–RBD4 was highly increased in the presence of GTPγS. Also, this interaction was highly specific because association of the GST–RBD4 with 27-kDa GTP-binding protein in the presence of GTPγS could be completely disrupted by the presence of excess of free (unfused) RBD4. In addition, GST by itself did not associate with any retinal proteins.

[20] W. Wray, T. Boulikas, V. Wray, and R. Hancock, *Anal. Biochem.* **118,** 197 (1981).

Bov. 27-kDa Protein: DPALAAQYEHDLEVAQTTA

Hum. Ran-GTPase: D_{99}PALAAQYEHDLEVAQTTA$_{117}$

FIG. 4. Comparison of amino acid sequence between bovine 27-kDa RBD4-binding protein and human Ran-GTPase. The sequence of 19 amino acids obtained from the bovine 27-kDa protein matched 100% the counterpart sequence of human Ran-GTPase.[22]

Purification and Identification of Retinal 27-kDa RBD4- and GTP-Binding Protein

To determine the identity of the retinal 27-kDa GTP-binding protein, analytical incubation reactions have been scaled up 432-fold and carried out in several 15-ml conical tubes. Retinal coprecipitates are fractionated in 10 SDS–PAGE gels, stained overnight with 0.5% (w/v) Coomassie Brilliant Blue R-250 in 30% (v/v) 2-propanol, and destained in 30% (v/v) 2-propanol. Under these conditions it is possible to visualize faintly the 27-kDa RBD4-binding protein. The 27-kDa bands are cut and electroeluted at 150 V into Centricon-10s (Millipore) in Laemmli SDS–electrophoresis buffer[16] that was thoroughly degassed. Samples are then concentrated, pulled together, and 5% (v/v) of the total purified protein is resolved in parallel with analytical reactions in SDS–PAGE (Fig. 3b). Silver staining of these gels confirms the high purity and correct electrophoretic mobility of the electroeluted protein.

The N-terminal sequence of the purified protein provided no sequence information. Thus, we carried out chemical digest of the purified protein. Peptide fragments are isolated by high-performance liquid chromatography (HPLC) for amino acid sequence analysis. To this end, SDS is removed from electroeluted and purified protein sample with 3 × 1-ml washes of 10 mM ammonium bicarbonate in a Centricon-10 followed by lyophilization of the sample in a microcentrifuge tube. Then, the sample is subjected to CNBr digestion in 70% formic acid for 18 hr at room temperature and in the dark as described by Stone et al.[21] Reagents are removed with N_2 and digested peptides applied to a C_{18} HPLC column and fractionated in a 0.1% trifluoroacetic acid/acetonitrile buffer. From the few peaks obtained (not shown), one fraction yielded 19 cycles of amino acid sequence. Comparision of this amino acid sequence with other proteins in the database showed that it was 100% identical to the human Ran-GTPase protein (Fig. 4).[22]

---

[21] K. Stone, M. Lopresti, J. Crawford, R. DiAngelis, and K. Williams, "A Practical Guide to Protein and Peptide Purification for Microsequencing," pp. 31–47. Academic Press, San Diego, 1989.

[22] G. Drivas, A. Shih, E. Coutavas, M. Rush, and P. D'Eustachio, *Mol. Cell. Biol.* **10,** 1793 (1990).

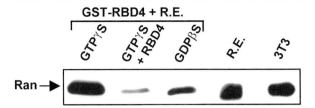

FIG. 5. Western blot analysis of GST–RBD4 coprecipitates of retinal reactions with a polyclonal antibody against Ran-GTPase. Binding of Ran to RBD4 was highly increased, decreased and abolished in the presence of GTPγS, GDPβS, and competitor protein (unfused RBD4), respectively. Ran was also present in 3T3 fibroblasts. R.E., Bovine retinal extracts; 3T3, SDS-solubilized homogenates of 3T3 fibroblasts.

To confirm that the purified bovine GTP-binding protein is Ran, we performed Western blot analysis with a polyclonal antibody against Ran-GTPase (gift from P. D'Eustachio) on coprecipitates of analytical binding reactions, and aliquots of CHAPS-solubilized retinal extracts and homogenates of 3T3 fibroblasts solubilized in SDS sample buffer. To this end, samples are resolved on 13.5% SDS–PAGE and blotted onto a polyvinylidene difluoride (Immobilon-P) membrane on a Genie electrophoretic blotter (Idea Scientific, Corvallis, OR) at 14 V for 2 hr in a transfer buffer (25 mM Tris, 190 mM glycine, 15% methanol).[23] Blots are blocked overnight at 4° in 1× blocking solution [10% glycerol, 1× PBS, pH 7.4, 0.5% nonfat dry milk (BioRad)][10,11] with gentle shaking. Blocked membranes are incubated with anti-Ran antibody at 1:1400 dilution in 1:5 diluted blocking solution for 2 hr at room temperature. Blots are washed three times for 10 min in a washing buffer (1× PBS, pH 7.4, 0.05% Tween 20).[10,11] Then, blots are incubated with goat anti-rabbit IgG (Kirkegaard & Perry Laboratories, Inc., Gaithersburg, MD) at 25 ng/ml in diluted blocking buffer for 1 hr, washed twice with washing buffer, and developed for 1 min in a chemiluminescent substrate (Kirkegaard & Perry Laboratories, Inc.).[10] As shown in Fig. 5, GST-RBD4 specifically associated with Ran-GTPase in a GTP-dependent fashion. This association could be virtually abolished in the presence of free RBD4. Ran was also present both in retinal extracts and homogenates of 3T3 fibroblasts.

Characterization of CLD Domain of RanBP2

We have extended structure–function studies on another RanBP2 domain, the cyclophilin-like domain (CLD) (Fig. 1).[1,12] This domain exhibited

[23] H. Towbin, T. Staehelin, and J. Gordon, *Proc. Natl. Acad. Sci. U.S.A.* **76**, 4350 (1979).

very low homology to cyclophilin proteins[1] and there was no clue of its putative function. We screened retinal extracts for proteins that associated specifically with CLD by the same methods previously described.

A series of GST-fusion constructs is prepared containing CLD by itself or combined with one or more of its flanking domains, RBD3 and W1W2 (Fig. 1).[12] Incubation of GST–CLD (2.2 μM) with retinal extracts (2–3 mg) leads to the association of CLD to a retinal protein with a molecular mass of 112 kDa (p112) that is visible on silver-stained gels (not shown).[12] This association can be disrupted by coincubation with excess of unfused CLD but not with other unrelated constructs.[12] Thus, the interaction is highly specific. Purification and sequence analysis of 112-kDa protein by methods already described identify this protein as the largest 112-kDa subunit (not shown)[24,25] of the 19S regulatory complex of the proteasome,[26] the major nonlysosomal proteolytic machine in the cell. To confirm the identity of p112 and investigate whether CLD mediates the interaction of RanBP2 with p112 and other components of the 19S regulatory complex in a tissue-specific fashion, we incubate GST–CLD with retinal and other tissue extracts prepared in the presence of proteasome inhibitors. Concentrations of tissue extracts are normalized to that of the retinal extracts. Western blot analysis of GST–CLD coprecipitates is carried out with an antibody (1:2500) against the whole purified 19S regulatory particle of the proteasome (gift from George DeMartino) and blots are developed for 5 min in the presence of a more sensitive chemiluminescent substrate, SuperSignal (Pierce, Rockford, IL).[12]

As shown in Fig. 6, a polyclonal antibody against the 19S regulatory cap of the proteasome, recognizes several non- and ATPase components of this particle across different mammalian tissues. Interestingly, there was also some variation in composition and level of expression of these components among different tissues tested (Fig. 6b). In retinal extracts, p112 and p58 associated with the CLD domain with very high affinity and, in contrast to our previous report,[18] this interaction was significantly enhanced by the presence of ATPγS. Interaction of these components with CLD could be disrupted by addition of unfused CLD to incubation reactions. Finally, association of p112 and p58 with CLD occurred in the retina, but surprisingly, not in other tissues such as kidney, liver, and spleen.

---

[24] G. DeMartino, C. Moomaw, O. Zagnitko, R. Proske, M. Chu-Ping, S. Afendis, J. Swaffield, and C. Slaughter, *J. Biol. Chem.* **269,** 20878 (1994).

[25] K. Yokota, S. Kagawa, Y. Shimizu, H. Akioka, C. Tsurumi, C. Noda, M. Fujimuro, H. Yokosawa, T. Fujiwara, E. Takahashi, M. Ohba, M. Yamasaki, G. DeMartino, C. Slaughter, A. Toh-e, and K. Tanaka, *Mol. Biol. Cell* **7,** 853 (1996).

[26] W. Baumeister, J. Walz, F. Zuhl, and E. Seemuller, *Cell* **92,** 367 (1998).

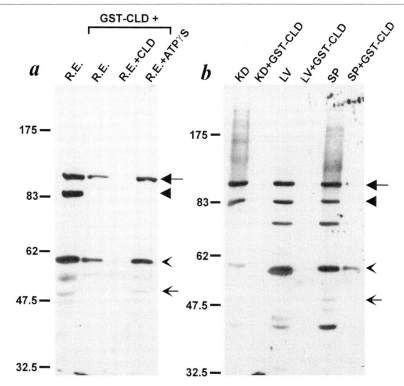

Fig. 6. The CLD domain mediates the interaction of RanBP2 with subunits of the 19S regulatory complex of the proteasome in a tissue-specific fashion. Western blot analysis of GST–CLD coprecipitates from (a) retinal and (b) other tissue extracts with a polyclonal antibody against PA700 (19S complex). PA700 antibody detected subunits of the 19S complex in different tissues. In particular, it recognized well the non-ATPase subunits, p112 (arrow), p97 (arrowhead) and p58 (open arrowhead). Also, some lower ATPase subunits, such as p48 (open arrow), could be detected. Expression profile of some of these subunits varied among different tissues. In retina, CLD associated very strongly with p112 and p58 (a, lane 2) and this interaction was significantly increased in the presence of ATP (a, lane 4). The interaction was highly specific because it was abolished in the presence of free CLD (a, lane 3). Moreover, CLD interaction with 19S regulatory subunits of the proteasome occurred in the retina but not kidney, liver, and spleen extracts (b). LV, Liver; KD, kidney; SP, spleen; prestained molecular weight markers (New England Biolabs) on the left side of each panel. (Note that the 83-kDa prestained marker has an abnormal mobility as it coruns with the 97-kDa regular marker.)

Conclusion

Identification of retinal components interacting specifically with distinct domains of RanBP2 has advanced our understanding of the molecular macroassembly architecture of RanBP2 and its role in integrating protein machines with key roles in protein biogenesis and kinesis. The screening of retinal proteins against different proteins or domains of interest by modified affinity-purification methods here presented has proved to be a powerful tool for identifying functional domains in RanBP2 and molecular partners interacting with these domains. This screening method is a valuable tool to identify and characterize specific protein–protein interactions and thus, unravel protein networks in photoreceptors and other cells.

Acknowledgments

I thank W. L. Pak (Purdue University), G. H. Travis (U.T.S.M.C.) for their support throughout the initial characterization of RanBP2. I thank A. Mahrenholz at Purdue University for excellent assistance in protein sequence analysis of Ran. I also thank D. Schick for excellent technical assistance, Y. Cai for help in preparation of some retinal binding reactions, J. Albanesi for 3T3 cells, P. D'Eustachio and G. DeMartino for anti-Ran/TC4 and anti-PA700 antibodies, respectively. This work was supported in part by Karl Kirchgessner Foundation, Fight for Sight, Inc., NATO/JNICT (Portugal) fellowship and grant, MCW institutional funds and NIH Grant EY11993 to P.A.F.

Section IV

Transducin and Regulators of G-Protein Signaling

[32] Intrinsic Biophysical Monitors of Transducin Activation: Fluorescence, UV–Visible Spectroscopy, Light Scattering, and Evanescent Field Techniques

By OLIVER P. ERNST, CHRISTOPH BIERI, HORST VOGEL, and KLAUS PETER HOFMANN

Introduction

To fully elucidate the mechanism of catalytic interaction of G proteins with their receptors, information on the temporal order, the kinetics, and the energetics of each of the reaction intermediates is required. Intrinsic monitors, such as light scattering, absorption, and fluorescence, exploit endogenous properties of proteins, and therefore offer the advantage of leaving the system under investigation undisturbed. Such assays currently available for G proteins include the well-known intrinsic fluorescence changes of a tryptophan residue located near the G protein's active center (Trp-207 in transducin). In addition, the visual system offers UV–Vis spectrophotometric assays in which the retinal chromophore itself serves as an intrinsic reporter group. These assays include monitoring the transducin sensitive formation of the active metarhodopsin-II (MII) conformation, as well as the photoregeneration of dark state rhodopsin from MII. A third technique is kinetic light scattering, which makes use of the unique activation-dependent solubility of the rod G protein (see also the chapter by Heck *et al.* in this volume[1]).

While these monitors are highly specialized, more generally applicable—although, with respect to the preparations required, more demanding—monitors have been developed in recent years and successfully tested with the visual system. This family of techniques is based on the evanescent field in the vicinity of a reflecting optical surface, and includes surface plasmon resonance and resonant mirror spectroscopy. This chapter describes applications and limitations of these techniques (Table I).

In Vitro Reaction Scheme

Intrinsic biophysical reporters allow one to monitor individual steps of the interaction between active rhodopsin (R*) and transducin (G_t), which is based on the following reaction sequence:

[1] M. Heck, A. Pulvermüller, and K. P. Hofmann, *Methods Enzymol.* **315** [22], 1999 (this volume).

TABLE I
INTRINSIC BIOPHYSICAL MONITORS OF G_t INTERACTION[a]

Method	Sample	Time resolution	Monitor assignment	Linearity with [R*]/fidelity
Fluorescence	Purified R, WMs, liposomes	Seconds	$G\alpha \cdot GTP$ (Trp-207 in $G_{t\alpha}$)	Low
UV–vis (extra MII)	WMs, liposomes	Seconds (low temperature)	MI/MII shift (deprotonated RSB)	High
UV–vis (photoregeneration)	Purified R	Measures $R^* \cdot G_t$ at flash	Photoregeneration of ground state (reprotonated RSB)	Low
Light scattering	ROS, WMs, liposomes	Real time (ms)	Gain or loss of mass	WMs: high Vesicles: low[c]
SPR	Supported membranes[b]	Seconds	Gain or loss of mass, lateral resolution in membranes	Low[c]
Resonant mirror	WMs[b]	Seconds	Gain or loss of mass	Low[c]

[a] R, Rhodopsin; R*, photoactivated rhodopsin; G_t, transducin; RSB, retinal–Schiff base; ROS, rod outer segments; WMs, washed ROS membranes, ROS devoid of peripheral and soluble proteins.
[b] Experiments with solubilized rhodopsin conceivable.
[c] Due to nondefined orientation of the receptor in the bilayer.

$$G_{tsol} \cdot GDP \rightleftharpoons G_{tmb} \cdot GDP \qquad (1)$$
$$G_{tmb} \cdot GDP + R^* \rightleftharpoons R^* \cdot G_{t[empty]} + GDP \qquad (2)$$
$$R^* \cdot G_{t[empty]} + GTP \rightarrow [R^* \cdot G_t \cdot GTP] \rightarrow R^* + G_{tsol} \cdot GTP \qquad (3)$$
$$G_{tsol} \cdot GTP \rightarrow G_{tsol} \cdot GDP + P_i \qquad (4)$$

Step 1 describes the equilibrium of the inactive GDP-bound G_t, which is distributed between a soluble ($G_{tsol} \cdot GDP$) and a membrane-bound fraction ($G_{tmb} \cdot GDP$) (see Ref. 2). In the micromolar range, binding of G_t to the membrane occurs within a few seconds with an apparent K_D of 10^{-5}–10^{-6} M.[3] Similar rates are measured in detergent solution where the rate of $R^* \cdot G_t$ formation may depend on the transition of G protein from detergent micelles to the solubilized rhodopsins[3] (see section on fluorescence).

Step 2 describes the formation of the nucleotide-free complex between R* and G_t ($R^* \cdot G_{t[empty]}$). Liberation of GDP by R^*–G_t association can be shown directly by measuring release of ^{32}P-labeled GDP.[4]

[2] P. A. Liebman, K. R. Parker, and E. A. Dratz, *Annu. Rev. Physiol.* **49,** 765 (1987).
[3] A. Schleicher and K. P. Hofmann, *J. Membr. Biol.* **95,** 271 (1987).
[4] O. P. Ernst, K. P. Hofmann, and T. P. Sakmar, *J. Biol. Chem.* **270,** 10580 (1995).

Step 3 describes GTP-induced dissociation of the $R^* \cdot G_t$ complex and formation of transducin in its active GTP-bound form. Under conditions *in vitro*, most of the transducin (mainly the α subunit) leaves the membrane ($G_{tsol} \cdot GTP$). Weak binding of GTP to the $R^* \cdot G_{t[empty]}$ complex induces a conformational change in G_t leading to higher affinity for the nucleotide. In this model[5] nucleotide and R^* mutually displace each other. In the presence of its effector, cGMP phosphodiesterase (PDE), G_t remains membrane bound due to its association with the effector. (For monitoring G_t–PDE interaction by light scattering see the contribution by Heck *et al.*[1] in this volume.)

Step 4 describes the hydrolysis of GTP due to the GTPase activity of G_t, which takes minutes at room temperature, but less than 1 sec in rod outer segment (ROS) preparations.[6]

Intrinsic $G_{t\alpha}$ Fluorescence

This assay is based on the intrinsic fluorescence change of $G\alpha$ subunits arising from the conformational change accompanying the formation of activated $G\alpha$, $G\alpha \cdot GTP$, or $G\alpha \cdot GDP \cdot AlF_4$, respectively (see Refs. 7 and 8 and references therein). In the case of transducin, the fluorescence sensor is Trp-207 in the $G_{t\alpha}$ subunit.[9]

Figure 1A shows the fluorescence increase of G_t due to addition of GTPγS, a nonhydrolyzable analog of GTP, to a mixture of photoactivated affinity-purified rhodopsin and G_t. Evaluation of the time course of fluorescence increase can be used to compare different R^* samples (e.g., rhodopsin mutants or complexes of opsin and all-*trans*-retinal (see Sachs *et al.*, this volume[10]) in their efficiency in G_t activation. In the case of substoichiometric addition of GTPγS, the relative change of fluorescence intensity (difference between plateau values, indicated by double arrows) allows convenient determination of the pool of activatable $G_{t\alpha}$ subunits in the sample. By titration with known amounts of GTPγS the concentration of activatable $G_{t\alpha}$ subunits of transducin preparations can be determined precisely.[11] In Fig. 1B transient fluorescence increase arising from $G_t \cdot GTP$ formation is seen which fades away due to the intrinsic GTPase activity of transducin.

[5] M. Kahlert and K. P. Hofmann, *Biophys. J.* **59**, 375 (1991).
[6] T. M. Vuong and M. Chabre, *Proc. Natl. Acad. Sci. U.S.A.* **88**, 9813 (1991).
[7] R. A. Cerione, *Methods Enzymol.* **237**, 409 (1994).
[8] T. Higashijima and K. M. Ferguson, *Methods Enzymol.* **195**, 321 (1991).
[9] E. Faurobert, A. Otto-Bruc, P. Chardin, and M. Chabre, *EMBO J.* **12**, 4191 (1993).
[10] K. Sachs, D. Maretzki, and K. P. Hofmann, *Methods Enzymol.* **315** [17], 1999 (this volume).
[11] K. Fahmy and T. P. Sakmar, *Biochemistry* **32**, 7229 (1993).

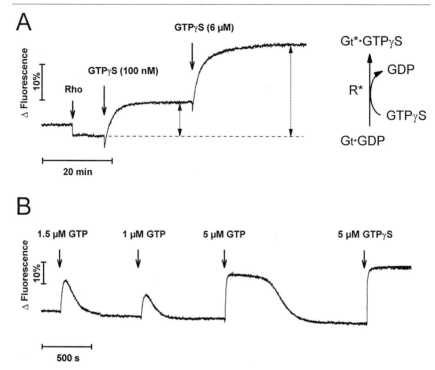

FIG. 1. Monitoring formation of $G_{t\alpha} \cdot GTP$ as a fluorescence increase after GTP or GTPγS uptake. Addition of 50 μl affinity-purified recombinant rhodopsin[4] (20 nM final concentration) to a stirred, continuously illuminated (543 nm, He/Ne Laser, Melles Griot) sample at 10° containing 1.5 ml 200 nM transducin in 10 mM Tris-HCl (pH 7.4), 100 mM NaCl, 2 mM $MgCl_2$, and 0.01% (w/v) dodecylmaltoside causes an initial decrease of fluorescence due to dilution of the sample. Subsequent additions of GTPγS (a nonhydrolyzable GTP analog; final concentrations indicated) lead to an increase of fluorescence intensity (excitation: 300 nm; emission: 345 nm) arising from $G_t \cdot GTP\gamma S$ formation (A). Titration with GTPγS can be used to determine the amount of activatable $G_{t\alpha}$ subunits. With GTP, the fluorescence increase is transient, reflecting G_t-catalyzed GTP hydrolysis (B). The measurement (excitation: 300 nm; emission: 340 nm) was performed at 20° with a sample (1.5 ml) containing 0.5 μM opsin (prepared from ROS),[10] 0.5 μM all-*trans*-retinal, 1 μM G_t, 50 mM bis-tris propane (BTP) (pH 7.5), 130 NaCl, 1 mM $MgCl_2$. For these measurements a SPEX fluorolog 2 (Instruments S.A., Inc., Edison, NJ) with two double monochromators was used.

By addition of higher concentrations of GTP the $G_t \cdot GTP$ complex accumulates, which is observed as a longer steady-state phase.

An optimum for excitation at 300 nm was determined from the difference of the fluorescence excitation spectra of the activated $G_{t\alpha}$ subunit ($G_{t\alpha} \cdot GDP \cdot AlF_4$), measured after addition of 10 μM $AlCl_3$, 5 mM NaF, and 100 μM GTP, versus the inactive subunit ($G_{t\alpha} \cdot GDP$; 1 μM). At that wavelength the shielding of excitation light due to absorption by added GTP or GTPγS nucleotides (λ_{max} 252 nm) is low. Fluorescence emission of $G_{t\alpha}$ in its inactive and active forms can be monitored most efficiently around 340–345 nm.

Despite other tryptophan residues (a total of two in the $G_{t\alpha}$ subunit, eight in the $G_{t\beta\gamma}$ complex, and five in rhodopsin) that increase background fluorescence, we obtained readouts of G_t activation even with high concentrations of rhodopsin (respective washed membranes) and G_t (e.g., 10 μM rhodopsin, 1 μM G_t). However, in these samples, the superimposed decay of R* into opsin and all-*trans*-retinal, seen as an exponential increase of fluorescence intensity (see Ref. 12; excitation: 295 nm; emission maximum: 330 nm) becomes visible and accounts for a major part of the baseline drift.

Receptor catalyzed nucleotide exchange in G_t can be followed fluorimetrically either on reconstituted samples containing G_t and disk membranes, rhodopsin-containing vesicles, or solubilized rhodopsin. When using these preparations, we must take into account that binding of G_t to the membrane (step 1 in the reaction sequence outlined earlier) can become the rate-limiting step. This can be seen dramatically with opsin embedded in disk membranes that is subjected to partial regeneration with 11-*cis*-retinal to rhodopsin. Assaying this photoactivated preparation for G_t activation shows that increasing amounts of retinal lead to a linear increase of the G_t activation rate up to a retinal : opsin ratio of 0.03, above which the membrane binding step becomes rate limiting.[13] When working with native disk membranes, with their high rhodopsin density, it is crucial to take into account this intrinsic limitation of the assay. An example is the quantitative determination of the activity of the opsin · all-*trans*-retinal complex, as compared to light-induced metarhodopsin-II.[10] Only under conditions where this membrane-binding step does not cause kinetic artifacts is the reaction limited by R*. For membrane preparations, this is achieved by either partial regeneration of opsins embedded in native disk membranes or by reconstitution of solubilized rhodopsin into vesicles at low molar ratios of rhodopsin : lipid.

When comparing the efficiency of solubilized rhodopsin mutants in activating G_t in detergent solution similar precautions have to be taken. Usually low rhodopsin concentrations (1–5 nM) and low detergent concentrations [0.01% (w/v) dodecylmaltoside (DM)] around the critical micellar concentration are used to avoid that the reaction is limited by the efficiency of collision of G_t with micelles containing R* rather than the actual formation of the R* · G_t complex. Doubling the DM concentration from 0.01 to

---

[12] D. L. Farrens and H. G. Khorana, *J. Biol. Chem.* **270**, 5073 (1995).
[13] S. Jäger, K. Palczewski, and K. P. Hofmann, *Biochemistry* **35**, 2901 (1996).

0.02% (w/v) already decreases the G_t activation rate fourfold in a filter-binding assay, which monitors the light-dependent guanine-nucleotide exchange by G_t.[14] It is necessary to be aware of these limitations because a moderate loss of receptor activity, e.g., by mutation, may escape this assay when performed under non-R*-limiting conditions.

UV–Vis Spectroscopy

Besides revealing different photointermediates, UV–vis spectroscopy can be applied for monitoring interaction of transducin with photoactivated rhodopsin in its interactive conformation. Two techniques, *extra MII* and *photoregeneration from the signaling state of the receptor,* are described next.

Extra MII

The extra MII effect arises from the ability of transducin and other rhodopsin-interacting proteins to stabilize the 380-nm absorbing meta intermediate of rhodopsin (MII) at the expense of the tautomeric forms MI (478 nm) and MIII (470 nm).[15,16] Among these other proteins are arrestin[17] and its splice variant p44,[18] and transducin-derived peptides.[19] Interestingly rhodopsin kinase does not show this extra MII effect, suggesting that rhodopsin kinase interacts with all meta intermediates of rhodopsin.[20] Formation of extra MII can either be measured by static UV–vis spectroscopy or in real time as an enhanced absorption change, measured at 380 nm after photolysis of rhodopsin with green flashes. The examples in Fig. 2 show such time-resolved measurements.

The amount of MII formed in excess of the normal, spontaneously formed MII is a stoichiometric measure of the complexes formed with the interacting transducin or arrestin.[16–18] In the case of G_t, it is the nucleotide-free $R^* \cdot G_{t[empty]}$ complex (no exogenous GDP added), which is reflected in the enhanced MII absorption. Presence of GTP causes dissociation of the $R^* \cdot G_{t[empty]}$ complex (step 3 in the reaction sequence given earlier), which is expressed in transient formation of extra MII due to decay of the (380-nm absorbing) $R^* \cdot G_{t[empty]}$ complex and fast return of the freed R^*

---

[14] M. Han, M. Groesbeek, S. O. Smith, and T. P. Sakmar, *Biochemistry* **37,** 538 (1998).
[15] D. Emeis and K. P. Hofmann, *FEBS Lett.* **136,** 201 (1981).
[16] D. Emeis, H. Kühn, J. Reichert, and K. P. Hofmann, *FEBS Lett.* **143,** 29 (1982).
[17] A. Schleicher, H. Kühn, and K. P. Hofmann, *Biochemistry* **28,** 1770 (1989).
[18] A. Pulvermüller, D. Maretzki, M. Rudnicka-Nawrot, W. C. Smith, K. Palczewski, and K. P. Hofmann, *Biochemistry* **36,** 9253 (1997).
[19] O. G. Kisselev, C. K. Meyer, M. Heck, O. P. Ernst, and K. P. Hofmann, *Proc. Natl. Acad. Sci. U.S.A.* **96,** 4898 (1999).
[20] A. Pulvermüller, K. Palczewski, and K. P. Hofmann, *Biochemistry* **32,** 14082 (1993).

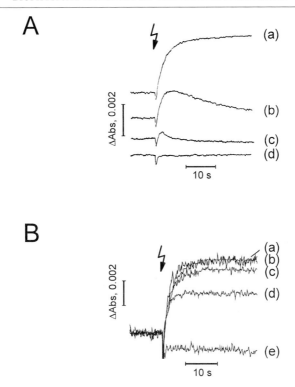

FIG. 2. Monitoring R* · G_t complex formation by time-resolved formation of extra MII by dual-wavelength UV–vis spectroscopy. Under the conditions of the experiment (4°, pH 7.5–8) flash photolysis of rhodopsin (10 μM) predominantly produces the MI photoproduct (λ_{max} 478 nm), which is in a dynamic equilibrium with a small proportion of MII (λ_{max} 380 nm, isosbestic point for MI/MII at λ = 417 nm). Traces represent readings of a 380- to 417-nm absorbance difference (ΔAbs). Heterotrimeric G_t or G_t mimetic peptides shift the MI $\rightleftharpoons$ MII equilibrium toward MII in a concentration-dependent manner and produce extra MII. (A) Flash-induced extra MII measured on samples containing 3 μM rhodopsin (WMs) and 0.4 μM G_t in the presence of increasing concentrations of GTP: (a) 0 μM, (b) 1.5 μM, (c) 7 μM, or (d) 100 μM, in 10 mM PIPES (pH 7.5), 130 mM KCl, 0.5 mM MgCl$_2$, 1 mM CaCl$_2$, 0.5 mM EDTA at 4°; the mole fraction of flash-activated rhodopsin (R*/R) was 4%; 1-cm cuvette path length. (B) Extra MII produced by 10 μM rhodopsin (WMs) and (a) 1 μM G_t, or (b) 1 mM of G_t mimetic peptides: high-affinity analog of the $G_{t\alpha}$ (340–350) sequence VLEDLKSCGLF, (c) $G_{t\gamma}$ (50–71) farnesylated at Cys-71, (d) $G_{t\alpha}$ (340–350), or (e) no peptide. Measurements were done in 100 mM HEPES (pH 7.9), 50 mM NaCl, 1 mM DTT, 1 mM MgCl$_2$, 1 mM EDTA at 1.5°. Cuvette path length was 2 mm. Rhodopsin (12%) was flash activated by 500 ± 20 nm light.

into the MI/MII equilibrium [Fig. 2A, traces (b) and (c)].[21] High concentrations of GTP do not allow any measurable accumulation of this complex, which is reflected in a trace comparable to the control measurement without G_t [Fig. 2A, trace (d)]. At low GTP concentrations, however, the dissociation of the complex is slow enough to make the decay of extra MII a real-time monitor of the $R^* \cdot G_t$ dissociation/G_t activation kinetics.[21,22] Figure 2B shows the effect of peptides derived from transducin's C-terminal sequence of the α and γ subunits.

Mandatory for measuring extra MII are conditions under which the MI $\rightleftharpoons$ MII equilibrium is not on the MII side. This is the case for native rhodopsin in disk membranes at low temperatures (typically 0–4°, measurable up to ca. 14°) and pH 7.5–8. Extra MII on solubilized rhodopsin cannot be performed because commonly used detergents like octylglucoside and dodecylmaltoside shift the equilibrium completely toward MII. This detergent effect can also be seen if small amounts of these detergents are added to the native lipid host. The ability to measure extra MII with solubilized rhodopsin samples can be restored in so-called "doped micelles."[23] However, a much more reliable and stable sample is obtained after removal of the detergent and reconstitution of rhodopsin into lipid bilayers composed of, e.g., egg lecithin or retina lipids[24] (see also the section on light scattering below).

Time-resolved extra MII, evoked by light flashes can be measured with modified dual-wavelength spectrometers (in our laboratory we use a Shimadzu UV300/UV3000 or home-built spectrometer[25]), which offer low preactivation by the measuring light. The sample is illuminated by a flash of green light (photoflash, equipped with a 500 ± 20 nm broadband interference filter, and a blue-green filter (BG 24, 2 mm, Schott, Mainz, Germany) in the light path of the measuring beam, to protect the photomultiplier tube (see Ref. 21 for details of the apparatus). To suppress light scattering artifacts caused by the membranous suspensions, cuvettes with short optical path (2 mm) are used, positioned next to the cathode of the photomultiplier tube, and the absorption difference $A_{380} - A_{417}$ is recorded (417 nm denotes the isosbestic point of the MI/MII equilibrium). The presence of noninteracting proteins does not disturb this assay.

[21] B. Kohl and K. P. Hofmann, *Biophys. J.* **52,** 271 (1987), and references therein.
[22] K. P. Hofmann, *Biochim. Biophys. Acta* **810,** 278 (1985).
[23] B. König, W. Welte, and K. P. Hofmann, *FEBS Lett.* **257,** 163 (1989).
[24] W. J. DeGrip, J. VanOostrum, and P. H. M. Bovee-Geurts, *Biochem. J.* **330,** 667 (1998).
[25] K. P. Hofmann, "Handbook of Experimental Pharmacology," Vol. 108/II, "GTPases in Biology II," p. 267. Springer-Verlag, Berlin, 1993.

Photoregeneration from MII State

This assay makes use of the fact that rhodopsin can be regenerated from the deprotonated meta intermediate MII directly by light (see Ref. 26 and references therein). The photoconversion of MII's all-*trans*-retinal yields ca. 50% 11-*cis*-retinal (and to a small degree 9-*cis*-retinal). This is followed by reprotonation of the Schiff base with a thermal transition of the protein moiety (50-ms kinetics), and an accompanying absorption shift (λ_{max} approximately 500 nm), the opsin shift characteristic for rhodopsin. The other half of MII's all-*trans*-retinal is photoconverted with fast kinetics (1 ms) into one or several 470-nm absorbing species, which may possibly be identified with 7-*cis* isoforms. The large absorption shift indicates that both photoconverted species are likely to contain protonated Schiff bases.

Interestingly, the slow phase, which reflects the formation of the rhodopsin ground state, can be blocked by G_t[26] or by G_t-derived peptides.[19] The complex $R^* \cdot G_{t[empty]}$ is photolyzed in the presence of G_t (in equilibrium with its endogenous GDP). Figure 3 shows that the slow component (regeneration of rhodopsin) is supressed with increasing amounts of G_t. If formation of $R^* \cdot G_{t[empty]}$ complex is prevented by the presence of additional GDP or GTP, G_t does not exert its effect on photoregeneration. Photoregeneration is therefore an assay to probe (1) the total amount of MII (spontaneously formed and extra MII) present in a photolyzed rhodopsin sample and thus (2) the capability of a protein or peptide to stabilize MII.

In practice, after quantitative photoconversion of rhodopsin to MII by continuous illumination with green light (1 min), the sample is subjected to a blue flash and the formation of the rhodopsin ground state (photoregeneration) is monitored at 500 or 530 nm. At the latter wavelength, the contribution of the fast 470-nm photoproduct to the signal is reduced. The slow component of the record (rhodopsin regeneration, λ_{max} approximately 500 nm) is then evaluated. For instrumental details, see Ref. 26.

Near-Infrared Light Scattering

Flashes of green light applied to rod outer segments evoke changes in the light scattering intensity of near-infrared light. These changes are demonstrated to be stoichiometric to the formation and dissociation of the rhodopsin–transducin complex (see Ref. 25 and references therein).

In this assay the measuring light (near-infrared, 840 nm) is scattered by rhodopsin embedded in membranous particles, i.e., disk membranes or

---

[26] S. Arnis and K. P. Hofmann, *Biochemistry* **34**, 9333 (1995).

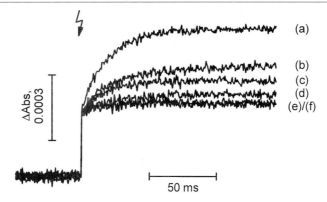

Fig. 3. Probing the complex between G_t or G_t mimetic peptides and R* by photolysis of MII. Rhodopsin in its active MII state (λ_{max} 380 nm) with an intact, deprotonated retinal Schiff base bond can be photolyzed by blue light resulting in the biphasic formation of photoproducts with protonated Schiff bases expressed in the opsin shift (a). The slow component reflects the photoregeneration of rhodopsin (λ_{max} 500 nm), whereas the fast arising product is characterized by a blue-shifted absorption maximum (λ_{max} 470 nm; see Ref. 26). Photoregeneration of rhodopsin (but not the formation of the 470-nm species) can be suppressed gradually by increasing amounts of G_t. Sample contained 1 μM rhodopsin and (a) 0, (b) 1.2, (c) 1.6, (d) 2.0, (e) 3.0, or (f) 4.0 μM G_t; 50 mM MES (pH 6), 130 mM NaCl, and 0.4 mM DM, at 18°. Traces are the sum of absorption changes from 20 flashes (412 ± 7 nm), recorded at 500 nm due to a small mole fraction of MII photolysis per flash (0.8%). Path length of the cuvette was 7 mm.

lipid vesicles. A description of theory and applications to proteins other than G_t is given in Ref. 1. The changes of light scattering intensity which are attributed to R*–G_t interaction are made possible by the distribution of G_t between the membranous surface and the aqueous phase (see step 1 in the reaction sequence given earlier).

In the so-called *binding signal* [Fig. 4, trace (a), increase of the scattering intensity] additional soluble G_t binds as a consequence of flash-induced R* formation to the membranes, thereby increasing the mass of the scattering particles. The binding signal does not show the real-time binding of soluble G_t to R* but reflects the redistribution of G_t between the soluble and membrane-bound state.[3] In the case of disk membrane preparations the binding signal has to be corrected by a reference signal (N-signal, rhodopsin signal) without G_t, which arises from the change of polarizability of the disk membranes, and not yet fully understood effects, linked to the formation of MII (see Ref. 1). This signal does not occur in the corresponding G_t-free measurement with rhodopsin reconstituted into lipid vesicles [Fig. 4, trace (c)]. In Fig. 4, trace (b), a competing reaction takes place: hydroxylamine

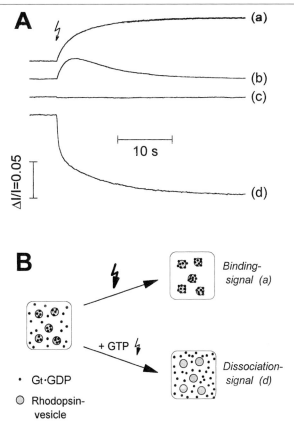

FIG. 4. Real-time monitoring of R*–G_t formation by near-infrared light scattering. Shown is the time course of normalized light scattering intensity originated from rhodopsin-containing vesicles (diameter 100–200 nm) measured at 16 ± 2° (details see text). (A) Flash activation of a sample containing 700 nM recombinant rhodopsin and (a) 4 μM G_t, (b) sample as in (a) with additional 100 mM NH$_2$OH, (c) control without G_t. The mole fraction of flash-activated rhodopsin (R*/R) was 55%. In trace (d) the flash activated only catalytic amounts of rhodopsin (R*/R = 0.2%) of a sample as in (a) with additional 1 mM GTP. Measuring conditions: 20°, 10 mM BTP (pH 7), 130 mM NaCl, 1 mM MgCl$_2$. I represents the intensity of the sample before the flash corrected for the scattering contribution of the cuvette with buffer alone. Specific optic and electronic devices are described in Ref. 1. (B) illustrates light-induced mass changes of the scattering membranous particles causing the binding and dissociation signal, respectively (details see text).

destroys R*, once formed, and removes it from the equilibrium by forming retinaloxime and opsin.

Figure 4, trace (d), shows a flash-induced decrease of light scattering intensity of a sample containing recombinant rhodopsin embedded in liposomes, G_t and GTP. This decrease is due to the activation of G_t by catalytic amounts of R* and subsequent dissociation from the membrane (step 3 in the reaction sequence above). This *dissociation signal* monitors the fate of G_t, which is bound to the membrane before the flash (step 1). Because the diffusion of active G_t from the membrane is very fast and the activation of G_t, which is moving up according to step 1 of the reaction sequence, is just canceled with the G_t lost by activation in the steady-state reaction, the dissociation signal measures in real time with <5-ms delay[27] the dissociation of the R* · G_t complex.

When using this technique for studying G_t interaction with mutant rhodopsins, the pigment has to be reconstituted into lipid vesicles.[4] For reconstitution of rhodopsin into lipid bilayers, techniques such as dilution,[28] dialysis, and detergent extraction[7,24] are described among others. From our experience, the following microdialysis method[4] works quite well for small amounts of rhodopsin. A mixture of solubilized rhodopsin (1.5–2 μM), purified by immunoaffinity chromatography in 10 mM BTP (pH 7.0), 130 mM NaCl, 1 mM MgCl$_2$, 1.5% (w/v) octylglucoside,[4] and a 5 mg/ml phosphatidylcholine (PC) suspension in deionized water (PC from fresh egg yolk was obtained from Fluka, Buchs, Switzerland) in a molar protein to lipid ratio (1 : 350) is dialyzed over 24 hr at 4–8° against 2 liters of detergent-free buffer in a Microdialyzer 500 system (Pierce, Rockford, IL; molecular weight cutoff 8000).

Evanescent Field Techniques

Surface Plasmon Resonance

Surface plasmon resonance (SPR) is an evanescent wave technique that is sensitive to the optical properties in the vicinity of metal surfaces.[29] The resonant angle can be measured as a dip in a reflection versus angle of incidence scan. Adsorption of organic material such as proteins or lipids from an aqueous buffer to the metal surface (i.e., SPR sensor chip) changes the resonant angle by altering the mean refractive index in the volume probed by the evanescent wave. Using Fresnel equations, the average area per adsorbed molecule can be calculated from the resonance angle shift. Commercially available instruments based on this technique (e.g., BIAcore, Uppsala, Sweden) are extensively used to study interactions of molecules immobilized to the chip with analytes in the volume.[30] More advanced SPR

---

[27] M. Heck and K. P. Hofmann, *Biochemistry* **32**, 8220 (1993).
[28] M. L. Jackson and B. J. Litman, *Biochim. Biophys. Acta* **812**, 369 (1985).
[29] E. Burstein, W. P. Chen, W. J. Chen, and A. Hartstein, *J. Vac. Sci. Technol.* **11**, 1004 (1974).
[30] M. Fivash, E. M. Towler, and R. J. Fisher, *Curr. Opin. Biotechnol.* **9**, 97 (1998).

instrumentation allows for time-resolved and laterally resolved SPR, which can be used to monitor different regions of patterned sensor chips simultaneously.[31] The lateral resolution of SPR is in the micrometer range and depends on the properties of the sensor chip and the wavelength of the light used for excitation.[32]

Surface Plasmon Resonance and Self-Assembled Monolayers. Self-assembled monolayers (SAMs) are used to modify the chemical properties of the metallic SPR sensor surfaces. SAMs form spontaneously when gold is exposed to solutions containing millimolar concentrations of linear alkyl chains with a terminal thiol group.[33] The ω-terminal methyl groups of the alkyl chains can be substituted by functional groups, which produce SAMs with defined chemical properties, e.g., hydrophilic [using $HS-(CH_2)_{11}-OH$] or hydrophobic [using $HS-(CH_2)_{11}-CH_3$].

Supported Lipid Bilayers Containing Rhodopsin. Supported lipid bilayers (SLBs) are artificial membranes formed on solid supports.[34,35] They "float" on a thin (ca. 1-nm) water film,[36] therefore transmembrane proteins can be incorporated that extend on both sides of the membrane. SLBs are produced by spreading lipid vesicles on oxide surfaces,[35] by hydration of lipid films deposited on the surface,[37] or by micellar dilution.[38] In the last case, a solution containing a detergent/lipid mixture is slowly diluted below the critical micellar concentration (cmc) of the detergent, leading to spontaneous formation of lipid bilayers on hydrophilic supports, or lipid monolayers on hydrophobic supports, respectively.

SPR was first used to follow the incorporation of detergent-solubilized rhodopsin into lipid bilayers formed by hydration and reported to measure conformational changes on rhodopsin photolysis.[37] In later experiments, the binding of transducin to this rhodopsin/lipid film was measured.[39] Alternatively, rhodopsin is incorporated into SLBs by micellar dilution (see section on Patterns of Rhodopsin Membranes).[40] Rhodopsin is solubilized in micromolar concentrations in 50 mM octylglucoside supplemented with

[31] W. Hickel and W. Knoll, *Thin Solid Films* **199,** 367 (1991).
[32] B. Rothenhäusler, J. Rabe, P. Korpiun, and W. Knoll, *Surf. Sci.* **137,** 373 (1984).
[33] C. D. Bain and G. M. Whitesides, *Angew. Chem.* **101,** 522 (1989).
[34] S. Heyse, T. Stora, E. Schmid, J. H. Lakey, and H. Vogel, *Biochim. Biophys. Acta* **1376,** 319 (1998).
[35] E. Sackmann, *Science* **271,** 43 (1996).
[36] S. J. Johnson, T. M. Bayerl, D. C. McDermott, G. W. Adam, A. R. Rennie, R. K. Thomas, and E. Sackmann, *Biophys. J.* **59,** 289 (1991).
[37] Z. Salamon, Y. Wang, M. F. Brown, H. A. Macleod, and G. Tollin, *Biochemistry* **33,** 13706 (1994).
[38] H. Lang, C. Duschl, and H. Vogel, *Langmuir* **10,** 197 (1994).
[39] Z. Salamon, Y. Wang, J. L. Soulages, M. F. Brown, and G. Tollin, *Biophys. J.* **71,** 283 (1996).
[40] S. Heyse, O. P. Ernst, Z. Dienes, K. P. Hofmann, and H. Vogel, *Biochemistry* **37,** 507 (1998).

egg phosphatidylcholine (molar protein to lipid ratio 1 : 100). Octylglucoside is especially suited for micellar dilution because it is uncharged and has a very high cmc (20–25 mM), which facilitates bilayer formation. Upon washing with buffer lacking detergent and under constant stirring, a mixed supported lipid/rhodopsin membrane forms that contains high amounts of rhodopsin. The reconstituted membrane can be used to monitor G_t–membrane interactions by SPR simultaneously in regions of low or high rhodopsin content (see Fig. 5). Membrane binding of solubilized G_t (step 1 in the reaction scheme) leads to a resonance angle shift. Kinetics and concentration dependency of this process can be followed readily. Transducin binds to the SLB membrane with an (apparent) affinity higher but comparable to that of native disk membranes.[40]

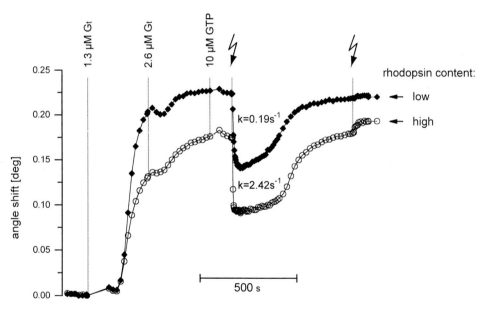

FIG. 5. SPR sensorgram of rhodopsin–transducin interaction on a patterned supported lipid bilayer containing rhodopsin formed by micellar dilution.[40] The formation of regions with low or high rhodopsin content is shown in Fig. 6. G_t was added in two subsequent identical doses to final concentrations of 1.3 and 2.6 μM, respectively. Addition of GTP (10 μM) does not change the resonance angle, whereas a light flash (λ = 530 nm) results in a fast desorption signal. The fast component of double-exponential fits reveals a 10-fold higher rhodopsin activity in the region with a high rhodopsin content (lipid bilayer) as in the region with low rhodopsin content (tethered lipid bilayer). Upon depletion of the GTP pool, the angle returns almost to the starting value. A further flash induces formation of R* · $G_{t[empty]}$ complexes, as evidenced by the slight resonance angle increment (region with high rhodopsin content).

In the presence of GTP, a light flash that activates a catalytic fraction of the immobilized rhodopsin leads to a fast desorption signal, comparable to the dissociation signal in the light scattering assay (see earlier comments), indicating the desorption of transducin from the membrane, followed by readsorption of transducin when the GTP pool is used up. Consistently, a subsequent light flash leads to the formation of R* · G_t complexes (Fig. 5, region with high rhodopsin content), as indicated by a slight increase of the resonance angle, analoguous to the binding signal in the light scattering assay [Fig. 4A, trace (a)]. Kinetic analysis of the parameters of this step revealed a catalytic turnover of 2.6 $\mu M_{GTP}^{-1} s^{-1}$ of the immobilized rhodopsin.[40] The rate of G_t desorption is significantly higher in the region with high rhodopsin content, whereas the return to the baseline due to GTP hydrolysis is, as a R* independent process, the same in both regions.

Patterns of Rhodopsin Membranes. As mentioned, laterally resolved SPR can be employed to monitor different regions of patterned sensors simultaneously.[31] This has the potential to investigate lateral processes, such as the diffusion of membrane-bound G protein. Patterning the SLB with respect to its rhodopsin content allows one to simultaneously investigate the interaction between transducin and, in the ideal case, membranes containing or lacking rhodopsin.[40]

Patterning of supported membranes can be achieved by patterning the underlying SAM. Two different possible approaches are shown in Fig. 6. In recent work,[40] a lithographic approach[41] was used (see Fig. 6, left-hand side): First, a hydrophilic SAM is created by immersing the gold surface of the sensor chip with millimolar concentrations of ω-carboxyundecanoic acid (CTA) in ethanol. Second, after rinsing, an electron microscopy grid mask is placed as a lithography mask on the chip, and the sample is illuminated for 1 hr with UV light. In the unprotected regions, the alkanethiols are oxidized to alkane sulfonates, which subsequently are washed away. Third, the resulting bare gold regions are filled by immersing the chip in a 0.1 mM solution of thiolipids in water/50 mM octylglucoside. Thiolipids[38] are synthetic glycerophospholipids connected on their headgroup to a flexible, hydrophilic linker with a terminal thiol group. They assemble on gold to form monolayers much like thioalkanes, mimicking one layer of a lipid bilayer. This results in a patterned SAM consisting of two regions: a hydrophilic, negatively charged region, and a hydrophobic region forming one leaflet of a tethered lipid bilayer.

On micellar dilution, a SLB forms on the hydrophilic regions of the SAM, whereas in the thiolipid regions, a fluid monolayer is deposited on the immobilized thiolipids. If rhodopsin is present during micellar dilution,

---

[41] J. Huang, D. A. Dahlgren, and J. C. Hemminger, *Langmuir* **10**, 626 (1994).

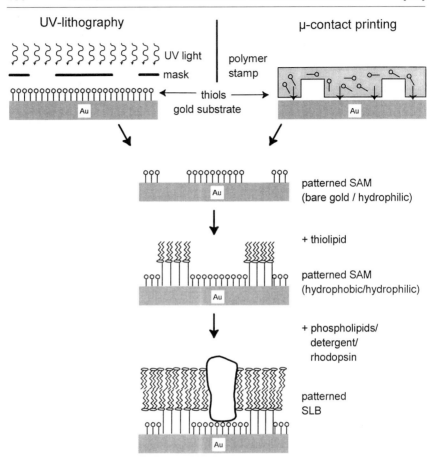

FIG. 6. Two possible methods to pattern SAMs. Left panel: A first SAM formed on the sensor chip is patterned by UV illumination through an appropriate mask, e.g., an electromicroscopy grid. The resulting sulfonates are washed away, and the bare gold regions are filled with the second thiol, e.g., a thiolipid. Right panel: A polymer stamp soaked with an ethanolic solution of the thiols forming the first SAM is brought in contact with the gold surface of the sensor chip. Self-assembly occurs only where the stamp touches the surface. The remaining regions are filled with a second thiol. If a pattern consisting of a hydrophilic SAM and a SAM formed by thiolipids is made, micellar dilution leads to a patterned lipid layer consisting of two regions: a "standard" SLB where both layers are fluid, and a tethered lipid bilayer, where the lower leaflet is covalently bound to the support. Transmembrane proteins will preferentially integrate into the first region.

it is preferentially incorporated into SLBs on hydrophilic regions, because the monolayers do not offer enough space. For transducin, however, both regions are identical with respect to the lipid layers, allowing direct comparison of G_t interaction with membranes containing high and low contents of rhodopsin. The lithographic approach resulted in a tenfold enrichment of rhodopsin in the bilayer regions of the supported membrane. Further enhancement of this contrast seems unlikely because of the unspecific adsorption of rhodopsin to the sensor chip.

Other approaches to pattern SAM have been developed.[42,43] An intriguing method is μcontact printing (see Fig. 6, right-hand side). This technique uses stamps that are produced by polymerization of dimethoxysiloxane on an appropriate silicon master. To produce the patterned SAM, the relief of the stamp is first wetted for some minutes with a millimolar ethanolic solution of an alkanethiol. The liquid film on the relief is then removed by an air stream, and the stamp is put on the sensor chip. Where contact occurs between the stamp and the substrate, a SAM is formed, whereas the other regions stay bare and can be filled by immersion of the chip in an appropriate self-assembly solution. With this method, structures in the micrometer range can be produced repeatedly and without microalignment.[44]

Resonant Mirror Spectroscopy

The resonant mirror technique,[45] as used in a commercially available instrument (IAsys, Affinity Sensors, Cambridge, United Kingdom), makes use of a thin monomode waveguide that is highly sensitive to changes in the surface refractive index. These changes, e.g., adsorption or desorption of proteins from the aqueous environment, are expressed in the changes of the propagation angle θ_{guide}. At the "resonance" angle waveguiding and propagation of the light along the surface of the device are maximal. In practice, 670-nm light from a laser diode is coupled via prism coupling into the waveguide, which forms the bottom surface of a stirred 200-μl cuvette, and reflected light from the inside of the resonant mirror is measured as a function of incident angle. In a resonant mirror sensorgram, the angle where the light propagates maximally is plotted versus time.

---

[42] A. Kumar, N. L. Abbott, E. Kim, H. A. Biebuyck, and G. M. Whitesides, *Acc. Chem. Res.* **28,** 219 (1995).

[43] Y. Xia and G. M. Whitesides, *Angewandte Chemie Int. Ed.* **37,** 550 (1998).

[44] C. Bieri, O. P. Ernst, S. Heyse, K. P. Hofmann, and H. Vogel, *Nature Biotechnol.*, in press (1999).

[45] R. Cush, J. M. Cronin, W. J. Stewart, C. H. Maule, J. Molloy, and N. J. Goddard, *Biosensors Bioelectron.* **8,** 347 (1993).

This technique gives sensorgrams similar to the SPR technique. In the example given in Fig. 7, rhodopsin in its native host membrane was immobilized. Affinity purified anti-biotin antibody (Vector Laboratories, Burlingame, CA, stabilizing bovine serum albumin was removed by gel filtration) was first coupled to the aminosilane surface of the sensor cuvette by using the homobifunctional reagent bis(sulfosuccinimidyl) suberate (Pierce, Rockford, IL) according to the manufacturer's protocol (not shown in Fig. 7). The resonance angle obtained by this cuvette filled with buffer is set to 0 arc-sec. For immobilization of washed disk membranes, biotinyl lipid (18:1 Biotinyl-PE, Avanti Polar Lipids, Alabaster, AL) was incorporated into the lipid bilayer by ultrasonication at a lipid:biotinyl lipid molar ratio of 100:1. Replacing the buffer in the stirred, antibody-coated cuvette by the membrane suspension leads to an increase of the resonance angle due

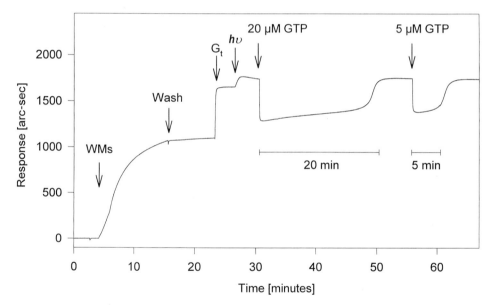

FIG. 7. Sensorgram of rhodopsin–transducin interaction monitored by the resonant mirror technique (IAsys device). Biotinyl lipid was incorporated into washed membranes (WMs) by ultrasonication at a lipid:biotinyl lipid ratio of 100:1. This membrane suspension (5 μM rhodopsin) in 10 mM BTP (pH 7.0), 130 mM NaCl, 1 mM MgCl$_2$, 1 mM DTT was added to the cuvette as indicated. Subsequent capture of the membranes via the biotin tag by an anti-biotin antibody coupled to the sensor surface was stopped by replacing the membrane suspension with 200 μl buffer (Wash). Binding of added transducin (50 μl, 1.75 μM final concentration) can be seen as an increase of the resonance angle, which is enhanced after illumination by light (>495 nm). GTP-induced desorption of G$_t$ from the immobilized membrane (decrease of the signal) and readsorption after GTP hydrolysis occurs after addition of 20 and 5 μM GTP (final concentrations) as indicated.

to immobilization of the membranes via biotin and the cognate antibody (Fig. 7). Before addition of G_t to the cuvette the suspension with unbound membranes was removed by aspiration and replaced with buffer. After binding of G_t to the immobilized membranes in the dark (step 1 in the reaction scheme given earlier) additional G_t is adsorbed due to light-induced formation of $R^* \cdot G_t$ (step 2, reaction scheme) as expressed in an increasing signal. Addition of GTP causes desorption of G_t due to GTP uptake (step 3, reaction scheme) and subsequent readsorption after GTP hydrolysis (step 1, reaction scheme). As seen, under the conditions of the experiment, the time for hydrolysis of the whole pool of bound G_t (length of the plateau) is linearly related to the amount of GTP added.

The drawback of some commercial sensing devices is that information about the chemical specification of the sensor chip is limited and that control measurements often have to be obtained from individually prepared samples. However, in an approach like the SPR method described here, the lateral resolution allows one to simultaneously monitor a reference area that faces the same sample during the whole experiment and therefore allows identification and correction of nonspecific interactions, which in some cases contribute a major part of the signal.

The long-term goal is to combine surface patterning with uniformly oriented incorporation of rhodopsin into native lipid bilayers, thereby yielding a sample region (with rhodopsin) next to an identical reference region (without rhodopsin). We have recently shown that such a system based on SPR can provide a prototypical sensor chip that may be equipped with a variety of ligand-activated receptors.[44]

[33] Fluorescent Probes as Indicators of Conformation Changes in Transducin on Activation

By CHII-SHEN YANG, NIKOLAI P. SKIBA, MARIA R. MAZZONI, TARITA O. THOMAS, and HEIDI E. HAMM

Introduction

Heterotrimeric guanine nucleotide binding proteins (G proteins) are composed of Gα and G$\beta\gamma$ subunits. During the signaling cycle, Gα controls the binding affinity of its partner proteins by undergoing local conformational changes in regions around the nucleotide binding pocket depending on whether GTP or GDP is bound. The regions that undergo conformational change on activation are defined as switch I, switch II, and switch III, and have been demonstrated by comparison of the crystal structures of Gα_t and Gα_{i1} in their active and inactive states.[1-4] Comparison of the crystal structures of the Gα_t and Gα_{i1} subunits in the activated versus inactivated states has greatly advanced our knowledge of the molecular principles of nucleotide binding, hydrolysis, and the nature of the conformational changes that define the activation state of Gα. However, structurally flexible regions, which often play an important role in protein function, are disordered in the crystal structure and remain undefined. For example, neither the C-terminal region of Gα_t, known to be important for receptor interaction,[5] nor the α2 helix of Gα_{i1}–GDP, known to be involved in effector and $\beta\gamma$ interaction, were visible in their crystal structures.[1-3] Thus, other biochemical and biophysical methods are required to define the structural features of these regions.

In this chapter, we have applied an approach utilizing specific fluorescent probes as reporters, to monitor the conformational changes of Gα_t. The method is based on targeted fluorescent labeling of Gα_t mutants at single surface-exposed cysteine residues. Gα_t has only two cysteine residues accessible for labeling with thiol-specific fluorescent reagents, located at positions

---

[1] J. P. Noel, H. E., Hamm, and P. B. Sigler, *Nature* **366,** 654 (1993).
[2] D. G. Lambright, J. P. Noel, H. E. Hamm, and P. B. Sigler, *Nature* **369,** 621 (1994).
[3] M. B. Mixon, E. Lee, D. E. Coleman, A. M. Berghuis, A. G. Gilman, and S. R. Sprang, *Science* **270,** 954 (1995).
[4] D. E. Coleman, A. M. Berghuis, E. Lee, M. E. Linder, A. G. Gilman, and S. R. Sprang, *Science* **269,** 1405 (1994).
[5] H. E. Hamm, D. Deretic, A. Arendt, P. A. Hargrave, B. Koenig, and K. P. Hofmann, *Science* **241,** 832 (1988).

210 and 347. The same cysteine residues are exposed in the $G\alpha_t/G\alpha_{i1}$ chimera (Chi6; for more details see Skiba et al.[6]), a functional derivative of $G\alpha_t$. In this chimeric protein region 216–294 of $G\alpha_t$ is replaced with the corresponding residues from $G\alpha_{i1}$. By substituting Cys-210 with Ser, or Cys-347 with Ser, we have generated $G\alpha$ subunits, which can be labeled at a single residue, Cys-347 (Chi6a) or C-210 (Chi6b), respectively.[7] After Cys-347 (Chi6a) or Cys-210 (Chi6b) have been specifically labeled with the fluorescent group Lucifer Yellow vinyl sulfone (LY) we can monitor the conformational changes that occur at the switch II and C-terminal regions of $G\alpha_t$ on its activation with aluminum fluoride (AlF_4^-).

Construction of Recombinant $G\alpha_t$ Variants

$G\alpha_t$ expressed in *Escherichia coli* is not functional, because it cannot assemble properly,[8] whereas $G\alpha_{i1}$ is fully functional. A strategy was found to "rescue" recombinant $G\alpha_t$ by replacement of specific regions that perturb proper folding with the corresponding regions of $G\alpha_{i1}$.[8] Using this approach a number of soluble and functional $G\alpha_t/G\alpha_{i1}$ chimeras have been generated.[6] One of these chimeras, Chi6, was chosen for targeted fluorescent labeling due to (1) its similarity to $G\alpha_t$ in nucleotide, receptor, and $G\beta\gamma$ binding properties,[8] and (2) the crystal structure of Chi6, in complex with $G\beta_1\gamma_1$ has been solved[9] and it exhibits a structure similar to $G\alpha_t$.

The construction of the mutants used for these studies are based on Chi6. Cys210Ser-Chi6 (Chi6a), Cys210Ser-Cys347Ser-Chi6 (Chi6ab), Ala3Cys-Chi6ab, and Val301Cys-Chi6ab were constructed using the QuikChange site-directed mutagenesis kit (Strategene, La Jolla, CA). The oligonucleotides for PCR (polymerase chain reaction) were designed as suggested by the manufacturer with two complementary oligonucleotides containing the desired mutation in the central region. Following PCR and *Dpn*I enzyme treatment for 1 hr at 37°, plasmid DNA was transformed into *E. coli* strainBL21 (DE3) (Novagen, Madison, WI). Recombinant proteins were expressed and purified as described elsewhere.[6]

Fluorescent Probe Labeling

The primary fluorescence probe used in this study was LY (Sigma, St. Louis, MO). We have also used 6-acryloyl-2-dimethylaminonaphthalene

[6] N. P. Skiba, T. O. Thomas, and H. E. Hamm, *Methods Enzymol.* **315** [34], 1999 (this volume).
[7] C. Yang, N. Skiba, M. Mazzoni, and H. Hamm, *J. Biol. Chem.* **274**, 2379 (1999).
[8] N. P. Skiba, H. Bae, and H. E. Hamm, *J. Biol. Chem.* **271**, 413 (1996).
[9] D. Lambright, J. Sondek, A. Bohm, N. Skiba, H. Hamm, and P. Sigler, *Nature* **379**, 311 (1996).

(acrylodan) and 5-(2-iodoacetyl amino)ethylaminonaphthalene-1-sulfonic acid (IAEDANS) from Molecular Probes Inc. (Eugene, OR) as thiol-specific fluorescent reagents of different size and charge. However, as discussed later, only LY resolved species with different labeling stoichiometry, as well as species with the same stoichiometry of labeling that differ only at positions of labeling.

Figure 1 shows the molecular structure and chemical reaction of LY with cysteine. Labeling of Gα with LY is performed immediately after samples are purified using Mono Q ion exchange high-performance liquid chromatography (HPLC). A stock solution of LY (10 mM) is prepared by dissolving the compound in methanol. The labeling reaction of protein (0.8–1.2 mg/ml) and fluorescent probe proceeds at a molar ratio of 1:5, Gα protein:LY for 30 min on ice. The reaction is stopped by addition of 4 mM 2-mercaptoethanol. Overnight dialysis using 50 mM Tris-HCl, pH 8.0, 50 mM NaCl, 5 mM MgCl$_2$ (buffer A) containing 25 μM GDP, 5 mM 2-mercaptoethanol and 0.1 mM phenylmethylsulfonyl fluoride (PMSF) removes excess, unreacted fluorescent probe. The labeled sample is further purified by HPLC using a 2-ml Protein Pak Q15 HR column (Millipore, Bedford, MA).

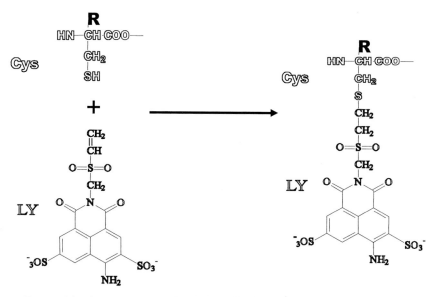

FIG. 1. Molecular structure and chemical modification of Lucifer Yellow to cysteine. LY is a polysulfonate with high polarity and water solubility. Chemical reduction of the disulfide present in the cysteine residue allows conjugation with the fluorescent label. Residues conjugated with LY have two additional negative charges that are part of the adduct, allowing charge differentiation of modified residues from unmodified residues.

Resolution of Labeled Species

HPLC purification on an anion-exchange resin (Millipore) equilibrated with buffer A, allows resolution of differentially labeled fluorescent proteins. The presence of LY adds negative charges to the protein, thus allowing separation by anion-exchange chromatography. After labeling, the reaction mixes were applied to a Waters HPLC AP-1 column (Waters Chromatography Division, Millipore) packed with Protein-Pak Q 15 HR (Millipore). Proteins were eluted from the column with buffer A using a gradient concentration of NaCl (from 0.1 to 1 M).

Figure 2 shows a typical elution profile of Chi6 labeled with LY protein. The first peak is unmodified protein, the second peak corresponds to Chi6 with Cys-210 modified by LY, the third peak is Chi6, Cys-347 modified by LY, and the fourth peak is Chi6 with both Cys-210 and Cys-347 modified by LY.

Determination of Stoichiometry of Labeled Proteins

To determine the stoichiometry, samples were collected from the various elution peaks and scanned from 260 to 550 nm with a 8452A Diode Array Spectrophotometer (Hewlett Packard, Palo Alto, CA). The readings of absorbance at 280 and 430 nm (for LY) were recorded. The stoichiometry (n) of LY labeling is calculated as a ratio of the concentration of LY and the concentration of the labeled protein in the HPLC eluted samples.[7] The relationship applied is:

$$n = C_{LY}/C_{Chi6-LY}$$

where C_{LY} is the concentration of LY and $C_{Chi6-LY}$ is the concentration of labeled Chi6 protein (or other labeled Gα chimeras). The concentration of LY (C_{LY}) is determined using the equation:

$$C_{LY}(M) = A_{430,LY}/\varepsilon_{430,LY}$$

where $A_{430,LY}$ is the absorbance of LY at 430 nm and $\varepsilon_{430,LY}$ is the molar extinction coefficient of LY at 430 nm, which is 12,400.[10] Because LY also absorbs at 280 nm, the concentration of total labeled protein ($C_{Chi6-LY}$) is determined by subtraction of LY absorbance at 280 nm from the total absorbance at 280 nm using the relationship:

$$C_{Chi6-LY} = [A_{280,Chi6-LY} - (A_{430,LY}\varepsilon_{280,LY}/\varepsilon_{430,LY})]/\varepsilon_{280,Chi6}$$

---

[10] W. Stewart, *Nature* **292**, 17 (1981).

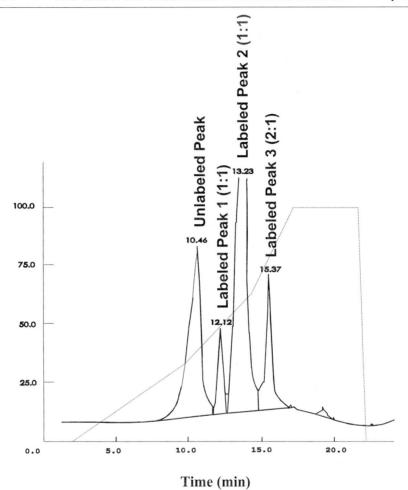

Fig. 2. HPLC purification of LY-labeled Chi6. Labeled Gα (Chi6) samples were purified by HPLC using a Protein-Pak Q 15 column. Unlabeled protein has the shortest retention time (10.46 min). Labeled peak 1 represents a 1:1 labeling stoichiometry of LY to protein with a retention time of 12.12 min. Labeled peak 2 represents a 1:1 labeling stoichiometry of LY to protein with a retention time of 13.23 min. Labeled peak 3 represents a 2:1 labeling stoichiometry of LY to protein with a retention time of 15.37 min. The y axis shows the salt concentration at which each protein with or without adduct was eluted.

where $A_{280,\text{Chi6}}$ is the absorbance of Chi6-LY at 280 nm and $\varepsilon_{280,\text{Chi6}}$ is the molar extinction coefficient of Chi6 at 280 nm, which is 43,100. The value of $\varepsilon_{280,\text{Chi6}}$ is calculated by applying the relationship:

$$\varepsilon_{280,\text{Chi6}} = 2\varepsilon_{280,\text{Trp}} + 13\varepsilon_{280,\text{Tyr}} + \varepsilon_{280,\text{GDP}}$$

where $\varepsilon_{280,Trp}$ is the molar extinction coefficient of Trp at 280 nm (5300), $\varepsilon_{280,Tyr}$ is the molar extinction coefficient of Tyr at 280 nm (1400), and the molar extinct coefficient of GDP at 280 nm, $\varepsilon_{280,GDP}$, is 9000. The relationship used to determine $\varepsilon_{280,Chi6}$ is based on the Trp, Tyr, and GDP content in Chi6 (Chi6 has 3 Trp, 13 Tyr, and 1 GDP). For the concentration of Chi6, the following relationship is used:

$$C_{Chi6}(M) = A_{280,Chi6}/\varepsilon_{280,Chi6}$$

Following stoichiometry determination all purified labeled proteins underwent overnight dialysis in buffer A with 50 μM GDP, 0.1 mM PMSF, and 5 mM 2-mercaptoethanol in 40% (v/v) glycerol and were stored at $-80°$.

Proteolytic Analysis of Labeled Peaks

Limited typtic digestion was performed to characterize the labeled sites in each of the samples collected from HPLC elution. The potential tryptic cleavage sites in the Chi6 protein[11,12] are Lys-18, Arg-204, and Arg-310, as illustrated at the bottom of Fig. 3. Protein samples were treated with trypsin (in buffer A) using a weight ratio of 16/6 (protein/trypsin) for 0, 5, 15, and 30 min before the stopping the reactions with N-α-p-tosyl-L-lysinechloromethyl ketone (TLCK).[11,12] The LY-labeled fragments are resolved on SDS–PAGE and observed by UV illumination. As observed in the upper left panel of Fig. 3, the double labeled peak 3 from Chi6-LY has fluorescent fragments of 38, 19, 15, and 5 kDa, suggesting the two labeled cysteines are within the region composed of amino acids 205–350, which contain the Cys-210 and Cys-347 residues. Further evidence of the specific labeling is demonstrated in the upper right panel of Fig. 3, as Chi6a, the protein on which Cys-347 alone is labeled, has labeled fragments of 38, 19, and 5 kDa after treatment with trypsin.

Functional Activity of Labeled Proteins

To check that LY-labeled Gα chimeras were functional, measurement of AlF$_4^-$-dependent conformational change was performed. In the presence of GDP and Mg^{2+}, AlF$_4^-$ mimics the γ-phosphate of Gα–GTP, thus, Gα–GDP–AlF$_4^-$ is considered the activated conformation. The AlF$_4^-$-dependent conformational change tests whether GDP is present in the guanine nucleotide binding pocket of Gα, as well as determining the ability of the protein to undergo an AlF$_4^-$-dependent activating conformational

---

[11] B. Fung and C. Nash, *J. Biol. Chem.* **258**, 10503 (1983).
[12] M. Mazzoni, J. Malinski, and H. Hamm, *J. Biol. Chem.* **266**, 14072 (1991).

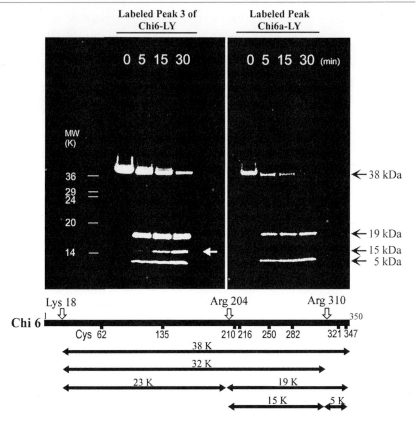

FIG. 3. Proteolytic analysis of labeled peak of Chi6 and Chi6a-LY. Labeled peak 3 of Chi6-LY (upper left panel) and labeled peak of Chi6a-LY (upper right panel) were treated with trypsin as described in the text. Each gel contains protein samples treated with trypsin for the indicated amount of time: 0, 5, 15, or 30 min. The reactions were stopped by addition of TLCK. UV illumination was used to visualize the LY-labeled fragments on the gel. The white arrow on the left panel identifies a 15-kDa labeled fragment that is not present in that of Chi6a-LY. Below the gel is a linear representation of Chi 6 in which the possible trypsin cleavage sites are identified by hollow arrows above Lys-18, Arg-204, and Arg-310, as well as the various fragments produced by cleavage at these sites.

change.[2,13,14] Tryptophan fluorescence (excitation at 280 nm and emission at 340 nm) changes were measured before and after addition of AlF_4^- (simultaneous addition of 10 mM NaF and 50 μM $AlCl_3$ to the test sample). A functionally intact Gα produces an approximately 40% increase in trypto-

[13] T. Higashijima, K. Ferguson, M. Smigel, and A. Gilman, *J. Biol. Chem.* **262,** 757 (1987).
[14] W. Phillips and R. Cerione, *J. Biol. Chem.* **263,** 15498 (1988).

phan fluorescence. This same procedure was applied to the kinetic measurements of fluorescence change in either conformational change or intermolecular interaction studies. All the fluorescent experiments are performed on an AMINCO Bowman Series 2 Luminescence Spectrometer running with OS/2 Warp 4.0.

Lucifer Yellow, Fluorescent Reporter of Conformational Change in Switch II Region

Cys-210 is located in the switch II region. We determined whether there was an AlF_4-dependent conformational change at Cys-210 using the C347S-Chi6 protein (Chi6b) labeled with LY. The fluorescence of Cys-210–LY (Chi6b) increased by 220% in response to AlF_4^- (Fig. 4), indicating that LY at Cys-210 is a sensitive reporter of conformational changes that occur in the switch II region on activation. This conformational change can be reversed by EDTA (data not shown), which chelates free Al^{3+} and causes

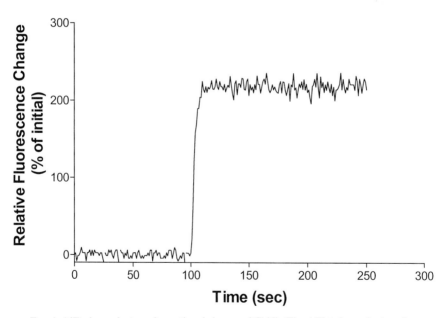

FIG. 4. AlF_4-dependent conformational change of Chi6b. The AlF_4^--dependent conformational change at Cys-210 was assessed in Chi6b-LY. 10 mM NaF and 30 μM $AlCl_3$ were added simultaneously at time 100 sec. The fluorescence increase on Gα activation with AlF_4^- is expressed as a percent of the initial fluorescence using the equation $F(\%) = (F-F_0)/F_0 \times 100$, where F_0 is the initial fluorescence and F is the fluorescence after addition of AlF_4^-. The initial fluorescence of Chi6b-LY is shown as 0%. A 220% increase in fluorescence was observed with the addition of AlF_4^- at 100 sec.

a complete dissociation of AlF_4^- from the nucleotide binding pocket, thus reversing LY fluorescence to its background state.

Lucifer Yellow, Fluorescent Reporter of Conformational Change in C Terminus of $G\alpha$

The C terminus of $G\alpha_t$ is not ordered in the crystal structure of the molecule.[1,2] To probe conformational changes that might occur in this region of $G\alpha_t$ in the different functional states, we targeted LY labeling to the Cys-347 residue using the C210S–Chi6 (Chi6a) protein. The functional and structural integrity of Chi6a–LY was confirmed by measuring its AlF_4^--dependent increase in intrinsic tryptophan fluorescence (Fig. 5, dashed

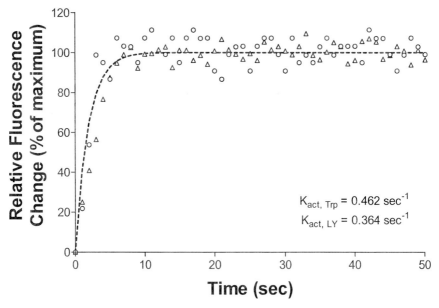

FIG. 5. AlF_4^--dependent fluorescence change monitored at 340 nm (Trp) and 520 nm (LY). The Trp fluorescence change (dashed line) of Chi6a-LY (Cys-347–LY) during activation by AlF_4^- (10 mM NaF and 50 μM $AlCl_3$) produces a K_{act} = 0.462. The LY fluorescence change (solid line) of Chi6a-LY (Cys-347–LY) during activation by AlF_4^- (10 mM NaF and 30 μM $AlCl_3$) produces a K_{act} = 0.0.364. The fluorescence is expressed as a percentage of maximum fluorescence change on addition of AlF_4^-. The best fits obtained for both data sets by using equation $F_t = F_0 + F_1(1 - e^{-K_{act}t})$ are shown where F_t is fluorescent emission at time t, F_0 is the initial fluorescence before addition of AlF_4^-, F_1 is maximal change in fluorescence, e is an exponent, and K_{act} is the rate constant for protein activation. Initial fluorescence was set to 0 and maximal fluorescence change is expressed as 100%. Both fluorescence change in LY and Trp can be reversed by addition of 10 mM EDTA (data not shown).

line). A substantial increase in the fluorescence of Chi6a-LY (Fig. 5, solid line; approximately 90% from the initial fluorescence) was detected in the presence of the activator, AlF_4^-. This observation indicates that on activation the conformation of the C terminus of $G\alpha$ undergoes a significant change. The rates of the fluorescence changes of Trp and LY were very similar (K_{act} = 0.462 and 0.364, respectively, Fig. 5), suggesting that on activation the C terminus of $G\alpha$ undergoes a fast conformational change that occurs after the slow conformational change in the switch II region. Alternatively, the change in conformation of the switch II region could result in a shorter distance between the switch II region and the C terminus of the molecular, thus leading to an increase of LY fluorescence.

Acrylodan at Cys^{-3} of $G\alpha_t$ Reporting Activation-Dependent Conformational Changes in N Terminus

Substitution of Ala-3 for Cys in the context of the Chi6ab chimera (Ala-3 Cys-Chi6ab) is an example of how this method can be used to create a unique labeling site for a fluorescent group in a region of choice. Since in the Chi6ab chimera both Cys-210 and Cys-347 were replaced with Ser, a cysteine introduced at position 3 is the only accessible residue for thiol-specific labeling reagents. Labeling of Ala3Cys-Chi6ab with LY did not result in any change in fluorescence on protein activation. However, when Ala3Cys-Chi6ab is fluorescently labeled with another fluorescence group, acrylodan, it showed a decrease of nearly 20% in the presence of AlF_4^-, indicating that environmental changes at the N terminus of the $G\alpha$ subunit do occur on activation (Fig. 6A). This mutant had a normal AlF_4^--dependent increase in intrinsic Trp fluorescence (Fig. 6B), indicating the functional and structural integrity of the protein. The decrease in acrylodan fluorescence noted for Chi6ab on its activation suggests that the fluorescent group moves toward a more hydrophobic environment.

In a control experiment we replaced Val-301 of Chi6ab, which is located in a structurally rigid region of α subunit, with Cys. The Val301Cys-Chi6ab chimera was specifically labeled with either LY or acrylodan. Addition of AlF_4^- to the Val301Cys-Chi6ab-LY or Val301Cys-Chi6ab-acrylodan labeled proteins did not result in any detectable fluorescence changes (data not shown). The absence of any fluorescence changes in this region, an area that is known to be conformationally rigid, highlights the conformational changes that occur at the N-terminal region of $G\alpha$.

Conclusions

We have developed an approach to monitor in real time the structural changes in $G\alpha$ that occur on activation using targeted fluorescent labeling

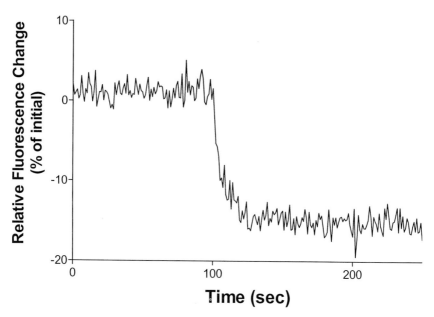

FIG. 6. A3C-Chi6ab labeled with acrylodan reports conformational change. (A) Ala3Cys-Chi6ab labeled with acrylodan shows a 20% fluorescence decrease on the addition of AlF_4^- at 100 sec. This decrease in fluorescence represents a conformational change in the N terminus of the protein since Ala3Cys is the only accessible thiol group in Ala3Cys-Chi6ab to be labeled. (B) Ala3Cys-Chi6ab showed an increase in AlF_4^--dependent Trp fluorescence, indicating the protein is functional and structural integrity of the protein. Basal fluorescence of Ala3Cys-Chi6ab was recorded and fluorescence at 100 sec. AlF_4^- was added to produce a 40% increase in fluorescence.

of single surface exposed cystein residues. Chi6, a functional derivative of $G\alpha_t$, can be specifically labeled with LY at Cys-210 and Cys-347. The protein species labeled either at different Cys residues or with different stoichiometry can be separated by HPLC on an anion-exchange column. Substitution of Chi6 Cys-210 or Cys-347 residues with Ser resulted in only one Cys being accessible for LY labeling, Cys-347 in the case of Chi6a or Cys-210 in the case of Chi6b. Change of both exposed cysteines (Cys210Ser and Cys347Ser; Chi6ab) resulted in a protein resistant for labeling with LY, as well as labeling by other fluorescent reagents such as acrylodan and IAEDANS. Substitution of an exposed residue with cysteine (Ala3Cys) in

B

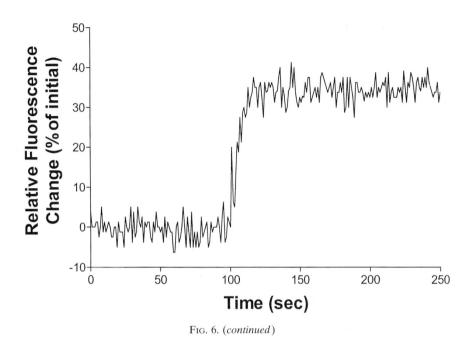

Fig. 6. (*continued*)

the Chi6ab context targets the fluorescent label to a new position of Gα. Therefore, substitution of an amino acid residue for cysteine in the context of the Cys-210/347 Ser mutant of Gα provides an opportunity to probe nearly any position of Gα for changes that occur in fluorescence following a given event such as activation or binding of another protein.

Site-specific Cys-directed fluorescent groups could be reporters for conformational changes in various regions of Gα. When Chi6b, a protein in which Cys-210 is labeled with LY, an activation-dependent conformational change in the switch II region of $G\alpha_t$ was observed. This region has a well-characterized conformational change that is known to occur on protein activation.[1,2] We have documented that an activation-dependent conformational change at the C terminus of Gα subunit also occurs using Chi6a, a protein in which Cys-347 is labeled with LY as a sensitive reporter. Cys-3 labeled with acrylodan is a molecular probe that reports environmental changes around the N terminus of $G\alpha_t$ on its activation. Another important application of directionally labeled Gα subunits is the ability to monitor

$G\alpha$ interactions with its partners, which include receptors, $G\beta\gamma$ subunits, and effectors.

Acknowledgment

This work was supported by grants EY06062 (H.E.H.) and EY10291 (H.E.H.), and a Distinguished Investigator Award from the National Alliance for Research on Schizophrenia and Depression (H.E.H.), and an American Heart Association Grant-In-Aid (N.P.S.).

[34] $G\alpha_t/G\alpha_{i1}$ Chimeras Used to Define Structural Basis of Specific Functions of $G\alpha_t$

By Nikolai P. Skiba, Tarita O. Thomas, and Heidi E. Hamm

Introduction

The α subunits of heterotrimeric G proteins function as molecular switches between the active and inactive state of signaling pathways depending on whether GTP or GDP is bound. There is high homology in amino acid sequences between $G\alpha$ subunits highlighting a set of their common functions including guanine nucleotide binding, GTP hydrolysis, and $\beta\gamma$ binding. The crystal structures of $G\alpha_t$, $G\alpha_{i1}$, and $G\alpha_s$ indicate that the geometry of these α subunits is nearly identical, suggesting the side chains of nonconservative amino acids exposed on the surface of $G\alpha$ subunits determine the specificity of interaction between a given $G\alpha$ and its effectors, receptors, or regulators of G-protein signaling (RGS) proteins. Multiresidue mutagenesis of nonconserved amino acids is a useful approach to define the structural determinants of specific functions for homologous proteins, like the α subunits of G proteins. This is achieved by replacing a region of $G\alpha$ with a corresponding region of a structurally homologous but functionally diverse protein. The resulting protein, which is a combination of different proteins, is defined as a chimera. A group of $G\alpha_s/G\alpha_{i1}$ chimeras used to map the adenylyl cyclase interaction sites of $G\alpha_s$ set the stage for the use of a chimeric approach to study structure–function relationships in G proteins.[1,2] As described here, the usefulness of $G\alpha$ chimeras is not limited to mapping effector binding sites. Rather the $G\alpha$ chimeras can be valuable tools to define the structural determinants of other $G\alpha_t$ functions

---

[1] S. Osawa, N. Dhanasekaran, C. W. Woon, and G. L. Johnson, *Cell* **63,** 697 (1990).
[2] C. Berlot and H. R. Bourne, *Cell* **68,** 911 (1992).

including nucleotide binding and interaction with RGS proteins in addition to interaction with cGMP phosphodiesterase (PDE).

$G\alpha_t$ and $G\alpha_{i1}$ constitute an ideal pair of molecules to study G-protein functions since they are unique functionally but similar structurally. The two proteins couple different receptors to different effectors,[3] have significantly different affinity for GDP,[4] and have specificity for RGS proteins.[5] Structurally they share 68% homology in amino acid sequence and have nearly identical geometry, according to their crystal structures.[6,7] Therefore, analysis of functional properties of chimeric polypeptides composed of different regions of $G\alpha_t$ and $G\alpha_{i1}$ provides an extremely valuable approach to define the molecular determinants of unique $G\alpha$ functions.

Studying $G\alpha_t$ interactions with effectors and RGS proteins in addition to probing its nucleotide binding properties requires relatively large amounts of pure protein. Such amounts of protein can easily be produced in bacteria. Therefore, mapping functional regions of parent proteins using a chimera approach usually requires functional expression of both protein components. However, unlike recombinant $G\alpha_{i1}$, which is functional,[8] $G\alpha_t$ expressed in *Escherichia coli* is misfolded and insoluble. Numerous attempts to express large amounts of functional $G\alpha_t$ in different expression systems were not successful. Thus, construction of $G\alpha_t/G\alpha_{i1}$ chimeras provides a strategy to "rescue" recombinant $G\alpha_t$ by systematic replacement of $G\alpha_t$ regions, which inhibit proper folding, with $G\alpha_{i1}$ regions that are permissive for proper folding. By systematic replacement of more and more $G\alpha_{i1}$ regions for the corresponding regions of $G\alpha_t$, we have constructed a set of $G\alpha_t/G\alpha_{i1}$ chimeras that is soluble, functional, and express at high levels in *E. coli*. The most $G\alpha_t$-like chimera contains only 11 amino acids different from native $G\alpha_t$. These chimeric proteins have been used to assign unique functional features of $G\alpha_t$ to their structural determinants.

Construction of $G\alpha_t/G\alpha_{i1}$ Chimeras

Chimeric genes were constructed by introduction of unique restriction enzyme sites, flanking target fragments, into $G\alpha_{i1}$ and $G\alpha_t$ cDNAs using polymerase chain reaction (PCR) amplification with corresponding oligonu-

---

[3] J. R. Hepler and A. G. Gilman, *Trends Biochem. Sci.* **17,** 383 (1992).
[4] N. P. Skiba, H. Bae, and H. E. Hamm, *J. Biol. Chem.* **271,** 413 (1996).
[5] N. P. Skiba, C.-S. Yang, T. Huang, H. Bae, and H. E. Hamm, *J. Biol. Chem.* **274,** 8770 (1999).
[6] J. P. Noel, H. E. Hamm, and P. B. Sigler, *Nature* **366,** 654 (1993).
[7] D. E. Coleman, A. M. Berghuis, E. Lee, M. E. Linder, A. G. Gilman, and S. R. Sprang, *Science* **265**(5177), 1405 (1994).
[8] M. E. Linder, D. A. Ewald, R. J. Miller, and A. G. Gilman, *J. Biol. Chem.* **265**(14), 8243 (1990).

cleotide primers followed by replacement of $G\alpha_{i1}$ fragments with corresponding $G\alpha_t$ cDNA fragments and vice versa. Figure 1 illustrates the general scheme of construction of two representative chimeras using PCR-based mutagenesis. Expression vectors pHis$_6$Gα_{i1} and pHis$_6$Gα_t,[4] which contain $G\alpha_{i1}$ and $G\alpha_t$ cDNAs, respectively, were used as templates for

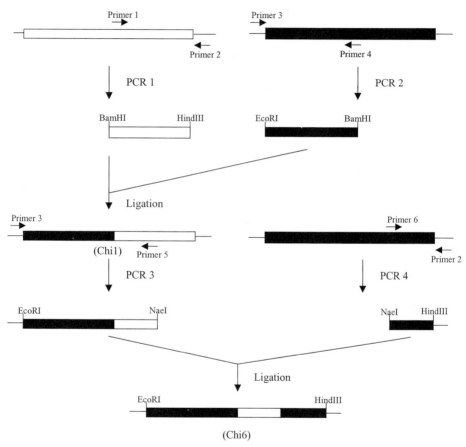

Primer 1: 5'-GAAATGGATCCACTGCTTGCAAGGCGTG
Primer 2: 5'-TTACTAAGCTTCTGCAGAGCTTAGAAG
Primer 3: 5'-AAAAGAATTCTAAGGAGGTTTAAC
Primer 4: 5'-AGTGGATCCACTTCTTGCGCTCTGAGC
Primer 5: 5'-ATACGCGCCGGCCTCTTCATATGTGTTTG
Primer 6: 5'-GGACGCCGGCAATTACATCAAGGTGC

FIG. 1. Construction of $G\alpha_t/G\alpha_{i1}$ chimeras using PCR. Filled bars indicate $G\alpha_t$ sequence. Open bars represent $G\alpha_{i1}$ originated sequence. Underlined primer sequence shows restriction enzyme sites.

PCR amplification. Gα cDNAs are preceded by a nucleotide sequence encoding a hexahistidine tag and are under the control of the bacteriophage T7 promoter.

To construct chimera 1, we first generated PCR fragments encoding the N-terminal half of $G\alpha_t$ and the C-terminal half of $G\alpha_{i1}$. Simultaneously, a *Bam*HI site was introduced into the fragment originating from $G\alpha_{i1}$ using primer 1 for PCR. The corresponding *Bam*HI site was present in $G\alpha_t$ cDNA. Two oligonucleotide primers (primer 2 and 3) were used for PCR reactions 1 and 2 (Fig. 1) and correspond to the 3' and 5' noncoding regions of Gα subunits, respectively, and thus are common for construction of many $G\alpha_t/G\alpha_{i1}$ chimeras. PCR products of reactions 1 and 2 were cut with *Bam*HI and *Hin*dIII or *Eco*RI and *Bam*HI, respectively, and simultaneously ligated with the large fragment of $pHis_6G\alpha_{i1}$ cut with *Eco*RI and *Hin*dIII. The resulting Chi1 contained the first N-terminal 216 amino acid residues of $G\alpha_t$ and the rest of the sequence originated from $G\alpha_{i1}$. Chi1 was further used as an intermediate construct to generate a set of derivative chimeras. The bottom part of Fig. 1 shows construction of Chi6 using Chi1 cDNA as a template for PCR. The PCR fragment encoding the first 295 amino acids of chimera 1 was generated using universal primer 3 and specific primer 5, which directed insertion of a *Nae*I site. The C-terminal portion of Chi6 originated from $G\alpha_t$, which is encoded by a DNA fragment resulting from PCR amplification of $G\alpha_t$ cDNA using primers 2 and 6. The *Eco*RI–*Nae*I DNA fragment was mixed with the *Nae*I–*Hin*dIII DNA fragment and simultaneously ligated with the large fragment of $pHis_6G\alpha_{i1}$ cut with *Eco*RI and *Hin*dIII, creating Chi6.

Figure 2 illustrates an example of exchange of short corresponding fragments of Gα subunits. The quadruple $G\alpha_{i1}$ mutant L232M/A235V/E238D/M240V changed the switch III region of $G\alpha_{i1}$ to that of $G\alpha_t$. Two regions of $G\alpha_{i1}$ cDNA were PCR amplified in separate reactions. The downstream primer 7 for synthesis of the 5'-terminal fragment directed not only substitution of the indicated 4 amino acids but also introduction of a *Sph*I site. The upstream PCR primer 8 for amplification of the 3'-terminal fragment directed introduction of the *Sph*I site as well. The resulting fragments were cut with *Eco*RI and *Sph*I (5'-terminal fragment) and *Sph*I and *Hin*dIII (3'-terminal fragment) and simultaneously ligated with the large fragment of $pHis_6G\alpha_{i1}$ digested with *Eco*RI and *Hin*dIII (Chi3). Chi3 was an intermediate construct used to generate Chi8 and Chi10 using the strategy described for construction of the Chi6.

Chimera G_i/G_tH is a derivative of His_6-$G\alpha_{i1}$ in which residues 60–177 of $G\alpha_{i1}$ encompassing the α-helical domain are substituted with the corresponding region of $G\alpha_t$, residues 56–173. The chimeric gene was constructed by introduction of unique restriction enzyme sites flanking the

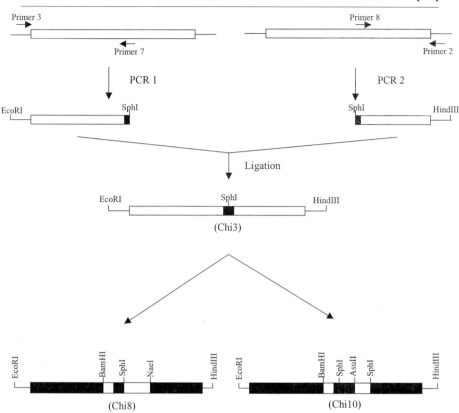

Primer 7: 5'-GTA<u>GCATGC</u>GGTTCA**C**TTCGTCATCCTCAACAAGAACCATGTC
Primer 8: 5'-ATGAACC<u>GCATGC</u>ATGAAAG

FIG. 2. Replacement of switch III region of $G\alpha_{i1}$ with the corresponding region of $G\alpha_t$ (Chi3). Filled bars indicate $G\alpha_t$ sequence. Open bars show $G\alpha_{i1}$ sequence. Underlined primer sequence shows SphI site. Bold letters are the points of mutations.

DNA fragments of $G\alpha_{i1}$ cDNA and $G\alpha_t$ cDNA, which encode the α-helical domain. A *Mlu*I restriction enzyme site (3' end of the fragment) was inserted in both $G\alpha_t$ and $G\alpha_{i1}$ cDNAs, whereas a *Bst*XI site, which is present in the $G\alpha_{i1}$ gene, was inserted only in $G\alpha_t$ cDNA (5' end of the fragment) using a QuikChange site-directed mutagenesis kit (Stratagene, La Jolla, CA) with corresponding oligonucleotide primer mutagenes. The *Bst*XI–*Mlu*I DNA fragment of $G\alpha_t$ was inserted into the $G\alpha_{i1}$ cDNA after cutting off the corresponding fragment of $G\alpha_{i1}$ with *Bst*XI and *Mlu*I restriction enzymes. Chimera Chi6/G$_i$H contains the α-helical domain of Gα_{i1} in the context of Chi6. The Gα_t α-helical domain of this chimera (residues 56–174) was replaced with the corresponding region of Gα_{i1} using the same approach as described for the construction of chimera G$_i$/G$_t$H.

The resulting plasmid DNAs were transformed into competent *E. coli* BL21(DE3) cells (Novagen, Madison, WI). BL21(DE3) strain provides an excellent background for protein expression, since it lacks *lon* and *ompT* proteases. This strain is also lysogenic for λDE3 locus, which carries the T7 RNA polymerase gene under the control of a *lacUV5* promoter. Therefore, this system provides a dual control for foreign gene expression that eliminates background expression, and thereby eliminates the potential toxic effect of expressing foreign proteins in a bacterial cell. Addition of isopropyl-β-D-thiogalactoside (IPTG) stimulates expression of T7 RNA polymerase, which activates expression of Gα_t/Gα_{i1} cDNAs under the control of a bacteriophage T7 promoter.

Expression and Purification of Chimeras

Cells harboring plasmids encoding all chimeras except Chi9 and Chi10 are grown in 1–3 liters of 2× YT medium in the presence of 100 μg/ml of ampicillin at room temperature up to an OD$_{600}$ of 0.5–0.8 and then induced with 30 μM IPTG for 16–20 hr. The cell pellet is resuspended in 1/20 of a cell culture volume of buffer containing 50 mM Tris-HCl, pH 8.0, 50 mM NaCl, 5 mM MgCl$_2$, 50 μM guanosine diphosphate (GDF), 0.1 mM phenylmethylsulfonyl fluoride (PMSF), and 5 mM 2-mercaptoethanol (buffer A). The cell suspension is disrupted by ultrasonication at 4°. The crude cell lysate is cleared by centrifugation at 100,000g for 60 min at 4°. The supernatant is collected and adjusted to 500 mM NaCl and 20 mM imidazole concentrations by adding ice-cold 8× binding buffer: 160 mM Tris-HCl, pH 8.0, 4 M NaCl, and 160 mM imidazole. The resulting mixture is loaded onto a 5-ml Ni^{2+}-nitrilotriacetic acid agarose resin column (Ni-NTA, His-Bond, Novagen) preequilibrated with the binding buffer. The column is washed with 10 volumes of 1× binding buffer and bound material was eluted with 20 mM Tris-HCl, pH 8.0, 500 mM NaCl, and 100 mM imidazole (buffer I-100). NaCl and imidazole are removed from the protein sample by overnight dialysis against buffer A in the presence of 20% glycerol using Spectra/Por membrane tubing (Spectrum Medical Industries, Inc). The protein samples are directly applied to a 20-ml Waters HPLC AP-1 column (Waters Chromatography Division, Millipore, Bedford, MA) packed with the Protein-Pak Q 15 HR anion-exchange resin (Millipore) equilibrated with buffer A free of GDP and 2-mercaptoethanol. This was followed by protein elution using a NaCl gradient in buffer A.

The concentration of chimeras in the eluates, which did not contain GDP, was immediately determined spectrophotometrically. GDP, 2-mercaptoethanol, and PMSF were added to the eluates to a concentration of 25 μM, 2 mM, and 0.1 mM, respectively, and samples were aliquoted and stored at $-80°$ for several months with no loss of functional activity. The final yield of highly soluble chimeras and $G\alpha_{i1}$ ranged from 3 to 5 mg of more than 95% pure protein/liter of bacterial culture.

Expression and Purification of Low-Soluble Chimeras Chi9 and Chi10

Cells (4–8 liters) containing the plasmids encoding Chi9 and Chi10 are grown and induced at conditions described earlier. Cell pellets are resuspended in 200 ml of ice-cold buffer A and insoluble material is centrifuged at 100,000g for 60 min at 4°. Supernatants are collected and adjusted to 500 mM NaCl and 20 mM imidazole concentrations by adding ice-cold 8× binding buffer. Ni-NTA resin (1 ml) equilibrated with 1× binding buffer was resuspended in 2 ml 1× binding buffer and added to the supernatant. The mix is gently shaken for 10 min and then centrifuged at 1000g for 5 min at 4°. The supernatant is discarded and the resin resuspended in 5 ml of ice-cold 1× binding buffer and then packed into a 3-ml column. The rest of the Ni-chelate affinity chromatography procedure is done at standard conditions as described earlier. Proteins eluted in I-100 buffer fraction are dialyzed against buffer A containing 20% glycerol and subjected to a second purification step using anion-exchange chromatography on a 2-ml Waters AP minicolumn, packed with Protein-Pak Q HR15 resin, as described earlier. Typical yield of Chi9 and Chi10 is 0.1–0.2 mg of 80–90% pure protein per 1 liter of bacterial culture.

Protein Concentrations

The concentrations of $G\alpha$ subunits are determined spectrophotometrically using calculated extinction coefficients. The molar extinction coefficient of $G\alpha$GDP at 280 nm, ε_{280}, is calculated by using the following equation: $\varepsilon_{280}[M^{-1}\,\text{cm}^{-1}] = 5500n_{\text{Trp}} + 1490n_{\text{Tyr}} + 7900$, where 5500, 1490, and 7900 are molar extinction coefficients of tryptophan, tyrosine, and GDP, respectively, at 280 nm.

Measured concentrations are corrected for amount of functional protein based on a fluorescent assay detecting AlF_4^--dependent increase in Trp fluorescence, which reflects conformational change on G protein activation. For a detailed description of this assay, see the section on AlF_4^--Dependent Conformational Change of $G\alpha$'s.

Maximal fluorescent change detected in this assay for best preparations of homogeneous $G\alpha_t$GDP, containing two Trp, was consistently 70%. We thus estimated that a 70% increase in Trp fluorescence corresponds to 100% of functionally active protein. Maximal fluorescent change for the best preparations of homogeneous $G\alpha_{i1}$, containing three Trp, was 55%. The concentrations of functional $G\alpha_t$GDP and chimeras Chi9 and Chi10 (two Trp), C_{func}, were calculated using the formula $C_{\text{func}} = n\%/70\%\ C_{\text{spec}}$, where $n\%$ is AlF_4^--dependent fluorescent change and C_{spec} is protein concentration determined spectrophotometrically. For calculation of concentrations of functional $G\alpha_{i1}$ and other chimeras (three Trp) the formula is $C_{\text{func}} = n\%/55\%\ C_{\text{spec}}$.

Solubility of $G\alpha_t/G\alpha_{i1}$ Chimeras

The insolubility of recombinant $G\alpha_t$ in bacterial lysates was "inherited" by a set of chimeras listed in Fig. 3. These chimeras were exclusively localized in the insoluble fractions apparently in the form of inclusion bodies. Western blotting analysis of soluble fractions revealed very low amounts of these proteins (less than 1–2% of total expressed protein), which were nonfunctional and aggregated. Analysis of the structures of insoluble chimeras indicated that they contained either region 216–227 or 271–294 of $G\alpha_t$. Avoiding the two "problem" regions, a second set of $G\alpha_t/G\alpha_{i1}$ chimeras was constructed. The distributions of recombinant proteins between soluble and insoluble fractions of the bacterial lysates were analyzed by Western blotting using an antipeptide antibody against the C-terminal 10 amino acids of $G\alpha_{i1}$. This antibody recognizes both $G\alpha_t$ and $G\alpha_{i1}$, which have only one amino acid difference in this region (E342N), as well as all of the $G\alpha_t/G\alpha_{i1}$ chimeras. Figure 4 shows schematically the structure of a representative set of soluble chimeras that we used to map effector, RGS9, and nucleotide binding sites of $G\alpha$ subunits. Another criterion of chimera solubility was the yield of recombinant protein from 1 liter of an initial cell culture. Because losses of recombinant protein during standard purification procedure were likely similar, the final yield of pure chimera closely reflects the initial amount of soluble polypeptide in the bacterial cell. Chimeras, listed in Fig. 4, except Chi9 and Chi10, were fully soluble (final yield 3–5 mg/ml). Chimeras 9 and 10, which contained region 237–270 of $G\alpha_t$, had significantly reduced solubility (final yield 0.1–0.25 mg/ml).

Large amounts of $G\alpha_t$-like proteins were expressed in functional form in *E. coli* by incorporating sequences from $G\alpha_{i1}$ into the $G\alpha_t$ context. Surprisingly, nearly 95% of $G\alpha_t$ can be incorporated into the chimeric protein, and still be permissive for proper folding. Two regions of $G\alpha_t$

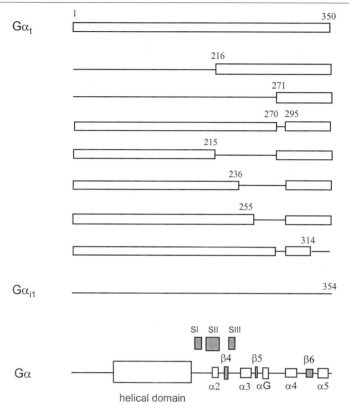

Fig. 3. Schematic structure of insoluble $G\alpha_t/G\alpha_{i1}$ chimeras. Numbers above the lines representing chimeras indicate the junction points of $G\alpha_t$ and $G\alpha_{i1}$ sequences according the $G\alpha_t$ nomenclature. The conformational switch regions SI, SII, and SIII are shown. The α-helical domain and the regions of secondary structure outside the α-helical domain are indicated on the bottom line.

located between residues 216–227 and 271–294, which differ by a total of 11 amino acids, caused improper folding of chimeras. Another problem region between residues 237 and 270 decreased solubility of chimeras, presumably by perturbing the folding pathway in a more subtle way.

Functional Activity of Soluble Chimeras

Because one of the recombinant parent proteins used for chimera construction, $G\alpha_t$, is insoluble, folding state and functional activity of the resulting soluble chimeras required careful evaluation in several alternative functional assays. Their ability to interact with receptor and $G\beta\gamma$ was

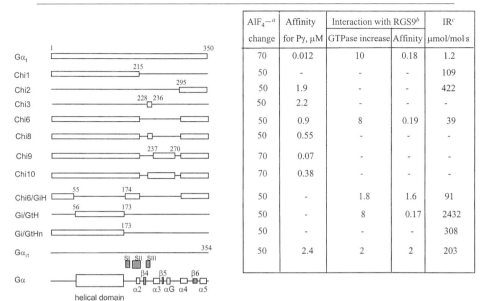

FIG. 4. Structure and functional properties of $G\alpha_t/G\alpha_{i1}$ chimeras. [a]Percent increase in tryptophan fluorescence in the presence of AlF_4^-. [b]GTPase increase represents fold increase of GTPase activity of $G\alpha$ in the presence of 10 μM of RGS9d. [c]Intrinsic rate of spontaneous nucleotide release (IR). Dashes indicate that functional activity was not determined.

determined in a GTPγS binding assay. The rate of GTP hydrolysis for chimeras, $G\alpha_t$, and $G\alpha_{i1}$ was measured by a single turnover GTPase assay. Finally, the ability of chimeras to undergo an activating conformational change was analyzed by an increase in intrinsic fluorescence of tryptophan on addition of AlF_4^-. These assays are described in detail next.

Measurement of Rhodopsin-Stimulated GTP Binding by Gα's Using GTPγS Binding Assay

One of the most common functional assays for Gα's is binding of the nonhydrolyzable analog of GTP–GTPγS as a result of its nucleotide exchange for GDP. As described later in detail, the intrinsic GDP–GTP exchange rate for $G\alpha_t$ is very low. This rate is higher for $G\alpha_{i1}$. Light-activated rhodopsin in the presence of $G\beta\gamma_t$ significantly stimulates the nucleotide exchange rate on both $G\alpha_t$ and $G\alpha_{i1}$.[4] Therefore, acceleration of the rate of GTPγS binding to Gα provides a measure of Gα's ability to interact with receptor and Gβγ.

Purified recombinant Gα subunits (4 μM) are preincubated with 4 μM of Gβγ$_t$ at room temperature for 15 min in 100 mM HEPES, pH 8.0, 1 mM EDTA, 10 mM MgSO$_4$ and 10 mM dithiothreitol (DTT) containing 5 μM [^{35}S]GTPγS (5000 cpm/pmol; DuPont, New England Nuclear, 1000 Ci/mmol). Nucleotide exchange reaction is initiated by adding 100 nM of light-activated rhodopsin. Samples (total volume 200 μl) were incubated from 0 to 60 min at room temperature. Aliquots (20 μl) were withdrawn at indicated times, passed through the Millipore Multiscreen-HA 96-well filtration plate, and washed 10 times with ice-cold wash buffer (200 μl per well) containing 20 mM Tris-HCl, pH 8.0, 100 mM NaCl, and 25 mM MgCl$_2$. Before loading samples, filters are presoaked by passing ice-cold wash buffer twice. Filters are dried for a few minutes at room temperature and punched out into glass 7-ml Kimble vials using the Millipore Multiscreen Puncher. Biosafe II (3 ml) scintillation fluid is added to each sample and counted by liquid scintillation spectrometry.

GTPase Assay

Single turnover GTPase reactions are performed under conditions described by Arshavsky and Bownds[9] with some modifications. Freshly illuminated urea-washed rod outer segment (ROS) membranes (final concentration 15 μM) are reconstituted with 1 μM of G$_t$ or 1 μM Gα and 1 μM Gβγ in 10 mM Tris-HCl, pH 7.5, 100 mM NaCl, 2 mM MgCl$_2$, 0.02 mM AMP-PNP, and incubated for 10 min in the dark at room temperature. The reaction is started by addition of RGS9 and 200 nM [γ-^{32}P]GTP ($\sim 10^5$ cpm/pmol) to the reconstituted membranes and quenched by the addition of 100 μl of 7% perchloric acid followed by addition of 500 μl of 50 mM potassium phosphate, pH 7.6, containing 10% (w/v) activated charcoal (Sigma, St. Louis, MO) to remove nucleotides. Reaction tubes are centrifuged for 10 min at 10,000g and free [^{32}P]P$_i$ in the supernatant was measured by scintillation counting.

AlF$_4^-$-Dependent Conformational Change of Gα's

AlF$_4^-$ mimics the γ-phosphate of GTP when it enters the nucleotide binding pocket and binds next to GDP.[10] Binding of AlF$_4^-$ activates GαGDP by establishing contacts with the γ-phosphate interacting residues leading to conformational changes in the switch I, II, and III regions.[11] A movement

---

[9] V. Y. Arshavsky and M. D. Bownds, *Nature* **357**, 416 (1992).
[10] B. Antonny and M. Chabre, *J. Biol. Chem.* **267**(10), 6710 (1992).
[11] J. Sondek, D. G. Lambright, J. P. Noel, H. E. Hamm, and P. B. Sigler, *Nature* **372**, 276 (1994).

of Trp-270 of Gα_t located in the switch II region to a more hydrophobic environment[12] causes an increase in intrinsic tryptophan fluorescence. Thus, the ability of the recombinant protein to undergo an AlF$_4^-$-dependent increase in Trp fluorescence tests the presence of GDP in the nucleotide binding site and structural changes leading to activation of the α subunit— features that reflect the structural and functional integrity of the protein.

Fluorescent measurements were performed on Aminco-Bowman Series 2 luminescence Spectrometer (SLM Aminco) at room temperature in 50 mM Tris-HCl, pH 8.0, 50 mM NaCl, 2 mM MgCl$_2$ using excitation at 280 nm and emission at 340 nm. Fluorescence of 200 nM of Gα–GDP was measured before and after addition of 10 mM NaF and 30 μM AlCl$_3$. The fluorescence increase on Gα activation with AlF$_4^-$ is expressed as a percent of the initial fluorescence: $\Delta F(\%) = (F - F_0)/F_0 \times 100$, where F_0 is an initial fluorescence and F is the fluorescence after addition of AlF$_4^-$.

Monitoring Spontaneous GTPγS Binding to Chimeras Using Fluorescent Assay: Regions of Gα_t That Define High-Affinity Nucleotide Binding

The rate of spontaneous nucleotide release is a property of Gα subunits that distinguishes Gα_t and Gα_{i1}. Gα_t binds GDP very tightly, therefore it has a very slow, not even reliably detectable rate of intrinsic nucleotide release. Gα_{i1} exhibits a significantly faster rate of intrinsic nucleotide exchange[4,13] that reflects a lower affinity for GDP compared to Gα_t. The different rates of GDP release for Gα_t and Gα_{i1} cannot be explained based on the stereochemistry of their complexes with GDP. The crystal structures of Gα_tGDP[12] and Gα_{i1}GDP[14] demonstrated that the side chains contacting the nucleotide and coordinating magnesium are identical for these two Gα subunits. Therefore, structural elements of Gα subunits distal to the nucleotide binding pocket play an important role in defining the strength of interaction with GDP and, correspondingly, the rate of intrinsic GDP release. To determine additional regions of Gα subunits that control affinity of nucleotide binding we used Gα_t/Gα_{i1} chimeras to evaluate their rates of GDP release in a GTPγS binding fluorescence assay. The principle underlying this assay is that the intrinsic fluorescence of tryptophan-207 of Gα_t or chimera, located in the switch II region, increases 50–70% on GDP dissociation followed by binding of GTPγS and activation of α subunit.

---

[12] D. G. Lambright, J. P. Noel, H. E. Hamm, and P. B. Sigler, *Nature* **369,** 621 (1994).
[13] T. O. Thomas, H. Bae, and H. E. Hamm, submitted.
[14] M. B. Mixon, E. Lee, D. E. Coleman, A. M. Berghuis, A. G. Gilman, and S. R. Sprang, *Science* **270**(5238), 954 (1995).

The crystal structures of $G\alpha_t$ in its GDP[12] and GTPγS[6] forms illustrates that Trp-207 moves into a more hydrophobic environment. This change in the environment of Trp-207 is the structural basis for the increase in intrinsic Trp fluorescence accompanying $G\alpha$ activation. Tryptophan fluorescence of $G\alpha$'s was measured on an Aminco-Bowman Series 2 Luminescence spectrometer at the constant temperature of 25°, in 50 mM Tris-HCl, 50 mM NaCl, and 5 mM $MgCl_2$ using excitation at 280 nm and emission at 340 nm. To minimize bleaching of the fluorophore during the time course of the experiment, the shutter was closed between readings.

The program that controls instrument functions during the course of the experiment is shown in Fig. 5. At the step "Add Protein & Nucleotides" $G\alpha$, GTPγS, and GDP were added to final concentrations of 100 nM, 10 μM, and 1 μM, respectively. The total recording time was 43.5 min divided into blocks as follows: to measure initial Trp fluorescence the first five recordings were taken every 10 sec; other regular recordings were taken every minute for 40 data points. To determine the maximal change of Trp fluorescence, the final 18 recordings were taken every 10 sec after addition of 10 mM NaF and 20 μM $AlCl_3$ at the step "Add 10 mM NaF and 20 μM $AlCl_3$ Simultaneously." The first 5 recordings taken every 10 seconds and final 5 recordings taken after addition of AlF_4^- were independently averaged and set as the initial (0%) and maximal (100%) fluorescence change. Normalized Trp fluorescence change plotted as a function of time is shown in Fig. 6. Data for the initial rates of fluorescence change were fit into a linear function of time. Alternatively, whole fluorescence time traces were analyzed using an exponential function $\Delta F\% = 100(1 - e^{-kt})$, where $\Delta F\%$ is percent of Trp fluorescent change and k is a rate constant of GTPγS binding (Fig. 6C).

The rate of spontaneous nucleotide release of $G\alpha_t$ was found to be very low (1.2 μmol/mol · sec, nearly undetectable). A similar very low rate of GDP release (4.9 μmol/mol · sec) was documented using a filter assay of [^{35}S]GTPγS binding to $G\alpha_t$ in the absence of receptor and $G\beta\gamma$.[4] The spontaneous GDP release rate of $G\alpha_{i1}$ determined by the fluorescent method was significantly higher (202 μmol/mol · sec) and was similar to that determined in the filter assay[4] (225 μmol/mol · sec). Placement of the N-terminal 216 amino acids of $G\alpha_t$ into a $G\alpha_{i1}$ context (Chi1) decreased the nucleotide exchange rate approximately twofold (109 μmol/mol · sec) indicating that the N terminus of $G\alpha_t$ possesses additional determinants of GDP binding. The C-terminal 55 amino acids of $G\alpha_t$ (Chi2) did not increase the nucleotide binding. Moreover this chimera exhibited a higher rate of nucleotide exchange (422 μmol/mol · sec), likely reflecting a role for both N-terminal and C-terminal regions of $G\alpha_t$ in high-affinity GDP binding. Indeed Chi6, which has both these regions of $G\alpha_t$ in $G\alpha_{i1}$ context, had

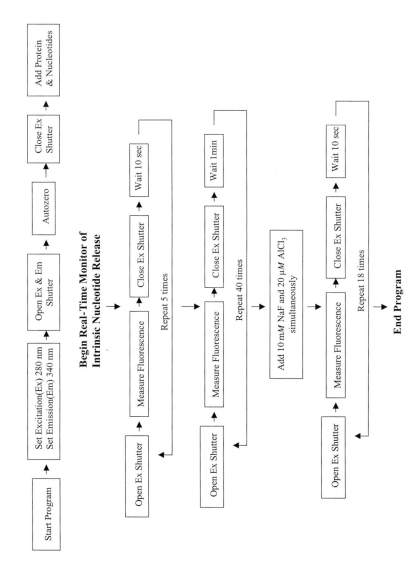

FIG. 5. Program describing sequence of commands for instrument control during real-time monitoring of intrinsic GDP release.

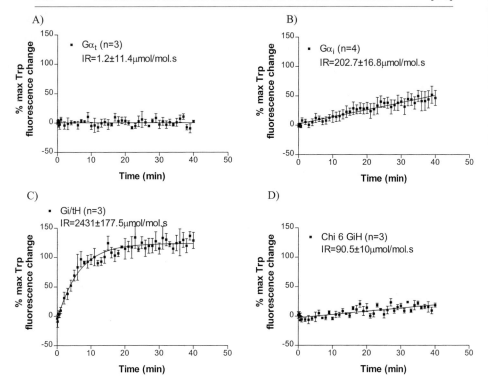

FIG. 6. Time traces of spontaneous nucleotide release for (A) $G\alpha_t$, (B) $G\alpha_{i1}$, (C) ChiGi/GtH, and (D) Chi6/GiH as measured by an increase in intrinsic tryptophan fluorescence. The intrinsic rate of spontaneous GDP release (IR) of $G\alpha$ subunits (100 nM) was measured in the presence of 10 μM GTPγS and 1 μM GDP. Data points represent the mean ± S.E.M. from three or four independent experiments (n) and expressed as a percent of maximal fluorescence change. The initial rates of fluorescent change (A, B, and D) were fit into a linear function of time. The time trace for G_i/G_tH (C) was analyzed using an exponential function $\Delta F\% = 100(1-e^{-kt})$, where $\Delta F\%$ is percent of maximal Trp fluorescence change and k is a rate constant of GTPγS binding.

a much higher affinity for GDP and, correspondingly, lower nucleotide exchange rate (39 μmol/mol · sec). To narrow down sites within the N-terminal 1–216 region of $G\alpha_t$ that are involved in nucleotide binding, we constructed an additional set of chimeras. Chimera G_i/G_tHn containing a shorter N-terminal region (1–173) of $G\alpha_t$ as compared to Chi1 displayed a high $G\alpha_{i1}$-like rate of nucleotide exchange (308 μmol/mol · sec), indicating that region 174–216 is important for GDP binding. Excluding an additional 55 amino acids from the very N terminus of $G\alpha_t$ (G_iG_tH) suprisingly had a hyperelevated nucleotide exchange rate (2432 μmol/mol · sec), probably reflecting destabilized GDP binding. Placement of the N-terminal 55 amino acids and the region 174–216 of Gα_t in a Chi2 context reduces the nucleotide exchange rate approximately fourfold (90 μmol/mol · sec) indicating that the presence of both these regions is necessary for tight GDP binding. Figure 10, shown later, illustrates these regions in a ribbon drawing of Gα_t. Analysis of the crystal structure of Gα_t and Gα_{i1} has revealed that in Gα_t Asp-26, the amino-terminal helix interacts with Asn-191 near the proximal end of the switch II region, an area known to play an important role in GDP binding. On the other hand, in Gα_{i1}, the corresponding residues Ala-30 and His-195 are further apart and do not form a molecular contact. Therefore, we speculate that the interaction between Asp-26 and Asn-191 of Gα_t stabilizes GDP binding, likely by stabilizing the residues in the switch I and II regions directly involved in nucleotide binding.

Receptor–G Protein Interaction

In a physiologic environment G_t and G_i interact with different receptors. G_t interacts with the photoreceptor rhodopsin,[15] whereas G_i interacts with a broad spectrum of receptors linking them to adenylyl cyclase and ion channel regulation.[16] Surprisingly, rhodopsin does not discriminate between G-protein heterotrimers containing Gα_t and Gα_{i1},[4] indicating that rhodopsin interacts with conserved regions of the Gα and $\beta\gamma$ subunits. This fact prevented us from using Gα_t/Gα_{i1} chimeras to specify the rhodopsin binding determinants of Gα_t. However, analysis of Gα_{i1} and Gα_t coupling to different receptors[17] has revealed that G_{i1} can interact with the 5-HT$_{1B}$ receptor and stabilize a high-affinity agonist binding state of the receptor, while G_t cannot. Testing the functional coupling abilities of the Gα_t/Gα_{i1} chimeras has been used to define the molecular determinants of selectivity in 5-HT$_{1B}$ receptor–Gα_{i1} interactions.[17]

Probing Effector–Gα_t Interaction: Binding of Chimeras to Inhibitory Subunit Pγ of cGMP Phosphodiesterase

Rhodopsin activated Gα_t interacts with cGMP PDE and activates it by binding to Pγ and displacing it from an inhibitory site on catalytic subunits P$\alpha\beta$.[18] Gα_{i1} can not activate PDE.[19,20] We have taken advantage of the

---

[15] L. Stryer, *J. Biol. Chem.* **266**, 10711 (1991).

[16] B. R. Conklin and H. R. Bourne, *Cell* **73**(4), 631 (1993).

[17] H. Bae, K. Anderson, L. A. Flood, N. P. Skiba, H. E. Hamm, and S. G. Graber, *J. Biol. Chem.* **272**(51), 32071 (1997).

[18] T. Wensel and L. Stryer, *Biochemistry* **29**, 2155 (1990).

[19] A. Otto-Bruc, T. M. Vuong, and B. Antonny, *FEBS Lett.* **343**(3), 183 (1994).

[20] M. Natochin, A. Granovsky, and N. O. Artemyev, *J. Biol. Chem.* **273**, 21808 (1998).

specific interaction of $G\alpha_t$ with PDE to map its effector binding sites using $G\alpha_t/G\alpha_{i1}$ chimeras.

Binding of $G\alpha_t/G\alpha_{i1}$ chimeras to Pγ was measured using the fluorescent assay developed earlier in our laboratory.[21] Pγ was labeled with the thiol-specific fluorescent reagent Lucifer Yellow (LY) at its single cysteine-68. LY attached to Cys-68 of Pγ is a sensitive reporter of Pγ binding to $G\alpha_t$. Fluorescence of Pγ-LY increases threefold at the maximal dose on addition of $G\alpha_t$–GTPγS. Fluorescence measurements were performed on a Perkin-Elmer (Norwalk, CT) LS5B spectrofluorimeter at room temperature in 50 mM Tris-HCl, pH 8.0, 50 mM NaCl, and 2 mM MgCl$_2$ using excitation at 430 nm and emission at 520 nm. Relative increase in fluorescence (F/F_0) of Pγ-LY (50 nM) in the presence of AlF$_4^-$ was measured after addition of increasing concentrations of $G\alpha_t$, $G\alpha_{i1}$, or chimera. Data were best fit to the following equation: $Y = 1 + (Y_{max} - 1)/1 + 10^{(\log EC_{50} - X)*H}$, where Y_{max} represents the maximal relative increase in fluorescence, X is the logarithm of $G\alpha$ concentration, and H is the Hill coefficient. K_d values for the affinity of $G\alpha$'s for Pγ-LY were determined based on calculated EC$_{50}$ parameters using the equation $K_d = EC_{50} - 1/2[P\gamma]_{tot}$.

Figure 7 demonstrates binding of $G\alpha_t$, $G\alpha_{i1}$, and chimeras to Pγ-LY in the presence of AlF$_4^-$. Binding of Pγ-LY to $G\alpha_t$ activated in two different ways, either by GTPγS or AlF$_4^-$ ($G\alpha_t^*$), yielded essentially the same affinity constants for Pγ-LY-$G\alpha_t^*$ complexes (K_d 12 nM). Thus, for further testing of chimeras binding to Pγ-LY, their rapid activation was produced by addition of AlF$_4^-$. $G\alpha_t$GDP was also capable of binding to Pγ-LY, however affinity was ~70-fold lower than in its activated form (K_d 0.88 μM). The binding of $G\alpha_{i1}$GDP to Pγ-LY was undetectable at concentrations up to 20 μM. However, on activation with AlF$_4^-$ it formed a low-affinity complex with Pγ-LY (K_d 2.4 μM, Fig. 5). Addition of first 216 N-terminal amino acids and last 55 C-terminal amino acids of $G\alpha_t$ to the $G\alpha_{i1}$ context (Chi6) resulted only in a modest 2.5-fold increase in affinity of the chimera to Pγ (K_d 0.9 μM) compared to $G\alpha_{i1}$. This indicates that earlier identified regions of effector interaction—switch II[20,22] and 293–314[23]—of $G\alpha_t$ present in this chimera contribute little to the binding of $G\alpha_t$ to Pγ. Insertion of the switch III region of $G\alpha_t$ in the Chi6 context (Chi8) resulted in an additional twofold increase of affinity for chimeric $G\alpha$ to Pγ, indicating an involvement of the switch III region in effector binding. A significant increase in affinity of chimera to Pγ was achieved by addition of the $G\alpha_t$ region 237–270 into

---

[21] N. O. Artemyev, H. M. Rarick, J. S. Mills, N. P. Skiba, and H. E. Hamm, *J. Biol. Chem.* **267**(35), 25067 (1992).

[22] E. Faurobert, A. Otto-Bruc, P. Chardin, and M. Chabre, *EMBO J.* **12**(11), 4191 (1993).

[23] H. M. Rarick, N. O. Artemyev, and H. E. Hamm, *Science* **256**(5059), 1031 (1992).

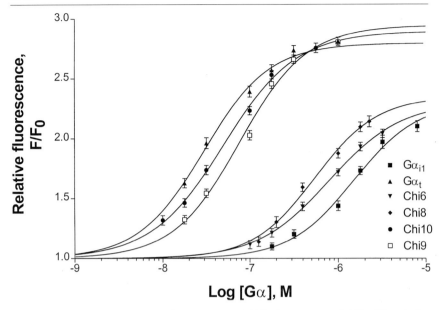

FIG. 7. Binding of $G\alpha_t$, $G\alpha_{i1}$, and chimeras to Pγ-LY in the presence of AlF$_4^-$. The relative increase in fluorescence (F/F_0) of Pγ-LY (50 nM) was measured after addition of increasing concentrations of indicated Gα subunits. The solid lines represent the best fit to a four-parameter logistic equation (see text for details) where the fluorescence at each point has been corrected for dilution resulted by addition of Gα. Data points represent the mean ± S.E. from four independent experiments.

Chi6 or Chi8 context (Chi9 and Chi10, respectively). These two chimeras both formed a high-affinity complex with Pγ-LY in the presence of AlF$_4^-$ (K_d 70 and 38 nM, respectively), just 3- to 6-fold lower than the affinity of wild-type $G\alpha_t^*$ for Pγ. The 237–270 region of $G\alpha_t$ consists of the α_3 helix and α_3/β_5 loop as well as conserved β_5 sheet and NKXD guanidine ring binding element. Comparison of $G\alpha_t$ and $G\alpha_{i1}$ sequences in this region shows that the main differences between the two proteins occur in the α_3 helix and α_3/β_5 loop. By ruling out the conserved regions, we can narrow down the most important determinant for high-affinity Pγ binding to residues 237–257 of $G\alpha_t$.

Interaction of Chimeras with RGS9: Molecular Determinants of Selective Interaction with RGS9

RGS proteins are potent accelerators of the intrinsic GTPase activity of G protein α subunits. These proteins are encoded by at least 19 genes and have been identified in mammalian tissues based on homology to the diagnostic RGS core domain of ~120 amino acid residues.[24] One of the members of this family, RGS9, is expressed predominantly in photoreceptor cells and is unique in its ability to act synergistically with a downstream effector to stimulate GTPase activity of $G\alpha_t$.[25] Interestingly, the core domain of RGS9 (RGS9d) is a more potent stimulator of GTPase activity of $G\alpha_t$ than $G\alpha_{i1}$. RGS9d stimulated $G\alpha_t$ GTPase activity by 10-fold and $G\alpha_{i1}$ GTPase by only twofold at a concentration of 10 μM.[5] To compare the functional effects of RGS9d with its affinity for different substrates, we have developed a sensitive fluorescence binding assay. As we reported earlier,[26,27] Cys-210 located at the distal end of the switch II region of $G\alpha_t$ can be labeled selectively with the thiol-specific reagent Lucifer Yellow. $G\alpha_t$ and Chi6 have only two cysteines accessible for modification with LY, located at positions 210 and 347. We have replaced Cys-347 of Chi6 with Ser. The resulting mutant (Chi6b) was labeled selectively at the only accessible Cys-210 with the fluorescent group. LY at Cys-210 is a reporter of the activating conformational change in the switch II region. The addition of AlF_4^- to the labeled protein resulted in a 200% increase in LY fluorescence (Fig. 8). The addition of RGS9 increased the fluorescence of Chi6b-LY by 108% at the maximal dose (2 μM RGS9d, Fig. 8).

We have used $G\alpha_t/G\alpha_{i1}$ chimeras in this fluorescence assay to determine their affinities for RGS9d. Binding of RGS9d to Chi6b-LY in the presence of different $G\alpha$'s was monitored by a fluorescence change of LY at Cys-210. Fluorescent measurements were performed on an Aminco-Bowman Series 2 Luminescence Spectrometer (SLM Aminco) at room temperature in 50 mM Tris-HCl, pH 8.0, 50 mM NaCl, 2 mM $MgCl_2$ using excitation at 430 nm and emission at 520 nm. To determine the affinity of RGS9d to $G\alpha$ substrates, we used competition between Chi6b-LY and different $G\alpha$'s for binding to RGS9d. The principle of a competition binding assay is that binding of unlabeled $G\alpha$ to RGS9 causes a displacement of fluorescently labeled Chi6b from its complex with RGS9d, resulting in a decrease in the initial fluorescence of the Chi6-LY-RGS9d complex. For competition measurements 50 nM Chi6b-LY activated with AlF_4^- was mixed with 100 nM RGS9d. A typical increase in fluorescence was 50–60% of the initial. Fluorescence of the Chi6b-LY-RGS9d complex was set to 100% (maximal fluorescence change). The decrease in fluorescence was monitored after

---

[24] D. M. Berman and A. G. Gilman, *J. Biol. Chem.* **273**(3), 1269 (1998).

[25] W. He, C. W. Cowan and T. G. Wensel, *Neuron* **20,** 95 (1998).

[26] C.-S. Yang, N. P. Skiba, M. R. Mazzoni, and H. E. Hamm, *J. Biol. Chem.* **274**(4), 2379 (1999).

[27] C.-S. Yang, N. P. Skiba, M. R. Mazzoni, T. O. Thomas, and H. E. Hamm, (1999), *Methods Enzymol.* **315** [33], 1999 (this volume).

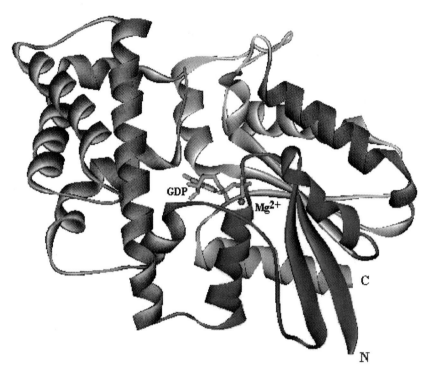

FIG. 10. Structural determinants of specific $G\alpha_t$ functions. Ribbon drawing of $G\alpha_t$–GDP (gray) with the determinants of GDP specificity (residues 26–55 and 174–215) highlighted in green, the determinants of RGS9 specificity (residues 56–173) highlighted in pink, and the determinants of effector specificity (residues 237–270) highlighted in brown. GDP is indicated in light blue and Mg^{2+} is seen as a dark blue sphere. The N and C termini of the molecule are indicated.

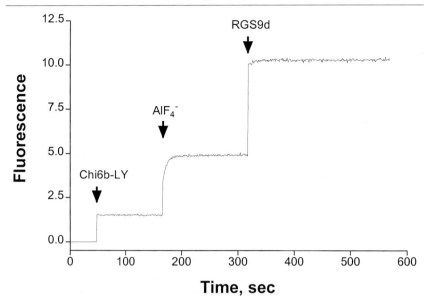

FIG. 8. Binding of RGS9d to Chi6b-LY. Sequential fluorescence increase of Chi6-LY (50 nM) on addition of 10 mM NaF and 30 μM AlCl₃, followed by addition of 2 μM RGS9d. The AlF₄⁻ dependent increase of Chi6-LY was 200 ± 20%. The maximal fluorescence increase of activated chimera in the presence of 2 μM RGS9d was 104 ± 5%. The fluorescence trace represents one of five independent similar experiments.

the addition of increasing concentrations of $G\alpha_t$, $G\alpha_{i1}$, or chimera and normalized to the percent of maximal fluorescence. The initial fluorescence of Chi6b-LY activated with AlF_4^- was set to 0%. The fluorescence change, after addition of $G\alpha$, is expressed as a percent of maximal change and plotted against the $G\alpha$ concentrations using the four-parameter logistic equation: $Y = Y_0 + (Y_{max} - Y_0)/1 + 10^{(\log EC_{50} - X)*H}$, where Y_0 and Y_{max} represent, respectively, the initial and maximal values of the binding function, which describes the normalized fluorescence at a given concentration of interacting protein; X is the logarithm of $G\alpha$ concentration; and H is the Hill coefficient. K_d values for binding of $G\alpha_t$, $G\alpha_{i1}$, and chimeras were determined from their competition curves using the equation $K_d = EC_{50}/(1 + [L]/K_d c)$, where [L] = 50 nM—concentration of competing, labeled $G\alpha$, Chi6b-LY; $K_d c$ is the K_d of Chi6b-LY-RGS9d complex, which is 190 nM.[5]

The K_d of $G\alpha_t$GDP–AlF_4^- complex calculated from the competition curve (Fig. 9A) was 185 nM. $G\alpha_t$GTPγS binds to RGS9d fivefold weaker (K_d 900 nM) compared to $G\alpha_t$ in its nucleotide transition state form. Replacement of the 216–294 region of $G\alpha_t$ with the corresponding region of

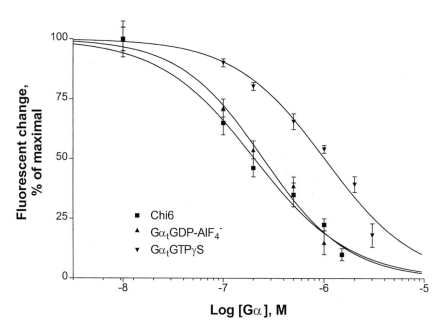

A

- ■ Chi6
- ▲ Gα$_t$GDP-AlF$_4^-$
- ▼ Gα$_t$GTPγS

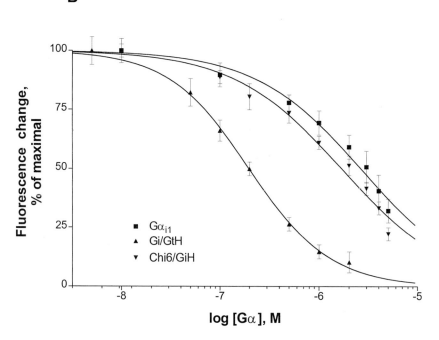

B

- ■ Gα$_{i1}$
- ▲ Gi/GtH
- ▼ Chi6/GiH

Gα$_{i1}$ (Chi6), a region containing the switch III region, does not significantly change the affinity of Chi6 to RGS9d (K_d 174 nM, Fig. 9A). This indicates that the least conserved switch region, region III, seemingly does not play a role in providing specificity of RGS9–Gα interaction. Gα$_{i1}$ was also able to compete with Chi6b-LY for binding to RGS9d. The affinity of Gα$_{i1}$ for RGS9d calculated from the competition curve (K_d 2 μM, Fig. 9B) was more than 10-fold lower than that of Gα$_t$. Decreased affinity of Gα$_{i1}$ for RGS9 and decreased ability of RGS9 to stimulate Gα$_{i1}$ GTPase activity, as compared to Gα$_t$, indicate that RGS9 interaction with Gα$_t$ is more specific than its interaction with Gα$_{i1}$.

To determine the region of Gα$_t$ that is responsible for the increased affinity to RGS9d we used Gα$_t$/Gα$_{i1}$ chimeras with exchanged helical domains in the fluorescent competition assay (Fig. 9B). Replacement of the α-helical domain of Gα$_{i1}$ with the corresponding region of Gα$_t$ (chimera Gi/GtH) resulted in more than a 10-fold increase in the affinity for RGS9d (K_d 170 nM) compared with Gα$_{i1}$. On the other hand, the reciprocal chimera where the α-helical domain of Gα$_t$ was replaced with the corresponding domain of Gα$_{i1}$ (chimera Chi6/GiH) exhibited decreased affinity for RGS9d (K_d 1.6 μM, Fig. 9B) compared with Gα$_t$ or Chi6, but was similar to Gα$_{i1}$. These data demonstrate that the determinants of Gα$_t$ specificity toward RGS9 reside in the α-helical domain. This region is highlighted in a ribbon and coil schematic of Gα$_t$ (Fig. 10, see color insert).

Conclusions

Here we demonstrate a variety of useful applications of Gα$_t$/Gα$_{i1}$ chimeras to study the structural determinants of Gα$_t$ functions. First, we were able to solve the folding problem of Gα$_t$ by replacing its problem regions (residues 216–228 and 271–294) with the corresponding regions of Gα$_{i1}$, which is able to undergo functional self-assembly on expression in bacterial cells. Surprisingly, nearly 95% of the insoluble protein (Gα$_t$) can be incorporated into a chimeric protein and still maintain its solubility and functionality. By targeting and replacing specific regions within Gα$_t$ with correspond-

---

FIG. 9. Binding of Gα$_t$, Gα$_{i1}$, and chimeras to RGS9d. (A) Competition between Gα$_t$GDP-AlF$_4^-$, Gα$_t$GTPγS, or Chi6 and Chi6B-LY for binding to RGS9d. The fluorescence of the complex of RGS9d (100 nM) with Chi6b-LY (50 nM) in the presence of AlF$_4^-$ before and after the addition of increasing concentrations of Gα$_t$GDP, Gα$_t$GTPγS and Chi6. (B) Competition between Gα$_{i1}$ and chimeras for binding to RGS9d. The fluorescence of Chi6b-LY-AlF$_4^-$ (50 nM) in the presence of RGS9d (100 nM) was measured before and after the addition of increasing concentrations of Gα$_{i1}$, ChiG$_i$/G$_t$H or Chi6/GiH. Data points represent the mean percent fluorescence change ± S.E. (n = 3).

ing regions of $G\alpha_{i1}$ we have defined the structural determinants of several $G\alpha$ functions that are distinct for the parent proteins. Analysis of the rate of intrinsic nucleotide exchange for different chimeras highlights the important role of intramolecular contacts between N terminus and switch I/II regions of $G\alpha_t$ in stabilizing and strengthening GDP binding. Using another set of chimeras we have defined that molecular determinants for $G\alpha$ specificity toward RGS9 reside in the α-helical domain of $G\alpha_t$. And finally, analysis of binding of chimeras to the target site of $G\alpha_t$ effector, Pγ, has revealed that the $G\alpha_t$ region encompassing α helix 3 and α_3/β_5 loop contributes most to interaction with Pγ.

Acknowledgments

We thank Jennifer Connor for preparation of bovine retinal $G\alpha_t$ and $G\beta\gamma$, Chii-Shen Yang for preparation of Chi6b labeled with LY, and Theresa Vera and Annette Gilchrist for critical reading of the manuscript. This work was supported by NIH grants EY06062 and EY10291 to H.E.H. and American Heart Association Grant-In-Aid to N.P.S.

[35] Enzymology of GTPase Acceleration in Phototransduction

By CHRISTOPHER W. COWAN, THEODORE G. WENSEL, and VADIM Y. ARSHAVSKY

Introduction

GTPase Cycle in Phototransduction

The photoreceptor G protein, transducin (G_t), acts as a molecular switch that carries information between the light-activated receptor, rhodopsin (R*), and the effector, cGMP phosphodiesterase (PDE) (reviewed in Refs. 1 and 2). The cycle of G_t activation and inactivation includes at least seven individual steps whose understanding is essential for designing experiments described in this chapter:

1. Until activated, G_t exists almost exclusively as the heterotrimeric form with GDP bound to its α subunit, $G_t\alpha\beta\gamma$–GDP. GDP dissocia-

---

[1] M. Chabre and P. Deterre, *Eur. J. Biochem.* **179,** 255 (1989).
[2] E. N. Pugh and T. D. Lamb, *Biochim. Biophys. Acta BioEnergetics* **1141,** 111 (1993).

tion from this complex is extremely low, with a rate constant of less than 10^{-4} s^{-1}.[3,4] In the absence of R* this step is the rate-limiting one for the whole cycle.

2. On light activation, R* binds to $G_t\alpha\beta\gamma$–GDP and triggers rapid GDP release from $G_t\alpha$. On disk membranes this event happens on a timescale of a few milliseconds,[5,6] but in dilute suspensions, this encounter can be much slower.

3. $G_t\alpha$ then rapidly binds GTP in a second-order reaction with a rate constant of $\sim 10^7\ M^{-1}$ s^{-1},[5] which is essentially instantaneous at millimolar intracellular GTP concentrations. However, GTP binding could become rate limiting in experiments utilizing low concentrations of GTP or R*–G_t.

4. $G_t\alpha$–GTP dissociates from R* and $G_t\beta\gamma$. At this stage, $G_t\alpha$ has the ability to activate PDE. The primary target of transducin is the γ subunit of PDE (PDEγ) whose inhibitory action on the PDE catalytic α and β subunits is removed by $G_t\alpha$–GTP.

5. GTP bound to $G_t\alpha$ is hydrolyzed to GDP and P_i. The intrinsic rate of transducin GTPase is slow, ~ 0.05 s^{-1}.[7,8] As discussed later, in photoreceptors this rate is significantly accelerated by a coordinated action of at least two proteins, PDEγ and the complex between the ninth member of the regulators of G-protein signaling family (RGS9) and the long splice variant of type 5 G protein β subunit (Gβ_{5L}). Dim flash response kinetics[9] and direct measurements[9a] suggest that the rate of transducin GTPase likely exceeds 7 s^{-1} when maximally accelerated.

6. P_i is released from the $G_t\alpha$–GDP–P_i complex. This reaction is clearly not rate limiting at low to moderate hydrolysis rates,[7] but its kinetics at maximal GTPase acceleration have yet to be determined.

7. At this final step $G_t\alpha$–GDP reassociates with $G_t\beta\gamma$ so it can be activated by R* in the next cycle. This "recycling" step can be quite slow in dilute suspensions. Of particular concern for assays of GTPase acceleration is that proteins which accelerate GTP hydrolysis can

---

[3] L. Ramdas, R. M. Disher, and T. G. Wensel, *Biochemistry* **30,** 11637 (1991).
[4] A. B. Fawzi and J. K. Northup, *Biochemistry* **29,** 3804 (1990).
[5] M. Kahlert and K. P. Hofmann, *Biophys. J.* **59,** 375 (1991).
[6] T. M. Vuong, M. Chabre, and L. Stryer, *Nature* **311,** 659 (1984).
[7] V. Y. Arshavsky, M. P. Antoch, and P. P. Philippov, *FEBS Lett.* **224,** 19 (1987).
[8] J. K. Angleson and T. G. Wensel, *Neuron* **11,** 939 (1993).
[9] S. H. Tsang, M. E. Burns, P. D. Calvert, P. Gouras, D. A. Baylor, S. P. Goff, and V. Y. Arshavsky, *Science* **282,** 117 (1998).
[9a] C. W. Cowan, unpublished observations (1998).

also inhibit recycling. The clearest example is PDEγ, which binds to $G_t\alpha$–GDP and slows recycling.[10,11]

GTPase Acceleration in Phototransduction

Photoreceptor recovery from the light response requires that $G_t\alpha$–GTP hydrolyze its bound GTP.[12–14] However, isolated $G_t\alpha$ hydrolyzes GTP too slowly to account for the rapid recovery rates observed *in vivo*.[7,8] A solution for this problem has been found by showing that photoreceptors contain a mechanism responsible for activating the slow intrinsic rate of $G_t\alpha$–GTPase to the rates sufficient for photoresponse recovery on the physiologic timescale. Two proteins in rod outer segments (ROS) have been shown to accelerate $G_t\alpha$–GTPase: RGS9[15,16] and PDEγ.[9,17–19] Makino et al.[20] have demonstrated that RGS9 in photoreceptors exists as a tight complex with $G\beta_{5L}$.

The role of RGS9 is to provide transducin with the RGS homology domain, which most likely stabilizes $G_t\alpha$ in the conformation favorable for the GTP hydrolysis. This became evident from crystallographic studies with the transition state analog GDP–AlF$_4$ bound either to $G_t\alpha$,[21] or to $G_t\alpha$ complexed with the RGS9 homolog, RGS4.[22] This idea is also supported by the finding that RGS domains bind preferentially to the GDP–AlF$_4^-$ forms of several different $G\alpha$ subunits.[23] This stabilization and GTPase acceleration can be observed with RGS domains alone, including the RGS domain of RGS9.[16]

[10] D. F. Morrison, J. M. Cunnick, B. Oppert, and D. J. Takemoto, *J. Biol. Chem.* **264**, 11671 (1989).
[11] A. Yamazaki, M. Yamazaki, S. Tsuboi, A. Kishigami, K. O. Umbarger, L. Hutson, W. T. Madland, and F. Hayashi, *J. Biol. Chem.* **268**, 8899 (1993).
[12] W. A. Sather and P. B. Detwiler, *Proc. Natl. Acad. Sci. U.S.A.* **84**, 9290 (1987).
[13] T. D. Lamb and H. R. Matthews, *J. Physiol.* **407**, 463 (1988).
[14] M. S. Sagoo and L. Lagnado, *Nature* **389**, 392 (1997).
[15] W. He, C. W. Cowan, and T. G. Wensel, *Neuron* **20**, 95 (1998).
[16] C. W. Cowan, R. N. Farris, I. Sokal, K. Palczewski, and T. G. Wensel, *Proc. Natl. Acad. Sci. U.S.A.* **95**, 5351 (1998).
[17] V. Y. Arshavsky and M. D. Bownds, *Nature* **357**, 416 (1992).
[18] V. Y. Arshavsky, C. L. Dumke, Y. Zhu, N. O. Artemyev, N. P. Skiba, H. E. Hamm, and M. D. Bownds, *J. Biol. Chem.* **269**, 19882 (1994).
[19] J. K. Angleson and T. G. Wensel, *J. Biol. Chem.* **269**, 16290 (1994).
[20] E. R. Makino, J. W. Handy, T. Li, and V. Y. Arshavsky, *Proc. Natl. Acad. Sci. U.S.A.* **96**, 1947 (1999).
[21] J. Sondek, D. G. Lambright, J. P. Noel, H. E. Hamm, and P. B. Sigler, *Nature* **372**, 276 (1994).
[22] J. J. G. Tesmer, D. M. Berman, A. G. Gilman, and S. R. Sprang, *Cell* **89**, 251 (1997).
[23] N. Watson, M. E. Linder, K. M. Druey, J. H. Kehrl, and K. J. Blumer, *Nature* **383**, 172 (1996).

PDEγ itself does not activate transducin GTPase, but it enhances the catalytic action of RGS9. The degree of this potentiation observed in physiologically intact photoreceptors is approximately sevenfold.[9] The mechanism of PDEγ action is an area still under active investigation, as is the possible involvement of additional proteins such as $G\beta_{5L}$.

Two Kinetic Approaches for Measuring Transducin GTPase

The major complication in measuring the rate at which GTP bound to $G_t\alpha$ is hydrolyzed to GDP and P_i is that reactions of the G_t activation/inactivation cycle, different from GTPase itself, may become rate limiting for GTP hydrolysis under various experimental conditions. This consideration is particularly important for designing the most traditionally used *multiple-turnover method* consisting of monitoring the steady-state rate of $G_t\alpha$–GTP hydrolysis during cycles of transducin activation and inactivation (for detailed methods see Refs. 24–26). The major characteristic feature of this approach is that GTP has to be present in amounts larger than transducin. The use of this assay requires establishing conditions where GTP hydrolysis is the rate-limiting step. Such conditions are easily achieved in a reconstituted system containing purified G_t and ROS membranes free of any components regulating the rates of $G_t\alpha$–GTPase or GTP uptake. Four essential requirements have to be met. First, GTP should be present at saturating concentration. This should be tested by demonstrating that the rate of transducin GTPase remains constant when GTP concentration is reduced by two- to threefold. Second, the rate of GTP binding to transducin should not be limited by the amount of R* in the assay. This can be controlled by demonstrating that a severalfold increase in the membrane concentration does not lead to an increased rate of transducin GTPase. Third, the rate of G_t recycling should not be limited by the rate of $G_t\alpha$–GDP reassociation with $G_t\beta\gamma$. This should be controlled by demonstrating that the GTPase rate constant does not increase on an increase in transducin concentration. Finally, the production of GDP and P_i has to be a linear function of time until a substantial fraction of added GTP is hydrolyzed.

This method is rather straightforward; however, it not useful in experiments addressing the function of G_t–GTPase activating proteins. As discussed earlier, at least some of them may not only accelerate GTP hydrolysis but may also interfere with G_t recycling. For the same reason, the multiple turnover method could yield ambiguous results when performed with sus-

---

[24] V. Y. Arshavsky, M. P. Gray-Keller, and M. D. Bownds, *J. Biol. Chem.* **266**, 18530 (1991).
[25] F. Pages, P. Deterre, and C. Pfister, *J. Biol. Chem.* **267**, 22018 (1992).
[26] F. Pages, P. Deterre, and C. Pfister, *J. Biol. Chem.* **268**, 26358 (1993).

pensions of ROS containing endogenous GTPase activating proteins (GAPs).

The best way to avoid this problem is to use the *single-turnover methodology*. The design of the single-turnover GTPase assay restricts $G_t\alpha$–GTP primarily to only one round of GTP hydrolysis, removing the potential rate-limiting process of G_t recycling. Single-turnover conditions are generated in one of two ways: (1) addition of GTP at a much lower concentration than that of G_t under conditions where uptake is very rapid relative to hydrolysis and recycling,[8,24] and (2) rapid separation of activated $G_t\alpha$–GTP from R*.[27,28]

Alternative indirect approaches to monitoring hydrolysis under single-turnover conditions include measuring tryptophan fluorescence changes associated with $G_t\alpha$–GTPase,[27,29] light scattering changes accompanying the cycle of transducin activation/inactivation,[30,31] and dynamics of PDE stimulation by $G_t\alpha$–GTP.[32] These methods have the advantage of providing a continuous readout, and in the case of PDE activity sufficient sensitivity to follow hydrolysis at very low R* levels. However, they require performing additional control experiments in order to verify that the signal changes observed are due exclusively to GTP hydrolysis.

The major emphasis of this chapter is on describing a single-turnover method for measuring $G_t\alpha$–GTPase that is relatively simple and reliable and uses ^{32}P-labeled GTP as the substrate. We first provide a detailed discussion of all principle steps of the assay, including initiating the reaction by GTP binding to $G_t\alpha$, quenching the reaction, and determining the amount of hydrolyzed GTP. We then describe how this method could be utilized for studying the regulation of transducin GTPase by various GAPs. At the end, we list two detailed protocols outlining different modifications of this technique.

Initializing Reaction by GTP Binding to $G_t\alpha$

There are two central ideas behind the single-turnover method for measuring the rate of transducin GTPase. First, the reaction is initiated by addition of GTP in amounts much less than G_t present in the system, so only one turnover of GTP hydrolysis is allowed. Second, the reaction is

---

[27] B. Antonny, A. Otto-Bruc, M. Chabre, and T. M. Vuong, *Biochemistry* **32**, 8646 (1993).
[28] A. Otto-Bruc, B. Antonny, and T. M. Vuong, *Biochemistry* **33**, 15215 (1994).
[29] P. M. Guy, J. G. Koland, and R. A. Cerione, *Biochemistry* **29**, 6954 (1990).
[30] H. Kuhn, N. Bennett, M. Michel-Villaz, and M. Chabre, *Proc. Natl. Acad. Sci. U.S.A.* **78**, 6873 (1981).
[31] R. Wagner, N. Ryba, and R. Uhl, *FEBS Lett.* **234**, 44 (1988).
[32] T. M. Vuong and M. Chabre, *Proc. Natl. Acad. Sci. U.S.A.* **88**, 9813 (1991).

performed under conditions where the rate of GTP binding to G_t is significantly faster than the rate of the subsequent GTP hydrolysis, so that the hydrolytic rate can be directly calculated from the exponential fit of the time course of GTP hydrolysis. An important feature of the single-turnover conditions when [GTP] ≪ [G_t–R*] is that the time constant of GTP binding to transducin is not dependent on concentration of GTP, but rather is dependent on the concentration of the G_t–R* complex. Therefore, demonstrating that variations in GTP concentration do not result in changes of the GTP hydrolysis rate is not a relevant control for showing that the requirement for GTP binding to be faster than GTP hydrolysis is fulfilled. In most cases, reliable determinations of the catalytic rate constant of transducin-bound GTP hydrolysis could be performed when the concentration of G_t–R* is 2 μM or higher.

Several approaches are used to show that the requirement for the GTP binding rate to exceed the rate of GTPase hydrolysis is met in a particular experimental design. The most simple control is to show that an increase in the G_t–R* concentration in the reaction mixture does not result in an increase in the overall rate of GTP hydrolysis. Unfortunately, this approach is limited to experiments performed with purified proteins and cannot be used with ROS membranes containing the GTPase activating factors. A more commonly used alternative consists of the GTP binding rate measurement by the pulse-chase technique. The GTPase reaction is started by mixing G_t–R* with [γ-^{32}P]GTP, and then at least a 100-fold excess of nonlabeled GTPγS (or GTP) is added at various times to prevent further binding of labeled GTP. The reaction should then be stopped at the time sufficient for the hydrolysis of all [γ-^{32}P]GTP bound at the time of the chase (5 min is sufficient). The amount of total GTP hydrolyzed should be then plotted as a function of time before the chase and the rate of GTP binding can be calculated from the exponential fit of the data.

On performing the GTP binding assay it is crucial to realize that GTP binding to G_t is reversible in the presence of R*.[18,33] The binding equilibrium is significantly shifted toward the $G_t\alpha$–GTP complex formation, so in most cases this reversibility does not affect the observed rate of GTP hydrolysis. However, it has to be considered in the pulse-chase assays because some [γ-^{32}P]GTP bound to $G_t\alpha$ at the time of chase could be exchanged for the nonlabeled GTP instead of being hydrolyzed. As a result, the binding rate would be underestimated. As described in Ref. 18, the best way to avoid this problem is to minimize the R* content in the sample. The G_t–R* complex should be formed by bleaching the mixture of G_t with ROS membrane preparation on ice and then the mixture should be incubated for at

---

[33] J. S. Mills, N. O. Artemyev, N. P. Skiba, and H. E. Hamm, *Biophys. J.* **64**, A383 (1993).

least one hour at room temperature. The incubation results in a thermal decay of free R*, but not R* within the complex with G_t, because the formation of this complex stabilizes the R* conformation (reviewed in Ref. 1). Alternatively, the experiments can be performed in the dark with only a small (<10%) fraction of rhodopsin bleached.

A similar assay can be performed for a single time point such as the 5-sec point used for the preincubation assay described later. In this case, the fraction of GTP bound at the time of chase should be calculated by dividing the amount of GTP hydrolyzed after the chase by the amount of GTP hydrolyzed in a parallel sample where the chase has not been performed.

For assays in which convolution of uptake with hydrolysis significantly affects the time course, an accurate value of catalytic rate constant can still be derived by fitting the entire time course to a system of equations,[8] with GTP uptake kinetics calibrated separately. Note that such time courses may have phases that resemble single exponential decays that are greatly affected by the GTP binding rate constant.

In some experiments addressing the properties of the G_t–GTPase regulators, such as RGS9 and PDEγ, an uncertainty may exist as to whether these proteins affect only the rate of GTP hydrolysis or if they also influence the rate of GTP binding. In this case it is often convenient to include a preactivation time (e.g., 5 sec[8,19]) before addition of candidate GTPase accelerating or inhibiting factor to ensure that essentially all of the GTP is bound to G_t prior to addition of test samples. With this design, test samples added at 5 sec can affect GTP hydrolysis, but not uptake. The single-turnover GTPase assay may be performed with purified ROS containing the endogenous complement of transducin (typically 5–10% of the rhodopsin concentration) or with urea-washed ROS (uw-ROS), lacking the endogenous transducin and GAPs, reconstituted with purified G_t. When using purified ROS to activate G_t, consider that it contains all of the endogenous components of phototransduction, including the GAP, RGS9. Therefore, the kinetics of $G_t\alpha$–GTPase will be affected by the concentration of ROS membranes.[8,34,35] In contrast, the concentration of reconstituted uw-ROS does not affect $G_t\alpha$–GTPase kinetics since the endogenous GAP activity of RGS9 is inactivated.[19] For this reason, uw-ROS membranes are preferred for testing the intrinsic $G_t\alpha$–GTPase rate and for testing candidate proteins for effects on GAP activity. A relatively small amount of uw-ROS is re-

---

[34] E. A. Dratz, J. W. Lewis, L. E. Schaechter, K. R. Parker, and D. S. Kliger, *Biochem. Biophys. Res. Commun.* **146,** 379 (1987).

[35] V. Y. Arshavsky, M. P. Antoch, K. A. Lukjanov, and P. P. Philippov, *FEBS Lett.* **250,** 353 (1989).

quired for efficient and rapid activation of transducin in the GTPase assay. Typically 5–15 μM uw-ROS freshly illuminated and reconstituted with 0.5–1.0 μM G_t is sufficient for rapid activation if the complex of a nucleotide-free G_t complex with R* is preformed prior to the addition of GTP.

Several factors have been reported to affect the rate of GTP binding. The most important is Mg^{2+},[36] which stabilizes and coordinates $G_t\alpha$ residues in the nucleotide-binding pocket.[37,38] $MgCl_2$ (2–8 mM) is routinely included in the assays. The GTP binding rate at 8 mM Mg^{2+} is about twice as fast as the rate observed at 2 mM Mg^{2+}.[38a] Many G proteins have been shown to be able to bind but not to hydrolyze GTP in the absence of Mg^{2+}. This feature has been utilized in developing an alternative single-turnover method applicable for $G_i\alpha$ and $G_o\alpha$.[23,39] Their approach is based on prebinding of GTP in the absence of Mg^{2+}, followed by initiation of the GTPase reaction with $MgCl_2$. Unfortunately, this method could not be used for transducin. Unlike the $G_{i/o}$ proteins, G_t does not readily exchange the bound GDP in the absence of activated receptor, R*, requiring several hours for nucleotide exchange.[3,4] Therefore, in the absence of R* the rate constant for GTP hydrolysis always exceeds that for nucleotide exchange. Another problem is that GTP hydrolysis by G_t can proceed in the presence of EDTA in excess over Mg^{2+}.[36,40]

Monitoring Time Course of GTP Hydrolysis and Quenching Reactions

Additional assay parameters found to influence $G_t\alpha$–GTPase kinetics are high salt concentration[8] and Cl^-,[27] both of which inhibit GTP hydrolysis. For this reason, the salt concentration in most $G_t\alpha$–GTPase assays is kept under ~100 mM. Raising the assay temperature from 23° to 37° increases the $G_t\alpha$–GTPase approximately two- to threefold.[7,8] Assays have been slowed by carrying them out at 4°.[41] Ca^{2+}, which effects the recovery of the light response in ROS, does not appear to influence $G_t\alpha$–GTPase as indicated by lack of an effect of 1 mM EGTA[8] or with 5–500 nM Ca^{2+}.[24]

The kinetic time course of transducin GTPase, in the absence of accelerating factors, occurs on a timescale sufficiently slow for assays at room temperature ($k_{cat} = 0.05$ sec^{-1}). For the accelerated $G_t\alpha$–GTPase reactions occurring on a timescale faster than ~5 sec, it is difficult to quench the

---

[36] G. Yamanaka, F. Eckstein, and L. Stryer, *Biochemistry* **24,** 8094 (1985).
[37] J. P. Noel, H. E. Hamm, and P. B. Sigler, *Nature* **366,** 654 (1993).
[38] D. G. Lambright, J. P. Noel, H. E. Hamm, and P. B. Sigler, *Nature* **369,** 621 (1994).
[38a] V. Y. Arshavsky, unpublished observations (1998).
[39] D. M. Berman, T. M. Wilkie, and A. G. Gilman, *Cell* **86,** 445 (1996).
[40] N. Bennett and Y. DuPont, *J. Biol. Chem.* **260,** 4156 (1985).
[41] T. Wieland, C.-K. Chen, and M. I. Simon, *J. Biol. Chem.* **272,** 8853 (1997).

reaction accurately at the intended time. To overcome this problem, a tape recorder may be used to verbally record the times of the reaction start and quench. The record could be then later replayed to determine the accurate time (the method was originally introduced by M. D. Bownds). Another way to improve the temporal resolution of the method is to conduct the GTPase reaction at low temperature.

For assays in which uptake is clearly much faster than hydrolysis, the hydrolysis time course can be fit by standard nonlinear least squares methods according to $P_i(t) = GTP_0[1 - \exp(-kt)]$, where $P_i(t)$ represents the amount of inorganic phosphate released on the GTP hydrolysis at time t after addition, GTP_0 represents the total amount of added GTP, and k is the first-order hydrolysis rate constant.

When testing a large number of samples for GAP activity, it can be useful to perform the assay using a single time point performed in duplicate or triplicate for each sample. Rate constants can be estimated from extent of hydrolysis by assuming an exponential time course according to $k = -\ln\{[GTP_0 - P_i(t)]/GTP_0\}/t$, where t is the time of quenching. If a preactivation method is used, t represents the time *after* addition of test sample at time t_{pre} (rather than after initial GTP addition), $P_i(t)$ represents the difference between total P_i produced and that present at t_{pre}, and GTP_0 represents the GTP remaining at t_{pre}. The single time point method with or without preactivation does not provide accurate determinations of the rate constants, but does provide a high-throughput method for comparing activity in numerous samples. Note that a relative uncertainty of only 5% in the determination of $P_i(t)$ can give rise to an uncertainty of 50% in the calculated value of k if, for example, 90% of GTP is hydrolyzed in 10 sec, while a similar uncertainty at 40% hydrolysis gives rise to a 10% uncertainty in k.

For most of the $G_t\alpha$–GTPase assays described in the literature, the reaction is quenched by the addition of acid. The choice of acid is determined by the GTPase detection method. For detection of hydrolysis of [γ-^{32}P]GTP, trichloroacetic acid (TCA) or perchloric acid is frequently used,[15,24] whereas for hydrolysis of [α-^{32}P]GTP, trifluoroacetic acid (TFA) has been used[8] because its volatility minimizes interference with chromatographic separations. The addition of acid to quench the reaction has two major effects: (1) rapid denaturation of transducin and other proteins involved in the GTP hydrolysis and (2) release of bound radioactive nucleotides for subsequent separation and quantification.

Separating Hydrolyzed from Unhydrolyzed GTP

For assays utilizing [γ-^{32}P]GTP, two major techniques are used for separating the hydrolysis product [^{32}P]P_i from unhydrolyzed [γ-^{32}P]GTP:

(1) incubation with activated charcoal, which binds guanine (and other) nucleotides, but not P_i (or pyrophosphate) or (2) molybdate precipitation of $[^{32}P]P_i$, which does not precipitate unhydrolyzed $[\gamma\text{-}^{32}P]GTP$.[42] In most cases, the charcoal binding technique is recommended to be used due to its simplicity, reliability, and rapid processing time. To efficiently bind and separate unhydrolyzed $[\gamma\text{-}^{32}P]GTP$, a sufficient quantity of charcoal must be used. The binding capacity of the charcoal suspension used in the assay should be tested to be sufficient for complete binding of GTP at amounts used in the experiment. The charcoal should be supplemented by Na_2HPO_4 to occupy any weak P_i binding sites on the charcoal and the test tube walls. The radioactivity in the supernatant from the charcoal suspension is then determined by scintillation counting.

The molybdate precipitation technique employs the quenching of the $G_t\alpha$–GTPase with perchloric acid, followed by addition of ammonium molybdate, triethylamine hydrochloride (TEA), and KH_2PO_4.[43] The procedure yields a yellow precipitate that can be separated from the unhydrolyzed $[\gamma\text{-}^{32}P]GTP$ by standard filtration of the precipitate using glass fiber filters. Sugino and Miyoshi[42] report that the precipitation results with perchloric acid are also obtained with H_2SO_4, HCl, or TCA.

Alternatively, GTP hydrolysis can be measured using $[\alpha\text{-}^{32}P]GTP$ by following its conversion to $[\alpha\text{-}^{32}P]GTP$ by chromatography, e.g., on polyethyleneimine cellulose thin-layer chromatography (TLC) plates.[8] This assay unambiguously identifies the substrate, reaction product, and impurities in the starting material. However, it is slow and tedious due to the relatively long time required for sample loading and capillary migration of the nucleotides. The TLC-separated hydrolysis products may then be quantified by scanning with a phosphorimager or by cutting out the hydrolysis products and scintillation counting of radioactivity.

General Controls

Since a small fraction of the $[\gamma\text{-}^{32}P]GTP$ or $[\alpha\text{-}^{32}P]GTP$ is always hydrolyzed in commercial stocks (usually <5%), a control reaction has to be performed in each experiment. The reaction mixture containing G_t–R* should be first "quenched" by acid and then mixed with $[^{32}P]GTP$. The amount of radioactivity remaining in the supernatant of the charcoal mix is then subtracted from all GTPase reactions containing G_t–R*, essentially removing GTP hydrolysis unrelated to $G_t\alpha$–GTPase.

[42] Y. Sugino and Y. Miyoshi, *J. Biol. Chem.* **239**, 2360 (1964).
[43] T. D. Ting and Y. K. Ho, *Biochemistry* **30**, 8996 (1991).

Another problem with commercial [^{32}P]GTP stocks is that they contain some radioactive material that can bind to charcoal but is not GTP. To accommodate for this problem the actual amount of [^{32}P]GTP in the stock has to be determined in a control sample. This sample containing normal amounts of G_t–R* should be incubated with [^{32}P]GTP for the time sufficient for a complete GTP hydrolysis, such as for 10 min. The amount of [^{32}P]P_i released on complete GTP hydrolysis in this sample less the amount of [^{32}P]P_i present in the stock (see the previous paragraph) should be used as the amount of radioactivity corresponding to 100% of the added GTP in all calculations. If this amount differs more than 15% from the total amount of radioactivity added to the sample, the [^{32}P]GTP stock should be replaced.

During the GTPase reaction, some of the GTP may be hydrolyzed by $G_t\alpha$–independent mechanisms, such as nonspecific GTP hydrolysis by contaminating ATPases. For this reason, a nonhydrolyzable ATP analog, AMP–PNP, is often included in assays to inhibit nonspecific GTPase activity in the sample. Because the ability of G proteins to discriminate between GTP and ATP by far exceeds that of most other GTP hydrolyzing enzymes, AMP–PNP does not significantly affect the $G_t\alpha$–GTPase kinetics. The easiest way to measure the rate of the nontransducin GTPase activity in ROS is to perform a control measurement in complete darkness. For reconstituted systems this activity could be assessed from separate measurements of GTPase activity in G_t and R* preparations. It is important to note that in the single-turnover experiments the actual impact of the non–G_t–GTPase activity is always smaller than in these controls. This is because the amount of GTP used in the single-turnover assays is small, and the binding of this GTP to G_t eliminates the possibility of its binding to other GTPase that could take place in the absence of G_t activation.

Testing Proteins That Regulate $G_t\alpha$–GTPase

One of the major applications of the assay described in this chapter is the testing of proteins or other molecules that might regulate $G_t\alpha$–GTPase. Many proteins of the RGS family enhance the rate of GTP hydrolysis by most G proteins of the G_i and G_q subfamilies (reviewed in Ref. 44). Several RGS proteins, including RET-RGS1,[45] RGS16,[41,46,47] RGS1,[23] RGS4, GAIP,[48] and RGS9,[15,16] accelerate the GTP hydrolysis by transducin. The

[44] D. M. Berman and A. G. Gilman, *J. Biol. Chem.* **273,** 1269 (1998).
[45] E. Faurobert and J. B. Hurley, *Proc. Natl. Acad. Sci. U.S.A.* **94,** 2945 (1997).
[46] C.-K. Chen, T. Wieland, and M. I. Simon, *Proc. Natl. Acad. Sci. U.S.A.* **93,** 12885 (1996).
[47] M. Natochin, A. E. Granovsky, and N. O. Artemyev, *J. Biol. Chem.* **272,** 17444 (1997).
[48] E. R. Nekrasova, D. M. Berman, R. R. Rustandi, H. E. Hamm, A. G. Gilman, and V. Y. Arshavsky, *Biochemistry* **36,** 7638 (1997).

most common technique employed in these studies is the single-turnover $G_t\alpha$–GTPase assay in a reconstituted system containing uw-ROS and purified G_t. For those proteins that could not be solubilized without detergent, $G_t\alpha$–GTPase assay could be performed with certain detergents present. They include n-octylglucoside[16,49] and lauryl sucrose.[20] It is important to note that a detergent may not affect the intrinsic $G_t\alpha$–GTPase rate, but may inhibit the potency of the GAP protein tested.

In addition to the GAP effects of RGS proteins, PDEγ dramatically enhances the rate of $G_t\alpha$–GTPase in the presence of RGS9.[15–19] PDEγ over 3.0 μM is usually saturating for the effect,[18] but larger amounts may be used at unusually high transducin concentrations in the assays. Native PDEγ, purified from holo-PDE, His-tagged recombinant PDEγ, expressed in bacteria, and recombinant PDEγ without affinity tag, expressed in bacteria all accelerate $G_t\alpha$–GTPase and inhibit the catalytic subunits of PDE in a similar way.[15,17–19]

The time of RGS protein or PDEγ (unless >20 μM PDEγ is used) addition to the assay is not essential for monitoring the GTPase activating effects. They could be added prior to the addition of GTP or at 5 sec following GTP addition, as in the preactivation method. However, the preactivation assay is strongly recommended to be used first in the experiments analyzing the effects of putative GTPase regulators. This strategy allows one to eliminate possible artifacts originating from the altered rates of GTP binding to G_t.

Standard and Preactivation Methods for Single-Turnover Measurements of G_t–GTPase Activity

Whereas both versions of the method described in this section could be performed with both ROS suspensions and reconstituted systems, the traditional method is outlined in application to the ROS suspension and the preactivation method in application to the reconstituted system.

Standard Method (Modified from Ref. 24)

Equipment

Vortex mixer
Electronic timer

---

[49] V. Y. Arshavsky, M. P. Antoch, A. M. Dizhoor, S. V. Rakhilin, and P. P. Philippov, in "Retinal Proteins" (Y. A. Ovchinnikov, ed.), p. 497. VNU Science Press, Utrecht, 1987.

Tape recorder
Microcentrifuge
Stirring plate
Liquid scintillation counter

Solutions

> *Assay buffer:* 10 mM Tris-HCl, pH 7.8, 100 mM NaCl, 8 mM MgCl$_2$, and 1 mM dithiothreitol (DTT)
> *ROS stock:* ROS suspension in the assay buffer (It should be bleached on ice shortly before the experiment.)
> *GTP stock:* [γ-^{32}P]GTP prepared by mixing nonradioactive GTP solution with [γ-^{32}P]GTP (NEN) at the amount of ~10^5 dpm per sample [This solution also contains 1 mM AMP–PNP (5'-adenylylimidodiphosphate).]
> *Quenching solution:* 6% perchloric acid
> *Charcoal suspension:* 0.1 g of activated charcoal/ml of 50 mM potassium phosphate buffer, pH 7.5)
> *Ecoscint liquid scintillation cocktail* (Fisher Scientific, Pittsburgh, PA)

Assay

A typical experiment contains six or seven time points distributed over the anticipated time course of the reaction. For each sample time point, a 20-μl reaction mixture containing at least 40 μM ROS (by rhodopsin) is placed in a 1.5-ml microcentrifuge tube. The use of lower ROS concentrations is not recommended because the rate of GTP binding might influence the value of the GTPase rate constant determined from the assay. If required, ROS could be supplemented by GAPs. Four additional samples are prepared: two "blanks" and two "totals." For each time point, the reaction mixture is continually vortexed while 20 μl of [γ-^{32}P]GTP is added to initiate the assay. The concentration of GTP is not essential, as long as it is at least fivefold lower than the G_t content in the sample. The tape recorder should continue throughout the experiment. The addition of GTP should be quick and, on addition, the experimenter should say "now." Some training might be required to learn how to say "now" exactly at the time of the GTP addition. At the predetermined time, 100 μl of 6% perchloric acid is added to vortexed mixture to quench the reaction. Again, the experimenter says "now" simultaneously with the addition. The "total" samples are incubated for 5–10 min to achieve a complete hydrolysis of all added GTP. The "blank" samples are quenched by acid prior to the addition of GTP. Each quenched sample is placed on ice and on accomplishment of the whole set of the samples, 700 μl of the charcoal solution is added to each of them. The charcoal suspension is constantly stirred in a beaker throughout the experiment. The samples are vortexed three times during 20 min on ice. The tubes are centrifuged at 12,000g for 4 min at room temperature. Supernatants (500 μl) are removed and added to 10 ml of Ecoscint scintillation cocktail, and the [^{32}P]P$_i$ radioactivity is determined by liquid scintillation counting. The actual time points are determined with an electronic timer after playing the tape recorded during the experiment. In principle, it is possible to make reliable measurements at 0.7- to 0.8-sec time points, but it requires participation of two persons: one initiating the reaction and another quenching it. The counts in two "blank" samples are averaged and subtracted from the counts in all other samples. The GTPase rate constant is determined by single exponential fit of the data by using any appropriate software. The counts in the "total" points are treated as the total amount of GTP in the samples.

Preactivation Method

Equipment

    Small magnetic stirring plate
    Sterile, plastic microtiter plates (96-well size)
    5- × 2-mm Microspinbars (1 per assay well)
    Beckman TL-100 Ultracentrifuge and 1.5-ml polyallomer ultracentrifuge tubes
    Electronic timer
    1.5-ml Microcentrifuge tubes (1 per assay time point or assay condition)
    Microcentrifuge

Solutions

    Urea-washed bovine ROS membranes: 60–120 μM prepared in the dark as previously described[36] and stored in 1 × GAPN buffer
    Transducin: 15–30 μM
    His$_6$-PDEγ: 10–30 μM expressed and affinity-purified as previously described[15] (PDEγ is stable for many months when stored dry −80°, and for many hours on ice in pure water. In neutral pH buffers it tends to lose activity over time.)
    1 × GAPN buffer: 10 mM Tris-HCl, pH 7.4, 100 mM NaCl, 2 mM MgCl$_2$, and solid phenylmethylsulfonyl fluoride (PMSF)
    GTPγS (Guanosine 5'-O-(3-thiotriphosphate): a concentrated stock (200 μM), made up in 1 mM DTT and Milli-Q H$_2$O, 0.2-μm NC filtered, and stored at −80°

[γ-^{32}P]GTP: 1 μM stocks, prepared in advance with 4.99 ml 1 μM GTP (Sigma, St. Louis, MO, lithium salt) in 1 × GAPN buffer and 10 μl of 10 mCi/ml of [γ-^{32}P]GTP (Amersham-Pharmacia, Piscataway, NJ; >5000 Ci/mmol), filtered and aliquoted into ~500-μl aliquots and stored at −20°

GTP: (1–20 mM) in 1 × GAPN, filtered

AMP–PNP: 1-mM aliquots stored at −20°

Trichloroacetic acid (TCA): 50% (w/v), stock solution, prepared in advance and stored at 4°

Charcoal suspension: 5% (w/v) activated charcoal, 5 mM NaH$_2$PO$_4$

Scintillation cocktail

Assay

This example uses initial concentrations of 10 μM uw-ROS, 1 μM G$_t$, and 0.2 μM [γ-^{32}P]GTP (all diluted twofold on addition of GAP test samples), to test GAP activity of 2 μM RGS9d[15] in the presence or absence of 300 nM His$_6$-PDEγ. uw-ROS (15 μM) and G$_t$ (1.5 μM) are premixed in GAPN buffer and 30-μl aliquots pipetted into microtiter plate wells containing stir fleas, with one set of control samples having buffer only, and one having 10 μM GTPγS in addition to the uw-ROS and G$_t$. In this example 11 samples, including controls, are assayed in triplicate. The assay is initiated by adding 20 μl of 0.5 μM [γ-^{32}P]GTP, followed at 5 sec by addition of 50 μl GAP test sample. The GAP test samples are prepared in GAPN buffer containing one of the following, with the indicated concentrations twice the final concentration in the assay, with or without 0.6 μM His$_6$-PDEγ added just before the assay: (1) buffer only; (2) ROS, 80 μM, *not* urea washed; (3) 4 μM RGS9d; or (4) 200 μM GTPγS (uptake control). Five seconds after addition of GAP test sample, 50 μl 50% TCA is added. One additional control sample has the same composition as the one containing unwashed ROS and PDEγ, but everything including [γ-^{32}P]GTP is mixed at the start and allowed to incubate 30 min before adding TCA. After TCA quench, the samples are added to charcoal and assayed for [^{32}P]P$_i$ essentially as described earlier for the standard method. The data are analyzed as described earlier. Briefly, the triplicate data sets are averaged, average background GTP hydrolysis is subtracted, then the fraction of total GTP hydrolyzed is determined relative to the 30-min incubation control (100% hydrolysis). An estimate of the rate constant can be calculated according to the equation given earlier.

[36] Mutational Analysis of Functional Interfaces of Transducin

By MICHAEL NATOCHIN and NIKOLAI O. ARTEMYEV

Introduction

The visual cascade is an informative model system[1] for studying signaling via heterotrimeric G proteins mainly because the major signaling components, rhodopsin (R), transducin (G_t), and cGMP phosphodiesterase (PDE), can be readily purified in large quantities from rod outer segment (ROS) membranes. The crystal structure of $G_t\alpha$ opened up an avenue for indepth analysis of its structure–function relationships.[2] However, further understanding of molecular mechanisms of transducin signaling has been hindered by the lack of an efficient expression system for $G_t\alpha$. Although a functional expression of $G_t\alpha$ using the Sf9 (*Spodoptera frugiperda*)/baculovirus system has been demonstrated,[3] low yields of active $G_t\alpha$ probably prohibited extensive use of this approach. Only a relatively small number of $G_t\alpha$ mutants expressed in the insect cells have been characterized.[3,4] In contrast, *in vitro* translation of $G_t\alpha$ allowed for extensive screening of $G_t\alpha$ mutants.[5] The drawback of the *in vitro* translation system is related to limitations in functional testing. For example, assays for the ability of $G_t\alpha$ mutants to bind and activate PDE that require relatively high concentrations of recombinant $G_t\alpha$ are not feasible with *in vitro* translated $G_t\alpha$ mutants. Attempts to functionally express $G_t\alpha$ in *Eschericha coli* have been unsuccessful, although this expression system works very well for other Gα subunits such as $G_i\alpha$ and $G_s\alpha$. The strategy of expression of $G_t\alpha/G_i\alpha$ chimeric proteins in *E. coli*[6] helps to overcome the disadvantages of the Sf9/baculovirus and *in vitro* translation systems by enabling expression and purification of a large number of mutants in significant quantities. A chimeric $G_t\alpha/G_i\alpha$ protein ($G_{ti}\alpha$), Chi8,[6] was initially utilized as a template for scanning mutational analysis of the effector interface of $G_t\alpha$.[7] Later, a more

---

[1] S. Yarfitz and J. B. Hurley, *J. Biol. Chem.* **269**, 14329 (1994).
[2] J. P. Noel, H. E. Hamm, and P. B. Sigler, *Nature* **366**, 654 (1993).
[3] E. Faurobert, A. Otto-Bruc, P. Chardin, and M. Chabre, *EMBO J.* **12**, 4191 (1993).
[4] Q. Li and R. A. Cerione, *J. Biol. Chem.* **272**, 21673 (1997).
[5] R. Onrust, P. Herzmark, P. Chi, P. D. Garcia, O. Lichtarge, C. Kingsley, and H. R. Bourne, *Science* **275**, 381 (1997).
[6] N. P. Skiba, H. Bae, and H. E. Hamm, *J. Biol. Chem.* **271**, 413 (1996).
[7] M. Natochin, A. E. Granovsky, and N. O. Artemyev, *J. Biol. Chem.* **273**, 21808 (1998).

transducin-like chimeric $G_t\alpha/G_i\alpha$ protein ($G_t\alpha^*$) containing only 16 $G_i\alpha$ residues was generated for the purpose of mutagenesis (Fig. 1).[8] In this chapter, we describe procedures for site-directed mutagenesis, expression, purification, and functional evaluation of mutants with substitutions of $G_t\alpha$ residues in the chimeric $G_t\alpha$ and $G_t\alpha^*$ subunits.

Targeting of Specific Regions and Residues of $G_t\alpha$ for Mutational Analysis of Its Effector Interface

Existing biochemical evidence on transducin interaction with effector, PDE, provides for the targeting of several selected regions for scanning mutagenesis. A region corresponding to the α_4 helix and the α_4-β_6 loop of $G_t\alpha$ was implicated as a major effector domain.[9–11] Since the affinity of $G_t\alpha$ for PDE is significantly enhanced in the active GTP-bound form, the conformation sensitive domains, called switches, are likely to be involved in the effector binding. The failure of the $G_t\alpha$ mutant W207F to activate PDE provided the first indication that switch II might represent an important effector domain of $G_t\alpha$.[3] The crystal structure of $G_s\alpha$ complexed with catalytic domains of adenylyl cyclase furnishes conclusive proof for the involvement of switch II in binding to the effector enzyme.[12] Another potential effector region of $G_t\alpha$ corresponding to the α_3 helix and the α_3-β_5 loop is implicated by the functional analysis of $G_t\alpha/G_i\alpha$ chimeras.[6] Based on the evidence summarized above, three regions of $G_t\alpha$, switch II, α_3-β_5, and α_4-β_6, have been selected for mutational analysis to delineate effector-interacting residues of $G_t\alpha$. The Ala-scanning approach is chosen to study the α_4-β_6 region and conservative residues within switch II (Fig. 1). Nonconserved $G_t\alpha$ switch II residues have been replaced by the corresponding residues from $G_s\alpha$ (Fig. 1). Substitutions of the effector-interacting residues would be expected to cause a decrease in the $G_{ti}\alpha$ or $G_t\alpha^*$ affinities for Pγ. The interaction between the α_3-β_5 region and PDE is $G_t\alpha$ specific. Therefore, only the residues that are different between $G_i\alpha$ and $G_t\alpha$ within the α_3-β_5 region of $G_{ti}\alpha$ have been replaced by corresponding $G_t\alpha$ residues. Substitutions of the $G_t\alpha$ effector-specific residues for $G_i\alpha$ residues in $G_{ti}\alpha$ would lead to an increased affinity for Pγ.

Analysis of solvent accessibility of residues within selected regions per-

---

[8] M. Natochin, A. E. Granovsky, K. G. Muradov, and N. O. Artemyev, *J. Biol. Chem.* **274**, 7865 (1999).
[9] H. M. Rarick, N. O. Artemyev, and H. E. Hamm, *Science* **256**, 1031 (1992).
[10] N. O. Artemyev, J. S. Mills, K. R. Thornburg, D. R. Knapp, K. L. Schey, and H. E. Hamm, *J. Biol. Chem.* **268**, 23611 (1993).
[11] Y. Liu, V. Y. Arshavsky, and A. E. Ruoho, *J. Biol. Chem.* **271**, 26900 (1996).
[12] J. J. G. Tesmer, R. K. Sunahara, A. G. Gilman, and S. R. Sprang, *Science* **278**, 1907 (1997).

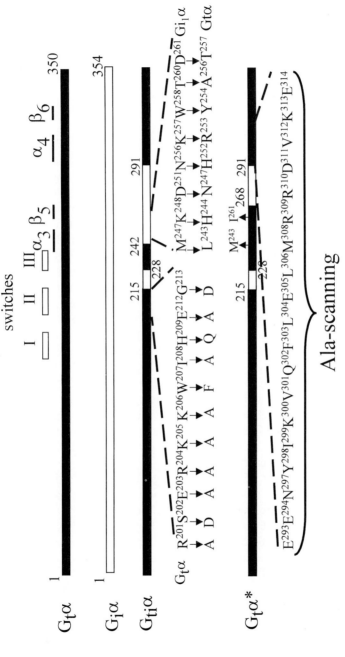

FIG. 1. Schematic representation of chimeric $G_{ti}\alpha$ and $G_t\alpha^*$ subunits. Substitutions of residues in the switch II, α_3-β_5, and α_4-β_6 regions are shown.

mits a further reduction in the number of mutations to be made in order to characterize a specific protein interface. Higher solvent accessibility of a residue increases the likelihood for its participation in intermolecular interactions. Solvent-exposed residues of $G_t\alpha$ can be identified using coordinates of the crystal structures of $G_t\alpha$ in different conformations[2,13,14] and RasMol (version 2.6) software, which is available for downloading from the Internet (*http://www.umass.edu/microbio/rasmol/*).

Site-Directed Mutagenesis of Chimeric Transducins

Mutagenesis of the $G_t\alpha$ switch II residues is performed using the vector for expression of His$_6$-tagged $G_t\alpha/G_{i1}\alpha$ chimera, $G_{ti}\alpha$, as a template for polymerase chain reaction (PCR) amplifications. The solvent-exposed amino acid residues from the switch II region are chosen for site-directed mutagenesis (Fig. 1). The unique *Bam*HI restriction site (overlaps codons for $G_t\alpha$ residues Trp^{207}-Ile^{208}-His^{209}) is conveniently located in the middle of the switch II region of $G\alpha$ subunits. It allows for a simple PCR-based substitution of any $G_t\alpha$ switch II residue that is N terminal to Trp-207 or C terminal to His-209. For PCR amplifications, a forward primer can include a unique *Nco*I site with the Met start codon (*Nco*I primer), and a pairing reverse primer may contain a unique *Hin*dIII site located ~50 bp downstream from the stop codon (*Hin*dIII primer).

The following procedure can be employed to substitute residues that coincide with a restriction site. To mutate Trp-207 and Ile-208, the vector is prepared by digesting pHis$_6$-G$_{ti}\alpha$ with *Bam*HI followed by removal of the 5' protruding ends with mung bean nuclease. After inactivation of the nuclease, the vector is digested with *Nco*I. Reverse PCR primers carrying mutations are designed to substitute deleted *Bam*HI-originated 5' protruding ends with mutant sequences. The PCR products are digested with *Nco*I and ligated into the vector.

Analogously to the PCR-based mutagenesis of the switch II region, substitutions within the α_3-β_5 region can be made by taking advantage of existing restrictions sites such as the *Sph*I site (at $G_t\alpha$ Met-238), or the newly introduced silent *Spe*I site (at $G_t\alpha$-Thr^{258}-Ser^{259}). Eight residues of $G_{ti}\alpha$ within the α_3-β_5 region that are different between $G_t\alpha$ and $G_i\alpha$ are replaced by the corresponding $G_t\alpha$ residues (Fig. 1). The only remaining residue, $G_t\alpha$–Val-261, is conservatively substituted in $G_{i1}\alpha$ (Ile-265), and has limited solvent exposure. Therefore, it is excluded from the mutagenesis. To reduce a number of mutants and simplify initial functional screening,

---

[13] D. G. Lambright, J. P. Noel, H. E. Hamm, and P. B. Sigler, *Nature* **369**, 621 (1994).
[14] J. Sondek, D. G. Lambright, J. P. Noel, H. E. Hamm, and P. B. Sigler, *Nature* **372**, 276 (1994).

the eight α_3-β_5 residues are first substituted in three cluster mutants, $M^{247}K^{248}D^{251} \rightarrow L^{243}H^{244}N^{247}$ (LHN), $N^{256}K^{257}W^{258} \rightarrow H^{252}R^{253}Y^{254}$(HRY), and $T^{260}D^{261} \rightarrow A^{256}T^{257}$ (AT). The first triple mutation, LHN, is introduced using PCR with a mutant forward primer carrying the *Sph*I site (at $G_t\alpha$–Met-238) and the *Hin*dIII-primer. The PCR product is cut with *Sph*I and *Hin*dIII and inserted into the pHis$_6$G$_{ti}\alpha$ vector. To aid in cloning of the HRY and AT mutants, a silent *Spe*I site (A/CTAGT coding for $G_t\alpha$–Thr^{258} Ser^{259}) is created in pHis$_6$G$_{ti}\alpha$ using a QuikChange kit (Stratagene, La Jolla, CA). The "whole-plasmid DNA" PCR approach utilized in the QuikChange and other similar protocols represents a practical method for creating silent restriction sites anywhere in the cDNA of the protein of interest. The pHis$_6$G$_{ti}\alpha$–*Spe*I vector is then used for PCR-directed mutagenesis with the 5′ *Nco*I primer and 3′ primers containing HRY and AT mutations and the *Spe*I site.

Based on the functional analysis of the "cluster" mutants, individual mutations, $M^{247} \rightarrow L^{243}$(L), $K^{248} \rightarrow H^{244}$(H), $D^{251} \rightarrow N^{247}$(N), have been generated using PCR-directed mutagenesis and the *Sph*I and *Hin*dIII restriction sites. Incidentally, a substitution $K^{248} \rightarrow H^{244}$ produced an internal restriction site *Sph*I (GC ATG/C), which was eliminated by introduction of a synonymous codon for $G_t\alpha$–Ser-242 (AGC $\rightarrow$ AGT). To determine a cumulative effect of the $G_t\alpha$ effector residues His-244 and Asn-247, a double $K^{248}D^{251} \rightarrow H^{244}N^{247}$ (HN) mutant has been made by PCR using the LHN template and the *Sph*I and *Hin*dIII restriction sites.

A collection of different G$_{ti}\alpha$ mutants that has been accumulated during analysis of the α_3-β_5 region permits simple, one- or two-step PCR procedures for incorporation of the α_3-β_5 G$_t\alpha$ residues into G$_{ti}\alpha$ in any desired combination. An example of such a procedure outlines the generation of transducin-like, effector-competent, chimeric protein G$_t\alpha$*. A plasmid containing cDNA coding for the HN mutant is PCR amplified with the *Nco*I primer and the reverse primer carrying AT mutations. After subcloning of the PCR product into the pHis6G$_{ti}\alpha$–*Spe*I vector, the HN-AT containing cDNA is PCR amplified with a reverse primer carrying HRY-AT mutations. As a result, all G$_i\alpha$ residues in the α_3-β_5 region of G$_{ti}\alpha$, except for Met-247 (Leu-243 in G$_t\alpha$) and Ile-265 (Val-261 in G$_t\alpha$) are replaced by G$_t\alpha$ residues (Fig. 1).

Appropriate restriction sites may not be available, and a cDNA sequence may not allow creation of suitable silent site(s). Scanning mutational analysis of the G$_t\alpha$ α_4 helix and α_4-β_6 loop (amino acids 293–314) illustrates a PCR-based strategy for mutagenesis that does not rely on restriction sites. Forward primers carrying mutations are used for the first round of PCR amplification with a reverse *Hin*dIII primer. This amplification produces ~180- to 200-bp products. Purified PCR fragments are then used as reverse primers in combination with a forward primer carrying the *Bam*HI site. Resulting PCR products (~500 bp) are purified and subcloned into pHis$_6$G$_t\alpha$* using the *Bam*HI and *Hin*dIII sites. Despite the two rounds of PCR amplification, the incident mutation rate is very low when *Pfu* DNA polymerase (Stratagene, La Jolla, CA) is used. In fact, only one misincorporation has been detected on automated DNA sequencing of the 19 α_4-β_6 mutants. Following ligation of PCR products into appropriately digested vectors, ligation mixtures are used directly to transform *Escherichia coli* DH5α cells. Plasmid DNAs are purified using standard protocols and sequenced to verify mutations. *E. coli* BL21(DE3) cells are then transformed for protein expression.

Expression and Purification of Chimeric Transducin Mutants

E. coli BL21(DE3) cells harboring a gene for the T7 RNA polymerase in their chromosomal DNA are well suited for efficient protein expression under the highly specific T7 promoter.[15] Cells transformed with a vector for mutant G$_{ti}\alpha$ or G$_t\alpha$* are grown overnight at 37° in 50 ml Luria–Bertani (LB) medium in the presence of 100 μg/ml ampicillin. The night culture is used for inoculation of 0.5–1.0 liter of 2× TY medium. When the culture reaches OD$_{600}$ of ~0.5, the temperature is decreased to 25° and 30 μM isopropylthiogalactoside (IPTG) is added. Following a 16-hr incubation, cells are pelleted and frozen at −70°. To lyse the cells, the pellet is resuspended in 25–30 ml of 50 mM Tris-HCl (pH 8.0) buffer containing 50 mM NaCl, 5 mM MgSO$_4$, 50 μM GDP, 0.1 mM phenylmethylsulfonyl fluoride (PMSF) and 5 mM 2-mercaptoethanol. The cell suspension is sonicated on ice with the Sonic Dismembrator 550 (Fisher Scientific, Pittsburgh, PA) for 5 min in 30-sec intervals with 30-sec pauses. The dismembrator power output is set on 5. The resulting crude cell lysate is cleared by centrifugation at 100,000g for 60 min at 4°. In many instances inclusion bodies are observed in the pellets. However, the presence of inclusion bodies or their amounts do not appear to correlate with yields and activity of soluble recombinant proteins. After addition of 500 mM NaCl and 20 mM imidazole, the supernatant is loaded on a 1-ml column containing His-Bind resin (Novagen, Madison, WI) charged with NiSO$_4$ in accordance with the manufacturer's recommendations. The resin is washed from nonspecifically bound proteins with 10–20 volumes of 20 mM Tris-HCl (pH 8.0) buffer, containing 500 mM NaCl and 20 mM imidazole. Specifically bound proteins are eluted with the buffer containing 100 mM imidazole. The eluate is supplemented with

---

[15] F. W. Studier, A. H. Rosenberg, J. J. Dunn, and J. W. Dubendorff, *Methods Enzymol.* **185**, 60 (1990).

GDP (final concentration 50 μM) and dialyzed overnight against 1000× volume of 50 mM Tris-HCl (pH 8.0) buffer containing 50 mM NaCl, 5 mM MgSO$_4$, 0.1 mM PMSF, 5mM 2-mercaptoethanol, 50 μM GDP, and 50% (v/v) glycerol. Proteins are stored at $-20°$, or for a long term at $-80°$. The purity of mutant G$_{ti}\alpha$ or G$_t\alpha$* after the affinity purification is ~60–90% as estimated by Coomassie-stained gel electrophoresis. Expression of most mutants yields similar amounts of soluble protein (~5 mg/liter of culture). If necessary, further purification of the recombinant proteins is accomplished by ion-exchange high-performance liquid chromatography (HPLC) on a Mono Q HR 5/5 column (Pharmacia, Piscataway, NJ). The column is equilibrated with 30 mM Tris-HCl (pH 8.0) buffer containing 2 mM MgSO$_4$. The proteins are eluted using a 0–500 mM gradient of NaCl over 30 min. G$_{ti}\alpha$ (G$_t\alpha$*) elutes at ~350 mM NaCl. Typically, the purity of G$_{ti}\alpha$ (G$_t\alpha$*) or their mutants after this step is >90%.

Evaluation of Correct Folding of Transducin Mutants

To properly evaluate and interpret results of mutational analysis of chimeric G$_t\alpha$, confirmation of correct folding is absolutely essential. The first indication that a particular mutation may have led to the misfolding of a protein could come from estimates of expression levels. We monitor the yields of soluble protein for all G$_{ti}\alpha$ (G$_t\alpha$*) mutants. A significant reduction in the level of expression of soluble mutant warrants very careful attention and evaluation of its structural integrity with as many assays as possible. Several mutants—G$_{ti}\alpha$M$^{247} \rightarrow$ L^{243}, G$_t\alpha$*Y298A, I299A, and F303A—have notably reduced (5- to 10-fold) levels of expression. Decreased expression levels for these mutants correlate well with the low degree of proteolytic protection in the trypsin protection assay described later. Also noteworthy, the G$_t\alpha$ residues at mutated positions, Leu-243, I299A, and F303A, participate in the network of intramolecular interactions between the α_3 and α_4 helices of G$_t\alpha$, which may provide an explanation to the observed properties. Several methods such as a trypsin protection assay, tryptophan fluorescence monitoring, R*-induced GTPγS binding, and measurements of GTPase activity can be utilized for complete evaluation of the functional folding of transducin mutants.

Trypsin Protection Assay

The basic well-established criterion for correct folding of the α subunits of heterotrimeric G proteins, including G$_t\alpha$, is the ability to undergo a conformational change to the active state. Activation of G$_t\alpha$–GDP on exchange with GTPγS or binding of AlF$_4^-$ leads to (1) protection of the switch II region from cleavage with trypsin[16] and (2) to an increase in intrinsic tryptophan ($G_t\alpha$ W207) fluorescence.[3] The rate of receptor-independent GDP–GTP exchange on $G_t\alpha$ is very slow, unlike a significant spontaneous nucleotide exchange on $G_i\alpha$ or other $G\alpha$ subunits. Chimeric $G_{ti}\alpha$ and $G_t\alpha^*$ exhibit basal nucleotide exchange rates similar to those of $G_t\alpha$.[6,8] The AlF_4^- complex bypasses the requirement of rhodopsin for $G_t\alpha$ activation. It causes a conformational change on $G_t\alpha$ by binding to $G_t\alpha$–GDP and mimicking the γ-phosphate of GTP. Therefore, the trypsin protection assay in the presence and in the absence of AlF_4^- represents a simple, first choice assay for correct folding of mutant $G_{ti}\alpha$ ($G_t\alpha^*$).

In a typical trypsin protection assay, the recombinant proteins are incubated for 10 min at 25° in 20 mM HEPES (pH 8.0) buffer containing 100 mM NaCl, 50 μM GDP, and 10 mM $MgSO_4$ with or without addition of 30 μM $AlCl_3$ and 10 mM NaF. The proteins are then treated with trypsin (25μg/ml) at a 20:1 protein: trypsin molar ratio for 15–20 min at 25°. The reactions are stopped with addition of sodium dodecyl sulfate–polyacrylamide gel electrophoresis (SDS–PAGE) sample buffer and immediate heat treatment for 5 min at 100° followed by analysis with SDS gel electrophoresis.

The trypsin protection pattern for $G_{ti}\alpha$ shown in Fig. 2A is characteristic for properly folded $G_{ti}\alpha$ mutants. However, this assay is not universal. While the protection of a $G_t\alpha$ mutant from tryptic cleavage does indicate proper folding, the opposite conclusion, i.e., that a failure of the protection means incorrect folding, may not be accurate. Four switch II mutants, R201A, E203A, R204A, and W207F, exhibit a low degree of proteolytic protection (10% or less) in the trypsin protection assay in the presence of AlF_4^-. However, based on other assays of $G_t\alpha$ structural integrity, such as intrinsic tryptophan fluorescence enhancement on activation, R* and $G_t\beta\gamma$-dependent GTPγS binding, and GTPase activity, all four $G_t\alpha$ mutants are correctly folded. Side chains of Arg-201, Arg-204, and Trp-207 in the $G_t\alpha$–GTPγS or $G_t\alpha$–GDP–AlF_4^- form ordered interactions with the switch III/α_3 region,[2,14] which are likely involved in preventing the tryptic cleavage at Arg-204 or Lys-205.

Tryptophan Fluorescence Assay

The Trp fluorescence measurement provides a critical additional assay for verification of the mutant's ability to undergo a conformational change. An important benefit of this assay is the capacity to determine the kinetics of activation for different $G_t\alpha$ mutants. Time traces of AlF_4^- induced in-

---

[16] B. K.-K. Fung and C. Nash, *J. Biol. Chem.* **258**, 10503 (1983).

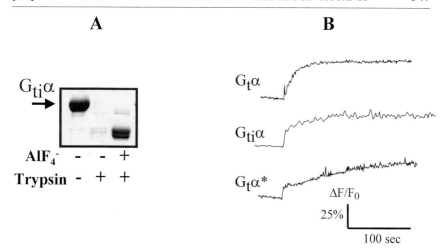

FIG. 2. (A) Trypsin protection assay for $G_{ti}\alpha$. A Coomassie-stained SDS gel shows $G_{ti}\alpha$ treated with trypsin (25 µg/ml) at a 20:1 $G_{ti}\alpha$:trypsin molar ratio for 15 min at 25° in the absence or presence of AlF_4^-. (B) Time courses of AlF_4^--induced enhancement in the intrinsic tryptophan fluorescence of $G_t\alpha$, $G_{ti}\alpha$, and $G_t\alpha^*$. Time traces of tryptophan fluorescence of $G_t\alpha$, $G_{ti}\alpha$, and $G_t\alpha^*$ (200 nM) recorded with excitation at 280 nm and emission at 340 nm. At indicated times, a mixture of $AlCl_3$ and NaF is injected into the cuvette to final concentrations of 30 µM and 10 mM, respectively.

creases in the tryptophan fluorescence of $G_{ti}\alpha$ and its mutants (100–200 nM) can be recorded on an AB2 fluorescence spectrophotometer (Spectronic Instruments, Rochester, NY) in a stirred cuvette with 1 ml of 20 mM HEPES buffer (pH 7.5) containing 100 mM NaCl and 4 mM $MgSO_4$ using excitation at 280 nm and emission at 340 nm. Stock solutions of $AlCl_3$ and NaF are mixed immediately prior to injection of 10 µl of the mixture into a cuvette (final concentrations 30 µM $AlCl_3$ and 10 mM NaF).

Representative time traces of AlF_4^--induced increases in the tryptophan fluorescence of $G_t\alpha$, $G_{ti}\alpha$, and $G_t\alpha^*$ are shown in Fig. 2B. Typical Trp fluorescence increases of $G_{ti}\alpha$ and it mutants (>90% purity) on addition of AlF_4^- are ~35% with kinetics that are intermediate between fast activation of $G_i\alpha$[7] and a more slow activation of $G_t\alpha$. The maximal Trp fluorescence enhancement is strongly dependent on the purity of $G_{ti}\alpha$ mutants and drops significantly when the preparations are less than 90% pure. Two out of four mutants with a low degree of trypsin protection, R201A and R204A, demonstrate normal Trp fluorescence enhancement on activation, thus underscoring the necessity to verify folding of $G_t\alpha$ mutants using the latter assay if they fail in the trypsin protection assay. Evidently, the Trp fluorescence assay cannot be utilized for mutations of Trp-207. Another rare example of when this assay is unable to recognize a conformational change in a correctly folded mutant in seen in the substitution R204A. The R204A mutant has a very low degree of fluorescent enhancement (<5%),[7] and yet it has normal R*-dependent GTPγS binding and GTPase activity. In the crystal structure of $G_t\alpha$–GDP–AlF_4^-, side chains of Arg-204 contact Trp-207.[14] If this interaction is responsible for the Trp fluorescence change, it would explain the lack of notable fluorescence increase in the R204A mutant.

[^{35}S]GTPγS Binding Assay

A GTPγS binding assay is a simple and informative test for proper folding of $G_{ti}\alpha$ mutants. This assay attests to the mutant competency to interact with $G_t\beta\gamma$ and R*, to exchange GTPγS for GDP, and to undergo an activational conformational change. An additional benefit from this assay is the opportunity to estimate accurately the concentrations of functionally active proteins in preparations of different mutants. This is especially important when a preparation may contain both functional and unfolded recombinant protein. The limitation of this assay for evaluation of a $G_{ti}\alpha$ mutant's structural integrity is that the GTPγS binding properties might be affected if the mutated residue is involved in the $G_t\beta\gamma$ or R* binding interface.

We recommend the following procedure for quick analysis of total GTPγS binding by $G_{ti}\alpha$ mutants. $G_t\beta\gamma$ and urea-washed ROS membranes (uw-ROS) are prepared according to previously published protocols.[17,18] $G_{ti}\alpha$ or mutants (400 nM), 2 μM $G_t\beta\gamma$, and uROS membranes (10 μM R*) are mixed in 100 μl of 20 mM HEPES (pH 8.0) buffer containing 100 mM NaCl, 10 mM dithiothreitol (DTT), and 8 mM MgSO$_4$. The binding reaction is initiated with addition of 100 μl (∼200,000–500,000 cpm) of 10 μM [^{35}S]GTPγS (New England Nuclear, Boston, MA). Following incubation for 20 min at 25°, the binding mixtures are passed through wet Whatman cellulose nitrate filters (0.45 μm). The filters are washed three times with 1.5 ml of 20 mM Tris-HCl (pH 8.0) buffer containing 100 mM NaCl and 10 mM MgSO$_4$ and transferred to scintillation vials. Radioactivity is counted after the filters are fully dissolved in scintillation cocktail. Nonspecific GTPγS binding is determined in the absence of $G_{ti}\alpha$ and usually represents less than 1% of the total binding.

---

[17] G. Yamanaka, F. Eckstein, and L. Stryer, *Biochemistry* **24**, 8094 (1985).
[18] C. Kleuss, M. Pallast, S. Brendel, W. Rosenthal, and G. Schultz, *J. Chromatogr.* **407**, 281 (1987).

Analysis of Binding between Mutant $G_{ti}\alpha$ or $G_t\alpha^*$ and Pγ Subunit

One of the simplest and most sensitive assays to monitor the interaction between $G_t\alpha$ and Pγ utilizes Pγ labeled with environmentally sensitive fluorescent probes. The Pγ molecule contains only one cysteine residue at position 68. The Cys residue represents a particularly convenient site for introduction of a fluorescent probe. First, cysteine residues can be selectively labeled with a large number of commercially available SH-reactive molecular probes. Second, PγCys-68 is positioned within the Pγ region that is involved in the interaction with $G_t\alpha$.[19] Two environmentally sensitive probes have been extensively used for labeling Pγ, Lucifer Yellow vinyl sulfone (4-amino-N-[3-(vinylsulfonyl)phenyl]naphthalimide 3,6-disulfonate) (LY)[20] and 3-(bromoacetyl)-7-diethylaminocoumarin (BC).[21] For studies of the effector interface of $G_t\alpha$, labeling with BC offers several advantages over the use of LY. Binding of $G_t\alpha$–GTPγS or $G_t\alpha$–GDP–AlF$_4^-$ to PγLY and to PγBC results in maximal fluorescence increases (F/F_0) of ~2- to 3- and 6- to 8-fold, respectively.[20,21] Furthermore, the affinities of these $G_t\alpha$ forms for PγBC (2–4 nM)[21] are significantly higher than those for PγLY (20–30 nM).[6,20] The inactive, GDP-bound $G_t\alpha$ also displays a stronger interaction with PγBC (K_d 75 nM)[21] as compared to PγLY (K_d 880 nM).[6] The elevated fluorescence increases of PγBC combined with the higher affinities are especially critical when testing the interaction of $G_{ti}\alpha$ or its mutants with the effector. $G_{ti}\alpha$ lacks effector-specific $G_t\alpha$ residues within the α_3 helix and consequently binds relatively weakly to PγBC (K_d values of 48 nM for $G_{ti}\alpha$–GDP–AlF$_4^-$ and 980 nM for $G_{ti}\alpha$–GDP).[7] Nonetheless, these affinities are well within the range of concentrations that allows an accurate assessment of the K_d values, whereas the affinity of $G_{ti}\alpha$–GDP for PγLY is so low ($K_d > 20$ μM) that a K_d cannot be accurately determined.[6]

Fluorescence Assay of Binding of $G_{ti}\alpha$, $G_t\alpha^*$, or Mutants to PγBC

The recombinant Pγ subunit is expressed in *E. coli*, purified, and labeled with BC at PγCys-68 as described in this volume.[22] The fluorescence measurements are performed on a F-2000 fluorescence spectrophotometer (Hitachi, Naperville, IL) or an AB2 fluorescence spectrophotometer (Spec-

---

[19] N. P. Skiba, N. O. Artemyev, and H. E. Hamm, *J. Biol. Chem.* **270**, 13210 (1995).
[20] N. O. Artemyev, H. M. Rarick, J. S. Mills, N. P. Skiba, and H. E. Hamm, *J. Biol. Chem.* **267**, 25067 (1992).
[21] N. O. Artemyev, *Biochemistry* **36**, 4188 (1997).
[22] A. E. Granovsky, K. G. Muradov, and N. O. Artemyev, *Methods Enzymol.* **315** [42], 2000 (this volume).

tronic Instruments, Rochester, NY) in 1 ml of 20 mM HEPES buffer (pH 7.5) containing 100 mM NaCl and 4 mM MgSO$_4$. When the AlF$_4^-$-activated conformations of G$_{ti}\alpha$ or G$_t\alpha$* are examined, immediately prior to addition of the recombinant proteins, AlCl$_3$ and NaF are added to the assay buffer to final concentrations of 30 μM and 10 mM, respectively. G$_{ti}\alpha$, G$_t\alpha$*, or mutants are then incubated for 1–2 min, which allows them to assume the active conformation. Fluorescence of PγBC is monitored at equilibrium before and after additions of increasing concentrations of recombinant G$_t\alpha$'s using excitation at 445 nm and emission at 495 nm. The concentration of PγBC is determined using ε_{445} 53,000 (see Fig. 3).

The K_d values are calculated by fitting the data to the equation:

$$\frac{F}{F_0} = 1 + \frac{[(F/F_0)_{max} - 1]X}{K_d + X} \quad (1)$$

where F_0 is the basal fluorescence of PγBC, F is the fluorescence after addition of recombinant G$_t\alpha$, $(F/F_0)_{max}$ is the maximal relative increase of fluorescence, and X is the concentration of free G$_{ti}\alpha$, G$_t\alpha$*, or mutant. The

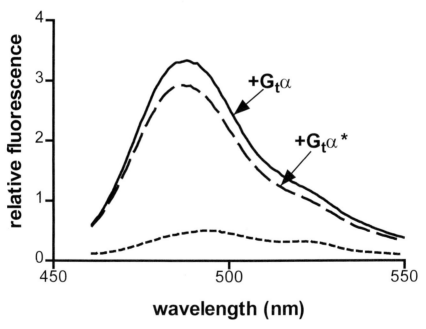

FIG. 3. Binding of AlF$_4^-$-activated G$_t\alpha$ and G$_t\alpha$* to PγBC. The emission spectra of PγBC (10 nM) alone and with addition of 100 nM G$_t\alpha$ or G$_t\alpha$* are recorded on an AB2 fluorescence spectrophotometer in a stirred cuvette with 1 ml of 20 mM HEPES buffer (pH 7.5) containing 100 mM NaCl, 4 mM MgSO$_4$, 30 μM AlCl$_3$, and 10 mM NaF using excitation at 445 nm.

X value is determined as $[G_t\alpha]_{total} - [G_t\alpha]_{bound} = [G_t\alpha]_{total} - [P\gamma BC](F - F_0)/(F_{max} - F_0)$; where $[P\gamma BC]$ is the concentration of PγBC in the assay, typically 10 nM.

Activation of Holo-PDE by Mutant G$_t\alpha$

Binding between G$_t\alpha$ and Pγ represents a simple and informative model assay to monitor the transducin–effector interaction. However, to activate PDE *in vivo*, transducin interacts with Pγ subunits that are complexed with the PDE catalytic subunits.[1] Additional interactions between G$_t\alpha$ and the PDE catalytic subunits might be involved in PDE activation. Although there is a correlation between the G$_t\alpha$ mutants' affinities for PγBC and their ability to stimulate PDE activity,[7] this correlation may not be universal. For example, if residues involved in the putative G$_t\alpha$/P$\alpha\beta$ interaction are mutated, this will not be detected by the PγBC binding assay. Furthermore, in the PγBC binding assay, AlF$_4^-$ is used to activate mutants to avoid light scattering caused by addition of R*-containing ROS membranes. However, despite the fact that the AlF$_4^-$ and GTPγS-activated G$_t\alpha$ interact with PγBC comparably, addition of AlF$_4^-$ in contrast to GTPγS does not cause appreciable activation of PDE in suspensions of ROS membranes. Therefore, the PDE activation test complements the PγBC binding assay and allows for a more comprehensive examination of the effector properties of G$_t\alpha$ mutants.

To test the capacity of G$_{ti}\alpha$ mutants to stimulate PDE activity, the holoenzyme is reconstituted with uROS membranes and G$_t\beta\gamma$ in the presence of GTPγS. For these experiments, holo-PDE can be extracted from ROS membranes and purified as previously described.[23] uw-ROS, G$_t\alpha$–GDP, and G$_t\beta\gamma$ are isolated according to established protocols.[17,18] The assay mixture (20 μl total volume) includes PDE (0.2 nM), 2 μM G$_{ti}\alpha$ or its mutants, 2 μM G$_t\beta\gamma$, and uROS membranes (10 μM rhodopsin) in 20 mM HEPES (pH 8.0) buffer containing 100 mM NaCl and 8 mM MgSO$_4$. GTPγS (10 μM) is added to the mixture, which is then incubated for 10 min at 25°. cGMP hydrolysis is initiated with the addition of 100 μM [^{3}H]cGMP and allowed to proceed for 10 min. The reaction is stopped by heating the samples for 2 min at 100°. The samples are cooled and treated with 0.1 u of bacterial alkaline phosphatase for 20 min at 37°. AG 1-X2 cation exchange resin (Bio-Rad Laboratories Hercules, CA) (1 ml of a 20% bed volume suspension) is added and the samples are incubated for 10 min with periodic vortex mixing. Aliquots (0.6 ml) are withdrawn from the

---

[23] N. O. Artemyev and H. E. Hamm, *Biochem. J.* **283**, 273 (1992).

supernatants containing [^{3}H]cGMP, mixed with 5 ml of scintillation cocktail and counted.

Analysis of Interaction between $G_{ti}\alpha$ Mutants and RGS Proteins

Recent findings have established that proteins termed regulators of G-protein signaling (RGS) serve as GTPase-activating proteins (GAPs) for $G_i\alpha$, $G_q\alpha$, and $G_{12}\alpha$ families and attenuate G-protein-mediated signal transduction.[24] A photoreceptor-specific member of the RGS family, RGS9, is a GAP for transducin.[25] The Pγ subunit potentiates the GAP function of RGS9, thus providing a mechanism for the effector role in turn-off of the visual signal.[25] Our mutational analysis of $G_{ti}\alpha$ has identified effector residues within the $G_t\alpha$ switch II region.[7] Switch II is also an important site for binding to RGS proteins.[26] This finding and the synergistic modulation of $G_t\alpha$ GTPase activity by RGS9 and Pγ have served as catalysts for a comparative analysis of the effector and RGS interfaces on $G_t\alpha$.

Binding of $G_t\alpha$ Mutants to RGS Protein

The ability of $G_{ti}\alpha$ mutants to bind RGS proteins can be examined using precipitation of mutants by glutathione agarose beads containing immobilized glutathione S-transferase (GST)–RGS protein. The RGS domain of RGS9 (RGS9d) is inactive when expressed in *E. coli* as GST fusion protein.[27] His$_6$-tagged RGS9d can be functionally refolded.[25] However, it cannot be used in the precipitation experiments using Ni^{2-}-bound resin because the $G_{ti}\alpha$ mutants are expressed as His$_6$-tagged proteins as well. Moreover, the affinity of RGS9d for $G_t\alpha$ is significantly lower than that of RGS16.[25,28] Therefore, we utilize RGS16 as a model protein to probe the binding of $G_{ti}\alpha$ mutants with RGS proteins. GST–RGS16 protein is expressed in *E. coli* and purified as previously described.[28] To obtain a resin with immobilized GST–RGS16, glutathione agarose from Sigma (~20% slurry) is mixed with GST–RGS16 (3–5 mg/ml) in 20 mM HEPES buffer (pH 7.6) containing 100 mM NaCl, 8 mM MgSO$_4$, and 6 mM 2-mercaptoethanol. The resin suspension is incubated for 20 min at 25° with gentle vortex mixing, pelleted, and washed three times with a 20× volume of buffer to remove unbound proteins. Typically, $G_{ti}\alpha$ or its mutants (10 μg) are added

---

[24] D. M. Berman and A. G. Gilman, *J. Biol. Chem.* **273**, 1269 (1998).
[25] W. He, C. W. Cowan, and T. G. Wensel, *Neuron* **20**, 95 (1998).
[26] J. J. G. Tesmer, D. M. Berman, A. G. Gilman, and S. R. Sprang, *Cell* **89**, 251 (1997).
[27] M. Natochin and N. O. Artemyev, unpublished results (1998).
[28] M. Natochin, A. E. Granovsky, and N. O. Artemyev, *J. Biol. Chem.* **272**, 17444 (1997).

to 100 μl of GST–RGS16 containing glutathione agarose (20% slurry, retains ~10 μg of GST–RGS16) in 20 mM HEPES buffer (pH 7.6) containing 100 mM NaCl, 8 mM MgSO$_4$ with or without addition of 30 μM AlCl$_3$ and 10 mM NaF. The mixtures are then incubated for 20 min at 25°, the agarose beads are spun down, washed three times with 1 ml of the buffer, and the bound proteins are eluted with a sample buffer for SDS–PAGE. Coomassie-stained gels can be scanned and analyzed to make quantitative assessments of the amounts of precipitated G$_t$α mutants. Under the procedural conditions described earlier GST–RGS16 precipitates nearly stoichiometric amounts of G$_t$α or G$_{ti}$α in the presence of AlF$_4^-$. G$_{ti}$α mutants with impaired RGS binding can be readily identified by a reduction in the amount of the precipitated mutant.

Single-Turnover GTPase Assay

A measurement of the RGS protein's capacity to activate GTP hydrolysis by G$_t$α (or mutants) is a central test in examining the effects of G$_t$α mutations on the G$_t$α–RGS interaction. RGS proteins accelerate the rate of GTP hydrolysis (k_{cat}) by Gα subunits but do not affect GDP–GTP exchange induced by activated G-protein-coupled receptors.[24] Hence, a single-turnover GTPase assay, rather than a steady-state assay, is required.

Commonly, our single-turnover GTPase activity measurements are carried out in suspensions of uROS membranes (5 μM rhodopsin) in 20 mM HEPES buffer (pH 8.0) containing 120 mM NaCl and 8 mM MgSO$_4$. The uROS membranes are reconstituted with G$_t$α, G$_{ti}$α, or mutants (2 μM) and G$_t$βγ (1 μM) with or without addition of increasing concentrations of RGS16. They are then bleached with ambient light during incubation for 5 min at 25°. The GTPase reactions are initiated by addition of 200 nM [γ-^{32}P]GTP (~4 × 10^5 dpm/pmol, Amersham, Piscataway, NJ) to a total reaction volume of 20 μl. The individual reactions are quenched at 6–8 time points for a single GTP hydrolysis time course of 0–60 sec by addition of 100 μl of 7% perchloric acid. A suspension of charcoal Norit A (Fisher, Pittsburgh, PA) (700 μl, 10% w/v) is added to the reaction mixtures to precipitate nucleotides. Following incubation for 20 min at 25° with an occasional vortex mixing, the charcoal is pelleted (4 min, 10,000g) and aliquots (500 μl) are withdrawn from the supernatants for ^{32}P$_i$ liquid scintillation counting. The GTPase rate constants were calculated by fitting the experimental data to an exponential function: %GTP hydrolyzed = $100(1 - e^{-kt})$, where k is a rate constant for GTP hydrolysis. The calculated GTPase rate constants are plotted as a function of RGS protein concentration on a logarithmic scale and the EC$_{50}$ values for the RGS

GAP effects are calculated by fitting the data to the sigmoidal dose–response curve with variable slope according to the following equation:

$$k = k_0 + \frac{k_{max} - k_0}{1 + 10^{[\log EC_{50} - X]*H}}$$

where k_0 is the basal GTPase rate constant of $G_{ti}\alpha$ mutant, k_{max} is the maximal, RGS-stimulated GTPase rate constant, X is the logarithm of RGS protein concentration, and H is the Hill slope. We found a good correlation between the EC_{50} values for stimulation of GTPase activity of $G_{ti}\alpha$ mutants by RGS16 and the ability of GST–RGS16 to precipitate these mutants using the binding assay described earlier.[7]

Conclusions

The high-yield functional expression on chimeric $G_t\alpha/G_i\alpha$ proteins in *E. coli* coupled with the ease of affinity purification helps to overcome the limitations of previously developed expression systems for $G_t\alpha$ and facilitates a study of a large number of $G_t\alpha$ mutants. Such a system is especially appreciated at a time when the crystal structures of $G_t\alpha$ may guide the targeting of regions and residues for substitutions. Mutational analysis of $G_t\alpha$ residues based on chimeric $G_t\alpha/G_i\alpha$ proteins has already been proven to be an effective tool in studying the structure and function of transducin. As a result, critical PDE-interacting residues of $G_t\alpha$ have been identified and the structure–function relationships between the effector and RGS interfaces have been examined.[7] In the course of these studies a transducin-like chimeric protein, $G_t\alpha^*$, has been generated, which is efficiently expressed in *E. coli* and capable of high-affinity interaction with PDE.[8] $G_t\alpha^*$ is also fully competent of interaction with $G_t\beta\gamma$ and R* and can be further utilized to investigate these functions of transducin.

Acknowledgments

Our research is supported by the National Institutes of Health Grant EY-10843 and the American Heart Association Grant-in-Aid 9750334N.

Section V

Photoreceptor Protein Phosphatases

[37] Isolation and Properties of Protein Phosphatase Type 2A in Photoreceptors

By MUHAMMAD AKHTAR, ALASTAIR J. KING, and NINA E. M. MCCARTHY

Introduction

Bovine rhodopsin is composed of 11-*cis*-retinal linked to the protein opsin via a Schiff base linkage[1-3] involving Lys-296.[4-8] Following light stimulation the *cis* double bond is isomerized to a *trans* geometry resulting in straightening of the polyene chain, which promotes conformational changes in the protein structure. These changes are sensed by the chromophore as a succession of rapidly produced species, with characteristic absorption spectra, until a relatively long-lived intermediate (λ_{max} 389 nm), metarhodopsin-II, is formed. The $t_{1/2}$ of the decay[9] of this intermediate at 20° is about 3 min, and <1 min at 37°. Metarhodopsin-II represents the activated form of the visual receptor (symbolized as Rho*). At a certain juncture after the visual transmission has been initiated, Rho* is inactivated by processes that include its phosphorylation catalyzed by rhodopsin kinase.[10] The phosphorylation occurs at the C-terminal region of Rho*,[11-14] and under forcing conditions *in vitro* a large number of serine and threonine residues are modified. However, a consensus has now begun to emerge that the most sensitive sites[14-18] of phosphorylation, and presumably relevant *in vivo*, are predominantly[17,18] Ser-343 and Ser-338, then Ser-334.[18]

---

[1] D. Bownds and G. Wald, *Nature* **205**, 254 (1965).
[2] M. Akhtar, P. T. Blosse, and P. B. Dewhurst, *Life Sci.* **4**, 1221 (1965).
[3] M. Akhtar, P. T. Blosse, and P. B. Dewhurst, *Biochem. J.* **110**, 693 (1968).
[4] D. Bownds, *Nature* **216**, 1178 (1967).
[5] E. Mullen and M. Akhtar, *FEBS Lett.* **182**, 261 (1982).
[6] Y. A. Ovchinnikov, *FEBS Lett.* **148**, 179 (1982).
[7] E. A. Dratz and P. A. Hargrave, *Trends Biochem. Sci.* **8**, 125 (1983).
[8] J. B. C. Findlay, M. Brett, and D. J. C. Pappin, *Nature* **293**, 314 (1981).
[9] N. E. M. McCarthy, Ph.D. Thesis, University of Southampton, UK (1998).
[10] C.-K. Chen and J. B. Hurley, *Methods Enzymol.* **315** [27], 1999 (this volume).
[11] H. Kühn and J. H. McDowell, *Biophys. Struct. Mech.* **3**, 199 (1977).
[12] G. J. Sale, P. Towner, and M. Akhtar, *Biochem. J.* **175**, 421 (1978).
[13] H. Shichi and R. L. Somers, *J. Biol. Chem.* **253**, 7040 (1978).
[14] P. A. Hargrave, S. L. Fong, J. H. McDowell, M. T. Mas, D. R. Curtis, J. K. Wang, R. Juszcak, and D. P. Smith, *Neurochem. Int.* **1**, 231 (1980).
[15] J. H. McDowell, J. P. Nawrocki, and P. A. Hargrave, *Biochemistry* **32**, 4968 (1993).
[16] D. I. Papac, J. E. Oatis Jr., R. K. Crouch, and D. R. Knapp, *Biochemistry* **32**, 5930 (1993).

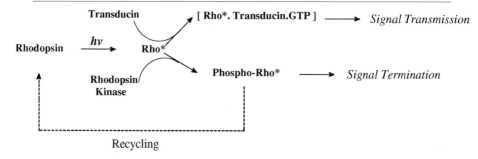

Fig. 1. Involvement of photoactivated rhodopsin (Rho*) in signal transmission and signal termination processes. The recycling process occurs in three steps as mentioned in the text.

The role of Rho* in the initiation and subsequent termination of the visual signal is shown in Fig. 1. The *in vivo* recycling of phospho-Rho* (Fig. 1) should require the presence of a protein phosphatase.[19] An enzyme fulfilling such a role has been isolated from rod outer segments (ROS) and classified as a type 2A protein phosphatase.[20–23] For a detailed characterization of the ROS phosphatase, the purification was carried out using 1040 bovine retinas (750 g wet weight) which were processed in eight independent batches, each consisting of 130 retinas.[23] The protocol described here is also suitable for small-scale preparation with 100–150 retinas, which when processed up to the Mono Q step (Table I) give a partially purified enzyme suitable for most enzymologic studies. It can be further extended for the isolation of a 500- to 2000-fold purified enzyme by combining extracts from eight batches and adding two further chromatographic steps (gel filtration and affinity chromatography).

Experimental

Buffers

Buffer A: 100 mM Tris-HCl, 0.1 mM EDTA, 2 mM $MgCl_2$, 0.1% (v/v) 2-mercaptoethanol, 1 mM benzamidine hydrochloride, 0.2 mM

---

[17] N. Pullen and M. Akhtar, *Biochemistry* **33**, 14536 (1994).
[18] H. Ohguro, M. Rudnicka-Naawrot, J. Buczylko, X. Zhao, J. A. Taylor, K. A. Walsh, and K. Palczewski, *J. Biol. Chem.* **271**, 5215 (1996), and references cited therein to the earlier work of this group.
[19] The earliest studies that pointed to the presence of a phosphatase in ROS were by Kühn; Miller and Paulsen; Weller *et al.*; and Thompson and Findlay and these are cited in Ref. 23.
[20] K. Palczewski, P. A. Hargrave, J. H. McDowell, and T. S. Ingebritsen, *Biochemistry* **28**, 415 (1989).

TABLE I
PURIFICATION OF TYPE 2A PHOSPHATASES FROM BOVINE RETINAS

Extract	Total volume (ml)	Total units	(%)[a]	Total protein[b] (μg)	Specific activity (units/mg)	Purification (x-fold)
Crude	289	1387	(100)	83232	16.7	1
Heparin-agarose	473	2639	(190)	36894	71.5	4
Mono Q, peak 1, Fig. 2B (catalytic subunit)	11	594	(43)	3960	150	9
Mono Q, Peak 2, Fig. 2B (holoenzyme)	10	340	(25)	2550	133	8
Superose-12, peak ▲, Fig. 2C (catalytic subunit)	2	148	(11)	346	428	26
Superose-12, peak ●, Fig. 2C (holoenzyme)	1.5	74	(8)	530	140	8
Thiophosphorylase a Sepharose (catalytic subunit)	9.5	107	(8)	3	35,660	2136
Thiophosphorylase a Sepharose (holoenzyme)	15	154	(11)	17	9058	542

[a] The phosphatase activity was determined using [^{32}P]phosphoopsin as substrate in the presence of 500 μg/ml protamine sulfate (one phosphatase unit corresponded to the release of 1 nmol inorganic phosphate/min at 30°). Other details are as given in Fig. 2.
[b] Protein was determined by the method of Bradford using Bio-Rad microassay reagent.

phenylmethylsulfonyl fluoride (PMSF), pH 7.0 (This buffer was used to make up the various sucrose solutions used in the flotations and gradient centrifugation steps.)

Buffer B (high-salt buffer): 500 mM NaCl, 20 mM Tris-HCl, 0.1 mM EDTA, 5% (v/v) glycerol, 1 mM benzamidine hydrochloride, 0.1 mM PMSF, 0.1% (v/v) 2-mercaptoethanol, pH 7.0

Buffer C: 20 mM Tris-HCl, 0.1 mM EDTA, 5% (v/v) glycerol, 1 mM benzamidine hydrochloride, 0.1 mM PMSF, 0.1% (v/v) 2-mercaptoethanol, pH 7.0

Buffer D: 140 mM NaCl, 20 mM Tris-HCl, 0.1 mM EDTA, 5% (v/v) glycerol, 1 mM DL-dithiothreitol, pH 7.4

Buffer E: 100 mM potassium orthophosphate, 2 mM MgCl$_2$, 0.1 mM EDTA, 0.1% (v/v) 2-mercaptoethanol, pH 7.0

---

[21] C. Fowles, P. Cohen, and M. Akhtar, *Biochemistry* **28**, 9385 (1989).
[22] K. Palczewski, J. H. McDowell, S. Jakes, T. S. Ingebritsen, and P. A. Hargrave, *J. Biol. Chem.* **264**, 15770 (1989).
[23] A. J. King, N. Andjelkovic, B. A. Hemmings, and M. Akhtar, *Eur. J. Biochem.* **225**, 383 (1994).

Preparation of Rod Outer Segments

All operations are carried out at 4° and up to the extraction of the ROS, light is avoided and the experiments are performed under Kodak (Rochester, NY) OC safelight filters. Dark-adapted retinas (130) are suspended in buffer A containing 45% (w/v) sucrose (65 ml; 2 retinas/ml), shaken vigorously for 2 min, and the dense tissue suspension subjected to flotation by centrifugation for 40 min at 10,000g. The walls of the tubes are gently scraped with a spatula to release the attached ROS material and the supernatant removed. The pellet is resuspended in buffer A containing 45% (w/v) sucrose and processed as above. The two ROS-containing supernatants from the flotation step are combined and diluted with 0.5 volume of buffer A (without sucrose) and centrifuged for 60 min at 40,000g. The sedimented ROS pellet is resuspended in buffer A containing 34% (w/v) sucrose (13 ml; 10 retina per ml) and layered onto six centrifuge tubes, each containing a three-step discontinuous sucrose gradient generated from 34% (5 ml), 36% (5 ml), and 38% (2 ml) sucrose in buffer A. The tubes are centrifuged, using a swing-out rotor, for 90–120 min at 130,000g when the ROS are resolved in two distinct bands. The first (band 1) is at the interface of the layered solution and 34% (w/v) sucrose and the second (band 2) settles at the interface of 34% and 36% (w/v) sucrose. These bands are separately removed from the gradient with a syringe, diluted with 0.3 volume of buffer A, centrifuged for 25 min at 40,000g to collect the ROS pellets, which are stored at $-70°$ until required. The distribution of rhodopsin in the two bands differs between experiments, depending on how harshly the retinas are shaken in the initial stage. In the particular experiments considered here, more than 80% of rhodopsin was in band 1 (60–80 mg with an A_{280}/A_{500} of 2.2–3.0) and the remaining in band 2 (12–20 mg).

Preparation of Phosphoopsin

The band 2 ROS suspension (1 mg of rhodopsin/ml of buffer E) is hand-homogenized and transferred to a stoppered glass round-bottom flask, and [γ-^{32}P]ATP (50,000–180,000 dpm/nmol) is added to a final concentration of 3 mM ATP. Samples are sonicated for 1 min, incubated for 5 min at 30° and then the light-dependent phosphorylation process initiated by illumination with continuous white light (150-W photoflood lamp at a distance of 30 cm). After the desired incubation period, samples are centrifuged for 20 min at 40,000g and the pellet is washed three times with buffer E and three times with water. The [^{32}P]phosphoopsin pellet is stored at $-20°$.

Phosphatase Assay

For the standard phosphatase assay,[21] 1 nmol of well-washed [^{32}P]phosphoopsin in buffer A (10–20 µl), 10–20 µl phosphatase extract, and 10–20 µl buffer A (in which various activating agents, inhibitors, etc., could be added) are mixed, giving a final incubation volume of 30–60 µl. They are incubated at 30° for 30 min. Samples (10 µl) are removed at various time intervals and, following precipitation with 200 µl trichloroacetic acid, the radioactivity released into the supernatant is measured by scintillation counting (using Optiphase 'Hisafe' 3 liquid scintillation cocktail, Fisher Chemicals, Loughborough, England).

The dephosphorylation of phosphopeptides is illustrated using a peptide that contains the sequence of the C terminal of rhodopsin, ^{338}SKTETSQ-VAPA^{348}, predominantly phosphorylated at Ser-343. Phosphopeptide (30–60 pmol), calculated from the specific activity of ^{32}P, assuming that the phosphopeptides contain 1 mol phosphate/mol peptide, is suspended in 10–20 µl buffer A or AnalaR water, mixed with the phosphatase extract (10–20 µl), and made up to a final volume of 30–60 µl with buffer A. The mixture is incubated at 30°, from which aliquots (15 µl) are removed at various time intervals, quenched with 200 µl of trichloroacetic acid (TCA), and mixed with 600 µl of 1.25% (w/v) ammonium molybdate in 1.1 M HCl containing 50 μM KH$_2$PO$_4$. The samples are extracted with 900 µl of 1:1 mixture of benzene:2-methylpropan-3-ol and the organic layer subjected to the determination of radioactivity by scintillation counting.

Unit

One phosphatase unit corresponds to the release of 1 nmol inorganic phosphate/min, at 30°.

Extraction and Purification of ROS Phosphatases[23]

The band 1 ROS (60–80 mg of rhodopsin), from each of the eight batches, are suspended in 3–4 ml of buffer A containing 100 µg/ml soybean trypsin inhibitor, 10 µg/ml each of aprotinin, leupeptin, and pepstatin, and then homogenized in a Potter-Elvehjem (Jencons, Leighton Buzzard, UK) 5-ml glass manual homogenizer using several strokes of the pestle. The sample is transferred to a 10-ml flat-bottom glass vial, sonicated in an ice–water cooled sonication bath (Decon Ultrasonics Ltd., Hove, Sussex, England) for 6 min and then centrifuged at 240,000g for 30 min. The supernatant containing the released protein is saved and the ROS pellet suspended in high-salt buffer B (3–4 ml) and subjected to the extraction protocol given earlier. The two extracts are combined and diluted with buffer C to give 100 mM NaCl concentration. The sample is made up to 1 mM benzamidine hydrochloride, 0.1 mM PMSF, 5% (v/v) glycerol and had a final volume of 35 ml. The extracts from eight such experiments are combined (total volume ~ 300 ml) and in three batches of ~100 ml each applied to a 30-ml column of heparin-agarose that had been preequilibrated in buffer C (containing 100 mM NaCl). After washing with 3–6 column volumes of the running buffer C (containing 100 mM NaCl), the adsorbed proteins are eluted with 3–6 column volumes of high-salt buffer B (Fig. 2A). In small-scale experiments, extracts from a single experiment, originating from 60 to 80 mg of ROS, are separated as described but using a 10-ml heparin-agarose column.

Figure 2A shows that when phosphoopsin is used as the substrate and the fractions assayed in the presence of protamine sulfate to stimulate the activity, 94% of the phosphoopsin phosphatase activity is in flow-through peak 1 and only 6% in the 500 mM NaCl elute. Using phosphorylase a as the substrate and performing incubations in the presence and absence of either protamine sulfate or inhibitor-2, it is established that the peak 1 and 2 phosphatase activities are due to type 2A and type 1 protein phosphatases, respectively.[21] This conclusion is confirmed using the authentic catalytic subunits of protein phosphatase 1 and 2A. We mention in passing that using phosphorylase a, which is a substrate for both the type 1 and type 2A phosphatases, the dephosphorylation activities in peaks 1 and 2 are in the ratio of 1.5 : 1.0. The data show that both protein phosphatases are present in ROS but it is type 2A protein phosphatase that is most effective in the dephosphorylation of phosphoopsin.

The flow-through fractions from three heparin-agarose columns are pooled, concentrated from ~430 to 27 ml in a 350-ml Amicon (Danvers, MA) YM30 concentrator at 4° using N_2 pressure and split into three batches. Each portion of about 9 ml was run down a 1-ml Mono Q column and preequilibrated with buffer C, using a Pharmacia FPLC (fast protein liquid chromatography) system. The column is developed with a 30-ml gradient of 0–500 mM NaCl in buffer C with a final wash with 10 ml of 1 M NaCl in the same buffer. The phosphoopsin phosphatase activity elutes in two sharp peaks at 250–290 mM (peak 1, Fig. 2B) and 350–400 mM NaCl (peak 2, Fig. 2B).

Gel filtration is the next chromatographic step performed, and it serves two purposes. First, it allows the approximate sizing of each phosphatase species and, second, the fractions are desalted for further purification by affinity chromatography. The fractions in peaks 1 and 2, from the three Mono Q runs (Fig. 2B), are concentrated from their combined volume of

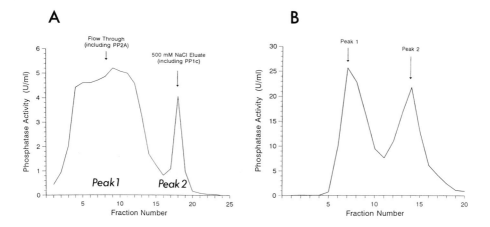
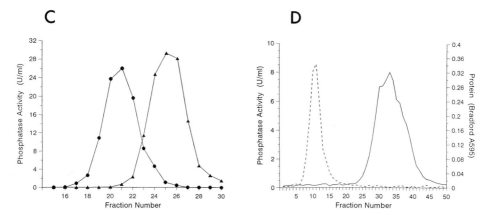

FIG. 2. (A) A 100-ml portion of the ROS extract chromatographed on a heparin-agarose column using step elution. (B) The flow-through peak from three runs of the type in (A) was concentrated to 27 ml and a portion (9 ml) separated by Mono Q FPLC. (C) The fractions in peaks 1 and 2 from three runs of the type in (B) were combined and separately concentrated to 200–500 µl and chromatographed on Superose-12: the elution profile of Mono Q peak 1 (▲) and peak 2 (●). (D) The fractions from each Superose-12 FPLC run were pooled, brought to 70 mM NaCl, and applied to a 9-ml column of thiophosphorylase a Sepharose. The phosphatase activity was eluted by a NaCl gradient: Protein (dashed line), as measured by Bradford assay and phosphatase activity (solid line). Part (D) shows the elution profile of the Superose-12 peak containing the holoenzyme. In these experiments, the phosphatase activity was measured in a final volume of 30 µl using 10-µl aliquots of each fraction, [^{32}P]phosphoopsin (1 nmol) and protamine sulfate at a concentration of 500 µg/ml. The incubations were performed at 30° for 30 min and the release of [^{32}P]phosphate measured by scintillation counting.

about 10 ml down to 200–500 µl in Amicon Centricon-10 cells. These are separately applied to a 20-ml Superose-12 FPLC column preequilibrated and developed in buffer D at a rate of 0.4 ml/min. The peak 1 phosphatase from Mono Q (peak 1, Fig. 2B) elutes from Superose-12 in fraction 25 (▲, Fig. 2C), corresponding to a molecular mass of 34 kDa. The elution profile is identical to that found for the authentic type 2A catalytic subunit of M_r 36 (the subunit is often referred to as the C subunit). The phosphoopsin phosphatase activity in the concentrate from Mono Q peak 2 (Fig. 2B), on the other hand, is found to elute at a M_r of about 100,000 (●, Fig. 2C) attributed to a holoenzyme form of protein phosphatase 2A.

The active fractions from each Superose-12 FPLC run (▲ and ●, Fig. 2C) are separately pooled, diluted with buffer C to reduce the NaCl concentration to 70 mM, and passed down a 9-ml column of thiophosphorylase a Sepharose[24] that has been equilibrated with buffer C at a flow rate of 0.5 ml/min. The column is developed with a linear 60-ml gradient from 0 to 600 mM NaCl in buffer C using a gravity-fed gradient former. Affinity chromatography removes large amounts of unwanted proteins in the flow-through fraction (for example, see Fig. 2D) and results in the elution of the putative catalytic subunit at a concentration of 130 mM NaCl (not shown), as is found with an authentic type 2A catalytic subunit from rabbit muscle. When the same procedure is applied to the holoenzyme-containing fractions (the 100-kDa peak, ●, Fig. 2C), the phosphatase activity elutes in a rather broad peak at 130–150 mM NaCl (—, Figure 2D). This behavior is reminiscent of rabbit muscle protein phosphatase $2A_1$.[24,25] The summary of the purification procedure is outlined in Table I.

Characterization of ROS Protein Phosphatases

For sodium dodecyl sulfate–polyacrylamide gel electrophoresis (SDS–PAGE) analysis and immunologic characterization, the fractions containing the phosphatase activity from the thiophosphorylase a column are pooled, dialyzed overnight against 2 liters of 5 mM Tris-HCl, pH 7.2, and the desalted protein solution freeze-dried. Following resuspension in AnalaR water the samples containing the catalytic subunit (22 ng; lane 1, Fig. 3) and the holoenzyme (80 ng; lane 2, Fig. 3) are subjected to 12% SDS–PAGE analysis. Although these samples are not pure, the main band in lane 1 corresponds to a M_r of 36,000 expected for the catalytic subunit of phosphatase 2A. Lane 2 also contains a band in this position and five other bands.

---

[24] H. Y. L. Tung, S. Alemany, and P. Cohen, *Eur. J. Biochem.* **148**, 251 (1985).

[25] P. Cohen, S. Alemany, B. A. Hemmings, T. J. Resink, P. Stalfors, and H. Y. L. Tung, *Methods Enzymol.* **159**, 390 (1988).

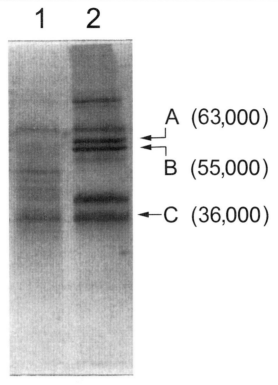

FIG. 3. SDS–PAGE (12%) analysis of ROS phosphatases. The affinity chromatography-purified catalytic subunit (lane 1) or the holoenzyme (lane 2) was electrophoresed and visualized either by Coomassie blue/silver staining or subjected to Western blot analysis. The band at 36,000 (C) in the two lanes cross-reacted with several antibodies raised against the catalytic subunit of protein phosphatase 2A. The 55,000 (B) and 63,000 (A) bands in lane 2 cross-reacted only with antibodies raised against the α subtype 55,000 and 65,000 regulatory subunit of protein phosphatase 2A, respectively.

Further characterization of the proteins in the samples of lane 1 and 2 is performed immunologically,[23] using antibodies produced by Hemmings and co-workers.[26,27] The 36,000 bands in the two samples cross-react with several specific antibodies generated against the rabbit muscle catalytic subunit of type 2A phosphatase, whereas the 63,000 (marked A) and 55,000 (marked B) bands in the holoenzyme sample (lane 2, Fig. 3) cross-react with antibod-

[26] P. Hendtix, R. E. Mayer-Jaekel, P. Cron, J. Goris, J. Hofsteenge, W. Merlevede, and B. A. Hemmings, *J. Biol. Chem.* **268,** 15267 (1993).
[27] P. Hendtix, R. E. Mayer-Jaekel, P. Cron, J. Goris, J. Hofsteenge, W. Merlevede, and B. A. Hemmings, *J. Biol. Chem.* **268,** 7330 (1993).

ies raised against the 65,000 and 55,000 regulatory polypeptides of subtype α. (These are referred to in the literature as the A and B subunits, respectively.) No cross-reaction with any of the proteins in the two samples is observed with sera[26,27] directed against the following proteins: catalytic subunit of protein phosphatase 1; the β regulatory polypeptide of type 2A phosphatase having M_r of 65,000a and 55,000; or the 72,000 and 130,000 regulatory subunit of protein phosphatase $2A_3$.

Cumulatively these results characterize the 36,000 species as the catalytic subunit of type 2A protein phosphatase (subunit C) and suggest that the higher M_r phosphatase contains the catalytic subunit complexed with the α isoforms of each of the 65,000 (A subunit) and 55,000 (B subunit) regulatory subunits (comprehensive reviews[25,28] on protein phosphatases are available). Such a composition is reminiscent of the trimeric form of the holoenzyme (ABC or $PP2A_1$) or the inactive tetramer ($AB'C_2$). The M_r determined by gel filtration experiments is lower than expected for any of the holoenzyme forms of the phosphatases that contain A as well as B regulatory subunits. The discrepancy is either due to the underestimation of the M_r by the gel permeation method utilized here or the trimeric enzyme is in equilibrium with a dimeric species, notably AC (for details on the dissociation/association behavior of the trimeric enzyme and its subunits, see Ref. 29). Notwithstanding some uncertainty on the specific issue, the immunologic detection of all three subunits in the holoenzyme fraction is consistent with the notion that, in its native state, the active form of ROS phosphoopsin phosphatase is the trimeric enzyme designated as the ABC form of protein phosphatase 2A, i.e., $PP2A_1$. Support for this view is provided by the work of Palczewski et al.[22] who showed, in bovine ROS, the presence of several subunit forms of type 2A phosphatase by ion-exchange chromatography. We have also found that the flow-through peak 1 (Fig. 2A) when directly subjected to Superose-12 chromatography, bypassing the Mono Q step, gives the phosphatase activity predominantly in a peak corresponding to positions of proteins with $M_r > 100,000$.

Effect of Inhibitors and Activators

The activities of the two ROS phosphatase species are stimulated by protamine sulfate (Fig. 4) and other polyamines, which are known to activate type 2A phosphatases from other tissues.[25,27] Similarly, okadaic acid, a well-established inhibitor of the class of phosphatases was found to inhibit the activities of the two ROS phosphatase species by more than 90% at

---

[28] P. Cohen, *Annu. Rev. Biochem.* **58**, 453 (1989).
[29] C. Kamibayashi, R. Estes, C. Slaughter, and M. C. Mumby, *J. Biol. Chem.* **266**, 13251 (1991).

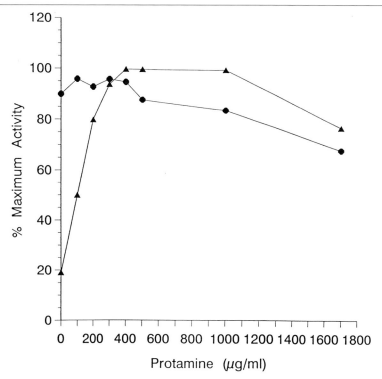

FIG. 4. The dephosphorylation of protamine-treated phosphoopsin in the presence of various concentrations of protamine sulfate. [^{32}P]Phosphoopsin samples (50 nmol) were treated either with 1 mg/ml of protamine sulfate (1.5 ml) or buffer A only, then extensively washed in buffer A. Portions (1 nmol) of the pretreated samples were incubated in a final volume of 30 μl for 30 min at 30° with the catalytic subunit of ROS phosphatase in the presence of various amounts of protamine sulfate. The ^{32}P release was measured as in Fig. 2. Samples treated with protamine (bull;); with buffer A (▲).

concentrations of <1 nM, whereas Ca^{2+}, Mn^{2+}, and related inorganic cations as well as simple anions and nucleotides at concentrations of up to 2 mM were without any effect. Note, however, that the presence of a Ca^{2+}-stimulated phosphatase in ROS, which acts on phosphoopsin, has been reported recently.[30] In the present study and earlier reports[20–23] such an activity could not be detected. If a Ca^{2+}-stimulated phosphatase does exist in ROS, its activity must be rather low and obscured by the overwhelming presence of the type 2A phosphatases in our preparations.

[30] M. A. Kutuzov and N. Bennett, *Eur. J. Biochem.* **238,** 613 (1996).

Substrate Specificity

Phosphoopsin, phosphorhodopsin and thus by implication phospho-Rho* were equally good substrates for the ROS phosphatase catalytic subunit as well as the holoenzyme. The broad specificity leaves open the issue as to what is the true substrate for the enzyme *in vivo*, and hence the precise reaction sequence through which rhodopsin is recycled from the bleached intermediate. The recycling process (Fig. 1) requires three different reactions involving (1) the dephosphorylation process, (2) the hydrolysis of the Schiff base to release all-*trans*-retinal, and (3) reaction with 11-*cis*-retinal to produce the 500-nm chromophore. These reactions can operate by three alternative routes (see Ref. 21), but in view of the properties of the ROS phosphatase described earlier no clear choices between these is possible at present.

A spinoff from our studies on the mechanism of action of rhodopsin kinase was the discovery that peptides corresponding to the C-terminal region of rhodopsin could be conveniently phosphorylated by rhodopsin kinase in the presence of Rho*.[31,32] Using such a protocol several phosphopeptides were prepared and it was found that, in the absence of protamine sulfate, these were dephosphorylated by the catalytic subunit as well as the holoenzyme two to five times the rate found for phosphoopsin.[23] Interestingly, however, the dephosphorylation of the phosphopeptides was *not* stimulated by protamine sulfate and other polyamines, traditionally known to activate type 2A phosphatases toward their protein substrates. The maximum rate of dephosphorylation of the best peptide substrate (^{338}SKTETSQ-VAPA^{348} phosphorylated at Ser-343), in the absence of protamine sulfate, was twice that of phospho-opsin in the presence of protamine sulfate (see Ref. 23).

Mechanism of Activation of ROS Type 2A Phosphatases by Protamine

The observation of the lack of effect of protamine sulfate on the dephosphorylation of phosphopeptides provided the stimulus to critically examine the mechanism through which type 2A phosphatase-catalyzed dephosphorylation of phosphoproteins is stimulated by basic proteins. We took advantage of the particulate nature of phosphoopsin containing ROS which were preincubated with protamine sulfate and then the unbound reagent conveniently removed by centrifugation and further washings. It was found that the protamine-pretreated phosphoopsin (which contained 0.7 mol of prot-

---

[31] C. Fowles, R. Sharma, and M. Akhtar, *FEBS Lett.* **238,** 56 (1988).
[32] N. G. Brown, C. Fowles, R. P. Sharma, and M. Akhtar, *Eur. J. Biochem.* **208,** 659 (1992).

amine per mole of opsin) was maximally dephosphorylated by the two ROS phosphatases without requiring the presence of exogenously added protamine. The parallel experiment conducted with an appropriate control sample of phosphoopsin showed the expected dose-dependent stimulation of phosphate release (Fig. 4).

The cumulative observations may be rationalized by assuming that the protamine treatment somehow modifies phosphoopsin so that the phosphoryl groups of the substrate become exposed for an effective interaction with the phosphatase, thus facilitating hydrolysis. This inference may be embodied in a hypothetical model (Fig. 5), which assumes that the conversion of a neutral hydroxyl group of an amino acid into a dianionic functionality, following phosphorylation (A → B), promotes conformational changes, drawing suitable cationic sites within the protein toward the vicinity of the phosphate group (Fig. 5C). In the presence of a high concentration of a

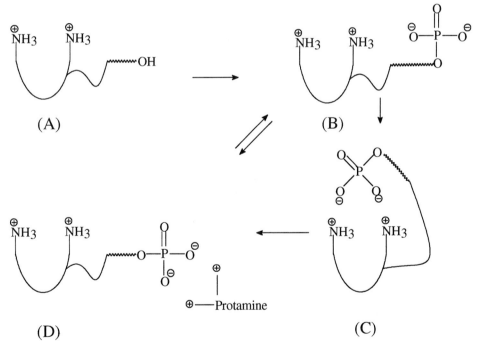

FIG. 5. A hypothetical mechanism for the stimulation of the activity of type 2A phosphatase by polyamines. The model assumes that following phosphorylation (A → B) a conformational change (B → C) allows phosphoproteins to adopt a salt-bridged resting state in which the phosphoryl group is inaccessible to the phosphatase. The function of protamine is to shift the equilibrium toward the open structure (B or D).

suitable basic species, the equilibrium is shifted from a closed structure, of the type shown in Fig. 5C, into an open conformation (Fig. 5B or 5D) in which the phosphate group is exposed for effective interaction with the phosphatase. The fact that the small phosphopeptides are efficiently hydrolyzed without requiring the presence of protamine is explained by invoking that with these substrates the ground state is populated by a large number of rapidly equilibrating structural forms from which the species containing unbridged structures are readily scavenged by the phosphatase. The hypothesis thus envisages that the stimulation of the activity of the two ROS phosphatases (catalytic subunit and holoenzyme) by basic amines may be a substrate-directed phenomenon as illustrated in the model of Fig. 5. For further comments on the model see Ref. 23.

Acknowledgments

This research was supported by the Wellcome Trust, Medical Research Council, and Ulverscroft Foundation.

[38] Purification and Characterization of Protein Phosphatase Type 2C in Photoreceptors

By SUSANNE KLUMPP and DAGMAR SELKE

Introduction

Increasing appreciation of the importance of protein phosphatases has generated considerable interest in methods for their separation, detection, and characterization. Phosphatases catalyzing the dephosphorylation of phosphoserine- and phosphothreonine-containing proteins are composed of two structurally distinct families (old nomenclature is given in parentheses)[1,2]: The PPP family, which includes PPP1 (PP1), PPP2 (PP2A), PPP3 (PP2B or calcineurin), etc.; and the PPM family of Mg^{2+}- or Mn^{2+}-dependent phosphatases including PPM1 (PP2C) enzymes. The presence of five distinct PP2C genes (α, β, γ = FIN13, and δ-Wip1) has been reported in mammalian cells.[3-7] PP2Cα and PP2Cβ are the major and best characterized isoforms

[1] T. S. Ingebritsen and P. Cohen, *Science* **221,** 331 (1983).
[2] P. T. W. Cohen, *Adv. Prot. Phosphatases* **8,** 371 (1994).
[3] C. H. McGowan, D. G. Campbell, and P. Cohen, *Biochim. Biophys. Acta* **930,** 279 (1987).
[4] C. H. McGowan and P. Cohen, *Eur. J. Biochem.* **166,** 713 (1987).
[5] S. M. Travis and M. J. Welsh, *FEBS Lett.* **412,** 415 (1997).

with a molecular mass of 43–48 kDa (75% identity in primary structure).[8] Mouse PP2Cβ has been found to have five distinct isoforms, which are splicing variants originating from a single pre-mRNA.[9] The crystal structure of recombinant human 2Cα is known.[10] PP2C isoforms are ubiquitously expressed.

The major differences of PP2C compared to the other Ser/Thr protein phosphatases include (1) strict requirement for Mg^{2+} or Mn^{2+} ions for activity[11]; (2) inhibition by Ca^{2+} ions[12,13]; (3) insensitivity toward okadaic acid, tautomycin, microcystin, etc., and insensitivity toward low molecular weight, acidic, and thermostable inhibitor proteins affecting PP1 and PP2A—thus, compounds to specifically enhance or inhibit PP2C activity are not available; and (4) monomeric structure: regulatory or targeting subunits referring localization and substrate specificity to the catalytic entity are unknown. With regard to regulation of PP2C activity, phosphorylation of the α isozyme by casein kinase II has been reported.[14,15] Both α and β isozymes were found to be sensitive to activation by unsaturated fatty acids.[16]

Determination of Phosphatase Type 2C Activity

PP2C activity is measured in the presence of magnesium ions using ^{32}P-labeled casein as a substrate. Phosphorylation of bovine casein is performed slightly modifying the procedure of McGowan and Cohen.[17] Casein (10 mg) (Sigma, St. Louis, MO) and 2 mg cAMP-dependent protein kinase (Sigma) are incubated in the presence of 50 μM cAMP, 10 mM magnesium

---

[6] M. A. Guthridge, P. Bellosta, N. Tavoloni, and C. Basilico, *Mol. Cell. Biol.* **17,** 5485 (1997).

[7] M. Fiscella, H. L. Zhang, S. Fan, K. Sakaguchi, S. Shen, W. E. Mercer, G. F. Van de Woude, P. M. O'Connor, and E. Appella, *Proc. Natl. Acad. Sci. U.S.A.* **94,** 6048 (1997).

[8] D. J. Mann, D. G. Campbell, C. H. McGowan, and P. T. W. Cohen, *Biochim. Biophys. Acta* **1130,** 100 (1992).

[9] S. Kato, T. Terasawa, T. Kobayashi, M. Ohnishi, Y. Sasahara, K. Kusuda, Y. Yanagawa, A. Hiraga, Y. Matsui, and S. Tamura, *Arch. Biochem. Biophys.* **318,** 387 (1995).

[10] A. K. Das, N. R. Helps, P. T. W. Cohen, and D. Barford, *EMBO J.* **15,** 6798 (1996).

[11] M. D. Pato and R. S. Adelstein, *J. Biol. Chem.* **258,** 7055 (1983).

[12] M. D. Pato and E. Kerc, *Mol. Cell. Biochem.* **101,** 31 (1991).

[13] S. Klumpp, D. Selke, D. Fischer, A. Baumann, F. Müller, and S. Thanos, *J. Neurosci. Res.* **51,** 328 (1998).

[14] T. Kobayashi, S. Kanno, T. Terasawa, T. Murakami, M. Ohnishi, K. Ohtsuki, A. Hiraga, and S. Tamura, *Biochem. Biophys. Res. Commun.* **195,** 484 (1993).

[15] T. Kobayashi, K. Kusuda, M. Ohnishi, H. Wang, S. Ikeda, M. Hanada, Y. Yanagawa, and S. Tamura, *FEBS Lett.* **430,** 222 (1998).

[16] S. Klumpp, D. Selke, and J. Hermesmeier, *FEBS Lett.* **437,** 229 (1998).

[17] C. H. McGowan and P. Cohen, *Methods Enzymol.* **159,** 416 (1988).

acetate, 200 μM adenosine triphosphate (ATP), and 37 MBq [γ-^{32}P]ATP (110 TBq/mmol). The phosphorylation reaction (1 ml) is incubated for 6–8 hr at 30° (incorporation rate 5–10%) and terminated by addition of 100 μl containing 100 mM Na$_4$P$_2$O$_7$, 100 mM EDTA, pH 7. The centrifugation supernatant (10,000g for 5 min) is applied to Sephadex G-50 superfine (1.2 × 26 cm; 18 ml/hr) equilibrated and run in 50 mM Tris-HCl, pH 7, 0.1 mM EDTA, and 0.1% (v/v) 2-mercaptoethanol to separate the labeled protein from unincorporated nucleotides. [^{32}P]Casein fractions are pooled and stored at 4° (freeze–thawing would favor hydrolysis). Prior to being used as a substrate for PP2C activity, an aliquot of this stock solution is withdrawn and adjusted to 3 × 10^6 cpm/ml (equivalent to 3 nmol ^{32}P/ml, Cerenkov radiation) by the addition of substrate buffer [50 mM Tris-HCl, pH 7, 0.1 mM EGTA, 1 mg/ml bovine serum albumin (BSA), 0.1% 2-mercaptoethanol]. Then 30 μl 100 mg/ml BSA is added to 1 ml of this solution, to ensure that trichloroacetic acid (TCA) precipitation of the substrate remaining after phosphatase treatment is quantitative. PP2C assays are performed in 30-μl reactions at 30° for 10 min (10 μl substrate to start the reaction, 10 μl enzyme diluted in substrate buffer, 5 μl divalent cation solution, 5 μl for other reagents). Final concentrations for a PP2C standard assay are 33 mM Tris-HCl, pH 7, 0.07% (v/v) 2-mercaptoethanol, 20 mM magnesium acetate, 1.3 mg/ml BSA, and 1 μM [^{32}P]casein with ~5 × 10^4 cpm/assay tube (measured in the presence of 1 ml scintillant). Reactions are terminated by the addition of 200 μl 20% TCA and placed on ice for 5 min. After centrifugation at 10,000g (5 min), 200 μl of the supernatant is analyzed for ^{32}P content. Radioactive P$_i$ is converted and extracted as phosphomolybdic complex to discriminate phosphatase from unspecific protease activities.[18] Care must be taken to ensure that activity measurements are kept within the linear range of time and protein (maximally 25% release of P$_i$).

Separation of Retinal PP2C from Other Serine Threonine Protein Phosphatases

To distinguish dephosphorylation of [^{32}P]casein carried by either PP2A or PP2C, assays need to be performed in duplicate, one in the absence and one in the presence of magnesium ions (e.g., 0.1 mM EDTA versus 20 mM Mg^{2+}). Alternatively, 100 nM okadaic acid can be added to the Mg^{2+}-containing reactions. This tumor promotor completely blocks PP2A activity without affecting PP2C. One does not have to worry about eliminating PP1

---

[18] J. F. Antoniw and P. Cohen, *Eur. J. Biochem.* **68,** 45 (1976).

activity when working with casein as a substrate for PP2C: [^{32}P]casein is not dephosphorylated by PP1.

Stepwise Purification by Column Chromatography

Enzyme preparations are carried out at 4° with buffers containing 0.02% NaN$_3$ and 0.1% (v/v) 2-mercaptoethanol. The isolation of PP2Cα and PP2Cβ as described here yields 2–6 μg PP2C proteins with a specific activity of 30–90 nmol P$_i$/min/mg.[13] Eighty bovine retinas are shaken for 90 sec in 80 ml sucrose medium [20 mM Tris-HCl, pH 7.5, 600 mM sucrose, 10 mM glucose, 0.2 mM EDTA, 0.1 mM phenylmethylsulfonyl fluoride (PMSF), 1 mM benzamidine]. The soluble extract (48,000g, 30 min at 4°, followed by 100,000g, 1 hr) is applied to DEAE-Sephacel (2.5 × 10 cm; 80 ml/hr) equilibrated in buffer A (20 mM Tris-HCl, pH 7, 0.1 mM EDTA). The column is washed with buffer A containing 0.1 M NaCl prior to elution of PP2C with buffer A containing 0.5 M NaCl. PP2C activity containing eluate is diluted 1:5 with buffer B (20 mM Tris-HCl, pH 7, 0.1 mM EDTA, 5% glycerol) and applied to heparin-Sepharose CL-6B equilibrated in buffer B (2.5 × 10 cm; 120 ml/hr). PP2C activity is recovered in the heparin-Sepharose flow-through and concentrated by ultrafiltration. PP2C is remarkably insensitive to short-term exposure of 1 M NaCl (but not ammonium sulfate), thus enabling hydrophobic interaction chromatography. NaCl is added to yield a 1 M solution prior to application onto phenyl-Sepharose CL-6B (1 × 2.6 cm) equilibrated in buffer C (20 mM Tris-HCl, pH 7, 1 mM EDTA) containing 1 M NaCl. PP2C activity is eluted with buffer C containing 20% glycerol (1 ml/min) and concentrated on Mono Q HR5/5 run with buffer C containing 5% glycerol (0.5 ml/min) using addition of 0.5 M NaCl for elution. Chromatography on Superdex 75 preparation grade 26/60 (1 ml/min) is performed in buffer D (20 mM Tris-HCl, pH 7, 1 mM EDTA, 5% glycerol, 40 mM NaCl). MgCl$_2$ (3 mM) is added to fractions containing PP2C activity. Isozymes are separated on a 1-ml Blue-Sepharose column (Affi-Gel Blue in a HR 5/5 column or HiTrap) equilibrated in buffer D containing 3 mM MgCl$_2$ (0.75 ml/min). A linear 60-ml gradient (2–0 mM Mg^{2+}) is required to separate PP2Cβ (eluting at 1.5 mM Mg^{2+}) from PP2Cα (0.5 mM Mg^{2+}).[19] Purification is summarized in Table I.

Batch Procedure for Rapid Enrichment

A simplified method has been developed to highly enrich PP2C proteins quickly.[16] One buffer is used throughout: 20 mM Tris-HCl, pH 7, 1 mM

---

[19] S. Klumpp and D. Selke, *Meth. Mol. Biol.* **93,** 213 (1998).

TABLE I
PURIFICATION OF PP2C FROM BOVINE RETINAS

Fraction	Volume (ml)	Total protein (mg)	Total activity (nmol/min)	Specific activity (nmol/min/mg)	Purification (-fold)	Recovery (%)
Supernatant	170	654	73	0.11	1	100
DEAE-Sephacel	625	284	74	0.26	2.4	102
Heparin-Sepharose	482	135	40	0.29	2.6	54
Ultrafiltration	44	124	20	0.16	1.5	27
Phenyl-Sepharose	31	34	18.5	0.54	4.9	25
Mono Q	4	11	12.0	1.10	10	16
Superdex 75	15	0.39	1.5	3.85	35	2.1
Blue-Sepharose						
Peak I (2Cβ)	13	0.006	0.17	28	264	0.2
Peak II (2Cα)	10	0.002	0.18	90	818	0.2

EDTA, 5% glycerol. The concentrations of NaCl for washing and elution are the same as those described for the column procedure. The beads (DEAE-Sephacel and heparin-Sepharose) are used with gentle rocking. Binding and elution times are 1 hr, respectively. Beads are pelleted by centrifugations using a swing-out rotor at 700g for 1 min at 4°.

The soluble retinal extract is applied to 50-ml DEAE-Sephacel. Heparin-Sepharose (10 ml) was sufficient for the next step. PP2C in the supernatant fraction is concentrated by Centriplus (Amicon, Danvers, MA) and further enriched using Mono Q HR 5/5 (0.7 M NaCl for elution). Gel filtration, as the final step, is performed as described earlier, yielding highly enriched PP2C free from other phosphatases. For assays dealing with the cation dependence and fatty acid stimulation of PP2C, EDTA in the ultimate column buffer was reduced to 0.1 mM.

Purification of Recombinant PP2C Isozymes

PP2Cα and PP2Cβ have been cloned from a bovine cDNA library.[20] They share high identity with PP2C isozymes from other mammalian species and tissues. For example, a comparison of PP2C from rat liver with the enzyme from bovine retina shows 5 and 26 substitutions in the amino acid sequence for α and β, respectively.

PP2C isozymes can be heterologously expressed.[13] His-tagged proteins are purified by affinity chromatography on Ni^{2+}-NTA agarose. Enzyme activity of purified recombinant PP2C decreases on long-term storage (gone

---

[20] D. Selke, S. Klumpp, B. Kaupp, and A. Baumann, *Meth. Mol. Biol.* **93,** 243 (1997).

within 6 months at $-80°$). Enzyme activity in crude bacterial extracts, in contrast, is fairly stable (at least for 1 year). Therefore, a procedure has been developed to purify PP2C from small volumes: 300 μl lysate onto 50 μl Ni^{2+}-NTA agarose plugged into a Pasteur pipette, equilibrated in buffer E (20 mM Tris-HCl, pH 7.9, 5 mM imidazole, 0.5 M NaCl), and run by gravity. After washing in buffer E containing 60 mM imidazole, PP2C is eluted using buffer F (10 mM Tris-HCl, pH 7.9, 0.5 M imidazole, 0.25 M NaCl). Recovery of PP2C isozymes is 70%, respectively. A 20-ml culture of *Escherichia coli* normally yields 0.5 mg PP2C protein with a specific activity of 3–15 nmol P$_i$/min/mg. Expression of PP2Cα routinely results in only about half of the amount of protein compared to PP2Cβ. Activity of both isozymes is strongly inhibited by imidazole (IC$_{50}$ 250 mM), therefore, immediate and at least fivefold dilution (substrate buffer) is recommended.

General Comments on Purification and Analytics

Separation of PP2C from PP1 is possible using chromatography on heparin-Sepharose: PP2C lacks affinity to heparin-Sepharose, thus, it can be quantitatively recovered in the flow-through fractions. Separation of PP2C from PP2A is achieved by gel filtration chromatography (Superdex 75 or Sephacryl S-200): the catalytic entities of PP2C α and β (~45 kDa) elute in clear distance to the high molecular weight PP2A holoenzymes.

Separation of PP2C α and β isozymes can be accomplished by two distinct methods. Isozymes from rabbit liver are nicely separated on chromatography using ion-exchange resins.[4] This does not work for the retinal PP2C isozymes. Due to a few species- and tissue-specific amino acid substitutions, retinal PP2Cα and PP2Cβ—in contrast to PP2C isozymes from rabbit liver—end up with almost identical pI values. Separation of retinal PP2C isozymes is possible, however, on chromatography on Blue-Sepharose[19]: both isozymes are retained by Cibachron Blue provided Mg^{2+} ions are present. Decreasing the divalent cation concentration consequently elutes first β, then α.

In general, it is recommended that long-lasting dialysis steps be avoided. For short-term storage, e.g., overnight interruption during the purification process, freezing is best. Purified PP2C should be stored in aliquots at $-80°$. Repetitive freeze–thawing of specifically recombinant PP2Cα should be avoided. Activity and protein "vanish." Adsorption to plastic vials might be the reason. This phenomenon is not observed with PP2Cβ. Analyzing PP2C isozymes on sodium dodecyl sulfate–polyacrylamide gel electrophoresis (SDS–PAGE) bears two unexpected findings: (1) Unusual migration behavior: according to the amino acid composition, the molecular mass of PP2Cα is less compared to PP2Cβ (e.g., 382 versus 387 amino acids for

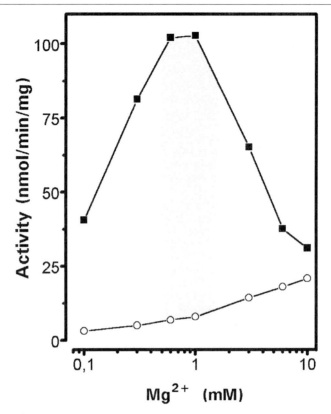

FIG. 1. Mg^{2+} dependence of PP2C. Activity of recombinant PP2Cα was assayed under standard reaction conditions (○; 12.5 ng protein/tube) or in the presence of 0.5 mM oleic acid (■; 6.3 ng protein/tube). The shaded area marks the activation in the physiologic concentration range of free Mg^{2+} (0.5–1.5 mM).

the enzymes from bovine retina). On denaturing electrophoresis, however, PP2Cα runs as a protein band *above* PP2Cβ. (2) For most proteins, silver staining is ~10-fold more sensitive than Coomassie staining. Detection of PP2C isozymes by silver staining seems to be impaired.

Characterization of Native and Recombinant PP2C Isozymes

Mechanical disintegration of retinas (Dounce homogenizer) and subsequent centrifugation (100,000g for 60 min) revealed that PP2C activity distributes equally well between the soluble and particulate fractions. Addition of detergents did not release further PP2C activity.

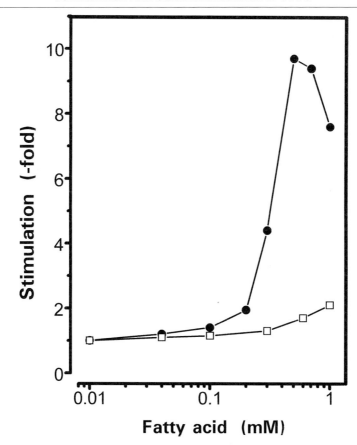

FIG. 2. Differential subcellular fatty acid composition affecting PP2C activity. Recombinant PP2Cα (6.3 ng/tube; specific activity 8 nmol P_i/min/mg) was assayed in the presence of 0.7 mM Mg^{2+} under otherwise standard reaction conditions. Fatty acids added were docosahexaenoic acid (●, 22:6) known to be concentrated in the photoreceptor outer segments, and palmitic acid (□, 16:0) enriched in the neighboring pigment epithelium cells.

Except for the effect of fatty acids (*vide infra*), enzymatic properties of recombinant 2Cα and 2Cβ are almost indistinguishable from one another and from their corresponding native counterparts. The optima for temperature and pH were 45°–55° (activation energy 70–90 kJ/mol) and pH 7.5–8 (Tris-HCl). A feature exhibited by all PP2C characterized to date is the requirement of unphysiologic high Mg^{2+} concentrations for activity. A recent report describes that addition of unsaturated fatty acids results in high PP2C activity measurable at physiologic Mg^{2+} concentrations.[16] Figure 1 shows the Mg^{2+} dependence of PP2C in the absence versus presence of oleic acid (18:1). Activity increases 10-fold and is highest at physiologic free Mg^{2+} levels (0.5–1.5 mM). Under those conditions, Ca^{2+} ions are inhibitory in the micromolar range.[16]

The amount of fatty acids and the fatty acid composition are known to vary within cell compartments and from membrane to membrane. In vertebrate retinas, a striking concentration gradient exists between docosahexaenoic acid (22:6; major component in photoreceptor outer segments) and palmitic acid (16:0; found in the adjacent pigment epithelium cells) with important functional consequences. Supply of the photoreceptors with retinoids is based on the differential subcellular distribution of those long-chain fatty acids: Docosahexaenoic acid—but not palmitic acid—induces release of 11-*cis*-retinal from a 140-kDa interphotoreceptor matrix retinoid-binding protein.[21] Within vertebrate retinas, PP2Cα and PP2Cβ isozymes are highly enriched in the photoreceptor outer segments.[13] Figure 2 shows that docosahexaenoic acid—but not palmitic acid—increases the activity of recombinant PP2Cα up to 10-fold. The half-maximal activating concentration for docosahexaenoic acid was 350 μM. Stimulation of recombinant PP2Cβ and native PP2C was somewhat less. PP2C and docosahexaenoic acid are not only colocalized in rod outer segments, but are both associated with retinal degeneration processes.[13]

Conclusions

PP2C, a "black sheep" phosphatase (no activator, no inhibitor) had been considered to be 10-fold less compared to PP1 and PP2A. The rise in less-quantity amount of activity by fatty acids now brings PP2C onto stage as well.

Fatty acids and Ca^{2+} ions are established players in retinal signal transduction processes. We currently hypothesize regulation of PP2C at two levels: (1) Its basal activity is depending on the presence of unsaturated fatty acids. A long-term change in the fatty acid distribution and concentration as exemplified by apoptosis or retinitis pigmentosa will dramatically affect PP2C activity. (2) On top of that fatty acid sustained PP2C activity, shifts in the Ca^{2+} ion concentration would account for short-term regulation of PP2C causing immediate changes in enzyme activity.

Acknowledgment

This work was supported by the Deutsche Forschungsgemeinschaft (Kl 601/8-1).

---

[21] Y. Chen, L. A. Houghton, J. T. Brenna, and N. Noy, *J. Biol. Chem.* **271**, 20507 (1996).

[39] Photoreceptor Serine/Threonine Protein Phosphatase Type 7: Cloning, Expression, and Functional Analysis

By Xizhong Huang, Mark R. Swingle, and Richard E. Honkanen

Introduction

Reversible phosphorylation regulates the biological activity of many proteins involved in the phototransduction cascade of both vertebrate and invertebrate photoreceptors, with phosphorylation influencing the sensitivity and kinetics of both the excitatory and recovery processes.[1,2] In photoreceptors, as observed in all eukaryotic cells, protein phosphorylation occurs on serine, threonine, and tyrosine residues. The phosphorylation reactions are catalyzed by protein kinases, serine/threonine protein kinases, and tyrosine protein kinases. Protein dephosphorylation is catalyzed by three families of protein phosphatases: serine/threonine protein phosphatases, which catalyze the dephosphorylation of phosphoserine and phosphothreonine residues; tyrosine phosphatases, which are specific for phosphotyrosine; and dual-specificity phosphatases, which dephosphorylate phosphoserine, phosphothreonine, or phosphotyrosine residues on suitable protein substrates.

The serine/threonine protein phosphatases (PPases) are classified based on their biochemical characteristics, their sensitivities to specific inhibitors, and a limited amount of substrate specificity that can be demonstrated *in vitro*. Accordingly, four subtypes, PP1, PP2A, PP2B, and PP2C, have been established.[3,4] More recent studies, however, indicate that the primary amino acid sequences of PP1, PP2A, and PP2B are related, while PP2C is structurally distinct and belongs to a different gene family.[5,6] In humans, three isoforms of PP1, two isoforms of PP2A, and three structurally related

---

[1] W. S. Stark, D. Hunnius, J. Mertz, and D. M. Chen, *in* "Degenerative Disease of the Retina" (R. E. Anderson, M. M. LaVail, and J. G. Hollyfield, eds.), p. 217. Plenum Press, New York, 1995.

[2] P. Ferreira and W. Pak, *in* "Degenerative Disease of the Retina" (R. E. Anderson, M. M. LaVail, and J. G. Hollyfield, eds.), p. 263. Plenum Press, New York, 1995.

[3] P. Cohen, *Annu. Rev. Biochem.* **58,** 453 (1989).

[4] S. Shenolikar and A. C. Nairn, *Adv. Sec. Mess. Phosphopro. Res.* **23,** 1 (1991).

[5] P. T. W. Cohen, N. D. Brewis, V. Hughes, and D. J. Mann, *FEBS Lett.* **268,** 355 (1990).

[6] G. Walter and M. Mumby, *Biochim. Biophys. Acta.* **1155,** 207 (1993).

PPases, designated as PP4,[7] PP5,[8] and PP6,[9] have been identified. PP4 and PP6 are closely related to PP2A, sharing 65% and 57% identity at the level of their primary amino acid sequence, respectively. PP5 is more distantly related to PP1 and PP2A, containing a catalytic domain common to the PP1/PP2A/PP2B/PP4/PP6 family of enzymes and a unique N-terminal domain with four 34-amino-acid "tetratricopeptide" (TPR) repeat motifs.[8]

This chapter focuses on the most recently identified family of human serine/threonine protein phosphatases, PP7,[10] (GenBank accession number AF027977). PP7 has a catalytic core that contains all 53 amino acids that are absolutely conserved in the PP1–PP6 family of PPases, including all 16 amino acids predicted to form the site of catalytic activity based on the crystal structure analysis of PP1.[11] However, PP7 has unique N- and C-terminal regions, with the extended C-terminal region possessing five putative high-affinity calcium binding domains referred to as EF-hand and EF-hand-like motifs.[10] The human gene encoding PP7, also referred to as PPEF,[12] and a highly homologous gene, referred to as PPEF-2,[13] have been identified (accession numbers X97867 and AF023456, respectively). The expression of these PPases appears to be limited to brain, retina, and sensory neurons. Structurally, the amino acid sequence for PP7 shares the greatest homology with the gene product encoded by the *Drosophila* retinal degeneration C gene[14] (rdgC; accession number M89628). A direct comparison of PP7 (AF027997) and rdgC (M89628) indicates that they share 42.1% identity and 54.3% similarity.[10] However, almost nothing is known about the roles of PP7 in vertebrate photoreceptors. Therefore, to aid in the illumination of the physiologic and potential pathologic roles of PP7, the remainder of this chapter describes protocols and procedures useful for the study of PP7.

---

[7] N. D. Brewis, A. J. Street, A. R. Prescott, and P. T. W. Cohen, *EMBO J.* **12,** 987 (1993).

[8] W. Becker, H. Kentrup, S. Klumpp, J. E., Schultz, and H. G. Joost, *J. Biol. Chem.* **269,** 22586 (1994).

[9] H. Bastians and H. Ponstingl, *J. Cell Sci.* **109,** 2865 (1996).

[10] X. Huang and R. E. Honkanen, *J. Biol. Chem.* **273,** 1462 (1998).

[11] J. Goldberg, H. B. Huang, Y. G. Kwon, P. Greengard, A. C. Nairn, and J. Kuriyan, *Nature* **376,** 745 (1995).

[12] E. Montini, E. I. Rugarli, E. Van de Vosse, G. Andolfi, M. Mariani, A. A. Puca, G. G. Consalez, J. T. den Dunne, A. Ballabio, and B. Franco, *Hum. Mol. Genet.* **6,** 1137 (1997).

[13] P. M. Sherman, H. Sun, J. P. Macke, J. Williams, P. M. Smallwood, and J. Nathans, *Proc. Natl. Acad. Sci. U.S.A.* **94,** 11639 (1998).

[14] F. R. Steel, T. Washburn, R. Rieger, and J. E. O'Tousa, *Cell* **69,** 669(1992).

Procedures

Expression of Recombinant PP7 in Insect Cells

Production of cDNA

Polymerase chain reaction (PCR)-mediated amplification of cDNA produced from poly(A)$^+$ enriched RNA is one of the most sensitive methods of detecting the expression of mRNA encoding a protein in tissue, cultured cells, or isolated photoreceptors. The identification of cDNA encoding PP7 is achieved using PCR with degenerate primers designed to complement conserved regions in known PPases using the procedures that follow.

Total RNA from either cell cultures (i.e., Y-79 cells) or human retina is prepared using a monophasic solution of guanidine isothiocyanate and phenol, TRIZOL Reagent (Life Technologies, Gaithersburg, MD) according to the methods of the manufacturer. Poly(A)$^+$ RNA is then obtained by affinity chromatography with an oligo(dT) spin column [Poly(A) Spin mRNA isolation kit, New England Biolabs, Beverly, MA]. First strand human cDNA is generated from poly(A)$^+$ RNA by incubating 1–5 μg of poly(A)$^+$ RNA with 1 μg of oligo(dT$_{12-18}$) primer and 200 units of reverse transcriptase (SuperScript II RNase H-reverse transcriptase; Life Technologies).

Identification and Cloning of PP7 cDNA

All of the known human PPases have four highly conserved regions [GD(YF)VDRG, RGNHE DILLWSDP, (FW)SA(PS)NY]. To characterize the PPases expressed in human retina, degenerate oligonucleotide primers complementary to several of three conserved regions were designed and employed in PCR-mediated amplification of cDNA produced from human retina and Y-79 cells (Fig. 1). A typical PCR reaction was conducted for 35 cycles (1 min at 95°, 45 sec each at 40° and 45°, 1 min at 72°) in a solution containing 50 mM KCl, 2 mM MgCl$_2$, 0.2 mM dNTP, 10 mM Tris-HCl, pH 8.3, and 100 ng of each appropriate primer.

The PCR products produced with the degenerate primer pairs were cloned into *Sma*I digested pBlueScript (Stratagene, La Jolla, CA) and used to transform highly competent *Escherichia coli*. Approximately 500 isolated colonies were then analyzed by restriction digestion and agarose gel electrophoresis. Approximately 130 promising clones (i.e., clones with different size inserts) were sequenced, and each sequence obtained was then compared to the sequences of the known PPases using the FASTA computer search [Genetics Computer group (GCG), Madison, WI]. Sequences identi-

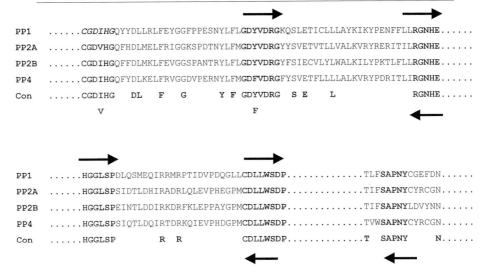

FIG. 1. PCR-based strategy used to identify novel retinal PPases. Degenerate oligonucleotide primers were made in the sense and antisense orientation as indicated by arrows and employed in the amplification of cDNA produced from human retina or Y-79 cells. Consensus (con) regions are indicated in bold, and the nucleotide sequences of individual human PPases are provided in Table I.

cal to three isoforms of PP1 (PP1α, PP1δ, PP1γ), PP2A (PP2Aα), PP4, PP5, and PP7 were identified (Table I). In addition, several clones with little or no homology to known PPases were identified. To distinguish PCR artifacts from clones that encode proteins, a BLAST computer search (GCG) was performed comparing sequences in the expressed sequence tag database (dbEST) to each sequence obtained with PCR. The search identified one ~875-bp fragment from human fetal brain (GenBank accession number H18854), and this human brain clone (ID51064) was obtained from Research Genetics, Inc. (Huntsville, AL; I.M.A.G.E. Consortium Lawrence Livermore National Laboratory, Livermore, CA). Sequencing of clone 51064 confirmed that it is identical to the PCR-generated DNA fragment amplified from human retinal cDNA, and further analysis revealed homology (sharing 37.4% similarity) with the *Drosophila* retinal degeneration C gene product (rdgC).

To obtain the full-length cDNA encoding the human rdgC homolog, clone 51064 was digested with *Hin*dIII. The ~600-bp fragment produced was then radiolabeled ($>2 \times 10^9$ cpm/μg) using random decamer oligonucleotide primers and exonuclease-free Klenow enzyme (DECAprime II, Ambion, Austin, TX). The probes with high specific activity were then used

TABLE I
NUCLEOTIDE SEQUENCES ENCODING HIGHLY CONSERVED REGIONS OF HUMAN SERINE THREONINE PROTEIN PHOSPHATASES

Nucleotide[a]	Amino acid[b] and nucleotide sequence of conserved regions contained in human serine/threonine protein phosphatases		
	GD(YF)VDRG	RGNHE	FSA(PS)NY
PP1α	GGGGACTATGTGGACAGGGG-	CCGTGGGAACCACGAGTGTG-	CTCAGCTCCCAACTACTGTGG-
PP1δ	GGAGATTATGTGGACAGAGG-	AGAGGAAACCATGAGTGTGC-	TCAGCCCCAAATTACTGTGGC-
PP1γ	GGGGACTATGTGGACAGGGA-	AGAGGGAACCATGAATGTGC-	TCTGCGCCAATTATTGCGGA-
PP2Aα	GGAGATTATGTTGACAGAGG-	CGAGGGAATCATGAGAGCAG-	TTCAGTGCTCCAAACTATTGT-
PP4	GGGGACTTTGTTGGACCGTGG-	CGGGGCAACCATGAGAGTCG-	TCGGGCACCCAACTACTGCTA-
PP5	GGTGACTTTGTGGACCGAGG-	CGAGGCAACCACGAGACAGA-	TTCTCTGCCCCAACTACTGC-
PP7	GGTGACTTTGTAGATCGAGG-	CAGAGGGAACCACGAAGATT-	TTTTCTGCTTCTAATTATTAT-

[a] Nucleotide sequence encoding conserved regions of individual human PPases were identified by sequence analysis of PCR products produced by the amplification of human retinal or Y-79 cDNA with degenerate primers encoding the indicated conserved regions in PP1, PP2A, PP2B, and PP4 (Fig. 1). All nucleotide sequences are from the sense strand and provided in the 5'-3' orientation.
[b] Identified by PCR amplification of human cDNA from retina or Y-79 cells.

to screen ~4 × 10⁶ clones from a human retina cDNA library constructed in λgt10. Four positive clones were isolated and sequenced completely. One of the λ clones contained an ~2.8-kb insert with an open reading frame that encodes a 653-amino-acid protein. This open reading frame encodes all of the conserved regions contained in the PP1–PP6 family of phosphatases, and shares 42.1% identity with a putative PPase encoded by *Drosophila* rdgC (54.3% similarity when conservative substitutions are considered). In adherence with the traditional nomenclature system used for human PPase, this PPase was given the name PP7.

Construction of PP7 Expression Vector

Human PP7 is expressed in Sf21 cells using the Bac-to-Bac baculovirus expression system (Life Technologies). To construct a PP7 donor plasmid, the full-length open reading frame encoding PP7 is amplified with PCR employing a nested primer pair, (RH 108; 5'-AACTCTGCAGTTAGC-CAAGGTTGGTGACATCAGG-3' and RH112; 5'-AGCCGGATCCAT-GGGATGCAGCAGTTCTT C-3'). These primers hybridize to the last 24 and the first 20 bases of the PP7 coding region in the antisense and sense orientation, respectively. To facilitate cloning, RH108 and RH112 contain restriction sites (*Bam*HI and *Pst*I) added to the 5' ends. The PCR reaction is conducted for 35 cycles (1 min at 94°, 55 sec at 55° and 1 min at 72° in a solution containing 50 mM KCl, 2 mM MgCl$_2$, 0.2 mM dNTP, 10 mM Tris-HCl, pH 8.3, and 100 ng of each primer. The PCR product is gel purified, digested with *Bam*HI/*Pst*I and ligated into *Bam*HI/*Pst*I digested pFastBac-HTb (Life Technologies). The PP7–pFastBac construct contains the bacterial transposon Tn7 with the polyhedrin promoter from *Autographa californica* nuclear polyhedrosis virus (AcNPV), the entire coding region of PP7 with six consecutive in-frame His⁺ residues attached to the N-terminal region of PP7, and a simian virus 40 (SV40) poly(A) signal inserted between the left and right arms of Tn7 (Fig. 2). After sequencing PP7–pFastBac to confirm the fidelity of the construct, PP7–PFastBac (1 ng in 5 μl) is then used to transform *E. coli* (DH10Bac), which contain a baculovirus shuttle vector[15] (bacmid; bMON14272) and a helper plasmid (pMON7124). The bacmid bMON14272 contains a low copy number mini-F replicon, a segment of DNA encoding the lacZ protein that has a short in-frame segment containing the attachment site for Tn7 (mini-att Tn7), and a region that confers resistance to kanamycin. *E. coli* DH10Bac contain a chromosomal *lacZ* deletion, which can be complemented by *lacZ* contained in the bacmid. The helper plasmid pMON7124 provides Tn7 transpo-

---

[15] V. A. Luckow, S. C. Lee, G. F. Barry, and P. O. Olins, *J. Virol.* **67**, 4566 (1993).

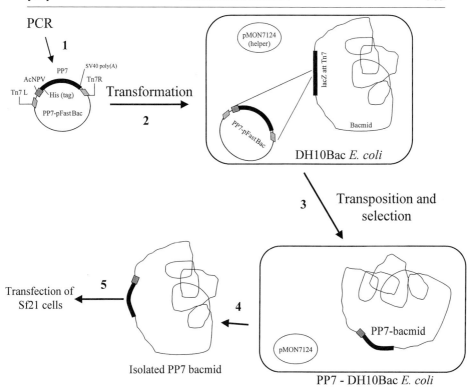

FIG. 2. Production of a baculovirus shuttle vector for the expression of recombinant human PP7 in insect cells. (1) Human retinal cDNA is amplified using PP7 specific primers, and the product is subcloned into the pFastBac-Htb expression cassette downstream of the polyhedrin promoter from *A. californica* nuclear polyhedrosis virus (p-AcNPV). (2) The PP7–pFastBac plasmid produced is then used to transform *E. coli* containing a baculovirus expression shuttle vector that provides resistance to kanamycin and complements the LacZ genomic deficiency of DH10Bac *E. coli* (Bacmid). (3). In the presence of transposition proteins, which are provided in *trans* by a helper plasmid (pMon 7124), the mini-Tn7 element of the PP7–pFastBac plasmid transposes to the mini-att Tn7 target site on the bacmid. Colonies containing recombinant bacmids are identified by the disruption of the *lacZ* gene, which produces a white phenotype when grown on media containing X-Gal. (4) The bacmid DNA from clones containing human PP7 is then used to transfect Sf21 cells and generate recombinant baculovirus particles. The recombinant virus particles are then used to infect Sf21 cells for the expression of human PP7.

sition functions in *trans*, which allows the transposition of a mini-Tn7 element from the donor plasmid (PP7–pFastBac) to the mini-att Tn7 attachment site on the bacmid. Transposition of the mini-Tn7 from the PP7–pFastBac construct to the bacmid interrupts the lacZ protein, which disrupts the expression of the lacZ protein in DH10Bac *E. coli*. Therefore, colonies containing recombinant bacmids adopt a white phenotype when grown on Luria–Bertani (LB) plates containing a chromogenic substrate (i.e., X-Gal). To identify clones in which PP7 has been incorporated into the bacmid, several individual white colonies are collected. Conformation that the isolated colonies contained bacmids transformed with PP7 is obtained with PCR analysis employing PP7 specific primers (RH 108 and RH 112). PP7–bacmid DNA is then isolated from overnight culture of a single, isolated bacterial colony according to the methods provided by Life Technologies. Bacmid DNA can be stored at $-20°$; however, repeated freeze-thawing should be avoided.

Transfection and Expression of PP7 in Spodoptera Frugiperda (Sf21) Cells

Insect cells in mid-log phase growth from a 3- to 4-day-old suspension culture with viability of $>97\%$ are seeded at 9×10^5 cells per well (35 mm) in 2 ml of Sf-900 medium (Life Technologies) containing penicillin (50 units/ml) and streptomycin (50 μg/ml). After allowing the cells at least 2 hr to attach, they are washed once with 2 ml of Sf-900 medium without antibiotics. Following aspiration of the wash media, 1 ml of transfection medium containing dilute PP7–bacmid and CellFECTIN (Life Technologies) is gently added to the cells. (Dilute PP7–bacmid is made by adding 5 μg of PP7–bacmid DNA to 100 μl of Sf-900 medium without antibiotics and mixing. Dilute CellFECTIN is made by adding 6 μl freshly mixed CellFECTIN to 100 μl of Sf-900 media without antibiotics. The transfection media is made by mixing the dilute CellFECTIN, dilute PP7–bacmid, and 800 μl of Sf-900 media without antibiotics. The transfection medium is allowed to stand for at least 15 min, but not more than 45 min, at room temperature prior to transfection.) The cells are incubated in transfection medium for 5 hr at $27°$. The transfection medium is then replaced with 2 ml of Sf-900 media containing antibiotics. Recombinant baculovirus is harvested 72 hr later by removing 2 ml of the medium from infected cultures. The medium is clarified by centrifugation for 5 min at 500g, and the virus-containing supernatant is transferred to a fresh tube. The initial stock [$\sim 9 \times 10^5$ pfu(plaque-forming units)] is reamplified and used to infect three 125-mm flasks of Sf21 cells at a multiplicity of infection of one. The cells are harvested 48 hr after the infection, collected by centrifugation at 1000 rpm, sonicated in 20 mM Tris-HCl (pH 7.9) containing 5 mM imidazole, 0.5 M NaCl, and 2 mM phenylmethylsulfonyl fluoride (PMSF) (binding buffer) and subjected to $4°$ ultracentrifugation at 400,000g for 20 min. The supernatant is passed through a 0.45-μm filter and applied to an 1 ml Ni$^+$ charged chelating His-tag column (Novagene, Madison, WI). The column is washed with 10 volumes of binding buffer and then with 6 volumes of 20 mM Tris-HCl (pH 7.9) containing 60 mM imidazole and 0.5 M NaCl (washing buffer). The recombinant PP7 is eluted from the column with the addition of 3 column volumes of 20 mM Tris-HCl (pH 7.9) containing 1 M imidazole and 0.5 M NaCl (elution buffer). Fractions (0.5 ml) are collected, and 10-μl aliquots from each fraction are tested for PPase activity using pNPP as a substrate (see later section). Fractions containing PPase activity are pooled, concentrated and subjected to fast protein liquid chromatography (FPLC) gel filtration on a Superose-12 HR10/30 column equilibrated with 20 mM Tris-HCl (pH 7.9) containing 5 mM imidazole and 0.5 M NaCl. The protein containing fractions are identified by absorbance at 280 nM, and fractions containing PP7 are identified by assaying for PPase activity. Active fractions are then pooled and subjected to a second round of affinity purification employing a Ni$^+$ charge chelating His-tag column as described earlier. The purity of each active fractions is assessed by Sodium dodecyl sulfate–polyacrylamide gel electrophoresis (SDS–PAGE) and coomassie blue or silver staining. This procedure produces a single major band that migrates with a similar rate to an ~75-kDa molecular mass standard.[10]

Assessment of PPase Activity

Preparation of Phosphoprotein Substrates

Phosphohistone, with a specific activity $\geq 5 \times 10^6$ dpm/nmol incorporated phosphate, is prepared by the phosphorylation of bovine brain histone with 3′, 5′-cAMP-dependent protein kinase from rabbit muscle in the presence of [γ-^{32}P]ATP. For the phosphorylation reaction, approximately 22 mg of histone type 2AS (H775 from Sigma) is dissolved in 80 mM Tris base (pH 6.8), containing 20 mM MgCl$_2$, to achieve a solution with a final concentration of 20 mg histone/ml. One milligram of protein kinase A (PKA) from rabbit muscle (PKA purified to an extent similar to the peak II kinase activity obtained after the second DEAE cellulose column described by Beavo et al.[16]; P-3891; Sigma) is dissolved in 320 μl of 50 mM Tris HCl (pH 6.8) containing 1 mM dithiothreitol (DTT) and 0.1 mM EDTA. The following are then added to a 15-ml polypropylene tube, in order: 1 ml of water, 100 μl of a 400 μM cAMP stock, 1 ml of the freshly prepared histone Tris-MgCl$_2$ solution (20 mg histone/ml), 62.4 μl of a 10 mM adenosine triphosphate (ATP) stock, 1 mCi [^{32}P]ATP (>6000 Ci/mmol, crude fraction from NEN) and water to achieve final volume 3.70 ml. The solution is mixed, and the phosphorylation reaction is initiated by the addition of PKA (300 μl from above). The final volume is 4 ml, and the

---

[16] J. A. Beavo P. J. Bechtel, and E. G. Krebs, *Methods Enzymol.* **38**, 299 (1974).

phosphorylation reaction is allowed to continue for 3.5 hr at 30°. Aliquots (2 × 5 μl) are removed for use in the calculation of specific activity, and the solution is incubated in a 30° shaking water bath for 3.5 hr. The reaction is terminated by the addition of 1.3 ml of ice-cold 100% trichloroacetic acid (TCA). After placing the tube in ice for 10 min, the precipitated phosphohistone is collected by centrifugation at 1500g for 5 min. The supernatant is discarded, and the histone pellet is redissolved in 4 ml of 1.0 M Tris-HCl (pH 8.5) to initiate washing. TCA (1.3 ml of 100% w/v) is added to precipitate the phosphohistone a second time. The tubes are placed on ice for at least 10 min, and the precipitated phosphohistone is collected by centrifugation at 1500g for 7 min. This washing procedure is repeated four additional times. Failure to adequately resuspend or thoroughly remove the supernatant after centrifugation will result in increased background in the phosphatase assay. To remove the TCA, the precipitated histone produced by the last wash is resuspended in 4 ml of ethanol : ethyl ether (1 : 4; v : v) and collected by centrifugation at 1800g for 10 min at room temperature. This ethanol : ether wash is repeated one time as described and then two additional times with 4 ml acidified ethanol : ethyl ether (1 : 4; 0.1 N HCl). The resuspension is facilitated by gentile mixing with small glass rods (heat-sealed Pasteur pipette), taking great care to suspend all of the pellet. The histone is collected by centrifugation at 1800g for 10 min at room temperature after each ethanol/ether wash. After the last ethanol/ether wash, the histone pellet is allowed to air dry overnight. The next day, the dried pellet is dissolved in 4 ml of 5 mM Tris HCl, pH 7.4. (This takes ~1 hr with vortexing every 5–10 min.) After the histone is dissolved, it is aliquoted and can be stored at −20°. Specific activity is calculated as follows:

1. ATP incorporation ratio R

$$R = \frac{\text{cpm in incubation mixture (incubation volume/aliquot counted)}}{\text{cpm in redissolved pellet (final volume/aliquot counted)}}$$

2. Concentration of incorporated phosphate; C_P; the initial concentration of ATP is 160 μM:

$$C_P = (160\ \mu M)\ (R)\ (\text{incubation volume/final volume})$$

Measurement of Protein Phosphatase Activity

The determination of protein phosphatase activity against phosphoproteins is determined by the quantification of ^{32}P liberated from phosphohistone, prepared according to the procedures described earlier. The phosphatase assay (80 μl final volume) is conducted in 50 mM Tris-HCl (pH 7.4), 0.5 mM DTT, 0.1 mM EDTA, phosphoprotein (2 μM PO$_4$), and the indicated amount of divalent cations (assay buffer; final concentration). The assay is conducted in a 1.5-ml microcentrifuge tube at 30° for 10 min, and the assay is initiated by adding 30 μl of phosphohistone substrate (dissolved in 5 mM Tris-HCl, pH 7.4) to PP7 diluted with 4× assay buffer and water. A typical reaction contains 20 μl 4× assay buffer, 5–10 μl PP7 (2.5 μM), 4 μl 100 mM MgCl$_2$, and water to bring the volume to 50 μl). The dephosphorylation reaction is stopped by the addition of 100 μl of 1 N H$_2$SO$_4$ containing 1 mM K$_2$H PO$_4$. [^{32}P]Phosphate liberated by the enzymes is then extracted as a phosphomolybdate complex and measured according to the methods of Killilea et al.[17] With this method, free phosphate is extracted by adding 20 μl of ammonium molybdate [7.5% (w/v) in 1.4 N H$_2$SO$_4$] and 250 μl of isobutanol:benzine (1:1, v/v) to each tube. The tubes are mixed vigorously for ~10 sec (vortex set at maximum speed), followed by centrifugation at 14,000g for 2 min. This produces a biphasic solution, and radioactivity (phosphate liberated from the phosphohistone) in the upper phase is quantified by counting 100-μl aliquots in a scintillation counter.

For inhibition studies, inhibitors are dissolved in dimethylformamide (DMF) and added to the enzyme mixture 10 min before the initiation of the reaction with the addition of substrate. Controls receive solvent alone, and, in all experiments, the amount of enzyme is diluted to ensure that the samples are below the titration end point. The titration end point is defined as the concentration of enzyme after which further dilution no longer affects the IC$_{50}$ of the inhibitor and represents a point where the concentration of enzyme used in the assay no longer approaches that of the inhibitor. This ensures that IC$_{50}$ represents the potency of the inhibitor alone and is not representative of a combination of potency of the inhibitor and titration artifacts of the assay system.[18] Preliminary assays must also be performed to ensure the dephosphorylation reaction is linear with respect to enzyme concentration and time.

Measurement of PP7 Activity with P-Nitrophenyl Phosphate (pNPP)

The dephosphorylation of pNPP is measured spectrophotometrically following the change in absorbance at 410 nM that occurs when the phosphate is liberated from pNPP. The assay (300-μl final volume) is performed at 30° in 40 mM Tris-HCl (pH 8.1 at 30°) containing 20 mM KCl, 20 mM MgCl$_2$ and 2 mM DL-dithiothreitol (assay buffer). To minimize pipetting errors, a master mix of 1.5× assay buffer containing substrate (pNPP) is

---

[17] S. D. Killilea, J. H. Aylward, R. L., Mellgren, and E. Y. C. Lee, *Arch. Biochem. Biophys.* **191,** 638 (1978).
[18] A. H. Walsh, A. Cheng, and R. E. Honkanen, *FEBS Lett.* **416,** 230 (1997).

made for each set of assays by adding (per reaction) 30 μl of 10× assay buffer (400 mM Tris-HCl, pH 8.1, 200 mM KCl, 200 mM MgCl$_2$, and 20 mM DL-dithiothreitol), 170−x μl of water and x μl of 0.5 M pNPP (x being the volume of pNPP stock solution required to produce the desired substrate concentration, e.g., 12 μl of 0.5 M pNPP would yield 20 mM pNPP per 300-μl reaction). To assay pNPP phosphatase activities, each reaction is started by adding 100 μl of PP7 (diluted from the PP7 stock immediately prior to use) to a cuvette containing 200 μl of the 1× buffer/substrate master mix. The sample is mixed by pipetting up and down several times, and blanks are made by substituting buffer or water for the enzyme solution. The rate of P$_i$ released is determined indirectly by measuring the change in absorbance at 410 nM produced by the formation of p-nitrophenol. Typically, A_{410} readings are taken on each sample at 10-sec intervals for 20 min using a spectrophotometer (e.g., Beckman DU460). The reaction rates are calculated by linear regression analysis of the data to obtain the best straight-line fit using the basic rate equation: rate = $(A_2 − A_1)/(t_2 − t_1)$. Initial and final times for these calculations are selected from data points contained in the linear portion of each data set collected over a 20-min period using ~0.5–2.0 μg of PP7. With both pNPP and phosphohistone, the concentrations of enzymes are adjusted to ensure that the dephosphorylation of substrate is <10% of the total available substrate and the reaction is linear with respect to enzyme concentration and time.

Functional Analysis of Recombinant PP7

Substrates for PP7

The physiologic substrates for PP7 have not been established. Recombinant human PP7 expressed in insect cells is soluble and has activity against pNPP and histone phosphorylated by cAMP-dependent protein kinase. Presently, pNPP is the preferred substrate because the dephosphorylation of pNPP (20 mM) is linear for ~20 min using a wide range of PP7 (0.5–10 μM). Unlike PP1 and PP2A, recombinant PP7 has no detectable activity against phosphorylase a employing standard assay conditions. Recombinant PP7 dephosphorylates histone phosphorylated by cyclic AMP-dependent PKA at a rate of 0.35 ± 0.05 nmol/min/mg of protein. This compares to a rate of 1925 ± 41 and 254 ± 15 for PP2A and PP1, respectively, when measured under identical conditions. Therefore, phosphohistone with a specific activity of $>1 \times 10^6$ dpm/nmol should be used when histone is employed as a substrate in studies with PP7.

Effect of pH and Divalent Cations on Activity of PP7

The activity of PP7 is sensitive to pH, with little activity observed under acid (pH < 6) or alkaline (pH > 10.5) conditions. PP7 has maximal activity at pH ~8.1. Under all conditions tested, the activity of PP7 against phosphohistone or pNPP is dependent on the presence of divalent cations (Fig. 3). Unlike PP1, PP2A, PP4, and PP5, which are active in the absence of divalent cations, almost no PP7 activity is observed when the divalent cations are omitted and EDTA is included in the assay buffer (blank, Fig. 3). Of the cations tested, the addition of 2–10 mM Mg^{2+} to the assay buffer produces the greatest increase in basal activity. Mg^{2+}-dependent PP7 activity is further increased by Ca^{2+}. However, calcium does not activate PP7 in the absence of Mg^{2+} or Mn^{2+}. When assayed in the presence of 2–10 mM Mg^{2+} or Mn^{2+}, the addition of calcium results in a dose-dependent increase in the activity, with ~250 μM Ca^{2+} producing half-maximal stimulation.[10] Maximal stimulation (approximately threefold) occurs with the addition of calcium at concentrations between 400 and 500 μM. No additional increase in PP7

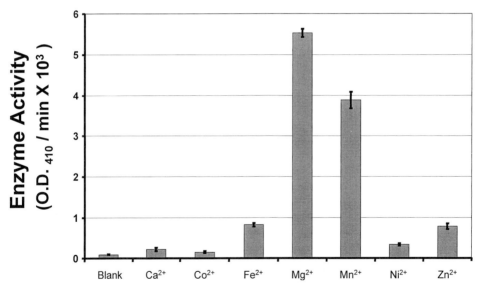

FIG. 3. Effect of divalent cations on the activity of purified recombinant PP7. PP7 activity (2 μg/ml) was measured using pNPP (20 μM) as a substrate in the presence of the indicated cations (10 mM). The initial reaction rate within the first 10 min of the assay was recorded using a Beckman DU-640 spectrophotometer, and the data are expressed as reaction rates (enzyme activity = OD_{410}/min × 10^3). The data are expressed as the mean ± SD ($n = 5$) and were obtained with at least two separate preparations of PP7.

activity is observed on the addition of >5 mM Ca^{2+}. The addition of calmodulin to the assay, which activates PP2B in the presence of calcium, has no detectable effect on the Ca^{2+}-mediated activation of PP7.

Effect of PPase Inhibitors on Activity of PP7

Treatment of PP7 with okadaic acid (1 μM), microcystin-LR (100 nM), calyculin A (100 nM), or cantharidin (20 μM), which completely inhibit the activity of PP1, PP2A, PP4, and PP5 at the concentrations indicated, has no apparent effect on PP7. Thus, like PP2B and PP2C, PP7 is insensitive to these natural toxins.

Assessment of PP7 Expression

RT-PCR Detection of PP7 in Retina and Y-79 Cells

Poly (A)$^+$ RNA is prepared from tissue or Y-79 cells as described earlier. cDNA is then produced by incubating 1 μg poly (A)$^+$ RNA in a solution (20 μl total volume) containing 50 mM Tris-HCl (pH 8.3), 75 mM KCl, 3 mM MgCl$_2$, 10 mM DTT, 0.5 mM dNTP mix, 100 ng of random primer, and 200 units reverse transcriptase (RT) (Superscript II; Life Technologies) at 42° for 50 min. After normalizing the concentration of cDNA in the samples, aliquots containing equal amounts of cDNA (~5 μl) are employed as template in PCR reactions containing PP7 or GAPDH (employed as a positive control to ensure equal amounts of cDNA) specific primer pairs. The PCR reactions are carried out in a volume of 25 μl, which contained 20 mM Tris-HCl (pH 8.4), 50 mM KCl, 15 mM MgCl$_2$, 0.2 mM dNTP mix, 100 ng each of the primers, and 100 units *Taq* DNA polymerase (Promega, Madison, WI) on a PTC-100 programmable thermal cycler (MJ Research Inc., Watertown, MA). The reaction mixture is denatured at 94° for 5 min, and the amplification reaction consists of 25 (for GAPDH) or 38 (for PP7) cycles of denaturation (94°, 30 sec), annealing (55°, 50 sec), and extension (72°, 2 min) with a final extension at 72° for 10 min. The sequences for PP7 and GAPDH specific primers pairs are 5'-CAGAGCATGAATGGGAACAGA-3',5'-ACATGGCACGAAATTCAACC-3', and 5'-TGAAGGTCGGAGTCAACGGATTTGGT-3', 5'-CATGTGGGCCATGAGGTCCACCAC-3', respectively.[10]

Assessment of PP7 in Genomic DNA

PCR with nested primers contained in exon 17 (5-GGAAGAATTTCGTGCCATGT-3; sense; nucleotides 1999–2018; and 5-TTAGCAAGGTTGGTGACATCAGG-3; antisense; nucleotides 2173–2196; accession number

AF027977) amplify a 199-bp intronless fragment of PP7 from human genomic DNA that is not amplified in rodent DNA. PCR reactions (25 μl) are conducted in 50 mM Tris-HCl (pH 9.0) containing 50 mM NaCl, 1.5 mM MgCl$_2$, 0.1% Triton X-100, 200 μM dNTP, 100 ng of each primer, 0.25 unit *Taq* DNA polymerase, and 100 ng template DNA. DNA amplification is performed on a thermal cycler (Perkin-Elmer 2400, Norwalk, CT) with 30 cycles of denaturation (94°, 30 sec), annealing (55°, 30 sec), and extension (72°, 1 min), with an initial denaturation (95°, 5 min) and a final extension (72°, 10 min). The splice junction sequences for the other exons are provided by Montini *et al.*[12]

Acknowledgments

This work was supported by a grant from the National Institutes of Health (NCI; CA60750), and the Lions/University of South Alabama Eye Research Institute.

Section VI

Photoreceptor Phosphodiesterase and Guanylyl Cyclase

[40] Purification and Assay of Bovine Type 6 Photoreceptor Phosphodiesterase and its Subunits

By TERRY A. COOK and JOSEPH A. BEAVO

Introduction

The 3',5'-cyclic nucleotide phosphodiesterase (PDE) 6 family is one of a growing number of characterized PDE gene families. To date, ten PDE families have been reported,[1] although it is quite likely that more will soon be discovered. They share the ability to hydrolyze cyclic nucleotides, but differ in a number of their properties, including substrate specificity, kinetics, tissue distribution, and regulation. Although the PDE 6 family probably has the most well-defined physiologic role, the importance of other PDEs in regulating processes as diverse as insulin release, T-cell proliferation, and vasodilation has more recently been recognized.

A large body of literature describing experiments using type 6 PDE exists. However, a comprehensive collection of methods used in the study of this enzyme has never been published. This chapter provides a brief review of the function and features of type 6 PDE, followed by detailed methods for the purification and assay of the holoenzymes and their various subunits. Particular emphasis is placed on techniques our laboratory has used extensively that often contain modifications from the published versions. Our hope is that this will facilitate research into this interesting and important enzyme.

Characteristics of Holoenzyme

PDE 6, along with the other components of the phototransduction cascade, is found at very high concentration in the outer segments of photoreceptor cells. Because bovine rod outer segments (ROS) can be quickly purified and retain their ability to respond to light, their signal transduction mechanism has been well studied. When light hits the transmembrane protein, rhodopsin, 11-*cis*-retinal embedded in the protein is isomerized to all-*trans*-retinal. This causes a conformational change in rhodopsin that allows it to interact with the G protein transducin, which can then exchange guanosine diphosphate (GDP) for guanosine 5'-triphosphate (GTP). When in its GTP-bound state, transducin is active and can interact with the gamma

---

[1] S. H. Soderling, S. J. Bayuga and J. A. Beavo, *Proc. Natl. Acad. Sci. U.S.A.* **95,** 8991 (1998).

subunits of type 6 PDE. By shifting the position of the gamma subunits with respect to the catalytic site,[2] transducin activates the PDE. PDE 6 catalyzes the hydrolysis of cGMP into 5'GMP. The consequent reduction in cGMP in the cell causes cGMP gated channels in the plasma membrane to close, resulting in the hyperpolarization of the photoreceptor and a reduction in neurotransmitter release. Thus the light signal is passed to the next cell. Once transducin hydrolyzes GTP to GDP, it no longer interacts with the gamma subunits and the PDE becomes less active. cGMP is resynthesized by guanylate cyclase and the cGMP gated channels open again.

The type 6 PDEs are among the few PDEs whose physiologic importance is well appreciated. The finding that these PDEs are the primary light-regulated modulator of cGMP in the ROS led to a flurry of interest in the kinetics, structure, and function of the enzyme. As a result, these basic parameters are well understood. Type 6 PDEs have a catalytic subunit structure that is typical of all PDEs investigated thus far. The catalytic subunits form dimers, and each catalytic subunit contains a conserved sequence for catalysis. The catalytic subunits are unusual among PDEs in that they have a very limited tissue distribution—in mammals, they are found at very high levels in the photoreceptor cells of the eye, but little expression in other tissues has been detected. Purified trypsin activated type 6 PDEs have K_m values of about 15–20 μM, although in intact rods its apparent affinity may be lower—about 100 μM.[3] They have the highest V_{max} values of any known PDE: measured values of turnover number range from about 600–4200 cGMP/molecule PDE/sec,[3] though it is unlikely that this rate is ever reached *in vivo* since the average concentration of free cGMP in the ROS is only about 10 μM. Recent research on this enzyme has focused on more subtle aspects of its behavior, from the regulation of the interactions of its subunits to the function of cGMP binding to its noncatalytic subunits.

Type 6 PDEs are usually purified as an oligomer that contains two catalytic subunits and various accessory subunits. The rod isozyme contains catalytic subunits from two different genes, α and β, while only α', the product of a third gene, is expressed in cones (see Fig. 1). Evidence suggests that the majority of the PDE found in rods is an α/β heterodimer, although it is likely that small amounts of $\alpha\alpha$ or $\beta\beta$ homodimers may also be present.[4] The functional significance of the different catalytic subunit combinations is not known. The cone enzyme is an $\alpha'\alpha'$ homodimer. Structure–function studies of the catalytic subunits have been hindered by difficulties with

---

[2] A. E. Granovsky, M. Natochin, and N. O. Artemyev, *J. Biol. Chem.* **272**, 11686 (1997).
[3] E. N. J. Pugh and T. D. Lamb, *Biochim Biophys Acta* **1141**, 111 (1993).
[4] N. O. Artemyev, R. Surendran, J. C. Lee, and H. E. Hamm, *J. Biol. Chem.* **271**, 25382 (1996).

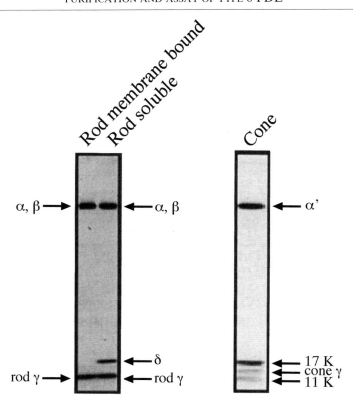

FIG. 1. Sodium dodecyl sulfate–polyacrylamide gel electrophoresis (SDS–PAGE) of cone and rod enzymes after purification using the ROS-1 antibody. The membrane-bound form of the rod enzyme has two types of catalytic subunits, α and β, with apparent molecular masses of 88 and 84 kDa, respectively, as well as the 11-kDa γ subunit. The soluble form is identical to the membrane-bound form except for the additional δ subunit. The cone enzyme is purified as a single type of catalytic subunit (α') with an apparent monomer molecular mass of 94 kDa, along with a 17-kDa subunit (cone δ), a 13-kDa subunit (cone γ), and an 11-kDa subunit (most likely rod γ). Because the gamma subunit does not stain well with standard Coomassie stains, these gels were stained using the silver staining method of Blum et al.[33] (Data redrawn from the dissertation thesis of Peter Gillespie, 1987.)

expressing significant amounts of active recombinant catalytic subunits. However, relatively simple procedures have been developed for purification of this enzyme from tissue (see Fig. 2).

The rod and cone enzymes share the ability to bind cGMP at noncatalytic sites on the α and β subunits. Bovine rod enzyme binds cGMP much more tightly than the cone enzyme. The cone enzyme has a single class of binding sites (based on binding affinity), whereas the rod enzyme has high- and low-affinity sites. The function of noncatalytic binding is not clear because

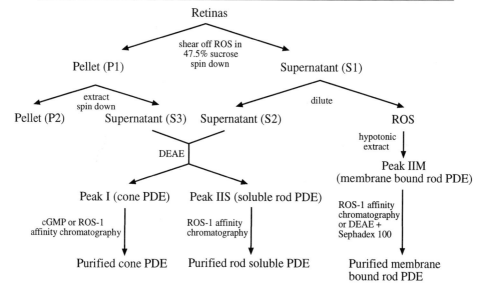

FIG. 2. Purification scheme for cone and rod isozymes. (Adapted from Peter Gillespie, dissertation thesis, 1987.)

it seems to have little effect on the enzyme's catalytic properties under the conditions tested thus far. However, cGMP binding can affect the interaction of the inhibitory gamma subunit with the catalytic subunits,[5] suggesting that the noncatalytic sites may be important for regulation of the shutoff of the enzyme.

It is interesting that most of the enzymes of the phototransduction cascade share the ability to bind to membranes. These interactions may be permanent, as is the case with the transmembrane protein rhodopsin, or may be transient, as is true of rhodopsin kinase. A large fraction of rod PDE 6 is bound to the membrane. The C termini of the catalytic subunits are modified by isoprenylation and methylation, and these modifications appear to be required for the interaction between the enzyme and the membrane to occur.[6] However, when PDE 6 is purified from bovine retinal extract, about 30% of the rod activity is found in the soluble fraction.[7] The delta subunit of PDE 6 is found on the soluble form of the rod enzyme. The cone enzyme appears to be primarily soluble although this may depend

[5] V. Y. Arshavsky, C. L. Dumke, and M. D. Bownds, *J. Biol. Chem.* **267**, 24501 (1992).
[6] N. Qin and W. Baehr, *J. Biol. Chem.* **269**, 3265 (1994).
[7] P. G. Gillespie, R. K. Prusti, E. D. Apel, and J. A. Beavo, *J. Biol. Chem.* **264**, 12187 (1989).

on the state of the retinas (frozen or fresh) when the preparation is done. It is not known whether the cone enzyme normally contains the delta subunit.[8]

γ and δ Subunits

The catalytic subunits of PDE 6 copurify with smaller molecular weight subunits, γ and δ. The γ subunit inhibits the catalytic activity of the enzyme by interacting with the cGMP binding pocket of the catalytic site.[2] It also increases the rate of GTP hydrolysis by transducin,[9] and is required for correct shutoff of the signal transduction cascade.[10] A number of posttranslational modifications have been reported to occur on the γ subunit. For instance, *in vitro* phosphorylation of the subunit by both protein kinase C (PKC) and protein kinase A (PKA) have been reported and have been demonstrated to affect the activity of the gamma subunit.[11–14] The effect of these modifications has not yet been demonstrated in the intact photoreceptor, but it is conceivable that the activity of the inhibitor, and thus of the holo-PDE, might be regulated in such a manner.

The δ subunit of type 6 PDE was identified when PDE from the soluble fraction of ROS was purified.[7] Prior to this purification, soluble PDE was usually washed off of the membranes and discarded. After the cloning of the delta subunit and its expression in *spodoptera frugiperda* (Sf 9) cells, it was shown that the recombinant protein could remove membrane-bound rod PDE into the supernatant.[8] It is unclear whether the cone form of type 6 PDE also contains a δ subunit and, if so, whether there are separate isoforms of this subunit in rods and cones. Unfortunately, due to the difficulty in obtaining pure cone preparations from retina, these questions may be difficult to answer.

Data have suggested that referring to the δ subunit as a subunit of PDE 6 may be somewhat of a misnomer. It has been shown to interact with other proteins using the yeast two-hybrid system.[15] It also contains a consensus sequence for interaction with postsynaptic density protein, disks large, zona occludens (PDZ) domains. Unlike type 6 PDEs, the δ subunit is expressed

---

[8] S. K. Florio, R. P. Prusti, and J. A. Beavo, *J. Biol. Chem.* **271**, 24036 (1996).
[9] F. Pages, P. Deterre, and C. Pfister, *J. Biol. Chem.* **268**, 26358 (1993).
[10] S. H. Tsang, M. E. Burns, P. D. Calvert, P. Gouras, D. A. Baylor, S. P. Goff, and V. Y. Arshavsky, *Science* **282**, 117 (1998).
[11] I. P. Udovichenko, J. Cunnick, K. Gonzalez, and D. J. Takemoto, *J. Biol. Chem.* **269**, 9850 (1994).
[12] I. P. Udovichenko, J. Cunnick, K. Gonzales, and D. J. Takemoto, *Biochem. J.* **295**, 49 (1993).
[13] S. Tsuboi, H. Matsumoto, and A. Yamazaki, *J. Biol. Chem.* **269**, 15016 (1994).
[14] F. Hayashi, *FEBS Lett.* **338**, 203 (1994).
[15] A. M. Marzesco, T. Galli, D. Louvard, and A. Zahraoui, *J. Biol. Chem.* **273**, 22340 (1998).

in a number of tissues outside of the retina.[15a] These data suggest that the δ subunit may be a mediator of protein–protein interactions between prenylated proteins and perhaps nonprenylated proteins as well.

Materials and Methods. Part 1: Purification of PDE 6 Holoenzyme and Subunits

Purification of Membrane-Bound Rod PDE

Purification of the membrane-bound rod isoform of the PDE (see Fig. 2) is simplified by its reversible interaction with membranes. In isotonic buffer, this PDE is found in the insoluble fraction, so it can be purified away from soluble ROS proteins. On treatment with hypotonic buffers, the PDE becomes soluble and then can be purified by using anion exchange and gel filtration chromatography. The following procedure is a modification of the original purification procedure.[16] It is reported here for a 50-retina preparation, but the procedure can be scaled up. Generally we extract the retinas under dark or dim red light conditions to decrease the percent of activated PDE (which may lose its γ subunits more easily) in the final preparation. This may not be necessary for some purposes.

Solutions

> ROS buffer (make 500 ml): 20 mM 4-morpholinopropanesulfonic acid (MOPS), pH 7.2, 30 mM NaCl, 60 mM KCl, 2 mM MgCl$_2$, 1 mM DL-dithiothreitol (DTT) 0.1 mg/ml 4-(2-aminoethyl)benzenesulfonyl fluoride hydrochloride (AEBSF)
>
> Hypotonic buffer (make 500 ml): 10 mM tris(hydroxymethyl)aminomethane (Tris), pH 7.5, 1 mM MgCl$_2$, 1 mM DTT, 0.1 mg/ml AEBSF
>
> Retinas: Dark-adapted frozen retinas can be obtained from Lawson (Lincoln, NE) or Hormel (Austin, MN). For some purposes, fresh tissue may be more appropriate. If fresh tissue is used, retinas should be dissected in the dark or under dim red light prior to the first step of this procedure.

Procedure

> Steps 3–11 are performed in dark or under dim red light.
>
> 1. Thaw 50 frozen retinas on ice in the dark (overnight at 4° is also convenient).

[15a] T. A. Cook, unpublished data (1997).
[16] W. Baehr, M. J. Devlin, and M. L. Applebury, *J. Biol. Chem.* **254**, 11699 (1979).

2. Add 75 ml ROS buffer containing 47.5% sucrose (w/v) to a 250-ml beaker. This leaves ample room for 50 retinas (about 50 ml of tissue).
3. In the dark or under dim red light, add the retinas to the beaker. Stir vigorously on a stir plate with a magnetic stir bar for 10 min to detach ROS from retinas.
4. Divide the retinal homogenate into 30 ml Oak Ridge centrifuge tubes (Nalgene, NUNC, Milwaukee, WI). Centrifuge at $4000g$ in an angle rotor for 2 hr, $4°$.
5. Carefully pour off and save the supernatant (S1), which should contain ROS membranes. If ROS (pink or orange in light; appear whitish under red light) are visibly adhering to the sides of the tube, scrape them into the supernatant. The pellet (P1) contains some cone PDE and can be extracted for purification of the cone enzyme if desired.
6. Dilute S1 with 1 volume of ROS buffer without sucrose.
7. To isolate ROS, centrifuge the diluted S1 at $16,000g$ for 1 hr, $4°$.
8. Pour supernatant (S2) off of the ROS. S2 contains soluble rod and cone PDE and can be saved for purification of this isozyme.
9. Wash the ROS by resuspending the pellets in 10 ml ROS buffer. Centrifuge at $16,000g$ for 5 min at $4°$.
10. Pour off the supernatant. Extract membrane-bound PDE by resuspending ROS in 0.5 ml hypotonic buffer per retina and homogenizing with a loose fitting Dounce homogenizer.
11. Pellet the ROS at $16,000g$, 5 min at $4°$. Save the supernatant. Repeat extraction. In our hands, three extractions usually remove most of the PDE. Usually about 10% of the PDE remains irreversibly membrane bound. The cause of this tight binding is not known.

Extracted PDE is then subjected to chromatography for further purification.

12. Batch load the hypotonic extract onto 0.5 ml diethyl aminoethyl (DEAE)-cellulose resin/retina for 2–4 hr (or overnight), tumbling, slowly, at $4°$.
13. Load the resin onto a column with an inside diameter of about 1 in. If desired, a 5–10 ml layer of DEAE-cellulose that has not been incubated with the PDE may be placed at the bottom of the column. This often improves the resolution of the activity peaks.
14. Wash the resin with 5 column volumes hypotonic buffer + 20 mM NaCl. Alternately, the resin may be washed more quickly prior to loading it on the column by using a scintered glass filter. We routinely use a flow rate of 0.5 ml/min for all steps except the wash, which can be done at a faster flow rate.

15. Elute the column with a gradient of 20–500 mM NaCl in ROS buffer. This is made by adding the appropriate amount of NaCl to the previously made ROS buffer. Total gradient volume should be at least three times the volume of the column.
16. Assay the fractions for PDE activity (see section on activity assays). There should be one large peak of PDE activity (see Fig. 3).
17. Fractions containing PDE activity may also contain transducin. This should be visible on SDS–PAGE. PDE may be further purified, if necessary, by using ROS-1 affinity chromatography (see later process) or by using a Sephadex G-100 gel filtration column.
18. The enzyme can be stored for several months at $-20°$ in the buffer from the column in 50% (v/v) glycerol.

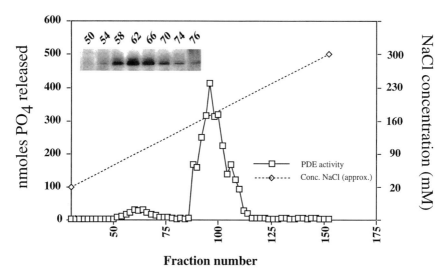

FIG. 3. DEAE cellulose profile of cone and rod-soluble PDE. Soluble proteins from bovine retina were separated over DEAE cellulose basically as described in text. In this experiment, S1 + P1 extract (S3) from 50 retinas was loaded onto 30 ml DEAE cellulose. This was poured onto a column of ~1-in. diameter. Resin was washed and then proteins were eluted with a gradient (200 ml) of 20–300 mM NaCl run at ~0.6 ml/min. Dotted line represents beginning (as determined by increase in protein coming off column) and end of gradient. Actual concentration of NaCl will be lower than calculated from this line. Fractions were tested for PDE activity. Typically, the cone peak only reaches about 10% of the maximum PDE activity in the rod peak. Membrane-bound PDE elutes at about the same conductance as the soluble form of the enzyme, but there should be no cone peak. Fractions were then run on SDS–PAGE and blotted with an antibody specific for the cone enzyme (α' I) to demonstrate that the first smaller peak of PDE activity contains the cone form of the enzyme (inset).

Partial Purification of Soluble Rod PDE and Cone PDE

Soluble rod PDE and cone PDE can be purified from S2 (see Figs. 2 and 3). This procedure is a modification of the procedure from Gillespie et al.[17]

Procedure

1. Dilute S2 (see step 8, above) with 4 volumes of hypotonic buffer, to bring the salt concentration down to about 20 mM.
2. Extract P1 pellet from step 5, above, by homogenizing it with a Polytron in 50 ml ROS buffer. Centrifuge at 100,000g at 4° for 45 min. Mix supernatant (S3) with 4 volumes hypotonic buffer.
3. Batch load S2 and S3 onto DEAE-cellulose as in step 12, above.
4. Wash and elute as in steps 14–16, above, except use 20–300 mM NaCl gradient.
5. There should be two peaks of PDE activity—the cone enzyme elutes first, and then the soluble rod enzyme (see Fig. 3). At this point the PDE is still so impure that it is difficult to distinguish using SDS–PAGE, so appropriate further purification procedures must be performed to isolate the enzyme. Either enzyme may be purified using ROS-1 affinity chromatography. The cone form of the enzyme may also be purified using cGMP affinity chromatography.

ROS-1 Affinity Chromatography

ROS-1 is a monoclonal antibody that recognizes both soluble and membrane-bound forms of rod PDE as well as the cone form.[18] Its affinity for these enzymes is lowered if they are missing their γ subunits, so this procedure cannot be used to isolate active PDE. The antibody can be coupled to Sepharose using a variety of methods. Most commonly, our laboratory uses cyanogen bromide activated Sepharose 4B coupling gel (Pharmacia, Piscataway, NJ) following manufacturer's instructions. PDE binds tightly to this antibody at physiologic pH and high salt, but can be eluted by using high pH. For a 50-retina preparation, 1–2 ml of Sepharose coupled to 3–6 mg purified ROS-1 is sufficient to bind all of the PDE. From this, we can obtain 80–100 μg soluble rod PDE. ROS-1 is available by contacting the authors.

This procedure can also be used to simplify purification of the membrane-bound form of the enzyme. Apply the extracted PDE from step 11 to the ROS-1 column after adding NaCl to 300 mM to the PDE extract.

---

[17] P. G. Gillespie and J. A. Beavo, *J. Biol. Chem.* **263**, 8133 (1988).
[18] R. L. Hurwitz, M. A. H. Bunt, and J. A. Beavo, *J. Biol. Chem.* **259**, 8612 (1984).

Solutions

   Binding buffer (make 50 ml): 10 mM Tris, pH 7.5, 300 mM NaCl, 1 mM MgCl$_2$
   Elution buffer (make 10 ml): 25 mM glycine, pH 10.7, 2 mM MgCl$_2$, 200 mM NaCl
   Neutralizing buffer (make 2 ml): 0.5 M Tris, pH 7, 2 mM MgCl$_2$, 1 mM DTT, 10 mg/ml AEBSF

Procedure

1. Add NaCl to the PDE-containing solution for a final NaCl concentration of 300 mM.
2. Wash the ROS-1 column with 5 volumes binding buffer.
3. Circulate the PDE solution onto the column at 0.5 ml/min. We recirculate this several times at 4° to ensure maximal binding.
4. Wash the column with 5 volumes binding buffer.
5. Elute the PDE with 10 ml elution buffer. This can be done at 0.5 ml/min with a pump or can also be done via gravity flow.
6. Collect 1-ml fractions into tubes containing 100 μl neutralizing buffer.
7. Measure PDE activity in fractions. *Note:* PDE has little catalytic activity at high pH. Therefore, on elution of PDE from the ROS-1 column, eluate should be collected in neutralizing buffer and incubated at 4°. After 30 min, the PDE should have regained most of its activity.

cGMP Affinity Chromatography

The cGMP binding properties of the cone enzyme have been used to simplify its purification.[17] The first peak of PDE activity from the DEAE-cellulose fractionation of soluble retinal proteins is from the cone enzyme (see Fig. 3). This peak can be further purified using cGMP-Sepharose. This procedure does not work for the rod isoform, possibly because the affinity of this enzyme for cGMP is so high that the endogenous cGMP remains too tightly bound to capture the enzyme on cGMP resin.

cGMP-Coupled Resin Preparation

The cGMP affinity column is prepared by coupling cGMP to epoxy-activated Sepharose (can be obtained from a number of sources including Sigma, St. Louis, MO). The conditions used for the preparation of this resin can greatly affect the binding capacity of the resin for cone PDE (for details, see Ref. 17). The following procedure has given us the best yield of the procedures tried thus far. A fresh column should be prepared for each preparation because the capacity of the resin for the enzyme often decreases significantly after the first use.

Solutions

Coupling buffer (make 150 ml): 200 mM NaCl, 100 mM sodium phosphate, pH 11

Prewash buffer 1 (make 50 ml): 0.1 M sodium acetate, pH 4, 0.5 M NaCl

Prewash buffer 2 (make 50 ml): 0.1 M sodium borate, 0.5 M NaCl

Wash buffer (make 100 ml): 10 mM Tris, pH 7.5, 1 mM ethylenediamine-N,N,N',N'-tetraacetic acid (EDTA), 250 mM NaCl, 1 mM DTT, 0.2 mM phenylmethylsulfonyl fluoride (PMSF)

Procedure

1. Swell 10 ml resin in distilled H_2O, then wash the resin with 10 volumes distilled H_2O.
2. Suspend the resin in 2 volumes coupling buffer, pH 11, containing 20 mM cGMP. Incubate 36 hr at room temperature.
3. Wash the resin with 4–5 volumes of coupling buffer without cGMP.
4. Block unreacted sites by incubation with 1 M ethanolamine in coupling buffer overnight.
5. Prewash the resin with prewash buffer 1 followed by prewash buffer 2. Cycle these buffers for a total of three washes with each buffer.
6. Wash the resin with several volumes of wash buffer. The column can be stored for several months in wash buffer with 0.02% NaN_3 added.

Preparation of the Sample and Chromatography

Solutions

cGMP elution buffer (make 50 ml with cGMP; make 10 ml without cGMP): 10 mM Tris, pH 7.5, 1 mM EDTA, 50 mM NaCl, 5 mM cGMP, 1 mM DTT

Wash buffer (make 100 ml): 10 mM Tris, pH 7.5, 1 mM EDTA, 250 mM NaCl, 1 mM DTT

DEAE cellulose elution buffer (make 10 ml): 10 mM Tris, pH 7.5, 250 mM NaCl, 1 mM $MgCl_2$, 1 mM DTT

Procedure

1. Treat pooled fractions from the first peak of PDE activity (cone peak) with PMSF (0.5 mM) and then warm to 30° for 2 hr. This treatment allows more of the PDE activity to bind to the affinity column.

2. Adjust pool to 2 mM EDTA + 250 mM NaCl.
3. Preequilibrate the cGMP and cAMP columns in wash buffer.
4. Run the solution sequentially over a 1-ml 8-(6-aminohexylamino)-cAMP-Sepharose column (Pharmacia), which will remove protein kinase G and the regulatory subunit of PKA, and a 10-ml cGMP-epoxy Sepharose column. This step should be done slowly (less than 20 ml/hr).
5. Wash the cGMP-epoxy Sepharose column with 5 volumes wash buffer.
6. Before eluting the PDE from the resin, place a 0.5-ml DEAE-cellulose column between the cGMP-Sepharose column and the fraction collector to concentrate the PDE and to remove free cGMP.
7. Warm the cGMP-Sepharose column to room temperature and pump 1 volume of elution buffer onto the column at ~3 ml/min. Stop flow through the column for 30 min to allow the PDE to dissociate from the column.
8. Resume flow at ~1.2 ml/min. PDE eluted from the column should bind to the DEAE cellulose column at this salt concentration, but cGMP will flow through.
9. Uncouple the DEAE-cellulose column from the cGMP-Sepharose column. Wash the DEAE-cellulose column with 5 volumes cGMP elution buffer without cGMP.
10. Elute the PDE from the DEAE-cellulose column by slow (0.1 ml/min) elution with DEAE-cellulose elution buffer.

Purification of γ Subunit

The original purification methods for the gamma subunit from rod outer segments take advantage of its heat and acid stability.[19,20] We have purified small amounts of the small molecular weight subunits from purified PDE by using reversed-phase high-performance liquid chromatography (RP-HPLC).[17] The γ subunit separates well on a C_4 column and can be reconstituted. If large amounts of the protein are desired, the γ subunit can be expressed in bacteria as a fusion protein, and can be isolated using standard methods.[21–23]

[19] J. B. Hurley and L. Stryer, *J. Biol. Chem.* **257**, 11094 (1982).
[20] I. L. Dumler and R. N. Etingof, *Biochim. Biophys. Acta* **429**, 474 (1976).
[21] R. L. Brown and L. Stryer, *Proc. Natl. Acad. Sciences U.S.A.* **86**, 4922 (1989).
[22] S. E. Hamilton, R. K. Prusti, J. K. Bentley, J. A. Beavo, and J. B. Hurley, *FEBS Lett.* **318**, 157 (1993).
[23] F. M. Ausubel, R. Brent, R. E. Kingston, D. D. Moore, J. G. Seidman, J. A. Smith, and K. Struhl (eds.), "Current Protocols in Molecular Biology." Wiley Interscience, Boston, 1996.

Method

Purified PDE in DEAE-cellulose elution buffer is injected directly onto a C_4 column equilibrated in 20% acetonitrile and 0.096% trifluoroacetic acid (TFA). For small preps, a 3-cm C_4 reversed-phase column (such as from Pierce) is convenient, because it is inexpensive so a column can be dedicated to purification of a single subunit. Larger preps can be done using a standard 30-cm C_4 column.

Subunits are eluted using a 30-min gradient from 16% acetonitrile, 0.096% TFA in distilled H_2O to 64% acetonitrile, 0.084% TFA in distilled H_2O. Flow rate for the 3-cm column is 0.3 ml/min; the larger column can be run at about 1 ml/min. OD_{260} is measured to determine where the protein peaks are.

After fractionation of the gradient, solvents can be removed by lyophilizing the fractions. Then the γ-subunit peak can be reconstituted in a small volume of 1 M sodium bicarbonate.

Because the γ subunit is denatured in the course of this purification, it is important to make sure that it regains activity after reconstitution. The percent of the subunit that is active may depend on a number of factors, including the purification method used as well as the speed and temperature at which the preparation is done. If an affinity of the gamma subunit is to be reported, it is important to know how much active protein is present in the sample. Activity of the γ subunit can be tested by adding it back to trypsin-activated PDE. If the gamma subunit is active, an inhibition of PDE activity should occur. See Hamilton *et al.*[22] for details.

Purification of δ Subunit

The δ subunit can also be purified from the holoenzyme by using HPLC, but it cannot be renatured after lyophilization. However, it has been expressed both in Sf9 cells and as a glutathione *S*-transferase (GST) fusion protein. The fusion protein retains the ability to remove PDE from ROS membranes[23a,24] and can be easily purified using standard methods.[23]

To purify the untagged recombinant δ subunit from Sf9 cells,[8] it is possible to take advantage of its small size compared to most other proteins expressed in the cells. Since fusion protein expression is generally easier, faster, and more convenient than expression and purification from Sf9 cells, only a brief description of this method is given here; for details of Sf9 cell culture and purification, see Florio *et al.*[8]

[23a] T. A. Cook, unpublished data (1997).
[24] N. Li and W. Baehr, *FEBS Lett.* **440,** 454 (1998).

Sf9 cells overexpressing the δ subunit are spun down and homogenized in low-salt buffer (20 mM Tris, 1 mM EDTA, pH 7.4). Membranes are pelleted, and the extract is loaded onto DEAE-cellulose and eluted with a gradient from 20 to 300 mM NaCl in the aforementioned buffer. Fractions containing δ subunit (as determined by SDS–PAGE or Western blots) are pooled. This pool is then run through Amicon (Danvers, MA) Centricon-100 and Centricon-50 concentrators; the δ subunit flows through both of these, but few other Sf9 cell proteins do. The resulting δ subunit can be concentrated in Ultrafree-4 or Ultrafree-10 centrifugal filter devices (Millipore, Bedford, MA) with a molecular weight cutoff of 10,000. This preparation typically yields δ subunit that is 80–95% pure.

The δ subunit purified from either Sf9 cells or as a GST fusion protein has a tendency to aggregate after about 2 weeks at 4°. More dilute solutions (<1 mg/ml) do not aggregate as quickly. If frozen, it aggregates immediately. Although ideal storage conditions have not been completely worked out, it retains activity for at least a month when stored at −20° in 50% glycerol in buffers with a pH of about 7.5.

Materials and Methods. Part 2: Activity Assays for Type 6 PDE

Several different methods can be useful for measurement of type 6 PDE activity. Its activity can be measured using a tritium-based assay[25,26] or other methods appropriate for measuring the activity of other PDEs. Two methods that are unique to the photoreceptor enzyme are presented here. Because the enzyme has such high V_{max}, a colorimetric assay, known as the phosphate release assay, can be used for most studies. In this assay, cGMP is hydrolyzed by PDE, then the resulting GMP is acted on by a phosphatase found in snake venom. The free phosphate from this reaction is then reacted with molybdate ion to form a blue product that can be measured spectrophotometrically. This assay can be extremely useful because it can be conveniently done in a 96-well plate, it is fast, and it uses no radioactivity.

Activity of type 6 PDE can also be measured over time in permeablized isolated bovine ROS. In this assay, the production of GMP is measured by using a pH meter to detect the hydrogen ion produced when cGMP is hydrolyzed. Therefore, this assay is known as the pH assay. The advantage of this procedure is that it can be used to measure the rate of cGMP hydrolysis in response to a physiologic stimulus over time, instead of product accumulation at fixed time points.

[25] J. A. Beavo, J. G. Hardman, and E. W. Sutherland, *J. Biol. Chem.* **246**, 3841 (1971).
[26] T. J. Martins, M. C. Mumby, and J. A. Beavo, *J. Biol. Chem.* **257**, 1973 (1982).

Phosphate Release Assay[17,26]

All reagents for this assay (with the exception of the complete molybdate reagent) can be made ahead of time, aliquotted, and stored. Trypsin, trypsin inhibitor, cGMP, and purified snake venom are all stored at $-20°$. Developing reagent stock can be stored at room temperature but the ascorbic acid should be added immediately prior to use of the reagent. The SDS solution should be stored at room temperature.

The PDE can also be activated by active transducin [for instance, bound to guanosine 5'-O(3-thiotriphosphate) (GTPγS)] or by the addition of histones. The mechanism by which histones activate the enzyme is not known.

The range of this assay is 0–60 nmol phosphate released.

Solutions

10× assay buffer: 200 mM Tris, pH 7.5, 100 mM MgCl$_2$

Developing reagent stock: 0.4 N H$_2$SO$_4$, 0.2% ammonium molybdate (w/v)

Purified snake venom: The snake venom used for this assay must be purified to remove endogenous phosphates. This can be done by dialysis or by using a desalting column.

Procedure

1. Place the PDE samples in a total volume of 50 μl into wells of 96-well plates. We have found that NUNC microwell plates work well. Take care to use the appropriate dilutions to keep the phosphate released by the PDE within the range of the assay (0–60 nmol phosphate released).
2. Plate out the appropriate standards and controls.
3. Add 10 μl of 10× sample buffer to each well.
4. To activate the PDE, add 10 μl of 0.2 mg/ml L-1-tosylamide-2-phenylethyl chloromethyl ketone (TPCK) treated trypsin to each well that contains PDE. Incubate on ice for 5–10 min. (A time course for trypsin activation of PDE should be performed to ensure the correct activation time.)
5. To stop the activation, add 10 μl of 10 mM 1.25 mg/ml soybean trypsin inhibitor.
6. Transfer the plate to a 30° water bath. Begin the hydrolysis reaction by addition of substrate. We typically use 10 μl of 10 mM cGMP as substrate, unless very concentrated samples must be measured. In this case we sometimes use cAMP as substrate to slow the rate of the reaction.
7. Three minutes before the end of the period for incubation of the PDE with substrate, add 10 μl of purified snake venom (4 mg/ml).

8. Stop the reaction by addition of 50 μl 6% (w/v) SDS.
9. Prepare the developing reagent by adding 2% ascorbic acid.
10. Add 150 μl of developing reagent to each well. Incubate for 20 min at 30°.
11. Read absorbance at 650 nm using a plate reader.

Cautions

Because free phosphate is a relatively common component of many solutions, it is important to do the appropriate controls for this assay. To make sure that reagents have not been contaminated with phosphate, running a reaction without any PDE sample is always useful. To test whether the sample containing the PDE itself has any endogenous phosphate, the sample can be treated with SDS prior to the addition of any of the other assay components. In addition, type 6 PDE may be activated by endogenous proteases. To control for this possibility, PDE samples without trypsin, or with trypsin inhibitor added before the trypsin, are also run.

pH Assay

This assay[27] allows the experimenter to visualize the rate of cGMP hydrolysis in response to a flash of light (see Fig. 4). In the presence of adenosine 5'-triphosphate (ATP), the rate of hydrolysis should eventually return to dark levels. However, if ATP is left out of the mixture, cGMP hydrolysis should continue until either GTP or cGMP is depleted from the mixture, because rhodopsin will remain in its active state. The following protocol is based on that used by Gray-Keller *et al.*[28] For more information, see detailed papers on this assay.[29,30]

The preparation of ROS is modified from Papermaster and Dreyer.[31] All steps are performed in the dark or under dim red light. If a darkroom centrifuge is not available, Oak Ridge tubes (Nalgene) can be wrapped in aluminum foil for centrifugation steps.

Solutions

34% Sucrose (make 75 ml): 5 mM Tris, pH 7.4, 34% (w/v) sucrose, 65 mM NaCl, 2 mM MgCl$_2$, 1 mg/ml AEBSF

[27] R. Yee and P. A. Liebman, *J. Biol. Chem.* **253**, 8902 (1978).
[28] M. P. Gray-Keller, P. B. Detwiler, J. L. Benovic, and V. V. Gurevich, *Biochemistry* **36**, 7058 (1997).
[29] P. A. Liebman and A. T. Evanczuk, *Methods Enzymol.* **81**, 532 (1982).
[30] M. W. Kaplan and K. Palczewski, *Methods Neurosci.* **15**, 205 (1993).
[31] D. S. Papermaster and W. J. Dreyer, *Biochemistry* **13**, 2438 (1974).

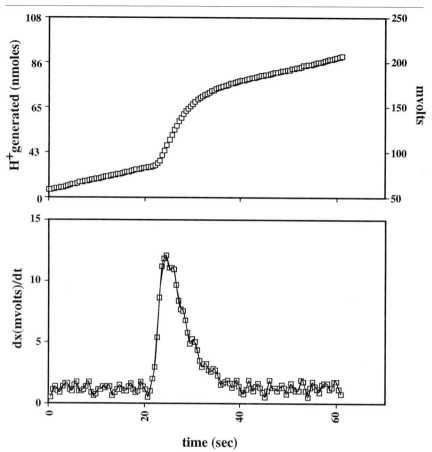

FIG. 4. pH assay data. *Top:* Raw data. The pH assay was performed as described in the methods section. Note the slow, constant level of cGMP hydrolysis in the dark (time 0–20 sec). At about 20 sec, a 20-msec flash of light was given. The rate of cGMP hydrolysis quickly increased and eventually returned to the dark rate of hydrolysis. *Bottom:* Derivative of the data at top. This shows the rate of cGMP hydrolysis at any given point. It can also be used to measure the point at which activity has returned to the dark level. Raw data were taken from the pH meter using an inexpensive Serial Box Interface for the Macintosh from Vernier Software (Portland, OR). Data were analyzed using the Data Logger program, also from Vernier.

1.105 density sucrose (make 25 ml): 27.4% (w/v) sucrose, 10 mM Tris, pH 7.4, 1 mM MgCl$_2$

1.115 density sucrose (make 25 ml): 30% (w/v) sucrose, 10 mM Tris, pH 7.4, 1 mM MgCl$_2$

1.135 density sucrose (make 25 ml): 35.4% (w/v) sucrose, 10 mM Tris, pH 7.4, 1 mM MgCl$_2$

Procedure

1. Place the 34% sucrose solution into a screw-cap container that has enough additional room for all the retinas (each retina has about 1 ml of volume). Keep this container on ice.
2. Dissect the retinas from 50 fresh bovine eyes in the dark using infrared (IR) goggles or under dim red light. Place them into the ice-cold 34% sucrose.
3. Place the lid on the container and shake it vigorously for 1 min to shear the ROS from the retinas.
4. Place the resultant homogenate into Oak Ridge tubes. Centrifuge at 4° for 4 min at 1250g.
5. Decant the supernatant and dilute it with 2 volumes of 10 mM Tris, pH 7.4. Spin down ROS by centrifuging at 4° for 4 min at 1250g.
6. Remove the supernatant. Suspend the pellet in 20 ml of 1.105 density sucrose. Homogenize gently (2×) with a loose Dounce homogenizer to break up clumps.
7. Layer onto the sucrose density gradient using the prepared 1.115 and 1.135 density solutions. Spin in a swinging bucket rotor for 30 min at 100,000g and 4°.
8. Remove the thick band of ROS from the 1.135/1.115 interface. This should look white in red light or with IR goggles.
9. Dilute band 1:1 with pH assay buffer. Spin down, 10 min, 4°, 16,000g.
10. Remove supernatant. Pellet can be resuspended in a small amount of buffer, aliquotted, and stored at −20° or −80° in foil-wrapped tubes. Protein concentration can be measured with a Bradford assay and used to estimate rhodopsin content (rhodopsin is about 75% of the preparation). Alternately (and more accurately), a spectrophotometric reading can be used to determine the amount of unbleached rhodopsin in the preparation. To do this, aliquot ROS (20 μl) into 980 μl of 10 mM 3-[3-(chloroamidopropyl)dimethylammonio]-1-propane sulfonate (CHAPS), 25 mM hydroxylamine in the dark. Mix well and measure the OD of this solution at 498 nm. Expose the cuvette to bright light for 3 min to bleach the rhodopsin, and then measure the OD$_{498}$ again. Because bleached rhodopsin will not absorb, the difference in the absorbances can be used to determine the amount of unbleached rhodopsin in the original sample. The molar absorbance of rhodopsin at 498 nm is 39,000.

Solutions

> pH assay buffer: 140 mM potassium aspartate, 7 mM KCl, 5 mM NaCl, 5 mM N-(2-hydroxyethyl)piperazine-N'-(2-ethanesulfonic acid) (HEPES), 1 mM ethylene glycol-bis(β-aminoethyl ether)-N,N,N',N'-tetraacetic acid (EGTA), 3.3 mM MgCl$_2$, 0.986 mM CaCl$_2$. Adjust pH to 8.0 using KOH. Unchelated calcium in this buffer is calculated to be about 600 nM, the level of calcium in ROS in the dark.[32]
>
> Nucleotides: All nucleotides are made up in pH assay buffer. The pH of these solutions should be carefully checked to ensure that it is as close as possible to pH 8.

Procedure

It is best to do this assay in complete darkness using IR goggles, because the ROS will respond to dim red light.

1. In light, mix together buffer, ATP, and GTP in a clear glass reaction vial. This vial should have a very small stir bar or stir vane so that the reaction can be constantly stirred. In our laboratory, the reaction is done in 400 μl, using 8 μl of 100 mM ATP (final concentration is 2 mM) and 4 μl 50 mM GTP (final concentration is 0.5 mM).
2. Turn off lights. In dark, triturate the ROS by drawing it into an insulin syringe (28-gauge, needle attached) 10 times.
3. Dispense the appropriate amount of ROS into the reaction vial. We use 10–20 μM rhodopsin.
4. Dispense the appropriate amount of cGMP into the reaction vial. We use 40 μl of 50 mM cGMP. Because this solution is rather viscous when kept on ice, thus making it difficult to measure in the dark, we leave the solution at room temperature while performing the assays.
5. As quickly as possible, put the pH electrode into the vial. (pH meter can be attached to a chart recorder or a computer to record data.) Normally there is a low rate of dark cGMP hydrolysis. If this rate is high, new ROS should be made. This mixture should be stirred the entire time the pH is being measured.
6. Expose the ROS to a brief light flash. pH should change rapidly at first, then return to its original dark rate (see Fig. 4). Usually, several flashes can be given per vial, but care must be taken not to deplete

---

[32] M. P. Gray-Keller and P. B. Detwiler, *Neuron* **17,** 323 (1996).
[33] H. Blum, H. Beier, and H. J. Gross, *Electrophoresis* **8,** 93 (1987).

the cGMP or the rate of hydrolysis will be limited by substrate concentration.

Functional Assay of the δ Subunit

The δ subunit should remove PDE from washed ROS membranes. The following is a procedure used in our laboratory to test the activity of the delta subunit.

Solutions

ROS buffer: 20 mM MOPS, pH 7.2, 60 mM KCl, 30 mM NaCl, 2 mM MgCl$_2$

Procedure

1. Wash the appropriate amount of ROS membranes in ROS buffer to remove soluble PDE. We use about 2 mg/ml ROS protein in the assays, and dilute this 1:25 when testing PDE activity. The wash is performed by triturating ROS in ROS buffer through an insulin syringe, 10×, on ice.
2. Pellet ROS membranes, 10 min, 90,000g, room temperature. Repeat wash and centrifugation for a total of three washes.
3. Resuspend the ROS in ROS buffer + 1 mg/ml AEBSF (Boehringer Mannheim).
4. Incubate the ROS with purified δ subunit for 1 hr at 4° We incubate the ROS with the δ subunit in a 35-μl volume in 1.7- or 0.65-ml Eppendorf tubes. Incubation time and temperature can be varied. The 2 μM δ subunit should remove significant amounts of PDE activity from the membrane.
5. Pellet ROS membranes for 15 min at 14,000g, 4°.
6. Remove supernatant fraction from pellet. Resuspend pellet in ROS buffer. Respin supernatant fractions. Assay both supernatant and pellet fractions for PDE activity.

Conclusions

We have presented methods for purification of each of the known isozymes of type 6 PDE from tissue, as well as purification methods for individual subunits and for assay of these preparations. Creative application of these methods will hopefully lead to valuable insight into the mechanisms of signal transduction in the photoreceptor.

[41] Transcriptional Regulation of the cGMP Phosphodiesterase β-Subunit Gene

By LEONID E. LERNER and DEBORA B. FARBER

Introduction

The β subunit of rod-specific cGMP phosphodiesterase (EC 3.1.4.17; β-PDE) is an important component of the catalytic core of the enzyme. Mutations in the β-PDE gene (PDE6B) have been associated with inherited retinal degenerations in a number of species and are the most common known cause of human autosomal recessive retinitis pigmentosa (ARRP).[1-8] However, PDE6B mutations are responsible for disease only in 5–6% of all studied cases of ARRP. Undoubtedly, appropriate levels of transcription of the β-PDE gene are crucial for the integrity and function of the photoreceptors. Poor expression of the β-PDE gene causes an abnormally low phosphodiesterase activity leading to high levels of cGMP and degeneration of rods. Eventually, the genes encoding proteins that play key roles in the regulation of the β-PDE gene expression will become additional candidates for ARRP and other related diseases.

As the field of transcriptional regulation of gene expression evolved, it became clear that genes are differentially expressed according to their interplay with particular sets of transcription factors. Therefore, cell-specific expression of a gene is likely to be regulated by a unique set of transcription factors specific for a particular cell type, rather than a single cell-specific transcription factor. These transcription factors bind avidly and selectively to short sequences of DNA (4–12 bp in length). If a gene contains a DNA

---

[1] C. Bowes, T. Li, M. Danciger, L. C. Baxter, M. L. Applebury, and D. B. Farber, *Nature* **347,** 677 (1990).
[2] S. J. Pittler and W. Baehr, *Proc. Natl. Acad. Sci. U.S.A.* **88,** 8322 (1991).
[3] D. B. Farber, J. S. Danciger, and G. Aguirre, *Neuron* **9,** 349 (1992).
[4] M. L. Suber, S. J. Pittler, N. Qin, G. C. Wright, V. Holcombe, R. H. Lee, C. M. Craft, R. N. Lolley, W. Baehr, and R. L. Hurwitz, *Proc. Natl. Acad. Sci. U.S.A.* **90,** 3968 (1993).
[5] M. Danciger, J. Blaney, Y. Q. Gao, D. Y. Zhao, J. H. Heckenlively, S. G. Jacobson, and D. B. Farber, *Genomics* **30,** 1 (1995).
[6] M. Danciger, V. Heilbron, Y. Q. Gao, D. Y. Zhao, S. G. Jacobson, and D. B. Farber, *Mol. Vis.* **2,** http://www.emory.edu/molvis (1996).
[7] M. E. McLaughlin, M. A. Sandberg, E. L. Berson, and T. P. Dryja, *Nature Genet.* **4,** 130 (1993).
[8] M. Bayes, M. Giordano, S. Balcells, D. Grinberg, L. Vilageliu, I. Martínez, C. Ayuso, J. Benítez, M. A. Ramos-Arroyo, P. Chivelet, T. Solans, D. Valverde, S. Amselem, M. Goossens, M. Biget, R. González-Duarte, and C. Besmond, *Hum. Mutant.* **5,** 228 (1995).

binding site for a particular factor, as long as the site is appropriately positioned relative to other regulatory components of the gene, it may be responsive to that factor *in vivo*. We have performed a scrupulous mutational analysis as well as studies of protein–DNA interactions on the 5'-flanking region of the human β-PDE gene in order to localize and characterize its regulatory *cis*-acting elements and to demonstrate their interactions with nuclear *trans*-acting factors.

This chapter focuses on specific techniques that have been employed to characterize the molecular mechanisms that control expression of the β-PDE gene at the transcriptional level. We describe detailed protocols for the preparation of unidirectional nested deletion mutants, site-specific substitution mutagenesis, transient transfections in Y-79 or WERI-Rb-1 human retinoblastoma cell lines, cotransfections of a reporter construct carrying a fragment of the human β-PDE gene 5'-flanking region and one or more expression construct(s) containing a cDNA for a transcription factor, the preparation of nuclear extracts from Y-79 or WERI-Rb-1 human retinoblastoma cells, DNase I footprinting, and gel mobility shift assays.

Methods for Functional Analysis of *cis*-acting Elements in β-PDE 5'-Flanking Region

The potential for transcriptional activation of a regulatory sequence can be quantitatively assayed *in vitro*. This is achieved by transfecting an appropriate cell line maintained in culture with a construct containing a reporter gene driven by the regulatory sequence of interest and measuring the level of the reporter gene expression.

A number of different reporter genes have been used in transient transfection assays. These include chloramphenicol acetyltransferase (CAT),[9] β-galactosidase,[10] luciferase,[11] and β-globin.[12] We have used the GeneLight reporter vectors pGL2-Basic and pGL2-Control (Promega Corp., Madison, WI), which contain the coding region of the firefly *Photinus pyralis* luciferase gene. Luciferase activity can be easily quantified using a rapid and sensitive assay that allows the analysis of a large number of samples. There is no endogenous luciferase activity in eukaryotic cells. Therefore, the luciferase activity measured following transfection of an appropriate cell line with a pGL2-based construct reflects the level of expression of the luciferase reporter gene. A disadvantage of an assay that depends on the

[9] C. M. Gorman, L. F. Moffat, and B. H. Howard, *Mol. Cell. Biol.* **2**, 1044 (1982).
[10] C. V. Hall, P. E. Jacob, G. M. Ringold, and F. Lee, *J. Mol. Appl. Genet.* **2**, 101 (1983).
[11] J. R. de Wet, K. V. Wood, M. DeLuca, D. R. Helinski, and S. Subramani, *Mol. Cell. Biol.* **7**, 725 (1987).
[12] B. J. Knoll, S. T. Zarucki, D. C. Dean, and B. W. O'Malley, *Nucleic Acids Res.* **11**, 6733 (1983).

measurement of reporter enzyme activity is that it does not directly measure the transcriptional activation reflected by the levels of mRNA. It has been accepted, however, that the level of luciferase activity correlates well with the steady-state level of its mRNA. Direct analysis of mRNA levels may be performed by primer extension or RNase protection assays, which can also be used to confirm the correct initiation of transcription. However, this approach is less sensitive and substantially more time consuming.

When a transient transfection assay using a pGL2-based construct is employed, the luciferase activity measured in cell lysates must be normalized for both transfection efficiency and general effects on transcription. This is achieved by introducing an internal control plasmid. A suitable control plasmid that we have used contains the bacterial *lacZ* gene driven by the simian virus 40 (SV40) early promoter, the pSV-β-galactosidase control vector (Promega). Luciferase activity is determined by measuring the light generated by the luciferin substrate in the presence of adenosine triphosphate (ATP) and Mg^{2+}. The reaction is carried out in a luminometer cuvette and the peak light emission is recorded using a luminometer (e.g., Monolight 2010, Analytical Luminescence, Ann Arbor, MI). A simple protocol for the preparation of cell extracts suitable for luciferase analysis is described later.

A number of different methods have been developed for the introduction of DNA into eukaryotic cells. The method chosen depends to a large extent on the particular cell line to be transfected. For Y-79 human retinoblastoma cells, we have successfully used the calcium phosphate-mediated transfection method,[13] which has the additional advantage of being relatively inexpensive compared to other approaches such as lipofection. Y-79 retinoblastoma cells have been shown to express both cone- and rod-specific genes including the β-PDE gene.[14]

Construction of Reporter Vectors Containing β-PDE 5'-Flanking Sequences

Unidirectional nested deletion constructs containing various lengths of the 5'-flanking region of the human β-PDE gene are generated by polymerase chain reaction (PCR) using sequence-specific primers.[15] The 3' primer, complementary to the +34 to +53 bp sequence (relative to the transcription start site of the β-PDE gene) adjacent to the translation initiation codon, is common to all of the constructs and contains a *Bgl*II linker. The 5'

---

[13] M. Wigler, S. Silverstein, L. S. Lee, A. Pellicer, Y. C. Cheng, and R. Axel, *Cell* **11**, 223 (1977).
[14] A. Di Polo and D. B. Farber, *Proc. Natl. Acad. Sci. U.S.A.* **92**, 4016 (1995).
[15] A. Di Polo, L. E. Lerner, and D. B. Farber, *Nucleic Acids Res.* **25**, 3863 (1997).

primers vary in order to generate different length fragments, but each primer contains an *Nhe*I linker. PCR products are directionally subcloned into the pGL2-Basic vector. Inserts are fully sequenced in both directions to ensure 100% identity with the original template.

Site-specific substitution constructs are produced by introducing A-C and G-T transversion mutations into specific sequences of the β-PDE 5'-flanking region. We have employed two alternative approaches in order to generate transversions:

1. PCR amplification of the β-PDE 5'-flanking region using primers containing the desired mutation. The length of the primers varies depending on the position of the mutation(s). The 3' primer complementary to the β-PDE 5'-flanking sequence extends upstream from the +53 position (adjacent to the translation initiation codon) and contains a *Bgl*II linker. The 5' primer extends downstream from the −72 position, which may vary if different lengths of the β-PDE 5'-flanking region are to be tested, and contains an *Nhe*I linker. Either the 3' or the 5' primer or both contain a desired mutation. For example, the 5' primer used to generate the −69/−64 bp mutation in the context of the −72 to +53 bp fragment of the human β-PDE 5'-flanking region is as follows (the *Nhe*I linker is shown in lowercase and the mutated nucleotides are underlined): 5'-gggctagcGAGG-TCTGAAGCTGACCC-3'. It is important to design the primer so that the mismatched nucleotides are located far enough from its 3' end to ensure adequate annealing between the template and the 3' end of the primer where extension occurs. The amplified sequence containing nucleotide substitutions is directionally subcloned into the pGL2-Basic vector and sequenced in both directions.

2. Recombinant PCR to introduce a mutation into a sequence located far from the nearest restriction site.[16] This approach is employed when the synthesis of a sufficiently long primer encompassing both the mutation and the restriction site is difficult. Briefly, two sets of primers are synthesized. One set of complementary primers contains the desired site-specific mutation based on the sequence of the β-PDE 5'-flanking region. For example, in order to generate the −7/−6 bp mutation in the context of the −72 to +53 bp fragment we have used the following primers (the mutated nucleotides are underlined): 5'-GCTGATGACAGTTTTTCCTGGGAGTCC-3' and 5'-GGACTCCCAGGAAAAACTGTCATCAGC-3'. The other pair of primers is designed so that one is a "sense" primer complementary to a sequence upstream from the planned mutation and the other is an

---

[16] R. Higuchi, *in* "PCR Protocols: A Guide to Methods and Applications" (M. Innis, D. Gelfand, D. Sninsky and T. White, eds.), p. 177. Academic Press, San Diego, 1990.

"antisense" primer complementary to a sequence downstream from the mutation. We have used the GLprimer1 and GLprimer2 (Promega) complementary to the pGL2 vector sequences flanking the inserted fragment.

Two PCR reactions are set up, each using a mutated primer and one of the GLprimers, and the pGL2-Basic vector containing a fragment of the β-PDE 5'-flanking region as template. The reactions yield two sequences that partially overlap creating a short double-stranded region containing the mutation and a long protruding end on each side. The protruding ends are complementary to the flanking sequences of the pGL2 vector containing unique restriction sites (*Nhe*I upstream and *Bgl*II downstream, respectively). The two PCR products are purified using agarose gel electrophoresis and annealed. Using cloned *Pfu* DNA polymerase (Stratagene), each of the two sequences is extended utilizing the protruding end of the opposite sequence as template. The double-stranded product containing the desired mutation is then digested with *Nhe*I and *Bgl*II restriction endonucleases, purified and directionally subcloned into the pGL2-Basic vector as described earlier. Inserts are sequenced in both directions to ensure 100% identity with the original template. In our hands, the efficiency of the recombinant PCR is very high (>85%).

A detailed protocol for the secondary extension reaction is as follows:

1. In a 0.5-ml microfuge tube, mix 50 ng of each of the purified products of the primary PCR with 10 μl of 10× *Pfu* reaction buffer, 1 μl of 20 mM dNTP mix, 2.5 U of *Pfu* DNA polymerase, and doubly distilled H$_2$O to 50 μl total volume. Add three drops of mineral oil to avoid evaporation.
2. Denature for 4 min at 94°.
3. Perform 30 cycles of denaturation at 94° for 1 min, annealing at 48° for 1.5 min, and extension at 72° for 1.5 min in a thermal cycler. It is important to adjust the extension time according to the length of the fragment to be amplified.
4. Incubate for additional 10 min at 72° (1 cycle) to allow extension to go to completion.

For efficient and consistent transfection results in Y-79 retinoblastoma cells, it is very important to prepare highly purified plasmid DNA. We obtain best results with the Endofree Maxi kit (Qiagen, Valencia, CA) designed to remove bacterial endotoxin.

Transient Transfections of Y-79 or WERI-Rb-1 Human Retinoblastoma Cells Using Calcium Phosphate Precipitation

1. Prepare the following solutions:
   - 2× HEPES-buffered saline/Na$_2$HPO$_4$, pH 7.0 (HBS/P): 45 mM

HEPES (tissue culture grade), 280 mM NaCl, and 2.8 mM Na$_2$HPO$_4$. Adjust pH accurately to 7.0 with NaOH, filter sterilize. This buffer may be stored at 4° for several months; recheck pH after prolonged storage.
- Phosphate-buffered saline (PBS), pH 7.4: 140 mM NaCl, 2.7 mM KCl, 4.5 mM Na$_2$HPO$_4$, and 1.5 mM KH$_2$PO$_4$. Adjust pH to 7.4 with HCl, filter sterilize.
- 2.5 M CaCl$_2$ (tissue culture grade), filter sterilize.
- 10× TE buffer, pH 7.0: 10 mM Tris-HCl and 1 mM EDTA, filter sterilize. This buffer may be stored at 4° for several months.
- 0.2 mg/ml Poly(D-lysine) in PBS, filter sterilize, store at 4°.
- 5 μg/ml Fibronectin (Sigma, St. Louis, MO) in PBS, filter sterilize, store at 4°.

2. Initiate Y-79 retinoblastoma cell cultures (ATCC, Rockville, MD, HTB 18) at least 2 weeks before transfection. Cells maintained in culture for a long period of time may not produce optimal results, therefore we prefer to use cell cultures less than 1.5 months old. Propagate the cells in suspension in RPMI 1640 (Gibco-BRL, Gaithersburg, MD) supplemented with 15% fetal bovine serum (FBS). Feed the cells the day before plating for transfection.

3. Plate cells 24 hr before transfection. Under sterile conditions, coat 60-mm-diameter tissue culture plates with 0.4 ml of poly(D-lysine) (spread over the surface with brisk rotational movements). Replace the lid and wait for 10 min. Add 0.2 ml of fibronectin solution and spread over the surface. Replace the lid and wait for 30–60 min. Meanwhile, count the cells and dilute them with RPMI 1640 supplemented with 15% FBS to 1.5 × 10^6 cells/ml. Aspirate fibronectin/poly(D-lysine) and wash the dishes briefly with 3 ml of serum-free RPMI 1640 followed by aspiration. Plate 3 ml of cell suspension per plate (approximately 4.5 × 10^6 cells/plate). If larger dishes are used, the number of cells plated and the volumes given should be adjusted accordingly.

4. The following day, gently aspirate the medium (including any dead cells) without drying the cells, and replace with 4 ml of Dulbecco's modified Eagle's medium (DMEM)/F12 medium (50/50 mix, Cellgro, Mediatech Inc., Herndon, VA) supplemented with 15% FBS (RPMI 1640 has a high Ca^{2+} content, which may interfere with calcium phosphate-mediated transfection). Place the cells back into the humidified incubator for 3 hr.

5. Prepare the following mixture in a 14-ml Falcon tube: 1.35 ml of 10× TE buffer, pH 7.0; 15 μg of pSV-β-galactosidase vector (5 μg

per plate) and 30 µg of the appropriate pGL2 construct (10 µg per plate) in a total volume of 45 µl of TE buffer; 0.15 ml of 2.5 M CaCl$_2$.
6. Add 1.5 ml of 2× HBS/P, pH 7.0, and mix thoroughly by pipetting up and down. Do not vortex. This transfection mixture (3 ml) is used for three plates (transfection of each construct is carried out in triplicate).
7. Remove the plates from the incubator. Add 1.0 ml of transfection mixture to each plate and spread over the cells by grid movements (forward–backward and left–right). Do not use rotational movements because the precipitate tends to spread around the periphery of the plate.
8. Place the plates into the humidified incubator overnight. The following morning, a fine-grained precipitate will have formed. It is readily visualized under the microscope around the cells and attached to the surface of the cells. It is important to examine all the plates for the amount of precipitate formed, which should be comparable, as well as to make sure that the cells remain attached to the plate. If there are plates with significantly less precipitate or with many cells detached, those plates should be disregarded.
9. Gently aspirate the medium from the plates to avoid cell detachment and damage. Cells may be carefully washed with 5 ml of serum-free medium for up to three times but this may increase the cell loss from the plate. Feed the cells with 5 ml RPMI 1640 supplemented with 15% FBS, incubate for 24 hr and harvest.

Essentially the same protocol can be used for transient transfection of WERI-Rb-1 human retinoblastoma cells.

Preparation of Cell Lysate and Measurement of Luciferase Activity

1. Luciferase assay mixture (LAM): 20 mM Tricine, 1 mM MgCO$_3$, 2.7 mM MgSO$_4$, 0.1 mM EDTA, 30 mM dithiothreitol (DTT), 0.3 mM coenzyme A, 0.5 mM luciferin and 0.5 mM ATP; adjust pH to 7.8 with 1 M HCl. Store 1-ml aliquots at $-20°$ in complete darkness (may be wrapped in aluminum foil).
2. Prepare the cell lysis buffer by adding 1 volume of 5× Reporter Lysis Buffer (Promega) to 4 volumes of doubly distilled H$_2$O. Vortex.
3. Aspirate the medium from the plates and wash the cells twice with 4 ml of PBS at room temperature. Thoroughly remove all PBS after the final wash by tilting the plate (residual PBS may interfere with the cell lysis buffer and prevent complete cell lysis).

4. Pipette 0.1 ml of cell lysis buffer onto the cell monolayer and spread it over the surface by tilting the plate. Leave the plate horizontal for 15 min at room temperature. The cells will be lysed (if viewed under the microscope, only intact nuclei are visible).
5. Scrape the plate thoroughly with a cell scraper and transfer the cell lysate into a precooled microfuge tube. Keep on ice. Centrifuge for 2 min at 4° at 16,000g to pellet the cell debris, and transfer the supernatant into a clean 1.5-ml tube. Store on ice.
6. Pipette 20 μl of the extract obtained from different plates into appropriately labeled polystyrene luminometer cuvettes. Bring the cuvettes and LAM alongside the luminometer and turn it on.
7. Add 0.1 ml of LAM (thawed at room temperature in the dark, e.g., in a drawer) to one cuvette at a time, briefly hand-vortex by finger tapping once on the lower portion of the cuvette, insert the cuvette into the luminometer, and measure the luciferase activity. It is important to perform this step in a timely fashion consistent for all the cuvettes because once LAM is added to the extract, the light-emitting reaction begins. One can perform this sequence to a count of one to four. Alternatively, if the luminometer is equipped with a dispenser, LAM can be injected automatically.

LAM alone and extracts from mock-transfected cells provide negative controls (between 90 and 150 light units). Luciferase activity produced by transfection of Y-79 cells with the pGL2-Control vector can be used as a positive control and the values obtained usually range between 80,000 and 110,000 light units.

Determination of Galactosidase Activity and Quantification of Transcription Levels

1. Prepare the following solutions:
   β-Galactosidase assay buffer, pH 7.0 (buffer Z): 60 mM Na$_2$HPO$_4 \cdot$ 7H$_2$O, 40 mM NaH$_2$PO$_4 \cdot$ H$_2$O, 10 mM KCl, 1 mM MgSO$_4 \cdot$ 7H$_2$O, and 50 mM 2-mercaptoethanol.[17] Do not autoclave. This buffer is stable at 4°.
   4 mg/ml o-Nitrophenyl-β-D-galactopyranoside (ONPG) in buffer Z, store in 1 ml aliquots at $-20°$.
   1 M Na$_2$CO$_3$ in doubly distilled H$_2$O.
2. Pipette 40 μl of 4 mg/ml ONPG stock solution into appropriately labeled glass tubes.

[17] J. Sambrook, E. F. Fritsch, and T. Maniatis, "Molecular Cloning: A Laboratory Manual," Vol. 3, p. 16.66. Cold Spring Harbor Laboratory Press, New York, 1989.

3. Add 40 μl of cell extract followed by 120 μl of buffer Z. Mix.
4. Incubate at 37° until yellow color develops (approximately 1.5–3 hr).
5. Add 100 μl of 1 M Na_2CO_3 to each tube and determine the OD at 420 nm.

Relative luciferase activity (*RLA*) is calculated as follows:

RLA = Luciferase activity (light units)/β-galactosidase activity
$(OD_{420}$ units) × 1000

A typical example of relative luciferase activity calculated following transient transfections of Y-79 retinoblastoma cells is shown in Fig. 1. Constructs containing different lengths of the 5'-flanking region of the human β-PDE gene were subcloned into the pGL2-Basic vector and tested in transient transfections (Figs. 1A and 1B). In addition, the results of transfections using site-specific nucleotide substitution mutants of the 5'-flanking region of the β-PDE gene are illustrated in Fig. 1B.

For cotransfection experiments, when two or more different DNA constructs are transfected simultaneously (e.g., a reporter construct and an expression construct), the protocol is essentially the same as above except that an equal amount of plasmid used as vector for the expression construct(s) is added to the control transfection mixture. For example, when cotransfecting 10 μg of a construct containing a fragment of the 5'-flanking region of the β-PDE gene (a reporter construct) and 1 μg of plasmid containing the cDNA for the CRX transcription factor (an expression construct),[18,19] the control transfection contains 10 μg of the β-PDE construct and 1 μg of the empty plasmid used as vector for the CRX cDNA.

Methods for Assaying Protein–DNA Interactions at β-PDE 5'-Flanking Region

Gel Mobility Shift Assay

Gel mobility shift assay (GMSA) is a sensitive and convenient technique for the detection of sequence-specific protein–DNA interactions. It is particularly useful for the analysis of low abundance transcription factor–DNA complexes formed in the context of the β-PDE 5'-flanking region.

---

[18] S. Chen, Q. L. Wang, Z. Nie, H. Sun, G. Lennon, N. G. Copeland, D. J. Gilbert, N. A. Jenkins, and D. J. Zack, *Neuron* **19,** 1017 (1997).
[19] T. Furukawa, E. M. Morrow, and C. L. Cepko, *Cell* **91,** 531 (1997).

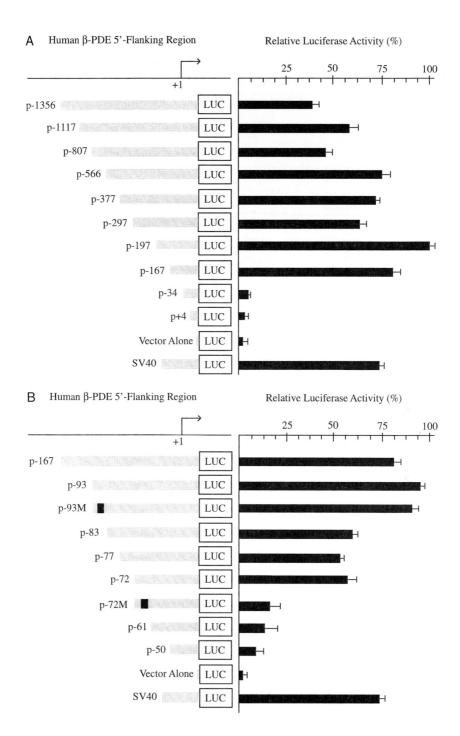

Preparation of Nuclear Protein Extracts from Y-79 or WERI-Rb-1 Human Retinoblastoma Cells

We use a modification of the method previously described by Naeve et al.[20] The advantage of this method over that reported by Dignam et al.[21] is that the present procedure can be accomplished in 1–1.5 hr thus minimizing the potential nuclear protein loss. Essentially the same protocol is used for the nuclear extract preparation from Y-79 or WERI-Rb-1 retinoblastoma cells.

1. Prepare the following solutions:
   Buffer A: 10 mM HEPES, pH 7.9, 10 mM KCl, 0.1 mM EDTA, 0.1 mM EGTA, 1 mM DTT, 0.75 mM spermidine, 0.15 mM spermine, 100 μM phenylmethylsulfonyl fluoride (PMSF), 1 μg/ml aprotenin, 1 μg/ml leupeptin, and 1 μg/ml pepstatin.
   Buffer B: 50 mM HEPES, pH 7.9, 10 mM KCl, 0.2 mM EDTA, 0.2 mM EGTA, 1 mM DTT, 0.75 mM spermidine, 0.15 mM spermine, 100 μM PMSF, 1 μg/ml aprotenin, 1 μg/ml leupeptin, 1 μg/ml pepstatin, and 67% (w/v) sucrose.
   Buffer C: 20 mM HEPES, pH 7.9, 100 mM KCl, 0.2 mM EDTA, 0.2 mM EGTA, 2 mM DTT, 100 μM PMSF, 1 μg/ml aprotenin, 1 μg/ml leupeptin, 1 μg/ml pepstatin, and 20% (v/v) glycerol.
   Buffers are stored at $-20°$ without PMSF, DTT, and sucrose, which are added immediately before use.
2. Retinoblastoma cells grown in suspension for less than 1.5 months are transferred to a 15-ml conical tube and the packed cell volume (PCV) determined after the cells have been pelleted by centrifugation at 1,000g for 5 min at 4°. Cells are resuspended in a small volume

---

[20] G. S. Naeve, Y. Zhou, and A. S. Lee, *Nucleic Acids Res.* **23**, 475 (1995).
[21] J. D. Dignam, R. M. Lebovitz, and R. G. Roeder, *Nucleic Acids Res.* **11**, 1475 (1983).

FIG. 1. (A) Relative luciferase activity of constructs containing different lengths of the human β-PDE 5'-flanking region from -1356 to $+4$ bp on transient transfection into Y-79 human retinoblastoma cells. Plasmids (15 μg) were cotransfected with the control *lacZ* gene driven by the SV40 promoter (5 μg). Luciferase activity was normalized to the corresponding β-galactosidase activity for each sample and expressed as percent activity of the construct p-197. Values represent the average of at least three transfections and standard deviation bars are shown. (B) Relative luciferase activity produced by constructs containing unidirectional nested deletions of the upstream end of the human β-PDE 5'-flanking region (from -167 to -50) on transient transfection into Y-79 human retinoblastoma cells. Plasmids p-93M and p-72M contain point mutations in the E box and the AP-1 consensus sequences, respectively. Luciferase activity was normalized to the corresponding β-galactosidase activity for each sample and expressed as percent activity of the construct p-197. Values represent the average of at least three transfections and standard deviation bars are shown.

of cold PBS and transferred to a 50-ml conical tube (it may be more convenient to split the cells into two tubes if PCV is greater than 0.5 ml).
3. The cells are gently washed with 30× PCV of cold PBS, resuspended in 5× PCV of cold buffer A (hypotonic) and allowed to swell on ice for 10 min. Cells are centrifuged at 1000g for 3 min at 4°.
4. The pellet is carefully resuspended in 2× the original PCV of cold buffer A and transferred into a Dounce homogenizer. Cells are broken with 5 strokes of a tight pestle.
5. Immediately, 0.3× PCV of cold buffer B is added and mixed with 5 strokes of a loose pestle, and the mixture is transferred to a 1.5-ml microfuge tube. Nuclei are separated from lysed cells by centrifuging at 16,000g for 30 sec at 4°.
6. The supernatant is carefully removed. The viscous nuclear pellet is resuspended in 0.66× PCV of buffer C and subjected to three cycles of rapid freeze–thawing (60 sec in dry ice/acetone and 90 sec in a water bath preheated to 37°) in order to lyse the nuclei. Lysed nuclei are centrifuged at 16,000g for 10 min at 4°.
7. The supernatant is collected, aliquotted, and quick-frozen. Nuclear extract may be stored at −80° for several months (avoid temperature fluctuations). Protein concentration is determined using the Bradford assay.[22]

Synthesis and Purification of DNA Probe

We generally use synthetic oligonucleotides for making the probe for GMSA, although restriction fragments from a recombinant plasmid containing the sequence of interest may be used as well, particularly when a longer probe (e.g., 50–250 bp) is desired.

1. Two complementary oligonucleotides identical to a short sequence in the human β-PDE 5′-flanking region are synthesized. For example, we have used oligonucleotides identical to the region from −72 to −58 bp, which contains the consensus sequence for the AP-1 response element, TGAGTCA: 5′-GAGTGAGTCAGCTGA-3′.
2. For optimal results, purify the oligonucleotides using a 12–15% polyacrylamide gel containing 6 M urea. (We run one oligonucleotide per gel to avoid cross-contamination.) Briefly, the bands are sliced out and minced with a razor blade (use a new blade for each oligonucleotide) and crushed with a pipette tip in a microfuge tube. DNA is eluted overnight in 1 ml of a solution containing 0.1% (w/v) sodium dodecyl sulfate (SDS), 0.5 M ammonium acetate, and 10 mM magne-

---

[22] M. M. Bradford, *Anal. Biochem.* **72**, 248 (1976).

sium acetate at 37° in a shaker incubator. Polyacrylamide gel pieces are pelleted by centrifugation for 2 min at 12,000g at 4° and the supernatant is collected followed by ethanol/sodium acetate precipitation.

Labeling of DNA Probe

1. Unlike restriction fragments, synthetic oligonucleotides do not have a 5'-phosphate and therefore can be directly labeled using T4 polynucleotide kinase and [γ-^{32}P]ATP. We prefer to label single-stranded oligonucleotides which are subsequently annealed.

   Prepare reaction mixture:

Oligonucleotide (10 pmol/μl)	1.0 μl
10× T4 polynucleotide kinase buffer	2.0 μl
[γ-^{32}P]ATP, 6000 Ci/mmol (10 pmol)	5.0 μl
Doubly distilled H$_2$O	11.4 μl

   In general, ng of oligonucleotide = pmol of oligonucleotide × 0.33 × N (number of bases in the oligonucleotide)

   Add T4 polynucleotide kinase (10 units), mix well, and incubate for 45 min at 37°. Stop the reaction by heating to 68° for 10 min.

2. We always use [γ-^{32}P]ATP received on the day when the labeling reaction is performed. The higher the specific activity, the less the amount of the probe needed for the detection of the DNA-binding protein of interest. If the specific activity of the probe is low and many oligonucleotide molecules remain unlabeled, more probe should be used per reaction. However, this increases the competitive binding of the protein to unlabeled molecules therefore decreasing its detectability in GMSA. For example, if the protein of interest is present in the nuclear extract in scarce amounts (e.g., a rare transcription factor) it will preferentially interact with the excess of unlabeled oligonucleotides therefore getting sequestered. This will prevent the formation of a retarded band in GMSA. To achieve the highest specific activity of the probe, the concentration of the oligonucleotide in the labeling reaction should be decreased threefold to 3 pmol, the amount of [γ-^{32}P]ATP should be increased threefold to 15 μl, and the volume of H$_2$O decreased to 1.4 μl (compare to the standard reaction described earlier). In this way up to 90% of oligonucleotide molecules will be labeled compared to only 50% labeled when the standard reaction is performed using equimolar concentrations of the oligonucleotide and [γ-^{32}P]ATP. However, about 90% of [γ-^{32}P]ATP molecules will be wasted.[23]

[23] J. Sambrook, E. F. Fritsch, and T. Maniatis, "Molecular Cloning: A Laboratory Manual," Vol. 2, p. 11.31. Cold Spring Harbor Laboratory Press, New York, 1989.

3. After radiolabeling is complete, annealing of the complementary oligonucleotides is achieved by mixing them in equimolar amounts, heating to 75° for 5 min in a large water bath, and allowing to cool slowly to room temperature. The probe is purified by centrifugation through a 50-μl Sephadex G-25 microcentrifuge spin column. The efficiency of [γ-^{32}P]ATP transfer can be calculated by scintillation counting 1 μl of the probe on a piece of filter paper before and after the removal of unincorporated radiolabel by the spin column. To evaluate the quality of the radiolabeled probe, it is electrophoresed on a native 12–15% polyacrylamide gel followed by autoradiography. The labeled DNA probe is stored at $-20°$ for no more than 5 days to avoid radiolysis.

Protein–DNA Binding Reaction and Gel Electrophoresis of Formed Complexes

In general, the strength of protein–DNA interactions *in vitro* can be influenced by a number of parameters including monovalent (Na$^+$ and K$^+$) and divalent (Mg^{2+}) cations, pH, nonionic detergents, protein concentration, the type and concentration of the competitor DNA, and the temperature of the binding reaction. For example, low salt (<150 mM KCl) tends to favor protein–DNA interactions. Low amounts of nonionic detergent (0.5–0.05% Nonidet P-40), a carrier protein (e.g., 50 ng/μl bovine serum albumin) and a polyamine such as spermidine (2 mM) can stabilize certain protein–DNA complexes. Thus, systematic alteration of the binding reaction composition and binding conditions ensures optimal detection of protein–DNA complexes. In addition, different parameters of gel electrophoresis can alter the dissociation rate and the electrophoretic mobility of protein–DNA complexes. In general, low percentage and cross-linking of the polyacrylamide gel and low ionic strength of the electrophoresis buffer enhance the detection of the complexes in GMSA.

1. Prepare 5× GMSA buffer (10 mM Tris-HCl, pH 7.5, 50 mM KCl, 0.2 mM EDTA, 1 mM MgCl$_2$, 4% (v/v) glycerol, 1 mM DTT, 0.05% (v/v) Nonidet P-40, and 1 mM PMSF) from stock solutions:

1 M Tris-HCl, pH 7.5	10 μl
1 M KCl	50 μl
100 mM EDTA	2 μl
500 mM MgCl$_2$	2 μl
5% Nonidet P-40	10 μl
Glycerol	40 μl
1 M DTT	1 μl
100 mM PMSF	10 μl

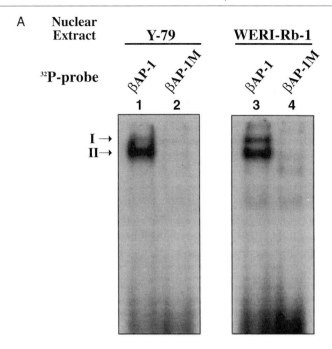
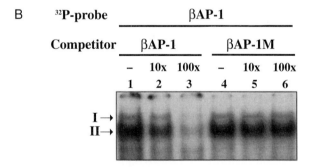

FIG. 2. (A) Two shifted bands (I and II) are detected when the labeled βAP-1 probe is incubated with either Y-79 or WERI-Rb-1 cell nuclear extract. No shifted bands are visualized when the labeled βAP-1M is used as a probe. (B) Competition gel shift experiments using the labeled βAP-1 probe and Y-79 retinoblastoma nuclear extract. The intensity of both shifted bands I and II decreases progressively on addition of increasing amounts of unlabeled βAP-1 competitor (lanes 1–3), but shows no inhibition of complex formation with excess unlabeled βAP-1M (lanes 4–6).

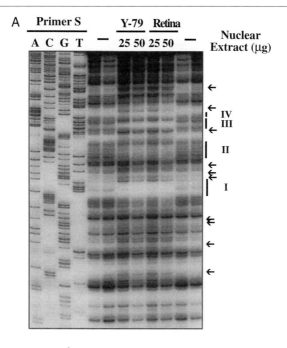

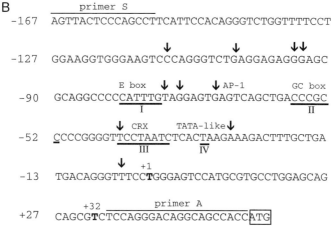

FIG. 3. (A) DNase I footprinting analysis performed with the human β-PDE proximal 5'-flanking region. The sense probe spanning nucleotides −167 to +53 was generated by PCR using primers S and A [shown in (B)]. Binding reactions contained 25 or 50 μg of either Y-79 retinoblastoma or pig retina crude nuclear extracts. A 1 : 50 and 1 : 10 dilution of DNase I was used, respectively, for digestion of nonprotein control and protein-bound probe. Protection areas are designated by vertical lines and numbered consecutively [compare to (B)]; hypersensitivity bands are indicated by arrows. A dideoxynucleotide chain termination sequencing

Add doubly distilled H_2O to a total volume of 200 µl. Use 4 µl of 5× GMSA buffer per 20 µl binding reaction (see below).

2. The DNA probe (1–5 × 10⁵ cpm) identical to the sequence of the human β-PDE promoter containing the consensus AP-1 response element (βAP-1, 5'-GAG<u>TGAGTCA</u>GCTGA-3') and the probe containing transversions in the AP-1 consensus sequence (βAP-1M, 5'-GAG<u>GTCTTCA</u>GCTGA-3') are incubated with 5–20 µg of either Y-79 or WERI-Rb-1 retinoblastoma nuclear extract in the presence of 2 µg of poly(dI-dC) (Pharmacia) as nonspecific DNA competitor in GMSA buffer. The binding reaction is performed at room temperature for 20 min in a total volume of 20 µl. Meanwhile, a native polyacrylamide gel (39:1 acrylamide:bisacrylamide, w/w) containing 0.25× TBE buffer (22.5 mM Tris–borate and 0.5 mM EDTA, pH 8.0) is preelectrophoresed for 60–90 min at 150–200 V (the current should drop from 20–25 mA to less than 10 mA). Protein–DNA complexes are resolved at room temperature on a 5–8% (w/v) polyacrylamide gel, depending on the size of the protein–DNA complexes, at 8–80 V · hr. In our experience, the complexes appear to be stable when electrophoresed for the appropriate time at 45–200 V. Following electrophoresis, the gel is dried and autoradiographed (Fig. 2).

DNase I Footprinting Assay

This technique allows the analysis of simultaneous interactions between different transcription factors and several binding sites on the same DNA molecule. Another advantage of this technique is that the sites of protein–DNA interactions can be accurately localized.

Preparation of DNA Probe

The quality of the DNA probe is fundamental for the success of this type of experiment. The specific activity of the probe must be high

---

reaction using primer S was run in parallel with the footprinting reactions as a marker. (B) Sequence of the proximal 5'-flanking region of the human β-PDE gene (from −167 to +53 bp) showing the sites of protein–DNA interactions identified by DNase I footprinting. Protected areas are underlined and consecutively numbered; hypersensitivity bands are shown as arrows. Sequences homologous to previously described regulatory elements, E box, AP-1 element, GC box, CRX element, and TATA-like element are labeled. Sense and antisense primers (S and A, respectively) used to generate the DNA probe are overlined. The two transcription start sites of this gene are shown as +1 and +32 and the translation initiation codon is boxed.

and the probe must not be older than 4–5 days to avoid radiolysis. For the β-PDE promoter footprinting assay, a DNA probe comprising the sequence from −167 to +53 bp of the human β-PDE promoter is generated by PCR. The 5′ primer, complementary to positions −167 to −153 (5′-AGTTACTCCCAGCCT-3′), is labeled with T4 polynucleotide kinase (Pharmacia) and [γ-^{32}P]ATP (6000 Ci/mmol, New England Nuclear, Boston, MA). The 3′ primer, complementary to +34 to +53 bp (5′-GGTGGCTGCCTGTCCCTGGA-3′), is not labeled. Both the 5′ and 3′ primers (10 pmol of each) are used to amplify pBamHI (100 ng), which contains 4.8 kb of the human β-PDE 5′-flanking region.[24] The amplified fragment is purified by electrophoresis on a denaturing 7% polyacrylamide gel and used for DNase I footprinting. The labeled DNA probe is stored at −20°.

DNase I Footprinting

The footprinting reaction is performed in three steps: proteins are allowed to bind to the DNA probe, the protein–DNA complexes are partially digested with DNase I, and the products of digestion are analyzed by gel electrophoresis. Incidentally, DNase I cleavage is not completely random. Therefore, free DNA probe is digested and resolved alongside the protein-bound probe on a polyacrylamide gel to control for sequence-specific cleavage.

1. Prepare the following solutions:
   10× Footprinting buffer: 100 mM HEPES, pH 7.9, 1 mM EDTA, glycerol, 20% polyvinyl alcohol, 500 mM KCl.
   Stop buffer: 20 mM EDTA, 1% SDS, 0.2 M NaCl, and 250 μg/ml yeast tRNA.
2. The binding reaction is performed in a total volume of 50 μl containing 5 μl of 10× footprinting buffer, 0.5 mM DTT, 1 μg poly(dI-dC), labeled DNA probe (approximately 5×10^4 cpm), and 25–50 μg of Y-79 retinoblastoma or pig retina crude nuclear extract as described elsewhere.[24] For consistent results, a master mix containing all of the components except nuclear extract is aliquotted into microfuge tubes. Add nuclear extract to each tube and incubate on ice for 30 min. Meanwhile, dilute the DNase I stock (1 mg/ml, Worthington Biochemical, Lakewood, NJ). We have used a 1:5 or 1:10 dilution of DNase I for the protein-bound probe and a 1:50 or 1:100 dilution for the control probe alone (dilutions are determined empirically). After the incubation period, 50 μl of a 10 mM MgCl$_2$/5 mM CaCl$_2$

[24] A. Di Polo, C. B. Rickman, and D. B. Farber, *Invest. Ophthalmol. Vis. Sci.* **37**, 551 (1996).

mixture is added, followed immediately by digestion with DNase I (mix gently with the pipette tip) for exactly 1 min.
3. Promptly terminate the reaction by the addition of 90 μl of stop buffer.
4. Nucleic acids are recovered by phenol–chloroform (1:1) extraction followed by ethanol precipitation. Pellets are dissolved in 99% formamide with tracking dyes, and DNA is resolved on a 7% (w/v) sequencing gel. The gel is dried and autoradiographed. A typical example of a DNase I footprinting gel is shown in Fig. 3. Protein–DNA interactions are evidenced as areas protected from DNase I digestion or as intense hypersensitivity bands. The latter may result from changes in DNA conformation caused by the binding of a nuclear protein, which makes this particular site more susceptible to DNase I digestion.

Acknowledgments

The development of the protocols described in this chapter was supported by NIH grants EY02651 (D.B.F.) and EY00367 (L.E.L.). D.B.F. is the recipient of a Senior Scientific Investigators Award from Research to Prevent Blindness.

[42] Inhibition of Photoreceptor cGMP Phosphodiesterase by Its γ Subunit

By ALEXEY E. GRANOVSKY, KHAKIM G. MURADOV, and NIKOLAI O. ARTEMYEV

Introduction

Rod cGMP phosphodiesterase (PDE) is the effector enzyme in the visual transduction cascade. In dark-adapted rod photoreceptors, the activity of PDE catalytic α and β subunits (Pαβ) is blocked by two identical inhibitory γ subunits (Pγ). Light stimulation of photoreceptors leads to enzyme activation by the GTP-bound transducin-α molecules (GtαGTP), which bind to Pγ subunits and displace them from the catalytic core.[1-3] Identification of the sites involved in the Pγ/Pαβ interface and insights into mechanisms of the PDE activity inhibition are essential for understanding

---

[1] M. Chabre and P. Deterre, *Eur. J. Biochem.* **179,** 255 (1989).
[2] S. Yarfitz and J. B. Hurley, *J. Biol. Chem.* **269,** 14329 (1994).
[3] L. Stryer, *Proc. Natl. Acad. Sci. U.S.A.* **93,** 557 (1996).

the molecular basis of visual signaling. Site-directed mutagenesis and synthetic peptide studies have identified two major regions of Pγ that interact with Pαβ, the central polycationic region, Pγ-24–45, and the C terminus of Pγ.[4–8] Approximately 5–7 C-terminal amino acid residues of Pγ are central to the inhibition of PDE activity.[4,7,8] The role of the polycationic region is to enhance the affinity of the Pγ/Pαβ interaction. Progress in structure–function studies of Pαβ has lagged behind, in part due to the lack of an efficient expression system for the catalytic subunits. For a long time the Pγ interaction sites on Pαβ remained largely obscure. A combination of a photo-cross-linking approach with fluorescence assays to probe the Pγ/Pαβ interface has yielded new valuable information about the inhibitory sites on Pαβ and the mechanism of PDE inhibition by the Pγ subunit.[9,10] In this chapter we describe these cross-linking and fluorescence approaches and procedures to the analysis of inhibition of PDE by Pγ.

Cloning Strategies for Pγ Mutants Designed for Labeling with Fluorescent and Cross-Linking Probes

The Pγ molecule contains a single cysteine residue, PγCys-68. This residue is localized outside the two regions of Pγ that interact with Pαβ, and its substitution of PγCys-68 for Ser does not affect its ability to inhibit PDE activity. The advantage of using PγCys-68 for introduction of Cys-reactive fluorescent or cross-linking probes is that modification at this position would not likely interfere with the Pγ/Pαβ interaction. Pγ fluorescently labeled at Cys-68 has been employed in emission anisotropy[11] or energy transfer assays of Pγ binding to Pαβ.[12] The shortcomings of labeling PγCys-68 to study Pγ/Pαβ interaction are the lack of its utility for the use of environmentally sensitive fluorescent probes and potential low efficiency in cross-linking experiments. Both of these approaches rely on the proximity of a probe to the Pαβ polypeptide chains. The fluorescence of Pγ labeled

---

[4] V. M. Lipkin, I. L. Dumler, K. G. Muradov, N. O. Artemyev, and R. N. Etingof, *FEBS Lett.* **234**, 287 (1988).
[5] N. O. Artemyev and H. E. Hamm, *Biochem. J.* **283**, 273 (1992).
[6] D. J. Takemoto, D. Hurt, B. Oppert, and J. Cunnick, *Biochem. J.* **281**, 637 (1992).
[7] R. L. Brown, *Biochemistry* **31**, 5918 (1992).
[8] N. P. Skiba, N. O. Artemyev, and H. E. Hamm, *J. Biol. Chem.* **270**, 13210 (1995).
[9] N. O. Artemyev, M. Natochin, M. Busman, K. L. Schey, and H. E. Hamm, *Proc. Natl. Acad. Sci. U.S.A.* **93**, 5407 (1996).
[10] A. E. Granovsky, M. Natochin, and N. O. Artemyev, *J. Biol. Chem.* **272**, 11686 (1997).
[11] T. G. Wensel and L. Stryer, *Biochemistry* **29**, 2155 (1990).
[12] A. L. Berger, R. A. Cerione, and J. W. Erickson, *J. Biol. Chem.* **272**, 2714 (1997).

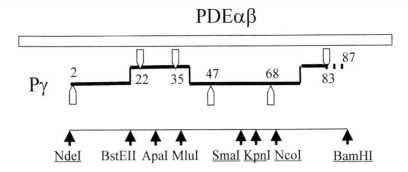

FIG. 1. Localization of molecular probes, BC or MBP, on mutant Pγ polypeptides. Restriction sites utilized in the cloning of Pγ mutants for fluorescence and photo-cross-linking studies are indicated. (Sites unique for the pET-11a vector are underlined.)

at Cys-68 with either of two probes, Lucifer Yellow vinyl sulfone (LY) or 3-(bromoacetyl)-7-diethylaminocoumarin (BC) (Fig. 1), is unchanged on binding to Pαβ.[10] Likewise, a cross-linking probe away from an interaction site is likely to produce inefficient cross-linking. Therefore, proper positioning of the probe is critical for successful application of fluorescent or cross-linking techniques. Favorable modification sites can be identified by placing a single cysteine residue at different positions in the Pγ polypeptide chain via site-directed mutagenesis.

The synthetic gene of Pγ assembled by Brown and Stryer[13] represents a convenient template for site-directed mutagenesis of Pγ because it contains a large number of restriction sites throughout the Pγ cDNA. For expression purposes this gene has been PCR amplified with primers containing NdeI and BamHI sites and subcloned into the pET-11a vector

---

[13] R. L. Brown and L. Stryer, *Proc. Natl. Acad. Sci. U.S.A.* **86,** 4922 (1989).

(Novagen, Madison, WI).[14] The resulting pET-11a-Pγ vector has been widely used for Pγ mutagenesis.[8,14]

Prior to introduction of a Cys residue at a selected position, PγCys-68 must be substituted. A substitution of Cys-68 for Ser can be made using *Nco*I restriction site (codons for Pro^{69}-Trp^{70}) and PCR-directed mutagenesis with a reverse primer containing an *Nco*I site and the mutation. A forward PCR primer may contain one of the upstream Pγ cDNA restriction sites such as the *Nde*I site, which includes a start codon (Fig. 1). The *Nde*I/*Nco*I digested PCR fragment is then inserted into the *Nde*I/*Nco*I fragment of the Pγ expression vector. The PγCys^{68}-Ser mutant gene serves as a basic construct for introduction of Cys. Similarly to the Cys^{68}-Ser substitution, mutations within the N terminus of Pγ can be introduced by polymerase chain reaction (PCR) amplifications using a 5′-primer carrying the desired mutation and a *Nde*I site, and a 3′-primer containing a *Nco*I or *Bam*HI site. In addition to *Nco*I and *Bam*HI sites, unique restriction site *Sma*I (codons for Pro^{55}-Gly^{56}) and *Kpn*I (codons for Gly^{61}-Thr^{62}) allow the placement of Cys at practically any position within the C-terminal portion of Pγ by means of PCR-directed mutagenesis (Fig. 1).

A more elaborate procedure is required to incorporate a Cys residue N terminally or within the polycationic region of Pγ, Pγ-24–45. The available restriction sites, *Bst*EII (Pro^{20}-Val^{21}-Thr^{22}), *Apa*I (Lys^{24}-Gly^{25}-Pro^{26}), and *Mlu*I (Thr^{35}-Arg^{36}), are not unique (Fig. 1), and the pET-11a vector contains a single copy of each site. A potential cloning strategy includes partial digestion of the pET-11a-Pγ vector with one of these enzymes. The linearized plasmid is isolated and subsequently cut at one of the unique restriction sites, *Nde*I, *Sma*I, *Kpn*I, *Nco*I, or *Bam*HI (Fig. 1). A plasmid DNA of appropriate size is separated and used for subcloning of a PCR-generated DNA fragment carrying a mutation and flanked with the chosen restriction sites. Alternatively, the Pγ cDNA can be subcloned from the pET-11a vector into a vector that does not contain *Bst*EII, *Apa*I, and *Mlu*I sites. Afterward, mutations can be generated by PCR-directed mutagenesis as outlined earlier. Because the *Bst*EII, *Apa*I, and *Mlu*I sites are in proximity to each other, mutations can also be made using two annealed complementary mutant oligonucleotides with protruding ends that are compatible with selected restriction enzymes.

Pγ mutants, each with a single Cys residue evenly spaced along the Pγ polypeptide chain, have been generated using the pET-11a-PγCys^{68}-Ser construct.[15] Residues Asn-2, Thr-22, Thr-35, and Val-47 are selected for

---

[14] V. Z. Slepak, N. O. Artemyev, Y. Zhu, C. L. Dumke, L. Sabacan, J. Sondek, H. E. Hamm, M. D. Bownds, and V. Y. Arshavsky, *J. Biol. Chem.* **270**, 14319 (1995).

[15] A. E. Granovsky, R. McEntaffer, and N. O. Artemyev, *Cell Biochem. Biophys.* **28**, 115 (1998).

substitutions by Cys (Fig 1). Such mutations provide for the placement of a probe not only at the N terminus of Pγ, but also N terminally to, in the middle of, and C terminally to the polycationic region, Pγ-24–45, which has been implicated in the interaction with Pαβ. In addition, to place a probe at the C terminus of Pγ, four C-terminal residues of Pγ have been truncated and replaced with Cys (Pγ83Cys). This mutant is designed based on the evidence that the C-terminal residues of Pγ (Ile^{86}–Ile^{87}) are critical for PDE inhibition.[8] They presumably have a complementary hydrophobic binding cavity on Pαβ. If a fluorescent probe occupies the binding pocket on Pαβ instead of the Pγ C terminus, an enhancement in probe quantum yield is likely.

Cloning Procedures for Pγ Mutants

Typically, PCR primers carrying a mutation are designed to allow a mutant codon to be flanked by at least 10 unmodified bases. PCR reactions are performed using a RoboCycler® PCR amplifier (Stratagene, La Jolla, CA). Sizes of PCR fragments for Pγ mutants usually vary from 150 to 250 bp. The PCR reaction mixtures contain 10–50 ng of pET-11a-PγCys^{68}-Ser as a template, 500 ng of each 5′ and 3′ primer, 0.2 mM dNTP mix, 5 μl of 10× PCR buffer, 2.5 units of *Taq* polymerase (Perkin-Elmer, Norwalk, CT) and doubly distilled H_2O to a final volume of 50 μl. After a 2-min dwell time at 95°, 30 cycles of PCR are performed as follows: 94°/50 sec, 56°/50 sec, and 72°/35 sec. The yields of purified PCR products are generally 2–4 μg. If several restriction sites (enzymes) are suitable for the mutagenesis, the preference is given to the enzymes that are optimally active in the same restriction buffer. The PCR product (1 μg) is incubated with 10 units of each restriction enzyme for 3–4 hr to ensure complete digestion. If simultaneous digestion is not possible, a PCR fragment is first cut with one enzyme, precipitated using 0.7 volumes of 2-propanol, dissolved in doubly distilled H_2O, and digested with a second enzyme. A DNA fragment is subcloned into the pET-11a-Pγ vector digested with appropriate restriction enzymes. Because the sizes of excised fragments are small (150–250 bp), it is very difficult to adequately separate fully and incompletely digested (linearized) plasmid. Therefore, the digested vector is additionally treated with shrimp alkaline phosphatase (SAP) (1 unit/0.5 μg DNA) prior to ligation with a PCR fragment. In control experiments, the digested and SAP-treated vector is tested using ligation in the absence of the PCR fragment.

For the introduction of mutations by a pair of annealed complementary mutant oligonucleotides, equimolar amounts (0.2 nmol/10 μl) of two oligonucleotides are mixed and incubated at 95° for 1 min. Afterward, the mix is incubated at room temperature for 20 min. The pET-11a-Pγ vector is prepared by cutting with selected restriction enzymes and treatment with SAP. The oligonucleotide duplex and the vector are mixed at a molar ratio of 20:1 and ligated overnight at 25°. The control ligation mix contains no duplex oligonucleotide, and normally yields no colonies after transformation.

Expression and Isolation of Pγ or Its Mutants

BL21(DE3) cells carrying the pET-11a-Pγ plasmid or mutant plasmid are grown at 37° overnight in LB medium (1 liter: 10 g tryptone, 5 g yeast extract, 5 g NaCl, 80 μl 12.5 M NaOH) containing 50 μg/ml ampicillin. The overnight culture is diluted (1:250) with 2× TY medium (1 liter: 16 g tryptone, 10 g yeast extract, 5 g NaCl) containing 50 μg/ml ampicillin and is grown at 37° until it reaches an OD_{600} of ~0.8. The incubation temperature is then reduced to 30° and Pγ expression is induced by addition of 0.5 mM isopropylthiogalactoside (IPTG). The induction is allowed to proceed for 3–4 hr. Afterward, the cells are spun down and washed two times with 50 mM Tris-HCl buffer (pH 7.5) and resuspended in 50 mM Tris-HCl buffer (pH 7.5) containing 20 mM NaCl, 5 mM EDTA, 1 mM dithiothreitol (DTT), and the protease inhibitors phenylmethylsulfonyl fluoride (PMSF) (1 mM) and pepstatin A (20 μg/ml). The cells are disrupted by sonication with 30-sec pulses (4 min total sonication time) using a Model 550 Sonic Dismembrator (Fisher, Pittsburgh, PA), followed by centrifugation at 100,000g for 30 min. The supernatant is loaded on an SP Fast Flow Sepharose column (Pharmacia, Piscataway, NJ) equilibrated with 50 mM Tris-HCl buffer (pH 7.5) containing 20 mM NaCl, 5 mM EDTA, and 1 mM DTT. The bound proteins are eluted at a flow rate of 0.5 ml/min using a 20–400 mM NaCl gradient. Pγ elutes at ~300 mM NaCl. Additional purification of Pγ is achieved by reversed-phase high-performance liquid chromatography (RP-HPLC) on a C_4 column (Microsorb-Rainin, Hesperia, CA, or 214TP54-Vydac, Woburn, MA) with a 0–80% gradient of acetonitrile in 0.1% trifluoroacetic acid (TFA)/H_2O. Pγ elutes at ~45% acetonitrile. Purified Pγ is lyophilized, dissolved in 20 mM HEPES buffer (pH 7.5), and stored at −80° until use. This procedure yields up to 20 mg of >95% pure Pγ or mutant per liter of culture.

Labeling of Pγ Mutants with Fluorescent Probe, BC

To label Pγ or its mutants containing a single Cys with BC, a stock solution of the probe (5 mM) is freshly prepared by dissolving BC in N',N-dimethylformamide in a light-protected tube. Typically, a two- to threefold molar excess of BC is added from the stock solution to 100–200 μM Pγ or mutant in 20 mM HEPES buffer (pH 7.5). The mixture is incubated for 30 min at room temperature, and the reaction is stopped by addition of 2-mercaptoethanol (5 mM final concentration). The BC-labeled Pγ or mutant is then passed through a PD-10 column (Pharmacia) equilibrated with 20 mM HEPES buffer (pH 7.5) containing 100 mM NaCl to remove excess BC. The final purification step includes (RP-HPLC) on a protein C$_4$ column (214TP54-Vydac) using a 0–80% gradient of acetonitrile/0.1% TFA to separate labeled Pγ from the unlabeled protein. Using ε_{445} 53,000 for BC, the molar ratio of BC to Pγ in the purified preparations is typically greater than 0.8 mol/mol. Under the conditions described, a cysteine-less PγCys^{68}-Ser is not modified by BC. The BC-labeled Pγ and mutants are referred to in the following section according to the position of the modified Cys residue as Pγ2BC, Pγ22BC, Pγ35BC, Pγ47BC, Pγ68BC, and Pγ83BC.

Effects of Mutations and Labeling on Pγ's Ability to Inhibit tPDE

Prior to analysis of the Pγ/P$\alpha\beta$ interaction in fluorescence assays, the ability of the labeled and unlabeled Pγ mutants to inhibit PDE activity is tested. The PDE inhibition test assists in the evaluation of fluorescence binding data, since it represents a singular assay for determination of P$\alpha\beta$ affinities to unlabeled Pγ mutants or BC-labeled mutants that do not change their fluorescence. Labeling of Pγ mutants with BC may possibly produce no or opposite effects on the affinity for P$\alpha\beta$. The probe at certain positions may disrupt specific interactions between the two proteins formed by residues adjacent to the labeled cysteine residue. In contrast, by forming additional hydrophobic interactions with the P$\alpha\beta$ chains, the probe could potentially enhance the affinity of the Pγ/P$\alpha\beta$ interaction.

P$\alpha\beta$ for the PDE inhibition assay can be prepared using limited proteolysis of the holoenzyme with trypsin and purification of trypsinized PDE (tPDE) as described previously.[5,16] Tryptic proteolysis removes intrinsic Pγ subunits and small farnesylated and geranylgeranylated C-terminal fragments of Pα and Pβ, respectively.[17] Proton evolution[18,19] and [^{3}H]cGMP hydrolysis assays can be used to measure PDE activity.[20] The analysis of PDE inhibition has demonstrated that Pγ2BC and Pγ68BC inhibit tPDE as effectively as the unlabeled proteins with K_i values of ~0.1–0.2 nM.[15] Mutations, Thr^{22}-Cys, Thr^{35}-Cys, and Val^{47}-Cys, modestly increase the K_i

---

[16] J. B. Hurley and L. Stryer, *J. Biol. Chem.* **257**, 11094 (1982).
[17] P. Catty and P. Deterre, *Eur. J. Biochem.* **199**, 263 (1991).
[18] P. A. Liebman and A. T. Evanczuk, *Methods in Enzymol.* **81**, 532 (1982).
[19] M. W. Kaplan and K. Palczewski, in "Methods in Neurosciences" (P. A. Hargrave, ed.), p. 205. Academic Press, San Diego, 1993.
[20] W. J. Thompson and M. M. Appleman, *Biochemistry* **10**, 311 (1971).

value for tPDE inhibition (~1 nM), but labeled mutants Pγ22BC, and Pγ35BC, and Pγ47BC inhibit tPDE with comparable K_i values.[15] The C-terminally truncated mutant Pγ83Cys has a significantly diminished interaction with P$\alpha\beta$. This mutant maximally inhibits only ~50% of tPDE activity with a K_i of ~13 nM. After modification with BC, the mutant is capable of full inhibition of tPDE with a K_i of ~3 nM.[10] The more potent and complete inhibition suggests that the probe partially substitutes for the Pγ C terminus and binds to the inhibitory pocket on P$\alpha\beta$.

Fluorescence Assay of Binding between BC-Labeled Pγ Mutants and P$\alpha\beta$

The assay is based on monitoring the fluorescence change resulting from binding of BC-labeled Pγ mutants to the P$\alpha\beta$ subunits.[10] The fluorescence measurements are performed on an F-2000 fluorescence spectrophotometer (Hitachi, Naperville, IL) in 1 ml of 20 mM HEPES buffer (pH 7.5), containing 100 mM NaCl and 4 mM MgCl$_2$. Fluorescence of a BC-labeled Pγ mutant (10 nM) is monitored at equilibrium before and after additions of increasing concentrations of P$\alpha\beta$ using excitation at 445 nm and emission at 495 nm. Concentrations of BC-labeled polypeptides are determined using ε_{445} of 53,000.

The K_d values for BC-Pγ mutant binding to P$\alpha\beta$ are calculated by fitting the data to the following equation:

$$\frac{F}{F_0} = 1 + \frac{\left[\frac{F}{F_0}_{max} - 1\right] X}{K_d + X} \tag{1}$$

where F_0 is the basal fluorescence of a BC-Pγ mutant, F is the fluorescence after additions of tPDE, $F/F_{0\,max}$ is the maximal relative increase of fluorescence, and X is the concentration of free tPDE.

Unlabeled Pγ mutants or competitive inhibitors of PDE such as zaprinast can be used to compete with BC-Pγ mutants for the binding to P$\alpha\beta$. This competition leads to dissociation of the BC-Pγ mutant from P$\alpha\beta$ and a decrease in the BC fluorescence. The IC$_{50}$ values can be calculated by fitting the data to the one site competition equation with variable slope:

$$\frac{F}{F_0} = 1 + \frac{\frac{F}{F_0}_{max} - 1}{1 + 10^{[X - \log IC_{50}]H}} \tag{2}$$

where X is the concentration of the competing ligand (Pγ mutant, zaprinast, cGMP analogs) and H is the Hill slope. Fitting of the data is performed

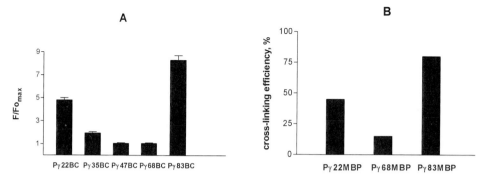

FIG. 2. Correlation of fluorescence changes with the Pγ/Pαβ cross-linking yields for labeled Pγ mutants. (A) Maximal fluorescence increases (F/F_o) of the BC-labeled Pγ mutants on binding to Pαβ (emission 495 nm, excitation 445 nm). (B) Typical yields of Pγ/Pαβ cross-linked products seen for the MBP-labeled Pγ mutants.

with nonlinear least squares criteria using GraphPad Prizm software. The dissociation constants ($K_{1/2}$) for the ligand/Pαβ interaction can be calculated from the IC_{50} values using the Cheng and Prusoff equation[21] describing competitive displacement, $K_{1/2} = IC_{50}/(1 + [BC-P\gamma m]/K_d)$, where IC_{50} is the concentration of ligand that reduces the relative fluorescence increase by 50% [from Eq. (2)], [BC-Pγm] is the total concentration of BC-labeled Pγ mutant, and K_d is the dissociation constant for the BC-Pγm/Pαβ complex [from Eq. (1)]. If the K_d and $K_{1/2}$ values are comparable to the total concentration of Pαβ in the assay, then equations derived by Linden[22] are appropriate for the calculation of $K_{1/2}$.

Analysis of Binding of BC-Labeled Pγ Mutants to Pαβ

Addition of tPDE (Pαβ) to Pγ2BC, Pγ47BC, and Pγ68BC does not alter their fluorescence.[15] Fluorescence of Pγ35BC is moderately (1.9-fold) enhanced on binding to Pαβ (Fig. 2A). Pγ22BC and Pγ83BC display large increases in probe fluorescence on binding to Pαβ.[15] Fluorescence of Pγ22BC and Pγ83BC is maximally enhanced by 4.8-fold and 8.3-fold, respectively (Fig. 2A). These data are consistent with existing biochemical evidence[4-8] and indicate that the C terminus of Pγ and the N-terminal portion of the polycationic region represent major points of contact between Pγ and Pαβ. The N-terminal region of Pγ and the region Pγ-47–68 are not intimately involved in the Pγ/Pαβ interaction. An agreement between data

[21] Y.-C. Cheng and W. H. Prusoff, *Biochem. Pharmacol.* **22**, 3099 (1973).
[22] J. Linden, *J. Cyc. Nuc. Res.* **8**, 163 (1982).

obtained from the fluorescence assay and the existing biochemical evidence on the Pγ/Pαβ interface suggests that the "scanning" probe approach represents a useful tool in identification of protein–protein interaction sites. Furthermore, the fluorescence assay of interaction between Pγ83BC and Pαβ has offered critical insights into the mechanism of Pαβ inhibition by Pγ subunits.[10] This assay reports binding of the Pγ C terminus to Pαβ that leads to PDE inhibition. Pγ83BC binds to Pαβ ($K_d \sim 4$ nM) less tightly than Pγ ($K_d < 100$ nM), which permits competition between Pγ83BC and other Pαβ ligands that have moderate affinity for the catalytic subunits. Competition experiments between Pγ83BC and zaprinast or cGMP analogs have demonstrated that Pγ inhibits Pαβ by blocking its catalytic site.[10]

Labeling of Pγ Mutants with Photoreactive Cross-Linking Probe, MBP

4-(N-Maleimido)benzophenone (MBP) is chosen as the photoreactive cross-linker. This probe is excited by UV light to a triplet biradical state. In the excited state, MBP is highly reactive toward tertiary C–H bonds and unreactive toward water.[23] Therefore, the MBP-labeled protein is often capable of producing high-yield cross-linking. The labeling of Pγ and mutants at a Cys residue with MBP is more efficient in the presence of an organic solvent such as N',N-dimethylformamide or acetonitrile. However, high concentrations of solvent may lead to reduced selectivity of MBP toward Cys. Acetonitrile (35–45% v/v) can be added to Pγ mutant (200–300 μM) in 20 mM HEPES (pH 7.5) buffer. Alternatively, a fraction, which contains mutant Pγ, after RP-HPLC purification on a C$_4$ column can be utilized immediately after chromatography. The pH value in the fraction is adjusted to 7.5 using 200 mM HEPES buffer (pH 8.4). Next, MBP is added from a 10-mM stock solution in acetonitrile to a final concentration of 500 μM. The reaction is allowed to proceed for 20 min at 25°, and is terminated with addition of 5 mM 2-mercaptoethanol. The MBP-labeled Pγ mutants are separated from free MBP on a PD-10 column (Pharmacia) equilibrated with 20 mM HEPES buffer (pH 7.5) containing 100 mM NaCl. All operations are carried out in dim light, and all fractions are collected into light-protected tubes. Ordinarily, the efficiency of the MBP-labeling is better than 70–75% based on the absorption spectra of the unlabeled and labeled proteins. The concentrations of the MBP-labeled proteins are determined using ε_{260} of 23,000 for MBP.

---

[23] G. Dorman and G. D. Prestwich, *Biochemistry* **33**, 5661 (1994).

Cross-Linking of Pγ or Its Mutants to Pαβ

For cross-linking experiments, trypsinized PDE (Pαβ) is mixed at a final concentration of 2 μM with 10 μM MBP-labeled Pγ or mutant in a polypropylene microcentrifuge tube and irradiated for 4 min at a distance of 4 cm with a Transilluminator UV lamp (UVP, Inc., Upland, CA). The polypropylene tube cuts off short-wavelength UV light (<300 nm) thus preventing protein damage. The longer wavelength UV light (300–350 nm), which is necessary for the excitation of benzophenone derivatives, passes through. Following photolysis, the cross-linked products are analyzed by sodium dodecyl sulfate–polyacrylamide gel electrophoresis (SDS–PAGE). Pαβ cross-linked with Pγ can be further purified by ion-exchange HPLC on a MonoQ HR 5/5 column (Pharmacia).[9] This purification step removes uncross-linked Pγ mutant and partially separates the Pγ/Pαβ cross-linked products from remaining free Pαβ.

Analysis of Pγ/Pαβ Interaction Using Photo-Cross-Linking

Pγ and its mutants with a Cys residue at different positions have been labeled with MBP and cross-linked to Pαβ. An interesting correlation is observed between the fluorescence change of BC-labeled mutant on binding to Pαβ and the yields of cross-linking. The Pγ68BC fluorescence is unchanged in the complex with Pαβ, and the cross-linking yields for Pγ68MBP do not exceed 15–20% (Fig. 2).[9] The two BC-labeled Pγ mutants, Pγ22BC and Pγ83BC, exhibit a large increase in fluorescence when bound to Pαβ,[10,15] and when labeled with MBP, efficiently form cross-linked products with the catalytic subunits.[9,24] Yields of 45 and 80% have been achieved for cross-linking of Pγ22BC and Pγ83BC to Pαβ, respectively (Fig. 2). The addition of unlabeled Pγ blocks this cross-linking. The cross-linking data suggest that an increase in fluorescence of a fluorescently labeled Pγ mutant is indicative of the potential efficiency of cross-linking between Pαβ and the Pγ mutant labeled with a photoprobe.

The specific, high yield of cross-linking between the Pγ83MBP and Pαβ permitted identification of the Pαβ site that binds the Pγ C terminus.[9] The site is within the PDE catalytic domain and corresponds to residues 751–763 of Pα. This sequence is unique for photoreceptor PDEs and is adjacent to the NKXD motif, a consensus sequence for binding the GTP-guanine ring in G proteins, and thus a potential binding site for cGMP. Combined with

---

[24] A. E. Granovsky and N. O. Artemyev, unpublished results (1998).

the evidence from the fluorescence assays, which indicate that Pγ blocks cGMP binding to the PDE catalytic site, the cross-linking results favor participation of the NKXD motif in the interaction with cGMP.

Acknowledgments

Our research is supported by the National Institutes of Health grant EY-10843 and the American Heart Association Grant-in-Aid 9750334N.

[43] Kinetics and Regulation of cGMP Binding to Noncatalytic Binding Sites on Photoreceptor Phosphodiesterase

By RICK H. COTE

Introduction

The photoreceptor phosphodiesterase (PDE; EC 3.1.4.35) is the central effector enzyme of the phototransduction pathway in the outer segments of rod and cone photoreceptors. Of the 10 families of phosphodiesterases discovered to date, the photoreceptor enzyme (classified as PDE6) is the only one known to be regulated by interaction with a heterotrimeric G protein, transducin. Activation of transducin following photolysis of the visual pigment stimulates cyclic nucleotide hydrolysis at the catalytic sites of PDE, resulting in a subsecond decrease in cytoplasmic cGMP levels in the outer segment. The rapid drop in cGMP levels during visual excitation induces dissociation of cGMP from allosteric binding sites on the cGMP-gated ion channel in the plasma membrane of the cell, leading to closure of the channels and hyperpolarization of the membrane.[1-5]

The inactivation of PDE must be as precisely controlled as its activation in order for rod and cone photoreceptors to respond rapidly to changes in light stimulation. One way of controlling PDE inactivation is by regulating the extent and lifetime of activation of the transducin α_t–GTP subunit that binds to and activates PDE. In addition to upstream mechanisms that control the rate and extent of formation of α_t–GTP, it is now appreciated

---

[1] C. Pfister, N. Bennett, F. Bruckert, P. Catty, A. Clerc, F. Pagès, and P. Deterre, *Cell. Signal.* **5**, 235 (1993).
[2] E. N. Pugh, Jr., and T. D. Lamb, *Biochim. Biophys. Acta* **1141**, 111 (1993).
[3] K.-W. Yau, *Invest. Ophthalmol. Vis. Sci.* **35**, 9 (1994).
[4] M. D. Bownds and V. Y. Arshavsky, *Behav. Brain Sci.* **18**, 415 (1995).
[5] E. J. M. Helmreich and K. P. Hofmann, *Biochim. Biophys. Acta* **1286**, 285 (1996).

that the inhibitory γ subunit of PDE (Pγ) can act in concert with a regulator of G-protein signaling (RGS), RGS9, to shorten the lifetime of α_t–GTP by accelerating its intrinsic GTPase activity.[6–8] Furthermore, other processes (e.g., posttranslational modifications of Pγ[9] or changes in membrane localization of PDE[10]) may also act to reduce the extent and lifetime of PDE activation during the recovery from a light stimulus and/or during light adaptation.

The rod PDE holoenzyme consists of a catalytic dimer (P$\alpha\beta$) to which the inhibitory Pγ subunits bind. In addition to containing the catalytic domain where cyclic nucleotides are hydrolyzed, the α and β subunits also contain high-affinity noncatalytic cGMP binding sites.[11] The occurrence of noncatalytic cGMP binding sites in photoreceptor PDE is a feature shared with other PDE isoforms. For the case of PDE2, cGMP binding to these noncatalytic sites allosterically stimulates catalysis at the active site of the enzyme.[12] For PDE5 and photoreceptor PDE, the role of these noncatalytic sites is less clear. However, it has been demonstrated that occupancy of the noncatalytic sites with cGMP increases the binding affinity of Pγ for P$\alpha\beta$,[13] and conversely, binding of Pγ to P$\alpha\beta$ confers high-affinity binding of cGMP to the noncatalytic sites.[14–16] This reciprocal relationship between Pγ binding and cGMP binding may be important for Pγ to function with RGS9 as an accelerator of transducin GTPase. The noncatalytic sites may also serve additional allosteric functions by altering the conformation of the α and β tertiary and quaternary structure, thereby changing the interaction of the P$\alpha\beta$ dimer with (1) the transducin α_t–GTP subunit, (2) the two distinct domains of interaction of Pγ with P$\alpha\beta$, and/or (3) the δ subunit that has been implicated in promoting release of PDE from its membrane-associated state.

[6] C. W. Cowan, R. N. Fariss, I. Sokal, K. Palczewski, and T. G. Wensel, *Proc. Natl. Acad. Sci. U.S.A.* **95,** 5351 (1998).
[7] W. He, C. W. Cowan, and T. G. Wensel, *Neuron* **20,** 95 (1998).
[8] S. H. Tsang, M. E. Burns, P. D. Calvert, P. Gouras, D. A. Baylor, S. P. Goff, and V. Y. Arshavsky, *Science* **282,** 117 (1998).
[9] S. Tsuboi, H. Matsumoto, K. W. Jackson, K. Tsujimoto, T. Williams, and A. Yamazaki, *J. Biol. Chem.* **269,** 15016 (1994).
[10] S. K. Florio, R. K. Prusti, and J. A. Beavo, *J. Biol. Chem.* **271,** 1 (1996).
[11] A. Yamazaki, I. Sen, M. W. Bitensky, J. E. Casnellie, and P. Greengard, *J. Biol. Chem.* **255,** 11619 (1980).
[12] T. J. Martins, M. C. Mumby, and J. A. Beavo, *J. Biol. Chem.* **257,** 1973 (1982).
[13] V. Y. Arshavsky, C. L. Dumke, and M. D. Bownds, *J. Biol. Chem.* **267,** 24501 (1992).
[14] A. Yamazaki, F. Bartucci, A. Ting, and M. W. Bitensky, *Proc. Natl. Acad. Sci. U.S.A.* **79,** 3702 (1982).
[15] R. H. Cote, M. D. Bownds, and V. Y. Arshavsky, *Proc. Natl. Acad. Sci. U.S.A.* **91,** 4845 (1994).
[16] A. Yamazaki, V. A. Bondarenko, S. Dua, M. Yamazaki, J. Usukura, and F. Hayashi, *J. Biol. Chem.* **271,** 32495 (1996).

This chapter describes experimental approaches that have been developed to study the regulation of amphibian PDE, with an emphasis on the role of the noncatalytic cGMP binding sites. Other articles in this volume[16a,b] describe procedures for the study of bovine PDE. There are several advantages to working with the amphibian system, including the ability to obtain fully dark-adapted and highly purified rod outer segments (ROS) at defined times of the circadian photoperiod; sufficient quantities of phototransduction proteins for biochemical studies; the wealth of information on the electrophysiologic responses of amphibian rod photoreceptors during excitation, recovery, and adaptation, and the ability to correlate biochemical and electrophysiologic responses to illumination in the same cell preparation.

Isolation of Purified Rod Outer Segments from Frog Retinas

Solutions

Frog Ringer's solution consists of (in mM): 105 NaCl, 2.0 MgCl$_2$, 2.0 KCl, 1.0 CaCl$_2$, 10 HEPES (hemisodium salt), pH 7.5, 232 mOs. Just before use, 5.0 mM glucose is added from a 100× stock.

Percoll (Sigma, St. Louis, MO) is first adjusted to physiologic ionic strength by the addition of 1 volume 10× Ringer's solution to 9 volumes of Percoll, and then dialyzed 3 hr in Spectra/Por tubing (molecular weight cutoff 6000–8000) against 10 volumes of Ringer's solution. The Ringer's solution is then replaced and dialysis repeated overnight at 4°. Dialysis adjusts the osmolality and the pH to that of the Ringer's solution, and removes free polyvinylpyrrolidone in the original Percoll suspension. On addition of 5 mM glucose, this isosmotic Percoll solution (defined as 100% Percoll) is used for subsequent dilutions.

Volumetric dilutions of 100% isosmotic Percoll are made with Ringer's to prepare 5, 30, 42, and 60% Percoll solutions. The density of the diluted Percoll solutions is easily checked by measuring the refractive index (Ringer's $\eta = 1.3340$, 100% isosmotic Percoll $\eta = 1.3510$).

Isolation of Retinas

Bullfrogs (*Rana catesbeiana*) 10–15 cm in length are obtained from Niles Biologicals (Sacramento, CA) at least 2 weeks prior to use in experiments. The animals are housed in stainless steel tanks with continuously flowing 20° water. Inside the light-tight tank is a light controlled by a timer

---

[16a] A. E. Granovsky, K. G. Musakov, and N. O. Artemyev, *Methods Enzymol.* **315** [40], 1999 (this volume).

[16b] T. A. Cook and J. A. Beavo, *Methods Enzymol.* **315** [40], 1999 (this volume).

to permit a 12-hr light–12-hr dark daily cycle. This serves to entrain the circadian oscillations known to occur in the retina and also to define the diurnal disk membrane renewal process.[17,18] During the "daytime," the illumination is controlled in 15-min periods: 5 min of dark, 5 min of continuous light, and 5 min of 1-Hz flashing light. The animals are fed twice weekly 3 g of pureed commercial dog food supplemented with 1% (w/w) AIN-76 Vitamin Mixture (Purina Test Diets, St. Louis, MO) from which the water-soluble vitamins are omitted.

Dissection of frog retina is performed 1–6 hr before the onset of "daytime." All operations are performed in a darkroom using infrared illumination (Kodak, Rochester, NY, #11 filters) and a Model 6100M infrared-imaging converter (Electrophysics, Nutley, NJ). The enucleated eye is opened by a shallow incision around the circumference of the eyeball, just anterior to the sclera. The cornea, lens, and attached retina are lifted out of the eyecup with forceps, and the retina is then teased from its points of attachment and placed in ~0.7 ml 5% Percoll–Ringer's solution in a glass depression slide. This approach minimizes contamination of the retina with the pigment epithelium layer.

Purification of Frog Rod Outer Segments

The two isolated retinas are held by their edges with a forceps and gently shaken together in 0.8 ml of 5% Percoll in a 1.5-ml siliconized microcentrifuge tube. As the solution becomes cloudy with dissociated photoreceptors, the retinas are transferred to another tube with fresh 5% Percoll, and the agitation of the retinas is gradually increased. This process is repeated three or four times until the yield of dissociated cells declines and the retinas begin to fall apart. Finally, the resulting pieces of retina are vigorously drawn up and expelled into 0.8 ml of 5% Percoll using a wide orifice 1000-μl micropipette tip to further disrupt the retinal pieces.

Once the visible particles in each of the tubes have settled, the suspension of ROS is applied to the centrifuge tube containing the discontinuous Percoll gradient. The particles and retinal pieces from each tube are pooled and resuspended in 5% Percoll. The retinal particles are further disrupted by gentle vortexing or pipetting, and the supernatant suspension is added to the gradient tube. This procedure is repeated until the supernatant solution does not appear cloudy.

Discontinuous Percoll density gradient centrifugation[19,20] permits the separation of ROS from other constituents of the crude suspensions ob-

[17] S. Basinger, R. Hoffman, and M. Matthes, *Science* **194,** 1074 (1976).
[18] G. M. Cahill and J. C. Besharse, *Neuron* **10,** 573 (1993).
[19] R. H. Cote, M. S. Biernbaum, G. D. Nicol, and M. D. Bownds, *J. Biol. Chem.* **259,** 9635 (1984).
[20] M. S. Biernbaum and M. D. Bownds, *J. Gen. Physiol.* **85,** 83 (1985).

tained from shaking and disrupting the retinas. As a density medium for amphibian ROS purification, Percoll offers several advantages over sucrose[21]: the isosmotic Percoll medium permits purification of osmotically intact cells; lower levels of contamination by other retinal neurons are seen; and an ~10-fold faster separation time is possible when preformed gradients are used.

Discontinuous gradients are typically prepared by carefully layering 60% Percoll, then 30% Percoll, and finally 0.5 ml of 5% Percoll in a 12- to 15-ml polycarbonate tube so that a sharp interface exists between each layer. The distance between the 5/30% and 30/60% Percoll interfaces should be ~1 cm for optimal separation and recovery of the purified ROS band. The ROS suspensions are layered on top of the 5% Percoll cushion, and centrifuged (8000–12,000g for 10 min at 4°) in a fixed-angle rotor. Following centrifugation, the band at the 30/60% Percoll interface contains a mixture of osmotically intact and "leaky" ROS. Because frog PDE is a peripheral membrane protein and does not dissociate from the disk membrane in Ringer's solution, the yield of enzyme is enhanced by including leaky ROS. In instances where osmotically intact ROS are required, an additional 42% Percoll layer ($\eta = 1.3413$) can be added to the discontinuous Percoll gradient to separate leaky ROS (30/42% Percoll interface) from intact ROS (42/60% Percoll interface). In either instance, microscopic examination of the bands collected with a 15-gauge syringe needle generally show <1% contamination of the purified ROS suspension with spherical cells, cone photoreceptors, or visible cellular debris. There is significant day-to-day variation, however, in the fraction of ROS that retain the ellipsoid portion of the inner segment; typically ≤30% of the ROS are attached to the ellipsoid region. (Methods are available to enrich for frog ROS still attached to the inner segment.[20,22])

The Percoll present in the ROS suspension is removed by diluting the Percoll concentration to ≤20% with Ringer's, gently sedimenting the ROS, and discarding the Percoll-containing supernatant. The ROS pellet is typically resuspended in a pseudo-intracellular medium or in a Tris buffer for subsequent use.

Gaining Access to the Cytosol

Several methods can be used to disrupt the plasma membrane of Percoll-purified ROS to permit biochemical assays, including sonication, freeze–thawing, forceful passage through a narrow-gauge syringe needle, electro-

---

[21] A. Yamazaki, N. Miki, and M. W. Bitensky, *Methods Enzymol.* **81**, 526 (1982).
[22] R. H. Cote, G. D. Nicol, S. A. Burke, and M. D. Bownds, *J. Biol. Chem.* **264**, 15384 (1989).

permeabilization, and homogenization.[21,23–25] The most consistent and reliable results in our laboratory have been obtained using homogenization with a Potter–Elvehjem tissue grinder closely following the procedure of Dumke et al.[25] The resuspended ROS are placed in an ice-cold, 2-ml glass tissue grinder, and a custom-made nylon pestle [attached to a Talboy Model 134-1 stirrer (VWR Scientific Products, West Chester, PA)] is used to homogenize the ROS; the stirrer is operated at 50% maximum speed. The number of strokes needed to completely disrupt cellular morphology at the light microscopic level must be empirically determined for each mortar and pestle, and the suspension is kept in an ice water bath throughout to minimize heating. The commercially available Teflon pestles do not fit tightly enough to completely homogenize ROS, thus necessitating use of a custom-made nylon pestle. The complications that result if stacks of disk membranes remain in the homogenate[25] require careful attention to the complete disruption of the ROS membrane system.

Following homogenization, the rhodopsin concentration is determined by difference spectroscopy[26] to provide an initial estimate of the PDE concentration. As much as 1 mg of rhodopsin can be recovered from one frog when "leaky" and intact ROS are combined. The homogenate is incubated in the dark at 22° for 30 min to degrade endogenous cGMP nucleotides that would interfere with measurements of [³H]cGMP binding, as well as GTP that could otherwise activate transducin on light exposure.

Determination of PDE Holoenzyme and Pγ Concentration in Frog ROS

Use of ROS Homogenates as Source of Frog PDE

Because frog photoreceptor PDE is the only protein in amphibian ROS that binds to cGMP with high affinity ($K_D < 1~\mu M$),[11,15,23] we are able to carry out many of our experiments with ROS homogenates. Likewise, because no other PDE isoforms have been detected in ROS, hydrolytic activity determinations can be made using ROS homogenates without concern for contaminating activities from other phosphodiesterases. Using unfractionated ROS homogenates allows PDE regulation to be stud-

---

[23] R. H. Cote and M. A. Brunnock, *J. Biol. Chem.* **268**, 17190 (1993).
[24] V. J. Coccia and R. H. Cote, *J. Gen. Physiol.* **103**, 67 (1994).
[25] C. L. Dumke, V. Y. Arshavsky, P. D. Calvert, M. D. Bownds, and E. N. Pugh, Jr., *J. Gen. Physiol.* **103**, 1071 (1994).
[26] D. Bownds, A. Gordon-Walker, A. C. Gaide Huguenin, and W. Robinson, *J. Gen. Physiol.* **58**, 225 (1971).

ied under relatively physiologic conditions with the full complement of phototransduction proteins and under conditions where light activation of nucleotide-supplemented ROS homogenates approached the light sensitivity of the excitation pathway *in vivo*.

Purification to homogeneity of the amphibian photoreceptor PDE holoenzyme is possible[13,21] but not routinely undertaken in our laboratory, primarily because of the instability of the purified enzyme once removed from the membrane. This contrasts with bovine rod PDE, which has been successfully purified and stored for extended periods.[27]

Determination of PDE Content of ROS Homogenates

An initial estimate of the PDE concentration that is accurate to within ±20% can be obtained by measuring the rhodopsin concentration of the ROS homogenate, along with a knowledge of the molar ratio of rhodopsin to PDE (270:1).[25] Frog-to-frog variations in the rhodopsin:PDE molar ratio are likely to be the major factor limiting the accuracy of this estimate. The PDE concentration in ROS homogenates can also be estimated by measuring the maximum amount of [^{3}H]cGMP bound (B_{max}) when the cGMP concentration is sufficient to saturate all of the high-affinity binding sites, and using a binding stoichiometry of 2 mol cGMP bound per mole of PDE holoenzyme.[15,23,28] A third approach is to measure the rate of cGMP hydrolysis of transducin-activated PDE in a ROS homogenate (>4 μM rhodopsin, with 10 mM cGMP), and to calculate the enzyme concentration using the published turnover number, k_{cat} = 4400 cGMP per PDE holoenzyme per second.[25] Finally, the PDE concentration in ROS can be estimated by first proteolytically cleaving the Pγ subunits to activate the enzyme,[29] and then titrating the amount of purified, recombinant Pγ needed to stoichiometrically restore full inhibition of the enzyme (see later section). The latter two methods that rely on activity measurements correlate very well (<10% variation) with estimates of PDE concentration from [^{3}H]cGMP B_{max} binding value.

Determination of Pγ Concentration in ROS Homogenates

Because the Pγ subunit is central to the regulation of PDE activity and cGMP binding, it is important to estimate its concentration accurately. The expression and purification of the recombinant bovine rod Pγ subunit[30] is described in detail elsewhere in this volume.[16a] Following purification of

[27] W. Baehr, M. J. Devlin, and M. L. Applebury, *J. Biol. Chem.* **254,** 11669 (1979).
[28] P. G. Gillespie and J. A. Beavo, *Proc. Natl. Acad. Sci. U.S.A.* **86,** 4311 (1989).
[29] J. B. Hurley and L. Stryer, *J. Biol. Chem.* **257,** 11094 (1982).
[30] R. L. Brown and L. Stryer, *Proc. Natl. Acad. Sci. U.S.A.* **86,** 4922 (1989).

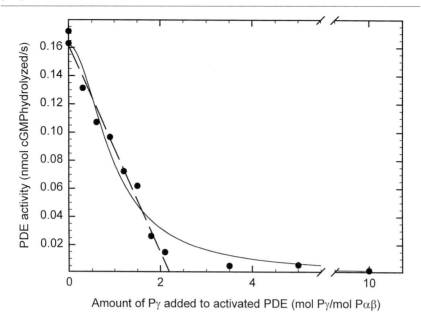

FIG. 1. Titration of trypsinized frog PDE with Pγ. Trypsinized frog PDE (Pαβ concentration, 5 nM) was incubated with the indicated concentrations of recombinant bovine Pγ for 6 min at room temperature. To determine the extent of inhibition of PDE activity, 10 mM cGMP (final concentration) was added, and portions quenched at three time points. The extent of cGMP hydrolysis was determined with the colorimetric phosphate assay. The curve is the fit to a logistic equation where the IC_{50} value is 0.96 Pγ/Pαβ ($r^2 = 0.972$). The dotted line is the regression line ($r^2 = 0.997$) for data points up to only 2.1 Pγ/Pαβ and intersects the x axis at 2.18 Pγ/Pαβ. The results demonstrate that the Pγ is 90–95% active.

>95% pure recombinant Pγ by reversed-phase high-performance liquid chromatography (HPLC), spectroscopic measurements in 45% acetonitrile, 0.1% trifluoroacetic acid in conjunction with amino acid analyses of identical samples provided an experimental determination of the extinction coefficient for Pγ of $\varepsilon_{277} = 7550$ OD · M^{-1} under these conditions. We have also determined the biological activity of purified Pγ by testing its ability to inhibit activated frog PDE under conditions where there is a linear relationship between Pγ added and inhibition (i.e., [PDE] ≫ K_D for Pγ binding).[31] We find that the addition of 2.0 moles of Pγ per mole of PDE will titrate >90% of the hydrolytic activity of activated frog PDE (Fig. 1), indicating that the purified Pγ is close to 100% active.

We have also used purified, recombinant Pγ as a standard for quantitative immunoblot analysis of the total Pγ content in frog ROS homogenates.

[31] S. E. Hamilton, R. K. Prusti, J. K. Bentley, J. A. Beavo, and J. B. Hurley, *FEBS Lett.* **318**, 157 (1993).

Frog ROS homogenates prepared as described in the previous section were centrifuged in a Beckman Airfuge (5 min at 130,000g at room temperature) to sediment ROS membranes from cytosolic proteins. Equivalent portions of the original ROS homogenate, membrane, and supernatant fractions, as well as known amounts of Pγ were subjected to sodium dodecyl sulfate–polyacrylamide gel electrophoresis (SDS–PAGE). After electrophoretic transfer to a nitrocellulose membrane, the blot is probed with a rabbit polyclonal antipeptide antibody, UNH9710 (amino acids 63–87 of bovine Pγ). Detection of anti-Pγ antibody is performed using a horseradish peroxidase-linked secondary antibody directed to the primary antibody, followed by luminescent visualization of the antibody complex.[32] As shown in Fig. 2, all of the Pγ originally present in ROS homogenates cosediments with PDE in the ROS membrane fraction, with no detectable Pγ immunoreactivity present in the supernatant fraction. The molar stoichiometry of Pγ to Pαβ has been determined to be 1.8 ± 0.2 mol Pγ per mole of Pαβ ($n = 14$). These results demonstrate that frog PDE has the same Pγ-subunit stoichiometry as had been determined for bovine rod PDE and that no excess Pγ is present in the cytoplasm of frog ROS.

Preparing Various Forms of Activated Frog PDE

Solutions

Pseudo-intracellular medium is a buffer that partially mimics the composition of the ROS cytosol and is routinely used for resuspending ROS pellets for homogenization. It consists of (in mM): 77 KCl, 35 NaCl, 2.0 MgCl$_2$, 1.0 CaCl$_2$, 1.18 EGTA ([Ca^{2+}$_{free}$] = 240 nM), 10 HEPES, pH 7.5. Just before use, the solution is supplemented with final concentrations of 1.0 mM dithiothreitol (DTT), 0.5 μg/ml leupeptin, 0.2 mM Pefabloc, and 0.7 μg/ml pepstatin.

As an alternative, we have also used a Tris-based buffer consisting of 100 mM Tris, 10 mM MgCl$_2$, 0.5 mM EDTA. This solution is supplemented on the day of the experiment with 0.5 mg/ml bovine serum albumin (BSA), as well as DTT and protease inhibitors (as used earlier).

Nonactivated PDE

Following homogenization and nucleotide depletion of the ROS homogenate, the preparation can be exposed to room light for the subsequent

---

[32] S. Gallagher, in "Current Protocols in Protein Science" (J. E. Coligan, B. M. Dunn, H. L. Ploegh, D. W. Speicher, and P. T. Wingfield, eds.), pp. 10.10.1–10.10.12. J. Wiley & Sons, New York, 1998.

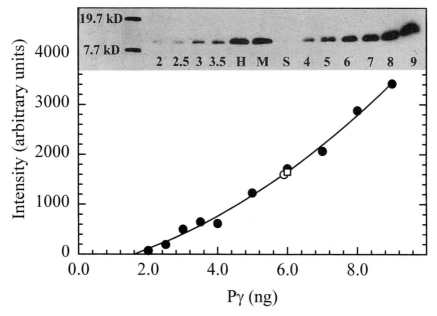

FIG. 2. Determination of the Pγ content in frog ROS. Samples of frog ROS homogenate (H, o), as well as an equivalent amount of fractionated membrane (M, □) and supernatant (S) components, were loaded on a 15% SDS–PAGE gel, along with known concentrations of Pγ (numbers below each standard lane indicate the nanograms of Pγ loaded.). The membrane and homogenate samples each contained 3.0 pmol nonactivated PDE holoenzyme, whereas the supernatant sample did not contain any detectable PDE (as judged by an activity assay). The Pγ content of unknown samples was determined by comparing to the quadratic curve fit ($r = 0.997$) of the known Pγ samples (●). In this experiment, the homogenate and membrane samples contained 5.9 and 6.0 ng Pγ, respectively, whereas the amount in the supernatant was below the detection limits of the assay (≤2 ng Pγ). The data are representative of one of four similar experiments.

experiment without concern for activation of PDE. When adjusted to rhodopsin concentration of 4 μM, this "nonactivated" PDE preparation binds 1.8–2 mol cGMP per mole of PDE holoenzyme, and has an enzyme activity that is ~2% of the fully transducin-activated rate. When stored at 4°, the cGMP binding stoichiometry gradually declines over several days, presumably due to PDE denaturation and/or loss of Pγ by proteolysis.

Transducin-Activated PDE

To prepare transducin-activated PDE, the poorly hydrolyzable analog of GTP, GTPγS, is added to the ROS homogenate in a two- to fivefold molar excess over the transducin concentration (assuming a molar ratio of 1:10 transducin:rhodopsin) in order to activate all transducin. The rhodopsin concentration must be kept ≥4 μM in order to prevent dissociation of membrane-bound transducin and a loss of maximal PDE activation.[25]

Extraction of Pγ from PDE to Prepare Membrane-Associated Pαβ Dimers

ROS membranes containing PDE depleted of most (~70%) of its bound Pγ can be prepared by extracting Pγ as a complex with the activated α_t–GTP subunit of transducin.[15,33] Nucleotide-depleted ROS homogenates (20 μM rhodopsin) in pseudo-intracellular medium are first exposed to light at 4°, and then incubated with 10 μM guanosine 5'-triphosphate (GTP) for 0.5 min. On diluting the preparation 10-fold with GTP-containing buffer, the homogenate is immediately centrifuged at 40,000g at 4° for 30 min. The membrane pellet is washed to remove residual GTP, and the ROS membranes are resuspended and homogenized in pseudo-intracellular medium (lacking GTP). PDE activity measurements and quantitative immunoblot analysis of the Pγ content of these Pγ-depleted membranes indicate that 50–80% of the total Pγ is removed by this treatment in any given experiment. Repeating the extraction procedure with additional GTP has been found to be mostly ineffective, although changing the ionic strength during Pγ extraction has been reported to affect Pγ release from ROS membranes.[34] The presence of a significant fraction of Pαβγ and/or Pαβγ$_2$ complicates the interpretation of experiments using this preparation for studies of PDE regulation by Pγ and its noncatalytic sites.

Limited Proteolysis of PDE with Trypsin to Prepare Pαβ Catalytic Dimers

Brief exposure of PDE holoenzyme to trypsin is known to release the inhibitory constraint of Pγ on frog[35] and bovine[29] PDE. To prepare trypsinized PDE, we use ROS homogenates (adjusted to 40 μM rhodopsin) that are prepared in the Tris homogenization buffer from which protease inhibitors are omitted. TPCK-treated trypsin (100 μg/ml final concentration) is added for 10 min at 4°, then the reaction halted with a sixfold excess

[33] A. Yamazaki, F. Hayashi, M. Tatsumi, M. W. Bitensky, and J. S. George, *J. Biol. Chem.* **265**, 11539 (1990).
[34] A. Yamazaki, M. Yamazaki, V. A. Bondarenko, and H. Matsumoto, *Biochem. Biophys. Res. Commun.* **222**, 488 (1996).
[35] N. Miki, J. M. Baraban, J. J. Keirns, J. J. Boyce, and M. W. Bitensky, *J. Biol. Chem.* **250**, 6320 (1975).

of soybean trypsin inhibitor. As shown in Fig. 1, inhibition of trypsinized frog PDE is nearly complete when 2 mol Pγ per mole of P$\alpha\beta$ is added back to this P$\alpha\beta$ preparation; this suggests that the primary locus of action of trypsin under these conditions is on Pγ, not on the catalytic subunits. Unlike the case of bovine rod PDE where trypsin first releases PDE from its membrane-associated state prior to full degradation of the Pγ subunits,[36] the frog enzyme shows a similar time course for cleavage of the C-terminal site of membrane attachment and hydrolysis of the Pγ subunit. Although trypsin proteolysis is much more efficient in removing bound Pγ than the Pγ extraction from membrane-associated PDE described in the previous section, there remains the concern that some aspects of the structure and/or function of the P$\alpha\beta$ dimer may be altered by this treatment.

Assays for PDE Hydrolytic Activity

The high catalytic constant for activated photoreceptor PDE combined with a relatively large value of the K_m for substrate[2] permit the use of less sensitive assays for cyclic nucleotide hydrolysis in addition to the standard radiotracer assay.[37] One method relies on the measurement of protons generated as a product of the PDE catalytic mechanism; in a minimally pH-buffered solution, changes in pH recorded with a pH microelectrode provide a continuous assay of the rate of cyclic nucleotide hydrolysis. This method has been described in a previous volume of this series,[38] and critical factors for use of this assay have been discussed thoroughly.[39] One little appreciated advantage of continuous monitoring of PDE activity is that the kinetic parameters can be determined from both initial velocities and from the complete reaction progress curve.[40] In our laboratory we routinely use both the radiotracer assay and a colorimetric assay to quantify PDE activity.

PDE Activity Assay Using Radiolabeled Substrates

Solutions and Materials

Buffer A: 20 mM Tris, pH 6.8, at room temperature.
Buffer B: buffer A + 0.5 M NaCl.
0.1 M HCl.
0.1 M Trizma base will result in a final pH of 7.5 ± 0.5 when mixed with an equal volume of acid.

[36] T. G. Wensel and L. Stryer, *Prot. Struct. Funct. Genet.* **1**, 90 (1986).
[37] R. L. Kincaid and V. C. Manganiello, *Methods Enzymol.* **159**, 457 (1988).
[38] P. A. Liebman and A. T. Evanczuk, *Methods Enzymol.* **81**, 532 (1982).
[39] M. W. Kaplan and K. Palczewski, *Meth. Neurosci.* **15**, 205 (1993).
[40] A. E. I. Barkdoll, E. N. Pugh, Jr., and A. Sitaramayya, *J. Neurochem.* **50**, 839 (1988).

2.5 mg/ml *Crotalus atrox* snake venom (Sigma, St. Louis, MO) in water. Addition of 100 μM isobutylmethyl xanthine (IBMX) is sometimes warranted with cAMP as substrate if the venom preparation has cAMP PDE activity.

Polystyrene chromatography columns (Evergreen Scientific, Los Angeles, CA).

DEAE-Sephadex A-25 resin is initially swelled overnight in a 10-fold excess (w/v) of buffer B. Decant the liquid to remove fines, and add sufficient buffer B to give 6 parts settled resin to 10 parts resin plus buffer. Add 1 ml of resuspended resin to the column to yield 0.6 ml settled resin. Equilibrate columns with 8 ml of buffer B, then 8 ml buffer A before use.

Activity Assay

The reaction is started by mixing 9 volumes of the PDE sample with a 10× cyclic nucleotide reaction mixture containing ~10^5 dpm of [^{3}H]cGMP or [^{3}H]cAMP per sample plus the appropriate amount of unlabeled nucleotide to give the desired concentration. In addition to the experimental samples, control samples should also be run (i.e., blanks lacking enzyme and "total hydrolysis" samples in which all substrate is converted to product). Portions ($\leq$50 μl) are withdrawn and quenched in 100 μl of 0.1 M HCl.

The samples are neutralized with 0.1 M Trizma, then the 5'-nucleoside monophosphate product is quantitatively converted to adenosine/guanosine by addition of 25 μg of snake venom, and incubation at 37° for 5–15 min (empirically determined for each batch of venom). The nucleoside is separated from unreacted substrate by passage through the DEAE-Sephadex A-25 column equilibrated in buffer A; cyclic nucleotide is retained, while the nucleoside elutes on washing the column with four 0.5-ml washes of buffer A. The eluate is mixed with 4 ml Ultima Gold XR (Packard, Meriden, CT) and counted in a Packard TR2300 scintillation counter. The columns are regenerated by washing in 8 ml buffer B, followed by storage in buffer A containing 0.1% sodium azide.

The radiotracer assay has unparalleled sensitivity (<0.1 pmol of product can be detected reliably), and is the method of choice for samples containing low enzyme activities or at low substrate concentrations. However, the need for radioactivity and the labor-intensive chromatography step are drawbacks to this method.

Colorimetric Determination of PDE Activity Based on Phosphate Production

This method relies on the stoichiometric production of inorganic phosphate in a coupled enzyme assay of photoreceptor PDE and snake venom

5'-nucleotidase: cyclic nucleotide → 5'-nucleoside monophosphate → nucleoside + inorganic phosphate.[41] Colorimetric quantitation of the inorganic phosphate produced using a 96-well microplate reader permits rapid assay of large numbers of samples. In the following protocol, it is assumed that the PDE reaction is carried out in a separate tube, and portions removed and quenched in acid in the 96-well plate.

Solutions and Materials

0.1 M HCl.
0.5 M Trizma base (tested with HCl to ensure that the neutralized solution has a pH of 7.5 ± 0.5).
2.5 mg/ml snake venom (see earlier section). If the snake venom has significant contamination of phosphate, it can be removed by gel filtration of an initial preparation of 3.75 mg/ml snake venom using a PD-10 column (Pharmacia).
Phosphate standard solution: 1.0 mM KH$_2$PO$_4$.
Molybdate solution: 0.4 N H$_2$SO$_4$, 0.2% (w/v) ammonium molybdate, 2.0% (w/v) sodium dodecyl sulfate (SDS), and 2% (w/v) ascorbic acid (added just before assay).
96-Well plates, untreated (Corning, New York).
Repetitive dispensing pipettes and 8-channel pipettors.
Microplate reader with dual-wavelength option, reading at 700–750 nm with a 450-nm reference wavelength (to reduce light scattering artifacts).

Protocol

Substrate is added to a PDE-containing sample, and ≤20-μl portions are removed at various times and quenched with 50 μl of 0.1 M HCl in a microplate well. The quenched sample should contain between 0.5 and 60 nmol hydrolyzed nucleotide in order to be accurately determined. The acidified samples are neutralized with 0.2 volume of 0.5 M Trizma. (The volumes of sample, acid, and base can be varied, as long as it is verified that the acid quenching is instantaneous and that the addition of base restores the pH to neutrality.) The samples are then treated with 10 μl snake venom per earlier section. Standards are prepared by addition of 1–70 nmol of the 1 mM KH$_2$PO$_4$ solution in water to the same final volume as the experimental samples.

Add 150 μl of the molybdate solution and incubate at 37° for 20 min. The results of the standard phosphate samples are fit to a second-order

[41] P. G. Gillespie and J. A. Beavo, *Mol. Pharmacol.* **36,** 773 (1989).

polynomial equation, and used to interpolate the concentration of phosphate in the unknowns.

Variations

If the PDE activity is substantially less than the 5'-nucleotidase activity in the snake venom preparation, simultaneous incubation of PDE and snake venom is possible.[41] In this instance, the reaction can be quenched with 2% SDS instead of acid, and the neutralization step avoided. The SDS should be omitted from the molybdate solution. Also, verify that the snake venom preparation has undetectable levels of PDE activity, especially if using cAMP as the substrate.

To determine the PDE concentration in a large number of samples (e.g., column chromatography fractions), 10 µl of PDE can be added to wells containing 10 µl of a 5× PDE assay buffer (1× concentrations: 20 mM Tris, pH 7.5, 10 mM MgCl$_2$, 0.5 mg/ml BSA). The samples can be activated by addition of 10 µl of 100 µg/ml TCPK-treated trypsin for a sufficient time to optimally activate the PDE, then treated with 10 µl of 0.6 mg/ml soybean trypsin inhibitor. Then 10 µl of cGMP (10 mM final concentration is typical) is added to all wells, and the reaction quenched with acid. The remaining steps follow the standard protocol.

Comments

The phosphate assay agrees to within 10% with the radiotracer assay described in the previous section. Although the colorimetric assay is more than a thousandfold less sensitive than the radiotracer assay, often the assay conditions and/or amount of enzyme can be adjusted to permit use of this economic, rapid, and nonradioactive procedure.

Membrane Filtration Assays for Measurements of cGMP Binding to Noncatalytic cGMP Binding Sites on PDE

Principles

Membrane filtration techniques for quantitating the binding of cyclic nucleotides to their receptor proteins[42] are widely used for characterizing the regulation of cyclic nucleotide-dependent protein kinases and cGMP-binding phosphodiesterases (i.e., PDE2, PDE5, and PDE6). It is often the method of choice because of its high sensitivity, excellent partitioning of

---

[42] A. G. Gilman, *Proc. Natl. Acad. Sci. U.S.A.* **1,** 305 (1970).

free from bound ligand, and its low extent of nonspecific binding. The major drawbacks to membrane filtration are the nonequilibrium nature of the separation process and its restriction to studying relatively high-affinity ligand–receptor interactions.

To accurately use membrane filtration to estimate the equilibrium binding parameters (K_D, B_{max}) of a ligand for its receptor, several criteria need to be satisfied: (1) the binding reaction must closely approach equilibrium prior to separating free from bound ligand; (2) the separation of bound from free ligand must not perturb the equilibrium during filtration; (3) the extent of nonspecific binding of ligand to the membrane must be determined precisely; and (4) the appropriate binding model must be selected to interpret the binding data.

Determining Approach to Equilibrium

Accurate calculation of the binding parameters demands that the binding reaction be essentially (e.g., >97%) at equilibrium. This can be empirically determined by monitoring the time course of increase in binding. However, because the approach to equilibrium slows down as ligand concentration decreases, the time course should be performed at the lowest ligand concentration to be used. Alternatively the time to attain 97% of equilibrium (T_e) can be shown[43] to depend on the association (k_{+1}) and dissociation (k_{-1}) rate constants:

$$T_e = \frac{3.5}{k_{+1}[L] + k_{-1}} \qquad (1)$$

where [L] is the free ligand concentration. Note that at low ligand concentrations ($k_{+1}[L] \ll k_{-1}$), the approach to equilibrium can be estimated if k_{-1} is known.

Estimating Extent of Dissociation of Bound Ligand during Separation of Free From Bound Ligand

The major constraint limiting the usefulness of membrane filtration is the rate of dissociation of the ligand–receptor complex during the separation procedure. It is the dissociation rate constant that determines how much ligand will dissociate during the time it takes to filter and rinse the receptor–ligand complex.[44] The amount of time needed to complete filtration and

---

[43] G. A. McPherson, in "Receptor Pharmacology and Function" (M. Williams, R. A. Glennon, and P. B. M. W. M. Timmermans, eds.). Marcel Dekker, New York, 1988.
[44] J. P. Bennett, Jr., and H. I. Yamamura, in "Neurotransmitter Receptor Binding," 2nd ed. (H. I. Yamamura, ed.), Vol. 2, p. 61. Raven Press, New York, 1985.

rinsing of a sample with the loss of ≤10% of the bound ligand ($T_{10\%}$) can be calculated from $T_{10\%} = 0.14/k_{-1}$. Later in this section an alternative filtration protocol using ammonium sulfate is presented to specifically address this issue. This concern can also be addressed by using another method (see later section) that does not perturb the equilibrium of receptor–ligand binding.

Quantitating Contribution of Nonspecific Ligand Binding to the Filter Membrane

Usually, membrane filtration techniques offer a high ratio of specific to nonspecific ligand binding. Nonspecific binding is a general term comprised of several components: ligand binding to the filter, incomplete washing of unbound ligand during filtration, and true nonspecific binding of ligand to the sample containing the receptor. The first component can be examined by performing filtration and rinsing of the radiolabeled ligand solution in the absence of the receptor. The efficacy of removing unbound ligand can be assessed and optimized by varying the number of rinses and the volume of the rinse solution. The last component to total nonspecific binding can be judged by incubating the receptor with the labeled ligand to which a large excess of unlabeled ligand (typically, a concentration ≥100 times the K_D) has been added. In our experience, nonspecific binding of radiolabeled cGMP in the presence of a large excess of cold cGMP is identical in the presence or absence of ROS membranes, and never exceeds 1% of the bound cGMP.

Analysis of Ligand Binding Data

Although graphical presentations of transformed data (e.g., Scatchard plot) remain helpful for visual inspection of ligand binding results, graphic approaches to determining binding parameters have major shortcomings and should be avoided.[45] Instead, nonlinear curve-fitting techniques should be employed with exact models of the predicted binding reaction(s). The development of specialized computer programs using iteratively weighted, nonlinear regression analysis to evaluate equilibrium and kinetic radioligand binding data (e.g., LIGAND,[46] KELL[47]) provides several important advantages: no transformation of the raw data (or their associated errors) is introduced; the exact binding model is used; statistical comparison of different models (e.g., one versus two sites, cooperativity, etc.) can be

[45] R. J. Leatherbarrow, *Trends Biochem. Sci.* **15**, 455 (1990).
[46] P. J. Munson, *Methods Enzymol.* **92**, 543 (1983).
[47] G. A. McPherson, *J. Pharmacol. Meth.* **14**, 213 (1985).

carried out to determine the most likely model; the weighting model that is most appropriate for the errors associated with different y-axis values can be selected; and the free ligand concentration is calculated by an iterative procedure that takes into account potential ligand depletion.[46,48] The collection of programs marketed as KELL (Biosoft, Ferguson, MO) are derived from the original LIGAND program and are routinely used in our laboratory for kinetic and equilibrium binding studies of the noncatalytic sites of PDE.

Filter Binding Assay to Detect High-Affinity Noncatalytic cGMP Binding Sites on PDE

The following filter binding protocol is useful for measuring cGMP binding to high-affinity (i.e., $K_D < \mu M$) sites on amphibian PDE to determine the K_D and the maximum binding stoichiometry (B_{max}) under equilibrium conditions, as well as the dynamics of cGMP association and dissociation.

The first consideration in performing a radiolabeled cGMP binding experiment with PDE is to ensure that all endogenous cGMP has been destroyed and all binding sites are unoccupied. Although 30-min incubation of frog ROS homogenates is sufficient to deplete the cGMP,[23] the considerably slower cGMP dissociation rate for bovine rod PDE[28] has until recently hampered studies of the noncatalytic sites on the mammalian rod photoreceptor enzyme.[49]

A second consideration arises from the fact that the ligand for the noncatalytic sites is also the substrate at the active site of PDE, necessitating that enzyme activity be abolished to prevent destruction of cGMP. PDE5-selective inhibitors such as E4021 and zaprinast are far more effective than EDTA in this regard, with K_I values for the activated enzyme of 1.7 and 32 nM for E4021 and zaprinast, respectively.[41,50] The efficacy of inhibition is confirmed by determining the extent of [^{3}H]cGMP hydrolysis under the actual binding assay conditions. Because of the high discrimination of the noncatalytic sites for binding cGMP compared with cGMP analogs or PDE inhibitors,[50,51] competition between cGMP and these compounds for binding to the noncatalytic sites is extremely weak.

[48] E. C. Hulme and N. J. M. Birdsall, in "Receptor-Ligand Interactions: A Practical Approach" (E. C. Hulme, ed.), p. 63. Oxford University Press, Oxford, UK, 1992.
[49] H. Mou, H. J. Grazio, T. A. Cook, J. A. Beavo, and R. H. Cote, *J. Biol. Chem.* **274**, 18813 (1999).
[50] M. R. D'Amours, A. E. Granovsky, N. O. Artemyev, and R. H. Cote, *Mol. Pharmacol.* **55**, 508 (1999).
[51] M. C. Hebert, F. Schwede, B. Jastorff, and R. H. Cote, *J. Biol. Chem.* **273**, 5557 (1998).

Solutions and Materials

> Pseudo-intracellular medium (see previous section).
>
> Wash buffer is identical to the pseudo-intracellular medium, except it lacks DTT and protease inhibitors.
>
> 100 mM zaprinast (Sigma) stock solution is dissolved in 1-methyl-2-pyrrolidinone.
>
> The exact composition of the radiolabeled ligand solution depends on the nature of the experiment. In general, the concentration of cGMP is varied by adding a fixed amount of [^{3}H]cGMP (NEN, Dupont, Boston, MA; ~10^4–10^5 dpm per sample to be filtered) to various amounts of unlabeled cGMP (Sigma). The concentration of unlabeled cGMP is verified by its UV absorption spectrum (ε_{254} = 1.29 × 10^4 OD · M^{-1} at pH 7). Zaprinast is typically added to a final concentration of 100 μM.
>
> Filter disks: 25-mm diameter, 0.45-μm pore size, mixed cellulose esters MF-Millipore (Bedford, MA) membrane filters.
>
> The vacuum filtration device (Hoefer 10-place manifold; Amersham/Pharmacia, Piscataway, NJ) is attached to a vacuum source that permits 1 ml of standard wash buffer to be filtered in ~1 sec.
>
> Ultima Gold (Packard) scintillation fluid is used to measure radioactivity eluted from the filters.

General Procedure for Filter Binding Assay

Following nucleotide depletion of the PDE-containing ROS homogenates (4 μM rhodopsin) for 30 min at room temperature in pseudo-intracellular medium, the cGMP binding reaction is initiated by addition of 10-fold concentrated [^{3}H]cGMP solution containing zaprinast. Immediately before filtration of the ROS sample, the membrane filter is prewet with 1 ml ice-cold wash buffer. ROS (25-μl portions) are directly pipetted to the filter, and immediately rinsed with three 1-ml portions of wash buffer. This process should be completed within ≤4 sec.

It is not advised to add the PDE sample to a tube containing wash buffer and then filter the diluted sample. Diluting the PDE will alter the equilibrium binding of Pγ to P$\alpha\beta$ and thereby lower the cGMP binding affinity to those enzyme molecules lacking bound Pγ.

Nonspecific binding is measured by supplementing the [^{3}H]cGMP solution with a 1000-fold excess of unlabeled cGMP prior to adding to the PDE sample. Aliquots of the original [^{3}H]cGMP ligand solution and the final incubation mixture are pipetted into scintillation vials to determine the actual specific activity of the cGMP solution and the total DPMs applied to the filter.

Validation of Method

Values for the K_D and B_{max} for the high-affinity noncatalytic sites on nonactivated frog PDE using the standard membrane filtration assay agree well with two equilibrium methods described later, namely, ultrafiltration[23] and a centrifugal assay.[52] For general use, the membrane filtration method is preferred for several reasons: greater reproducibility, lower experimental error, and ability to process large numbers of samples. The major limitations of the filter binding method result from the loss of cGMP bound to lower affinity sites during the filtration process. This drawback prevents the filtration method from being able to detect a second class of cGMP binding sites present in amphibian ROS,[23,52] as well as obscuring the transition from high- to low-affinity cGMP binding that results following Pγ release from P$\alpha\beta$.[15]

Alternative Filter Binding Protocol Using 96-Well Multiscreen Assay Plate

For high-throughput screening of cGMP binding to PDE,[51] we have modified the preceding filter binding protocol to use Multiscreen (Millipore, MAHA N45) filtration plates containing the same mixed cellulose ester membrane as earlier. Each well is prewet with wash buffer in order to obtain a good seal of the plate with the Multiscreen vacuum manifold. A 20-μl portion is added, then immediately rinsed with three 100-μl washes. A disadvantage to this method is the slower time for filtration of the sample and wash solutions ($\sim$10 sec). The filters are punched from the base of the plate into 7-ml scintillation vials. Water is added, the vial vortexed, then 3 ml of Ultima Gold scintillation fluid is added and the vials treated as earlier.

Use of Ammonium Sulfate to Stabilize Bound cGMP during Membrane Filtration

In an attempt to reduce the likelihood of dissociation of bound cGMP from PDE during the membrane filtration procedure, we have recently utilized an ammonium sulfate solution to stabilize bound cGMP. This approach has been used before to precipitate soluble binding proteins as well

---

[52] R. S. Forget, J. E. Martin, and R. H. Cote, *Anal. Biochem.* **215**, 159 (1993).

as for stabilizing protein-bound nucleotide.[53] This method is particularly well suited for extending the operating range of the filtration assay to lower affinity binding sites where ligand exchange is more rapid and nucleotide dissociation will otherwise occur during filtration and washing. In addition, the ammonium sulfate stop solution can be used to more precisely define the exact duration of the association or dissociation reaction in kinetic studies.

Solutions

> Saturated (100%) ammonium sulfate solution is prepared by adding 353.3 g ammonium sulfate to 500 ml distilled water. After dissolving completely, cool to 4° to achieve complete saturation.
> 200 mM Tris, pH 7.5.
> Buffered ammonium sulfate (BAS) is made by adding 1 volume of 200 mM Tris to 19 volumes of saturated ammonium sulfate (final 95%).
> A 2% SDS solution is used to solubilize the radioactivity trapped on the filter.
> The scintillation fluid used is Ultima Gold XR (Packard), which has a high sample load capacity.

Procedure

Incubate the PDE-containing samples with [^{3}H]cGMP as earlier. Meanwhile, chill tubes containing 200 μl of BAS to 4° in an ice water bath. When ready to stop the binding reaction, add a 20-μl portion of PDE to the BAS solution and vortex immediately. Once stabilized with BAS, the samples can remain on ice for several minutes prior to filtering.

Immediately prior to filtration, prewet the 25-mm filter disks with 2 ml BAS (no vacuum), and apply vacuum after filter is thoroughly wetted. Then apply the entire sample to the filter, and wash with three 1-ml portions of BAS. The filters are then placed in vials with 2 ml 2% SDS, and shaken for 10 min. Add 3.5 ml Ultima Gold XR, mix well, and count.

Comments

There is no evidence that addition of BAS affects the dynamics of [^{3}H]cGMP exchange with the noncatalytic sites of PDE except to prevent already bound nucleotide from dissociating. Addition of a 1000-fold excess of unlabeled cGMP to the BAS to prevent additional [^{3}H]cGMP association shows no detectable change in the amount of cGMP bound over a 10-min

[53] S. O. Doskeland and D. Ogreid, *Methods Enzymol.* **159**, 147 (1988).

period. Over a longer period of time, bound [³H]cGMP stabilized with BAS dissociates its bound cGMP at a linear rate of 0.1% per min. Inclusion of cold cGMP has a minor effect on the stability of the cGMP–PDE complex (0.2% dissociation per minute).

The use of BAS results in several-fold higher levels of nonspecific binding of radiolabel to the filters compared with the standard filtration procedure. The absolute levels of nonspecific binding remain well below 1% of the total disintegrations per minute applied to the filter.

Ammonium sulfate concentrations as low as 50% are effective in stabilizing cGMP bound to PDE for short periods of time. If samples are to be filtered shortly after adding the PDE sample to BAS, then lower concentrations of ammonium sulfate can be used.

Dissociation Kinetics of cGMP Release from Noncatalytic Sites

Although equilibrium binding parameters provide important information about the extent of occupancy of the noncatalytic cGMP binding sites on PDE in the resting state, an understanding of the role of these binding sites in shaping the photoresponse also requires knowledge of the rate of cGMP exchange to and from the noncatalytic sites. In addition, small (<10-fold) differences between the K_D values for the two high-affinity sites on PDE not easily detected using equilibrium methods can be more readily resolved using kinetic approaches.

Release of bound cGMP from the high-affinity noncatalytic sites on PDE can be evaluated by either a rapid drop in the free cGMP concentration ("concentration jump") or by isotopic dilution ("cold chase"). Concentration jump experiments are classically performed by volumetric dilution of the solution containing an equilibrium mixture of free ligand and the receptor–ligand complex, resulting in net dissociation of bound ligand as the new equilibrium state is approached. Volumetric dilution of cGMP–PDE complexes is not feasible, since the Pγ–P$\alpha\beta$ binding equilibrium is also affected,[36,54] which then affects the cGMP binding affinity to the noncatalytic sites. Calvert et al.[55] have circumvented this problem in a novel manner by inducing cGMP dissociation following addition of the PDE1 (calmodulin-dependent) isoform to rapidly destroy all free cGMP and to induce net dissociation (and destruction) of bound cGMP. The advantage of this "enzymatic concentration jump" method is that it more closely mimics the conditions likely to pertain during the light activation phase of the photoresponse. However, establishing the proper conditions

---

[54] M. R. D'Amours and R. H. Cote, *Biochem. J.* **340**, 863 (1999).

and controls to perform these experiments is more difficult than for the cold chase protocol.

Cold Chase Protocol for Measuring [³H]cGMP Release from Noncatalytic Sites on PDE

This approach relies on greatly reducing the specific activity of the free [³H]cGMP by adding a large (>100-fold) molar excess of unlabeled cGMP. When a bound cGMP molecule dissociates from PDE, an unlabeled cGMP will reassociate, leading to progressive decline in bound radiolabeled ligand over time.

PDE is prepared as described earlier, and the binding of 1 μM [³H]cGMP (sufficient to saturate the high-affinity sites) allowed to approach equilibrium. Just prior to initiating dissociation of the bound [³H]cGMP, portions of the incubated PDE sample are assayed to determine the maximum extent of binding. Then 1 mM unlabeled cGMP (final concentration) is added at time zero, and samples filtered at various times thereafter. Nonspecific binding controls are included in which the 1 mM cGMP is premixed with the radiolabeled cGMP before incubation with PDE. The filters are processed as described earlier for the general filtration protocol.

For the case of nonactivated PDE in which the dissociation kinetics typically follow a single exponential decay, <10 time points covering 10–90% dissociation can provide sufficient data for accurate estimation of k_{-1}. The biphasic kinetics seen with transducin-activated PDE,[15] for example, require twice as many data points to adequately define the fast and slowly dissociating components. The cold chase method described here has been shown to give equivalent results to the enzymatic concentration jump method for determining the dissociation rate constant of cGMP from nonactivated frog PDE.[55]

Equilibrium Methods to Determine cGMP Binding to Lower Affinity cGMP Binding Sites

Various methods are available to detect binding of ligands to receptors without significantly perturbing the equilibrium of the binding reaction, including equilibrium dialysis, gel filtration, centrifugation, and ultrafiltration.[56] These equilibrium methods are required for relatively weak ligand–receptor interactions where dissociation of bound ligand will occur during the process of separating free from bound ligand. They are also useful as

---

[55] P. D. Calvert, T. W. Ho, Y. M. LeFebvre, and V. Y. Arshavsky, *J. Gen. Physiol.* **111**, 39 (1998).
[56] E. C. Hulme, "Receptor-Ligand Interactions: A Practical Approach." Oxford University Press, Oxford, 1992.

an adjunct to nonequilibrium methods to confirm the validity of the binding parameters. Disadvantages of the equilibrium methods described later include relatively high levels of nonspecific binding, lower sensitivity, and difficulty in performing kinetic studies.

Centrifugal Ultrafiltration Method for Measuring cGMP Binding to ROS Homogenates

In this method, cGMP bound to specific binding sites is partitioned from free cGMP using a permeable membrane that permits the ligand to pass through while retaining the receptor.[57,58]

Procedure

Nucleotide-depleted ROS homogenates are incubated with 0.04–20 μM [^{3}H]cGMP as described earlier for the membrane filtration assay. In addition, [^{14}C]sorbitol is added as an internal volume marker. A 140-μl portion is added to the top chamber of an Ultrafree-MC ultrafiltration unit (30,000 nominal molecular weight cutoff, polysulfone membrane; Millipore). The unit is centrifuged (5 min at 6000g at 4°) in a fixed-angle centrifuge rotor. Defined volumes of the filtrate, retentate, and initial reaction mixture are analyzed by dual-label scintillation counting to determine ^{3}H and ^{14}C disintegrations per minute in each fraction. Nonspecific binding controls are treated identically, except that a >100-fold excess of unlabeled cGMP is added to the [^{3}H]cGMP solution before mixing with the ROS sample.

Comments

When this method was used to measure cGMP binding affinity to PDE in the submicromolar range, the K_D and B_{max} values obtained agreed reasonably well with those from membrane filtration experiments.[23] This equilibrium method has an operating range greater than that of membrane filtration; experiments carried out at cGMP concentrations up to 20 μM revealed a second class of cGMP binding sites ($K_D = 8\ \mu M$) in ROS homogenates that was not detected by filtration.[23]

Centrifugal Separation Assay Using Silicone Oil

Simple sedimentation assays have been employed with membrane-associated receptors to separate free from bound ligand by centrifugation.

[57] H. Paulus, *Anal. Biochem.* **32**, 91 (1969).
[58] J. A. Sophianopoulos, S. J. Durham, A. J. Sophianopoulos, H. L. Ragsdale, and W. P. J. Cropper, *Arch. Biochem. Biophys.* **187**, 132 (1978).

The resulting pellet is often washed with buffer, in which case low-affinity ligands will dissociate. The alternative, to assay the pellet without washing, results in substantial nonspecific binding due to free ligand present in the pellet volume.

We developed the use of centrifugal separation through silicone oil,[52] in which the aqueous phase containing unbound ligand is stripped away from the membrane-associated receptors by passage through a layer of silicone oil. This method relies on the fact that ROS membranes have a greater density than the silicone oil through which they pass; in contrast, the aqueous solution (containing unbound ligand) remains above the oil layer following centrifugation. The advantage of using the silicone oil layer is that nonspecific entrapment of unbound radiolabeled ligand is greatly reduced compared with a simple centrifugation of membranes from supernatant. Also, passage of the ligand–receptor complex through the oil does not appear to perturb the binding equilibrium. One limitation of this method is that it is applicable only to membrane-associated receptors; soluble receptor proteins will remain in the upper aqueous layer with free ligand and not be detected.

Solutions

Dow-Corning silicone oils are obtained from William F. Nye, Inc. (New Bedford, MA). To obtain the proper density, Dow Corning 550 fluid (density 1.07 g/ml) and Dow Corning 220 fluid (1.0 cS, density 0.818 g/ml) were mixed to achieve a final density of 1.02 g/ml.

Verifying Quantitative Partitioning of Membranes and Free Ligand

It is important to ascertain that the sedimentation conditions (oil density, centrifugal force) result in complete sedimentation of the membrane-associated receptor through the silicone oil layer and into the pellet. We assay the recovery of rhodopsin in the pellet as a measure of quantitative sedimentation of ROS membranes. We have observed that the manner in which the ROS membranes are prepared (see earlier section) can affect the membrane density.

It is also important to verify that the radiolabeled ligand is effectively excluded from entering the oil layer. ^{14}C-Labeled compounds that do not bind to ROS membranes (e.g., sorbitol) can be added both as volume markers and to quantify the ability of small molecules to enter into the oil layer or to cosediment with the membranes in the pellet.

Centrifugal Separation Protocol for ROS Membranes

Silicone oil (150 μl, 1.020 g/ml) is added to a 175-μl polyethylene Beckman Airfuge tube, and briefly spun to sediment the oil from the side walls of the tube. (A 5-μl cushion of 50% glycerol can be added before the oil to enhance the ease of recovery of the pellet.) ROS homogenates are incubated with [^{3}H]cGMP (0.04–20 μM) as earlier. Portions (25 μl) are then layered on top of the silicone oil layer. Centrifugation is carried out at room temperature for 1 min at 130,000g in the Beckman Airfuge. After removing the samples from the centrifuge, the tubes are immersed in a dry-ice–ethanol bath to freeze the aqueous layer. The tubes are cut on a diagonal through the oil layer, and the frozen top layer is discarded. The contents of the bottom half of the tube (including residual oil) are transferred to scintillation vials. The pellets at the bottom of the tubes are resuspended with 50–100 μl of detergent (2% SDS or 50 mM hexadecyltrimethylammonium chloride) and added to the vials. Scintillation fluid is added and the radioactivity sedimenting through the oil layer quantified. Nonspecific entrapment of unbound [^{3}H]cGMP can be assessed by incubating an identical sample with a large excess of unlabeled ligand.

Comments

This centrifugal separation assay has been used with submicromolar levels of [^{3}H]cGMP to determine the K_D (73 ± 19 nM) and B_{max} (0.005 ± 0.001 mol cGMP/mol rhodopsin) for the high-affinity class of noncatalytic sites on frog PDE.[52] These values are in excellent agreement with membrane filtration measurements,[23] although it is noted that the error associated with the centrifugal separation procedure is greater than that observed with filter binding measurements.

When the ligand concentration is extended up to 20 μM, the centrifugal separation procedure is able to detect a second class of lower affinity binding sites, in general agreement with the ultrafiltration method discussed earlier. Depending on the type of ROS preparation the K_D falls in the range of 5–12 μM.[23,52] The relatively large error in estimating the binding parameters for this lower affinity class of sites is primarily due to the low specific radioactivity of the [^{3}H]cGMP solutions at the highest cGMP concentrations. Use of [^{32}P]cGMP in future work would overcome this limitation.

Conclusions

The experimental approaches described in this chapter have advanced our understanding of the roles that the noncatalytic cGMP binding sites play in the visual transduction pathway of frog rod outer segments. Under conditions where PDE is not activated and Pγ is associated with the Pαβ dimer, the high-affinity class of cGMP binding sites will be completely occupied. Furthermore, the slow rate of cGMP dissociation from the high-affinity sites rules out any dynamic changes in binding site occupancy on the timescale of visual excitation. Thus, the high-affinity sites on nonactivated PDE effectively sequester much of the total cGMP content in the rod outer segment and effectively reduce the cytoplasmic free cGMP concentration.[23] On activation of PDE by transducin, the high-affinity cGMP binding sites on PDE undergo changes in binding affinity and dissociation kinetics,[14,15] but the regulatory significance of these changes is not fully understood at present. One likely scenario is that allosteric interactions between Pγ and the noncatalytic binding sites may regulate the ability of Pγ to act in concert with RGS9 to control the lifetime of activated PDE during the photoresponse (see Introduction section).

Much less is known about the locus and physiologic significance of the moderate-affinity binding sites in ROS, which can only be detected using equilibrium binding assays. This second class of sites may play a role in dynamically buffering the free cGMP concentration in the outer segment.[2,23] The moderate-affinity cGMP binding sites may also represent a second noncatalytic cGMP binding domain on PDE, based on the identification in the primary sequence of two conserved putative cGMP binding domains per PDE catalytic subunit.[59] It remains for future investigations to determine precisely how cGMP binding to distinct classes of binding sites in the photoreceptor encode information that is used to regulate the amplitude, duration, and/or state of adaptation of the phototransduction cascade.

Acknowledgments

I would like to thank past and present members of my laboratory, as well as valued colleagues, who have contributed to the work described in this article. This research has been supported by the National Institutes of Health (National Eye Institute grant EY-05798) and the New Hampshire Agricultural Experiment Station (Scientific Contribution #2008).

---

[59] H. Charbonneau, R. K. Prusti, H. Letrong, W. K. Sonnenburg, P. J. Mullaney, K. A. Walsh, and J. A. Beavo, *Proc. Natl. Acad. Sci. U.S.A.* **87,** 288 (1990).

[44] Purification and Autophosphorylation of Retinal Guanylate Cyclase

By Jeffrey P. Johnston, Jennifer G. Aparicio, and Meredithe L. Applebury

Introduction

Photoreceptor signaling in the vertebrate retina is mediated by cGMP. Light-induced changes in the G-coupled receptor signaling cascade decrease the cGMP levels. The lower levels in turn decrease the ability of this second messenger to gate the cGMP-dependent cation channel in the photoreceptor surface membrane.[1,2] The intracellular level of cGMP is controlled by the opposing actions of two key phototransduction enzymes, cGMP phosphodiesterase (PDE-VI) and guanylate cyclase (GC). This article addresses the properties and regulation of the latter.

GC synthesizes cGMP; its activity restores the dark current and is involved in the process of visual adaptation.[3,4] The major guanylate cyclase found in rod outer segments (ROS) has been isolated from bovine, toad, and frog ROS membranes and is an oligomeric enzyme comprised of a 115-kDa integral membrane protein subunit.[5–7] Cloning of the retinal GCs identified two isoforms that function in photoreceptors. GC-E (RetGC-1, ROS-GC1)[8–11] is expressed in retina and the pineal. GC-F (RetGC-2, ROS-GC2)[10,12,13] is expressed exclusively in the retina. Both GC-E and

---

[1] E. E. Fesenko, S. S. Kolesnikov, and A. L. Lyubarsky, *Biochim. Biophys. Acta* **856,** 661 (1986).
[2] K. W. Yau and K. Nakatani, *Nature* **317,** 252 (1986).
[3] L. Stryer, *J. Biol. Chem.* **266,** 10711 (1991).
[4] E. N. Pugh, T. Duda, A. Sitaramayya, and R. K. Sharma, *Biosci. Rep.* **17,** 429 (1997).
[5] F. Hayashi and A. Yamazaki, *Proc. Natl. Acad. Sci. U.S.A.* **88,** 4746 (1991).
[6] K.-W. Koch, *J. Biol. Chem.* **266** 8634 (1991).
[7] J. A. Aparicio and M. L. Applebury, *Prot. Expr. Purif.* **6,** 501 (1995).
[8] A. W. Shyjan, F. J. de Sauvage, N. A. Gillett, D. V. Goeddel, and D. G. Lowe, *Neuron* **9,** 727 (1992).
[9] R. M. Goraczniak, T. Duda, A. Sitaramayya, and R. K. Sharma, *Biochem. J.* **302,** 455 (1994).
[10] R. Yang, D. C. Foster, D. L. Garbers, and H. J. Fulle, *Proc. Natl. Acad. Sci. U.S.A.* **92,** 602 (1995).
[11] J. P. Johnston, F. Farhangfar, J. G. Aparicio, S. H. Nam, and M. L. Applebury, *Gene* **193,** 219 (1997).
[12] D. G. Lowe, A. M. Dizhoor, K. Liu, Q. Gu, M. Spencer, R. Laura, L. Lu, and J. B. Hurley, *Proc. Natl. Acad. Sci. U.S.A.* **92,** 5535 (1995).
[13] R. Goraczniak, T. Duda, and R. K. Sharma, *Biochem. Biophys. Res. Commun.* **234,** 666 (1997).

GC-F have been localized to the outer segments of rod photoreceptors, and they both exist as homomers.[14]

The deduced amino acid sequences of GC-E and GC-F indicate that these proteins are members of a family of membrane-bound GCs. The structures of this type of GC have four functional domains: an extracellular domain (ECD), a single transmembrane domain, a kinase homology domain (KHD), and a C-terminal catalytic domain (CAT) (Fig. 1). The ECD of some GCs bind peptide hormones, resulting in activation of the cyclase activity (reviewed by Drewett and Garbers.[15]). To date, no ligand has been identified for the photoreceptor guanylate cyclases. The KHD shows sequence similarity to several serine/threonine and tyrosine kinases, and is most homologous to the receptor tyrosine kinase.[16] The catalytic domain has homology to the catalytic core of adenylate cyclase and mutation of key amino acids within the CAT can transform the enzyme from a guanylate cyclase to an adenylate cyclase.[17,18]

Given the complexity of the GC protein structure, it is not surprising that its activity is regulated by several different mechanisms. Three guanyl-

---

[14] R. B. Yang and D. L. Garbers, *J. Biol. Chem.* **272,** 13738 (1997).
[15] J. G. Drewett and D. L. Garbers, *Endocr. Rev.* **15,** 135 (1994).
[16] S. Singh, D. G. Lowe, D. S. Thorpe, H. Rodriguez, W. J. Kuang, L. J. Dangott, M. Chinkers, D. V. Goeddel, and D. L. Garbers, *Nature* **334,** 708 (1988).
[17] C. L. Tucker, J. H. Hurley, T. R. Miller, and J. B. Hurley, *Proc. Natl. Acad. Sci. U.S.A.* **95,** 5993 (1998).
[18] R. K. Sunahara, A. Beuve, J. J. Tesmer, S. R. Sprang, D. L. Garbers, and A. G. Gilman, *J. Biol. Chem.* **273,** 16332 (1998).
[19] M. Itakura, M. Iwashina, T. Mizuno, T. Ito, H. Hagiwara, and S. Hirose, *J. Biol. Chem.* **269,** 8314 (1994).
[20] J. T. Stults, K. L. O'Connell, C. Garcia, S. Wong, A. M. Engel., D. L. Garbers, and D. G. Lowe, *Biochemistry* **33,** 11372 (1994).
[21] N. McNicoll, J. Gagnon, J. J. Rondeau, H. Ong, and A. DeLean, *Biochemistry* **35,** 12950 (1996).
[22] K. W. Koch, P. Stecher, and R. Kellner, *Eur. J. Biochem.* **222,** 589 (1994).
[23] S. S. Taylor, D. R. Knighton, J. Zheng, J. M. Sowadski, C. S. Gibbs, and M. J. Zoller, *Trends Biochem. Sci.* **18,** 84 (1993).
[24] S. R. Hubbard, L. Wei, L. Ellis, and W. A. Hendrickson, *Nature* **372,** 746 (1994).
[25] S.-H. Hu, M. W. Parker, J. Y. Lei, M. C. J. Wilce, G. M. Benian, and B. E. Kemp, *Nature* **369,** 581 (1994).
[26] E. M. Wilson and M. Chinkers, *Biochemistry* **34,** 4696 (1995).
[27] J. J. Tesmer, R. K. Sunahara, A. G. Gilman, and S. R. Sprang, *Science* **278,** 1907 (1997).
[28] G. Zhang, Y. Liu, A. E. Ruoho, and J. H. Hurley, *Nature* **386,** 247 (1997).
[29] R. P. Laura and J. B. Hurley, *Biochemistry* **37,** 11264 (1998).
[30] T. Duda, R. M. Goraczniak, N. Pozdnyakov, A. Sitaramayya, R. K. Sharma, *Biochem. Biophys. Res. Comm.* **242,** 118 (1998).

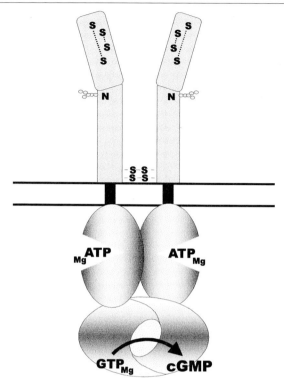

FIG. 1. Model of the photoreceptor membrane-bound guanylate cyclase GC-E or GC-F. GC-E and GC-F exist as individual homodimers, stabilized by intersubunit disulfide bonds[7,14]; studies with gel filtration suggest the proteins exist as homooligomers ranging from dimers to tetramers.[7] The extracellular domain has six cysteine residues conserved in GC-E and GC-F. Drawing on knowledge of other members of the superfamily, this N-terminal region is considered to be folded in an immunoglobulin-like domain stabilized by disulfide bonds.[19,20] This region specifies the ligand binding that regulates the activation of the cytoplasmic cyclase domain.[21] GC-E and GC-F differ significantly in this region and no ligand has been identified for either enzyme. GC-E is glycosylated,[22] putatively at a consensus sequence indicated. On the cytoplasmic side of the membrane, a kinase homology domain is positioned adjacent to the membrane. An ATP binding pocket is proposed based on amino acid homology with other protein kinases whose crystal structures have been determined.[23–25] Not represented is a short hinge segment at the C terminus of the KHD that connects to the catalytic core of the enzyme and may be involved in subunit dimerization.[26] The cytoplasmic terminus of guanylate cyclase is formed by a pair of catalytic domains (CAT). Based on structures of adenylate cyclases,[27,28] with which guanylate cyclases are homologous, the active site would be formed along the interface of the two subunits. In a homodimer, two active sites formed about an inverted axis of symmetry might be present. Whether both would be accessible to substrate remains to be explored. Regulation by GCAP-1 and GCAP-2 is conferred by interaction of the C terminus of the KHD and possibly the C terminus of the CAT domain.[29] CD-GCAP appears to interact with the C terminus of the CAT domains.[30]

ate cyclase activating proteins (GCAPs), GCAP-1,[31–33] GCAP-2,[34,35] and CD-GCAP (S100-β),[36] confer calcium sensitivity to the ROS membrane GCs. The details of this mode of regulation are presented elsewhere in this volume (see Chapters 45, 46, and 48).

The cyclase activity of ROS membrane GCs is modulated by adenosine triphosphate (ATP).[37–40] Biochemical analyses using nonhydrolyzable analogs of ATP and 8-N$_3$[α-^{32}P]ATP labeling suggest that ATP plays an allosteric role. The effects are noncompetitive with guanosine triphosphate (GTP) and are probably caused by ATP binding to a site within the KHD.[37] This domain also shows an ATP-Mg–dependent autophosphorylating activity and is capable of phosphorylating exogenous substrates, as discussed in more detail later.

In addition, GC activity is modulated by the heterologous kinases PKA and PKC.[41,42] Both GC-E and GC-F contain several phosphorylation consensus sites. Identification of the sites that are phosphorylated *in vivo*, the kinases that have physiologic roles, and the effects of phosphorylation on GC activity have yet to be established. Phosphorylation regulates the activities of other membrane GCs. GC-A and GC-B are phosphorylated at multiple sites within their KHDs. Phosphorylation of these proteins is required for activation of cyclase activity by the natriuretic peptides, and dephosphorylation is associated with receptor desensitization.[43,44] The role of phosphorylation in regulating these enzymes may prove insightful for understanding modulation of ROS membrane GC activity.

[31] W. A. Gorczyca, M. P. Gray-Keller, P. B. Detwiler, and K. Palczewski, *Proc. Natl. Acad. Sci. U.S.A.* **91,** 4014 (1994).
[32] K. Palczewski, I. Subbaraya, W. A. Gorczyca, B. S. Helekar, C. C. Ruiz, H. Ohguro, J. Huang, X. Zhao, J. W. Crabb, R. S. Johnson, K. A. Walsh, M. P. Gray-Keller, P. B. Detwiler, and W. Baehr, *Neuron* **13,** 395 (1994).
[33] S. Frins, W. Bonigk, F. Muller, R. Kellner, and K. W. Koch, *J. Biol. Chem.* **271,** 8022 (1996).
[34] A. M. Dizhoor, D. G. Lowe, E. V. Olshevskaya, R. P. Laura, and J. B. Hurley, *Neuron* **12,** 1345 (1994).
[35] A. M. Dizhoor, E. V. Olshevskaya, W. J. Henzel, S. C. Wong, J. T. Stults, I. Ankoudinova, J. and B. Hurley, *J. Biol. Chem.* **270,** 25200 (1995).
[36] N. Poznyakov, A. Yoshida, N. G. F. Cooper, A. Margulis, T. Duda, R. K. Sharma, and A. Sitaramayya, *Biochemistry* **34,** 14279 (1995).
[37] J. G. Aparicio, and M. L. Applebury, *J. Biol. Chem.* **271,** 27083 (1996).
[38] W. A. Gorczyca, J. P. Van Hooser, and K. Palczewski, *Biochemistry* **33,** 3217 (1994).
[39] A. Sitaramayya, T. Duda, and R. K. Sharma, *Mol. Cell. Biochem.* **148,** 139 (1995).
[40] A. Sitaramayya, R. B. Marala, S. Hakki, and R. K. Sharma, *Biochemistry* **30,** 6742 (1991).
[41] G. Wolbring and P. P. M. Schnetkamp, *Biochemistry* **34,** 4689 (1995).
[42] G. Wolbring and P. P. M. Schnetkamp, *Biochemistry* **35,** 11013 (1996).
[43] L. R. Potter and T. Hunter, *Mol. Cell Biol.* **18,** 2164 (1998).
[44] L. R. Potter and T. Hunter, *J. Biol. Chem.* **273,** 15533 (1998).

Characterization of the ROS membrane GC activity and the mechanisms of its regulation require a purified preparation of protein for *in vitro* studies. This chapter summarizes a simple and efficient protocol for the affinity purification of GC-E from bovine rod outer segment membranes. The purity and stability of the protein has allowed us to biochemically characterize GC-E[7] and discover an inherent autophosphorylating protein kinase activity.[37] The issues important for successful purification and a detailed protocol are presented. In addition, we describe the protocols used to assess the kinase activity of the KHD.

Purification of Bovine Retinal GC-E

Several procedures for the purification of GC from ROS membranes have been reported.[5–7,45] The following, more recent protocol describes a one-column affinity purification of ROS membrane GC from fresh bovine retina.[7] This method generates a preparation where GC-E is the major component. Peptide sequencing of the purified protein identifies it as GC-E[11]; no additional peptides were isolated that are consistent with the presence of GC-F. Moreover, no GC-F is detected on a Western blot of the purified GC using an antibody against the N-terminus of GC-F (a generous gift of K. Palczewski). Sodium dodecyl sulfate–polyacrylamide gel electrophoresis (SDS–PAGE) on several different percentage gels shows no other major band. A significant amount of ROS membrane GC activity remains resistant to detergent solubilization,[5–7,45] and may represent a pool of photoreceptor guanylate cyclase associated with the cytoskeleton.[46,47a] GC-F has not yet been isolated biochemically, thus its role and abundance are unknown.

The details of purification are presented next and summarized in flowchart form in Fig. 2. Important steps in the protocol include stringent washing of the ROS membranes prior to solubilization, use of dodecylmaltoside (DDM) for solubilization of GC activity, and establishment of effective conditions for stabilizing activity during GTP-agarose chromatography. This protocol produces a relatively large yield of enzyme with high activity. The simplicity of this procedure and the ability to work in ambient light are additional advantages.

---

[45] S. Hakki and A. Sitaramayya, *Biochemistry* **29,** 1088 (1990).
[46] D. Fleischman and M. Denisevich, *Biochemistry* **18,** 5060 (1979).
[47a] D. Fleischman, M. Denisevich, D. Raveed, and R. G. Pannbacker, *Biochim. Biophys. Acta* **630,** 176 (1980).

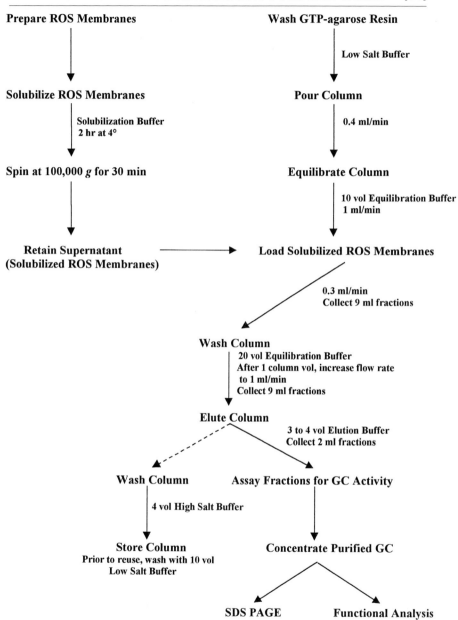

FIG. 2. Flowchart of GC-E purification from bovine retina.

Guanylate Cyclase Assay

Several assays of GC activity are available in this volume.[38,41,47b–d,48–50] We have chosen the method of Saloman,[51] which measures the conversion of [α-^{32}P]GTP to [^{32}P]cGMP, to monitor the purification of ROS membrane GC-E. This assay allows simultaneous processing of large numbers of samples. Moreover, the inclusion of [^{3}H]cGMP to monitor degradation of cGMP by PDE and recovery during chromatographic separation of the nucleotides provides more reproducibility throughout the purification scheme.

Our modifications[7] of the assay are described briefly: A standard reaction with a final volume of 100 μl contains 2 mM MnCl$_2$, 2 mM Na$_2$GTP, 100 mM NaCl, 20 mM HEPES, pH 7.6, 1 mM 3-isobutyl-1-methylxanthine (IBMX), and 1 mM 8 Br-cGMP with ~10^6 cpm [α-^{32}P]GTP and ~10^4 cpm [^{3}H]cGMP. To start the reaction, enzyme is added to the assay buffer and incubated at 37° for 10 min. The reaction is quenched by addition of 10 μl of 2.5% SDS, 5 mM GTP, 50 mM Tris-HCl, pH 7.6, and incubated at 100° for 3 min. The volume is adjusted to 200 μl with H$_2$O. GTP is separated from cGMP using Dowex (50X4-200) and neutral alumina columns in tandem as originally described, except that the imidazole is buffered to pH 7.5. We routinely use 1.6 ml of Dowex and 0.3 g of alumina, but the column size and buffer volumes for washing and eluting the columns should be empirically checked for each new lot of resin used.

Preparation of ROS Membranes

ROS membranes are prepared using a modified procedure of Papermaster and Dreyer[52] as previously described.[7,53] This protocol has been developed using retina isolated from fresh bovine eyes obtained from a slaughterhouse on the day of the purification. We have also tried this protocol using frozen bovine retina obtained commercially, but the purity and the yield of GC is consistently lower.

Two important modifications of the original procedure contribute to

---

[47b] W. A. Gorczyca, *Methods Enzymol.* **315** [45], 2000 (this volume).
[47c] J. B. Hurley and A. M. Dizhoor, *Methods Enzymol.* **315** [46], 2000 (this volume).
[47d] G. Wolbring and P. P. M. Schnetkamp, *Methods Enzymol.* **315** [47], 2000 (this volume).
[48] A. Sitaramayya, L. Lombardi, and A. Margulis, *Visual Neurosci.* **10,** 991 (1993).
[49] J. A. d. Nesbitt, W. B. Anderson, Z. Miller, I. Pastan, T. R. Russell, and D. Gospodarowicz, *J. Biol. Chem.* **251,** 2344 (1976).
[50] A. L. Steiner, A. S. Pagliara, L. R. Chase, and D. M. Kipnis, *J. Biol. Chem.* **247,** 1114 (1972).
[51] Y. Saloman, *Adv. Cyclic Nucleotide Res.* **10,** 35 (1979).
[52] D. S. Papermaster and W. J. Dreyer, *Biochemistry* **13,** 2438 (1974).
[53] D. A. Nicoll and M. L. Applebury, *J. Biol. Chem.* **264,** 16207 (1989).

the stability and purity of the final GC preparation. First, the buffers (10 mM Tris-HCl, pH 7.6, 65 mM NaCl, 2 mM MgCl$_2$ with appropriate sucrose concentrations) used for homogenization of the retinas, sucrose density centrifugation, and washing of the isolated membranes contain 1 mM dithiothreitol (DTT), 20 μM NGDA (nordihydroguaiaretic acid, an inhibitor of lipid oxidation and rhodopsin polymerization), and protease inhibitors [1 μg/ml each of pepstatin, leupeptin, aprotinin, 10 μg/ml phenylmethylsulfonyl fluoride (PMSF) and 100 μM EDTA. 4-(2-aminoethyl)-benzenesulfonyl fluoride hydrochloride (AEBSF) (Boehringer Mannheim, Indianapolis, IN) at 0.1–1 mg/ml may substitute for PMSF]. Second, the procedure for washing the ROS membranes has been modified. The band I ROS membranes isolated from the sucrose gradient are diluted in two volumes of an "isotonic" buffer (10 mM Tris-HCl, pH 7.6, 2 mM MgCl$_2$, 65 mM NaCl). The membranes are pelleted and washed twice with a hypotonic/GTP/hydroxylamine buffer (10 mM Tris-HCl, pH 7.4, 0.1 mM EDTA, 40 μM GTP, 10 mM hydroxylamine). Finally, the membranes are washed twice with a hypertonic buffer (50 mM Tris-HCl, pH 7.6, 1 M KCl). The resulting membranes, termed *stripped* ROS membranes, retain 36% of the total retinal guanylate cyclase activity. For best results, washing of ROS membranes should be performed immediately after isolation. Stripped ROS membranes may be stored at −80° in 25% glycerol in isotonic buffer.

Solubilization of ROS Membranes

The key to purification of GC-E is to effect solubilization of ROS membranes in a detergent that maintains GC activity and minimizes the formation of protein aggregates. The ability of several nonionic detergents to solubilize GC-E from ROS membranes was tested.[7]

DDM solubilizes a significant fraction of GC activity, but only 60% of the total protein resulting in an increased activity. The presence of DDM also results in a 40% increase in total soluble and insoluble GC activity. Correcting for the detergent activation of GC suggests that ~65% of the GC activity is solubilized. Detergent activation of other membrane guanylate cyclases has been described and reviewed elsewhere.[54] Activation may result from exposure of a previously unavailable pool of membrane GC. However, the soluble, sodium nitroprusside stimulated GC is also activated by detergent. Thus, the activation of GC activity by detergents is due, at least in part, to induced conformational changes of the GC protein. This phenomenon stresses that activity is not a linear measurement of the amount of GC present during the various stages of purification.

[54] A. White and P. J. Lad, *Methods Enzymol.* **195,** 363 (1991).

Buffers

3× solubilization buffer: 90 mM DDM, 60 mM HEPES, pH 7.6, 150 mM NaCl, 6 mM MnCl$_2$, 3 mM DTT, 3 μg/ml each leupeptin, pepstatin, and aprotinin, 60 μM NGDA. Filter through a 0.45-μm filter and degas.

Protocol

All procedures are carried out at 4° using prechilled buffers and tubes. Thaw stripped ROS membranes on ice (~45 mg ROS at 3–4.5 mg/ml). To a 100-ml plastic centrifuge bottle add 1 volume H$_2$O, 1 volume 3× solubilization buffer, and 1 volume stripped ROS membranes. Rinse the tube that contained the stripped ROS membranes with a small volume of 3× solubilization buffer and add it to the solubilization mixture. Mix gently for 2 hr. Centrifuge the solubilization mixture at 100,000g for 30 min and load the supernatant onto the GTP-agarose column.

GTP-Agarose Chromatography

The affinity purification of GC-E is achieved by binding the enzyme to GTP-agarose in the presence of Mn^{2+}. We have used MnCl$_2$ as a cation for substrate binding since binding in the presence of MgCl$_2$ was variable. Several factors influence the success of the protocol including characterization of the binding capacity of the GTP-agarose resin, and the stability of GC activity during chromatography. These issues are discussed following the protocol so that the reader can optimize the procedure.

Buffers

Low-salt wash buffer: 50 mM NaCl, 10 mM HEPES, pH 7.6.

Equilibration buffer: 50 mM NaCl, 2 mM MnCl$_2$, 10 mM HEPES, pH 7.6, 2.5 mM DDM, 0.025% phosphatidylcholine, 20% glycerol, 1 mM DTT, 1 μg/ml each leupeptin, pepstatin, and aprotinin, 20 μM NGDA. Approximately 400 ml is needed for a single GTP column. Filter through a 0.45-μm filter and degas.

Elution buffer: 120 mM NaCl, 6 mM EGTA, 10 mM HEPES, pH 7.6, 2.5 mM DDM, 0.025% phosphatidylcholine, 20% glycerol, 1 mM DTT, 1 μg/ml each leupeptin, pepstatin, and aprotinin, 20 μM NGDA. Filter through a 0.45-μm filter and degas. Approximately 50 ml of elution buffer is needed for a single GTP column.

High-salt wash buffer: 1 M NaCl, 10 mM HEPES, pH 7.6.

Protocol

1. Wash ~12.5 ml of GTP-agarose (Sigma 11 atom spacer arm) several times with low-salt buffer to remove the glycerol and salt in which the GTP agarose is stored. Suspend the GTP agarose in low-salt wash buffer and pour a 1.5-cm-diameter × 6.5-cm column (Bio-Rad, Hercules, CA, Econo column with flow adapter). Pack the column using a flow rate of 0.4 ml/min. When the resin has settled, attach the flow adapter and wash the column with 3 volumes of low-salt wash buffer at 1 ml/min.
2. Wash the column with 10 column volumes of equilibration buffer at 1 ml/min.
3. Load the solubilized ROS membranes (30–45 ml) at 0.3 ml/min. Collect 9-ml fractions.
4. Wash the column with 20 volumes of equilibration buffer. After one column volume, increase the flow rate to 1 ml/min. Collect 9-ml fractions.
5. Elute the column with 3–4 volumes of elution buffer. Collect 2-ml fractions.
6. To reuse the column, wash with 4 volumes of high-salt wash buffer. The column may be stored in this buffer in the presence of 0.02% sodium azide. Prior to reuse, wash the column with 10 volumes of low-salt buffer.

Assay 30 μl of each fraction for GC activity. Greater than 90% of the GC activity elutes as a large peak. A small second peak, containing <10% of the GC activity, is discarded. Concentrate the fractions containing the first peak of GC activity one fraction at a time in a CentriPrep-30 unit (Amicon, Beverly, MA). Add additional fractions as the volume is reduced. Care should be taken to minimize the surface area to which the guanylate cyclase containing solution is exposed since the small quantity of protein is easily lost to nonspecific binding on surfaces. Continue the concentration until all fractions have been combined and the volume of the pooled fractions is about 600 μl. Finally, assay 1 μl of the concentrated sample for guanylate cyclase activity and protein. Evaluate the purity by SDS–PAGE. Make the concentrated GC-E 30% glycerol and store in aliquots at $-20°$.

Notes on GTP-Agarose Chromatography

Individual batches of GTP-agarose behave differently in their binding and elution properties. For GTP-agarose (Sigma) with an 11 atom spacer arm, optimal conditions for binding were found to be 50 mM NaCl and 2 mM MnCl$_2$. Efficient elution was obtained in buffer containing 120 mM

NaCl and 6 mM EGTA. However, a new batch of GTP-agarose should be assessed for binding capacity and the optimal salt concentrations for binding and elution.

Stabilization of GC-E activity during chromatography is required for successful purification. GC-E solubilized in DDM is stable in the presence of native lipids and 1 mM DTT. However, only 1% of the activity is recovered after elution of the GTP-agarose column with buffers containing only 10 mM HEPES, pH 7.6, 50 mM NaCl, and 0.1% DDM. Addition of 20% glycerol, 0.025% phosphatidylcholine, or 1 mM DTT to the column buffers individually improved the recovery of the GC-E activity.[7] The addition of all three resulted in a 17.5-fold increase in the recovery of GC-E activity.[7] The phosphatidylcholine that we routinely use (Sigma) is prepared from soybean and is only 40% phosphatidylcholine. Thus, the stabilizing effect of this lipid on GC activity may be due to other lipids present as contaminants in the phosphatidylcholine preparation.

Other Methods

Protein concentrations are determined by the method of Kaplan and Pederson[55] using BSA as a standard.

Results

A summary of the purification is given in Table I. Roughly 0.25% of the ROS membrane protein, ~120 μg of GC, may be purified from 100 fresh retina. SDS–PAGE shows that the enzyme consists of a single 115-kDa subunit and constitutes >95% of the total protein in the sample as detected by Coomassie staining. Studies of the quaternary structure indicate that the enzyme exists as a multimer, ranging from dimers to tetramers. Typical preparations show specific activities ranging between 1850 and 5879 nmol cGMP/min/mg using MnGTP as substrate. The K_m and V_{max} for the physiologic substrate MgGTP are 1.07 ± 0.20 mM and 3262 ± 514 nmol/min/mg, respectively. The activity is stable for 2–4 weeks at 4° and at least 2 years at −20°.

Measurement of Guanylate Cyclase Kinase Activities

The protocol described for the purification of GC-E from fresh bovine retina allowed us to characterize the activity of the kinase homology domain of this complex enzyme. The role of this domain has been somewhat elusive,

---

[55] R. S. Kaplan and P. L. Pederson, *Anal. Biochem.* **150,** 97 (1985).

TABLE I
Guanylate Cyclase Purification from Bovine Rod Outer Segments[a]

Sample	Activity				Specific activity
	nmol cGMP		Protein		nmol cGMP
	min	percent	mg	percent	min mg
Rod outer segments (ROS)	[b]	—	183[c]		—
Hypotonic/GTP wash	5.85	0.76	88.0	42.0	0.30
KCl wash (1 M)	0.27	0.03	19.5	18.8	0.03
Stripped ROS membranes	773	100	9.0	100	16.1
DDM solubilized ROS	912	118	48.0	87.1	21.8
Insoluble fraction	276	35.7	41.8	22.3	25.1
GTP column			10.7		
Flow-through and wash	74.3	9.6	42.5	88.5	1.75
Eluate pool 1	467	60.4	0.12	0.24	4030
Eluate pool 2	74.6	9.7	0.02	0.05	3240

[a] ROS were isolated from 100 retina yielding stripped ROS membranes following hypotonic/GTP and hypertonic (1 M KCl) washes. These membranes were solublized in DDM and purified by GTP affinity chromatography. Fractions containing the first 90% of the guanylate cyclase activity were pooled for eluate 1. The remaining fractions containing guanylate cyclase activity were pooled for eluate 2. All assays were carried out using MnGTP as the substrate. Data reported are from a single purification.

[b] The measurements of guanylate cyclase activity in these fractions are not reliable due to a large amount of degradation of cGMP by PDE.

[c] For tabulating activity and yield, data were normalized to stripped ROS membranes because it is the first step in which guanylate cyclase activity can be measured without large statistical deviation.

outside of recent reports that indicate the C terminus of the domain binds or influences GCAP binding.[29] The roles of phosphorylation, ATP binding, or kinase function are difficult to assess in nonpurified extracts. Heterologous kinases such as PKC or PKA[41,42] which are generally present in cell extracts, are known to regulate the activity of the cyclase domain, but the mechanisms are unknown.

The catalytic domains of the family of protein kinases contain 36 characteristic amino acids that contribute to the active site or stabilize its conformation.[56,57] The KHD of GC-E contains 24 of these characteristic amino acids suggesting that an ATP binding site is present. ATP, or its nonhydrolyzable analogs, play an allosteric role in conferring higher cyclase activity. ATP stimulates the cyclase activity about twofold.[31,37] Moreover, GC-E is labeled by 8-$N_3[\alpha$-^{32}P]ATP at a site independent of GTP binding, suggesting a functional ATP binding site in the kinase domain.[37] This effect is analo-

[56] S. K. Hanks and A. M. Quinn, *Methods Enzymol.* **200**, 38 (1991).
[57] S. S. Taylor, D. R. Knighton, J. Zheng, L. F. Ten Eyck, and J. M. Sowadski, *Ann. Rev. Cell Biol.* **8**, 429 (1992).

gous to the absolute requirement of ATP for stimulation of GC-A or GC-B cyclase activity by their extracellular ligands.[58,59]

The kinase domain has not been considered to possess a protein kinase activity since an aspartate residue that is important for catalysis of phosphotransfer at the active site in PKA (D166)[56,57] is absent in the kinase domains of membrane-bound GC enzymes.[8,9,12] However, in the presence of Mg[γ-^{32}P]ATP, we noticed that phosphate was incorporated into the purified GC-E, which suggested that GC-E has a protein kinase activity. To rule out the presence of contaminating heterologous kinases, incorporation of phosphate into GC-E, or an exogenous substrate (see later section), was examined in the presence of cAMP, cGMP, Ca^{2+} PS/PMA, or the nonspecific kinase inhibitor staurosporine. The lack of effect of any of these factors and a level of inhibition by staurosporine that is distinct from that for known kinases indicated that the incorporation could not be attributed to protein kinase A, protein kinase G, protein kinase C, Ca^{2+}-dependent kinases, or opsin kinase.[37] Thus, the incorporation of phosphate is consistent with a GC-E kinase activity leading to autophosphorylation. This activity could be intramolecular or intermolecular.

To further assess the kinase domain activity, we sought a substrate that could be used for routine assays. Myelin basic protein (MBP), a substrate for many kinases, proved most effective; β-casein, histones, and protamines are also phosphorylated, but less effectively.[37] Use of MBP as a substrate reduces the amount of enzyme needed for assay, enables one to vary substrate concentrations, and thereby allows characterization of ATP binding, kinase activity, and the effects of other modes of regulation on kinase activity.

The protocols for assay of the purified GC autophosphorylation and assay for kinase activity using MBP as a substrate are presented next. These simple methods provide a foundation on which more sophisticated studies should be built to elucidate the allosteric and regulatory role of the kinase homology domain of the photoreceptor membrane-bound guanylate cyclases.

Autophosphorylation Protocol

Autophosphorylation is assessed by adding labeled ATP to purified GC-E and quantifying the amount of label incorporated into the isolated 115-kDa GC-E subunit. The reaction is carried out in relatively high concen-

[58] H. Kurose, T. Inagami, and M. Ui, *FEBS Lett.* **219**, 375 (1987).
[59] C. H. Chang, K. P. Kohse, B. Chang, M. Hirata, B. Jiang, J. E. Douglas, and F. Murad, *Biochim. Biophys. Acta* **1052**, 159 (1990).

trations of GTP to block any spurious incorporation of label into the active site of the cyclase domain. The reactions are run at ~30 μM ATP, which is significantly below the K_m of 80 μM for ATP. Although the reaction proceeds faster using higher ATP concentrations, large amounts of isotope are required to maintain an effective specific activity. The amount of purified guanylate cyclase assayed has ranged from 0.05 to 1 μg. More labeling is observed with higher cyclase concentrations, but the volume added must be kept to a minimum; something in the GTP column elution buffer reduces the stoichiometry of phosphorylation. Reaction times should be optimized for each GC-E preparation. They have ranged from 5 min up to 1 hr, but are generally carried out for 20 min to 1 hr (Fig. 3).

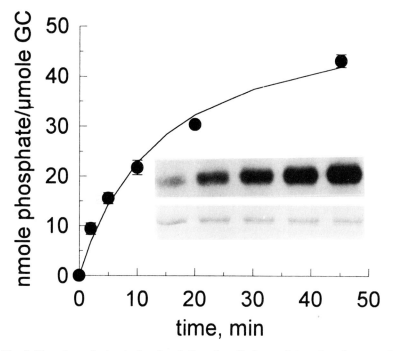

FIG. 3. Time-dependent autophosphorylation of purified guanylate cyclase. Incorporation of phosphate into GC-E is time dependent and reaches saturation indicative of a specific site of phosphoryl transfer. Kinase reactions included 0.3 μg of purified guanylate cyclase and were incubated for the times indicated. Following autoradiography, the guanylate cyclase band was excised and counted for ^{32}P. The stoichiometry of phosphorylation was calculated using the specific activity of the [γ-^{32}P]ATP. The data plotted are the average of duplicate reactions. The inset shows an autoradiogram (top) and the corresponding Coomassie stained gel (bottom) of the guanylate cyclase protein.

Reagents and Buffers

100 mM ATP stocks are made in H$_2$O, adjusted to pH 7.0 with NaOH, and the concentration determined using ε_{max} = 15,400 M^{-1} cm^{-1} at 252 nm, pH 7.0. Aliquots of the ATP stock are stored at $-20°$. Repeated freeze–thaw of the stock solutions should be avoided.

100 mM GTP stocks are prepared like the ATP stocks, but ε_{max} = 13,700 M^{-1} cm^{-1} at 252 nm, pH 7.0.

5× reaction buffer: 100 mM MOPS, pH 7.1, 500 mM NaCl, 40 mM MgCl$_2$, 5 mM NaGTP, 5 mM DTT, and protease inhibitors (aprotinin, pepstatin, leupeptin each at 5 μg/ml and PMSF at 50 μg/ml). AEBSF (Boehringer Mannheim) at 0.5–5.0 mg/ml may substitute for PMSF. This buffer is made from individual stock solutions just prior to use.

10× [γ-^{32}P]ATP: 300 μM (~5000 cpm/pmol).

3× quench solution: 50 mM Tris-HCl, pH 6.8, 3% SDS, 30% glycerol, 15 mM ATP, 30 mM EDTA, 90 mM DTT, bromphenol blue.

Assay

The reactions for autophosphorylation are run in a volume of 25 μl in 20 mM MOPS, pH 7.1, 100 mM NaCl, 5 mM MgCl$_2$, 1 mM NaGTP, 30 μM [γ-^{32}P]ATP, 1 mM DTT, 10 μg/μl PMSF, and 1 μg/ml of protease inhibitors. Mix 5 μl of 5× reaction buffer, 3 to 5 μl purified GC-E (0.3 μg), and bring the volume to 23.5 μl with H$_2$O. Start the reaction by the addition of 2.5 μl of 10× [γ-^{32}P]ATP. Incubate the samples at room temperature for varying lengths of time (see earlier discussion). Stop the reactions by addition of 12.5 μl of 3× quench solution. Boil the samples for 3 min. Resolve unincorporated ATP from the phosphorylated GC by carring out SDS–PAGE on a 9% gel. The incorporated ^{32}P may be quantitated either by excision of stained bands and scintillation counting, autoradiography, and densitometry, or most easily by use of a phosphorimager. For scintillation counting, stain the gel with Coomassie and dry the gel onto a piece of cellophane membrane. Cut out the 115-kDa band and quantitate ^{32}P by scintillation counting.

Measuring GC-E Kinase Activity Using MBP as Substrate

Using an exogenous substrate to assess kinase activity enhances the level of labeling, since the kinase activity is catalytic and substrate need not be limiting. The reactions with MBP are run in the same manner as described earlier for autophosphorylation, except that 0.9 μg of MBP and 0.05 μg of purified GC (100 : 1 molar excess of MBP) are added per reaction.

The reaction volumes are reduced to 10 μl to save material and ease loading on an SDS gel. NaGTP may be omitted with no effect.

Reagents and Buffers

The reaction buffer and ATP stock solutions are the same as described earlier for the autophosphorylation assay, except that NaGTP may be omitted from the 5× reaction buffer.

MBP (Sigma) stock solution is dissolved in H_2O at 0.9 mg/ml, aliquotted, and stored at $-20°$.

Assay

Kinase assays are run in a volume of 10 μl. Mix 2 μl of 5× reaction buffer with 1 μl purified GC-E (0.05 μg) and 1 μl MBP (0.9 μg), then bring to 9 μl with H_2O. Start the reaction by the addition of 1 μl of 10× [γ-^{32}P]ATP. Incubate the reactions at room temperature for 5 min. Add 5 μl of 3× quench solution to stop the reaction. Boil the samples for 3 min. As described earlier, resolve the unincorporated [^{32}P]ATP from the phosphorylated proteins by SDS–PAGE on a 9% (or higher) gel. The incorporation of ^{32}P into the 18.4-kDa MBP band is quantitated as described earlier.

This assay may be used to assess kinetic constants for substrate, inhibitors, and cofactors. For example, to measure the K_m of ATP, the concentration is adjusted to vary from 0 to 250 μM. Thus, several 10× ATP solutions are required, ranging in concentration from 0 to 2.5 mM. The specific activity of [γ-^{32}P]ATP will vary and is accounted for in the data analysis.

Results and Discussion

Use of the protocols for assessment of autophosphorylation and a kinase assay with exogenous MBP as a substrate provided the first biochemical evidence that a member of the membrane guanylate cyclase family has an intrinsic kinase activity. For both autophosphorylation and MBP phosphorylation, the incorporation of labeled phosphate is time dependent and approaches saturation in the presence of limiting substrate (see Fig. 3). For autophosphorylation, the stoichiometry of phosphorylation was calculated to be 0.05 mol of phosphate/mol of GC. The meaning of this low value is unclear, since the endogenous levels of phosphorylation of these preparations are unknown. Phosphoamino acid analysis using the procedure described by Boyle et al.,[60] indicates that the autophosphorylation occurs on

---

[60] W. J. Boyle, P. Van der Geer, and T. Hunter, *Methods Enzymol.* **201**, 110 (1991).

a serine residue(s).[37] Studies are ongoing to identify the residues phosphorylated and to assess the role of autophosphorylation versus that by protein kinase C or A.

Protein kinases are bisubstrate enzymes that utilize both MgATP and a phosphate acceptor protein as substrates. The velocity of the reaction is dependent on both the concentration and affinity of each substrate. Using MBP as an exogenous protein substrate for the guanylate cyclase kinase *in vitro*, the properties of kinase activity of the guanylate cyclase have been examined.[37] The kinase activity exhibits simple Michaelis–Menten kinetic behavior showing a single ATP binding affinity. The K_m for ATP is 81 μM and the V_{max} is 2.1 ± 0.2 nmol phosphate transferred/min/mg guanylate cyclase. The turnover number of the kinase activity is 0.004. Several issues need to be clarified about this rather low turnover number. The value was obtained for preparations that were purified by optimizing cyclase activity, not kinase activity. Solubilization and the presence of a detergent may affect the kinase activity. We have reconstituted the purified GC into phosphatidylcholine vesicles and observed that the protein retains its kinase activity, although the stoichiometry of phosprylation and kinetic constants of the kinase activity have not been measured under these conditions. The identification of a physiologic substrate and the use of a preparation that is more akin to the protein's native environment may prove beneficial in elucidating the details of the kinase activity. If the photoreceptor GC enzymes do respond to ligand binding by their extracellular domains, this basal level of kinase activity might be altered. Likewise, other regulatory influences such as GCAPs or other kinases may change the rate of phosphorylation.

[45] Use of Nucleoside α-Phosphorothioates in Studies of Photoreceptor Guanylyl Cyclase: Purification of Guanylyl Cyclase Activating Proteins

By WOJCIECH A. GORCZYCA

Introduction

Guanylyl cyclases (GC), which catalyze conversion of guanosine 5'-triphosphate (GTP) to cyclic guanosine 3',5'-cyclic monophosphate (cGMP), are found in two forms: cytosolic (soluble GC) and membrane-bound (particulate GC). Soluble and particulate cyclases have different structures and are activated in different ways.[1] Photoreceptor-specific guanylyl cyclase, also termed retinal GC (RetGC), belongs to the family of particulate guanylyl cyclases and shares with them an extensive sequence similarity. The mechanism of activation, however, appears to be quite different for RetGC than for other members of the family. RetGC is regulated in a calcium-dependent manner and is insensitive to peptide ligands that activate other particulate cyclases. The activity of the enzyme is low at high calcium concentrations (over 500 nM) and significantly (5–20 times) increases at low calcium concentrations (below 100 nM).[2–4] Two isoforms of RetGC (named RetGC1 and RetGC2 in humans) have been identified in mammalian retina to date[5–9]; however, the relative abundance of each isoform in the outer segments of photoreceptor cells is unknown. Ca^{2+}-dependent regulation of RetGC activity is mediated by Ca^{2+}-binding proteins referred to as GCAPs (guanylyl cyclase activating proteins).[10,11] Two GCAPs (GCAP1 and GCAP2) have been isolated so far from vertebrate retina.[10–15] Both GCAPs are soluble, highly acidic (pK_a 4.1–4.4), and hydrophobic proteins. Their electrophoretic (SDS–PAGE) mobility depends on calcium ions. The apparent molecular mass of GCAP1 is 20 kDa in the presence of Ca^{2+} and

---

[1] D. L. Garbers and D. G. Lowe, *J. Biol. Chem.* **269**, 30741 (1994).

[2] R. N. Lolley and E. Racz, *Vision Res.* **22**, 1481 (1982).

[3] K.-W. Koch and L. Stryer, *Nature* **334**, 64 (1988).

[4] Y. Koutalos, K. Nakatani, T. Tamura, and K.-W. Yau, *J. Gen. Physiol.* **106**, 863 (1995).

[5] A. W. Shyjan, F. J. de Sauvage, N. A. Gillette, D. V. Goeddel, and D. G. Lowe, *Neuron* **9**, 727 (1992).

[6] R. M. Goraczniak, T. Duda, A. Sitaramayya, and R. K. Sharma, *Biochem. J.* **302**, 455 (1994).

[7] D. G. Lowe, A. M. Dizhoor, K. Liu, Q. Gu, M. Spencer, R. Laura, L. Lu, and J. B. Hurley, *Proc. Natl. Acad. Sci. U.S.A.* **92**, 5535 (1995).

[8] R.-B. Yang, D. C. Foster, D. L. Garbers, and H.-J. Fülle, *Proc. Natl. Acad. Sci. U.S.A.* **92**, 602 (1995).

[9] R. M. Goraczniak, T. Duda, and R. K. Sharma, *Biochem. Biophys. Res. Commun.* **234**, 666 (1997).

[10] W. A. Gorczyca, M. P. Gray-Keller, P. B. Detwiler, and K. Palczewski, *Proc. Natl. Acad. Sci. U.S.A.* **91**, 4014 (1994).

[11] K. Palczewski, I. Subbaraya, W. A. Gorczyca, B. S. Helekar, C. C. Ruiz, H. Ohguro, J. Huang, X. Zhao, J. W. Crabb, R. S. Johnson, K. A. Walsh, M. P. Gray-Keller, P. B. Detwiler, and W. Baehr, *Neuron* **13**, 395 (1994).

[12] A. M. Dizhoor, D. G. Lowe, E. V. Olshevskaya, R. P. Laura, and J. B. Hurley, *Neuron* **12**, 1345 (1994).

[13] W. A. Gorczyca, A. S. Polans, I. Surgucheva, I. Subbaraya, W. Baehr, and K. Palczewski, *J. Biol. Chem.* **270**, 22029 (1995).

[14] A. M. Dizhoor, E. M. Olshevskaya, W. J. Henzel, S. C. Wong, J. T. Stults, I. Ankoudinova, and J. B. Hurley, *J. Biol. Chem.* **270**, 25200 (1995).

[15] S. Frins, W. Bönigk, F. Müller, R. Kellner, and K.-W. Koch, *J. Biol. Chem.* **271**, 8022 (1996).

~25 kDa in the presence of EGTA.[13,16] GCAP2 appears as a band of 18–19 kDa in the presence of Ca^{2+} or 24 kDa when Ca^{2+} is chelated.[13] Both proteins have been cloned and their amino acid sequences deduced.[11,13–15] Bovine GCAP1 and GCAP2 are predicted to consist of 205 amino acids (calculated molecular mass 23.5 kDa)[11,15] and 204 amino acids (calculated molecular mass 23.7 kDa),[13,14] respectively. GCAP1 and GCAP2 share approximately 50% homology at the amino acid level. In accordance with the physiologic role postulated for GCAPs, they contain in their structures three EF-hand motifs that are potential sites of Ca^{2+} binding. The presence of a fourth, residual EF-hand motif (located at the N terminus) indicates that GCAPs belong to a subfamily of neuronal calcium-binding proteins that originates from a common calmodulin-like ancestor.[16] GCAPs are cotranslationally myristoylated at the N-terminal glycine.[11,17] Both proteins regulate activity of RetGC by direct interaction with its cytoplasmic kinase-like domain.[18,19] GCAP1 and GCAP2 have been localized by immunohistochemical techniques primarily in the photoreceptor layer of retina.[13,14,20,21] GCAP2 has also been detected in other cells of vertebrate retina.[20] Functionally active GCAP1 and GCAP2 have been successfully expressed in bacteria[15,17,21] and in eukaryotic cells.[13,17] Recently, GCAP1 has been shown to be involved in degenerative diseases of the retina.[22,23]

Since cyclic GMP is the central messenger molecule in visual transduction,[24] elucidation of the mechanisms regulating its synthesis in photoreceptor cells is necessary. A reliable assay for guanylyl cyclase activity in crude membranous material is crucial to studies of RetGC and to the identification of factors affecting its activity.

The main difficulty in the determination of RetGC activity is the rapid hydrolysis of cGMP by photoreceptor phosphodiesterase, which is activated, even with very weak illumination, in rod outer segment (ROS)

---

[16] A. S. Polans, W. Baehr, and K. Palczewski, *Trends Neurosci.* **19,** 547 (1996).

[17] E. V. Olshevskaya, R. E. Hughes, J. B. Hurley, and A. M. Dizhoor, *J. Biol. Chem.* **272,** 14327 (1997).

[18] T. Duda, R. Goraczniak, I. Surgucheva, M. Rudnicka-Nawrot, W. A. Gorczyca, K. Palczewski, A. Sitaramayya, W. Baehr, and R. K. Sharma, *Biochemistry,* **35,** 8478 (1996).

[19] R. P. Laura, A. M. Dizhoor, and J. B. Hurley, *J. Biol. Chem.* **271,** 11646 (1996).

[20] N. Cuenca, S. Lopez, K. Howes, and H. Kolb, *Invest. Ophthalmol. Vis. Sci.* **39,** 1243 (1998).

[21] A. Otto-Bruc, R. N. Fariss, F. Haeseleer, J. Huang, J. Buczylko. I. Surgucheva, W. Baehr, A. H. Milim, and K. Palczewski, *Proc. Natl. Acad. Sci. U.S.A.* **94,** 4727 (1997).

[22] S. L. Semple-Rowland, W. A. Gorczyca, J. Buczylko, B. S. Helekar, C. C. Ruiz, I. Subbaraya, K. Palczewski, and W. Baehr, *FEBS Lett.* **385,** 47 (1996).

[23] I. Sokal, N. Li, I. Surgucheva, M. J. Warren, A. M. Payne, S. S. Bhattacharya, W. Baehr, and K. Palczewski, *Mol. Cell* **2,** 129 (1998).

[24] E. N. Pugh, Jr., and T. D. Lamb, *Biochim. Biophys. Acta* **1141,** 111 (1993).

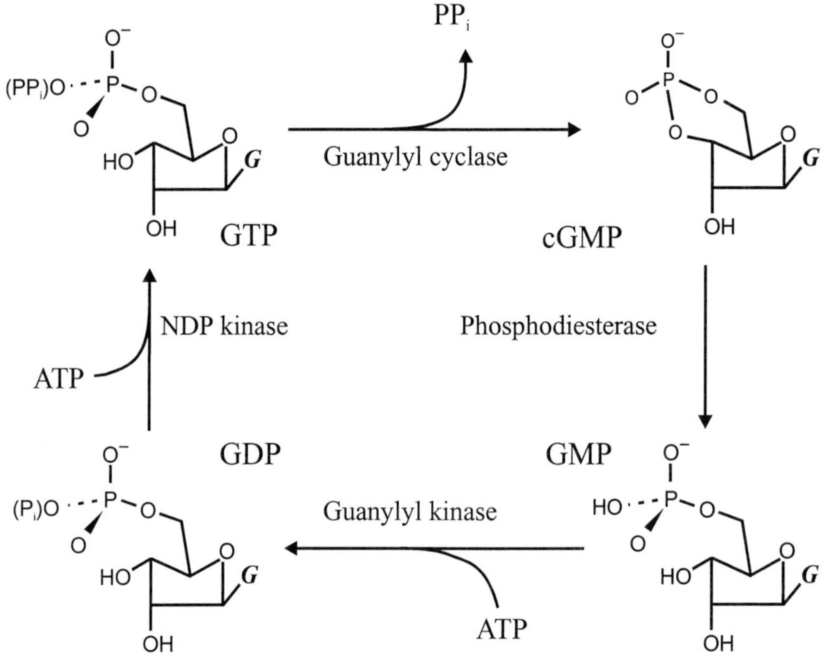

FIG. 1. The cycle of guanosine nucleotides in rod photoreceptor cells. In rod cells GTP is converted to cGMP by photoreceptor guanylyl cyclase, then cGMP is hydrolyzed by light-activated phosphodiesterase to GMP, which is subsequently phosphorylated by guanylyl kinase to GDP and finally by nucleoside diphosphate kinase to GTP.[25,26] G is an abbreviation for guanine.

homogenates. Moreover, the product of hydrolysis (GMP) is further metabolized to GDP and GTP (Fig. 1).[25,26] For these reasons, assays of RetGC activity in ROS need darkness or red light and application of PDE inhibitors.

One can easily avoid these inconveniences by taking advantage of the stereospecific metabolism of α-thionucleotides. Nucleoside phosphorothioates have been useful in many stereochemical analyses of enzymatic reactions. Enzymes that accept α-thionucleotides as substrates were found to be highly stereoselective with respect to the α phosphorus configuration.[27] Since the synthesis and resolution of S_p and R_p stereoisomers of cAMPS

[25] P. R. Kavipurapu, D. B. Farber, and R. N. Lolley, *Exp. Eye Res.* **34,** 181 (1982).
[26] W. A. Gorczyca, J. P. Van Hooser, and K. Palczewski, *Biochemistry* **33,** 3217 (1994).
[27] F. Eckstein, *Annu. Rev. Biochem.* **54,** 367 (1985).

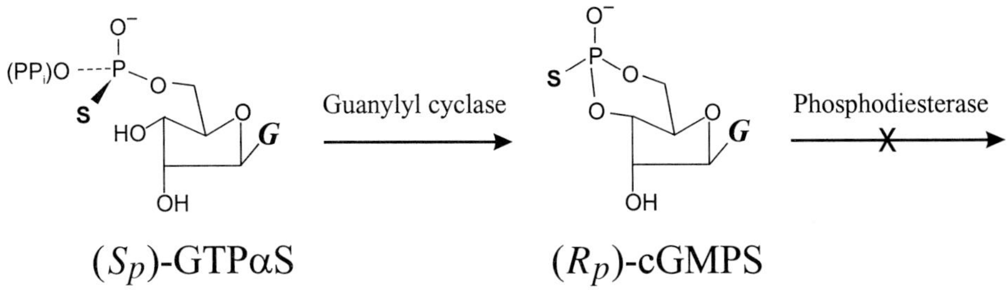

FIG. 2. Stereospecificity of photoreceptor guanylyl cyclase. The photoreceptor guanylyl cyclase catalyzes conversion of the S_p stereoisomer of GTPαS into the R_p stereoisomer of cGMPS,[31] which is an inhibitor of the photoreceptor phosphodiesterase.[32] G is an abbreviation for guanine.

(adenosine 3′,5′-cyclic thiomonophosphate) have been described,[28] stereochemical studies of adenylyl cyclase have become possible. It has been shown that the S_p stereoisomer of ATPαS (adenosine 5′-O-[1-thiotriphosphate]) is a substrate of brain adenylyl cyclase and is converted by the enzyme into the R_p stereoisomer of cAMPS.[29] Soluble and particulate guanylyl cyclases have also been demonstrated to utilize α-thionucleotide as substrates. The enzymes converted (S_p)-GTPαS (the S_p stereoisomer of guanosine 5′-O-[1-thiotriphosphate]) into (R_p)-cGMPS (the R_p stereoisomer of guanosine 3′,5′-cyclic thiomonophosphate).[30,31] These findings, together with the observation of Zimmerman et al.[32] that (R_p)-cGMPS is poorly hydrolyzed by PDE, have enabled development of a new assay for photoreceptor guanylyl cyclase[26] that has proved useful in its studies and purification of its activators.[10,13]

Assay of Photoreceptor Guanylyl Cyclase Activity

(S_p)-GTPαS is converted by RetGC into (R_p)-cGMPS about 10–20 times slower than GTP,[31] and (R_p)-cGMPS is hydrolyzed by PDE ~20,000 times slower than cGMP[32] (Fig. 2).

Using radioactive (S_p)-GTPα[^{35}S], one can simply determine GC activity in ROS homogenates by measurement of the reaction product radioactivity.

[28] J. Baraniak, R. W. Klimas, K. Lesniak, and W. J. Stec, J. Chem. Soc., Chem. Commun. 940 (1979).
[29] F. Eckstein, P. J. Romaniuk, W. Heideman, and D. R. Storm, J. Biol. Chem. 256, 9118 (1981).
[30] P. D. Senter, F. Eckstein, A. Mülsch, and E. Böme, J. Biol. Chem. 258, 6741 (1983).
[31] K.-W. Koch, F. Eckstein, and L. Stryer, J. Biol. Chem. 265, 9659 (1990).
[32] A. L. Zimmerman, G. Yamanaka, F. Eckstein, D. A. Baylor, and L. Stryer, Proc. Natl. Acad. Sci. U.S.A. 82, 8813 (1985).

The product is separated from the substrate (high excess) either with an anion exchange column or using neutral alumina gel. In comparison with other assay methods, this assay is fast, reproducible, and needs no darkness or PDE inhibitors. Guanylyl cyclase activity can be measured in homogenates of ROS or homogenates of washed ROS, which do not contain most of the soluble proteins.

Materials

(S_p)-GTPαS and (S_p)-GTPα[^{35}S] were obtained from DuPont-New England Nuclear (Boston, MA). Other nucleotides, EGTA, EDTA, IBMX (3-isobutyl-1-methylxanthine), dipyridamole, sodium nitroprusside, and BTP (1,3-bis[tris-(hydroxymethyl)methylamino]propane) were obtained from the Sigma-Aldrich Chemical Co., St. Louis, MO. Neutral alumina (ICN alumina) was obtained from ICN Biomedicals, Inc. (Costa Mesa, CA). The Mono Q HR 5/5 column is a product of Amersham Pharmacia Biotech.

Substrate Stock Solution

First, 250 μCi of (S_p)-GTPα[^{35}S] (20 μl; 1000–1500 Ci/mmol) is added to 1200 μl of 50 mM HEPES containing 60 mM KCl, 20 mM NaCl, and 25 μmol of "cold" (S_p)-GTPαS. Such a solution has a specific activity ~20,000 cpm/nmol at day 0. Since $t_{1/2}$ for ^{35}S is 87.4 days and the compound is stable when stored at $-80°$, the stock solution can be used without significant loss of assay sensitivity for at least 3 months. Stock solution (4 μl) added to the sample results in a final substrate concentration of 1.3 mM. In most cases, a more dilute stock solution (final volume 2 ml) that gives a 0.8 mM final concentration of the substrate is sufficient. The stock solution is divided into 100-μl aliquots and stored at $-80°$.

Buffers

The minimal assay buffer consists of 50 mM HEPES, pH 7.8, 60 mM KCl, 20 mM NaCl, 10 mM MgCl$_2$, and 0.4 mM EGTA, with CaCl$_2$ added to obtain the desired free calcium concentration (45 nM at 0.16 mM CaCl$_2$ and 1 μM at 0.375 mM CaCl$_2$).

The concentration of free calcium in each assay buffer can be adjusted by adding EGTA and CaCl$_2$ in amounts that are calculated using available computer programs, for example, CHELATOR 1.0.[33] Such calculations should take into account all other factors (other ion concentrations, concen-

---

[33] T. J. Schoenmakers, G. J. Visser, G. Flik, and A. P. R. Theuvenet, *Bio Techniques* **12**, 870 (1992).

trations of nucleotides, ionic strength, pH, and temperature) that influence the final concentration of free calcium in the prepared solution.

Taking into account that 50 μl of the total sample volume of 64 μl is destined for the buffer and assayed material, prepare stock solutions (concentrated 2×) of the following buffers, which are mostly used during studies of RetGC:

2× "*Low-calcium*" *buffer* (45 nM free calcium final concentration in the sample): (1) 100 mM HEPES, pH 7.8, (2) 120 mM KCl, (3) 40 mM NaCl, (4) 20 mM $MgCl_2$, (5) 1.024 mM EGTA, and (6) 0.205 mM $CaCl_2$.

2× "*High-calcium*" *buffer* (1 μM free calcium final concentration in the sample): (1)–(5) as for 2× low-calcium buffer and (6) 0.96 mM $CaCl_2$.

Methods

ROS Homogenate

All steps are performed on ice or at 4° under dim red light. Bovine retinas are dissected from fresh bovine eyeballs obtained from the local slaughterhouse. ROS are prepared from 100 retinas using a discontinuous sucrose gradient according to the procedure described by Papermaster[34] and homogenized using a glass/glass homogenizer (6–8 strokes) in 4 ml of 50 mM HEPES containing 60 mM KCl and 20 mM NaCl. The concentration of rhodopsin in the suspension thus obtained (ROS homogenate) is determined by the method described by McDowell[35] and adjusted to be 8 mg/ml. When stored at −80°, the ROS homogenate is suitable for use for at least 3 months. To avoid repeated freezing and thawing, the ROS homogenate should be divided into 0.5-ml aliquots and kept in polypropylene (Eppendorf) tubes wrapped tightly with aluminum foil to protect from light exposure.

Washed ROS Homogenate

The ROS membranes are washed to remove soluble activators of RetGC. ROS from 100 bovine retinas, isolated as stated earlier, is homogenized in 35 ml of distilled H_2O containing 20 μg/ml of leupeptin (10 mM HEPES, pH 7.5, containing 20 mM NaCl can be also used). The homogenate is centrifuged at 30,000g for 30 min at 4°. The supernatant can be discarded or saved for further isolation of GC activator(s) or other soluble proteins. The collected pellet is homogenized and centrifuged again as stated earlier.

[34] D. S. Papermaster, *Methods Enzymol.* **81,** 48 (1982).
[35] J. H. McDowell, *Methods Neurosci.* **15,** 123 (1993).

The second supernatant is discarded, the pellet is resuspended in 50 mM HEPES containing 60 mM KCl and 20 mM NaCl, and the concentration of rhodopsin is adjusted to 8 mg/ml as earlier. This suspension is termed *washed ROS homogenate* and is used during purification of activators of RetGC for monitoring the enzyme activity stimulation by purified material. Washed ROS homogenate can also be stored for a long time at −80° if aliquotted and protected from light as mentioned earlier.

Separation of Substrate and Product

Two methods have been developed for separating (R_p)-cGMPS from (S_p)-GTPαS: Mono Q chromatography and adsorption on neutral alumina.

Mono Q Chromatography

3′,5′-Cyclic mono-, 5′-mono-, 5′-di-, and 5′-triphosphates can be separated using a Mono Q HR 5/5 column and high-performance liquid chromatography (HPLC) or fast protein liquid chromatography (FPLC) systems. The column is equilibrated with 5 mM BTP, pH 7.5, and nucleotides are eluted with a linear gradient of NaCl (0–300 mM) over 20 min with a flow rate of 1 ml/min. Absorbance of the eluted material is monitored at 260 nm. Under the indicated conditions, cGMP, GMP, guanosine diphosphate (GDP), and GTP elute at 8, 9, 14, and 18 min, respectively, and the adenosine nucleotides adenosine monophosphate (AMP), cAMP, adenosine diphosphate (ADP), and adenosine triphosphate (ATP) elute at 8, 9, 13, and 17 min, respectively. Under the same conditions, (R_p)-cGMPS and (S_p)-GTPαS have retention times significantly longer than those of cGMP and GTP. (R_p)-cGMPS appears as a peak at 16 min and (S_p)-GTPαS is retained on the column and has to be eluted with a higher salt concentration. When radioactive substrate is used, the product and substrate are easily traced by measurements of radioactivity eluted from the column.

Adsorption on Neutral Alumina

This very simple method is convenient in assays dealing with many samples. The method is based on earlier observations that alumina at neutral or slightly alkaline pH binds 5′-nucleotides but not cyclic nucleotides.[36] The binding capacity of the gel is high; 150 mg of alumina binds up to 2 μmol of 5′-nucleotides.[26] After mixing 1 μmol of nucleotide with 150 mg of neutral alumina, the unbound fraction for different nucleotides was found to be: (R_p)-cGMPS > 98%, cGMP > 98%, GDP < 1%, GMP < 1%, ATP < 1%, GTP ∼ 1%, and (S_p)-GTPαS ∼ 1%. Therefore,

[36] A. A. White and T. V. Zenser, *Anal. Biochem,* **41,** 372 (1971).

it is easy to separate (R_p)-cGMPS from (S_p)-GTPαS by adding the mixture of nucleotides to neutral alumina suspended in buffer at neutral pH (0.5 ml of 200 mM Tris-HCl, pH 7.4, containing 50 mM EDTA was found to be suitable). After vigorous shaking (Vortex) the mixture is centrifuged. The alumina, with tightly bound substrate, forms a pellet, and the product remains in the supernatant. Two-milliliter polypropylene (Eppendorf) tubes containing 150 mg of neutral alumina are prepared a few days before the assay and filled with 0.5 ml of 200 mM Tris-HCl, pH 7.4, containing 50 mM EDTA immediately before the separation.

Assay Procedure

Samples are prepared in duplicate as described later. Accuracy during preparation of samples is very important, and use of a dedicated set of pipettes that has previously been checked for accuracy is recommended. The following components are added, in order, to a 1.5-ml polypropylene tube:
  25 μl 2× Assay buffer (low or high calcium, or other)
  4 μl Substrate stock solution
  25 μl Assayed material or H$_2$O
  10 μl ROS or washed ROS homogenate.
The samples are vortexed and kept at 30° in a water bath for 10–15 min. The reaction is stopped by adding 15 μl of 0.4 N HCl to each sample. The samples are vortexed and centrifuged (Eppendorf microcentrifuge) for 4 min at 14,000 rpm. The supernatant (50 μl) is transferred immediately to prepared tubes that contain alumina. Since the alumina is suspended in 0.5 ml of 200 mM Tris-HCl, pH 7.4, containing 50 mM EDTA, the transferred supernatant is neutralized and any residual GC activity is quenched. The samples are vortexed for 8 min and centrifuged as earlier. Supernatant (350 μl) from each tube is transferred to vials containing scintillation cocktail, and radioactivity is measured to determine the amount of (R_p)-cGMP[^{35}S] formed during the reaction. Control samples (to determine the background radioactivity) are treated exactly as test samples, except that 0.4 N HCl is added to the reaction mixture before the ROS homogenate (zero time of reaction). The total time of assay for 24 samples (11 experimental points plus control) is estimated to be approximately 2 hr. This estimated time includes preparation of samples (30 min), reaction (15 min), separation of products (25–30 min), and measurement of radioactivity in the scintillation β-counter (1 hr).

Calculations

To determine the total (R_p)-cGMP[^{35}S] formed in the reaction, the background radioactivity is subtracted from the radioactivity of the samples, and the result is multiplied by a factor of 2.48, because only fractions of samples are transferred during manipulations. The activity of the RetGC is expressed in nmol/min/mg of rhodopsin.

Properties of Photoreceptor Guanylyl Cyclase

The presence of Mg^{2+} or Mn^{2+} is obligatory for RetGC activity. During development of this assay, 10 mM $MgCl_2$ was found to be optimal for the enzyme activity, and the optimal pH was established to be 7.8–7.9. The assay is linear up to 15 min and is only slightly sensitive to light.[26]

The property that distinguishes the photoreceptor GC from other particulate guanylyl cyclases is the dependence of its activity on calcium ion concentration. This calcium-dependent regulation of RetGC is clearly shown in the assay with (S_p)-GTPα[^{35}S] as a substrate (Fig. 3). The calcium sensitivity of RetGC is lost by washing a soluble factor[3] from ROS membranes and restored by adding back GCAP1.

Adenosine nucleotides (except AMP) and their nonhydrolyzable analogs, when used in concentrations up to 0.5 mM, slightly stimulate RetGC activity in accordance with earlier observations that have been made for ATP.[37] Guanosine nucleotides containing three or more phosphates are competitive inhibitors of retinal GC. Polyphosphates also inhibit RetGC activity (Fig. 4).

Sodium nitroprusside (SNP) has no stimulatory effect on GC activity in ROS homogenates, indicating that contribution of soluble GC to cGMPS formation is negligible (Fig. 5). Interestingly, PDE inhibitors IBMX (3-isobutyl-1-methylxanthine) and dipyridamole[38] inhibit retinal GC with IC_{50} values of 5.5 and 0.17 mM, respectively (Fig. 5). IBMX has also been shown to affect other cyclases,[39] and investigators should use it to inhibit PDE hydrolytic activity with caution. It was also found that some commonly used compounds decrease activity of the enzyme. An important observation is that EGTA in concentrations higher than 1.5 mM inhibits RetGC with an IC_{50} of about 3.5 mM. Organic solvents also strongly inhibit RetGC activity (acetonitrile, for example, has an IC_{50} of 3%).

Comments

Because other guanylyl cyclases can use (S_p)-GTPαS as a substrate,[30] the preceding assay can be employed in their studies as well. The assay

---

[37] N. Krishnan, R. T. Fletcher, G. J. Chader, and G. Krishna, *Biochim. Biophys. Acta* **523**, 506 (1978).

[38] P. G. Gillespie and J. A. Beavo, *Mol. Pharmacol.* **36**, 773 (1989).

[39] R. M. Graeff, T. F. Walseth, and N. D. Goldberg, *Neurochem. Res.* **12**, 551 (1987).

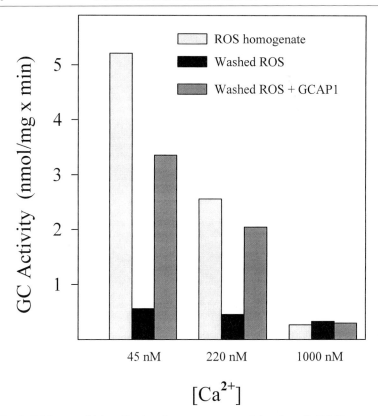

FIG. 3. Calcium sensitivity of bovine photoreceptor guanylyl cyclases. In ROS homogenates (light gray bars) guanylyl cyclase is active at low calcium concentration (45 nM) and loses its activity at high calcium concentration (1000 nM). This calcium dependence of cyclase activity may be destroyed by washing ROS membranes with low ionic strength buffer (black bars). Since washing of ROS membranes removes the soluble factor that mediates calcium-dependent activation of GC, the high activity of GC and its calcium sensitivity is restored by adding GCAP1 to washed ROS membranes (dark gray bars). The half-maximal activation of ROS guanylyl cyclase occurs at ~240 nM of free calcium.[10]

has also been used with GTPα[^{32}P] as a substrate in systems reconstituted with recombinant RetGC and recombinant GCAP.[13]

Isolation of GCAP1 and GCAP2 from Bovine Retina

The assay just described can be used for monitoring RetGC activity stimulation at different stages in the purification of guanylyl cyclase activating proteins from bovine retina. RetGC activity is assayed at low calcium

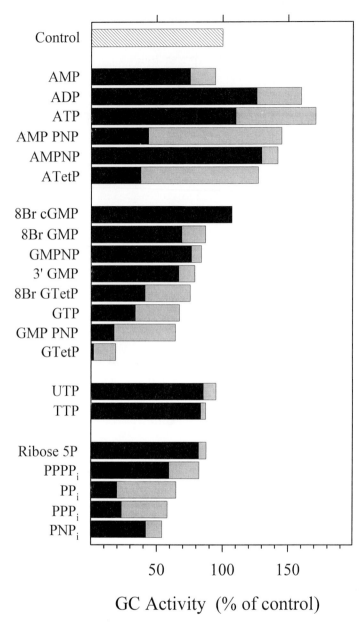

FIG. 4. Activators and inhibitors of photoreceptor guanylyl cyclase. The effects of the indicated compounds on the catalytic activity of photoreceptor guanylyl cyclase were determined under standard assay conditions (ROS homogenates, at 45 nM free calcium concentration). The gray and black bars represent GC activity in the presence of 0.4 mM (0.2 mM for adenosine nucleotides) or 2 mM concentrations of tested compounds, respectively. ATetP and GTetP are abbreviations for adenosine and guanosine tetrakisphosphate, respectively.

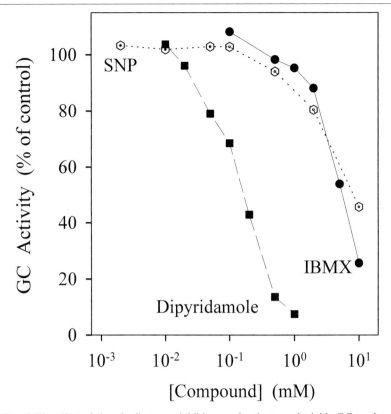

FIG. 5. The effect of phosphodiesterase inhibitors and activators of soluble GC on photoreceptor guanylyl cyclase activity. IBMX, dipyridamole (inhibitors of PDE), and sodium nitroprusside (SNP; a stimulator of soluble forms of cyclases) are added at indicated concentrations to the assay solution containing ROS homogenate. The assay is performed at low calcium conditions (45 nM free calcium).

in washed ROS homogenate that does not contain endogenous GCAP activity. To ensure that stimulatory activity is related to the presence of GCAPs, RetGC activity is controlled at high calcium concentration and SDS-PAGE analysis is performed.

GCAP1 can be isolated either from ROS extract or from retinal extract. GCAP2 is isolated from retinal extract. Two methods have been developed to isolate GCAPs: conventional column chromatography[10,14] and immunoaffinity purification, a fast and efficient method that exploits highly specific antibodies raised against both GCAPs.[13] The purification of GCAP1 from ROS extract by sequential column chromatography and the purification of GCAP1 and GCAP2 from retinal extract by immunoaffinity chromatography are described next.

Materials

DEAE-Sepharose Fast Flow, CNBr-activated Sepharose 4B, and protein A-Sepharose are products of Amersham Pharmacia Biotech. Nitrocellulose is a product of Schleicher & Schuell (Keene, NH). Immobilon P was obtained from Millipore (Bedford, MA). All chemicals are of reagent grade.

Antibodies

Polyclonal Antibody UW14 (pAbUW14). The antibody, immunoreactive with bovine GCAP1, is raised in rabbit against bacterially expressed truncated bovine GCAP1 (GCAP180, Met^{26}-Pro^{205}).[13] The antibody specifically recognizes GCAP1 and does not cross-react with bovine GCAP2. Antiserum UW14 can be used in Western blots at an optimal dilution of 1:10,000 and is suitable for identification of GCAP1.

pAbGS31. Rabbit antibody GS31 is raised against bovine GCAP1 C-terminal peptide (Asp^{189}-Asp^{205}-Lys).[13] Because the C terminus of GCAP1 differs from the C terminus of GCAP2, this antibody specifically recognizes only bovine GCAP1. The optimal dilution of antiserum GS31 in Western blots is 1:2000. The antibody has been chosen for separation of GCAP1 from GCAP2.[13]

pAb850. Antibody 850 is raised in rabbit against recombinant (*Escherichia coli* expressed) bovine GCAP2, in which six N-terminal residues are replaced by a hexahistidine (His_6) tag.[21] It does not cross-react with bovine GCAP1. Antiserum 850 can be used in Western blots at a dilution 1:20,000 or more (even up to 200,000) and has been found suitable for isolation of GCAP2 from retinal extract and separation from GCAP1.

Monoclonal Antibody G2 (MAbG2). Raised against truncated bovine GCAP1 (GCAP180) in BALB/c mice,[11] G2 has high specific activity against GCAP1 (in Western blots it works well at a concentration of 0.2 μg/ml). It is, however, cross-reactive with bovine GCAP2.[13] This antibody can be used for isolation of either GCAP2 or GCAP1, or for simultaneous isolation of both GCAPs from retinal extract.

Methods

Purification of GCAP1 from ROS Extract by Column Chromatography

The purification of GCAP1 from bovine ROS extract comprises the following steps: preparation of bovine ROS extract, chromatography over

DEAE-Sepharose Fast Flow, HPLC over hydroxylapatite, and reversed-phase HPLC over C_4.

Preparation of Bovine ROS Extract

The first steps of purification are performed in dim red light. All solutions, containers, centrifuges, etc., are kept at 4°. ROS, obtained from 50 retinas, is homogenized in a glass/glass homogenizer (6–8 strokes) with 20 ml of distilled water containing 0.5 mM benzamidine and 20 μg/ml of leupeptin. The homogenate is clarified by centrifugation at 30,000g for 30 min at 4°. To remove the remaining impurities, the collected supernatant is centrifuged again at 100,000g for 30 min at 4° and the resulting supernatant (the ROS extract), after dialysis for 8 hr against distilled water, is subjected to DEAE chromatography as described next.

Chromatography over DEAE-Sepharose Fast Flow

This step and subsequent steps can be performed in daylight or room light. The ROS extract is loaded at a flow rate of 0.5 ml/min onto a small column (HR5/5; Amersham Pharmacia Biotech) filled with DEAE-Sepharose Fast Flow previously equilibrated with 5 mM BTP, pH 7.5, containing 50 mM NaCl. After all the extract is loaded, the column is washed at a flow rate of 1 ml/min with 5 mM BTP, pH 7.5, containing 100 mM NaCl until the absorbance at 280 nm is below 0.01. Bound proteins are eluted with a linear NaCl gradient (100–350 mM NaCl in 5 mM BTP, pH 7.5) at a flow rate of 1 ml/min. Fractions (1 ml) are collected. Elution of proteins is monitored by absorbance at 280 nm. The ability of eluted fractions to stimulate RetGC activity is checked at low calcium (45 nM) conditions. Fractions that stimulate RetGC activity (fractions 12–22), eluted at 220 mM NaCl, are combined and then subjected to hydroxylapatite chromatography.

HPLC over Hydroxylapatite

The combined fractions are loaded onto a hydroxylapatite column (7.5 × 100 mm; Pentax Column SH-07110m. Asahi Optical Co., Ltd., Japan) equilibrated with 10 mM BTP, pH 7.5, containing 100 mM NaCl. Fractions (1 ml) are collected at a flow rate of 1 ml/min. Fractions containing GCAP activity are eluted from the column with an increasing concentration of KH_2PO_4 (0–60 mM) and decreasing concentration of NaCl (100–0 mM) in 10 mM BTP, pH 7.5. Stimulatory activity is eluted at 30 mM KH_2PO_4 and 50 mM NaCl in three fractions (usually between fractions 17–21) containing GCAP1 (20-kDa protein as determined by SDS–PAGE analysis). These fractions are pooled and concentrated to 0.5 ml using a Speed-Vac. Acetonitrile is added to a final concentration of 15% and the sample is subjected to reversed-phase HPLC.

Reversed-Phase HPLC over C_4

The sample is loaded onto a C_4 column (4.6 × 150 mm; W-Porex 5 C4; (Phenomenex, Torrence, CA) equilibrated with 30% acetonitrile in 5 mM BTP, pH 7.5. GCAP1 is eluted with a linear gradient of acetonitrile (30–60%) in BTP, pH 7.5, at a flow rate of 1.5 ml/min. Fractions of 0.75 ml are collected. All eluted fractions are analyzed by SDS–PAGE (12.5% gel, 20 μl of fraction/lane), and their ability to stimulate RetGC is tested. Because the presence of acetonitrile strongly inhibits RetGC activity, it is recommended to withdraw only a small volume (2 μl) from each fraction or, alternatively, to lyophilize 150 μl of each fraction and then resuspend it in 30 μl of H_2O. GCAP1 is eluted as a peak containing 4–5 fractions (fractions 8–12). In SDS–PAGE GCAP1 appears as the main, 20-kDa band.[10] It is slightly contaminated by a low molecular weight protein identified as the γ subunit of transducin. Sometimes a minor band with a molecular weight of about 19 kDa appears in fractions 13–15 eluted at higher acetonitrile concentration.[10] The final yield of GCAP1 obtained by this method is up to 10 μg of protein from 50 retinas.

Immunoaffinity Purification of GCAP1 and GCAP2 from Retinal Extract

Immunoaffinity chromatography is a relatively simple, fast, and efficient method of purification of GCAPs from retinal extracts. It is possible to obtain highly purified proteins in one step with much higher yield than with conventional chromatography. The procedure includes preparation of immunoadsorbents, preparation of retinal extract, and purification of GCAP1 and GCAP2.

Preparation of Immunoadsorbents

In general, antibodies that are used as immunoadsorbents should be highly specific and not cross-react with other proteins, although sometimes cross-reactivity can be advantageous. The specificity of antibodies can be determined by Western blot analysis of purified GCAPs (0.2–0.5 μg/lane), ROS extract (2–4 μl/lane), or retinal extract (1–2 μl/lane) separated by SDS–PAGE (12.5% gel).

Monoclonal antibodies G2 are isolated from ascites fluid, either by standard affinity chromatography on protein A-Sepharose or by ammonium sulfate precipitation followed by ion-exchange chromatography using DEAE-Sepharose Fast Flow and a NaCl step gradient (50 and 100 mM NaCl). Both methods give highly purified IgGs that retain high reactivity against bovine GCAP1, but are also weakly reactive against bovine GCAP2. Polyclonal antibodies 850 are purified in a similar way (protein A affinity chromatography is recommended). Polyclonal antibodies GS31 are affinity purified using the synthetic C-terminal peptide (Asp^{189}-Asp^{205}-Lys) of GCAP1 coupled to Sepharose 4B.[13]

Three different immunoadsorbents are prepared from the three antibodies by coupling with CNBr-activated Sepharose 4B according to the procedure recommended by the manufacturer. Antibodies G2 and 850 are coupled with high density (6–8 mg IgG/ml of gel and 5 mg IgG/ml of gel, respectively). Antibody GS31 is substituted with lower density (<1 mg/ml of gel) because of low yield during affinity purification of this antibody. Prepared immunoadsorbents (2 ml) are loaded into small columns (Pharmacia C10/10 columns with flow adapters are very convenient) and equilibrated with 10 mM HEPES, pH 7.4, containing 50 mM NaCl.

Preparation of Retinal Extract

The extract from bovine retinas is obtained essentially in the same way as that from ROS. The first steps of purification are performed in dim red light and 4°. Fifty retinas (fresh or stored at −20°) are homogenized using a Teflon/glass homogenizer and then a glass/glass homogenizer (6–8 strokes each time) with 50 ml of distilled H_2O containing 0.5 mM benzamidine and 20 μg/ml of leupeptin. The homogenate is clarified by centrifugation at 30,000g for 30 min at 4°. To remove impurities, the collected supernatant is centrifuged again at 100,000g for 30 min at 4°, and the resulting supernatant (the retinal extract) is used as a source of GCAPs.

Purification of GCAP1 and GCAP2

Two methods were developed to separate GCAP1 from GCAP2 by immunoaffinity chromatography of retinal extracts: chromatography on pAbGS31-Sepharose4B followed by chromatography on MAbG2-Sepharose 4B or, alternatively, chromatography on pAb850-Sepharose 4B followed by chromatography on MAbG2-Sepharose 4B.

pAbGS31 + MAbG2 Immunoaffinity Chromatography. This method was used to show that both GCAPs are present in the same retinal extract and that each of them is able to stimulate RetGC activity separtely.[13] All steps are performed at 4°. Retinal extract obtained from 25 retinas (25 ml) is loaded at a flow rate of 6 ml/hr onto a pAbGS31-Sepharose column (2 ml). After the extract is loaded, the column is washed with 10 mM HEPES, pH 7.4, containing 50 mM NaCl, and GCAP1 is specifically eluted with a 1 mM concentration of the C-terminal peptide (Asp^{189}-Asp^{205}-Lys). The column is washed with 0.1 M glycine hydrochloride, pH 2.5, to remove adsorbed proteins and then reequilibrated with 10 mM HEPES, pH 7.4, containing 50 mM NaCl. The extract that passed through the column during the first loading is loaded again and the procedure is repeated four times to ensure that all GCAP1 is adsorbed. Eluted GCAP1 is analyzed by

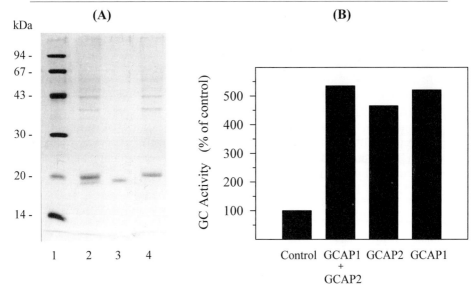

FIG. 6. Separation of bovine GCAP1 and GCAP2 from retinal extract by tandem immunoaffinity chromatography. (A) A Coomassie-stained, 12.5% SDS–polyacrylamide gel showing fractions eluted from immunoaffinity columns. The extract, obtained from bovine retinas, was applied first onto a pAb850-Sepharose column, and unbound material was directly loaded onto an MAbG2-Sepharose column (connected in tandem). Columns were washed, disconnected, and then GCAPs were eluted from each column with low pH buffer. GCAP2 is eluted from pAb850-Sepharose (lane 3) and GCAP1 is eluted from MAbG2-Sepharose (lane 4). Retinal extract was also directly applied onto an MAbG2-Separose column, and both GCAPs were bound and then eluted (lane 2). On each lane of the gel, 30 μl from the GCAP fractions were loaded. Lane 1 contains molecular weight markers. (B) Stimulatory activity of eluted fractions. Activity of RetGC was determined in washed ROS homogenates at low calcium conditions in the presence of 0.3 μg of GCAP1, GCAP2, or both GCAPs. Basal activity of washed membranes (control) was determined to be 0.43 nmol/min/mg of rhodopsin.

SDS–PAGE and its stimulatory activity is tested in the RetGC activity assay.[13] In the next step, the extract that now contains only GCAP2 is loaded onto a MAbG2-Sepharose column (2 ml) at a flow rate of 6 ml/hr. After the extract is loaded, the column is washed with 10 mM HEPES, pH 7.4, containing 200 mM NaCl (20 ml) and then with 10 mM HEPES, pH 7.4. GCAP2 is eluted with 0.1 M glycine hydrochloride, pH 2.5. Collected fractions (0.5 ml) are immediately neutralized with 0.15 ml of 1 M Tris-HCl, pH 8.4. Twenty microliters from each fraction is subjected to SDS–PAGE (12.5% gel) and stimulatory activity is tested in washed ROS homogenates using a GC activity assay in low calcium buffer. In Coomasie-stained gels GCAP2 appears as an approximately 18-kDa protein in fractions 8–10; the same fractions stimulate RetGC activity.

pAb850 + MAbG2 Immunoaffinity Chromatography. Retinal extract obtained from 50 retinas (50 ml) is loaded overnight at a flow rate of 3 ml/hr onto a 2-ml pAb850-Sepharose column that is connected in tandem with an identical column filled with MAbG2-Sepharose. In this manner, the extract passes first through the column that adsorbs GCAP2 and then through the column that adsorbs GCAP1. After all extract is loaded, the columns are washed (still connected) until the absorbance at 280 nm is below 0.02. The columns are then disconnected and washed separately with 10 mM HEPES, pH 7.4, containing 200 mM NaCl (20 ml for each column) at a flow rate of 18 ml/hr followed by 10 mM HEPES, pH 7.4 (4–6 ml), at the same flow rate. GCAP2 is eluted from pAb850-Sepharose and GCAP1 is eluted from the MAbG2-Sepharose with 0.1 M glycine hydrochloride, pH 2.5, at a flow rate of 15–20 ml/hr. Eluted fractions (1 ml) are immediately neutralized with 35 μl of 1.5 M Tris-HCl, pH 8.8, subjected to SDS–PAGE analysis (Fig. 6), and checked for stimulation of RetGC activity in washed ROS. Contaminating proteins (~36- and 40-kDa bands) can be removed using C_4 column RP-HPLC under the chromatography conditions described earlier in this chapter. The final yield of GCAP1 purified by this method is about 70 μg of protein obtained from 50 retinas, and the yield of GCAP2 is about 30 μg of protein purified from the same quantity of retinas.

Comments

The preceding methods can also be applied to purification of recombinant GCAPs. Recombinant GCAP1 (expressed in *E. coli*) is recovered from bacterial extracts in good yield using C_4 RP-HPLC as described earlier.[13] Affinity chromatography can be performed successfully with GCAP1 and GCAP2 expressed in insect cells and in bacteria.[13]

Acknowledgments

I thank Greg Garwin for help in preparing the manuscript. This work was supported by the Polish Committee for Scientific Research (KBN grant 4 P05A 082 14). In part, experiments described in this contribution were carried out at the University of Washington, Department of Ophthalmology, in the laboratory of Dr. Kris Palczewski and supported by an NIH grant from the National Eye Institute, EY 08061.

[46] Heterologous Expression and Assays for Photoreceptor Guanylyl Cyclases and Guanylyl Cyclase Activating Proteins

By James B. Hurley and Alexander M. Dizhoor

Introduction

Vertebrate photoreceptors recover quickly following photoexcitation induced by a short flash of light. They also adapt to light so that they can respond to variations in intensity even in the presence of background illumination.[1–3] Ca^{2+} appears to play an important role in recovery and adaptation. Free intracellular Ca^{2+} in dark-adapted rod photoreceptors has been estimated to be ~500 nM and on illumination it falls to ~50 nM.[4–6] This change appears to be necessary for light adaptation since conditions that prevent changes in intracellular free Ca^{2+} also prevent light adaptation.[7,8] One of the ways in which Ca^{2+} may control recovery and light adaptation in vertebrate photoreceptors is by regulating photoreceptor membrane guanylyl cyclases (RetGCs) via Ca^{2+}-binding proteins.[1–3]

Guanylyl cyclase (GC) activity in photoreceptor membranes is derived from two membrane guanylyl cyclases referred to as RetGC-1 and RetGC-2. (Because they were discovered by independent laboratories the human forms are referred to as RetGC-1 and RetGC-2, the rat counterparts are referred to as GC-E and GC-F, and the bovine counterparts as ROS-GC1 and ROS-GC2.) RetGCs are unusual membrane guanylyl cyclases because they do not respond to the types of extracellular peptide ligands that activate other closely related membrane GCs, e.g., the natriuretic peptide receptors.[9] An important and unusual property of RetGCs is that they are active only at low concentrations of free Ca^{2+}.[10] Biochemical and electrophysiologic

---

[1] K. W. Yau, *Invest. Ophthalmol. Vis. Sci.* **35**, 9 (1994).
[2] J. L. Miller, A. Picones, and J. I. Korenbrot, *Curr. Opin. Neur.* **4**, 488 (1994).
[3] E. N. Pugh, Jr., T. Duda, A. Sitaramayya, and R. K. Sharma, *Biosci. Rep.* **17**, 429 (1998).
[4] G. M. Ratto, R. Payne, W. G. Owen, and R. Y. Tsien, *J. Neurosci.* **8**, 3240 (1988).
[5] M. P. Gray-Keller and P. B. Detwiler, *Invest. Ophthalmol. Vis. Sci.* **35**, 1486 (1994).
[6] A. P. Sampath, H. R. Matthews, M. C. Cornwall, and G. L. Fain, *J. Gen. Physiol.* **111**, 53 (1998).
[7] H. R. Matthews, R. L. W. Murphy, G. L. Fain, and T. D. Lamb, *Nature* **334**, 67 (1988).
[8] K. Nakatani and K.-W. Yau, *Nature* **334**, 69 (1988).
[9] D. L. Garbers and D. L. Lowe, *J. Biol. Chem.* **269**, 30741 (1994).
[10] K.-W. Koch and L. Stryer, *Nature* **334**, 64 (1988).

evidence suggests that lowered Ca^{2+} levels produced in response to illumination stimulate RetGCs to synthesize cGMP.[3]

Koch and Stryer[10] reported in 1988 that a soluble protein was required to impart Ca^{2+} sensitivity to GC activity in photoreceptor membranes. Later, two soluble Ca^{2+}-binding proteins with this activity, GCAP-1 and GCAP-2, were identified and purified.[11,12,12a] GCAP-1 was isolated from a retinal fraction containing photoreceptor outer segment membranes and it was purified based on its ability to activate GC in washed photoreceptor outer segment membranes.[12] GCAP-2 was isolated from a heat-stable fraction of soluble retinal proteins and it too was purified to homogeneity by the same criterion but using a different purification scheme.[11] Both GCAP-1 and GCAP-2 have three distinct EF hands. A myristoylation signal is found at the N termini of GCAP-1 and GCAP-2 and both proteins are fatty acylated at their N termini.[13,14]

Specific antibodies to both GCAPs have been raised and immunoaffinity procedures were developed to isolate native GCAPs directly from retinal extracts.[15,16] For example, a highly specific antibody, ΔNp24, raised against a recombinant fragment of GCAP-2 has been used to purify from a crude retinal extract functional homogenous GCAP-2 that is free of GCAP-1 contamination.[15]

GCAP-1 and GCAP-2 are both present in photoreceptors,[15–18] but studies reporting their precise localization have been, unfortunately, contradictory and confusing. In addition to staining photoreceptors, some anti-GCAP-1 and GCAP-2 antibodies stain retinal amacrine or ganglion cells weakly, but it is not clear whether this represents a low level of GCAP expression or weak cross-reactivity of the antibodies with structurally re-

---

[11] A. M. Dizhoor, D. G. Lowe, E. V. Olshevskaya, R. P. Laura, and J. B. Hurley, *Neuron* **12**, 1345 (1994).

[12] W. A. Gorczyca, M. P. Gray-Keller, P. B. Detwiler, and K. Palczewski, *Proc. Natl. Acad. Sci. U.S.A.* **91**, 4014 (1994).

[12a] W. A. Gorczyca *Methods Enzymol.* **315** [45], 1999 (this volume).

[13] K. Palczewski, I. Subbaraya, W. A. Gorczyca, B. S. Helekar, C. C. Ruiz, H. Ohguro, J. Huang, X. Zhao, J. W. Crabb, and R. S. Johnson, *Neuron* **13**, 395 (1994).

[14] E. V. Olshevskaya, R. E. Hughes, J. B. Hurley, and A. M. Dizhoor, *J. Biol. Chem.* **272**, 14327 (1997).

[15] A. M. Dizhoor, E. V. Olshevskaya, W. J. Henzel, S. C. Wong, J. T. Stults, I. Ankoudinova, and J. B. Hurley, *J. Biol. Chem.* **270**, 25200 (1995).

[16] W. A. Gorczyca, A. S. Polans, I. G. Surgucheva, I. Subbaraya, W. Baehr, and K. Palczewski, *J. Biol. Chem.* **270**, 22029 (1995).

[17] K. Howes, J. D. Bronson, N. Li, K. Zhang, M. Lee, I. Subbaraya, J. Frederick, A. Surguchov, H. Kolb, K. Palczewski, and W. Baehr, *Invest. Ophthalmol. Vis. Sci.* **38**, A93 (1997).

[18] A. Otto-Bruc, R. N. Fariss, F. Haeseleer, J. Huang, J. Buczylko, I. Surgucheva, W. Baehr, A. H. Milam, and K. Palczewski, *Proc. Natl. Acad. U.S.A.* **94**, 4727 (1997).

lated protein(s).[17] GCAP-1 and GCAP-2 have both been detected in outer and inner segments of photoreceptors. GCAP-1 was detected in outer segments, cell bodies, and synaptic termini of both cones and rods,[16,18] and highly specific GCAP-1 antibodies detected heavy expression of GCAP-1 in cones and much weaker expression in rods.[17,19] GCAP-2 has been detected primarily in outer and inner segments of rods using a highly specific polyclonal antibody, DNP24, preadsorbed with recombinant GCAP-1.[15,19] A similar result has been reported using a monoclonal antibody to detect GCAP-2 in mouse retinas.[17] Studies using another highly specific anti-GCAP-2 antibody preadsorbed with excess recombinant GCAP-1 has revealed that GCAP-2 is heavily expressed in outer segments of bovine rods.[19] However, note that there is some disparity in studies of GCAP-2 localization that have been published. For example, some antibodies against GCAP-2 label predominantly cones.[18] In general, however, most studies agree that GCAP-1 and GCAP-2 are both expressed in photoreceptors and are both present in outer segments of photoreceptors. The relative distribution of GCAP-1 and -2 between rods and cones and within compartments of photoreceptors may depend on which antibody preparation is used, on the animal species being analyzed, and on how extensively different methods of fixation affect epitope accessibility and immunoreactivity.[17]

Major progress leading to our present understanding of the role and molecular mechanisms of cGMP synthesis in photoreceptors has been achieved during the last few years. One reason for this progress is that methods have been developed to produce guanylyl cyclases and their regulatory proteins in heterologous expression systems and to assay their activities. *Escherichia coli*, insect cell, and mammalian cell cultures have been used to express photoreceptor membrane guanylyl cyclases and guanylyl cyclase activating proteins (GCAPs). Here we describe some of the expression systems and biochemical assays that we have found to be particularly useful for these studies.

Methods for Assaying Photoreceptor GCs

Photoreceptor membrane homogenates are rich in GC activity. However, analysis of GC activity in photoreceptor homogenates is complicated by the presence of a highly active phosphodiesterase (PDE) that is stimulated by light. The PDE is fully activated even by very dim light, so special precautions must be taken to prevent cGMP hydrolysis.

[19] S. Kachi, E. V. Olshevskaya, Y. Nishizava, N. Watanabe, A. Yamazaki, A. Dizhoor, and J. Usukura, *Exp. Eye. Res.* **68,** 465 (1999).

Investigators studying photoreceptor guanylyl cyclase have developed three different types of approaches to avoid interference from cGMP hydrolysis when assaying GC activity in photoreceptor membranes:

1. GC activity can be measured under infrared illumination using fully dark-adapted photoreceptor membranes. To suppress basal PDE activity, PDE inhibitors, such as zaprinast, dipyridamole or isobutylmethylxanthine, are also used. In GC assays that are supplemented with purified GCAPs PDE activity is further reduced by first washing the photoreceptor membranes with a low-salt solution.[11] [α-^{32}P] GTP is used as a substrate and cGMP formed from it isolated by thin-layer chromatography (TLC). It is important to monitor recovery of cGMP by including a known amount of [^{3}H]cGMP as an internal standard.[15] This method is time consuming because of the TLC analysis, but it is accurate, produces low nonspecific background, requires small reaction volumes (25–50 μl), and is inexpensive.

2. GC activity can also be measured using a radiolabeled thio derivative of GTP, (S_P)-GTPαS as a substrate.[20] GC activity inverts the configuration of the thiophosphate of this substrate to produce (R_P)-cGMPS, which is a poor substrate for the photoreceptor PDE. Because of the reduced rate of product hydrolysis the reaction can be performed under dim illumination. The radiolabeled product is isolated using alumina oxide. The substrate is expensive but this technique is fast, simple, and accurate.

3. A third type of GC assay monitors production of pyrophosphate (PP$_i$) produced during conversion of GTP to cGMP. This assay can be performed under normal illumination.[21] The activity of exogenous inorganic pyrophosphatase enzymatically coupled to NADH oxidation is monitored by an optical method. This method requires large reaction volumes but it allows GC activity to be measured *in vitro* continuously in real time.

An additional type of measurement, cGMP radioimmunoassay, has also been used to monitor recombinant RetGC activity.[22] This method requires low levels of PDE activity and is not suitable for ROS preparations but it can be performed using commercial kits; and its high sensitivity is helpful for assaying low levels of activity.

Heterologous Expression of RetGCs

Recombinant RetGC-1 and RetGC-2 (or their homologs from other animal species) can be expressed in active form using mammalian cell

[20] W. A. Gorczyca, J. P. Van Hooser, and K. Palczewski, *Biochemistry* **33**, 3217 (1994).
[21] G. Wolbring, and P. P. Schnetkamp, *Biochemistry* **34**, 4689 (1995).
[22] A. W. Shyjan, F. J. de Sauvage, N. A. Gillett, D. V. Goeddel, and D. G. Lowe, *Neuron* **9**, 727 (1992).

cultures. Successful bacterial expression systems have not yet been reported. RetGC-1 was first expressed in a stable cell line using an expression vector that harbored ouabain resistance.[22] More recently, functional RetGC-1 and RetGC-2 were produced in HEK293 and COS-7 cells[11,23–25] or in insect cells using eukaryotic expression vectors.[26] We describe here in detail a protocol for expressing RetGCs in human transformed embryonic kidney cells HEK293.

Expression Vector

RetGC-1 or RetGC-2 cDNA is inserted into pRCCMV (Invitrogen) under control of a strong viral promoter. This vector also encodes neomycin resistance that allows selection for a stable RetGC expressing cell line. However, we have found that expression of RetGCs in transiently transfected cells gives sufficiently high yields of GC activity.

Cell Culture

HEK293 cells (available from ATCC, Rockville, MD) are propagated in a CO_2 incubator at 5% (v/v) CO_2 in 10 ml of RPMI 1640 medium supplemented with 10% heat-inactivated calf serum, glutamine, antibiotics, and antimycotic buffered with HEPES at about pH 7.2. (All reagents are available from Life Technologies/GIBCO-BRL Gaithersburg, MD.)

Several days before transfection (depending on the rate of cell growth) the cells from each 100-mm culture dish are gently rinsed with sterile phosphate-buffered saline (PBS) and treated with 0.05% (w/v) trypsin, 0.5 mM EDTA in sterile HEPES-buffered saline solution (Life Technologies) for 1 min. The cell suspension from one plate is then split into three to five new dishes with 10-ml fresh media. The cell are then grown until 50–60% confluence, at which point they can be used for transfection.

Transfection

The pRCCMV RetGC cDNA expression plasmid is purified from *E. coli* using QIAGEN or Promega (Madison, WI) plasmid purification

[23] D. G. Lowe, A. M. Dizhoor, K. Liu, Q. Gu, M. Spencer, R. Laura, L. Lu, and J. B. Hurley, *Proc. Natl. Acad. Sci. U.S.A.* **92,** 5535 (1995).
[24] R.-B. Yang, D. C. Foster, D. L. Garbers, and H.-J. Fulle, *Proc. Natl. Acad. Sci. U.S.A.* **92,** 602 (1994).
[25] R. Goraczniak, T. Duda, and R. K. Sharma *Biochem. Biophys. Res. Commun.* **234,** 666 (1997).
[26] I. Sokal, N. Li, I. Surgucheva, M. J. Warren, A. M. Payne, S. S. Bhattacharya, W. Baehr, and K. Palczewski. *Molec. Cell* **2,** 129 (1998).

kits. We typically use 15–20 μg of DNA per standard 100-mm culture dish. Solutions for calcium phosphate DNA transfection can be made or can be purchased from commercial sources (for example, Life Technologies/ GIBCO). Every new batch of reagents should be tested for optimal transfection.

Approximately 2 hr before transfection, the growth medium is replaced with fresh medium buffered with HEPES/KOH to pH 7.1–7.2. For each transfection, DNA is diluted in 440 μl of deionized sterile water, thoroughly mixed with 62 μl of 2 M $CaCl_2$, and then added dropwise into 0.5 ml of 2× HEPES-buffered saline (50 mM HEPES, 280 mM NaCl, 1.5 mM Na_2HPO_4, pH 7.1) while vortexing gently at room temperature. The resulting slightly opaque suspension of precipitated DNA is then added dropwise to each plate so that it is evenly distributed over the cell culture. Sedimented DNA precipitated on cells can be clearly observed in approximately 1 hr under a microscope. After incubation of the cells for an additional 5 hr in a CO_2 incubator at 37°, the medium is changed and the cells allowed to grow an additional 48 hr before harvesting.

Membrane Preparations

The harvested cells are washed with PBS to remove growth medium, covered with 10 mM Tris (pH 7.5) containing 2 mM $MgCl_2$, 1 mM adenosine triphosphate (ATP) and 10 μg/ml each leupeptin and aprotinin and 0.1 mM phenylmethylsulfonyl fluoride (PMSF), scraped from the plates and disrupted by 15 strokes in a Dounce homogenizer on ice. The homogenate is then centrifuged at approximately 2000g for 5 min at 4° to remove most of the nuclear material, and then membranes from the supernatant are collected at 100,000g for 1 hr at 4°. Membranes should be aliquotted, frozen in liquid nitrogen and stored at −70° until used in the assay.

Heterologous Expression of GCAPS

Both eukaryotic and prokaryotic expression systems have been used successfully to produce functional recombinant GCAPs. Originally GCAP-1 was expressed in insect cell culture using a baculovirus expression vector,[16] and GCAP-2 was expressed in transfected HEK293 cells.[15] A bacterial expression system was later developed that allows efficient production of large quantities of both GCAP-2 and GCAP-1 in *E. coli*.[14,15]

We describe here an efficient and reliable method for producing large quantities of functional GCAPs in *E. coli*. The method is versatile because it allows large quantities of GCAPs to be produced with or without an

N-terminal myristoyl residue. Expression of GCAP-2 and GCAP-1 using this system has also been described in previous publications.[14,27,28]

Expression Vectors

The vectors for expression of GCAPs in *E. coli* were constructed on the basis of pET11 expression vectors (Novagen). The GCAP cDNA was amplified from a bovine retinal cDNA library using *Pfu* polymerase (Stratagene, La Jolla, CA) and ligated into the *Nco*I and *Bam*HI restriction endounuclease sites of pET11d. The cDNA insert contained the initiator ATG as part of the *Nco*I site and a stop codon preceeding the *Bam*HI site. Expression of GCAP from this vector is under control of the T7 polymerase promoter so that it is regulated by the *lac* operator. High-level GCAP expression is therefore induced by the presence of isopropylthiogalactoside (IPTG). The vector also encodes resistance to ampicillin for selection to maintain transformed bacterial clones.

Bacterial Strains

We have successfully used *E. coli* strains that carry T7 polymerase incorporated as a prophage, but in our experience the strain of choice is BLR(DE3)pLysS (Novagen). We initially used BL21 (DE3)pLysS, as well, but with this strain the level of expression for some forms of GCAPs tends to decline after several generations, perhaps due to the tendency of this strain to lose its DE3 prophage.

Transformation

Chemically treated competent bacterial cells can be purchased from Novagen, or directly prepared using conventional techniques described in common molecular biology handbooks. We prefer to use electroporation as a method of transformation because of its high efficiency. The expression vector is first mixed either with freshly prepared or previously stored ($-80°$) electrocompetent cells in a final volume of 50–80 μl at 0°: The mixture is then immediately subjected to electroporation at 1500 V using a standard 1-mm cuvette. The transformed cells are incubated in 1 ml of standard LB or SOC media for 30 min at 37° then plated on LB agar containing 100 mg/ml ampicillin. After overnight incubation at 37° the Amp^r colonies are tested for expression of GCAPs upon growing them in LB media to OD$_{600}$ 0.3–0.5 and induction with 1 mM IPTG.

[27] A. M. Dizhoor and J. B. Hurley, *J. Biol. Chem.* **271**, 19346 (1996).
[28] A. M. Dizhoor, S. G. Boikov, E. V. Olshevskaya, *J. Biol. Chem.* **273**, 17311 (1998).

Expression

To produce large quantities of GCAPs, we typically grow 500 ml of bacterial culture with 50 μg/ml ampicillin in an environmental shaker at 37° until it reaches OD_{600} of 0.3–0.5. IPTG is added as an aliquot of 1 M stock solution to a final concentration of 1 mM and the incubation continues for another 3–5 hr. Cells are then collected by centrifugation at 8000g for 20 min at 4°, and washed twice in TE buffer (20 mM Tris-HCl, 1 mM EDTA). For each wash the pellet is resuspended in 25–50 ml of TE buffer and disrupted by several cycles of ultrasonication in ice followed by centrifugation. The insoluble material from the homogenate is collected by centrifugation at 10,000g for 15 min at 4°. Although solubility may vary for different GCAPs and GCAP mutants, GCAPs are generally found in the insoluble fraction of inclusion bodies. This provides substantial purification of the expressed insoluble proteins. The final pellet is then dissolved in an appropriate volume of TE buffer containing freshly deionized 6–8 M urea and 15 mM 2-mercaptoethanol for 1 hr at 0°. The insoluble material should be removed by centrifugation at 10,000g for 30 min at 4°. The clarified supernatant is then dialyzed against 500 volumes of TE buffer containing 2-mercaptoethanol. The dialysis should be repeated once more to ensure removal of urea. Mass spectrometry analysis has shown us that it is also important to avoid prolonged exposure of the proteins to oxidizing environments at this step and during storage; cysteine residues of the urea-solubilized GCAPs become very sensitive to oxidation and can even be modified with mercaptoethanol residues. The denatured protein precipitate that forms during dialysis is removed by centrifugation at 10,000g for 30 min and the cleared supernatant containing soluble GCAPs is used for further purification. If the conditions of bacterial growth and protein expression are optimal, the purity of GCAPs at this step can be as high as 70%.

Purification

For most biochemical applications an additional step of purification using gel filtration is usually both necessary and sufficient. Various gel-permeation chromatography media can be used for this purpose. We found HiPrep 26/60 (Pharmacia, Piscataway, NJ) columns packed with Sephacryl S-100 to be most convenient for this step. Cleared supernatant from the previous step is concentrated under argon or nitrogen pressure using an Amicon (Danvers, MA) YM10 membrane, loaded on a column equilibrated with an appropriate buffer solution [for example, 10 mM Tris-HCl (pH 7.5) and 1 mM EDTA] and eluted at 0.5–1 ml/min. Peak fractions collected from the column typically contain GCAP of 90% or greater purity.

Biochemical Assay of GCAP Activity

To assay the ability of a GCAP preparation to stimulate photoreceptor membrane GC outer segment membranes must first be washed to remove endogenous GCAPs. Both GCAP-1 and -2 tend to associate with membranes at normal ionic strength,[14,16] so three to four washes with a low-salt buffer are required to prepare rod outer segment (ROS) membranes for reconstitution experiments.[11,20] Tissue culture cells, such as HEK293 or insect cells expressing recombinant RetGCs or their mutants, can also be used as a source of RetGC.[11,23,26,29–33] An advantage to these preparations is that they are free of endogenous GCAPs. On the other hand, their disadvantage is that the GC specific activity of recombinant cyclases is usually much lower than that of native OS membranes.

To assay activation of RetGCs by GCAPs, a 5-μl aliquot containing recombinant GCAP-2 or GCAP-1 is added to 3 μg of washed dark-adapted bovine OS membranes in 50 mM MOPS-KOH (pH 7.5), 60 mM KCl, 8 mM NaCl, 10 mM MgCl$_2$, 2 mM Ca^{2+}/EGTA buffer, 10 μM each of cGMP PDE inhibitors dipyridamole and zaprinast, 1 mM ATP, 1 mM guanosine triphosphate (GTP), 4 mM cGMP, 1 μCi [α-^{32}P]GTP, 0.1 μCi [^{3}H]cGMP, and washed bovine outer segment membranes (3.5 μg rhodopsin). The final reaction volume is 25 μl. The reaction mixtures are then incubated for 12 min at 30° under infrared illumination to avoid activation of light-dependent phosphodiesterase. To terminate the reaction, it is heated for 2 min at 95°, chilled on ice, centrifuged for 5 min at 10,000g and analyzed by TLC on polyethylenemine cellulose plastic-backed plates with fluorescent background (Merck). After development in 0.2 M LiCl, cGMP spots are visualized under UV illumination, cut, eluted with 1 ml of 2 M LiCl, shaken with 10 ml of an Ecolume scintillation mixture (ICN Costa Mesa, CA) and both ^{3}H and ^{32}P radioactivity are counted using a liquid scintillation counter. [^{3}H]cGMP in this assay is used as the internal standard to ensure the absence of cGMP hydrolysis by PDE during the course of the reaction.

Myristoylation of GCAPs

Escherichia coli cells do not posses N-myristoyltransferase (NMT) activity. Therefore, unlike native GCAPs isolated from retinas, GCAPs ex-

---

[29] R. P. Laura, A. M. Dizhoor, and J. B. Hurley, *J. Biol. Chem.* **271**, 11646 (1996).
[30] T. Duda, R. Goraczniak, I. Surgucheva, M. Rudnicka-Nawrot, W. A. Gorczyca, K. Palczewski, A. Sitaramayya, W. Baehr, and R. K. Sharma, *Biochemistry* **35**, 8478 (1996).
[31] A. Krishnan, R. M. Goraczniak, T. Duda, and R. K. Sharma *Mol. Cell. Biochem.* **178**, 251 (1998).
[32] R. P. Laura, and J. B. Hurley, *Biochemistry* **37**, 11264 (1998).
[33] C. L. Tucker, J. H. Hurley, T. R. Miller, and J. B. Hurley, *Proc. Natl. Acad. Sci. U.S.A.* **95**, 5993 (1998).

pressed in *E. coli* strains will not be myristoylated at their N termini unless additional steps are taken. Myristoylation has only a minor effect on the activity of recombinant GCAP-2,[14] but it substantially increases the activity of recombinant GCAP-1.[34,35] Myristoylated GCAP-2 or GCAP-1 can be produced in *E. coli* transformed with a pET vector encoding yeast NMT obtained from Dr. J. Gordon (Washington University, St. Louis, MO). This vector encodes kanamycin resistance, so after transformation with the GCAP expression vector, the bacterial strains must be maintained in the presence of both ampicillin and kanamycin, otherwise expression of NMT may be lost. To produce myristoylated GCAPs, we use the same protocol as described earlier, except that the bacteria must be grown in culture media containing both ampicillin and 25 μg/ml kanamycin. Myristic acid (Sigma) as a donor of myristoyl residues must be added to the media from a 100 mg/ml stock ethanol solution to a final concentration of 50 μg/ml 30 min prior to induction of protein synthesis with IPTG. The N-terminal sequence of GCAP-1 is poorly recognized by yeast NMT. We replaced Asp-6 for Ser in GCAP-1 to make it a more suitable substrate for yeast NMT. This substitution does not inactivate GCAP-1 and it allows more efficient production of myristoylated GCAP-1.[28,35] We have used mass spectrometry to analyze the extent of myristoylation of GCAPs produced by this method. Typically, coexpression of GCAPs with NMT in *E. coli* results in GCAP preparations that are more than 90% myristoylated.

Conclusions

The methods described in this paper have been used for a variety of purposes including defining the roles of N-terminal fatty acylation,[14] the importance of the three EF hand Ca^{2+}-binding sites,[27] and the biochemical effects of a naturally occurring mutation in GCAP-1.[28] In addition, the ability to express various recombinant forms of RetGCs has made it possible to study the roles of the various domains of the cyclases and the mechanisms by which they are regulated by GCAPs.

Acknowledgments

J. B. H. is supported by the Howard Hughes Medical Institute and NIH grant EY06641 and A. M. D. is supported by NIH grant EY11522 and by Career Award from Research to Prevent Blindness, Inc.

---

[34] A. Otto-Bruc, J. Buczylko, I. Surgucheva, I. Subbarayya, M. Rudnicka-Nawrot, J. W. Crabb, A. Arendt, P. A. Hargrave, W. Baehr, and K. Palczewski, *Biochemistry* **36**, 4295 (1997).

[35] D. Krylov, G. Niemi, A. Dizhoor, and J. B. Hurley, *J. Biol. Chem.* **274**, 10833 (1999).

[47] Spectrophotometric Determination of Retinal Rod Guanylyl Cyclase

By GREGOR WOLBRING and PAUL P. M. SCHNETKAMP

Introduction

Retinal rod outer segment guanylyl cyclase (ROS-GC) is a key enzyme in the process of visual transduction in vertebrate rod photoreceptors. ROS-GC catalyzes the conversion of guanosine triphosphate (GTP) into pyrophosphate and cyclic guanosine monophosphate (cGMP); cGMP has been well established as the second messenger in the process of vertebrate phototransduction. The critical role of ROS-GC in rod vision is accentuated by the fact that mutations in ROS-GC cause severe retinal dystrophy in humans[1] and in chicken.[2] ROS-GC belongs to the guanylyl cyclase receptor family, which consists of two main groups, NO-stimulated soluble GCs and particulate and peptide-regulated GCs containing a single transmembrane spanning domain.[3] ROS-GC belongs to the particulate group, but it is unique because it requires an accessory protein named GCAP that binds to the cytosolic domain of ROS-GC[4]; GCAP confers the well-established calcium sensitivity to ROS-GC.[5]

Most (ROS-)GC assays rely on radiolabeled guanine nucleotides and, in the case of ROS-GC, special care needs to be taken that part of the cGMP formed is not hydrolyzed by the potent rod cGMP phosphodiesterase (PDE). This requires working in total darkness, the inclusion of PDE inhibitors, inclusion of radiolabeled cGMP to measure the amount of cGMP hydrolyzed during the assay time, or by the use of hydrolysis-resistant GTP analogs.[6–8] The assay described here sidesteps the problem of the potent and

---

[1] I. Perrault, J. M. Rozet, P. Calvas, S. Gerber, A. Camuzat, H. Dollfus, S. Chatelin, E. Souied, I. Ghazi, C. Leowski, M. Bonnemaison, P. D. Le, J. Frezal, J. U. Dufier, A. Munnich, and J. Kaplan, *Nature Gen.* **14**, 461 (1996).
[2] S. L. Semple-Rowland, N. R. Lee, J. P. Van Hooser, K. Palczewski, and W. Baehr, *Proc. Natl. Acad. Sci. U.S.A.* **95**, 1271 (1998).
[3] D. L. Garbers and D. G. Lowe, *J. Biol. Chem.* **269**, 30741 (1994).
[4] K. Palczewski, I. Subbaraya, W. A. Gorczyca, B. S. Helekar, C. C. Ruiz, H. Ohguro, J. Huang, X. Zhao, J. W. Crabb, R. S. Johnson, K. A. Walsh, M. P. Gray-Keller, P. B. Detwiler, and W. Baehr, *Neuron* **13**, 395 (1994).
[5] K. W. Koch, *Rev. Physiol. Biochem. Pharmacol.* **125**, 149 (1994).
[6] H. G. Lambrecht, and K. W. Koch, *EMBO J.* **10**, 793 (1991).
[7] W. A. Gorczyca, J. P. Van-Hooser, and K. Palczewski, *Biochemistry* **33**, 3217 (1994).
[8] A. Sitaramaya, T. Duda, and R. K. Sharma, *Mol. Cell. Biochem.* **148**, 139 (1995).

light-activated rod PDE by measuring pyrophosphate rather than cGMP. Pyrophosphate is measured colorimetrically by a commercially available kit (Sigma St. Louis, MO), which makes it possible to obtain real-time traces of ROS-GC activity. We have previously applied this assay to examine the effect of kinases on ROS-GC activity,[9,10] and we have taken advantage of the lack of interference by the light-activated PDE to examine the effect of light and transducin subunits on ROS-GC.[11]

General Methods

Purification of ROS Membranes

Bovine eyeballs are obtained from a local abattoir and stored in a light-tight container. Retinas are dissected within 2 hr of the death of the animals. All procedures with retinas or rod outer segments (ROS) are carried out in darkness or under dim red light illumination. Retinas are dissected and intact, Ca^{2+}-depleted bovine ROS are purified as described previously.[12] Briefly, retinas are collected in ice-cold 600 mM sucrose, 5% Ficoll 400, 50 mM NaCl, 1 mM EGTA, and 20 mM HEPES (pH 7.4). Retinas are vortexed for 30 sec, filtered through a Nylon screen, and applied to the top of a Ficoll 400 gradient (4–16%, w/w) in the above medium. After centrifugation (90 min, 26,000 rpm, at 4°), ROS are collected, diluted with 3 volumes of washing medium [600 mM sucrose, 20 mM HEPES (pH 7.4), and 100 μM EDTA], and sedimented. Subsequently, ROS are further purified on a linear sucrose gradient (20–40%, w/w) containing 20 mM HEPES (pH 7.4) and 100 μM EDTA, washed with washing medium, sedimented, and resuspended in the washing medium (containing 2.5% Ficoll 400) to a final rhodopsin concentration of about 200 μM. ROS are either used the same day, or aliquotted in portions of 200 μl and stored at $-20°$ in a light-tight container. All sucrose and sucrose–Ficoll 400 solutions are passed over a mixed-bed ion-exchange resin column before addition of salts and pH or calcium buffer.

Preparation of Stripped ROS Membranes

Frozen ROS are thawed and diluted with 40 volumes of ice-cold 20 mM HEPES (pH 7.4); after 5 min, ROS membranes are sedimented for

---

[9] G. Wolbring and P. P. M. Schnetkamp, *Biochemistry* **34,** 4689 (1995).
[10] G. Wolbring and P. P. M. Schnetkamp, *Biochemistry* **35,** 11013 (1996).
[11] G. Wolbring, W. Baehr, K. Palczewski, and P. P. M. Schnetkamp, *Biochemistry* **38,** 2616 (1999).
[12] P. P. M. Schnetkamp, *J. Physiol.* **373,** 25 (1986).

45 min at 40,000g and 4°, the pellet is resuspended in 300 mM sucrose, 2.5% Ficoll 400, and 10 mM HEPES (pH 7.4) to a final rhodopsin concentration of 200 μM rhodopsin, and stored frozen.

GCAP1 Purification

Frozen *spodoptera frugiperda* (Sf9) cells infected with the baculovirus containing the His-tagged GCAP1 cDNA were a gift from Dr. K. Palczewski. Frozen cells (5 ml) are thawed and homogenized on ice with two 5-sec Polytron bursts. Further sample preparation and purification of His-tagged GCAP1 is carried out with the His Trap procedure according to the manufacturer's instructions (Pharmacia Biotech, Baie d'Urfé, Quebec). GCAP1 is used directly without further dialysis.

Principle of Assay

Pyrophosphate is determined according to the following coupled reactions:

$$\text{Pyrophosphate} + \text{F-6-P} \xrightarrow{\text{fructose 6-phosphate kinase, pyrophosphate-dependent}} \text{F-1,6-DP} + \text{inorganic phosphate}$$

$$\text{F-1,6-DP} \xrightarrow{\text{aldolase}} \text{GAP} + \text{DHAP}$$

$$\text{GAP} \xrightarrow{\text{triosephosphate isomerase}} \text{DHAP}$$

$$2\text{ DHAP} + 2\text{ H}^+ + 2\text{ }\beta\text{-NADH} \xrightarrow{\text{glycerophosphate dehydrogenase}} 2 \text{ glycerol-3-phosphate} + 2\text{ }\beta\text{-NAD}^+$$

where F-6-P is D-fructose 6-phosphate; F-1,6-DP, D-fructose 1,6-diphosphate; GAP, D-glyceraldehyde-3-phosphate; and DHAP, dihydroxyacetone phosphate. This reaction scheme is based on a bacterial enzyme (fructose 6-phosphate kinase) that catalyzes the pyrophosphate-dependent phosphorylation of fructose 6-phosphate in contrast to the eukaryotic fructose 6-phosphate kinase that catalyzes adenosine triphosphate (ATP)-dependent phosphorylation of fructose 6-phosphate. In this reaction scheme, 2 mol of NADH are oxidized to nicotinamide adenine dinucleotide (NAD$^+$) per mole of pyrophosphate consumed.

Instrumentation

The pyrophosphate assay couples the measurement of pyrophosphate to the oxidation of β-NADH (see preceding paragraph). ROS-GC is a particulate enzyme that copurifies with ROS membranes and is assayed in a suspension of ROS or ROS membranes. Real-time measurements of ROS-GC activity, therefore, need to be carried out by measuring β-NADH oxidation at 340 nm in a turbid suspension of ROS or ROS membranes under constant stirring, preferably in a thermostatted cuvette house. The turbidity of the suspension adds a significant amount of apparent light absorption due to light scattering, while the swirling eddies caused by the constant stirring add a time-dependent component to the light scattering. In our earlier studies, ROS-GC activity was measured in an SLM-Aminco DW2C dual-wavelength spectrophotometer (SLM Instruments, Urbana, IL). With this instrument light loss due to light scattering is minimized by placing the cuvette directly in front of the photomultiplier, while changes in apparent absorption due to swirling eddies were eliminated by the dual-wavelength mode. In the dual-wavelength mode, changes in light scattering are eliminated or minimized by comparing at 60-Hz light absorption at the measuring wavelength with that at a reference wavelength, in our case 340 and 410 nm, respectively. The result is low-noise recordings of light absorption in turbid samples such as suspension of cells. Unfortunately, dual-wavelength spectrophotometers appear to be no longer commercially available, which led us to explore possible alternatives. Many of the new generation of relatively inexpensive spectrophotometers are equipped with the capability of rapid wavelength scans. We have tested the Cary 50, which has a maximum wavelength scan rate of 24,000 nm/min, and, therefore, can sample $\Delta A_{340}-A_{410}$ at a frequency of 3 Hz.

Application to Intact Ros Permeabilized with Saponin

ROS-GC Assay in Intact ROS: Interfering Enzyme Activities

We have measured ROS-GC activity in intact bovine ROS with a full complement of cytosolic constituents after permeabilizing the plasma membrane by the addition of 0.025% saponin. The ROS-GC/PP$_i$ assay is carried out in a volume of 2 ml containing 75 mM NaCl, 100 μM Pefabloc SC, 400 μM 1,2-bis(2-aminophenoxy)ethane-N,N,N',N'-tetraacetic acid (BAPTA), 0.025% saponin, and components of the pyrophosphate detection kit (Sigma, St. Louis, MO), in final concentration: 15 mM imidazole hydrochloride (pH 7.4), 0.03 mM EDTA, 0.066 mM MnCl$_2$, 0.006 mM CoCl$_2$, 2 mM MgCl$_2$, 0.26 mM β-NADH, 4 mM fructose 6-phosphate, 1 mg bovine serum albumin (BSA), 1 mg sugar stabilizer, 0.33 units of PP$_i$-dependent fructose 6-phosphate kinase, 4.95 units of aldolase (EC 4.1.2.13), 3.33 units of glycerophosphate dehydrogenase (EC 1.1.1.8), and 33.3 units of triosephosphate isomerase (EC 5.3.1.1).

Enzymatic activities in the ROS cytosol could conceivably interfere with or feed into the enzymatic cascade of the PP_i assay kit; we have observed two such cases. First, we test the assay by addition of 10 nmol of exogenous PP_i to a suspension of saponin-permeabilized ROS in the presence of the PP_i assay kit, and observe no signal (Fig. 1). Added PP_i is rapidly hydrolyzed by a pyrophosphatase endogenous to ROS, which appears to compete rather effectively for substrate with the components of the PP_i detection kit. When Mn^{2+} is added to the incubation medium at increasing concentrations, a progressive increase in the amplitude of the PP_i-induced signal of NADH oxidation is observed until at a concentration of 2.5 mM Mn^{2+} the full amplitude is recovered by complete inhibition of endogenous pyrophosphatase (Fig. 1). Alternatively, 15 mM NaF is equally effective as an inhibitor of ROS pyrophosphatase.[9] The PP_i-induced signal is also used as a convenient internal calibration.

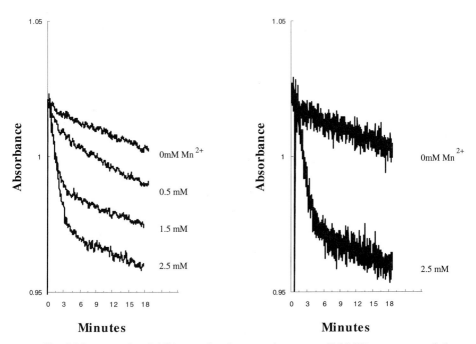

FIG. 1. Manganese ions inhibit pyrophosphatase endogenous to ROS. PP_i assay was carried out in a suspension of intact ROS as described in the text (with a final rhodopsin concentration of 1.2 μM). The ROS plasma membrane was permeabilized with 0.025% saponin, and 1 mM potassium citrate was added to the medium to inhibit fructose 6-phosphate kinase. Time traces of recordings of $\Delta A_{340} - A_{410}$ are illustrated. PP_i assay was initiated by addition of 5 μM PP_i. $MnCl_2$ was added as indicated. Right-hand side: no filtering; left-hand side: with filtering as described in text. ROS-GC traces illustrated here and in all other illustrations except for Fig. 3 were recorded in the Cary 50 as described under Instrumentation. Temperature: 25°.

The traces illustrated in Fig. 1 are recorded with the simple Cary 50 spectrophotometer as described in the Instrumentation section. The right-hand panel (Fig. 1) illustrates two of the traces without application of any filtering and a rather significant noise level is observed when compared with traces obtained with a true dual-wavelength spectrophotometer 9 (see also Fig. 3). The high noise level is most likely caused by swirling eddies in the suspension under constant stirring; the frequency of the eddies is higher than the 3-Hz sampling frequency of the spectrophotometer. In another contribution to this volume we describe a simple filtering method applied to time traces obtained with a fluorometer or spectrophotometer and imported into Microsoft Excel.[13] Briefly, the value of $\Delta A_{340} - A_{410}$ for each time point is averaged with five neighboring values on either side on the time axis; because our sampling frequency is 3 Hz, averaging is carried out over a time interval of 3 sec. We apply this noise reduction method to all the ROS-GC traces illustrated here, and its benefits can be easily appreciated by comparing the unfiltered traces in the right-hand panel of Fig. 1 with those in the left-hand panel, which are filtered.

Mammalian fructose 6-phosphate kinase does not utilize pyrophosphate, but ATP or other nucleotide triphosphates such as GTP. The ROS cytosol contains glycolytic enzymes,[14,15] and addition of ATP or GTP together with the PP_i detection kit results in NADH oxidation that could be attributed erroneously to ROS-GC activity or ATP-dependent modulation of ROS-GC activity. The ATP-dependent NADH oxidation by ROS fructose 6-phosphate kinase is illustrated in Fig. 2; 1.5 mM citrate completely inhibits ROS fructose 6-phosphate kinase activity. Some care should be taken, however, because fructose 6-phosphate kinase is a key regulatory point in glycolysis, and many allosteric effectors of this enzyme have been described. For example, we notice that ammonium ions, often a constituent of protein preparations, could reverse the inhibition by citrate and stimulate fructose 6-phosphate kinase activity (Fig. 2).

Application to Washed or Stripped ROS Membranes

ROS-GC Activity in Washed ROS Membranes

The necessity to use citrate and fluoride or Mn^{2+} to inhibit enzymes present in the ROS cytosol that interfere with the pyrophosphate assay

---

[13] C. B. Cooper, R. T. Szerencsei, and P. P. M. Schnetkamp, *Methods Enzymol.* **315** [56], 1999 (this volume).

[14] R. Lopez-Escalera, X. Li, R. T. Szerencsei, and P. P. M. Schnetkamp, *Biochemistry* **30**, 8970 (1991).

[15] S. C. Hsu. and R. S. Molday, *J. Biol. Chem.* **266,** 21745 (1991).

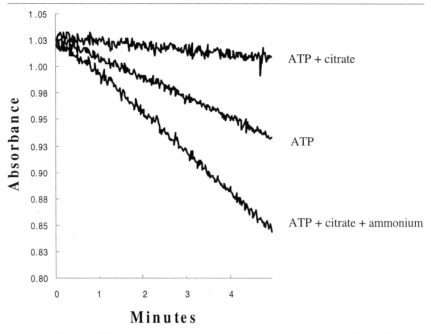

FIG. 2. Citrate inhibits fructose 6-phosphate kinase endogenous to ROS. ROS-GC assay was carried out in a suspension of intact ROS as described in the text (with a final rhodopsin concentration of 1.2 μM). The ROS plasma membrane was permeabilized with 0.025% saponin. Time traces of recordings of $\Delta A_{340}-A_{410}$ are illustrated. Fructose 6-phosphate kinase was measured at a concentration of 500 μM ATP. Potassium citrate (1.5 mM) or NH$_4$Cl (20 mM) were present as indicated. Temperature: 25°.

makes it desirable to remove most soluble proteins from ROS without removing GCAP and, thus, remove most interfering enzymatic activities. Washing with calcium-free salt solutions containing Mg^{2+} resulted in an irreversible loss of calcium-sensitive ROS-GC activity; in contrast, when Mn^{2+} replaced Mg^{2+}, little loss in calcium-sensitive ROS-GC activity was observed indicating little or no loss of endogenous GCAP (Fig. 3). In contrast, both pyrophosphatase activity and fructose 6-phosphate kinase endogenous to ROS were removed completely. BAPTA was used in the preceding medium as a Ca^{2+} chelator that binds neither Mn^{2+} nor Mg^{2+}. It should be pointed out that EGTA, often used as a Ca^{2+} chelator that does not bind Mg^{2+}, does bind Mn^{2+}, and could not be used in our application. We have previously used the washed ROS preparation to study the effects of exogenous and endogenous kinases on ROS-GC activity.[9,10]

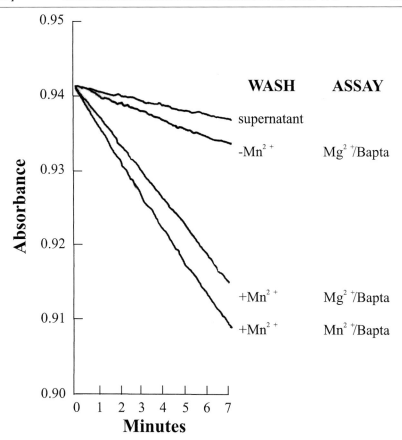

FIG. 3. Washing in the presence of manganese retains full ROS-GC activity. Intact ROS were frozen, thawed, and diluted to a final rhodopsin concentration of 5 μM in 100 mM NaCl, 25 mM HEPES (pH 7.4), and 1 mM MnCl$_2$ or MgCl$_2$ as indicated. ROS were sedimented in a microcentrifuge at 14,000 rpm, and resuspended in the standard assay medium containing 400 μM BAPTA and 1 mM MnCl$_2$ or 1 mM MgCl$_2$ as indicated; final rhodopsin concentration was 1.1 μM. Time traces of recordings of $\Delta A_{340} - A_{410}$ are illustrated. ROS-GC activity was measured at a concentration of 500 μM GTP. ROS-GC traces were recorded in a SLM-Aminco DW2C dual-wavelength spectrophotometer. Temperature: 25°.

ROS-GC Activity in Stripped ROS Membranes Reconstituted with Exogenous GCAP1

To further delineate the interactions between ROS-GC and GCAP with other ROS constituents it would be highly desirable to use a reconstituted system of purified components. Figure 4 illustrates ROS-GC activity of ROS membranes stripped from soluble and peripheral proteins as a function

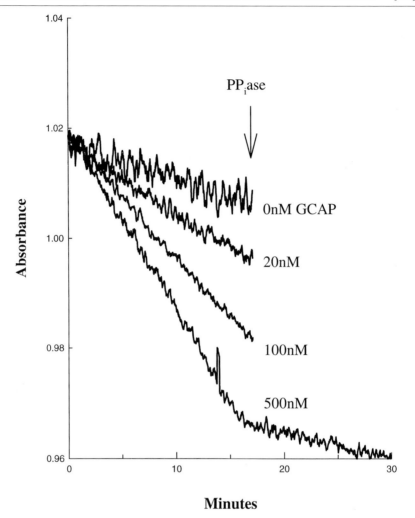

FIG. 4. Reconstitution of ROS-GC with stripped ROS membranes and recombinant GCAP1. Stripped ROS membranes were incubated in a medium containing 50 mM potassium acetate, 4 mM MgCl$_2$, 400 μM BAPTA, pryophosphate kit, and the indicated concentration of recombinant GCAP1. Time traces of recordings of $\Delta A_{340} - A_{410}$ are illustrated. ROS-GC activity was measured at a concentration of 500 μM GTP. Between 20 and 25 min after the start of the trace (indicated by the arrow), 0.4 unit of inorganic pyrophosphatase was added. Temperature: 25°.

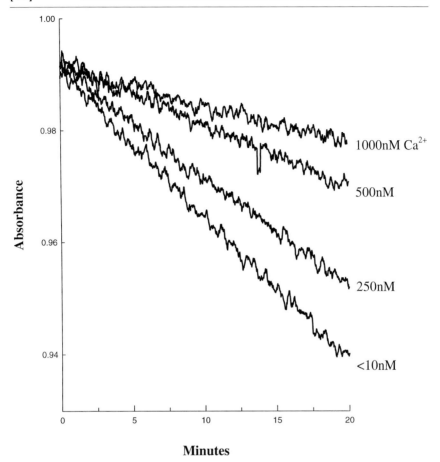

FIG. 5. Calcium dependence of reconstituted ROS-GC activity. Stripped ROS membranes were incubated in a medium containing 50 mM KOAc, 4mM MgCl$_2$, 400 μM BAPTA, pryrophosphate kit, and 500 nM GCAP1. Time traces of recordings of $\Delta A_{340}-A_{410}$ are illustrated. ROS-GC activity was measured at a concentration of 500 μM GTP. CaBAPTA was added to reach the indicated final free Ca^{2+} concentrations. Temperature: 25°.

of added recombinant GCAP1; little ROS-GC activity was observed until GCAP1 was added. As a control we added excess exogenous pyrophosphatase to hydrolyze pyrophosphate produced by ROS-GC before it can enter the enzymatic cascade of the PP$_i$ assay kit. Excess exogenous pyrophosphatase reduced the GCAP1-dependent ROS-GC rate to the background rate observed without added GCAP1. We used inhibition of the reconstituted ROS-GC activity by submicromolar calcium concentrations as a second test to validate our enzyme-coupled ROS-GC assay as applied to reconstituted

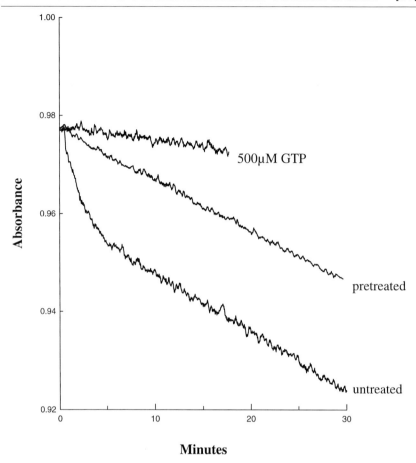

FIG. 6. Initial rates of GTP-induced ROS-GC activity. ROS-GC activity was measured in stripped ROS membranes reconstituted with recombinant GCAP1 as described in the legend of Fig. 4. Time traces of recordings of $\Delta A_{340}-A_{410}$ are illustrated. Time traces were initiated by addition of 500 μM GTP as indicated, from an untreated GTP stock or from a GTP stock treated with the pyrophosphate kit. Upper trace: no GCAP1 present. Temperature: 25°.

system. Figure 5 illustrates that ROS-GC activity in stripped ROS membranes reconstituted with recombinant GCAP1 could be fully inhibited when free Ca^{2+} was raised to 1 μM.

In most of our experiments GTP-induced ROS-GC rates, as judged from NADH oxidation in our enzyme-coupled assay, were linear for at least 30 min after a rapid initial phase of NADH oxidation had subsided; the ROS-GC rates reported here represent this linear portion of the GTP-induced NADH oxidation. The initial rapid phase of GTP-induced NADH oxidation was due to a minor pyrophosphate contamination in the GTP stock solution since it was eliminated by pretreatment of this stock solution with the pyrophosphate kit (Fig. 6).

Conclusions

In this contribution we described the application of a simple colorimetric pyrophosphate assay to the study of ROS-GC. The virtues of the assay are that ROS-GC is measured in real-time traces and without the need for radiolabeled guanine nucleotides. We identified pyrophosphatase and fructose 6-phosphate kinase as two enzymatic activities endogenous to ROS that could interfere with our assay. Fluoride or manganese ions and citrate were found to be useful as inhibitors of pyrophosphatase and fructose 6-phosphate kinase, respectively, without interfering with calcium-dependent ROS-GC activity. Washed ROS (to remove soluble proteins) and stripped ROS membranes (to remove all but integral membrane proteins) were prepared free of interfering enzyme activities. In the case of stripped ROS membranes, we were able to demonstrate calcium-dependent ROS-GC activity on reconstitution of the membranes with recombinant GCAP1 (Figs. 3 and 4). Optical recordings of a turbid suspension of cells or cell membranes are best carried out in a dual-wavelength spectrophotometer equipped with a thermostatted cuvette house under constant stirring. (Figure 3 illustrates time traces of ROS-GC activity obtained with a dual-wavelength spectrophotometer.) Unfortunately, dual-wavelength spectrophotometers are expensive and appear to be no longer commercially available. We discussed here how to modify the rapid wavelength scanning capability of an inexpensive spectrophotometer (Cary 50) for obtaining time traces of ROS-GC activity and to apply to the time traces simple data averaging in Microsoft Excel (Fig. 1).

Acknowledgments

This research was supported by an operating grant from the Canadian Retinitis Pigmentosa Foundation (P.P.M.S.). P.P.M.S. is a medical scientist of the Alberta Heritage Foundation for Medical Research. G.W. was a recipient of a Fellowship from the Canadian Retinitis Pigmentosa Foundation. We thank Robert T. Szerencsei and Shardelle Brown for preparing ROS.

[48] Calcium-Dependent Activation of Membrane Guanylate Cyclase by S100 Proteins

By Ari Sitaramayya, Nikolay Pozdnyakov, Alexander Margulis, and Akiko Yoshida

Introduction

Membrane guanylate cyclases (GCs) are regulated either by peptide hormones or intracellular proteins.[1] In the latter case all the proteins known to date appear to be calcium-binding proteins. In *Paramecium*, a unicellular organism, calcium influx results in increased cyclic guanosine monophosphate (cGMP) formation.[2] This increase appears to be due to activation of a membrane GC by calmodulin.[3] Membrane GCs of *Tetrahymena* and crayfish are also activated by calmodulin.[4,5] A different type of membrane GC exists in the rod outer segments (ROS) of vertebrate retinal photoreceptor cells. This enzyme (ROS-GC) is activated by a light-induced decrease in the intracellular calcium concentration.[1,6–8] Two calcium-binding proteins capable of mediating this activation have been characterized and named guanylate cyclase activating proteins or GCAPs.[9,10] GCAP binds ROS-GC and activates it at calcium concentrations lower than about 500 nM.[10–12] At higher calcium concentrations, GCAP binds calcium and dissociates from GC, resulting in a lower rate of cGMP production.

This chapter describes the purification of a protein from vertebrate retina, which activates ROS-GC in a calcium-dependent manner. It differs

---

[1] E. N. Pugh, Jr., T. Duda, A. Sitaramayya, and R. K. Sharma, *Biosci. Reports* **17,** 429 (1997).
[2] J. E. Schultz and U. Schade, *J. Membr. Biol.* **109,** 251 (1989).
[3] J. E. Schultz and S. Klumpp, *FEMS Microbiol. Lett.* **13,** 303–306 (1982).
[4] S. Kakiuchi, K. Sobue, R. Yamazaki, S. Nagao, S. Umeki, Y. Nozawa, M. Yazawa, and K. Yagi, *J. Biol. Chem.* **256,** 19 (1981).
[5] W. Riediger, H.-G. Hergenhahn, and D. Sedlmeier, *Int. J. Biochem.* **21,** 333 (1989).
[6] A. L. Hodgkin and B. J. Nunn, *J. Physiol.* **403,** 439 (1988).
[7] K.-W. Koch and L. Stryer, *Nature* **334,** 64 (1988).
[8] M. P. Gray-Keller and P. B. Detwiler, *Neuron,* **13,** 849 (1994).
[9] W. A. Gorczyca, M. Gray-Keller, P. B. Detwiler, and K. Palczewski, *Proc. Natl. Acad. Sci. U.S.A.* **91,** 4014 (1994).
[10] A. M. Dizhoor, D. G. Lowe, E. V. Olshevskaya, R. P. Laura, and J. B. Hurley, *Neuron,* **12,** 1345 (1994).
[11] W. A. Gorczyca, A. S. Polans, I. G. Surgucheva, I. Subbaraya, W. Baehr, and K. Palczewski, *J. Biol. Chem.* **270,** 22029 (1995).
[12] T. Duda, R. Goraczniak, I. Surgucheva, M. Rudnicka-Nawrot, W. A. Gorczyca, K. Palczewski, A. Sitaramayya, W. Baehr, and R. K. Sharma, *Biochemistry,* **35,** 8478 (1996).

from GCAPs in that it activates cyclase at micromolar calcium concentrations, which do not support activation by the latter. Characterization of the protein shows that it belongs to the S100 family of calcium-binding proteins.

Methods

Preparation of ROS-GC

ROS membranes are used as the source of ROS-GC. They are prepared from fresh bovine eyes brought to the laboratory from a local slaughterhouse in a light-tight container and dark-adapted for 3 hr. All further operations are conducted under infrared light with the aid of an image converter. Retinas are extracted from the eyes and ROS purified according to the protocol of Schnetkamp et al.[13] ROS are frozen, thawed, washed three times in 10 mM Tris, pH 7.5, to remove endogenous proteins that might regulate cyclase activity, suspended in 5 mM isobutylmethylxanthine (IBMX) containing 5 mM dithiothreitol (DTT), and stored at $-80°$.

Assay of ROS-GC

GC activity is measured under infrared light in a 40-μl assay mixture containing 500 μg/ml of ROS membrane protein, 40 mM HEPES, pH 7.4, 1 mM guanosine triphosphate (GTP), 6 μCi of [α-^{32}P]GTP, 2 mM cGMP, 0.2 μCi of [^{3}H]cGMP, 15 mM MgCl$_2$, 2.5 mM IBMX, and 10 mM DTT. The tubes containing the membranes are preincubated for 5 min at 37° or 25° and the reaction initiated with the addition of substrate. After 10 min the reaction is terminated with the addition of 20 μl of 150 mM EDTA containing GTP, cGMP, and GMP, all at 2 mM. The tubes are held at 95° for 5 min followed by centrifugation and analysis of [^{32}P]cGMP and [^{3}H]cGMP in the supernatant by thin-layer chromatography.[14] [^{3}H]cGMP added to the assay mixture serves as a monitor for phosphodiesterase activity. Usually 98% of the added cyclic nucleotide remains intact for the duration of the assay.

The effect of various calcium concentrations is tested on the basal GC activity and the ability of protein preparations to regulate it. For routine measurements a "low" calcium concentration is obtained by the addition of 2 mM EGTA to the assay, and a "high" calcium concentration by the

---

[13] P. P. M. Schnetkamp, A. A. Klompmakers, and F. J. M. Daemen, *Biochim. Biophys. Acta* **552**, 379 (1979).
[14] A. Sitaramayya, L. Lombardi, and A. Margulis, *Visual Neurosci.* **10**, 991 (1993).

addition of 1 mM CaCl$_2$ in the absence of calcium chelators. To obtain precise free-calcium concentrations either dibromo-BAPTA–CaCl$_2$ or EGTA–CaCl$_2$ buffers are used.[15] Free-calcium concentration in the buffer is calculated using Maxchelator[16] and verified by measurement using a Corning calcium electrode (Corning Inc., Corning, NY).

Preparation of Crude Retinal Extract

Crude retinal extract is prepared by homogenizing 100 frozen retinas in 50 ml of 10 mM Tris, pH 7.5, containing 5 mM MgCl$_2$, 0.1 mM phenylmethylsulfonyl fluoride (PMSF), 50 μg/ml of benzamidine, and 10 μg/ml each of leupeptin, aprotinin, and trypsin inhibitor. The homogenate is centrifuged at 100,000g for 30 min, and concentrated solutions of Tris (pH 8.0) and CaCl$_2$ are added to the supernatant to bring their final concentrations to 50 and 5 mM, respectively. The material is then distributed, 3 ml each, into 5-ml capacity polypropylene tubes, which are then placed in a 75° water bath. After 3 min, the tubes are cooled on ice, and centrifuged for 30 min at 27,000g. The resulting supernatant is dialyzed overnight at 4° against 5 liters of dialysis buffer (10 mM Tris, pH 8.0), centrifuged at 27,000g for 30 min and 4° to remove particulate material, and the resulting supernatant stored at 4°.

Purification of Calcium-Dependent Guanylate Cyclase Activator Protein (CD-GCAP)

The crude extract from 100 retinas is loaded on a 1- × 6-cm column of DEAE-Sepharose CL-6B that has been equilibrated at 4° in the dialysis buffer. The column is washed with 100 ml of the same buffer at a flow rate of 8.5 ml/hr and eluted with an 80-ml linear gradient of 0–500 mM NaCl. Fractions of approximately 2 ml are collected and 15 μl of each is tested for influence on cyclase activity at high calcium concentration. The active fractions are pooled and solid NaCl is added to bring the final concentration to 2.5 M. The preparation is warmed to room temperature, loaded on a 1- × 5-cm column of phenyl-Sepharose CL-4B equilibrated at room temperature in 10 mM Tris, pH 7.5, containing 2.5 M NaCl, washed with the same buffer at a flow rate of 20 ml/hr until the absorbance monitored at 280 nm returned to baseline, and eluted with a linear gradient consisting of 40 ml of equilibration buffer and 40 ml of the buffer without NaCl. Fractions of approximately 2 ml are collected, desalted on Centricon-10, and tested

---

[15] S. M. Harrison and D. M. Bers, *Biochim. Biophys. Acta* **925**, 133 (1987).
[16] D. M. Bers, C. W. Patton, and R. Nuccitelli, *Methods Cell Biol.* **40**, 3 (1994).

for cyclase-stimulating activity at high calcium concentration. The active fractions are pooled and concentrated to about 600 μl. The buffer is exchanged on Centricon-10 to 50 mM Tris, pH 6.8, containing 50 mM NaCl, and the material is then subjected to high-performance liquid chromatography (HPLC) in the same buffer at room temperature on a 0.8- × 60-cm gel filtration column, Biosep-SEC-2000 (Phenomenex, Torrance, CA). The flow rate is 0.4 ml/min and fractions of 0.2 ml are collected and analyzed for influence on GC activity. Fractions containing cyclase-stimulatory activity (usually 3–4) are pooled and stored at 4°.

Zinc Affinity Chromatography of Purified Activator

Two hundred thirty-five μg of the activator (CD-GCAP), purified by the preceding protocol, is subjected to zinc affinity chromatography as described by Baudier et al.[17] Briefly, the protein is incubated at 4° for 24 hr with 15 ml of 1 mM ZnSO$_4$ in 50 mM Tris, pH 7.5, containing 10 mM mercaptoethanol, and loaded on a 1- × 6-cm phenyl-Sepharose column equilibrated at 4° in 50 mM Tris, pH 7.5, 10 mM mercaptoethanol, 0.3 M NaCl, and 0.25 M ZnSO$_4$. The column is washed at 20 ml/hr with 20 ml of equilibration buffer, followed by elution with 20 ml of 50 mM Tris, pH 7.5, 10 mM mercaptoethanol, and 2 mM EDTA. The protein in the eluate is filtered and washed on Centricon-10 to remove EDTA and ZnSO$_4$, and the buffer exchanged to 50 mM Tris, pH 6.8, containing 50 mM NaCl and 10 mM mercaptoethanol. The preparation is stored at 4°.

Recovery of Activator after Preparative SDS–PAGE

The activator purified as given is mixed with an equal volume of 2× electrophoresis sample buffer, loaded in equal volume into all the 15 lanes of an 8- × 10-cm sodium dodecyl sulfate (SDS)–15% polyacrylamide gel, and subjected to electrophoresis according to Laemmli.[18] After electrophoresis, the location of the activator is detected by staining the first and last lanes of the gel with Coomassie Brilliant Blue. Once the location of the protein is known, a 5-mm-wide band of the entire gel containing the protein is cut out and extracted overnight into 10 volumes of 10 mM Tris, pH 7.5. SDS is removed from the extract by filtration and washing on Centricon-10, and the concentrated preparation is stored at 4°.

---

[17] J. Baudier, C. Holtzscherer, and D. Gerad, *FEBS Lett.* **148**, 231 (1982).
[18] U. K. Laemmli, *Nature,* **227**, 680 (1970).

Immunoreactivity of Activator with Antibodies against S100 Proteins

Cyclase activator and S100 proteins are electrophoresed and transferred to nitrocellulose membrane. The membrane is fixed for 45 min at room temperature in 0.2% glutaraldehyde[19] in TTBS (20 mM Tris, pH 7.5, 0.5 M NaCl, and 0.05% Tween 20), and blocked overnight with 5% dry milk in TTBS. It is then washed and incubated for 60 min with monoclonal antibodies against S100A1 (also called S100α) or S100β (Sigma Chemical Company, St. Louis, MO) diluted in 1% bovine serum albumin dissolved in TTBS. The membrane is washed, incubated for 30 min with biotinylated secondary antibody, washed again, and incubated with avidin–horseradish peroxidase complex. Following washing in TTBS devoid of Tween 20, the blot is developed with 4-chloro-1-naphthol.

Results

Basal Activity

Under the conditions of assay used, ROS-GC activity in the washed membranes was linear with protein concentration and incubation time and varied between preparations from 0.7 to 3.0 nmol of cGMP formed per minute per milligram of protein. This activity was referred to as basal activity and was not significantly influenced by calcium concentrations in the tested range.

Crude Retinal Extract Activated ROS-GC Both at Low and High Calcium Concentrations

As shown in Fig. 1, crude extract from retina stimulated GC activity of washed ROS membranes both at low and high calcium concentrations. At low calcium concentration the stimulated activity was about 250% of the basal and at high calcium, about 400%. Because proteins that mediate the activation of GC in low calcium have already been identified (GCAPs),[9–12] we attempted to identify factors responsible for activation of the enzyme at high calcium concentration.

---

[19] L. J. Eldik and S. R. Wolchok, *Biochem. Biophys. Res. Commun.* **124,** 752 (1984).

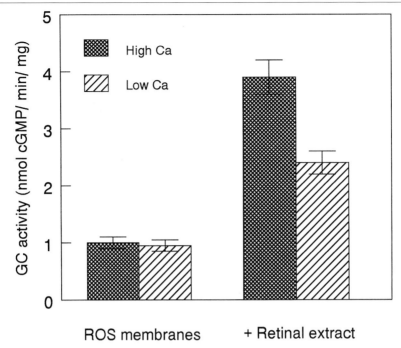

FIG. 1. Activation of ROS-GC by crude retinal extract at both low and high calcium concentrations. Cyclase activity of the washed ROS membranes was measured in the absence or presence of 10 μl of crude retinal extract. Data are shown as mean ± S.D. from three independent experiments. Activation of the cyclase at low calcium concentrations was probably due to one of the known GCAPs. Activation at high calcium concentration suggested that the extract might also contain a novel regulator of GC.

Purification of Calcium-Dependent Guanylate Cyclase Activator Protein

Chromatography of the retinal extract on DEAE-Sepharose resulted in the elution of a single peak of cyclase-stimulatory activity. It eluted at approximately 300 mM NaCl. When the active fractions were pooled and adjusted to 2.5 M salt concentration required for chromatography on phenyl-Sepharose, the preparation became completely inactive. The inactivation was, however, reversible on desalting either by dialysis or filtration through Centricon-10. Fractions eluted from the phenyl-Sepharose column therefore had to be desalted. Determination of the cyclase-stimulatory activity after desalting revealed a broad peak of activity in fractions eluting between 2.0 and 1.1 M NaCl. The highest activity was found in fractions eluting at about 1.5 M NaCl. In the next step, cyclase-stimulating activity was found to elute from the gel-filtration column as a single peak matching the elution of a major protein peak. Application of the procedure given here to the extract from 100 retinas yielded about 150 μg of the activator protein.

Properties of the Activator

Based on the calibration of the gel-filtration column with standard proteins, the molecular mass of the activator was estimated at 40,000 Da (Fig. 2A). SDS–PAGE in a 10–20% gradient polyacrylamide gel, followed by staining with Coomassie Brilliant Blue showed that the purified activator contained a single protein band with a molecular mass of about 6–7 kDa (Fig. 2B). Electrophoresis of the aged activator preparations, particularly in the absence of reducing agents, showed a band at 21 kDa in addition to

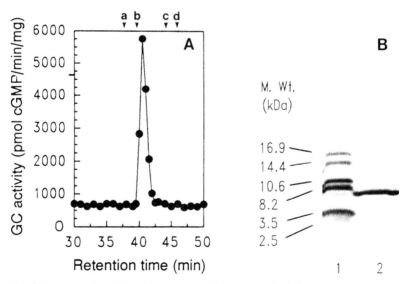

FIG. 2. Determination of the molecular mass of the activator by (A) gel-filtration chromatography and (B) SDS–PAGE. In (A), cyclase-stimulatory activity of the proteins eluting from the gel-filtration column (Biosep-SEC-2000) was determined. The activator eluted as a single peak with a retention time of 40.5 min. Based on the retention times of standard proteins, marked a–d (a, 37.8 min for the 66-kDa bovine serum albumin; b, 39.6 min for the 44-kDa ovalbumin; c, 44 min for the 29-kDa carbonic anhydrase; and d, 46 min for the 19-kDa myoglobin), the molecular mass of the activator was calculated as 40 kDa. (B) Shows the SDS–PAGE of 4 μg of purified protein (lane 2) from the gel-filtration column. Lane 1 shows standard proteins with their molecular masses marked on the left. Based on the mobility of the standard proteins, the activator has an apparent mass of 6–7 kDa.

the 6- to 7-kDa band (as seen later in Fig. 4). The 21-kDa band was apparently a dimer or a trimer. Mass spectrometric analysis revealed that the mass of the activator was 10,580 Da, suggesting that the protein's electrophoretic mobility as a 6- to 7-kDa band was anomalous. It was therefore likely that the appearance of the 21-kDa band in SDS-PAGE, and the elution of the activator from gel-filtration column as a 40-kDa protein, were due to self-association as dimer and tetramer, respectively.

The activator was stable to treatment with SDS.[20] The 6- to 7-kDa protein extracted from SDS-PAGE gel, and regenerated, retained its ability to stimulate ROS-GC in a calcium-dependent fashion.

Treatment of the activator with pronase, a nonspecific protease (from *Streptomyces griseus*), completely destroyed its ability to activate cyclase.

Although heat treatment of the crude retinal extract was used as the first step in the purification of the activator, purified activator in dilute protein solution was completely inactivated when heated for 3 min at 75°.

The purified protein activated ROS-GC in a dose-dependent and saturable fashion. When assayed at 1.0 mM calcium concentration, maximal activation, a 24-fold increase in cyclase activity was observed in the presence of 160 μg/ml of the activator protein.

Calcium Dependence of Activation of Cyclase by Purified Protein

During the purification of the activator, its activity was monitored in assays containing 1 mM CaCl$_2$. To determine the calcium dependence of the activity, dibromo-BAPTA-CaCl$_2$ buffers were used to obtain the desired free calcium concentrations. Figure 3 shows the activation of cyclase at different calcium concentrations by a supersaturating concentration (240 μg/ml) of the activator. Half-maximal activation was observed at 1.5 μM free calcium.[21] Based on the calcium dependence of the activation, the protein was referred to as CD-GCAP (calcium-dependent guanylate cyclase activator protein).

Characterization of CD-GCAP: Identity with S100β

A survey of the literature revealed that a protein that has neurite extension properties and runs as a 6.5-kDa protein in SDS-PAGE was

---

[20] N. Pozdnyakov, A. Yoshida, N. G. F. Cooper, A. Margulis, T. Duda, R. K. Sharma, and A. Sitaramayya, *Biochemistry* **34,** 14279 (1995).

[21] N. Pozdnyakov, R. Goraczniak, A. Margulis, T. Duda, R. K. Sharma, A. Yoshida, and A. Sitaramayya, *Biochemistry* **36,** 14159 (1997).

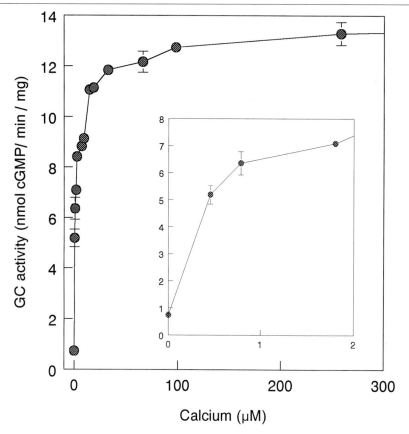

FIG. 3. Calcium dependence of the stimulation of ROS-GC by the retinal activator. Cyclase activity was measured in the presence of 240 μg/ml of the activator at different free-calcium concentrations. *Inset:* activation at 0–2 μM calcium. Error bars indicate variation in the triplicate measurements.

later identified as S100β, a 10.5-kDa protein and a member of the S100 family of calcium-binding proteins.[22] This suggested that CD-GCAP, which has a similar anomalous electrophoretic mobility, could also be a member of the S100 family. To test this possibility, the reactivity of CD-GCAP with antibodies against two S100 proteins, S100A1 and S100β, was tested by

---

[22] D. Kligman and D. Marshak, *Proc. Natl. Acad. Sci. U.S.A.* **82,** 7136 (1985).

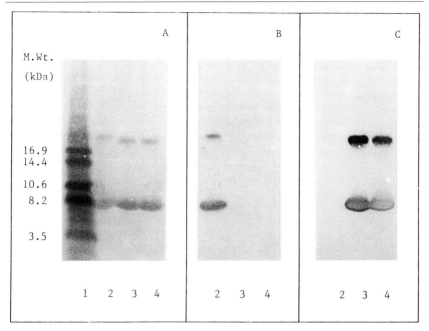

FIG. 4. Immunoreactivity of the retinal GC activator (CD-GCAP). Commercially obtained S100A1 (lane 2) and S100β (lane 3), and CD-GCAP purified from retina (lane 4) were electrophoresed and transferred to nitrocellulose membrane. (A) One set of samples was stained with amido black, (B) another was reacted with antibodies against S100A1; and (C) a third with antibodies against S100β. CD-GCAP reacted with S100β antibodies but not with those of S100A1. All three of the proteins showed two bands at 6–7 and 21 kDa, which represent the monomer and dimer, respectively. Lane 1 shows standard proteins with their molecular masses marked on the left.

Western blotting. As shown in Fig. 4, CD-GCAP was found to react with antibodies against S100β, but not with those against S100A1.[21] Partial sequencing of the protein revealed that its primary structure is identical with that of S100β.[21]

Commercially obtained S100β and S100A1 (from Sigma Chemical Company, St. Louis, MO, and Calbiochem-Novabiochem Corp., San Diego, CA) also activated ROS-GC in a calcium-dependent manner.[23] At saturating concentrations, S100A1 increased the cyclase activity to about 300% of

[23] A. Margulis, N. Pozdnyakov, and A. Sitaramayya, *Biochem. Biophys. Res. Commun.* **218**, 243 (1996).

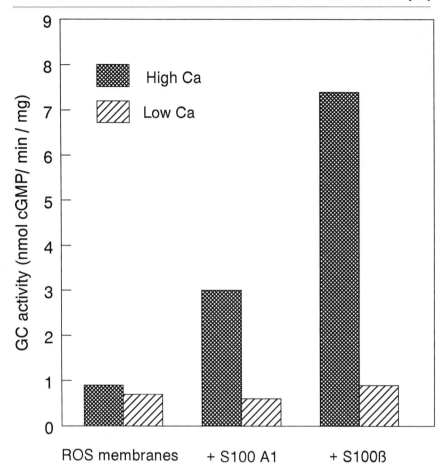

FIG. 5. Calcium-dependent activation of ROS-GC by S100A1 and S100β. GC activity was assayed in washed ROS membranes in the absence and presence of 147 μg/ml of S100 proteins obtained commercially. The assays were carried out at both low and high calcium concentrations. Cyclase activity was stimulated by S100 proteins only at high calcium concentrations and it was found more responsive to S100β than S100A1.

control compared to about 700% of control by S100β (Fig. 5). Calmodulin, which activates guanylate cyclases in *Paramecium, Tetrahymena,* and crayfish, had no influence on the ROS cyclase activity.

Since CD-GCAP isolated in our laboratory has the same primary structure as that of S100β and reacted with its antibodies, it was surprising that commercially obtained S100β was consistently less active than CD-GCAP. Commercial preparations of S100β also required a nearly 20-fold higher concentration of calcium for cyclase activation than CD-GCAP.[21,23] Since it appeared that commercial S100β preparations were purified by zinc affinity chromatography,[17] we tested the effect of putting purified CD-GCAP through the same procedure on its activation of cyclase and calcium sensitivity.[21] It was observed that CD-GCAP subjected to zinc affinity chromatography activated cyclase to a lesser degree than CD-GCAP and it also required a higher calcium concentration for the activation. Its properties were essentially similar to those of commercial S100β.[21]

Discussion

The results shown here demonstrate that retinal extract activates ROS-GC in both low and high calcium and that the activation at high calcium is due to a single protein, CD-GCAP, whose properties identify it as S100β. We found that S100A1, another S100 protein, also activates the ROS-GC though to a lesser extent than S100β. S100β purified by our protocol (CD-GCAP) has higher cyclase-stimulatory activity and higher calcium sensitivity than the protein purified by zinc affinity chromatography, suggesting that tightly bound zinc might leave the protein in a conformation less effective in calcium-dependent activation of cyclase.

The presence of S100 proteins, A1 and β, in bovine retina was demonstrated by other investigators using molecular biological[24] and immunohistochemical[25] methods. Whether these proteins activate ROS-GC under physiologic conditions would depend on their presence in the proximity of the enzyme and whether the concentration of calcium in the ROS supports the activation. Rambotti et al.[25] found in a recent immunohistochemical study that both S100β and S100A1 are localized in the ROS of bovine photoreceptors. They also confirmed the calcium-dependent activation of ROS-GC by these proteins. The present study shows that the calcium concentration in the dark-adapted ROS, about 250–500 nM,[1] is sufficient for a three- to fivefold activation of the cyclase by S100β (CD-GCAP). It is therefore conceivable that S100 proteins play an important role in the cGMP synthesis of dark-adapted ROS.

Immunohistochemical studies have demonstrated the presence of ROS-GC-like cyclases in retinal synaptic layers.[26] Both ROS-GC1 and S100β were shown to be present in the pineal gland and possibly involved in the adrenergic activation of cGMP synthesis.[27] S100β-mediated activation

---

[24] T. Duda, R. M. Goraczniak, and R. K. Sharma, *Biochemistry* **35,** 6263 (1996).
[25] M. G. Rambotti, I. Giambanco, A. Spreca, and R. Donato, *Neuroscience* **92,** 1089 (1999).
[26] N. Cooper, L. Liu, A. Yoshida, N. Pozdnyakov, A. Margulis, and A. Sitaramayya, *J. Mol. Neurosci.* **6,** 211 (1995).
[27] V. Venkataraman, T. Duda, and R. K. Sharma, *FEBS Lett.* **427,** 69 (1998).

of a membrane cyclase was demonstrated in the Muller cells of bovine retina.[25] It therefore appears that membrane guanylate cyclases similar to the ROS enzyme may be present in other retinal and nonretinal locations where they may be activated by S100 proteins in response to an increase in calcium concentration.

Acknowledgment

This study was supported by National Eye Institute grant EY 07158.

[49] Characterization of Guanylyl Cyclase and Phosphodiesterase Activities in Single Rod Outer Segments

By YIANNIS KOUTALOS and KING-WAI YAU

Introduction

In darkness, cGMP binds and opens nonselective cation channels on the surface membrane of rod and cone outer segments, maintaining an inward current and hence a membrane depolarization.[1] Light absorption by the visual pigment elevates cGMP phosphodiesterase activity to increase cGMP hydrolysis.[2] The consequent decrease in cytoplasmic cGMP concentration causes the cGMP-gated channels to close, producing a membrane hyperpolarization as the light response.[3] The dynamic changes in cGMP level during illumination result from a continuously changing difference between the rates of cGMP synthesis and hydrolysis. Thus, the guanylyl cyclase and the phosphodiesterase activities are key determinants of the amplitude and kinetics of the light response. At the same time, negative-feedback modulations of these two enzyme activities as a result of light-induced Ca^{2+} changes have strong effects on the light sensitivity of the cell.[4–6]

[1] K.-W. Yau and D. A. Baylor, *Annu. Rev. Neurosci.* **12**, 289 (1989).
[2] L. Stryer, *Annu. Rev. Neurosci.* **9**, 87 (1986).
[3] K.-W. Yau, *Invest. Ophthalmol. Vis. Sci.* **35**, 9 (1994).
[4] H. R. Matthews, R. L. W. Murphy, G. L. Fain, and T. D. Lamb, *Nature* **334**, 67 (1988).
[5] K. Nakatani and K.-W. Yau, *Nature* **334**, 69 (1988).
[6] Y. Koutalos and K.-W. Yau, *TINS* **19**, 73 (1996).

Although the guanylyl cyclase and the cGMP phosphodiesterase activities have been measured biochemically (see elsewhere in this volume), it is useful to repeat these measurements under near-physiologic conditions. In particular, these measurements would be useful for calculations of the light response of an intact photoreceptor and for assessing the relative importance of the various Ca^{2+}-feedback pathways in setting light sensitivity.[6-9]

The methods described here allow one to measure the two enzyme activities from single rod outer segments (ROS).[8,9] While the descriptions are for rods, the same approach is, in principle, applicable to cones.

Method

A single ROS is sucked halfway into a glass micropipette for recording membrane current. The outer segment is then truncated so that cGMP or guanosine triphosphate (GTP) can be dialyzed into its interior through the open end[8,9] (Fig. 1A).

Phosphodiesterase Activity

Consider a truncated ROS into which cGMP is dialyzed from bath solution (Fig. 1B). On sudden removal of bath cGMP, the cGMP inside the outer segment will dissipate because of two processes: (1) hydrolysis by the endogenous phosphodiesterase activity, and (2) diffusion out of the outer segment. Under these conditions, it can be shown[9] that the cGMP concentration inside the outer segment declines as a single exponential with time, with a rate constant of $(\beta + r)$, where β is the phosphodiesterase activity, and is given by V_{max}/K_m (where V_{max} and K_m are the maximal velocity and the Michaelis constant of the enzyme, respectively), and r is a parameter related to the longitudinal diffusion coefficient, D, of cGMP in the outer segment, and is given by $\pi^2 D/4L^2$, where L is the length of the truncated rod outer segment.[9] The term for β holds when the enzyme operates in its linear range, which is valid for cGMP concentrations significantly below the K_m of 70–95 μM for the enzyme.[10,11] At low cGMP concentrations, the cGMP-activated current varies as the nth power of cGMP concentration, where n (≈ 2) is the Hill coefficient for activation of the

---

[7] K. Nakatani, Y. Koutalos, and K.-W. Yau, *J. Physiol.* **484,** 69 (1995).
[8] Y. Koutalos, K. Nakatani, T. Tamura, and K.-W. Yau, *J. Gen. Physiol.* **106,** 863 (1995).
[9] Y. Koutalos, K. Nakatani, and K.-W. Yau, *J. Gen. Physiol.* **106,** 891 (1995).
[10] T. G. Wensel and L. Stryer, *Proteins* **1,** 90 (1986).
[11] C. L. Dumke, V. Y. Arshavsky, P. D. Calvert, M. D. Bownds, and E. N. Pugh, Jr., *J. Gen. Physiol.* **103,** 1071 (1994).

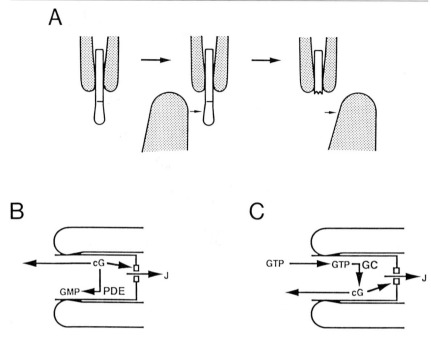

FIG. 1. Schematic diagrams showing how the cGMP phosphodiesterase and guanylyl cyclase activities can be measured from a truncated ROS. (A) Procedure of truncating a ROS. (B) By rapidly removing bath cGMP after dialyzing it into the outer segment, the phosphodiesterase activity can be measured from the rate of decline of the cGMP-activated current. (C) By dialyzing GTP, the substrate for guanylyl cyclase, into the outer segment and measuring the cGMP-activated current in steady state, the cyclase activity can be calculated. cG, cGMP; PDE, phosphodiesterase; GC, guanylyl cyclase; J, cGMP-activated current.

current by cGMP. Thus, the current will also decline exponentially with time, with a rate constant of $n(\beta + r)$. The terms containing β and r can be separated by performing the experiment first in the absence, then in the presence, of 3-isobutyl-1-methylxanthine (IBMX), which inhibits the phosphodiesterase activity (hence, $\beta = 0$). The experiment to measure phosphodiesterase activity is carried out in both darkness and light in order to capture the dark and light-activated enzyme activities, respectively. On the other hand, the experiment to measure cGMP diffusion (the parameter r) by using IBMX to inhibit the phosphodiesterase activity is carried out in darkness because nascent phosphodiesterase activity produced by an absorbed photon is not instantaneously inhibited by IBMX. Finally, the Hill coefficient, n, can be obtained separately in the same or another outer segment by measuring the steady-state relation between activated current and bath cGMP concentration in the presence of IBMX[8] (so that the steady-state cGMP concentration in the outer segment is the same as in the bath solution). By repeating the experiment at different buffered concentrations of Ca^{2+} in the bath solution, the dependence of cGMP phosphodiesterase activity on Ca^{2+} can also be measured.

Guanylyl Cyclase Activity

In this case, GTP is dialyzed from the bath solution into a truncated ROS, to be converted into cGMP by guanylyl cyclase (Fig. 1C). In the steady state, the rate of cGMP synthesis by the enzyme is balanced by the rate of cGMP dissipation which, in the presence of IBMX, is simply due to diffusion out of the outer segment (see earlier section). For a given GTP concentration in the bath, the "effective" steady-state cGMP concentration in the outer segment, $[cG]_i$, can be derived from the steady-state activated current based on the relation between activated current and cGMP concentration mentioned earlier. The rate of cGMP synthesis by guanylyl cyclase is then given by $r[cG]_i$, where r is the diffusion parameter introduced earlier. By repeating the experiment at different buffered concentrations of Ca^{2+} in the bath solution, the dependence of guanylyl cyclase activity on Ca^{2+} concentration can be measured.

Preparation

Larval tiger salamander (*Ambystoma tigrinum*), the marine toad (*Bufo marinus*), or the bullfrog (*Rana catesbeiana*) are suitable animal species because their ROS are large (≥ 6 μm in diameter) and, hence, suitable for suction pipette recording. The eyes are removed from a dark-adapted, pithed animal in dim red light, and the retinas isolated in infrared light with the help of infrared light viewers. Isolated rod cells are obtained by finely chopping a piece of retina on Sylgard elastomer (Dow Corning, Midland, MI), under Ringer's solution, with a razor blade.[12] A single ROS, preferably still attached to the inner segment, is sucked partially and tip-first into a fire-polished, snug-fitting glass micropipette. The micropipette is coated with tri-*n*-butylchlorosilane to allow smooth passage of the outer segment into it. The suction pipette is filled with the appropriate solution (see next section) and connected to a current-to-voltage converter to measure the membrane current through the part of the outer segment inside the pipette. The outer segment is then truncated by shearing off its basal part outside the pipette with a glass microprobe similar in size to the suction

---

[12] K. Nakatani and K.-W. Yau, *J. Physiol.* **395**, 695 (1988).

pipette.[13] In most cases, the truncated end remains open so that the interior of the outer segment can be dialyzed with bath solution. The electrical potential between the pipette interior and the bath is held at 0 mV, because the leak current will otherwise be very large due to the relatively low seal resistance between the outer segment plasma membrane and the wall of the pipette tip. Thus, ion flux is driven exclusively by the concentration gradient across the outer segment membrane.

Solutions

Amphibian Ringer's solution contained 110 mM NaCl, 2.5 mM KCl, 1.6 mM MgCl$_2$, 1 mM CaCl$_2$, 5 mM TMA (tetramethylammonium)-HEPES, and 5 mM glucose, pH 7.55.

For measurement of phosphodiesterase activity, either of two solution pairs can be used[9]: (1) a pipette solution containing Na$^+$ and a bath solution for dialysis containing choline (which is membrane impermeant), so the membrane current is inward (flowing from the pipette interior into the outer segment) at 0 mV; or (2) a pipette solution containing choline and a bath solution containing K$^+$, so the current is outward at 0 mV. In either case, the pipette solution contains 0.5 mM buffered free Mg^{2+} and a low (≤ 1 μM) buffered free Ca^{2+} concentration, and the bath solution contained one of several buffered free Ca^{2+} concentrations together with 0.5 mM buffered free Mg^{2+}. Any Na/Ca-K exchange activity on the outer segment membrane,[14,15] which potentially affects the Ca^{2+} concentration in the outer segment, is eliminated because of the absence of K$^+$ (case 1) or Na$^+$ (case 2) in both pipette and bath solutions. Both cations are required for the exchange activity.[14,15] When present, IBMX is at 0.5 mM. cGMP is at 1 or 3 mM except for cases in which a steady-state relation between activated current and cGMP concentration is to be measured, in which case lower cGMP concentrations are also used. Adenosine triphosphate (ATP) and GTP are at 0.2 and 0.1 mM, respectively, in the bath solution. ATP is for supporting rhodopsin phosphorylation by rhodopsin kinase to bring about normal termination of phototransduction, and GTP is required for transducin activation.[3] At 0.1 mM GTP, the rate of its conversion into cGMP by guanylyl cyclase is negligible compared to the exogenous cGMP supplied.[8]

For measurement of guanylyl cyclase activity, the same solution pairs can be used.

---

[13] K. Nakatani and K.-W. Yau, *J. Physiol.* **395,** 731 (1988).
[14] K.-W. Yau and K. Nakatani, *Nature* **311,** 661 (1984).
[15] L. Cervetto, L. Lagnado, R. J. Perry, D. W. Robinson, and P. A. McNaughton, *Nature* **337,** 740 (1989).

Results

cGMP Phosphodiesterase Activity

Figure 2A shows currents activated by 3 mM cGMP at different Ca^{2+} concentrations from a truncated salamander ROS, with a pipette-choline/bath-K^+ solution pair. A steady light was also present throughout. The arrow indicates the time of truncation of the outer segment. The decays

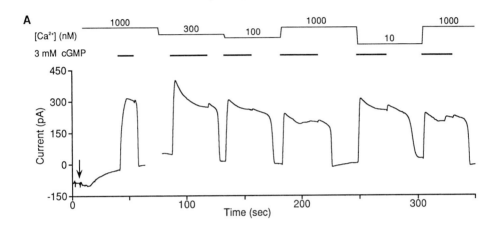

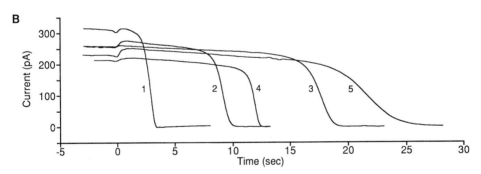

FIG. 2. Measurement of light-stimulated phosphodiesterase activity as a function of Ca^{2+} concentration. (A) Membrane current recorded from a truncated tiger salamander ROS in the presence of a steady light [27.8 photons (520 nm) μm^{-2} sec^{-1}]. The arrow indicates the time of truncation. The current rise soon after the truncation was due to a junction current resulting from bath solution change. (B) Expanded records of the decays of the cGMP-activated currents from (A). The traces have been synchronized at the time of switching bath solution to 0 cGMP (time 0). Ca^{2+} concentrations were (1) 1 μM, (2) 300 nM, (3) 100 nM, (4) 1 μM, and (5) 10 nM. [From Y. Koutalos, K. Nakatani, and K.-W. Yau, *J. Gen. Physiol.* **106,** 891 (1995).]

of the cGMP-activated currents on removing bath cGMP are shown in Fig. 2B, synchronized at the time of removing bath cGMP (time 0). The rate of current decay increased with increasing Ca^{2+} concentration, indicating increased phosphodiesterase activity. In each trace, the latency for the current decline is not necessarily correlated with the decline rate, because the cGMP concentration inside the outer segment had not necessarily reached steady state before bath cGMP was removed. The decline time courses of the currents in Fig. 2B, after normalization of the initial values, are plotted in semilogarithmic scales in Fig. 3A. The late decline phase of each current trace is exponential with time, as expected. With the same outer segment, the experiment was subsequently repeated in darkness and also the presence of 0.5 mM IBMX, so that the contribution from cGMP diffusion to the current decline can be determined (Fig. 3B). The phosphodiesterase activity calculated from these measurements is plotted as a function of free Ca^{2+} concentration in Fig. 3C. This activity includes the basal rate in darkness, which can also be independently measured.[9]

Guanylyl Cyclase Activity

Figure 4A shows membrane currents elicited from a salamander ROS by 0.5 mM GTP at different buffered Ca^{2+} concentrations, in conditions of a pipette-choline/bath-K^+ solution pair. The GTP-elicited current, normalized by the saturated current activated by 1 mM cGMP, is plotted against Ca^{2+} concentration in Fig. 4B. These currents can be converted into "effective" internal cGMP concentration, $[cG]_i$, using the steady-state dose–response relation between current and cGMP concentration obtained separately, and plotted in Fig. 4C (right ordinate). With the constant r measured separately (see earlier discussion), the guanylyl cyclase activity can be calculated from $r[cG]_i$, and is plotted on the left ordinate in Fig. 4C. The results demonstrate an inhibition of the enzyme activity by Ca^{2+}. Strictly speaking, the use of the expression $r[cG]_i$ to calculate guanylyl cyclase activity is only valid for constant cGMP concentration in the outer segment. In actuality, there is a cGMP diffusion gradient along the outer segment. Nonetheless, calculations based on diffusion theory[8] indicate that, as long as the parameter r is larger than the ratio V_{max}/K_m for the enzyme (i.e., diffusion is sufficiently fast to keep up with the activity of the enzyme), the expression $r[cG]_i$ does provide a reasonable approximation of the cyclase activity. This is the case for the rod guanylyl cyclase.

Comments

In addition to affecting phosphodiesterase and guanylyl cyclase activities, Ca^{2+} also changes the half-activation constant, $K_{1/2}$, for the relation

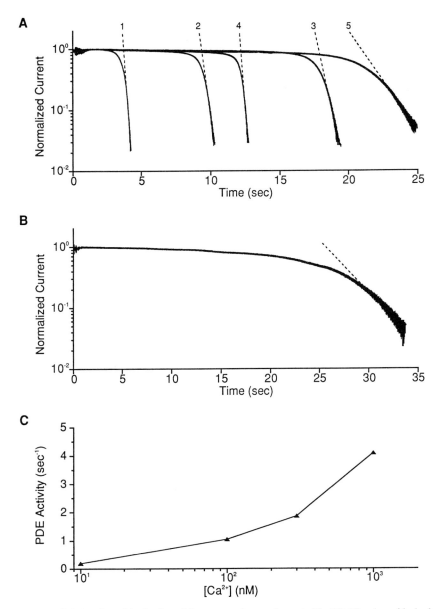

FIG. 3. (A) Semilogarithmic plot of the current decays shown in Fig. 2B. The slope (dashed line) of the linear part of each trace gives the rate of current decay. (B) Measurement of the rate of cGMP loss through diffusion. Same outer segment as in (A). This is essentially the same kind of experiment as in Figs. 2 and 3A, but it was carried out in darkness and in the presence of 0.5 mM IBMX, which inhibited the phosphodiesterase activity. (C) Calculated phosphodiesterase activity plotted as a function of free Ca^{2+} concentration. [From Y. Koutalos, K. Nakatani, and K.-W. Yau, *J. Physiol.* **106,** 891 (1995).]

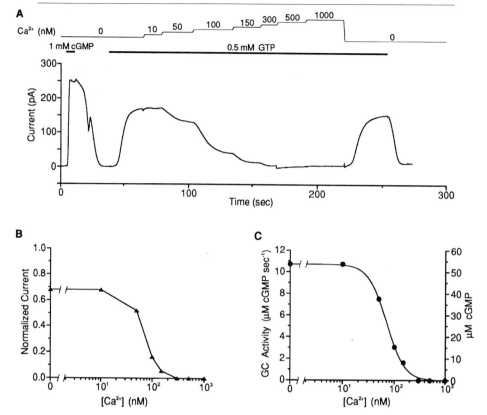

FIG. 4. (A) Effect of Ca^{2+} on the current elicited with 0.5 mM GTP from a truncated salamander ROS. (B) Current (normalized with respect to the saturated cGMP-activated current) from (A) plotted against Ca^{2+} concentration. (C) Currents from (B) converted to guanylyl cyclase activity. The two ordinates give the cyclase activity in "effective" cGMP concentration (right) and in the actual biochemical rate expressed as micromolar cGMP sec^{-1} (left). [From Y. Koutalos, K. Nakatani, T. Tamura, and K.-W. Yau, *J. Gen. Physiol.* **106**, 863 (1995).]

between activated current and cGMP concentration.[7] However, this effect does not interfere with the preceding measurements of the cGMP phosphodiesterase activity because the parameter $K_{1/2}$ does not appear explicitly in the expression, $n(\beta + r)$; the Hill coefficient, n, likewise stays approximately constant with changes in Ca^{2+} concentration.[7]

The effect of Ca^{2+} on channel gating does, in principle, affect the measurement of guanylyl cyclase activity. In practice, however, this turns out not to be a major concern as well, for the reason given later.

The truncated outer segment provides a convenient preparation for rapidly changing the solution inside the outer segment. By the same token, there is a gradual washout of soluble factors from the outer segment. In the experiments to measure phosphodiesterase activity, the washout of rhodopsin and arrestin appears to be minor, at least for the time period of a typical experiment (several to 10 minutes), as judged from the observation that the light-stimulated phosphodiesterase activity does not increase with time as would otherwise be expected.[9] On the other hand, the Ca^{2+} dependence of the enzyme does slowly disappear, especially after dialysis with low (10 nM) Ca^{2+}. To minimize such washout, it is helpful to make measurements as rapidly as possible after truncation of the outer segment, and to use low Ca^{2+} solutions as late in the experiment as possible. As for the guanylyl cyclase measurements, washout actually helps because, in these experiments, the truncated ROS is dialyzed first with low Ca^{2+} so that the Ca^{2+} effect on the channel has already disappeared. The cyclase activity and its dependence on Ca^{2+} do not appear to be largely affected by the low Ca^{2+} treatment.[8]

Measurement of Enzyme Activities in Cone Outer Segments

The preceding measurements should in principle be applicable to cone outer segments as well. Because the cone outer segment is very small, it is more difficult to draw the outer segment only partially into a suction pipette and to keep it stable during truncation. Nonetheless, it is possible, at least for tiger salamander, to draw the entire cone outer segment together with part of the inner segment into the suction pipette, and truncate the cell at the level of the inner segment.[16] Whether the slender ciliary stalk connecting the outer and inner segments would present a diffusion barrier, and whether the tightly packed mitochondria in the ellipsoid region would be a problem for proper dialysis, however, remain to be examined.

Conclusions

The measurements described here for the cGMP-phosphodiesterase and guanylyl cyclase activities agree broadly with the biochemical measurements (see elsewhere in this volume). The single-cell experiments have the advantages that very little biological material is required for the measurements, and that the conditions, including the concentrations of proteins,

---

[16] K. Nakatani and K.-W. Yau, *Invest. Ophthalmol. Vis. Sci.* **27**, 300A (1986).

are much closer to the physiologic situation. The same methods are applicable to other cell structures with an elongated geometry, such as olfactory cilia.[17]

Acknowledgment

This work was supported in part by a grant from the National Eye Institute.

[17] C. Chen, T. Nakamura, and Y. Koutalos, *Biophys. J.* **76,** 2861 (1999).

Section VII

Cyclic Nucleotide-Gated Channels

[50] Covalent Tethering of Ligands to Retinal Rod Cyclic Nucleotide-Gated Channels: Binding Site Structure and Allosteric Mechanism

By JEFFREY W. KARPEN, MARIALUISA RUIZ, and R. LANE BROWN

Introduction

Light stimulation of retinal rods triggers a membrane hyperpolarization that modulates the release of neurotransmitter at the synapse.[1] This electrical response is generated by the closure of cyclic nucleotide-gated (CNG) cation channels in the plasma membrane of the outer segment.[2,3] These channels function as cGMP-sensitive electrodes that track light-induced changes in the intracellular concentration of cGMP. The native rod channel is thought to be a tetramer composed of homologous 63-kDa α and 240-kDa β subunits.[4–7] Each subunit contains six transmembrane domains followed by a cyclic nucleotide-binding domain in the carboxy-terminal tail.[4,8] Significant activation requires the binding of three or four molecules of cGMP.[9–11]

In addition to playing a central role in phototransduction, CNG channels have been embraced by biophysicists as a model for the study of protein allostery.[12] CNG channels are well suited to this role because they can be studied readily using high-resolution patch-clamp techniques down to the level of a single protein molecule, and they do not desensitize. We have developed a novel method for studying the activation mechanism of the rod CNG channel by combining the biochemical technique of photoaffinity

---

[1] K.-W. Yau, *Invest. Ophthalmol. Vis. Sci.* **35,** 9 (1994).
[2] E. E. Fesenko, S. S. Kolesnikov, and A. L. Lyubarsky, *Nature* **313,** 310 (1985).
[3] J. T. Finn, M. E. Grunwald, and K.-W. Yau, *Annu. Rev. Physiol.* **58,** 395 (1996).
[4] U. B. Kaupp, T. Niidome, T. Tanabe, S. Terada, W. Bönigk, W. Stühmer, N. J. Cook, K. Kangawa, H. Matsuo, T. Hirose, T. Miyata, and S. Numa, *Nature* **342,** 762 (1989).
[5] T.-Y Chen, Y.-W. Peng, R. S. Dhallan, B. Ahamed, R. R. Reed, and K.-W. Yau, *Nature* **362,** 764 (1993).
[6] H. G. Körschen, M. Illing, R. Seifert, F. Sesti, A. Williams, S. Gotzes, C. Colville, F. Müller, A. Dosé, M. Godde, L. Molday, U. B. Kaupp, and R. S. Molday, *Neuron* **15,** 627 (1995).
[7] D. T. Liu, G. R. Tibbs, and S. A. Siegelbaum, *Neuron* **16,** 983 (1996).
[8] R. L. Brown, R. Gramling, R. J. Bert, and J. W. Karpen, *Biochemistry* **34,** 8365 (1995).
[9] L. W. Haynes, A. R. Kay, and K.-W. Yau, *Nature* **321,** 66 (1986).
[10] A. L. Zimmerman and D. A. Baylor, *Nature* **321,** 70 (1986).
[11] ML. Ruiz and J. W. Karpen, *Nature* **389,** 389 (1997).
[12] W. N. Zagotta and S. A. Siegelbaum, *Annu. Rev. Neurosci.* **19,** 235 (1996).

labeling with patch-clamp recording. In the first portion of this chapter, we detail the biochemical methods used to photoaffinity label rod CNG channels. This work demonstrated that both subunits bind cGMP and led to the identification of amino acid residues in the binding pocket.[8,13] These results also provided a strong foundation for functional studies. In the second half, we detail the methods for covalently attaching ligands to channels in excised patches and for extracting mechanistic information from both multichannel and single-channel records. The ability to attach cGMP moieties to the channel's binding sites one at a time allows dissection of the mechanism of activation with unprecedented resolution.[11,14]

Design and Chemical Synthesis of APT–cGMP

Photoaffinity labeling is a classical biochemical technique typically used to identify proteins that bind small molecules and to localize the interaction sites.[15] A photosensitive group is attached to the ligand of interest, and the resulting derivative is incubated with target proteins. Photolysis with ultraviolet light generates electron-deficient carbon or nitrogen species that can insert into any protein bond within reach. If it does not immediately react within the binding site, the optimal photoaffinity probe is rapidly deactivated by solvent. The cGMP derivative, 8-azido-cGMP (8-N_3-cGMP), had been used to photoaffinity label cGMP-dependent protein kinase,[16] but early attempts to identify the rod CNG channel using this probe were unsuccessful. In light of this, we designed a new photoaffinity analog of cGMP, 8-*p*-azidophenacylthio-cGMP (APT–cGMP).[13] To construct this probe, we attached a phenylazide moiety to the C8 position of the cGMP guanine ring via a thioether linkage (Fig. 1A). Hydrophobic substitutions at this position were known to create potent activators of the rod channel.[17] The success of APT–cGMP is likely due to the longer, more flexible linker, which affords it a greater selection of targets than 8-N_3-cGMP. Although the protocol used in previous studies was not optimal for labeling of the CNG channel, we also failed to detect any labeling with 8-N_3-cGMP under conditions nearly identical to those that produced strong labeling with APT–cGMP. Benzophenone derivatives are often used as photoaffinity probes. We have tested a benzophenone derivative created by the reaction

---

[13] R. L. Brown, W. V. Gerber, and J. W. Karpen, *Proc. Natl. Acad. Sci. U.S.A.* **90,** 5369 (1993).
[14] J. W. Karpen and R. L. Brown, *J. Gen. Physiol.* **107,** 169 (1996).
[15] J. Brunner, *Annu. Rev. Biochem.* **62,** 483 (1993).
[16] R. L. Geahlen, B. E. Haley, and E. G. Krebs, *Proc. Natl. Acad. Sci. U.S.A.* **76,** 2213 (1979).
[17] R. L. Brown, R. J. Bert, F. E. Evans, and J. W. Karpen, *Biochemistry* **32,** 10089 (1993).

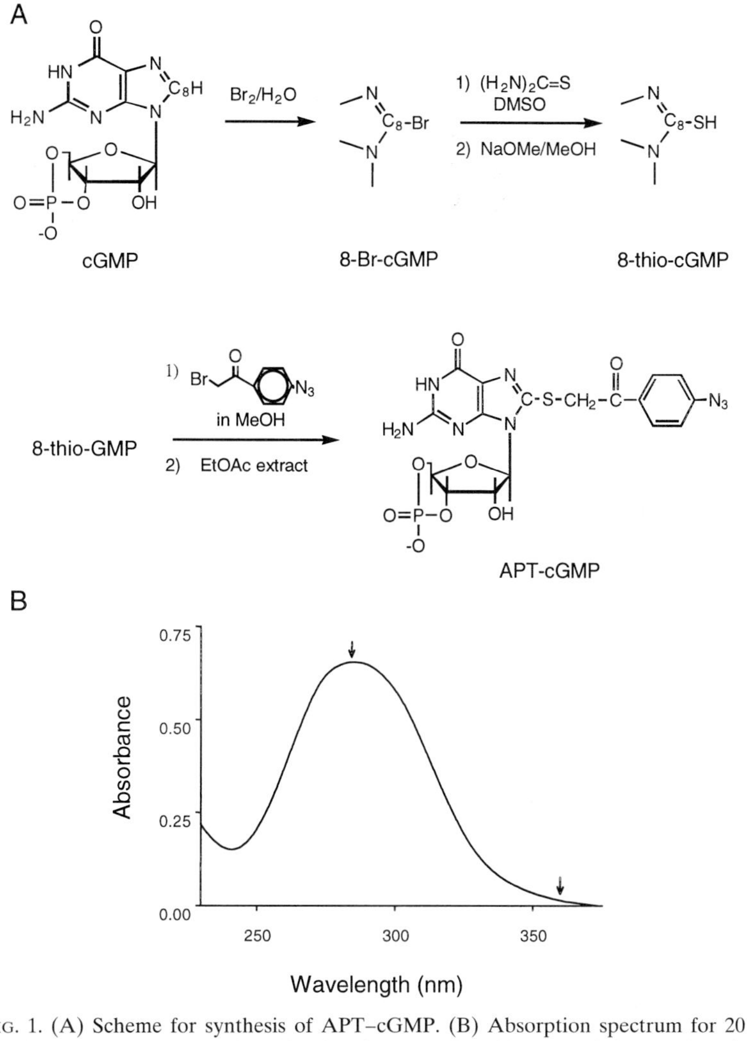

FIG. 1. (A) Scheme for synthesis of APT–cGMP. (B) Absorption spectrum for 20 μM APT–cGMP. The two arrows show the absorbance peak at 284 nm and the wavelength (360 nm) at which the compound is photolyzed.

of 8-thio-cGMP and iodoacetamidobenzophenone (BP–cGMP). In preliminary experiments with this compound, however, we were able to achieve only a limited amount of permanent activation of *tax-4*, a CNG channel from *Caenorhabditis elegans*, using conditions that gave robust covalent activation with APT–cGMP.

Nonradioactive APT–cGMP

The basic scheme for synthesizing 8-*p*-azidophenacylthio-cGMP is outlined in Fig. 1A. Three hundred milligrams of cGMP (sodium salt; Sigma, St. Louis, MO) is dissolved in 28 ml of 250 mM ammonium acetate (pH 3.9 with acetic acid) and brominated by adding 6 ml of a 1:100 bromine/water solution. This is allowed to react for 2 hr at room temperature before addition of a second 3-ml aliquot of the bromine/water reagent. Formation of 8-Br-cGMP is monitored by thin-layer chromatography (TLC) on silica gel (kieselgel 60 F254; EM Science, Gibbstown, NJ) developed with 1-butanol/acetic acid/water (5:3:2). Products are detected by UV shadowing under 254-nm illumination. The 8-Br-cGMP is diluted in water and dried repeatedly in a Speed-Vac (Savant, Holbrook, NY) to remove residual bromine and ammonium acetate. The yield of this reaction is typically greater than 90%.

To synthesize 8-thio-cGMP, 300 mg of 8-Br-cGMP is dissolved in 5 ml of redistilled dimethyl sulfoxide (DMSO) containing 800 mg of thiourea. This reaction mixture is divided into five screw-cap microcentrifuge tubes and heated to 120° for 20 hr. Reaction progress is monitored by UV shadowing on poly(ethyleneimine) (PEI)-cellulose plates (0.1 mm; Macherey-Nagel, Easton, PA) developed with 200 mM LiCl, 50 mM triethanolamine, pH 7.9. On completion, the reaction mixture is diluted to 50 ml with methanol, and the intermediate isothiuronium salt is decomposed by the addition of 0.5 g of sodium methoxide. The crude product is dried, resuspended in about 100 ml of water, and the solution neutralized with acetic acid. The 8-thio-cGMP is purified by anion-exchange chromatography on a 1.7- × 25-cm column of Whatman (Clifton, NJ) DE-52 matrix. The column eluate is monitored at 274 nm. After washing the column with five volumes of water, the 8-thio-cGMP is eluted with a step of 1 M ammonium acetate, pH 7.0. The fractions containing the product are pooled and dried repeatedly from water to remove ammonium acetate. The yield of this step is typically 40–50%. In some cases 8-thio-cGMP is further purified by reversed-phase high-performance liquid chromatography (RP-HPLC).

In the final step, 8-thio-cGMP is alkylated by the addition of *p*-azidophenacyl bromide (ICN, Costa Mesa, CA) to form APT–cGMP. One hundred milligrams of 8-thio-cGMP is dissolved in 5 ml of methanol, and a twofold molar excess of *p*-azidophenacyl bromide is added in the dark. This mixture is shielded from light and incubated at room temperature for 6–12 hr. Reaction progress is monitored by silica TLC as described earlier. All subsequent manipulations are carried out under red light or dim room light conditions.

In our original experiments, APT–cGMP was purified by preparative layer chromatography on silica gel (25 mg/plate; 20 × 20 × 0.2 cm) using the solvent system described earlier. After development, the center of the plate was shielded with aluminum foil, and the product located by exposing the edges of the plate to UV light. The silica gel containing the product was scraped from the plate and extracted with 4 × 25-ml aliquots of methanol to recover the APT–cGMP. The overall yield of APT–cGMP after purification was typically 10–20%. We have found subsequently that extraction of the reaction mix with an excess of ethyl acetate after a twofold dilution with water efficiently removes the unreacted *p*-azidophenacyl bromide. This method reduces labor considerably and achieves a higher yield of APT–cGMP. The product is further purified by RP-HPLC using a C_{18} column eluted with a gradient of aqueous methanol containing 5 mM ammonium acetate, pH 5. The UV detector is shut off, samples of the collected fractions are tested for absorbance at 280 nm, and the peak fractions are pooled. After drying, the purified APT–cGMP is redissolved in water and quantified by absorbance at 284 nm (Fig. 1B, $\varepsilon = 32{,}000\ M^{-1}\ cm^{-1}$). The identity of APT–cGMP is confirmed by fast atom bombardment-mass spectrometry as a negative ion with the expected *m/e* ratio of 535. It is stored frozen and protected from light.

^{32}P-Labeled APT–cGMP

The preparation of ^{32}P-labeled APT–cGMP follows the general scheme just presented with a few modifications. Because of the expense of [^{32}P]cGMP, we find it more cost effective to use [^{32}P]GTP as starting material. Eight millicuries of [^{32}P] GTP (100 Ci/mmol) are converted to cGMP by incubation with guanylyl cyclase (GC) from sea urchin sperm (*Strongylocentrotus purpuratus*). In brief, sperm is collected as it is expelled after injection of the body cavity with 0.5 ml of 0.5 M KCl. After sedimentation, the sperm is extracted with a Lubrol-containing buffer [25 mM triethanolamine, pH 7.4, 2 mM dithiothreitol (DTT), 1% Lubrol WX] to solubilize the GC. The GC is partially purified by GTP-agarose affinity chromatography.[18] For the formation of cGMP, 20 μg of GC is incubated with [^{32}P]GTP in 700 μl of buffer containing 50 mM triethanolamine, pH 7.9, 8 mM theophylline, and 5 mM MnCl$_2$ for 3 hr at room temperature. Conversion is tested on PEI-cellulose.

The bromination reaction is a scaled-down version (50-fold lower volume) of that described earlier except that 20 mM unlabeled 5′-GMP is added as a scavenger to prevent overbromination of cGMP. After the reaction is complete, the 5′-GMP is converted to uncharged guanosine by treatment with 5′-nucleotidase (12.5 units from *Crotalus atrox*; Sigma) to fa-

[18] D. L. Garbers, *J. Biol. Chem.* **251,** 4071 (1976).

cilitate separation. This reaction is performed in 500 mM glycine, pH 9.0, at 37° for 24–48 hr. 8-Br-[^{32}P]cGMP is purified by anion-exchange HPLC eluted with a 5–500 mM ammonium acetate gradient, pH 5.0, with acetic acid.

The remainder of the synthesis is a scaled-down version of the procedure described earlier. Reaction progress on TLC plates is monitored by autoradiography. The entire synthesis procedure takes about 1 week to complete.

Photoaffinity Labeling of Rod Channels in Biochemical Preparations

Partial Purification of Rod CNG Channels, Labeling, and Electrophoretic Separation of Subunits

Rod outer segments (ROS) are obtained from frozen bovine retinas (J. A. and W. L. Lawson, Lincoln, NE) and isolated by centrifugation through a sucrose gradient.[19] The typical preparation starts with 300 retinas, which yield ~ 150 mg rhodopsin. The rod CNG channel protein is partially purified according to the procedure of Cook et al.[20] The ROS are washed with hypotonic saline to strip peripheral membrane proteins before solubilization in CHAPS detergent. To remove rhodopsin, the solubilized membranes are chromatographed on the anion-exchange matrix DEAE-Fractogel (EM Science, Gibbstown, NJ) and eluted in 0.75 M KCl. This detergent-solubilized channel preparation is used for labeling without further purification or reconstitution into liposomes.

Labeling of channels with 5 μM APT-[^{32}P]cGMP is carried out in quartz cuvettes in a volume of 3–6 ml in the presence and absence of 1 mM 8-Br-cGMP. The preparation is illuminated for 10–30 min from a distance of 0.5 m with long-wavelength UV light from a 200-W mercury lamp (Spectral Energy, Westwood, NJ) screened with both a 360-nm bandpass filter (Schott UG1, Duryea, PA) and an IR-blocking filter (Oriel 51950, Stratford, CT). These filters allow photolysis of APT–cGMP (see Fig. 1B), but block wavelengths of light absorbed by aromatic amino acid residues, and prevent heating of the sample. The preparation is diluted 40-fold in 1% CHAPS and concentrated to 0.5 ml in a Centriprep-30 (Amicon, Danvers, MA) to remove salt and unincorporated nucleotide. The labeled sample is diluted with 2 volumes of gel sample-loading buffer containing 10% sodium dodecyl sulfate (SDS), 7 M urea, 195 mM Tris-HCl, pH 8.8, 20% (w/v) glycerol, 20 mM DTT, 5% (v/v) 2-mercaptoethanol, and 0.01% bromphenol blue, and electrophoresed on a 20- × 20- × 0.15-cm SDS–9% polyacrylamide gel to

---

[19] A. Cavaggioni and R. T. Sorbi, *Proc. Natl. Acad. Sci. U.S.A.* **78**, 3964 (1981).
[20] N. J. Cook, W. Hanke, and U. B. Kaupp, *Proc. Natl. Acad. Sci. U.S.A.* **84**, 585 (1987).

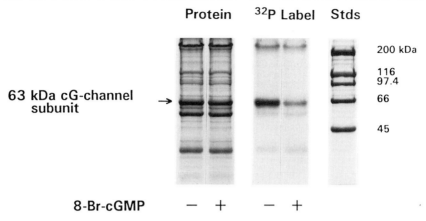

FIG. 2. Specific labeling of partially purified native bovine rod cGMP-gated channel with APT-[^{32}P]cGMP. Paired lanes from an SDS–9% polyacrylamide gel show labeling without (−) and with (+) 8-Br-cGMP (100 μM) added as a competitive inhibitor. Protein lanes show staining with Coomassie Blue and the ^{32}P Label lanes are an autoradiogram of the same gel. The α subunits are the 63-kDa bands (arrow), and β subunits run at about 240 kDa (top bands). Molecular weight standards are shown at the right. [Reproduced with permission from R. L. Brown, W. V. Gerber, and J. W. Karpen, *Proc. Natl. Acad. Sci. U.S.A.* **90,** 5369 (1993). Copyright (1993) National Academy of Sciences, U.S.A.]

purify subunits for subsequent analysis. Note that the rod CNG channel proteins are difficult to resolubilize when the standard method of trichloroacetic acid (TCA) precipitation is used to remove salt and concentrate the sample. An example of an analytical gel of the labeled preparation is shown in Fig. 2. APT-[^{32}P]cGMP labeled both the α and β subunits. The effective competition by 8-Br-cGMP demonstrates that labeling is largely specific to the cGMP-binding sites. We estimate the stoichiometry of labeling to be about 0.5–2%.

Electroelution, CNBr Cleavage, and Electrophoresis of Labeled Peptides

Protein bands on preparative gels are identified by negative staining in 0.3 M copper chloride for less than 5 min. The bands containing the two subunits are excised from the gel, destained in 0.25 M EDTA, 0.25 M Tris-HCl, pH 9, and chopped into 5-mm squares. The proteins are electroeluted in a Centrilutor apparatus (Amicon) at 100 V for 4 hr in 20 mM 3-[cyclohexylamino]-1-propanesulfonic acid (CAPS), pH 11, 0.1% SDS. The samples are twice concentrated to 50 μl (after dilution in 0.1% SDS) in Centricon-30 units.

Subunits are cleaved in 0.25 ml of 0.25 M CNBr and 70% formic acid for 36 hr in the dark. Samples are repeatedly dried from water to remove CNBr and formic acid.

The subunit fragments are dissolved in 50 mM Tris-HCl, pH 6.8, 4% SDS, 12% (w/v) glycerol, 2% 2-mercaptoethanol, and 0.01% bromphenol blue, and electrophoresed on a Tricine–SDS–polyacrylamide gel.[21] To improve resolution, we used 15-cm-long resolving gels, either 16.5% (α subunit) or 13% (β subunit) total acrylamide with 6% and 3% cross-linker, topped by a 2-cm spacer gel (10% total acrylamide). Running the gels slowly at 50-mA constant current resulted in sharper protein bands. Electrophoretic separation of CNBr digests for both α and β subunits revealed striking, specifically labeled bands of 7.5 and 8.2 kDa, respectively.[8] A gel and autoradiogram for the β-subunit digest is shown in Fig. 3.

Sequence Analysis of Labeled Peptides

For sequence analysis, the peptide gels are electroblotted onto PVDF membrane (Immobilon) in 20 mM CAPS, pH 11, 20% methanol (15-W constant power for 2 hr). Protein bands are visualized by staining in 0.1% Coomassie Brilliant Blue R-250, excised and subjected to Edman degradation in an Applied Biosystems (Foster City, CA) 470A protein sequencer. The product of each cycle is collected and identified by RP-HPLC on an Applied Biosystems analyzer. Sequence analysis of the labeled fragment from the α subunit reveals that it is a 66-amino-acid peptide derived from CNBr cleavage at methionine-514. The molecular weight indicates that the fragment extends through methionine-580.[8] This fragment is contained entirely within the cGMP-binding site predicted on the basis of sequence homology with protein kinases and the bacterial catabolite gene activator (CAP) protein.[4] Sequence analysis of the labeled peptide from the β subunit[6] indicates that the APT–cGMP attaches to the homologous region of this protein (tyrosine-1128–methionine-1211).[8]

For the identification of labeled residues, radioactive peptides are identified by autoradiography and subjected to Edman degradation. The product of each cycle is extracted from the blot using n-butyl chloride, collected, and analyzed by liquid scintillation counting. This results in a peak of radioactivity in cycles 9–15, with the maximum in cycle 12. A comparison between this and the sequence of the corresponding fragment revealed labeling of amino acids valine-524, valine-525, and alanine-526 in the α subunit. Although this procedure is successful, extraction of radioactive products is inefficient using standard solvents. Improved yields might be

---

[21] H. Schägger and G. von Jagow, *Anal. Biochem.* **166**, 368 (1987).

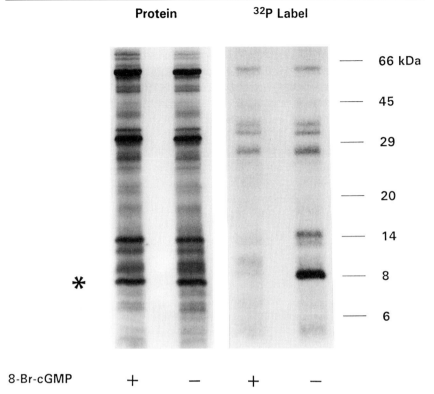

FIG. 3. Identification of the peptide fragment from the β subunit specifically labeled by APT-[^{32}P]cGMP. Paired lanes from a Tricine SDS-13% polyacrylamide gel show CNBr fragments of β subunits labeled in the presence (+) and absence (−) of 8-Br-cGMP. Protein lanes show Coomassie staining of all protein fragments, and the ^{32}P Label lanes show an autoradiogram of the same gel. The peptide fragment indicated by the asterisk was electroblotted to polyvinylidene difluoride (PVDF) membrane and subjected to sequence analysis. [Reproduced with permission from R. L. Brown, R. Gramling, R. J. Bert, and J. W. Karpen, *Biochemistry*, **34,** 8365 (1995). Copyright (1995) American Chemical Society.]

obtained by solid-phase sequencing in which peptides are covalently coupled to the sequencing membrane.[22] This method would enable the use of harsher extraction solvents (e.g., TFA), which may be more efficient in eluting amino acids modified with a polar photoaffinity reagent.

---

[22] J. M. Coull, D. J. C. Pappin, J. Mark, R. Aebersold, and H. Köster, *Anal. Biochem.* **194,** 110 (1991).

Covalent Activation of CNG Channels

In patch-clamp recordings, APT–cGMP can be used to covalently activate CNG channels. In these experiments, an excised, inside-out patch is perfused with a nearly saturating concentration of APT–cGMP (20 μM), so that almost all of the binding sites are occupied. The patch is then exposed to UV light for various time intervals. As in the biochemical experiments, photolysis leads to the formation of a covalent bond with the protein. In this way, cGMP is locked into the binding sites, and after all free ligand is washed away, the channel remains permanently activated. With this approach, the contribution of individual cGMP-binding events to channel gating can be assessed, without the ambiguities that arise from the continuous binding and unbinding of ligands.

Macroscopic Currents

Native rod channels are studied in excised patches from isolated ROS of tiger salamander (*Ambystoma tigrinum*) or leopard frog (*Rana pipiens*) retina. Experiments are performed in visible light at room temperature. Following rapid decapitation, the brain and spinal cord are pithed, the eyes removed and hemisected, and the retinas isolated in Ringer's solution containing (in mM) 111 NaCl, 2.5 KCl, 1 CaCl$_2$, 1 MgCl$_2$, 10 glucose, 0.02 EDTA, 3 HEPES, pH 7.6. A small piece of retina is transferred to the experimental chamber (~100 μl) and teased with fine needles to release ROS. Homomultimeric channels composed of the bovine rod α subunit[4] are heterologously expressed by injecting *in vitro* transcribed RNA into *Xenopus* oocytes, using established methods.[23]

Electrodes are fabricated from borosilicate glass pipettes and fire-polished to resistances of 2–5 MΩ for rods and 0.5–1 MΩ for oocytes. The pipette contains the following control solution (in mM) 130 NaCl, 0.02 EDTA, and 2 HEPES, pH 7.6. For dose–response relations, cGMP is added to the control solution at different concentrations and perfused onto the patch. Currents are elicited by 50-mV pulses from a holding potential of 0 mV, filtered at 2 kHz, and sampled at 10 kHz. After a dose–response relation is determined, the patch is perfused with APT–cGMP and exposed to UV light for varying time intervals. The illumination conditions are as described earlier for biochemical experiments. The patch is washed for at least 15 min to remove all free ligand. The persistent current that remains is a result of channels with covalently attached ligands. This is verified by

---

[23] A. Colman, *in* "Transcription and Translation: A Practical Approach" (B. D. Hames and S. J. Higgins, eds.), p. 271. I.R.L. Press, Washington, D.C., 1984.

blocking the currents with 5 mM Mg^{2+} in the bath. The persistent current increases with longer exposures to UV light as shown in Fig. 4 for native channels composed of α and β subunits, and for expressed channels composed of the α subunit alone. These data indicate that the degree of covalent activation can be controlled by the length of UV exposure time, and that the two subunits of the native channel, as well as α subunits from different species, label with similar efficiency.

Macropatches contain hundreds to thousands of channels, and submaximal exposures result in mixed populations of channels with 1, 2, 3, or 4

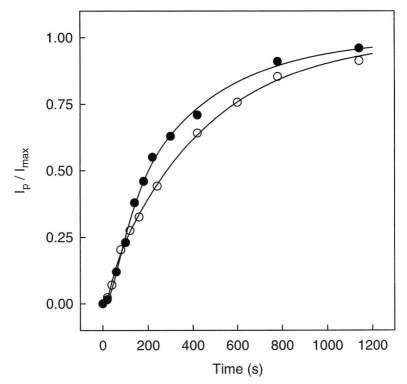

FIG. 4. Time course of covalent activation of cGMP-gated channels. Filled circles, native CNG channels from frog ROS; open circles, homomultimeric channels formed from bovine rod α subunits heterologously expressed in *Xenopus* oocytes. I_p/I_{max} is the fractional persistent current that remained after exposure to APT–cGMP and UV light for the indicated time intervals. Persistent current was measured after patches were washed for 15 min in control solution to remove free nucleotide. I_{max} values were measured in saturating cGMP before exposure and were 706 pA (native) and 5210 pA (expressed). Solid lines were drawn by eye. [Reproduced from J. W. Karpen and R. L. Brown, *J. Gen. Physiol.* **107,** 169 (1996) by copyright permission of the Rockefeller University Press.]

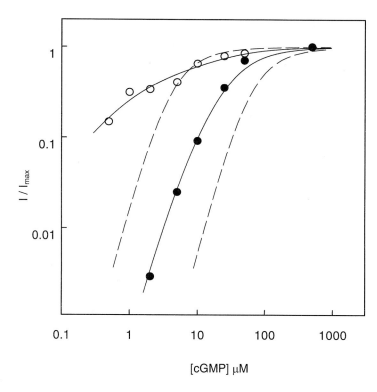

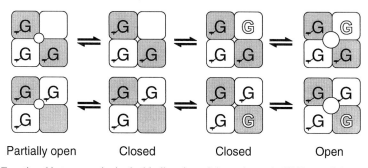

FIG. 5. Functional heterogeneity in the binding sites of the native rod cGMP-gated channel from frog retina. (A) I/I_{max} is the fractional current measured in response to free cGMP before covalent activation (filled circles) and atop a persistent current (from covalent activation) that was 87% of the maximal cGMP-induced current (open circles). All channels that were not fully liganded were assumed to be channels missing only a single ligand. The solid line through the open circles is a best fit to the data assuming two classes of binding sites in equal abundance

ligands covalently attached. We have assumed that the binding sites are labeled at random because (1) the sites are nearly saturated with APT–cGMP, which ensures that labeling does not depend on binding site affinity; and (2) the two subunits of the native channel label with similar efficiency (Fig. 4). Under these conditions, the fraction of channels in each liganded state is given by a simple binomial distribution. From this, we predict that a patch with a persistent current ≥80% of the maximal cGMP-induced current contains almost exclusively channels that are either fully liganded or have three ligands attached. Thus, we can measure the functional properties of the last binding site by adding free cGMP to the patch. Figure 5 shows dose–response relations for native channels from frog rods before covalent activation (filled circles) and atop a persistent current that was 87% of maximum (open circles). Surprisingly, the dose–response relation for the last binding event (open circles) could not be fit with a single binding site model, which predicts a relation that is much too steep; instead the fit shown in Fig. 5 assumes that channels missing only a single ligand fall into two different, equally abundant classes with apparent dissociation constants of 1.2 and 19 μM. These results indicate that there is a functional heterogeneity in the binding sites of the native rod channel (Fig. 5B). Further evidence suggests this is due, at least in part, to the different properties of the α and β subunits.[14] When the two apparent affinities are incorporated into a four-site model of the channel,[14] assuming measured values for the channel-opening equilibrium constants, a good fit to the original dose–response relation (Fig. 5A, filled circles) is obtained. In contrast, a good fit could not be obtained using only one of the apparent affinities (dashed curves in Fig. 5A), suggesting that neither the high nor the low affinity site alone is created by the process of covalent activation.

---

with the following equation: $I/I_{max} = 0.5$ [cGMP]/([cGMP] + K_1) + 0.5 [cGMP]/([cGMP] + K_2), where K_1 and K_2 are the apparent dissociation constants of the two classes (1.2 and 19 μM). The solid line through the filled circles is from a four-site model of channel activation that incorporates the two apparent affinities [see (B)]. The dashed curves show the predicted relations when only one of the apparent affinities is incorporated. (B) Model showing how heterogeneity of the last binding step may arise. The channel is shown with four subunits, two of one binding-site class and two of another (open and shaded), that have different intrinsic affinities for cGMP. cGMP moieties (filled G) are covalently tethered to all but one site, leaving an equal mixture of the two types of binding sites unoccupied. Free cGMP (open G) can bind to the unoccupied binding sites. Channels with only three ligands bound can open partially, and channels with four ligands bound open maximally. [Adapted from J. W. Karpen and R. L. Brown, *J. Gen. Physiol.* **107**, 169 (1996) by copyright permission of The Rockefeller University Press.]

Single-Channel Currents

APT-cGMP can also be used to covalently attach cGMP moieties to single channels isolated in excised membrane patches. In these experiments, the conformational changes of a single protein can be observed in every possible liganded state. In a number of respects, this is the most powerful application of the approach of tethering ligands, because it enables a complete dissection of the allosteric mechanism of channel opening.

We have performed this assay on homomultimeric channels formed from bovine rod α subunits expressed in *Xenopus laevis* oocytes. Isolation of single channels is easiest when 0.05–0.25 ng of cRNA are injected into oocytes. The concentrations used vary depending on the translation efficiency of the cRNA. In this case, the channel cDNA includes 178 bp of 5' and 460 bp of 3' untranslated regions present in the bovine channel gene, and is subcloned into the SP72 vector (Promega, Madison, WI). The oocytes are incubated at 16° for 3–5 days before recording. Electrodes are fabricated from borosilicate glass pipettes, coated with Sylgard, and fire-polished to resistances of about 10–20 MΩ. The solutions are the same as described earlier, except they also contain 1 mM EGTA and the pipette solution includes 500 μM niflumic acid to reduce background channel activity. Channel activity is recorded with 10-sec pulses to ± 50 mV, with 15 sec at 0 mV in between. Records are filtered at 5 kHz and digitized at 88 kHz (Neurocorder DR-484 PCM unit, Cygnus Technology, Delaware Water Gap, PA) for storage on VHS tape. The record is replayed for analysis, filtered at either 1 or 5 kHz, and sampled at either 5 or 25 kHz.

The protocol for covalent attachment of cGMP moieties to the channel is performed as described earlier with exposure times ranging from 10 to 180 sec. To determine how many ligands are tethered to the channel, dose–response relations are compared before and after exposure as shown in Fig. 6. Here, I/I_{max} is the increase in current measured in response to addition of free cGMP. Covalent attachment of ligand causes the dose–response relations to become shallower (lower Hill coefficient) because there are fewer unoccupied binding sites available for binding free cGMP. The $K_{1/2}$ value is also lower because the channel is half-maximally activated at lower concentrations of cGMP. (It is best to determine control dose–response relations 10 min postexcision in order to rule out spontaneous shifts in $K_{1/2}$.)[24] The dose–response relations fall into four obvious groups that are fit with Hill coefficients of 2.9, 1.9, 1.4, and 1.1, suggesting that the relations represent channels with 0, 1, 2, and 3 ligands covalently attached.

---

[24] E. Molokanova, B. Trivedi, A. Savchenko, and R. H. Kramer, *J. Neurosci.* **17**, 9068 (1997).

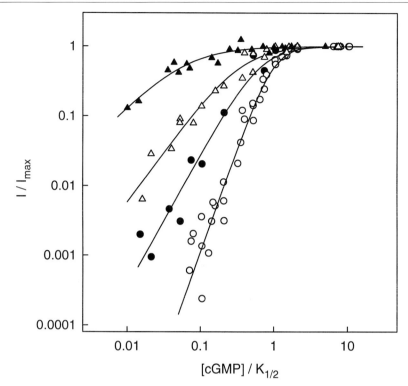

FIG. 6. Shifted and shallower dose–response relations indicate the number of ligands covalently attached to CNG channels. Superimposed dose–response relations from 10 data sets of single-channel recordings before (open circles) and after covalent attachment of ligand. I/I_{max} is the fractional increase in current in response to free cGMP. cGMP concentrations are expressed relative to each channel's $K_{1/2}$ (the concentration that gives a half-maximal response) in order to align the control relations so subsequent shifts can be compared between patches. Solid lines are fits of the Hill equation: $I/I_{max} = [cGMP]^n/([cGMP]^n + K_{1/2}^n)$ where the progressive decrease in the Hill coefficients (n) indicates the covalent attachment of one (filled circles), two (open triangles), and three (filled triangles) ligands. [Reproduced from ML. Ruiz and J. W. Karpen, *Nature* **389**, 389 (1997) by copyright permission of Macmillan Magazines, Ltd.]

Fully liganded channels are not shown, because the addition of free cGMP causes no increase in current.

These tentative assignments of the number of ligands attached are supported by obvious changes in channel behavior, as shown in Fig. 7. We found the fractional currents through these channels to be about 10^{-5}, unliganded; 10^{-5}, one ligand; 10^{-2}, two ligands; 0.33, three ligands; and 1.00, four ligands. Clearly, the channel is very flexible even though no ligands

One Ligand

Two Ligands

Three Ligands

Four Ligands

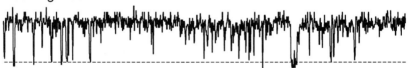

FIG. 7. Single-channel recordings portray dynamic flexibility of the protein with ligands locked into the binding sites. Traces are from channels with one, two, three, and four ligands covalently attached, as indicated by the dose–response relations in Fig. 6. Each trace is 305 ms of channel behavior recorded at +50 mV. Openings are in the upward direction, and the dashed lines show the closed current amplitude.

are binding or unbinding: it exhibits bursting behavior, multiple conductance states, and multiple kinetic states. This is the first time single-channel behavior has been correlated to the precise number of ligands bound.

Amplitude histograms show that the channel freely moves between three conductance states no matter how many ligands are attached.[11] However, subconductance openings are preferred in partially liganded channels, whereas the fully open state is favored in the fully liganded channel. This behavior allows us to rule out simple mechanisms of opening including the

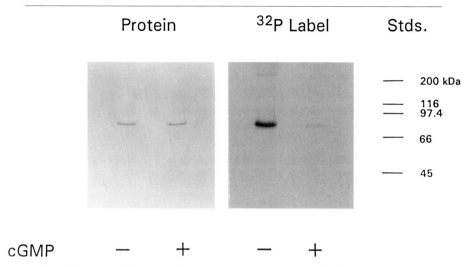

FIG. 8. Labeling cGMP-dependent protein kinase with APT-[^{32}P]cGMP. Protein lanes (SDS-9% polyacrylamide gel) show Coomassie staining of cGMP-dependent kinase after photolysis in the presence (+) or absence (−) of 2 mM cGMP (added as a competitive inhibitor). ^{32}P Label lanes show an autoradiogram of the same gel. Each lane contained 1 μg of protein.

sequential (KNF)[25] and concerted (MWC)[26] models previously proposed for this channel. Further kinetic analyses of dwell-time histograms and adjacent states have provided a basis for constructing a connected state model that can simulate the complex opening behavior of rod cGMP-gated channels.[27]

Future Applications and Conclusions

In addition to experiments on the retinal channel, APT–cGMP promises to be useful in dissecting the activation mechanisms of the different varieties of CNG channels as well. Along these lines, we have demonstrated that APT–cGMP can covalently activate CNG channels formed from the α subunit of the mouse olfactory CNG channel and the *tax-4* subunit from *C. elegans*.[28]

The utility of APT-derivatized cyclic nucleotides is not confined to CNG channels. We have also demonstrated that APT-[^{32}P]cGMP specifically labels cGMP-dependent protein kinase (Fig. 8). The labeling is carried out

[25] D. E. Koshland, Jr., G. Nemethy, and D. Filmer, *Biochemistry* **5**, 365 (1966).
[26] J. Monod, J. Wyman, and J.-P. Changeux, *J. Mol. Biol.* **12**, 88 (1965).
[27] ML. Ruiz and J. W. Karpen, *J. Gen. Physiol.* **113**, 873 (1999).
[28] ML. Ruiz, J. W. Karpen, and K.-W. Yau, unpublished observations (1998).

with ~15 µg kinase (type 1α from bovine lung, Promega, Madison, WI) and 1 µM APT-[^{32}P]cGMP in a buffer containing 125 mM NaCl, 8 mM MgCl$_2$, 5 mM potassium phosphate, pH 6.8, 10 mM 2-mercaptoethanol, and 5% sucrose. The mixture is photolyzed for 30 min using the lamp and filters described earlier. The labeling is confined almost entirely to the binding sites, as it is virtually eliminated by the presence of 2 mM cGMP.

In summary, biochemical labeling experiments with APT–cGMP provided the first direct evidence that both the α and β subunits of the rod CNG channel bind cGMP. This type of experiment assumes increased importance with the recent cloning of *ether-a-go-go* and related genes. These channels would appear to contain cyclic nucleotide-binding domains on the basis of sequence homology, however, there is little and, in some cases, no evidence to indicate that these sites actually bind cGMP or cAMP. Perhaps the most novel and compelling use of APT–cGMP has been to covalently activate CNG channels in patch-clamp experiments. In functional studies on the rod channel, we have shown that the binding sites of the native channel behave in a heterogeneous fashion, mostly due to the differential binding properties of the two subunits. In patch-clamp recordings from single channels, covalent attachment of one ligand at a time has allowed us to study the functional behavior of the channel in every liganded state. This has given us an unparalleled look at the activation mechanism of an allosteric protein.

Acknowledgments

This work was supported by National Eye Institute grants EY09275 (to J.W.K.) and EY11397 (to R.L.B.), and individual National Research Service Awards EY06425 (to R.L.B.) and EY06713 (to ML.R.).

[51] Using State-Specific Modifiers to Study Rod cGMP-Activated Ion Channels Expressed in *Xenopus* Oocytes

By SHARONA E. GORDON

Introduction

Ion channels activated by the direct binding of cyclic guanosine monophosphate (cGMP) are the targets of the enzyme cascade initiated when rhodopsin absorbs a photon. The relatively high cGMP concentration in the dark opens channels, through which Na^+ and Ca^{2+} enter to depolarize the rod. The decrease in cGMP concentration in the light causes some of the channels to close, decreasing the Na^+ and Ca^{2+} influx and hyperpolarizing the rod.

The properties of these ion channels have been extensively studied in both native rods and in exogenous expression systems such as the oocytes of *Xenopus laevis*. Many years of study have taught us much about the role of these channels in visual transduction. However, the contribution of modulation to the function of the channels, the molecular mechanisms for such modulations, and conformational events that couple cGMP binding to opening of the pore remain obscure. In this chapter, we cover some of the methods used in studying activation of the channels by cGMP and other state-dependent forms of modulation. It is through studying such state-dependent interactions that we gain insight into the dynamics of channel function and how they allow the channel to fulfill its role in visual transduction. Before examining state-dependent interactions, we examine some of the general methods used for expressing and studying the channels in *Xenopus* oocytes.

Expression in Oocytes

Robust Expression

Nature seems to have designed *Xenopus* oocytes just for our use in expressing cyclic nucleotide-gated channels. The methods involved are not different from expressing other ion channels[1] and thus are not discussed here in detail, but the expression levels can be much higher than observed for other types of ion channels. For example, we have observed patch currents using bovine rod channel α subunits that exceed 100 nA. The advantage of such high expression is that channels containing mutations that decrease expression by even several orders of magnitude can still be studied.

To get such large currents in a patch, a modified version of the giant patch technique can be used.[2] Because cyclic nucleotide-gated channels are blocked by divalent cations, the bath and patch pipette solutions used are usually free of divalent cations. This makes getting giant patches a bit more difficult, but using mineral oil-based seal glue,[2] seals with patch pipette inner diameters of up to 15–20 μm can be obtained regularly. The seal glue does not seem to alter any of the properties of the channels examined

---

[1] T. M. Shih, R. D. Smith, L. Toro, and A. L. Goldin, *Methods Enzymol.* **293,** 529 (1998).
[2] D. W. Hilgemann and C. C. Lu, *Methods Enzymol.* **293,** 267 (1998).

thus far including voltage dependence, apparent affinity for cyclic nucleotide, cyclic nucleotide specificity, modulation by Ni^{2+}, or dephosphorylation by the endogenous oocyte phosphatase.[3]

Common Artifacts

Although robust expression is clearly desirable, there are both general and cyclic nucleotide-gated channel-specific limitations to the size of usable currents. The general limitation is due to the significant voltage drop across the patch pipette when passing large currents (V_{SR}). For example, using a 200-kΩ pipette and measuring 50 nA of current would give $V_{SR} \sim 10$ mV. Because the open probability of the channels is nearly voltage independent, V_{SR} effects on channel gating would be minimal and need not be considered. Significant V_{SR} errors, however, make measuring a dose–response curve for activation by cGMP nearly impossible because as the cGMP concentration is varied, the size of the current—and thus the true voltage across the membrane—also varies. Therefore, unless the patch-clamp amplifier's circuitry is used to correct for series resistance, care should be taken to keep V_{SR} to less than 10% of the applied voltage.

Voltage-dependent decreases in the current due to ion accumulation and depletion[4] also may limit the size of useful currents. This is because when cellular debris restricts diffusion in the patch, large outward currents move the permeant ion through the channels faster than it can reequilibrate with its concentration in bulk solution. Similarly, large inward currents cause the permeant ion to briefly accumulate in the patch. The local, temporary depletion or accumulation of the permeant ion in the patch decreases the driving force for that ion and is seen as a sag in the current during voltage steps. The hallmark of this effect is a smaller, mirror-image current that occurs when the voltage is returned to the holding potential. This is due simply to the built-up concentration gradient of the permeant ion across the membrane and goes away as the concentration in the patch reequilibrates with that in the bulk solution. The amount of ion depletion and accumulation is not just dependent on the size of the current; it is also dependent on the architecture of the patch. Luckily, the more open architecture of giant patches seems to minimize its effects.

Another artifact to watch out for is that of blocking ions in solution.[5] Two sources have recently been described. One is the presence of metal

---

[3] E. Molokanova, B. Trivedi, A. Savchenko, and R. H. Kramer, *J. Neurosci.* **17,** 9068 (1997).
[4] A. L. Zimmerman, J. W. Karpen, and D. A. Baylor, *Biophys. J.* **54,** 351 (1988).
[5] J. I. Crary, S. E. Gordon, and A. L. Zimmerman, *Vis. Neurosci.* **15,** 1189 (1998).

ions in solution when using metal in the perfusion pathway (for example, metal syringe needles) along with solutions containing high concentrations (several hundred micromolar or more) of EGTA. This is an effect that has been previously described for other ion channels and is thought to be due to a stripping off of metal from the needle or other object by the EGTA.[6] It appears as a voltage-dependent block of the channels that can be eliminated by switching to a solution that has never been exposed to metal or to high EGTA concentrations.

A second source of voltage-dependent block is the plastic used in the perfusion pathway.[5] Substances used in the plastic manufacturing process, such as tinuvin 770, can leach out of the plastic over several minutes. Some plastic sterile syringes used for solution reservoirs have been particularly implicated. To avoid this problem, solutions should not be kept in plastic syringes for long periods of time. This appears to be an ubiquitous problem, because block by plastics-derived agents has been described for other ion channels as well.[7-9]

Expressing Heteromultimeric Channels

The recent cloning of second subunits for both rod[10] and olfactory[11,12] channels has led to attempts to reconstitute properties of native cyclic nucleotide-gated channels by expressing RNA coding for both types of subunits in the same oocyte. The goal for expression in these cases is to ensure that all or almost all of the channels have both kinds of subunits. When expressing both kinds of subunits, two classes of functional channels are possible: homomultimeric channels that contain only the first subunit and heteromultimeric channels that contain both the first and second subunits. The olfactory and rod second subunits do not form functional cyclic nucleotide-activated channels on their own, so homomultimeric channels composed of the second subunit are not observed under normal conditions. To reduce the number of homomultimeric channels, an excess of RNA

---

[6] B. Sakmann and E. Neher, in "Single-recording," Vol. xxii, p. 503. Plenum Press, New York, 1983.
[7] H. Glossmann, S. Hering, A. Savchenko, W. Berger, K. Friedrich, M. L. Garcia, M. A. Goetz, J. M. Liesch, D. L. Zink, and G. J. Kaczorowski, *Proc. Natl. Acad. Sci. U.S.A.* **90,** 9523 (1993).
[8] R. L. Papke, A. G. Craig, and S. F. Heinemann, *J. Pharmacol. Exp. Ther.* **268,** 718 (1994).
[9] T. O. Reuhl, M. Amador, and J. A. Dani, *Brain Res. Bull.* **25,** 433 (1990).
[10] T. Y. Chen, Y. W. Peng, R. S. Dhallan, B. Ahamed, R. R. Reed, and K. W. Yau, *Nature* **362,** 764 (1993).
[11] J. Bradley, J. Li, N. Davidson, H. A. Lester, and K. Zinn, *Proc. Natl. Acad. Sci. U.S.A.* **91,** 8890 (1994).
[12] E. R. Liman and L. B. Buck, *Neuron* **13,** 611 (1994).

coding for the second subunit should be used. For rod channels, a 1:4 ratio of subunit 1:subunit 2 appears to be sufficient.

Possible contamination by homomultimeric channels can be assayed in several ways:

1. The potency of cyclic adenosine monophosphate (cAMP) as an agonist. For rod homomultimeric channels, cAMP activates about 1% of the cGMP-activated current,[13,14] but activates about 10–15% of the current for heteromultimeric channels. For olfactory channels, cAMP is a full agonist for both homo- and heteromultimeric channels but has about an eightfold lower $K_{1/2}$ (concentration that activates half the maximal current) in heteromultimeric compared to homomultimeric channels.[11,12]
2. The voltage dependence of activation. For both olfactory and rod channels, heteromultimeric channels are more steeply voltage dependent and have increased rectification compared to homomultimeric channels, particularly at low concentrations of cyclic nucleotide.
3. The differential pharmacology of pore block. For example, rod heteromultimeric channels have a higher affinity for l-*cis*-diltiazem and are less susceptible to the snake venom pseudechetoxin compared to homomultimeric channels.[10,15]
4. The form of the dose–response relation for activation by cyclic nucleotide. Olfactory heteromultimeric channels have a smaller Hill coefficient than do homomultimeric channels.
5. Olfactory heteromultimeric channels desensitize when expressed in oocytes (but not in cultured mammalian cells).[11,12]

cGMP as State-Specific Modulator of Channel Function

In the absence of cyclic nucleotide, the open probability of rod channels is very low (10^{-5} or lower).[16,17] Binding of cGMP to the cyclic nucleotide-binding sites of all four channel subunits increases the open probability to a value near 1. This coupling of cGMP binding to channel opening is a consequence of a very simple idea—that the affinity of the channels for cGMP is much higher when the channels are open than when they are closed. This open state-specific binding of cGMP is a restatement of the thermodynamic requirement that if channels open better when cGMP is bound then cGMP binds better when channels are open. There are, then,

---

[13] S. E. Gordon and W. N. Zagotta, *Neuron,* **14,** 177 (1995).
[14] E. H. Goulding, G. R. Tibbs, and S. A. Siegelbaum, *Nature* **372,** 369 (1994).
[15] R. L. Brown, T. L. Haley, K. A. West, and J. W. Crabb, *Biophys. J.* **76,** A7 (1999).
[16] G. R. Tibbs, E. H. Goulding, and S. A. Siegelbaum, *Nature* **386,** 612 (1997).
[17] M. L. Ruiz and J. W. Karpen, *Nature* **389,** 389 (1997).

at least two affinities of the channels for cGMP we must consider: the initial, low-affinity binding of cGMP to the closed channel and the subsequent, higher affinity binding of cGMP to open channels. Thus, cGMP is the first state-dependent modulatory agent that we consider.

Many mutations, including the many rod channel/olfactory channel chimeras that have been studied, alter the apparent affinity of the channels for cGMP. How can we tell whether a particular mutation affects the initial binding steps or the subsequent opening conformational change? One trick that we have is to use ligands other than cGMP. Partial agonists, such as cAMP or cyclic inosine monophosphate (cIMP), activate the channels more poorly than cGMP (open probability of ~0.01 for cAMP and ~0.5 for cIMP, at saturating concentrations) and can be useful tools in determining whether the effects of a mutation or treatment are on the initial binding steps or the subsequent opening steps.[13,14,18] Because at saturating concentrations the cyclic nucleotide-binding sites are already maximally occupied (i.e., the initial binding events have already occurred), mutations that affect the open probability at saturating cAMP/cIMP concentrations are not acting by altering the initial binding of the cyclic nucleotides to the closed channels. Rather, such mutations act by altering the energetics of the subsequent opening conformational changes to which initial binding is coupled. Furthermore, cyclic nucleotide-specific changes produced by a mutation suggest that the mutated residue interacts with the cyclic nucleotide as part of the conformational change in the cyclic nucleotide-binding domain. Looking for changes in open probability at saturating concentrations of partial agonists is thus an excellent tool for isolating mutations and treatments that target the later steps in channel activation.

General Considerations for State-Specific Channel Modification

The higher affinity of open channels for cGMP arises because the presence of cGMP in the binding site produces a conformational change in the binding site itself. The conformational change of the binding site around the cGMP provides the energy that drives the opening conformational change in the rest of the protein, including the pore. The binding site, then, in addition to other regions of the channel, is in a different conformation when the channels are open than when they are closed. One approach to identifying specific amino acids, both in the binding site and the rest of the protein, that move as part of the opening conformational change is to use channel modifying reagents. Useful channel modifications range from simple binding of protons or other ions to specific amino acids to covalent

[18] M. D. Varnum, K. D. Black, and W. N. Zagotta, *Neuron* **15,** 619 (1995).

attachment of functional groups to catalysis of interprotein bonds or even proteolysis. If an amino acid or region of the channel is modified in a state-dependent manner then it, like the cyclic nucleotide-binding site and the pore, "looks" different when the channels are open compared to when they are closed. State-dependent modification can thus be a tool for determining if a particular amino acid or region is part of the conformational change that is coupled to cGMP binding.

As stated earlier, the implication that changes in conformation of the channel change its affinity for a ligand applies to modulatory agents other than cyclic nucleotides. Any modulator that acts by altering the conformational change within or outside the cyclic nucleotide-binding site will, by definition, display state dependence. Stated the other way around, any modulator that binds better to open channels will cause the channels to open better when it is bound. Conversely, any modulator that binds better to closed channels will cause channels to close better when it is bound. Thus, the state dependence of binding or modification leads directly to the modulator's potentiation or inhibition of channel function. Localizing the particular amino acid residues or channel region to which a modulator binds or which a modulator modifies identifies a site that moves or changes environment as part of the opening conformational change.

The change in affinity for a modulator when a channel opens or closes means that the rate of modification and/or reversal of modification will depend on whether it is applied to open or to closed channels. The high affinity of one state of the channel compared to the other can come from an increase in the on rate and/or a decrease in the off rate. This is observed as a faster rate of modification by a potentiating modifier when channels are open and/or a faster rate of reversal of modification when channels are closed. Conversely, inhibiting modifiers will produce their effects at a faster rate when channels are closed and/or these inhibitory effects will be reversed more quickly when channels are open. Thus, deciding whether to apply a modulator in the presence or absence of high cGMP is an important determinant of the rate of the modification reaction. It should also be considered that part of the opening conformational change my have already occurred in an intermediate state, such as the ligand bound but closed state observed in the presence of saturating concentrations of cAMP. In such a case the rate of modification of cAMP-bound, closed channels would resemble that of cGMP-bound, open channels. This type of analysis can yield insights into the stage of activation during which the modified residue undergoes the change in conformation or environment.

We now proceed to examine several different channel modulators: cysteine-specific reagents, Ni^{2+}, tetracaine, and Ca^{2+}/calmodulin. These very different reagents have diverse mechanisms and properties. Some of them, such as the cysteine-specific reagents, can modify multiple sites in the primary sequence of the channels, whereas others, such as tetracaine, likely act at a unique site. Some bind reversibly to the channel, whereas other irreversibly modify through covalent modification. What they all share in common is their state dependence. Other state-dependent channel modifiers that we will not cover here include serine/threonine phosphorylation, tyrosine phosphorylation, diacylglycerol, PIP_2, and protonation. The principles that guide our understanding of modulation by cGMP, Ni^{2+}, tetracaine, and Ca^{2+}/calmodulin apply to these modifiers as well.

State-Specific Modification of Cysteines

The broad availability of cysteine-modifying reagents has made them a reagent of choice for structure–function studies of many proteins. Reversible and irreversible reagents are available with a wide variety of chemical characteristics (charge, size, hydrophobicity, fluorescent properties, further reactivity, etc.). There are seven endogenous cysteines in the bovine rod α subunit and more in the beta subunit. For the α subunit, one cysteine that is modified preferentially in the open state (C481, producing channel potentiation)[19] and another that is modified preferentially in the unliganded state (C505, producing channel inhibition)[20] have been identified.

Many cysteine-specific reagents (such as N-ethyl malaimide [NEM] and the methane thiosulfanate [MTS] reagents) are not stable once in solution and care should be taken to use them the same day they are made (for some the reagents should be made fresh for each experiment) and to protect them from light. Thimerosal is another light-sensitive cysteine-specific reagent, although it is stable for up to several days, even at room temperature. For those reagents that are not very water soluble, dimethyl sulfoxide (DMSO) is the preferred solvent for making concentrated stocks. Final DMSO concentrations of up to 1% have been used without confounding effects on the channels or on seal stability. Those cysteine-specific reagents that participate in mixed disulfides with the channel, such as thimerosal, can usually be removed with either dithiothreitol (DTT) or glutathione. High concentrations of these reducing agents (several millimolar and above), however, seem to disrupt the membrane or the seal.

Another class of cysteine-specific reagents does not itself covalently link with the cysteine, but rather catalyzes its modification by other amino acids or parts of the protein. Several general oxidants fall in this category including hydrogen peroxide, iodide, and copper(II) phenanthroline. Al-

[19] S. E. Gordon, M. D. Varnum, and W. N. Zagotta, *Neuron* **19,** 431 (1997).
[20] K. Matulef, G. E. Flynn, and W. N. Zagotta, *Biophys. J.* **76,** A337 (1999).

though not firmly established, the mechanism of this catalysis is thought to involve production of a highly reactive free radical from molecular oxygen in solution. These reactions are, in general, much slower (minutes to hours) than those of the reagents that directly modify cysteines because they require at least one extra step. Also, to maintain specificity, only very low concentrations can be used. For example, 1.5 mM copper(II)/5 mM phenanthroline will very quickly (seconds to minutes) convert cysteines to higher oxidation states (sulfenic, sulfinic, and sulfonic acids) and with exposure over several hours will cause near-complete proteolysis. Hydrogen peroxide in concentrations as low as 0.1% will also convert cysteines to higher oxidation states.

When using copper(II) phenanthroline to catalyze the formation of disulfide bonds, the following procedure can be used. A high concentration cupric sulfate stock (e.g., 1 mM) should be made in water and is stable many weeks when stored at 4°. A high concentration phenanthroline stock (e.g., 50 mM) should be made in dry ethanol and should be stable several days when stored at 4°. These stocks should be diluted to final concentrations of 1.5 μM cupric sulfate and 5 μM phenanthroline in the aqueous buffer of choice and should be used for only 1 day. DTT and glutathione can be used to reduce intraprotein disulfide bonds, although particularly stable disulfide bonds may reduce quite slowly or not at all.

State-Specific Binding of Ni^{2+}

Many proteins are known to bind divalent transition metals, such as Zn^{2+}, Cd^{2+}, and Ni^{2+}, with high affinity. The usual amino acid ligands are histidines, cysteines, aspartic acid, and glutamic acid. The binding of a divalent transition metal to an individual amino acid residue occurs with only low affinity. For example Ni^{2+} binds to free imidazole in solution with a K_D in the millimolar range.[21] High-affinity binding is produced by the coordination of the metal by multiple amino acids with a specific spacing and geometry.

For cyclic nucleotide-gated channels, two high-affinity binding sites for divalent transition metals have been identified. They have been studied best using Ni^{2+}. Ni^{2+} is the divalent transition metal best suited for these studies because the concentrations needed for the state-dependent high-affinity binding are lower than those that produce voltage-dependent channel block.[22] The binding site for Ni^{2+} in rod channel α subunits is a single histidine residue (H420). In *open* channels, two of these same histidine

[21] L. G. Sillén and A. E. Martell, in "Stability Constants," Suppl. 1. Alden Press, Oxford, 1971.
[22] J. W. Karpen, R. L. Brown, L. Stryer, and D. A. Baylor, *J. Gen. Physiol.* **101**, 1 (1993).

residues on adjacent subunits come together to coordinate a Ni^{2+} with an affinity in the range of tens of nanomolar.[13] The same state-dependent adage used earlier applies here: because Ni^{2+} binds better to open channels, channels open better when Ni^{2+} is bound, producing potentiation. The rat olfactory cyclic nucleotide-gated channel α subunit has a different histidine, only three amino acids apart in the primary sequence (H396, analogous to the 417 position of the rod channel). H396 binds Ni^{2+} with high affinity when the channels are *closed,* giving channels that close better when Ni^{2+} is bound, producing inhibition.[23] It is intriguing that two amino acids so close to one another in the primary sequence would have opposite state dependencies for Ni^{2+} coordination. The simplest interpretation of these data is that the region containing the histidines moves as part of activation bringing one set of histidines (420) into alignment between subunits when channels are open and the other (396/417) into alignment when the channels close.

Because of its ability to effectively stabilize the open state of the channel (increase open probability), and the nucleotide-nonselective nature of the potentiation,[13] Ni^{2+} has been used as a tool for studying partial agonists and the effects of mutations on channel function. Such a tool is necessary because some mutations decrease open probability so much that the currents are difficult to resolve. Similarly, some cyclic nucleotides, such as cAMP, activate the channels with such a low open probability that significant channel current may be difficult to obtain. By increasing the open probability in these cases—by as much as 50- to 100-fold—mutant channels that in the absence of Ni^{2+} have too low an open probability to be studied properly and partial agonists that are similarly difficult to study can be "boosted" to a level that makes examination of their properties possible.

The concentration of Ni^{2+} used in experiments is chosen to balance the kinetics and affinity with which it binds to the channels with the concentrations that produce voltage-dependent pore block. For both potentiation of the rod channel and inhibition of the olfactory channel, 1 μM Ni^{2+} is sufficient to produce the effects but can take several minutes to reach steady state. Increasing the concentration to 10 μM greatly speeds the kinetics of onset with the drawback of more significant pore block (10–15% at +100 mV). One compromise is to expose the patch to the higher concentration of Ni^{2+} for a few minutes and then switch back to the lower concentration for measuring currents. Both a high concentration $NiCl_2$ stock made in water and the Ni^{2+} solutions mentioned earlier are stable for weeks at room temperature.

---

[23] S. E. Gordon and W. N. Zagotta, *Neuron* **14**, 857 (1995).

Because standard chelating compounds, such as EDTA and EGTA, have too high an affinity for Ni^{2+}, they must be excluded from the Ni^{2+}-containing solutions. This is a problem both because it prevents chelation of contaminant Ca^{2+} in the solution and because it prevents chelation of contaminant heavy metals from both the water and the $NiCl_2$ (see later discussion). The presence of contaminant Ca^{2+} in solutions causes activation of a Ca^{2+}-activated Cl^- channel endogenous to oocytes. The magnitude of the interference from Ca^{2+}-activated Cl^- channels varies from batch to batch of oocytes and, to a lesser extent, from oocyte to oocyte within a batch. The current due to Ca^{2+}-activated Cl^- channels can range from a few picoamps to several nanoamps and is inwardly rectifying in the presence of symmetrical NaCl. High concentrations of niflumic acid in the patch pipette can reduce these currents significantly. A high concentration (e.g., 150 mM) niflumic acid stock can be made in DMSO. It should be diluted to 500 μM in the pipette solution. The pH of the pipette solution must then be rechecked and adjusted, because the niflumic acid can acidify it significantly. Because 500 μM is near the limit of solubility of niflumic acid in water, lengthy shaking may be required to get it all into solution. The solution should then be filtered with a 0.2-μm pore size syringe filter immediately before use. Both the niflumic acid stock and the niflumic acid-containing pipette solution are stable from a few days to 1 week. After this time, a decrease in seal stability can occur.

Whenever using a new lot of $NiCl_2$, a negative control should be performed to ascertain its purity. The best negative control is to apply 10 μM Ni^{2+} to rod channels in which the Ni^{2+}-binding histidine has been mutated to another amino acid (such as H420Q). This is essential, because contaminant cations such as Al^{3+} in the $NiCl_2$ powder can produce potentiation of the rod channel similar to that produced by Ni^{2+} by acting at an unknown site distinct from H420. Without such a control, it is not possible to determine whether the results observed are really due to Ni^{2+} or to the contaminant cation. Significant potentiation due to such contaminants has been observed in lots of $NiCl_2$ purchased from commercial sources.

State-Specific Binding within Pores

Tetracaine is a local anesthetic that blocks CNG channels in a voltage-dependent manner. Its binding in the pore occurs with high affinity—but only when channels are closed. In fact, the affinity of open channels for tetracaine is nearly 1000-fold lower than the affinity of closed channels for tetracaine.[24] The entire state dependence of binding can be attributed to

---

[24] A. A. Fodor, S. E Gordon, and W. N. Zagotta, *J. Gen. Physiol.* **109**, (1997).

a single glutamic acid residue within the pore (E363); mutations in which E363 is neutralized abolish the state dependence of block and the affinity of the channels for tetracaine (whether open or closed) is close to the low affinity of wild-type open channels.[25]

Because of its ability to "sense" whether a channel is closed, tetracaine case be used as a sensitive reporter of the tendency of a given channel to be closed. This is similar to the way in which Ni^{2+}, as an open channel-specific modifier, can be used as a reporter of a channel's tendency to be open. Ni^{2+}, however, is a poor tool for channels like the olfactory channel, which have open probabilities near 1 at saturating cGMP, because the Ni^{2+}-induced increases in open probability for those channels tend to be quite low. In these cases, it is especially useful to have an independent tool such as tetracaine to probe conformational changes in the pore.

The use of tetracaine requires few special procedures. A 5- to 10-mM stock can be made in any aqueous buffer and is stable for weeks. Diluted solutions in aqueous buffers last weeks as well. One caution to be noted is that when applying tetracaine to the channels it can take a few minutes for its inhibitory effects to reach steady state. Therefore, the current should be monitored during each tetracaine application to ensure that measurements are made at steady state.

State-Dependent Modification by Ca^{2+}/Calmodulin

The Ca^{2+} concentration in rods varies directly with activity of the cGMP-activated ion channels, allowing Ca^{2+} to be an effective feedback regulator in visual transduction. One of its targets appears to be the channels themselves. In the related olfactory cyclic nucleotide-activated channels, Ca^{2+}, in combination with calmodulin, binds to a stretch of about 30 amino acids in the amino-terminal region of the alpha subunits.[26,27] The Ca^{2+}/calmodulin binding site in rod channels appears to be in the amino-terminal region of the β subunits, although a potential binding site in the carboxyl-terminal region of the β subunits has been identified as well.[28,29] For both of these channel types, binding of Ca^{2+}/calmodulin is state dependent: binding of Ca^{2+}/calmodulin occurs with higher affinity to closed channels so that channels close better when Ca^{2+}/calmodulin is bound.

---

[25] A. A. Fodor, K. D. Black, and W. N. Zagotta, *J. Gen. Physiol.* **110,** 591 (1997).
[26] T. Y. Chen and K. W. Yau, *Nature* **368,** 545 (1994).
[27] M. Liu, T. Y. Chen, B. Ahamed, J. Li, and K. W. Yau, *Science* **266,** 1348 (1994).
[28] M. E. Grunwald, W. P. Yu, H. H. Yu, and K. W. Yau, *J. Biol. Chem.* **273,** 9148 (1998).
[29] D. Weitz, M. Zoche, F. Muller, M. Beyermann, H. G. Korschen, U. B. Kaupp, and K. W. Koch, *EMBO J.* **17,** 2273 (1998).

The case of the olfactory channel, in which the mechanism of Ca^{2+}/calmodulin inhibition is better understood, reveals two competing types of state-dependent binding. The first type is the interaction of Ca^{2+}/calmodulin with the amino-terminal region and is inhibitory.[27] The second involves the role of the amino-terminal region in the absence of Ca^{2+}/calmodulin. The 30-amino-acid stretch appears to function as an autoexcitatory domain, binding to a receptor region in the carboxyl-terminal region preferentially when channels are open.[27,30] The mechanism by which Ca^{2+}/calmodulin binding inhibits the channels, then, is clear. By capturing the amino-terminal region it prevents it from interacting with the carboxyl-terminal region. In this way, Ca^{2+}/calmodulin and the carboxyl-terminal region exhibit competitive binding to the amino-terminal region, both with respect to binding site and the state of the channel they bind to best.

Calmodulin is an extraordinarily highly conserved protein. Therefore, it does not seem to matter from which species or which tissue it is derived. High concentrations (<500 nM) have been shown to cause a small voltage-dependent block of the channels,[31] although this effect may well be caused by a contaminant in the preparation rather than by calmodulin itself. The Ca^{2+} concentration should be buffered carefully in these experiments using a method such as that developed by Fabiato and Fabiato.[32] For experiments with rod channels, a free Ca^{2+} concentration of 1 μM is adequate, whereas for experiments with olfactory channels, a free Ca^{2+} concentration of about 50 μM is typically used.

The inhibition by Ca^{2+}/calmodulin is much more pronounced in olfactory channels than in rod channels.[26,33] Furthermore, at least when in their native rod membrane, endogenous calmodulin or another Ca^{2+}-binding protein with similar properties may already be bound to the channels.[31] In such cases, adding exogenous calmodulin may be ineffective because its binding sites are already occupied. The endogenous inhibitor can then be removed using solutions very low in Ca^{2+} (no added Ca^{2+} and millimolar concentrations of EGTA). Following the low-Ca^{2+} treatment, exogenous calmodulin is expected to have its full effect.

Conclusions

Studying state-dependent binding and modification can be a powerful method for examining how cyclic nucleotide binding is coupled to the

---

[30] M. D. Varnum and W. N. Zagotta, *Science* **278,** 110 (1997).
[31] S. E. Gordon, J. Downing-Park, and A. L. Zimmerman, *J. Physiol.* **486,** 533 (1995).
[32] A. Fabiato and F. Fabiato, *J. Physiol.* **75,** 463 (1979).
[33] Y. T. Hsu and R. S. Molday, *Nature* **361,** 76 (1993).

opening conformational change throughout the primary sequence of the protein. The functional consequences of state-dependent binding and modification—potentiation for open state-specific reagents and inhibition for closed state-specific reagents—provide an accessible and reliable assay for these types of interactions. Some state-specific modifiers can also be used as tools to probe the effects of other manipulations such as mutations and nonphysiologic agonists.

[52] Identification and Characterization of Calmodulin Binding Sites in cGMP-Gated Channel Using Surface Plasmon Resonance Spectroscopy

By KARL-WILHELM KOCH

Introduction

Cyclic nucleotide-gated (CNG) channels are nonselective cation channels that are gated by cAMP and cGMP. They serve as downstream targets of cGMP and cAMP signaling cascades in rod and cone phototransduction and in olfaction.[1,2] CNG channels are composed of two homologous subunits (α and β or 1 and 2) that form heterooligomeric complexes. When α subunits are heterologously expressed, they can form functional channels on their own, whereas β subunits alone do not form a functional channel.[3,4] The ligand sensitivity of CNG channels in rod photoreceptor cells and in olfactory neurons is modulated by Ca^{2+}-calmodulin (Ca^{2+}-CaM).[5,6] In rod CNG channels, the $K_{1/2}$ of cGMP increases 2-fold when Ca^{2+}-CaM binds to a binding site on the β subunit.[4] The effect of Ca^{2+}-CaM on the $K_{1/2}$ for cAMP of the olfactory channels is much larger and can result in a 50-fold decrease in cAMP affinity.[6,7] The binding site for CaM in the olfactory channel is located at the cytoplasmic N-terminal segment of the α subunit.[8]

---

[1] U. B. Kaupp, *Curr. Opin. Neurobiol.* **5**, 434 (1995).
[2] J. T. Finn, M. E. Grunwald, and K.-W. Yau, *Annu Rev. Physiol.* **58**, 395 (1996).
[3] H. G. Körschen, M. Illing, R. Seifert, F. Sesti, A. Williams, S. Gotzes, C. Colville, F. Müller, A. Dose, M. Godde, L. Molday, U. B. Kaupp, and R. S. Molday, *Neuron* **15**, 627 (1995).
[4] T.-Y. Chen, M. Illing, L. L. Molday, Y.-T. Hsu, and R. S. Molday, *Proc. Natl. Acad. Sci. U.S.A.* **91**, 11757 (1994).
[5] T.-Y. Hsu and R. S. Molday, *Nature* **361**, 76 (1993).
[6] T.-Y. Chen and K.-W. Yau, *Nature* **368**, 545 (1994).
[7] S. Balasubramanian, J. W. Lynch, and P. H. Barry, *J. Membr. Biol.* **152**, 13 (1996).
[8] M. Liu, T.-Y. Chen, B. Ahamed, J. Li, and K.-W. Yau, *Science* **266**, 1348 (1994).

FIG. 1. Domain structure of the β subunit of the rod CNG channel. It consists of a GARP part and a β' part. Only regions located in the β' part are shown. The two CaM-binding sites are represented by black boxes (CaM1 and CaM2); the six membrane spanning regions and the cGMP-binding site are drawn as gray boxes. Locations of overlapping peptides P1–P11 are indicated.

Two CaM-binding sites were recently identified in the β subunit of the rod CNG channel.[9,10] One site, denoted CaM1, is located in the N terminus, whereas the other site, denoted CaM2 is located in the C terminus of the β subunit (Fig. 1). Electrophysiologic recordings on deletion mutants have demonstrated that the CaM-binding site in the N terminus is necessary for CaM-mediated control of cGMP affinity, whereas channels with a deleted C-terminal CaM-binding site displayed an almost unchanged CaM sensitivity.[9,10] Several biochemical and biophysical techniques including overlay binding assays, gel shift assays, fluorescence emission experiments, and surface plasmon resonance (SPR) spectroscopy were employed to characterize these binding sites and to determine the apparent affinities for CaM.[9,10] By these different approaches the two CaM-binding sites were localized in a stretch of 20–26 amino acids. Synthetic peptides derived from both regions displayed a moderate to low affinity for CaM (range of K_D values between 0.24 and 10 μM), which is two to three orders of magnitude lower than the apparent affinity for CaM of the native or the heterologously expressed CNG channel.[5–11] These discrepancies have been extensively discussed elsewhere[9,10]; it was suggested that high-affinity binding sites in the native channel could originate from the concerted binding of two interacting N-terminal domains or that other regions in the channel protein assist in binding CaM.

SPR spectroscopy is an evanescent wave biosensor technology that monitors an interaction of two or more molecules in real time. Details concerning the physical principle of detection and the optical configuration

---

[9] M. E. Grunwald, W.-P. Yu, H.-H. Yu, and K.-W. Yau, *J. Biol. Chem.* **273,** 9148 (1998).
[10] D. Weitz, M. Zoche, F. Müller, M. Beyermann, H. G. Körschen, U. B. Kaupp, and K.-W. Koch, *EMBO J.* **17,** 2273 (1998).
[11] P. J. Bauer, *J. Physiol.* **494,** 675 (1996).

of SPR biosensors in commercial and noncommercial systems have been described in several reviews.[12-14] SPR biosensors are sensitive to changes in mass on the sensor surface, therefore they detect changes in refractive index. The technique can be used to measure interactions of biomolecules including peptides, proteins, nucleic acids, glucans, and phospholipid vesicles. One binding partner is immobilized on a surface of a sensor chip, whereas the other binding partner is present in the mobile phase and is carried in a flow of buffer through a miniature flow cell. Any binding event on the surface of the sensor chip leads to a change in refractive index at the surface layer and is monitored by a detector (e.g., diode array). Time-dependent changes in refractive index are recorded as sensorgrams. Sensorgrams not only provide information about binding or nonbinding, they also contain information about the kinetics and the affinity of the studied interaction.

SPR spectroscopy was recently applied to identify and characterize CaM-binding sites in different target proteins of CaM including calcineurin,[15] nitric oxide synthase isoforms,[16,17] and the CNG channel from rod photoreceptors and olfactory neurons.[10] We describe here the experimental procedures and the analysis of SPR data to investigate hypothetical CaM-binding regions in the β subunit of the rod CNG channel.

Experimental Procedures

Buffers and Materials

Running buffer: 150 mM NaCl, 10 mM HEPES, pH 7.4, 3 mM CaCl$_2$, 2 mM EGTA, 10 mM MgCl$_2$, 0.005% (v/v) Tween 20

Regeneration buffer: Same as running buffer without 3 mM CaCl$_2$.

Designation, position, and sequences of synthetic peptides and channel constructs are exactly as in Ref. 10. Overlapping peptides P1–P3 representing CaM1 and P4–P11 containing CaM2 correspond to A^{667}-D^{708} and to L^{1133}-G^{1277} in the β subunit of the bovine rod CNG channel.

---

[12] P. B. Garland, *Quart. Rev. Biophys.* **29,** 91 (1996).
[13] P. Schuck, *Annu. Rev. Biophys. Biomol. Struct.* **26,** 541 (1997).
[14] J. H. Lakey and E. M. Raggett, *Curr. Opin. Struct. Biol.* **8,** 119 (1998).
[15] E. Takano, M. Hatanaka, and M. Maki, *FEBS Lett.* **352,** 247 (1994).
[16] M. Zoche, M. Bienert, M. Beyermann, and K.-W. Koch, *Biochemistry* **35,** 8742 (1996).
[17] M. Zoche, M. Beyermann, and K.-W. Koch, *Biol. Chem.* **378,** 851 (1997).

Overview of Experimental Design

Sensorgrams were recorded with a BIAcore 1000 system using the BIAcore control software 1.2. One binding partner was immobilized on the surface of a sensor chip (CM5, BIAcore). In the present project only chips with a carboxymethylated (CM) dextran surface were used. CM-dextran surfaces allow the covalent coupling of molecules via amine ($-NH_2$), sulfhydryl ($-SH$), aldehyde ($-CHO$) or carboxyl ($-COOH$) groups. The other binding partner was dissolved in the running buffer and was injected over the miniature flow cell by an automated liquid handling system. Sensorgrams can be used to determine apparent association (k_a) and dissociation (k_d) rate constants and equilibrium binding constants (K_A and K_D) with the assistance of available software packages (BIAevaluation 3.0 BIAcore; Clamp by Myszka and Morton[18]).

Immobilization by Ligand Thiol Coupling

A peptide or a protein can be covalently coupled via an intrinsic thiol group (i.e., a cysteine) by exchange with a reactive disulfide group on the surface of the dextran layer of a sensor chip. Peptides corresponding to putative CaM-binding sites contained an additional cysteine at the N terminus that allowed a site-directed coupling to the activated dextran layer.[10,16,17] The CM-dextran layer was first activated by 50 mM N-hydroxysuccinimide (NHS) and 200 mM N-ethyl-N'-[(dimethylamino)propyl]carbodiimide (EDC) for 7 min at a flow rate of 5 μl/min (injection volume 35 μl). The activated dextran matrix was derivatized by 80 mM 2-(2-pyridinyldithio) ethaneamine hydrochloride (PDEA) in 0.1 M sodium borate buffer (pH 8.5) for 4 min at a flow rate of 5 μl/min (injection volume 20 μl). Peptides were immobilized by injection of 35 μl of a peptide solution (5–50 μg/ml in 50 mM sodium formate buffer, pH 4.0). Reactive PDEA-modified disulfide groups were deactivated by 50 mM L-cysteine in 1 M NaCl for 3 min at a flow rate of 5 μl/min. Sensorgrams recorded during the immobilization of cysteine modified peptides can be seen in Refs. 16 and 17. CaM-binding peptides were immobilized at densities between 0.4 and 2 ng/mm^2. Surface densities can be varied by changing the time of activation and derivatization or the contact time of the applied sample. Other parameters that influence the surface concentration are the pH and the ionic strength.

---

[18] D. G. Myszka and T. A. Morton, *Trends Biochem. Sci.* **23,** 149 (1998).

Immobilization by Amine Coupling

Activation of the CM-dextran layer was achieved by a mixture of NHS/EDC as described earlier for the ligand thiol method. Proteins or peptides can be covalently linked to the activated matrix by uncharged primary amine groups (lysine residues or free N termini). For example, a maltose binding protein fusion construct of the N terminus[10] of the rod CNG β subunit was injected into the flow cell for 8 min at a concentration of 10–100 μg/ml in 10 mM sodium acetate buffer, pH 4.0. Afterward, a solution of 1 M ethanolamine, pH 8.5, was injected for 8 min to deactivate excessive reactive groups.

Recording of SPR Titration Series

The binding partner in the mobile phase, i.e., CaM or a CaM-binding peptide, was dissolved in the running buffer at varying concentrations of 10, 50, 100, 250, 500, 750, 1000, 2000, 5000, 7500, and 10,000 nM. CaM or peptide was injected for 4–8 min (20–40 μl) at a flow rate of 5 μl/min. To minimize changes in refractive index that are mainly caused by differences in buffer compositions (bulk refractive index change) the sample in the mobile phase should be diluted into the running buffer from a concentrated stock. Alternatively, the sample can be dialyzed against the running buffer before the SPR experiment. Complex formation between the immobilized target and CaM was monitored by a change in refractive index and recorded as a sensorgram [resonance units, (RU) as a function of time]. Dissociation of the formed complex was initiated by injecting running buffer without CaM or peptide. After each cycle, the sensor chip was regenerated by injection of 5–15 μl of regeneration buffer (containing 2 mM EGTA without CaCl$_2$). The same experiment can be performed at different flow rates (1–100 μl/min). Higher flow rates above 30 μl/min are generally recommended when data are further used to determine rate constants (see later section). Lower flow rates are sometimes necessary to reach a binding equilibrium during the time of sample application.

A set of sensorgrams recorded with different concentrations of CaM over immobilized CNG-peptide P1 ([667]ATSTASQNSAIINDRLQELVKL-FKER[692]) is shown in Fig. 2 in the upper trace. Each sensorgram in Fig. 2 consists of four parts: a prerunning phase without CaM (open bar), an association phase (black bar) when CaM was present in the running buffer, a dissociation phase (open bar) after removal of CaM, and a regeneration phase (gray bar) when the flow cell was flushed with EGTA. CaM bound reversibly and in a concentration-dependent manner. A comparison of several sets of recordings revealed that signals saturated between 5 and

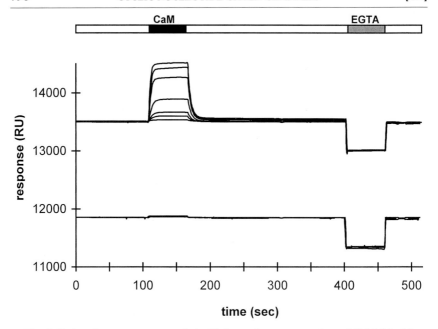

FIG. 2. Series of sensorgrams recorded with increasing concentrations of CaM (black bar; 10, 50, 100, 250, 500, 750, and 1000 nM). Two different flow cells of the same sensor chip are directly compared: sensorgrams displayed in the upper trace were obtained with peptide P1 immobilized on the sensor chip surface; sensorgrams displayed in the lower trace were obtained with a nonmodified CM surface and served as a control. Both flow cells were flushed with running buffer (open bar) and were routinely regenerated with EGTA (gray bar) after each injection of CaM. Sensorgrams of one flow cell were superimposed. Each flow cell of a single sensor chip has a different starting point, but sensorgrams recorded on the same flow cell started at the same RU (upper trace, 13,500 RU; lower trace, 11,856 RU; differences were 0.1% at maximum), which indicates that the complex of peptide and CaM completely dissociated by the regeneration pulse. Flow rate was set to 40 μl/min. Binding signals are recorded as relative changes in RU. Before data are evaluated, the baseline of every sensorgram is Y-transformed (set to 0).

10 μM CaM when P1 was immobilized. Sensorgrams in Fig. 2 (upper trace) also show that a plateau amplitude is reached during the time CaM was supplied in the running buffer. Plateau amplitudes can be used to determine R_{eq} values (see section on Analysis of Data).

The lower trace in Fig. 2 shows a set of control runs recorded with the same concentrations of CaM over a control surface (see later section). CaM in the running buffer caused a small change in refractive index of 20–30 RU, which was similar on different control surfaces. Both sets of recordings were performed on the same sensor chip and are directly compared in Fig. 2. Data were further evaluated by methods of analysis described later.

Surfaces containing peptide P1 as well as other peptides were rather robust and could withstand several cycles of binding and regeneration. For instance, seven different series of CaM titration as in Fig. 2 gave almost equal results with respect to the determined constants.

Changing Configuration on Sensor Chip Surface

It is good practice in the performance of SPR experiments to reverse the geometry on the sensor chip. Equilibrium binding constants should be rather similar irrespective of the geometry on the chip surface. If large differences in the binding constants are observed, this could indicate, for example, that the chemistry of surface immobilization has affected one of the two binding partners. Another test for effects caused by the immobilization chemistry is to immobilize one binding partner in different configurations. CaM was successfully immobilized by three different methods[10]: (1) spinach CaM was immobilized by a unique Cys at position 26 (thiol coupling), (2) bovine CaM was coupled via NH_2 groups (amine coupling), and (3) biotinylated CaM was bound to immobilized streptavidin. Several constructs of the rod CNG β subunit were tested on binding to differently immobilized CaM. Binding affinities were similar for each of the three configurations indicating that immobilization of CaM did not alter binding to its targets.

Controls

Peptides and proteins can nonspecifically bind to dextran surfaces. In addition, slight differences in buffer composition between the running buffer and the sample buffer could lead to a change in refractive index, which is typically seen as a small rectangular step response. Therefore, control sensorgrams should be recorded by flowing the sample over a control surface. The best choice of a control surface is a peptide or protein similar to the one that is used in the binding experiment, but that does not show any interaction with the sample in the mobile phase. However, this cannot be realized in every case. Thus, control runs can alternatively be performed on an activated CM-dextran layer that was only deactivated by cysteine or ethanolamine or on a nonactivated dextran layer.

Analysis Of Data

Qualitative Determination of Binding and Nonbinding

SPR spectroscopy can be used as a screening tool to identify a hypothetical interaction partner. If a search for a CaM-binding domain in a region

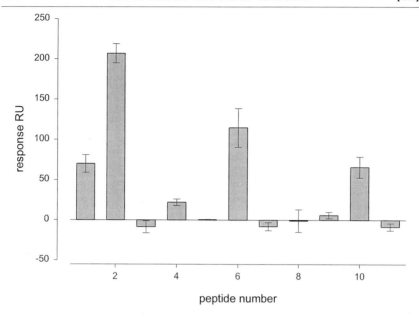

FIG. 3. Peptide screening with peptide P1–P11 using immobilized CaM. Each peptide was dissolved at a concentration of 1 μM in running buffer and injected over immobilized CaM at a flow rate of 40 μl/min. Maximum response amplitudes were recorded and corrected for nonspecific binding by subtracting the amplitude obtained by flushing the corresponding peptide over a control surface. Amplitudes of control runs did not exceed 50 RU, negative responses resulted from slightly higher amplitudes in control runs. Data are the mean ± S.D. of at least two determinations.

of a CaM target is envisaged, the following strategy could be applied. CaM is immobilized and peptides representing potential CaM-binding domains are supplied at a concentration of 1 μM in the running buffer. The plateau amplitudes of recorded sensorgrams are determined and the corresponding control runs are subtracted. Such an approach is performed with peptides P1–P11 and immobilized spinach CaM. A summary of corrected plateau amplitudes for each of peptides P1–P11 is shown in Fig. 3. Plateau amplitudes of control runs were 20–30 RU at maximum. Reliable binding signals were above 50 RU and were observed with peptides P1, P2, P6, P10; other peptides did not show significantly large amplitudes. Identical results were obtained using the reversed geometry (i.e., when peptides were immobilized) as described by Weitz et al.[10] Results from screening experiments are helpful to identify a domain as an interaction site. These results can serve as a starting point for a more thorough quantitative analysis.[19–22]

[19] D. J. O'Schannessy, M. Brigham-Burke, K. K. Soneson, P. Hensley, and I. Brooks, *Methods Enzymol.* **240,** 323 (1994).

Quantitative Analysis of Sensorgrams

A sensorgram is a change of resonance units RU with time and reflects the formation of a complex

$$A + B \rightleftharpoons AB \quad (1)$$

The rate of formation of complex AB with time t is

$$d[AB]/dt = k_a[A]([B]_0 - [AB]) - k_d[AB] \quad (2)$$

or as a change of resonance units with time

$$dRU/dt = k_aC(RU_{max} - RU_t) - k_dRU_t \quad (3)$$

where dRU/dt is the rate by which a complex of two components A and B (e.g., CaM and a CaM-binding peptide) is formed. RU_{max} represents the total binding capacity on the surface of the sensor chip and is equivalent to the number of immobilized binding sites B at $t = 0([B]_0)$. RU_t is the number of occupied binding sites at time t (equivalent to $[AB]$). $RU_{max} - RU_t$ is the number of free binding sites on the chip surface at time t (or $[B]_0 - [AB]$); k_a and k_d are association and dissociation rate constants, respectively. C is the concentration of the binding partner in the mobile phase ($[A]$).

At equilibrium $dRU/dt = 0$, Eq. (4)

$$k_aC(RU_{max} - RU_t) - k_dRU_t = 0 \quad (4)$$

can be arranged to

$$k_aCRU_{max} - k_aCRU_t = k_dRU_t \quad (5)$$
$$CRU_{max} - CRU_t = k_d/k_aRU_t \quad (6)$$
$$RU_t/C = K_ARU_{max} - K_ARU_t \quad (7)$$

At equilibrium RU_t is expressed as R_{eq}. A Scatchard plot analysis (R_{eq}/C against R_{eq}) can be used to obtain K_A (or K_D) from the slope of the plot.

Integration of the rate Equation [Eq. (3)] yields:

$$RU_t = \frac{Ck_aRU_{max}}{Ck_a + k_d}(1 - e^{-[(Ck_a+k_d)t-t_0]}) \quad (8)$$

The dissociation phase of a sensorgram can be described as

$$dRU/dt = -k_dRU \quad (9)$$

---

[20] D. J. O'Shannessy and D. J. Winzor, *Anal. Biochem.* **236**, 275 (1996).
[21] D. G. Myszka, *Curr. Opin. Biotechnol.* **8**, 50 (1997).
[22] P. Schuck, *Curr. Opin. Biotechnol.* **8**, 498 (1997).

or in integrated form

$$RU_t = RU_0 e^{-k_d(t-t_0)} \tag{10}$$

Primary data of a sensorgram RU versus t during the association phase can be directly fit to Eqs. (8) and (10) assuming a 1:1 interaction model.[19]

Sensorgrams obtained with CNG peptides or CaM as immobilized binding partner were analyzed in a Scatchard plot to derive K_D values [Eq. (7)]. Interaction of P1 with CaM exhibited a rather high K_D of 1.5 ± 0.46 μM irrespective of the geometry on the chip surface ($K_D = 2 \mu M$). A representative titration series with CaM in the mobile phase and immobilized P1 is shown in Fig. 2. The affinity of CaM to P2 was determined previously[10]; it is about sixfold higher than the affinity of CaM to P1. Peptide P3 did not show any binding to CaM (Fig. 3 and Ref. 10). Thus, the binding of CaM to the N terminus of the rod CNG β subunit is confined to the boundaries set by P2.

Among the eight overlapping peptides, P4–P11, that span the entire sequence of a part of the C-terminal region in the β subunit (Fig. 1), only P6 and P10 gave signals of a considerable amplitude (Fig. 3 and Ref. 10).

A titration series with peptide P10 in the mobile phase and immobilized CaM is shown in Fig. 4A. The peptide bound reversibly and in a concentration-dependent manner. Signals saturated between 0.5 and 1 μM. A Scatchard analysis of the data in Fig. 4A is shown in Fig. 4B and revealed a K_D value of 86 nM. Sensorgrams shown in Fig. 4A were also evaluated using the curve-fitting program BIAevaluation 3.0. The complete set of binding curves was fit to a single set of rate constants (global fitting) by application of a 1:1 binding model; only R_{max} values were fit locally. A K_D of 47 nM was determined by this approach, which is similar to the K_D obtained from the Scatchard plot. Calculated R_{max} values differed maximally by 10%.

Several titration series with P6 in the mobile phase and immobilized CaM confirmed that P6 interacts with CaM. Data analysis using Scatchard plots exhibited a low affinity of CaM for the site represented by P6 and a very large scattering of the derived K_D values (2.1 ± 1.7 μM). Fusion constructs that contain P6 but lack P10 do not bind CaM.[10] Therefore, low-affinity binding of CaM to P6 is probably not of biological significance.

Interpretation of SPR Data

Kinetics observed in sensorgrams can be rather complex even if a simple 1:1 Langmuir binding model is a plausible interaction model (e.g., CaM binding to a peptide). Deviation from a single exponential association and dissociation process is mainly caused by[13,20–22] (1) the immobilization chemistry (see section on Changing Configuration on Sensor Chip Surface),

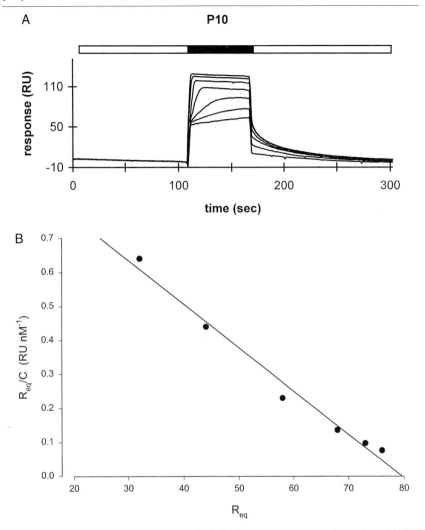

FIG. 4. (A) Series of sensorgrams recorded with increasing concentrations of peptide P10 in the running buffer (black bar: 10, 50, 100, 250, 500, 750, and 1000 nM). All sensorgrams were superimposed and the baseline was set to 0. The flow rate was 40 μl/min. (B) Scatchard plot for binding of P10 to CaM (R_{eq}/C versus R_{eq}). Data analysis was performed with the sensorgrams shown in (A). Maximum amplitudes were read and amplitudes obtained in a control run were subtracted to yield R_{eq}. Maximum amplitudes in control runs did not exceed 50 RU. Determination of K_D from the slope yielded 86 nM.

IQ-Motif   *IQ* x x x *KG* x x x *R*

P2 ^{676}AIINDR**L**QEL**V**KL**F**KERTEK**V**KEKLI^{701}
         1   5   8   15

P10 ^{1236}GGRGGR**L**ALLRAR**L**KELAA**L**EAALEAAARQ^{1261}
          1   8   14

FIG. 5. Amino acid sequences of CaM-binding sites CaM1 and CaM2 represented by peptides P2 and P10. Numbers represent the position of amino acids in the β subunit of the rod CNG channel according to Ref. 3. Conserved positions of hydrophobic amino acids are shaded in gray. A consensus sequence of an IQ motif is shown above the peptide sequences.

(2) a heterogeneous population of targets on the chip surface, (3) mass transfer limitation and rebinding, and (4) steric hindrance.

An inhomogeneity on the chip surface could result in a curvilinear Scatchard plot, which was observed when peptide P10 was immobilized. Furthermore, binding of CaM to P10 occurred with considerably lower affinity (EC$_{50}$ value of 560 nM)[10] than binding of P10 to immobilized CaM (K_D = 86 nM; see earlier section and Fig. 4). This indicates that immobilization of P10 has led to two different populations of P10 on the surface, and that a significant portion of P10 has a reduced affinity for CaM when attached to the surface. The curvilinear Scatchard plot became linear when a larger version of peptide P10[23] or a C-terminal fusion construct containing P10 and adjacent regions[10] were immobilized instead. Furthermore, the affinity increased to a K_D of 207 nM when the C-terminal construct was coupled to the dextran layer. These results show that influences of the immobilization chemistry can be avoided if a larger molecule is used for covalent binding. If the covalent attachment is site directed by the noninteracting part of the molecule, the actual binding regions remain flexible and can adopt a high-affinity conformation.

Mass transfer limitations, rebinding, and steric hindrances can be minimized by using low surface densities, higher flow rates, and/or the addition of a binding competitor during the dissociation phase.[13,20–22] The extent of mass transfer limitation can be ascertained by fitting the data with parameters that include a mass transport coefficient (BIAevaluation 3.0). However, binding constants that are derived purely from curve fitting should be

---

[23] D. Weitz, Doctoral Thesis, University of Cologne (1998).

considered with caution.[22,23] Whenever possible, they should be compared with binding constants determined from a Scatchard plot.

Conclusion

SPR spectroscopy is a useful tool for identifying CaM-binding sites in the β subunit of the rod CNG channel. The CaM2 site represented by the synthetic peptide P10 (Figs. 1 and 5) contains the *1-8-14* recognition motif[24,25] for Ca^{2+}-dependent CaM binding (numbers indicate the position of hydrophobic amino acids). However, there is no evidence that CaM binds to the CaM2 site in functional channels. The CaM1 site represented by peptide P2 is not a conventional target site for CaM. It lacks a hydrophobic amino acid at position 14 (a valine is at position 15 instead) and contains neither the *1-5-10* nor the *1-8-14* motif.[25] The motif is partly similar to the IQ motif,[25] a consensus site for Ca^{2+}-independent binding of CaM (Fig. 5). However, the interaction of CaM with CaM1 is strictly Ca^{2+} dependent, which is untypical for a conventional IQ-binding motif.

Acknowledgment

This work was supported by grants from the Deutsche Forschungsgemeinschaft.

[24] A. Crivici and M. Ikura, *Annu. Rev. Biophys. Biomol. Struct.* **24,** 85 (1995).
[25] A. R. Rhoads and F. Friedberg, *FASEB J.* **11,** 331 (1997).

[53] Determination of Fractional Calcium Ion Current in Cyclic Nucleotide-Gated Channels

By STEPHAN FRINGS, DAVID H. HACKOS, CLAUDIA DZEJA, TSUYOSHI OHYAMA, VOLKER HAGEN, U. BENJAMIN KAUPP, and JUAN I. KORENBROT

Introduction

Cyclic nucleotide-gated ion channels (CNG channels) select cations over anions, discriminate poorly among monovalent cations, and are also permeable to divalent cations. In physiologic ionic solutions, therefore, both monovalent and divalent cations permeate the channels and any one ion species carries only a fraction of the total current. We describe a method to determine the fraction of the ionic current carried by Ca^{2+} ions both in native photoreceptors and in cells expressing recombinant CNG channels.

The method requires the simultaneous measurement of total membrane currents, through electrical means, and Ca^{2+} flux through photometric measurements. The method is based on concepts and procedures first developed by Neher and collaborators in order to measure the fractional Ca^{2+} currents in glutamate and acetylcholine receptor channels.[1,2] Variants of this method have been used to examine fractional Ca^{2+} currents in recombinant and native glutamate receptors,[3-6] CNG channels,[7,8] capsaicin-activated channels in dorsal root ganglion neurons,[9] and in transduction channels in inner ear hair cells.[10] We have combined measurements of membrane current and cytoplasmic Ca^{2+} with the use of caged cyclic nucleotides to rapidly activate the CNG channels. Implementation of the method in studies of intact photoreceptors, in addition, offers technical challenges that may be important to consider in investigations of any cells in which ion channels are not uniformly distributed throughout the cell's surface. We first briefly summarize the theoretical basis for the experiments as outlined by Neher and collaborators.[1,2,11] We then describe the use of caged cyclic nucleotides for channel activation, the recording hardware, and data analysis. Finally, we present several experimental caveats that must be considered in the successful application of this method.

Method to Determine Fractional Ca^{2+} Current in CNG Channels

The fraction of the ionic current carried by Ca^{2+} ions in CNG channels, P_f, is determined by simultaneously measuring total membrane current, with patch-clamp electrodes, and Ca^{2+} influx with the Ca^{2+}-sensitive fluorescent dye, Fura-2. Figure 1 illustrates the recording configuration: A cell expressing CNG channels is bathed in normal physiologic saline solution and is voltage clamped using a tight-seal electrode in the whole-cell configuration. Because the electrode is filled with a solution containing caged cGMP (see later section for details on this molecule) and Fura-2, these

[1] R. Schneggenburger, Z. Zhou, A. Konnerth, and E. Neher, *Neuron* **11,** 133 (1993).
[2] Z. Zhou and E. Neher, *Pflügers Arch.,* **425,** 511 (1993).
[3] N. Burnasheve, Z. Zhou, E. Neher, and B. Sakmann, *J. Physiol.,* **485,** 403 (1995).
[4] O. Garaschuk, R. Schneggenburger, C. Schirra, F. Tempia, and A. Konnerth, *J. Physiol.,* **491,** 757 (1996).
[5] R. Schneggenburger, *Biophys. J.* **70,** 2165 (1996).
[6] F. Tempia, M. Kano, R. Schneggenburger, C. Schirra, O. Garaschuk, T. Plant, and A. Konnerth, *J. Neurosci.,* **15,** 456 (1996).
[7] S. Frings, R. Seifert, M. Godde, and U. B. Kaupp, *Neuron,* **15,** 169 (1995).
[8] C. Dzeja, V. Hagen, U. B. Kaupp, and S. Frings. 1999, *EMBO J.* **18,** 131 (1999).
[9] H. U. Zeilhofer, M. Kress, and D. Swandulla, *J. Physiol.,* **503,** 67 (1997).
[10] A. J. Ricci and R. Fettiplace, *J. Physiol.* **506,** 159 (1998).
[11] E. Neher and G. J. Augustine, *J. Physiol.* **450,** 273 (1992).

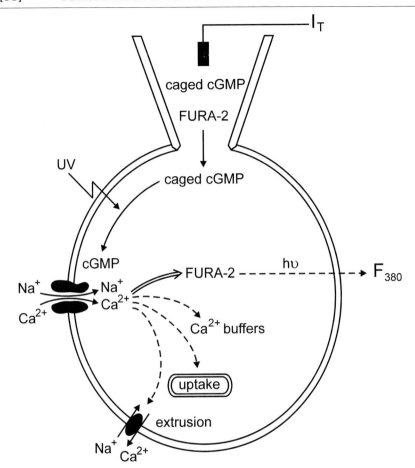

FIG. 1. Recording configuration for measurements of fractional Ca^{2+} current in CNG channels. A cell expressing CNG channels is voltage clamped and loaded with caged cGMP and Fura-2 using a patch pipette. Irradiation with an UV flash liberates cGMP, which activates the channels. The patch electrode records the consequent inward current, I_T, which is carried by monovalent cations and Ca^{2+}. Ca^{2+} binds to Fura-2 and changes the fluorescence emission of the dye, F_{380}, which is monitored by a photon counter. I_T and F_{380} can be used to calculate the fractional Ca^{2+} current, $P_f = I_{Ca}/I_T$. [Reprinted with permission from C. Dzeja, V. Hagen, U. B. Kaupp, and S. Frings. *EMBO J.* **18,** 131 (1999). Copyright (1999) Oxford University Press.]

molecules diffuse into the cell and, after some minutes, their concentrations in the cytoplasm and the electrode are equal. After this equilibrium is reached, an intense flash of UV light generates free cyclic nucleotide, channels open, and current flows across the membrane (the total current I_T).

Because the channels exhibit only limited ion selectivity, I_T is composed of contributions from several ion species: Na$^+$ (I_{Na}), K$^+$ (I_K), Mg^{2+} (I_{Mg}), and Ca^{2+} (I_{Ca}):

$$I_T = I_{Na} + I_K + I_{Ca} + I_{Mg}$$

Fura-2 binds Ca^{2+} with high affinity (K_d^{Ca} = 220 nM in the presence of 100 mM KCl and 1 mM Mg^{2+}) and the Ca^{2+}-free and Ca^{2+}-bound forms of the dye fluoresce with distinct fluorescence excitation spectra.[12] The Ca^{2+}-free form fluoresces with maximum intensity when excited at about 380 nm and this intensity declines as Ca^{2+} is bound.[12] When holding the membrane voltage at −70 mV, channel activation by uncaging of cGMP causes Ca^{2+} to enter the cell; the inflowing Ca^{2+} is rapidly bound by the Ca^{2+}-free form of Fura-2, resulting in a decrease in the intensity of emitted fluorescence excited by 380-nm light (F_{380}).

The extent of decline in F_{380} is proportional to free Ca^{2+} concentration, yet it measures the total amount of Ca^{2+} entering the cell if, and only if, a number of experimental conditions are strictly met. Establishing that these conditions are met must be part of the successful application of the analysis detailed here.

If a membrane current carries Ca^{2+} into the cell, then Δn_{in}, the change in the total mole amount of intracellular Ca^{2+} caused by the current over a time interval Δt, is given by

$$\Delta n_{in} = \frac{1}{2F} \int_0^{\Delta t} I_{Ca}(t)\, dt \qquad (1)$$

where F is Faraday's constant. If Fura-2 is used to monitor cytoplasmic Ca^{2+}, intracellular Ca^{2+} may distribute among several competing pools: (1) free Ca^{2+}, (2) Ca^{2+} bound to cytoplasmic buffers (CaB), (3) Ca^{2+} bound to Fura-2 (CaF), and (4) Ca^{2+} pumped into cellular organelles or out of the cell (P). Thus,

$$\Delta n_{in} = \Delta n_{free} + \Delta n_{CaF} + \Delta n_{CaB} + \Delta n_P \qquad (2)$$

where Δn_{free}, Δn_{CaF}, Δn_{CaB}, and Δn_P represent the changes in mole amount of free Ca^{2+}, Ca^{2+} bound to Fura-2, Ca^{2+} bound to cytoplasmic buffers, and Ca^{2+} pumped (or sequestered) out of the cytoplasm, respectively. Neher reasoned, and proved rigorously,[11] that under certain experimental conditions Eq. (2) can be simplified because all the inflowing Ca^{2+} is bound to Fura-2. These conditions are as follows: (1) Measurements must be limited to a few seconds after the change in current. The rapid rate of Ca^{2+} binding by Fura-2 (few milliseconds) allows a few seconds to be sufficiently long

---

[12] G. Grynkiewicz, M. Poenie, and R. Y. Tsien, *J. Biol. Chem.* **260**, 3440 (1985).

for binding to the dye to be complete, but still short relative to the time necessary to pump Ca^{2+} into organelles or out of the cell; this condition makes Δn_P negligible. (2) Measurements must be executed with high Fura-2 concentrations, 1–5 mM. Under this condition, the indicator dye, which is a very powerful buffer, outcompetes the endogenous Ca^{2+} buffers, rendering Δn_{CaB} negligible. Also, at high concentrations of Fura-2, all entering Ca^{2+} is bound by the dye and Δn_{free} becomes negligible. Under these experimental conditions, then:

$$\Delta n_{in} = \Delta n_{CaF} = \frac{1}{2F} \int_0^{\Delta t} I_{Ca}(t)\, dt \tag{3}$$

The Ca^{2+} binding capacity of Fura-2, κ_F is defined as:

$$\kappa_F = \frac{d[CaF]}{d[Ca^{2+}]} = \frac{K_D [F]_{tot}}{(K_D + [Ca^{2+}])^2} \tag{4}$$

where $[Ca^{2+}]$, $[CaF]$, and $[F]_{tot}$ are the concentrations of free Ca^{2+}, bound Ca^{2+}, and total Fura, respectively. K_D is the dissociation constant of the Fura–Ca^{2+} complex. With this definition, considering a small incremental elevation in $[Ca^{2+}]$ and under the experimental conditions that hold true for Eq. (3), then:

$$\Delta n_{in} = V_c \Delta [Ca^{2+}](1 + \kappa_F) \tag{5}$$

where V_c is the accessible cell volume. If we define the emitted intensity of Fura-2 fluorescence excited at 380 nm (F_{380}), as the sum of the fluorescence of its two molecular forms, one Ca^{2+} free, S_F, and the other Ca^{2+} bound, S_{CaF}, then:

$$\Delta F_{380} = \frac{\Delta n_{in}(S_{CaF} - S_F)\kappa_F}{1 + \kappa_F} = V_c \Delta[Ca^{2+}](S_{CaF} - S_F)\kappa_F \tag{6}$$

Equation (6) indicates that at high concentrations of Fura-2 the change in fluorescence is proportional to the total amount of Ca^{2+} entering the cell, as the ion shifts the fraction of Fura-2 between its free and bound states. Combining Eqs. (3), (5), and (6) yields:

$$\frac{\Delta F_{380}}{\int_0^{\Delta t} I_{Ca}(t)\, dt} = (S_{CaF} - S_F)\frac{\kappa_F}{1 + \kappa_F} \tag{7}$$

Equation (7) indicates that the ratio of fluorescence change to the time integral of the current reflects the shift in Fura-2 states and, therefore, measures the total amount of Ca^{2+} that enters the cell as a carrier of membrane current. Let us define f as the ratio of the change in fluorescence,

ΔF_{380}, and the integral of the membrane current:

$$f = \frac{\Delta F_{380}}{\int_0^{\Delta t} I_T(t)\, dt} \quad (8)$$

If Ca^{2+} were the exclusive current carrier, f takes the value f_{max}.

$$f_{max} = \frac{\Delta F_{380}}{\int_0^{\Delta t} I_{Ca}(t)\, dt} \quad (9)$$

The use of Eqs. (8) and (9) is valid if (1) Ca^{2+} influx through the CNG channels is the only source of Ca^{2+} that contributes to ΔF_{380} (e.g., Ca^{2+} release from cellular stores, leak conductances, or flux through other ion channels does not occur); (2) Ca^{2+} ions are not lost to intracellular buffers or Ca^{2+} transport systems (pumps, exchangers); (3) Fura-2 is not saturated by Ca^{2+}; and (4) F_{380} and I_T are recorded from the same part of the cell.

The value of the fraction of membrane current carried by Ca^{2+}, P_f, is simply

$$P_f = \frac{f}{f_{max}} \quad (10)$$

Activation of CNG Channels Using Photolysis of Caged Cyclic Nucleotides

Rapid and controlled channel activation in the intact cell is a requisite for P_f measurements and can be achieved by using photolabile derivatives of the channel ligands cGMP or cAMP—so-called caged cyclic nucleotides (for reviews, see Adams and Tsien,[13] Corrie and Trentham,[14] Nerbonne,[15] and Kaupp et al.[16]). These compounds are rendered biologically inactive by addition of a protecting group ("cage") that is linked by a photolabile ester bond to the phosphate group of the respective cyclic nucleotide. The bond is cleaved on absorption of UV light, releasing cyclic nucleotides, a photolysis by-product, and a proton. The caging groups of commercially available caged cGMP and caged cAMP are NPE [1-(2-nitrophenyl)ethyl] or DMNB (4,5-dimethoxy-2-nitrobenzyl). NPE-caged compounds display

---

[13] S. R. Adams and R. Y. Tsien, *Ann. Rev. Physiol.* **55**, 755 (1993).
[14] J. E. T. Corrie and D. R. Trentham, in "Bioorganic Photochemistry" (H. Morrison, ed.), Vol. 21, p. 243. Wiley, New York, 1993.
[15] J. Nerbonne, *Curr. Opin. Neurobiol.* **6**, 379 (1996).
[16] U. B. Kaupp, C. Dzeja, S. Frings, J. Bendig, and V. Hagen. *Methods Enzymol.* **291**, 415 (1998).

a higher quantum yield of photolysis than DMNB derivatives (the quantum yield Φ indicates the fraction of caged molecules that undergoes photolysis on absorption of one photon) and release during photolysis a by-product (2-nitrosoacetophenone) that is less toxic than the by-product of DMNB compounds (4,5-dimethoxy-2-nitrosobenzaldehyde). On the other hand, DMNB-caged compounds have faster release kinetics[15] and are used for applications where rapid concentration jumps of cyclic nucleotides are required. The rate of photolysis is >3000 sec^{-1} in DMNB-caged cGMP,[17] but only 4.2 sec^{-1} in NPE-caged cAMP.[18]

Application of caged cyclic nucleotides for P_f measurements requires several considerations concerning their chemical properties. Briefly, the relatively poor solubility of the compounds (typically 50–150 μM in intracellular solution) limits the concentration range of free cyclic nucleotides that can be released by irradiation. This imposes a constraint on experiments with rod and cone photoreceptors where concentration jumps to 50–500 μM cGMP would be desirable because of the relatively low ligand sensitivity of the cGMP-gated channels in intact cells. Furthermore, photoreleased cGMP is rapidly hydrolyzed in photoreceptors due to the high activity of phosphodiesterase (PDE) in these cells. Analysis of P_f data is, consequently, restricted to a brief time segment after photolysis. However, the PDE sensitivity of cGMP also offers two beneficial effects: (1) The commercially available batches of caged cGMP often contain a sizable fraction of free cyclic nucleotide (10–20%). In a whole-cell experiment, the endogenous PDE removes these contaminations very efficiently, so that each photolysis experiment starts with a very low concentration of channel ligand, i.e., with all channels closed. If caged cGMP is used for experiments with excised membrane patches that do not retain PDE activity, free cyclic nucleotides have to be removed by high-performance liquid chromatography (HPLC). (2) Because PDE removes cGMP from the cytosol after each flash, it is possible to perform repeated photorelease experiments in the same cell.

For P_f measurements of CNG channels with low ligand sensitivity, 8-bromo-substituted derivatives of cGMP and cAMP are very useful, since several CNG channels display 5- to 10-fold higher sensitivity to 8-Br-cGMP and 8-Br-cAMP compared to the nonsubstituted compounds.[19] The synthesis and application of caged 8-Br-cGMP and caged 8-Br-cAMP are de-

---

[17] J. W. Karpen, A. L. Zimmerman, L. Stryer, and D. A. Baylor, *Proc. Natl. Acad. Sci. U.S.A.* **85,** 1287 (1988).

[18] J. F. Wootton and D. R. Trentham, in "Photochemical Probes in Biochemistry" (P. E. Nielsen, ed.), Vol. 272, p. 277. Kluwer Academic, The Netherlands, 1989.

[19] A. L. Zimmerman, G. Yamana, F. Eckstein, D. A. Baylor, and L. Stryer. *Proc. Natl. Acad. Sci. U.S.A.* **82,** 8813 (1985).

TABLE I
Photochemical Properties of Caged Derivatives of 8-Br-cAMP and 8-Br-cGMP

Cyclic nucleotide ester	Solubility (μM)[a]	Half-life in aqueous buffer $t_{1/2}(h_r)$[b]	Long wavelength absorption maximum (nm)	Quantum yield Φ[c]
DMNB-caged 8-Br-cGMP				
Axial isomer	100	50	346 ($\varepsilon = 5800$)	0.0049
Equatorial isomer	10	5	346 ($\varepsilon = 5800$)	0.0052
DMNB-caged 8-Br-cAMP				
Axial isomer	120	60	345 ($\varepsilon = 5900$)	0.050
Equatorial isomer	80	8	345 ($\varepsilon = 5900$)	0.045
NPE-caged 8-Br-cGMP				
Axial isomer	35	300	264 ($\varepsilon = 19,400$)	0.33
Equatorial isomer	120	80	264 ($\varepsilon = 19,400$)	0.27
NPE-caged 8-Br-cAMP				
Axial isomer	225	>600	265 ($\varepsilon = 19,500$)	0.48
Equatorial isomer	300	95	265 ($\varepsilon = 19,500$)	0.49
MCM-caged 8-Br-cGMP				
Axial isomer	40	>400	328 ($\varepsilon = 13,400$)	0.20
Equatorial isomer	20	>400	325 ($\varepsilon = 13,400$)	0.10
MCM-caged 8-Br-cAMP				
Axial isomer	20	>400	326 ($\varepsilon = 13,500$)	0.14
Equatorial isomer	40	>400	326 ($\varepsilon = 13,500$)	0.09

[a] In 5% acetonitrile/0.01 M HEPES/KOH buffer, pH 7.2, containing 0.12 M KCl at room temperature. Saturation concentrations 1 hr after dissolution.
[b] Buffer as in footnote a.
[c] Φ values were determined at λ_{exc} 333 nm by the relative method [H. J. Kuhn, S. E. Braslavsky, and R. Schmidt, *Pure Appl. Chem.* **61**, 187 (1989)] using the standard *E,E*-1,4-diphenyl-1,3-butadiene. Data for DMNB and NPE compounds from Hagen *et al.*[20,21]; for MCM compounds: V. Hagen, unpublished.

scribed elsewhere,[20,21] and their photochemical properties are summarized in Table I. Several important features of these compounds are determined by the protecting group used for caging. In particular, the half-life in aqueous solution (the time until 50% of the ligand is solvolytically uncaged) is longest in the MCM [(7-methoxycoumarin-4-yl) methyl] derivatives. This is important because the 8-bromo-substituted derivatives are resistant to hydrolysis by PDE and may accumulate within the cell on solvolysis. MCM derivatives also show a high quantum yield of photolysis in the near-UV range, which is used for the uncaging flash (see later section). Despite their

[20] V. Hagen, C. Dzeja, S. Frings, J. Bendig, E. Krause, and U. B. Kaupp, *Biochemistry* **35**, 7762 (1996).
[21] V. Hagen, C. Dzeja, J. Bendig, I. Baeger, and U. B. Kaupp, *J. Photochem. Photobiol. B* **42**, 71 (1998).

limited solubility (Table I), MCM-caged cyclic nucleotides are, therefore, very useful substances to generate substantial concentration steps of free channel ligand. Higher concentrations can be reached by using the DMNB- and NPE-caged compounds, because their aqueous solubility is higher. However, the shorter half-life in aqueous solution causes a progressive increase of free 8-Br-cGMP and makes it necessary to prepare fresh pipette solutions every 2–3 hr to avoid channel activation during the loading phase. In summary, caged cGMP and caged cAMP can be used for transient activation of CNG channels as well as for repetitive experiments in an individual cell. The caged forms of 8-Br-cGMP and 8-Br-cAMP release ligands with higher biological efficiency that are resistant to hydrolysis by PDE. The choice of the appropriate caging group makes it possible to design the flash-induced concentration step with regard to time course and amplitude.

Instrumentation and Data Collection

A recording setup for P_f measurements is illustrated in Fig. 2. Light from two different light sources, Xe and Hg, is collected by a fiber optic light guide, directed into the epifluorescence light path of an inverted microscope (symbolized by the boxed-in area in Fig. 2), and focused onto cells held in a recording chamber on the microscope stage. The light from the Xe lamp passes through a filter wheel and is used for the excitation of Fura-2, while the Hg lamp produces light for the uncaging flash, which is controlled by shutter 1. Mercury lamps are well suited for uncaging experiments because their radiation spectrum includes an intense peak at around 365 nm, close to the absorption maxima of DMNB- and MCM-caged compounds (Table I). Uncaging can also be obtained with a Xe arc flash of sufficient power (200 J in 300 μs; Xe flash in Fig. 2). Cells in the recording chamber are viewed through a 40× oil-immersion objective that transmits light in the near-UV region. A light switch reflects the transmitted light into a video camera, allowing a pinhole aperture in the light path to be positioned under video control around a single cell. This ensures that only light from the investigated cell is collected and analyzed. After positioning the pinhole, the cell is patch clamped in the whole-cell configuration; a broad bandpass green filter (520–560 nm) placed in the microscope illumination light path prevents photolysis of caged compounds. While caged compounds and Fura-2 diffuse into the cell, the light from the microscope objective is directed toward a photomultiplier mounted on one of the microscope's ports. Fluorescence emission intensity is measured with the photomultiplier using photon counting instrumentation. The excitation intensity is sufficient to generate a reliable fluorescence signal, but does not

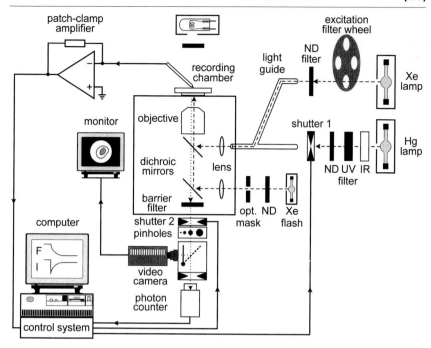

FIG. 2. Schematic representation of the recording setup as described in the text. Light sources are a 100-W Xe lamp attached to an excitation-filter wheel (Rainbow, Life Science Resources, Cambridge, UK) with three excitation filters (340HT15, 365HT25, and 380HT15; Omega Optical, Brattleboro, VT), and a 100-W Hg lamp equipped with an elliptical reflector and an infrared water filter (A1000, AMKO, Tornesch, Germany). Harmful UV light of λ <330 nm is removed by a UV filter (WG335, AMKO) before the light reaches the electronically controlled shutter 1 (Compur, München, Germany). A Y-shaped fiber optic UV light guide (AMKO) directs the light into an inverted microscope (Diaphot 300, Nikon, Düsseldorf, Germany) via an epifluorescence adapter and is reflected into the objective (CF Fluor 40×, n.a. 1.3, Nikon) by a dichroic mirror (400DCLP, Omega Optical). Maximal light intensity of the Hg lamp, measured at the aperture of the objective, is 15–20 mW at λ = 365 nm. It is critical that the fluorescence illumination intensity in the plane of the cells be homogenous in density (photons $\mu m^{-2}\ sec^{-1}$) to avoid errors arising from the fact that cells may be located anywhere throughout the objective's field of view. For measurements with photoreceptors light from a Xe flash lamp (Strobex, Chadwick-Helmuth, El Monte, CA) is focused into the objective through an optical mask. The fluorescence emitted from the cell passes through a barrier filter (510WB40, Omega Optical) and is directed toward the side port of the microscope. Shutter 2 (Uniblitz, Vincent Associates, Rochester, NY) controls access to the fluorescence recording system (PhoCal, Life Science Resources) that consists of a wheel with pinholes of various apertures (Nikon) and a light switch (Mikroflex, PFX-35, Nikon) to which a video camera (TM 500, Pulnix, Sunnyvale, CA) and a photon counter (QL30F/RFI, Thorn EMI, Middlesex, UK) are connected. A patch-clamp amplifier (L/M-EPC 7, List-Medical, Darmstadt, Germany) records the whole-cell current. The control system includes software for simultaneous recording of current and fluorescence (PhoCal/PhoClamp, Life Science Resources).

itself cause detectable uncaging (approximately 5×10^7 photons $\mu m^{-2} sec^{-1}$ at 380 nm). To induce photolysis of caged compounds, shutter 2 closes to protect the photon counter from the flash light, while shutter 1 opens for the required duration; subsequently shutter 2 reopens to allow recording of the fluorescence signal. The simultaneous current and fluorescence traces are recorded and used to calculate P_f

To measure P_f in intact, isolated photoreceptors two specific modifications of the instrument just described are critically important. First, spectral filters and/or dichroic mirrors are placed in the microscope illuminator to allow all microscopic manipulations to be carried out under infrared (IR) illumination, using appropriate video cameras and monitors. This is important because cell viability is compromised if isolated photoreceptors are continuously exposed to visible light. Second, the optical pathways that bring the fluorescence excitation light and the uncaging light into the epifluorescence attachment of the inverted microscope must be separated. Thus, an optical fiber brings in the fluorescence excitation light, but the uncaging light is brought in through its own pathway and they finally combine in the microscope objective (see Xe flash attachment in Fig. 2). This is important because photoreceptors, unlike HEK293 or other cells expressing recombinant channels, are structurally polarized; CNG channels are the only channels present in the plasma membrane of the outer segment, but they are also present in the inner segment membrane, albeit at much lower density. Also, the cytoplasm in the inner segment inevitably contains caged nucleotides and Fura-2. Because cyclic nucleotides generated in the inner segment diffuse into the outer segment (see later discussion for evidence), if uncaging is not restricted to the outer segment alone, then two waves of cyclic nucleotide occur in the outer segment; the first one generated by local uncaging of nucleotide and the second one generated by diffusion from the inner to the outer segment.

Because cyclic nucleotides diffuse from the inner to the outer segment, it is also true that Ca^{2+} may diffuse from the outer to the inner segment. This intracellular flow implies that Ca^{2+} entering the outer segment through the CNG channels may be sequestered away from the outer segment by simply diffusing into the inner segment. To meet the theoretical demands of the experimental design, therefore, the cGMP-dependent changes in the concentration of Ca^{2+}-bound Fura-2 must be measured in both inner and outer segments. This aim is achieved by adjusting the mask in front of the photomultiplier to include all of the photoreceptor cell, and by using an adjustable mask in the fluorescence excitation path (see Xe flash attachment in Fig. 2) in order to illuminate both inner and outer segment.

To quantitate the intensity of fluorescence emission, cell fluorescence signals are normalized using standard fluorescent microbeads (Fluoresbrite carboxy BB, 4.5 μm, Polyscience, Warrington, PA). Bead fluorescence is excited and emission intensity measured under identical conditions to those used to measure cell fluorescence. The mean fluorescence intensity of a single microbead is defined as a bead unit (BU) and is measured on the day of each experiment. This normalization accounts for fluctuations of excitation light intensity and allows comparison of results obtained under different experimental conditions.[1,2] A possible source for inaccuracy in P_f measurements arises from the fact that the fluorescence of microbeads used for normalization of F_{380} slowly fades with time. Beads should be replaced at frequent intervals.

To start an experiment, the cell under investigation is clamped at a negative membrane voltage, Xe illumination is switched to 380 nm, and data acquisition hardware and software record current and fluorescence. Delivery of the uncaging UV flash causes an inward current (I_T) and a concomitant decline in F_{380}. Figure 3A illustrates results of an experiment on a human embryonic kidney (HEK)293 cell expressing bovine olfactory CNG channels. The fluorescence trace shows a small gap around the time of the flash, caused by the closing of shutter 2 (see Fig. 2). For many applications, the amount of photoreleased cyclic nucleotide should be as large as possible. This can be achieved by using bright or prolonged flashes. However, both intensity and duration of the flash light must be limited because photobleaching of the dye can create problems with the analysis of fluorescence data. Light intensity can be reduced by ND filters in the Hg light path. The maximal safe flash duration and intensity can be determined by investigating the bleaching of Fura-2 loaded into cells that have not been transfected. Figure 3B (upper trace) shows an example in which photobleaching of Fura-2 in a HEK293 cell is apparent as a slight loss in fluorescence intensity caused by uncaging UV flashes of 200- and 500-ms duration. To measure F_{380} without inducing photolysis of caged compounds, fluorescence excitation intensity must be appropriately dim. The lower trace in Fig. 3B shows that dimming the Xe light with an ND filter (OD of 2) prevented channel activation while the residual light was sufficient to measure the free Ca^{2+} concentration. Photorelease of cGMP induced by a flash of the Hg lamp caused activation of CNG channels in this cell (arrow).

Data Analysis

For analysis, the F_{380} signal is first converted into bead units (Fig. 3C, negative F_{380} values are used for convenience of display). Offsets are subtracted from both electrical and photometric signals, so that the baseline data preceding the uncaging flash has a mean value of zero. However, the absolute value of the holding current must be established in order to assess

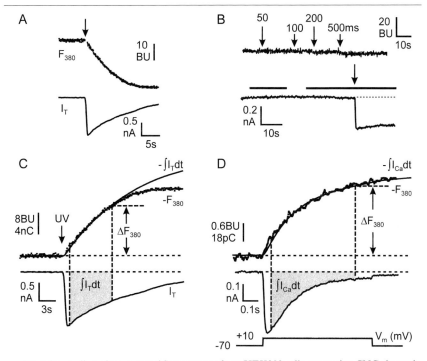

FIG. 3. Recording of current and fluorescence from HEK293 cells expressing CNG channels. (A) Example of simultaneous fluorescence (F_{380}) and current (I_T) recordings from a cell expressing bovine olfactory CNG channels, following release of cGMP from 75 μM NPE-caged cGMP by a 500-ms flash at −70mV. Fluorescence intensity is given in bead units (BU). Pipette solution (mM): 145 KCl, 8 NaCl, 1 MgCl$_2$, 2 Mg-ATP (adenosine triphosphate), 0.3 Na$_2$-GTP (guanosine triphosphate), 0.02 EGTA, 2 Fura-2-K$_5$, 10 HEPES/KOH at pH 7.2. Bath solution (mM): 120 NaCl, 3 KCl, 0.5 CaCl$_2$, 50 glucose, 10 HEPES/NaOH at pH 7.4. (B) *Upper trace:* Recording of F_{380} during application of UV flashes of the indicated duration. Bleaching of Fura-2 is detectable as a slight loss of the fluorescence signal at 200- and 500-ms flashes. *Lower trace:* Current recording at −70 mV with 150 μM NPE-caged cGMP from a cell expressing bovine cone CNG channels. During irradiation with Xe light (380 nm, black bar) through an ND filter (OD of 2), no current activation is detected. This light intensity allows fluorimetric measurement of [Ca^{2+}]$_i$ without channel activation. A 100-ms light flash (arrow) induces photorelease of cGMP and activates channels. (C) Recording of I_T and F_{380} at −70 mV from a HEK293 cell expressing olfactory CNG channels. Solutions as in (A) with 75 μM NPE-caged cGMP in the pipette solution. The arrow marks the application of a 500-ms flash of 0.6 mW. (D) Recording of I_{Ca} and F_{380} from a cell expressing BIII Ca^{2+} channels. A depolarizing voltage-pulse (*lower trace*) induces an inactivating Ca^{2+} current (*middle trace*) that leads to a change of F_{380} (*upper trace*), proportional to the integral of I_{Ca}. Pipette solution (mM): 130 CsCl, 20 tetraethylammonium chloride (TEA)-Cl, 2 MgCl$_2$, 2 Na$_2$-ATP, 0.2 Na$_2$-GTP, 1 Fura-2-K$_5$, 0.02 EGTA, 10 HEPES/CsOH at pH 7.2. Bath solution (mM): 120 NaCl, 10 TEA-Cl, 1 MgCl$_2$, 2 CaCl$_2$, 10 HEPES/NaOH at pH 7.4.

the quality of the experimental data. Leaky seals allow unaccounted Ca^{2+} flux, a violation of the theoretical demands of this analysis. In general, input impedance (cell membrane resistance plus seal resistance) must be $\geq 5 \times 10^8$ Ω. The integral of the inward current, $-\int I_T dt$ is computed and is then superimposed on the fluorescence intensity data. Scaling of the ordinate permits excellent superposition of the integral of the current data and the fluorescence intensity data (Fig. 3C, upper traces) over the first 5–15 sec following uncaging. This proportionality holds only as long as the theoretical presumptions of the analysis apply, i.e., where Eqs. (3)–(7) are valid. The change in fluorescence within this time segment, ΔF_{380}, and the corresponding current integral, $\int I_T dt$, are used to calculate the f value according to Eq. (8).

To calculate P_f [Eq. (10)], f_{max}, the value of f when Ca^{2+} carries 100% of the membrane current, must be measured in the same cell types and the same instrument. Two strategies can be used to measure f_{max}: (1) cells can be made to express Ca^{2+} selective channels, and (2) experimental conditions can be defined under which Ca^{2+} is the exclusive charge carrier through the CNG channels. Figure 3D illustrates an example of the first strategy in which f_{max} was measured in a HEK293 cell expressing voltage-gated Ca^{2+} channels (we used the BIII N-type Ca^{2+} channel from rabbit brain[7,22]). After loading with Fura-2, channels were activated by a depolarizing voltage step (Fig. 3D, lower trace), current and fluorescence were recorded and analyzed as described earlier. The integral of the current and F_{380} are proportional to each other over the entire 800-ms activation voltage pulse (Fig. 3D, upper traces), indicating that F_{380} accurately monitors the number of Ca^{2+} ions entering the cell through Ca^{2+} channels.

Measurement of f_{max} in a detached rod outer segment isolated from the retina of tiger salamanders is illustrated in Fig. 4. Methods of cell isolation and electrophysiologic recording can be found elsewhere.[23] cGMP-gated channels are the only channels present in the plasma membrane of the outer segment in both rods and cones.[23–25] In this experiment, the solution filling the rod outer segment consisted of (in mM) choline gluconate (80), choline aspartate (20), KCl (10), and HEPES (10), pH 7.2. The extracellular solution consisted of (in mM) choline chloride (165), KCl (2.5), $CaCl_2$ (1), and HEPES (10), pH 7.4. Under these ionic conditions, K^+ and Ca^{2+} are

---

[22] Y. Fujita, M. Mynlieff, R. T. Dirksen, M.-S. Kim, T. Niidome, J. Nakai, T. Friedrich, N. Iwabe, T. Miyata, T. Furuichi, D. Furutama, K. Mikoshiba, Y. Mori, and K. G. Beam, *Neuron*, **10**, 585 (1993).
[23] S. Hestrin and J. I. Korenbrot, *J. Gen. Physiol.* **90**, 527 (1987).
[24] D. A. Baylor and T. D. Lamb, *J. Physiol.* **328**, 19 (1982).
[25] J. L. Miller and J. I. Korenbrot, *J. Gen. Physiol.* **101**, 933 (1993).

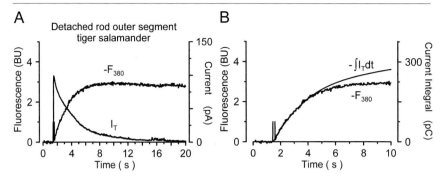

FIG. 4. Determination of f_{max} in a detached rod outer segment from tiger salamander. (A) Recording of I_T and ΔF_{380} at -35 mV. The composition of the solutions bathing the cell and filling the electrode are detailed in the text. The cell was filled with 2 mM Fura-2 and 50 μM DMNB-caged 8-Br-cGMP. The composition of the ionic solutions was such that uncaging the cyclic nucleotide caused an inward current carried exclusively by Ca^{2+}. (B) The change in F_{380} is superimposed on the integral of the same current illustrated in (A). The ratio of the change in fluorescence to the current integral over the first 3 sec following the uncaging flash defines the value of f_{max} for the experimental setup.

the only possible charge carriers.[26] By conducting the measurement at a holding voltage of -35 mV, the equilibrium potential for K^+ at the concentrations selected, Ca^{2+} is the only net current carrier and, therefore, any electrical current measured can only be carried by Ca^{2+}. Figure 4A shows the membrane current elicited by photorelease of 8-Br-cGMP and the fluorescence emission intensity excited with 380-nm light, F_{380}. The superposition of current integral and fluorescence trace (Fig. 4B) demonstrates that the two signals are proportional over the first 3 sec of the recording, allowing calculation of f_{max} according to Eq. (9).

Experimental Caveats

The determination of P_f described here is accurate only if executed under a restricted set of experimental conditions devised to transform the complex reality of a biological cell into simpler circumstances under which (1) the cell can be described to consist only of two homogeneous compartments, the inside and the outside, (2) Ca^{2+} flux is exclusively inward, and (3) all entering Ca^{2+} is bound by Fura-2. Successful application of the method demands confirmation that the assumed experimental conditions have indeed been met. We have found, as others have before, that these experimental conditions are relatively easy to meet in measurements of

---

[26] C. Picco and A. Menini, *J. Physiol.* **460**, 741 (1993).

transfected mammalian cultured cells and detached rod outer segments (dROS). Rod outer segments can be detached from the rest of the cell by simple mechanical means and remain able to phototransduce, if provided with the necessary biochemical substrates.[23] Both transfected cells and dROS are anatomically homogeneous, CNG channels are essentially the only channels in the plasma membrane, and active Ca^{2+} removal mechanisms (ATP dependent or Na^+ gradient dependent) are relatively sluggish. In contrast, experimental demands are more difficult to meet in measurements of intact photoreceptors because these cells are structurally polarized, cGMP-gated channels are also found in the inner segment membrane, and active Ca^{2+} removal can be extremely rapid. We briefly detail experimental conditions that must be met and methods to test whether they are met.

1. *The composition of the cytoplasm must be in full equilibrium with the solution filling the tight-seal electrode.* Cell filling can be monitored by continuously measuring the cell's fluorescence intensity excited at 380 nm. As Fura-2 flows between the pipette-filling solution and the cytosol, this signal should increase and reach a steady value at equilibrium. The data shown in Fig. 5A illustrate the typical time course of loading an HEK293 cell studied with a pipette solution containing 2 mM Fura-2. Equilibration is reached within 10 min after establishing the whole-cell configuration. In detached rod outer segments equilibration is reached in about 5 min in cells isolated from tiger salamander retina and about 10 min in those isolated from Gecko retina. These time intervals are consistent with diffusion-limited equilibration of small molecules (molecular mass 500–800 Da) in the rod outer segment.[27]

Equilibration between the outer segment cytoplasm and the electrode-filling solution is particularly slow in cone photoreceptors because of their unfavorable geometry. Unlike rods, viable cone outer segments cannot be detached from the inner segment, and the volume of the inner segment exceeds that of the outer segment. In tiger salamander cones, for example, the inner segment volume can be 10-fold larger than that of the outer segment. This poses a problem in photometric measurements because small optical masks must be used to isolate signals from the outer segment from those originating in the inner segment. Small masks, however, create experimental uncertainties because of the unavoidable light scattering. On the other hand, the outer segment phototransduces and this feature can be used to assess its cytoplasmic loading. The electrode-filling solution in P_f experiments contains Fura-2 and caged nucleotides, but not free cyclic nucleotides, or triphosphate nucleotides, which are required to sustain

---

[27] A. Olson and E. N. Pugh, Jr., *Biophys. J.* **65**, 1335 (1993).

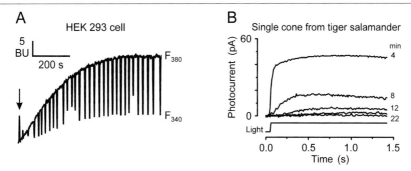

FIG. 5. (A) Time course of cytoplasmic loading in a HEK293 cell monitored by Fura-2 fluorescence. The arrow indicates whole-cell breakthrough. The fluorescence emission was recorded during alternating excitation wth constant photon flux at $\lambda_{exc} = 340$ nm and $\lambda_{exc} = 380$ nm. F_{380} alone is sufficient to reveal that cytoplasmic loading reached equilibrium after about 10 min. The ratio F_{340}/F_{380} reveals the cytoplasmic free Ca^{2+} concentration $[Ca^{2+}]_i$. (B) Time course of cytoplasmic loading of the outer segment of a single cone from tiger salamander. Light-dependent current activated with steps of light of constant intensity (at 380 nm, 6×10^7 photons μm^{-2} sec^{-1}) was measured at various times following whole-cell breakthrough, defined as $t = 0$. As the cytoplasm fills with Fura-2 and caged cGMP by diffusion, it also loses free cGMP and triphosphate nucleotides necessary for phototransduction. When the cytoplasmic content is in equilibrium with the contents of the patch electrode, the cone ceases phototransduction. With an electrode sealed onto the inner segment membrane, equilibrium was reached after about 22 min. The electrode was filled with (in mM) K$^+$-gluconate (70), K$^+$-aspartate (20), KCl (20), NaCl (9), MOPS (15), pH 7.25. The cell was bathed with (in mM) NaCl (100), KCl (2), NaHCO$_3$ (5), NaH$_2$PO$_4$ (1), CaCl$_2$ (1), MgCl$_2$ (1), glucose (10), HEPES (10), pH 7.4.

phototransduction. Thus, as cytoplasm and electrode come to equilibrium, the outer segment loads with Fura-2 and caged nucleotides, while it is depleted of free cyclic nucleotides and triphosphate nucleotides, thus losing its ability to phototransduce. If loading is assessed under conditions that maintain the dark-adapted state of a cone, then the ability to respond to illumination (phototransduce) will be lost as the outer segment and electrode contents equilibrate with each other. When equilibrium is reached, the cone will be unresponsive to light. Figure 5B illustrates the results of such measurements in a tiger salamander cone. Despite the very small size of the outer segment in these cells, about 3–5 μm at the base and 6–8 μm length, full equilibration required about 20 min, as made evident by the slow loss of photoresponsiveness.

2. *Cytoplasmic Fura-2 must be in its Ca^{2+}-free state and it must bind, at first, virtually every inflowing Ca^{2+} ion.* The cytoplasmic free Ca^{2+} concentration at the end of the loading phase should be below 100 nM, thus ensuring that Fura-2 is present mostly in its Ca^{2+}-free form and, therefore, able to bind inflowing Ca^{2+}. Ca^{2+} concentration can be determined by measuring

the ratio of the fluorescence intensities excited by both 340 nm (F_{340}) and 380 nm (F_{380}).[12] The F_{340}/F_{380} ratio is a function of Ca^{2+} concentration,[12] and this dependence can be defined for any cell using standard calibration procedures (Fig. 6A).

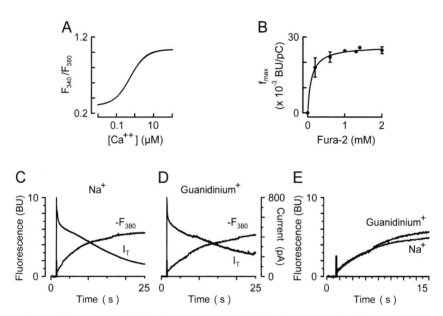

FIG. 6. Necessary controls for P_f measurements. (A) The free cystosolic Ca^{2+} concentration $[Ca^{2+}]_i$ must not exceed 100 nM. Calibration curve for the calculation of $[Ca^{2+}]_i$ from the fluorescence ratio F_{340}/F_{380} for a Fura-2 concentration of 2 mM in HEK293 cells. (B) The Fura-2 concentration must be high enough to outcompete other Ca^{2+} buffers. HEK293 cells expressing BIII Ca^{2+} channels were loaded with 2 mM Fura-2 and f_{max} values were measured at different times during loading. The cytosolic Fura-2 concentration at each time was derived from the fraction of the maximal fluorescence (measured after complete loading) at λ_{exc} = 365 nm. Values of f_{max} saturate at Fura-2 concentrations $\geq$ 1 mM, indicating that all inflowing Ca^{2+} will initially bind to the dye. The maximal value was 26.9 × 10^{-3} ± 3.2 × 10^{-3} BU/pC (n = 13). Data represent mean ± S.D. from 5 cells. The solid line is a Michaelis–Menten fit with a $K_{1/2}$ of 0.1 mM. [Reprinted with permission from S. Frings, R. Seifert, M. Godde, and U. B. Kaupp. Neuron **15**, 169 (1995). Copyright (1996) Cell Press.] (C–E) The flux of Ca^{2+} across the plasma membrane must be unidirectional. ΔF_{380} and I_T were measured in the same single cone from striped bass first (C) in the presence of Ringer's with normal Na^+ (106 mM) and then (D) in the presence of Ringer's in which all Na^+ was isosmotically replaced by guanidinium$^+$ (106 mM). Because Na^+ and guanidinium$^+$ are equally permeable through the cGMP-gated channels, the currents measured on uncaging cGMP were essentially the same. (E) ΔF_{380} was also the same under both experimental conditions, but only for the first 5–6 sec. Thereafter, the signals diverged, indicating that in cone outer segments the active extrusion of Ca^{2+} (via the Na^+/Ca^{2+}–K^+ exchanger, while in the presence of Na^+) can be neglected only for the first 4.5–5 sec following channel activation.

Fura-2 binds all entering Ca^{2+} if, and only if, the dye is present at a concentration that exceeds the buffering capacity of any other Ca^{2+}-binding system in the cell. The dye concentration necessary and sufficient to meet this goal is established experimentally by determining the value of f as a function of cytoplasmic Fura-2 concentration. The value of f rises at first, but becomes independent of concentration when Fura-2 outcompetes any endogenous Ca^{2+}-binding system. Figure 6B illustrates an experiment that serves to find the appropriate Fura-2 concentration: f_{max} values were measured in a HEK293 cell expressing voltage-gated Ca^{2+} channels at various Fura-2 concentrations. The value of f_{max} increased with [Fura-2] up to ≤ 1 mM, but became independent of [Fura-2] at higher concentrations. Thus, Fura-2 must be used at intracellular concentrations of 1–2 mM to ensure that ΔF_{380} is a quantitative measure of total Ca^{2+}.

3. *Ca^{2+} flux must be unidirectional, that is, active Ca^{2+} efflux must not be significant in the time interval while f is being measured.* Active Ca^{2+} removal can be mediated by ATP- and/or Na^+-gradient-dependent mechanisms. In HEK293 and detached rod outer segments, Ca^{2+} removal is relatively slow, as measured by the recovery of F_{380} after cessation of Ca^{2+} influx. In cone outer segments, on the other hand, Ca^{2+} removal by a Na^+/Ca^{2+}–K^+ exchanger is much more rapid than in the other cells, partly because of the large surface-to-volume ratio of the cone outer segment.[28,29] This high ratio arises from the fact that in cone outer segments the plasma membrane folds continuously, forming a closely spaced stack of disks that extends across the width of the outer segment. The disks have a repeat distance of about 30–50 nm and they are open to the extracellular space. In the single cones of both tiger salamander and striped bass and under physiologic conditions Ca^{2+} is cleared from the outer segment with a time constant of 20–30 ms. To test that active Ca^{2+} removal does not interfere with the determination of P_f, f should be measured in the same cell before and after blocking any potential active Ca^{2+} transport. If active Ca^{2+} transport is not significant, as demanded by theory, then f should have the same value in the presence and absence of active transport. Figures 6C–E illustrate results of one such experiment in single cone from striped bass. In the same cell, f was first measured in normal Ringer's solution (Fig. 6C) and then again with the cell bathed in a Ringer's solution in which Na^+ was isosmotically replaced by guanidinium$^+$ (Fig. 6D). This cation is nearly as permeable as Na^+, but it does not replace Na^+ in the Na^+/Ca^{2+}–K^+ exchanger, which, therefore, is inactive in the presence of guanidinium$^+$. The cGMP-dependent membrane currents under Na^+ or guanidinium$^+$ are nearly identical,

---

[28] S. Hestrin and J. I. Korenbrot, *J. Neurosci.* **10,** 1967 (1990).
[29] R. J. Perry and P. A. McNaughton, *J. Physiol.* **433,** 561 (1991).

as expected since their permeability is essentially the same. The time courses of F_{380} signals under both ions are initially the same, but diverge after 5.5–6 sec (Fig. 6E). This indicates that active Ca^{2+} transport is indeed negligible only at early times. Thus, the value of f in single cones is accurate only when measured over the first 4.5–5 sec following channel activation.

4. *The intracellular space sampled photometrically must be the same space sampled electrically and intracellular compartmentalization must not occur in the time interval while f is being measured.* HEK293 cells and dROS are structurally homogeneous. Intact photoreceptors, both rods and cones, in contrast, are structurally polarized. CNG channels are the only channels in the outer segment, which is at one end of the cell, while the inner segment onto which a tight-seal electrode is positioned is at the other end. This imposes an experimental problem since cGMP diffuses within the cell. Figure 7 presents evidence for this fact. In a single cone of the striped bass we measured the membrane current generated by flashes of UV light delivered to restricted sections of the intact cell using small optical masks. When cGMP was generated at the end of the inner segment, the current was small and slow in time course and it exhibited a delay relative to the

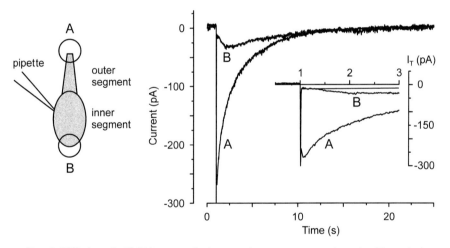

Fig. 7. Diffusion of cGMP between the inner and outer segments of a striped bass single cone. In the same cell, we measured currents activated by identical uncaging flashes of UV light delivered either onto (A) the outer segment or (B) the inner segment. Uncaging in the outer segment produced rapid and large current without detectable time delay. Uncaging in the inner segment also generated currents, but they were smaller and slower. More importantly they exhibited a significant time delay relative to the presentation of the uncaging flash (inset). The delay likely reflects the diffusion into the outer segment of cGMP produced in the inner segment. The delay, about 400 ms, is expected for the diffusion-limited travel of small molecules over a distance of about 20 μm. The cone inner segment is about 25 μm long.

presentation of the UV flash. On the other hand, when cGMP was generated within the outer segment, the current was larger and faster and, most significantly, exhibited no delay relative to the uncaging flash. That is, cGMP generated in the inner segment diffused into the outer segment to activate the channels located there, but diffusion imposed a delay in channel activation. Just as cGMP diffuses from the inner to the outer segment, it must be true that Ca^{2+} can diffuse from the outer to the inner segment. Therefore, any Ca^{2+} that enters the outer segment can, and does, diffuse into the inner segment. That is, the inner segment sequesters Ca^{2+} away from the outer segment. To determine f accurately under these conditions, Ca^{2+} concentration must be measured in the cytoplasm of both inner and outer segment, while uncaging is best restricted to the outer segment alone. This critical demand of the experimental protocol can be easily executed with the use of appropriate masks placed both on the fluoroscence excitation and fluorescence emission optical pathways.

[54] Modulation of Rod cGMP-Gated Cation Channel by Calmodulin

By MARIA E. GRUNWALD and KING-WAI YAU

Introduction

The hyperpolarizing responses of rod and cone photoreceptors to light are produced by the closing of cGMP-gated, nonselective cation channels.[1–3] In darkness, these channels are kept open by cGMP, which binds to them and serves as an activating ligand. In the light, the cytoplasmic cGMP level falls due to light-stimulated cGMP phosphodiesterase (PDE) activity, causing the channels to close.[4,5] This channel closure leads to a decrease in the cytoplasmic Ca^{2+} concentration in the outer segment, which triggers negative-feedback mechanisms to produce light adaptation.[6] In rods, one of these mechanisms is an inhibition of the cGMP-gated channel by Ca^{2+},

---

[1] K.-W. Yau and D. A. Baylor, *Annu. Rev. Neurosci.* **12,** 289 (1989).
[2] J. T. Finn, M. E. Grunwald, and K.-W. Yau, *Annu. Rev. Physiol.* **58,** 395 (1996).
[3] W. N. Zagotta and S. A. Siegelbaum, *Annu. Rev. Neurosci.* **19,** 235 (1996).
[4] L. Stryer, *Annu. Rev. Neurosci.* **9,** 87 (1986).
[5] K.-W. Yau, *Invest. Ophthalmol. Vis. Sci.* **35,** 9 (1994).
[6] Y. Koutalos and K.-W. Yau, *TINS* **19,** 73 (1996).

acting through a soluble factor.[7,8] This modulation of the channel by Ca^{2+} is not very strong, consisting of a maximum increase in the channel's half-activation constant ($K_{1/2}$) by only a factor of 2 (Ref. 7, but also see Ref. 8). Accordingly, its contribution to light adaptation is small compared to other mechanisms, in particular the negative modulation of guanylyl cyclase by Ca^{2+} (Ref. 6). However, the way in which Ca^{2+} modulates the channel is very interesting. There is now fairly good evidence that Ca^{2+} acts through Ca^{2+}/calmodulin,[9–11] which binds to the channel, though the involvement of an additional, unidentified Ca^{2+}-binding protein interacting with the same binding site has also been suggested.[10] In cones, an effect of Ca^{2+}/calmodulin on the native cGMP-gated channel is still not entirely clear, and may vary with animal species[12–14]; again, the involvement of an unidentified factor other than calmodulin has been proposed.[13,14]

A cyclic nucleotide-gated cation channel highly homologous to the rod and cone channels is known to mediate olfactory transduction.[15,16] Unlike the rod and cone channels, however, the olfactory channel is strongly modulated by Ca^{2+}/calmodulin, with its $K_{1/2}$ for activation by cAMP (the second messenger for olfactory transduction) being increased by up to 20-fold.[17] Unlike in photoreceptors, this modulation appears to be the predominant mechanism for sensory adaptation in olfactory receptor cells, at least on a short timescale.[18]

For the rod channel, the calmodulin-binding site mediating the physiologic effect has been localized to the cytoplasmic N terminus of the β subunit,[19–21] a subunit that is present in the native channel complex but that cannot form functional channels by itself.[22–24] For the olfactory channel,

---

[7] K. Nakatani, Y. Koutalos, and K.-W. Yau, *J. Physiol.* **484,** 69 (1995).
[8] M. S. Sagoo and L. Lagnado, *J. Physiol.* **497,** 309 (1996).
[9] Y.-T. Hsu and R. S. Molday, *Nature* **361,** 76 (1993).
[10] S. E. Gordon, J. Downing-Park, and A. L. Zimmerman, *J. Physiol.* **486,** 533 (1995).
[11] P. J. Bauer, *J. Physiol.* **494,** 675 (1996).
[12] L. W. Haynes and S. C. Stotz, *Vis. Neurosci.* **14,** 233 (1997).
[13] D. H. Hackos and J. I. Korenbrot, *J. Gen. Physiol.* **110,** 515 (1997).
[14] T. I. Rebrik and J. I. Korenbrot, *J. Gen. Physiol,* **112,** 537 (1998).
[15] T. Nakamura and G. H. Gold, *Nature* **325,** 442 (1987).
[16] R. S. Dhallan, K.-W. Yau, K. A. Schrader, and R. R. Reed, *Nature* **347,** 184 (1990).
[17] T.-Y. Chen and K.-W. Yau, *Nature* **368,** 545 (1994).
[18] T. Kurahashi and A. Menini, *Nature* **385,** 729 (1997).
[19] M. E. Grunwald, W. P. Yu, H.-H. Yu, and K.-W. Yau, *J. Biol. Chem.* **273,** 9148 (1998).
[20] D. Weitz, M. Zoche, F. Müller, M. Beyermann, H. G. Körschen, U. B. Kaupp, and K.-W. Koch, *EMBO J.* **17,** 101 (1998).
[21] Y.-T. Hsu and R. S. Molday, *J. Biol. Chem.* **269,** 29765 (1994).
[22] T.-Y. Chen, Y.-W. Peng, R. S. Dhallan, B. Ahamed, R. R. Reed, and K.-W. Yau, *Nature* **362,** 764 (1993).
[23] T.-Y. Chen, M. Illing, L. L. Molday, Y.-T. Hsu, K.-W. Yau, and R. S. Molday, *Proc. Natl. Acad. Sci. U.S.A.* **91,** 11757 (1994).

on the other hand, the calmodulin-binding site resides on the cytoplasmic N terminus of the α subunit.[25] Interestingly, the cone channel α subunit (of human, for example) also has a *bona fide* calmodulin-binding site on its cytoplasmic N terminus, but a homomeric channel formed by this subunit does not show functional modulation by Ca^{2+}/calmodulin.[26] In this chapter, the focus will be on the rod channel, although brief mention will also be made of the olfactory channel, for which much more is known about the mechanism of modulation by calmodulin.

Inspection of Amino Acid Sequence for Calmodulin-Binding Motifs

There are consensus motifs for a calmodulin-binding site, including the *1-5-10, 1-8-14,* and IQ motifs.[27] Generally, a binding site has both hydrophobic and positively charged residues (often forming an amphiphilic helix) to interact with hydrophobic and negatively charged residues, respectively, on calmodulin.[27–29] However, there are many instances in which the binding site on a target protein does not conform to these motifs very well and is therefore difficult to recognize. At the same time, there may be multiple binding sites, not all of which are functionally relevant. For the rod channel β subunit, there are sites on the cytoplasmic C terminus that bind Ca^{2+}/calmodulin with micromolar affinities in binding assays; however, when these sites were deleted, the effect of Ca^{2+}/calmodulin on the channel persisted.[19,20,29,30] The cytoplasmic N terminus does not contain any obvious calmodulin-binding site, but in a physiologic assay on deletion mutants (see later section), a region upstream of the first transmembrane domain was identified that is required for the Ca^{2+}/calmodulin effect.[19,20] The site also binds calmodulin in biochemical assays.[19,20]

Binding of Calmodulin to Heterologously Expressed Proteins

Previously, in a gel-overlay assay, calmodulin has been shown to bind to what subsequently turned out to be the native bovine rod channel β subunit.[9,21] A gel-overlay experiment can also be carried out on the channel

---

[24] H. G. Körschen, M. Illing, R. Seifert, F. Sesti, A. Williams, S. Gotzes, C. Colville, F. Müller, A. Dosé, M. Godde, L. Molday, U. B. Kaupp, and R. S. Molday, *Neuron* **15,** 627 (1995).

[25] M. Liu, T.-Y. Chen, B. Ahamed, J. Li, and K.-W. Yau, *Science* **266,** 13648 (1994).

[26] M. E. Grunwald, J. Lai, and K.-W. Yau, *Biophys. J.* **74,** A125 (1998).

[27] A. R. Rhoads and F. Friedberg, *FASEB J.* **11,** 331 (1997).

[28] M. Ikura, G. M. Clore, A. M. Gronenborn, G. Zhu, C. B. Klee, and A. Bax, *Science* **256,** 632 (1992).

[29] A. Crivici and M. Ikura, *Annu. Rev. Biophys. Biomol. Struct.* **24,** 85 (1995).

[30] R. S. Molday, *Curr. Opin. Neurobiol.* **6,** 445 (1996).

protein heterologously expressed in human embryonic kidney (HEK) 293 cells (see Electrophysiological Studies section for expression procedure). This experiment, which works well on the olfactory channel α subunit,[25] has failed to demonstrate calmodulin binding to the heterologously expressed and immunoprecipitated human rod channel β subunit, possibly reflecting weak binding or failed renaturation of the protein.[19] Another means of study is to use a fusion protein constructed from glutathione s-transferase (GST) and a relevant part of the channel protein, and expressed in bacteria.[19] This method has the advantage that milligrams of the fusion protein can be obtained. The N and C termini of the β subunit were examined, based on amino-sequence inspection as described earlier. These termini were amplified by polymerase chain reaction (PCR) using primers with flanking *Bam*HI and *Eco*RI restriction sites, and the PCR fragments were subcloned into the pGEX-2T expression vector (Amersham Pharmacia Biotech, Uppsala, Sweden). The resulting constructs were transformed into *Escherichia coli* BL21 cells, and the fusion proteins were isolated and purified using the bulk GST purification module from Amersham Pharmacia Biotech.

For gel-overlay assays, the purified fusion proteins are run on sodium dodecyl sulfate–polyacrylamide gel electrophoresis (SDS–PAGE), transferred to nitrocellulose or polyvinylidene difluoride (PVDF) membrane in 10 mM 3-(cyclohexylamino)-1-propanesulfonic acid (CAPS), pH 10.8, or Towbin buffer[31] containing 2–10% methanol. After transfer, the blots are probed with biotinylated calmodulin or calmodulin coupled to alkaline phosphatase in the presence of 1 mM Ca^{2+} (or 5 mM EGTA as control). The synthesis of calmodulin-coupled alkaline phosphatase and the detection procedure are both according to Walker *et al.*[32] In assays with biotinylated calmodulin, the membrane is blocked in a buffer containing 150 mM NaCl, 10 mM Tris-HCl, pH 7.5, 1 mM CaCl$_2$ (or 5 mM EGTA as control), 0.1% antifoam A, and 5% nonfat dry milk for 30 min. Biotinylated calmodulin (Biomedical Technologies) was added to give a final concentration of 1 μg/ml, followed by incubation for 1–2 hr at room temperature. After extensive washing in the same buffer without additives, the membrane is incubated with avidin and horseradish peroxidase (ABC system, Vector Laboratories, Burlingame, CA) and developed using the ECL enhanced chemiluminescence system (Amersham Pharmacia Biotech). To confirm that calmodulin binds to the appropriate protein and to verify the amount of protein on the blot, a Western blot analysis can be carried out on the same blot with

---

[31] H. Towbin, T. Staehelin, and J. Gordon, *Proc. Natl. Acad. Sci. U.S.A.* **76**, 4350 (1979).

[32] R. G. Walker, A. J. Hudspeth, and P. G. Gillespie, *Proc. Natl. Acad. Sci. U.S.A.* **90**, 2807 (1993).

an antibody, after stripping calmodulin off the blot with 1% SDS, 1 mM EDTA in Tris-buffered saline. Figure 1B (left lane) shows the binding signal from biotinylated calmodulin with a GST fusion protein of the partial N terminus of the rod channel β subunit. The binding disappears when the putative calmodulin-binding site is removed (Fig. 1B, right lane). Somewhat surprisingly, a control experiment using 5 mM EGTA instead of 1 mM CaCl$_2$ indicated that calmodulin still bound to the fusion protein.[19] This seemingly Ca^{2+}-independent binding of calmodulin was probably an artifact from the experimental conditions because the functional modulation of the α/β-heteromeric channel complex requires Ca^{2+} (Ref. 19) as does the binding of calmodulin to peptides corresponding to the binding site (see later section). Contradictory observations on the Ca^{2+} dependence of calmodulin binding under different conditions have been reported for other proteins.[33] Sometimes the gel-overlay experiment has to be interpreted with caution because, as pointed out earlier, the protein is not necessarily completely renatured after blotting.

Other methods, not described here, allow one to check the interaction of a fusion protein with calmodulin, such as using Sepharose beads (or column) coupled to glutathione (or an antibody) to pull down the GST fusion protein together with calmodulin. Conversely, calmodulin immobilized to Sepharose can be used to pull down the fusion protein. These methods are potentially superior to the gel-overlay assay because they examine calmodulin binding in solution; on the other hand, interactions detectable by the gel-overlay method may be too weak to be detected by these other assays.

Binding Studies with Peptides

A synthetic peptide corresponding to the binding site provides a useful tool for verifying the binding to calmodulin and for measuring the affinity constant.

Gel-Shift Assay

This assay gives a qualitative indication of the affinity of calmodulin for the binding site.[19] Figure 1C shows such an experiment for two peptides (MEG1 and MEG2) that span the putative binding site on the N terminus on the rod channel β subunit.[19] For comparison, two other peptides corresponding, respectively, to the calmodulin-binding site on the olfactory channel α subunit (KY9)[25] and that on an N-methyl-D-aspartate (NMDA) recep-

---

[33] P. D. Wes, M. Yu, and C. Montell, *EMBO J.* **15**, 5839 (1996).

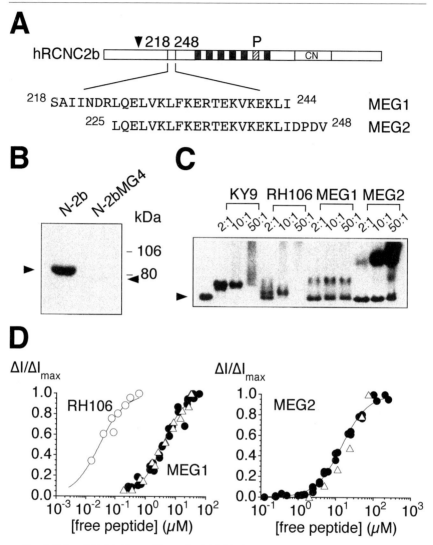

Fig. 1. Calmodulin binding to N-terminal GST fusion proteins and peptides. (A) Schematic diagram of the β subunit (incomplete at its N terminus; see Ref. 19) showing the locations of the two N-terminal peptides. The arrowhead indicates the end of the GARP region.[19] Below the diagram, the sequences of the peptides MEG1 and MEG2 are shown with numbers indicating the first and last residues. (B) Calmodulin overlay of N-terminal fusion proteins probed with biotinylated calmodulin. N-2b, wild-type N-terminal fusion protein; N-2bMG4, the same fusion protein with residues 225–248 deleted (see Fig. 2). (c) Gel-shift experiments with peptides. The leftmost lane contains calmodulin only. The arrowhead indicates the position of unbound calmodulin, and the numbers indicate peptide/calmodulin mole ratios; 2 mM Ca^{2+} present. Protein visualization by Coomassie Blue staining. See text for details.

tor channel subunit (RH106)[34] were included. Calmodulin (375 pmol) was incubated with the peptides in several mole ratios in a buffer containing 10 mM HEPES/NaOH (pH 7.2) and 2 mM CaCl$_2$ for 30 min at room temperature. The calmodulin peptide complexes were then resolved by nondenaturing gel electrophoresis on 15% gels according to the standard procedures for SDS–PAGE except that SDS was omitted and 2 mM CaCl$_2$ added. The protein bands were visualized by Coomassie Blue staining. The leftmost lane in Fig. 1C has no peptides, showing the position of the unbound-calmodulin band. KY9 has high affinity for Ca^{2+}/calmodulin[25] and completely shifts the calmodulin already in a peptide/calmodulin mole ratio of 2:1. RH106 has a lower affinity for Ca^{2+}/calmodulin[34] and produces only a partial shift when the peptide/calmodulin mole ratio is 2:1. MEG1 and MEG2 show even lower affinities for Ca^{2+}/calmodulin, with a band of pure calmodulin remaining for a peptide/calmodulin mole ratio of 50:1. In a control experiment in which 5 mM EGTA replaced 2 mM Ca^{2+}, no shift in the position of the calmodulin band was observed in any of the lanes (not shown), indicating that the binding of calmodulin to the site requires Ca^{2+}.

Fluorescence Measurements

This method allows quantitative measurement of the dissociation constant between Ca^{2+}/calmodulin and the binding site.[19,25] The method makes use of the change in spectrum and intensity of the fluorescence of a protein when the molecular environment of the fluorescent group changes, such as when the protein is bound to another protein or a peptide.[35] Dansylcalmodulin[36] (Sigma, St. Louis, MO) at a given concentration was incubated with a peptide of increasing concentration in a buffer containing 50 mM Tris-HCl, pH 7.3, 150 mM NaCl, and 0.5 mM CaCl$_2$. The emission spectrum at 400–600 nm was recorded on a spectrophotometer using an excitation wavelength of 340 nm, the bandwidth being 10–15 nm for both excitation and emission. The increase in fluorescence at 480 nm (the exact wavelength

---

[34] M. D. Ehlers, S. Zhang, J. P. Bernhardt, and R. L. Huganir, *Cell* **84,** 745 (1996).
[35] R. L. Kincaid, M. L. Billingsley, and M. Vaughan, *Methods Enzymol.* **159,** 605 (1988).
[36] R. L. Kincaid, M. Vaughan, J. C. Osborne, Jr., and V. A. Tkachuk, *J. Biol. Chem.* **257,** 10638 (1982).

(D) Fluorescence measurements of the interaction between the peptides and dansylcalmodulin; 0.5 mM Ca^{2+} present. The normalized fluorescence increase (representing the fraction of bound calmodulin) is plotted against the concentration of free peptide. Curves are fit according to the Hill equation. See text for details. [From M. E. Grunwald, W.-P. Yu, H.-H. Yu, and K.-W. Yau, *J. Biol. Chem.* **273,** 9148 (1998).]

is unimportant, as long as the change in fluorescence intensity is substantial at the chosen wavelength) was used to assay for the concentration of dansylcalmodulin bound to the peptide. Assuming a 1:1 stoichiometry of binding between Ca^{2+}/calmodulin and peptide, the fraction of dansylcalmodulin, f_b, bound to peptide is given by $f_b = (I_m - I_f)/(I_b - I_f)$, where I_f is the dansylcalmodulin fluorescence with no peptide present, I_b is the fluorescence when all dansylcalmodulin is peptide-bound, and I_m is the fluorescence of intermediate mixtures.[25] The dissociation constant, K_d, between Ca^{2+}/calmodulin and the peptide can be derived from a plot between the fractional increase in fluorescence and the calculated concentration of free peptide.[25] The results are independent of the purity of the dansylcalmodulin, provided that calmodulin and dansylcalmodulin behave identically in the binding experiments. Figure 1D shows these measurements for the peptides MEG1 and MEG2, together with RH106 for comparison. The plots give a K_d of 3.6 μM for MEG1, 14.4 μM for MEG2, and 29 nM for RH106. The Hill coefficient is 1.2 for MEG1, 1.3 for MEG2, and 1.0 for RH106. The Hill coefficient being higher than unity for MEG1 and MEG2 may indicate that the assumption of a 1:1 stoichiometry between peptide and calmodulin is not entirely valid, or that the experimental design does not agree with the *in vivo* conditions (see Comments section).

The choice of dansylcalmodulin concentration in the preceding measurements depends on the value of K_d. If the calmodulin concentration is much higher than the K_d value, all added peptide will bind to calmodulin, as a result of which the fraction of bound dansylcalmodulin will simply increase linearly with the peptide concentration until saturation.[25] Thus, under these conditions, simply reading off the half-saturated signal would give a faulty K_d value. In the experiments of Fig. 1D, two dansylcalmodulin concentrations were used, 100 nM (open triangles) and 300 nM (filled circles), both considerably lower than the K_d value and therefore fine. Without prior information about the K_d value, a more foolproof, but more laborious, way to measure its value would be to always have equal concentrations of dansylcalmodulin ($[CaM]_T$) and peptide ($[Pep]_T$) in the solution, and to covary them. Suppose $[CaM]_T = [Pep]_T = nK_d$; it can be shown (again assuming 1:1 stoichiometry between calmodulin and peptide) that the fraction of bound dansylcalmodulin (f_b) is given by $f_b = [(2n + 1) - (4n + 1)^{1/2}]/2n$. When n equals 1, f_b equals 0.382. Thus, the value of K_d can be read from a complete plot of f_b against $[CaM]_T$ (or $[Pep]_T$) at the point where f_b is 0.382.

Electrophysiologic Studies

To verify that the calmodulin-binding site in question mediates the modulation of the channel, it is necessary to carry out a functional assay and to show loss of modulation when the binding site is deleted or disrupted. cDNAs in the pCIS expression vector and coding for various deletion mutants of the β subunit were individually cotransfected with the cDNA coding for the wild-type channel α subunit into HEK293 cells (American Type Culture Collection, Rockville, MD) using the calcium phosphate precipitate method. The HEK cells were cultured at 37° in media containing 10% fetal calf serum and penicillin/streptomycin in a humidified atmosphere with 5% (v/v) CO_2. At 48–72 hr after transfection, inside-out membrane patches were excised from the transfected cells and tested for an effect of Ca^{2+}/calmodulin on the cGMP-activated current using the patch-clamp recording technique.[17,19,25,37] For zero-Ca^{2+} conditions, the pipette and bath solutions both contained 140 mM NaCl, 5 mM KCl, 2 mM EGTA, and 10 mM HEPES/NaOH, pH 7.4. In experiments involving Ca^{2+}/calmodulin, the bath solution contained calmodulin at a specified concentration (typically 250 nM) and 50 μM buffered free Ca^{2+} (achieved by replacing the EGTA with 2 mM nitrilotriacetic acid and 704 μM $CaCl_2$, Ref. 17). The affinity of Ca^{2+} for calmodulin is sufficiently high that, at 50 μM Ca^{2+}, all of the calmodulin is complexed with Ca^{2+}. To verify that any loss of calmodulin effect associated with a β-subunit mutant was not due to failure of functional expression of the mutant protein, the cGMP-activated current was always tested for blockage at +60 mM by 10 μM L-*cis*-diltiazem,[19,38] a pharmacologic blocker that requires the presence of the β subunit for its potency.

Figure 2 shows such an experiment for various deletion mutants of the rod channel β subunit lacking different amino acid segments on the N terminus.[19] The mutants MG4 and BA104, which both lack the calmodulin-binding site, no longer show the inhibition by Ca^{2+}/calmodulin. In contrast, the other mutants retain the effect of Ca^{2+}/calmodulin.

A competition experiment can also be carried out by premixing calmodulin with an excess amount of the peptide before applying it to the membrane patch.[25] In this case, the inhibition of the cGMP-activated current is not expected to take place, because all of the Ca^{2+}/calmodulin is already bound to the peptide.

Comments

In experiments on the native rod channel[21] or the heterologously expressed channel complex composed of α and β subunits,[19,20] it was found that half-maximal inhibition of the channel occurred at a few nanomolars calmodulin. The large discrepancy between this apparent high affinity and

---

[37] B. Sakmann and E. Neher, eds., "Single-Channel Recording," 2nd ed. Plenum Press, New York (1995).

[38] K.-W. Koch and U. B. Kaupp, *J. Biol. Chem.* **260**, 6788 (1985).

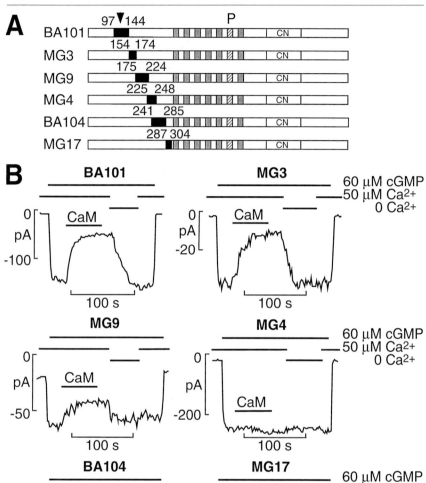

the 1000-fold lower affinity (see later discussion) measured with a peptide for the calmodulin-binding site (see earlier section) can arise from several possibilities. The first is that the peptide is structurally unable to reproduce the binding properties of the whole protein. The second is that other regions on the channel protein facilitate or stabilize the binding of calmodulin to the binding site. There is precedent for such a scenario.[39–42] Finally, the dansyl group adduct to calmodulin may somehow interfere with the binding, though this does not appear to be a problem for other binding sites (see, for example, Ref. 25).

Another tool for studying protein interactions is based on surface plasmon resonance spectroscopy[43] (e.g., the BIAcore system). When applied to the interaction between calmodulin and a peptide or a fusion protein containing the N terminus of the rod channel β subunit, the derived K_d is of the order of 10-fold lower than that measured with dansylcalmodulin described earlier.[20] The reason for this discrepancy is unclear. In any case, however, the discrepancy between the affinity constant suggested by the peptide experiments and that suggested by the electrophysiologic experiments would remain at 100- to 1000-fold.

Olfactory Channel

Exactly how Ca^{2+}/calmodulin inhibits the rod channel after binding to the N terminus of the β subunit is still unclear. This question is not easy to solve, because the degree of modulation is weak and because the presence

---

[39] C. T. Craescu, A. Bouhss, J. Mispelter, E. Diesis, A. Popescu, M. Chiriac, and O. Barzu, *J. Biol. Chem.* **270,** 7088 (1995).
[40] H. Munier, F. J. Blanco, B. Precheur, E. Diesis, J. L. Nieto, C. T. Craescu, and O. Baszu, *J. Biol. Chem.* **268,** 1695 (1993).
[41] M. Dasgupta, T. Honeycutt, and D. K. Blumenthal, *J. Biol. Chem.* **264,** 17156 (1989).
[42] D. Ladant, *J. Biol. Chem.* **263,** 2612 (1988).
[43] A. Szabo, L. Stolz, and R. Granzow, *Curr. Opin. Struct. Biol.* **5,** 699 (1995).

FIG. 2. Effects of N-terminal deletions in the rod channel β subunit on the Ca^{2+}/calmodulin modulation of the heterologously expressed α/β-heteromeric channel. (A) Locations of various N-terminal deletions of the β subunit (with an incomplete N terminus, see Ref. 19) are indicated by black boxes. The numbers indicate the first and last residues of each deleted domain. (B) Macroscopic currents induced by 60 μM cGMP at -60 mV from excised, inside-out membrane patches of HEK293 cells transfected with cDNAs coding for the wild-type rod channel α subunit and the respective β-subunit mutants. Calmodulin concentration is 250 nM. [From M. E. Grunwald, W.-P. Yu, H.-H. Yu, and K.-W. Yau, *J. Biol. Chem.* **273,** 9148 (1998).]

of both α and β subunits is required for the effect to take place. The situation is simpler for the olfactory channel, because, as pointed out earlier, the modulation is much more pronounced and the calmodulin-binding site resides on the α subunit, which forms functional channels by itself. In this case, it appears that a region containing the calmodulin-binding site on the N terminus contributes to the high open probability of the channel.[25] When Ca^{2+}/calmodulin binds, the influence of this domain on channel gating is removed, thus causing the open probability of the channel to decrease.[25] Because channel gating is kinetically coupled to the step of cGMP binding to the channel, the result is an increase in the half-activation constant, $K_{1/2}$, for the cGMP-activated channel.[25] It now appears that the N terminus of the olfactory channel physically interacts with the C terminus through the calmodulin-binding site (which thus serves a dual purpose); on binding to the site, Ca^{2+}/calmodulin disrupts the N–C terminal interaction to lower the channel's open probability.[44]

Conclusions

It is clear from the preceding account that the study of the calmodulin modulation of ion channels requires a number of experimental approaches. Cyclic nucleotide-gated channels were the first to be demonstrated to bind calmodulin, but other ion channels have now been found to have the same property.[34,45–48] More examples will undoubtedly follow. The sequence of experiments to identify the calmodulin-binding site and to characterize the molecular mechanism of the calmodulin effect is likely to vary with individual ion channels, depending on the subtleties of the binding site as well as the nature and degree of the modulation.

---

[44] M. D. Varnum and W. N. Zagotta, *Science* **278,** 110 (1997).
[45] Y. Saimi and C. Kung, *FEBS* **350,** 155 (1994).
[46] C. G. Warr and L. E. Kelly, *Biochem. J.* **314,** 497 (1996).
[47] X.-M. Xia, B. Fakler, A. Rivard, G. Wayman, T. Johnson-Pais, J. E. Keen, T. Ishii, B. Hirschberg, C. T. Bond, S. Lutsenko, J. Maylie, and J. P. Adelman, *Nature* **395,** 503 (1998).
[48] B. Z. Peterson, C. D. DeMaria, and D. T. Yue, *Neuron* **22,** 549 (1999).

Section VIII

$Na^+/Ca^{2+}-K^+$ Exchanger and ABCR Transporter

[55] Purification and Biochemical Analysis of cGMP-gated Channel and Na^+/Ca^{2+}-K^+ Exchanger of Rod Photoreceptors

By ROBERT S. MOLDAY, RENÉ WARREN, and TOM S. Y. KIM

Introduction

The cGMP-gated channel and the Na^+/Ca^{2+}-K^+ exchanger play a central role in phototransduction and light adaptation in vertebrate rod and cone photoreceptor cells by controlling the flow of cations across the outer segment plasma membrane.[1-4] In dark-adapted photoreceptors, Na^+ and Ca^{2+} enter the outer segment through cGMP-gated channels in the plasma membrane that are kept in their open state by bound cGMP. This influx of cations, coupled with the efflux of K^+ through voltage-gated channels in the photoreceptor inner segment, constitutes the dark current and maintains the photoreceptor cell in a depolarized state. Low intracellular Na^+ concentration is sustained by the extrusion of Na^+ through the Na^+,K^+-ATPase transporter in the plasma membrane of the photoreceptor inner segment. Steady-state intracellular Ca^{2+} concentration of ~400 nM is maintained by the balanced efflux of Ca^{2+} by the Na^+/Ca^{2+}-K^+ exchanger in the outer segment. Photoexcitation of rhodopsin and activation of the visual cascade culminates in a decrease in cGMP levels, the closure of cGMP-gated channels, and a transient hyperpolarization of the cell. Continual removal of Ca^{2+} by the Na^+/Ca^{2+}-K^+ exchanger leads to the decrease in intracellular Ca^{2+} in the outer segment, a process that is important in photorecovery and adaptation.[5,6]

The cGMP-gated channel of rod photoreceptors consists of two distinct subunits referred to as α and β that assemble into a heterotetrameric complex.[7-11] Each subunit contains six putative membrane spanning segments designated as S1–S6, a cyclic nucleotide-binding domain near the C

---

[1] E. E. Fesenko, S. S. Kolesnikov, and A. L. Lyubarsky, *Nature* **313,** 310 (1985).
[2] P. P. M. Schnetkamp, *Prog. Biophys. Molec. Biol.* **54,** 1 (1989).
[3] K.-W Yau, *Invest. Ophthalmol. Vis. Sci.* **35,** 9 (1994).
[4] K.-W Yau and D. A, Baylor, *Annu. Rev. Neurosci.* **12,** 289 (1989).
[5] U. B. Kaupp and K.-W. Koch, *Annu. Rev. Physiol.* **54,** 153 (1992).
[6] A. Polans, W. Baehr, and K. Palczewski, K. *TINS* **19,** 547 (1996).
[7] U. B. Kaupp, T. Niidome, T. Tanabe, S. Terada, W. Bönigk, W. Stühmer, N. J. Cook, K. Kangawa, H. Matsuo, T. Hirose, T. Miyata, and S. Numa, *Nature* **343,** 762 (1989).
[8] T.-Y. Chen, Y-W. Peng, R. S. Dhallan, B. Ahamed, R. R. Reed, and K.-W. Yau, *Nature* **362,** 764 (1993).

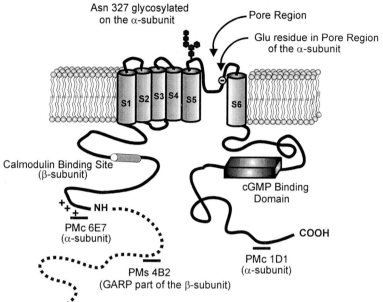

Fig. 1. Current topological model for the rod cGMP-gated channel subunits showing structural features and selected monoclonal antibody binding sites. Structural features found in both the α and β subunits include six transmembrane segments (S1–S6), a pore region between S5 and S6, and a cGMP-binding domain near the C terminus. In addition, the α subunit is glycosylated at Asn-327, has a negatively charged glutamic acid residue in its pore region, and contains 92 amino acid at the N terminus that is missing in channel from ROS. The locations of the binding sites for monoclonal antibodies PMc 1D1 and PMc 6E7 on the α subunit are indicated. The β subunit contains the binding site for calmodulin and an extended N terminus referred to as GARP. The location for monoclonal antibody PMs 4B2 within the GARP region is shown.

terminus, a voltage sensor-like motif comprising the S4 segment, and a pore region positioned between transmembrane segments S5 and S6 (Fig. 1). The latter two structural features and the subunit membrane topology are characteristic features of voltage-gated cation channels. The α subunit

[9] H. G. Körschen, M. Illing, R. Seifert, F. Sesti, A. Williams, S. Gotzes, C. Colville, F. Müller, A. Dosé, M. Godde, L. L. Molday, U. B. Kaupp, and R. S. Molday, Neuron 15, 627 (1995).
[10] U. B. Kaupp, Curr. Opin. Neurobiol. 5, 434 (1995).
[11] R. S. Molday, Curr. Opin. Neurobiol. 6, 445 (1996).

can self-assemble into a functional homotetrameric channel, whereas the β subunit cannot form a functional channel on its own.[7–9] Coexpression of the α and β subunits produces a heterotetrameric channel with conductance properties that closely resemble the native channel of rod photoreceptors.[8,9]

The rod Na^+/Ca^{2+}-K^+ exchanger consists of a high molecular weight sialoglycopolypeptide that exists as a homodimer in the plasma membrane of ROS.[12–15] Primary structural analysis and labeling studies indicate that it contains two multispanning membrane domains that are connected by a large cytoplasmic loop.[16,17] A large, highly glycosylated extracellular segment precedes the first membrane domain (Fig. 2).

Both the cGMP-gated channel and the Na^+/Ca^{2+}-K^+ exchanger are present in relatively large quantities in outer segments of rod photoreceptors.[12,13,18,19] As a result, sufficient quantities of purified channel and exchanger can be obtained for structural and functional analyses. In this article, we describe several procedures that have been developed to isolate the cGMP-gated channel and Na^+/Ca^{2+}-K^+ exchanger from bovine rod outer segments (ROS) for molecular characterization and functional reconstitution.

Isolation of Bovine ROS Membranes

Buffers

Homogenization buffer: 20% (w/v) sucrose, 20 mM Tris–acetate, pH 7.4, 10 mM glucose, 1 mM MgCl$_2$

30% Sucrose solution: 30% (w/v) sucrose, 20 mM Tris-acetate, pH 7.4, 10 mM glucose, 1 mM MgCl$_2$

50% sucrose solution: 50% (w/v) sucrose, 20 mM Tris-acetate, pH 7.4, 10 mM glucose, 1 mM MgCl$_2$

Hypotonic lysis buffer: 10 mM HEPES–KOH, pH 7.4, 1 mM ethylenediaminetetraacetic acid (EDTA), 1 mM dithiothreitol (DTT)

---

[12] N. J. Cook and U. B. Kaupp, *J. Biol. Chem.* **263**, 11382 (1988).

[13] D. A. Nicoll and M. L. Applebury, *J. Biol. Chem.* **264**, 16207 (1989).

[14] D. M. Reid, U. Friedel, R. S. Molday, and N. J. Cook, *Biochemistry* **29**, 1601 (1990).

[15] A. Schwarzer, T. S. Y. Kim, V. Hagen, R. S. Molday, and P. J. Bauer, *Biochemistry* **36**, 13667 (1997).

[16] H. Reilander, A. Achilles, U. Friedel, G. Maul, F. Lottspeich, and N. J. Cook. *EMBO J.* **11**, 1689 (1992).

[17] T. S, Y, Kim, D. M. Reid, and R. S. Molday, *J. Biol. Chem.* **273**, 16561 (1998).

[18] N. J. Cook, W. Hanke, and U. B. Kaupp, *Proc. Natl. Acad. Sci. U.S.A.* **84**, 585 (1987).

[19] N. J. Cook, L. L. Molday, D. Reid, U. B. Kaupp, and R. S. Molday, *J. Biol. Chem.* **264**, 6996 (1989).

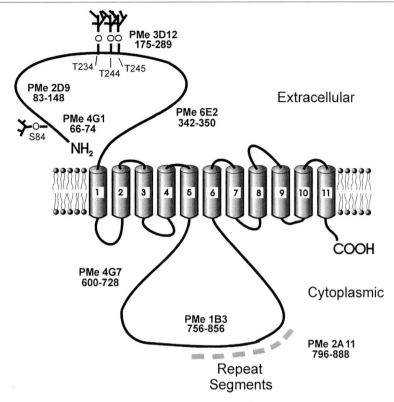

FIG. 2. Current topological model for the rod $Na^+/Ca^{2+}-K^+$ exchanger showing structural features and the location of various monoclonal antibodies. Two membrane domains containing multiple transmembrane segments are separated by a large, negatively charge intracellular loop containing repeat segments. A large extracellular domain at the N terminus contains multiple O-linked sialo-oligosaccharides (possible serine and threonine attachment sites are indicated). The model is based on sequence analysis and biochemical and immunochemical labeling studies.[16,17]

Procedures

ROS are isolated from bovine retina under dim red light at 4° by continuous sucrose gradient centrifugation.[20,21] Typically, retinas from 100 dark-adapted bovine eyes are immersed in 40 ml homogenizing buffer and gently agitated for 1 min to break off the outer segments. The suspension

[20] R. S. Molday and L. L. Molday, *J. Cell Biol.* **105,** 2589 (1987).
[21] R. S. Molday and L. L. Molday, *Methods Neurosci.* **15,** 131 (1993).

is passed through a Teflon screen to remove the retina tissue and layered on six 20-ml continuous sucrose gradients made in polycarbonate tubes. The sucrose gradients are prepared from the 30% (w/v) and 50% (w/v) sucrose solutions utilizing a conventional gradient mixing apparatus. Centrifugation is performed in a Beckman (Fullerton, CA) SW 28 rotor at 25,000 rpm (82,500g) for 45 min at 4°. The pink band from each tube is carefully collected with a syringe, pooled, diluted with 5 volumes homogenizing solution and centrifuged at 13,000 rpm (20,000g) in a Sorvall (Newtown, CT) SS-34 rotor for 20 min. The ROS pellet is washed in 40 ml homogenizing buffer by centrifugation and the final pellet is resuspended in 8–10 ml homogenizing buffer to obtain a final ROS concentration of 8–10 mg/ml protein as measured by the bicinchoninic acid (BCA) assay (Pierce, Rockford, IL). The ROS are either used immediately or stored in light-tight vials at −80°. A yield of 80–100 mg of ROS protein is typically obtained from 100 bovine retinas.

Membranes are isolated by hypotonic lysis of ROS followed by centrifugation. Typically, 20 mg unbleached ROS are mixed with 10 volumes hypotonic lysis buffer, maintained on ice for 30 min, and then centrifuged in a Sorvall SS-34 rotor at 13,000 rpm for 10 min. The membrane pellet is resuspended in the same buffer and the centrifugation step is repeated. This washing procedure is repeated twice in order to remove soluble and weakly associated proteins from the membranes. The final pellet is resuspended in 10 mM HEPES-KOH, pH 7.4, in the original volume (∼2 ml) to obtain a protein concentration of 8–10 mg/ml protein. This procedure is particularly applicable for processing relatively large amounts of ROS (>10 mg). The procedure can be scaled down to process smaller samples. A Beckman Optima TL centrifuge with a fixed-angle rotor is used for the centrifugation steps.

Solubilization of ROS Membranes

Buffer

CHAPS solubilization buffer: 10 mM HEPES-KOH, pH 7.4, 1 mM DTT, 10 mM CaCl$_2$, 0.15 M KCl, 18 mM CHAPS, 2 mg/ml asolectin (soybean phosphatidylcholine, type IV-S; Sigma, St. Louis, MO) and protease inhibitors [0.1 mM diisopropylfluorophosphate, 5 μg/ml aprotinin, 1 μg/ml leupeptin, and 2 μ-g/ml pepstatin or 20 μM Pefabloc SC (Boehringer Mannheim, Indianapolis, IN)]

Procedures

Cook et al.[22] have described optimal conditions for solubilization of ROS membranes in CHAPS detergent. In our laboratory, we slowly add unbleached ROS membranes (~8–10 mg/ml) to the CHAPS solubilization buffer at 4° with constant stirring under dim red light to yield a final protein concentration of 1–1.5 mg/ml. The solution is stirred for 30 min and subsequently centrifuged at 27,000g or greater for 30 min to remove any residual aggregated material. Only a small pellet should be observed. The supernatant is then exposed to room lighting and used directly for purification or functional reconstitution as described later.

Purification of the Channel and Exchanger

Buffers

CHAPS column buffer: 10 mM HEPES-KOH, pH 7.4, 1 mM DTT, 1 mM CaCl$_2$, 0.15 M KCl, 12 mM CHAPS, and 2 mg/ml asolectin (~40% L-α phosphatidylcholine from soybean, Type IV-S; Sigma P-3644)

DEAE elution buffer: 10 mM HEPES-KOH, pH 7.4, 1 mM DTT 10 mM CaCl$_2$, 0.7 M KCl, 12 mM CHAPS, and 2 mg/ml asolectin

Red-dye elution buffer: 10 mM HEPES-KOH, pH 7.4, 1 mM DTT, 10 mM CaCl$_2$, 1.8 M KCl, 12 mM CHAPS, and 2 mg/ml asolectin

DEAE and Red-Dye Chromatography

The cGMP-gated channel and Na$^+$/Ca^{2+}-K$^+$ exchanger were first isolated from CHAPS-solubilized ROS membranes using a series of chromatographic methods.[12,18,23] These procedures are carried out at 4° under normal lighting conditions. In the first step, CHAPS-solubilized ROS membranes (10–20 mg protein) are applied to 2 ml of DEAE-Fractogel TSK 650 S (Supelco, Bellefonte, PA) preequilibrated with CHAPS column buffer containing 10 mM CaCl$_2$. The column is washed at a flow rate of 0.5 ml/min with the same buffer until the 280-nm absorbance returns to baseline levels. Most of the rhodopsin and 220-kDa rim protein (ABCR) are present in this unbound or flow-through fraction. The column is then eluted at higher ionic strength with DEAE elution buffer. The resulting fraction contains both the channel and exchanger along with a number of other

---

[22] N. J. Cook, C Zeilinger, K-W. Koch, and U. B. Kaupp, *J. Biol. Chem.* 261, 17033 (1986).
[23] N. J. Cook, *Methods Neurosci.* **15,** 271 (1993).

proteins. This DEAE fraction can be used either directly for functional reconstitution studies or for separation of the channel from the exchanger using AF red dye chromatography as follows. The pooled DEAE elution fraction is applied to 2 ml of AF Red-Fractogel TSK (E. Merck, Darmstadt, Germany) equilibrated in DEAE elution buffer. The flow-through fraction contains the exchanger as the major protein; a number of minor proteins can also be visualized by sodium dodecyl sulfate–polyacrylamide gel electrophoresis (SDS–PAGE). After the column is thoroughly washed with DEAE elution buffer, the channel is eluted with red dye elution buffer. In their initial study, Cook and co-workers[18] reported a 110-fold purification of functionally active channel after AF Red-Fractogel chromatography with a 32% recovery. In our experience, there is some variation in activity of the channel after AF Red-Fractogel chromatography, possibly due to inconsistencies in the AF Red-Fractogel matrix.

The exchanger present in the flow-through fraction from the AF red dye column can be directly used for functional studies or further purified by Concanavalin A (ConA)-Sepharose chromatography.[12,23] A 65-fold purification with a 41% yield was obtained for the exchanger isolated after AF Red-Fractogel chromatography. The purification was further increased by a factor of 1.7 after ConA-Sepharose chromatography, but the yield was reduced by half.

Calmodulin Affinity Chromatography

Calmodulin binds strongly to the β subunit of the rod cGMP-gated channel in a calcium-dependent manner.[11,24–29] This property has been used to obtain a highly pure preparation of the rod GMP-gated channel using calmodulin affinity chromatography.[25] The procedures are typically carried out at 4° under normal lighting conditions. A solution of CHAPS-solubilized ROS (20–30 mg protein) is applied to a column containing 2.5 ml of calmodulin-Sepharose (Amersham-Pharmacia Biotech, Piscataway, NJ) preequilibrated at 4° with CHAPS column buffer. The solution is passed through the column at a flow rate of about 0.5 ml/min and the column is then washed with 10 volumes of CHAPS column buffer to ensure the complete

[24] Y.-T Hsu and R. S. Molday, *Nature* **361**, 76 (1993).
[25] Y.-T Hsu and R. S. Molday, *J. Biol. Chem.* **269**, 29765 (1994).
[26] T.-Y. Chen, M. Illing, L. L. Molday, Y.-T. Hsu, K.-W. Yau, and R. S. Molday, *Proc. Natl. Acad. Sci. U.S.A.* **91**, 11757 (1994).
[27] P. J. Bauer, *J. Physiol.* **494**, 675 (1996).
[28] M. E. Grunwald, W-P. Yu, H. H. Yu, and K.-W. Yau, *J. Biol. Chem* **273**, 9148 (1998).
[29] D. Weitz, M. Zoche, F. Muller, M. Beyermann, H. G. Körschen, U. B. Kaupp, and K.-W. Koch, *EMBO J.* **17**, 2273 (1998).

removal of unbound protein. CHAPS column buffer in which the CaCl$_2$ is replaced with 2 mM EDTA or EGTA is used to elute the channel from the column. The fractions are pooled and the CaCl$_2$ concentration is adjusted to 2 mM. A 30–40% recovery is typically obtained by this method. The channel is the major protein observed by SDS–PAGE. Blots labeled with calmodulin, however, reveal the presence of other calmodulin-binding proteins that are not present in channel preparations isolated by immunoaffinity methods.[24,25]

Ricin-Agarose Chromatography

The rod Na$^+$/Ca^{2+}-K$^+$exchanger contains a high content of O-linked sialo-oligosaccharides.[14,17] After removal of the terminal sialic acid residues with neuraminidase, galactose residues are exposed that serve as binding sites for *ricinus communis agglutinin*, a lectin derived from castor beans. This property can be used to obtain highly enriched preparations of the exchanger from bovine ROS for functional studies.[14] In this procedure, 20 mg ROS are digested with 0.2 units of *Arthrobacter ureafaciens* neuraminidase (Boehringer Mannheim) for at least 2 hr at 4° prior to isolation and solubilization of ROS membranes as described earlier. The CHAPS-solubilized ROS membranes are applied to a 3-ml ricin-agarose column (RCA$_{60}$-agarose or RCA$_{120}$-agarose; Sigma). After washing the column with 10 volumes of CHAPS column buffer to remove unbound protein, the exchanger is eluted in the CHAPS solubilization buffer containing 0.2 M galactose. In this preparation, the 230-kDa exchanger appears as the major protein, but several other ricin-binding proteins are typically observed. Because the exchanger is highly susceptible to digestion by protease, it is important to use freshly dissected ROS and protease inhibitors to minimize degradation.

Immunoaffinity Methods

A number of monoclonal antibodies have been generated to distinct sequences (typically 8–10 amino acids) of the channel or exchanger (Figs. 1 and 2). Many of these antibodies bind strongly to CHAPS or Triton X-100 solubilized channel or exchanger and, therefore, have application in immunoprecipitation and/or immunoaffinity purification studies.[25,30–34]

[30] L. L. Molday, N. J. Cook, U. B. Kaupp, and R. S. Molday, *J. Biol. Chem.* **265**, 18690 (1990).
[31] R. S. Molday, L. L. Molday, A. Dosé, I. C. Lewis, M. Illing, N. J. Cook, E. Eismann, and U. B. Kaupp, *J. Biol. Chem.* **266**, 21917 (1991).
[32] R. S. Molday and L. L. Molday, *Vision Res.* **38**, 1315 (1998).
[33] R. S. Molday, *Invest. Ophthalmol. Vis. Sci.* **39**, 2493 (1998).
[34] R. S. Molday and L. L. Molday, *Methods Enzymol.* **294**, 246 (1998).

Here, we describe the use of the monoclonal antibody PMc 1D1 to immunoprecipitate the channel from solubilized ROS membranes, and PMc 6E7 to purify the channel from ROS for functional reconstitution.

Monoclonal antibodies are isolated from either hybridoma culture fluid or ascites fluid by ammonium sulfate precipitation followed by DEAE chromatography or alternatively by protein G or protein A-agarose chromatography as described by Harlow and Lane.[35] The antibody is dialyzed against borate buffer (10 mM sodium borate/0.1 M NaCl, pH 8.4) overnight at 4° and its concentration is determined from the absorbance at 280 nm (1 mg/ml antibody gives an absorbance of 1.3). Typically, 4 mg of antibody is coupled to 2 ml of packed CNBr-activated Sepharose (Pharmacia) in 10 mM sodium borate/0.1 M NaCl buffer, pH 8.4, for at least 4 hr at 4°. The reaction is stopped by the addition of Tris buffered saline (TBS; 10 mM Tris-HCl, 150 mM NaCl, pH 7.4) containing 50 mM glycine. The matrix is then washed thoroughly with TBS and stored at 4° in the presence of 1 mM sodium azide. Because the coupling efficiency is typically 85–95%, the matrix contains ~1.7–2.0 mg protein per milliliter of packed beads.

Immunoprecipitation

Immunoprecipitation experiments are generally carried out when the goal is simply to separate the channel complex from other ROS proteins for analysis by SDS–PAGE and Western blotting, rather than for functional studies.[30] Typically, a solution of CHAPS-solubilized ROS membranes (200–500 μl) is added to 50 μl of PMc 1D1-Sepharose preequilibrated with CHAPS column buffer in a spin column (Ultrafree-MC 0.45-μm filter unit; Millipore, Bedford, MA). The matrix is gently agitated for 30 min and collected on the filter after centrifugation in a mini tabletop centrifuge (10 sec at ~2000g). The unbound or flow-through fraction is collected in an Eppendorf tube. The matrix is then resuspended in 0.4 ml of CHAPS column buffer and centrifuged as earlier. This washing procedure is repeated a total of 4–6 times and then the protein is eluted with 2% SDS for analysis by SDS–PAGE and Western blotting. Centrifugation steps are carried out at low speed and for short times in order to avoid drying the matrix. DTT is generally excluded from the CHAPS solubilization and column buffer to avoid the loss of antibody from the column. Asolectin can also be omitted for these studies. Triton X-100 or other mild detergents can be used in place of CHAPS. For example, in some studies we use 1% Triton X-100 to solubilize the ROS membranes and 0.1% Triton X-100 in the column buffer in place of CHAPS.

[35] E. Harlow and D. Lane, "Antibodies: A Laboratory Manual," p. 298. Cold Spring Harbor Laboratory, Cold Spring Harbor, New York, 1988.

Using this procedure, we have immunoprecipitated the channel complex from detergent-solubilized ROS membranes with the PMc 1D1 and PMc 6E7 monoclonal antibodies against different epitopes within the α subunit and PMs 4B2, PMs 5E11, and PMb 3C9 antibodies against various regions of the β subunit. Monoclonal antibodies PMe 2A11 and PMe 1B3 have been used to immunoprecipitate the bovine rod exchanger.[14,17]

Immunoaffinity Chromatography

Immunoaffinity chromatography has been used to purify the channel and exchanger from CHAPS-solubilized ROS for functional reconstitution.[25] For these studies, synthetic peptides corresponding to the epitope of the antibody are used to competitively elute the protein from the antibody-Sepharose matrix. Synthetic peptides of 8–12 amino acids in length are typically used; elution is most effective when the affinity of the antibody for the peptide and protein is similar.

Immunoaffinity purification of the channel is typically carried out by applying 20 ml of CHAPS-solubilized ROS membranes (~20 mg protein) to 3 ml of the PMc 6E7-Sepharose preequilibrated with CHAPS column buffer at 4°. After incubation for 30 min, the solution is eluted from the column at a flow rate of about 0.5 ml/min and the flow-through is collected as the unbound fraction. The column is then washed in the same buffer until the 280-nm absorbance returns to baseline. The column matrix is then incubated with 3 ml of CHAPS column buffer containing 0.1 mg/ml of the 6E7 competing peptide for at least 60 min, and the channel is collected as the "bound" fraction. This elution procedure can be repeated again to maximize the recovery of the channel or exchanger. The column matrix can be regenerated by washing the column first with CHAPS column buffer and then with 0.1 M acetic acid to remove bound peptide. Finally, the column is washed and stored in CHAPS column buffer containing 0.1% sodium azide. In our experience, there is some loss in efficiency of the column when it is reused, possibly due to irreversibly bound protein and/or inactivation of the antibody. The recovery of the channel varies with the antibody used, but typical recoveries range from 30% to more than 70%. The procedure can readily be scaled down using spin columns as described earlier. Immunoaffinity purification of the channel on a PMc 6E7-Sepharose column is illustrated in Fig. 3 and purification of the exchanger on a PMe 2A11 is shown in Fig. 4. This procedure has been effectively used for the purification of other ROS membrane proteins (rhodopsin, peripherin/rds-rom-1 complex, and ABCR) for which well-characterized monoclonal antibodies have been generated.[33]

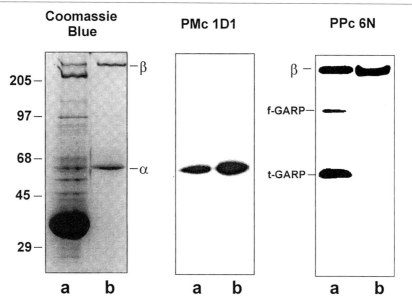

FIG. 3. Immunoaffinity purification of the rod cGMP-gated channel as analyzed by SDS–PAGE and Western blotting. Bovine ROS (lane a) were solubilized in CHAPS and passed through a PMc 6E7-Sepharose column. After removal of the unbound protein, the channel (lane b) was eluted with 0.1 mg of the 6E7 competing peptide. The left panel shows an SDS gel stained with Coomassie Blue, middle panel is a Western blot labeled with PMc 1D1 against the α subunit, and right panel is a Western blot labeled with the polyclonal antibody PPc 6N against the N terminus of the β subunit. In ROS, the PPc 6N antibody labels the two GARP variants (f-GARP and t-GARP) in addition to the β subunit of the channel.

Analysis by SDS–PAGE and Western Blotting

Detection of the channel and exchanger in ROS and during purification is carried out by SDS–PAGE and Western blotting. Protein sample is mixed with an equal volume of SDS cocktail consisting of 125 mM Tris-HCl, pH 6.8, 4% SDS, 8% 2-mercaptoethanol, 40% sucrose, and ~0.02% bromphenol blue. The sample is run without prior heating on a 6 or 8% polyacrylamide gel using the Laemmli buffer system.[36] Approximately 10–20 μg of ROS membrane proteins and 1–10 μg of fractions obtained during purification are applied per lane. For Western blotting, proteins fractionated by SDS–PAGE are transferred electrophoretically onto Immobilon-P membranes (Millipore) in 25 mM Tris-HCl, 192 mM glycine, pH 8.3, and 5% methanol using a BioRad (Richmond, CA) Trans-blot SD

[36] U. K. Laemmli, *Nature* **227,** 680 (1970).

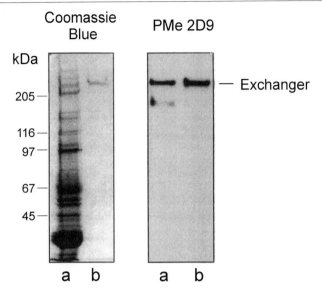

FIG. 4. Immunoaffinity purification of the Na$^+$/Ca^{2+}-K$^+$ exchanger as analyzed by SDS–PAGE and Western blotting. Bovine ROS (lane a) were solubilized in CHAPS and passed through a PMe 2A11 column. After removal of the unbound protein, the exchanger (lane b) was eluted with 0.1 mg of the 2A11 peptide. The left panel shows an SDS gel stained with Coomassie Blue and the right panel shows a Western blot labeled with monoclonal antibody PMe 2D9. The band below the 230-kDa exchanger in ROS is a proteolytic fragment of the exchanger.

semidry transfer apparatus. Transfer is carried out at 300 mA for 40 min. The blot is blocked with PBST buffer (0.01 M sodium phosphate, 0.15 M NaCl, 0.1% Tween-20) containing 0.5% nonfat evaporated milk for 30 min at room temperature. After washing in PBST, the blot is incubated with hybridoma cell supernatant diluted 1:15 with PBST for 30–60 min. The blot is washed again in PBST buffer and incubated with the secondary antibody (sheep anti-mouse immunoglobulin diluted 1:5000 in PBST) for 30–60 min at room temperature. The blot is finally washed with PBST and antibody labeling is detected using enhanced chemiluminescence (ECL). Sheep anti-mouse immunoglobulin and ECL kit is purchased from Amersham-Pharmacia Biotech.

By SDS–PAGE, the α subunit of the channel migrates as a 63- to 68-kDa polypeptide and the β subunit as a 240-kDa polypeptide (Fig. 3). The mass of the α subunit from ROS is significantly smaller than the 79.6-kDa size predicted from its amino acid sequence and observed in heterologous cell expression studies.[31] This difference is attributed to the absence of the first 92 N-terminal amino acids of the α subunit from ROS as measured by N-terminal amino acid sequencing. The β subunit of the bovine rod channel has an apparent molecular mass of 240 kDa, significantly higher than the 155-kDa size predicted from its sequence.[9] The abnormally slow migration of the β subunit by SDS–PAGE has been attributed to the high content of proline and glutamic acid residues present in the N-terminal glutamic acid-rich region called GARP.[9,37]

In addition to being part of the β subunit of the channel, GARP is also expressed in ROS in two additional spliced forms.[37] The full-length GARP form (f-GARP) contains 590 amino acids and is identical in sequence to the GARP part of the β subunit except for a 19-amino-acid C-terminal extension. The truncated GARP form (t-GARP) contains the first 291 amino acids of f-GARP and is capped by an 8-amino-acid C-terminal segment. Like the β subunit, these GARP variants migrate anomalously on SDS polyacrylamide gels with apparent molecular mass of ~140 kDa for f-GARP and ~60 kDa for t-GARP (Fig. 3). These two GARP forms are not associated with the purified channel. Their role in ROS remains to be determined.

The Na^+/Ca^{2+}-K^+ exchanger of rod photoreceptors consists of a single polypeptide chain of 1199 amino acids with a calculated molecular mass of 130 kDa.[16] By SDS–PAGE, however, the polypeptide migrates with an apparent molecular mass of 230 kDa (Fig. 4). The high content of O-linked sialo-oligosaccharides is partially responsible for the abnormally high molecular mass observed by SDS–PAGE.[17] Digestion of the exchanger with neuraminidase and O-glycosidase reduces the apparent molecular mass to 180 kDa. The high content of aspartic acid residues within the large cytoplasmic loop also contributes to the slow migration of the exchanger on SDS polyacrylamide gels.[16]

In purified channel preparations, the α and β subunits are observed as the most intensely stained bands. In some preparations, a faint band just below the 240-kDa β subunit, has been observed. Western blots indicate that this minor band is the rod Na^+/Ca^{2+}-K^+ exchanger.[32] These results indicate that the channel interacts weakly with the exchanger in detergent solutions. Previously, Bauer and Drechsler[38] have suggested that the channel and exchanger are associated in ROS plasma membrane on the basis of ion flux assays.

---

[37] C. Colville and R. S. Molday, *J. Biol. Chem.* **271**, 32968 (1996).
[38] P. J. Bauer and M. Drechsler, *J. Physiol.* **451**, 109 (1992).

Functional Reconstitution Assays

Buffers

Reconstitution buffer: 10 mM HEPES-KOH, pH 7.4, 0.1 M KCl, 1 mM DTT, 2 mM CaCl$_2$, and 10 mM CHAPS and 18 mg/ml asolectin sonicated to obtain a homogeneous suspension
Dialysis buffer: 10 mM HEPES-KOH, pH 7.4, 0.1 M KCl, and ±2 mM CaCl$_2$
Arsenazo III buffer: 10 mM HEPES-KOH, pH 7.4, 0.1 M KCl, 75 μM Arsenazo III

Reconstitution into Liposomes

The channel and exchanger have to be reconstituted into liposomes for analysis of ion transport activities.[12,18] Soybean phospholipid, generally referred to as asolectin, is particularly suitable for reconstitution because these lipids form small, relatively uniform unilamellar vesicles after removal of the detergent. Typically, the channel and exchange activity is determined by monitoring the release of Ca^{2+} from liposomes on the addition of cGMP or NaCl, respectively. Arsenazo III is used as a Ca^{2+} indicator.

Typically, 1 ml of CHAPS-solubilized protein (solubilized ROS membranes; 1–1.5 mg protein) or a fraction containing either the purified channel or exchanger (~1–50 μg protein) is added to 1 ml of reconstitution buffer and the solution is placed in a dialysis bag (Spectro/Por MWCO 12–14,000, 10-mm width; Spectrum Medical Industries, Houston, TX). The solution is dialyzed at 4° against 1 liter of dialysis buffer containing 2 mM CaCl$_2$. Dialysis is carried out over a period of 2 days with two changes of the dialysis buffer. Finally, the solution is dialyzed at 4° for an additional 6–12 hr against Ca-free dialysis buffer to generate a transmembrane Ca^{2+} gradient. Alternatively, the external Ca^{2+} can be rapidly removed by passing the reconstituted vesicles through a Chelex-100 (BioRad) ion-exchange column. For analysis of Na$^+$/Ca^{2+}-K$^+$ exchange activity it is important that all buffers used in the reconstitution procedures are devoid of Na$^+$ ions. The protein concentration of the purified channel and exchanger is measured using a modified Bradford assay as described by Read and Northcote.[39]

Ca^{2+} Flux Measurements

The efflux of Ca^{2+} from liposomes containing the channel or exchanger is measured spectrophotometrically using the metallochromic dye, Arsenazo

[39] S. M. Read and D. H. Northcote, *Anal. Biochem.* **116,** 53 (1981).

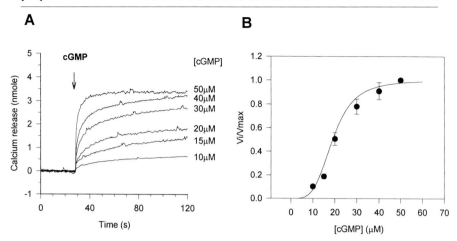

FIG. 5. Cyclic GMP-dependent Ca^{2+} efflux from liposomes reconstituted with the purified cGMP-gated channel. The cGMP-gated channel was purified from CHAPS-solubilized bovine ROS membranes on a calmodulin-Sepharose column and reconstituted into Ca^{2+} containing liposomes. (A) Time course for the efflux of Ca^{2+} from the vesicles on addition of different cGMP concentrations (arrow). Ca^{2+} release was measured in a dual-wavelength spectrophotometer using Arsenazo III as a Ca^{2+} indicator. (B) The dependence of the initial velocity (V_i) normalized to the velocity at saturating cGMP concentration (V_{max}) as a function of cGMP concentration. Data represent the mean from three experiments. A Michaelis constant (K_m) of 19 μM and a Hill coefficient (n) of 3.8 were used to fit the data using the relationship $V_i/V_{max} = [cGMP]^n / ([cGMP]^n + K_m^n)$.

III.[18] Typically, 0.3 ml reconstituted liposomes is added to 1.7 ml Arsenazo III buffer in a cuvette with stirring. After recording a baseline in a dual-wavelength spectrophotometer set at 650–730 nm, cGMP-dependent Ca^{2+} efflux is measured by monitoring the absorbance increase on addition of 4 μl of cGMP to the cuvette. The final concentration of cGMP is in the range of 1–150 μM. Maximal rate of Ca^{2+} release is generally observed at a cGMP concentration above 100 μM and half-maximum rate is observed at about 20 μM cGMP. The Na-dependent Ca^{2+} exchange activity is measured by adding a concentrated NaCl solution to the curette to give a final Na^+ concentration of 5–100 mM. For exchange activity, the liposome solutions also contain 2 μM carbonyl cyanide p-trifluoromethoxyphenylhydrazone (FCCP) and 2 μM valinomycin. Under these conditions, exchange activity is measured under symmetrical KCl concentrations. Enhanced exchange activity can be obtained using nonsymmetrical KCl. In this case the reconstituted solution is diluted into an Arsenazo III solution containing a reduced KCl concentration.[40]

---

[40] U. Friedel, G. Wolbring, P Wohlfart, and N. J. Cook, *Biochim. Biophys. Acta* **1061**, 247 (1991).

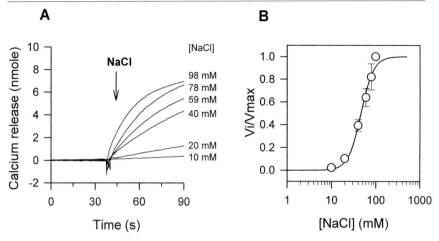

FIG. 6. Na$^+$-dependent Ca^{2+} efflux from liposomes reconstituted with the Na$^+$/Ca^{2+}-K$^+$ exchanger. ROS membranes were solubilized in CHAPS and reconstituted into liposomes Ca^{2+} containing liposomes. (A) Time course for the efflux of Ca^{2+} from the vesicles on addition of different concentrations of NaCl (arrow). Ca^{2+} release was measured in a dual-wavelength spectrophotometer using Arsenazo III as a Ca^{2+} indicator. (B) The dependence of the initial velocity (V_i) normalized to the velocity at 100 mM NaCl (V_{max}) as a function of NaCl concentration. Data represent the mean from three experiments. A Michaelis constant (K_m) of 46 mM and a Hill coefficient of 3.1 were used to fit the data using the relationship $V_i/V_{max} = [Na+]^n/([Na+]^n + K_m^n)$.

The initial rates of Ca^{2+} efflux are obtained by taking a tangent of the absorbance curve at the time of cGMP or NaCl addition (zero time). Absorbance values are converted to nmoles Ca^{2+} by calibration with a 1 nmol solution of CaCl$_2$. Figures 5 and 6 show the rate of release of Ca^{2+} from liposomes containing the channel and exchanger, respectively, and the dependence of the rate of release on cGMP and NaCl concentration.

Conclusions

A number of conventional and affinity chromatography methods are now available to purify the cGMP-gated channel and Na$^+$/Ca^{2+}-K$^+$ exchanger from bovine ROS membranes for structural and functional studies. The purified channel consists of an α subunit having an apparent molecular mass of 63 kDa and a β subunit that migrates as a 240-kDa protein by SDS–PAGE. These subunits are present in similar quantities suggesting that the native rod channel is composed of a $\alpha_2\beta_2$ heterotetramer. The exchanger, on the other hand, appears to be composed of a single, highly glycosylated polypeptide that migrates as a 230-kDa polypeptide by SDS–PAGE. Cross-linking studies, however, indicate that the exchanger likely exists as a dimer in ROS membranes. Although some information is available about the topological organization of the channel and exchanger in ROS membranes, high-resolution structural analyses of these membrane proteins remain to be performed. The methods described in this chapter should prove useful in obtaining sufficient quantities of highly pure proteins for structural studies. The channel and exchanger are typically isolated in CHAPS detergent. This detergent can be easily removed and in the presence of phospholipid the channel and exchanger can be reconstituted into lipid vesicles. Ion flux measurements confirm that both the reconstituted channel and exchanger exhibit most functional properties observed for these proteins in native ROS membranes. The reconstituted channel and exchanger should prove useful in further evaluating the functional and regulatory properties of these ion transporters and possible interactions with other ROS proteins.

Acknowledgments

This work was supported by grants from the National Eye Institute (EY 2422) and the Medical Research Council of Canada.

[56] Spectrofluorometric Detection of $Na^+/Ca^{2+}-K^+$ Exchange

By CONAN B. COOPER, ROBERT T. SZERENCSEI, and PAUL P. M. SCHNETKAMP

Introduction

Proteins that extrude calcium across the cell plasma membrane are of critical importance to all cells since they maintain cytosolic free calcium at a low value of about 100 nM while extracellular calcium is in the 1–2 mM range. Two classes of proteins contribute to calcium export across the plasma membrane: an adenosine triphosphate (ATP)-dependent calcium pump and sodium-coupled calcium transporters. Furthermore, two classes of sodium-coupled calcium transporters have been described to date: Na-Ca exchangers (NCX) are found in most tissues and operate at a 3 Na:1 Ca stoichiometry.[1] Potassium-dependent Na-Ca exchangers (NCKX), however, appear to have a more limited tissue distribution. NCKX1 was first

[1] K. D. Philipson and D. A. Nicoll, *Int. Rev. Cyt.* **137C,** 199 (1993).

described in the outer segments of retinal rod photoreceptors (ROS).[2] The NCKX1 cDNA was cloned from bovine retina and abundant transcript was shown to be present in retina, but no transcript was detected in cardiac or skeletal muscle, brain, kidney, or liver.[3] Functional studies suggest the presence of a potassium-dependent Na-Ca exchanger in platelets.[4] Recently, a second NCKX2 isoform was cloned from rat brain and was shown to be expressed in various regions of the brain.[5] NCKX2 is a considerably smaller protein of 670 amino acids compared with the mammalian rod NCKX1 proteins, which range in length between 1013 and 1216 amino acids.[6] To date, in situ NCKX activity has only been described in significant detail for retinal ROS, where a stoichiometry of 4 Na:1 Ca + 1 K was found (reviewed in Refs. 2 and 7). Compared to most cells with typical values of resting free Ca^{2+} of 50–100 nM, the outer segments of retinal rod photoreceptors maintain an unusually high cytosolic free-Ca^{2+} concentration of 500–700 nM in darkness due to continuous Ca^{2+} influx via the light-sensitive and cGMP-gated channels.[8,9] A few drugs such as tetracaine and L-cis-diltiazem,[2] and 3′,5′-dichlorobezamil[10] were found to inhibit NCKX activity in ROS but only at rather high concentrations; moreover, all three drugs were considerably more potent as inhibitors of the cGMP-gated channels in ROS when compared with their effect on NCKX activity, suggesting that pharmacologic intervention is not a suitable method for detecting NCKX activity. In this chapter, we describe the use of fluorescent calcium-indicating dyes to detect NCX or NCKX activity in retinal ROS or in other cells. In particular, we discuss the logistics of detecting NCX or NCKX activity in cell lines transfected with NCX or NCKX cDNAs.

Retinal ROS are a rather simple system in terms of calcium homeostasis since the ROS plasma membrane has been shown to contain only two cation transporters: the cGMP-gated channels and the Na^+/Ca^{2+}-K^+ exchanger. No evidence has been found to suggest the presence of other ion channels or ion pumps.[11,12] Moreover, NCKX protein is present at a rather high

[2] P.P.M. Schnetkamp, Prog. Biophys. Mol. Biol. **54**, 1 (1989).
[3] H. Reilander et al. EMBO J. **11**, 1689 (1992).
[4] M. Kimura, A. Aviv, and J. P. Reeves, J. Biol. Chem. **268**, 6874 (1993).
[5] M. Tsoi et al., J. Biol. Chem. **273**, 4155 (1998).
[6] C. B. Cooper, R. J. Winkfein, R. T. Szerencsei, and P. P. M. Schnetkamp, Biochemistry **38**, 6276 (1999).
[7] L. Lagnado and P. A. McNaughton, J. Membr. Biol. **113**, 177 (1990).
[8] M. P. Gray-Keller and P. B. Detwiler, Neuron **13**, 849 (1994).
[9] A. P. Sampath, H. R. Matthews, M. C. Cornwall, and G. L. Fain, J. Gen. Physiol. **111**, 53 (1998).
[10] G. D. Nicol, P. P. M. Schnetkamp, Y. Saimi, E. J. Cragoe, Jr., and M. D. Bownds, J. Gen. Physiol. **90**, 651 (1987).
[11] L. Lagnado, L. Cervetto, and P. A. McNaughton, Proc. Natl. Acad. Sci. U.S.A. **85**, 4548 (1988).
[12] P. P. M. Schnetkamp, R. T. Szerencsei, and D. K. Basu, J. Biol. Chem. **266**, 198 (1991).

concentration and can mediate changes in total intracellular calcium of up to 0.5 mM/sec in small mammalian ROS.[13] Mammalian or insect cell lines transfected with NCX or NCKX cDNAs present a more challenging problem as other components of the calcium homeostasis machinery provide for tight regulation of intracellular free calcium. In particular, it is difficult or impossible to inhibit the plasma membrane calcium pump and to thus impose sustained elevated cytosolic calcium levels that would permit measurement of sodium-dependent calcium extrusion. Here, we describe manipulation of transmembrane ion gradients by switching extracellular solutions and/or the application of cation ionophores to manipulate intracellular sodium concentration to detect NCX or NCKX actvity. To do so, we use two key properties shared by NCX and NCKX: (1) the absolute ion selectivity of NCX/NCKX for sodium ions with no other alkali cation including lithium, being able to replace sodium, and (2) the fact that NCX and NCKX are bidirectional and will mediate both calcium influx and calcium efflux dependent on the transmembrane sodium gradient.[13] Finally, we use the potassium dependence unique to NCKX to distinguish between NCX and NCKX activity.

General Methods

Solutions

Sucrose or a mixture of sucrose-Ficoll 400 is the main component of medium used to isolate, purify, and store retinal ROS. The cation contamination of the sucrose and sucrose-Ficoll solutions is minimized by passing them over a mixed-bed ion-exchange resin prior to addition of buffers and salts; this has the added benefit of removing bacteria and fungi and the solutions can be maintained free of bacterial or fungal contamination for extended periods of time on refrigeration. The NCKX Na$^+$/Ca^{2+}-K$^+$ exchanger transports both sodium and potassium ions, which necessitates careful control of both sodium and potassium concentration in the solutions applied. Hence, HEPES used throughout our experiments to control solution pH is adjusted to the desired pH by addition of arginine or Tris rather than KOH or NaOH.

Purification of Bovine ROS with Intact Plasma Membrane

Bovine eyes are obtained from a local abattoir and intact ROS are purified from freshly dissected bovine retinas by centrifugation on a mixed

[13] P. P. M. Schnetkamp, *J. Physiol.* **373**, 25 (1986).

sucrose-Ficoll 400 gradient as described previously.[14] ROS are isolated in a calcium-depleted/sodium-rich form by inclusion of 50 mM NaCl and 1 mM EGTA throughout the entire isolation and purification procedure.[13] Retinas are dissected and ROS purified and maintained in darkness or under dim red illumination until loading with the fluorescent calcium-indicating dye. ROS containing 150 μM rhodopsin are bleached with white light for 10 sec (to reach photochemical equilibrium between the all-*trans* and 11-*cis* conformations of retinal) and then incubated for 20–30 min with 40 μM fluo-3AM at room temperature in a medium containing 400 mM sucrose, 20 mM HEPES (pH 7.4), and 50 μM EDTA. Subsequently, extracellular fluo-3 and unhydrolyzed fluo-3AM are removed by placing the ROS suspension on top of 35 ml of ice-cold 600 mM sucrose, 20 mM HEPES (pH 7.4), and 50 μM EDTA, followed by centrifugation of fluo-loaded ROS through the sucrose solution in a fixed-angle rotor (12 min at 3220g). The ROS pellet is resuspended in ice-cold 600 mM sucrose, 20 mM HEPES (pH 7.4), 2.5% Ficoll 400, and 50 μM EDTA, and stored in a refrigerated light-tight container.

Maintenance of Cells and Dye Loading

Transformed Chinese hamster ovary (CHO) cells are maintained in Isacov's modified Dulbecco's medium (IMDM) supplemented with 10% fetal bovine serum (FBS), and grown in culture flasks at 37°, 5% CO_2. Cells in 150-mm^2 flasks at 60–80% confluence are suspended by washing with phosphate buffered saline (PBS) followed by treatment with 0.05% trypsin, 0.5 mM EDTA. Trypsin is inactivated by at least a 10 times dilution into regular medium containing FBS. For cells intended to be measured in suspension, the cells are sedimented at 1500g for 4 min, then resuspended and incubated in a buffer containing 150 mM NaCl, 3 mM KCl, 1.5 mM $CaCl_2$, 20 mM HEPES (pH 7.4), 250 μM sulfinpyrazone (to inhibit Ca^{2+}-indicating dye from being pumped out), and 25 μM fluo-3AM. After a 30-min loading, the buffer is diluted 5× with buffer not containing dye. Cells are sedimented, then washed in a buffer containing 150 mM LiCl, 20 mM HEPES (pH 7.4), 200 μM EGTA, and 250 μM sulfinpyrazone. Cells are again sedimented and then resuspended in a small volume of 150 mM LiCl, 20 mM, 20 mM HEPES (pH 7.4), 200 μM EGTA, 250 μM sulfinpyrazone, and held on ice until, and during, use. For cells intended to be used in single-cell imaging, suspended cells are seeded onto coverslips in 100-mm-diameter plates and grown to 30–40% confluence over 1–2 days. Coverslips with attached CHO cells are transferred to 12-well plates for loading with

[14] P. P. M. Schnetkamp and F. J. M. Daemen, *Methods Enzymol.* **81,** 110 (1982).

fluorescent Ca^{2+}-indicating dye. Fura-2AM (50 nmol) is dissolved in 10 μl 10% pluronic F-127 in dimethyl sulfoxide (DMSO), and then suspended in 5–10 ml serum-free IMDM for a final fura-2 concentration of 5–10 μM. Coverslips are washed once with 500 μl serum-free IMDM and then bathed in 500 μl fura-2 loading medium at 23° for 15–45 min. After dye loading, coverslips are washed once with serum-free IMDM, then bathed in serum-free IMDM until use.

Instrumentation

Fluorescence Measurements of Cells or ROS in Suspension

Fluorescence measurements of cells in suspension are carried out in an SLM-Aminco series 2 luminescence spectrometer (SLM Instruments, Urbana, IL). The following features of a fluorescence spectrometer should be considered. First, measurement of NCX or NCKX activity in cells requires successive additions to the external medium while continuously measuring fluorescence in the cells or ROS in suspension. Hence, the fluorometer should contain a thermostatted cuvette holder equipped with a magnetic stirrer and an opening port through which additions (e.g., NaCl, KCl, $CaCl_2$) can be made from the top. When a pipette tip was lowered through the opening port into the cuvette and into the light beam, a sharp spike in fluorescence was observed due to light that bounced off the pipette tip into the emission monochromator; this spike was used as a convenient marker for additions to the cuvette. In all our experiments with suspension of ROS or cells, we have used fluo-3 as a convenient dye to indicate cytosolic free calcium; fluo-3 undergoes at least a 100-fold increase in fluorescence on binding Ca^{2+}.[15,16] Two steps were taken to minimize apparent fluorescence due to light scattered into the emission monochromator by the suspended cells or ROS: (1) the excitation wavelength was lowered from the maximal value of 500 to 480 nm (with the emission monochromator set at 530 nm and a bandwidth of 8 nm); (2) a 495-nm long-pass filter was placed between the cuvette and the emission monochromator to reduce the amount of light scattered into the emission monochromator by two- to threefold.

Fluorescence Measurements in Single Cells

For single-cell Ca imaging with the fluorescent Ca-indicating dye, fura-2 (Molecular Probes, Eugene, OR), an Axiovert-135 inverted microscope

---

[15] A. Minta, J. P. Y. Kao, and R. Y. Tsien, *J. Biol. Chem.* **264,** 8171 (1989).
[16] R. P. Haugland, "Handbook of Fluorescent Probes and Research Chemicals," 6th ed. Molecular Probes Inc., Eugene, Oregon, 1996.

(Zeiss, Don Mills, Ontario) was equipped with FLUAR 20×/0.75 and 40×/1.3 objectives, designed for optimal transmission of light in the 300- to 400-nm range. An aluminum PH-3 perfusion platform (Warner Instruments, Hamden, CT) was designed to fit the stage of the Axiovert without need for adaptation. The platform can be heated via current applied to two attached resistors, and a thermistor (temperature feedback sensor) fits into a bore in the side of the platform; both are controlled by a TC-344A heater controller unit (Warner Instruments). Coverslips, sealed by vacuum grease, fit the recess on the bottom side of a plastic perfusion chamber, which itself fits into a clamp on the heated perfusion platform. Solutions for perfusion are heated by an SH-27A in-line heating unit (Warner Instruments), which consists of a heating wire wrapped around a core tube, through which solutions pass. The heater controller maintains a steady solution temperature by means of a built-in thermistor in the in-line unit. A thermometer is also incorporated into the wiring for the platform heating, which allows for direct measurement of the in-chamber solution temperature. Excitation light is provided by a Delta-Ram High Speed Illuminator (Photon Technologies International, Monmout Junction, NJ) which consists of a 75-W xenon arc and a computer-controlled random-access wavelength monochromator. The light passes from the monochromator to the microscope input via a 1-m quartz fiber optic bundle. Wavelengths are further selected prior to cell illumination by a dichroic filter (Chroma Technology Corporation, Battleboro, VT) mounted in a sliding apparatus under the objectives. Emission light passes through a long-pass filter (Chroma Technology Corporation) in the filter-slider, and is then directed to either the microscope oculars for direct visualising, or to an intensified CCD (Photon Technologies International, Monmouth Junction, NJ). Image Master v1.4 software (Photon Technologies International) allows direct control of the camera, illumination, and data acquisition. For single-cell imaging, coverslips are loaded into a heated microscope platform chamber, and viewed, at 400× magnification, through an inverted fluorescent microscope with attached intensified camera (ICCD). Cells are excited at alternating 340- and 380-nm wavelength light. Then 510-nm emission images are collected, via the ICCD, by integrated software, which is also used to calculate ratios of 340/380 fura-2 emission fluorescence.

Calcium Measurements in Bovine ROS

In a series of previous studies from our laboratory we used fluo-3 to examine, in detail, the functional properties of rod NCKX1[17-20] and the

---

[17] P. P. M. Schnetkamp, X. B. Li, D. K. Basu, and R. T. Szerencsei, *J. Biol. Chem.* **266**, 22975 (1991).

role of NCKX1 in Ca^{2+} homeostasis and calcium compartmentalization in ROS.[21,22] Changes in fluo-3 fluorescence indicate changes in free cytosolic Ca^{2+}, at best a qualitative indicator for NCKX flux activity. Ca^{2+}-depleted ROS contain no measurable calcium and a very low cytosolic free Ca^{2+} concentration (<10 nM), thus enabling us to initiate calcium uptake via reverse Na^+/Ca^{2+}-K^+ exchange and to correlate increases in free cytosolic calcium with changes in total calcium content (as measured with ^{45}Ca uptake). For short time intervals (<5 min) and for free Ca^{2+} concentrations of <1 μM a linear relationship was observed between free cytosolic Ca^{2+} and total ROS Ca^{2+},[21] which enables us to convert changes in fluo-3 fluorescence into quantitative Ca^{2+} fluxes.

Calibration of Fluo-3 in Intact Bovine ROS

We previously determined that fluo-3, hydrolyzed by esterases, endogenous to bovine ROS, is located exclusively in the ROS cytosol with no dye detectable in the intradiskal space.[17] The calcium sensitivity of fluo-3 in the ROS cytosol is calibrated with the use of the calcium ionophore A23187 to equilibrate extracellular calcium (controlled by calcium buffers) and cytosolic calcium; an in situ K_D value of 500 nM was obtained, slightly higher than the value of 400 nM obtained for fluo-3 in buffered salt solutions. Because A23187 is a calcium-proton exchanger, it is important to use high salt concentrations to minimize the internal Donnan potential, which could otherwise cause a large and salt-dependent acidification of the ROS cytosol, and a concomitant accumulation of Ca^{2+} when A23187 is added.[22a] Figure 1 illustrates fluorescence measurements of Ca^{2+}-depleted and fluo-loaded ROS in a collage of time traces obtained with separate cuvettes, which summarize the calibration procedure. Ca^{2+}-depleted and fluo-3-loaded ROS suspended in a buffered sucrose solution containing EDTA show little fluorescence in addition to light scattered into the emission monochromator. We use two procedures that should increase cytosolic Ca^{2+} sufficiently to saturate fluo-3: (1) the Ca^{2+} ionophore A23187 and (2) the channel ionophore gramicidin to control intracellular sodium and drive Ca^{2+} in via

---

[18] P. P. M. Schnetkamp, D. K. Basu, X. B. Li, and R. T. Szerencsei, *J. Biol. Chem.* **266,** 22983 (1991).

[19] P. P. M. Schnetkamp, R. T. Szerencsei, J. E. Tucker, and P. Van den Elzen, *Am. J. Physiol.* (*Cell Physiol.*) **269,** c1147 (1995).

[20] P. P. M. Schnetkamp, J. E. Tucker, and R. T. Szerencsei, *Am. J. Physiol.* (*Cell Physiol.*) **269,** c1153 (1995).

[21] P. P. M. Schnetkamp and R. T. Szerencsei, *J. Biol. Chem.* **268,** 12449 (1993).

[22] P. P. M. Schnetkamp, *J. Biol. Chem.* **270,** 13231 (1995).

[22a] P. P. M. Schnetkamp, unpublished observations (1996).

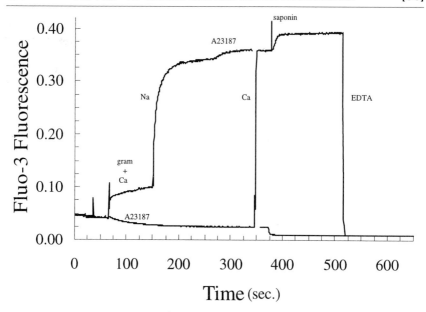

FIG. 1. Calibration of fluo-3 in Ca^{2+}-depleted intact bovine ROS in suspension. This figure comprises a collage, showing the calibration procedure in ROS, of three separate cuvettes with initial suspension medium containing 600 mM sucrose, 10 mM KCl, 20 mM HEPES (pH 7.4), 50 μM EDTA. First and second cuvettes: 1 μM A23187 is added at 60 sec, followed at 340 sec by the addition of 100 μM CaCl$_2$ to the first cuvette, and at 360 sec, 0.01% saponin to the second cuvette. Third cuvette: 100 nM gramicidin was added, followed at 60 sec by the addition of 100 μM CaCl$_2$; after the initial jump in fluorescence due to external fluo-3, at 160 sec, 50 mM NaCl was added; 1 μM A23187 was added at 260 sec, followed by 0.01% saponin at 360 sec. EDTA (1 mM) was added to cuvettes 1 and 2 at 480 sec. Temperature: 25°.

reverse Na^+/Ca^{2+}-K^+ exchange. The first addition spike (see earlier discussion) indicates addition of 100 nM of the channel alkali cation ionophore gramicidin, which did not cause any change in fluorescence. To the first cuvette, addition of 1 μM A23187 causes only a minor decrease in fluorescence, indicating a very low initial free Ca^{2+} concentration in ROS; next, raising Ca^{2+} to 100 μM causes a rapid and large increase in fluorescence. To another cuvette, 100 μM Ca^{2+} is added first, and a small and instantaneous increase in fluorescence is observed due to external fluo-3. The trace continues with a gradual increase in fluorescence indicating slow Ca^{2+} leakage into the ROS cytosol until addition of 50 mM NaCl causes a rapid and large increase in fluorescence to a level very close to that observed with 100 μM Ca^{2+} and 1 μM A23187.

The sodium-induced rise in cytosolic Ca^{2+} is caused by Ca^{2+} influx via reverse Na-Ca exchange (driven by internal sodium that entered the ROS cytosol via gramicidin). No other alkali cation could replace sodium to elicit Ca^{2+} influx. Saponin (0.01% final concentration) is added to both cuvettes at 360 sec to release fluo-3 from ROS, which allows us to obtain values for maximal and minimal fluorescence. A small increase and decrease in fluorescence is observed in the presence of 100 μM Ca^{2+} and 50 μM EDTA, respectively. These saponin-induced changes in fluorescence appear to be caused by changes in fluorescence quenching by the ROS suspension (high Ca^{2+}) or by changes in the amount of excited light scattered into the emission monochromator (EDTA), respectively, since similar changes in fluorescence are observed when saponin is added to cuvettes containing ROS (not loaded with fluo-3), external fluo-3, and Ca^{2+} or EDTA, respectively.

Cation Dependence of Calcium Uptake in ROS: Detecting NCX or NCKX Activity

The most common way to test for NCX or NCKX activity in cells is to measure Na_{in}-dependent Ca^{2+} uptake. This test uses the bidirectional capability of both types of exchangers. On reversal of the transmembrane sodium gradient, Ca^{2+} is driven into the cell by reverse Na-Ca exchange, which can be measured as either ^{45}Ca uptake or as a rise in intracellular free Ca^{2+} with fluo-3 or other fluorescent calcium-indicating dyes. In most studies, cells are preloaded with sodium by incubating cells in a Ringer's solution containing ouabain to inhibit the plasma membrane Na-K pump, followed by measurement of calcium uptake in lithium or potassium medium with calcium uptake in sodium medium as a control. (High external sodium is thought to inhibit calcium uptake via reverse Na-Ca exchange by competing with calcium for the transport site on the exchanger.) In our studies, we have used sodium ionophores such as monensin and gramicidin to clamp and control the internal sodium, potassium, or lithium concentration. Although no alkali cation gradients are likely to exist in the presence of such ionophores, intracellular sodium is sufficient to drive calcium into the cell when a large inward calcium gradient is presented.

Figure 2 illustrates this experimental paradigm with an experiment in which the cation dependence of Ca^{2+} influx into the ROS cytosol was examined. Fluo-3-loaded and Ca^{2+}-depleted ROS are preincubated for 1 min with 100 nM gramicidin in a buffered sucrose solution containing 75 mM LiCl and 50 μM EDTA; the added gramicidin allows intracellular sodium to be replaced by lithium (sodium released from ROS is diluted to below 0.5 mM in the large external volume). Addition of 150 μM $CaCl_2$ (marked by the addition spike, see earlier discussion) results in an instantaneous, small rise in fluorescence, which is followed by a very slow time-dependent rise in fluo-3 fluorescence. The instantaneous rise in fluorescence

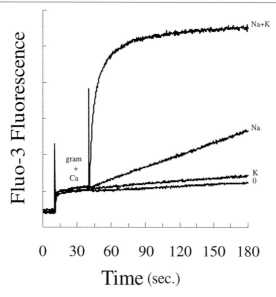

FIG. 2. Cation dependence of Ca^{2+} influx into Ca^{2+}-depleted intact bovine ROS. Composite of four separate cuvettes. In all cuvettes the initial suspension contained 400 mM sucrose, 20 mM HEPES (pH 7.4), 50 μM EDTA, 75 mM LiCl, and 100 nM gramicidin, which was added 15 sec prior to the start of each trace; 100 μM CaCl$_2$ was added at the first spike on all traces, with another addition at the second spike on each trace as follows: 0, no addition; K, 10 mM KCl; Na, 50 mM NaCl; and Na+K, 50 mM NaCl plus 10 mM KCl. Temperature: 25°.

is due to external fluo-3, most likely released from ROS during the sedimentation/resuspension process used to remove unhydrolyzed fluo-3AM. About 40 sec after the start of the traces, the second addition spike indicates the addition of 10 mM KCl (K), 50 mM NaCl (Na), the combined addition of 10 mM KCl and 50 mM NaCl (Na+K), or no further additions (0). Addition of KCl causes no change in the very slow rate of calcium influx compared with the trace where no further additions are made; while addition of NaCl causes a slight increase in calcium influx. In contrast, the combined addition of NaCl and KCl causes a very rapid and large rise in fluo-3 fluorescence due to a rapid Ca^{2+} influx that required the presence of both sodium and potassium, indicative for NCKX. Very similar results are obtained when the channel ionophore gramidicin is replaced by the shuttle ionophore, monensin, although a higher concentration of monenisn was required (1 μM). The potassium-independent calcium influx observed in the presence of NaCl alone probably presents a minor component of potassium-independent Na-Ca exchange via the $Na^+/Ca^{2+}-K^+$ exchanger observed before.[23]

[23] P. P. M. Schnetkamp and R. T. Szerencsei, *J. Biol. Chem.* **266**, 189 (1991).

Effect of Rhodopsin Bleaching on Ca^{2+} Measurements

Disks within intact bovine ROS have been shown to accumulate Ca^{2+} and to release Ca^{2+} under certain conditions.[21,24] Furthermore, indirect evidence suggests that light may cause Ca^{2+} release from disks in dark-adapted toad rods.[25] Therefore, we examined the effect of rhodopsin bleaching on Ca^{2+} release from disks. A significant time-dependent increase in fluo-3 fluorescence was observed when dark-adapted and fluo-3-loaded ROS were placed in the fluorometer, and when bleaching of rhodopsin and measurement of fluo-3 fluorescence were initiated at the same time by opening the shutter of the excitation monochromator. Further control experiments, however, showed that the time-dependent increase in fluo-3 fluorescence was associated with changes in fluorescence quenching by the ROS suspension. To illustrate this effect, fluo-3-loaded ROS were incubated with 1 μM A23187 and a saturating Ca^{2+} concentration of 100 μM (to eliminate changes in fluorescence due to changes in free Ca^{2+}). Subsequently, the ROS suspension was placed in the fluorometer, the excitation shutter was opened, and the time-dependent fluorescence changes were recorded as a function of the rhodopsin concentration (Fig. 3). Fluorescence rose instantly (i.e., within the sampling time) to an initial value, followed by a time-dependent component whose relative amplitude increased dramatically as the rhodopsin concentration was increased from 1 to 15 μM. This implies that the observed light-induced fluorescence changes are not due to changes in Ca^{2+}, but rather reflect a change in fluorescence quenching by ROS, which precludes meaningful measurements of light-induced changes in ROS Ca^{2+}. In view of the timescale involved, it would appear that the metarhodopsin-II–metarhodopsin-III transition is accompanied by large changes in fluo-3 fluorescence quenching.

NCX or NCKX Activity in (Transfected) Cells

Detecting NCX or NCKX Activity in Cells or Cells Transfected with NCX or NCKX cDNAs

The presence of the Na^+/Ca^{2+}-K^+ exchanger in ROS has been well established, and cDNAs have been cloned of bovine,[3] human,[26] and dolphin rod NCKX1.[6] We have adopted the preceding protocol for detecting NCX/NCKX activity on transfecting cells with the appropriate cDNA. Using CHO or HEK293 cells loaded with fluo-3 we determined that neither cell

---

[24] P. P. M. Schnetkamp, *Cell Calcium* **18,** 322 (1995).
[25] G. L. Fain and W. H. Schröder, *J. Physiol.* **389,** 361 (1987).
[26] J. E. Tucker, R. J. Winkfein, C. B. Cooper, and P. P. M. Schnetkamp, *IOVS* **39,** 435 (1998).

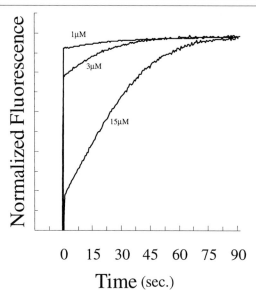

Fig. 3. Time-dependent fluorescence changes as a function of rhodopsin concentration in suspensions of intact bovine ROS. Initial suspensions were as in Fig. 1. Prior to the start of each trace, 1 μM A23187 and saturating 100 μM CaCl$_2$ were added. At time zero, the excitation shutter was opened, and fluorescence changes were observed. Suspensions contained 1, 3, or 15 μM rhodopsin as indicated in the figure. Values were normalized to the peak fluorescence obtained for each rhodopsin concentration. Temperature: 25°.

type contained an endogenous NCX/NCKX activity when a suspension of such cells was assayed with our protocol (i.e., addition of gramicidin did not cause any increase in cytosolic free calcium when sufficient sodium and calcium were present to drive reverse Na$^+$/Ca^{2+}-K$^+$ exchange).

Next, we tested our experimental paradigm with CHO cells, stably transfected with bovine NCX1 (Fig. 4). CHO cells transfected with NCX1 are loaded with fluo-3 and suspended in 150 mM LiCl medium containing 200 μM EGTA. The first addition spike at about 40 sec indicates the addition of 2.5 μM gramicidin, which does not cause any change in fluo-3 fluorescence; at about 110 sec, 1 mM CaCl$_2$ is added, which causes an instantaneous and small increase in fluorescence due to external fluo-3 with little further increase in fluorescence over time (trace labeled "Ca"). In a separate cuvette, 50 mM NaCl is added to the same CHO cell suspension (trace labeled "Na"), which causes a large and time-dependent increase in fluorescence, indicating a rise in cytosolic free Ca^{2+} due to reverse Na-Ca exchange; the sodium-dependent rise in cytosolic free Ca^{2+} does not require potassium nor is it modified when 20 mM KCl is present (not illustrated).

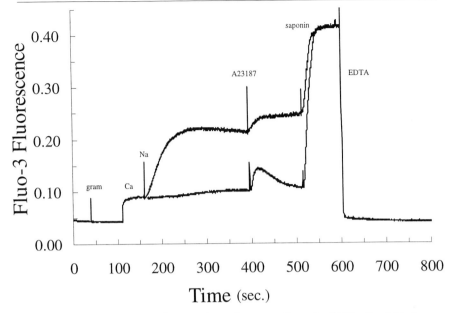

Fig. 4. Detection of Na$^+$/Ca^{2+}-K$^+$ exchange in mammalian cells. CHO cells, stably transformed with the bovine cardiac Na-Ca exchanger, and loaded with fluo-3 as described in the text, were suspended in a cuvette containing 150 mM LiCl, 20 mM HEPES (pH 7.4), 200 μM EGTA. Gramicidin (2.5 μM) was added, at the first spike on the trace, 75 sec prior to the addition of 1 mM CaCl$_2$ indicated on the trace by the first jump in fluorescence. At 160 sec, there was either no addition or a 50 mM NaCl addition for the top trace in the figure. At 400 sec, 1 μM A23187 was added, followed at 510 sec by the addition of 0.01% saponin. Minimal fluorescence was obtained by the addition of 4 mM EDTA at 600 sec. Temperature: 25°.

For calibration purposes we add 1 μM A23187, which causes a small further increase in fluo-3 fluorescence.

Remarkably, the A23187-induced increase in fluorescence observed in the cuvette with no sodium present is transient and did not reach the fluorescence level observed in the cuvette with sodium present despite the fact the external free Ca^{2+} concentration was 800 μM. We believe these findings show that A23187 (even at 1 μM) does not significantly increase the plasma membrane permeability for Ca^{2+} and thus cause Ca^{2+} influx into CHO cells; instead, A23187 appears to enter the cell and release Ca^{2+} from intracellular stores. Subsequent permeabilization of the plasma membrane by addition of 0.01% saponin causes a large increase in fluorescence that reaches the same level in both cuvettes, indicating saturation of fluo-3 fluorescence. Finally, addition of 4 mM EDTA causes a large drop in fluorescence to a level below that observed at the start of the experiment.

The sodium-induced rise in fluorescence is about half that observed with saponin; this could be interpreted to indicate either that fluo-3 is only half saturated (free Ca^{2+} concentration of about 500 nM) or that only 50% of the CHO cells express a functional NCX1 exchanger. This issue is addressed later in the section on digital Ca^{2+} imaging in single cells.

We examined HEK293 cells or CHO cells, transfected with cDNAs from either human or bovine rod NCKX1, but observed no NCKX function using the experimental paradigms described earlier. We recently cloned the cDNA of dolphin NCKX1, and observed robust potassium-dependent Na-Ca exchange activity when transfected into HEK293 cells and when analyzed either with single-cell digital calcium imaging (see later section) or with fluo-3 measurements of an ensemble of cells in suspension.[6] CHO cells permanently transfected with dolphin NCKX1 were loaded with fluo-3 and analyzed for Na-Ca exchange activity by collapsing the sodium gradient across the plasma membrane with gramicidin. Addition of gramicidin caused a rise in cytosolic free Ca^{2+} to indicate Na-Ca exchange activity (Fig. 5). Compared with a similar experiment in ROS we noted several differences: (1) a much higher concentration of gramicidin was required (2.5 μM compared with 100 nM); (2) a 30-sec delay between addition of gramicidin and the rise in cytosolic Ca^{2+} was observed in CHO cells (a similar observation was made for CHO cells transformed with NCX1 cDNA); and (3) monensin could not replace gramicidin even when a high concentration of 25 μM was used. We believe these findings suggest that the plasma membrane of HEK293 or CHO cells is very rigid compared to that of ROS, and does not permit proper function of shuttle type ionophores such as A23187 and monensin. A channel-type ionophore like gramicidin did permit equilibration of alkali cation gradients once the ionophore was properly inserted in the membrane (the lag phase observed in Fig. 5; no lag phase was observed when Ca^{2+} was added after preincubation for 1 min with gramicidin; see also Fig. 4).

Simple Method for Noise Reduction

Often, the fluo-3 fluorescence traces obtained with cells transfected with NCX/NCKX cDNAs were noisy due to the limited number of cells available and due to the limited level of fluo-3 loading (excessive fluo-3 loading is undesirable in most cells to avoid fluo-3 accumulation in intracellular organelles). The software of the fluorometer provides data, consisting of a series time points and associated emission measurements, that were transferred into a Microsoft Excel 97 spreadsheet. An equation for each datum is entered into the column adjacent to the data set to be noise-filtered, which calculates the average value over the data preceding and following

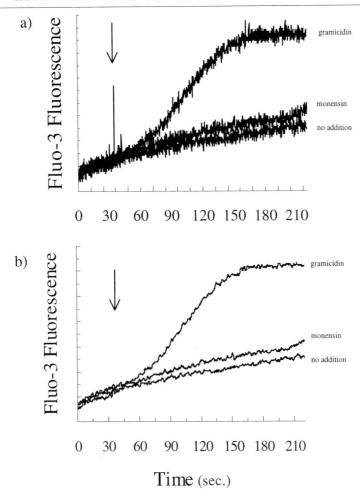

FIG. 5. Measurement of Na^+/Ca^{2+}-K^+ exchange using Na^+ ionophores. CHO cells stably transformed with dolphin NCKX1 and loaded with fluo-3 as described in the text were suspended in a solution containing 50 mM NaCl, 100 mM LiCl, 200 μM EGTA. Prior to the start of the trace, to each cuvette was added 10 mM KCl and 1 mM $CaCl_2$. In three separate cuvettes, at 45 sec, additions were made as follows: no addition; 25 μM monensin; 2.5 μM gramicidin. Temperature: 25°. In (a) no noise reduction has been applied to the trace; (b) shows the result of filtering, by a datum-averaging method described in the text, over a period of 2 sec for each datum.

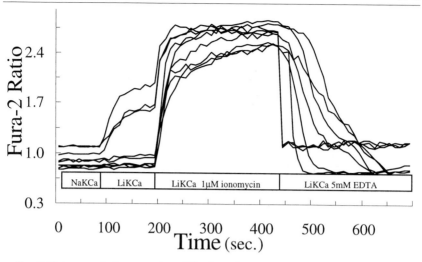

FIG. 6. Heterogeneity in expression of Na-Ca exchanger in stably transformed CHO cells. Each trace represents the change in fura-2 ratio in response to the indicated solution changes as measured in single NCX1-transformed CHO cells, loaded with fura-2 as described in the text. Changes in fura-2 ratio represent cytosolic Ca^{2+} flux. Cells on coverslips were perfused as described in the text. Initial solution contained 150 mM NaCl, 5 mM KCl, 100 μM $CaCl_2$. The first solution switch replaced the NaCl with 150 mM LiCl. At 200 sec 1 μM ionomycin was added to the solution bath. At 440 sec, 5 mM EDTA was added. Temperature: 25°.

the datum for a specified time period. Each datum is averaged by the same equation for the same time period centered over each datum. Smoothing of the resultant curve is increased as the time period over which averages are taken is increased. For data acquired at a frequency of 5 Hz, averaging over 2 sec, or five points either side of each datum, results in a reasonable reduction of noise without averaging-out kinetic resolution. On very fast kinetics, resolution may be lost if filtering is carried out over too long a time period.

Digital Calcium Imaging in Single Cells

Digital Ca^{2+} imaging of single CHO cells attached to a coverslip was used to address the issue raised earlier of cell heterogeneity. The 340/380 ratio of fura-2 fluorescence is illustrated in Fig. 6 as a measure of free cytosolic Ca^{2+} as described in the Instrumentation section. Cells perfused with sodium-containing solution displayed 340/380 ratios of around 1, indicating a low intracellular free Ca^{2+} concentration. Reversal of the sodium gradient by replacing sodium in the medium with lithium caused a sharp increase in the 340/380 ratio in some cells, while other cells showed no

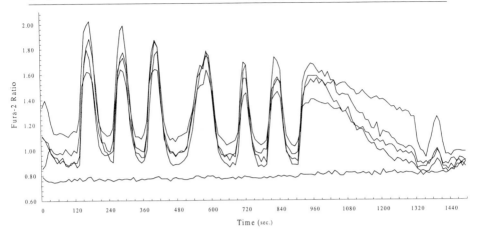

Fig. 7. Quantitative analysis using single-cell imaging. The flat trace represents the response of some cells in the field which were not affected by solution changes and which are presumed not to express NCKX1 protein. Each successive increase in fluorescence was elicited by a solution switch from 150 mM NaCl, 5 mM KCl, 100 μM CaCl$_2$ to a solution with 150 mM LiCl replacing the NaCl. On the seventh switch to LiCl, the cells were perfused in LiCl until most had run down cytosolic Ca^{2+} to starting level. The small "blip" at 1350 sec was a solution switch the same as the previous seven switches from NaCl to LiCl buffer, but with little or no response after the previous run down.

response. Subsequent addition of 1 μM of the calcium ionophore ionomycin caused a rapid and large rise in the 340/380 ratio in all cells that could be reversed by removing calcium from the perfusion medium and adding EDTA. From this experiment we conclude that the CHO cells stably transfected with NCX1 cDNA display considerable heterogeneity in terms of functional NCX1 expression, perhaps due to different expression levels at different stages of the cell cycle. A practical consequence is that a simple conversion of fluorescence into a scale of free cytosolic Ca^{2+} is not possible when fluorescence is measured in an ensemble of cells in suspension.

When we measured NCX/NCKX activity in single cells, either CHO cells transformed with NCX/NCKX cDNA or HEK293 cells transiently transfected with NCX/NCKX cDNA, functional expression levels varied considerably from one experiment to the next.[6] To obtain quantitative information about NCX/NCKX function, e.g., a dependence on potassium concentration, a series of potassium concentrations have to be presented to the same cell. Figure 7 illustrates the feasibility of this approach. HEK293 cells were transfected with dolphin NCKX1 cDNA and cytosolic free Ca^{2+} was imaged when the cells were subjected to seven series of consecutive solution switches in which the sodium gradient was replaced by lithium and vice versa.

Acknowledgments

This research was supported by operating grants from the Medical Research Council of Canada to P.P.M.S., and to the MRC-Group on Ion Channels and Transporters. P.P.M.S. is an Alberta Heritage Foundation for Medical Research Medical Scientist.

[57] Purification and Characterization of ABCR from Bovine Rod Outer Segments

By JINHI AHN *and* ROBERT S. MOLDAY

Introduction

The 220-kDa Rim protein (now known as ABCR) is an abundant rod photoreceptor membrane protein localized to the rim and incisures of rod outer segment disks.[1-3] The bovine and human ABCR genes were cloned independently using two methods: (1) screening a retinal cDNA expression library with an antibody against the 220-kDa Rim protein followed by rescreening with oligonucleotide probes[1,4] and (2) a genetic approach designed to isolate the gene linked to Stargardt disease.[5] In addition to Stargardt disease, mutations in the ABCR gene have recently been implicated in several other human retinal degenerative diseases including specific forms of autosomal recessive retinitis pigmentosa,[6] cone-rod dystrophy,[7] and age-related macular degeneration.[8] The cDNA for the mouse ABCR protein has also been cloned as an initial step in generating ABCR knockout mice.[9]

---

[1] M. Illing, L. L. Molday, and R. S. Molday, *J. Biol. Chem.* **272,** 10303 (1997).
[2] D. S. Papermaster, B. G. Schneider, M. A. Zorn, and J. P. Kraehenbuhl, *J. Cell Biol.* **78,** 415 (1978).
[3] H. Sun and J. Nathans, *Nat. Genet.* **17,** 15 (1997).
[4] I. Nasonkin, M. Illing, M. R. Koehler, M. Schmid, R. S. Molday, and B. H. F. Weber, *Hum. Genet.* **102,** 21 (1998).
[5] R. Allikmets, N. Singh, H. Sun, N. F. Shroyer, A. Hutchinson, A. Chidambaram, B. Gerrard, L. Baird, D. Stauffer, A. Peiffer, A. Rattner, P. Smallwood, Y. Li, K. L. Anderson, R. A. Lewis, J. Nathans, M. Leppert, M. Dean, and J. R. Lupski, *Nat. Genet.* **15,** 236 (1997).
[6] A. Martinez-Mir, E. Paloma, R. Allikmets, C. Ayuso, T. del Rio, M. Dean, L. Vilageliu, R. Gonzalez-Duarte, and S. Balcells, *Nat. Genet.* **18,** 11 (1998).
[7] F. P. Cremers, D. J. van de Pol, M. van Driel, A. I. den Hollander, F. J. van Haren, N. V. Knoers, N. Tijmes, A. A. Bergen, K. Rohrschneider, A. Blankenagel, A. J. Pinckers, A. F. Deutman, and C. B. Hoyng, *Hum. Mol. Genet.* **7,** 355 (1998).
[8] R. Allikmets, N. F. Shroyer, N. Singh, J. M. Seddon, R. A. Lewis, P. S. Bernstein, A. Peiffer, N. A. Zabriskie, Y. Li, A. Hutchinson, M. Dean, J. R. Lupski, and M. Leppert, *Science* **277,** 1805 (1997).
[9] S. M. Azarian and G. H. Travis, *FEBS Lett.* **409,** 247 (1997).

ABCR is a member of the superfamily of adenosine triphosphate (ATP)-binding cassette proteins or ABC transporters.[10,11] These proteins typically contain two ATP-binding cassettes, also called nucleotide-binding domains, and two membrane domains that reside on the same or different subunits. ABC transporters typically utilize energy derived from ATP hydrolysis to transport a substrate across or out of biological membranes. The transport function of ABC proteins is often mirrored by the activation of ATPase activity by the substrate.[12,13] Like P-glycoprotein and CFTR (cystic fibrosis transmembrane conductance regulator), ABCR consists of a single polypeptide chain divided into two tandemly arranged, homologous halves, each of which contains an ATP-binding cassette preceded by a membrane domain containing multiple transmembrane segments. Photoaffinity labeling studies have confirmed that ABCR binds 8-azido-ATP and this labeling reaction is effectively inhibited by either ATP or guanosine triphosphate (GTP).[1]

To determine the possible role of ABCR as an active transporter, it is important to isolate ABCR from ROS membranes for analysis of its basal and substrate-stimulated ATPase activity and, more importantly, its ATP-dependent transport of putative substrates. In this chapter, we describe procedures developed to (1) purify ABCR from detergent-solubilized ROS membranes, (2) reconstitute purified ABCR into lipid vesicles, and (3) assay for basal and retinoid-stimulated ATPase activity of solubilized and reconstituted ABCR. We also discuss methods used to express and purify the two nucleotide-binding domains (NBDs) of ABCR as fusion proteins for analysis of their ATPase activities.

Purification of the ABCR by Immunoaffinity Chromatography

Preparation of ROS and the Rim 3F4 Monoclonal Antibody Affinity Column

Rod outer segments (ROS) are isolated from frozen dark-adapted bovine retinas using a continuous sucrose density gradient as published previously[14] and stored in light-tight vials at $-80°$ at a concentration of 8–10 mg/ml protein in 20% sucrose, 0.01 M Tris-HCl buffer, pH 7.4.

---

[10] C. F. Higgins, *Annu. Rev. Cell Biol.* **8**, 67 (1992).
[11] J. M. Croop, *Methods Enzymol.* **292**, 101 (1998).
[12] A. B. Shapiro and V. Ling, *J. Biol. Chem.* **269**, 3745 (1994).
[13] X. B. Chang, Y. X. Hou, and J. R. Riordan, *J. Biol. Chem.* **272**, 30962 (1997).
[14] R. S. Molday and L. L. Molday, *J. Cell Biol.* **105**, 2589 (1987).

The Rim 3F4 monoclonal antibody recognizes a 9-amino-acid segment (Y-D-L-P-L-H-P-R-T) close to the C terminus of bovine ABCR.[1] This antibody cross-reacts with ABCR from most mammalian rod photoreceptors including human, rat, mouse, and dog. The antibody is purified from mouse ascites fluid by precipitation with 50% saturated ammonium sulfate at 4° followed by DEAE-Sephacel chromatography[15] and coupled to cyanogen bromide-activated Sepharose 2B in 0.02 M sodium borate buffer, pH 8.4, at a concentration of 2 mg protein per milliliter of packed beads. The immunoaffinity matrix is stored at 4° in 20 mM Tris-HCl, pH 7.4, 140 mM NaCl, 0.02% NaN$_3$.

Column Buffer

The column buffer consists of 50 mM Na-HEPES, pH 7.5, 0.1 M NaCl, 10 mM 3-[(cholamidopropyl)-dimethylammonio]-1-propanesulfonate (CHAPS), 1 mg/ml soybean phospholipid (~40% L-α-phosphatidylcholine, Type IV-S; Sigma, St. Louis, MO), 10% (v/v) glycerol, 3 mM MgCl$_2$ and 1 mM dithiothreitol (DTT). This buffer is prepared just prior to use by mixing the soybean phospholipid with Na-HEPES/NaCl/CHAPS solution, sonicating the suspension in a water bath for 10 min to disperse the lipids, stirring the solution until the lipid is completely dissolved, and sonicating the solution for another 5 min. Glycerol, MgCl$_2$, and DTT are then added to the buffer.

Immunoaffinity Purification Procedure

Just prior to use, 100 μl of Rim 3F4-Sepharose 2B is transferred to a spin column (Ultrafree-MC 0.45-μm filter unit; Millipore, Bedford, MA) and washed with 12 volumes (3 × 0.4 ml) of column buffer by low-speed centrifugation in a mini tabletop centrifuge (10 sec spin at 2000g). ABCR is purified from dark-adapted ROS in the dark or under dim red light at 4° as follows: ROS (1 mg rhodopsin) are centrifuged at 86,000g for 10 min in a Beckman (Palo Alto, CA) Optima TL Ultracentrifuge (40,000 rpm in a TLA100.4 rotor). The pellet is resuspended in 3 ml of 10 mM Na-HEPES, pH 7.5, and recentrifuged. This wash step is repeated and the final membrane pellet is resuspended in 150 μl 10 mM Na-HEPES, pH 7.5. This procedure results in the hypotonic lysis of ROS and removal of soluble and weakly associated membrane proteins.[16,17] The ROS membranes are

---

[15] E. Harlow and D. Lane, "Antibodies: A Laboratory Mannual," p. 298. Cold Spring Harbor Laboratory Press, Cold Spring Harbor, New York, 1988.
[16] H. Kuhn, *Nature* **283,** 587 (1980).
[17] R. S. Molday and L. L. Molday, *Methods Neurosci.* **15,** 131 (1993).

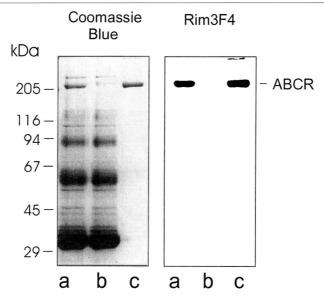

FIG. 1. Immunoaffinity purification of the ABCR on a Rim 3F4-Sepharose 2B column. Lane a: 15 μg CHAPS detergent-solubilized ROS membranes; lane b: flow-through or unbound fraction from the column; and lane c: the bound fraction eluted with the 3F4 peptide. The Coomassie Blue stained gel (left panel) shows that the 220-kDa ABCR is a major component of ROS membranes and can be isolated using immobilized Rim 3F4 antibody. The intense bands migrating at 36, 60, and 90 kDa in lanes a and b represent rhodopsin monomers, dimers, and trimers, respectively. The corresponding Western blot (right panel) shows that ABCR is present in solubilized ROS (a) and the peptide eluate (c), but is absent in the flow-through fraction (b).

solubilized by the dropwise addition of the membrane suspension to 1.35 ml of column buffer containing an additional 9 mg of CHAPS detergent (final CHAPS concentration of 18 mM) and the solution is stirred for 20 min. Residual insoluble material is removed by centrifugation at 86,000g for 10 min. The supernatant is then added to 100 μl Rim 3F4-Sepharose 2B beads (equilibrated with column buffer) in an Eppendorf tube and the mixture is gently agitated for 1 hr. The mixture is then transferred in batches to the spin column and the beads are collected on the filter by low-speed centrifugation in a mini centrifuge at room temperature. The flow-through or unbound fraction contains solubilized ROS membrane proteins that do not bind the Rim 3F4 antibody. Complete binding of ABCR to the immunoaffinity matrix is verified by comparing the solubilized ROS membrane fraction that has been applied to the column (Fig. 1, lane a) with the flow-through fraction (Fig. 1, lane b) by sodium dodecyl sulfate–

polyacrylamide gel electrophoresis (SDS–PAGE) and Western blotting. The latter should be depleted of ABCR as measured by Western blots labeled with the Rim 3F4 monoclonal antibody. The beads are washed by adding 0.4 ml of column buffer to the matrix followed by low-speed centrifugation. This washing procedure is repeated a total of six times. The bound ABCR is finally eluted by gently shaking the beads for 15 min in 60 μl column buffer containing 0.2 mg/ml of 3F4 synthetic peptide (Y-D-L-P-L-H-P-R-T-G) under normal room light. This elution step is repeated two more times and the fractions are pooled to yield 180 μl of eluted fraction. Approximately 30–40% of the ABCR is recovered by this procedure. Up to 80% of ABCR can be recovered when longer elution times and/or larger elution volumes are employed. The purity of ABCR is assessed by SDS–PAGE and Western blotting (Fig. 1, lane c). ABCR migrates as the predominant Coomassie Blue stained polypeptide having an apparent molecular mass of 220 kDa. A less intensely stained polypeptide having a molecular mass of 45 kDa is observed in some preparations. This band has been identified as the p44 spliced variant of arrestin.[18]

SDS–PAGE, Western Blots, and Protein Determination

For SDS–PAGE, 20 μl of protein sample is mixed with 10 μl of 3× SDS buffer [188 mM Tris-HCl, pH 6.8, 6% SDS, 12% (v/v) 2-mercaptoethanol, 30% (v/v) glycerol, 0.03% bromphenol blue], loaded on a 7.5% polyacrylamide gel and electrophoresed using the Laemmli buffer system.[19] Varying amounts of bovine serum albumin (0.1–2.0 μg) are loaded in adjacent wells. Gels are stained with Coomassie Brilliant Blue, destained, then scanned with an LKB densitometer. The amount of ABCR in the sample is determined from a standard curve constructed from the relative intensities of the albumin standards. Similar protein content is obtained using the Bradford assay.[20] Rhodopsin, 11-*cis*-retinal, all-*trans*-retinal, and all-*trans*-retinol concentrations are determined spectrophotometrically using molar extinction coefficients of 40,000 M^{-1} cm^{-1} ($\lambda = 500$ nm),[21] 24,935 M^{-1} cm^{-1} ($\lambda = 380$ nm, in ethanol), 42,880 M^{-1} cm^{-1} ($\lambda = 383$ nm, ethanol) and 52,770 M^{-1} cm^{-1} ($\lambda = 325$ nm, ethanol), respectively.[22]

For Western blotting, proteins are electrophoretically transferred onto Immobilon-P membranes (Millipore) in 25 mM Tris-HCl, 192 mM glycine,

---

[18] L. L. Molday and R. S. Molday, unpublished observations (1999).
[19] U. K. Laemmli, *Nature* **227**, 680 (1970).
[20] S. M. Read and D. H. Northcote, *Anal. Biochem.* **116**, 53 (1981).
[21] M. L. Applebury, D. M. Zuckerman, A. A. Lamola, and T.M. Jovin, *Biochemistry* **13**, 3448 (1974).
[22] R. Hubbard, P. K. Brown, and D. Bownds, *Methods Enzymol.* **18C**, 615 (1971).

pH 8.3, containing 10% methanol.[23] Transfer is carried out in a BioRad (Hercules, CA) Trans-blot SD semidry transfer apparatus at 300 mA for 40 min. The membranes are blocked with phosphate-buffered saline containing 0.1% Tween-20 (PBST) and 1.0% nonfat dry milk for 30 min at room temperature and incubated for 30 min at room temperature with the Rim 3F4 hybridoma cell supernatant diluted 1:15 with PBST. The membranes are washed with PBST and incubated with sheep anti-mouse Ig-peroxidase (diluted 1:5000 in PBST) for 30 min at room temperature. The membranes are finally washed with PBST and developed with an ECL Kit per the manufacturer's instructions (Amersham-Pharmacia Biotech, UK).

Measurement of ATPase Activity

Solutions

>ATP solution: A 10× ATP solution is prepared by adding [α-^{32}P]ATP (3000 Ci/mmol; NEN) to unlabeled ATP in column buffer or assay buffer to obtain an activity of 0.2 μCi/μl and an ATP concentration of 0.2–10 mM.
>Retinoid solution: A 10× retinoid solution is prepared by dissolving the retinoid in ethanol and subsequently diluting the solution with column buffer or assay buffer to reduce the ethanol to 5%. A 5% ethanol solution in the same buffer without the retinoid is used as a control.
>Column buffer: 50 mM Na-HEPES, pH 7.5, 0.1 M NaCl, 10 mM CHAPS, 1 mg/ml soybean phospholipid (~40% L-α-phosphatidylcholine, Type IV-S; Sigma), 10% (v/v) glycerol, 3 mM MgCl$_2$ and 1 mM DTT prepared as described earlier.
>Assay buffer: 50 mM Na-HEPES, pH 7.5, 0.1 M NaCl, 1 mM DTT, 3 mM MgCl$_2$ and 10% (v/v) glycerol.

Assay Procedure

ATP hydrolysis is determined by quantifying the production of [α-^{32}P]ADP from [α-^{32}P]ATP by thin-layer chromatography (TLC).[24] The reaction is carried out in a total volume of 10 μl. Purified ABCR (40–100

---

[23] H. Towbin, T. Staehelin, J. Gordon, *Proc. Natl. Acad. Sci. U.S.A.* **76**, 4350 (1979).
[24] C. Li, M. Ramjeesingh, W. Wang, E. Garami, M. Hewryk, D. Lee, J. M. Rommens, K. Galley, and C. E. Bear, *J. Biol. Chem.* **271**, 28463 (1996).

ng) in 8 μl of column buffer (for detergent-solubilized samples) or assay buffer (for reconstituted samples) is added to a 0.5-ml Eppendorf tube. Then 1 μl of the retinoid solution, ethanol control, or buffer is added and the sample is kept on ice for 10 min. A sample without enzyme is included as a blank. One microliter of the ATP solution is added and the reaction is started by immersing the tube in a 37° water bath. The reaction is stopped at the designated time, usually 30 min, by the addition of 4 μl of 10% SDS. After brief centrifugation in a microfuge, 1 μl of the sample is spotted onto a polyethyleneimine-cellulose TLC plate (12 × 24 cm; Aldrich) and allowed to dry at room temperature. The plate is then placed in a tank containing 0.5 M LiCl, 1 M formic acid until the solvent front migrates to 1 cm from the top. The plate is air dried, wrapped in plastic film, and exposed to a storage phosphor screen for 3 hr. The exposed screen is scanned in a Molecular Dynamics PhosphorImager and spots are quantitated using IPLab Gel analysis software. The ATPase activity is expressed as nanomoles of ADP formed per min, where nmol ADP = (intensity of ADP spot)/ (intensity of ADP spot + intensity of ATP spot) × initial nmol of ATP. The linear response of the PhosphorImager is tested by spotting different amounts of radiolabel and plotting the intensity versus known amount of radioactivity. Only intensities within this linear range are used for quantitation. Each reaction assay is run in duplicate. With purified ABCR, the [α-^{32}P]ADP that is formed is not hydrolyzed further because no other radioactive species are detected on the TLC plate. However, crude ATPase-containing mixtures, such as ROS membranes, hydrolyze ADP further to produce other ^{32}P-containing compounds. This must be taken into account when calculating the amount of ATP hydrolyzed.

GTPase activity is measured by replacing the [α-^{32}P]ATP with [α-^{32}P]GTP and using a full sheet of polyethyleneimine-cellulose (24 × 24 cm) to separate GDP from GTP.

ATPase Activity of the CHAPS-Solubilized ABCR

A typical time course for the production of ADP from ATP by purified, CHAPS-solubilized ABCR as measured by the TLC method is shown in Fig. 2. ADP production deviates from linearity, especially at ATP concentrations below 200 μM. As a result, initial rates have to be obtained by extrapolation of time course plots to zero time. The decrease in rate of ADP production with time appears to be due to the instability of CHAPS-solubilized ABCR at 37°. We have found that there is a complete loss of ATPase activity when ABCR is incubated at 37° for 30 min in the absence of ATP prior to the start of the ATPase reaction. When ABCR is incubated at 37° in the presence of 1 mM ATP, a 50% reduction in the rate of ATP

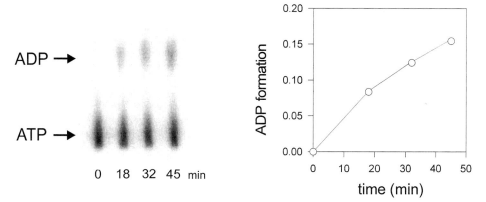

FIG. 2. Measurement of ATPase activity using TLC to separate [α-^{32}P]ADP from [α-^{32}P]ATP. Purified ABCR (50 ng) was incubated with 0.5 mM ATP (0.04 μCi ^{32}P/nmol) in a 10-μl reaction volume. At the indicated times, 4 μl of 10% SDS was added to stop the reaction and 1 μl of the reaction mixture was applied to the TLC plate shown in the left panel. The positions of ADP and ATP spots on the PhosphorImager scan are indicated by arrows. The amount of ADP produced relative to the amount of ATP added initially to the reaction was quantitated and plotted versus time (right panel).

hydrolysis is observed after 30 min. Furthermore, we have observed that the basal activity of the CHAPS-solubilized ABCR is preserved after it has been maintained on ice for 24 hr, but the enzyme is no longer stimulated by retinal. In constrast, purified ABCR reconstituted into phospholipid vesicles is very stable. Both the basal and retinal-stimulated ATPase activity is maintained when reconstituted ABCR is kept on ice for up to 4 days and the time course for ATP hydrolysis at 37° is constant for at least 60 min.

The ATPase activity of immunoaffinity purified, CHAPS-solubilized ABCR exhibits typical Michaelis–Menten kinetics. The K_m and V_{max} for ATP is 82 ± 14 μM and 200 ± 40 nmol/min/mg protein, respectively. ABCR also exhibits GTPase activity that is essentially indistinguishable from its ATPase activity. This is consistent with previous 8-azido ATP photoaffinity labeling experiments, which showed that ATP and GTP bind the ABCR with similar affinities.[1]

The effect of various agents on the ATPase activity of the purified CHAPS-solubilized ABCR has been investigated. As illustrated in Fig. 3, the ATPase activity of ABCR in the presence of soybean phospholipids is stimulated up to twofold by 25 μM 11-*cis*- or all-*trans*-retinal. Half maximal stimulation of the ATPase activity is observed at 10 μM all-*trans*-retinal. Retinol shows only limited stimulation at 25 μM. ATPase activity of ABCR is strongly dependent on the presence of phospholipids and sulfhydryl reducing agent. Basal and retinal-stimulated ATPase activities are reduced

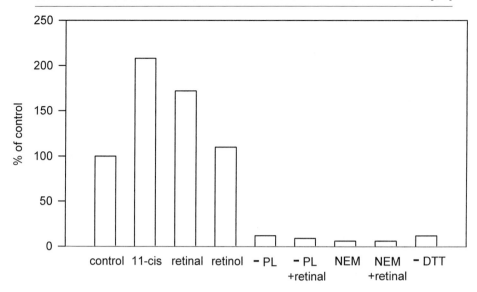

FIG. 3. The effect of various reagents on the ATPase activity of CHAPS-solubilized ABCR. Purified ABCR was incubated with 1 mM ATP (0.02 μCi ^{32}P/nmol) for 30 min at 37° and the reaction product was separated as shown in Fig. 2. Under these standard conditions, 163 ± 24 nmol ATP was hydrolyzed/min/mg protein. This is defined as 100% activity. Samples were preincubated with 25 μM 11-*cis*-retinal (11-*cis*), all-*trans*-retinal (retinal), all-*trans*-retinol (retinol), or 10 mM N-ethylmaleimide (NEM) for 10 min on ice before adding ATP. To measure the effect of dithiothreitol (DTT) or lipids (PL) on ATPase activity, ABCR was purified and assayed in the absence of these reagents.

by more than 90% in the absence of soybean phospholipid and DTT. N-ethylmaleimide (NEM), a sulfhydryl reactive agent, also inhibits ATPase activity. Previously, phospholipids and sulfhydryl reducing agents have been reported to be essential for maintaining the function of P-glycoprotein and multi-drug resistance protein (MRP) during the solubilization and purification of these ABC transporters.[13,25]

ATPase activities of various fractions obtained during the purification of ABCR have been measured and are listed in Table I. The ATPase activity of ABCR constitutes only a small amount of the total ATPase activity of isolated ROS preparations. The activity of ABCR recovered from the immunoaffinity column comprises 1.2% of the ATPase activity of ROS and about 2.3% of the activity of CHAPS-solubilized ROS membranes applied to the immunoaffinity column. Because the recovery of ABCR

[25] R. Callaghan, G. Berridge, D. R. Ferry, and C. F. Higgins, *Biochim. Biophys. Acta* **1328**, 109 (1997).

TABLE I
PURIFICATION OF ATPase ACTIVITY FROM ROD OUTER SEGMENT MEMBRANES[a]

| Fraction | Volume (ml) | Protein[b] (mg) | ATPase activity | | | | Recovery (%) | |
| | | | Basal | | +Retinal | | | |
			S.A.[c]	Total (nmol/min)	S.A.[c]	Total (nmol/min)	Protein	Activity[d]
ROS	1.0	0.91	87	79	62	56	100	100
Soluble proteins	2.0	0.11	36	4.0	34	3.7	12	7
ROS membrane	0.14	0.86	42	36	22	19	95	34
CHAPS extract	1.6	0.94	44	41	23	22	103	39
Flow-through	1.6	0.90	40	36	20	18	99	32
Eluate	0.18	.005	191	0.96	290	1.4	0.55	2.5

[a] ROS were washed in hypotonic buffer by centrifugation. The supernatant contains soluble proteins that associate with ROS membranes. There was little ATPase activity present in the soluble protein fraction. The ROS membrane fraction (pellet from wash) was solubilized in column buffer containing 18 mM CHAPS and contained most of the ATPase activity. The CHAPS extract was incubated with Rim 3F4-Sepharose 2B to purify the ABCR. Most of the ATPase was found in the flow-through fraction, however, the specific activity was higher for the purified ABCR and only this fraction showed retinal-stimulated ATPase activity.
[b] Determined by Bradford method.
[c] S.A. = specific activity in nmol/min/mg.
[d] Retinal-stimulated activity.

from the immunoaffinity column is estimated to be about 30% in this study, ABCR may contribute to as much as 7% of the total ATPase activity of solubilized ROS membranes and about 2% of the total protein.

The ATPase activity of the unbound (ABCR-depleted) fraction from the immunoaffinity column has been compared with the activity of the purified ABCR. Whereas the ATPase activity of the purified ABCR is stimulated by all-*trans*-retinal, the activity of the unbound fraction is actually reduced in the presence of a large excess of retinal (200 μM). Furthermore, unlike ABCR, the ATPase activity of the unbound fraction does not require phospholipid or DTT to maintain enzyme activity.

Reconstitution of the Purified ABCR in Soybean Phospholipid Vesicles

Solutions

Reconstitution buffer: 50 mM Na-HEPES, pH 7.5, 0.1 M NaCl, 20 mg/ml soybean phospholipids, 1 mM EDTA, 1 mM DTT

Dialysis buffer: 20 mM Na-HEPES, pH 7.5, 0.1 M NaCl, 1 mM DTT
Assay buffer: 50 mM Na-HEPES, pH 7.5, 0.1 M NaCl, 1 mM DTT, 3 mM MgCl$_2$, 10% (v/v) glycerol

Procedure

ABCR, like other membrane proteins, can be reconstituted into lipid vesicles by the addition of excess phospholipid and removal of detergent. In initial studies, we have employed the procedure used by Cook *et al.*[26] to reconstitute the cGMP-gated channel into soybean phospholipid (asolectin) vesicles. The immunoaffinity purified ABCR is added to an equal volume of reconstitution buffer containing the phospholipid and the mixture is stirred for 15 min at 4°. The mixture is then placed in a dialysis bag (Spectra/ Por 4 Membrane MWCO 12-14,000) and dialyzed for 24 hr against 500 ml of dialysis buffer. During this period, the dialysis buffer is changed three times to ensure complete removal of the CHAPS. Finally, the reconstituted ABCR is dialyzed for 12 hr against 500 ml of assay buffer.

ATPase Activity of ABCR Reconstituted into Soybean Phospholipid Vesicles

As shown in Fig. 4, the ATPase activity of ABCR reconstituted into soybean phospholipid vesicles displays typical Michaelis–Menten kinetics. Double-reciprocal plots give K_m and V_{max} values for ATP of 25 ± 9 μM and 50 ± 24 nmol/min/mg, respectively. The K_m value is comparable to the value of 33 μM ATP reported by Sun *et al.*[27] for ABCR reconstituted into brain polar lipid. Our V_{max} value of 50 nmol/min/mg for ABCR reconstituted in soybean phospholipid, however, is significantly higher than the value of 1.3 nmol/min/mg measured for ABCR in brain polar lipid. In the presence of 20 μM all-*trans*-retinal, the K_m and V_{max} for ABCR reconstituted in soybean phospholipids increase to 50 ± 20 μM ATP and 110 ± 60 nmol/min/mg, respectively. These values differ significantly from the K_m and V_{max} values of 725 μM and 29 nmol/min/mg observed for ABCR in brain polar lipid in the presence of 80 μM retinal.[27] Evidently, the lipid environment strongly influences the kinetics and retinal stimulation of ABCR. We are now carrying out a more detailed study to examine the effects of specific lipids on the basal and retinal-stimulated ATPase activity of purified, reconstituted ABCR.

---

[26] N. J. Cook, C. Zeilinger, K. Koch, and U. B. Kaupp, *J. Biol. Chem.* **261,** 17033 (1986).
[27] H. Sun, R. S. Molday, and J. Nathans, *J. Biol. Chem.* **274,** 8269 (1999).

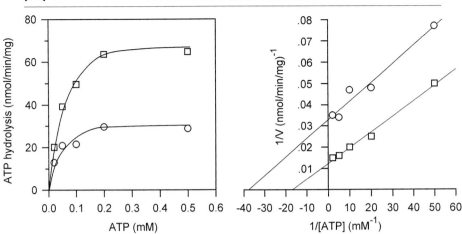

FIG. 4. ATPase activity as a function of ATP concentration in the absence (circles) and presence of all-*trans*-retinal (squares). Purified ABCR reconstituted into soybean phospholipids was incubated with various concentrations of ATP (0.2 μCi ^{32}P/tube) and the reaction was monitored over a 30-min period. The initial rate of ATP hydrolysis was plotted versus ATP concentration (left panel). Double-reciprocal plots (right panel) were used to calculate V_{max} and K_m values for basal and retinal-stimulated ATPase activities.

Expression of ABCR Nucleotide-Binding Domains as MBP Fusion Proteins

The NBDs of several mammalian ABC proteins have been expressed in bacteria as glutathione *S*-transferase (GST) or maltose binding protein (MBP) fusion proteins.[28-30] Relatively large quantities of these fusion proteins have been purified as soluble proteins for analysis of their ATP-binding and ATPase activities. We have expressed both the N- and C-terminal NBDs of ABCR as GST and MBP fusion proteins in *Escherichia coli*. The MBP fusion proteins are highly soluble, whereas most of the GST fusion proteins are expressed as insoluble proteins in inclusion bodies. We describe here the methods we have employed to express and purify MBP fusion proteins.

[28] H. Baubichon-Cortay, L.G. Baggetto, G. Dayan, and A. Di Pietro, *J. Biol. Chem.* **269**, 22983 (1994).
[29] S. Sharma and D. R. Rose, *J. Biol. Chem.* **270**, 14085 (1995).
[30] Y. H. Ko and P. L. Pedersen, *J. Biol. Chem.* **270**, 22093 (1995).

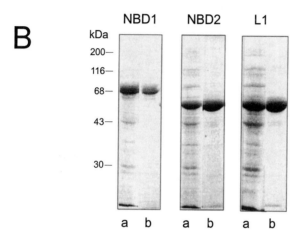

FIG. 5. (A) Linear representation of ABCR showing the regions of the protein used to construct MBP fusion proteins. The shaded boxes represent the protein domains of the human ABCR, which were used to generate the NBD1, NBD2, Loop1, and NBD1del fusion proteins. The amino acid boundaries are indicated above each shaded box: for the Walker A deletion mutant, NBD1del, amino acids 963–970 (GHNGAGKT) were deleted. (B) Coomassie Blue stained SDS gel showing the expression and purification of MBP–NBD1 (~70 kDa), MBP–NBD2 (~58 kDa), and MBP–Loop1 (L1, ~58 kDa). Lane a, total cell lysate (equivalent to ~20 μl culture); lane b, purified fusion protein. Positions of molecular weight markers are indicated.

Construction of Expression Plasmids

Constructs encoding the two nucleotide-binding domains, NBD1 and NBD2, of the human ABCR[4,5] are prepared by amplifying the cDNA sequences encompassing the NBDs (amino acids 853–1105 for NBD1 and amino acids 1951–2109 for NBD2) and inserting these fragments into the EcoRI site of the pMAL vector (New England BioLabs). As negative controls, a 159-amino-acid segment, Loop1 (amino acids 80–238), upstream of NBD1 and predicted to be present as an extracellular loop of ABCR[1] and the NBD1 missing the 8-amino-acid Walker A consensus sequence (NBD1del) are also expressed as MBP fusion proteins (see Fig. 5A). The following PCR primers are used to amplify the corresponding DNA coding regions of ABCR:

NBD1:	forward	TAGAATTCACTCGCTTGGTACCTTGATC
	reverse	GAGGAATTCTACAGGAGCAGATCCCAG
NBD2:	forward	GGGAATTCCAGCCCAGCAGTGGAC
	reverse	GATGAATTCTCACAGCATGCGGCGTGC
Loop1:	forward	GTGAATTCTCCCTGTTTTCAAAGC
	reverse	GTGAATTCTATTCACTGTAGGGTGCC
NBD1del:	forward	CGAGAACCAGATCACCGCATTCCTGACCACCTTGTCCATCC
	reverse	CCGACAGCTTTCTCTGCATGCC

Polymerase chain reaction (PCR) conditions are as follows: 3 min at 94°, 25 cycles of 45 sec at 94°, 45 sec at 55°, 90 sec at 72°, followed by 7 min at 72°. The first three PCR products (Loop1, NBD1, NBD2) are digested with *Eco*RI, purified with the Geneclean kit (Bio101, Vista, CA), and ligated to the *Eco*RI site of pGEX4T3 (Pharmacia). The PCR product NBD1del is digested with *Sph*I and *Bsm*I, purified with Geneclean, and ligated to the *Sph*I and *Bsm*I sites of pGEX4T3-NBD1. To create pMAL expression plasmids, the ABCR domains are excised from the pGEX4T3 constructs with *Eco*RI and inserted into the *Eco*RI site of pMALcxh (pMAL with a modified polylinker, which puts the insert in frame with MBP). Alternatively, the PCR products for Loop1, NBD1, and NBD2 can be cloned directly into the *Eco*RI site of pMALcxh.

Expression and Purification of Fusion Proteins

Escherichia coli DH5α cells are transformed and grown at 37° in 50 ml of Luria–Bertani (LB) medium containing 50 mg/liter ampicillin to an absorbance at 600 nm of ~0.5. Fusion protein expression is induced with 0.3 mM isopropyl β-D-thiogalactopyranoside for 3 hr at 30°. The cells are pelleted and suspended in 10 ml buffer B (20 mM Tris, pH 7.5, 200 mM NaCl, 1 mM DTT) containing protease inhibitors [1 mM EDTA, 1 μg/ml aprotinin, 0.5 μg/ml leupeptin, 0.8 μg/ml pepstatin, 0.1 mg/ml Pefabloc (Boehringer-Mannheim)]. The cell suspension is passed through an Aminco French pressure cell (1200-psi gauge pressure at high setting) and centrifuged at 27,000*g* for 20 min. The supernatant is slowly passed through a 2-ml amylose resin column (equilibrated in buffer B). This is repeated twice to allow binding of the MBP fusion proteins to the column. The column is washed with 10 volumes of buffer B and eluted in buffer B containing 10 mM maltose.

Figure 5B shows that the fusion proteins are highly expressed in transformed *E. coli* cells after induction (lane a). The protein that is eluted from the amylose column is highly pure (lane b); however, minor bands are still visible when the SDS gel is overloaded with protein.

ATPase Activity of MBP Fusion Proteins

ATPase activities of the four affinity isolated fusion proteins (NBD1, NBD2, Loop1, and NBD1del) are determined using a colorimetric method.[31] The ATPase activities of NBD1 and NBD2 (K_m for ATP of ~0.3 mM; V_{max} ~ 30 nmol/min/mg) were identical to those of Loop1 and NBD1del. The latter two domains either do not contain a NBD (Loop1) or contain a disrupted domain (NBD1del) and therefore should not catalyze the hydrolysis of ATP. Because we measure similar ATPase activities for all four fusion proteins, we conclude that most, if not all, of the ATPase activity present in the fusion protein preparations likely arises from a contaminating bacterial ATPase, possibly a chaperone protein, that is present in low concentration but copurifies with the fusion protein. We have carried out several additional studies. The ATPase activity of these fusion proteins is decreased by less than 50% in the presence of NEM, unlike the ABCR protein purified from rod cells, which loses more than 90% of its ATPase activity with NEM. Furthermore, 8-azido photoaffinity labeling studies indicate that the fusion proteins are not specifically labeled with this reagent in contrast to ABCR.

These studies indicate that the NBD1 and NBD2 domains can be expressed as soluble MBP proteins. However, most, if not all of the protein does not bind or hydrolyze ATP, suggesting that it is not folded into an active conformation.

Conclusions

ABCR, an abundant protein in ROS disk membranes, can be readily purified by immunoaffinity chromatography using Rim 3F4-Sepharose columns. This procedure has the advantage that it is relatively simple, efficient, and mild. The purified ABCR exhibits basal and retinal-stimulated ATPase activity in both its detergent solubilized and reconstituted state. The presence of both phospholipid and DTT is required for activity. The ATPase kinetics and stimulatory properties of ABCR are also strongly influenced by its lipid environment. Because the ATPase activity of ABCR is relatively low (compared to P-glycoprotein), sensitive radioisotope ATPase assays need to be used for enzymatic determinations. The ATPase assay described here, measuring the production of [α-^{32}P]ADP by TLC, is highly sensitive

---

[31] S. Chifflet, A. Torriglia, R. Chiesa, and S. Tolosa, *Anal. Biochem.* **168,** 1 (1988).

and reproducible, but requires the use of a PhosphorImager for quantification of the product.

Studies reported here and by Sun et al.[27] indicate that retinal activates the ATPase of ABCR in a highly specific manner, suggesting the possibility that ABCR is a retinal-specific transporter. Experiments are now under way to determine if ABCR reconstituted into lipid vesicles is capable of transporting retinal from membranes in an ATP dependent manner.

Acknowledgments

This work was supported by grants from the Steinbach Fund, the National Eye Institute (EY 02422), and the Medical Research Council of Canada. We also wish to thank Hui Sun and Jeremy Nathans for helpful discussions and Laurie Molday, Michelle Illing, Tom Kim, and Jason Wong for assistance with a number of the preliminary studies. We also thank Dr. Rosalie Crouch for 11-*cis*-retinal.

[58] ABCR: Rod Photoreceptor-Specific ABC Transporter Responsible for Stargardt Disease

By HUI SUN and JEREMY NATHANS

Introduction

Stargardt disease (MIM #248200; also called fundus flavimaculatus) is an autosomal recessive disorder that produces central vision loss and a progressive bilateral atrophy of the macular region of the retina and retinal pigment epithelium (RPE).[1] A distinctive feature of Stargardt disease is the accumulation of fluorescent, short-wave absorbing material in the RPE, which appears in postmortem samples as lipofuscin-like deposits. The gene responsible for Stargardt disease codes for a member of the ABC transporter family, ABCR.[2] The identification of ABCR occurred via two independent routes: by purification and characterization of an abundant high molecular weight ROS membrane protein,[3–5] and by a combination of high-

---

[1] P. A. Blacharski, in "Retinal Dystrophies and Degenerations" (D. A. Newsome, ed.), p. 135. Raven Press, New York, 1988.
[2] R. Allikmets, N. Singh, H. Sun, N. F. Shroyer, A. Hutchinson, A. Chidambaram, B. Gerrard, I. Baird, D. Stauffer, A. Peiffer, A. Rattner, P. M. Smallwood, K. L. Anderson, R. A. Lewis, J. Nathans, M. Leppert, M. Dean, and J. R. Lupski, *Nat. Genet.* **15,** 236 (1997).
[3] M. Illing, L. L. Molday, and R. S. Molday, *J. Biol. Chem.* **272,** 10303 (1997).
[4] S. M. Azarian and G. H. Travis, *FEBS Lett.* **409,** 247 (1997).

throughput cDNA sequencing and positional cloning.[2] The spectrum of mutations identified thus far in Stargardt disease patients suggests that these individuals carry at least one partially functional allele.[2,6–8] By contrast, some cases of autosomal recessive retinitis pigmentosa are associated with homozygosity of frameshift or splice site mutations in the ABCR gene, which are likely to result in complete loss of ABCR function.[9,10] An association between a subset of subjects with age-related macular degeneration and variant alleles of the ABCR gene has been reported,[11] but interpretation of the data has been controversial.[12]

In this chapter we discuss methods and results pertaining to localization and functional reconstitution of ABCR.[2,13,14] A recent review of the genetics of ABCR can be found in Ref. 15.

Materials and Methods

Reagents and Equipment

Reagents and their sources are as follows: Extracti-Gel detergent removal resin (Pierce, Rockford, IL); CNBr-activated Sepharose 4 Fast Flow beads (Pharmacia, Kalamazoo, MI); dark-adapted bovine retinas (Schenk

---

[5] J. L. Thomson, H. Brzeski, B. Dunbar, J. V. Forrester, J. E. Fothergill, and C. A. Converse, *Curr. Eye Res.* **16,** 741 (1997).

[6] J.-M. Rozet, S. Gerber, E. Souied, I. Perrault, S. Chatelin, I. Ghazi, C. Leowski, J.-L. Dufier, A. Munnich, and J. Kaplan, *Eur. J. Hum. Genet.* **6,** 291 (1998).

[7] I. Nasonkin, M. Illing, M. R. Koehler, M. Schmid, R. S. Molday, and B. H. F. Weber, *Hum. Genet.* **102,** 21 (1998).

[8] R. A. Lewis, N. F. Shroyer, N. Singh, R. Allikmets, A. Hutchinson, Y. Li, J. R. Lupski, and M. Dean, *Am. J. Hum. Genet.* **64,** 422 (1999).

[9] A. Martinez-Mir, E. Paloma, R. Allikmets, C. Ayoso, T. del Rio, M. Dean, R. Gonzalez-Duarte, and S. Balcells, *Nat. Genet.* **18,** 11 (1998).

[10] F. P. M. Cremers, D. J. R. van de Pol, M. van Driel, A. I. den Hollander, F. J. J. van Haren, N. V. A. M. Knoers, N. Tijmes, A. A. B. Bergen, K. Rohrschneider, A. Blankenagel, A. J. L. G. Pinckers, A. F. Deutman, and C. B. Hoyng, *Hum. Molec. Genet.* **7,** 355 (1998).

[11] R. Allikmets, N. F. Shroyer, N. Singh, J. M. Seddon, R. A. Lewis, P. Bernstein, A. Peiffer, N. Zabriskie, Y. Li, A. Hutchinson, M. Dean, J. R. Lupski, and M. Leppert, *Science* **277,** 1805 (1997).

[12] T. P. Dryja, C. E. Briggs, E. Berson, P. J. Rosenfeld, and M. Abitbol, *Science* **279,** 1107 (1998).

[13] H. Sun and J. Nathans, *Nat. Genet.* **17,** 15 (1997).

[14] H. Sun, R. S. Molday, and J. Nathans, *J. Biol. Chem.* **274,** 8269 (1999).

[15] M. A. van Driel, A. Maugeri, B. J. Klevering, C. B. Hoyng, and F. P. M. Cremers, *Ophthalmol. Genet.* **19,** 117 (1998).

Packing Company, Stanwood, WA); activated charcoal, caprylic acid, crude sheep brain PE, all-*trans*-retinal, actinomycin D, amiodarone, cholchicine, nifedipine, trifluoperazine, verapamil, vinblastine, vincristine, CHAPS, progesterone (Sigma, St. Louis, MO); β-octyl-glucoside and *n*-dodecyl-beta-D-maltoside (CalBiochem, San Diego, CA); porcine brain polar lipids and egg PC (Avanti Polar Lipids, Alabaster, AL). Lipid sonication was performed using the G12SP1 Special Ultrasonic Cleaner (Laboratory Supplies Co., Inc., Hicksville, NY).

Conjugation of Rim 3F4 Antibody to Sepharose Beads

Monoclonal antibody Rim 3F4, which recognizes a linear epitope near the C terminus of bovine ABCR,[3] was purified from ascites fluid by caprylic acid precipitation,[16] and conjugated to CNBr-activated Sepharose 4 Fast Flow beads (Pharmacia) at a ratio of 1 mg of antibody per milliliter of resin. The resin was discarded after each use to improve reproducibility in the purification of ABCR.

Purification of ABCR from Bovine ROS

All purification procedures prior to the elution step are performed under red dim light at 4° to minimize aggregation of photobleached rhodopsin. Purified ROS[17] from 4–8 bovine retinas are diluted into greater than 10 volumes 1× phosphate-buffered saline (PBS), pelleted, and solubilized in 1.4 ml of ROS/CHAPS buffer (10% glycerol, 0.75% CHAPS, 50 mM HEPES, pH 7.0, 0.5 mg/ml crude brain PE and 0.5 mg/ml egg PC, 140 mM NaCl, 3 mM MgCl$_2$, 5 mM 2-mercaptoethanol). The sample is rotated at 4° in the dark for 1 hr to completely solubilize the ROS membranes. Insoluble material is removed by microcentrifugation at 16,000 rpm for 5 min and the supernatent added to 200 μl of Rim 3F4-Sepharose and rotated in the dark at 4° for 3 hr. The resin is recovered by microcentrifugation at 3000 rpm and washed 3 × 1 min and then 2 × 20 min (with gentle rotation) using ice-cold 1 ml ROS/CHAPS buffer. Three successive elutions are performed at 4° using 80 μl each of 0.2 mg/ml of a peptide (H$_2$N-NETYDL-PLHPRTAG-COOH) containing the Rim 3F4 epitope[3] in ROS/CHAPS buffer for a total of 45 min. An irrelevant peptide, JN50 (H$_2$N-MLRNNLGNSSDC-CONH$_2$), is used as a control for the specificity of elution.

[16] E. Harlow and D. Lane, "Antibodies: A Laboratory Manual." Cold Spring Harbor Laboratory Press, Cold Spring Harbor, New York, 1998.
[17] D. S. Papermaster, *Methods Enzymol.* **81**, 48 (1982).

Reconstitution of ABCR into Proteoliposomes

Immunoaffinity purified ABCR (290 μl) in ROS/CHAPS buffer is added to a premixed solution containing 30 μl 15% octylglucoside and 90 μl 50 mg/ml sonicated porcine brain polar lipid in 25 mM HEPES, 140 mM NaCl, 10% glycerol. The sample is mixed, incubated on ice for 30 min, and then rapidly diluted into 2 ml of reconstitution buffer (25 mM HEPES, pH 7.0, 140 mM NaCl, 1 mM EDTA, 10% glycerol) at room temperature and incubated at room temperature for an additional 2 min. To remove detergent, the sample is passed (at room temperature) over a 2-ml Extracti-Gel (Pierce) column that has been thoroughly prewashed with the reconstitution buffer. The flow-through is stored on ice. Light microscopic examination of the reconstituted preparation stained with 1,1'-dioctadecyl-3,3,3',3'-tetraethylindocarbocyanine (dil) shows a heterogeneous distribution of vesicle sizes.

ATPase Assays

ATPase assays are performed using ABCR that has been purified from ROS and reconstituted into liposomes on the same day. ATPase assays are performed in 300 μl reconstitution buffer supplemented with MgCl$_2$ to a final concentration of 3 mM. Appropriate dilutions and/or mixtures of test compounds are prepared in ethanol and added to 0.1% of the volume of the ATPase reaction; control reactions receive the same volume of ethanol alone. The reactions are started by the addition of 30 μl of 10× adenosine triphosphate (ATP) mix containing 50 mM 2-mercaptoethanol, 500 uM ATP with 50 μCi/ml of 3000 Ci/mmol [γ-^{32}P]ATP, and incubated at 37° for 2 hr. For experiments in which the ATP concentration varies, the [^{32}P]ATP is held constant at 5 μCi/ml. At the starting and final time points, triplicate samples of 50 μl are added to 200 μl of 10% activated charcoal in 10 mM HCl, vortexed vigorously for 2 min, and microcentrifuged to pellet the charcoal. The resulting supernatant (50 μl) is counted. For experiments to differentiate the effects of retinal isomers, the ATPase assay is manipulated under dim red light and incubated in the dark. All other ATPase assays are performed in room light. Data points in Lineweaver–Burk plots are fit to a straight line by a least squares criterion; no attempt is made to fit the data to more complex curves.

Handling and Storage of Retinoids

All procedures are performed under dim red light or in the dark. Retinoids are dissolved in ethanol and for each experiment their integrity is monitored by determining an absorption spectrum. Retinoids are handled in glass vials to minimize absorption to plastic. For storage, aliquots are dried under argon and maintained in the dark at $-80°$.

In situ Hybridization

In situ hybridization with digoxigenin-labeled riboprobes is performed as described.[18] For mouse and rat, whole eyes are frozen and sectioned; macaque retinas are obtained following cardiac perfusion with PBS/4% paraformaldehyde. An extra incubation of 30 min in PBS with 1% Triton X-100 is applied to the macaque retina sections immediately after the acetylation step.

Immunostaining

Macaque retinas are harvested following cardiac perfusion with PBS/4% paraformaldehyde and overnight fixation of isolated eye cups in the same buffer. Fresh bovine retinas are fixed in PBS/4% paraformaldehyde for 3 hr at 4°. After cryoprotection in 30% sucrose and snap freezing in optimal cutting temperature (OCT) compound, 10-μm frozen sections are cut and preincubated for 1 hr in PBS containing 5% normal goat serum/0.3% Triton X-100, and then incubated in the same buffer containing the primary antibody overnight. Biotinylated goat anti-rabbit secondary antibody and Texas Red-conjugated streptavidin (Vector Laboratories, Burlingame, CA) or avidin-horseradish peroxidase (Extravidin-peroxidase; Sigma) are used to visualize the ABCR immunostaining. Affinity purified anti-ABCR1156-1258 is used for the experiment shown in Fig. 1 (see color insert). An identical pattern of immunostaining is observed with ABCR425-570. The cGMP-gated channel is visualized with mouse monoclonal PMc1D1[19] and fluorescein-conjugated goat anti-mouse secondary antibody (Vector Laboratories).

Results

In situ Hybridization and Immunolocalization of ABCR mRNA and Protein

In situ hybridization of rodent and primate ocular tissues reveals ABCR transcripts in the retina but not in the RPE. Within the retina the

---

[18] N. Schaeren-Wiemers and A. Gerfin-Moser, *Histochemistry* **100,** 431 (1993).
[19] N. J. Cook, Molday, L. L., D. Reid, U. B. Kaupp, and R. S. Molday, *J. Biol. Chem.* **264,** 6996 (1989).

hybridization signal is present only in rods.[2] To localize ABCR by immunostaining, rabbit antisera are raised against four nonoverlapping 100–150 amino acid segments of human ABCR that are produced in *Escherichia coli* as bacteriophage T7 gene 10 fusion proteins.[13] Following affinity purification with four analogous fusions to the *E. coli* maltose binding protein (MBP), three of the four purified antibodies are found to detect only the predicted ~250-kDa band on immunoblots. This band is highly enriched in purified bovine ROS relative to total retina, and by quantitative immunoblotting with the MBP fusion proteins as internal standards it appears to be present in ROS membranes at a mole ratio of 1 : 120 to rhodopsin.

In agreement with the *in situ* hybridization, immunostaining of primate retina reveals ABCR only in rod photoreceptors where it is localized to outer segments (Fig. 1B). Double immunolabeling of bovine retina with rabbit anti-ABCR and a mouse MAb against the cGMP-gated channel shows that ABCR is present predominantly or exclusively in the disk membrane (Figs. 1C–E). A close examination of the immunostaining pattern using confocal microscopy shows greatest ABCR immunoreactivity at the edges of the disks. Similar results have been obtained by Molday and colleagues using immunoelectron microscopy.[3] The molecular weight and subcellular distribution of mammalian ABCR strongly suggest that it is identical to the high molecular weight RIM protein from frog outer segments that was characterized immunochemically by Papermaster and colleagues 20 years ago.[20] The rod outer segment (ROS) localization of ABCR suggests that ABCR transports a molecule that plays a specialized role in photoreceptor homeostasis, and the localization of ABCR to the disk membranes implies that the transport event is between the lumenal and cytosolic faces of the disk. Because outer segment turnover occurs via RPE phagocytosis, aberrant accumulation of one or more compounds in the outer segment could plausibly explain the progressive accumulation of material within the RPE in Stargardt disease.

Purification and Reconstitution of ABCR

For functional reconstitution, the ABCR purification strategy is essentially that of Illing *et al.*[3] modified by the use of a mixture of CHAPS, PC, and PE (ROS/CHAPS buffer) for the solubilization and chromatography steps. In preliminary experiments, we observed that the ATPase activity

---

[20] D. S. Papermaster, B. G. Schneider, M. A. Zorn, and J. P. Kraehenbuhl, *J. Cell Biol.* **78**, 415 (1978).

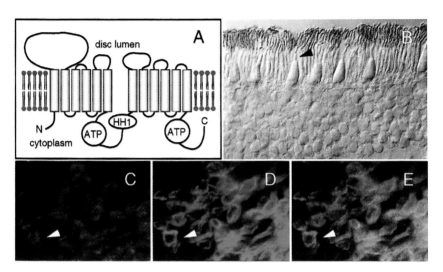

Fig. 1. Structure and localization of ABCR. (A) Schematic diagram of ABCR in the ROS disk membrane. ATP, ATP-binding cassettes; HH1, hydrophobic segment found in ABCR and closely related ABC transporters; N, amino terminus; C, carboxy terminus. (B) Immunoperoxidase labeling of ABCR in the macaque retina. Arrowhead indicates one of the cone outer segments. The large cone inner segments are clearly evident among the slender and more numerous rod inner segments. (C–E) Confocal images (0.4-μm optical sections) of bovine retina cut through the outer segment region parallel to the plane of the retina and double stained for ABCR (Texas Red; C) and the cGMP-gated channel (fluorescein; D), a plasma membrane protein. (E) The merged images. ABCR immunoreactivity is present in the interior of the outer segment and is most intense just inside the plasma membrane.

from the ABCR preparation was higher if solubilization and purification were performed with CHAPS than with either octylglucoside or dodecylmaltoside.

In brief, the ABCR purification involves (1) purifying bovine ROS on a sucrose step gradient, (2) solubilizing the ROS in ROS/CHAPS buffer, (3) binding ABCR to MAb Rim 3F4-conjugated Sepharose, and (4) eluting ABCR with a synthetic peptide corresponding to the Rim 3F4 epitope. All but the last of these steps are carried out under dim red light to minimize denaturation and nonspecific trapping of rhodopsin.

Figure 2A shows the results of the purification procedure. The specificity of the immunoaffinity purification step was assessed by halving a sample of Rim 3F4-Sepharose to which ABCR had been bound and carrying out parallel elutions with either the Rim 3F4 peptide or an irrelevant peptide, JN50. Coomassie staining and Western blotting shows that the ABCR polypeptide, with an apparent molecular mass of approximately 250 kDa, is eluted with the Rim 3F4 peptide but is barely detectable following elution with the JN50 peptide. The extensive washing of the column-bound material should remove all low molecular weight compounds initially present in the ROS membrane sample.

Reconstitution of ABCR into brain lipid membranes is accomplished by gently mixing immunoaffinity purified ABCR in ROS/CHAPS buffer with an excess of sonicated porcine brain polar lipids (referred to hereafter as "brain lipid"). Efficient removal of CHAPS with a high yield of ABCR ATPase activity is most effectively achieved by absorption to Extracti-Gel resin (Pierce). This method of detergent removal is rapid and efficient, and results in minimal losses of ABCR, most likely because the proteolipid complexes are effectively excluded by the low molecular weight pore size of the Extracti-Gel resin. In developing this protocol we have avoided ultracentrifugation of the reconstituted ABCR proteolipid. This circumvents problems associated with inefficient recovery of membranes in the presence of 10% glycerol, used here as a protein stabilizer, and with dispersing a compact proteolipid pellet. Several other methods for detergent removal and proteolipid formation were found to be less effective for reconstitution of ABCR: rapid dilution followed by ultracentrifugation[21] resulted in inefficient removal of detergent and poor recovery of proteoliosomes; gel filtration followed by ultracentrifugation[22,23] resulted in loss of ABCR ATPase activity; and detergent removal by binding to Bio-beads (BioRad,

---

[21] S. V. Ambudkar and P. C. Maloney, *J. Biol. Chem.* **261,** 10079 (1986).
[22] A. Shapiro and V. Ling, *J. Biol. Chem.* **269,** 3745 (1994).
[23] S. Dunn and R. P. Thuynsma, *Biochem.* **33,** 757 (1994).

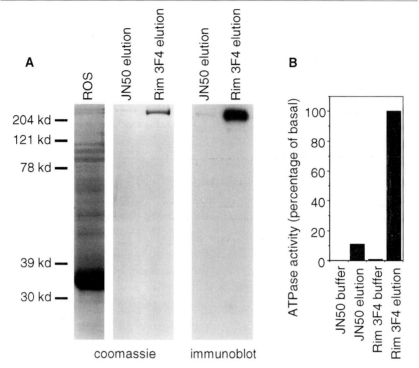

FIG. 2. Purification of ABCR and identification of an associated ATPase activity. (A) Proteins were resolved by SDS–PAGE (10% acrylamide) and stained with Coomassie Blue (left) or immunoblotted and probed with affinity purified rabbit anti-ABCR antibodies (right). The ROS lane shows the starting material for immunoaffinity purification. Proteins eluted from the Rim 3F4 affinity column by the JN50 or Rim 3F4 peptides are shown in the right pair of panels. ABCR migrates in SDS–PAGE with an apparent molecular mass of approximately 250 kDa. (B) ATPase assays in ROS/CHAPS buffer were performed with the material eluted by the JN50 or Rim 3F4 peptides or with the pure peptides in ROS/CHAPS buffer as controls. The basal activity level refers to the ATPase activity of material eluted with Rim 3F4.

Hercules, CA) followed by ultracentrifugation[24] efficiently removed detergent but was accompanied by significant adsorption of ABCR protein to the beads.

Purified ABCR solubilized in ROS/CHAPS buffer or reconstituted in brain lipid posesses considerable ATPase activity (Fig. 2B; and see Fig. 5 later). This activity appears to derive from ABCR because no other polypeptides are detected in this preparation by Coomassie staining and

[24] P. von Dippe and D. Levy, *J. Biol. Chem.* **265,** 14812 (1990).

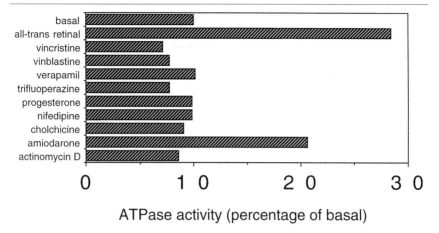

FIG. 3. Effect of 20 μM all-*trans*-retinal or each of nine P-glycoprotein substrates or sensitizers on the ATPase activity of purified and reconstituted ABCR. ATPase activity is shown as a percent of basal activity, the level observed in the absence of added compounds.

because the ATPase activity is far lower in the sample eluted with the irrelevant peptide JN50.

Stimulation of ABCR ATPase Activity by Low Molecular Weight Compounds

As one approach to identifying substrates or allosteric regulators of ABCR, various structurally diverse compounds were tested for the ability to stimulate or inhibit the ATPase activity of purified and reconstituted ABCR. Figure 3 shows the results of an initial test with all-*trans*-retinal, a candidate substrate, or nine P-glycoprotein substrates. The compounds were tested at 20 μM, a concentration at which many P-glycoprotein substrates and sensitizers enhance ATP hydrolysis by P-glycoprotein.[22,25–27] Most compounds had little or no effect on ATPase activity, and none of the 43 tested thus far significantly inhibit ATPase activity. However, 7 of the 43 compounds—three retinal isomers (all-*trans*-retinal, 11-*cis*-retinal, and 13-*cis*-retinal), amiodarone, digitonin, dehydroabietylamine, and 2-*tert*-butylanthroquinone—reproducibly activate the ATPase activity two- to

---

[25] S. V. Ambudkar, I. H. Lelong, J. P. Zhang, C. O. Carderelli, M. M. Gottesman, and I. Pastan, *Proc. Natl. Acad. Sci. U.S.A.* **89**, 8472 (1992).

[26] B. Sarkadi, E. M. Price, R. C. Boucher, U. A. Germann, and G. A. Scarborough, *J. Biol. Chem.* **267**, 4854 (1992).

[27] I. L. Urbatsch, M. K. Al-Shawi, and A. E. Senior, *Biochemistry* **33**, 7069 (1994).

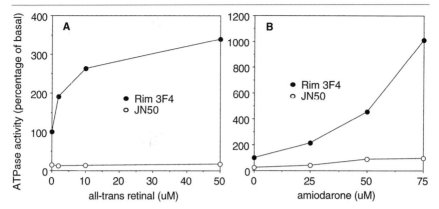

FIG. 4. Concentration dependence of (A) all-*trans*-retinal and (B) amiodarone in stimulating the ATPase activity of purified and reconstituted ABCR. To assess the specificity of the ATPase reaction parallel elution and reconstitution procedures were performed using either the Rim 3F4 peptide or the JN50 peptide as indicated.

fivefold (the fold activation depending on the compound), a degree of activation similar to that reported for purified and reconstituted P-glycoprotein in the presence of the most potent P-glycoprotein substrates and sensitizers.[22,27,28]

The specificity of ATPase stimulation can be assessed by comparing preparations that had been identically reconstituted following elution from the Rim 3F4 immunoaffinity column with either the Rim 3F4 peptide or the irrelevant peptide JN50 (Fig. 4). In each case, ATPase stimulation is observed only in the sample that was eluted with the Rim 3F4 peptide. The concentration dependence of ATPase activation by each of the stimulatory compounds has a characteristic shape: all-*trans*-retinal produces a dose-dependent ATPase activation that is half maximal at 10–15 μM and shows a simple Michaelis–Menten dose dependence (Fig. 4); amiodarone activates the ATPase beginning at approximately 10–20 μM and shows no evidence of saturation up to 50 μM (Fig. 4). The dose–response curves suggest a mechanistic distinction between all-*trans*-retinal, which appears to act at a single site or class of sites without cooperativity, and amiodarone, which appears to act through a cooperative mechanism that involves two or more sites.

[28] J. F. Rebbeor and A. E. Senior, *Biochim. Biophys. Acta* **1369**, 85 (1998).

Factors Affecting the All-*Trans*-Retinal-Dependent Stimulation of ATP Hydrolysis

The ATPase activity of purified and reconstituted ABCR in the presence or absence of all-*trans*-retinal exhibits a broad pH optimum and the reaction continues linearly for at least 180 min at 37°. The basal and all-*trans*-retinal-stimulated ATPase activities of ABCR are sensitive to lipid environment, a property shared with P-glycoprotein.[29] Crude sheep brain PE, brain polar lipid, and *E. coli* polar lipid support all-*trans*-retinal-stimulated ATPase, but egg PC and soybean PC do not. In contrast to the reconstituted sample, purified ABCR solubilized in ROS/CHAPS buffer shows no stimulation by all-*trans*-retinal, suggesting that the precise arrangement of protein-lipid contacts is a critical factor in the ability of retinal to activate the ATPase activity. The lack of all-*trans*-retinal stimulation may be related to the high basal V_{max} observed for ABCR in a mixed CHAPS/lipid environment relative to that in lipid. A similar difference has been observed between detergent-solubilized and reconstituted P-glycoprotein in ATPase stimulation by both substrates and sensitizers.[22]

An analysis of the ATPase activity of ABCR in the presence of varying ATP concentration shows simple Michaelis–Menten behavior for ATP regardless of whether ABCR is solubilized in ROS/CHAPS buffer (Figs. 5A and 5C) or reconstituted in brain lipid (Figs. 5B and 5D). In ROS/CHAPS buffer, the K_m for ATP is 278 μM and the V_{max} is 27 nmol ATP/min/mg ABCR, assuming that all of the ABCR is functional. For comparison, highly purified P-glycoprotein in detergent has a K_m for ATP of 940 μM and a V_{max} of 321 nmol ATP/min/mg.[22] Reconstituted ABCR in the absence or presence of 80 μM all-*trans*-retinal shows a K_m for ATP of 33 or 725 μM and a V_{max} of 1.3 or 29 nmol ATP/min/mg ABCR, respectively. At the 1–3 mM ATP concentration present in the outer segment, the reaction velocity will be close to V_{max}. In these calculations we assume that all of the ABCR present in the reaction (quantitated by Western blotting) is functionally reconstituted and has equivalent access to ATP. Because these assumptions are unlikely to apply, the calculated V_{max} represents a lower limit of the true V_{max}, a consideration that makes any quantitative comparison between solubilized and reconstituted preparations problematic.

In the Lineweaver–Burk plot shown in Fig. 5D the parallel shift of the curve of 1/V versus 1/[ATP] upon addition of all-*trans*-retinal indicates that all-*trans*-retinal acts as an "uncompetitive activator" analogous in effect

---

[29] I. L. Urbatsch and A. E. Senior, *Arch. Biochem. Biophys.* **316**, 135 (1995).

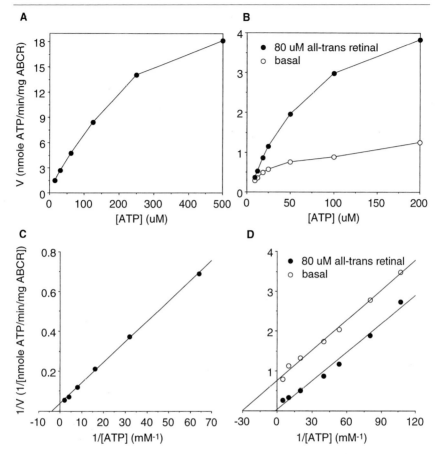

FIG. 5. K_m for ATP and V_{max} for ATP hydrolysis determined for purified ABCR in ROS/CHAPS buffer or following reconstitution in the presence or absence of all-*trans*-retinal. (A, C) The dependence of ATP hydrolysis on ATP concentration by purified ABCR in ROS/CHAPS buffer shows simple Michaelis–Menten behavior. (B, D) The dependence of ATP hydrolysis on ATP concentration by purified and reconstituted ABCR in the absence or presence of 80 μM all-*trans*-retinal. Panels (C) and (D) show the Lineweaver–Burk plots derived from the data shown in panels (A) and (B), respectively.

but opposite in sign to the action of a classical uncompetitive enzyme inhibitor.[30,31] The simplest interpretation of this kinetic data is that all-*trans*-retinal binds to and alters the ABCR–ATP intermediate at the rate-limiting step in the ATP hydrolytic pathway.

[30] A. Lehninger, "Biochemistry," p 183. Worth Publishers, New York, 1975.
[31] I. H. Segel, "Biochemical Calculations," p. 246. Wiley, New York, 1976.

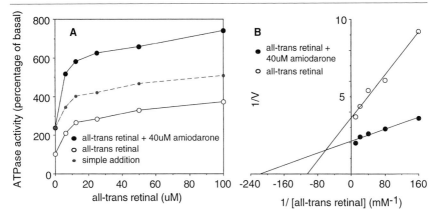

FIG. 6. (A) Dose–response curves for all-*trans*-retinal stimulation of ATP hydrolysis with or without 40 μM amiodarone. (B) Lineweaver–Burk plot of the data from (A) showing that, in the presence of 50 μM ATP, 40 μM amiodarone increases the V_{max} for ATP by 1.7-fold and decreases the $K_{apparent}$ for all-*trans*-retinal from 9.6 to 4.6 μM. The vertical axis is shown with arbitrary units because the efficiency of reconstitution and the fraction of ABCR proteins oriented with the ATPase sites facing the extravesicular space are unknown. For this analysis, the value of the basal ATPase activity has been subtracted from the values in the presence of all-*trans*-retinal, and the ATPase activity referable to amiodarone alone has been subtracted from the values in the presence of all-*trans*-retinal plus amiodarone.

Synergistic Activation of ABCR by All-*Trans*-Retinal and Amiodarone

Further evidence that all-*trans*-retinal interacts with ABCR in a manner that differs from the interactions of nonretinoid compounds with ABCR comes from the observation that ABCR ATPase is activated in a greater than additive manner by mixtures of all-*trans*-retinal and the stimulatory nonretinoid compounds. For example, Fig. 6A compares the degree of activation in the presence of varying concentrations of all-*trans*-retinal at a fixed concentration of amiodarone versus that calculated for simple additivity. The effect of amiodarone on the apparent affinity for all-*trans*-retinal is shown in Fig. 6B. The Lineweaver–Burk plot of the relationship between the rate of ATP hydrolysis at 50 μM ATP and the concentration of all-*trans*-retinal shows that a 1.7-fold increase in V_{max} produced by the addition of 40 μM amiodarone is associated with a twofold increase in the apparent affinity, a decrease in $K_{apparent}$ from 9.6 to 4.6 μM of all-*trans*-retinal. (We refer to this value as an apparent affinity or $K_{apparent}$ rather than a true K_m because the binding and release of all-*trans*-retinal may be complex and the reaction was not performed at saturating ATP concentration.)

As revealed by a Lineweaver–Burk analysis, amiodarone and all-*trans*-retinal alter the K_m of ATP and the V_{max} for ATP hydrolysis via distinct mechanisms (Fig. 7). The principal effect of amiodarone (at 40 μM) is to increase V_{max} with only a modest change in K_m for ATP, as measured from the intercept of the best fitting straight line and the horizontal axis, which corresponds to $-1/K_m$. Amiodarone addition in the absence of all-*trans*-retinal lowers the K_m for ATP by approximately two-fold (Fig. 7B), whereas amiodarone addition in the presence of 100 μM all-*trans*-retinal raises the K_m for ATP approximately two-fold (Fig. 7C). Importantly, these changes are accompanied by a two fold decrease in the ratio K_m/V_{max} (the slope of the best fitting straight line), an effect that is analogous, but opposite in sign, to that of classical noncompetitive enzyme inhibitors.[30,31] Amiodarone can therefore be considered a "noncompetitive activator." The simplest model to account for the kinetic behavior of amiodarone is one in which ABCR is assumed to exist in an equilibrium mixture of enzymatically active and inactive species. By analogy with the diminution in the number of active enzymes resulting from the action of a noncompetitive inhibitor, amiodarone binding appears to shift the equilibrium between active and inactive enzymes to favor the active species.

In contrast to amiodarone, the effect of all-*trans*-retinal, either alone or in combination with amiodarone, is to increase the V_{max} for ATP hydrolysis and the K_m for ATP by the same factor, thereby leaving the ratio K_m/V_{max} (the slope of the best fitting straight line) unchanged (Fig. 7D). This effect on kinetic parameters is analogous but opposite in sign to that displayed by classical uncompetitive enzyme inhibitors,[30,31] and all-*trans*-retinal can therefore be considered an "uncompetitive activator." The simplest model that accounts for this kinetic behavior is one in which all-*trans*-retinal interacts specifically with an intermediate in the ATP hydrolysis pathway. By analogy with the specific trapping of enzyme-substrate intermediates produced by the action of a classical uncompetitive enzyme inhibitor, the binding of all-*trans*-retinal creates an accelerated pathway for ATP hydrolysis. The data can also be formalized along the lines of a double-displacement ("ping-pong") kinetic scheme in which the progress of one substrate, ATP, through the reaction pathway is accelerated by the binding of the second component, all-*trans*-retinal, at an intermediate point in the pathway.[30,31]

The two different modes of interaction between ABCR and amiodarone or all-*trans*-retinal predict that their combined effects will by multiplicative rather than additive, and therefore accounts in a simple way for their ability to act simultaneously and synergistically on ABCR. The simple Michaelis–Menten behavior and uncompetitive mode of ATPase activation exhibited by all-*trans*-retinal strongly suggest that this compound is a transport substrate. By contrast, the cooperative behavior and the noncompeti-

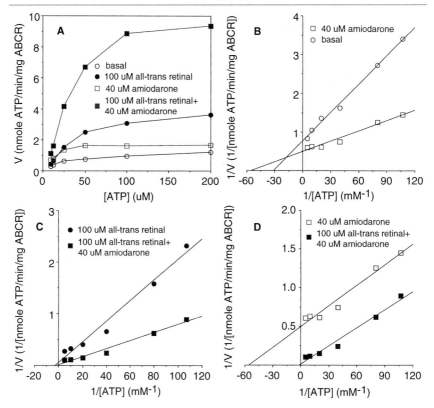

FIG. 7. The effect of all-*trans*-retinal, amiodarone, or a combination of the two on the K_m for ATP and V_{max} of ATP hydrolysis for purified and reconstituted ABCR. (A) ATP hydrolysis at different ATP concentrations. In these calculations of reaction velocity we assume that all of the ABCR present in the reaction is functional and has access to ATP. (B–D) Lineweaver–Burk plots comparing the effect of (B) no addition versus amiodarone, (C) all-*trans*-retinal versus all-*trans*-retinal plus amiodarone, and (D) amiodarone versus amiodarone plus all-*trans*-retinal. Concentrations of amiodarone and all-*trans*-retinal are indicated in each panel. In panels (A–C), it is apparent that amiodarone addition produces a shallower best-fitting line to the data points, i.e., it lowers the slope (K_m/V_{max}), whereas all-*trans*-retinal addition produces a downward shift of the best-fitting line, i.e., the slope (K_m/V_{max}) is unchanged. Comparison of panel (B) with (C) and (D) indicates that amiodarone and all-*trans*-retinal act independently. Based only on the intercepts of the fitted lines, the K_m for ATP and the V_{max} for ATP hydrolysis in the presence of the indicated compounds are as follows: 33 μM and 1.3 nmol ATP/min/mg (no addition), 18 μM and 2 nmol ATP/min/mg (amiodarone), 400 μM and 20 nmol ATP/min/mg (all-*trans*-retinal), 666 μM and 50 nmol ATP/min/mg (all-*trans*-retinal plus amiodarone). In this experiment and the one shown in Fig. 5D, the K_m and V_{max} values should be considered less reliable when obtained only from the intercepts of those fitted lines that lie close to the origin. More reliable values are obtained from the intercepts of the fitted lines that lie further from the origin and from the slope (K_m/V_{max}) of the fitted lines.

tive mode of ATPase activation exhibited by amiodarone suggest that this compound is an allosteric effector.

Discussion

Retinal Stimulation of ABCR ATPase

The finding that all-*trans*-retinal (and other geometric isomers of retinal) stimulates the ATPase activity of ABCR can be explained by any of three mechanisms: (1) all-*trans*-retinal exerts a general effect on the lipid environment that alters the conformation of ABCR; (b) all-*trans*-retinal binds to ABCR as an allosteric effector of the ATPase cycle but is not itself a substrate for transport; and (3) all-*trans*-retinal is a substrate for transport and its presence accelerates ATP hydrolysis by a conformational coupling between the transport domain and one or both ATP-binding domains.

The first mechanism is unlikely for three reasons. First, it does not account for the specificity of ATPase stimulation by all-*trans*-retinal relative to the other compounds tested; second, it seems implausible that retinal should exert an effect on bulk membrane properties in a reaction in which lipid is present at 1–1.5 mg/ml and retinal is present at 10 μM (equivalent to 3.3 μg/ml); and third, a quantitative analysis of the concentration dependence of all-*trans*-retinal stimulation indicates that it acts with simple Michaelis–Menten behavior.

The second mechanism, if correct, would imply a novel strategy for modulating the activity of a ROS protein, i.e., sensing the concentration of free retinal or a related retinoid. Under this assumption, ABCR activity might be modulated by the level of all-*trans*-retinal released by photactivated rhodopsin or of free 11-*cis*-retinal imported from the RPE for rhodopsin regeneration. Arguing against a simple allosteric mechanism is the effect of all-*trans*-retinal on the K_m for ATP and the V_{max} for ATP hydrolysis, which indicates that all-*trans*-retinal acts via an uncompetitive interaction with ABCR.

The third possible mechanism, that various geometric isomers of retinal and/or other retinoids are transported by ABCR in a reaction that is coupled to ATP hydrolysis, can only be definitively distinguished from an allosteric mechanism by directly demonstrating ATP-dependent vectorial transport of these compounds in a reconstituted membrane system. At present, the strongest evidence in favor of this hypothesis is the observation that all-*trans*-retinal acts without cooperativity at a single class of binding site(s) on ABCR and exhibits uncompetitive stimulation of ATPase activity.

This kinetic behavior implies that all-*trans*-retinal binds to an intermediate in the ATPase reaction pathway and that this binding accelerates a rate-limiting step in ATP hydrolysis and/or release of the hydrolysis products. In the discussion that follows, we assume that ABCR plays a role in retinoid transport in the ROS.

A Model for ABCR in the Visual Cycle

If ABCR transports one or more retinoids, it could act by flipping PE–all-*trans*-retinal adducts from the lumenal to the cytosolic face of the disk membrane, move free all-*trans*-retinal from the lipid phase of the disk membrane to a juxta-membrane location, or possibly reorient all-*trans*-retinal in the bilayer. In the first case, ABCR activity would resemble the PC flippase activity of mouse mdr-2 (Ref. 32) and human MDR3 (Ref. 33), whereas in the latter cases, ABCR activity would resemble the drug extrusion activity of P-glycoprotein, especially as envisioned by the "hydrophobic vacuum cleaner model" in which P-glycoprotein is proposed to extract hydrophobic compounds from the lipid phase and deliver them to the extracellular space.[34] In each case, the effect of ABCR activity would be to more efficiently deliver all-*trans*-retinal to all-*trans*-retinol dehydrogenase, leading to lower photoreceptor noise, more rapid recovery following illumination, and decreased accumulation of all-*trans*-retinal or its adducts within the disk membrane (Fig. 8A). In these models of ABCR action, the vectorial movement of all-*trans*-retinal is topologically equivalent to importing a substrate from the extracellular space, a direction of transport opposite to the direction of lipid flippase activity demonstrated for mouse mdr2 (Ref. 32) and the human MDR1 and MDR3 P-glycoproteins[33] and opposite to the direction of drug extrusion mediated by P-glycoprotein. Substrate import by ABC transporters is commonly observed in bacteria, but has thus far not been reported in multicellular organisms.[35] The stimulation of ATPase activity by 11-*cis*-retinal and to a lesser extent by all-*trans*-retinol suggests that ABCR may also play a role in the movement of these compounds.

---

[32] S. Ruetz and P. Gros, *Cell* **77**, 1071 (1994).
[33] A. van Helvoort, A. J. Smith, H. Sprong, I. Fritzsche, A. H. Schinkel, P. Borst, and G. van Meer, *Cell* **87**, 507 (1996).
[34] Y. Raviv, H. B. Pollard, E. P. Bruggemann, I. Pastan, and M. M. Gottesman, *J. Biol. Chem.* **265**, 3975 (1990).
[35] C. F. Higgins, *Ann. Rev. Cell Biol.* **8**, 67 (1992).

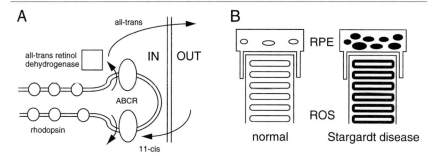

Fig. 8. Possible models for ABCR action in the visual cycle and for the pathogenesis of Stargardt disease. (A) Model of the ROS showing the edge of one outer segment disk and the adjacent plasma membrane. A hypothetical path of all-*trans*-retinal and all-*trans*-retinol movement following rhodopsin photobleaching is indicated by arrows in the upper half of the figure. In this model all-*trans*-retinal is released into the disk membrane and is transported or presented to all-*trans*-retinol dehydrogenase on the cytosolic face of the disk by ABCR. ABCR may also facilitate the export of all-*trans*-retinol by extracting it from the disk membrane as postulated for the action of P-glycoprotein and its hydrophobic substrates by the "hydrophobic vacuum cleaner" model.[34] Ultimately, all-*trans*-retinol is exported across the plasma membrane to the adjacent RPE. A hypothetical path of 11-*cis*-retinal movement is indicated by arrows in the lower half of panel (A). If 11-*cis*-retinal partitions into the disk membrane, ABCR may facilitate its removal from the lumen of the disk and/or its binding to opsin to regenerate rhodopsin. Whether retinal is released from or attached to opsin with a specific orientation or location relative to the membrane is currently unknown. (B) Model for the accumulation of short-wave absorbing material in Stargardt disease. The distal region of a ROS and the adjacent RPE cell is shown. Left, in the normal retina there is minimal accumulation of retinoid derivatives in the disk membrane and therefore only a slow accumulation of these or related derivatives in the RPE. Right, in the Stargardt disease retina, retinoids accumulate in the disk membrane (represented by a heavy outline of the disk membrane), and these or related derivatives accumulate in the RPE in phagolysosomes containing ingested ROS (represented by filled cytoplasmic inclusions).

Implications for the Pathogenesis of Stargardt Disease and Other Retinopathies

The model outlined here suggests that Stargardt disease and other retinopathies caused by mutations in ABCR arise from defects in retinoid metabolism (Fig. 8B). This idea was originally proposed as an explanation for the delayed recovery of rod sensitivity observed in Stargardt disease patients.[36] The persistence of this delay in rod recovery despite high levels of dietary vitamin A[37] distinguishes it from classical night blindness of dietary origin,[38] and more specifically suggests a model in which the defect

[36] G. A. Fishman, J. S. Farbman, and K. R. Alexander, *Ophthalmol.* **98,** 957 (1991).
[37] A. M. Glenn, G. A. Fishman, L. D. Gilbert, and D. J. Derlacki, *Retina* **14,** 27 (1994).
[38] J. E. Dowling and G. Wald, *Proc. Natl. Acad. Sci. U.S.A.* **44,** 648 (1958).

in Stargardt disease is not related to the availability of 11-*cis*-retinal, but rather to a defect in removing the product of photobleaching, all-*trans*-retinal.

The model outlined recently may also be relevant to recent structural analyses of the fluorescent, short-wave absorbing compounds that accumulate with age in the human RPE. These analyses show that the major component is a diretinal-ethanolamine derivative, A2-E.[39,40] A2-E has been hypothesized to form from the condensation of all-*trans*-retinal and PE in the outer segment and then to accumulate over time in the RPE as a residual product of ROS digestion.[39] Although it is not known definitively whether A2-E comprises the fluorescent, short-wave absorbing material that accumulates to high levels in the RPE of Stargardt disease patients, one *in vivo* study suggests that this compound or a similar one is present at high concentration in these patients.[41,42]

---

[39] G. E. Eldred and M. R. Lasky, *Nature* **361,** 724 (1993).
[40] N. Sakai, J. Decataur, K. Nakanishi, and G. E. Eldred, *J. Am. Chem. Soc.* **118,** 1559 (1996).
[41] F. C. Delori, C. K. Dorey, G. Staurenghi, O. Arend, D. G. Goger, and J. J. Weiter, *Invest. Ophthalmol. Vis. Sci.* **36,** 718 (1995).
[42] F. C. Delori, G. Staurenghi, O. Arend, C. K. Dorey, D. G. Goger, and J. J. Weiter, *Invest. Ophthalmol. Vis. Sci.* **36,** 2327 (1995).

Author Index

Numbers in parentheses are footnote reference numbers and indicate that an author's work is referred to although the name is not cited in the text.

A

Abbott, N. L., 487
Abdulaev, N. G., 3, 5(8), 11(8), 13, 59, 60, 92, 269, 273(15)
Abdulaeva, G., 447
Abitbol, M., 880
Achilles, A., 833, 834(16), 843(16)
Adam, G. W., 483
Adams, M., 227
Adams, S. R., 802
Adamus, G., 56, 57, 63, 71, 75, 414, 453
Adelman, J. P., 828
Adelstein, R. S., 571
Adesnik, M., 78
Adler, M., 114
Adman, E. T., 453
Aebersold, R., 763
Aebi, U., 93, 99(31), 456
Afendis, S., 466
Agarwal, N., 3, 10(2), 11, 13, 30, 44, 57(2, 26), 272, 275(32), 278(32), 291
Agre, P., 93
Aguirre, G., 617
Ahamed, B., 755, 783, 784(27), 785, 786(8), 818, 819, 820(25), 821(25), 823(25), 824(25), 827(25), 828(25), 831, 833(8)
Ahn, J., 864
Aho, A. C., 159
Akeson, A., 389
Akhtar, M., 197, 557, 557(17), 558, 558(21, 23), 559, 561(21, 23), 562(21), 564(23), 567(21, 23), 568, 568(23), 570(23)
Akino, T., 349, 350, 350(11)
Akioka, H., 466
Akita, H., 117, 147, 158(19), 162(19), 219, 227
Akong, M., 3
Alb, J. G., Jr., 79

Albeck, A., 163, 171, 200, 203
Albert, A. D., 29, 107, 111, 113(16), 115(15)
Alderfer, J. L., 29, 111, 113(16), 115(15)
Alemany, S., 564, 566(25)
Alexander, K. R., 896
Alfano, R. R., 144
Allen, L. F., 258
Allen, N. E., 45
Allikmets, R., 864, 876(5), 879, 880, 880(2), 884(2)
al-Obeidi, F., 389
Alpern, M., 190
Al-Shawi, M. K., 887, 888(27)
Altenbach, C., 104, 106, 112(30, 31), 113, 115, 115(30, 31), 122, 179, 269, 287, 290(14), 292(46), 385
Amador, M., 775
Ambudkar, S. V., 885, 887
Amersham Pharmacia Biotech, 460, 466(18)
Ames, J. B., 405
Amons, R., 389
Amos, L. A., 93, 103(36)
Andersen, W. B., 679
Anderson, K., 364, 517
Anderson, K. L., 864, 876(5), 879, 880(2), 884(2)
Anderson, S., 441
Andersson, L., 73
Andjelkovic, N., 558, 558(23), 559, 561(23), 564(23), 567(23), 568(23), 570(23)
Andolfi, G., 580, 593(12)
Andrews, H. N., 106
Angleson, J. K., 232, 238, 525, 526, 526(8), 530(8, 19), 531(8), 532(8), 533(8), 535(19)
Ankoudinova, I., 405, 676, 690, 691(14), 701(14), 709, 710(15), 711(15), 713(15)
Antoch, M. P., 525, 526(7), 530, 531(7), 535

Antoniw, J. F., 572
Antonny, B., 512, 517, 528
Aoyama, H., 92
Aparicio, J. G., 673, 675(7), 676, 677(7, 37), 679(7), 680(7), 683(7), 684(37), 685(37), 689(37)
Apel, E. D., 600
Apparsundaram, S., 258
Appella, E., 570(7), 571
Applebury, M. L., 139, 213, 377(37), 384(37), 602, 617, 652, 673, 675(7), 676, 677(7), 679, 679(7), 680(7), 683(7), 685(37), 689(37), 833, 868
Appleman, M. M., 641
Archavsky, V. Y., 524, 525, 526(7), 531(7)
Arend, O., 897
Arendt, A., 56, 60, 63, 75, 79, 88(27), 108, 111, 115(15), 254, 363, 364(8), 376(8), 388, 414, 437, 447, 453, 490, 717
Argos, P., 295, 316(40), 324
Arnis, S., 194, 269, 287(12), 292(12), 377, 380, 385, 479, 480(26)
Arshavsky, V. Y., 75, 512, 525, 526, 526(9), 527, 527(9), 529(18), 530, 531, 531(24), 532(24), 534, 535, 535(17, 18, 20), 600, 601, 638, 646, 647, 651, 651(15), 652(13, 15, 25), 656(15, 25), 665(15), 667(55), 668, 672(15), 743
Artamonov, I. D., 296, 297(9), 309, 349
Artemyev, N. O., 517, 518, 526, 529, 529(18), 534, 535(18), 539, 540, 546(8), 548(7), 549, 549(7), 551, 551(7), 552, 552(7), 554(7), 598, 635, 636, 637(10), 638, 638(8), 639(8), 641(5), 642(10, 15), 643(4, 5, 8, 15), 644(10), 645, 645(9), 648, 652(16a), 663
Asato, A. E., 298
Asenjo, A. B., 117, 204, 322
Asseon-Batres, M. A., 238, 239(8)
Asson-Batres, M. A., 252, 264(5)
Aton, B., 198
Auer, M., 93
Augustine, G. J., 798, 800(11)
Ausubel, F. M., 443, 608, 609(23)
Axel, R., 619
Aylward, J. H., 589
Ayoso, C., 880
Ayuso, C., 864
Azarian, S. M., 864
Azuma, M., 50

B

Baasov, T., 200
Baccanari, D., 389
Badalov, P. R., 3, 11(6)
Bae, H., 364, 491, 503, 504(4), 511(4), 513, 513(4), 514(4), 517, 517(4), 520(5), 539, 540(6), 546(6), 549(6)
Baeger, I., 804
Baehr, W., 139, 213, 405, 600, 602, 609, 617, 652, 676, 690, 691, 691(11, 13), 693(13), 699(13), 701(13), 702(11, 13, 21), 705(13), 707(13), 709, 710(17, 18), 712, 713(16), 716, 716(16, 26), 717, 718, 719, 730, 734(11, 12), 831
Baggetto, L. G., 875
Bahia, D., 388
Bailey, J. E., 20
Bain, C. D., 483
Baird, I., 879, 880(2), 884(2)
Baird, L., 864, 876(5)
Balashov, S., 200, 201(25), 202, 204, 206
Balashov, S. P., 147
Balasubramanian, P., 389
Balasubramanian, S., 785, 786(7)
Balcells, S., 617, 864, 880
Balch, W. E., 77, 83(6)
Baldwin, D., 389
Baldwin, J. M., 69, 91, 93, 97, 103(20), 104(2, 8), 112, 118, 126, 128(38), 129(17), 179, 185(1, 3), 268
Ballabio, A., 580, 593(12)
Ballesteros, J. A., 112
Balogh-Nair, V., 117, 147, 158(19), 162(19), 227
Banin, E., 79
Bankaitis, V. A., 78, 79
Baraban, J. M., 656
Baraniak, J., 693
Barbas, C. R., 399, 402(28)
Barboy, N., 158
Barford, D., 571
Barkdoll, A. E. I., 657
Barlow, H. B., 159
Barlow, R. B., 159, 160, 161(60), 162(54, 60), 163, 199
Barlow, R. B., Jr., 159
Barrett, R., 389
Barry, B., 198
Barry, G. F., 16, 584

Barry, P. H., 785, 786(7)
Bartucci, F., 647, 652(14)
Barzu, O., 827
Basilico, C., 570(6), 571
Basinger, S., 649
Bastians, H., 580
Basu, D. K., 848, 852, 852(18), 853, 853(17)
Baubichon-Cortay, H., 875
Baudier, J., 733, 741(17)
Bauer, P. J., 786, 818, 833, 837, 843
Baumann, A., 571, 573(13), 574, 574(13), 578(13)
Baumeister, W., 93, 99(31), 466
Bax, A., 110, 819
Baxter, L. C., 617
Bayerl, T. M., 483
Bayes, M., 617
Bayley, H., 59
Baylor, D. A., 159, 199, 268, 270, 281(25), 285, 286(25), 294, 525, 526(9), 527(9), 601, 647, 693, 742, 755, 774, 780, 803, 810, 817, 831
Bayuga, S. J., 597
Bazan, N. G., 77, 78(3)
Beach, J. M., 307, 377
Beam, K. G., 810
Bear, C. E., 869
Beauve, A., 674
Beavo, J. A., 587, 597, 600, 601, 605, 606(17), 608, 608(17), 609(8, 22), 610, 647, 648, 652, 653, 659, 660(41), 663, 663(41), 672, 698
Bechtel, P. J., 587
Beck, M., 123, 181, 182(20, 21), 183(20), 184, 185(20), 189, 206
Becker, W., 580
Beckmann, E., 93
Beck-Sickinger, A. G., 12
Behbehani, M., 199
Beier, H., 599(33), 615
Bell, S., 388
Bellosta, P., 570(6), 571
Belsham, G. J., 16
Benckhuijsen, W., 389
Bendig, J., 804
Benian, G. M., 674, 675(25)
Bennet, N., 307
Bennett, J. P., Jr., 661
Bennett, M. K., 78

Bennett, N., 270, 329, 333(6), 343, 377, 528, 531, 567, 646
Benovic, J. L., 411, 419, 421, 422, 423, 424, 424(4, 5), 427(4), 429(5), 433, 433(4, 5), 434(4, 19), 435, 435(5, 11), 436, 436(11), 437, 453, 455, 612
Bentley, J. K., 608, 609(22), 653
Berg, P., 37
Bergen, A. A., 864, 880
Berger, A. L., 636
Berghuis, A. M., 363, 372(2), 490, 503, 513
Bergsma, D. J., 91
Berkower, C., 59
Berlose, J.-P., 114
Berlot, C., 502
Berman, D. M., 520, 526, 531, 534, 552, 553(24)
Bernard, G. D., 217
Bernhardt, J. P., 823, 828(34)
Bernstein, M., 144
Bernstein, P., 864, 880
Berova, N., 219, 222, 226, 227(14), 229(14)
Berridge, G., 872
Bers, D. M., 732
Berson, E. L., 60, 79, 208, 617, 880
Bert, R. J., 755, 756, 756(8), 762(8), 763
Besharse, J. C., 57, 649
Beyermann, M., 786, 787, 787(10), 789(10), 792(10), 794(10), 796(10), 818, 825(20), 827(20), 837
Bezanilla, F., 278, 291(44), 292(44)
Bhattacharya, S. S., 417, 691, 712, 716(26)
Bhawasar, N., 111, 115(15)
Bibi, E., 59
Biebuyck, H. A., 487
Bienert, M., 787
Bieri, C., 244, 330, 342(14), 344(12), 377, 471, 487
Biernbaum, M. S., 649, 650(20)
Bifone, A., 158
Bigay, J., 346
Billeter, M., 111
Billingsley, M. L., 823
Bindig, J., 802
Binghan, E. L., 190
Birdsall, N. J. M., 663
Birge, R. R., 117, 143, 147, 148, 151, 151(26, 27), 152(27, 68), 153(27), 154(21, 27, 28), 155(21), 158, 158(19), 159, 160, 160(51),

161(60), 162(19, 51, 54, 60, 69), 163, 189, 199, 268, 269(4)
Bitensky, M. W., 350, 647, 650, 651(11, 21), 652(14, 21), 656
Blacharski, P. A., 879
Black, K. D., 777, 782, 783
Blackshear, P. J., 371
Blanco, F. J., 114, 827
Blaney, J., 617
Blankenagel, A., 864, 880
Blau, H. M., 39
Blin, N., 364
Blobel, G., 69, 456
Blosse, D., 197
Blosse, P. T., 557
Blum, H., 599(33), 615
Blume, A., 329
Blumenthal, D. K., 827
Blumer, K. J., 526, 531(23), 534(23)
Blundell, T. L., 114
Bohm, A., 363, 372(6), 376(6), 491
Boikov, S. G., 714
Boland, L. M., 273
Bollum, F. J., 419
Böme, E., 692, 693, 698(30)
Bond, C. T., 828
Bond, R. A., 258
Bondarenko, V. A., 647, 656
Bönigk, W., 676, 690, 691(15), 755, 762(4), 764(4), 831, 833(7)
Bonnermaison, M., 718
Bonnycastle, L., 399, 402(28)
Borgijin, J., 56
Borhan, B., 226, 227(14), 229(14)
Borst, P., 895
Boucher, F., 154
Boucher, R. C., 887
Bouhss, A., 827
Boulikas, T., 463
Bourne, H. R., 106, 112, 122, 179, 364, 388, 388(14, 15), 389, 502, 517, 539
Bovee, P. H. M., 194
Bovee-Geurts, P. H. M., 26, 27, 28, 28(24), 29(22, 24), 123, 193, 194, 196, 478
Bowes, C., 617
Bowlin, T., 389
Bowmaker, J. K., 29
Bownds, D., 70, 197, 557, 868
Bownds, M. D., 512, 526, 529(18), 535(17, 18), 600, 638, 646, 647, 649, 650, 650(20), 651,

651(15), 652(13, 15, 25), 656(15, 25), 665(15), 672(15), 743, 848
Boyce, J. J., 656
Boyle, W. J., 688
Boytos, C., 389
Bradbury, A., 440
Bradford, M. M., 39, 371, 460, 461(17), 628
Bradley, J., 775, 776(11)
Braell, W. A., 77, 83(6)
Braiman, M. S., 186
Brann, M. R., 258
Braslavsky, S. E., 144, 145, 145(4, 5, 8, 9), 147(12), 148, 148(14), 154(14, 23), 157
Braun, W. J., 111
Braunk, W., 111
Braunschweiler, L., 110
Bredberg, D. L., 238, 239(8), 252, 264(5)
Brendel, S., 350, 548, 551(18)
Brenna, J. T., 578
Brent, R., 443, 608, 609(23)
Breton, J., 118
Brett, M., 557
Brewis, N. D., 579, 580
Briggs, C. E., 880
Brigham-Burke, M., 792, 794(19)
Brolley, D., 404, 406
Bronson, J. D., 709, 710(17)
Brooks, I., 792, 794(19)
Brown, M. F., 483
Brown, N. G., 568
Brown, P. K., 8, 198, 294, 377, 868
Brown, R. L., 608, 636, 637, 643(7), 652, 755, 756, 756(8), 761, 762(8), 763, 765, 767, 767(14), 776, 780
Brownds, M. D., 527, 531(24), 532(24)
Bruckert, F., 346, 646
Brueggemann, L., 31, 268, 292
Bruel, C., 405
Bruggemann, E. P., 895, 896(34)
Brunger, A. T., 78
Brunissen, A., 114
Brunner, J., 756
Brunner, M., 77, 78(7)
Brunnock, M. A., 651, 652(23), 665(23), 669(23), 671(23), 672(23)
Brzeski, H., 879(5), 880
Brzovic, P. S., 405
Buck, L. B., 775, 776(12)
Buckholtz, R. G., 3
Buczylko, J., 70, 217, 238, 239(8), 240, 241,

243, 249(7), 250, 252, 264, 264(5), 334, 335, 339, 342(28), 437, 455, 557(18), 558, 691, 702(21), 709, 710(18), 717
Buda, F., 158
Bui, J. D., 142
Bünemann, M., 388, 404(12)
Bunt, M. A. H., 605
Burke, S. A., 650
Burnasheve, N., 798
Burnette, W. N., 44, 212
Burns, M. E., 525, 526(9), 527(9), 601, 647
Burrows, G. G., 130, 137(2), 142(2)
Burstein, E. S., 258, 482
Busman, M., 636, 645, 645(9)
Buss, V., 227
Butler, J., 389

C

Cafiso, D. S., 270
Cahill, G. M., 649
Calhoon, R. D., 160, 180
Callaghan, R., 872
Callender, R. H., 117, 144, 152(67, 68), 158, 161, 162(61), 163, 197, 198
Calvas, P., 718
Calvert, P. D., 525, 526(9), 527(9), 601, 647, 651, 652(25), 656(25), 667(55), 668, 743
Campbell, D. G., 570, 571
Camuzat, A., 718
Cao, H., 78
Cao, T., 60
Carderelli, C. O., 887
Careaga, C. L., 140
Caron, M. G., 59, 258, 259(20)
Carr, S. A., 241, 243
Carroll, K. S., 117
Caska, T. A., 93
Casnellie, J. E., 647, 651(11)
Cattaneo, A., 440
Catty, P., 346, 641, 646
Cavaggioni, A., 760
Cavanagh, H. D., 70
Cepko, C. L., 625
Cerione, R. A., 241, 256, 473, 482(7), 496, 528, 539, 636
Cervetto, L., 746, 848
Ceska, T. A., 99
Cha, K., 405

Chabre, M., 118, 329, 330, 333(6), 346, 473, 512, 518, 524, 525, 528, 530(1), 539, 540(3), 546(3), 635
Chader, G. J., 698
Chan, C., 78
Chan, T., 117
Chang, B., 685
Chang, C. H., 685
Chang, X., 865, 872(13)
Changeux, J.-P., 771
Charbonneau, H., 672
Chardin, P., 473, 518, 539, 540(3), 546(3)
Charlton, C. A., 39
Chase, A. M., 267
Chase, L. R., 679
Chassaing, G., 114
Chatelin, S., 718, 880
Chavko, M., 462
Chen, A. H., 223, 226(13)
Chen, C., 36, 272, 752
Chen, C.-K., 404, 405, 406(8), 409(8), 531, 534, 534(41)
Chen, D. M., 579
Chen, H.-B., 39, 322
Chen, N., 200
Chen, S., 625
Chen, T.-Y., 755, 775, 776(10), 783, 784(26), 785, 786(6, 8), 818, 819, 820(25), 821(25), 823(25), 824(25), 825(17), 827(25), 828(25), 831, 833(8), 837
Chen, W. J., 482
Chen, W. P., 482
Chen, Y., 578
Cheng, A., 589
Cheng, Y.-C., 619, 643
Cherry, R. J., 168
Chervitz, S. A., 130
Chi, P., 364, 539
Chidambaram, A., 864, 876(5), 879, 880(2), 884(2)
Chiesa, R., 878
Chifflet, S., 878
Chinkers, M., 674, 675(26)
Chinn, J., 389
Chiriac, M., 827
Chiu, W., 93
Chomczynski, P., 321
Chorev, M., 112
Chuman, L., 456
Chu-Ping, M., 466

Cideciyan, A. V., 79
Clerc, A., 646
Clore, G. M., 819
Cobbs, W. H., 347
Coccia, V. J., 651
Cohen, G. B., 51, 57, 57(34), 185, 190, 204(2, 3), 208, 209(1), 238, 249(3), 250, 252, 257(11), 258, 258(11), 263, 263(11)
Cohen, P., 558(21), 559, 561(21), 562(21), 564, 566, 566(25), 567(21), 570, 571, 572, 575(4), 579, 580
Coleman, D. E., 363, 372(2, 7), 490, 503, 513
Coleman, M., 195
Colman, A., 764
Colville, C., 755, 762(6), 785, 818(24), 819, 831(9), 832, 833(9), 843, 843(9)
Cone, R. A., 167, 270, 275(18)
Conklin, B. R., 364, 388(14), 389, 517
Conklin, R. B., 388(15), 389
Conrad, S., 456
Consalez, G. G., 580, 593(12)
Conti, F., 59
Converse, C. A., 81, 153, 879(5), 880
Convert, O., 114
Conway, B. P., 60
Cook, N. J., 755, 760, 762(4), 764(4), 831, 833, 833(7), 834(16), 836, 836(12, 18), 837(12), 838, 838(14), 839(30), 840(14), 842(31), 843(16, 18), 845, 845(18), 874, 883
Cook, T. A., 597, 601, 609, 663
Cooper, A., 148, 152(31), 153, 153(25), 154(25, 30, 32), 157(32)
Cooper, C. B., 723, 845, 848, 857, 857(6), 860(6), 863(6)
Cooper, N., 741
Cooper, N. G. F., 676, 737
Cooper, T. M., 147, 148, 151, 151(26, 27), 152(27), 153(27), 154(21, 27, 28), 155(21), 157
Copeland, N. G., 625
Cork, T. A., 648
Corless, J. M., 97
Cornille, F., 114
Cornwall, M. C., 219, 232, 239, 264, 267(33), 708, 848
Corrie, J. E. T., 802
Corson, D. W., 126, 239, 264, 267(33)
Costa, T., 259, 260(27)
Cote, R. H., 646, 647, 649, 650, 651, 651(15), 652(15, 23), 656(15), 663, 665, 665(15, 23, 51), 667, 669(23), 670(52), 671(23, 52), 672(15, 23)
Cotecchia, S., 258, 259, 259(20), 260(27)
Coull, J. M., 763
Coulson, A. R., 34, 36(15), 209, 448
Courtin, J., 158
Coutavas, E., 456, 464
Cowan, C. W., 232, 520, 524, 525, 526, 532(15), 534(15), 535(15), 537(15), 538(15), 552, 647
Cowen, C. W., 238
Cowley, G. S., 60
Crabb, J. W., 241, 243, 405, 437, 676, 690, 691(11), 702(11), 709, 717, 718, 776
Craescu, C. T., 827
Craft, C., 417, 617
Cragoe, E. J., Jr., 848
Craig, A. G., 775
Craig, L., 389
Craig, W. S., 3
Crary, J. I., 774, 775(5)
Crawford, J., 464
Creemers, A. F. M., 26
Cregg, J. M., 3
Cremers, F. P., 864, 880
Crivici, A., 797, 819
Cron, P., 565, 566(26, 27)
Cronin, J. M., 487
Cronin, T. W., 217
Croop, J. M., 865
Cropper, W. P. J., 669
Crosby, J. R., 78
Crouch, R. K., 126, 200, 201(25), 219, 238, 239, 239(8), 249(7), 250, 252, 264, 264(5), 267(33), 557
Cruze, J., 3
Cuatrecasas, P., 136
Cuenca, N., 691
Cull, M. G., 390, 392(27)
Cunnick, J., 526, 601, 636, 643(6)
Curtis, D. R., 295, 316(40), 324, 557
Cush, R., 487
Cwirla, S., 389
Cypess, A. M., 181, 255

D

Dabauvalle, M.-C., 456
Dacey, D. M., 404

Daemen, F. J. M., 731, 850
Dahlgren, D. A., 485
D'Amours, M. R., 663, 667
Danciger, M., 617
Dangott, L. J., 674
Dani, J. A., 775
Daniel, K., 59
Darszon, A., 292
Das, A. K., 571
Das, D., 29
Dasgupta, M., 827
Davenport, C. M., 44, 57(27), 60, 291
Davey, J., 83
Davidson, F. F., 3
Davidson, J. S., 112
Davidson, N., 775, 776(11)
Davies, A., 106
Davies, M. V., 209
Davis, A., 389
Davis, D. G., 110
Davis, G. R., 3
Davis, L. C., 108
Davis, T. R., 18
Dawes, J., 70
Dayan, G., 875
Dayhoff, M. O., 320
Dean, D. C., 618
Dean, M., 864, 876(5), 879, 880, 880(2), 884(2)
DeCaluwé, G. L. C., 27
DeCaluwé, G. L. J., 14, 20(8), 21(8), 92
DeCaluwé, L. L. J., 123, 194
Decataur, J., 897
DeDecker, B. S., 117
DeGrip, W. J., 3, 10(5), 12, 13, 14(3), 26, 26(3), 27, 28, 28(24), 29, 29(22, 24), 56, 92, 96, 110, 123, 193, 194, 196, 380, 478
deGroot, H. J. M., 26, 158
Deisenhofer, J., 92
Delange, F., 26, 193, 196
DeLean, A., 674, 675(21)
Delori, F. C., 897
del Rio, T., 864, 880
DeLuca, M., 618
DeMaria, C. D., 828
DeMartino, G., 466
Demin, V. V., 92
Dencher, N. A., 385
den Dunne, J. T., 580, 593(12)
Deng, H., 117, 161, 162(61), 198
Deng, W. P., 32, 272

den Hollander, A. I., 864, 880
Denisevich, M., 677
Dennis-Sykes, C. A., 45
Denny, M., 298
Deretic, D., 77, 78(3), 79, 79(1, 2), 80, 80(1), 81(1, 30), 83(1, 2), 84(2), 85(1, 2, 4, 30), 86(1, 2, 4, 5), 87(1, 2, 4), 88(27), 363, 364(8), 376(8), 388, 453, 490
Derguini, F., 223, 226(13), 264
Derlacki, D. J., 896
de Sauvage, F. J., 673, 685(8), 690, 711, 712(22)
Deterre, P., 346, 524, 527, 530(1), 601, 635, 641, 646
Detwiler, P. B., 405, 526, 612, 615, 676, 684(31), 690, 691(11), 693(10), 699(10), 701(10), 702(11), 704(10), 708, 709, 718, 730, 734(9), 848
D'Eustachio, P., 464
Deutman, A. F., 864, 880
Devine, C. S., 406
Devlin, M. J., 602, 652
de Vries, R., 389
de Wet, J. R., 618
Dewhurst, P., 197, 557
Dhallan, R. S., 755, 818, 831, 833(8)
Dhanasekaran, N., 502
DiAngelis, R., 464
Dice, A., 363, 364(12), 388, 404(11)
Dickerson, C. D., 412, 415(5), 416, 418
Dienes, Z., 483, 484(40), 485(40)
Diesis, E., 827
Dignam, J. D., 627
Dilley, R. A., 329
Dineen, B., 363, 364(12), 388, 404(11)
Dingus, J., 369
Dinur, U., 158
Dion, S. B., 421, 433, 435(11), 436(11)
Di Pietro, A., 875
Di Polo, A., 619, 634
Dirksen, R. T., 810
Disher, R. M., 525, 531(3)
Dixon, J., 457, 459(13), 460(13)
Dizhoor, A. M., 75, 404, 405, 406, 535, 673, 676, 685(12), 690, 691, 691(14), 701(14), 708, 709, 710, 710(15), 711(11, 15), 712, 712(11), 713(14, 15), 714, 714(14), 716, 716(11, 14, 23), 717, 717(14, 27, 28), 730, 734(10)
Dollfus, H., 718

Donato, R., 741, 742(25)
Donner, K., 159
Donoso, L. A., 431
Dorey, C. K., 897
Dorey, M., 111, 115(15)
Dorman, G., 644
Dorner, A. J., 209
Dosé, A., 755, 762(6), 785, 818(24), 819, 831(9), 832, 833(9), 838, 842(31), 843(9)
Doskeland, S. O., 666
Douglas, J. E., 685
Doukas, A., 144, 198
Dower, W., 389
Dowling, J. E., 896
Downing, K. H., 93, 99, 103(20)
Downing-Park, J., 784, 818
Dratz, E. A., 97, 367, 472, 530, 557
Drechsler, M., 843
Drewett, J. G., 674
Dreyer, W. J., 70, 96, 612, 679
Drijfhout, J., 389
Drivas, G., 464
Druckman, S., 163
Druey, K. M., 526, 531(23), 534(23)
Dryja, T. P., 60, 79, 208, 617, 880
Dua, S., 647
Dubendorff, J. W., 544
Duchesne, M., 114
Duda, T., 673, 674, 675(30), 676, 685(9), 690, 691, 708, 712, 716, 718, 730, 734(12), 737, 739(21), 741, 741(1, 21)
Duffin, D., 389
Dufier, J.-L., 880
Dufier, J. U., 718
Dumke, C. L., 526, 529(18), 535(18), 600, 638, 647, 651, 652(13, 25), 656(25), 743
Dumler, I. L., 3, 11(6), 608, 636, 643(4)
Dunbar, B., 879(5), 880
Dunham, T. D., 122
Dunn, J. J., 544
Dunn, S., 885
Dunphy, W. G., 77, 83(6)
Dunwiddie, T. V., 363, 364(12), 388, 404(11)
Dupont, Y., 270, 343, 531
Durham, S. J., 669
Duronio, R. J., 406
Duschl, C., 483, 485(38)
Dutton, D. P., 130, 137(2), 142(2)
Dyson, H. J., 107
Dzeja, C., 797, 798, 800(11), 802, 804

E

Ebrey, T., 117, 147, 158, 161, 162(61), 196, 197, 198, 199, 200, 201, 201(25), 202, 204, 205, 205(38), 206, 206(18), 207(18), 230
Eckstein, F., 371, 531, 537(36), 548, 551(17), 692, 693, 698(30), 803
Edwards, P. C., 93, 101(23), 106(23)
Ehlers, M. D., 823, 828(34)
Einterz, C. M., 148, 157, 158(24), 162(69), 163, 165, 168, 171, 175, 176(12), 355
Eismann, E., 838, 842(31)
Ekholm, J., 448
Elagina, R. B., 79
Eldik, L. J., 734
Eldred, G. E., 897
Ellenberg, J., 88
Elling, C. E., 112
Ellis, L., 674, 675(24)
El-Sayed, M. A., 145, 146, 147, 147(13, 15)
Emeis, D., 180, 329, 331, 332(4), 333(17), 343, 348, 352(3), 353(3), 377, 379(1), 476, 481(15)
Engel, A., 93, 99(30, 31), 674, 675(20)
Engelhard, M., 186, 201
Engelman, D. M., 59, 69(5)
Erdjument-Bromage, H., 77, 78(7)
Erickson, J. W., 636
Ericsson, L. H., 76, 405
Ermolaeva, M., 364
Ernst, O. P., 178, 243, 244, 249, 330, 332, 342(14), 344(12), 377, 384, 471, 472, 474(4), 476, 479(19), 482(4), 483, 484(40), 485(40), 487
Ernst, R. R., 110
Estes, R., 566
Etingof, R. N., 608, 636, 643(4)
Evanczuk, A. T., 386, 612, 641, 657
Evans, F. E., 756
Ewald, D. A., 503
Exum, S., 258, 259(20)

F

Fabiato, A., 784
Fabiato, F., 784
Fahmgi, K., 350
Fahmy, K., 123, 159, 160, 178, 181, 184, 186, 187, 188(37, 46, 49, 50), 189, 190(46), 191,

192(49, 50), 193, 194, 194(43), 195, 206, 243, 256, 260, 269, 287(12), 292(9, 12), 385, 473
Fain, G. L., 232, 708, 742, 848, 857
Fakler, B., 828
Falke, J. J., 130, 140, 140(1)
Fan, S., 570(7), 571
Fann, Y. C., 164
Farahbakhsh, Z. T., 104, 106, 112(31), 113, 115, 115(31), 179, 179(11), 180, 269, 290(13)
Farber, D. B., 617, 619, 634, 692, 696(25)
Farfel, Z., 364, 388(14, 15), 389
Farhangfar, F., 673
Fariss, R. N., 647, 691, 702(21), 709, 710(18)
Farrens, D. L., 104, 106, 112(30, 31), 113, 115, 115(30, 31), 122, 179, 269, 290(14), 475
Farris, R. N., 526, 534(16), 535(16)
Faruqi, A. R., 106
Fasick, J. I., 209, 321, 324(28)
Fasshauer, D., 78
Faurobert, E., 473, 518, 534, 539, 540(3), 546(3)
Fawzi, A. B., 525, 531(4)
Felsenstein, J., 319
Feng, Y., 200, 201(25)
Ferguson, K. M., 241, 473, 482(8), 496
Ferreira, P., 456, 457(10, 11), 460(10–12), 461(10–12), 462(10–12), 465(10, 11), 466(12), 579
Ferreira, P. A., 21, 455, 456(1), 458(1), 465(1)
Ferretti, L., 60, 135, 209, 321
Ferry, D. R., 872
Fesenko, E. E., 673, 755, 831
Fettiplace, R., 798
Fields, S., 440
Filmer, D., 771
Findlay, J. B. C., 557
Findsen, L. A., 117, 147, 158, 158(19), 162(19)
Finlay, B. B., 440
Finn, J. T., 755, 785, 817
Fiscella, M., 570(7), 571
Fischer, D., 571, 573(13), 574(13), 578(13)
Fisher, R. J., 482
Fishman, G. A., 896
Fivash, M., 482
Flaherty, K. M., 406
Flanagan, C., 112
Fleishman, D., 677
Fletcher, J., 364

Fletcher, R. T., 698
Flik, G., 696
Flood, L. A., 364, 517
Florio, S. K., 601, 609(8), 647
Flynn, G. E., 779
Fodor, A. A., 782, 783
Fong, C., 388
Fong, S. L., 295, 316(40), 324, 557
Ford, R. C., 93, 99(30)
Forget, R. S., 665, 670(52), 671(52)
Forquet, F., 346
Forrester, J. V., 879(5), 880
Foster, D. C., 673, 690, 712
Foster, K. W., 238, 239(5)
Fothergill, J. E., 879(5), 880
Fowles, C., 558(21), 559, 561(21), 562(21), 567(21), 568
Fox, G. C., 158
Frabman, J. S., 896
Franco, B., 580, 593(12)
Frank, R., 70, 78
Franke, R. R., 57, 116, 134, 158, 163, 181, 186, 187, 187(32), 206, 252, 253, 437
Franklin, P., 223, 226(13)
Frederick, J., 709, 710(17)
Frezel, J., 718
Friedberg, F., 797, 819
Friedel, U., 833, 834(16), 838(14), 840(14), 843(16), 845
Friedlander, M., 69
Friedman, N., 163, 171, 183, 189(26), 201, 203, 205(32)
Friedrich, T., 810
Fringeli, U. P., 192
Frings, S., 797, 798, 800(11), 802, 804, 810(7), 814
Frins, S., 676, 690, 691(15)
Fritsch, E. F., 34, 313, 322(7), 442, 459, 462(15), 624, 629
Fritzsche, I., 895
Fujimuro, M., 466
Fujita, Y., 810
Fujiwara, T., 466
Fujiyoshi, Y., 93
Fukada, F., 348, 349(7), 357(7), 361(7)
Fukada, Y., 117, 128(11), 195, 221, 294, 294(6), 296, 297(9), 298, 301, 307, 308, 308(25), 309, 348, 349, 350, 350(11)
Fukuda, Y., 197
Fukui, M., 456

Fülle, H.-J., 673, 690, 712
Fung, B., 495
Fung, B. K.-K., 9, 68, 364, 365, 365(22), 367(22), 371, 372, 376(27), 546
Furuichi, T., 810
Furukawa, T., 625
Furutama, D., 810

G

Gaertner, W., 145, 148(14), 154(14), 157
Gagnon, J., 674, 675(21)
Gaide Huguenin, A. C., 651
Gallagher, S., 654
Galley, K., 869
Galli, T., 601
Gallop, M., 389
Ganter, U. M., 126, 183, 186(22), 189, 189(23)
Gao, Y. Q., 617
Garami, E., 869
Garaschuk, O., 798
Garbay, C., 114
Garbers, D. L., 673, 674, 674(14), 675(20), 690, 708, 712, 718, 759
Garcia, A., Jr., 20
Garcia, C., 674, 675(20)
Garcia, P. D., 364, 388, 539
Garland, P. B., 787
Garnovskaya, M. N., 3, 11(6)
Garrison, J., 364
Gärtner, W., 144, 145, 145(9), 147(12), 148, 154(23), 183, 189(23)
Garwin, G. G., 267
Gassner, B., 274, 279(37)
Gat, Y., 201, 205(31)
Gates, C. M., 389, 390, 392(27)
Gautam, N., 164, 364, 369
Gawinowicz, M., 230
Geahlen, R. L., 756
Gebhard, R., 158
Geluk, A., 389
Gensch, T., 145, 148(14), 154(14), 157
George, J. S., 656
Gerad, D., 733, 741(17)
Gerber, S., 718, 880
Gerber, W. V., 756, 761
Gerchman, S.-E., 59, 69(5)
Gerfin-Moser, A., 883
Germann, U. A., 887

Geromanos, S., 77, 78(7)
Gerrard, B., 864, 876(5), 879, 880(2), 884(2)
Gershengorn, M. C., 258
Ghazi, I., 718, 880
Giambanco, I., 741, 742(25)
Gibbs, C. S., 674, 675(23)
Gibson, T. J., 313
Gierschik, P., 365
Gies, D. R., 36, 51, 51(17), 298
Gilbert, D. J., 625
Gilbert, L. D., 896
Gilchrist, A., 363, 364(12), 388, 404(11, 12)
Gillespie, P. G., 599, 600, 605, 606(17), 608(17), 652, 659, 660(41), 663(41), 698, 820
Gillett, N. A., 673, 685(8), 711, 712(22)
Gillette, N. A., 690
Gilman, A. G., 241, 363, 365, 372(2, 3, 7), 490, 496, 503, 513, 520, 526, 531, 534, 540, 552, 553(24), 660, 674, 675(27)
Gilson, H., 205
Gilula, N. B., 93, 97(26)
Giordano, M., 617
Glaeser, R. M., 97
Glenn, A. M., 896
Glossmann, H., 775
Gluzman, Y., 31, 32(7), 62, 271
Gmachl, M., 78
Goddard, N. J., 487
Godde, M., 755, 762(6), 785, 798, 810(7), 814, 818(24), 819, 831(9), 832, 833(9), 843(9)
Godfrey, R. E., 168
Goebl, M., 79
Goeddel, D. V., 673, 674, 685(8), 690, 711, 712(22)
Goff, S. P., 525, 526(9), 527(9), 601, 647
Goger, D. G., 897
Gold, G. H., 818
Goldberg, J., 580
Goldberg, N. D., 698
Goldin, A. L., 773
Goldsmith, P., 388
Golub, G. H., 355
Gonzalez, K., 601
Gonzalez-Duarte, R., 864, 880
Gonzalez-Fernandez, F., 60
Good, W., 60
Goodman, J. L., 144, 147(7)
Goraczniak, R., 673, 674, 675(30), 685(9), 690,

691, 712, 716, 730, 734(12), 737, 739(21), 741, 741(21)
Gorczyca, W. A., 405, 676, 679(38), 684(31), 689, 690, 691, 691(11, 13), 692, 693(10, 13, 26), 698(26), 699(10, 13), 701(10, 13), 702(11, 13), 704(10), 705(13), 707(13), 709, 711, 713(16), 716, 716(16, 20), 718, 730, 734(9, 11, 12)
Gordon, J., 375, 405, 406, 465, 820, 869
Gordon, S. A., 772
Gordon, S. E., 774, 775(5), 776, 777(13), 779, 781, 781(13), 782, 784, 818
Gordon-Walker, A., 651
Goris, J., 565, 566(26, 27)
Gorman, C. M., 36, 51, 51(17), 298, 618
Gorny, M. K., 44
Gospodarowicz, D., 679
Gottesman, M. M., 887, 895, 896(34)
Gottlieb, H., 200
Gottlieb, M., 462
Gotzes, S., 755, 762(6), 785, 818(24), 819, 831(9), 832, 833(9), 843(9)
Goudreau, N., 114
Goulding, E. H., 776, 777(14)
Gouras, P., 525, 526(9), 527(9), 601, 647
Govindjee, R., 147, 200, 201(25)
Gowan, B. E., 106
Graber, S. G., 364, 517
Graeff, R. M., 698
Graham, R. M., 181, 253
Gramling, R., 755, 756(8), 762(8), 763
Granados, R. R., 18
Granovsky, A. E., 517, 534, 539, 540, 546(8), 548(7), 549, 549(7), 551(7), 552, 552(7), 554(7), 598, 635, 636, 637(10), 638, 642(10, 15), 643(15), 644(10), 645, 663
Granzow, R., 827
Gray, K.-M. P., 612, 615
Gray-Keller, M. P., 405, 527, 531(24), 532(24), 676, 684(31), 690, 691(11), 693(10), 699(10), 701(10), 702(11), 704(10), 708, 709, 718, 730, 734(9), 848
Grazio, H. J., 663
Green, N. M., 93
Greene, N. M., 367
Greengard, P., 580, 647, 651(11)
Gregerson, D. S., 431
Grellmann, K. H., 160
Grigorieff, N., 99
Grinberg, D., 617

Groesbeek, M., 124, 127(34), 220, 255, 264, 476
Gronenborn, A. M., 819
Gronovsky, A. E., 648, 652(16a)
Gros, P., 895
Gross, H. J., 599(33), 615
Groves, J. D., 59
Grunwald, M. E., 755, 783, 785, 786, 817, 818, 819, 819(19), 820(19), 821(19), 822(19), 823, 823(19), 825(19), 827, 837
Grynkiewicz, G., 800, 814(12)
Gstaunthaler, G., 274, 279(37)
Gu, Q., 673, 685(12), 690, 712, 716(23)
Guan, K., 457, 459(13), 460(13)
Guntert, P., 111
Gurevich, V. V., 3, 11(6), 419, 421, 422, 423, 424, 424(4, 5), 427(4), 429(5), 433, 433(4, 5), 434(4, 18, 19), 435, 435(5, 11), 436, 436(11), 437, 453, 455, 612
Guthridge, M. A., 570(6), 571
Gutman, C., 240
Gutman, M., 385
Gutmann, C., 217, 335
Guy, P. M., 256, 528

H

Hackos, D. H., 797, 818
Haeseleer, F., 453, 691, 702(21), 709, 710(18)
Hagen, V., 797, 798, 800(11), 802, 804, 833
Hagiwara, H., 674, 675(19)
Haig, C., 267
Hakki, S., 676, 677
Haley, B. E., 756
Haley, T. L., 776
Hall, C. V., 618
Hall, S. W., 422
Hamill, O. P., 274, 281(39), 290(39)
Hamilton, S. E., 608, 609(22), 653
Hamm, H. E., 350, 363, 364, 364(8, –12), 365, 365(23), 367(11, 23), 369, 371(23), 372(6, 11, 23), 373(1), 376(1, 6, 8, 29), 388, 389, 397(19), 404(11, 12, 19), 437, 453, 490, 491, 493(7), 495, 496(2), 498(1, 2), 501(1, 2), 502, 503, 504(4), 511(4), 512, 513, 513(4), 514(4, 6, 12), 517, 517(4), 518, 520, 520(5), 526, 529, 529(18), 531, 534, 535(18), 539, 540, 540(6), 542, 542(2), 546(2, 6, 14), 548(14), 549, 549(6), 551,

598, 636, 638, 638(8), 639(8), 641(5), 643(5, 8), 645, 645(9)
Hamm, H. H., 363
Hammer, D. A., 18
Han, M., 116, 117, 119, 123(21), 124, 126, 126(33), 127(34), 179, 185(6), 208, 220, 226, 227(17, 18), 245, 249, 251, 252, 255, 257, 257(8), 258(8, 9, 19), 260(19), 261(9), 263(8, 9, 19), 264, 269, 291, 292(7), 476
Hanada, M., 571
Hanahan, D., 448
Hanck, D. A., 273, 274(36), 278(36)
Hancock, R., 463
Handy, J. W., 526, 535(20)
Hanke, W., 760, 833, 834(18), 836(18), 845(18)
Hanks, S. K., 684, 685(56)
Hansen, J., 108
Hanson, P. I., 78
Hardman, J. G., 610
Hargrave, P. A., 8, 12, 56, 60, 63, 70, 71, 75, 76, 79, 88(27), 91, 92, 97, 98, 102(46), 103, 104(8), 108, 111, 115(15), 179, 185(1), 240, 254, 295, 316(40), 324, 363, 364(8), 376(8), 388, 414, 422, 437, 447, 453, 455, 490, 557, 558, 558(22), 559, 566(22), 567(20, 22), 717
Harlow, E., 839, 866, 881
Harrick, N. J., 192
Harris, G., 456
Harrison, S. M., 732
Hartstein, A., 482
Harzmark, P., 388(14), 389
Hashimoto, S., 122
Hatanaka, M., 787
Hatta, G., 363, 364(9), 388
Haugland, R. P., 851
Havelka, W. A., 93, 95
Hawkins, R. E., 444
Hayashi, F., 526, 601, 647, 656, 673, 677(5)
Hayashi, N., 456
Hayashida, T., 456
Haynes, L. W., 755, 818
Hazelbauer, G. L., 130, 137(2), 142(2)
Hde-DeRuyscher, R., 440
He, W., 520, 526, 532(15), 534(15), 535(15), 537(15), 538(15), 552, 647
He, Y., 186
Heasley, L., 389
Heaton, R. J., 453
Heberle, J., 192, 385
Hebert, M. C., 663, 665(51)
Hecht, S., 199, 251, 267
Heck, M., 241, 249, 329, 330, 332, 332(11), 335(11), 346, 347(11), 386, 471, 476, 479(19), 480(1), 482
Heckenlively, J., 60, 617
Hefti, A., 93, 99(30, 31)
Heibel, G. E., 144, 145(4)
Heihoff, K., 144, 145(5, 8)
Heilbron, V., 617
Heinemann, S. F., 775
Heinkel, G., 389
Helekar, B. S., 405, 676, 690, 691, 691(11), 702(11), 709, 718
Helinski, D. R., 618
Helmreich, E. J., 238
Helmreich, E. J. M., 646
Helps, N. R., 571
Hemminger, J. C., 485
Hemmings, B. A., 558, 558(23), 559, 561(23), 564, 564(23), 565, 566(25–27), 567(23), 568(23), 570(23)
Henderson, R., 92, 93, 95, 97, 99, 103(20, 36), 104(52), 106, 106(52), 108, 112(2), 128
Hendler, R. W., 172
Hendrickson, W. A., 674, 675(24)
Hendtix, P., 565, 566(26, 27)
Henley, J. R., 78
Henn, C., 93, 99(30)
Hennessey, J. C., 60
Henry, E. R., 355, 357(24)
Hensley, P., 792, 794(19)
Henzel, W. J., 405, 676, 690, 691(14), 701(14), 709, 710(15), 711(15), 713(15)
Hepler, J. R., 503
Hergenhahn, H.-G., 730
Hermesmeier, J., 571, 573(16), 577(16), 578(16)
Hersh, D., 78
Herzmark, P., 364, 539
Hestrin, S., 270, 712(23), 810, 815
Heuckeroth, R. O., 406
Heuser, J. E., 78
Hewryk, M., 869
Heyduk, T., 75
Heymann, J. A. W., 95
Heymann, J. B., 93
Heymann, J. S. W., 95
Heyn, M. P., 270, 292(26)
Heyse, S., 483, 484(40), 485(40), 487

AUTHOR INDEX

Hickel, W., 483, 485(31)
Hicks, D., 63
Hideg, K., 269, 290(13)
Hiemstra, H., 389
Higashijima, T., 241, 473, 482(8), 496
Higgins, C. F., 865, 872, 895
Higgins, D. G., 313
Higuchi, R., 620
Hildebrandt, J. D., 369
Hilgemann, D. W., 773
Hiraga, A., 571
Hirai, T., 93
Hirata, M., 685
Hirose, S., 674, 675(19)
Hirose, T., 755, 762(4), 764(4), 831, 833(7)
Hiroshima, T., 222
Hirschberg, B., 828
Hirschberg, K., 88
Hirsh, J. A., 435
Hisatomi, O., 298
Hitz, B., 108
Ho, T. W., 667(55), 668
Ho, Y. K., 126, 533
Hobbs, J. N., Jr., 45
Hodges, R. S., 453
Hodgkin, A. L., 730
Hoenger, A., 93, 99(30)
Hoffman, R., 649
Hoffmann, W. F., 331, 336
Hofmann, K. P., 70, 108, 178, 180, 183, 189(26), 194, 238, 239(6, 8), 240, 241, 241(6), 243, 243(6), 244, 244(6), 245(6), 247, 247(6), 249, 252, 257(6), 264(5), 269, 287(12), 292(12), 329, 330, 331, 331(7), 332, 332(4, 7, 11, 18), 333(17), 334, 335, 335(11, 18), 336, 336(18), 339, 339(7, 18), 341, 342(14, 28), 343, 343(7), 344(7, 12), 346, 347(11), 348, 352(3, 4), 353(3, 4), 363, 364(8), 376(8), 377, 379(1), 384, 385, 386, 388, 421, 433, 435(12, 13), 436(12, 13), 437, 453, 471, 472, 473, 474(4, 10), 475, 476, 478, 479, 479(19), 480(1, 3, 26), 481(15, 46), 482, 482(4), 483, 484(40), 485(40), 487, 490, 525, 646
Hofrichter, J., 355, 357(24)
Hofsteenge, J., 565, 566(26, 27)
Hogness, D. S., 32, 60, 272, 276(29), 313, 411
Holcombe, V., 617
Holden, F., 389
Hollister, J. R., 13

Holloway, R. A., 148, 151(27), 152(27), 153(27), 154(27), 157
Holtzscherer, C., 733, 741(17)
Holzwarth, A. R., 145, 147(12)
Hom, J., 455, 456(1), 458(1), 465(1)
Honeycutt, T., 827
Hong, K., 371, 413
Honig, B., 108, 144, 158, 162(71), 163, 205
Honkanen, R. E., 579, 580, 587(10), 589, 592(10)
Hood, D. C., 79
Horiuchi, K., 300
Horwitz, J. S., 165
Hosey, M. M., 388, 404(12), 421, 433, 435(11), 436(11)
Hou, Y., 865, 872(13)
Houghton, L. A., 578
Howard, B. H., 618
Howell, K. E., 78
Howes, K., 691, 709, 710(17)
Hoyng, C. B., 864, 880
Hruby, V., 389
Hsu, K.-W., 818
Hsu, S. C., 723
Hsu, T.-Y., 785, 786(5)
Hsu, Y.-T., 784, 785, 818, 819(9, 21), 825(21), 837, 838(24, 25)
Hu, S., 223, 226(13)
Hu, S.-H., 674, 675(25)
Huang, H. B., 580
Huang, J., 405, 485, 676, 690, 691, 691(11), 702(11, 21), 709, 710(18), 718
Huang, K.-S., 59
Huang, L., 117, 161, 162(61), 198
Huang, M., 51
Huang, T., 503, 520(5)
Huang, X., 579, 580, 587(10), 592(10)
Huang, Y., 79
Hubbard, L. M., 158
Hubbard, R., 159, 198, 377, 868
Hubbard, S. R., 674, 675(24)
Hubbell, W. L., 104, 106, 112(30, 31), 113, 115, 115(30, 31), 122, 179, 179(11), 180, 269, 270, 287, 290(13, 14), 292(46), 371, 385, 413
Hubbell, W. W., 179
Huber, L. A., 77, 86(5)
Huddleston, M. J., 241, 243
Hudson, T. H., 366
Hudspeth, A. J., 820

Hug, S. J., 148, 157, 158(24), 165, 167, 171, 175, 355
Huganir, R. L., 823, 828(34)
Hughes, R. E., 405, 691, 709, 713(14), 714(14), 716(14), 717(14)
Hughes, V., 579
Hulme, E. C., 663, 668
Hunnius, D., 579
Hunt, D. M., 29
Hunt, T., 424, 425(7), 428(7)
Hunter, T., 676, 688
Hurley, J. B., 9, 68, 365, 371, 404, 405, 406, 406(8), 409(8), 534, 539, 551(1), 608, 609(22), 635, 641, 652, 653, 656(29), 673, 674, 675(29), 676, 684(29), 685(12), 690, 691, 691(14), 701(14), 708, 709, 710(15), 711(11, 15), 712, 712(11), 713(14, 15), 714, 714(14), 716, 716(11, 14, 23), 717, 717(14, 27), 730, 734(10)
Hurley, J. H., 674, 675(28), 676, 716
Hurt, D., 636, 643(6)
Hurtley, S. M., 83
Hurwitz, R. L., 605, 617
Hutchinson, A., 864, 876(5), 879, 880, 880(2), 884(2)
Hutson, L., 526
Huttner, W. B., 78, 79, 79(14), 81(26), 83(26)
Hyde, D. R., 79
Hyde, E. Q., 158
Hyden, C., 159
Hyek, M. F., 258

I

Ikeda, S., 571
Ikura, M., 405, 797, 819
Illing, M., 755, 762(6), 785, 818, 818(24), 819, 831(9), 832, 833(9), 837, 838, 842(31), 843(9), 864, 865(1), 871(1), 876(1, 4), 879, 880, 881(3), 884(3)
Imai, H., 293, 294, 294(7, 8), 295, 296, 299, 300(8), 301, 302, 303(19), 304, 305, 306(16), 308(8), 310, 347, 348, 349(5, 7), 350(1), 353(5), 357(7), 361(7), 363(8)
Imami, N. R., 455
Imamoto, Y., 222, 294, 294(7), 296, 301, 302, 303, 303(19), 304, 305, 348, 349(5, 7), 353(5), 357(7), 361(7)
Imanoto, Y., 117, 128(11)

Imasheva, E., 200, 201(25)
Inagami, T., 685
Ingebritsen, T. S., 70, 558, 558(22), 559, 566(22), 567(20, 22), 570
Inglese, J., 405, 406(8), 409(8)
Iniguez-Lluhi, J. A., 363, 372(7)
Ishida, S., 388(14), 389
Ishii, T., 828
Ishima, R., 405
Israel, D. I., 209
Itakura, M., 674, 675(19)
Ito, M., 221, 222
Ito, T., 674, 675(19)
Ivanov, I. E., 78
Iwabe, N., 810
Iwasa, T., 122, 199, 206(18), 207(18)
Iwashina, M., 674, 675(19)
Iwata, S., 92

J

Jackson, K. W., 647
Jackson, M. L., 482
Jackson, R. J., 424, 425(7), 428(7)
Jackson-Machelski, E., 406
Jacob, P. E., 618
Jacob-Mosier, G., 364
Jacobs, G. H., 117
Jacobs, J., 389
Jacobson, S. G., 60, 79, 617
Jaeger, F., 206
Jäger, F., 123, 183, 184, 187, 188(37), 189(26), 269
Jäger, S., 168, 177(10), 178, 178(10), 183, 189(26), 238, 239(6, 8), 241(6), 243(6), 244(6), 245(6), 247(6), 249, 252, 257(6), 264(5), 377, 384, 475
Jahn, R., 78
Jakes, S., 70, 558(22), 559, 566(22), 567(22)
Jan de Vries, K., 78
Jansen, M., 329
Janssen, J. J. M., 3, 10(5), 13, 14, 14(3), 20(8), 21(8), 26(3), 27, 29(22), 56
Jap, B. K., 93, 97, 99(31)
Jarvis, D. L., 13, 20
Jastorff, B., 663, 665(51)
Jaszczack, E., 316(40), 324
Jenkins, N. A., 625
Jiang, B., 685

Jin, J., 239, 264, 267(33)
Joha, H., 274, 279(37)
Johnson, G. L., 139, 213, 366, 389, 413, 415, 502
Johnson, J. D., 371
Johnson, K., 457, 459(14)
Johnson, R. S., 76, 405, 676, 690, 691(11), 702(11), 709, 718
Johnson, S. J., 483
Johnson, T. D., 258
Johnson-Pais, T., 828
Johnston, J. P., 673
Jonas, R., 201
Jones, D., 389
Jones, D. T., 320
Jones, S. M., 78
Joost, H. G., 580
Jovin, T. M., 868
Jubb, J. S., 97
Juillerat, M. A., 59
Julius, D., 364, 388(15), 389
Junnarkar, M. R., 144
Jurman, M. E., 273
Juszcak, R., 557
Juszczak, E., 295

K

Kaback, H. R., 59
Kachi, S., 710
Kachurin, A. M., 92
Kagawa, S., 466
Kahlert, M., 108, 247, 330, 437, 453, 473, 525
Kai, K., 115
Kaiser, C., 3
Kakitani, H., 162(71), 163, 226
Kakitani, T., 144, 162(71), 163, 226
Kakiuchi, S., 730
Kalsow, C. M., 431
Kamibayashi, C., 566
Kamps, K. M. P., 97
Kandor, H., 189, 196(42)
Kandori, H., 117, 129, 195, 196, 294(6), 296, 308
Kangawa, K., 755, 762(4), 764(4), 831, 833(7)
Kanno, S., 571
Kano, M., 798
Kao, J., 164
Kao, J. P. Y., 851

Kaplan, E., 159, 163
Kaplan, J., 718, 880
Kaplan, M., 612, 641, 657
Kaplan, R. S., 683
Karnaukhova, E., 219
Karnauknova, E., 226, 227(14), 229(14)
Karnik, S. S., 39, 60, 135, 209, 321, 322
Karpen, J. W., 755, 756, 756(8, 11), 761, 762(8), 763, 765, 767, 767(14), 769, 770(11), 771, 774, 776, 780, 803
Katial, A., 417
Kato, S., 571
Katsuta, Y., 222
Katz, A., 364
Katz, B. A., 440
Katz, B. M., 239, 264, 267(33)
Kaufman, R. J., 3, 8(1), 10(1), 13, 30, 50(1), 62, 66(21), 181, 209, 216(11), 322
Kaupp, B., 574
Kaupp, U. B., 755, 760, 762(4, 6), 764(4), 785, 786, 787(10), 789(10), 792(10), 794(10), 796(10), 797, 798, 800(11), 802, 804, 810(7), 814, 818, 818(24), 819, 825, 825(20), 827, 827(20), 831, 831(9, 10), 832, 833, 833(7, 9), 834(18), 836, 836(12, 18), 837, 837(12), 838, 839(30), 842(31), 843(9), 845(18), 874, 883
Kaushal, S., 10
Kavipurapu, P. R., 692, 696(25)
Kawamura, S., 313, 322, 324(35)
Kawar, Z. S., 13
Kay, A. R., 755
Kay, B. K., 440
Kayada, S., 298
Keegstra, W., 97
Keen, J. E., 828
Kefalov, V. J., 219
Kehrl, J. H., 526, 531(23), 534(23)
Keirns, J. J., 656
Kellaris, P. A., 3
Kelle, R., 26
Kelleher, D., 389, 415
Kellner, R., 674, 675(22), 676, 690, 691(15)
Kelly, L. E., 828
Kemp, B. E., 674, 675(25)
Kenakin, T. P., 258
Kentrup, H., 580
Kenyon, K. R., 70
Kerc, E., 571
Kerker, M., 330, 331(14)

Kersting, U., 274, 279(37)
Khorana, G. H., 209, 216(11)
Khorana, H. B., 321
Khorana, H. G., 3, 8, 8(1), 10, 10(1, 3), 11(11), 13, 30, 39, 50(1), 55, 56(36), 57, 57(36), 59, 60, 62, 66(21), 91, 104, 106, 112(30, 31), 113, 115, 115(30, 31), 116, 122, 134, 135, 158, 163, 179, 179(11), 180, 181, 186, 187, 187(32), 195, 206, 209, 252, 253, 268, 269, 270, 278, 287, 290(14), 292(26, 46), 322, 367, 385, 405, 417, 421, 437, 475
Kibelbek, J., 307, 377
Kikkawa, M., 199, 206(18), 207(18)
Kikkawa, S., 122
Killilea, S. D., 589
Kim, C. M., 421, 433, 435(11), 436(11)
Kim, E., 487
Kim, H., 92
Kim, J.-M., 287, 292(46), 371, 385
Kim, M.-S., 810
Kim, T. S. Y., 831, 833, 834(17), 838(17), 840(17), 843(17)
Kimura, M., 314, 848
Kincaid, R. L., 657, 823
King, A. J., 557, 558, 558(23), 559, 561(23), 564(23), 567(23), 568(23), 570(23)
Kingsley, C., 364, 539
Kingston, R. E., 443, 608, 609(23)
Kipnis, D. M., 679
Kishigami, A., 526
Kisselev, O. G., 164, 332, 364, 476, 479(19)
Kistler, J., 93, 99(30)
Kito, Y., 50
Kitts, P. A., 16
Klaassen, C. H. W., 12, 26, 27, 28(24), 29(24), 196
Klee, C. B., 819
Klee, W., 365
Klein, D. C., 404
Kleuss, C., 350, 548, 551(18)
Klevering, B. J., 880
Klevitt, R. E., 405
Kliger, D. S., 148, 157, 158(24), 164, 165, 167, 168, 171, 173, 175, 175(18), 176(12, 20), 177(10), 178, 178(10), 249, 269, 306, 351, 355, 357(21, 25), 377, 384, 530
Kligman, D., 738
Klimas, R. W., 693
Klinge-Roode, E. C., 16
Klompmakers, A. A., 731

Klumpp, S., 570, 571, 573, 573(13, 16), 574, 574(13), 575(19), 577(16), 578(13, 16), 580, 730
Knapp, D. R., 540, 557
Knapp, H. M., 162(69), 163
Knighton, D. R., 674, 675(23), 684, 685(57)
Knoers, N. V., 864, 880
Knoll, B. J., 618
Knoll, W., 483, 485(31)
Knospe, V., 431
Knox, B. E., 3, 238, 239(5), 245, 252, 257(10), 258(10), 278, 323, 417
Ko, Y. H., 875
Kobashi, K., 132, 140(7)
Kobayashi, T., 571
Kobilka, B. K., 59, 112(29), 113
Kobilka, T. S., 59
Koch, A. L., 329
Koch, K.-W., 673, 674, 675(22), 676, 677(6), 690, 691(15), 693, 698(3), 708, 709(10), 718, 730, 785, 786, 787, 787(10), 789(10), 792(10), 794(10), 796(10), 818, 825, 825(20), 827, 827(20), 831, 836, 837
Kochendoerfer, G. G., 117, 122(12), 164, 221
Kochersperger, L., 389
Kodama, A., 221
Koehler, M. R., 864, 876(4), 880
Koenig, B., 363, 364(8), 376(8), 388, 453, 490
Kohl, B., 478
Kohse, K. P., 685
Kojima, D., 129, 189, 196(42), 204, 294(8), 295, 296, 298, 300(8), 301, 308(8), 310
Kokame, K., 349, 350(11)
Koland, J. G., 256, 528
Kolb, H., 691, 709, 710(17)
Kolesnikov, S. S., 673, 755, 831
Kolster, K., 227
Konig, B., 108, 437, 453, 478
Konnerth, A., 798, 808(1)
Kono, M., 112, 130, 131(5), 134, 134(5), 135(5), 140(9), 141, 200, 201(25)
Konvicka, K., 112
Korenbrot, J. I., 270, 708, 712(23), 797, 810, 815, 818
Korf, H. W., 404
Korpiun, P., 483, 485(31)
Körschen, H. G., 755, 762(6), 785, 786, 787(10), 789(10), 792(10), 794(10), 796(10), 818, 818(24), 819, 825(20), 827(20), 831(9), 832, 833(9), 837, 843(9)

Koshland, D. E., Jr., 130, 140(1), 142, 771
Kostenis, E., 364
Köster, H., 763
Koutalos, Y., 197, 198, 205, 690, 742, 743, 744(8), 746(7, 8), 747, 748(8), 749, 750, 750(7), 751(8, 9), 752, 817, 818
Kozak, M., 322, 429
Kraehenbuhl, J. P., 864, 884
Kraemer, D., 456
Kragl, U., 26
Kramer, R. H., 768, 774
Krause, A., 330
Krause, E., 804
Kräutle, O., 183, 189(26)
Krebs, A., 93, 101, 101(23), 106(23)
Krebs, E. G., 587, 756
Krebs, M. P., 270, 292(26)
Krebs, W., 446
Kress, M., 798
Kreusch, A., 92
Kreutz, W., 336, 481(46), 487
Krishna, G., 698
Krishnan, A., 716
Krishnan, N., 698
Kroll, M., 388
Krueger, S., 273, 274(36), 278(36)
Krupnick, J. G., 435, 437, 453
Krylov, D., 717
Kuang, W. J., 674
Kühlbrandt, W., 93, 99(32)
Kühn, H., 60, 70, 180, 240, 307, 329, 333(6), 334, 343(22), 348, 352(3), 353(3), 377, 379(1), 421, 422, 433, 435(12), 436(12), 446, 476, 528, 557, 866
Kuhne, W., 251
Kuma, K., 456
Kuman, N. M., 93, 97(26)
Kuman, S., 405
Kumar, A., 110, 487
Kumar, S., 320, 404
Kung, C., 828
Kunkel, T. A., 322
Kunze, D., 388
Kurahashi, T., 818
Kurakin, A. V., 440
Kuriyan, J., 580
Kurose, H., 685
Kushner, S. R., 442
Kusuda, K., 571
Kutuzov, M. A., 567

Kuwata, O., 202, 204, 205
Kwon, Y. G., 580
Kyle, J. W., 273, 274(36), 278(36)

L

Lacapere, J. J., 93, 99(33)
Lad, P. J., 680
Ladant, D., 827
Laemmli, U. K., 373(44), 374, 375(44), 416, 459, 460(16), 462(16), 464(16), 733, 843, 868
Lagnado, L., 526, 746, 818
Lagnado, M. P., 848
Lai, J., 819
Lake, J. A., 314
Lakey, J. H., 483, 787
Lam, V., 112(29), 113
Lamb, T. D., 232, 239, 267, 524, 526, 598, 646, 657(2), 691, 708, 742, 810
Lambrecht, H. G., 718
Lambright, D. G., 350, 363, 372(6), 376(6), 490, 491, 496(2), 498(2), 501(2), 512, 513, 514(12), 526, 531, 542, 546(14), 548(14)
Lamola, A. A., 106, 868
Landro, J. A., 59
Lane, D., 839, 866, 881
Lang, H., 483, 485(38)
Langen, R., 115
Lanyi, J. K., 145, 147, 147(13), 154, 155(35), 205
Larrivee, D., 456
Larsen, L. O., 159
Lasky, M. R., 897
Laura, R. P., 673, 674, 675(29), 676, 684(29), 685(12), 690, 691, 709, 711(11), 712, 712(11), 716, 716(11, 23), 730, 734(10)
Lavielle, S., 114
Lawrence, A. F., 147, 154(21), 155(21)
Lawton, R., 364
Lazarevic, M. B., 363, 364(9), 388
Le, P. D., 718
Leatherbarrow, R. J., 662
Lebert, M. R., 130, 137(2), 142(2)
Lebioda, L., 437
Leblanc, R. M., 154
Lebovitz, R. M., 627
Lee, A. S., 627
Lee, C., 364

Lee, C.-P., 195
Lee, D., 869
Lee, E., 363, 372(2, 7), 490, 503, 513
Lee, E. Y. C., 589
Lee, F., 618
Lee, G. F., 130, 137(2), 142(2)
Lee, J. C., 598
Lee, L. S., 619
Lee, M., 117, 709, 710(17)
Lee, N., 322
Lee, N. R., 718
Lee, R. H., 617
Lee, S. C., 16, 584
Lee, S. S., 131
Lee, S. S. J., 60, 63(13), 64(13)
LeFebvre, Y. M., 667(55), 668
Leff, P., 258, 259, 260(28)
Lefkowitz, R. J., 59, 258, 259, 259(20), 260(27), 405, 406(8), 409(8)
Lehninger, A., 890, 892(30)
Lei, J. Y., 674, 675(25)
Leibrock, C. S., 232, 239, 267
Lelong, I. H., 887
Lennon, G., 625
Leong, E., 399, 402(28)
Leowski, C., 718, 880
Lepault, J., 93, 103(20)
Leppert, M., 864, 876(5), 879, 880, 880(2), 884(2)
Lerner, L. E., 617, 619
Lerner, R. A., 107
Lerro, K., 230
Lesniak, K., 693
Lester, H. A., 775, 776(11)
Letrong, H., 672
Leunissen, J., 389
Levis, R. A., 291
Levy, D., 93, 99(33), 886
Lewis, I. C., 838, 842(31)
Lewis, J. W., 148, 157, 158(24), 164, 165, 167, 168, 171, 173, 175, 175(18), 176(12), 177(10), 178, 178(10), 249, 269, 306, 351, 355, 357(21, 25), 377, 384, 530
Lewis, R. A., 864, 876(5), 879, 880, 880(2), 884(2)
Li, A., 388, 404(12)
Li, C., 869
Li, J., 775, 776(11), 783, 784(27), 785, 786(8), 819, 820(25), 821(25), 823(25), 824(25), 827(25), 828(25)

Li, N., 609, 691, 709, 710(17), 712, 716(26)
Li, Q., 539
Li, T., 526, 535(20), 617
Li, X., 723
Li, X. B., 852, 852(18), 853, 853(17)
Li, Y., 864, 876(5), 880
Li, Z. Y., 79
Liang, J., 204, 205(38)
Liao, M.-J., 59
Licari, P. J., 20
Lichtarge, O., 106, 112, 122, 179, 364, 539
Liebman, P. A., 193, 269, 377, 381, 386, 472, 612, 641, 657
Lien, T., 230
Light, D. R., 114
Liman, E. R., 775, 776(12)
Lin, C. M., 130
Lin, J.-H., 114
Lin, S. W., 116, 117, 119, 120(20), 124, 126(33), 128(11), 197, 208, 245, 249, 252, 257, 257(8), 258(8, 19), 260(19), 263(8, 19), 291
Lin, T. Y., 388
Lindau, M., 270
Linden, J., 363, 364(12), 388, 404(11), 643
Linder, M. E., 363, 372(2), 503, 526, 531(23), 534(23)
Lindorfer, M., 364
Ling, V., 865, 885, 887(22), 888(22), 889(22)
Lipkin, V. M., 636, 643(4)
Lippincott-Schwartz, J., 88
Lisman, J. E., 161
Litman, B. J., 72, 95, 97, 307, 377, 482
Liu, D. T., 755
Liu, J., 226, 364
Liu, K., 673, 685(12), 690, 712, 716(23)
Liu, L., 741
Liu, M., 783, 784(27), 785, 786(8), 819, 820(25), 821(25), 823(25), 824(25), 827(25), 828(25)
Liu, R. S. H., 298
Liu, X., 8, 11(11), 26, 195, 196
Liu, Y., 273, 540, 674, 675(28)
Livingston, R., 160
Livnah, N., 200, 204, 205(38)
Logunov, I., 201
Logunov, S. L., 145, 146, 147, 147(13, 15)
Lolley, R. N., 617, 690, 692, 696(25)
Lombardi, L., 679, 731
Lomize, A. L., 179, 185(5)

London, E., 59
Longstaff, C., 160, 180, 240
Lopez, S., 691
Lopez-Escalera, R., 723
Loppnow, G., 198
Lopresti, M., 464
Lottspeich, F., 833, 834(16), 843(16)
Lou, J., 219, 226, 227(14), 229(14)
Louvard, D., 601
Lowe, D. G., 673, 674, 675(20), 676, 685(8, 12), 690, 709, 711, 711(11), 712, 712(11, 22), 716(11, 23), 718, 730, 734(10)
Lowe, D. L., 708
Lozier, R. H., 173
Lu, C. C., 773
Lu, L., 673, 685(12), 690, 712, 716(23)
Lu, Z., 8, 11(11)
Luckow, V. A., 16, 584
Ludwig, B., 92
Luecke, H., 205
Lugtenburg, J., 26, 158, 191
Lui, T., 388
Lukashev, E., 200, 201(25)
Lukjanov, K. A., 530
Lupica, C. R., 363, 364(12)
Lupski, J. R., 864, 876(5), 879, 880, 880(2), 884(2)
Lupski, M. J., 864
Lussier, L. S., 163
Lustig, K. D., 364, 388(15), 389
Lutsenko, S., 828
Lynch, J. W., 785, 786(7)
Lyubarsky, A. L., 673, 755, 831

M

Macke, J. P., 60, 321, 324(29), 580
MacKenzie, D., 38, 44(21), 136, 209, 254, 273, 322, 324(32)
Macleod, H. A., 483
MacNichol, E. F., 239, 264, 267(33)
Macosko, J. C., 78
Madden, K. R., 3
Maddison, D. R., 320
Maddison, W. P., 320
Madland, W. T., 526
Maeda, A., 117, 129, 189, 195, 196, 196(42), 205, 206, 299
Maggio, R., 59

Mah, T. L., 168, 177(10), 178, 178(10), 249, 377, 384
Maki, M., 787
Makino, C. L., 270, 281(25), 285, 286(25)
Makino, E. R., 526, 535(20)
Malinski, J. A., 346, 364, 365(23), 367(23), 371, 371(23), 372(23), 495
Maloney, P. C., 885
Mancini, M., 77, 86(5)
Manganiello, V. C., 657
Mangini, N. J., 126
Maniatis, T., 34, 313, 322(7), 442, 459, 462(15), 624, 629
Mann, D. J., 571, 579
Mäntele, W., 183
Manutis, M., 456
Marala, R. B., 676
Maretzki, D., 329, 330, 341, 433, 435(13), 436(13), 473, 474(10), 476
Margulis, A., 676, 679, 730, 731, 737, 739, 739(21), 741, 741(21, 23)
Mariani, M., 580, 593(12)
Mark, J., 763
Marr, K., 148, 154(22)
Marshak, D., 738
Marshall, G. R., 164
Martell, A. E., 780
Marti, S., 206
Marti, T., 186, 195
Martin, E. L., 363, 364(10), 389, 390, 392(27), 397(19), 404(19)
Martin, J. E., 665, 670(52), 671(52)
Martinez, I., 617
Martinez-Mir, A., 864, 880
Martins, T. J., 610, 647
Marty, A., 274, 281(39), 290(39)
Marzesco, A. M., 601
Mas, M. T., 557
Masthay, M. B., 147, 154(21), 155(21)
Mathews, G., 199
Mathews, R., 198
Mathies, R. A., 117, 122(12), 128(11), 164, 191, 197, 198, 221, 269, 350
Matsuda, T., 195, 308, 348, 349, 349(7), 350(11), 357(7), 361(7)
Matsui, Y., 571
Matsumoto, H., 266, 300, 601, 647, 656
Matsuo, H., 755, 762(4), 764(4), 831, 833(7)
Mattheakis, L., 389
Matthes, M., 649

Matthews, G., 159, 268
Matthews, H. R., 526, 708, 742, 848
Matthews, R. G., 377
Matulef, K., 779
Maugeri, A., 880
Maul, G., 833, 834(16), 843(16)
Maule, A. J., 16
Maule, C. H., 487
Maumenee, I. H., 60
Mauzerall, D., 144, 147(10)
Mayer-Jaekel, R. E., 565, 566(26, 27)
Maylie, J., 828
Maze, M., 112(29), 113
Mazzoni, M. R., 363, 364, 364(11, 12), 365, 365(23), 367(11, 23), 369, 371, 371(23), 372(11, 23), 376(29), 388, 404(11), 490, 491, 493(7), 495, 520
McAleer, W. J., 45
McCarthy, N. E. M., 557
McCaslin, D. R., 97
McConnell, D. G., 329
McCray, G., 36, 51, 51(17), 298
McDermott, D. C., 483
McDowell, J. H., 12, 56, 60, 63, 70, 71, 76, 79, 88(27), 95, 96(40), 108, 111, 115(15), 254, 274, 295, 316(40), 324, 414, 422, 437, 447, 453, 455, 557, 558, 558(22), 559, 566(22), 567(20, 22), 695
McEntaffer, R., 638, 642(15), 643(15)
McGee, T. L., 60
McGowan, C. H., 570, 571, 575(4)
McIntire, W. E., 369
McKay, D. B., 406
McKee, T. D., 112, 130, 131(5), 134(5), 135(5)
McKenzie, D., 63
McLaughlin, M. E., 617
McLaughlin, S., 371
McMinn, T. R., 258
McNaughton, P. A., 746, 815, 848
McNew, J. A., 78
McNicoll, N., 674, 675(21)
McNiven, M. A., 78
McPherson, G. A., 661, 662
Mead, D., 298
Melia, T. J., Jr., 232, 238
Mellgren, R. L., 589
Menick, D., 200, 201(25)
Menini, A., 811, 818
Mercer, W. E., 570(7), 571
Merkx, M., 27, 29(22), 193

Merlevede, W., 565, 566(26, 27)
Mertz, J., 579
Merutka, G., 107
Metz, G., 201
Meyboom, A., 329, 330
Meyer, C. K., 332, 377, 476, 479(19)
Michaelis, S., 59
Michel, H., 92
Michel-Villaz, M., 270, 307, 329, 333(6), 528
Mierke, D. F., 112
Miki, N., 650, 651(21), 652(21), 656
Mikoshiba, K., 810
Milam, A. H., 71, 79, 404, 691, 702(21), 709, 710(18)
Milano, C. A., 258
Millar, R. P., 112
Miller, C. M., 88
Miller, J., 70
Miller, J. L., 367, 708, 810
Miller, L. K., 16
Miller, R., 447
Miller, R. J., 503
Miller, T. R., 674, 716
Miller, W. J., 45
Miller, Z., 679
Milligan, G., 365, 369, 388
Milligan, S., 79
Mills, J. S., 518, 529, 540, 549
Min, K. C., 181, 189, 255
Minkova, M., 208, 245, 252, 257(8), 258(8), 263(8)
Minta, A., 851
Mir, I., 13
Misra, S., 200
Mitchell, D. C., 97, 307, 377
Mitra, A. K., 93
Mitsuoka, K., 93
Mixon, M. B., 490, 513
Miyata, T., 456, 755, 762(4), 764(4), 810, 831, 833(7)
Miyoshi, Y., 533
Mizobe, T., 112(29), 113
Mizukami, T., 302, 303(19), 306, 348, 349(5, 7), 353(5), 357(7), 361(7)
Mizuno, T., 674, 675(19)
Moffat, L. F., 618
Mogi, T., 186
Mohanna Rao, J. K., 316(40), 324
Mohler, W. A., 39
Molday, L. L., 453, 755, 762(6), 785, 818,

818(24), 819, 831(9), 832, 833, 833(9), 834, 837, 838, 839(30), 842(31), 843(9, 32), 864, 865, 865(1), 866, 868, 871(1), 876(1), 879, 881(3), 883, 884(3)
Molday, R. S., 3, 8(1), 10(1), 13, 30, 38, 44(21), 50(1), 62, 63, 66(21), 136, 181, 209, 216(11), 254, 273, 322, 324(32), 453, 723, 755, 762(6), 784, 785, 786(5), 818, 818(24), 819, 819(9, 21), 825(21), 831, 831(9, 11), 832, 833, 833(9), 834, 834(17), 837, 837(11), 838, 838(14, 17, 24, 25), 839(30), 840(14, 17, 33), 842(31), 843, 843(9, 17, 32), 864, 865, 865(1), 866, 868, 871(1), 874, 876(1, 4), 879, 879(27), 880, 881(3), 883, 884(3)
Mollaghababa, R., 3, 270, 292(26)
Molokanova, E., 768, 774
Moltke, S., 270, 292(26)
Monod, J., 771
Montal, M., 292
Montell, C., 821
Montini, E., 580, 593(12)
Moomaw, C., 466
Moore, D. D., 608, 609(23)
Moore, D. M., 443
Moore, J. E., 314
Mori, Y., 810
Morisaki, H., 78
Morita, E. A., 139, 213
Morrison, D. F., 108, 126, 526
Morrow, E. M., 625
Morser, J., 114
Morton, R., 196
Morton, T. A., 788
Mosberg, H. I., 179, 185(5)
Moss, J., 388
Mosser, G., 93, 99(33)
Mou, H., 663
Mullaney, P. J., 672
Mullenberg, C. G., 118
Müller, F., 571, 573(13), 574(13), 578(13), 676, 690, 691(15), 755, 762(6), 785, 786, 787(10), 789(10), 792(10), 794(10), 796(10), 818, 818(24), 819, 825(20), 827(20), 831(9), 832, 833(9), 837, 843(9)
Mülsch, A., 692, 693, 698(30)
Mumby, M. C., 566, 579, 610, 647
Munier, H., 827
Munnich, A., 718, 880
Munson, P. J., 662, 663(40)

Murad, F., 685
Muradov, K. G., 540, 546(8), 549, 635, 636, 643(4)
Murakami, T., 571
Murata, K., 93
Murphy, R. L. W., 708, 742
Murray, L. P., 117, 147, 148, 151(26, 27), 152(27), 153(27), 154(27), 157, 158(19), 162(19, 69), 163
Musakov, K. G., 648, 652(16a)
Mynlieff, M., 810
Myszka, D. G., 788, 792(21), 793, 794(21), 796(21)

N

Naarendorp, F., 190
Nachliel, E., 385
Naeve, G. S., 627
Nagao, S., 730
Nagaraja, R., 45
Nagata, T., 189, 196(42)
Nagita, T., 129
Nagle, J. F., 173
Nairn, A. C., 579, 580
Nakagawa, M., 122, 196, 199, 206(18), 207(18)
Nakai, J., 810
Nakamura, T., 752, 818
Nakanishi, K., 117, 147, 158(19), 162(19), 219, 221, 222, 223, 226, 226(13), 227, 227(14), 229(14), 230, 264, 897
Nakashima, R., 92
Nakatani, K., 673, 690, 708, 742, 743, 744(8), 745, 746, 746(7, 8), 747, 748(8), 749, 750, 750(7), 751, 751(8, 9), 818
Nakayama, T., 456, 457(10, 11), 460(10, 11), 461(10, 11), 462(10, 11), 465(10, 11)
Nakayama, T. A., 21
Nam, S. H., 673
Nash, C., 364, 365(22), 367(22), 495, 546
Nasi, E., 3, 278
Nasonkin, I., 864, 876(4), 880
Nassal, M., 60, 135, 209, 321
Nathans, J., 3, 10(2), 11, 13, 30, 32, 44, 50, 56, 57, 57(2, 3, 26, 27, 32), 60, 116(2), 117, 185, 186, 206, 269, 272, 275(32), 276(29), 278(32), 291, 298, 313, 321, 324(29), 411, 415, 580, 864, 874, 876(5), 879, 879(27), 880, 880(2), 884(2)

Natochin, M., 517, 534, 539, 540, 546(8), 548(7), 549(7), 551(7), 552, 552(7), 554(7), 598, 636, 637(10), 642(10), 644(10), 645, 645(9)
Natochin, M. Yu., 3, 11(6)
Navon, S. E., 365, 372, 376(27)
Nawrocki, J. P., 70, 76, 557
Nazawa, Y., 730
Needels, M., 389
Needleman, R., 205
Neer, E. J., 365, 366(25), 367(25)
Nef, P., 405
Neher, E., 274, 281(39), 290(39), 775, 798, 800(11), 808(1, 2), 825
Nei, M., 313, 315, 320
Neitz, J., 117
Neitz, M., 117
Nekrasova, E. R., 534
Nemethy, G., 771
Nerbonne, J., 802, 803(15)
Nesbitt, J. A. D., 679
Nessenzveig, D. R., 258
Nestel, U., 92
Neubert, T. A., 405
Neubig, R., 364
Newton, A. C., 367
Nichimura, S., 196
Nicklen, S., 34, 36(15), 209, 448
Nickoloff, J. A., 32, 272
Nicol, G. D., 649, 650, 848
Nicoll, D. A., 679, 833, 845
Nie, Z., 625
Nielsen, S. M., 112
Niemi, G., 404, 406, 717
Nieto, J. L., 827
Niidome, T., 755, 762(4), 764(4), 810, 831, 833(7)
Nir, I., 3, 10(2), 30, 57(2), 272, 275(32), 278(32)
Nishii, K., 456
Nishimoto, T., 456
Nishimura, S., 117, 195
Nishizava, Y., 710
Nitecki, D. E., 114
Noda, C., 466
Noda, M., 59
Nodes, B. R., 219
Noel, J. P., 350, 363, 373(1), 376(1), 490, 496(2), 498(1, 2), 501(1, 2), 503, 512, 513, 514(6, 12), 526, 531, 539, 542, 542(2), 546(2, 14), 548(14)

Noren, C., 399, 402(28)
Noren, K., 399, 402(28)
Norisuye, T., 331
Northcote, D. H., 844, 868
Nothrup, J. K., 525, 531(4)
Noy, N., 578
Nuccitelli, R., 732
Numa, S., 59, 755, 762(4), 764(4), 831, 833(7)
Nunn, B. J., 730

O

Oatis, J. E., Jr., 557
Oberleithner, H., 274, 279(37)
O'Connell, K. L., 674, 675(20)
O'Connor, P. M., 570(7), 571
Oesterhelt, D., 95, 186
Ogreid, D., 666
Ohashi, M., 78
Ohba, M., 466
Ohba, T., 456
Ohgura, H., 690, 691(11), 702(11)
Ohguro, H., 71, 76, 217, 308, 350, 405, 557(18), 558, 676, 709, 718
Ohkita, Y. J., 196
Ohnishi, M., 571
Ohtsuki, K., 571
Ohyama, T., 797
Okada, T., 294(6), 296, 308, 348, 349(7), 357(7), 361(7)
Okano, T., 296, 297, 297(9), 298, 309, 349
Okano, Y., 204
Okayama, H., 36, 272
Olins, P. O., 16, 406, 584
Oliva, G., 114
Olofsson, A., 93, 99(33)
Olshevskaya, E. M., 690, 691(14), 701(14)
Olshevskaya, E. V., 405, 676, 690, 691, 709, 710, 710(15), 711(11, 15), 712(11), 713(14, 15), 714, 714(14), 716(11, 14), 717(14), 730, 734(10)
Olson, A., 812
O'Malley, B. W., 618
Ong, H., 674, 675(21)
Ono, T., 456
Onorato, J. J., 421, 433, 435(11), 436(11)
Onrust, R., 364, 388, 539
Oppert, B., 526, 636, 643(6)

Oprian, D. D., 3, 8(1), 10(1), 13, 30, 50(1, 4), 51, 57, 57(34), 60, 62, 66(21), 112, 116(3), 117, 122(12), 123, 130, 131(5), 134, 134(5), 135, 135(5), 139, 140(8, 9), 141, 158, 162(44), 181, 185, 186, 190, 204, 204(2, 3), 205, 206, 208, 209, 209(1), 210, 211, 212, 213(12), 216(11, 16, 18), 238, 249(3), 250, 252, 257(11), 258, 258(11), 263, 263(11), 291, 321, 322, 406
Orcutt, B. C., 320
O'Reilly, D. R., 16
Orsini, M. J., 424
Ort, D. R., 154, 156(34)
Osawa, S., 388, 389, 411, 412, 415(5), 416, 418, 436, 437, 502
Osborne, J. C., Jr., 823
Oseroff, A., 197
O'Shannessy, D. J., 792, 792(20), 793, 794(19, 20), 796(20)
Osterhelt, D., 93
Ostermeier, C., 92
Ostroy, S. E., 377
O'Tousa, J. E., 79, 456, 459(2), 580
Ottenhoff, T., 389
Otto-Bruc, A., 473, 517, 518, 528, 539, 540(3), 546(3), 691, 702(21), 709, 710(18), 717
Ottolenghi, M., 158, 163, 171, 197, 198, 201, 203, 203(16), 205(32)
Oura, T., 294(8), 295, 296, 300(8), 301, 308(8), 310
Ovchinnikov, Y. A., 60, 557
Owen, W. G., 708
Ozaki, M., 456

P

Pages, F., 527, 601, 646
Pagliara, A. S., 679
Pak, W., 21, 270, 275(18), 455, 456, 456(1), 457(10), 458(1), 459(2), 460(10), 461(10), 462(10), 465(1, 10), 579
Pakula, A. A., 130
Palczewski, K., 56, 63, 70, 71, 75, 76, 217, 238, 239(6, 8), 240, 241, 241(6), 243, 243(6), 244(6), 245(6), 247(6), 249(7), 250, 252, 257(6), 264, 264(5), 267, 331, 332(18), 334, 335, 335(18), 336(18), 339, 339(18), 341, 342(28), 405, 406, 411, 414, 433, 435(13), 436(13), 437, 453, 455, 475, 476, 526, 534(16), 535(16), 557(18), 558, 558(22), 559, 566(22), 567(20, 22), 612, 641, 647, 657, 676, 679(38), 684(31), 690, 691, 691(11, 13), 692, 693(10, 13, 26), 698(26), 699(10, 13), 701(10, 13), 702(11, 13, 31), 704(10), 705(13), 707(13), 709, 710(17, 18), 711, 712, 713(16), 716, 716(16, 20, 26), 717, 718, 719, 730, 734(9, 11, 12), 831
Palings, I., 191
Pallast, M., 350, 548, 551(18)
Palmer, R., 60
Paloma, E., 864, 880
Palvzewski, K., 691
Pande, A., 163, 198
Pande, J., 163
Pannbacker, R. G., 677
Pante, N., 456
Papac, D. I., 557
Papermaster, D. S., 3, 10(2), 11, 13, 30, 44, 57(2, 26), 77, 78(3), 79(1), 80, 80(1), 81, 81(1, 30), 83(1), 85(1, 4, 30), 86(1, 4, 5), 87(1, 4), 96, 272, 275(32), 278(32), 291, 334, 612, 679, 694, 864, 881, 884
Papke, R. L., 775
Pappin, D. J. C., 557, 763
Parker, F., 114
Parker, J. M. R., 453
Parker, K. R., 472, 530
Parker, M. W., 674, 675(25)
Parkes, J. H., 193, 269, 377, 381
Parlati, F., 78
Parmley, S. P., 444
Parodi, L. A., 173
Parr, G. R., 59
Parson, W. W., 154, 156(34)
Pastan, I., 679, 887, 895, 896(34)
Patel, N., 195
Pato, M. D., 571
Pattabhi, V., 114
Patton, C. W., 732
Paulus, H., 669
Payne, A. M., 691, 712, 716(26)
Payne, R., 708
Pedersen, P. L., 875
Pederson, P. L., 683
Peiffer, A., 864, 876(5), 879, 880, 880(2), 884(2)
Pelletier, S. L., 322
Pellicer, A., 619

Pelligrini, M., 112
Peng, Y.-W., 755, 818, 831, 833(8)
Pepperberg, D. R., 126, 219, 247, 330
Perez-Sala, D., 126, 183, 186(22)
Perlman, J. I., 219
Perrault, I., 718, 880
Perry, R. J., 746, 815
Peteanu, L. A., 269
Peters, K. S., 144, 147(7), 148, 154(22)
Peters-Bhatt, E., 389
Peterson, B. Z., 828
Pfaller, W., 274, 279(37)
Pfister, C., 346, 527, 601, 646
Phair, R. D., 88
Philippov, P. P., 75, 404, 405, 525, 526(7), 530, 531(7), 535
Philipson, K. D., 845
Phillips, W., 496
Phillips, W. J., 241, 256
Phizicky, E. M., 440
Picco, C., 811
Picones, A., 708
Pierce, B. M., 117, 147, 158(19), 162(19)
Pinckers, A. J. L. G., 880
Pines, M., 365
Pinkers, A. J., 864
Pirenne, M. H., 199, 251
Pistorius, A. M., 110
Pistorius, A. M. A., 193
Pittler, S. J., 617
Plant, T., 798
Poenie, M., 800, 814(12)
Pogozheva, I. D., 179, 185(5)
Poirier, M. A., 78
Polans, A. S., 437, 690, 691, 691(13), 693(13), 699(13), 701(13), 702(13), 705(13), 707(13), 709, 713(16), 716(16), 730, 734(11), 831
Pollard, H. B., 895, 896(34)
Ponder, J. W., 164
Ponstingl, H., 580
Popescu, A., 827
Popot, J.-L., 59, 69(5)
Popp, M. P., 3, 5(8), 11(8), 13
Porath, J., 73
Portier, M. D., 27, 28(24), 29(24)
Porumb, T., 405
Posner, B. A., 363, 372(7)
Possee, R. D., 16
Potter, L. R., 676

Pozdnyakov, N., 674, 675(30), 676, 730, 737, 739, 739(21, 23), 741, 741(21)
Pratt, D., 160
Precheur, B., 827
Prescott, A. R., 580
Prestwich, G. D., 644
Price, E. M., 887
Proctor, W. R., 363, 364(12)
Promin, A., 364
Pronin, A. N., 369
Proske, R., 466
Prussoff, W. H., 643
Prusti, R. K., 600, 608, 609(22), 647, 653, 672
Prusti, R. P., 601, 609(8)
Ptasienski, J., 421, 433, 435(11), 436(11)
Puca, A. A., 580, 593(12)
Pugh, E. N., Jr., 347, 524, 598, 646, 651, 652(25), 656(25), 657, 657(2), 673, 691, 708, 730, 741(1), 743, 812
Puig, J., 447
Puleo Scheppke, B., 77, 79(2), 83(2), 84(2), 85(2), 86(2), 87(2)
Pullen, N., 557(17), 558
Pulsifer, L., 365
Pulvermüller, A., 70, 238, 240, 241, 243, 249, 252, 329, 331, 332(18), 334, 335, 335(18), 336(18), 339, 339(18), 341, 342(28), 433, 435(13), 436(13), 471, 476, 480(1)

Q

Qin, N., 600, 617
Quinn, A. M., 684, 685(56)

R

Raap, J., 26
Rabe, J., 483, 485(31)
Racz, E., 690
Radding, C., 198
Radlwimmer, F. B., 312, 322, 324(35)
Rae, J. L., 291
Rafferty, C. N., 118, 189, 329
Raggett, E. M., 787
Ragsdale, H. L., 669
RajBhandary, U. L., 195
Rakhilin, S. V., 535

Raman, D., 411, 437
Rambotti, M. G., 741, 742(25)
Ramdas, L., 525, 531(3)
Ramjeesingh, M., 869
Ranck, J. L., 93, 99(33)
Randall, L., 456, 459(2)
Rando, R. R., 126, 160, 180, 183, 186(22), 240
Ransom, N., 77, 86(5)
Rao, J. K. M., 295
Rao, V. R., 51, 57(34), 190, 208, 258
Rarick, H. M., 518, 540, 549
Rasenick, M. M., 363, 364(9)
Rath, P., 26, 123, 193, 194, 196, 198
Rattner, A., 864, 876(5), 879, 880(2), 884(2)
Ratto, G. M., 708
Raveed, D., 677
Raviv, Y., 895, 896(34)
Ray, S., 404, 406
Read, S. M., 844, 868
Rebbeor, J. F., 888
Rebrik, T. I., 818
Reed, R. R., 755, 818, 831, 833(8)
Rees, S., 388
Reeves, P. J., 3, 10(3), 55, 56(36), 57(36)
Regan, J. W., 59
Reichert, J., 180, 329, 332(4), 348, 352(3), 353(3), 377, 379(1), 476
Reid, D. M., 833, 834(17), 838(14, 17), 840(14, 17), 843(17), 883
Reilander, H., 833, 834(16), 843(16), 848, 857(3)
Reinsch, C., 355
Reithmeier, R. A. F., 118
Rennie, A. P., 483
Rens-Domiano, S., 363, 364(10), 389, 397(19), 404(19)
Resek, J. F., 179(11), 180
Resink, T. J., 564, 566(25)
Reuhl, T. O., 775
Reuter, T., 232, 239
Rhoads, A. R., 797, 819
Ricci, A. J., 798
Richards, F. M., 111
Richards, J. E., 57, 190
Richter, H.-T., 205
Rickman, C. B., 634
Ridge, K. D., 3, 5(8), 8, 10, 11(8, 11), 13, 59, 60, 63(13), 64(13), 106, 115, 131, 179, 269, 273(15), 417

Riediger, W., 730
Rieger, R., 580
Rigaud, J. L., 93, 99(33)
Rim, J., 117, 211, 406
Ringold, G. M., 618
Riordan, J. R., 865, 872(13)
Rivard, A., 828
Robert, M., 346
Roberts, G. C. K., 111
Roberts, J. D., 322
Robertson, S., 431
Robinson, D. W., 746
Robinson, P. R., 57, 123, 139, 185, 204(2, 3), 207, 208, 209, 209(1), 212, 216(18), 238, 249(3), 250, 252, 257(11), 258, 258(11), 263, 263(11), 291, 321, 324(28)
Robinson, W., 651
Robishaw, J. D., 365
Rockman, H. A., 258
Rodman, H., 162(71), 163
Rodriguez, H., 674
Rodriguez de Turco, E. B., 77, 78(3)
Roeber, J. F., 366
Roeder, R. G., 627
Roep, B., 389
Roepman, R., 456, 460(12), 461(12), 462(12), 465(12), 466(12)
Rohr, M., 144, 145, 145(9), 147(12)
Rohrschneider, K., 864, 880
Romens, J. M., 869
Rondeau, J. J., 674, 675(21)
Roosien, J., 3, 10(5), 16, 56
Roques, B. P., 114
Rose, D. R., 875
Rosenberg, A. H., 544
Rosenfeld, P. J., 60, 880
Rosenick, M. M., 388
Rosenthal, W., 350, 548, 551(18)
Ross, E. M., 241
Rossien, J., 13, 14(3), 26(3)
Roth, R., 78
Rothberg, J. E., 144
Rothenhäusler, B., 483, 485(31)
Rothman, J. E., 77, 78, 78(7), 83(6)
Rothschild, K. J., 26, 123, 186, 193, 194, 195, 196
Rousso, I., 201, 205(32)
Royo, M., 112
Rozet, J.-M., 718, 880
Rudnicka-Nawrot, M., 217, 341, 433, 435(13),

436(13), 476, 557(18), 558, 691, 716, 717, 730, 734(12)
Rudzki, J. E., 144, 147(7)
Rueter, T., 159
Ruetz, S., 895
Rugarli, E. I., 580, 593(12)
Ruiz, C. C., 405, 676, 690, 691, 691(11), 702(11), 709, 718
Ruiz, M. L., 755, 756(11), 769, 770(11), 771, 776
Ruoho, A. E., 540, 674, 675(28)
Ruppel, H., 270
Rush, M., 464
Russell, S. J., 444
Russell, T. R., 679
Rustandi, R. R., 534
Ryba, N., 528

S

Saari, J. C., 238, 239(8), 249(7), 250, 264, 267
Sabacan, L., 638
Sabatini, D. D., 78
Sacchi, N., 321
Sachs, K., 473, 474(10)
Sackmann, E., 483
Sagoo, M. S., 526, 818
Saibal, H. R., 106
Saimi, Y., 828, 848
Saito, T., 308, 350
Saitou, N., 315
Sakaguchi, K., 570(7), 571
Sakai, M., 222
Sakai, N., 897
Sakamoto, T., 421
Sakmann, B., 274, 281(39), 290(39), 775, 798, 825
Sakmar, T. P., 30, 39, 57, 91, 106, 112, 116, 117, 119, 120(20), 122, 123, 123(21), 124, 126(33), 127(34), 134, 158, 159, 160, 163, 178, 179, 181, 182(21), 184, 185(6, 20), 186, 187, 187(32), 188(37, 46, 49, 50), 189, 190(46), 191, 192(49, 50), 193, 194, 194(43), 199, 206, 208, 220, 243, 245, 249, 251, 252, 253, 255, 256, 257, 257(8), 258(8, 9, 19), 260, 260(19), 261(9), 263(8, 9, 19), 264, 269, 287(12), 291, 292(7, 9, 12), 322, 332, 350, 385, 388, 437, 472, 473, 474(4), 476, 482(4)

Salamero, J., 78
Salamon, Z., 483
Sale, G. J., 557
Saloman, Y., 679
Samama, P., 259, 260(27)
Sambrook, J., 34, 313, 322(7), 442, 459, 462(15), 624, 629
Sampath, A. P., 708, 848
Sampogna, R., 205
Sandberg, M. A., 60, 79, 617
Sandorfy, C., 163
Sanger, F., 34, 36(15), 209, 448
Santerre, R. F., 45
Sarkadi, B., 887
Sasahara, Y., 571
Sasaki, J., 195, 196, 205
Sassenrath, G., 230
Sastry, L., 230
Satchwell, M. F., 31, 272, 273(31), 276(31), 278(31), 287(31), 289(31)
Sather, W. A., 526
Sato, M. H., 114
Savchenko, A., 768, 774
Sawyer, T., 389
Scarborough, G. A., 93, 887
Scavarda, N., 456, 459(2)
Schaad, N. C., 404
Schade, U., 730
Schaechter, L. E., 530
Schaeren-Wiemers, N., 883
Schaffner, K., 144, 145(8)
Schägger, H., 762
Schatz, P. J., 363, 364(10), 389, 390, 392(27), 397(19), 404(19)
Schepers, T., 437, 453
Schertler, G. F. X., 69, 91, 93, 94, 97, 98, 99, 99(18), 101(23), 102(18, 46), 103, 103(18), 104(8, 52), 106, 106(18, 23, 52), 118, 128, 129(17), 179, 185(1)
Schertler, G. R. X., 108, 112(2)
Schey, K. L., 369, 540, 636, 645, 645(9)
Scheybani, T., 93, 99(31)
Schick, D., 456, 460(12), 461(12), 462(12), 465(12), 466(12)
Schick, G. A., 148, 151(27), 152(27), 153(27), 154(27), 157
Schiltz, E., 92
Schimmel, P., 59
Schinkel, A. H., 895
Schirmer, R. H., 118

Schirra, C., 798
Schlaer, S., 199
Schleicher, A., 329, 331(7), 332(4, 7), 339(7), 343(7), 344(7), 386, 421, 433, 435(12), 436(12), 472, 476, 480(3)
Schloot, N., 389
Schmerl, S., 79, 88(27)
Schmid, E., 483
Schmid, E. D., 126, 183, 186(22)
Schmid, M., 864, 876(4), 880
Schmidt, H. H., 148, 151(26)
Schmidt, R. J., 45
Schnapp, B. J., 39, 277
Schneggenburger, R., 798, 808(1)
Schneider, B. G., 11, 44, 57(26), 291, 864, 884
Schnetkamp, P. P. M., 580, 676, 679(41, 42), 684(41, 42), 711, 718, 719, 722(9), 723, 724(9, 10), 731, 831, 845, 848, 849, 850(13), 852, 852(18–20), 853, 853(17), 856, 857, 857(6, 21), 860(6), 863(6)
Schneuwly, S., 456
Schoenlein, R. W., 269
Schoenmakers, T. J., 696
Schouten, A., 16
Schrader, K. A., 818
Schroder, W. H., 857
Schubert, C., 435
Schuck, P., 787, 792(22), 793, 794(13, 22), 796(13, 22), 797(22)
Schulenberg, P. J., 144, 145(9)
Schulten, K., 144, 201
Schults, G., 350
Schultz, G., 548, 551(18)
Schultz, J. E., 580, 730
Schulz, G. E., 92, 118
Schwartz, R. M., 320
Schwartz, T. W., 112
Schwarzer, A., 833
Schwede, F., 663, 665(51)
Schweitzer, G., 145, 147(12)
Schwemer, J., 198
Scott, B. L., 97
Scott, J., 389, 399, 402(28)
Scott, K., 190
Scott, K. M., 57
Sealfon, S. C., 112
Seavello, S., 456
Seddon, J., 864, 880
Sedlmeier, D., 730
Seemuller, E., 466

Segel, I. H., 890, 892(31)
Seidman, J. G., 443, 608, 609(23)
Seifert, R., 755, 762(6), 785, 798, 810(7), 814, 818(24), 819, 831(9), 832, 833(9), 843(9)
Seki, T., 456
Sekoguti, Y., 50
Selke, D., 570, 571, 573, 573(13, 16), 574, 574(13), 575(13, 16, 19), 577(16), 578(13)
Semple-Rowland, S. L., 691, 718
Sen, I., 647, 651(11)
Senecal, J.-L., 456
Senior, A. E., 887, 888, 888(27), 889
Senter, P. D., 692, 693, 698(30)
Serrano, L., 114
Sesti, F., 755, 762(6), 785, 818(24), 819, 831(9), 832, 833(9), 843(9)
Shank, C. V., 269
Shapiro, A., 865, 885, 887(22), 888(22), 889(22)
Sharma, R. K., 673, 674, 675(30), 676, 685(9), 690, 691, 708, 712, 716, 718, 730, 734(12), 737, 739(21), 741, 741(1, 21)
Sharma, R. P., 568
Sharma, S., 875
Shaw, G. P., 71
Sheets, M. F., 273, 274(36), 278(36)
Sheikh, S. P., 106, 112, 122, 179
Shen, J., 389
Shen, S., 570(7), 571
Shenolikar, S., 579
Sherman, P. M., 580
Sheves, M., 163, 171, 183, 189(26), 197, 198, 200, 201, 203, 203(16), 204, 205(31, 32, 38)
Shi, W., 412, 415(5), 416, 418
Shichi, H., 118, 189, 557
Shichida, Y., 117, 128(11), 129, 189, 196, 196(42), 197, 204, 221, 222, 293, 294, 294(6–8), 295, 296, 298, 299, 300(8), 301, 302, 303, 303(19), 304, 305, 306(16), 308, 308(8), 310, 347, 348, 349, 349(5, 7), 350, 350(1), 353(5), 357(7), 361(7), 363(8)
Shieh, B.-H., 456
Shieh, T., 117, 179, 185(6), 269, 292(7)
Shih, A., 464
Shih, T. M., 773
Shimizu, Y., 466
Shimonishi, Y., 349, 350(11)
Shin, Y. K., 78
Shinohara, T., 417

Shinzawaitoh, K., 92
Shirokova, E. P., 3, 11(6)
Shlaer, S., 251
Shmukler, B. E., 3, 11(6)
Shopsin, B., 78
Shortridge, R., 456
Shrager, R. I., 172
Shroyer, N. F., 864, 876(5), 879, 880, 880(2), 884(2)
Shukla, P., 31, 268, 271, 275(27), 277(27), 281(27), 286, 286(27), 287(45), 289(27, 45), 292(45)
Shuler, M. L., 18
Shyjan, A. W., 673, 685(8), 690, 711, 712(22)
Siebert, F., 123, 126, 178, 180, 181, 182(21), 183, 184, 185(20), 186, 186(22), 187, 188(37, 46, 49, 50), 189, 189(23, 26), 190(46), 191, 192(49, 50), 193, 201, 206, 260, 269, 292(9), 350, 378, 481(46), 487
Siegel, R., 3
Siegelbaum, S. A., 755, 776, 777(14), 817
Siemiatkowski-Juszcak, E. D., 60
Sieving, P. A., 57, 60, 190
Siggia, E. D., 88
Sigler, P. B., 273(1), 276(1), 350, 363, 372(6), 376(6), 435, 490, 491, 496(2), 498(1, 2), 501(1, 2), 503, 512, 513, 514(6, 12), 526, 531, 539, 542, 542(2), 546(2, 14), 548(14)
Sigworth, F. J., 274, 281(39), 290(39)
Silbaugh, T. H., 159
Sillén, L. G., 780
Silverstein, S., 619
Simon, J. D., 144
Simon, J. P., 78
Simon, M. I., 130, 364, 365, 531, 534, 534(41)
Simonds, W., 388
Simons, K., 77, 86(5)
Sinelnikova, V. V., 405
Singer, S., 389
Singh, N., 864, 876(5), 879, 880, 880(2), 884(2)
Singh, S., 674
Sitaramayya, A., 70, 657, 673, 674, 675(30), 676, 677, 679, 685(9), 690, 691, 708, 716, 718, 730, 731, 737, 739, 739(21), 741, 741(1, 21, 23)
Siu, J., 81
Skehel, J. J., 78
Skiba, N. P., 363, 364, 372(6), 376(6), 490, 491, 493(7), 502, 503, 504(4), 511(4), 513(4), 514(4), 517, 517(4), 518, 520, 520(5),
526, 529, 529(18), 535(18), 539, 540(6), 546(6), 549, 549(6), 636, 638(8), 639(8), 643(8)
Slaughter, C., 466, 566
Slepak, V. Z., 638
Slice, L. W., 406
Smallwood, P. M., 580, 864, 876(5), 879, 880(2), 884(2)
Smigel, M. D., 241, 496
Smiley, B. L., 3
Smith, A. J., 895
Smith, B. L., 93
Smith, D., 457, 459(14)
Smith, D. P., 557
Smith, G. E., 16
Smith, G. P., 437, 444
Smith, H. G., 95
Smith, J. A., 443, 608, 609(23)
Smith, L., 431
Smith, P. H., 294
Smith, R. D., 773
Smith, S. O., 117, 124, 126, 126(33), 127(34), 158, 179, 185(6), 208, 220, 226, 227(17, 18), 245, 249, 252, 255, 257, 257(8), 258(8, 9, 19), 260(19), 261(9), 263(8, 9, 19), 264, 269, 291, 292(7), 476
Smith, W. C., 3, 5(8), 11(8), 13, 341, 433, 435(13), 436(13), 437, 476
Smyk-Randall, E., 60
Snoek, G., 78
Snook, C. F., 114
Sobue, K., 730
Soderling, S. H., 597
Sokal, I., 453, 526, 534(16), 535(16), 647, 691, 712, 716(26)
Sollner, T. H., 77, 78, 78(7)
Somers, R. L., 557
Sonar, S., 195
Sondek, J., 363, 372(6), 376(6), 491, 512, 526, 542, 546(14), 548(14), 638
Soneson, K. K., 792, 794(19)
Song, L., 145, 147(13)
Sonnenburg, W. K., 672
Soparkar, S., 389
Sophianopoulos, A. J., 669
Sophianopoulos, J. A., 669
Sorbi, R. T., 760
Souied, E., 718, 880
Soulages, J. L., 483
Southern, P. J., 37

Sowadski, J. M., 674, 675(23), 684, 685(57)
Spalding, T. A., 258
Spalink, J. D., 270, 287(20)
Spencer, M., 404, 673, 685(12), 690, 712, 716(23)
Spiegel, A., 365, 388
Sports, C. D., 411, 436
Sprang, S. R., 363, 372(2, 3, 7), 490, 503, 513, 526, 540, 552, 674, 675(27)
Spreca, A., 741, 742(25)
Sprong, H., 895
Srebro, R., 199
Stadel, J. M., 91
Staehelin, T., 375, 465, 820, 869
Stahlman, M., 70
Stalfors, P., 564, 566(25)
Stamnes, M., 456
Stark, W. S., 579
Stauffer, D., 864, 876(5), 879, 880(2), 884(2)
Staurenghi, G., 897
Stec, W. J., 693
Stecher, P., 674, 675(22)
Steel, F. R., 580
Stefani, E., 278, 291(44), 292(44)
Steigner, W., 274, 279(37)
Steinberg, G., 163, 198, 203(16), 204, 205(38)
Steiner, A. L., 679
Stephenson, R., 456, 459(2)
Stern, L. J., 186
Sterne-Marr, R., 411, 433, 435(11), 436(11)
Sternmarr, R., 421
Sternweis, P. C., 241
Stevens, P. A., 329, 330
Stewart, W. J., 487, 493
Stieve, H., 270, 287(20)
Stillman, B. W., 31, 32(7), 271
Stillman, C., 3
Stoddard, B. L., 142
Stokes, D. L., 93
Stolz, L., 827
Stone, E. M., 79
Stone, K., 464
Stora, T., 483
Stotz, S. C., 818
Strassburger, J. M., 145, 148, 148(14), 154(14, 23), 157
Street, A. J., 580
Strong, L. A., 161
Struhl, K., 443, 608, 609(23)
Strühmer, W., 59

Struthers, M., 130, 134, 140(9)
Stryer, L., 9, 68, 330, 371, 404, 405, 406, 517, 525, 531, 537(36), 548, 551(17), 608, 635, 636, 637, 641, 652, 656(29), 657, 667(36), 673, 690, 693, 698(3), 708, 709(10), 730, 742, 743, 780, 803, 817
Stuart, J. A., 158
Stubbs, G. W., 95
Studier, F. W., 544
Stühmer, W., 755, 762(4), 764(4), 831, 833(7)
Stults, J. T., 405, 674, 675(20), 676, 690, 691(14), 701(14), 709, 710(15), 711(15), 713(15)
Subbaraya, I., 405, 676, 690, 691, 691(11, 13), 693(13), 699(13), 701(13), 702(11, 13), 705(13), 707(13), 709, 710(17), 713(16), 716(16), 717, 718, 730, 734(11)
Suber, M. L., 617
Subo, H., 59
Subramani, S., 618
Subramaniam, S., 206
Sugino, Y., 533
Sullivan, J. M., 31, 268, 271, 272, 273(31), 275, 275(27), 276(31), 277, 277(27, 40), 278(31), 281(27), 283, 284(40), 286, 286(27, 40, 42), 287(31, 45), 289(27, 31, 45), 291(40), 292, 292(45)
Sullivan, R., 388
Summers, M. D., 16
Sun, C., 321, 324(29)
Sun, H., 112, 580, 625, 864, 874, 876(5), 879, 879(27), 880, 880(2), 884(2)
Sun, Y., 388(14), 389
Sunahara, R. K., 363, 372(3), 540, 674, 675(27)
Sung, C.-H., 11, 44, 57(26, 27), 291
Surendran, R., 598
Surgucheva, I., 690, 691, 691(13), 693(13), 699(13), 701(13), 702(13, 21), 705(13), 707(13), 709, 710(18), 712, 713(16), 716, 716(16, 26), 717, 730, 734(11, 12)
Surguchov, A., 709, 710(17)
Surya, A., 238, 239(5), 245, 252, 257(10), 258(10)
Suryanarayana, S., 112(29), 113
Sutherland, E. W., 610
Sutton, R. B., 78
Suzuki, H., 59
Suzuki, T., 50, 152(67), 163
Swaffield, J., 466
Swandulla, D., 798

Swanson, R., 3, 278
Swanson, R. J., 139, 213
Swingle, M. R., 579
Sykes, B. D., 111
Szabo, A., 827
Szarian, S. M., 879
Szerencsei, R. T., 723, 845, 848, 852, 852(18–20), 853, 853(17), 856, 857(6, 21), 860(6), 863(6)
Sztul, E. S., 78
Szundi, I., 168, 173, 175(18), 177(10), 178, 178(10), 249, 377, 384

T

Tachibanaki, S., 294(8), 295, 296, 299, 300(8), 301, 302, 306(16), 308(8), 310, 347, 348, 349(7), 357(7), 361(7), 363(8)
Takahashi, E., 466
Takano, E., 787
Takao, T., 349, 350(11)
Takemoto, D. J., 108, 526, 601, 636, 643(6)
Takemoto, L. J., 108
Takeuchi, H., 122
Tallent, J. R., 158, 160, 161(60), 162(60), 163
Tamura, S., 571
Tamura, T., 690, 743, 744(8), 746(8), 750, 751(8)
Tan, Q., 219, 226, 227(14), 229(14)
Tanabe, T., 755, 762(4), 764(4), 831, 833(7)
Tanaka, K., 466
Tanaka, T., 405
Tanis, S. P., 227
Taniuchi, H., 59
Tanner, M. J. A., 59
Tansik, R., 389
Tate, E., 389
Tatsumi, M., 350, 656
Tavan, P., 144
Tavoloni, N., 570(6), 571
Taylor, J., 364
Taylor, J. A., 217, 557(18), 558
Taylor, S. S., 406, 674, 675(23), 684, 685(57)
Taylor, W. R., 270, 281(25), 285, 286(25), 320
Tempia, F., 798
Tempst, P., 77, 78(7)
Ten Eyck, L. F., 684, 685(57)

Tensel, T. G., 346
Teplow, D. B., 365
Terada, S., 755, 762(4), 764(4), 831, 833(7)
Terakita, A., 129, 189, 196(42), 293, 294(8), 295, 296, 299, 300(8), 301, 302, 306(16), 308, 308(8), 310, 347, 348, 349(7), 357(7), 361(7), 363(8)
Terasawa, T., 571
Terstegen, F., 227
Tesmer, J. J. G., 363, 372(3), 526, 540, 552, 674, 675(27)
Thanh, H. L., 163
Thanos, S., 571, 573(13), 574(13), 578(13)
Thaw, C. N., 258
Theuvenet, A. P. R., 696
Thill, G. P., 3
Thomas, D., 313
Thomas, R. K., 483
Thomas, T. O., 490, 491, 502, 513, 520
Thompson, D. A., 3, 278
Thompson, J. D., 313
Thompson, W. J., 641
Thomson, J. L., 879(5), 880
Thorgeirsson, E. T., 306
Thorgeirsson, T. E., 148, 157, 158(24), 175, 176(20), 177(20), 355, 357(25)
Thornburg, K. R., 540
Thornton, J. M., 320
Thorpe, D. S., 674
Thurmond, R. L., 3, 10(3), 55, 56(36), 57(36), 287, 292(46), 385
Thuysnma, R. P., 885
Tibbs, G. R., 755, 776, 777(14)
Tiberi, M., 258
Tijmes, N., 880
Tijmjes, N., 864
Ting, A., 647, 652(14)
Ting, T. D., 126, 533
Tittor, J., 186
Tkachuk, V. A., 823
Tocqué, B., 114
Toh-e, A., 466
Tokunaga, F., 298
Tollin, G., 483
Tolosa, S., 878
Tomizaki, T., 92
Tomson, F. L., 447
Tooze, S. A., 78, 79, 79(14), 81(26), 83(26)
Torney, D. C., 350
Toro, L., 773

Torregrossa, R., 3
Torriglia, A., 878
Towbin, H., 375, 465, 820, 869
Towler, E. M., 482
Towner, P., 557
Toyoshima, C., 93
Travis, G., 456, 457(10, 11), 460(10, 11), 461(10, 11), 462(10, 11), 465(10, 11)
Travis, G. H., 21, 864, 879
Travis, S. M., 570
Trehan, A., 298
Trenham, D. R., 802
Trentham, D. R., 803
Trippe, C., 77, 79(2), 83(2), 84(2), 85(2), 86(2), 87(2)
Trissl, H. W., 270, 292
Trivedi, B., 768, 774
Tsang, S., 264
Tsang, S. H., 525, 526(9), 527(9), 601, 647
Tschopp, J. F., 3
Tsien, R. Y., 708, 800, 802, 814(12), 851
Tsoi, M., 848
Tsuboi, S., 526, 601, 647
Tsuda, M., 122, 196, 199, 204, 205(38), 206(18), 207(18), 230
Tsujimoto, K., 647
Tsukida, K., 221
Tsukihara, T., 92
Tsurumi, C., 466
Tsygannik, I. N., 97
Tucker, C. L., 674, 716
Tucker, J. E., 852(19, 20), 853, 857
Tung, H. Y.L., 564, 566(25)

U

Udovichenko, I. P., 601
Uhl, R., 336, 528
Ui, M., 685
Umbarger, K. O., 526
Umeki, S., 730
Unger, V. M., 69, 91, 93, 97(26), 99, 104(8), 118, 129(17), 179, 185(1)
Unson, C., 388
Unwin, P. N. T., 93, 103(36)
Urbatsch, I. L., 887, 888(27), 889
Usmany, M., 16
Usukura, J., 647, 710

V

Vahrenhorst, R., 227
Vallury, V., 126
Van Amsterdam, J. R., 365, 366(25), 367(25)
Van Breemen, J. F. L., 97
Van Bruggen, E. F. J., 97
Van de Hulst, H. C., 331
van den Berg, E. M. M., 191
van de Pol, D. J., 864, 880
Van der Elzen, P., 852(19), 853
Van der Geer, P., 688
van de Ven, W. J. M., 3, 10(5), 13, 14(3), 26(3), 56
Van de Vosse, E., 580, 593(12)
Van de Woude, G. F., 570(7), 571
VanDort, M., 364
van Driel, M. A., 864, 880
van Groningen-Luyben, W. A. H. M., 3, 10(5), 13, 14(3), 26(3), 56
van Haren, F. J., 864, 880
van Helvoort, A., 895
Van Hooser, J. P., 71, 267, 676, 679(38), 692, 693(26), 698(26), 711, 716(20), 718
Van Hooser, P., 70, 238, 241, 243, 252, 334
VanLent, J. W. M., 16
van Meer, G., 895
van Meijgaarden, K., 389
Van Oostrum, J., 14, 20(8), 21(8), 28, 92, 478
van Veelen, P., 389
Varma, A., 258
Varnum, M. D., 777, 779, 784, 828
Varo, G., 154, 155(35)
Vaughan, M., 388, 823
Velicelebi, G., 3
Venkataraman, V., 741
Venturini, S., 399, 402(28)
Vihtelic, T. S., 79
Vilageliu, L., 617, 864
Villa, C., 93, 99, 101(23), 104(52), 106(23, 52), 108, 112(2), 128
Visser, G. J., 696
Vissers, P. M. A. M., 27, 28(24), 29, 29(24)
Vlak, J. M., 3, 10(5), 13, 14(3), 16, 26(3), 56
Vocelle, D., 163
Vogel, H., 244, 330, 342(14), 344(12), 377, 471, 483, 484(40), 485(38, 40), 487
Vogel, Z., 59
Vogt, T., 111, 115(15)
von Dippe, P., 886

von Heijne, G., 118
von Jagow, G., 762
Vought, B. W., 143
Voyno-Yasenetskaya, T., 388(14), 389
Vrabec, T., 431
Vuong, T. M., 330, 473, 517, 525, 528

W

Waddell, W. H., 230
Wagner, R., 528
Wagstrom, C., 389
Wahsburn, T., 580
Wald, G., 8, 198, 239, 294, 377, 557, 896
Walker, R. G., 820
Wall, M. A., 363, 372(7)
Wallace, B. A., 114
Wallace-Williams, S. E., 26, 196, 355, 357(25)
Wallace-Williams, S. T., 306
Walseth, T. F., 698
Walsh, A. H., 589
Walsh, K. A., 76, 217, 404, 405, 557(18), 558, 672, 676, 690, 691(11), 702(11), 718
Walter, G., 579
Waltho, J. P., 107
Walz, J., 466
Walz, T., 93
Wang, D. N., 93, 117, 122(12)
Wang, H., 571
Wang, J., 223, 226(13)
Wang, J. K., 295, 316(40), 324, 557
Wang, Q. L., 625
Wang, W., 869
Wang, X., 59
Wang, Y., 483
Wang, Z., 117, 122(12), 204, 205
Warner, J., 168, 176(12)
Warr, C. G., 828
Warren, G., 78, 83
Warren, M. J., 691, 712, 716(26)
Warren, R., 831
Warshel, A., 158
Wasley, L. C., 209
Watanabe, M., 363, 364(9), 388
Watanabe, N., 710
Watson, N., 526, 531(23), 534(23)
Wayman, G., 828
Weber, B. H. F., 864, 876(4), 880

Weber, T., 78
Wechesser, J., 92
Wei, L., 674, 675(24)
Weidlich, O., 186
Weidman, P. J., 85
Weigel DiFranco, C., 79
Weinstein, H., 112
Weiss, E. R., 388, 389, 411, 412, 415(5), 416, 418, 436, 437
Weiss, M. S., 92
Weiss, U., 78
Weiter, J. J., 897
Weitz, C. J., 3, 10(2), 13, 30, 57, 57(2), 185, 269, 272, 275(32), 278(32)
Weitz, D., 786, 787(10), 789(10), 792(10), 794(10), 796, 796(10), 797(23), 818, 825(20), 827(20), 837
Welsh, M. J., 570
Welte, W., 478
Welter, W., 92
Wensel, T. G., 232, 238, 371, 517, 520, 524, 525, 526, 526(8), 530(8, 19), 531(3, 8), 532(8), 533(8), 534(16), 535(16, 19), 552, 636, 647, 657, 667(36), 743
Wes, P. D., 821
Wess, J., 59, 364, 388
Wessling-Resnick, M., 139, 213, 413
West, J. L., 230
West, K. A., 776
West, R. E., 388
Westermann, B., 78
Wetlaufer, D. B., 59
Wetzel, M. G., 57
White, A. A., 680, 696
White, B. H., 404
Whiteheart, S. W., 77, 78(7)
Whitehorn, E., 389
Whitesides, G. M., 483, 487
Wickham, T. J., 18
Wieland, T., 531, 534, 534(41)
Wigler, M., 619
Wilce, M. C. J., 674, 675(25)
Wilden, U., 240, 334, 422
Wiley, D. C., 78
Wilken, N., 456
Wilkie, S. E., 29
Wilkie, T. M., 531
Willemen, S., 389
Williams, A., 755, 762(6), 785, 818(24), 819, 831(9), 832, 833(9), 843(9)

Williams, D. S., 367
Williams, J., 580
Williams, K., 464
Williams, T., 647
Wilson, D. R., 440
Wilson, E. M., 674, 675(26)
Wilson, S., 91
Winkfein, R. J., 848, 857, 857(6), 860(6), 863(6)
Winslow, J. W., 365, 366(25), 367(25)
Winter, G., 444
Winter, W. M., 85
Winzor, D. J., 792(20), 793, 794(20), 796(20)
Wirtz, K., 78
Wise, R. J., 209
Wishart, D. S., 111
Wistow, G. J., 417
Wohlfart, P., 845
Wolbring, G., 676, 679(41), 684(41), 711, 718, 719, 722(9), 724(9, 10), 845
Wolchok, S. R., 734
Wolf, L. G., 365
Wolley, G. A., 114
Wong, J. K., 377
Wong, S. C., 405, 674, 675(20), 676, 690, 691(14), 701(14), 709, 710(15), 711(15), 713(15)
Wood, H. A., 18
Wood, K. V., 618
Wood, S. P., 114
Woodard, C., 388
Woodfork, K., 364
Woody, R. W., 222
Woon, C., 389
Woon, C. W., 502
Wootton, J. F., 803
Wray, V., 463
Wray, W., 463
Wright, G. C., 617
Wright, P. E., 107
Wrighton, N., 389
Wu, J., 456
Wüthrich, K., 110, 111
Wyman, J., 771

X

Xia, D., 92
Xia, X.-M., 828
Xia, Y., 487
Xian, J. Z., 92
Xiao, W., 78
Xu, J.-Y., 44

Y

Yach, K., 106
Yagi, K., 730
Yahagi, N., 59
Yamaguchi, H., 92
Yamamoto, T., 230
Yamamoto, Y., 197
Yamamura, H. I., 661
Yamana, G., 803
Yamanaka, G., 371, 531, 537(36), 548, 551(17), 693
Yamane, T., 106
Yamasaki, M., 466
Yamashita, E., 92
Yamazaki, A., 350, 526, 601, 647, 650, 651 (11, 21), 652(14, 21), 656, 673, 677(5), 710
Yamazaki, M., 526, 647, 656
Yamazaki, R., 730
Yamazaki, Y., 205
Yanagawa, Y., 571
Yang, A.-S., 108
Yang, C.-S., 490, 491, 493(7), 503, 520, 520(5)
Yang, K., 104, 112(30, 31), 113, 115, 115(30, 31), 122, 179, 269, 290(14)
Yang, R., 673
Yang, R.-B., 674(14), 690, 712
Yang, T., 57, 204(3), 208, 263
Yang, Z., 320
Yanofsky, S., 389
Yao, K. W., 199
Yao, L. L., 60, 131
Yaono, R., 92
Yarfitz, S., 539, 551(1), 635
Yasuda, H., 364
Yau, K.-W., 159, 268, 294, 646, 673, 690, 708, 742, 743, 744(8), 745, 746, 746(3, 7, 8), 747, 748(8), 749, 750, 750(7), 751, 751(8, 9), 755, 771, 783, 784(26, 27), 785, 786, 786(6, 8), 817, 818, 819, 819(19), 820(19, 25), 821(19, 25), 822(19), 823, 823(19, 25), 824(25), 825(17, 19), 827, 827(25), 828(25), 831, 833(8), 837

Yazawa, M., 730
Yeager, M., 93, 95, 97(26)
Yeagle, P. L., 29, 107, 111, 113(16), 115(15)
Yee, R., 612
Yellen, G., 273
Yokosawa, H., 466
Yokota, K., 466
Yokoyama, N., 456
Yokoyama, R., 312, 313(5), 323
Yokoyama, S., 312, 313, 313(1, 5), 315, 319(4), 320, 322, 323, 324, 324(35), 325(2)
Yomosa, S., 226
Yonekura, K., 93
Yonemoto, W., 406
Yoshida, A., 676, 730, 737, 739(21), 741, 741(21)
Yoshikawa, S., 92
Yoshizawa, T., 117, 128(11), 196, 197, 204, 221, 222, 266, 294, 294(6, 7), 296, 297(9), 298, 299, 300, 301, 302, 303, 304, 305, 307, 308, 308(25), 309, 348, 349, 350, 350(11)
Yu, C. A., 92
Yu, G., 324
Yu, H., 112, 130, 131(5), 134, 134(5), 135(5), 140(8, 9), 141, 331
Yu, H.-H., 783, 786, 818, 819(19), 820(19), 821(19), 822(19), 823, 823(19), 825(19), 827, 837
Yu, L., 92
Yu, M., 821
Yu, W.-P., 783, 786, 818, 819(19), 820(19), 821(19), 822(19), 823, 823(19), 825(19), 827, 837
Yuan, C., 202, 204, 205
Yue, D. T., 828
Yun, J., 364
Yunfei, C., 456, 460(12), 461(12), 462(12), 465(12), 466(12)
Yurkova, E. V., 92

Z

Zabriskie, N., 864, 880
Zack, D. J., 625
Zagnitko, O., 466
Zagotta, W. N., 755, 776, 777, 777(13), 779, 781, 781(13), 782, 783, 784, 817, 828
Zahraoui, A., 601
Zakuor, R. A., 322
Zam, Z. S., 56, 63, 71, 414, 453
Zarucki, S. T., 618
Zeilhofer, H. U., 798
Zeilinger, C., 836, 874
Zemelman, B. V., 78
Zemlin, F., 93, 103(20)
Zenser, T. V., 696
Zhand, L., 92
Zhang, C.-F., 147, 154(21), 155(21), 189
Zhang, D., 144, 147(10)
Zhang, G., 674, 675(28)
Zhang, H., 230, 313, 315
Zhang, H. L., 570(7), 571
Zhang, J., 320
Zhang, J. P., 887
Zhang, K., 709, 710(17)
Zhang, L., 411, 436
Zhang, P. J., 93
Zhang, S., 823, 828(34)
Zhao, D. Y., 617
Zhao, X., 217, 405, 557(18), 558, 676, 690, 691(11), 702(11), 709, 718
Zheng, J., 674, 675(23), 684, 685(57)
Zhou, W., 112
Zhou, Y., 627
Zhou, Z., 798, 808(1, 2)
Zhu, G., 819
Zhu, Y., 526, 529(18), 535(18), 638
Zhukovsky, E. A., 116(3), 117, 123, 139, 158, 162(44), 186, 206, 208, 209, 209(1), 212, 213(12), 216(18), 252, 258, 291
Zidovetzki, R., 147, 154(21), 155(21)
Zimmerman, A. L., 693, 755, 774, 775(5), 784, 803, 818
Zinn, K., 775, 776(11)
Zipp, A., 106
Zoche, M., 786, 787, 787(10), 789(10), 792(10), 794(10), 796(10), 818, 825(20), 827(20), 837
Zolla-Pazner, S., 44
Zoller, M. J., 674, 675(23)
Zorn, M. A., 864, 884
Zozulya, S., 3, 11(6), 405, 406
Zscherp, C., 192
Zuckerman, D. M., 868
Zuhl, F., 466

Zuidema, D., 16
Zuker, C. S., 456
Zulauf, M., 93, 99(30, 31)
Zvyaga, T. A., 3, 11(6), 106, 112, 122, 123, 179, 181, 184, 188(49), 191, 192(49), 206, 255
Zwick, M., 389, 399, 402(28)
Zylka, M. J., 39, 277

Subject Index

A

ABCR
  ATPase activity
    activators and inhibitors, overview, 871–872, 887–888
    all-*trans*-retinal activation
      dose response, 888–889
      factors affecting activation, 889–890
      mechanisms, 894–895
      synergistic activation with amiodarone, 891–892, 894
    amiodarone activation
      dose response, 888
      synergistic activation with all-*trans*-retinal, 891–892, 894
    charcoal-binding assay, 882
    kinetic parameters, 871
    nucleotide-binding domains expressed as maltose binding protein fusion proteins, 878
    reconstituted protein, 874, 886
    thin-layer chromatography assay
      incubation conditions, 869–870
      linearity, 870
      product separation, 870
      sensitivity, 878–879
      solutions, 869
  ATP-binding cassette, 865
  gene cloning, 864
  localization in eye
    immunostaining, 883–884
    in situ hybridization, 883–884
  mutation in disease, 864, 879–880
  nucleotide-binding domain expression as maltose binding protein fusion proteins
    affinity chromatography, 877–878
    expression vector construction, 876–877
    induction of *Escherichia coli*, 877
    solubility, 875
  purification from bovine retina
    gel electrophoresis and Western blot analysis, 868–869
    immunoaffinity chromatography
      antibody preparation, 866
      binding and elution, 867–868, 881, 885
      buffers, 866
      column preparation, 881
      sample preparation and loading, 866–867, 884–885
      rod outer segment isolation, 865
      stability, 870–871
      yield, 872–873
  reconstitution
    porcine brain lipids, 882, 885–886
    soybean phospholipid vesicles, 873–874
  retinal flipping function, 895
  Stargardt disease pathogenesis, 896–897
  structure, 865
Absorption spectroscopy, rhodopsin
  advantages and applications, 164
  all-*trans*-retinal activation of opsin, 244–245
  bovine opsin expressed in *Pichia pastoris*, 8
  constitutively active opsin mutants, 216–217
  fragment reconstitution studies, 66–68
  kinetic range of photoreactions, 166
  meta-rhodopsin intermediates, 302–303, 305–308
  primary photolysis
    laser sources, 165
    signal/noise ratio optimization, 165
  Schiff base
    bleach measurement, 171
    flash lamps as light sources, 166–167
    flow cells, 168–169
    global fitting of time-dependent spectra, 171–173
    indicator dye monitoring, 177–178
    isospectral intermediate analysis, 177

935

kinetic resolution of activation steps, 176
kinetic scheme determination, 173, 175
pK_a determination from absorbance spectra with model chromophores, 201, 203
probe beam polarization, 167–168
protonated form, 196–198
signal averaging, 168–169, 171
time-resolved low-temperature absorption spectroscopy of rhodopsin intermediates interacting with transducin
  absorption maxima of intermediates, 361–362
  applications and overview, 350–351
  collected spectra, 353, 355
  experimental setup, 351–352
  GTP analog studies, 352, 359, 361, 363
  intermediates interacting with transducin, 348–349, 358–359, 361–363
  photoconverted pigment, quantification, 352–353
  rhodopsin purification, 349
  spectral analysis, 355, 357–358
  transducin purification, 349–350
transducin activation assays
  extra metarhodopsin-II, 476, 478
  photoregeneration from metarhodopsin-II state, 479
tryptophans in transmembrane segment
  data collection, 119–120
  difference spectroscopy analysis of mutants, 120, 122
  chromophore photoisomerization sensing, 123–124
wavelengths of maximal absorption, identification of critical amino acids
  ancestral pigment sequence elucidation, 320–321
  phylogenetic tree construction
    neighbor-joining method, 315, 319–320
    programs, 313–314
    unweighted pair-group method with arithmetic mean, 314–315
  site-directed mutagenesis in functional testing
    absorbance spectroscopy, 324
    applications, 325
    bovine A292S, 324
  cloning of genes, 321–322
  mutagenesis, 322–323
  purification, 324
  regeneration, 323–324
  RNA isolation, 321
  wild-type pigment absorption, 312, 323–324
ADRP, see Autosomal dominant retinitis pigmentosa
All-*trans*-retinal
  ABCR interactions
    ATPase activation
      dose response, 888–889
      factors affecting activation, 889–890
      mechanisms, 894–895
      synergistic activation with amiodarone, 891–892, 894
    flipping function, 895
  activation of constitutively active opsin mutants
    basal activity correlation with *trans*-retinal activation, 258–260
    immunoaffinity chromatography, 254
    membrane preparation, 253–254
    modeling of activation, 258–260
    retinal analog studies of activation, 264, 266
    Schiff base formation in mutants, 260–263
    transducin activation
      filter-binding assay, 255–256
      intrinsic tryptophan fluorescence assay, 256
    transient transfection in COS-1 cells, 253
    wild-type activation, 257
  activation of opsin
    absorption spectroscopy, 244–245
    activity level, 238, 249, 252, 257–258
    arrestin binding, 244
    mechanism of activation, 245
    opsin preparations
      membrane-bound preparation, 239–240
      methylation of membranes, 240
      phosphorylated opsin, 240
    pH dependence, 249–250, 257
    physiological role, 238–239, 266–267
    rhodopsin kinase binding, 244, 249–250

SUBJECT INDEX

transducin activation assay by intrinsic tryptophan fluorescence, 241, 243–245, 247–249
visual desenditization implications, 266–267
agonist functions, 257–260, 264
reduction by retinol dehydrogenase, 238
Amiodarone, ABCR ATPase activation
dose response, 888
synergistic activation with all-*trans*-retinal, 891–892, 894
Arrestin
binding of opsin/all-*trans*-retinal complex, 244
direct binding assay for rhodopsin
dissociation constant determination, 435–436
incubation conditions, 434–435
radiolabeled arrestin preparation, 433–434
troubleshooting, 435–436
variations of assay, 436
expression and purification of recombinant protein from *Escherichia coli*
ammonium sulfate precipitation, 430
anion-exchange chromatography, 431
cells
growth, 429
lysis, 429–430
expression vector construction, 428–429
gel electrophoresis, 428
heparin-Sepharose chromatography, 430–431
troubleshooting, 432–433
Western blot analysis, 431–432
expression in cell-free translation system
gel electrophoresis and fluorography, 427
radiolabeling, 423–424
stock solutions, 425
transcription, 423–425
translation reaction, 424–426
troubleshooting, 427–428
yield calculation, 426–429
gel electrophoresis assay of rhodopsin binding
arrestin preparation and metabolic radiolabeling, 417, 419

binding reaction, 419, 421
dissociation constant determination, 421–422
gel electrophoresis, 421
rhodopsin preparation and phosphorylation, 417
kinetic light scattering of rhodopsin complexes
arrestin preparation, 335
modeling of binding reaction, 342
peptide competition assay for rhodopsin ligands, 453–455
phage display of rhodopsin ligands
advantages in ligand screening, 439–440
data analysis, 448–451
data confirmation, 451–452
insert preparation, 440–442
library evaluation, 442–443
M13 coat protein utilization, 438–439
panning
controls, 445–446, 448
DNA preparation and sequencing, 447–448
elution of bound phage, 444–445, 447
phage culture, 446–447
rounds, 443–444, 447
substrate coating on dishes, 444
vectors, 440
purification, 240
rhodopsin binding
overview, 70–71, 238, 411, 422–423, 437
phosphorylation requirement, 422–423, 433–435
selectivity profile, 433–434
ARRP, *see* Autosomal recessive retinitis pigmentosa
ATR, *see* All-*trans*-retinal
Attenuated total reflection, *see* Fourier transform infrared spectroscopy
Autosomal dominant retinitis pigmentosa, rhodopsin mutations, 79
Autosomal recessive retinitis pigmentosa, cyclic GMP phosphodiesterase mutations, 617
8-*p*-Azidophenacylthio-cyclic GMP, *see* Photoaffinity labeling

B

Bacteriorhodopsin
  energy storage studies with laser-induced optoacoustic spectroscopy, 147–148
  photocycle energetics, 154, 156
Baculovirus–insect cell recombinant expression systems
  cyclic GMP phosphodiesterase δ subunit, expression and purification, 609–610
  $G_t\alpha$ expression, 539
  insect cell lines, 14–15
  posttranslational modification advantages, 13–14
  promoters, 15
  protein phosphatase type 7, expression of human protein
    complementary DNA production, 581
    identification and cloning of gene, 581–582, 584
    transfection and expression in Sf21 cells, 586–587
    vector construction, 584–586
  recombinant baculovirus production
    homologous recombination, 15–16
    transposon-mediated exchange, 16–20
  rhodopsin expression
    bovine protein, optimization of expression, 25
    functional characterization, overview of assays, 29
    infection duration optimization, 23–24
    insect cell adaptation to serum-free and protein-free culture conditions, 22–23
    large-scale production, 25
    media, 21–22
    multiplicity of infection
      effects on yield, 20–21
      optimization, 23–24
    purification
      immunoaffinity chromatography, 26
      lectin affinity chromatography, 26–27
      nickel affinity chromatography of histidine-tagged protein, 27
    reconstitution into proteoliposomes, 28
    regeneration of pigment, 24–25
    stable-isotope labeling, 25–26
    yield relationship to functional protein, 20–21
  rhodopsin kinase expression and recoverin affinity purification, 409–410, 409–410
  Sf9 cell culture, 16
Bromocresol purple, see Proton uptake, rhodopsin activation

C

Caged compounds, see Cyclic nucleotide-gated channel
Calcium, see also Cyclic nucleotide-gated channel; Sodium/calcium-potassium exchanger, photoreceptors
  fractional ionic current determination for cyclic nucleotide-gated channel
    caged cyclic nucleotide photolysis
      8-bromo cyclic nucleotide features, 803–805
      caging groups, 802–805
      degradation following uncaging, 803
      endogenous cyclic GMP, 803
      solubility limitations, 803–805
    calculations, 800–802, 808, 810–811
    caveats
      assumptions, 811–812
      cytoplasmic equilibrium with pipette solution, 812–813
      Fura-2 controls, 813–815
      sampling space uniformity, 816–817
      unidirectionality of calcium flux, 815–816
    data acquisition, 808
    Fura-2 measurements of calcium, 798, 800–802, 807–808, 810–811, 813–815
    instrumentation, 805, 807
    overview, 797–798
    recording configuration, 798–799
  guanylate cyclase regulation, see Guanylate cyclase, photoreceptors
  light adaptation role, 708
Calcium-dependent guanylate cyclase activator protein
  guanylate cyclase assay of activation
    basal activity, 734
    calcium concentration manipulation and activation, 731–732, 734, 737

comparison with other S100 proteins, 739–741
radiolabeled GTP assay, 731
rod outer segment preparation, 731
immunoreactivity with anti-S100 antibodies, Western blot analysis, 734, 737–739
purification from bovine extract
    anion-exchange chromatography, 732, 735
    crude extract preparation, 732
    gel electrophoresis
        analytical, 736–737
        preparative, 733, 737
    gel filtration chromatography, 733, 736
    heat stability, 737
    hydrophobic interaction chromatography, 732, 735–736
    zinc affinity chromatography, 733, 741
Calmodulin
    cyclic nucleotide-gated channel binding
        calcium concentration buffering, 784
        deletion mutants of channels, electrophysiologic studies, 824–825
        dissociation constants of protein and peptide, 786
        effects on ligand binding, 785
        fluorescence binding assays with dansylcalmodulin, 823–824, 827
        gel-overlay assay with heterologously-expressed channels, 819–821
        gel shift assay with channel peptides, 821, 823
        magnitude of response
            channels versus channel peptides, 825, 827
            vision versus olfaction, 818, 827–828
        olfactory versus rod channel inhibition, 784
        sites of interaction, 783, 785–786, 797, 818–819
        state-dependent modification, 783–784
        surface plasma resonance spectroscopy
            buffers and materials, 787
            comparison of binding constants with other techniques, 827
            controls, 791
            immobilization by amine coupling, 789

immobilization by ligand thiol coupling, 788
            kinetic data interpretation, 794, 796
            mass transfer limitations, 786
            overview of experimental design, 788
            principle, 786–787
            qualitative analysis, 791–792
            quantitative analysis, 793–794
            recording of titration series, 789–791
            Scatchard plot analysis, 796–797
            sensor chip geometry, 791
        guanylate cyclase activation in non-mammals, 730, 740
CD, see Circular dichroism
CD-GCAP, see Calcium-dependent guanylate cyclase activator protein
Circular dichroism, rhodopsin peptides, initial analysis for nuclear magnetic resonance determination of structure, 110
CNG channel, see Cyclic nucleotide-gated channel
Cone visual pigment
    absorption spectroscopy, 299
    classification, 312
    comparison with rhodopsin, overview, 219–220, 293–294
    meta intermediates
        metarhodopsin-II decay rate and Glu-122 role, 294, 311–312
        thermal behavior
            absorption spectroscopy studies, 302–303, 305–308
            cooling of samples, 302–303
            transducin activation assay, 308–309, 311
    preparation from chicken
        buffers, 294, 296
        concanavalin A affinity chromatography, 296–297
        ion-exchange chromatography, 296–298
        transient transfection in HEK293S cells, 298–299
    regeneration time
        assay with absorption spectroscopy, 299–300, 302
        comparison to rhodopsin, 294, 302
Cresol red, see Proton uptake, rhodopsin activation
Cross-linking, cyclic GMP phosphodiesterase Pγ to P$\alpha\beta$

cross-linking reaction, 645
4-(N-maleimido)benzophenone labeling of Pγ mutants, 644
sites of cross-linking, 645–646
Cryogenic photocalorimetry
  efficiency factor, 153
  heat production measurement, 149
  instrumentation, 148–149
  partition functions, 153
  rhodopsin photobleaching energetics, 156, 158
  rhosopsin photoisomerization
    enthalpy changes, 148, 151, 153
    quantum yields and photochemical processes, 151
Crystallography, see Electron crystallography
Cyclic GMP phosphodiesterase
  assays for activity
    overview, 610
    pH assay
      advantages, 657
      incubation and flash conditions, 615–616
      rod outer segment preparation, 612–614
      solutions, 615
    phosphate release colorimetric assay, 611–612, 658–660
    radioassay with tritiated cyclic nucleotides, 657–658
    regulatory effects on GTPase, see Transducin
    single rod outer segment assay
      advantages, 751–752
      amphibian rod outer segment preparation, 745–746
      calcium-dependent measurements, 747–748, 750
      cone outer segment measurements, 751
      inhibitor measurements, 744
      overview, 743–745
      pipette solutions, 746
      washout minimization, 751
  β subunit gene
    DNase I footprinting analysis
      binding reaction, 634
      digestion reaction, 634–635
      gel electrophoresis, 635
      overview, 633
      probe preparation, 633–634
    gel mobility shift assay for transcription factor binding
      binding reaction, 630, 633
      gel electrophoresis, 633
      nuclear extract preparation, 627–628
      overview, 625
      probe labeling, 629–630
      probe synthesis and purification, 628–629
    mutation in autosomal recessive retinitis pigmentosa, 617
    reporter gene transfection assays for cis element analysis in 5'-flanking region
      cell lysate preparation, 523
      β-galactosidase assay, 624–625
      luciferase assay, 623–624
      overview, 618–619
      quantification of transcription levels, 625
      transient transfection with calcium phosphate, 621–623
      vector construction, 619–621
  cyclic GMP, noncatalytic binding
    dissociation kinetics
      cold chase experiment, 667–668
      concentration jump experiments, 667–668
    functions, 599–600, 647, 671–672
    low-affinity binding assays
      centrifugal separation using silicone oil, 669–671
      centrifugal ultrafiltration, 669
      overview of equilibrium methods, 668–669
    membrane filtration assays for high-affinity sites
      ammonium sulfate stabilization of bound ligand, 665–667
      data analysis, 662–663
      dissociation of bound ligand during separations, 661–662, 665–667
      endogenous cyclic GMP destruction, 663
      equilibrium state determination, 661
      high-throughput screening assay, 665
      incubation conditions, 664

SUBJECT INDEX

inhibition of enzyme during assay, 663
materials, 664
nonspecific binding quantification, 662, 664
principles, 660–661
radioactivity quantification, 665
validation, 665
reciprocal relationship with Pγ binding, 647
families, 597
$G\alpha_t/G\alpha_{i1}$ chimera mutant binding to Pγ subunit
activation assay of phosphodiesterase, 551–552
fluorescence assay
dissociation constant calculation, 550–551
incubation conditions, 550
principle, 549
probes, 549
non-mutant binding, 517–519
GTPase cycle in phototransduction, 524–526, 742
inhibition by Pγ subunit
assays
competition assays, 643–644
fluorescence assay, 642–644
nonfluorescence assay as control for mutation and labeling effects, 641–642
fluorescence labeling of mutant Pγ, 640–641
functional assay of activity removal from rod outer segment membranes, 616
interaction sites with Pαβ, 635–636, 645–646
mutant generation for assays
cysteine insertion sites, 636–638
expression and isolation from *Escherichia coli*, 640
polymerase chain reaction site-directed mutagenesis, 639–640
vector design, 637–639
photoreactive cross-linking to Pαβ
cross-linking reaction, 645
4-(N-maleimido)benzophenone labeling of Pγ mutants, 644
sites of cross-linking, 645–646

kinetic light scattering studies of rhodopsin/transducin interactions, phosphodiesterase signal
kinetic analysis, 346–347
overview, 345–346
phosphodiesterase preparation, 335
membrane interactions, 600–601
phototransduction role, 597–598, 646
preparations from frog retina
determination of content in rod outer segment homogenates
Pγ subunit, 652–654
phosphodiesterase, 651–652
nonactivated phosphodiesterase, 654–655
Pαβ dimers
catalytic dimer preparation with trypsin proteolysis, 656–657
membrane-associated dimers, 656
rod outer segment isolation
homogenization for assays, 650–651
Percoll density gradient centrifugation, 649–650
retina isolation, 648–649
solutions, 648
transducin-activated phosphodiesterase, 655–656
purification from bovine retina
anion-exchange chromatography, 603–604
cyclic GMP affinity chromatography, 606–608
δ subunit, 609–610
extraction, 603
γ subunit, 608–609
immunoaffinity chromatography, 605–606
overview, 602
partial purification from rods and cones, 605
rod outer segment membrane preparation, 602–603
solutions, 602
storage, 604
regulation in phototransduction, overview, 646–647
RGS9 enhancement, 527
substrate affinity in rods versus cones, 599–600
subunits

942   SUBJECT INDEX

δ subunit functions, 601–602
γ subunit functions, 601, 616, 635–636
overview, 598
tissue distribution of type 6, 598, 651
turnover numbers, 598
Cyclic nucleotide-gated channel
  calmodulin binding
    calcium concentration buffering, 784
    deletion mutants of channels, electrophysiologic studies, 824–825
    dissociation constants of protein and peptide, 786
    effects on ligand binding, 785
    fluorescence binding assays with dansyl-calmodulin, 823–824, 827
    gel-overlay assay with heterologously-expressed channels, 819–821
    gel shift assay with channel peptides, 821, 823
    magnitude of response
      channels versus channel peptides, 825, 827
      vision versus olfaction, 818, 827–828
    olfactory versus rod channel inhibition, 784
    sites of interaction, 783, 785–786, 797, 818–819
    state-dependent modification, 783–784
    surface plasma resonance spectroscopy
      buffers and materials, 787
      comparison of binding constants with other techniques, 827
      controls, 791
      immobilization by amine coupling, 789
      immobilization by ligand thiol coupling, 788
      kinetic data interpretation, 794, 796
      mass transfer limitations, 786
      overview of experimental design, 788
      principle, 786–787
      qualitative analysis, 791–792
      quantitative analysis, 793–794
      recording of titration series, 789–791
      Scatchard plot analysis, 796–797
      sensor chip geometry, 791
  cyclic GMP as state-specific modulator
    binding site identification, 777–778
    mutant studies, 777
    opening of channel, 776–777
    cysteinyl residues, reagents for state-specific modification, 779–780
  expression in *Xenopus* oocyte for patch clamp studies
    artifacts, 774–775
    heteromultimeric channels, 775–776
    robust expression, 773–774
  fractional calcium ionic current determination
    caged cyclic nucleotide photolysis
      8-bromo cyclic nucleotide features, 803–805
      caging groups, 802–805
      degradation following uncaging, 803
      endogenous cyclic GMP, 803
      solubility limitations, 803–805
    calculations, 800–802, 808, 810–811
    caveats
      assumptions, 811–812
      cytoplasmic equilibrium with pipette solution, 812–813
      Fura-2 controls, 813–815
      sampling space uniformity, 816–817
      unidirectionality of calcium flux, 815–816
    data acquisition, 808
    Fura-2 measurements of calcium, 798, 800–802, 807–808, 810–811, 813–815
    instrumentation, 805, 807
    overview, 797–798
    recording configuration, 798–799
  ion permeability, 797
  nickel binding
    affinities, 780
    concentration response, 781
    controls for study, 782
    niflumic acid inclusion in pipette solution, 782
    partial agonist studies, 781
    state-specific binding, 780–781
  patch clamp recordings following covalent activation with 8-*p*-azidophenacylthio-cyclic GMP
    conductance states, 770–771
    data acquisition, 764–765, 768
    functional heterogeneity in binding sites, 767
    ligand number assignment, 768–770, 772

macroscopic currents, 764–765, 767
  principle, 764
  single-channel currents, 768–771
photoaffinity labeling of rod channels
  applications with other proteins, 771–772
  8-*p*-azidophenacylthio-cyclic GMP
    phosphorous-32 labeling, 759–760
    purification, 758–759
    synthesis, 756–758
  cyanogen bromide cleavage of labeled subunits, 761–762
  gel electrophoresis
    electroelution, 761
    peptide fragments, 762
    subunit separation, 760–761
  labels
    labeling reaction, 760
    selection, 756–757
  partial purification of channels, 760
  principle, 756
  sequence analysis of labeled peptides, 762
phototransduction role, 755, 772–773, 785, 817–818, 831
purification from bovine rod outer segments with sodium/calcium-potassium exchanger
  anion-exchange chromatography, 836–837
  calmodulin affinity chromatography, 837–838
  dye affinity chromatography, 837
  gel electrophoresis and Western blot analysis, 841–843, 846–847
  immunoaffinity methods
    antibodies, 838–839
    immunoaffinity chromatography, 840
    immunoprecipitation, 839–840
  ricin-agarose chromatography, 838
  rod outer segment
    isolation, 833–835
    solubilization, 835–836
reconstitution into liposomes
  buffers, 844
  calcium flux measurements with Arsenazo III, 844–846
  incubation conditions, 844
structure, 755, 785, 831–833
tetracaine binding studies, 782–783

Cysteine scanning mutagenesis, *see* Split receptor, mapping tertiary interactions in rhodopsin

D

DNase I footprinting, cyclic GMP phosphodiesterase β subunit gene analysis
  binding reaction, 634
  digestion reaction, 634–635
  gel electrophoresis, 635
  overview, 633
  probe preparation, 633–634

E

Early receptor current, rhodopsin
  advantages and limitations of study, 290–292
  cones versus rods, 277
  early receptor potential origins, 270, 292
  electrical structure–function approach to R_2 signal, 286–287, 289
  fusion of cells, 273–274, 278–279
  giant HEK293S recordings
    E134Q mutant, 287, 289, 292
    transient transfection studies, 289–290
    wild-type, 284, 286
  immunocytochemistry, 273, 276–277
  instrumentation
    flash apparatus, 284
    microscope, 281, 283–284
    patch-clamp, 279, 281, 291
  materials for study, 271–272
  polarity, 270
  rationale for study, 270–271, 292–293
  regeneration of pigment, 274, 286–287, 291
  rod photoreceptor recordings, 284, 286
  signal enhancement, 277–279
  site-directed mutagenesis, 272
  transfection of HEK293S cells, 272–273, 275–278, 290–291
  whole-cell recording, 274–275
Electron crystallography
  comparison with other structure elucidation methods, 92–93

crystal requirements, 92–93
crystallization of rhodopsins
  induction by reconstitution
    added lipid reconstitution, 99, 101
    natural lipid reconstitution, 101, 106
    overview, 99
  induction by selective extraction of membranes, 96–97
  rod outer segment purification
    bovine, 96
    frog, 95
  overview, 93–94
  resolution, 106
  rhodopsin three-dimensional structures, 103–106
  screening for two-dimensional crystals
    optical diffraction, 102
    overview pictures, 102
    sample preparation for electron microscopy, 101–102
  structures solved with technique, 93
  vitrification of specimens for cryomicroscopy, 102
Electron microscopy
  crystallography, see Electron crystallography
  membrane morphology analysis in cell-free frog retinal system, 86, 88
Electron paramagnetic resonance, conformation studies of rhodopsin activation, 179–180
Electrophoretic mobility shift assay
  calmodulin binding to cyclic nucleotide-gated channel peptides, 821, 823
  cyclic GMP phosphodiesterase, β subunit gene analysis
    binding reaction, 630, 633
    gel electrophoresis, 633
    nuclear extract preparation, 627–628
    overview, 625
    probe
      labeling, 629–630
      synthesis and purification, 628–629
ELISA, see Enzyme-linked immunosorbent assay
Enzyme-linked immunosorbent assay
  rhodopsin fragments, 65–66
  rhodopsin peptide ligand screening
    crude lysates, 402–403
    overview, 399, 402
    purified maltose-binding protein fusion proteins, 403–404
EPR, see Electron paramagnetic resonance
ERC, see Early receptor current, rhodopsin

F

Fourier transform infrared spectroscopy
  difference spectroscopy of rhodopsin mutants
    active state characterization
      aspartate/glutamate mutants, 184
      dark state, 183
      H211N mutant, 184–185
      lipid vesicle versus detergent spectra, 183–184
      metarhodopsin-II, 181, 183
    attenuated total reflection sampling
      principle, 180, 192–193
      proton exchange studies, 193–195
    band assignment, 195
    molecular changes leading to activation, 185–186
    sample preparation, 180–181
    Schiff base environment studies
      Glu-113, mutation and role in stabilization/activation, 186–187, 189–190
      Gly-90 mutants, 190–192
      low-temperature studies, 191–192
      water molecule interactions, 188–189
  hydrogen-bonding network of water molecules in opsin mutants, 129
  rhodopsin activation studies, overview, 195–196
  stable-isotope labeling of opsins in baculovirus–insect cell recombinant expression system, 25–26
FTIR, see Fourier transform infrared spectroscopy

G

GC, see Guanylate cyclase, photoreceptors
GCAP, see Guanylate cyclase activating proteins

SUBJECT INDEX

G-protein
  conformational changes with receptor activation, 178–179, 364, 490
  Gα
    carboxyl-terminal interactions with receptors, 388, 490
    chimera applications, 502–503
    subunit structure, 363–364, 502
    superfamily of coupled receptors, 12, 91, 251–252
    transducin activation of rhodopsin, *see* Transducin
    transitory nature of receptor interactions, 389
Guanylate cyclase, photoreceptors
  activating proteins, *see also* Calcium-dependent guanylate cyclase activator protein; Guanylate cyclase activating proteins
    calmodulin activation in non-mammals, 730, 740
    overview of types, 674, 676, 690–691, 709
    S100A1, 739–741
    S100β, 739–742
  assays
    avoidance of rod outer segment light activation, 691–692, 710–711, 718
    cyclic GMP radioimmunoassay, 711
    (S_p)-GTPαS as substrate
      advantages, 692–693, 698–699, 711
      anion-exchange chromatography for product separation, 696
      calculations, 697–698
      incubation conditiomns, 697
      kinetic, 693
      materials and solutions, 694–695
      principle, 693–694
      rod outer segment preparation and washing, 695–696
      unreacted nucleotide adsorption on neutral alumina, 696–697
    kinase activity
      autophosphorylation assay, 685–687
      kinetics, 689
      myelin basic protein as substrate, 685, 687–688
    pyrophosphate production assay
      GCAP1 purification for assay, 720
      intact membrane assays and interfering enzymes, 721–723, 729
      overview, 711, 719, 729
      principle of coupled reaction assay, 720
      rod outer segment membranes, purification and stripping, 719–720
      spectrophotometer requirements and modification, 720–721, 729
      stripped membrane assay reconstituted with GCAP1, 725, 727–729
      washed membrane assay, 723–725
    radiolabeled GTP as substrate, 679, 731
    single rod outer segment assay
      advantages, 751–752
      amphibian rod outer segment preparation, 745–746
      calcium-dependent measurements, 748, 750
      cone outer segment measurements, 751
      overview, 743, 745
      pipette solutions, 746
      washout minimization, 751
  ATP regulation, 676, 684–685
  calcium regulation, 690, 698, 708–709, 717, 730
  comparison with soluble cyclases, 689–690, 718
  divalent cation dependence, 698
  domains, 674, 683–684
  expression of human protein in transfected HEK293 cells
    cell culture, 712
    membrane preparations, 713
    transfection, 712–713
    vector, 712
  inhibitors, 698
  isoforms, 673–674, 708
  mutation in disease, 718
  nomenclature, 708
  phosphorylation
    autophosphorylation
      evidence, 685
      preparation of autophosphorylated protein, 685–687
      sites, 688–689
      stoichiometry, 688
    kinases, 676, 684

sites, 676
purification of GC-E from bovine retina
  GTP affinity chromatography
    binding and elution, 682–683
    buffers, 681
    manganese dependence of binding, 681
    kinetic parameters, 683
  overview, 677–678
  rod outer segment membranes
    preparation, 679–680
    solubilization, 680–681
  yield, 683–684
Guanylate cyclase activating proteins
  assays, 699, 701, 716
  calcium binding, 690, 698, 708–709, 717
  GCAP1 reconstitution with guanylate cyclase in stripped membranes
    assay, 725, 727–729
    purification for assay, 720
  localization in retina, 691, 709–710
  overview of types, 674, 676, 690–691, 709
  purification from bovine retina
    assay, 699, 701
    GCAP1 purification from rod outer segments
      anion-exchange chromatography, 703
      extract preparation, 703
      hydroxylapatite chromatography, 703
      overview, 702–703
      reversed-phase high-performance liquid chromatography, 704
    immunoaffinity chromatography
      antibodies, 702, 709
      GCAP1, 705–707
      GCAP2, 706–707
      immunoadsorbent preparation, 704–705
      recombinant protein purification, 707
      retinal extract preparation, 705
    overview, 701–702
  purification of recombinant proteins from *Escherichia coli*
    gel filtration chromatography, 715
    inclusion body solubilization, 715
    induction, 715
    myristoylation, 716–717
    strains of bacteria, 714
    transformation, 714
    vectors, 714

S100, *see* Calcium-dependent guanylate cyclase activator protein

H

Horner reaction, seunthesis of 11-*cis*-retinal analogs, 234–235

I

IEF, *see* Isoelectric focusing
Immunostaining, ABCR localization in eye, 883–884
Isoelectric focusing, phosphorylated rhodopsins, 71–72, 76

K

Kinetic light scattering
  advantages in membrane protein–protein interaction studies, 330
  arrestin–rhodopsin complex studies
    arrestin preparation, 335
    modeling of binding reaction, 342
  instrumentation, 333–334
  light-induced binding of soluble proteins to rhodopsin disk membranes
    bimolecular analysis, 338–339
    centrifugation assay comparison, 336, 338, 343
    data acquisition, 336
    overview, 335–336
  rhodopsin kinase–rhodopsin complex studies
    bimolecular rate constant, 341–342
    dissociation constant determination, 339, 341
    rhodopsin kinase preparation, 335
  rhodopsin-containing membrane preparation, 334–335
  theory
    difference scattering curve, 331–332
    intensity of scattering, 332
    light scattering by vesicles, 330–331
    relative mass change versus relative intensity change, 332

SUBJECT INDEX

triggered scattering change, 331–332
transducin association/dissociation with rhodopsin
  dissociation signal
    kinetic analysis, 346–347
    overview, 344–345
  light-independent binding, 343
  modeling of binding reaction, 343–344
  nucleotide affinity, 342–343
  origin of signal, 329–330
  phosphodiesterase signal
    kinetic analysis, 346–347
    overview, 345–346
    phosphodiesterase preparation, 335
  rates and dissociation constant determination, 344
  sample preparation, 334–335
KLS, *see* Kinetic light scattering

L

Laser-induced optoacoustic spectroscopy
  energy storage studies
    bacteriorhodopsin, 147–148
    rhodopsin, 148, 154
  instrumentation, 145
  principles, 143–145
  reference compounds, 145–146
  rhodopsin photobleaching energetics, 156, 158
  temperature dependence of signal, 146
Light scattering, *see* Kinetic light scattering
LIOAS, *see* Laser-induced optoacoustic spectroscopy

M

4-(*N*-Maleimido)benzophenone, *see* Cross-linking
Membrane trafficking, rhodopsin
  carboxy-terminal domain role, 79, 88
  cell-free frog retinal system for post-trans-Golgi network trafficking assay
    applications, 77–79, 88
    electron microscopy analysis of membrane morphology, 86, 88

    polyacrylamide gel electrophoresis analysis of ATP-dependent transfer, 86
    post-Golgi carrier membrane formation, 81, 83–85
    pulse-labeling of proteins in cultured retina, 79–80
    rod outer segment removal and enrichment, 80–81
    sucrose density gradient fractionation of postnuclear supernatant, 85–86
    Western blot analysis, 86
  fractionation studies, 77
  opsin fragments, 69

N

NinaA, functions, 456
NMR, *see* Nuclear magnetic resonance
Nuclear magnetic resonance
  rhodopsin, structure determination of domains
    applications, 115
    assembly of structure, 107–108
    cytoplasmic surface structure elucidation, 112–113
    data acquisition, 110
    initial analysis, 109–110
    long-range constraints, 114–115
    peak assignment, 110–111
    peptides
      selection for analysis, 108–109
      structures, 111–112
    proline kinks, 114
    secondary structure analysis, 110, 114
    short-range interactions, 113–114
    validity of whole protein structure determination, 107–108
    stable-isotope labeling of opsins in baculovirus–insect cell recombinant expression system, 25–26

O

Opsin
  basal activity, 232, 238
  bovine opsin expression in *Pichia pastoris*

absorption spectroscopy characterization, 8
advantages of system, 10
chromophore formation, 8, 11
expression vector construction, 4–5, 10
G-protein activation, 9
glycosylation analysis, 7, 10
large-scale culture, 5–6
media, 4
purification
　buffers, 6–7
　immunoaffinity chromatography, 8
subcellular localization, 7
transformation by electroporation, 5
yield, 5, 11
constitutively active opsin mutants
　absorbance spectroscopy, 216–217
　all-*trans*-retinal activation
　　basal activity correlation with *trans*-retinal activation, 258–260
　　immunoaffinity chromatography, 254
　　membrane preparation, 253–254
　　modeling of activation, 258–260
　　retinal analog studies of activation, 264, 266
　　Schiff base formation in mutants, 260–263
　　transducin activation, 255–256
　　transient transfection in COS-1 cells, 253
　definition, 208
　Glu113Gln mutant, 218
　immunoaffinity chromatography, 211–212, 216
　Lys296Gly mutant, 218
　materials for functional assays, 208–209
　membrane preparation, 210–211
　mutagenesis of bovine gene, 209
　reconstitution into lipid vesicles, 212
　rhodopsin kinase assay, 217–218
　Schiff base complex preparation, 212
　transducin activation assay
　　filter binding assay, 213–214, 255–256
　　intrinsic tryptophan fluorescence assay, 256
　　pH dependence, 208, 214–215
　transducin purification, 213
　transient expression in COS-1 cells, 209–210
　Western blot analysis of expression, 212
generation from rhodopsin, 219, 236
recombinant expression systems, overview, 3, 13
regeneration with ligand, *see* Regeneration, rhodopsin

P

PDE, *see* Cyclic GMP phosphodiesterase
Peptide competition assay, rhodopsin ligands
　applications, 453
　arrestin binding competition, 454–455
　peptide synthesis, 453–454
Peptide-on-plasmid combinatorial library, rhodopsin ligands
　amplification of affinity purified plasmids, 395–396
　binding to other G-protein-coupled receptors, 404
　dark rhodopsin high-affinity sequences, 397–398
　higher affinity peptide selection, 398–399
　LacI lysate preparation, 396–397
　library construction
　　amplification of library, 392–393
　　electrocompetent cell preparation, 392
　　Escherichia coli strain, 390, 392
　　vector, 390
　light rhodopsin high-affinity sequences, 397–399
　maltose-binding protein fusion protein with clones
　　crude lysate preparation, 401
　　enzyme-linked immunosorbent assay
　　　crude lysates, 402–403
　　　overview, 399, 402
　　　purified fusion protein, 403–404
　　purification of fusion proteins, 401–402
　　rationale, 399, 402
　　subcloning into vectors, 399–400
　panning, 393–395
　transducin Gα carboxyl terminal ligands, overview, 388–390
Phage display, rhodopsin ligands

advantages in ligand screening, 439–440
data analysis with arrestin sequences, 448–451
data confirmation with arrestin sequences, 451–452
insert preparation, 440–442
library evaluation, 442–443
M13 coat protein utilization, 438–439
panning
  controls, 445–446, 448
  DNA preparation and sequencing, 447–448
  elution of bound phage, 444–445, 447
  phage culture, 446–447
  rounds, 443–444, 447
  substrate coating on dishes, 444
vectors, 440
Phosphodiesterase, see Cyclic GMP phosphodiesterase
Phosphorylation, rhodopsin
  arrestin binding to phosphorylated form, 70–71
  constitutively active opsin mutants as substrates, 217–218
  dephosphorylation, see Protein phosphatase type 2A, photoreceptors
  kinase, see Rhodopsin kinase
  preparation of phosphorylated rhodopsin
    anion-exchange chromatography, separation of forms, 75–76
    Chelex column chromatography, 73–75
    isoelectric focusing, 71–72, 76
    lectin affinity chromatography, 72–73
    phosphorylation conditions, 71, 240
    rod outer segment preparation, 71
Photoacoustic spectroscopy, see Laser-induced optoacoustic spectroscopy
Photoaffinity labeling, cyclic nucleotide-gated channels in rods
  applications with other proteins, 771–772
  8-p-azidophenacylthio-cyclic GMP
    phosphorous-32 labeling, 759–760
    purification, 758–759
    synthesis, 756–758
  cyanogen bromide cleavage of labeled subunits, 761–762
  gel electrophoresis
    electroelution, 761
    peptide fragments, 762
    subunit separation, 760–761
labels
  labeling reaction, 760
  selection, 756–757
partial purification of channels, 760
patch clamp recordings following covalent activation with 8-p-azidophenacylthio-cyclic GMP
  conductance states, 770–771
  data acquisition, 764–765, 768
  functional heterogeneity in binding sites, 767
  ligand number assignment, 768–770, 772
  macroscopic currents, 764–765, 767
  principle, 764
  single-channel currents, 768–771
principle, 756
sequence analysis of labeled peptides, 762
Photobleaching energetics, rhodopsin
  reaction scheme, 156
  enthalpy surface analysis, 156
  blue-shifted intermediate, 156, 158
  charge separation and energy storage, 158
  limitation of vision at high light intensity, 159
  measurement, see Cryogenic photocalorimetry; Laser-induced optoacoustic spectroscopy
  intermediates, 348
Photocalorimetry, see Cryogenic photocalorimetry
Phylogenetic analysis, rhodopsins
  classification of visual pigments, 312
  wavelengths of maximal absorption, identification of critical amino acids
    ancestral pigment sequence elucidation, 320–321
    phylogenetic tree construction
      neighbor-joining method, 315, 319–320
      programs, 313–314
      unweighted pair-group method with arithmetic mean, 314–315
    site-directed mutagenesis in functional testing
      absorbance spectroscopy, 324
      applications, 325
      bovine A292S, 324

cloning of genes, 321–322
    mutagenesis, 322–323
    purification, 324
    regeneration, 323–324
    RNA isolation, 321
    wild-type pigment absorption, 312, 323–324
PP2A, see Protein phosphatase type 2A, photoreceptors
PP2C, see Protein phosphatase type 2C, photoreceptors
PP7, see Protein phosphatase type 7, photoreceptors
Protein phosphatase type 2A, photoreceptors
    activators and inhibitors
        overview, 566–567
        protamine activation mechanism, 568–570
    assay
        incubation conditions, 561
        peptide dephosphorylation assay, 561
        phosphoopsin preparation, 560
        unit definition, 561
    classification of phosphatase types, 570–571, 579–580
    conserved sequences in humans, 582–583
    purification from bovine retina
        anion-exchange chromatography, 562
        buffers, 558–559
        extraction, 561–562
        gel filtration, 562, 564
        heparin-agarose chromatography, 562
        overview, 558–559
        rod outer segment preparation, 560
        thiophosphatase a sepharose chromatography, 564
    rhodopsin recycling role, 558, 568
    substrate specificity, 568
    Western blot analysis of subunits, 564–566
Protein phosphatase type 2C, photoreceptors
    activation by fatty acids, 578
    assay
        activity separation from other phosphatases, 572–573
        incubation conditions, 571–572
        product separation, 572
        radioactivity quantification, 572

classification of phosphatase types, 570–571, 579–580
enrichment from bovine retina, 573–574
magnesium dependence, 577–578
pH optima, 577
physiological significance, 578
purification
    bovine retina as starting material, 573, 575
    gel electrophoresis, 575–576
    recombinant isozymes, 574–575
    storage, 575
Protein phosphatase type 7, photoreceptors
    assay
        incubation conditions, 588–589
        inhibitor studies, 589
        p-nitrophenyl phosphate assay, 589–590
        phosphohistone substrate preparation, 587–588
        quantification of radioactivity, 589
    classification of phosphatase types, 570–571, 579–580
    conserved sequences in humans, 582–583
    divalent cation dependence, 591–592
    expression of human protein in baculovirus–insect cell system
        complementary DNA production, 581
        identification and cloning of gene, 581–582, 584
        transfection and expression in Sf21 cells, 586–587
        vector construction, 584–586
    genomic DNA screening with polymerase chain reaction, 592–593
    homology between species, 580
    pH optimum, 591
    phosphatase inhibitor sensitivity, 592
    reverse transcription–polymerase chain reaction assessment of expression, 592
    substrate specificity, 590
Proton uptake, rhodopsin activation
    bacteriorhodopsin comparison, 385
    overview, 198, 377–378
    pH indicator dye assays
        advantages, 378
        bromocresol purple, 378
        calibration of dye, 379, 381–383, 387
        cresol red, 378, 384

detergent-solubilized rhodopsin preparation, 380
general considerations, 378–379
instrumentation, 379
proton transfer results at different pH values, 383–384
pure pH signal determination, 381
sample preparation, 380–381
proton sources in activation, 386
proton transfer dynamics, 384–385

R

Ran-binding protein 2
glutathione S-transferase fusion protein with domain 4
affinity purification, 459–460
cell lysis, 459
cleavage with thrombin, 459–460
expression in *Escherichia coli*, 458–459
rationale for construction, 456–457
vector design, 456–458
screening for retinal binding proteins
functional domain elucidation, 468
gel electrophoresis and analysis, 462–463
proteasome 19s regulatory complex subunit binding to cyclophilin-like domain, 465–466
pull-down assay, 462
Ran-GTPase identification
gel electrophoresis and electroelution, 464
GTP dependence of binding, 463
sequencing, 464
Western blot analysis, 465
retinal extract preparation, 460–461
Recoverin
affinity purification of rhodopsin kinase
advantages, 405–406
column preparation, 407–408
isolation of calcium-dependent recoverin binding proteins, 408
recombinant kinase purification from baculovirus expression system, 409–410
recombinant recoverin expression and purification from *Escherichia coli*, 406–407

yield of kinase, 408–409
calcium binding, 404–405
N-terminal acylation, 405
Regeneration, rhodopsin, *see also* 11-*cis*-Retinal chromophore, rhodopsin
baculovirus–insect cell recombinant opsins, 24–25
bovine opsin, 241
bovine opsin expressed in *Pichia pastoris*, 8, 11
early receptor current studies, rhodopsin regeneration, 274, 286–287, 291
fragments of opsin, 66
regeneration time
assay with absorption spectroscopy, 299–300, 302
comparison between rhodopsin and cone pigments, 294, 302
retinal analogs, 237
Regulator of G-protein signaling, *see* RGS9
Resonant mirror spectroscopy, transducin activation of rhodopsin, assay, 487–489
11-*cis*-Retinal chromophore, rhodopsin
analog studies
activation of constitutively active opsin mutants, 264, 266
binding site generalizations, 220
binding studies, 228–229
circular dichroism
11,12-cyclopropylretinal, 226–228
11,12-dihydroretinal binding, 222–223
exciton-coupled spectroscopy, 222–228
11-*cis*-locked seven-membered ring, 11, 12 dihydrorhodopsins, 223–226
native rhodopsin spectra, 221–222
conformational analysis of cyclopropylretinal analogs, 229–230
9-methyl group importance, 264, 266
13-methyl group importance, 230–233
regeneration of opsins, 237
synthesis
cyclopropyl retinal analogs, 235
Horner reaction, 234–235
binding site
geometry, 179
spectral tuning with site-directed mutagenesis, 116–118

conformation analysis in photoisomerization, 220–233
discovery, 196
photobleaching energetics, *see* Photobleaching energetics, rhodopsin
photoisomerization reaction, overview, 12–13, 122–123, 148, 251, 269, 557
photoisomerization, sensing in transmembrane domains using glycine-121 mutants
  phenylalanine-261 interactions mediated by retinal, 128–129
  retinal C-9 methyl group interactions, analog studies, 126–127
  substitution effects on activity, 124, 126
planarity in binding site, 220–222
regeneration, *see* Regeneration, rhodopsin
removal, 219, 236
reverse agonist activity, 233
Schiff base
  absorption spectra of protonated form, 196–198
  butylamine Schiff base complex preparation, 235–236
  counterion protonation and thermal isomerization, 200–201
  counterion tuning, 268
  deprotonation, activation and noise reduction role, 160–161, 163, 198, 207, 269, 377
  distribution in vertebrate visual pigments, 197–198
  environment studies by Fourier transform infrared difference spectroscopy of rhodopsin mutants
    Glu-113, mutation and role in stabilization/activation, 186–187, 189–190
    Gly-90 mutants, 190–192
    low-temperature studies, 191–192
    water molecule interactions, 188–189
  ethylamine or propylamine Schiff base complex preparation, 212
pK_a
  bovine rhodopsin, 203, 207
  chloride ion effects on gecko pigment, 204–205
  counterion pK_a, 201
  determination from absorbance spectra with model chromophores, 201, 203
  environmental effects, 205–206
  octopus rhodopsin, 205–206
  P521, 204, 207
  P531, 205
  P568, 205
  rationale for high value, 198–201, 207
  thermal isomerization consequences and noise reduction, 199–201
proton exchange, *see* Proton uptake, rhodopsin activation
resonance Raman spectroscopy, 197
structure, 12, 91
thermal isomerization, 159–161
Retinitis pigmentosa, *see* Autosomal dominant retinitis pigmentosa
RGS9
  enhancement of cyclic GMP phosphodiesterase, 527
  $G\alpha_t/G\alpha_{i1}$ chimera interactions, 519–521, 523
  $G\alpha_t/G\alpha_{i1}$ chimera switch II mutant interactions
    glutathione *S*-transferase–RGS9 fusion protein pull-down assay, 552–553
    single-turnover GTPase assay, 553–554
  $G\beta_{5L}$ complex, 526
  GTPase, assay of regulatory effects, *see* Transducin
  transducin conformation stabilization for GTP hydrolysis, 526
Rhodopsin
  absorption spectroscopy, *see* Absorption spectroscopy, rhodopsin
  activation probability, 268
  apoprotein, *see* Opsin
  chromophore, *see* All-*trans*-retinal; 11-*cis*-retinal chromophore, rhodopsin
  cone pigment comparison, *see* Cone visual pigment
  crystallography, *see* Electron crystallography
  cysteine scanning mutagenesis, *see* Split receptor, mapping tertiary interactions in rhodopsin
  early receptor current, *see* Early receptor current, rhodopsin

fragment reconstitution studies
  absorption spectroscopy of complexes, 66–68
  chromophore formation, 66
  expression analysis by enzyme-linked immunosorbent assay, 65–66
  gene fragments, vector construction, 60–62
  glycosylation analysis, 64
  G-protein activation assays, 68–69
  immunoaffinity chromatography, 66
  membrane integration assay, 63–64
  noncovalent interactions, 69–70
  rationale, 59–60
  trafficking, 69
  transient transfection in COS-1 cells, 62–63
  Western blot analysis of expression, 63
G-protein activation, *see* Transducin
lifetimes of intermediates, 269
membrane topology, 12, 116
meta intermediates
  metarhodopsin-II decay rate and Glu-122 role, 294, 311–312
  thermal behavior
    absorption spectroscopy studies, 302–303, 305–308
    cooling of samples, 302–303
    transducin activation assay, 308–309, 311
noise, origin and energetics, 159–161, 163
peptide ligands, *see* Peptide competition assay, rhodopsin ligands; Peptide-on-plasmid combinatorial library, rhodopsin ligands; Phage display, rhodopsin ligands
phosphorylation, *see* Phosphorylation, rhodopsin
photobleaching energetics, *see* Photobleaching energetics, rhodopsin
phylogenetic analysis, *see* Phylogenetic analysis, rhodopsins
preparations
  chicken protein
    buffers, 294, 296
    concanavalin A affinity chromatography, 296–297
    ion-exchange chromatography, 296–298
    transient transfection in HEK293S cells, 298–299
  protein interaction studies
    membrane preparation, 413–415
    transient transfection in HEK293 cells, 411–413
proton exchange, *see* Proton uptake, rhodopsin activation
recombinant protein expression, *see* Baculovirus–insect cell recombinant expression systems; Yeast recombinant expression systems
regeneration, *see* Regeneration, rhodopsin
site-directed mutagenesis, *see* Site-directed mutagenesis, rhodopsin
split receptor, *see* Split receptor, mapping tertiary interactions in rhodopsin
stability of ground state, 268
stable transfection, *see* Site-directed mutagenesis, rhodopsin
structure, *see* Electron crystallography; Nuclear magnetic resonance
trafficking, *see* Membrane trafficking, rhodopsin
transient transfection, limitations, 30
Rhodopsin kinase
  affinity purification with recoverin
    advantages, 405–406
    column preparation, 407–408
    isolation of calcium-dependent recoverin binding proteins, 408
    recombinant kinase from baculovirus expression system, 409–410
    recombinant recoverin expression and purification from *Escherichia coli*, 406–407
    yield of kinase, 408–409
  assays of rhodopsin phosphorylation, 217–218, 237, 415–417
  binding of opsin/all-*trans*-retinal complex, 244, 249–250
  kinetic light scattering of rhodopsin complexes
    bimolecular rate constant, 341–342
    dissociation constant determination, 339, 341
    rhodopsin kinase preparation, 335
  overview of rhodopsin binding and activation, 70, 411, 437–438

posttranslational modification, 336
purification, 217, 240–241
sites of rhodopsin phosphorylation, 557
Rim protein, *see* ABCR
Rod outer segment
  ATPase activity, 872–873
  bleaching, 236
  electron crystallography of rhodopsin, rod outer segment purification
    bovine, 96
    frog, 95
  light activation avoidance in guanylate cyclase assays, 691–692, 710–711, 718
  preparation for assays
    cyclic GMP phosphodiesterase, pH assay, 612–614
    cyclic GMP phosphodiesterase, single rod outer segment assay
      advantages, 751–752
      amphibian rod outer segment preparation, 745–746
      calcium-dependent measurements, 747–748, 750
      cone outer segment measurements, 751
      inhibitor measurements, 744
      overview, 743–745
      pipette solutions, 746
      washout minimization, 751
    guanylate cyclase, 695–696, 731
    guanylate cyclase, single rod outer segment assay
      advantages, 751–752
      amphibian rod outer segment preparation, 745–746
      calcium-dependent measurements, 748, 750
      cone outer segment measurements, 751
      overview, 743, 745
      pipette solutions, 746
      washout minimization, 751
    sodium/calcium-potassium exchanger
      cation dependence of calcium uptake, 855–856
      fluo-3 measurements and calibration, 852–855
      preparation, 849–850
      rhodopsin bleaching effects on calcium measurements, 857

  preparation for protein purification
    ABCR, 865
    cyclic GMP phosphodiesterase
      bovine retina, 602–603
      frog retina, 648–651
    cyclic nucleotide-gated channel
      isolation, 833–835
      solubilization, 835–836
    GCAP1, 703
    GC-E, 679–681
    phosphorylated rhodopsin, 71
    protein phosphatase type 2A, 560
    sodium/calcium-potassium exchanger
      isolation, 833–835
      solubilization, 835–836
Rod pigment, *see* Rhodopsin
ROS, *see* Rod outer segment

S

S100, *see* Calcium-dependent guanylate cyclase activator protein
Schiff base, rhodopsin, *see* 11-*cis*-Retinal chromophore, rhodopsin
Site-directed mutagenesis
  cyclic GMP phosphodiesterase Pγ subunit for autoinhibition assays
    cysteine insertion sites, 636–638
    expression and isolation from *Escherichia coli*, 640
    polymerase chain reaction site-directed mutagenesis, 639–640
    vector design, 637–639
  Gα_t/Gα_{i1} chimeras
    alanine-scanning mutagenesis in switch II region, 540
    cyclic GMP phosphodiesterase Pγ subunit binding assays, 549–552
    GTPγS binding assay for folding analysis of mutants, 548
    polymerase chain reaction-based mutagenesis, 504–505, 542–544
    RGS interaction assays, 552–554
    target selection and rationale, 540, 542
    trypsin protection assay for folding analysis of mutants, 545–546
    tryptophan fluorescence assay with AlF$_4^-$ for folding analysis of mutants, 546–548

SUBJECT INDEX

Site-directed mutagenesis, rhodopsin
  chromophore binding site and spectral tuning, 116–118
  chromophore photoisomerization, sensing in transmembrane domains
    absorption spectroscopy analysis, 123–124
    glycine-121 interactions
      phenylalanine-261 interactions mediated by retinal, 128–129
      retinal C-9 methyl group interactions, analog studies, 126–127
      substitution effects on activity, 124, 126
    overview, 122–123
  constitutively active opsin mutants, see Opsin
  Fourier transform infrared difference spectroscopy of rhodopsin mutants
    active state characterization
      aspartate/glutamate mutants, 184
      dark state, 183
      H211N mutant, 184–185
      lipid vesicle versus detergent spectra, 183–184
      metarhodopsin-II, 181, 183
    attenuated total reflection sampling principle, 180, 192–193
    proton exchange studies, 193–195
    band assignment, 195
    molecular changes leading to activation, 185–186
    sample preparation, 180–181
    Schiff base environment studies
      Glu-113, mutation and role in stabilization/activation, 186–187, 189–190
      Gly-90 mutants, 190–192
      low-temperature studies, 191–192
      water molecule interactions, 188–189
  human gene mutagenesis
    absorption spectroscopy of mutants, 50–51, 57
    immunoaffinity chromatography, 49–50
    immunocytochemistry, subcellular localization and quantification, 38–39
    materials, 31–32
    plasmid preparation, 33–34, 36
    primer
      annealing, 34
      design and sequences, 32–33
      phosphorylation, 33
      selection primer, 34, 36
    slot immunoblot assay for quantification, 39, 44–45, 47–48
    vectors, 32, 36
  hydrogen-bonding network of water molecules in mutants, 129
  stable transfection of human mutants in HEK293S cells
    advantages, 30–31
    calcium phosphate transfection, 36–37
    expression levels, mutant dependence, 45, 48–49, 55
    fluorescent probing of expression, 56–57
    large-scale culture, 58
    materials, 31–32
    sampling frequency for expression analysis, 56
    selection of clones, 37–38, 51, 55
    supertransfection and secondary drug selection, 45, 48–49, 55–57
    vectors, 37, 51
  tryptophans in transmembrane segment
    absorption spectroscopy
      data collection, 119–120
      difference spectroscopy analysis of mutants, 120, 122
    conservation in G-protein-coupled receptors, 118
    targeting of specific tryptophans, 118–119
  wavelengths of maximal absorption, identification of critical amino acids
    ancestral pigment sequence elucidation, 320–321
    phylogenetic tree construction
      neighbor-joining method, 315, 319–320
      programs, 313–314
      unweighted pair-group method with arithmetic mean, 314–315
    site-directed mutagenesis for functional testing
      absorbance spectroscopy, 324
      applications, 325
      bovine A292S, 324
      cloning of genes, 321–322

mutagenesis, 322–323
purification, 324
regeneration, 323–324
RNA isolation, 321
wild-type pigment absorption, 312, 323–324
in Situ hybridization, ABCR localization in eye, 883–884
Slot immunoblot, quantification of human opsin mutants, 39, 44–45, 47–48
Sodium/calcium-potassium exchanger, photoreceptors
  inhibitors, 848
  ion selectivity, 849
  phototransduction role, 831
  purification from bovine rod outer segments with cyclic nucleotide-gated channel
    anion-exchange chromatography, 836–837
    calmodulin affinity chromatography, 837–838
    dye affinity chromatography, 837
    gel electrophoresis and Western blot analysis, 841–843, 846–847
    immunoaffinity methods
      antibodies, 838–839
      immunoaffinity chromatography, 840
      immunoprecipitation, 839–840
    ricin-agarose chromatography, 838
    rod outer segment
      isolation, 833–835
      solubilization, 835–836
  reconstitution into liposomes
    buffers, 844
    calcium flux measurements with Arsenazo III, 844–846
    incubation conditions, 844
  spectrofluorometric assay
    calcium dye loading, 850–851
    cell culture, 850
    cell transfection studies
      A23187-induced fluorescence, 859
      calcium measurements, 858–859
      cell lines, 857–858
      gramicidin effects, 860
      noise reduction, 860, 862
      single-cell imaging, 862–863
      transfection efficiency, 860

distinguishing activity from sodium/calcium exchanger by cation dependence, 849, 855–856
instrumentation
  single cell measurements, 851–852
  suspension preparations, 851
rod outer segment
  cation dependence of calcium uptake, 855–856
  fluo-3 measurements and calibration, 852–855
  preparation, 849–850
  rhodopsin bleaching effects on calcium measurements, 857
solutions, 849
stoichiometry of transport, 848
structure, 833, 846–847
tissue distribution of isoforms, 847–848
Split receptor, mapping tertiary interactions in rhodopsin
  cysteine scanning mutagenesis
    absorption spectroscopy of double mutants, 139
    number of mutants, 132
    transducin activation by double mutants, 139–140
  disulfide bond formation in cysteine mutants
    detection with Western blot analysis with concanavalin A, 132, 138
    induction of formation, 132–134, 137–138, 140
    kinetics of formation, 140–143
    native cysteine cross-linking, 141
  fragment design and structure, 130–132
  interpretation of results, 142
  principles and overview, 130, 134–135
  receptor preparation
    constructs, 135
    immunoaffinity chromatography, 136–137
    reconstitution, 136
    transfection and expression in COS-1 cells, 135–136
SPR, see Surface plasmon resonance
Stargardt disease, ABCR mutations and pathogenesis, 864, 879–880, 896–897
Surface plasmon resonance
  calmodulin binding to cyclic nucleotide-gated channels

buffers and materials, 787
controls, 791
immobilization by amine coupling, 789
immobilization by ligand thiol coupling, 788
kinetic data interpretation, 794, 796
mass transfer limitations, 786
overview of experimental design, 788
principle, 786–787
qualitative analysis, 791–792
quantitative analysis, 793–794
recording of titration series, 789–791
Scatchard plot analysis, 796–797
sensor chip geometry, 791
transducin activation of rhodopsin, assay
overview, 482–483
patterns of rhodopsin membranes, 485, 487
self-assembled monolayers, 483, 485, 487
supported lipid bilayers containing rhodopsin, 483–485

T

TGN, see Trans-Golgi network
Transducin
activation assays of rhodopsins
absorption spectroscopy assays
extra metarhodopsin-II, 476, 478
photoregeneration from metarhodopsin-II state, 479
bovine opsin expressed in *Pichia pastoris*, 9
comparison of intrinsic biophysical assays, 471–472
constitutively active opsin mutants, transducin activation assay
filter binding assay, 213–214
pH dependence, 208, 214–215
transducin purification, 213, 241
constitutively active opsin mutants, transducin activation assay with all-*trans*-retinal
filter-binding assay, 255–256
intrinsic tryptophan fluorescence assay, 256
cysteine double mutants, 139–140
intrinsic tryptophan fluorescence of

transducin in assay, 118–119, 241, 243–245, 247–249, 473–476
meta intermediate thermal behavior, comparison of rod and cone pigments, 308–309, 311
near-infrared light scattering assay, 479–482
opsin/all-*trans*-retinal complex activation by transducin, 241, 243–245, 247–249
reconstituted fragments of opsin, 68–69
resonant mirror spectroscopy, 487–489
surface plasmon resonance assay
overview, 482–483
patterns of rhodopsin membranes, 485, 487
self-assembled monolayers, 483, 485, 487
supported lipid bilayers containing rhodopsin, 483–485
expression of $G_t\alpha$ in baculovirus–insect cell system, 539
fluorescent labels as transducin conformation probes in rhodopsin activation
acrylodan probing of amino-terminal conformational changes, 499–501
cysteine-substituted mutants for labeling, 500–502
functional activity of labeled proteins, 495–497
$G\alpha_t/G\alpha_{i1}$ chimera construction in *Escherichia coli*, 491
high-performance liquid chromatography of labeled species, 493
lucifer yellow
carboxyl-terminus conformational changes, 498–499
labeling reaction, 491–492
sites of labeling, 500
stoichiometry of labeling, 492–495
switch II region conformational changes, 497–498
principle, 490–491
proteolytic analysis of labeled peaks, 495
$G\alpha_t/G\alpha_{i1}$ chimeras
applications, 523–524, 554
concentration determination, 508–509
construction in *Escherichia coli*, 491, 503–507

expression, 507–508, 544
functional assays
  conformational change induction by AlF$_4^-$, intrinsic tryptophan fluorescence, 512–513
  GTPase assay, 512
  GTP binding stimulation by rhodopsin, 511–512
  overview, 510–511
GDP dissociation with GTPγS binding
  binding reactions, 514
  GDP binding to Gα subunits, 513, 517
  intrinsic fluorescence assay, 513–514
  spontaneous release rates, 514, 516–517
homology of subunits, 503
low-solubility chimeras, expression and purification, 508
purification
  anion-exchange chromatography, 507–508, 545
  nickel chromatography of histidine-tagged proteins, 507–508, 544–545
receptor interactions
  binding to inhibitory subunit of phosphodiesterase, 517–519
  overview, 517
  RGS9 interactions, 519–521, 523
site-directed mutagenesis
  alanine-scanning mutagenesis in switch II region, 540
  cyclic GMP phosphodiesterase Pγ subunit binding assays, 549–552
  GTPγS binding assay for folding analysis of mutants, 548
  polymerase chain reaction-based mutagenesis, 504–505, 542–544
  RGS interaction assays, 552–554
  target selection and rationale, 540, 542
  trypsin protection assay for folding analysis of mutants, 545–546
  tryptophan fluorescence assay with AlF$_4^-$ for folding analysis of mutants, 546–548
solubility, 509–510

Gα$_t$ peptide binding to rhodopsin, *see* Peptide-on-plasmid combinatorial library, rhodopsin ligands
GTPase acceleration, *see* Cyclic GMP phosphodiesterase; RGS9
GTPase assays
  multiple-turnover approach, 527–528
  single-turnover approach
    controls, 529, 533–534
    data analysis, 530, 532, 537
    detection techniques, 528
    equipment, 535–537
    GTPase regulators, assay of effects, 530–531, 534–535, 553–554
    initialization by GTP binding to G$_t$α, 528–531, 536, 538
    magnesium in binding reaction, 531
    overview, 528
    preactivation assay, 537–538
    quenching, 532
    rhodopsin complex formation, 529–530
    salt concentration, 531
    separation of radiolabeled products, 532–533, 536–538
    solutions, 536–538
    standard assay, 535–537
    time scale of reactions, 531–532
GTPase cycle in phototransduction, 524–526
kinetic light scattering studies of association/dissociation with rhodopsin
  dissociation signal
    kinetic analysis, 346–347
    overview, 344–345
  light-independent binding, 343
  modeling of binding reaction, 343–344
  nucleotide affinity, 342–343
  origin of signal, 329–330
  phosphodiesterase signal
    kinetic analysis, 346–347
    overview, 345–346
    phosphodiesterase preparation, 335
  rates and dissociation constant determination, 344
  sample preparation, 334–335
limited proteolysis of G-proteins as conformational probes
  rationale, 364–365

antibody preparation, 369, 372
GDP binding effects, 367
GTP binding effects, 365–367
incubation conditions
   protein binding, 372–374
   proteolysis, 374
materials, 369, 371–372
protected cleavage sites, 376
proteolytic fragment identification
   reversed-phase high-performance liquid chromatography, 367, 369
   Western blot analysis, 367, 369, 375–376
rod outer segment preparation, 371
transducin preparation, 371
tryptic cleavage sites in Gα and G$\beta\gamma$, 365
V8 protease clevage sites, 367, 376
vesicle preparation, 371
posttranslational modification, 336
rhodopsin activation
   conformational changes with activation, 178–179, 364, 490
   overview, 70, 342–343, 471–473
   signal amplification, 422
   time-resolved low-temperature absorption spectroscopy of rhodopsin intermediates in activation
     absorption maxima of intermediates, 361–362
     applications and overview, 350–351
     collected spectra, 353, 355
     experimental setup, 351–352
     GTP analog studies, 352, 359, 361, 363
     intermediates interacting with transducin, 348–349, 358–359, 361–363
     photoconverted pigment, quantification, 352–353
     rhodopsin purification, 349
     spectral analysis, 355, 357–358
     transducin purification, 349–350
Trans-Golgi network, *see* Membrane trafficking, rhodopsin

W

Western blot
   ABCR, 868–869
   arrestin, 431–432
   calcium-dependent guanylate cyclase activator protein, 734, 737–739
   constitutively active opsin mutant expression, 212
   cyclic nucleotide-gated channel, 841–843, 846–847
   opsin fragments, analysis of expression, 63
   protein phosphatase type 2A subunits, 564–566
   Ran-GTPase, 465
   rhodopsin
     quantification of transient transfection in HEK293 cells, 413–415
     trafficking assay, 86
   sodium/calcium-potassium exchanger, 841–843, 846–847
   transducin fragments, 367, 369, 375–376

Y

Yeast recombinant expression systems, bovine opsin expression in *Pichia pastoris*
   absorption spectroscopy characterization, 8
   advantages of system, 10
   chromophore formation, 8, 11
   expression vector construction, 4–5, 10
   G-protein activation, 9
   glycosylation analysis, 7, 10
   large-scale culture, 5–6
   media, 4
   purification
     buffers, 6–7
     immunoaffinity chromatography, 8
   subcellular localization, 7
   transformation by electroporation, 5
   yield, 5, 11